An Introduction to Statistical Methods and Data Analysis

Fifth Edition

R. Lyman Ott

Michael Longnecker
Texas A&M University

DUXBURY

THOMSON LEARNING

Australia • Canada • Mexico • Singapore • Spain • United Kingdom • United States

DUXBURY

THOMSON LEARNING

Sponsoring Editor: *Carolyn Crockett*
Marketing Team: *Tom Ziolkowski, Ericka Thompson, Samantha Cabaluna*
Editorial Assistant: *Ann Day*
Project Editor: *Laurel Jackson*
Production Service: *Helen Walden*
Manuscript Editors: *Pam Rockwell, Helen Walden*
Permissions Editor: *Lillian Campobasso*

Interior Design: *Helen Walden*
Cover Design: *Denise Davidson*
Cover Photo: *PhotoDisc*
Print Buyer: *Vena Dyer*
Typesetting and Illustration: *Bi-Comp, Inc.*
Cover Printing: *Phoenix Color Corp., Inc.*
Printing and Binding: *R. R. Donnelley & Sons, Inc./Willard*

For more information about this or any other Duxbury product, contact:
DUXBURY
511 Forest Lodge Road
Pacific Grove, CA 93950 USA
www.duxbury.com
1-800-423-0563 (Thomson Learning Academic Resource Center)

Printed in the United States of America

10   9   8   7   6   5   4   3   2   1

Library of Congress Cataloging-in-Publication Data

Ott, Lyman.
    An introduction to statistical methods and data analysis / R. Lyman Ott, Michael Longnecker.—5th ed.
        p.   cm.
    Includes bibliographical references and index.
    ISBN 0-534-25122-6
    1. Mathematical statistics. I. Longnecker, Michael. II. Title.

QA276.O77 2001
519.5—dc21

99-040068

Contents

P A R T 1   Introduction   1

CHAPTER 1   What Is Statistics?   2

1.1 Introduction   2
1.2 Why Study Statistics?   6
1.3 Some Current Applications of Statistics   6
1.4 What Do Statisticians Do?   10
1.5 Quality and Process Improvement   12
1.6 A Note to the Student   14
1.7 Summary   14
Supplementary Exercises   14

P A R T 2   Collecting Data   17

CHAPTER 2   Using Surveys and Scientific Studies to Gather Data   18

2.1 Introduction   18
2.2 Surveys   19
2.3 Scientific Studies   27
2.4 Observational Studies   34
2.5 Data Management: Preparing Data for Summarization and Analysis   35
2.6 Summary   38

P A R T 3   Summarizing Data   39

CHAPTER 3   Data Description   40

3.1 Introduction   40
3.2 Calculators, Computers, and Software Systems   41
3.3 Describing Data on a Single Variable: Graphical Methods   43

3.4   Describing Data on a Single Variable: Measures of Central Tendency   69
3.5   Describing Data on a Single Variable: Measures of Variability   81
3.6   The Boxplot   96
3.7   Summarizing Data from More Than One Variable   101
3.8   Summary   109
Key Formulas   110
Supplementary Exercises   110

P A R T 4   Tools and Concepts   121

CHAPTER 4   **Probability and Probability Distributions   122**

4.1   How Probability Can Be Used in Making Inferences   122
4.2   Finding the Probability of an Event   125
4.3   Basic Event Relations and Probability Laws   128
4.4   Conditional Probability and Independence   131
4.5   Bayes' Formula   136
4.6   Variables: Discrete and Continuous   141
4.7   Probability Distributions for Discrete Random Variables   142
4.8   A Useful Discrete Random Variable: The Binomial   144
4.9   Probability Distributions for Continuous Random Variables   154
4.10   A Useful Continuous Random Variable: The Normal Distribution   157
4.11   Random Sampling   166
4.12   Sampling Distributions   171
4.13   Normal Approximation to the Binomial   182
4.14   Minitab Instructions   185
4.15   Summary   186
Key Formulas   187
Supplementary Exercises   187

P A R T 5   Analyzing Data: Central Values, Variances, and Proportions   191

CHAPTER 5   **Inferences about Population Central Values   192**

5.1   Introduction and Case Study   192
5.2   Estimation of μ 196
5.3   Choosing the Sample Size for Estimating μ 204
5.4   A Statistical Test for μ 207
5.5   Choosing the Sample Size for μ 219
5.6   The Level of Significance of a Statistical Test   224
5.7   Inferences about μ for a Normal Population, σ Unknown   228
5.8   Inferences about the Median   243
5.9   Summary   250
Key Formulas   251
Supplementary Exercises   253

CHAPTER 6   **Inferences Comparing Two Population Central Values   263**

6.1 Introduction and Case Study   263
6.2 Inferences about $\mu_1 - \mu_2$: Independent Samples   267
6.3 A Nonparametric Alternative: The Wilcoxon Rank Sum Test   287
6.4 Inferences about $\mu_1 - \mu_2$: Paired Data   299
6.5 A Nonparametric Alternative: The Wilcoxon Signed-Rank Test   308
6.6 Choosing Sample Sizes for Inferences about $\mu_1 - \mu_2$ 314
6.7 Summary   316
Key Formulas   317
Supplementary Exercises   319

CHAPTER 7   **Inferences about Population Variances   341**

7.1 Introduction and Case Study   341
7.2 Estimation and Tests for a Population Variance   344
7.3 Estimation and Tests for Comparing Two Population Variances   355
7.4 Tests for Comparing $t > 2$ Population Variances   365
7.5 Summary   373
Key Formulas   373
Supplementary Exercises   374

CHAPTER 8   **Inferences about More Than Two Population Central Values   379**

8.1 Introduction and Case Study   379
8.2 A Statistical Test about More Than Two Population Means: An Analysis of Variance   384
8.3 The Model for Observations in a Completely Randomized Design   394
8.4 Checking on the AOV Conditions   396
8.5 An Alternative Analysis: Transformations of the Data   403
8.6 A Nonparametric Alternative: The Kruskal–Wallis Test   410
8.7 Summary   414
Key Formulas   415
Supplementary Exercises   416

CHAPTER 9   **Multiple Comparisons   427**

9.1 Introduction and Case Study   427
9.2 Linear Contrasts   431
9.3 Which Error Rate Is Controlled?   438
9.4 Fisher's Least Significant Difference   440

9.5 Tukey's *W* Procedure   444
9.6 Student–Newman–Keuls Procedure   447
9.7 Dunnett's Procedure: Comparison of Treatments to a Control   450
9.8 Scheffé's *S* Method   452
9.9 Summary   458
Key Formulas   459
Supplementary Exercises   459

CHAPTER 10   Categorical Data   469

10.1 Introduction and Case Study   469
10.2 Inferences about a Population Proportion π 471
10.3 Inferences about the Difference between Two Population Proportions, $\pi_1 - \pi_2$ 482
10.4 Inferences about Several Proportions: Chi-Square Goodness-of-Fit Test   488
10.5 The Poisson Distribution   497
10.6 Contingency Tables: Tests for Independence and Homogeneity   501
10.7 Measuring Strength of Relation   510
10.8 Odds and Odds Ratios   516
10.9 Summary   520
Key Formulas   520
Supplementary Exercises   521

P A R T 6   Analyzing Data: Regression Methods and Model Building   529

CHAPTER 11   Linear Regression and Correlation   531

11.1 Introduction and Case Study   531
11.2 Estimating Model Parameters   540
11.3 Inferences about Regression Parameters   557
11.4 Predicting New *Y* Values Using Regression   567
11.5 Examining Lack of Fit in Linear Regression   576
11.6 The Inverse Regression Problem (Calibration)   582
11.7 Correlation   590
11.8 Summary   600
Key Formulas   602
Supplementary Exercises   603

CHAPTER 12   Multiple Regression and the General Linear Model   617

12.1 Introduction and Case Study   617
12.2 The General Linear Model   625
12.3 Estimating Multiple Regression Coefficients   627

12.4   Inferences in Multiple Regression   646
12.5   Testing a Subset of Regression Coefficients   657
12.6   Forecasting Using Multiple Regression   666
12.7   Comparing the Slopes of Several Regression Lines   670
12.8   Logistic Regression   675
12.9   Some Multiple Regression Theory (Optional)   683
12.10 Summary   687
Key Formulas   688
Supplementary Exercises   689

CHAPTER 13   **More on Multiple Regression   705**

13.1   Introduction and Case Study   705
13.2   Selecting the Variables (Step 1)   707
13.2   Formulating the Model (Step 2)   727
13.3   Checking Model Assumptions (Step 3)   758
13.4   Summary   782
Key Formulas   783
Supplementary Exercises   783

P A R T 7   **Design of Experiments and Analysis of Variance   829**

CHAPTER 14   **Design Concepts for Experiments and Studies   830**

14.1   Introduction   830
14.2   Types of Studies   831
14.3   Designed Experiments: Terminology   832
14.4   Controlling Experimental Error   836
14.5   Randomization of Treatments to Experimental Units   840
14.6   Determining the Number of Replications   845
14.7   Summary   849
Supplementary Exercises   849

CHAPTER 15   **Analysis of Variance for Standard Designs   853**

15.1   Introduction and Case Study   853
15.2   Completely Randomized Design with Single Factor   855
15.3   Randomized Complete Block Design   859
15.4   Latin Square Design   879
15.5   Factorial Treatment Structure in a Completely Randomized Design   891
15.6   Factorial Treatment Structure in a Randomized Complete Block Design   914
15.7   Estimation of Treatment Differences and Comparisons of Treatment Means   916
15.8   Summary   922

Key Formulas   923
Supplementary Exercises   924

CHAPTER 16   **The Analysis of Covariance   943**

16.1 Introduction and Case Study   943
16.2 A Completely Randomized Design with One Covariate   946
16.3 The Extrapolation Problem   959
16.4 Multiple Covariates and More Complicated Designs   962
16.5 Summary   970
Supplementary Exercises   971

CHAPTER 17   **Analysis of Variance for Some Fixed-, Random-, and Mixed-Effects Models   975**

17.1 Introduction and Case Study   975
17.2 A One-Factor Experiment with Treatment Effects Random: A Random-Effects Model   978
17.3 Extensions of Random-Effects Models   983
17.4 Mixed-Effects Models   992
17.5 Rules for Obtaining Expected Mean Squares   1000
17.6 Nested Sampling and the Split-Plot Design   1010
17.7 Summary   1019
Supplementary Exercises   1020

CHAPTER 18   **Repeated Measures and Crossover Designs   1025**

18.1 Introduction and Case Study   1025
18.2 Single-Factor Experiments with Repeated Measures   1029
18.3 Two-Factor Experiments with Repeated Measures on One of the Factors   1031
18.4 Crossover Designs   1040
18.5 Summary   1044
Supplementary Exercises   1045

CHAPTER 19   **Analysis of Variance for Some Unbalanced Designs   1051**

19.1 Introduction and Case Study   1051
19.2 A Randomized Block Design with One or More Missing Observations   1053
19.3 A Latin Square Design with Missing Data   1059
19.4 Balanced Incomplete Block (BIB) Designs   1063
19.5 Summary   1072

Key Formulas   1072
Supplementary Exercises   1074

CHAPTER 20   **Communicating and Documenting the Results of Analyses   1077**

20.1   Introduction   1077
20.2   The Difficulty of Good Communication   1078
20.3   Communication Hurdles: Graphical Distortions   1079
20.4   Communication Hurdles: Biased Samples   1082
20.5   Communication Hurdles: Sample Size   1083
20.6   Preparing Data for Statistical Analysis   1084
20.7   Guidelines for a Statistical Analysis and Report   1087
20.8   Documentation and Storage of Results   1088
20.9   Summary   1089
Supplementary Exercise   1089

Appendix: Statistical Tables   1090

References   1130

Index   1133

Preface

An Introduction to Statistical Methods and Data Analysis, Fifth Edition, is a textbook for advanced undergraduate and graduate students from a variety of disciplines. It is intended to prepare students to deal with solving problems encountered in research projects, decision making based on data, and general life experiences beyond the classroom and university setting. We presume students using this book have a minimal mathematical background (high school algebra) and no prior coursework in statistics. The first eleven chapters present the material typically covered in an introductory statistics course. However, we have provided case studies and examples that connect the statistics concepts to problems of a very practical nature. The remaining chapters cover regression modeling and design of experiments. We develop and illustrate the statistical techniques and thought process needed to design a research study or experiment and then analyze the data collected using an intuitive and proven four-step approach.

Major Features

Learning from Data We approach the study of statistics by using a four-step process to learn from data:

1. Design the data collection process.
2. Prepare the data for analysis.
3. Analyze the data.
4. Communicate the results of the data analysis.

Case Studies To demonstrate the relevance and importance of statistics in solving real-world problems, we introduce the major topic of most chapters using a case study. This approach aims to assist in overcoming the natural initial perception held by many people that statistics is "just another math course." Introducing major topics with case studies provides a focus of the central nature of applied statistics. We hope these case studies on a wide variety of research- and business-related topics will give the reader an enthusiasm for the broad applicability of statistics and the four-step statistical thought process that we have found and used throughout many years of teaching, consulting, and R&D management. The following case studies from the text are typical of those used throughout:

- **Percentage of calories from fat** Assessment and quantification of a person's diet, which is crucial in evaluating the degree of relationship between diet and diseases

- **Evaluation of consistency among property assessors**   A study to determine whether county property assessors differ systematically in their property value assessments
- **Effect of timing of the treatment of port-wine stains with lasers**   A prospective study to determine whether treatment at a younger age would yield better results than treatment at an older age
- **Evaluation of oil spill on the growth of flora**   A study of effects on plant growth one year after cleanup of a marsh contaminated by spilled oil

With each case study, we illustrate the use of the four-step, learning from data process. A discussion of sample size determination, graphical displays of the data, and a summary of the necessary ingredients for a complete report of the statistical findings of the study are included with many of the case studies.

Graphical Displays of Data

Throughout the text, we provide a variety of graphical displays that are essential to the evaluation of the critical assumptions underlying the application of such statistical procedures as normal probability plots, boxplots, scatterplots, matrix plots, and residual plots. Furthermore, we emphasize the summarization of data using graphical displays in order to provide tools the reader can use to demonstrate treatment differences. These types of plots provide a method for making a distinction between statistically different treatments and practically different treatments. For example, in the case study "Are Interviewers' Decisions Affected by Different Handicap Types?" we use side-by-side boxplots to summarize the differences in ratings across the handicap types. A normal probability plot is used to assess the normality assumption of the residuals. The graphs are shown here in Figures 9.1 and 9.2.

FIGURE 9.1

Boxplots of ratings by handicap (means are indicated by solid circles)

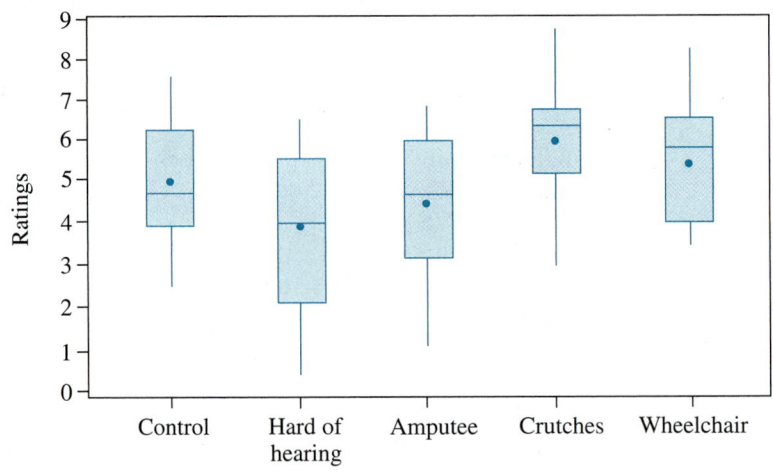

FIGURE 9.2

Normal probability plot of residuals

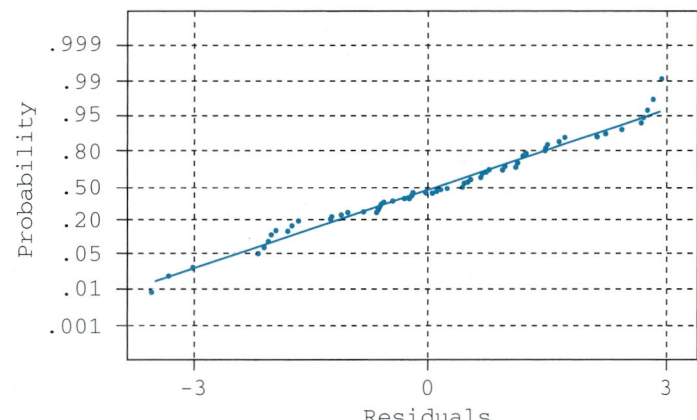

Average: -0.0000000   Anderson-Darling Normality Test
StDev: 1.63767   A-Squared: 0.384
N: 70   P-Value: 0.387

Examples and Exercises

We have further enhanced the practical nature of statistics by using examples and exercises from journal articles, newspapers, and our many consulting experiences. These provide students with further evidence of the practical uses of statistics in solving problems that are relevant to their everyday lives. Many new exercises and examples have been included in this edition; the number and variety of exercises will be a great asset to both the instructor and students in their study of statistics. In many of the exercises, we include computer output for the students to use as they solve the exercises. For example, in Exercise 9.7 on page 462, we provide in the SAS output an AOV table, mean separations using four different methods, and a residual analysis. The student is then asked a variety of questions that a researcher would ask when attempting to summarize the results of the study.

Topics Covered

This book can be used for either a one-semester or two-semester course and is divided into eight parts. The relationship among the eight parts of the text, the steps in learning from data, and the 20 chapters is illustrated in Table 1.1 on page 3. Parts 1 through 5 along with Chapters 11 and 20 would constitute a one-semester course:

Part 1   Introduction
Part 2   Collecting Data
Part 3   Summarizing Data
Part 4   Tools and Concepts
Part 5   Analyzing Data
Chapter 11   Linear Regression and Correlation
Chapter 20   Communicating and Documenting the Results of Analyses

The second semester of a two-semester course would then include model building and inferences in multiple regression analysis, logistic regression, design of experiments, and analysis of variance:

> **Part 6**   Analyzing Data: Regression Methods and Model Building
> **Part 7**   Design of Experiments and Analysis of Variance

Emphasis on Interpretation, Not Computation

The standard methodology in a first course in statistics is to define a particular statistical procedure and then carry out data analysis by using a computational form of the procedure. However, we find this to be a necessary impediment to the analysis of larger data sets. Furthermore, the students spend too much time in hand calculation with a form of the statistical procedure that is void of the intuitive nature of why the calculations are being done. Therefore, we provide examples and exercises that allow the student to study how to calculate the value of statistical estimators and test statistics using the definitional form of the procedure. After the student becomes comfortable with these aspects of the data the statistical procedure is reflecting, we then emphasize the use of computer software in making computations when analyzing larger data sets. We provide output from several major statistical packages: SAS, Minitab, Stata, Systat, JMP, and SPSS. We find that this approach gives the student the experience of computing the value of the procedure using the definition and hence, the student learns the basics behind each procedure. In most situations beyond the statistics course, the student should be using computer software in making the computations for both expedience and quality of calculation. In many exercises and examples, using the computer allows more time for emphasizing the interpretation of results of the computations rather than expending enormous time and effort in the actual computations.

We have demonstrated through examples and exercises the importance of the following aspects of hypothesis testing:

1. The statement of the research hypothesis through the summarization of the researcher's goals into a statement about population parameters.
2. The selection of the most appropriate test statistic, including sample size computations for many procedures.
3. The necessity of considering both Type I and Type II error rates (α and β) when discussing the results of a statistical test of hypotheses.
4. The importance of considering both statistical significance of a test result and the practical significance of the results. Thus, we illustrate the importance of estimating effect sizes and the construction of confidence intervals for population parameters.
5. The statement of the results of the statistical in nonstatistical jargon that goes beyond the statements "reject H_0" or "fail to reject H_0."

We have provided displays for tests of hypotheses that give the null and alternative hypotheses, the test statistic, rejection region, and the assumptions under which the test is valid. Alternative test statistics are suggested when the stated conditions are not realized in a study. For example, the display shown here is given in Chapter 5.

Summary of a Statistical Test for μ with a Normal Population Distribution (σ Known) or Large Sample Size n

Hypotheses:

Case 1. $H_0: \mu \le \mu_0$ vs. $H_a: \mu > \mu_0$ (right-tailed test)
Case 2. $H_0: \mu \ge \mu_0$ vs. $H_a: \mu < \mu_0$ (left-tailed test)
Case 3. $H_0: \mu = \mu_0$ vs. $H_a: \mu \ne \mu_0$ (two-tailed test)

T.S.: $z = \dfrac{\bar{y} - \mu_0}{\sigma/\sqrt{n}}$

R.R.: For a probability α of a Type I error,

Case 1. Reject H_0 if $z \ge z_\alpha$.
Case 2. Reject H_0 if $z \le -z_\alpha$.
Case 3. Reject H_0 if $|z| \ge z_{\alpha/2}$.

Note: These procedures are appropriate if the population distribution is normally distributed with σ known. In most situations, if $n \ge 30$, then the Central Limit Theorem allows us to use these procedures when the population distribution is nonnormal. Also, if $n \ge 30$, then we can replace σ with the sample standard deviation s. The situation in which $n < 30$ is presented later in this chapter.

Changes in the New Edition

- A case study in most chapters helps students appreciate the role applied statistics plays in solving practical problems. Emphasis is placed on illustrating the steps in the learning from data process.
- The text emphasizes the importance of the assumptions on which statistical procedures are based. We discuss and—through the use of computer simulations—illustrate the robustness or lack of robustness of many estimation and test procedures. The procedures needed to investigate whether a study satisfies the necessary assumptions are presented and illustrated in many examples and exercises. Furthermore, in many settings, we provide alternative procedures when the assumptions are not met.
- Emphasis is placed on interpreting results and drawing conclusions from studies used in exercises and examples.
- We encourage the use of the computer to make most calculations after an illustration of the computations using the definitional form of the procedure using a small data set.
- Most examples include a graphical display of the data. Computer use greatly facilitates the use of more sophisticated graphical illustrations of statistical results.
- We provide techniques for calculating sample sizes and the probability of Type II errors for the t test and F test.
- The exercises have been expanded and updated, and the examples and exercises span the various disciplines and include many practical, real-life problems. The exercises give the student experience in applying the steps in "learning from data."

- A new chapter on design concepts for experiments and studies discusses the important factors that need to be considered prior to collecting data. We emphasize how not considering all factors at the design stage can result in a study that fails to answer questions of importance to the researcher.
- Two chapters on linear regression and correlation have been combined into a single chapter.
- Although the last chapter is devoted to communicating and documenting the results of the data analyses, we have incorporated many of these ideas throughout the text by the use of the case studies.
- All data sets of any consequence from the exercises are available at the Web site http://www.duxbury.com.
- Answers to selected exercises are also available at the Web site.

Features Retained from Previous Editions

- We include many practical applications of statistical methods and data analysis from agriculture, business, economics, education, engineering, medicine, law, political science, psychology, environmental studies, and sociology.
- Exercises are grouped into Basic Techniques and Applications.
- Review exercises appear at the end of most chapters.
- We provide computer output from Minitab, SAS, Systat, JMP, Stata, and SPSS in numerous examples and exercises.
- Attention is paid to the underlying assumptions. Graphical procedures and test procedures are provided to determine whether assumptions have been violated.
- The first chapter provides a discussion of "What is Statistics?" We provide a discussion of why students should study statistics along with a discussion of several major studies that illustrate the use of statistics in the solution of real-world problems.
- A lengthy discussion on data management and preparation of data for analysis is included.

Ancillaries

- Student Solutions Manual (0-534-37123-X) contains select solutions for problems in the textbook.
- Solutions Manual (0-534-37121-3) provides instructors with the solutions for all the problems in the textbook.
- Test Bank (0-534-37122-1) contains test questions based on the topics covered in the textbook.
- The Ott/Longnecker Web Resource Center includes data sets, errata, and additional resources for both students and faculty. To access the resource center, go to www.duxbury.com and select "Online Book Companions."

Acknowledgments

Many people have made valuable, constructive suggestions for the development of the original manuscript and during the preparation of the subsequent editions.

Carolyn Crockett, our editor at Duxbury, has been a tremendous motivator throughout the writing of this edition. We are also deeply indebted to Chris Franklin of the University of Georgia for her thoughtful comments concerning changes needed for this edition and for her review of the revised chapters. We greatly appreciate the insightful and constructive comments from the following reviewers: Deborah J. Rumsey, Ohio State University; Larry J. Ringer, Texas A&M University; Mosuk Chow, The Pennsylvania State University; Christine Franklin, University of Georgia; and Darcy P. Mays, Virginia Commonwealth University. We give special acknowledgment to Felicita Longnecker, Michael's wife, for her help in preparing draft materials and proofreading and for her assistance in typing the first draft of this edition.

R. Lyman Ott
Michael Longnecker

1 What Is Statistics?

CHAPTER 1

What Is Statistics?

1.1 Introduction

1.2 Why Study Statistics?

1.3 Some Current Applications of Statistics

1.4 What Do Statisticians Do?

1.5 Quality and Process Improvement

1.6 A Note to the Student

1.7 Summary

1.1 Introduction

What is statistics? Is it the addition of numbers? Is it graphs, batting averages, percentages of passes completed, percentages of unemployment, and, in general, numerical descriptions of society and nature?

Statistics is a set of scientific principles and techniques that are useful in reaching conclusions about populations and processes when the available information is both limited and variable; that is, statistics is the science of *learning from data*. Almost everyone—including corporate presidents, marketing representatives, social scientists, engineers, medical researchers, and consumers—deals with data. These data could be in the form of quarterly sales figures, percent increase in juvenile crime, contamination levels in water samples, survival rates for patients undergoing medical therapy, census figures, or input that helps determine which brand of car to purchase. In this text, we approach the study of statistics by considering the four steps in learning from data: (1) designing the data collection process, (2) preparing data for analysis (summarization, models), (3) analyzing the data, and (4) reporting the conclusions obtained during data analysis.

The text is divided into eight parts. The relationship among the eight parts of the textbook, the four steps in learning from data, and the chapters is shown in Table 1.1. As you can see from this table, much time is spent discussing how to analyze data using the basic methods (for central values, variances, and proportions), regression methods, and analysis of variance methods. However, you must remember that for each data set requiring analysis, someone has developed a plan for collecting the data (Step 1). After preparing the data for analysis (Step 2) and analyzing the data (Step 3), someone has to communicate the results of the analysis (Step 4) in written or verbal form to the intended audience. All four steps are important in learning from data; data analysis, although time consuming, is only one of four steps. In fact, unless Step 1 is carried out properly, the goals of the experiment or study often will not be achieved because the data set will be incomplete or contain improper information. Throughout the text, we will try to keep you focused on the bigger picture of learning from data. Also, it would be good for you to refer to this table periodically as a reminder of where each chapter fits in the overall scheme of things.

TABLE 1.1

Organization of the text

Parts of the Textbook	Steps in Learning from Data	Chapters of the Textbook
1 Introduction		1 What Is Statistics?
2 Collecting Data	1	2 Using Surveys and Scientific Studies to Gather Data
3 Summarizing Data	2	3 Data Description
4 Tools and Concepts		4 Probability and Probability Distributions
5 Analyzing Data: Central Values, Variances, and Proportions	3	5 Inferences about Population Central Values
		6 Inferences Comparing Two Population Central Values
		7 Inferences about Population Variances
		8 Inferences about More Than Two Population Central Values
		9 Multiple Comparisons
		10 Categorical Data
6 Analyzing Data: Regression Methods and Model Building	3	11 Linear Regression and Correlation
		12 Multiple Regression and the General Linear Model
		13 Building Regression Models with Diagnostics
7 Analyzing Data: Design of Experiments and Analysis of Variance	3	14 Design Concepts for Experiments and Studies
		15 Analysis of Variance for Standard Designs
		16 The Analysis of Covariance
		17 Analysis of Variance for Some Fixed, Random, and Mixed Effects Models
		18 Repeated Measures and Crossover Designs
		19 Analysis of Variance for Some Unbalanced Designs
8 Communicating and Documenting the Results of Analyses	4	20 Communicating and Documenting the Results of Analyses

Before we jump into the study of statistics, let's consider three instances in which the application of statistics could help to solve a practical problem.

1. A lightbulb manufacturer produces approximately a half million bulbs per day. The quality control department must monitor the defect rate of the bulbs. It can accomplish this task by testing each bulb, but the cost would be substantial and would greatly increase the price per bulb. An alternative approach is to select 1,000 bulbs from the daily

production of 500,000 bulbs and test each of the 1,000. The fraction of defective bulbs in the 1,000 tested could be used to estimate the fraction defective in the entire day's production, provided that the 1,000 bulbs were selected in the proper fashion. We will demonstrate in later chapters that the fraction defective in the tested bulbs will probably be quite close to the fraction defective for the entire day's production of 500,000 bulbs.

2. To investigate the claim that people who quit smoking often experience a subsequent weight gain, researchers selected a random sample of 400 participants who had successfully participated in programs to quit smoking. The individuals were weighed at the beginning of the program and again one year later. The average change in weight of the participants was an increase of 5 pounds. The investigators concluded that there was evidence that the claim was valid. We will develop techniques in later chapters to assess when changes are truly significant changes and not changes due to random chance.

3. For a study of the effects of nitrogen fertilizer on wheat production, a total of 15 fields were available to the researcher. She randomly assigned three fields to each of the five nitrogen rates under investigation. The same variety of wheat was planted in all 15 fields. The fields were cultivated in the same manner until harvest, and the number of pounds of wheat per acre was then recorded for each of the 15 fields. The experimenter wanted to determine the optimal level of nitrogen to apply to *any* wheat field, but of course, she was limited to running experiments on a limited number of fields. After determining the amount of nitrogen that yielded the largest production of wheat in the study fields, the experimenter then concluded that similar results would hold for wheat fields possessing characteristics somewhat the same as the study fields. Is the experimenter justified in reaching this conclusion?

4. Similar applications of statistics are brought to mind by the frequent use of the *New York Times/CBS News, Washington Post/ABC News*, CNN, Harris and Gallup polls. How can these pollsters determine the opinions of more than 195 million Americans who are of voting age? They certainly do not contact every potential voter in the United States. Rather, they sample the opinions of a small number of potential voters, perhaps as few as 1,500, to estimate the reaction of every person of voting age in the country. The amazing result of this process is that the fraction of those persons contacted who hold a particular opinion will closely match the fraction in the total population holding that opinion at a particular time. We will supply convincing supportive evidence of this assertion in subsequent chapters.

These problems illustrate the four steps in learning from data. First, each problem involved designing an experiment or study. The quality control group had to decide both how many bulbs needed to be tested and how to select the sample of 1,000 bulbs from the total production of bulbs to obtain valid results. The polling groups must decide how many voters to sample and how to select these individuals in order to obtain information that is representative of the population of all voters. Similarly, it was necessary to carefully plan how many participants in the weight gain study were needed and how they were to be selected from the list of all such participants. Furthermore, what variables should the researchers have measured on each participant? Was it necessary to know each participant's age, sex, physical fitness, and other health-related variables, or was weight the only important variable? The results of the study may not be relevant to the general population if many of the participants in the study had a particular health condition. In the wheat experiments, it was important to measure both soil characteristics of the fields and environmental conditions, such as temperature and rainfall, to obtain results that could be generalized to fields not included in the study. The design of a study or experiment is crucial to obtaining results that can be generalized beyond the study.

Finally, having collected, summarized, and analyzed the data, it is important to report the results in unambiguous terms to interested people. For the lightbulb example, management and technical staff would need to know the quality of their production batches. Based on this information, they could determine whether adjustments in the process are necessary. Therefore, the results of the statistical analyses cannot be presented in ambiguous terms; decisions must be made from a well-defined knowledge base. The results of the weight gain study would be of vital interest to physicians who have patients participating in the smoking-cessation program. If a significant increase in weight was recorded for those individuals who had quit smoking, physicians may have to recommend diets so that the former smokers would not go from one health problem (smoking) to another (elevated blood pressure due to being overweight). It is crucial that a careful description of the participants—that is, age, sex, and other health-related information—be included in the report. In the wheat study, the experiments would provide farmers with information that would allow them to economically select the optimum amount of nitrogen required for their fields. Therefore, the report must contain information concerning the amount of moisture and types of soils present on the study fields. Otherwise, the conclusions about optimal wheat production may not pertain to farmers growing wheat under considerably different conditions.

population

sample

To infer validly that the results of a study are applicable to a larger group than just the participants in the study, we must carefully define the **population** (see Definition 1.1) to which inferences are sought and design a study in which the **sample** (see Definition 1.2) has been appropriately selected from the designated population. We will discuss these issues in Chapter 2.

DEFINITION 1.1

A **population** is the set of all measurements of interest to the sample collector. (See Figure 1.1.)

FIGURE 1.1
Population and sample

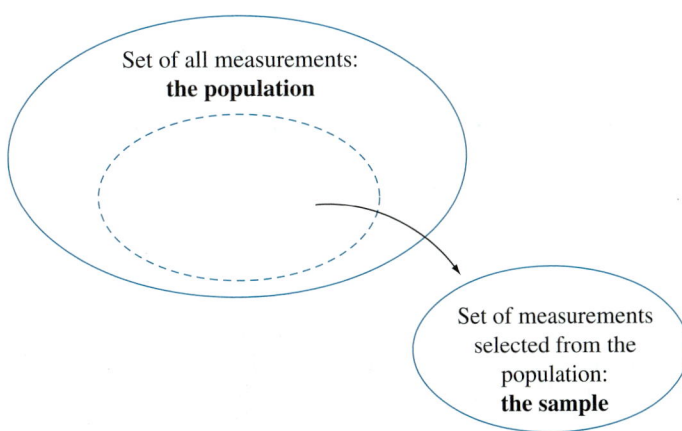

Set of all measurements:
the population

Set of measurements selected from the population:
the sample

DEFINITION 1.2   A **sample** is any subset of measurements selected from the population. (See Figure 1.1.)

1.2   Why Study Statistics?

We can think of two good reasons for taking an introductory course in statistics. One reason is that you need to know how to evaluate published numerical facts. Every person is exposed to manufacturers' claims for products; to the results of sociological, consumer, and political polls; and to the published results of scientific research. Many of these results are inferences based on sampling. Some inferences are valid; others are invalid. Some are based on samples of adequate size; others are not. Yet all these published results bear the ring of truth. Some people (particularly statisticians) say that statistics can be made to support almost anything. Others say it is easy to lie with statistics. Both statements are true. It is easy, purposely or unwittingly, to distort the truth by using statistics when presenting the results of sampling to the uninformed.

A second reason for studying statistics is that your profession or employment may require you to interpret the results of sampling (surveys or experimentation) or to employ statistical methods of analysis to make inferences in your work. For example, practicing physicians receive large amounts of advertising describing the benefits of new drugs. These advertisements frequently display the numerical results of experiments that compare a new drug with an older one. Do such data really imply that the new drug is more effective, or is the observed difference in results due simply to random variation in the experimental measurements?

Recent trends in the conduct of court trials indicate an increasing use of probability and statistical inference in evaluating the quality of evidence. The use of statistics in the social, biological, and physical sciences is essential because all these sciences make use of observations of natural phenomena, through sample surveys or experimentation, to develop and test new theories. Statistical methods are employed in business when sample data are used to forecast sales and profit. In addition, they are used in engineering and manufacturing to monitor product quality. The sampling of accounts is a useful tool to assist accountants in conducting audits. Thus, statistics plays an important role in almost all areas of science, business, and industry; persons employed in these areas need to know the basic concepts, strengths, and limitations of statistics.

1.3   Some Current Applications of Statistics

Acid Rain: A Threat to Our Environment

The accepted causes of acid rain are sulfuric and nitric acids; the sources of these acidic components of rain are hydrocarbon fuels, which spew sulfur and nitric oxide into the atmosphere when burned. Here are some of the many effects of acid rain:

- Acid rain, when present in spring snow melts, invades breeding areas for many fish, which prevents successful reproduction. Forms of life that depend on ponds and lakes contaminated by acid rain begin to disappear.

- In forests, acid rain is blamed for weakening some varieties of trees, making them more susceptible to insect damage and disease.
- In areas surrounded by affected bodies of water, vital nutrients are leached from the soil.
- Man-made structures are also affected by acid rain. Experts from the United States estimate that acid rain has caused nearly $15 billion of damage to buildings and other structures thus far.

Solutions to the problems associated with acid rain will not be easy. The National Science Foundation (NSF) has recommended that we strive for a 50% reduction in sulfur-oxide emissions. Perhaps that is easier said than done. High-sulfur coal is a major source of these emissions, but in states dependent on coal for energy, a shift to lower sulfur coal is not always possible. Instead, better scrubbers must be developed to remove these contaminating oxides from the burning process before they are released into the atmosphere. Fuels for internal combustion engines are also major sources of the nitric and sulfur oxides of acid rain. Clearly, better emission control is needed for automobiles and trucks.

Reducing the oxide emissions from coal-burning furnaces and motor vehicles will require greater use of existing scrubbers and emission control devices as well as the development of new technology to allow us to use available energy sources. Developing alternative, cleaner energy sources is also important if we are to meet NSF's goal. Statistics and statisticians will play a key role in monitoring atmosphere conditions, testing the effectiveness of proposed emission control devices, and developing new control technology and alternative energy sources.

Determining the Effectiveness of a New Drug Product

The development and testing of the Salk vaccine for protection against poliomyelitis (polio) provide an excellent example of how statistics can be used in solving practical problems. Most parents and children growing up before 1954 can recall the panic brought on by the outbreak of polio cases during the summer months. Although relatively few children fell victim to the disease each year, the pattern of outbreak of polio was unpredictable and caused great concern because of the possibility of paralysis or death. The fact that very few of today's youth have even heard of polio demonstrates the great success of the vaccine and the testing program that preceded its release on the market.

It is standard practice in establishing the effectiveness of a particular drug product to conduct an experiment (often called a *clinical trial*) with human participants. For some clinical trials, assignments of participants are made at random, with half receiving the drug product and the other half receiving a solution or tablet (called a *placebo*) that does not contain the medication. One statistical problem concerns the determination of the total number of participants to be included in the clinical trial. This problem was particularly important in the testing of the Salk vaccine because data from previous years suggested that the incidence rate for polio might be less than 50 cases for every 100,000 children. Hence, a large number of participants had to be included in the clinical trial in order to detect a difference in the incidence rates for those treated with the vaccine and those receiving the placebo.

With the assistance of statisticians, it was decided that a total of 400,000 children should be included in the Salk clinical trial begun in 1954, with half of them randomly assigned the vaccine and the remaining children assigned the placebo. No other clinical trial had ever been attempted on such a large group of participants. Through a public school inoculation program, the 400,000 participants were treated and then observed over the summer to determine the number of children contracting polio. Although fewer than 200 cases of polio were reported for the 400,000 participants in the clinical trial, more than three times as many cases appeared in the group receiving the placebo. These results, together with some statistical calculations, were sufficient to indicate the effectiveness of the Salk polio vaccine. However, these conclusions would not have been possible if the statisticians and scientists had not planned for and conducted such a large clinical trial.

The development of the Salk vaccine is not an isolated example of the use of statistics in the testing and developing of drug products. In recent years, the Food and Drug Administration (FDA) has placed stringent requirements on pharmaceutical firms to establish the effectiveness of proposed new drug products. Thus, statistics has played an important role in the development and testing of birth control pills, rubella vaccines, chemotherapeutic agents in the treatment of cancer, and many other preparations.

Applications of Statistics in Our Courts

Libel suits related to consumer products have touched each one of us; you may have been involved as a plaintiff or defendant in a suit or you may know of someone who was involved in such litigation. Certainly we all help to fund the costs of this litigation indirectly through increased insurance premiums and increased costs of goods. The testimony in libel suits concerning a particular product (automobile, drug product, and so on) frequently leans heavily on the interpretation of data from one or more scientific studies involving the product. This is how and why statistics and statisticians have been pulled into the courtroom.

For example, epidemiologists have used statistical concepts applied to data to determine whether there is a statistical ''association'' between a specific characteristic, such as the leakage in silicone breast implants, and a disease condition, such as an autoimmune disease. An epidemiologist who finds an association should try to determine whether the observed statistical association from the study is due to random variation or whether it reflects an actual association between the characteristic and the disease. Courtroom arguments about the interpretations of these types of associations involve data analyses using statistical concepts as well as a clinical interpretation of the data. Many other examples exist in which statistical models are used in court cases. In salary discrimination cases, a lawsuit is filed claiming that an employer underpays employees on the basis of age, ethnicity, or sex. Statistical models are developed to explain salary differences based on many factors, such as work experience, years of education, and work performance. The adjusted salaries are then compared across age groups or ethnic groups to determine whether significant salary differences exist after adjusting for the relevant work performance factors.

Estimating Bowhead Whale Population Size

Raftery and Zeh (1998) discuss the estimation of the population size and rate of increase in bowhead whales, *Balaena mysticetus*. The importance of such a study derives from the fact that bowheads were the first species of great whale for which commercial whaling was stopped; thus, their status indicates the recovery prospects of other great whales. Also, the International Whaling Commission uses these estimates to determine the aboriginal subsistence whaling quota for Alaskan Eskimos. To obtain the necessary data, researchers conducted a visual and acoustic census off Point Barrow, Alaska. The researchers then applied statistical models and estimation techniques to the data obtained in the census to determine whether the bowhead population had increased or decreased since commercial whaling was stopped. The statistical estimates showed that the bowhead population was increasing at a healthy rate, indicating that stocks of great whales that have been decimated by commercial hunting can recover after hunting is discontinued.

Ozone Exposure and Population Density

Ambient ozone pollution in urban areas is one of the nation's most pervasive environmental problems. Whereas the decreasing stratospheric ozone layer may lead to increased instances of skin cancer, high ambient ozone intensity has been shown to cause damage to the human respiratory system as well as to agricultural crops and trees. The Houston, Texas, area has ozone concentrations rated second only to Los Angeles that exceed the National Ambient Air Quality Standard. Carroll et al. (1997) describe how to analyze the hourly ozone measurements collected in Houston from 1980 to 1993 by 9 to 12 monitoring stations. Besides the ozone level, each station also recorded three meteorological variables: temperature, wind speed, and wind direction.

The statistical aspect of the project had three major goals:

1. Provide information (and/or tools to obtain such information) about the amount and pattern of missing data, as well as about the quality of the ozone and the meteorological measurements.
2. Build a model of ozone intensity to predict the ozone concentration at any given location within Houston at any given time between 1980 and 1993.
3. Apply this model to estimate exposure indices that account for either a long-term exposure or a short-term high-concentration exposure; also, relate census information to different exposure indices to achieve population exposure indices.

The spatial–temporal model the researchers built provided estimates demonstrating that the highest ozone levels occurred at locations with relatively small populations of young children. Also, the model estimated that the exposure of young children to ozone decreased by approximately 20% from 1980 to 1993. An examination of the distribution of population exposure had several policy implications. In particular, it was concluded that the current placement of monitors is not ideal if one is concerned with assessing population exposure. This project involved all four components of learning from data: planning where the monitoring stations should be placed within the city, how often data should be collected, and what variables should be recorded; conducting spatial–temporal graphing of the data; creating spatial–temporal models of the ozone data, meteorological data, and demographic data; and finally, writing a report that could assist local and federal officials in formulating policy with respect to decreasing ozone levels.

Opinion and Preference Polls

Public opinion, consumer preference, and election polls are commonly used to assess the opinions or preferences of a segment of the public for issues, products, or candidates of interest. We, the American public, are exposed to the results of these polls daily in newspapers, in magazines, on the radio, and on television. For example, the results of polls related to the following subjects were printed in local newspapers over a 2-day period:

- Consumer confidence related to future expectations about the economy
- Preferences for candidates in upcoming elections and caucuses
- Attitudes toward cheating on federal income tax returns
- Preference polls related to specific products (for example, foreign vs. American cars, Coke vs. Pepsi, McDonald's vs. Wendy's)
- Reactions of North Carolina residents toward arguments about the morality of tobacco
- Opinions of voters toward proposed tax increases and proposed changes in the Defense Department budget

A number of questions can be raised about polls. Suppose we consider a poll on the public's opinion toward a proposed income tax increase in the state of Michigan. *What was the population of interest to the pollster?* Was the pollster interested in all residents of Michigan or just those citizens who currently pay income taxes? *Was the sample in fact selected from this population?* If the population of interest was all persons currently paying income taxes, did the pollster make sure that all the individuals sampled were current taxpayers? *What questions were asked and how were the questions phrased?* Was each person asked the same question? Were the questions phrased in such a manner as to bias the responses? Can we believe the results of these polls? Do these results "represent" how the general public *currently* feels about the issues raised in the polls?

Opinion and preference polls are an important, visible application of statistics for the consumer. We will discuss this topic in more detail in Chapter 10. We hope that after studying this material you will have a better understanding of how to interpret the results of these polls.

1.4 What Do Statisticians Do?

What do statisticians do? In the context of learning from data, statisticians are involved with all aspects of designing a study or experiment, preparing the data for analysis using graphical and numerical summaries, analyzing the data, and reporting the results of their analyses. There are both good and bad ways to gather data. Statisticians apply their knowledge of existing survey techniques and scientific study designs or they develop new techniques to provide a guide to good methods of data collection. We will explore these ideas further in Chapter 2.

Once the data are gathered, they must be summarized before any meaningful interpretation can be made. Statisticians can recommend and apply useful methods for summarizing data in graphical, tabular, and numerical forms. Intelligent graphs and tables are useful first steps in making sense of the data. Also, measures of the average (or typical) value and some measure of the range or spread of the data help in interpretation. These topics will be discussed in detail in Chapter 3.

The objective of statistics is to make an inference about a population of interest based on information obtained from a sample of measurements from that population. The analysis stage of learning from data deals with making inferences. For example, a market research study reaches only a few of the potential buyers of a new product, but the probable reaction of the set of potential buyers (population) must be inferred from the reactions of the buyers included in the study (sample). If the market research study has been carefully planned and executed, the reactions of those included in the sample should agree reasonably well (but not necessarily exactly) with the population. We can say this because the basic concepts of probability allow us to make an inference about the population of interest that includes our best guess plus a statement of the probable error in our best guess.

We will illustrate how inferences are made by an example. Suppose an auditor randomly selects 2,000 financial accounts from a set of more than 25,000 accounts and finds that 84 (4.2%) are in error. What can be said about the set of 25,000 accounts? What inference can we make about the percentage of accounts in error for the population of 25,000 accounts based on information obtained from the sample of 2,000 accounts? We will show (in Chapter 10) that our best guess (inference) about the percentage of accounts in error for the population is 4.2%, and this best guess should be within $\pm.9\%$ of the actual unknown percentage of accounts in error for the population. The plus-or-minus factor is called the *probable error* of our inference. Anyone can make a guess about the percentage of accounts in error; concepts of probability allow us to calculate the (probable) error of our guess.

In dealing with the analyses of data, statisticians can apply existing methods for making inferences; some theoretical statisticians engage in the development of new methods with more advanced mathematics and probability theory. Our study of the methods for analyzing sample data will begin in Chapter 5, after we discuss the basic concepts of probability and sampling distributions in Chapter 4.

Finally, statisticians are involved with communicating the results of their analyses as the final stage in making sense of data. The form of the communication varies from an informal conversation to a formal report. The advantage of a more formal verbal presentation with visual aids or a study report is that the communication can use graphical, tabular, and numerical displays as well as the analyses done on the data to help convey the "sense" found in the data. Too often, this is lost in an informal conversation. The report or communication should convey to the intended audience what can be gleaned from the sample data, and it should be conveyed in as nontechnical terms as possible so there can be no confusion as to what is inferred. More information about the communication of results is presented in Chapter 20. We will identify the important components that should be included in the report while discussing case studies used to illustrate the statistical concepts in several of the chapters.

It is important to note that the ideas in the preceding discussion are relevant to everyone involved in a study or experiment. Degreed statisticians are somewhat rare individuals. Many organizations have no statisticians or only a few in their employment. Thus, in many studies, the design used in collecting the data, the summary and statistical analyses of the data, and the communication of the study results will be conducted by the individuals involved in the study with little or no support from a degreed statistician. In those cases where a statistician is an active member of the research team, it is still important for the other members of the group to have general knowledge of the concepts involved in a statistical design and data analysis. In fact, each member of the team brings an area of expertise and experience to the problems being addressed. Then within the context of the team dynamics, decisions will be made about the design of the study and how the results of the analyses will be communicated.

1.5   Quality and Process Improvement

One might wonder, at this stage, why we would bring up the subject of quality and process improvement in a statistics textbook. We do so to make you aware of some of the broader issues involved with learning from data in the business and scientific communities.

The post–World War II years saw U.S. business and the U.S. economy dominate world business, and this lasted for about 30 years. During this time, there was little attempt to change the ways things were done; the major focus was on doing things on a much grander scale, perfecting mass production. However, from the mid-1970s through today, many industries have had to face fierce competition from their counterparts in Japan and, more recently, from other countries in the Far East, such as China and Korea.

Quality, rather than *quantity,* has become the principal buying gauge used by consumers, and American industries have had a difficult time adjusting to this new emphasis. Unless there are drastic changes in the way many American industries approach their businesses, there will be many more casualties to the "quality" revolution.

The Japanese were the first to learn the lessons of quality. They readily used the statistical quality-control and process-control suggestions espoused by Deming (1981) and others and installed total quality programs. Through the organization—from top management down—they had a commitment to improving the quality of their products and procedures. They were never satisfied with the way things were and continually looked for new and better ways.

A number of American companies have now begun the journey toward excellence through the institution of a quality-improvement process. Listed below are ten basic requirements that provide the foundation for a successful quality-improvement process:

**Fundamental Requirements for a Successful
Quality-Improvement Process**

1. A focus on the customer as the most important part of the process
2. A long-term commitment by management to make the quality-improvement process part of the management system
3. The belief that there is room to improve
4. The belief that preventing problems is better than reacting to them
5. Management focus, leadership, and participation
6. A performance standard (goal) of zero errors
7. Participation by all employees, both as groups and as individuals
8. An improvement focus on the process, not the people
9. The belief that suppliers will work with you if they understand your needs
10. Recognition for success

Embedded in a companywide quality-improvement process or running concurrent with such a process is the idea of improving the work processes. For years, companies, in trying to boost and improve performance, have tried to speed up their processes, usually with additional people or technology but without addressing possible deficiencies in the work processes. In the ground-breaking book, *Reengineering the Corporation* (1993) by Michael Hammer and James Champy and in Hammer's later book, *Beyond Reengineering* (1996), Hammer and Champy addressed how a corporation could achieve radical improvement in quality, efficiency, and effectiveness by completely rethinking their business processes that have been maintained in a rapidly changing business and technology environment. If we define a task as a unit of work, and a process as a sequence of related tasks that create value for the customer, Hammer and Champy were offering corporations a way to refocus their change efforts in value-creating activities.

The case for change is compelling. Within almost every major business—apparel (e.g., Nike), chemicals (e.g., Dupont), computer equipment (e.g., Dell), computer software (e.g., Microsoft), electronics (e.g., General Electric), food (e.g., Nestlé), general merchandising (e.g., WalMart), network communications (e.g., Cisco), petroleum (e.g., Exxon Mobil), pharmaceuticals (e.g., Merck), and so on—the competitive position of the segment leader has been, is currently, or will soon be challenged. In many cases the industry leader has not kept pace with the dizzying changes occurring in the marketplace. Mergers proliferate with high expectations from management and shareholders for increased market share, cost synergies (reductions), and increased profitability. Unfortunately, the list of successful mergers (as defined by those meeting the initial case for action driving the merger) is pitifully small.

Something else is needed. Christopher Meyer, in his book, *Fast Cycle Time* (1993), makes the case that in an ever-changing marketplace, the competitor that can "consistently, reliably and profitably" provide the greatest value to the customer will "win." Meyer's basic premise is that a corporation must shorten its overall business cycle, which begins with identification of a customer's need and ends with the payment for a product delivered or service rendered. A company that can do this well *over time,* as needs and the competitive environment change, will win.

Whether a company focuses on business process improvement or fast cycle time, the foundation for change will be the underlying data about customer needs, current internal cycle time, and comparable benchmark data in the industry. "Winners" in the ongoing competition will be those who define what they're trying to do, establish ongoing data requirements to assess customer needs and current state of operations, rapidly implement recommended changes, and document their learning. These four points, which are very similar to the four steps in learning from data discussed earlier in the chapter, drive home the relevance of statistics (learning from data) to the business environment. A number of statistical tools and techniques that can help in these business improvement efforts are shown here.

Statistical Tools, Techniques, and Methods Used in Quality Improvement and Reengineering

- Histograms
- Numerical descriptive measures (means, standard deviations, proportions, etc.)
- Scatterplots
- Line graphs (scatterplots with dots connected)
- Control charts: $\bar{y}$ (sample mean), r (sample range), and s (sample standard deviation)
- Sampling schemes
- Experimental designs

The statistical tools and concepts listed here and discussed in this textbook are only a small component of a business process improvement or fast-cycle-time initiative. As you encounter these tools and concepts in various parts of this text, keep in mind where you think they may have application in business improvement efforts. Quality improvement, process redesign, and fast cycle time are clearly the focus of American industry for the 1990s in world markets characterized by increased competition, more consolidation, and increased specialization. These shifts will have impacts on us all, either as consumers or business participants, and it will be useful to know some of the statistical tools that are part of this revolution. Finally, in recent years the ideas and principles of quality control have been applied in areas outside of manufacturing. Service industries such as hotels, restaurants, and department stores have successfully applied the principles of quality control in their businesses. Many federal agencies—for example, the IRS, the Department of Defense, and the USDA—have adapted the principles of quality control to improve the performance of their agencies.

1.6   A Note to the Student

We think with words and concepts. A study of the discipline of statistics requires us to memorize new terms and concepts (as does the study of a foreign language). Commit these definitions, theorems, and concepts to memory.

Also, focus on the broader concept of making sense of data. Do not let details obscure these broader characteristics of the subject. The teaching objective of this text is to identify and amplify these broader concepts of statistics.

1.7   Summary

The discipline of statistics and those who apply the tools of that discipline deal with learning from data. Medical researchers, social scientists, accountants, agronomists, consumers, government leaders, and professional statisticians are all involved with data collection, data summarization, data analysis, and the effective communication of the results of data analysis.

Supplementary Exercises

Basic Techniques

Bio.   **1.1**   Selecting the proper diet for shrimp or other sea animals is an important aspect of sea farming. A researcher wishes to estimate the mean weight of shrimp maintained on a specific diet for a period of 6 months. One hundred shrimp are randomly selected from an artificial pond and each is weighed.

  a. Identify the population of measurements that is of interest to the researcher.
  b. Identify the sample.
  c. What characteristics of the population are of interest to the researcher?
  d. If the sample measurements are used to make inferences about certain characteristics of the population, why is a measure of the reliability of the inferences important?

Env.   **1.2**   Radioactive waste disposal as well as the production of radioactive material in some mining operations are creating a serious pollution problem in some areas of the United States. State health officials have decided to investigate the radioactivity levels in one suspect area. Two hundred points in the area are randomly selected and the level of radioactivity is measured at each point. Answer questions (a), (b), (c), and (d) in Exercise 1.1 for this sampling situation.

Soc.   **1.3**   A social researcher in a particular city wishes to obtain information on the number of children in households that receive welfare support. A random sample of 400 households is selected from the city welfare rolls. A check on welfare recipient data provides the number of children in each household. Answer questions (a), (b), (c), and (d) of Exercise 1.1 for this sample survey.

Pol. Sci.   **1.4**   Search issues of your local newspaper or news magazine to locate the results of a recent opinion survey.

a. Identify the items that were observed in order to obtain the sample measurements.
b. Identify the measurement made on each item.
c. Clearly identify the population associated with the survey.
d. What characteristic(s) of the population was (were) of interest to the pollster?
e. Does the article explain how the sample was selected?
f. Does the article include the number of measurements in the sample?
g. What type of inference was made concerning the population characteristics?
h. Does the article tell you how much faith you can place in the inference about the population characteristic?

Gov.   **1.5**   Because of a recent increase in the number of neck injuries incurred by high school football players, the Department of Commerce designed a study to evaluate the strength of football helmets worn by high school players in the United States. A total of 540 helmets were collected from the five companies that currently produce helmets. The agency then sent the helmets to an independent testing agency to evaluate the impact cushioning of the helmet and the amount of shock transmitted to the neck when the face mask was twisted.

a. What is the population of interest?
b. What is the sample?
c. What variables should be measured?
d. What are some of the major limitations of this study in regard to the safety of helmets worn by high school players? For example, is the neck strength of the player related to the amount of shock transmitted to the neck and whether the player will be injured?

Edu.   **1.6**   The faculty senate at a major university with 35,000 students is considering changing the current grading policy from A, B, C, D, F to a plus and minus system—that is, B+, B, B− rather than just B. The faculty is interested in the students' opinions concerning this change and will sample 500 students.

a. What is the population of interest?
b. What is the sample?
c. How could the sample be selected?
d. What type of questions should be included in the questionnaire?

2 Using Surveys and Scientific

Studies to Gather Data

Using Surveys and Scientific Studies to Gather Data

2.1 Introduction

2.2 Surveys

2.3 Scientific Studies

2.4 Observational Studies

2.5 Data Management: Preparing Data for Summarization and Analysis

2.6 Summary

2.1 Introduction

As mentioned in Chapter 1, the first step in learning from data is to carefully think through the objectives of the study (**think before doing**). The design of the data collection process is the crucial step in *intelligent data gathering.* The process takes a conscious, concerted effort focused on the following steps:

- Specifying the objective of the study, survey, or experiment
- Identifying the variable(s) of interest
- Choosing an appropriate design for the survey or scientific study
- Collecting the data

To specify the objective of the study you must understand the problem being addressed. For example, the transportation department in a large city wants to assess the public's perception of the city's bus system in order to increase usage of buses within the city. Thus, the department needs to determine what aspects of the bus system determine whether or not a person will ride the bus. The objective of the study is to identify factors that the transportation department can alter to increase the number of people using the bus system.

To identify the variables of interest, you must examine the objective of the study. For the bus system, some major factors can be identified by reviewing studies conducted in other cities and by brainstorming with the bus system employees. Some of the factors may be safety, cost, cleanliness of the buses, whether or not there is a bus stop close to the person's home or place of employment, and how often the bus fails to be on time. The measurements to be obtained in the study would consist of importance ratings (very important, important, no opinion, somewhat unimportant, very unimportant) of the identified factors. Demographic information, such as age, sex, income, and place of residence, would also be measured. Finally, the measurement of variables related to how frequently a person currently rides the buses would be of importance. Once the objectives are determined and the variables of interest are specified, you must select the most appropriate method to collect the data. Data collection processes include surveys, experiments, and the examination of existing data from business records, censuses, government records, and previous studies. The theory of sample surveys and the theory of experimental designs provide excellent methodology for data collection. Usually surveys are passive. The goal of the survey is to gather data on existing conditions, attitudes, or behaviors. Thus, the transportation department would need to construct a questionnaire and then sample current riders of the buses and persons who use other forms of transportation within the city.

Scientific studies, on the other hand, tend to be more active: The person conducting the study varies the experimental conditions to study the effect of the conditions on the outcome of the experiment. For example, the transportation department could decrease the bus fares on a few selected routes and assess whether the usage of its buses increased. However, in this example, other factors not under the bus system's control may also have changed during this time period. Thus, an increase in bus usage may have taken place because of a strike of subway workers or an increase in gasoline prices. The decrease in fares was only one of several factors that may have "caused" the increase in the number of persons riding the buses.

In most scientific experiments, as many as possible of the factors that affect the measurements are under the control of the experimenter. A floriculturist wants to determine the effect of a new plant stimulator on the growth of a commercially produced flower. The floriculturist would run the experiments in a greenhouse, where temperature, humidity, moisture levels, and sunlight are controlled. An equal number of plants would be treated with each of the selected quantities of the growth stimulator, including a control—that is, no stimulator applied. At the conclusion of the experiment, the size and health of the plants would be measured. The optimal level of the plant could then be determined, because ideally all other factors affecting the size and health of the plants would be the same for all plants in the experiment.

In this chapter, we will consider some of the survey methods and designs for scientific studies. We will also make a distinction between a scientific study and an observational study.

2.2 Surveys

Information from surveys affects almost every facet of our daily lives. These surveys determine such government policies as the control of the economy and the promotion of social programs. Opinion polls are the basis of much of the news reported by the various news media. Ratings of television shows determine which shows will be available for viewing in the future.

Who conducts surveys? We are all familiar with public opinion polls: the *New York Times/CBS News, Washington Post/ABC News,* Harris, Gallup for *Newsweek* and CNN polls. However, the vast majority of surveys are conducted for a specific industrial, governmental, administrative, or scientific purpose. For example, auto manufacturers use surveys to find out how satisfied customers are with their cars. Frequently we are asked to complete a survey as part of the warranty registration process following the purchase of a new product. Many important studies involving health issues are determined using surveys—for example, amount of fat in the diet, exposure to secondhand smoke, condom use and the prevention of AIDS, and the prevalence of adolescent depression.

The U.S. Bureau of the Census is required by the U.S. Constitution to enumerate the population every 10 years. With the growing involvement of the government in the lives of its citizens, the Census Bureau has expanded its role beyond just counting the population. An attempt is made to send a census questionnaire in the mail to every household in the United States. Since the 1940 census, in addition to the complete count information, further information has been obtained from representative samples of the population. In the 2000 census, variable sampling rates were employed. For most of the country, approximately five of six households were asked to answer the 14 questions on the short version of the form. The remaining households responded to a longer version of the form containing an additional 45 questions. Many agencies and individuals use the resulting information for many purposes. The federal government uses it to determine allocations of funds to states and cities. Businesses use it to forecast sales, to manage personnel, and to establish future site locations. Urban and regional planners use it to plan land use, transportation networks, and energy consumption. Social scientists use it to study economic conditions, racial balance, and other aspects of the quality of life.

The U.S. Bureau of Labor Statistics (BLS) routinely conducts more than 20 surveys. Some of the best known and most widely used are the surveys that establish the consumer price index (CPI). The CPI is a measure of price change for a fixed market basket of goods and services over time. It is a measure of inflation and serves as an economic indicator for government policies. Businesses tie wage rates and pension plans to the CPI. Federal health and welfare programs, as well as many state and local programs, tie their bases of eligibility to the CPI. Escalator clauses in rents and mortgages are based on the CPI. This one index, determined on the basis of sample surveys, plays a fundamental role in our society.

Many other surveys from the BLS are crucial to society. The monthly Current Population Survey establishes basic information on the labor force, employment, and unemployment. The consumer expenditure surveys collect data on family expenditures for goods and services used in day-to-day living. The Establishment Survey collects information on employment hours and earnings for nonagricultural business establishments. The survey on occupational outlook provides information on future employment opportunities for a variety of occupations, projecting to approximately 10 years ahead. Other activities of the BLS are addressed in the *BLS Handbook of Methods* (1982).

Opinion polls are constantly in the news, and the names of Gallup and Harris have become well known to everyone. These polls, or sample surveys, reflect the attitudes and opinions of citizens on everything from politics and religion to sports and entertainment. The Nielsen ratings determine the success or failure of TV shows. The Nielsen retail index furnishes up-to-date sales data on foods, cosmetics, pharmaceuticals, beverages, and many other classes of products. The data come from auditing inventories and sales in 1,600 stores across the United States every 60 days.

Businesses conduct sample surveys for their internal operations in addition to using government surveys for crucial management decisions. Auditors estimate account balances and check on compliance with operating rules by sampling accounts. Quality control of manufacturing processes relies heavily on sampling techniques.

Another area of business activity that depends on detailed sampling activities is marketing. Decisions on which products to market, where to market them, and how to advertise them are often made on the basis of sample survey data. The data may come from surveys conducted by the firm that manufactures the product or may be purchased from survey firms that specialize in marketing data.

Sampling Techniques

A crucial element in any survey is the manner in which the sample is selected from the population. If the individuals included in the survey are selected based on convenience alone, there may be biases in the sample survey, which would prevent the survey from accurately reflecting the population as a whole. For example, a marketing graduate student developed a new approach to advertising and, to evaluate this new approach, selected the students in a large undergraduate business course to assess whether the new approach is an improvement over standard advertisements. Would the opinions of this class of students be representative of the general population of people to which the new approach to advertising would be applied? The income levels, ethnicity, education levels, and many other socioeconomic characteristics of the students may differ greatly from the population of interest. Furthermore, the students may be coerced into participating in the study by their instructor and hence may not give the most candid answers to questions on a survey. Thus, the manner in which a sample is selected is of utmost importance to the credibility and applicability of the study's results.

simple random sampling   The basic design (**simple random sampling**) consists of selecting a group of *n* units in such a way that each sample of size *n* has the same chance of being selected. Thus, we can obtain a random sample of eligible voters in a bond-issue poll by drawing names from the list of registered voters in such a way that each sample of size *n* has the same probability of selection. The details of simple random sampling are discussed in Section 4.11. At this point, we merely state that a simple random sample will contain as much information on community preference as any other sample survey design, provided all voters in the community have similar socioeconomic backgrounds.

Suppose, however, that the community consists of people in two distinct income brackets, high and low. Voters in the high-income bracket may have opinions on the bond issue that are quite different from the opinions of low-income bracket voters. Therefore, to obtain accurate information about the population, we want to sample voters from each bracket. We can divide the population elements into two groups, or strata, according to income and select a simple random sample **stratified random sample**   from each group. The resulting sample is called a **stratified random sample.** (See Chapter 5 of Scheaffer et al., 1996.) Note that stratification is accomplished by using knowledge of an auxiliary variable, namely, personal income. By stratifying on high and low values of income, we increase the accuracy of our estimator. **ratio estimation**   **Ratio estimation** is a second method for using the information contained in an auxiliary variable. Ratio estimators not only use measurements on the response of interest but they also incorporate measurements on an auxiliary variable. Ratio estimation can also be used with stratified random sampling.

Although individual preferences are desired in the survey, a more economical procedure, especially in urban areas, may be to sample specific families, apartment buildings, or city blocks rather than individual voters. Individual preferences can then be obtained from each eligible voter within the unit sampled. This technique **cluster sampling**   is called **cluster sampling.** Although we divide the population into groups for both cluster sampling and stratified random sampling, the techniques differ. In stratified random sampling, we take a simple random sample within each group, whereas, in cluster sampling, we take a simple random sample of groups and then sample all items within the selected groups (clusters). (See Chapters 8 and 9 of Scheaffer et al., 1996, for details.)

Sometimes, the names of persons in the population of interest are available

systematic sample

in a list, such as a registration list, or on file cards stored in a drawer. For this situation, an economical technique is to draw the sample by selecting one name near the beginning of the list and then selecting every tenth or fifteenth name thereafter. If the sampling is conducted in this manner, we obtain a **systematic sample.** As you might expect, systematic sampling offers a convenient means of obtaining sample information; unfortunately, we do not necessarily obtain the most information for a specified amount of money. (Details are given in Chapter 7 of Scheaffer et al., 1996.)

The important point to understand is that there are different kinds of surveys that can be used to collect sample data. For the surveys discussed in this text, we will deal with simple random sampling and methods for summarizing and analyzing data collected in such a manner. More complicated surveys lead to even more complicated problems at the summarization and analysis stages of statistics.

The American Statistical Association (ASA, at Internet address: http://www.amstat.org or e-mail: asainfo@amstat.org) publishes a series of documents on surveys: *What Is a Survey? How to Plan a Survey, How to Collect Survey Data, Judging the Quality of a Survey, How to Conduct Pretesting, What Are Focus Groups?* and *More about Mail Surveys.* These documents describe many of the elements crucial to obtaining a valid and useful survey. They list many of the potential sources of errors commonly found in surveys with guidelines on how to avoid these pitfalls. A discussion of some of the issues raised in these brochures follows.

Problems Associated with Surveys

Even when the sample is selected properly, there may be uncertainty about whether the survey represents the population from which the sample was selected. Two of the major sources of uncertainty are nonresponse, which occurs when a portion of the individuals sampled cannot or will not participate in the survey, and measurement problems, which occur when the respondent's answers to questions do not provide the type of data that the survey was designed to obtain.

survey nonresponse

Survey nonresponse may result in a biased survey because the sample is not representative of the population. It is stated in *Judging the Quality of a Survey* that in surveys of the general population women are more likely to participate than men; that is, the nonresponse rate for males is higher than for females. Thus, a political poll may be biased if the percentage of women in the population in favor of a particular issue is larger than the percentage of men in the population supporting the issue. The poll would overestimate the percentage of the population in favor of the issue because the sample had a larger percentage of women than their percentage in the population. In all surveys, a careful examination of the nonresponse group must be conducted to determine whether a particular segment of the population may be either under- or overrepresented in the sample. Some of the remedies for nonresponse are

1. Offering an inducement for participating in the survey
2. Sending reminders or making follow-up telephone calls to the individuals who did not respond to the first contact

3. Using statistical techniques to adjust the survey findings to account for the sample profile differing from the population profile.

measurement problems

Measurement problems are the result of the respondent's not providing the information that the survey seeks. These problems often are due to the specific wording of questions in a survey, the manner in which the respondent answers the survey questions, and the fashion in which an interviewer phrases questions during the interview. Examples of specific problems and possible remedies are as follows:

1. *Inability to recall answers to questions:* The interviewee is asked how many times he or she visited a particular city park during the past year. This type of question often results in an underestimate of the average number of times a family visits the park during a year because people often tend to underestimate the number of occurrences of a common event or an event occurring far from the time of the interview. A possible remedy is to request respondents to use written records or to consult with other family members before responding.

2. *Leading questions:* The fashion in which an opinion question is posed may result in a response that does not truly represent the interviewee's opinion. Thus, the survey results may be biased in the direction in which the question is slanted. For example, a question concerning whether the state should impose a large fine on a chemical company for environmental violations is phrased as, "Do you support the state fining the chemical company, which is the major employer of people in our community, considering that this fine may result in their moving to another state?" This type of question tends to elicit a "no" response and thus produces a distorted representation of the community's opinion on the imposition of the fine. The remedy is to write questions carefully in an objective fashion.

3. *Unclear wording of questions:* An exercise club attempted to determine the number of times a person exercises per week. The question asked of the respondent was, "How many times in the last week did you exercise?". The word *exercise* has different meanings to different individuals. The result of allowing different definitions of important words or phrases in survey questions is to greatly reduce the accuracy of survey results. Several remedies are possible: The questions should be tested on a variety of individuals prior to conducting the survey to determine whether there are any confusing or misleading terms in the questions. During the training of the interviewer, all interviewers should have the "correct" definitions of all key words and be advised to provide these definitions to the respondents.

Many other issues, problems, and remedies are provided in the brochures from the ASA.

The stages in designing, conducting, and analyzing a survey are contained in Figure 2.1, which has been reproduced from an earlier version of *What Is a Survey?* in Cryer and Miller (1991). This diagram provides a guide for properly conducting a successful survey.

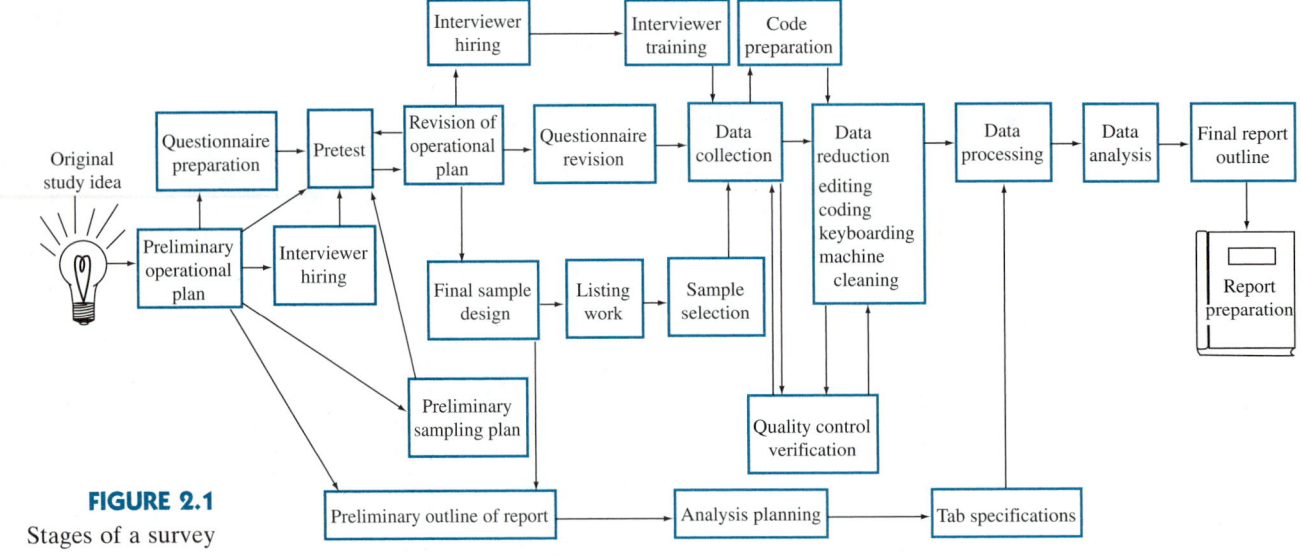

FIGURE 2.1

Stages of a survey

Data Collection Techniques

Having chosen a particular sample survey, how does one actually collect the data? The most commonly used methods of data collection in sample surveys are personal interviews and telephone interviews. These methods, with appropriately trained interviewers and carefully planned callbacks, commonly achieve response rates of 60% to 75% and sometimes even higher. A mailed questionnaire sent to a specific group of interested persons can achieve good results, but generally the response rates for this type of data collection are so low that all reported results are suspect. Frequently, objective information can be found from direct observation rather than from an interview or mailed questionnaire.

personal interviews   Data are frequently obtained by **personal interviews.** For example, we can use personal interviews with eligible voters to obtain a sample of public sentiment toward a community bond issue. The procedure usually requires the interviewer to ask prepared questions and to record the respondent's answers. The primary advantage of these interviews is that people will usually respond when confronted in person. In addition, the interviewer can note specific reactions and eliminate misunderstandings about the questions asked. The major limitations of the personal interview (aside from the cost involved) concern the interviewers. If they are not thoroughly trained, they may deviate from the required protocol, thus introducing a bias into the sample data. Any movement, facial expression, or statement by the interviewer can affect the response obtained. For example, a leading question such as "Are you also in favor of the bond issue?" may tend to elicit a positive response. Finally, errors in recording the responses can lead to erroneous results.

telephone interviews   Information can also be obtained from persons in the sample through **telephone interviews.** With the competition among telephone service providers, an interviewer can place any number of calls to specified areas of the country relatively inexpensively. Surveys conducted through telephone interviews are frequently less expensive than personal interviews, owing to the elimination of travel expenses.

The investigator can also monitor the interviews to be certain that the specified interview procedure is being followed.

A major problem with telephone surveys is that it is difficult to find a list or directory that closely corresponds to the population. Telephone directories have many numbers that do not belong to households, and many households have unlisted numbers. A few households have no phone service, although lack of phone service is now only a minor problem for most surveys in the United States. A technique that avoids the problem of unlisted numbers is random-digit dialing. In this method, a telephone exchange number (the first three digits of a seven-digit number) is selected, and then the last four digits are dialed randomly until a fixed number of households of a specified type are reached. This technique produces samples from the target population and avoids many of the problems inherent in sampling a telephone directory.

Telephone interviews generally must be kept shorter than personal interviews because responders tend to get impatient more easily when talking over the telephone. With appropriately designed questionnaires and trained interviewers, telephone interviews can be as successful as personal interviews.

**self-administered
questionnaire**

Another useful method of data collection is the **self-administered questionnaire,** to be completed by the respondent. These questionnaires usually are mailed to the individuals included in the sample, although other distribution methods can be used. The questionnaire must be carefully constructed if it is to encourage participation by the respondents.

The self-administered questionnaire does not require interviewers, and thus its use results in savings in the survey cost. This savings in cost is usually bought at the expense of a lower response rate. Nonresponse can be a problem in any form of data collection, but since we have the least contact with respondents in a mailed questionnaire, we frequently have the lowest rate of response. The low response rate can introduce a bias into the sample because the people who answer questionnaires may not be representative of the population of interest. To eliminate some of the bias, investigators frequently contact the nonrespondents through follow-up letters, telephone interviews, or personal interviews.

direct observation

The fourth method for collecting data is **direct observation.** If we were interested in estimating the number of trucks that use a particular road during the 4–6 P.M. rush hours, we could assign a person to count the number of trucks passing a specified point during this period, or electronic counting equipment could be used. The disadvantage in using an observer is the possibility of error in observation.

Direct observation is used in many surveys that do not involve measurements on people. The U.S. Department of Agriculture measures certain variables on crops in sections of fields in order to produce estimates of crop yields. Wildlife biologists may count animals, animal tracks, eggs, or nests to estimate the size of animal populations.

A closely related notion to direct observation is that of getting data from objective sources not affected by the respondents themselves. For example, health information can sometimes be obtained from hospital records, and income information from employer's records (especially for state and federal government workers). This approach may take more time but can yield large rewards in important surveys.

EXERCISES **Basic Techniques**

Soc.   **2.1** An experimenter wants to estimate the average water consumption per family in a city. Discuss the relative merits of choosing individual families, dwelling units (single-family houses, apartment buildings, etc.), and city blocks as sampling units.

Env.   **2.2** A forester wants to estimate the total number of trees on a tree farm that have diameters exceeding 12 inches. A map of the farm is available. Discuss the problem of choosing what to sample and how to select the sample.

Engin.   **2.3** A safety expert is interested in estimating the proportion of automobile tires with unsafe treads. Should he use individual cars or collections of cars, such as those in parking lots, in his sample?

H.R.   **2.4** An industry consists of many small plants located throughout the United States. An executive wants to survey the opinions of employees on the industry vacation policy. What would you suggest she sample?

Ag.   **2.5** A state department of agriculture wants to estimate the number of acres planted in corn within the state. How might one conduct such a survey?

Pol. Sci.   **2.6** A political scientist wants to estimate the proportion of adult residents of a state who favor a unicameral legislature. What could be sampled? Also, discuss the relative merits of personal interviews, telephone interviews, and mailed questionnaires as methods of data collection.

2.7 Discuss the relative merits of using personal interviews, telephone interviews, and mailed questionnaires as data collection methods for each of the following situations:
  a. A television executive wants to estimate the proportion of viewers in the country who are watching the network at a certain hour.
  b. A newspaper editor wants to survey the attitudes of the public toward the type of news coverage offered by the paper.
  c. A city commissioner is interested in determining how homeowners feel about a proposed zoning change.
  d. A county health department wants to estimate the proportion of dogs that have had rabies shots within the last year.

Soc.   **2.8** A Yankelovich, Skelly, and White poll taken in the fall of 1984 showed that one-fifth of the 2,207 people surveyed admitted to having cheated on their federal income taxes. Do you think that this fraction is close to the actual proportion who cheated? Why? (Discuss the difficulties of obtaining accurate information on a question of this type.)

Bus.   **2.9** Two surveys were conducted to measure the effectiveness of an advertising campaign for a low-fat brand of peanut butter. In one of the surveys, the interviewers visited the home and asked whether the low-fat brand was purchased. In the other survey, the interviewers asked the person to show them the peanut butter container when the interviewee stated he or she had purchased low-fat peanut butter.
  a. Do you think the two types of surveys will yield similar results on the percentage of households using the product?
  b. What types of biases may be introduced into each of the surveys?

Edu.   **2.10** *Time* magazine, in a late 1950s article, stated that "the average Yaleman, class of 1924, makes $25,111 a year," which, in today's dollars, would be over $150,000. *Time*'s estimate was based on replies to a sample survey questionnaire mailed to those members of the Yale class of 1924 whose addresses were on file with the Yale administration in late 1950.
  a. What is the survey's population of interest?
  b. Were the techniques used in selecting the sample likely to produce a sample that was representative of the population of interest?
  c. What are the possible sources of bias in the procedures used to obtain the sample?
  d. Based on the sources of bias, do you believe that *Time*'s estimate of the salary of a 1924 Yale graduate in the late 1950s is too high, too low, or nearly the correct value?

H.R. **2.11** A large health care corporation is interested in the number of employees who devote a substantial amount of time providing care for elderly relatives. The corporation wants to develop a policy with respect to the number of sick days an employee could use to provide care to elderly relatives. The corporation has thousands of employees, so it decides to have a sample of employees fill out a questionnaire.

  a. How would you define *employee*? Should only full-time workers be considered?
  b. How would you select the sample of employees?
  c. What information should be collected from the workers?

Bus. **2.12** The school of nursing at a university is developing a long-term plan to determine the number of faculty members that may be needed in future years. Thus, it needs to determine the future demand for nurses in the areas in which many of the graduates find employment. The school decides to survey medical facilities and private doctors to assist in determining the future nursing demand.

  a. How would you obtain a list of private doctors and medical facilities so that a sample of doctors could be selected to fill out a questionnaire?
  b. What are some of the questions that should be included on the questionnaire?
  c. How would you determine the number of nurses who are licensed but not currently employed?
  d. What are some possible sources for determining the population growth and health risk factors for the areas in which many of the nurses find employment?
  e. How could you sample the population of health care facilities and types of private doctors so as to not exclude any medical specialties from the survey?

2.3 Scientific Studies

The subject of experimental designs for scientific studies cannot be given much justice at the beginning of a statistical methods course—entire courses at the undergraduate and graduate levels are needed to get a comprehensive understanding of the methods and concepts of experimental design. Even so, we will attempt to give you a brief overview of the subject because much data requiring summarization and analysis arise from scientific studies involving one of a number of experimental designs. We will work by way of examples.

A consumer testing agency decides to evaluate the wear characteristics of four major brands of tires. For this study, the agency selects four cars of a standard car model and four tires of each brand. The tires will be placed on the cars and then driven 30,000 miles on a 2-mile racetrack. The decrease in tread thickness over the 30,000 miles is the variable of interest in this study. Four different drivers will drive the cars but the drivers are professional drivers with comparable training and experience. The weather conditions, smoothness of track, and the maintenance of the four cars will be essentially the same for all four brands over the study period. All extraneous factors that may affect the tires are nearly the same for all four brands. Thus, the testing agency feels confident that if there is a difference in wear characteristics between the brands at the end of the study, then this is truly a difference in the four brands and not a difference due to the manner in which the study was conducted. The testing agency is interested in recording other factors, such as the cost of the tires, the length of warranty offered by the manufacturer, whether the tires go out of balance during the study, and the evenness of wear across the width of the tires. In this example, we will only consider tread wear. There should be a recorded tread wear for each of the sixteen tires, four tires for each brand. The methods presented in Chapters 8 and 15 could be used to summarize and analyze the sample tread wear data in order to make comparisons (inferences) among the four tire brands. One possible inference of interest could be the selection of the brand having minimum tread wear. Can the best-performing tire brand in the sample data be expected to provide the best tread wear if the same study is repeated? Are the results of the study applicable to the driving habits of the typical motorist?

Experimental Designs

There are many ways in which the tires can be assigned to the four cars. We will consider one running of the experiment in which we have four tires of each of the four brands. First, we need to decide how to assign the tires to the cars. We could randomly assign a single brand to each car, but this would result in a design having the unit of measurement the total loss of tread for all four tires on the car and not the individual tire loss. Thus, we must randomly assign the sixteen tires to the four cars. In Chapter 15, we will demonstrate how this randomization is conducted. One possible arrangement of the tires on the cars is shown in Table 2.1.

TABLE 2.1

Completely randomized design of tire wear

Car 1	Car 2	Car 3	Car 4
Brand B	Brand A	Brand A	Brand D
Brand B	Brand A	Brand B	Brand D
Brand B	Brand C	Brand C	Brand D
Brand C	Brand C	Brand A	Brand D

In general, a **completely randomized design** is used when we are interested in comparing t "treatments" (in our case, $t = 4$, the treatments are brand of tire). For each of the treatments, we obtain a sample of observations. The sample sizes could be different for the individual treatments. For example, we could test twenty tires from Brands A, B, and C but only twelve tires from Brand D. The sample of observations from a treatment is assumed to be the result of a simple random sample of observations from the hypothetical population of possible values that could have resulted from that treatment. In our example, the sample of four tire-wear thicknesses from Brand A was considered to be the outcome of a simple random sample of four observations selected from the hypothetical population of possible tire-wear thicknesses for standard model cars traveling 30,000 miles using Brand A.

The experimental design could be altered to accommodate the effect of a variable related to how the experiment is conducted. In our example, we assumed that the effect of the different cars, weather, drivers, and various other factors was the same for all four brands. Now, if the wear on tires imposed by Car 4 was less severe than that of the other three cars, would our design take this effect into account? Because Car 4 had all four tires of Brand D placed on it, the wear observed for Brand D may be less than the wear observed for the other three brands because all four tires of Brand D were on the "best" car. In some situations, the objects being observed have existing differences prior to their assignment to the treatments. For example, in an experiment evaluating the effectiveness of several drugs for reducing blood pressure, the age or physical condition of the participants in the study may decrease the effectiveness of the drug. To avoid masking the effectiveness of the drugs, we would want to take these factors into account. Also, the environmental conditions encountered during the experiment may reduce the effectiveness of the treatment.

In our example, we would want to avoid having the comparison of the tire brands distorted by the differences in the four cars. The experimental design used to accomplish this goal is called a **randomized block design** because we want to "block" out any differences in the four cars to obtain a precise comparison of the four brands of tires. In a randomized block design, each treatment appears in every block. In the blood pressure example, we would group the patients according to the severity of their blood pressure problem and then randomly assign the drugs to the patients within each group. Thus, the randomized block design is similar to a stratified random sample used in surveys. In the tire wear example, we would use the four cars as the blocks and randomly assign one tire of each brand to each of the four cars, as shown in Table 2.2.

TABLE 2.2

Randomized block design of tire wear

Car 1	Car 2	Car 3	Car 4
Brand A	Brand A	Brand A	Brand A
Brand B	Brand B	Brand B	Brand B
Brand C	Brand C	Brand C	Brand C
Brand D	Brand D	Brand D	Brand D

Now, if there are any differences in the cars that may affect tire wear, that effect will be equally applied to all four brands.

What happens if the position of the tires on the car affects the wear on the tire? The positions on the car are right front (RF), left front (LF), right rear (RR), and left rear (LR). In Table 2.2, suppose that all four tires from Brand A are placed on the RF position, Brand B on RR, Brand C on LF, and Brand D on LR. Now, if the greatest wear occurs for tires placed on the RF, then Brand A would be at a great disadvantage when compared to the other three brands. In this type of situation we would state that the effect of brand and the effect of position on the car were confounded; that is, using the data in the study, the effects of two or more factors cannot be unambiguously attributed to a single factor. If we observed a large difference in the average wear among the four brands, is this difference due to differences in the brands or differences due to the position of the tires on the car? Using the design given in Table 2.2, this question cannot be answered. Thus, we now need two blocking variables: the "car" the tire is placed on and the "position" on the car. A design having two **Latin square design** blocking variables is called a **Latin square design.** A Latin square design for our example is shown in Table 2.3.

TABLE 2.3

Latin square design of tire wear

Position	Car 1	Car 2	Car 3	Car 4
RF	Brand A	Brand B	Brand C	Brand D
RR	Brand B	Brand C	Brand D	Brand A
LF	Brand C	Brand D	Brand A	Brand B
LR	Brand D	Brand A	Brand B	Brand C

Note that with this design, each brand is placed in each of the four positions and on each of the four cars. Thus, if position or car has an effect on the wear of the tires, the position effect and/or car effect will be equalized across the four brands. The observed differences in wear can now be attributed to differences in the brand of the car.

The randomized block and Latin square designs are both extensions of the completely randomized design in which the objective is to compare t treatments. The analysis of data collected according to a completely randomized design and the inferences made from such analysis are discussed further in Chapters 15 and 17. A special case of the randomized block design is presented in Chapter 6, where the number of treatments is $t = 2$ and the analysis of data and the inferences from these analyses are discussed.

Factorial Experiments

factors

Suppose that we want to examine the effects of two (or more) variables (**factors**) on a response. For example, suppose that an experimenter is interested in examining the effects of two variables, nitrogen and phosphorus, on the yield of a selected variety of corn. Also assume that we have three levels of each factor: 40, 50, and 60 pounds per plot for nitrogen; 10, 20, and 30 pounds per plot for phosphorus. For this study the experimental units are small, relatively homogeneous plots that have been partitioned from the acreage of a farm.

one-at-a-time approach

One approach for examining the effects of two or more factors on a response is called the **one-at-a-time approach.** To examine the effect of a single variable, an experimenter varies the levels of this variable while holding the levels of the other independent variables fixed. This process is continued until the effect of each variable on the response has been examined.

For example, suppose we want to determine the combination of nitrogen and phosphorus that produces the maximum amount of corn per plot. We would select a level of phosphorus, say 20 pounds, vary the levels of nitrogen, and observe which combination gives maximum yield in terms of bushels of corn per acre. Next, we would use the level of nitrogen producing the maximum yield, vary the amount of phosphorus, and observe the combination of nitrogen and phosphorus that produces the maximum yield. This combination would be declared the "best" treatment. The problem with this approach will be illustrated using the hypothetical yield values given in Table 2.4. These values would be unknown to the experimenter. We will assume that many replications of the treatments are used in the experiment so that the experimental results are nearly the same as the true yields.

TABLE 2.4

Hypothetical population yields (bushels per acre)

	Phosphorus		
Nitrogen	10	20	30
40	125	145	190
50	155	150	140
60	175	160	125

Initially, we run experiments with 20 pounds of phosphorus and the levels of nitrogen at 40, 50, and 60. We would determine that using 60 pounds of nitrogen with 20 pounds of phosphorus produces the maximum production, 160 bushels per acre. Next, we set the nitrogen level at 60 pounds and vary the phosphorus levels. This would result in the 10 level of phosphorus producing the highest yield, 175 bushels, when combined with 60 pounds of nitrogen. Thus, we would determine that 10 pounds of phosphorus with 60 pounds of nitrogen produces the maximum yield. The results of these experiments are summarized in Table 2.5.

TABLE 2.5

Yields for the experimental results

Phosphorus	20	20	20	10	30
Nitrogen	40	50	60	60	60
Yield	145	155	160	175	125

Based on the experimental results using the one-factor-at-a-time methodology, we would conclude that the 60 pounds of nitrogen and 10 pounds of phosphorus is the optimal combination. An examination of the yields in Table 2.4 reveals that the true optimal combination was 40 pounds of nitrogen with 30 pounds of phosphorus producing a yield of 190 bushels per acre. Thus, this type of experimentation may produce incorrect results whenever the effect of one factor on the response does not remain the same at all levels of the second factor. In this **interact** situation, the factors are said to **interact.** Figure 2.2 depicts the interaction between nitrogen and phosphorus in the production of corn. Note that as the amount of nitrogen is increased from 40 to 60 there is an increase in the yield when using the 10 level of phosphorus. At the 20 level of phosphorus, increasing the amount of nitrogen also produces an increase in yield but with smaller increments. At the 20 level of phosphorus, the yield increases 15 bushels when the nitrogen level is changed from 40 to 60. However, at the 10 level of phosphorus, the yield increases 50 bushels when the level of nitrogen is increased from 40 to 60. Furthermore, at the 30 level of phosphorus, increasing the level of nitrogen actually causes the yield to decrease. When there is no interaction between the factors, increasing the nitrogen level would have produced identical changes in the yield at all levels of phosphorus.

FIGURE 2.2

Yields from nitrogen–phosphorus treatments (interaction is present).

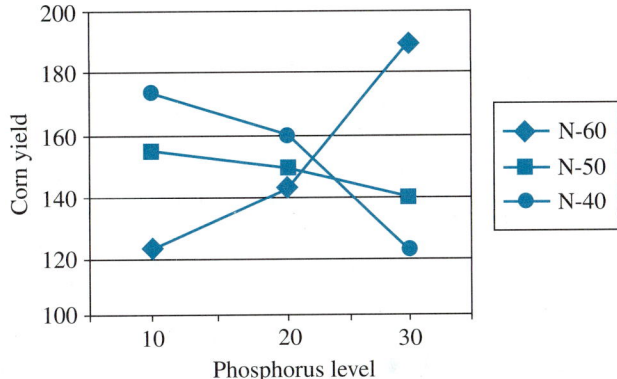

Table 2.6 and Figure 2.3 depict a situation in which the two factors do not interact. In this situation, the effect of phosphorus on the corn yield is the same for all three levels of nitrogen; that is, as we increase the amount of phosphorus, the change in corn yield is exactly the same for all three levels of nitrogen. Note that the change in yield is the same at all levels of nitrogen for a given change in phosphorus. However, the yields are larger at the higher levels of nitrogen. Thus, in the profile plots we have three different lines but the lines are parallel. When interaction exists among the factors, the lines will either cross or diverge.

TABLE 2.6

Hypothetical population yields (no interaction)

Nitrogen	Phosphorus		
	10	20	30
40	125	145	150
50	145	165	170
60	165	185	190

FIGURE 2.3

Yields from nitrogen–phosphorus treatments (no interaction)

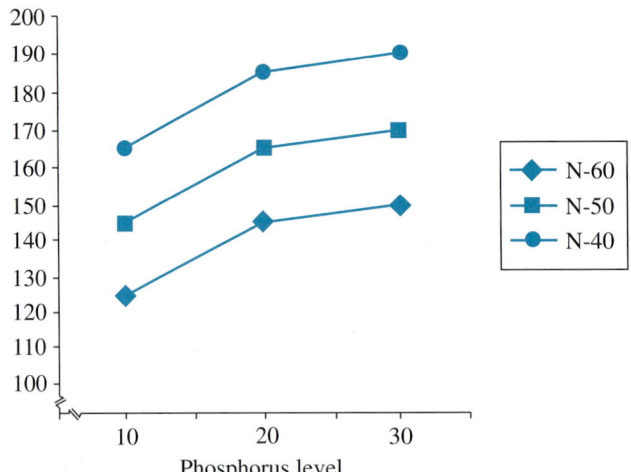

From Figure 2.3 we can observe that the one-at-a-time approach is appropriate for a situation in which the two factors do not interact. No matter what level is selected for the initial level of phosphorus, the one-at-a-time approach will produce the optimal yield. However, in most situations, prior to running the experiments it is not known whether the two factors will interact. If it is assumed that the factors *do not* interact and the one-at-a-time approach is implemented when in fact the factors *do* interact, the experiment will produce results that will often fail to identify the best treatment.

factorial experiment   **Factorial experiments** are useful for examining the effects of two or more factors on a response, whether or not interaction exists. As before, the choice of the number of levels of each variable and the actual settings of these variables is important. When the factor–level combinations are assigned to experimental units at random, we have a completely randomized design with treatments being the factor–level combinations.

DEFINITION 2.1   A **factorial experiment** is an experiment in which the response is observed at all factor–level combinations of the independent variables.

Using our previous example, we are interested in examining the effect of nitrogen and phosphorus levels on the yield of a corn crop. The nitrogen levels are 40, 50, and 60 pounds per plot and the phosphorus levels are 10, 20, and 30 pounds per plot. We could use a completely randomized design where the nine factor–level combinations (treatments) of Table 2.7 are assigned at random to the experimental units (the plots of land planted with corn).

It is not necessary to have the same number of levels of both factors. For

TABLE 2.7

Factor–level combinations for the 3 × 3 factorial experiment

Treatment	1	2	3	4	5	6	7	8	9
Phosphorus	10	10	10	20	20	20	30	30	30
Nitrogen	40	50	60	40	50	60	40	50	60

example, we could run an experiment with two levels of phosphorus and three levels of nitrogen, a 2 × 3 factorial experiment. Also, the number of factors can be more than two. The corn yield experiment could have involved treatments consisting of four levels of potassium along with the three levels of phosphorus and nitrogen, a 4 × 3 × 3 factorial experiment. Thus, we would have $4 \cdot 3 \cdot 3 =$ 36 factorial-level combinations or treatments. The methodology of randomization, analysis, and inferences for data obtained from factorial experiments in various experimental designs is discussed in Chapters 14, 15, 17, and 18.

More Complicated Designs

Sometimes the objectives of a study are such that we wish to investigate the effects of certain factors on a response while blocking out certain other extraneous sources of variability. Such situations require a block design with treatments from a factorial experiment and can be illustrated with the following example.

An investigator wants to examine the effectiveness of two drugs (A and B) for controlling heartworm in puppies. Veterinarians have conjectured that the effectiveness of the drugs may depend on a puppy's diet. Three different diets (Factor 1) are combined with the two drugs (Factor 2) and we have a 3 × 2 factorial experiment with six treatments. Also, the effectiveness of the drugs may depend on a transmitted inherent protection against heartworm obtained from the puppy's mother. Thus, four litters of puppies consisting of six puppies each were selected to serve as a blocking factor in the experiment because all puppies within a given litter have the same mother. The six factor–level combinations (treatments) were randomly assigned to the six puppies within each of the four litters. The design is shown in Table 2.8. Note that this design is really a randomized block design in which the blocks are litters and the treatments are the six factor–level combinations of the 3 × 2 factorial experiment.

TABLE 2.8

Block design for heartworm experiment

Puppy	Litter 1	2	3	4
1	A-D1	A-D3	B-D3	B-D2
2	A-D3	B-D1	A-D2	A-D2
3	B-D1	A-D1	B-D2	A-D1
4	A-D2	B-D2	B-D1	B-D3
5	B-D3	B-D3	A-D1	A-D3
6	B-D2	A-D2	A-D2	B-D1

Other more complicated combinations of block designs and factorial experiments are possible. As with sample surveys, though, we will deal only with the simplest experimental designs in this text. The point we want to make is that there are many different experimental designs that can be used in scientific studies for designating the collection of sample data. Each has certain advantages and disadvantages. We expand our discussion of experimental designs in Chapters

14–19, where we concentrate on the analysis of data generated from these designs. In those situations that require more complex designs, a professional statistician needs to be consulted to obtain the most appropriate design for the survey or experimental setting.

EXERCISES

Basic Techniques

2.13 Consider the yields given in Table 2.6. In this situation, there is no interaction. Show that the one-at-a-time approach would result in the experimenter finding the best combination of nitrogen and phosphorus—that is, the combination producing maximum yield. Your solution should include the five combinations you would use in the experiment.

2.14 The population values that would result from running a 2×3 factorial experiment are given in the following table. Note that two values are missing. If there is *no interaction* between the two factors, determine the missing values.

	Factor 2		
Factor 1	**I**	**II**	**III**
A	25	45	
B		30	50

2.4 Observational Studies

observational study

Before leaving the subject of sample data collection, we will draw a distinction between an **observational study** and a scientific study. In experimental designs for scientific studies, the observation conditions are fixed or controlled. For example, with a factorial experiment laid off in a completely randomized design, an observation is made at each factor–level combination. Similarly, with a randomized block design, an observation is obtained on each treatment in every block. These "controlled" studies are very different from observational studies, which are sometimes used because it is not feasible to do a proper scientific study. This can be illustrated by an example.

Much research and public interest centers on the effect of cigarette smoking on lung cancer and cardiovascular disease. One possible experimental design would be to randomize a fixed number of individuals (say, 1,000) to each of two groups—one group would be required to smoke cigarettes for the duration of the study (say, 10 years), while those in the second group would not be allowed to smoke throughout the study. At the end of the study, the two groups would be compared for lung cancer and cardiovascular disease. Even if we ignore ethical questions, this type of study would be impossible to do. Because of the long duration, it would be difficult to follow all participants and make certain that they follow the study plan. It would also be difficult to find nonsmoking individuals willing to take the chance of being assigned to the smoking group.

Another possible study would be to sample a fixed number of smokers and a fixed number of nonsmokers to compare the groups for lung cancer and for cardiovascular disease. Assuming one could obtain willing groups of participants, this study could be done in a *much shorter* period of time.

What has been sacrificed? Well, the fundamental difference between an observational study and a scientific study lies in the inferences(s) that can be drawn. For a scientific study comparing smokers to nonsmokers, assuming the two groups of individuals followed the study plan, the observed differences between the smoking and nonsmoking groups could be attributed to the effects of cigarette smoking because individuals were randomized to the two groups; hence, the groups were assumed to be comparable at the outset.

This type of reasoning does not apply to the observational study of cigarette smoking. Differences between the two groups in the observation could not necessarily be attributed to the effects of cigarette smoking because, for example, there may be hereditary factors that predispose people to smoking and cancer of the lungs and/or cardiovascular disease. Thus, differences between the groups might be due to hereditary factors, smoking, or a combination of the two. Typically, the results of an observational study are reported by way of a statement of association. For our example, if the observational study showed a higher frequency of lung cancer and cardiovascular disease for smokers relative to nonsmokers, it would be stated that this study showed that cigarette smoking was associated with an increased frequency of lung cancer and cardiovascular disease. It is a careful rewording in order not to infer that cigarette smoking *causes* lung cancer and cardiovascular disease.

Often, however, an observational study is the only type of study that can be run. Our job is to make certain that we understand the type of study run and, hence, understand how the data were collected. Then we can critique inferences drawn from an analysis of the study data.

2.5 Data Management: Preparing Data for Summarization and Analysis

In this section, we concentrate on some important data management procedures that are followed between the time the data are gathered and the time they are available in computer-readable form for analysis. This is not a complete manual with all tools required; rather, it is an overview—what a manager should know about these steps. As an example, this section reflects standard procedures in the pharmaceutical industry, which is highly regulated. Procedures may differ somewhat in other industries and settings.

We begin with a discussion of the procedures involved in processing data from a study. In practice, these procedures may consume 75% of the total effort from the receipt of the raw data to the presentation of results from the analysis. What are these procedures, why are they so important, and why are they so time consuming?

To answer these questions, let's list the major data-processing procedures in the cycle, which begins with receipt of the data and ends when the statistical analysis begins. Then we'll discuss each procedure separately.

Procedures in Processing Data for Summarization and Analysis

1. Receive the raw data source.
2. Create the database from the raw data source.
3. Edit the database.
4. Correct and clarify the raw data source.
5. Finalize the database.
6. Create data files from the database.

raw data source

1. *Receiving the raw data source.* For each study that is to be summarized and analyzed, the data arrive in some form, which we'll refer to as the **raw data source.** For a clinical trial, the raw data source is usually case report forms, sheets of $8\frac{1}{2} \times 11$-inch paper that have been used to record study data for each patient entered into the study. For other types of studies, the raw data source may be sheets of paper from a laboratory notebook, a magnetic tape (or other form of machine-readable data), hand tabulations, and so on.

data trail

It is important to retain the raw data source, because it is the beginning of the **data trail,** which leads from the raw data to the conclusions drawn from a study. Many consulting operations involved with the analysis and summarization of many different studies keep a log that contains vital information related to the study and raw data source. In a regulated environment such as the pharmaceutical industry, one may have to redo or reproduce data and data analyses based on previous work. Other situations outside the pharmaceutical industry may also require a retrospective review of what was done in the analysis of a study. In these situations, the study log can be an invaluable source of study information. General information contained in a study log is shown next.

Log for Study Data

1. Date received and from whom
2. Study investigator
3. Statistician (and others) assigned to study
4. Brief description of study
5. Treatments (compounds, preparations, etc.) studied
6. Raw data source
7. Response(s) measured and how measured
8. Reference number for study
9. Estimated (actual) completion date
10. Other pertinent information

Later, when the study has been analyzed and results have been communicated, additional information can be added to the log on how the study results were communicated, where these results are recorded, what data files have been saved, and where these files are stored.

2. *Creating the database from the raw data source.* For most studies that are scheduled for a statistical analysis, a machine-readable database is created. The steps taken to create the database and its eventual form vary from one operation to another, depending on the software systems to be used in the statistical analysis. However, we can give a few guidelines based on the form of the entry system.

When data are to be entered at a terminal, the raw data are first checked for legibility. Any illegible numbers or letters or other problems should be brought to the attention of the study coordinator. Then a coding guide that assigns column numbers and variable names to the data is filled out. Certain codes for missing values (for example, not available) are also defined here. Also, it is helpful to give a brief description of each variable. The data file keyed in at the terminal is

machine-readable database

referred to as the **machine-readable database.** A listing (printout) of the database should be obtained and checked carefully against the raw data source. Any errors should be corrected at the terminal and verified against an updated listing.

Sometimes data are received in machine-readable form. For these situations, the magnetic tape or disk file is considered to be the database. However, you must have a coding guide to "read" the database. Using the coding guide, obtain a listing of the database and check it *carefully* to see that all numbers and characters look reasonable and that proper formats were used to create the file. Any problems that arise must be resolved before proceeding further.

Some data sets are so small that it is not necessary to create a machine-readable data file from the raw data source. Instead, calculations are performed by hand or the data are entered into an electronic calculator. For these situations, check any calculations to see that they make sense. Don't believe everything you see; redoing the calculations is not a bad idea.

3. *Editing the database.* The types of edits done and the completeness of the editing process really depend on the type of study and how concerned you are about the accuracy and completeness of the data prior to the analysis. For example, in using a statistical software package (such as SAS or Minitab), it is wise to examine the minimum, maximum, and frequency distribution for each variable to make certain nothing looks unreasonable.

Certain other checks should be made. Plot the data and look for problems. Also, certain **logic checks** should be done depending on the structure of the data. If, for example, data are recorded for patients at several different visits, then the data recorded for visit 2 can't be earlier than the data for visit 1; similarly, if a patient is lost to follow-up after visit 2, there should not be any data for that patient at later visits.

For small data sets, the data edits can be done by hand, but, for large data sets, the job may be too time consuming and tedious. If machine editing is required, look for a software system that allows the user to specify certain data edits. Even so, for more complicated edits and logic checks it may be necessary to have a customized edit program written to machine edit the data. This programming chore can be a time-consuming step; plan for this well in advance of the receipt of the data.

4. *Correcting and clarifying the raw data source.* Questions frequently arise concerning the legibility or accuracy of the raw data during any one of the steps from the receipt of the raw data to the communication of the results from the statistical analysis. We have found it helpful to keep a list of these problems or discrepancies in order to define the data trail for a study. If a correction (or clarification) is required to the raw data source, indicate this on the form and make the appropriate change to the raw data source. If no correction is required, indicate this on the form as well. Keep in mind that the machine-readable database should be changed to reflect any changes made to the raw data source.

5. *Finalizing the database.* You may have been led to believe that all data for a study arrive at one time. This, of course, is not always the case. For example, with a marketing survey, different geographic locations may be surveyed at different times and, hence, those responsible for data processing do not receive all the data at one time. All these subsets of data, however, must be processed through the cycles required to create, edit, and correct the database. Eventually the study is declared complete and the data are processed into the database. At this time, the database should be reviewed again and final corrections made before beginning the analysis because for large data sets, the analysis and summarization chores take considerable human labor and computer time. It's better to agree on a final database analysis than to have to repeat all analyses on a changed database at a later date.

6. *Creating data files from the database.* Generally there are one or two sets of data files created from the machine-readable database. The first set, referred to as **original files,** reflects the basic structure of the database. A listing of the files

logic checks

original files

is checked against the database listing to verify that the variables have been read with correct formats and missing value codes have been retained. For some studies, the original files are actually used for editing the database.

work files   A second set of data files, called **work files,** may be created from the original files. Work files are designed to facilitate the analysis. They may require restructuring of the original files, a selection of important variables, or the creation or addition of new variables by insertion, computation, or transformation. A listing of the work files is checked against that of the original files to ensure proper restructuring and variable selection. Computed and transformed variables are checked by hand calculations to verify the program code.

If original and work files are SAS data sets, you should utilize the documentation features provided by SAS. At the time an SAS data set is created, a descriptive label for the data set, of up to 40 characters, should be assigned. The label can be stored with the data set imprinted wherever the contents procedure is used to print the data set's contents. All variables can be given descriptive names, up to 8 characters in length, which are meaningful to those involved in the project. In addition, variable labels up to 40 characters in length can be used to provide additional information. Title statements can be included in the SAS code to identify the project and describe each job. For each file, a listing (proc print) and a dictionary (proc contents) can be retained.

For files created from the database using other software packages, you should utilize the labeling and documentation features available in the computer program.

Even if appropriate statistical methods are applied to data, the conclusions drawn from the study are only as good as the data on which they are based—so you be the judge. The amount of time you should spend on these data-processing chores before analysis really depends on the nature of the study, the quality of the raw data source, and how confident you want to be about the completeness and accuracy of the data.

2.6   Summary

The first step in learning from data involves intelligent data gathering: specifying the objectives of the data-gathering exercise, identifying the variables of interest, and choosing an appropriate design for the survey or scientific study. In this chapter, we discussed various survey designs and experimental designs for scientific studies. Armed with a basic understanding of some design considerations for conducting surveys or scientific studies, you can address how to collect data on the variables of interest in order to address the stated objectives of the data-gathering exercise.

We also drew a distinction between observational and scientific studies in terms of the inferences (conclusions) that can be drawn from the sample data. Differences found between treatment groups from an observational study are said to be *associated with* the use of the treatments; on the other hand, differences found between treatments in a scientific study are said to be *due to* the treatments. In the next chapter, we will examine the methods for summarizing the data we collect.

PART

3

Summarizing

Data

3 Data Description

CHAPTER 3

Data Description

3.1 Introduction

3.2 Calculators, Computers, and Software Systems

3.3 Describing Data on a Single Variable: Graphical Methods

3.4 Describing Data on a Single Variable: Measures of Central Tendency

3.5 Describing Data on a Single Variable: Measures of Variability

3.6 The Boxplot

3.7 Summarizing Data from More Than One Variable

3.8 Summary

3.1 Introduction

In the previous chapter, we discussed how to gather data intelligently for an experiment or survey, Step 1 in learning from data. We turn now to Step 2, summarizing the data.

The field of statistics can be divided into two major branches: descriptive statistics and inferential statistics. In both branches, we work with a set of measurements. For situations in which data description is our major objective, the set of measurements available to us is frequently the entire population. For example, suppose that we wish to describe the distribution of annual incomes for all families registered in the 2000 census. Because all these data are recorded and are available on computer tapes, we do not need to obtain a random sample from the population; the complete set of measurements is at our disposal. Our major problem is in organizing, summarizing, and describing these data—that is, making sense of the data. Similarly, vast amounts of monthly, quarterly, and yearly data of medical costs are available for the managed health care industry, HMOs. These data are broken down by type of illness, age of patient, inpatient or outpatient care, prescription costs, and out-of-region reimbursements, along with many other types of expenses. However, in order to present such data in formats useful to HMO managers, congressional staffs, doctors, and the consuming public, it is necessary to organize, summarize, and describe the data. Good descriptive statistics enable us to make sense of the data by reducing a large set of measurements to a few summary measures that provide a good, rough picture of the original measurements.

In situations in which we are concerned with statistical inference, a sample is usually the only set of measurements available to us. We use information in the sample to draw conclusions about the population from which the sample was drawn. Of course, in the process of making inferences, we also need to organize, summarize, and describe the sample data.

For example, the tragedy surrounding isolated incidents of product tampering has brought about federal legislation requiring tamper-resistant packaging for certain drug products sold over the counter. These same incidents also brought about increased industry awareness of the need for rigid standards of product and packaging quality that must be maintained while delivering these products to the store shelves. In particular, one company is interested in determining the proportion of packages out of total production that are improperly sealed or have been damaged in transit. Obviously, it would be impossible to inspect all packages at all stores where the product is sold, but a random sample of the production could be obtained, and the proportion defective in the sample could be used to estimate the actual proportion of improperly sealed or damaged packages.

Similarly, in developing an economic forecast of new housing starts for the next year, it is necessary to use sample data from various economic indicators to make such a prediction (inference).

A third situation involves an experiment in which a food scientist wants to study the effect of two factors on the specific volume of bread loaves. The factors are type of fat and type of surfactant. (A surfactant is a substance that is mixed into the bread dough to lower the surface tension of the dough and thus produce loaves with increased specific volumes.) The experiment involves mixing a type of fat with a type of surfactant into the bread dough prior to baking the bread. The specific volume of the bread is then measured. This experiment is repeated several times for each of the fat–surfactant combinations. In this experiment, the scientist wants to make inferences from the results of the experiment to the commercial production of bread. In many such experiments, the use of proper graphical displays adds to the insight the scientist obtains from the data.

Whether we are describing an observed population or using sampled data to draw an inference from the sample to the population, an insightful description of the data is an important step in drawing conclusions from it. No matter what our objective, statistical inference or population description, we must first adequately describe the set of measurements at our disposal.

The two major methods for describing a set of measurements are graphical techniques and numerical descriptive techniques. Section 3.3 deals with graphical methods for describing data on a single variable. In Sections 3.4, 3.5, and 3.6, we discuss numerical techniques for describing data. The final topics on data description are presented in Section 3.7, in which we consider a few techniques for describing (summarizing) data on more than one variable.

3.2   Calculators, Computers, and Software Systems

Electronic calculators can be great aids in performing some of the calculations mentioned later in this chapter, especially for small data sets. For larger data sets, even hand-held calculators are of little use because of the time required to enter data. A computer can help in these situations. Specific programs or more general software systems can be used to perform statistical analyses almost instantaneously even for very large data sets after the data are entered into the computer from a terminal, a magnetic tape, or disk storage. It is not necessary to know computer programming to make use of specific programs or software systems for planned analyses—most have user's manuals that give detailed directions for their use. Others, developed for use at a terminal, provide program prompts that lead the user through the analysis of choice.

Many statistical software packages are available for use on computers. Three of the more commonly used systems are Minitab, SAS, and SPSS. Each is available in a mainframe version as well as in a personal computer version. Because a software system is a group of programs that work together, it is possible to obtain plots, data descriptions, and complex statistical analyses in a single job. Most people find that they can use any particular system easily, although they may be frustrated by minor errors committed on the first few tries. The ability of such packages to perform complicated analyses on large numbers of data more than repays the initial investment of time and irritation.

In general, to use a system you need to learn about only the programs in which you are interested. Typical steps in a job involve describing your data to the software system, manipulating your data if they are not in the proper format or if you want a subset of your original data set, and then calling the appropriate set of programs or procedures using the key words particular to the software system you are using. The results obtained from calling a program are then displayed at your terminal or sent to your printer.

If you have access to a computer and are interested in using it, find out how to obtain an account, what programs and software systems are available for doing statistical analyses, and where to obtain instruction on data entry for these programs and software systems.

Because computer configurations, operating systems, and text editors vary from site to site, it is best to talk to someone knowledgeable about gaining access to a software system. Once you have mastered the commands to begin executing programs in a software system, you will find that running a job within a given software system is similar from site to site.

Because this isn't a text on computer usage, we won't spend additional time and space on the mechanics, which are best learned by doing. Our main interest is in interpreting the output from these programs. The designers of these programs tend to include in the output everything that a user could conceivably want to know; as a result, in any particular situation, some of the output is irrelevant. When reading computer output look for the values you want; if you don't need or don't understand an output statistic, don't worry. Of course, as you learn more about statistics, more of the output will be meaningful. In the meantime, look for what you need and disregard the rest.

There are dangers in using such packages carelessly. A computer is a mindless beast, and will do anything asked of it, no matter how absurd the result might be. For instance, suppose that the data include age, gender (1 = female, 2 = male), religion (1 = Catholic, 2 = Jewish, 3 = Protestant, 4 = other or none), and monthly income of a group of people. If we asked the computer to calculate averages we would get averages for the variables gender and religion, as well as for age and monthly income, even though these averages are meaningless. Used intelligently, these packages are convenient, powerful, and useful—but be sure to examine the output from any computer run to make certain the results make sense. Did anything go wrong? Was something overlooked? In other words, be *skeptical.* One of the important acronyms of computer technology still holds; namely, GIGO: garbage in, garbage out.

Throughout the textbook, we will use computer software systems to do some of the more tedious calculations of statistics *after* we have explained how the calculations can be done. Used in this way, computers (and associated graphical and statistical analysis packages) will enable us to spend additional time on interpreting the results of the analyses rather than on doing the analyses.

3.3 Describing Data on a Single Variable: Graphical Methods

After the measurements of interest have been collected, ideally the data are organized, displayed, and examined by using various graphical techniques. As a general rule, the data should be arranged into categories so that *each measurement is classified into one, and only one, of the categories.* This procedure eliminates any ambiguity that might otherwise arise when categorizing measurements. For example, suppose a sex discrimination lawsuit is filed. The law firm representing the plaintiffs needs to summarize the salaries of all employees in a large corporation. To examine possible inequities in salaries, the law firm decides to summarize the 1997 yearly income rounded to the nearest dollar for all female employees into the following categories:

Income Level	Salary
1	less than $20,000
2	$20,000 to $39,999
3	$40,000 to $59,999
4	$60,000 to $79,999
5	$80,000 to $99,999
6	$100,000 or more

The yearly salary of each female employee falls into one, and only one, income category. However, if the income categories had been defined as

Income Level	Salary
1	less than $20,000
2	$20,000 to $40,000
3	$40,000 to $60,000
4	$60,000 to $80,000
5	$80,000 to $100,000
6	$100,000 or more

then there would be confusion as to which category should be checked. For example, an employee earning $40,000 could be placed in either category 2 or 3. To reiterate: If the data are organized into categories, it is important to define the categories so that a measurement can be placed into only one category.

When data are organized according to this guideline, there are several ways to display the data graphically. The first and simplest graphical procedure for data organized in this manner is the **pie chart.** It is used to display the percentage of the total number of measurements falling into each of the categories of the variable by partitioning a circle (similar to slicing a pie).

pie chart

The data of Table 3.1 represent a summary of a study to determine paths to authority for individuals occupying top positions of responsibility in key public-interest organizations. Using biographical information, each of 1,345 individuals was classified according to how she or he was recruited for the current elite position.

TABLE 3.1

Recruitment to top public-interest positions*

Recruitment From	Number	Percentage
Corporate	501	37.2
Public-interest	683	50.8
Government	94	7.0
Other	67	5.0

* Includes trustees of private colleges and universities, directors of large private foundations, senior partners of top law firms, and directors of certain large cultural and civic organizations.

Source: Thomas R. Dye and L. Harmon Zeigler, *The Irony of Democracy,* 5th ed. (Pacific Grove, CA: Duxbury Press, 1981), p. 130.

Although you can scan the data in Table 3.1, the results are more easily interpreted by using a pie chart. From Figure 3.1 we can make certain inferences about channels to positions of authority. For example, more people were recruited for elite positions from public-interest organizations (approximately 51%) than from elite positions in other organizations.

FIGURE 3.1

Pie chart for the data of Table 3.1

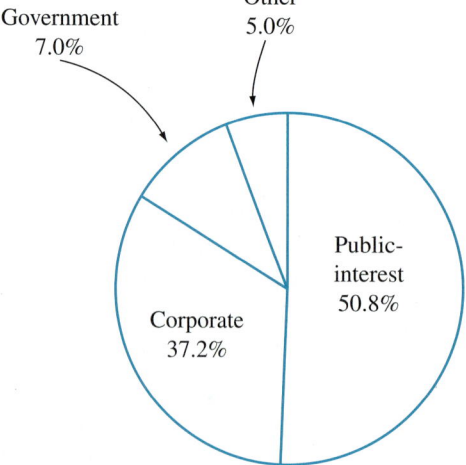

Other variations of the pie chart are shown in Figures 3.2 and 3.3. Clearly, from Figure 3.2, cola soft drinks have gained in popularity from 1980 to 1990 at the expense of some of the other types of soft drinks. Also, it's evident from Figure 3.3 that the loss of a major food chain account affected fountain sales for PepsiCo, Inc. In summary, the pie chart can be used to display percentages associated with each category of the variable. The following guidelines should help you to obtain clarity of presentation in pie charts.

FIGURE 3.2
Approximate market share of soft drinks by type, 1980 and 1990

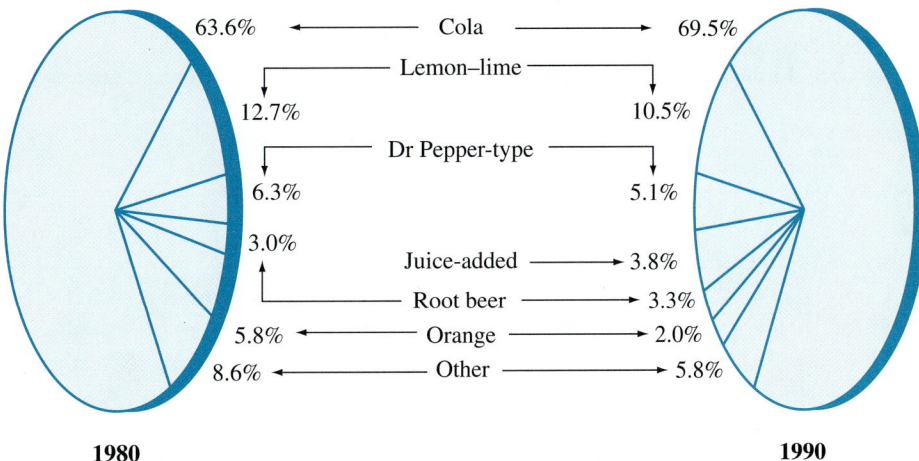

63.6%	← Cola →	69.5%
12.7%	Lemon–lime	10.5%
6.3%	Dr Pepper-type	5.1%
3.0%	Juice-added →	3.8%
	Root beer →	3.3%
5.8%	← Orange →	2.0%
8.6%	← Other →	5.8%

1980   **1990**

FIGURE 3.3
Estimated U.S. market share before and after switch in accounts*

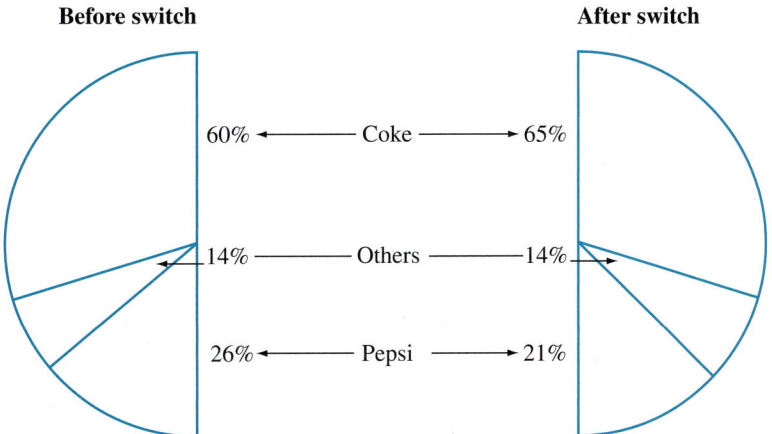

Before switch   **After switch**

60%	← Coke →	65%
14%	Others	14%
26%	← Pepsi →	21%

Guidelines for Constructing Pie Charts

1. Choose a small number (five or six) of categories for the variable because too many make the pie chart difficult to interpret.
2. Whenever possible, construct the pie chart so that percentages are in either ascending or descending order.

bar chart

A second graphical technique for data organized according to the recommended guideline is the **bar chart,** or bar graph. Figure 3.4 displays the number of workers in the Cincinnati, Ohio, area for the largest five foreign investors. The estimated total workforce is 680,000. There are many variations of the bar chart. Sometimes the bars are displayed horizontally, as in Figures 3.5(a) and (b). They can also be used to display data across time, as in Figure 3.6. Bar charts are relatively easy to construct if you use the guidelines given.

---

*A major fast-food chain switched its account from Pepsi to Coca-Cola for fountain sales.

FIGURE 3.4

Number of workers by major foreign investors

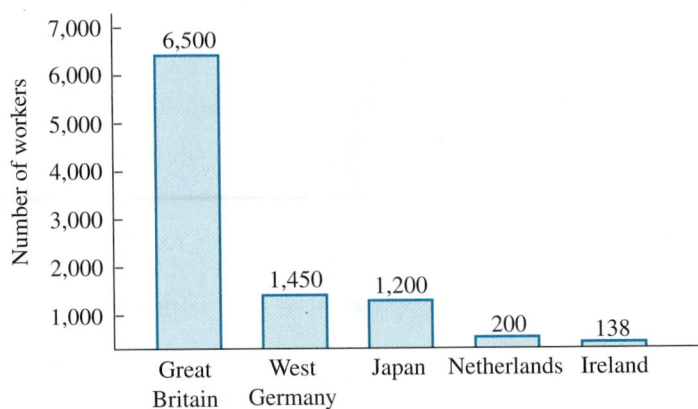

FIGURE 3.5

Greatest per capita consumption by country

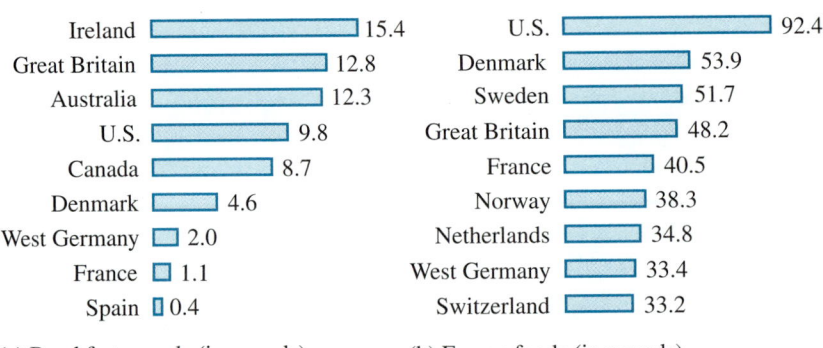

(a) Breakfast cereals (in pounds)   (b) Frozen foods (in pounds)

FIGURE 3.6

Estimated direct and indirect costs for developing a new drug by selected years

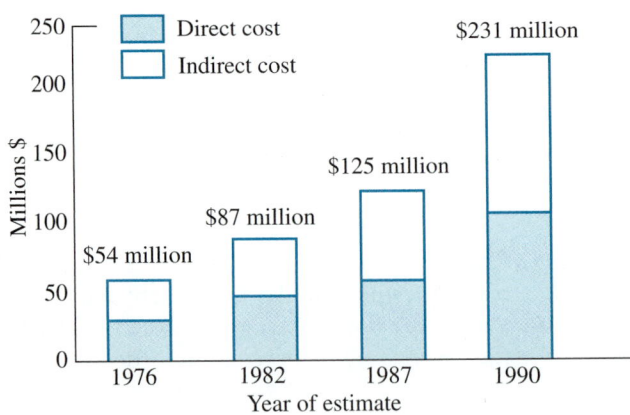

Guidelines for Constructing Bar Charts

1. Label frequencies on one axis and categories of the variable on the other axis.
2. Construct a rectangle at each category of the variable with a height equal to the frequency (number of observations) in the category.
3. Leave a space between each category to connote distinct, separate categories and to clarify the presentation.

frequency histogram, relative frequency histogram

The next two graphical techniques that we will discuss are the **frequency histogram** and the **relative frequency histogram.** Both of these graphical techniques are applicable only to quantitative (measured) data. As with the pie chart, we must organize the data before constructing a graph.

An animal scientist is carrying out an experiment to investigate whether adding an antibiotic to the diet of chicks will promote growth over the standard diet without the antibiotic. The scientist determines that 100 chicks will provide sufficient information to validate the results of the experiment. (In Chapter 5, we will present techniques for determining the proper sample size for a study to achieve specified goals.) From previous research studies, the average weight gain of chicks fed the standard diet over an 8-week period is 3.9 grams. The scientist wants to compare the weight gain of the chicks in the study to the standard value of 3.9 grams. To eliminate the effects of many other factors that could affect weight gain, the scientist rears and feeds the 100 chicks in the same building with individual feeders for each chick. The weight gains for the 100 chicks are recorded in Table 3.2.

TABLE 3.2

Weight gains for chicks (grams)

3.7	4.2	4.4	4.4	4.3	4.2	4.4	4.8	4.9	4.4
4.2	3.8	4.2	4.4	4.6	3.9	4.3	4.5	4.8	3.9
4.7	4.2	4.2	4.8	4.5	3.6	4.1	4.3	3.9	4.2
4.0	4.2	4.0	4.5	4.4	4.1	4.0	4.0	3.8	4.6
4.9	3.8	4.3	4.3	3.9	3.8	4.7	3.9	4.0	4.2
4.3	4.7	4.1	4.0	4.6	4.4	4.6	4.4	4.9	4.4
4.0	3.9	4.5	4.3	3.8	4.1	4.3	4.2	4.5	4.4
4.2	4.7	3.8	4.5	4.0	4.2	4.1	4.0	4.7	4.1
4.7	4.1	4.8	4.1	4.3	4.7	4.2	4.1	4.4	4.8
4.1	4.9	4.3	4.4	4.4	4.3	4.6	4.5	4.6	4.0

An initial examination of the weight gain data reveals that the largest weight gain is 4.9 grams and the smallest is 3.6 grams. Although we might examine the table very closely to determine whether the weight gains of the chicks are substantially greater than 3.9 grams, it is difficult to describe how the measurements are distributed along the interval 3.6 to 4.9. Are most of the measurements greater than 3.9, concentrated near 3.6, concentrated near 4.9, or evenly distributed along the interval? One way to obtain the answers to these questions is to organize the data in a **frequency table.**

frequency table

To construct a frequency table, we begin by dividing the range from 3.6 to 4.9 into an arbitrary number of subintervals called **class intervals.** The number of subintervals chosen depends on the number of measurements in the set, but we generally recommend using from 5 to 20 class intervals. The more data we have, the larger the number of classes we tend to use. The guidelines given here can be used for constructing the appropriate class intervals.

class intervals

Guidelines for Constructing Class Intervals

1. Divide the *range* of the measurements (the difference between the largest and the smallest measurements) by the approximate number of class intervals desired. Generally, we want to have from 5 to 20 class intervals.

2. After dividing the range by the desired number of subintervals, round the resulting number to a convenient (easy to work with) unit. This unit represents a common width for the class intervals.

3. Choose the first class interval so that it contains the smallest measurement. It is also advisable to choose a starting point for the first interval so that no measurement falls on a point of division between two subintervals, which eliminates any ambiguity in placing measurements into the class intervals. (One way to do this is to choose boundaries to one more decimal place than the data).

For the data in Table 3.2,

$$\text{range} = 4.9 - 3.6 = 1.3$$

Assume that we want to have approximately 10 subintervals. Dividing the range by 10 and rounding to a convenient unit, we have $1.3/10 = .13 \approx .1$. Thus, the class interval width is .1.

It is convenient to choose the first interval to be 3.55–3.65, the second to be 3.65–3.75, and so on. Note that the smallest measurement, 3.6, falls in the first interval and that no measurement falls on the endpoint of a class interval. (See Table 3.3.)

TABLE 3.3
Frequency table for the chick data

Class	Class Interval	Frequency f_i	Relative frequency f_i/n
1	3.55–3.65	1	.01
2	3.65–3.75	1	.01
3	3.75–3.85	6	.06
4	3.85–3.95	6	.06
5	3.95–4.05	10	.10
6	4.05–4.15	10	.10
7	4.15–4.25	13	.13
8	4.25–4.35	11	.11
9	4.35–4.45	13	.13
10	4.45–4.55	7	.07
11	4.55–4.65	6	.06
12	4.65–4.75	7	.07
13	4.75–4.85	5	.05
14	4.85–4.95	4	.04
Totals		$n = 100$	1.00

Having determined the class interval, we construct a frequency table for the data. The first column labels the classes by number and the second column indicates the class intervals. We then examine the 100 measurements of Table 3.2, keeping a tally of the number of measurements falling in each interval. The number of measurements falling in a given class interval is called the **class frequency.** These data are recorded in the third column of the frequency table. (See Table 3.3.)

class frequency

relative frequency

The **relative frequency** of a class is defined to be the frequency of the class divided by the total number of measurements in the set (total frequency). Thus, if we let f_i denote the frequency for class i and n denote the total number of measurements, the relative frequency for class i is f_i/n. The relative frequencies for all the classes are listed in the fourth column of Table 3.3.

histogram

The data of Table 3.2 have been organized into a frequency table, which can now be used to construct a *frequency histogram* or a *relative frequency* **histogram.** To construct a frequency histogram, draw two axes: a horizontal axis labeled with the class intervals and a vertical axis labeled with the frequencies. Then construct a rectangle over each class interval with a height equal to the number of measurements falling in a given subinterval. The frequency histogram for the data of Table 3.3 is shown in Figure 3.7(a).

FIGURE 3.7(a)

Frequency histogram for the chick data of Table 3.3

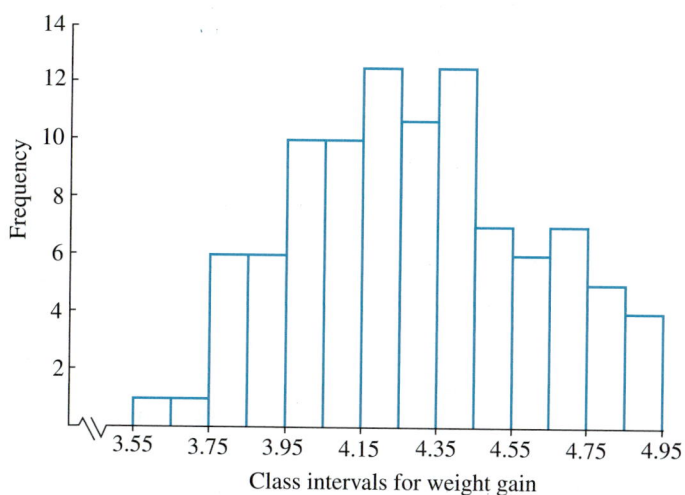

The relative frequency histogram is constructed in much the same way as a frequency histogram. In the relative frequency histogram, however, the vertical axis is labeled as relative frequency, and a rectangle is constructed over each class interval with a height equal to the class relative frequency (the fourth column of Table 3.3). The relative frequency histogram for the data of Table 3.3 is shown in Figure 3.7(b). Clearly, the two histograms of Figures 3.7(a) and (b) are of the same shape and would be identical if the vertical axes were equivalent. We will frequently refer to either one as simply a histogram.

FIGURE 3.7(b)

Relative frequency histogram for the chick data of Table 3.3

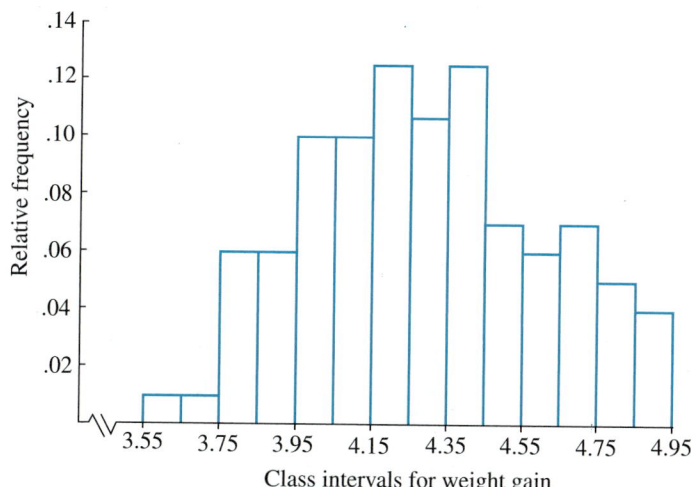

There are several comments that should be made concerning histograms. First, the distinction between bar charts and histograms is based on the distinction between *qualitative* and *quantitative* variables. Values of qualitative variables vary in kind but not degree and hence are not measurements. For example, the variable political party affiliation can be categorized as Republican, Democrat, or other, and, although we could label the categories as one, two, or three, these values are only codes and have no quantitative interpretation. In contrast, quantitative variables have actual units of measure. For example, the variable yield (in bushels) per acre of corn can assume specific values. *Pie charts and bar charts are used to display frequency data from qualitative variables; histograms are appropriate for displaying frequency data for quantitative variables.*

Second, the histogram is the most imporant graphical technique we will present because of the role it plays in statistical inference, a subject we will discuss in later chapters. Third, if we had an extremely large set of measurements, and if we constructed a histogram using many class intervals, each with a very narrow width, the histogram for the set of measurements would be, for all practical purposes, a smooth curve. Fourth, the fraction of the total number of measurements in an interval is equal to the fraction of the total area under the histogram over the interval. For example, if we consider the intervals containing weights greater than 3.9 for the chick data in Table 3.3, we see that there are exactly 86 of the 100 measurements in those intervals. Thus, .86, the proportion of the total measurements falling in those intervals, is equal to the proportion of the total area under the histogram over those intervals.

Fifth, if a single measurement is selected at random from the set of sample measurements, the chance, or **probability,** that it lies in a particular interval is equal to the fraction of the total number of sample measurements falling in that interval. This same fraction is used to estimate the probability that a measurement randomly selected from the population lies in the interval of interest. For example, from the sample data of Table 3.2, the chance or probability of selecting a chick with a weight gain greater than 3.9 grams is .86. The number .86 is an approximation of the proportion of all chickens fed the diet containing the antibiotic that would yield a weight gain greater than 3.9 grams, the value obtained from the standard diet.

Because of the arbitrariness in the choice of number of intervals, starting value, and length of intervals, histograms can be made to take on different shapes

probability

FIGURE 3.8
Histograms for the chick data

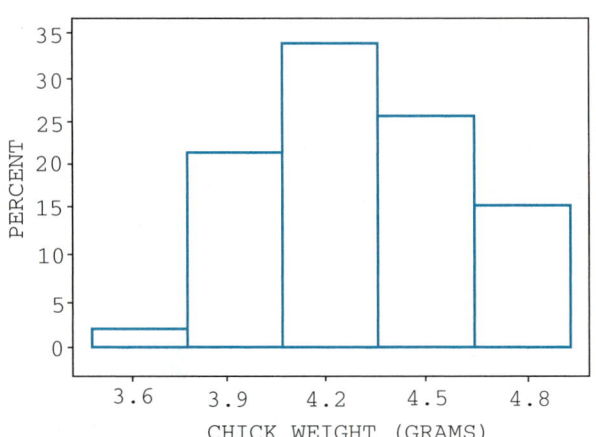

(a) Relative frequency histogram for chick data (5 intervals)

FIGURE 3.8

Histograms for the chick
data (*continued*)

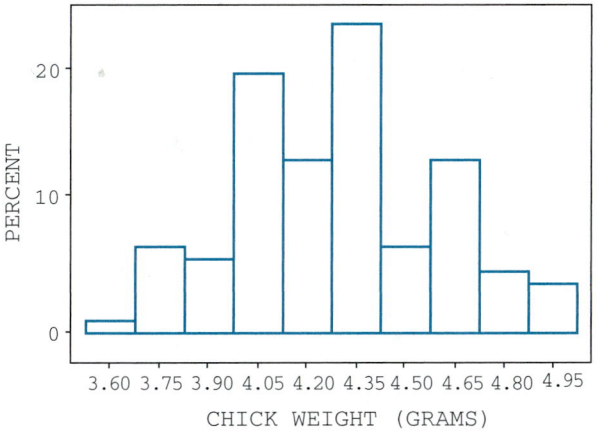

(b) Relative frequency histogram for chick data (10 intervals)

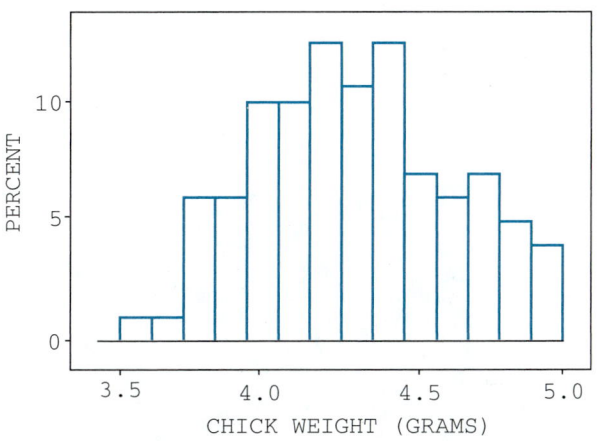

(c) Relative frequency histogram for chick data (14 intervals)

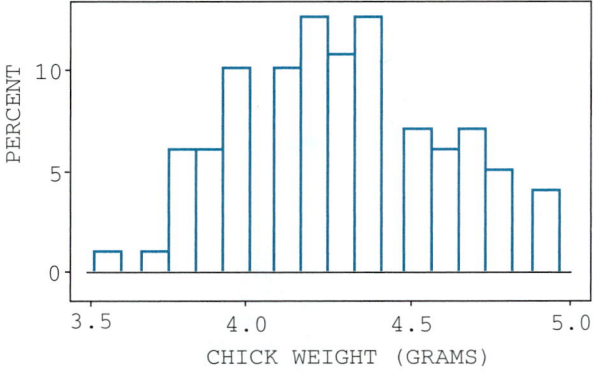

(d) Relative frequency histogram for chick data (18 intervals)

for the same set of data, especially for small data sets. Histograms are most useful
for describing data sets when the number of data points is fairly large, say 50 or
more. In Figures 3.8(a)–(d), a set of histograms for the chick data constructed using
5, 10, 14, and 18 class intervals illustrates the problems that can be encountered in

attempting to construct a histogram. These graphs were obtained using the Minitab software program.

When the number of data points is relatively small and the number of intervals is large, the histogram has several intervals in which there are no data values; see Figure 3.8(d). This results in a graph that is not a realistic depiction of the histogram for the whole population. When the number of class intervals is too small, most of the patterns or trends in the data are not displayed; see Figure 3.8(a). In the set of graphs in Figure 3.8, the histogram with 14 class intervals appears to be the most appropriate graph.

Finally, because we use proportions rather than frequencies in a relative frequency histogram, we can compare two different samples (or populations) by examining their relative frequency histograms even if the samples (populations) are of different sizes. When describing relative frequency histograms and comparing the plots from a number of samples, we examine the overall shape in the histogram. Figure 3.9 depicts many of the common shapes for relative frequency histograms.

unimodal   A histogram with one major peak is called **unimodal,** see Figures 3.9(b), (c), and (d). When the histogram has two major peaks, such as in Figures 3.9(e) and **bimodal**   (f), we state that the histogram is **bimodal.** In many instances, bimodal histograms

FIGURE 3.9

Some common shapes of distributions

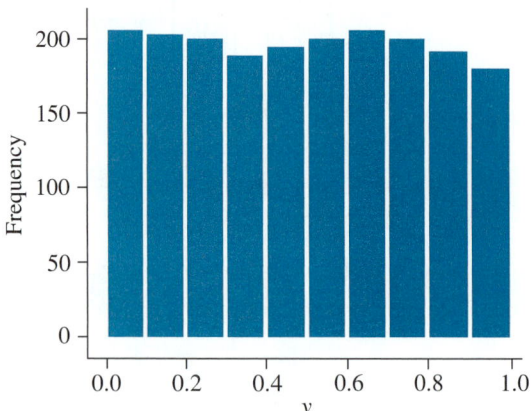

(a) Uniform distribution

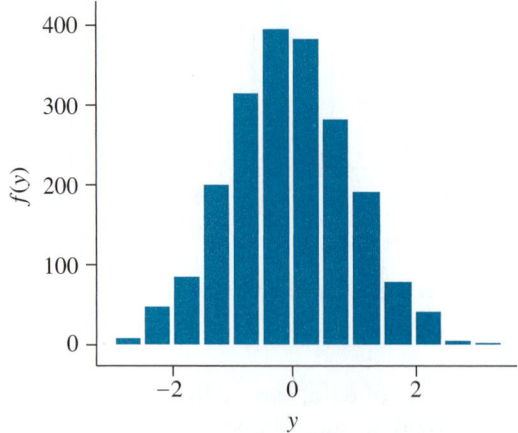

(b) Symmetric, unimodal (normal) distribution

FIGURE 3.9

Some common shapes of distributions (*continued*)

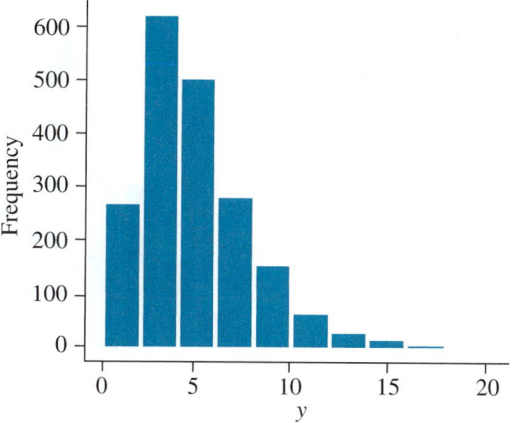

(c) Right-skewed distribution

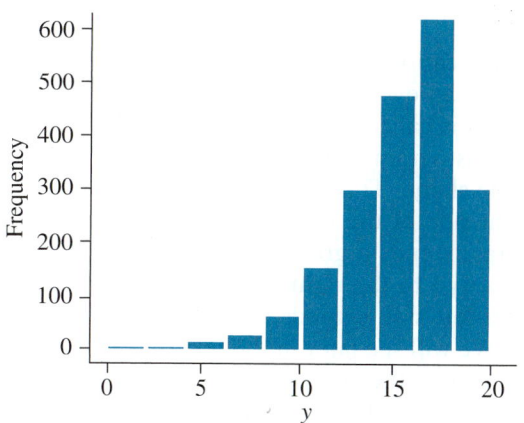

(d) Left-skewed distribution

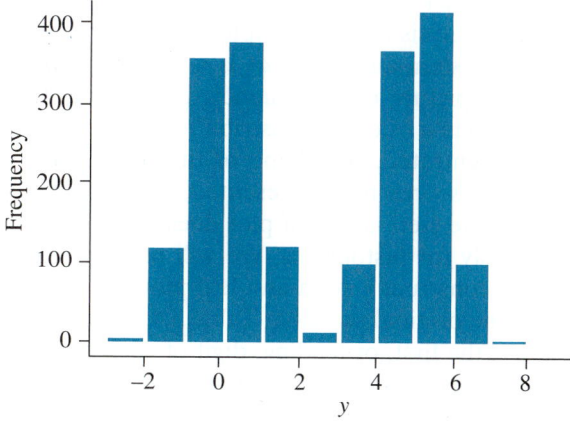

(e) Bimodal distribution

determines where a measurement is placed in a frequency table, the leading digit (stem of a score) determines the row in which a score is placed in a stem-and-leaf plot. The trailing digits for the score are then written in the appropriate row. In this way, each score is recorded in the stem-and-leaf plot, as in Figure 3.10 for the violent crime data.

We can see that each stem defines a class interval and the limits of each interval are the largest and smallest possible scores for the class. The values represented by each leaf must be between the lower and upper limits of the interval.

Note that a stem-and-leaf plot is a graph that looks much like a histogram turned sideways, as in Figure 3.10. The plot can be made a bit more useful by ordering the data (leaves) within a row (stem) from lowest to highest (Figure 3.11). The advantage of such a graph over the histogram is that it reflects not only frequencies, concentration(s) of scores, and shapes of the distribution but also the actual scores. The disadvantage is that for large data sets, the stem-and-leaf plot can be more unwieldy than the histogram.

FIGURE 3.11

Stem-and-leaf plot with ordered leaves

```
 1   89
 2   10 24 67 91 96 98
 3   36 41 52 54 74 76 88 93 93
 4   10 21 35 47 48 60 64 66 80 81 91 96 98
 5   04 05 08 16 26 29 37 57 59 61 62 62 62 63 70 71 78 85 92
 6   05 05 24 26 28 31 39 42 47 61 73 84 85 85 90 98
 7   03 06 18 19 20 31 35 39 51 58 71
 8   04 07 09 11 14 17 43 56 68 76 77 85
 9   28 71
10   20
```

Guidelines for Constructing Stem-and-Leaf Plots

1. Split each score or value into two sets of digits. The first or leading set of digits is the stem and the second or trailing set of digits is the leaf.
2. List all possible stem digits from lowest to highest.
3. For each score in the mass of data, write the leaf values on the line labeled by the appropriate stem number.
4. If the display looks too cramped and narrow, stretch the display by using two lines per stem so that, for example, leaf digits 0, 1, 2, 3, and 4 are placed on the first line of the stem and leaf digits 5, 6, 7, 8, and 9 are placed on the second line.
5. If too many digits are present, such as in a six- or seven-digit score, drop the right-most trailing digit(s) to maximize the clarity of the display.
6. The rules for developing a stem-and-leaf plot are somewhat different from the rules governing the establishment of class intervals for the traditional frequency distribution and for a variety of other procedures that we will consider in later sections of the text. Class intervals for stem-and-leaf plots are, then, in a sense slightly atypical.

The following stem-and-leaf plot is obtained from Minitab. The data consist of the number of employees in the wholesale and retail trade industries in Wisconsin measured each month for a 5-year period.

Data Display

```
Trade
    322   317   319   323   327   328   325   326   330   334
    337   341   322   318   320   326   332   334   335   336
    335   338   342   348   330   326   329   337   345   350
    351   354   355   357   362   368   348   345   349   355
    362   367   366   370   371   375   380   385   361   354
    357   367   376   381   381   383   384   387   392   396
```

Character Stem-and-Leaf Display

```
Stem-and-leaf of Trade      N = 60
Leaf Unit = 1.0

        31  789
        32  0223
        32  5666789
        33  00244
        33  556778
        34  12
        34  55889
        35  0144
        35  5577
        36  122
        36  6778
        37  01
        37  56
        38  01134
        38  57
        39  2
        39  6
```

Note that most of the stems are repeated twice, with the leaf digits split into two groups: 0 to 4 and 5 to 9.

The last graphical technique to be presented in this section deals with how certain variables change over time. For macroeconomic data such as disposable income and microeconomic data such as weekly sales data of one particular product at one particular store, plots of data over time are fundamental to business management. Similarly, social researchers are often interested in showing how variables change over time. They might be interested in changes with time in attitudes toward various racial and ethnic groups, changes in the rate of savings in the United States, or changes in crime rates for various cities. A pictorial method of presenting changes in a variable over time is called a **time series**. Figure 3.12 is a time series showing the percentage of white women age 30 to 34 who did not have any children during the years 1970 to 1986.

time series

Usually, time points are labeled chronologically across the horizontal axis (abscissa), and the numerical values (frequencies, percentages, rates, etc.) of the variable of interest are labeled along the vertical axis (ordinate). Time can be

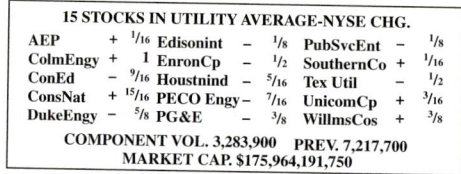

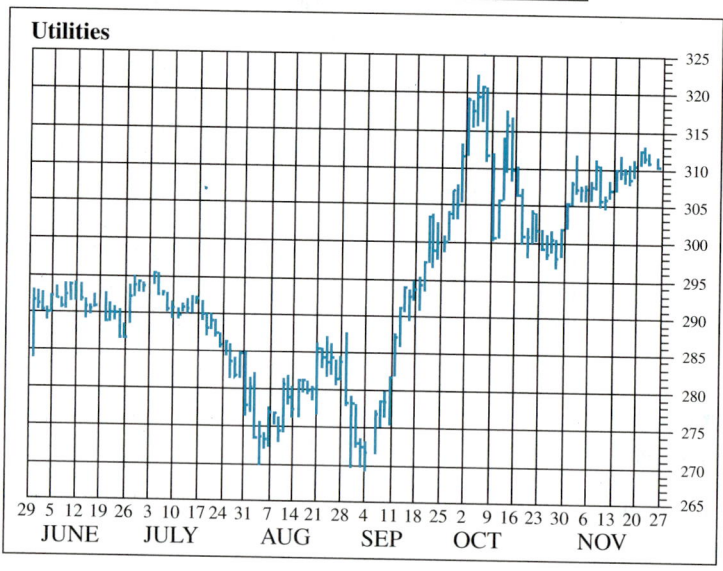

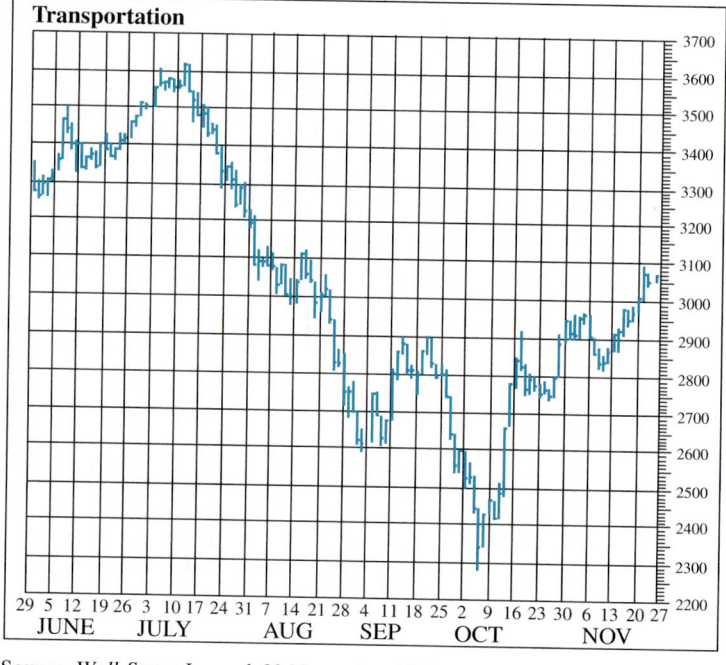

Source: *Wall Street Journal,* 30 November 1998

Sometimes it is important to compare trends over time in a variable for two or more groups. Figure 3.15 reports the values of two ratios from 1976 to 1988: the ratio of the median family income of African Americans to the median income of Anglo-Americans and the ratio of the median income of Hispanics to the median income of Anglo-Americans.

FIGURE 3.15

Ratio of African American and Hispanic median family income to Anglo-American median family income, 1976–1988

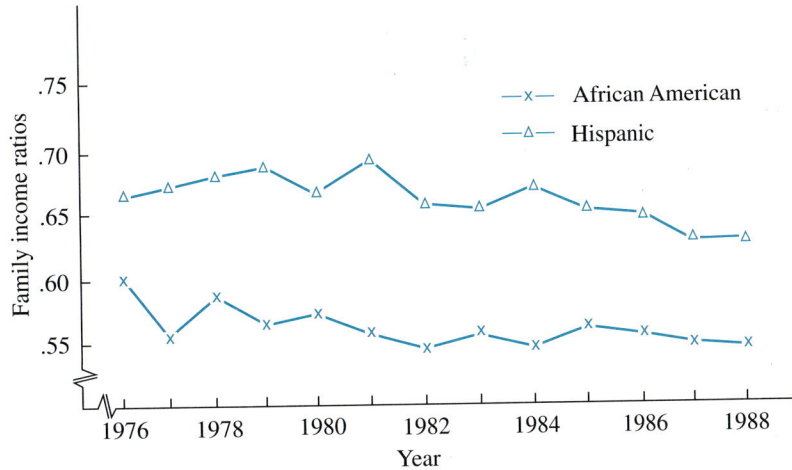

Median family income represents the income amount that divides family incomes into two groups—the top half and the bottom half. In 1987, the median family income for African Americans was $18,098, meaning that 50% of all African American families had incomes above $18,098, and 50% had incomes below $18,098. The median, one of several measures of central tendency, is discussed more fully later in this chapter.

Figure 3.15 shows that the ratio of African American or Hispanic to Anglo-American family income fluctuated between 1976 and 1988, but the overall trend in both ratios indicates that they declined over this time. A social researcher would interpret these trends to mean that the income of African American and Hispanic families generally declined relative to the income of Anglo-American families.

Sometimes information is not available in equal time intervals. For example, polling organizations such as Gallup or the National Opinion Research Center do not necessarily ask the American public the same questions about their attitudes or behavior on a yearly basis. Sometimes there is a time gap of more than 2 years before a question is asked again.

When information is not available in equal time intervals, it is important for the interval width between time points (the horizontal axis) to reflect this fact. If, for example, a social researcher is plotting values of a variable for 1985, 1986, 1987, and 1990, the interval width between 1987 and 1990 on the horizontal axis should be three times the width of that between the other years. If these interval widths were spaced evenly, the resulting trend line could be seriously misleading. Other examples of graphic distortion are discussed in Chapter 20.

Figure 3.16 presents the trend in church attendance among American Catholics and Protestants from 1958 to 1988. The width of the intervals between time points reflects the fact that Catholics were not asked about their church attendance every year.

Industry Group	Employment Growth Percentage, 1980–1990
Service	33.2
Manufacturing	25.0
Retail trade	17.9
Finance, insurance, real estate	6.6
Wholesale trade	4.8
Construction	4.6
Transportation	3.9
Government	2.7
Other	1.3

Construct a pie chart to display these data.

Soc.   **3.5**   From the same study described in Exercise 3.4, data were obtained on the job openings between 1980 and 1990. Use the data to construct a bar chart.

Occupational Groups	Percentage of Job Openings, 1980–1990
Clerical workers	20.9
Sales	7.3
Managers	9.5
Professional and technical	16.3
Laborers	3.7
Service workers	18.1
Operatives	13.1
Craft and kindred workers	11.1

Env.   **3.6**   The regulations of the board of health in a particular state specify that the fluoride level must not exceed 1.5 parts per million (ppm). The 25 measurements given here represent the fluoride levels for a sample of 25 days. Although fluoride levels are measured more than once per day, these data represent the early morning readings for the 25 days sampled.

.75	.86	.84	.85	.97
.94	.89	.84	.83	.89
.88	.78	.77	.76	.82
.72	.92	1.05	.94	.83
.81	.85	.97	.93	.79

a. Determine the range of the measurements.
b. Dividing the range by 7, the number of subintervals selected, and rounding, we have a class interval width of .05. Using .705 as the lower limit of the first interval, construct a frequency histogram.
c. Compute relative frequencies for each class interval and construct a relative frequency histogram. Note that the frequency and relative frequency histograms for these data have the same shape.
d. If one of these 25 days were selected at random, what would be the chance (probability) that the fluoride reading would be greater than .90 ppm? Guess (predict) what proportion of days in the coming year will have a fluoride reading greater than .90 ppm.

Gov.   **3.7** The National Highway Traffic Safety Administration has studied the use of rear-seat automobile lap and shoulder seat belts. The number of lives potentially saved with the use of lap and shoulder seat belts is shown for various percentages of use.

Percentage of Use	Lives Saved Wearing	
	Lap Belt Only	Lap and Shoulder Belt
100	529	678
80	423	543
60	318	407
40	212	271
20	106	136
10	85	108

Suggest several different ways to graph these data. Which one seems more appropriate and why?

3.8 Construct a frequency histogram for the data of Table 3.4.
   a. Compare the histogram to the stem-and-leaf plot of Figure 3.11.
   b. Describe the shape of the histogram using the standard terminology of histograms. Which graph is more informative?
   c. Explain how to design a histogram that would have essentially the same shape as the stem-and-leaf plot for the same data set.

Soc.   **3.9** Construct a relative frequency histogram for the data in the accompanying table. Describe the shape of the histogram using the standard terminology of histograms.

Per capita public welfare expenses, by number of states

Dollars	Number of States
50–74	3
75–99	6
100–124	14
125–149	11
150–174	2
175–199	5
200–224	2
225–249	5
250–274	1
275–299	1
Total	50

Soc.   **3.10** The 1994 annual per capita city taxes for the 24 largest cities in the United States are given in the following table.

2470	520	561	488	986	359
1305	512	467	270	360	451
4904	572	498	382	271	634
1682	784	298	643	947	686

Source: *Statistical Abstract of the United States, 1997*

Soc.   **3.20**   The following table presents the homeownership rates, in percentages, by state for the years 1985 and 1996. These values represent the proportion of homes owned by the occupant to the total number of occupied homes.

State	1985	1996	State	1985	1996
Alabama	70.4	71.0	Montana	66.5	68.6
Alaska	61.2	62.9	Nebraska	68.5	66.8
Arizona	64.7	62.0	Nevada	57.0	61.1
Arkansas	66.6	66.6	New Hampshire	65.5	65.0
California	54.2	55.0	New Jersey	62.3	64.6
Colorado	63.6	64.5	New Mexico	68.2	67.1
Connecticut	69.0	69.0	New York	50.3	52.7
Delaware	70.3	71.5	North Carolina	68.0	70.4
Dist. of Columbia	37.4	40.4	North Dakota	69.9	68.2
Florida	67.2	67.1	Ohio	67.9	69.2
Georgia	62.7	69.3	Oklahoma	70.5	68.4
Hawaii	51.0	50.6	Oregon	61.5	63.1
Idaho	71.0	71.4	Pennsylvania	71.6	71.7
Illinois	60.6	68.2	Rhode Island	61.4	56.6
Indiana	67.6	74.2	South Carolina	72.0	72.9
Iowa	69.9	72.8	South Dakota	67.6	67.8
Kansas	68.3	67.5	Tennessee	67.6	68.8
Kentucky	68.5	73.2	Texas	60.5	61.8
Louisiana	70.2	64.9	Utah	71.5	72.7
Maine	73.7	76.5	Vermont	69.5	70.3
Maryland	65.6	66.9	Virginia	68.5	68.5
Massachusetts	60.5	61.7	Washington	66.8	63.1
Michigan	70.7	73.3	West Virginia	75.9	74.3
Minnesota	70.0	75.4	Wisconsin	63.8	68.2
Mississippi	69.6	73.0	Wyoming	73.2	68.0
Missouri	69.2	70.2			

Source: U.S. Bureau of the Census, Internet site:
http://www.census.gov/ftp/pub/hhes/www/hvs.html.

   a. Construct a relative frequency histogram plot for the homeownership data given in the table for the years 1985 and 1996.
   b. What major differences exist between the plots for the two years?
   c. Why do you think the plots have changed over these 11 years?
   d. How could Congress use the information in these plots for writing tax laws that allow major tax deductions for homeownership?

3.21   Construct a stem-and-leaf plot for the data of Exercise 3.20.

3.22   Describe the shape of the stem-and-leaf plot and histogram for the homeownership data in Exercises 3.20 and 3.21, using the terms *modality, skewness,* and *symmetry* in your description.

Bus.   **3.23**   A supplier of high-quality audio equipment for automobiles accumulates monthly sales data on speakers and receiver–amplifier units for 5 years. The data (in thousands of units per month) are shown in a table. Plot the sales data. Do you see any overall trend in the data? Do there seem to be any cyclic or seasonal effects?

Year	J	F	M	A	M	J	J	A	S	O	N	D
1	101.9	93.0	93.5	93.9	104.9	94.6	105.9	116.7	128.4	118.2	107.3	108.6
2	109.0	98.4	99.1	110.7	100.2	112.1	123.8	135.8	124.8	114.1	114.9	112.9
3	115.5	104.5	105.1	105.4	117.5	106.4	118.6	130.9	143.7	132.2	120.8	121.3
4	122.0	110.4	110.8	111.2	124.4	112.4	124.9	138.0	151.5	139.5	127.7	128.0
5	128.1	115.8	116.0	117.2	130.7	117.5	131.8	145.5	159.3	146.5	134.0	134.2

Bus.   **3.24**   A machine-tool firm that produces a variety of products for manufacturers has quarterly records of total activity for the previous 8 years. The data, which are shown in the table, reflect activity rather than price, so inflation is irrelevant.

	Quarter			
Year	1	2	3	4
1	97.2	100.2	102.8	102.6
2	106.1	107.8	110.5	110.6
3	116.5	117.3	119.9	119.3
4	126.1	125.7	128.3	132.1
5	133.2	133.8	141.1	142.1
6	144.2	146.1	151.6	154.0
7	155.8	158.6	165.8	167.0
8	171.1	172.6	176.5	179.7

a. Plot the data against time (quarters 1–32).
b. Does there appear to be a clear trend? If so, what form of trend equation would you suggest?
c. Can you detect cyclic or seasonal features?

3.4   Describing Data on a Single Variable: Measures of Central Tendency

Numerical descriptive measures are commonly used to convey a mental image of pictures, objects, and other phenomena. There are two main reasons for this. First, graphical descriptive measures are inappropriate for statistical inference, because it is difficult to describe the similarity of a sample frequency histogram and the corresponding population frequency histogram. The second reason for using numerical descriptive measures is one of expediency—we never seem to carry the appropriate graphs or histograms with us, and so must resort to our powers of verbal communication to convey the appropriate picture. We seek several numbers, called *numerical descriptive measures,* that will create a mental picture of the frequency distribution for a set of measurements.

The two most common numerical descriptive measures are measures of **central tendency** and measures of **variability;** that is, we seek to describe the center of the distribution of measurements and also how the measurements vary about the center of the distribution. We will draw a distinction between numerical descriptive measures for a population, called **parameters,** and numerical descriptive measures for a sample, called **statistics.** In problems requiring statistical infer-

central tendency
variability

parameters
statistics

For these data, we have two measurements appearing three times each. Hence, the data are bimodal, with modes of 4.4 and 4.8. The median for the odd number of measurements is the middle score, 4.5.

grouped data median

The **median for grouped data** is slightly more difficult to compute. Because the actual values of the measurements are unknown, we know that the median occurs in a particular class interval, but we do not know where to locate the median within the interval. If we assume that the measurements are spread evenly throughout the interval, we get the following result. Let

L = lower class limit of the interval that contains the median

n = total frequency

cf_b = the sum of frequencies (cumulative frequency) for all classes before the median class

f_m = frequency of the class interval containing the median

w = interval width

Then, for grouped data,

$$\text{median} = L + \frac{w}{f_m}(.5n - cf_b)$$

The next example illustrates how to find the median for grouped data.

EXAMPLE 3.4

Table 3.5 is the frequency table for the chick data of Table 3.3. Compute the median weight gain for these data.

TABLE 3.5
Frequency table for the chick data, Table 3.3

Class Interval	f_i	Cumulative f_i	f_i/n	Cumulative f_i/n
3.55–3.65	1	1	.01	.01
3.65–3.75	1	2	.01	.02
3.75–3.85	6	8	.06	.08
3.85–3.95	6	14	.06	.14
3.95–4.05	10	24	.10	.24
4.05–4.15	10	34	.10	.34
4.15–4.25	13	47	.13	.47
4.25–4.35	11	58	.11	.58
4.35–4.45	13	71	.13	.71
4.45–4.55	7	78	.07	.78
4.55–4.65	6	84	.06	.84
4.65–4.75	7	91	.07	.91
4.75–4.85	5	96	.05	.96
4.85–4.95	4	100	.04	1.00
Totals	$n = 100$		1.00	

Solution   Let the cumulative relative frequency for class j equal the sum of the relative frequencies for class 1 through class j. To determine the interval that contains the median, we must find the first interval for which the cumulative relative frequency exceeds .50. This interval is the one containing the median. For these data, the interval from 4.25 to 4.35 is the first interval for which the cumulative relative frequency exceeds .50, as shown in Table 3.5, column 5. So this interval contains the median. Then

$$L = 4.25 \qquad f_m = 11$$
$$n = 100 \qquad w = .1$$
$$cf_b = 47$$

and

$$\text{median} = L + \frac{w}{f_m}(.5n - cf_b) = 4.25 + \frac{.1}{11}(50 - 47) = 4.28$$

Note that the value of the median from the ungrouped data of Table 3.2 on p. 47 is 4.3. Thus, the approximated value and the value from the ungrouped data are nearly equal. The difference between the two values for the sample median decreases as the number of class intervals increases.

mean The third, and last, measure of central tendency we will discuss in this text is the arithmetic mean, known simply as the **mean.**

DEFINITION 3.3 The **arithmetic mean,** or **mean,** of a set of measurements is defined to be the sum of the measurements divided by the total number of measurements.

When people talk about an "average," they quite often are referring to the mean. It is the balancing point of the data set. Because of the important role that the mean will play in statistical inference in later chapters, we give special symbols to the population mean and the sample mean. The *population mean* is denoted

μ
$\bar{y}$

by the Greek letter μ (read "mu"), and the *sample mean* is denoted by the symbol $\bar{y}$ (read "y-bar"). As indicated in Chapter 1, a population of measurements is the complete set of measurements of interest to us; a sample of measurements is a subset of measurements selected from the population of interest. If we let y_1, y_2, ..., y_n denote the measurements observed in a sample of size n, then the sample mean $\bar{y}$ can be written as

$$\bar{y} = \frac{\Sigma_i y_i}{n}$$

where the symbol appearing in the numerator, $\Sigma_i y_i$, is the notation used to designate a sum of n measurements, y_i:

$$\sum_i y_i = y_1 + y_2 + \cdots + y_n$$

The corresponding population mean is μ.

In most situations, we will not know the population mean; the sample will be used to make inferences about the corresponding unknown population mean. For example, the accounting department of a large department store chain is conducting an examination of its overdue accounts. The store has thousands of such accounts, which would yield a population of overdue values having a mean

By trimming the data, we are able to reduce the impact of very large (or small) values on the mean, and thus get a more reliable measure of the central value of the set. This will be particularly important when the sample mean is used to predict the corresponding population central value.

Note that in a limiting sense the median is a 50% trimmed mean. Thus, the median is often used in place of the mean when there are extreme values in the data set. In Example 3.5, the value $807.80 is considerably larger than the other values in the data set. This results in 10 of the 15 accounts having values less than the mean and only 5 larger. The median value for the 15 accounts is $61.61. There are 7 accounts less than the median and 7 accounts greater than the median. Thus, in selecting a typical overdue account, the median is a more appropriate value than the mean. However, if we want to estimate the total amount overdue in all 150 accounts, we would want to use the mean and not the median. When estimating the sum of all measurements in a population, we would not want to exclude the extremes in the sample. Suppose a sample contains a few extremely large values. If the extremes are trimmed, then the population sum will be grossly underestimated using the sample trimmed mean or sample median in place of the sample mean.

In this section, we discussed the mode, median, mean, and trimmed mean. How are these measures of central tendency related for a given set of measurements? The answer depends on the **skewness** of the data. If the distribution is mound-shaped and symmetrical about a single peak, the mode (M_o), median (M_d), mean (μ), and trimmed mean (TM) will all be the same. This is shown using a smooth curve and population quantities in Figure 3.17(a). If the distribution is skewed, having a long tail in one direction and a single peak, the mean is pulled in the direction of the tail; the median falls between the mode and the mean; and depending on the degree of trimming, the trimmed mean usually falls between

skewness

FIGURE 3.17

Relation among the mean μ, the trimmed mean TM, the median M_d, and the mode M_o

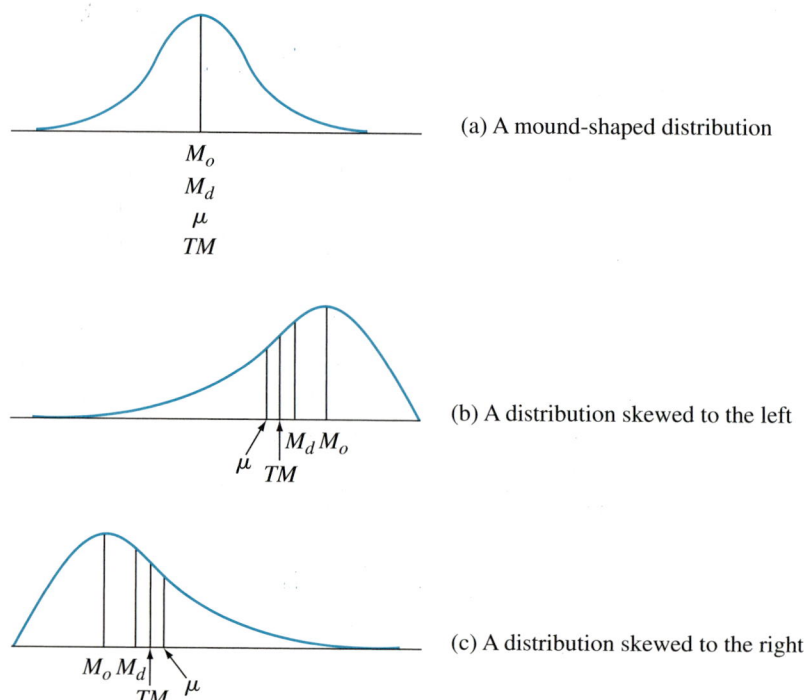

(a) A mound-shaped distribution

M_o
M_d
μ
TM

(b) A distribution skewed to the left

μ $M_d M_o$
    TM

(c) A distribution skewed to the right

$M_o M_d$
    TM μ

the median and the mean. Figures 3.17(b) and (c) illustrate this for distributions skewed to the left and to the right.

The important thing to remember is that we are not restricted to using only one measure of central tendency. For some data sets, it will be necessary to use more than one of these measures to provide an accurate descriptive summary of central tendency for the data.

Major Characteristics of Each Measure of Central Tendency

Mode

1. It is the most frequent or probable measurement in the data set.
2. There can be more than one mode for a data set.
3. It is not influenced by extreme measurements.
4. Modes of subsets cannot be combined to determine the mode of the complete data set.
5. For grouped data its value can change depending on the categories used.
6. It is applicable for both qualitative and quantitative data.

Median

1. It is the central value; 50% of the measurements lie above it and 50% fall below it.
2. There is only one median for a data set.
3. It is not influenced by extreme measurements.
4. Medians of subsets cannot be combined to determine the median of the complete data set.
5. For grouped data, its value is rather stable even when the data are organized into different categories.
6. It is applicable to quantitative data only.

Mean

1. It is the arithmetic average of the measurements in a data set.
2. There is only one mean for a data set.
3. Its value is influenced by extreme measurements; trimming can help to reduce the degree of influence.
4. Means of subsets can be combined to determine the mean of the complete data set.
5. It is applicable to quantitative data only.

Measures of central tendency do not provide a complete mental picture of the frequency distribution for a set of measurements. In addition to determining the center of the distribution, we must have some measure of the spread of the data. In the next section, we discuss measures of variability, or dispersion.

EXERCISES   **Basic Techniques**

3.25 Compute the mean, median, and mode for the following data:

```
11   17   18   10   22   23   15   17
14   13   10   12   18   18   11   14
```

3.26 Refer to the data in Exercise 3.25 with the measurements 22 and 23 replaced by 42 and 43. Recompute the mean, median, and mode. Discuss the impact of these extreme measurements on the three measures of central tendency.

3.27 Refer to Exercises 3.25 and 3.26. Compute a 10% trimmed mean for both data sets. Do the extreme values affect the 10% trimmed mean? Would a 5% trimmed mean be affected?

3.28 Determine the mode, median, and mean for the following measurements:

```
10   2   1    5
 1   5   7   10
 3   4   8   12
 5   6   8    9
```

3.29 Determine the mean, median, and mode for the data presented in the following frequency table:

Class Interval	Frequency
0–2	1
3–5	3
6–8	5
9–11	4
12–14	2

Applications

Engin.   **3.30** A study of the reliability of buses ["Large sample simultaneous confidence intervals for the multinominal probabilities on transformations of the cell frequencies," *Technometrics* (1980) 22:588] examined the reliability of 191 buses. The distance traveled (in 1000s of miles) prior to the first major motor failure was classified into intervals. A modified form of the table follows.

Distance Traveled (1,000 miles)	Frequency
0–20.0	6
20.1–40.0	11
40.1–60.0	16
60.1–100.0	59
100.1–120.0	46
120.1–140.0	33
140.1–160.0	16
160.1–200.0	4

a. Sketch the relative frequency histogram for the distance data and describe its shape.
b. Estimate the mode, median, and mean for the distance traveled by the 191 buses.
c. What does the relationship among the three measures of center indicate about the shape of the histogram for these data?
d. Which of the three measures would you recommend as the most appropriate representative of the distance traveled by one of the 191 buses? Explain your answer.

Med. **3.31** In a study of 1,329 American men reported in *American Statistician* [(1974) 28:115–122] the men were classified by serum cholesterol and blood pressure. The group of 408 men who had blood pressure readings less than 127 mm Hg were then classified according to their serum cholesterol level.

Serum Cholesterol (mg/100cc)	Frequency
0.0–199.9	119
200.0–219.9	88
220.0–259.9	127
greater than 259	74

a. Estimate the mode, median, and mean for the serum cholesterol readings (if possible).
b. Which of the three summary statistics is more informative concerning a typical serum cholesterol level for the group of men? Explain your answer.

Env. **3.32** The ratio of DDE (related to DDT) to PCB concentrations in bird eggs has been shown to have had a number of biological implications. The ratio is used as an indication of the movement of contamination through the food chain. The paper "The ratio of DDE to PCB concentrations in Great Lakes herring gull eggs and its use in interpreting contaminants data" [*Journal of Great Lakes Research* (1998) 24(1):12–31] reports the following ratios for eggs collected at 13 study sites from the five Great Lakes. The eggs were collected from both terrestrial- and aquatic-feeding birds.

DDE to PCB Ratio										
Terrestrial feeders	76.50	6.03	3.51	9.96	4.24	7.74	9.54	41.70	1.84	2.50	1.54
Aquatic feeders	0.27	0.61	0.54	0.14	0.63	0.23	0.56	0.48	0.16	0.18	

a. Compute the mean and median for the 21 ratios, ignoring the type of feeder.
b. Compute the mean and median separately for each type of feeder.
c. Using your results from parts (a) and (b), comment on the relative sensitivity of the mean and median to extreme values in a data set.
d. Which measure, mean or median, would you recommend as the most appropriate measure of the DDE to PCB level for both types of feeders? Explain your answer.

Med. **3.33** A study of the survival times, in days, of skin grafts on burn patients was examined in Woolson and Lachenbruch [*Biometrika* (1980) 67:597–606]. Two of the patients left the study prior to the failure of their grafts. The survival time for these individuals is some number greater than the reported value.

Survival time (days): 37, 19, 57*, 93, 16, 22, 20, 18, 63, 29, 60*

(The "*" indicates that the patient left the study prior to failure of the graft; values given are for the day the patient left the study.)

a. Calculate the measures of center (if possible) for the 11 patients.
b. If the survival times of the two patients who left the study were obtained, how would these new values change the values of the summary statistics calculated in (a)?

Engin. **3.34** A study of the reliability of diesel engines was conducted on 14 engines. The engines were run in a test laboratory. The time (in days) until the engine failed is given here. The study was terminated after 300 days. For those engines that did not fail during the study period, an asterisk is placed by the number 300. Thus, for these engines, the time to failure is some value greater than 300.

Failure time (days): 130, 67, 300*, 234, 90, 256, 87, 120, 201, 178, 300*, 106, 289, 74

   a. Calculate the measures of center for the 14 engines.
   b. What are the implications of computing the measures of center when some of the exact failure times are not known?

Ag. **3.35** Nitrogen is a limiting factor in the yield of many different plants. In particular, the yield of apple trees is directly related to the nitrogen content of apple tree leaves and must be carefully monitored to protect the trees in an orchard. Research has shown that the nitrogen content should be approximately 2.5% for best yield results. Note that some researchers report their results in parts per million (ppm); hence, 1% would be equivalent to 10,000 ppm.)

To determine the nitrogen content of trees in an orchard, the growing tips of 150 leaves are clipped from trees throughout the orchard. These leaves are ground to form one composite sample, which the researcher assays for percentage of nitrogen. Composite samples obtained from a random sample of 36 orchards throughout the state gave the following nitrogen contents:

2.0968	2.8220	2.1739	1.9928	2.2194	3.0926
2.4685	2.5198	2.7983	2.0961	2.9216	2.1997
1.7486	2.7741	2.8241	2.6691	3.0521	2.9263
2.9367	1.9762	2.3821	2.6456	2.7678	1.8488
1.6850	2.7043	2.6814	2.0596	2.3597	2.2783
2.7507	2.4259	2.3936	2.5464	1.8049	1.9629

   a. Round each of these measurements to the nearest hundredth. (Use the convention that 5 is rounded up.)
   b. Determine the sample mode for the rounded data.
   c. Determine the sample median for the rounded data.
   d. Determine the sample mean for the rounded data.

3.36 Refer to the data of Exercise 3.35 rounded to the nearest hundredth. Replace the fourth measurement (2.94) by the value 29.40. Compute the sample mean, median, and mode for these data. Compare these results to those you found in Exercise 3.35.

Gov. **3.37** Effective tax rates (per $100) on residential property for three groups of large cities, ranked by residential property tax rate, are shown in the following table.

Group 1	Rate	Group 2	Rate	Group 3	Rate
Detroit, MI	4.10	Burlington, VT	1.76	Little Rock, AR	1.02
Milwaukee, WI	3.69	Manchester, NH	1.71	Albuquerque, NM	1.01
Newark, NJ	3.20	Fargo, ND	1.62	Denver, CO	.94
Portland, OR	3.10	Portland ME	1.57	Las Vegas, NV	.88
Des Moines, IA	2.97	Indianapolis, IN	1.57	Oklahoma City, OK	.81
Baltimore, MD	2.64	Wilmington, DE	1.56	Casper, WY	.70
Sioux Falls, IA	2.47	Bridgeport, CT	1.55	Birmingham, AL	.70
Providence, RI	2.39	Chicago, IL	1.55	Phoenix, AZ	.68
Philadelphia, PA	2.38	Houston, TX	1.53	Los Angeles, CA	.64
Omaha, NE	2.29	Atlanta, GA	1.50	Honolulu, HI	.59

Source: Government of the District of Columbia, Department of Finance and Revenue, *Tax Rates and Tax Burdens in the District of Columbia: A Nationwide Comparison*, annual.

      a. Compute the mean, median, and mode separately for the three groups.
      b. Compute the mean, median, and mode for the complete set of 30 measurements.
      c. What measure or measures best summarize the center of these distributions? Explain.

3.38 Refer to Exercise 3.37. Average the three group means, the three group medians, and the three group modes, and compare your results to those of part (b). Comment on your findings.

3.5 Describing Data on a Single Variable: Measures of Variability

It is not sufficient to describe a data set using only measures of central tendency, such as the mean or the median. For example, suppose we are monitoring the production of plastic sheets that have a nominal thickness of 3 mm. If we randomly select 100 sheets from the daily output of the plant and find that the average thickness of the 100 sheets is 3 mm, does this indicate that all 100 sheets have the desired thickness of 3 mm? We may have a situation in which 50 sheets have a thickness of 1 mm and the remaining 50 sheets have a thickness of 5 mm. This would result in an average thickness of 3 mm, but none of the 100 sheets would have a thickness close to the specified 3 mm. Thus, we need to determine how dispersed are the sheet thicknesses about the mean of 3 mm.

      Graphically, we can observe the need for some measure of variability by examining the relative frequency histograms of Figure 3.18. All the histograms **variability** have the same mean but each has a different spread, or **variability,** about the mean. For illustration, we have shown the histograms as smooth curves. Suppose the three histograms represent the amount of PCB (ppb) found in a large number of 1-liter samples taken from three lakes that are close to chemical plants. The average amount of PCB, μ, in a 1-liter sample is the same for all three lakes. However, the variability in the PCB quantity is considerably different. Thus, the lake with PCB quantity depicted in histogram (a) would have fewer samples containing very small or large quantities of PCB as compared to the lake with

FIGURE 3.18
Relative frequency
histograms with different
variabilities but
the same mean

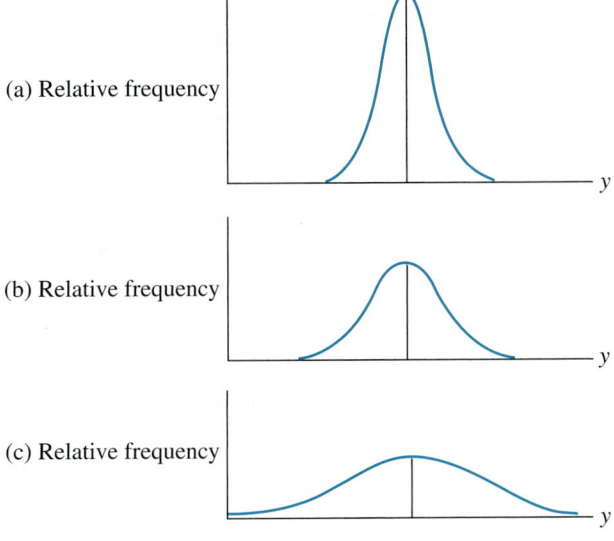

(a) Relative frequency

(b) Relative frequency

(c) Relative frequency

range

PCB values depicted in histogram (c). Knowing only the mean PCB quantity in the three lakes would mislead the investigator concerning the level of PCB present in all three lakes.

The simplest but least useful measure of data variation is the **range,** which we alluded to in Section 3.2. We now present its definition.

DEFINITION 3.4

The **range** of a set of measurements is defined to be the difference between the largest and the smallest measurements of the set.

EXAMPLE 3.7

Determine the range of the 15 overdue accounts of Example 3.5.

Solution   The smallest measurement is $4.88 and the largest is $807.80. Hence, the range is

$$807.80 - 4.88 = \$802.92$$

grouped data range

For **grouped data,** because we do not know the individual measurements, the **range** is taken to be the difference between the upper limit of the last interval and the lower limit of the first interval.

Although the range is easy to compute, it is sensitive to outliers because it depends on the most extreme values. It does not give much information about the pattern of variability. Referring to the situation described in Example 3.5, if in the current budget period the 15 overdue accounts consisted of 10 accounts having a value of $4.88, 3 accounts of $807.80, and 1 account of $11.36, then the mean value would be $165.57 and the range would be $802.92. The mean and range would be identical to the mean and range calculated for the data of Example 3.5. However, the data in the current budget period are more spread out about the mean than the data in the earlier budget period. What we seek is a measure of variability that discriminates between data sets having different degrees of concentration of the data about the mean.

percentile

A second measure of variability involves the use of **percentiles.**

DEFINITION 3.5

The **pth percentile** of a set of n measurements arranged in order of magnitude is that value that has at most $p\%$ of the measurements below it and at most $(100 - p)\%$ above it.

For example, Figure 3.19 illustrates the 60th percentile of a set of measurements. Percentiles are frequently used to describe the results of achievement test scores

FIGURE 3.19
The 60th percentile of a set of measurements

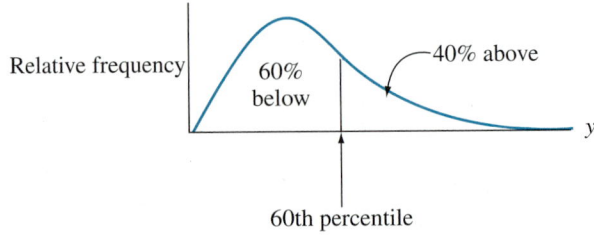

and the ranking of a person in comparison to the rest of the people taking an examination. Specific percentiles of interest are the 25th, 50th, and 75th percentiles, often called the *lower quartile*, the *middle quartile* (median), and the *upper quartile,* respectively (see Figure 3.20).

FIGURE 3.20

Quartiles of a distribution

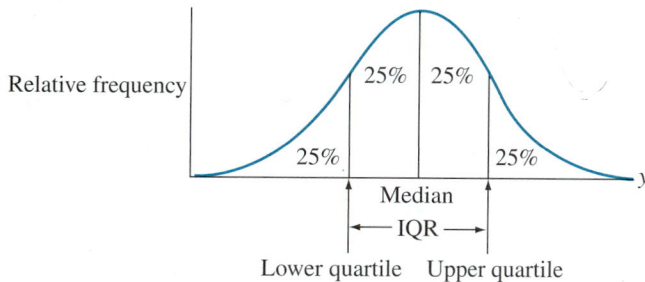

The computation of percentiles is accomplished as follows: Each data value corresponds to a percentile for the percentage of the data values that are less than or equal to it. Let $y_{(1)}, y_{(2)}, \ldots, y_{(n)}$ denote the ordered observations for a data set; that is,

$$y_{(1)} \le y_{(2)} \le \cdots \le y_{(n)}$$

The *i*th ordered observation, $y_{(j)}$, corresponds to the $100(j - .5)/n$ percentile. We use this formula in place of assigning the percentile $100j/n$ so that we avoid assigning the 100th percentile to $y_{(n)}$, which would imply that the largest possible data value in the population was observed in the data set, an unlikely happening. For example, a study of serum total cholesterol (mg/l) levels recorded the levels given in the following table for 20 adult patients. Thus, each ordered observation is a data percentile corresponding to a multiple of the fraction $100(j - .5)/n = 100(2j - 1)/2n = 100(2j - 1)/40$.

Observation (j)	Cholesterol (mg/l)	Percentile
1	133	2.5
2	137	7.5
3	148	12.5
4	149	17.5
5	152	22.5
6	167	27.5
7	174	32.5
8	179	37.5
9	189	42.5
10	192	47.5
11	201	52.5
12	209	57.5
13	210	62.5
14	211	67.5
15	218	72.5
16	238	77.5
17	245	82.5
18	248	87.5
19	253	92.5
20	257	97.5

The 22.5th percentile is 152 (mg/l). Thus, 22.5% of persons in the study have a serum cholesterol less than or equal to 152. Also, the median of the above data set, which is the 50th percentile, is halfway between 192 and 201; that is, median $= (192 + 201)/2 = 196.5$. Thus, approximately half of the persons in the study have a serum cholesterol level less than 196.5 and half greater than 196.5.

When dealing with large data sets, the percentiles are generalized to quantiles, where a quantile, denoted $Q(u)$, is a number that divides a sample of n data values into two groups so that the specified fraction u of the data values is less than or equal to the value of the quantile, $Q(u)$. Plots of the quantiles $Q(u)$ versus the data fraction u provide a method of obtaining estimated quantiles for the population from which the data were selected. We can obtain a quantile plot using the following steps:

1. Place a scale on the horizontal axis of a graph covering the interval (0, 1).
2. Place a scale on the vertical axis covering the range of the observed data, y_1 to y_n.
3. Plot $y_{(i)}$ versus $u_i = (i - .5)/n = (2i - 1)/2n$, for $i = 1, \ldots, n$.

Using the Minitab software, we obtain the plot shown in Figure 3.21 for the cholesterol data. Note that, with Minitab, the vertical axis is labeled $Q(u)$ rather than $y_{(i)}$. We plot $y_{(i)}$ versus u to obtain a quantile plot. Specific quantiles can be read from the plot.

FIGURE 3.21

Quantile plot of cholesterol data

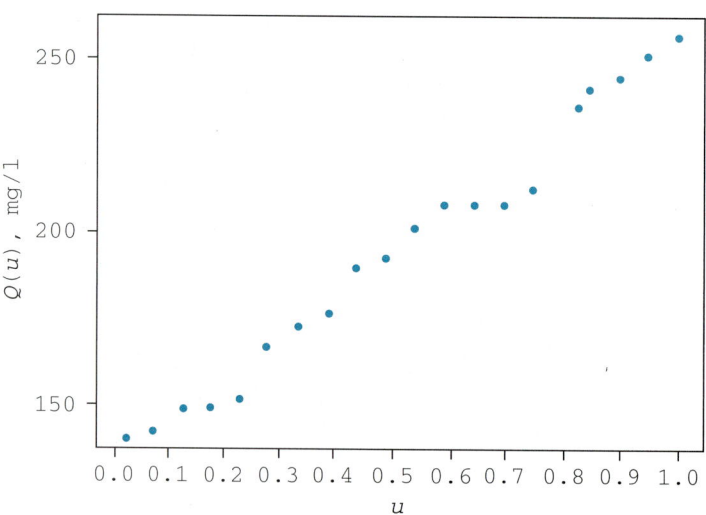

We can obtain the quantile, $Q(u)$, for any value of u as follows. First, place a smooth curve through the plotted points in the quantile plot and then read the value off the graph corresponding to the desired value of u.

To illustrate the calculations, suppose we want to determine the 80th percentile for the cholesterol data—that is, the cholesterol level such that 80% of the persons in the population have a cholesterol level less than this value, $Q(.80)$.

Referring to Figure 3.21, locate the point $u = .8$ on the horizontal axis and draw a perpendicular line up to the quantile plot and then a horizontal line over to the vertical axis. The point where this line touches the vertical axis is our

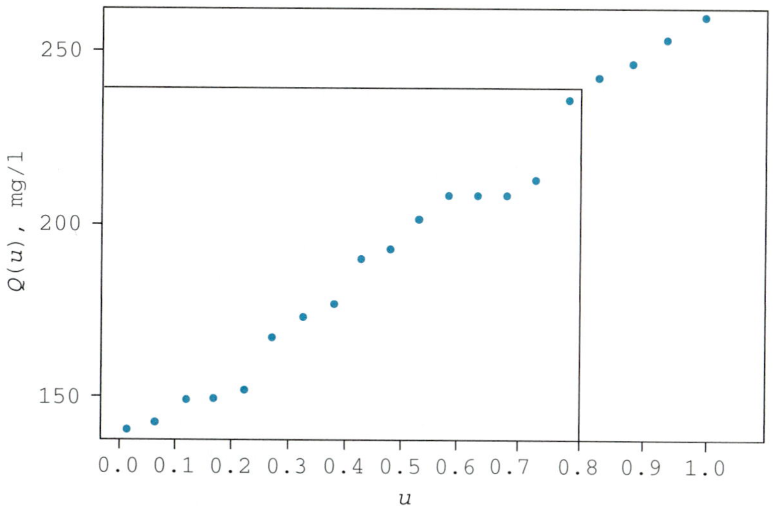

FIGURE 3.22

80th quantile of
cholesterol data

estimate of the 80th quantile. (See Figure 3.22.) Roughly 80% of the population have a cholesterol level less than 243.

When the data are grouped, the following formula can be used to approximate the percentiles for the original data. Let

P = percentile of interest

L = lower limit of the class interval that includes percentile of interest

n = total frequency

cf_b = cumulative frequency for all class intervals before the percentile class

f_p = frequency of the class interval that includes the percentile of interest

w = interval width

Then, for example, the 65th percentile for a set of grouped data would be computed using the formula

$$P = L + \frac{w}{f_p}(.65n - cf_b)$$

To determine L, f_p, and cf_b, begin with the lowest interval and find the first interval for which the cumulative relative frequency exceeds .65. This interval would contain the 65th percentile.

EXAMPLE 3.8

Refer to the chick data of Table 3.5. Compute the 90th percentile.

Solution Because the twelfth interval is the first interval for which the cumulative relative frequency exceeds .90, we have

$L = 4.65$

$n = 100$

$cf_b = 84$

$f_{90} = 7$

$w = .1$

divided by n in the definition of the sample variance s^2, the average sample variance computed from a large number of samples would be less than the population variance; hence, s^2 would tend to underestimate σ^2.

standard deviation   Another useful measure of variability, the **standard deviation,** involves the square root of the variance. One reason for defining the standard deviation is that it yields a measure of variability having the same units of measurement as the original data, whereas the units for variance are the square of the measurement units.

DEFINITION 3.8

The **standard deviation** of a set of measurements is defined to be the positive square root of the variance.

s
σ We then have s denoting the sample standard deviation and σ denoting the corresponding population standard deviation.

EXAMPLE 3.9

The time between an electric light stimulus and a bar press to avoid a shock was noted for each of five conditioned rats. Use the given data to compute the sample variance and standard deviation.

shock avoidance times (seconds):   5, 4, 3, 1, 3

Solution   The deviations and the squared deviations are shown here. The sample mean $\bar{y}$ is 3.2.

	y_i	$y_i - \bar{y}$	$(y_i - \bar{y})^2$
	5	1.8	3.24
	4	.8	.64
	3	$-.2$	.04
	1	-2.2	4.84
	3	$-.2$	.04
Totals	16	0	8.80

Using the total of the squared deviations column, we find the sample variance to be

$$s^2 = \frac{\Sigma_i(y_i - \bar{y})^2}{n - 1} = \frac{8.80}{4} = 2.2$$

We can make a simple modification of our formula for the sample variance to approximate the sample variance if only grouped data are available. Recall that in approximating the sample mean for grouped data, we let y_i and f_i denote the midpoint and frequency, respectively, for the ith class interval. With this notation, the sample variance for grouped data is $s^2 = \Sigma_i f_i (y_i - \bar{y})^2/(n - 1)$. The sample standard deviation is $\sqrt{s^2}$.

EXAMPLE 3.10

Refer to the chick data from Table 3.6 of Example 3.6. Calculate the sample variance and standard deviation for these data.

Solution   From Table 3.6, the sum of the $f_i(y_i - \bar{y})^2$ calculations is 9.4936. Using this value, we can approximate s^2 and s.

$$s^2 \cong \frac{1}{n-1} \sum_i f_i(y_i - \bar{y})^2 = \frac{1}{99}(9.4936) = 0.095895$$

$$s = \sqrt{0.095895} = 0.3097$$

If we compute s from the original 100 data values, the value of s to four decimal places is the same as is given above. The values of s computed from the original data and from the grouped data will generally be different. However, when the frequency table has a large number of classes, the approximation of s from the frequency table values will be very accurate.

We have now discussed several measures of variability, each of which can be used to compare the variabilities of two or more sets of measurements. The standard deviation is particularly appealing for two reasons: (1) we can compare the variabilities of *two or more* sets of data using the standard deviation, and (2) we can also use the results of the rule that follows to interpret the standard deviation of a single set of measurements. This rule applies to data sets with roughly a "mound-shaped" histogram—that is, a histogram that has a single peak, is symmetrical, and tapers off gradually in the tails. Because so many data sets can be classified as mound-shaped, the rule has wide applicability. For this reason, it is called the *Empirical Rule*.

EMPIRICAL RULE

Give a set of n measurements possessing a mound-shaped histogram, then

the interval $\bar{y} \pm s$ contains approximately 68% of the measurements

the interval $\bar{y} \pm 2s$ contains approximately 95% of the measurements

the interval $\bar{y} \pm 3s$ contains approximately 99.7% of the measurements

EXAMPLE 3.11

The yearly report from a particular stockyard gives the average daily wholesale price per pound for steers as $.61, with a standard deviation of $.07. What conclusions can we reach about the daily steer prices for the stockyard? Because the original daily price data are not available, we are not able to provide much further information about the daily steer prices. However, from past experience it is known that the daily price measurements have a mound-shaped relative frequency histogram. Applying the Empirical Rule, what conclusions can we reach about the distribution of daily steer prices?

Solution   Applying the Empirical Rule, the interval

$.61 \pm .07$ or $\$.54$ to $\$.68$

$$\text{approximate value of } s = \frac{\text{range}}{4}$$

Some people might wonder why we did not equate the range to $6s$, because the interval $\bar{y} \pm 3s$ should contain almost all the measurements. This procedure would yield an approximate value for s that is smaller than the one obtained by the preceding procedure. If we are going to make an error (as we are bound to do with any approximation), it is better to overestimate the sample standard deviation so that we are not led to believe there is less variability than may be the case.

EXAMPLE 3.12

The following data represent the percentages of family income allocated to groceries for a sample of 30 shoppers:

26	28	30	37	33	30
29	39	49	31	38	36
33	24	34	40	29	41
40	29	35	44	32	45
35	26	42	36	37	35

For these data, $\sum y_i = 1{,}043$ and $\sum (y_i - \bar{y})^2 = 1{,}069.3667$. Compute the mean, variance, and standard deviation of the percentage of income spent on food. Check your calculation of s.

Solution   The sample mean is

$$\bar{y} = \frac{\sum_i y_i}{30} = \frac{1{,}043}{30} = 34.77$$

The corresponding sample variance and standard deviation are

$$s^2 = \frac{1}{n-1} \sum (y_i - \bar{y})^2$$

$$= \frac{1}{29}(1{,}069.3667) = 36.8747$$

$$s = \sqrt{36.8747} = 6.07$$

We can check our calculation of s by using the range approximation. The largest measurement is 49 and the smallest is 24. Hence, an approximate value of s is

$$s \approx \frac{\text{range}}{4} = \frac{49 - 24}{4} = 6.25$$

Note how close the approximation is to our computed value.

Although there will not always be the close agreement found in Example 3.12, the range approximation provides a useful and quick check on the calculation of *s*.

The standard deviation can be deceptive when comparing the amount of variability of different types of populations. A unit of variation in one population might be considered quite small, whereas that same amount of variability in a different population would be considered excessive. For example, suppose we want to compare two production processes that fill containers with products. Process A is filling fertilizer bags, which have a nominal weight of 80 pounds. The process produces bags having a mean weight of 80.6 pounds with a standard deviation of 1.2 pounds. Process B is filling 24-ounce cornflakes boxes, which have a nominal weight of 24 ounces. Process B produces boxes having a mean weight of 24.3 ounces with a standard deviation of 0.4 ounces. Is process A much more variable than process B because 1.2 is three times larger than 0.4? To compare the variability in two considerably different processes or populations, we need to define another measure of variability. The **coefficient of variation** measures the variability in the values in a population relative to the magnitude of the population mean. In a process or population with mean μ and standard deviation σ, the coefficient of variation is defined as

coefficient of variation

$$CV = \frac{\sigma}{|\mu|}$$

provided $\mu \neq 0$. Thus, the coefficient of variation is the standard deviation of the population or process expressed in units of μ. The two filling processes would have equivalent degrees of variability if the two processes had the same CV. For the fertilizer process, the CV = 1.2/80 = .015. The cornflakes process has CV = 0.4/24 = .017. Hence, the two processes have very similar variability relative to the size of their means. The CV is a unit-free number because the standard deviation and mean are measured using the same units. Hence, the CV is often used as an index of process or population variability. In many applications, the CV is expressed as a percentage: $CV = 100(\sigma/|\mu|)\%$. Thus, if a process has a CV of 15%, the standard deviation of the output of the process is 15% of the process mean. Using sampled data from the population, we estimate CV with $100(s/|\bar{y}|)\%$.

EXERCISES

Basic Techniques

Engin.

3.39   Pushing economy and wheelchair propulsion technique were examined for eight wheelchair racers on a motorized treadmill in a paper by Goosey and Campbell [*Adapted Physical Activity Quarterly* (1998) 15: 36–50]. The eight racers had the following years of racing experience:

Racing experience (years): 6, 3, 10, 4, 4, 2, 4, 7

a. Verify that the mean years' experience is 5 years. Does this value appear to adequately represent the center of the data set?
b. Verify that $\Sigma_i(y - \bar{y})^2 = \Sigma_i(y - 5)^2 = 46$.
c. Calculate the sample variance and standard deviation for the experience data. How would you interpret the value of the standard deviation relative to the sample mean?

3.40   In the study described in Exercise 3.39, the researchers also recorded the ages of the eight racers.

Age (years): 39, 38, 31, 26, 18, 36, 20, 31

a. Generate a time series plot of the mercury concentrations and place lines for both sites on the same graph. Comment on any trends in the lines across the years of data. Are the trends similar for both sites?

b. Select the most appropriate measure of center for the mercury concentrations. Compare the center for the two sites.

c. Compare the variability in mercury concentrations at the two sites. Use the CV in your comparison and explain why it is more appropriate than using the standard deviations.

d. When comparing the center and variability of the two sites, should the years 1969–1972 be used for site 2?

3.6 The Boxplot

As mentioned earlier in this chapter, a stem-and-leaf plot provides a graphical representation of a set of scores that can be used to examine the shape of the distribution, the range of scores, and where the scores are concentrated. The **boxplot**
boxplot, which builds on the information displayed in a stem-and-leaf plot, is more concerned with the symmetry of the distribution and incorporates numerical measures of central tendency and location to study the variability of the scores and the concentration of scores in the tails of the distribution.

Before we show how to construct and interpret a boxplot, we need to introduce several new terms that are peculiar to the language of exploratory data analysis (EDA). We are familiar with the definitions for the first, second (median), and third quartiles of a distribution presented earlier in this chapter. The boxplot
quartiles uses the median and **quartiles** of a distribution.

We can now illustrate a *skeletal boxplot* using an example.

EXAMPLE 3.13

Use the stem-and-leaf plot in Figure 3.25 for the 90 violent crime rates of Table 3.4 on p. 55 to construct a skeletal boxplot.

FIGURE 3.25
Stem-and-leaf plot

```
 1   89
 2   10 24 67 91 96 98
 3   36 41 52 54 74 76 88 93 93
 4   10 21 35 47 48 60 64 66 80 81 91 96 98
 5   04 05 08 16 26 29 37 57 59 61 62 62 62 63 70 71 78 85 92
 6   05 05 24 26 28 31 39 42 47 61 73 84 85 85 90 98
 7   03 06 18 19 20 31 35 39 51 58 71
 8   04 07 09 11 14 17 43 56 68 76 77 85
 9   28 71
10   20
```

Solution When the scores are ordered from lowest to highest, the median is computed by averaging the 45th and 46th scores. For these data, the 45th score (counting from the lowest to the highest in Figure 3.25) is 571 and the 46th is 578, hence, the median is

$$M = \frac{571 + 578}{2} = 574.5$$

To find the lower and upper quartiles for this distribution of scores, we need to determine the 25th and 75th percentiles. We can use the method given on page

83 to compute $Q(.25)$ and $Q(.75)$. A quick method that yields essentially the same values for the two quartiles consists of the following steps:

1. Order the data from smallest to largest value.
2. Divide the ordered data set into two data sets using the median as the dividing value.
3. Let the lower quartile be the median of the set of values consisting of the smaller values.
4. Let the upper quartile be the median of the set of values consisting of the larger values.

In the example, the data set has 90 values. Thus, we create two data sets, one containing the $90/2 = 45$ smallest values and the other containing the 45 largest values. The lower quartile is the $(45 + 1)/2 = 23$rd smallest value and the upper quartile is the 23rd value counting from the largest value in the data set. The 23rd-lowest score and 23rd-highest scores are 464 and 719.

lower quartile, $Q_1 = 464$

upper quartile, $Q_3 = 719$

skeletal boxplot

These three descriptive measures and the smallest and largest values in a data set are used to construct a skeletal boxplot (see Figure 3.26). The **skeletal boxplot** is constructed by drawing a box between the lower and upper quartiles with a solid line drawn across the box to locate the median. A straight line is then drawn connecting the box to the largest value; a second line is drawn from the box to the smallest value. These straight lines are sometimes called whiskers, and the entire graph is called a **box-and-whiskers plot.**

box-and-whiskers plot

FIGURE 3.26

Skeletal boxplot for the data of Figure 3.25

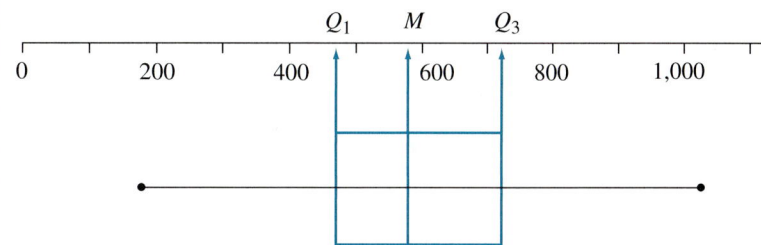

With a quick glance at a skeletal boxplot, it is easy to obtain an impression about the following aspects of the data:

1. The lower and upper quartiles, Q_1 and Q_3
2. The interquartile range (IQR), the distance between the lower and upper quartiles
3. The most extreme (lowest and highest) values
4. The symmetry or asymmetry of the distribution of scores

If we were presented with Figure 3.26 without having seen the original data, we would have observed that

$Q_1 \approx 475$

$Q_3 \approx 725$

$IQR \approx 725 - 475 = 250$

$M \approx 575$

most extreme values: 175 and 1,025

H.R.   **3.53**   Data on the age at the time of a job turnover and on the reason for the job turnover are displayed here for 250 job changes in a large corporation.

Reason for Turnover	Age (Years)				Total
	≤29	30–39	40–49	≥50	
Resigned	30	6	4	20	60
Transferred	12	45	4	5	66
Retired/fired	8	9	52	55	124
Total	50	60	60	80	250

Do a percentage comparison based on the row totals and use this to describe the data.

3.54   Refer to Exercise 3.53. What different summary would you get with a percentage comparison based on the column totals? Do this summary and describe your results.

Engin.   **3.55**   In the manufacture of soft contact lenses, the power (the strength) of the lens needs to be very close to the target value. In the paper "An ANOM-type test for variances from normal populations" [*Technometrics* (1997) 39: 274–283], a comparison of several suppliers is made relative to the consistency of the power of the lens. The following table contains the deviations from the target power value of lenses produced using materials from three different suppliers:

Supplier	Deviations from Target Power Value								
1	189.9	191.9	190.9	183.8	185.5	190.9	192.8	188.4	189.0
2	156.6	158.4	157.7	154.1	152.3	161.5	158.1	150.9	156.9
3	218.6	208.4	187.1	199.5	202.0	211.1	197.6	204.4	206.8

a. Compute the mean and standard deviation for the deviations of each supplier.
b. Plot the sample deviation data.
c. Describe the deviation from specified power for the three suppliers.
d. Which supplier appears to provide material that produces lenses having power closest to the target value?

Bus.   **3.56**   The federal government keeps a close watch on money growth versus targets that have been set for that growth. We list two measures of the money supply in the United States, M2 (private checking deposits, cash, and some savings) and M3 (M2 plus some investments), which are given here for 20 consecutive months.

Month	Money Supply (in Trillions of Dollars)	
	M2	M3
1	2.25	2.81
2	2.27	2.84
3	2.28	2.86
4	2.29	2.88

(*continued*)

(*continued*)

Month	Money Supply (in Trillions of Dollars)	
	M2	M3
5	2.31	2.90
6	2.32	2.92
7	2.35	2.96
8	2.37	2.99
9	2.40	3.02
10	2.42	3.04
11	2.43	3.05
12	2.42	3.05
13	2.44	3.08
14	2.47	3.10
15	2.49	3.10
16	2.51	3.13
17	2.53	3.17
18	2.53	3.18
19	2.54	3.19
20	2.55	3.20

a. Would a scatterplot describe the relation between M2 and M3?

b. Construct a scatterplot. Is there an obvious relation?

3.57 Refer to Exercise 3.56. What other data plot might be used to describe and summarize these data? Make the plot and interpret your results.

3.8   Summary

This chapter was concerned with graphical and numerical description of data. The pie chart and bar graph are particularly appropriate for graphically displaying data obtained from a qualitative variable. The frequency and relative frequency histograms and stem-and-leaf plots are graphical techniques applicable only to quantitative data.

Numerical descriptive measures of data are used to convey a mental image of the distribution of measurements. Measures of central tendency include the mode, the median, and the arithmetic mean. Measures of variability include the range, the interquartile range, the variance, and the standard deviation of a set of measurements.

We extended the concept of data description to summarize the relations between two qualitative variables. Here cross-tabulations were used to develop percentage comparisons. We examined plots for summarizing the relations between quantitative and qualitative variables and between two quantitative variables. Material presented here (namely, summarizing relations among variables) will be discussed and expanded in later chapters on chi-square methods, on the analysis of variance, and on regression.

Key Formulas

1. Median, grouped data

$$\text{Median} = L + \frac{w}{f_m}(.5n - cf_b)$$

2. Sample mean

$$\bar{y} = \frac{\Sigma_i y_i}{n}$$

3. Sample mean, grouped data

$$\bar{y} \cong \frac{\Sigma_i f_i y_i}{n}$$

4. Sample variance

$$s^2 = \frac{1}{n-1}\Sigma_i (y_i - \bar{y})^2$$

5. Sample variance, grouped data

$$s^2 \cong \frac{1}{n-1}\Sigma_i f_i(y_i - \bar{y})^2$$

6. Sample standard deviation

$$s = \sqrt{s^2}$$

7. Sample coefficient of variation

$$\text{CV} = \frac{s}{|\bar{y}|}$$

Supplementary Exercises

Env. **3.58** To control the risk of severe core damage during a commercial nuclear power station blackout accident, the reliability of the emergency diesel generators to start on demand must be maintained at a high level. The paper "Empirical Bayes estimation of the reliability of nuclear-power emergency diesel generators" [*Technometrics* (1996) 38: 11–23] contains data on the failure history of seven nuclear power plants. The following data are the number of successful demands between failures for the diesel generators at one of these plants from 1982 to 1988.

28	50	193	55	4	7	147	76	10	0	10	84	0	9	1	0	62
26	15	226	54	46	128	4	105	40	4	273	164	7	55	41	26	6

(*Note:* The failure of the diesel generator does not necessarily result in damage to the nuclear core because all nuclear power plants have several emergency diesel generators.)

    a. Calculate the mean and median of the successful demands between failures.

    b. Which measure appears to best represent the center of the data?

    c. Calculate the range and standard deviation, *s*.

    d. Use the range approximation to estimate *s*. How close is the approximation to the true value?

    e. Construct the intervals

$$\bar{y} \pm s \qquad \bar{y} \pm 2s \qquad \bar{y} \pm 3s$$

    Count the number of demands between failures falling in each of the three intervals. Convert these numbers to percentages and compare your results to the Empirical Rule.

    f. Why do you think the Empirical Rule and your percentages do not match well?

Edu. **3.59** The College of Dentistry at the University of Florida has made a commitment to develop its entire curriculum around the use of self-paced instructional materials such as videotapes, slide tapes, and syllabi. It is hoped that each student will proceed at a pace commensurate with his or her ability and that the instructional staff will have more free time for personal consultation in student–faculty interaction. One such instructional module was developed and tested on the first 50 students proceeding through the curriculum. The following measurements represent the number of hours it took these students to complete the required modular material.

16	8	33	21	34	17	12	14	27	6
33	25	16	7	15	18	25	29	19	27
5	12	29	22	14	25	21	17	9	4
12	15	13	11	6	9	26	5	16	5
9	11	5	4	5	23	21	10	17	15

    a. Calculate the mode, the median, and the mean for these recorded completion times.
    b. Guess the value of s.
    c. Compute s by using the shortcut formula and compare your answers to that of part (b).
    d. Would you expect the Empirical Rule to describe adequately the variability of these data? Explain.

Bus.   **3.60** The February 1998 issue of *Consumer Reports* provides data on the price of 24 brands of paper towels. The prices are given in both cost per roll and cost per sheet because the brands had varying numbers of sheets per roll.

Brand	Price per Roll	Number of Sheets per Roll	Cost per Sheet
1	1.59	50	.0318
2	0.89	55	.0162
3	0.97	64	.0152
4	1.49	96	.0155
5	1.56	90	.0173
6	0.84	60	.0140
7	0.79	52	.0152
8	0.75	72	.0104
9	0.72	80	.0090
10	0.53	52	.0102
11	0.59	85	.0069
12	0.89	80	.0111
13	0.67	85	.0079
14	0.66	80	.0083
15	0.59	80	.0074
16	0.76	80	.0095
17	0.85	85	.0100
18	0.59	85	.0069
19	0.57	78	.0073
20	1.78	180	.0099
21	1.98	180	.0011
22	0.67	100	.0067
23	0.79	100	.0079
24	0.55	90	.0061

    a. Compute the standard deviation for both the price per roll and the price per sheet.
    b. Which is more variable, price per roll or price per sheet?
    c. In your comparison in part (b), should you use s or CV? Justify your answer.

3.61 Use a scatterplot to plot the price per roll and number of sheets per roll.
    a. Do the 24 points appear to fall on a straight line?
    b. If not, is there any other relation between the two prices?
    c. What factors may explain why the ratio of price per roll to number of sheets is not a constant?

3.62 Construct boxplots for both price per roll and number of sheets per roll. Are there any "unusual" brands in the data?

Bio. **3.63** A study was conducted to determine urine flow of sheep (in milliliters/minute) when infused intravenously with the antidiuretic hormone ADH. The urine flows of 10 sheep are recorded here.

0.7   0.5   0.5   0.6   0.5   0.4   0.3   0.9   1.2   0.9

   a. Determine the mean, the median, and the mode for these sample data.
   b. Suppose that the largest measurement is 6.8 rather than 1.2. How does this affect the mean, the median, and the mode?

3.64 Refer to Exercise 3.63.
   a. Compute the range and the sample standard deviation.
   b. Check your calculation of s using the range approximation.
   c. How are the range and standard deviation affected if the largest measurement is 6.8 rather than 1.2? What about 68?

Env. **3.65** The paper "Conditional simulation of waste-site performance" [*Technometrics* (1994) 36: 129–161] discusses the evaluation of a pilot facility for demonstrating the safe management, storage, and disposal of defense-generated, radioactive, transuranic waste. Researchers have determined that one potential pathway for release of radionuclides is through contaminant transport in groundwater. Recent focus has been on the analysis of transmissivity, a function of the properties and the thickness of an aquifer that reflects the rate at which water is transmitted through the aquifer. The following table contains 41 measurements of transmissivity, T, made at the pilot facility.

9.354	6.302	24.609	10.093	0.939	354.81	15399.27	88.17	1253.43	0.75	312.10
1.94	3.28	1.32	7.68	2.31	16.69	2772.68	0.92	10.75	0.000753	
1.08	741.99	3.23	6.45	2.69	3.98	2876.07	12201.13	4273.66	207.06	
2.50	2.80	5.05	3.01	462.38	5515.69	118.28	10752.27	956.97	20.43	

   a. Draw a relative frequency histogram for the 41 values of T.
   b. Describe the shape of the histogram.
   c. When the relative frequency histogram is highly skewed to the right, the Empirical Rule may not yield very accurate results. Verify this statement for the data given.
   d. Data analysts often find it easier to work with mound-shaped relative frequency histograms. A transformation of the data will sometimes achieve this shape. Replace the given 41 T values with the logarithm base 10 of the values and reconstruct the relative frequency histogram. Is the shape more mound-shaped than the original data? Apply the Empirical Rule to the transformed data and verify that it yields more accurate results than it did with the original data.

3.66 Compute the mean, median, and standard deviation for the homeownership data in Exercise 3.20.
   a. Compare the mean and median for the 1996 data. Which value is most appropriate for this data set? Explain your answer.
   b. Is there a substantial difference between the summary statistics for the two years? What conclusions can you draw about the change in homeownership during the 11 years using these summary statistics?

3.67 Construct boxplots for the two years of homeownership data in Exercise 3.20.
   a. Has homeownership percentage changed over the years?
   b. Are there any states that have extremely low homeownership?

c. Are there any states that have extremely high homeownership?

d. What similarities exist for the states classified as having low homeownership? High ownership?

Soc.   **3.68** A random sample of 90 standard metropolitan statistical areas (SMSAs) was studied to obtain information on murder rates. The murder rate (number of murders per 100,000 people) was recorded, and these data are summarized in the following frequency table.

Class Interval	f_i	Class Interval	f_i
−.5–1.5	2	13.5–15.5	9
1.5–3.5	18	15.5–17.5	4
3.5–5.5	15	17.5–19.5	2
5.5–7.5	13	19.5–21.5	1
7.5–9.5	9	21.5–23.5	1
9.5–11.5	8	23.5–25.5	1
11.5–13.5	7		

Construct a relative frequency histogram for these data.

3.69 Refer to the data of Exercise 3.68.

a. Compute the sample median and the mode.

b. Compute the sample mean.

c. Which measure of central tendency would you use to describe the center of the distribution of murder rates?

3.70 Refer to the data of Exercise 3.68.

a. Compute the interquartile range.

b. Compute the sample standard deviation.

3.71 Using the homeownership data in Exercise 3.20, construct a quantile plot for both years.

a. Find the 20th percentile for the homeownership percentage and interpret this value for the 1996 data.

b. Congress wants to designate those states that have the highest homeownership percentage in 1996. Which states fall into the upper 10th percentile of homeownership rates?

c. Similarly identify those states that fall into the upper 10th percentile of homeownership rates during 1985. Are these states different from the states in this group during 1996?

Engin.   **3.72** Every 20 minutes a sample of 10 transistors is drawn from the outgoing product on a production line and tested. The data are summarized here for the first 500 samples of 10 measurements.

y_i	0	1	2	3	4	5	6	7	8	9	10
f_i	170	185	75	25	15	10	8	5	4	2	1

Construct a relative frequency distribution depicting the interquartile range. (*Note*: y_i in the table is the number of defectives in a sample of 10.)

3.73 Refer to Exercise 3.72.

a. Determine the sample median and the mode.

b. Calculate the sample mean.

c. Based on the mean, the median, and the mode, how is the distribution skewed?

3.74 Can the Empirical Rule be used to describe the set of measurements in Exercise 3.72? Justify your answer by referring to the relative frequency distribution.

Gov.   **3.75**   Per capita expenditure (dollars) for health and hospital services by state are shown here.

Dollars	f
45–59	1
60–74	4
75–89	9
90–104	9
105–119	12
120–134	6
135–149	4
150–164	1
165–179	3
180–194	0
195–209	1
Total	50

a. Construct a relative frequency histogram.
b. Compute approximate values for $\bar{y}$ and s from the grouped expenditure data.

3.76   Refer to the data of Table 3.4. Eliminate Philadelphia from the north and San Jose and Seattle from the west.
a. Compute $\bar{y}_i$ for the revised subgroups.
b. Combine the subgroup means ($\bar{y}_i$) to obtain the overall sample mean using the formula

$$\bar{y} = \frac{\Sigma_i n_i \bar{y}_i}{n}$$

where n_i is the number of observations in subgroup i.
c. Show that the sample mean computed in part (b) is identical to that obtained for the 87 measurements in part (a).

Engin.   **3.77**   The Insurance Institute for Highway Safety published data on the total damage suffered by compact automobiles in a series of controlled, low-speed collisions. The data, in dollars, with brand names removed are as follows:

361	393	430	543	566	610	763	851
886	887	976	1,039	1,124	1,267	1,328	1,415
1,425	1,444	1,476	1,542	1,544	2,048	2,197	

a. Draw a histogram of the data using six or seven categories.
b. On the basis of the histogram, what would you guess the mean to be?
c. Calculate the median and mean.
d. What does the relation between the mean and median indicate about the shape of the data?

Bus.   **3.78**   Production records for an automobile manufacturer show the following figures for production per shift (maximum production is 720 cars per shift):

| 688 | 711 | 625 | 701 | 688 | 667 | 694 | 630 | 547 | 703 | 688 | 697 | 703 |
| 656 | 677 | 700 | 702 | 688 | 691 | 664 | 688 | 679 | 708 | 699 | 667 | 703 |

a. Would the mode be a useful summary statistic for these data?
b. Find the median.

c. Find the mean.

d. What does the relationship between the mean and median indicate about the shape of the data?

3.79 Draw a stem-and-leaf plot of the data in Exercise 3.78. The stems should include (from highest to lowest) 71, 70, 69, ... Does the shape of the stem-and-leaf display confirm your judgment in part (d) of Exercise 3.78?

3.80 Refer to Exercise 3.79.
    a. Find the median and IQR.
    b. Find the inner and outer fences. Are there any outliers?
    c. Draw a boxplot of the data.

Soc.   **3.81** Data are collected on the weekly expenditures of a sample of urban households on food (including restaurant expenditures). The data, obtained from diaries kept by each household, are grouped by number of members of the household. The expenditures are as follows:

1 member:	67	62	168	128	131	118	80	53	99	68		
	76	55	84	77	70	140	84	65	67	183		
2 members:	129	116	122	70	141	102	120	75	114	81	106	95
	94	98	85	81	67	69	119	105	94	94	92	
3 members:	79	99	171	145	86	100	116	125				
	82	142	82	94	85	191	100	116				
4 members:	139	251	93	155	158	114	108					
	111	106	99	132	62	129	91					
5+ members:	121	128	129	140	206	111	104	109	135	136		

    a. Calculate the mean expenditure separately for each number of members.
    b. Calculate the median expenditure separately for each number of members.

3.82 Answer the following for the data in Exercise 3.81:
    a. Calculate the mean of the combined data, using the raw data.
    b. Can the combined mean be calculated from the means for each number of members?
    c. Calculate the median of the combined data using the raw data.
    d. Can the combined median be calculated from the medians for each number of members?

H.R.   **3.83** A company revised a long-standing policy to eliminate the time clocks and cards for nonexempt employees. Along with this change, all employees (exempt and nonexempt) were expected to account for their own time on the job as well as absences due to sickness, vacation, holidays, and so on. The previous policy of allocating a certain number of sick days was eliminated; if an employee was sick, he or she was given time off with pay; otherwise, he or she was expected to be working.

    In order to see how well the new program was working, the records of a random sample of 15 employees were examined to determine the number of sick days this year (under the new plan) and the corresponding number for the preceding year. The data are shown here:

Employee	This Year (new policy)	Preceding Year (old policy)
1	0	2
2	0	2
3	0	3
4	0	4
5	2	5

(continued)

(continued)

Employee	This Year (new policy)	Preceding Year (old policy)
6	1	2
7	1	6
8	3	8
9	1	5
10	0	4
11	5	5
12	6	12
13	1	3
14	2	4
15	12	4

a. Obtain the mean and standard deviation for each column.
b. Based on the sample data, what might you conclude (infer) about the new policies? Explain your reason(s).

3.84 Refer to Exercise 3.83. What happens to $\bar{y}$ and s for each column if we eliminate the two 12s and substitute values of 7? Are the ranges for the old and new policies affected by these substitutions?

Gov. **3.85** Federal authorities have destroyed considerable amounts of wild and cultivated marijuana plants. The following table shows the number of plants destroyed and the number of arrests for a 12-month period for 15 states.

State	Plants	Arrests
1	110,010	280
2	256,000	460
3	665	6
4	367,000	66
5	4,700,000	15
6	4,500	8
7	247,000	36
8	300,200	300
9	3,100	9
10	1,250	4
11	3,900,200	14
12	68,100	185
13	450	5
14	2,600	4
15	205,844	33

a. Discuss the appropriateness of using the sample mean to describe these two variables.
b. Compute the sample mean, 10% trimmed mean, and 20% trimmed mean. Which trimmed mean seems more appropriate for each variable? Why?

3.86 Refer to Exercise 3.85. Does there appear to be a relation between the number of plants destroyed and the number of arrests? How might you examine this question? What other variable(s) might be related to the number of plants destroyed?

Bus.   **3.87**   Monthly readings for the FDC Index, a popular barometer of the health of the pharmaceutical industry, are shown here. As can be seen, the index has several components—one each for pharmaceutical companies, diversified companies, and chain drugstores, and another for drug and medical supply wholesalers.

	Pharmaceuticals	**Diversified**	**Chain**	**Wholesaler**
January	123.1	154.6	393.3	475.5
February	122.4	146.0	407.6	504.1
March	125.2	169.2	405.0	476.6
April	136.1	156.7	415.1	513.3
May	149.3	177.0	418.9	543.5
June	145.7	158.1	443.2	552.6
July	162.4	156.6	419.1	526.2
August	168.0	178.6	404.0	516.3
September	155.6	170.4	391.8	482.1
October	177.0	162.9	410.9	484.0
November	196.6	182.4	459.8	522.6
December	195.2	195.4	431.9	536.8

a. Plot these data on a single graph.
b. Discuss trends within each component and any apparent relations among the separate components of the FDC Index.

3.88   Refer to Exercise 3.87. Compute the percent change for each month of each component of the index. (Assume that the percent changes in January were 12.3, −.7, 12.1, and 16.1, respectively, for the four components.) Plot these data. Are they more revealing than the original measurements were?

Bus.   **3.89**   The most widely reported index of the performance of the New York Stock Exchange (NYSE) is the Dow Jones Industrial Average (DJIA). This index is computed from the stock prices of 30 companies. When the DJIA was invented in 1896, the index was the average price of 12 stocks. The index was modified over the years as new companies were added and dropped from the index and was also altered to reflect when a company splits its stock. The closing New York Stock Exchange (NYSE) prices for the 30 components (as of May 1999) of the DJIA are given in the following table.

a. Compute the average price of the 30 stock prices in the DJIA.
b. Compute the range of the 30 stock prices in the DJIA.
c. The DJIA is no longer an average; the name includes the word "average" only for historical reasons. The index is computed by summing the stock prices and dividing by a constant, which is changed as stocks are added or removed from the index and when stocks split.

$$\text{DJIA} = \frac{\sum_{i=1}^{30} y_i}{C}$$

where y_i is the closing price for stock i, and $C = .211907$. Using the stock prices given, compute the DJIA for May 20, 1999.
d. The DJIA is a summary of data. Does the DJIA provide information about a population using sampled data? If so, to what population? Is the sample a random sample?

Components of DJIA

Company	Percent of DJIA	NYSE Stock Price (5/18/99)
Allied-Signal	2.640	60.8125
Alcoa	2.532	58.3125
American Express	5.357	123.3750
AT&T	2.570	59.1875
Boeing	1.948	44.8750
Caterpillar	2.559	58.9375
Chevron	4.060	93.5000
Citigroup	2.996	69.0000
Coca-Cola	2.950	67.9375
duPont	2.999	69.0625
Eastman Kodak	3.281	75.5625
Exxon	3.474	80.0000
General Electric	4.619	106.3750
General Motors	3.479	80.1250
Goodyear	2.578	59.3750
Hewlett-Packard	4.090	94.1875
IBM	2.380	232.5000
International Paper	2.380	54.8125
J.P. Morgan	6.049	139.3125
Johnson & Johnson	4.125	95.0000
McDonald's	1.739	40.0625
Merck	3.126	72.0000
Minnesota Mining	3.884	89.4375
Phillip Morris	1.745	40.1875
Procter & Gamble	4.198	96.6875
Sears, Roebuck	2.125	48.9375
Union Carbide	2.421	55.7500
United Technologies	2.640	60.8125
Wal-Mart Stores	1.935	44.5625
Walt Disney	1.294	29.8125

3.90 In Exercise 3.20 on p. 68 a relative frequency distribution was plotted for the homeownership data for the years 1985 and 1996.

   a. After examining these plots, do you think it would be appropriate to use the Empirical Rule to describe the data?

   b. Compute $\bar{y}$ and s for the year 1996. Compute the percentage of measurements falling in the intervals $\bar{y} \pm s$, $\bar{y} \pm 2s$, $\bar{y} \pm 3s$. Are these values consistent with the percentages provided by the Empirical Rule?

3.91 Refer to Exercise 3.90. Are there many extreme values affecting $\bar{y}$? Should this have been anticipated based on the data plot in Exercise 3.21? Compute the 10% trimmed mean for these data.

H.R. **3.92** As one part of a review of middle-manager selection procedures, a study was made of the relation between hiring source (promoted from within, hired from related business, hired from unrelated business) and the 3-year job history (additional promotion, same position, resigned, dismissed). The data for 120 middle managers follow.

Job History	Source			Total
	Within Firm	Related Business	Unrelated Business	
Promoted	13	4	10	27
Same position	32	8	18	58
Resigned	9	6	10	25
Dismissed	3	3	4	10
Total	57	21	42	120

a. Calculate job-history percentages within each source.

b. Would you say that there is a strong dependence between source and job history?

Env.   **3.93** A survey was taken of 150 residents of major coal-producing states, 200 residents of major oil- and natural-gas–producing states, and 450 residents of other states. Each resident chose a most preferred national energy policy. The results are shown in the following SPSS printout.

COUNT ROW PCT COL PCT TOT PCT	STATE			
	COAL	OIL AND GAS	OTHER	ROW TOTAL
OPINION	62	25	102	189
COAL ENCOURAGED	32.8	13.2	54.0	23.6
	41.3	12.5	22.7	
	7.8	3.1	12.8	
	3	12	26	41
FUSION DEVELOP	7.3	29.3	63.4	5.1
	2.0	6.0	5.8	
	0.4	1.5	3.3	
	8	6	22	36
NUCLEAR DEVELOP	22.2	16.7	61.1	4.5
	5.3	3.0	4.9	
	1.0	0.8	2.8	
	19	79	53	151
OIL DEREGULATION	12.6	52.3	35.1	18.9
	12.7	39.5	11.8	
	2.4	9.9	6.6	
	58	78	247	383
SOLAR DEVELOP	15.1	20.4	64.5	47.9
	38.7	39.0	54.9	
	7.3	9.8	30.9	
COLUMN	150	200	450	800
TOTAL	18.8	25.0	56.3	100.0

```
CHI SQUARE = 106.19406 WITH 8 DEGREES OF FREEDOM SIGNIFICANCE = 0.0000
CRAMER'S V = 0.25763
CONTINGENCY COEFFICIENT = 0.34233
LAMBDA = 0.01199 WITH OPINION DEPENDENT, = 0.07429 WITH STATE DEPENDENT.
```

a. Interpret the values 62, 32.8, 41.3, and 7.8 in the upper left cell of the cross tabulation. Note the labels COUNT, ROW PCT, COL PCT, and TOT PCT at the upper left corner.

b. Which of the percentage calculations seems most meaningful to you?

c. According to the percentage calculations you prefer, does there appear to be a strong dependence between state and opinion?

Bus. **3.94** A municipal workers' union that represents sanitation workers in many small midwestern cities studied the contracts that were signed in the previous years. The contracts were subdivided into those settled by negotiation without a strike, those settled by arbitration without a strike, and all those settled after a strike. For each contract, the first-year percentage wage increase was determined. Summary figures follow.

Contract Type	Negotation	Arbitration	Poststrike
Mean percentage wage increase	8.20	9.42	8.40
Variance	0.87	1.04	1.47
Standard deviation	0.93	1.02	1.21
Sample size	38	16	6

Does there appear to be a relationship between contract type and mean percent wage increase? If you were management rather than union affiliated, which posture would you take in future contract negotiations?

4 **Probability and Probability**

Distributions

CHAPTER 4

Probability and Probability Distributions

4.1 How Probability Can Be Used in Making Inferences

4.2 Finding the Probability of an Event

4.3 Basic Event Relations and Probability Laws

4.4 Conditional Probability and Independence

4.5 Bayes' Formula

4.6 Variables: Discrete and Continuous

4.7 Probability Distributions for Discrete Random Variables

4.8 A Useful Discrete Random Variable: The Binomial

4.9 Probability Distributions for Continuous Random Variables

4.10 A Useful Continuous Random Variable: The Normal Distribution

4.11 Random Sampling

4.12 Sampling Distributions

4.13 Normal Approximation to the Binomial

4.14 Minitab Instructions

4.15 Summary

4.1 How Probability Can Be Used in Making Inferences

We stated in Chapter 1 that a scientist uses inferential statistics to make statements about a population based on information contained in a sample of units selected from that population. Graphical and numerical descriptive techniques were presented in Chapter 3 as a means to summarize and describe a sample. However, a sample is not identical to the population from which it was selected. We need to assess the degree of accuracy to which the sample mean, sample standard deviation, or sample proportion represent the corresponding population values.

Most management decisions must be made in the presence of uncertainty. Prices and designs for new automobiles must be selected on the basis of shaky forecasts of consumer preference, national economic trends, and competitive actions. The size and allocation of a hospital staff must be decided with limited information on patient load. The inventory of a product must be set in the face of uncertainty about demand. Probability is the language of uncertainty. Now let us examine probability, the mechanism for making inferences. This idea is probably best illustrated by an example.

Newsweek, in its June 20, 1998 issue, asks the question, "Who Needs Doctors? The Boom in Home Testing." The article discusses the dramatic increase in medical screening tests for home use. The home-testing market has expanded beyond the two most frequently used tests, pregnancy and diabetes glucose monitoring, to a variety of diagnostic tests that were previously used only by doctors and certified laboratories. There is a DNA test to determine whether twins are fraternal or identical, a test to check cholesterol level, a screening test for colon cancer, and tests to determine whether your teenager is a drug user. However, the major question that needs to be addressed is how reliable are the testing kits? When a test indicates that a woman is not pregnant, what is the chance that the test is incorrect and the woman is truly pregnant? This type of incorrect result from a home test could translate into a woman not seeking the proper prenatal care in the early stages of her pregnancy.

Suppose a company states in its promotional materials that its pregnancy test provides correct results in 75% of its applications by pregnant women. We want to evaluate the claim, and so we select 20 women who have been determined by their physicians, using the best possible testing procedures, to be pregnant. The test is taken by each of the 20 women, and for all 20 women the test result is negative, indicating that none of the 20 is pregnant. What do you conclude about the company's claim on the reliability of its test? Suppose you are further assured that each of the 20 women was in fact pregnant, as was determined several months after the test was taken.

If the company's claim of 75% reliability was correct, we would have expected somewhere near 75% of the tests in the sample to be positive. However, none of the test results was positive. Thus, we would conclude that the company's claim is probably false. Why did we fail to state with certainty that the company's claim was false? Consider the possible setting. Suppose we have a large population consisting of millions of units, and 75% of the units are Ps for positives and 25% of the units are Ns for negatives. We randomly select 20 units from the population and count the number of units in the sample that are Ps. Is it possible to obtain a sample consisting of 0 Ps and 20 Ns? Yes, it is possible, *but* it is highly *improbable*. Later in this chapter we will compute the probability of such a sample occurrence.

To obtain a better view of the role that probability plays in making inferences from sample results to conclusions about populations, suppose the 20 tests result in 14 tests being positive—that is, a 70% correct response rate. Would you consider this result highly improbable and reject the company's claim of a 75% correct response rate? How about 12 positives and 8 negatives, or 16 positives and 4 negatives? At what point do we decide that the result of the observed sample is so improbable, assuming the company's claim is correct, that we disagree with its claim? To answer this question, we must know how to find the probability of obtaining a particular sample outcome. Knowing this probability, we can then determine whether we agree or disagree with the company's claim. Probability is the tool that enables us to make an inference. Later in this chapter we will discuss in detail how the FDA and private companies determine the reliability of screening tests.

Because probability is the tool for making inferences, we need to define probability. In the preceding discussion, we used the term *probability* in its everyday sense. Let us examine this idea more closely.

Observations of phenomena can result in many different outcomes, some of which are more likely than others. Numerous attempts have been made to give a precise definition for the probability of an outcome. We will cite three of these.

classical interpretation

The first interpretation of probability, called the **classical interpretation of probability,** arose from games of chance. Typical probability statements of this type are, for example, "the probability that a flip of a balanced coin will show 'heads' is 1/2" and "the probability of drawing an ace when a single card is drawn from a standard deck of 52 cards is 4/52." The numerical values for these probabilities arise from the nature of the games. A coin flip has two possible outcomes (a head or a tail); the probability of a head should then be 1/2 (1 out of 2). Similarly, there are 4 aces in a standard deck of 52 cards, so the probability of drawing an ace in a single draw is 4/52, or 4 out of 52.

outcome

event

In the classical interpretation of probability, each possible distinct result is called an **outcome;** an **event** is identified as a collection of outcomes. The probability of an event E under the classical interpretation of probability is computed by taking the ratio of the number of outcomes, N_e, favorable to event E to the total number N of possible outcomes:

$$P(\text{event } E) = \frac{N_e}{N}$$

The applicability of this interpretation depends on the assumption that all outcomes are equally likely. If this assumption does not hold, the probabilities indicated by the classical interpretation of probability will be in error.

relative frequency interpretation

A second interpretation of probability is called the **relative frequency concept of probability;** this is an empirical approach to probability. If an experiment is repeated a large number of times and event E occurs 30% of the time, then .30 should be a very good approximation to the probability of event E. Symbolically, if an experiment is conducted n different times and if event E occurs on n_e of these trials, then the probability of event E is approximately

$$P(\text{event } E) \approx \frac{n_e}{n}$$

We say "approximate" because we think of the actual probability $P(\text{event } E)$ as the relative frequency of the occurrence of event E over a very large number of observations or repetitions of the phenomenon. The fact that we can check probabilities that have a relative frequency interpretation (by simulating many repetitions of the experiment) makes this interpretation very appealing and practical.

The third interpretation of probability can be used for problems in which it is difficult to imagine a repetition of an experiment. These are "one-shot" situations. For example, the director of a state welfare agency who estimates the probability that a proposed revision in eligibility rules will be passed by the state legislature would not be thinking in terms of a long series of trials. Rather, the

subjective interpretation

director would use a **personal** or **subjective probability** to make a one-shot statement of belief regarding the likelihood of passage of the proposed legislative revision. The problem with subjective probabilities is that they can vary from person to person and they cannot be checked.

Of the three interpretations presented, the relative frequency concept seems to be the most reasonable one because it provides a practical interpretation of the probability for most events of interest. Even though we will never run the necessary repetitions of the experiment to determine the exact probability of an event, the fact that we can check the probability of an event gives meaning to the relative frequency concept. Throughout the remainder of this text we will lean heavily on this interpretation of probability.

EXERCISES

Applications

4.1 Indicate which interpretation of the probability statement seems most appropriate.

a. The National Angus Association has stated that there is a 60/40 chance that wholesale beef prices will rise by the summer—that is, a .60 probability of an increase and a .40 probability of a decrease.

b. The quality control section of a large chemical manufacturing company has undertaken an intensive process-validation study. From this study, the QC section claims that the probability that the shelf life of a newly released batch of chemical will exceed the minimal time specified is .998.

c. A new blend of coffee is being contemplated for release by the marketing division of a large corporation. Preliminary marketing survey results indicate that 550 of a random sample of 1,000 potential users rated this new blend better than a brand-name competitor. The probability of this happening is approximately .001, assuming that there is actually no difference in consumer preference for the two brands.

d. The probability that a customer will receive a package the day after it was sent by a business using an "overnight" delivery service is .92.

e. The sportscaster in College Station, Texas states that the probability that the Aggies will win their football game against the University of Florida is .75.

f. The probability of a nuclear power plant having a meltdown on a given day is .00001.

g. If a customer purchases a single ticket for the Texas lottery, the probability of that ticket being the winning ticket is 1/15,890,700.

4.2 Give your own personal probability for each of the following situations. It would be instructive to tabulate these probabilities for the entire class. In which cases did you have large disagreements?

a. The federal income tax will be eliminated.

b. You will receive an A in this course.

c. Two or more individuals in the classroom have the same birthday.

d. An asteroid will strike the planet earth in the next year.

e. A woman will be elected as vice president or president of the United States in the next presidential election.

4.2 **Finding the Probability of an Event**

In the preceding section, we discussed three different interpretations of probability. In this section, we will use the classical interpretation and the relative frequency concept to illustrate the computation of the probability of an outcome or event. Consider an experiment that consists of tossing two coins, a penny and then a dime, and observing the upturned faces. There are four possible outcomes:

TT: tails for both coins

TH: a tail for the penny, a head for the dime

HT: a head for the penny, a tail for the dime

HH: heads for both coins

What is the probability of observing the event exactly one head from the two coins?

This probability can be obtained easily if we can assume that all four outcomes are equally likely. In this case, that seems quite reasonable. There are $N = 4$ possible outcomes, and $N_e = 2$ of these are favorable for the event of interest, observing exactly one head. Hence, by the classical interpretation of probability,

$$P(\text{exactly 1 head}) = \frac{2}{4} = \frac{1}{2}$$

Because the event of interest has a relative frequency interpretation, we could also obtain this same result empirically, using the relative frequency concept. To demonstrate how relative frequency can be used to obtain the probability of an event, we will use the ideas of simulation. Simulation is a technique that produces outcomes having the same probability of occurrence as the real situation events. The computer is a convenient tool for generating these outcomes. Suppose we wanted to simulate 1,000 tosses of the two coins. We can use a computer program such as *SAS* or *Minitab* to simulate the tossing of a pair of coins. The program has a random number generator. We will designate an even number as *H* and an odd number as *T*. Since there are five even and five odd single-digit numbers, the probability of obtaining an even number is $5/10 = .5$, which is the same as the probability of obtaining an odd number. Thus, we can request 500 pairs of single-digit numbers. This set of 500 pairs of numbers will represent 500 tosses of the two coins, with the first digit representing the outcome of tossing the penny and the second digit representing the outcome of tossing the dime. For example, the pair (3, 6) would represent a tail for the penny and a head for the dime. Using version 13 of Minitab, the following steps will generate 1,000 randomly selected numbers from 0 to 9:

1. Select **Calc** from the toolbar
2. Select **Random Data** from list
3. Select **Integer** from list
4. Generate **20** rows of data
5. Store in column(s): **c1-c50**
6. Minimum value: **0**
7. Maximum value: **9**

The preceding steps will produce 1,000 random single-digit numbers that can then be paired to yield 500 pairs of single-digit numbers. (Most computer packages contain a random number generator that can be used to produce similar results.)

25	32	70	15	96	87	80	43	15	77	89	51	08	36	29	55	42	86	45	93	68	72	49	99	37
82	81	58	50	85	27	99	41	10	31	42	35	50	02	68	33	50	93	73	62	15	15	90	97	24
46	86	89	82	20	23	63	59	50	40	32	72	59	62	58	53	01	85	49	27	31	48	53	07	78
15	81	39	83	79	21	88	57	35	33	49	37	85	42	28	38	50	43	82	47	01	55	42	02	52
66	44	15	40	29	73	11	06	79	81	49	64	32	06	07	31	07	78	73	07	26	36	39	20	14
48	20	27	73	53	21	44	16	00	33	43	95	21	08	19	60	68	30	99	27	22	74	65	22	05
26	79	54	64	94	01	21	47	86	94	24	41	06	81	16	07	30	34	99	54	68	37	38	71	79
86	12	83	09	27	60	49	54	21	92	64	57	07	39	04	66	73	76	74	93	50	56	23	41	23
18	87	21	48	75	63	09	97	96	86	85	68	65	35	92	40	57	87	82	71	04	16	01	03	45
52	79	14	12	94	51	39	40	42	17	32	94	42	34	68	17	39	32	38	03	75	56	79	79	57
07	40	96	46	22	04	12	90	80	71	46	11	18	81	54	95	47	72	06	07	66	05	59	34	81
66	79	83	82	62	20	75	71	73	79	48	86	83	74	04	13	36	87	96	11	39	81	59	41	70
21	47	34	02	05	73	71	57	64	58	05	16	57	27	66	92	97	68	18	52	09	45	34	80	57
87	22	18	65	66	18	84	31	09	38	05	67	10	45	03	48	52	48	33	36	00	49	39	55	35
70	84	50	37	58	41	08	62	42	64	02	29	33	68	87	58	52	39	98	78	72	13	13	15	96
57	32	98	05	83	39	13	39	37	08	17	01	35	13	98	66	89	40	29	47	37	65	86	73	42
85	65	78	05	24	65	24	92	03	46	67	48	90	60	02	61	21	12	80	70	35	15	40	52	76
29	11	45	22	38	33	32	52	17	20	03	26	34	18	85	46	52	66	63	30	84	53	76	47	21
42	97	56	38	41	87	14	43	30	35	99	06	76	67	00	47	83	32	52	42	48	51	69	15	18
08	30	37	89	17	89	23	58	13	93	17	44	09	08	61	05	35	44	91	89	35	15	06	39	27

The summary of the simulation of the 500 tosses is shown in Table 4.1.

TABLE 4.1
Simulation of tossing a
penny and a dime 500 times

Event	Outcome of Simulation	Frequency	Relative Frequency
TT	(Odd, Odd)	129	129/500 = .258
TH	(Odd, Even)	117	117/500 = .234
HT	(Even, Odd)	125	125/500 = .250
HH	(Even, Even)	129	129/500 = .258

Note that this approach yields simulated probabilities that are nearly in agreement with our intuition; that is, intuitively we might expect these outcomes to be equally likely. Thus, each of the four outcomes should occur with a probability equal to 1/4, or .25. This assumption was made for the classical interpretation. We will show in Chapter 10 that in order to be 95% certain that the simulated probabilities are within .01 of the true probabilities, the number of tosses should be at least 7,500 and not 500 as we used previously.

If we wish to find the probability of tossing two coins and observing exactly one head, we have, from Table 4.1,

$$P(\text{exactly 1 head}) \approx \frac{117 + 125}{500} = .484$$

This is very close to the theoretical probability, which we have shown to be .5.

Note that we could easily modify our example to accommodate the tossing of an unfair coin. Suppose we are tossing a penny that is weighted so that the probability of a head occurring in a toss is .70 and the probability of a tail is .30. We could designate an H outcome whenever one of the random digits 0, 1, 2, 3,

4, 5, or 6 occurs and a T outcome whenever one of the digits 7, 8, or 9 occurs. The same simulation program can be run as before, but we would interpret the output differently.

<table>
<tr><td>**EXERCISES**</td><td>**Applications**</td></tr>
<tr><td>Edu.</td><td></td></tr>
</table>

EXERCISES

Applications

Edu.   **4.3**   Suppose an exam consists of 20 true-or-false questions. A student takes the exam by guessing the answer to each question. What is the probability that the student correctly answers 15 or more of the questions? [*Hint:* Use a simulation approach. Generate a large number (2,000 or more sets) of 20 single-digit numbers. Each number represents the answer to one of the questions on the exam, with even digits representing correct answers and odd digits representing wrong answers. Determine the relative frequency of the sets having 15 or more correct answers.]

Med.   **4.4**   The example in Section 4.1 considered the reliability of a screening test. Suppose we wanted to simulate the probability of observing at least 15 positive results and 5 negative results in a set of 20 results, when the probability of a positive result was claimed to be .75. Use a random number generator to simulate the running of 20 screening tests.

a. Let a two-digit number represent an individual running of the screening test. Which numbers represent a positive outcome of the screening test? Which numbers represent a negative outcome?

b. If we generate 2,000 sets of 20 two-digit numbers, how can the outcomes of this simulation be used to approximate the probability of obtaining at least 15 positive results in the 20 runnings of the screening test?

4.3   Basic Event Relations and Probability Laws

The probability of an event, say event A, will always satisfy the property

$$0 \leq P(A) \leq 1$$

that is, the probability of an event lies anywhere in the interval from 0 (the occurrence of the event is impossible) to 1 (the occurrence of an event is a "sure thing").

Suppose A and B represent two experimental events and you are interested in a new event, the event that **either A or B occurs.** For example, suppose that we toss a pair of dice and define the following events:

either *A* or *B* occurs

A: A total of 7 shows

B: A total of 11 shows

Then the event "either A or B occurs" is the event that you toss a total of either 7 or 11 with the pair of dice.

mutually exclusive

Note that, for this example, the events A and B are **mutually exclusive;** that is, if you observe event A (a total of 7), you could not at the same time observe event B (a total of 11). Thus, if A occurs, B cannot occur (and vice versa).

DEFINITION 4.1

Two events A and B are said to be **mutually exclusive** if (when the experiment is performed a single time) the occurrence of one of the events excludes the possibility of the occurrence of the other event.

The concept of mutually exclusive events is used to specify a second property that the probabilities of events must satisfy. When two events are mutually exclusive, then the probability that either one of the events will occur is the sum of the event probabilities.

DEFINITION 4.2

If two events, A and B, are mutually exclusive, the **probability** that either event occurs is $P(\text{either } A \text{ or } B) = P(A) + P(B)$.

Definition 4.2 is a special case of the union of two events, which we will soon define.

The definition of additivity of probabilities for mutually exclusive events can be extended beyond two events. For example, when we toss a pair of dice, the sum S of the numbers appearing on the dice can assume any one of the values $S = 2, 3, 4, \ldots, 11, 12$. On a single toss of the dice, we can observe only one of these values. Therefore, the values $2, 3, \ldots, 12$ represent mutually exclusive events. If we want to find the probability of tossing a sum less than or equal to 4, this probability is

$$P(S \leq 4) = P(2) + P(3) + P(4)$$

For this particular experiment, the dice can fall in 36 different equally likely ways. We can observe a 1 on die 1 and a 1 on die 2, denoted by the symbol $(1, 1)$. We can observe a 1 on die 1 and a 2 on die 2, denoted by $(1, 2)$. In other words, for this experiment, the possible outcomes are

(1, 1)	(2, 1)	(3, 1)	(4, 1)	(5, 1)	(6, 1)
(1, 2)	(2, 2)	(3, 2)	(4, 2)	(5, 2)	(6, 2)
(1, 3)	(2, 3)	(3, 3)	(4, 3)	(5, 3)	(6, 3)
(1, 4)	(2, 4)	(3, 4)	(4, 4)	(5, 4)	(6, 4)
(1, 5)	(2, 5)	(3, 5)	(4, 5)	(5, 5)	(6, 5)
(1, 6)	(2, 6)	(3, 6)	(4, 6)	(5, 6)	(6, 6)

As you can see, only one of these events, $(1, 1)$, will result in a sum equal to 2. Therefore, we would expect a 2 to occur with a relative frequency of $1/36$ in a long series of repetitions of the experiment, and we let $P(2) = 1/36$. The sum $S = 3$ will occur if we observe either of the outcomes $(1, 2)$ or $(2, 1)$. Therefore, $P(3) = 2/36 = 1/18$. Similarly, we find $P(4) = 3/36 = 1/12$. It follows that

$$P(S \leq 4) = P(2) + P(3) + P(4) = \frac{1}{36} + \frac{1}{18} + \frac{1}{12} = \frac{1}{6}$$

complement   A third property of event probabilities concerns an event and its **complement.**

DEFINITION 4.3

The **complement** of an event A is the event that A *does not* occur. The complement of A is denoted by the symbol $\overline{A}$.

Thus, if we define the complement of an event A as a new event—namely, "A does not occur"—it follows that

$$P(A) + P(\overline{A}) = 1$$

For an example, refer again to the two-coin-toss experiment. If, in many repetitions of the experiment, the proportion of times you observe event A, "two heads show," is 1/4, then it follows that the proportion of times you observe the event $\overline{A}$, "two heads do not show," is 3/4. Thus, $P(A)$ and $P(\overline{A})$ will always sum to 1.

We can summarize the three properties that the probabilities of events must satisfy as follows:

Properties of Probabilities

If A and B are any two mutually exclusive events associated with an experiment, then $P(A)$ and $P(B)$ must satisfy the following properties:

1. $0 \le P(A) \le 1$ and $0 \le P(B) \le 1$
2. $P(\text{either } A \text{ or } B) = P(A) + P(B)$
3. $P(A) + P(\overline{A}) = 1$ and $P(B) + P(\overline{B}) = 1$

union
intersection

We can now define two additional event relations: the **union** and the **intersection** of two events.

DEFINITION 4.4

The **union** of two events A and B is the set of all outcomes that are included in either A or B (or both). The union is denoted as $A \cup B$.

DEFINITION 4.5

The **intersection** of two events A and B is the set of all outcomes that are included in both A and B. The intersection is denoted as $A \cap B$.

These definitions along with the definition of the complement of an event formalize some simple concepts. The event $\overline{A}$ occurs when A *does not*; $A \cup B$ occurs when either A or B occurs; $A \cap B$ occurs when A *and* B occur.

The additivity of probabilities for mutually exclusive events, called the *addition law for mutually exclusive events,* can be extended to give the general addition law.

DEFINITION 4.6

Consider two events A and B; the **probability of the union** of A and B is

$$P(A \cup B) = P(A) + P(B) - P(A \cap B)$$

EXAMPLE 4.1

Events and event probabilities are shown in the Venn diagram in Figure 4.1. Use this diagram to determine the following probabilities:

a. $P(A), P(\overline{A})$
b. $P(B), P(\overline{B})$
c. $P(A \cap B)$
d. $P(A \cup B)$

FIGURE 4.1

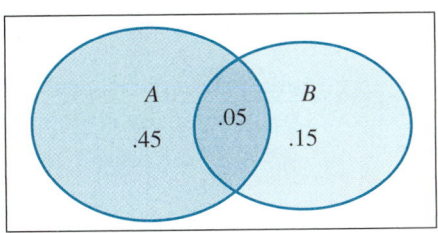

Solution   From the Venn diagram, we are able to determine the following probabilities:

a. $P(A) = .5$, therefore $P(\overline{A}) = 1 - .5 = .5$
b. $P(B) = .2$, therefore $P(\overline{B}) = 1 - .2 = .8$
c. $P(A \cap B) = .05$
d. $P(A \cup B) = P(A) + P(B) - P(A \cap B) = .5 + .2 - .05 = .65$

4.4 Conditional Probability and Independence

Consider the following situation: The examination of a large number of insurance claims, categorized according to type of insurance and whether the claim was fraudulent, produced the results shown in Table 4.2. Suppose you are responsible for checking insurance claims—in particular, for detecting fraudulent claims—and you examine the next claim that is processed. What is the probability of the event F, "the claim is fraudulent"? To answer the question, you examine Table 4.2 and note that 10% of all claims are fraudulent. Thus, assuming that the percentages given in the table are reasonable approximations to the true probabilities of receiving specific types of claims, it follows that $P(F) = .10$. Would you say that the risk that you face a fraudulent claim has probability .10? We think not, because you have additional information that may affect the assessment of $P(F)$. This additional information concerns the type of policy you were examining (fire, auto, or other).

Suppose that you have the additional information that the claim was associated with a fire policy. Checking Table 4.2, we see that 20% (or .20) of all claims are associated with a fire policy and that 6% (or .06) of all claims are fraudulent fire policy claims. Therefore, it follows that the probability that the claim is

TABLE 4.2
Categorization of insurance claims

| | Type of Policy (%) | | | |
Category	Fire	Auto	Other	Total %
Fraudulent	6	1	3	10
Nonfraudulent	14	29	47	90
Total	20	30	50	100

fraudulent, given that you know the policy is a fire policy, is

$$P(F|\text{fire policy}) = \frac{\text{proportion of claims that are fraudulent fire policy claims}}{\text{proportion of claims that are against fire policies}}$$

$$= \frac{.06}{.20} = .30$$

conditional probability   This probability, $P(F|\text{fire policy})$, is called a **conditional probability** of the event F—that is, the probability of event F given the fact that the event "fire policy" has already occurred. This tells you that 30% of all fire policy claims are fraudulent. The vertical bar in the expression $P(F|\text{fire policy})$ represents the phrase "given that," or simply "given." Thus, the expression is read, "the probability of the event F given the event fire policy."

unconditional probability   The probability $P(F) = .10$, called the **unconditional** or **marginal probability** of the event F, gives the proportion of times a claim is fraudulent—that is, the proportion of times event F occurs in a very large (infinitely large) number of repetitions of the experiment (receiving an insurance claim and determining whether the claim is fraudulent). In contrast, the conditional probability of F, given that the claim is for a fire policy, $P(F|\text{fire policy})$, gives the proportion of fire policy claims that are fraudulent. Clearly, the conditional probabilities of F, given the types of policies, will be of much greater assistance in measuring the risk of fraud than the unconditional probability of F.

DEFINITION 4.7   Consider two events A and B with nonzero probabilities, $P(A)$ and $P(B)$. The **conditional probability** of event A given event B is

$$P(A|B) = \frac{P(A \cap B)}{P(B)}$$

The conditional probability of event B given event A is

$$P(B|A) = \frac{P(A \cap B)}{P(A)}$$

This definition for conditional probabilities gives rise to what is referred to as the *multiplication law.*

DEFINITION 4.8   The **probability of the intersection** of two events A and B is

$$P(A \cap B) = P(A)P(B|A)$$
$$= P(B)P(A|B)$$

The only difference between Definitions 4.7 and 4.8, both of which involve conditional probabilities, relates to what probabilities are known and what needs to be calculated. When the intersection probability $P(A \cap B)$ and the individual probability $P(A)$ are known, we can compute $P(B|A)$. When we know $P(A)$ and $P(B|A)$, we can compute $P(A \cap B)$.

EXAMPLE 4.2

Two supervisors are to be selected as safety representatives within the company. Given that there are six supervisors in research and four in development, and each group of two supervisors has the same chance of being selected, find the probability of choosing both supervisors from research.

Solution   Let A be the event that the first supervisor selected is from research, and let B be the event that the second supervisor is also from research. Clearly, we want $P(A \text{ and } B) = P(A \cap B) = P(B|A)P(A)$.

For this example,

$$P(A) = \frac{\text{number of research supervisors}}{\text{number of supervisors}} = \frac{6}{10}$$

and

$$P(B|A) = \frac{\begin{array}{c}\text{number of research supervisors after}\\ \text{one research supervisor was selected}\end{array}}{\text{number of supervisors after one supervisor was selected}} = \frac{5}{9}$$

Thus,

$$P(A \cap B) = P(A)P(B|A) = \frac{6}{10}\left(\frac{5}{9}\right) = \frac{30}{90} = .333$$

Thus the probability of choosing both supervisors from research is .333, assuming that each group of two has the same chance of being selected.

Suppose that the probability of event A is the same whether event B has or has not occurred; that is, suppose

$$P(A|B) = P(A)$$

independent events   Then we say that the occurrence of event A is not dependent on the occurrence of event B or, simply, that A and B are **independent events.** When $P(A|B) \neq P(A)$, the occurrence of A depends on the occurrence of B, and events A and B **dependent events**   are said to be **dependent events.**

DEFINITION 4.9

Two events A and B are **independent events** if

$$P(A|B) = P(A) \qquad \text{or} \qquad P(B|A) = P(B)$$

(Note: You can show that if $P(A|B) = P(A)$, then $P(B|A) = P(B)$, and vice versa.)

Definition 4.9 leads to a special case of $P(A \cap B)$. When events A and B are independent, it follows that

$$P(A \cap B) = P(A)P(B|A) = P(A)P(B)$$

The concept of independence is of particular importance in sampling. Later in the text, we will discuss drawing samples from two (or more) populations to compare the population means, variances, or some other population parameters. For most of these applications, we will select samples in such a way that the observed values in one sample are independent of the values that appear in another **independent samples**   sample. We call these **independent samples.**

EXERCISES   ### Basic Techniques

4.5   A coin is to be flipped three times. List the possible outcomes in the form (result on toss 1, result on toss 2, result on toss 3).

4.6   In Exercise 4.5, assume that each one of the outcomes has probability 1/8 of occurring. Find the probability of
   a. A: observing exactly 1 head
   b. B: observing 1 or more heads
   c. C: observing no heads

4.7   For Exercise 4.6:
   a. Compute the probability of the complement of event A, event B, and event C.
   b. Determine whether events A and B are mutually exclusive.

4.8   Determine the following conditional probabilities for the events of Exercise 4.6.
   a. $P(A|B)$ **b.** $P(A|C)$ **c.** $P(B|C)$

4.9   Refer to Exercise 4.8. Are events A and B independent? Why or why not? What about A and C? What about B and C?

4.10   A die is to be rolled and we are to observe the number that falls face up. Find the probabilities for these events:
   a. A: observe a 6
   b. B: observe an even number
   c. C: observe a number greater than 2
   d. D: observe an even number and a number greater than 2

4.11   Refer to Exercise 4.10. Which of the events (A, B, and C) are independent? Which are mutually exclusive?

4.12   Consider the following outcomes for an experiment:

Outcome	1	2	3	4	5
Probability	.20	.25	.15	.10	.30

Let event A consist of outcomes 1, 3, and 5 and event B consist of outcomes 4 and 5.
   a. Find $P(A)$ and $P(B)$.
   b. Find P(both A and B occur).
   c. Find P(either A or B occurs).

4.13   Refer to Exercise 4.12. Does P(either A or B occurs) $= P(A) + P(B)$? Why or why not?

Applications

Edu.   **4.14**   A student has to have an accounting course and an economics course the next term. Assuming there are no schedule conflicts, describe the possible outcomes for selecting one section of the accounting course and one of the economics course if there are four possible accounting sections and three possible economics sections.

Engin.   **4.15**   The emergency room of a hospital has two backup generators, either of which can supply enough electricity for basic hospital operations. We define events A and B as follows:

event A: generator 1 works properly
event B: generator 2 works properly

Describe the following events in words:
   a. complement of A **b.** $B|A$ **c.** either A or B

H.R. **4.16** A survey of a number of large corporations gave the following probability table for events related to the offering of a promotion involving a transfer.

Promotion/ Transfer	Married Two-Career Marriage	Married One-Career Marriage	Unmarried	Total
Rejected	.184	.0555	.0170	.2565
Accepted	.276	.3145	.1530	.7435
Total	.46	.37	.17	

Use the probabilities to answer the following questions:
  a. What is the probability that a professional (selected at random) would accept the promotion? Reject it?
  b. What is the probability that a professional (selected at random) is part of a two-career marriage? A one-career marriage?

Bus. **4.17** An institutional investor is considering a large investment in two of five companies. Suppose that, unknown to the investor, two of the five firms are on shaky ground with regard to the development of new products.
  a. List the possible outcomes for this situation.
  b. Determine the probability of choosing two of the three firms that are on better ground.
  c. What is the probability of choosing one of two firms on shaky ground?
  d. What is the probability of choosing the two shakiest firms?

Soc. **4.18** A survey of workers in two manufacturing sites of a firm included the following question: How effective is management in responding to legitimate grievances of workers? The results are shown here.

	Number Surveyed	Number Responding "Poor"
Site 1	192	48
Site 2	248	80

Let A be the event the worker comes from Site 1 and B be the event the response is "poor." Compute $P(A)$, $P(B)$, and $P(A \cap B)$.

4.19 Refer to Exercise 4.16.
  a. Are events A and B independent?
  b. Find $P(B|A)$ and $P(B|\bar{A})$. Are they equal?

H.R. **4.20** A large corporation has spent considerable time developing employee performance rating scales to evaluate an employee's job performance on a regular basis, so major adjustments can be made when needed and employees who should be considered for a "fast track" can be isolated. Keys to this latter determination are ratings on the ability of an employee to perform to his or her capabilities and on his or her formal training for the job.

Workload Capacity	Formal Training None	Little	Some	Extensive
Low	.01	.02	.02	.04
Medium	.05	.06	.07	.10
High	.10	.15	.16	.22

The probabilities for being placed on a fast track are as indicated for the 12 categories of workload capacity and formal training. The following three events (A, B, and C) are defined:

A: an employee works at the high-capacity level
B: an employee falls into the highest (extensive) formal training category
C: an employee has little or no formal training and works below high capacity

a. Find $P(A)$, $P(B)$, and $P(C)$.
b. Find $P(A|B)$, $P(A|\bar{B})$, and $P(\bar{B}|C)$.
c. Find $P(A \cup B)$, $P(A \cap C)$, and $P(B \cap C)$.

Bus. **4.21** The utility company in a large metropolitan area finds that 70% of its customers pay a given monthly bill in full.

a. Suppose two customers are chosen at random from the list of all customers. What is the probability that both customers will pay their monthly bill in full?
b. What is the probability that at least one of them will pay in full?

4.22 Refer to Exercise 4.21. A more detailed examination of the company records indicates that 95% of the customers who pay one monthly bill in full will also pay the next monthly bill in full; only 10% of those who pay less than the full amount one month will pay in full the next month.

a. Find the probability that a customer selected at random will pay two consecutive months in full.
b. Find the probability that a customer selected at random will pay neither of two consecutive months in full.
c. Find the probability that a customer chosen at random will pay exactly one month in full.

4.5 Bayes' Formula

In this section, we will show how Bayes' Formula can be used to update conditional probabilities by using sample data when available. These "updated" conditional probabilities are useful in decision making. A particular application of these techniques involves the evaluation of diagnostic tests. Suppose a meat inspector must decide whether a randomly selected meat sample contains *E. coli* bacteria. The inspector conducts a diagnostic test. Ideally, a positive result (Pos) would mean that the meat sample actually has *E. coli,* and a negative result (Neg) would imply that the meat sample is free of *E. coli.* However, the diagnostic test is occasionally in error. The results of the test may be a **false positive,** for which the test's indication of *E. coli* presence is incorrect, or a **false negative,** for which the test's conclusion of *E. coli* absence is incorrect. Large-scale screening tests are conducted to evaluate the accuracy of a given diagnostic test. For example, *E. coli* (E) is placed in 10,000 meat samples, and the diagnostic test yields a positive result for 9,500 samples and a negative result for 500 samples; that is, there are 500 false negatives out of the 10,000 tests. Another 10,000 samples have all traces of *E. coli* (NE) removed, and the diagnostic test yields a positive result for 100 samples and a negative result for 9,900 samples. There are 100 false positives out of the 10,000 tests. We can summarize the results in the following table:

false positive
false negative

| Diagnostic | Meat Sample Status | |
Test Result	E	NE
Positive	9,500	100
Negative	500	9,900
Total	10,000	10,000

Evaluation of test results is as follows:

$$\text{True positive rate} = P(\text{Pos} \,|\, E) = \frac{9{,}500}{10{,}000} = .95$$

$$\text{False positive rate} = P(\text{Pos} \,|\, NE) = \frac{100}{10{,}000} = .01$$

$$\text{True negative rate} = P(\text{Neg} \,|\, NE) = \frac{9{,}900}{10{,}000} = .99$$

$$\text{False negative rate} = P(\text{Neg} \,|\, E) = \frac{500}{10{,}000} = .05$$

sensitivity
specificity

The **sensitivity** of the diagnostic test is the true positive rate, and the **specificity** of the diagnostic test is the true negative rate.

The primary question facing the inspector is to evaluate the probability of *E. coli* being present in the meat sample when the test yields a positive result; that is, the inspector needs to know $P(E \,|\, \text{Pos})$. Bayes' Formula answers this question, as the following calculations show. To make this calculation, we need to know the *rate* of *E. coli* in the type of meat being inspected. For this example, suppose that *E. coli* is present in 4.5% of all meat samples; that is, *E. coli* has prevalence $P(E) = .045$. We can then compute $P(E \,|\, \text{Pos})$ as follows:

$$P(E \,|\, \text{Pos}) = \frac{P(E \cap \text{Pos})}{P(\text{Pos})} = \frac{P(E \cap \text{Pos})}{P(E \cap \text{Pos}) + P(NE \cap \text{Pos})}$$

$$= \frac{P(\text{Pos} \,|\, E)P(E)}{P(\text{Pos} \,|\, E)P(E) + P(\text{Pos} \,|\, NE)P(NE)}$$

$$= \frac{(.95)(.045)}{(.95)(.045) + (.01)(1 - .045)} = .817$$

Thus, *E. coli* is truly present in 81.7% of the tested samples in which a positive test result occurs. Also, we can conclude that 18.3% of the tested samples indicated *E. coli* was present when in fact there was no *E. coli* in the meat sample.

EXAMPLE 4.3

A book club classifies members as heavy, medium, or light purchasers, and separate mailings are prepared for each of these groups. Overall, 20% of the members are heavy purchasers, 30% medium, and 50% light. A member is not classified into a group until 18 months after joining the club, but a test is made of the feasibility of using the first 3 months' purchases to classify members. The following percentages are obtained from existing records of individuals classified as heavy, medium, or light purchasers:

First 3 Months' Purchases	Group (%)		
	Heavy	**Medium**	**Light**
0	5	15	60
1	10	30	20
2	30	40	15
3+	55	15	5

If a member purchases no books in the first 3 months, what is the probability that the member is a light purchaser? (*Note*: This table contains "conditional" percentages for each column.)

The quantitative variable *y* is called a *random variable* because the value that *y* assumes in a given experiment is a chance or random outcome.

DEFINITION 4.10

When observations on a quantitative random variable can assume only a countable number of values, the variable is called a **discrete random variable.**

Examples of discrete variables are these:

1. Number of bushels of apples per tree of a genetically altered apple variety
2. Change in the number of accidents per month at an intersection after a new signaling device has been installed
3. Number of "dead persons" voting in the last mayoral election in a major midwest city

Note that it is possible to count the number of values that each of these random variables can assume.

DEFINITION 4.11

When observations on a quantitative random variable can assume any one of the uncountable number of values in a line interval, the variable is called a **continuous random variable.**

For example, the daily maximum temperature in Rochester, New York, can assume any of the infinitely many values on a line interval. It can be 89.6, 89.799, or 89.7611114. Typical continuous random variables are temperature, pressure, height, weight, and distance.

The distinction between **discrete** and **continuous random** variables is pertinent when we are seeking the probabilities associated with specific values of a random variable. The need for the distinction will be apparent when probability distributions are discussed in later sections of this chapter.

4.7 Probability Distributions for Discrete Random Variables

As previously stated, we need to know the probability of observing a particular sample outcome in order to make an inference about the population from which the sample was drawn. To do this, we need to know the probability associated with each value of the variable *y*. Viewed as relative frequencies, these probabilities

probability distribution

generate a distribution of theoretical relative frequencies called the **probability distribution** of *y*. Probability distributions differ for discrete and continuous random variables. For discrete random variables, we will compute the probability of specific individual values occurring. For continuous random variables, the probability of an interval of values is the event of interest.

The *probability distribution for a discrete random variable* displays the probability $P(y)$ associated with each value of *y*. This display can be presented as a table, a graph, or a formula. To illustrate, consider the tossing of two coins in

Section 4.2 and let y be the number of heads observed. Then y can take the values 0, 1, or 2. From the data of Table 4.1, we can determine the approximate probability for each value of y, as given in Table 4.3. We point out that the relative frequencies in the table are very close to the theoretical relative frequencies (probabilities), which can be shown to be .25, .50, and .25 using the classical interpretation of probability. If we had employed 2,000,000 tosses of the coins instead of 500, the relative frequencies for y = 0, 1, and 2 would be indistinguishable from the theoretical probabilities.

TABLE 4.3

Empirical sampling results for y: the number of heads in 500 tosses of two coins

y	Frequency	Relative Frequency
0	129	.258
1	242	.484
2	129	.258

The probability distribution for y, the number of heads in the toss of two coins, is shown in Table 4.4 and is presented graphically as a *probability histogram* in Figure 4.2.

TABLE 4.4

Probability distribution for the number of heads when two coins are tossed

y	P(y)
0	.25
1	.50
2	.25

FIGURE 4.2

Probability distribution for the number of heads when two coins are tossed

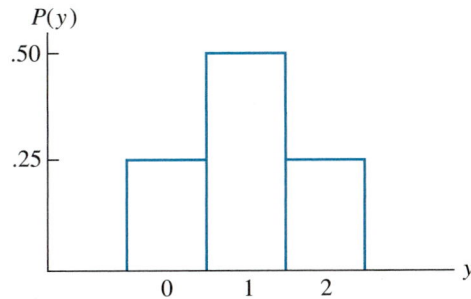

The probability distribution for this simple discrete random variable illustrates three important properties of discrete random variables.

Properties of Discrete Random Variables

1. The probability associated with every value of y lies between 0 and 1.
2. The sum of the probabilities for all values of y is equal to 1.
3. The probabilities for a discrete random variable are additive. Hence, the probability that y = 1 or 2 is equal to P(1) + P(2).

The relevance of the probability distribution to statistical inference will be emphasized when we discuss the probability distribution for the binomial random variable.

4.8 A Useful Discrete Random Variable: The Binomial

Many populations of interest to business persons and scientists can be viewed as large sets of 0s and 1s. For example, consider the set of responses of all adults in the United States to the question, "Do you favor the development of nuclear energy?" If we disallow "no opinion," the responses will constitute a set of "yes" responses and "no" responses. If we assign a 1 to each yes and a 0 to each no, the population will consist of a set of 0s and 1s, and the sum of the 1s will equal the total number of persons favoring the development. The sum of the 1s divided by the number of adults in the United States will equal the proportion of people who favor the development.

Gallup and Harris polls are examples of the sampling of 0, 1 populations. People are surveyed, and their opinions are recorded. Based on the sample responses, Gallup and Harris estimate the proportions of people in the population who favor some particular issue or possess some particular characteristic.

Similar surveys are conducted in the biological sciences, engineering, and business, but they may be called experiments rather than polls. For example, experiments are conducted to determine the effect of new drugs on small animals, such as rats or mice, before progressing to larger animals and, eventually, to human participants. Many of these experiments bear a marked resemblance to a poll in that the experimenter records only whether the drug was effective. Thus, if 300 rats are injected with a drug and 230 show a favorable response, the experimenter has conducted a "poll"—a poll of rat reaction to the drug, 230 "in favor" and 70 "opposed."

Similar "polls" are conducted by most manufacturers to determine the fraction of a product that is of good quality. Samples of industrial products are collected before shipment and each item in the sample is judged "defective" or "acceptable" according to criteria established by the company's quality control department. Based on the number of defectives in the sample, the company can decide whether the product is suitable for shipment. Note that this example, as well as those preceding, has the practical objective of making an inference about a population based on information contained in a sample.

The public opinion poll, the consumer preference poll, the drug-testing experiment, and the industrial sampling for defectives are all examples of a common, frequently conducted sampling situation known as a *binomial experiment*. The binomial experiment is conducted in all areas of science and business and only differs from one situation to another in the nature of objects being sampled (people, rats, electric lightbulbs, oranges). Thus, it is useful to define its characteristics. We can then apply our knowledge of this one kind of experiment to a variety of sampling experiments.

For all practical purposes the binomial experiment is identical to the coin-tossing example of previous sections. Here, n different coins are tossed (or a single coin is tossed n times), and we are interested in the number of heads observed. We assume that the probability of tossing a head on a single trial is π (π may equal .50, as it would for a balanced coin, but in many practical situations π will take some other value between 0 and 1). We also assume that the outcome for any one toss is unaffected by the results of any preceding tosses. These characteristics can be summarized as shown here.

DEFINITION 4.12   A **binomial experiment** is one that has the following properties:

1. The experiment consists of n identical trials.
2. Each trial results in one of two outcomes. We will label one outcome a success and the other a failure.
3. The probability of success on a single trial is equal to π and π remains the same from trial to trial.*
4. The trials are independent; that is, the outcome of one trial does not influence the outcome of any other trial.
5. The random variable y is the number of successes observed during the n trials.

EXAMPLE 4.5

An article in the March 5, 1998, issue of *The New England Journal of Medicine* discussed a large outbreak of tuberculosis. One person, called the index patient, was diagnosed with tuberculosis in 1995. The 232 co-workers of the index patient were given a tuberculin screening test. The number of co-workers recording a positive reading on the test was the random variable of interest. Did this study satisfy the properties of a binomial experiment?

Solution   To answer the question, we check each of the five characteristics of the binomial experiment to determine whether they were satisfied.

1. Were there n identical trials? Yes. There were $n = 232$ workers who had approximately equal contact with the index patient.
2. Did each trial result in one of two outcomes? Yes. Each co-worker recorded either a positive or negative reading on the test.
3. Was the probability of success the same from trial to trial? Yes, if the co-workers had equivalent risk factors and equal exposures to the index patient.
4. Were the trials independent? Yes. The outcome of one screening test was unaffected by the outcome of the other screening tests.
5. Was the random variable of interest to the experimenter the number of successes y in the 232 screening tests? Yes. The number of co-workers who obtained a positive reading on the screening test was the variable of interest.

All five characteristics were satisfied, so the tuberculin screening test represented a binomial experiment.

EXAMPLE 4.6

An economist interviews 75 students in a class of 100 to estimate the proportion of students who expect to obtain a "C" or better in the course. Is this a binomial experiment?

---

* Some textbooks and computer programs use the letter p rather than π. We have chosen π to avoid confusion with p-values, discussed in Chapter 5.

Solution   Check this experiment against the five characteristics of a binomial experiment.

1. Are there identical trials? Yes. Each of 75 students is interviewed.
2. Does each trial result in one of two outcomes? Yes. Each student either does or does not expect to obtain a grade of "C" or higher.
3. Is the probability of success the same from trial to trial? No. If we let success denote a student expecting to obtain a "C" or higher, then the probability of success can change considerably from trial to trial. For example, unknown to the professor, suppose that 75 of the 100 students expect to obtain a grade of "C" or higher. Then π, the probability of success for the first student interviewed, is $75/100 = .75$. If the student is a failure (does not expect a "C" or higher), the probability of success for the next student is $75/99 = .76$. Suppose that after 70 students are interviewed, 60 are successes and 10 are failures. Then the probability of success for the next (71st) student is $15/30 = .50$.

This example shows how the probability of success can change substantially from trial to trial in situations in which the sample size is a relatively large portion of the total population size. This experiment does not satisfy the properties of a binomial experiment.

Note that very few real-life situations satisfy perfectly the requirements stated in Definition 4.12, but for many the lack of agreement is so small that the binomial experiment still provides a very good model for reality.

Having defined the binomial experiment and suggested several practical applications, we now examine the probability distribution for the binomial random variable y, the number of successes observed in n trials. Although it is possible to approximate $P(y)$, the probability associated with a value of y in a binomial experiment, by using a relative frequency approach, it is easier to use a general formula for binomial probabilities.

Formula for Computing $P(y)$ in a Binomial Experiment

The probability of observing y successes in n trials of a binomial experiment is

$$P(y) = \frac{n!}{y!(n-y)!}\,\pi^y(1-\pi)^{n-y}$$

where

n = number of trials

π = probability of success on a single trial

$1 - \pi$ = probability of failure on a single trial

y = number of successes in n trials

$n! = n(n-1)(n-2)\cdots(3)(2)(1)$

As indicated in the box, the notation $n!$ (referred to as n factorial) is used for the product

$$n! = n(n-1)(n-2)\cdots(3)(2)(1)$$

For $n = 3$,

$$n! = 3! = (3)(3 - 1)(3 - 2) = (3)(2)(1) = 6$$

Similarly, for $n = 4$,

$$4! = (4)(3)(2)(1) = 24$$

We also note that 0! is defined to be equal to 1.

To see how the formula for binomial probabilities can be used to calculate the probability for a specific value of y, consider the following examples.

EXAMPLE 4.7

A new variety of turf grass has been developed for use on golf courses, with the goal of obtaining a germination rate of 85%. To evaluate the grass, 20 seeds are planted in a greenhouse so that each seed will be exposed to identical conditions. If the 85% germination rate is correct, what is the probability that 18 or more of the 20 seeds will germinate?

$$P(y) = \frac{n!}{y!(n - y)!} \pi^y (1 - \pi)^{n-y}$$

and substituting for $n = 20$, $\pi = .85$, $y = 18$, 19, and 20, we obtain

$$P(y = 18) = \frac{20!}{18!(20 - 18)!}(.85)^{18}(1 - .85)^{20-18} = 190(.85)^{18}(.15)^2 = .229$$

$$P(y = 19) = \frac{20!}{19!(20 - 19)!}(.85)^{19}(1 - .85)^{20-19} = 20(.85)^{19}(.15)^1 = .137$$

$$P(y = 20) = \frac{20!}{20!(20 - 20)!}(.85)^{20}(1 - .85)^{20-20} = (.85)^{20} = .0388$$

$$P(y \geq 18) = P(y = 18) + P(y = 19) + P(y = 20) = .405$$

The calculations in Example 4.7 entail a considerable amount of effort even though n was only 20. For those situations involving a large value of n, we can use computer software to make the exact calculations. An approach that yields fairly accurate results in many situations and does not require the use of a computer will be discussed later in this chapter.

EXAMPLE 4.8

Suppose that a sample of households is randomly selected from all the households in the city in order to estimate the percentage in which the head of the household is unemployed. To illustrate the computation of a binomial probability, suppose that the unknown percentage is actually 10% and that a sample of $n = 5$ (we select a small sample to make the calculation manageable) is selected from the population. What is the probability that all five heads of the households are employed?

Solution We must carefully define which outcome we wish to call a success. For this example, we define a success as being employed. Then the probability of success when one person is selected from the population is $\pi = .9$ (because the proportion unemployed is .1). We wish to find the probability that $y = 5$ (all five are employed) in five trials.

$$P(y = 5) = \frac{5!}{5!(5-5)!}(.9)^5(.1)^0$$

$$= \frac{5!}{5!0!}(.9)^5(.1)^0$$

$$= (.9)^5 = .590$$

The binomial probability distribution for $n = 5$, $\pi = .9$ is shown in Figure 4.3. The probability of observing five employed in a sample of five is shaded in the figure.

FIGURE 4.3
The binomial probability distribution for $n = 5$, $\pi = .9$

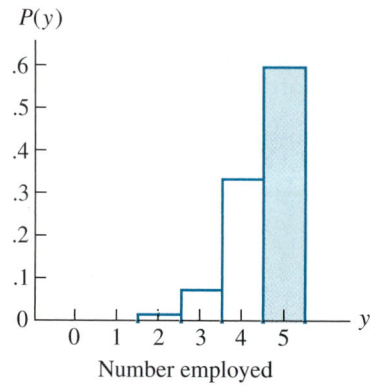

EXAMPLE 4.9

Refer to Example 4.8 and calculate the probability that exactly one person in the sample of five households is unemployed. What is the probability of one or fewer being unemployed?

Solution   Since y is the number of employed in the sample of five, one unemployed person would correspond to four employed ($y = 4$). Then

$$P(4) = \frac{5!}{4!(5-4)!}(.9)^4(.1)^1$$

$$= \frac{(5)(4)(3)(2)(1)}{(4)(3)(2)(1)(1)}(.9)^4(.1)$$

$$= 5(.9)^4(.1)$$

$$= .328$$

Thus, the probability of selecting four employed heads of households in a sample of five is .328, or, roughly, one chance in three.

The outcome "one or fewer unemployed" is the same as the outcome "4 or 5 employed." Since y represents the number employed, we seek the probability that $y = 4$ or 5. Because the values associated with a random variable represent mutually exclusive events, the probabilities for discrete random variables are additive. Thus, we have

$$P(y = 4 \text{ or } 5) = P(4) + P(5)$$

$$= .328 + .590$$

$$= .918$$

Thus, the probability that a random sample of five households will yield either four or five employed heads of households is .918. This high probability is consistent with our intuition: we could expect the number of employed in the sample to be large if 90% of all heads of households in the city are employed.

Like any relative frequency histogram, a binomial probability distribution possesses a mean, μ, and a standard deviation, σ. Although we omit the derivations, we give the formulas for these parameters.

Mean and Standard Deviation of the Binomial Probability Distribution

$$\mu = n\pi \quad \text{and} \quad \sigma = \sqrt{n\pi(1 - \pi)}$$

where π is the probability of success in a given trial and n is the number of trials in the binomial experiment.

If we know π and the sample size, n, we can calculate μ and σ to locate the center and describe the variability for a particular binomial probability distribution. Thus, we can quickly determine those values of y that are probable and those that are improbable.

EXAMPLE 4.10

We will consider the turf grass seed example to illustrate the calculation of the mean and standard deviation. Suppose the company producing the turf grass takes a sample of 20 seeds on a regular basis to monitor the quality of the seeds. If the germination rate of the seeds stays constant at 85%, then the average number of seeds that will germinate in the sample of 20 seeds is

$$\mu = n\pi = 20(.85) = 17$$

with a standard deviation of

$$\sigma = \sqrt{n\pi(1 - \pi)} = \sqrt{20(.85)(1 - .85)} = 1.60$$

Suppose we examine the germination records of a large number of samples of 20 seeds each. If the germination rate has remained constant at 85%, then the average number of seeds that germinate should be close to 17 per sample. If in a particular sample of 20 seeds we determine that only 12 had germinated, would the germination rate of 85% seem consistent with our results? Using a computer software program, we can generate the probability distribution for the number of seeds that germinate in the sample of 20 seeds, as shown in Figure 4.4.

Although the distribution is tending toward left skewness (see Figure 4.4), the Empirical Rule should work well for this relatively mound-shaped distribution. Thus, $y = 12$ seeds is more than 3 standard deviations less than the mean number of seeds, $\mu = 17$; it is highly improbable that in 20 seeds we would obtain only 12 germinated seeds if π really is equal to .85. The germination rate is most likely a value considerably less than .85.

Bus.   **4.34**   The weekly demand for copies of a popular word-processing program at a computer store has the probability distribution shown here.

y	P(y)
0	.06
1	.14
2	.16
3	.14
4	.12
5	.10
6	.08
7	.07
8	.06
9	.04
10	.03

a. What is the probability that three or more copies will be needed in a particular week?
b. What is the probability that the demand will be for at least two but no more than six copies?
c. If the store has eight copies of the program available at the beginning of each week, what is the probability the demand will exceed the supply in a given week?

Bio.   **4.35**   A biologist randomly selects ten portions of water, each equal to .1 cm³ in volume, from the local reservoir and counts the number of bacteria present in each portion. The biologist then totals the number of bacteria for the ten portions to obtain an estimate of the number of bacteria per cubic centimeter present in the reservoir water. Is this a binomial experiment?

Pol. Sci.   **4.36**   Examine the accompanying newspaper clipping. Does this sampling appear to satisfy the characteristics of a binomial experiment?

Poll Finds Opposition to Phone Taps

New York—People surveyed in a recent poll indicated they are 81% to 13% against having their phones tapped without a court order.

The people in the survey, by 68% to 27%, were opposed to letting the government use a wiretap on citizens suspected of crimes, except with a court order.

The survey was conducted for 1,495 households and also found the following results:

—The people surveyed are 80% to 12% against the use of any kind of electronic spying device without a court order.

—Citizens are 77% to 14% against allowing the government to open their mail without court orders.

—They oppose, by 80% to 12%, letting the telephone company disclose records of long-distance phone calls, except by court order.

For each of the questions, a few of those in the survey had no responses.

Env.   **4.37**   A survey is conducted to estimate the percentage of pine trees in a forest that are infected by the pine shoot moth. A grid is placed over a map of the forest, dividing the area into 25-foot by 25-foot square sections. One hundred of the squares are randomly selected and the number of infected trees is recorded for each square. Is this a binomial experiment?

Pol. Sci.   **4.38** A survey was conducted to investigate the attitudes of nurses working in Veterans Administration hospitals. A sample of 1,000 nurses was contacted using a mailed questionnaire and the number favoring or opposing a particular issue was recorded. If we confine our attention to the nurses' responses to a single question, would this sampling represent a binomial experiment? As with most mail surveys, some of the nurses did not respond. What effect might nonresponses in the sample have on the estimate of the percentage of all Veterans Administration nurses who favor the particular proposition?

Env.   **4.39** In an inspection of automobiles in Los Angeles, 60% of all automobiles had emissions that do not meet EPA regulations. For a random sample of ten automobiles, compute the following probabilities:

   a. All ten automobiles failed the inspection.
   b. Exactly six of the ten failed the inspection.
   c. Six or more failed the inspection.
   d. All ten passed the inspection.

Use the following Minitab output to answer the questions. Note that with Minitab, the binomial probability π is denoted by p and the binomial variable y is represented by x.

```
Binomial Distribution with n = 10 and p = 0.6

      x       P(X = x)      P(X <= x)
   0.00        0.0001        0.0001
   1.00        0.0016        0.0017
   2.00        0.0106        0.0123
   3.00        0.0425        0.0548
   4.00        0.1115        0.1662
   5.00        0.2007        0.3669
   6.00        0.2508        0.6177
   7.00        0.2150        0.8327
   8.00        0.1209        0.9536
   9.00        0.0403        0.9940
  10.00        0.0060        1.0000
```

4.40 Refer to Exercise 4.39.
   a. Compute the probabilities for parts (a) through (d) if $\pi = .3$.
   b. Indicate how you would compute $P(y \leq 100)$ for $n = 1,000$ and $\pi = .3$.

Bio.   **4.41** An experiment is conducted to test the effect of an anticoagulant drug on rats. A random sample of four rats is employed in the experiment. If the drug manufacturer claims that 80% of the rats will be favorably affected by the drug, what is the probability that none of the four experimental rats will be favorably affected? One of the four? One or fewer?

Soc.   **4.42** A criminologist claims that the probability of reform for a first-offender embezzler is .9. Suppose that we define *reform* as meaning the person commits no criminal offenses within a 5-year period. Three paroled embezzlers were randomly selected from the prison records, and their behavioral histories were examined for the 5-year period following prison release. If the criminologist's claim is correct, what is the probability that all three were reformed? At least two?

4.43 Consider the following experiment: Toss three coins and observe the number of heads y. Repeat the experiment 100 times and construct a relative frequency table for y. Note that these frequencies give approximations to the exact probabilities that $y = 0, 1, 2,$ and 3. (*Note*: These probabilities can be shown to be 1/8, 3/8, 3/8, and 1/8, respectively.)

4.44 Refer to Exercise 4.43. Use the formula for the binomial probability distribution to show that $P(0) = 1/8$, $P(1) = 3/8$, $P(2) = 3/8$, and $P(3) = 1/8$.

4.45   Suppose you match coins with another person a total of 1,000 times. What is the mean number of matches? The standard deviation? Calculate the interval ($\mu \pm 3\sigma$). (*Hint:* The probability of a match in the toss of a single pair of coins is $\pi = .5$.)

4.46   Refer to Exercise 4.39. Indicate how you could compute $P(y \leq 100)$ if $n = 1,000$ for $\pi = .6$.

Bus.   **4.47**   Over a long period of time in a large multinational corporation, 10% of all sales trainees are rated as outstanding, 75% are rated as excellent/good, 10% are rated as satisfactory, and 5% are considered unsatisfactory. Find the following probabilities for a sample of ten trainees selected at random:

    a. Two are rated as outstanding.
    b. Two or more are rated as outstanding.
    c. Eight of the ten are rated either outstanding or excellent/good.
    d. None of the trainees is rated as unsatisfactory.

Med.   **4.48**   A new technique, balloon angioplasty, is being widely used to open clogged heart valves and vessels. The balloon is inserted via a catheter and is inflated, opening the vessel; thus, no surgery is required. Left untreated, 50% of the people with heart-valve disease die within about 2 years. If experience with this technique suggests that approximately 70% live for more than 2 years, would the next five patients of the patients treated with balloon angioplasty at a hospital constitute a binomial experiment with $n = 5$, $\pi = .70$? Why or why not?

Med.   **4.49**   A prescription drug firm claims that only 12% of all new drugs shown to be effective in animal tests ever make it through a clinical testing program and onto the market. If a firm has 15 new compounds that have shown effectiveness in animal tests, find the following probabilities:

    a. None reach the market.
    b. One or more reach the market.
    c. Two or more reach the market.

4.50   Does Exercise 4.49 satisfy the properties of a binomial experiment? Why or why not?

Bus.   **4.51**   A random sample of 50 price changes is selected from the many listed for a large supermarket during a reporting period. If the probability that a price change is posted correctly is .93,

    a. Write an expression for the probability that three or fewer changes are posted incorrectly.
    b. What assumptions were made for part (a)?

4.9   Probability Distributions for Continuous Random Variables

Discrete random variables (such as the binomial) have possible values that are distinct and separate, such as 0 or 1 or 2 or 3. Other random variables are most usefully considered to be *continuous:* their possible values form a whole interval (or range, or continuum). For instance, the 1-year return per dollar invested in a common stock could range from 0 to some quite large value. In practice, virtually all random variables assume a discrete set of values; the return per dollar of a million-dollar common-stock investment could be $1.06219423 or $1.06219424 or $1.06219425 or .... However, when there are many possible values for a random variable, it is sometimes mathematically useful to treat the random variable as continuous.

Theoretically, then, a continuous random variable is one that can assume values associated with infinitely many points in a line interval. We state, without elaboration, that it is impossible to assign a small amount of probability to each value of y (as was done for a discrete random variable) and retain the property that the probabilities sum to 1.

   To overcome this difficulty, we revert to the concept of the relative frequency histogram of Chapter 3, where we talked about the probability of y falling in a given interval. Recall that the relative frequency histogram for a population containing a large number of measurements will almost be a smooth curve because the number of class intervals can be made large and the width of the intervals can be decreased. Thus, we envision a smooth curve that provides a model for the population relative frequency distribution generated by repeated observation of a continuous random variable. This will be similar to the curve shown in Figure 4.6.

FIGURE 4.6

Probability distribution for a
continuous random variable

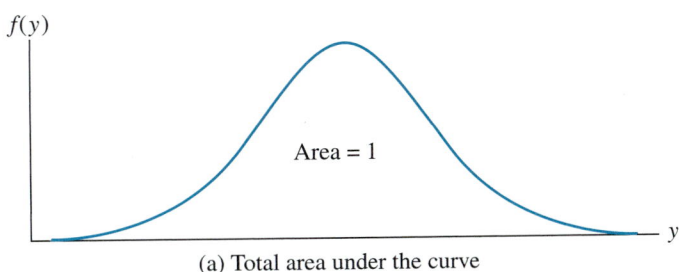

(a) Total area under the curve

(b) Probability

   Recall that the histogram relative frequencies are proportional to areas over the class intervals and that these areas possess a probabilistic interpretation. Thus, if a measurement is randomly selected from the set, the probability that it will fall in an interval is proportional to the histogram area above the interval. Since a population is the whole (100%, or 1), we want the total area under the probability curve to equal 1. If we let the total area under the curve equal 1, then areas over intervals are exactly equal to the corresponding probabilities.

   The graph for the probability distribution for a continuous random variable is shown in Figure 4.7. The ordinate (height of the curve) for a given value of y is denoted by the symbol $f(y)$. Many people are tempted to say that $f(y)$, like $P(y)$ for the binomial random variable, designates the probability associated with the

FIGURE 4.7

Hypothetical probability
distribution for student
examination scores

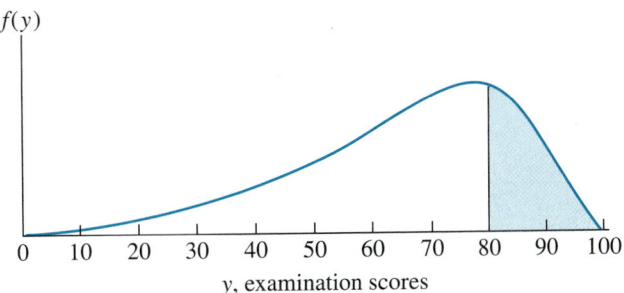

y, examination scores

continuous random variable *y*. However, as we mentioned before, it is impossible to assign a probability to each of the infinitely many possible values of a continuous random variable. Thus, all we can say is that *f(y)* represents the height of the probability distribution for a given value of *y*.

The probability that a continuous random variable falls in an interval, say between two points *a* and *b*, follows directly from the probabilistic interpretation given to the area over an interval for the relative frequency histogram (Section 3.3) and is equal to the area under the curve over the interval *a* to *b*, as shown in Figure 4.6. This probability is written $P(a < y < b)$.

There are curves of many shapes that can be used to represent the population relative frequency distribution for measurements associated with a continuous random variable. Fortunately, the areas for many of these curves have been tabulated and are ready for use. Thus, if we know that student examination scores possess a particular probability distribution, as in Figure 4.7, and if areas under the curve have been tabulated, we can find the probability that a particular student will score more than 80% by looking up the tabulated area, which is shaded in Figure 4.7.

Figure 4.8 depicts four important probability distributions that will be used extensively in the following chapters. Which probability distribution we use in a particular situation is very important because probability statements are determined by the area under the curve. As can be seen in Figure 4.8, we would obtain

FIGURE 4.8

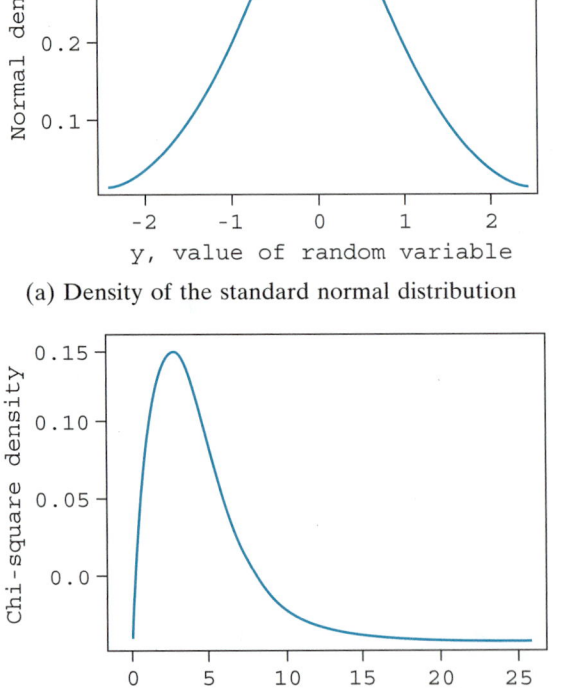

(a) Density of the standard normal distribution

(c) Density of the chi-square (df = 5) distribution

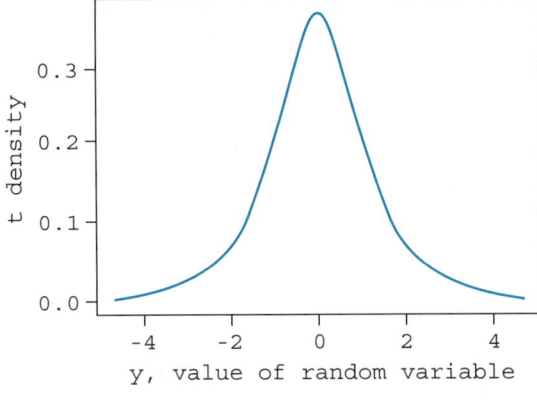

(b) Density of the *t*(df = 3) distribution

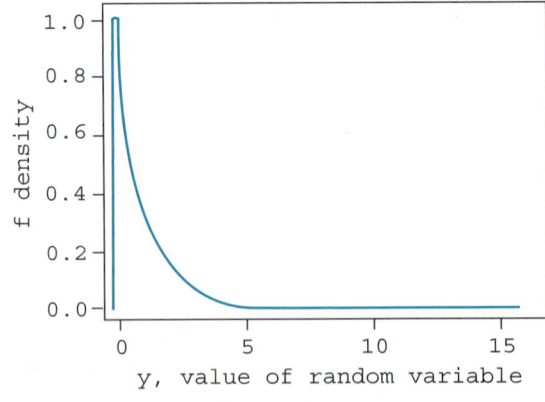

(d) Density of the *F* (df = 2, 6) distribution

very different answers depending on which distribution is selected. For example, the probability the random variable takes on a value less than 5.0 is essentially 1.0 for the probability distributions in Figures 4.8(a) and (b) but is 0.584 and 0.947 for the probability distributions in Figures 4.8(c) and (d), respectively. In some situations, we will not know exactly the distribution for the random variable in a particular study. In these situations, we can use the observed values for the random variable to construct a relative frequency histogram, which is a sample estimate of the true probability frequency distribution. As far as statistical inferences are concerned, the selection of the *exact* shape of the probability distribution for a continuous random variable is not crucial in many cases, because most of our inference procedures are insensitive to the exact specification of the shape.

We will find that data collected on continuous variables often possess a nearly bell-shaped frequency distribution, such as depicted in Figure 4.8(a). A continuous variable (the normal) and its probability distribution (bell-shaped curve) provide a good model for these types of data. The normally distributed variable is also very important in statistical inference. We will study the normal distribution in detail in the next section.

4.10   A Useful Continuous Random Variable: The Normal Distribution

normal curve

Many variables of interest, including several statistics to be discussed in later sections and chapters, have mound-shaped frequency distributions that can be approximated by using a **normal curve.** For example, the distribution of total scores on the Brief Psychiatric Rating Scale for outpatients having a current history of repeated aggressive acts is mound-shaped. Other practical examples of mound-shaped distributions are social perceptiveness scores of preschool children selected from a particular socioeconomic background, psychomotor retardation scores for patients with circular-type manic-depressive illness, milk yields for cattle of a particular breed, and perceived anxiety scores for residents of a community. Each of these mound-shaped distributions can be approximated with a normal curve.

Since the normal distribution has been well tabulated, areas under a normal curve—which correspond to probabilities—can be used to approximate probabilities associated with the variables of interest in our experimentation. Thus, the normal random variable and its associated distribution play an important role in statistical inference.

The relative frequency histogram for the normal random variable, called the *normal curve* or *normal probability distribution,* is a smooth bell-shaped curve. Figure 4.9(a) shows a normal curve. If we let y represent the normal random variable, then the height of the probability distribution for a specific value of y is represented by $f(y)$.* The probabilities associated with a normal curve form the basis for the Empirical Rule.

As we see from Figure 4.9(a), the normal probability distribution is bell shaped and symmetrical about the mean μ. Although the normal random variable y may theoretically assume values from $-\infty$ to $+\infty$, we know from the Empirical

---

* For the normal distribution, $f(y) = \dfrac{1}{\sqrt{2\pi}\,\sigma} e^{-(y-\mu)^2/2\sigma^2}$, where μ and σ are the mean and standard deviation, respectively, of the population of y-values.

FIGURE 4.9

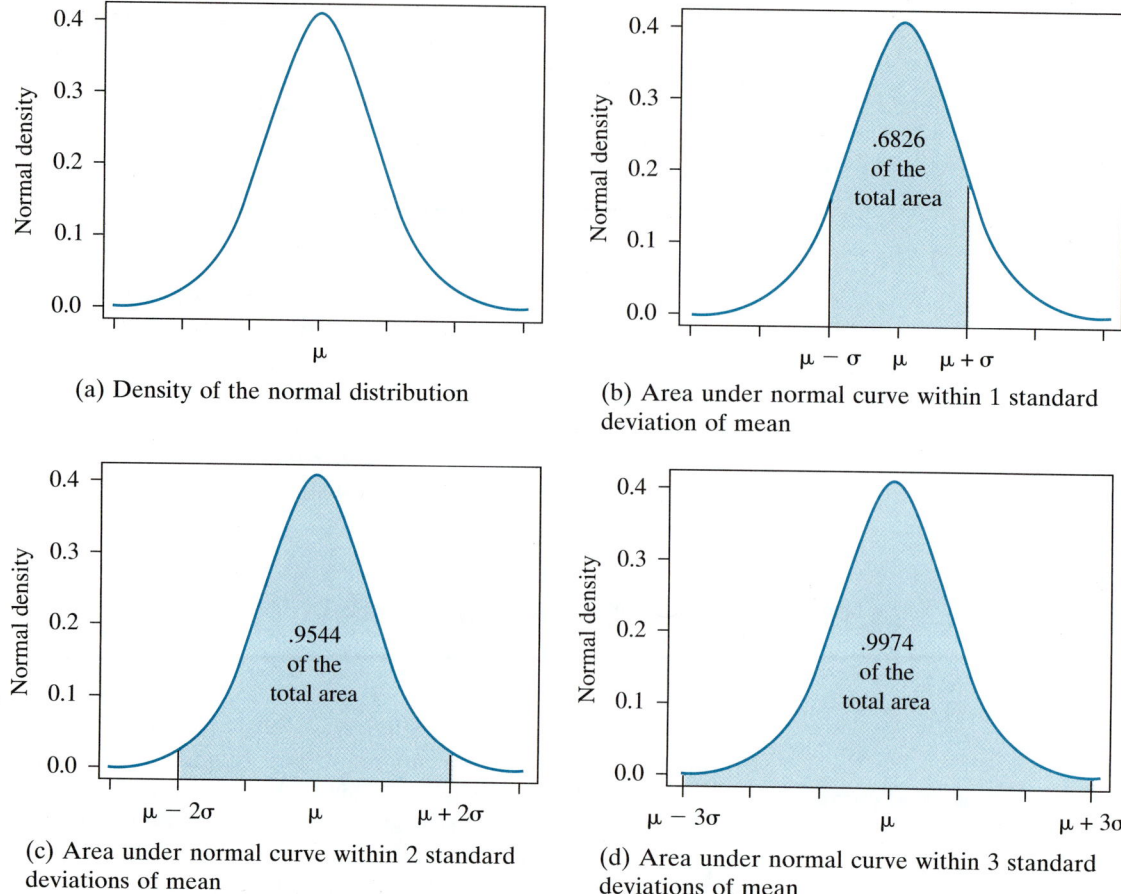

(a) Density of the normal distribution

(b) Area under normal curve within 1 standard deviation of mean

(c) Area under normal curve within 2 standard deviations of mean

(d) Area under normal curve within 3 standard deviations of mean

Rule that approximately all the measurements are within 3 standard deviations (3σ) of μ. From the Empirical Rule, we also know that if we select a measurement at random from a population of measurements that possesses a mound-shaped distribution, the probability is approximately .68 that the measurement will lie within 1 standard deviation of its mean (see Figure 4.9(b)). Similarly, we know that the probability is approximately .954 that a value will lie in the interval $\mu \pm 2\sigma$ and .997 in the interval $\mu \pm 3\sigma$ (see Figures 4.9(c) and (d)). What we do not know, however, is the probability that the measurement will be within 1.65 standard deviations of its mean, or within 2.58 standard deviations of its mean. The procedure we are going to discuss in this section will enable us to calculate the probability that a measurement falls within any distance of the mean μ for a normal curve.

Because there are many different normal curves (depending on the parameters μ and σ), it might seem to be an impossible task to tabulate areas (probabilities) for all normal curves, especially if each curve requires a separate table. Fortunately, this is not the case. By specifying the probability that a variable y lies within a certain number of standard deviations of its mean (just as we did in using the Empirical Rule), we need only one table of probabilities.

area under a normal curve   Table 1 in the Appendix gives the **area under a normal curve** to the left of a value y that is z standard deviations ($z\sigma$) away from the mean (see Figure 4.10). The area shown by the shading in Figure 4.10 is the probability listed in Table 1 in the Appendix. Values of z to the nearest tenth are listed along the left-hand column of the table, with z to the nearest hundredth along the top of the table.

FIGURE 4.10

Area under a normal curve as given in Appendix Table 1

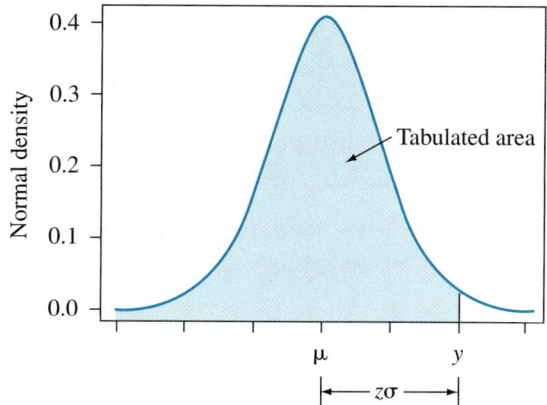

To find the probability that a normal random variable will lie to the left of a point 1.65 standard deviations above the mean, we look up the table entry corresponding to $z = 1.65$. This probability is .9505 (see Figure 4.11).

FIGURE 4.11

Area under a normal curve from μ to a point 1.65 standard deviations above the mean

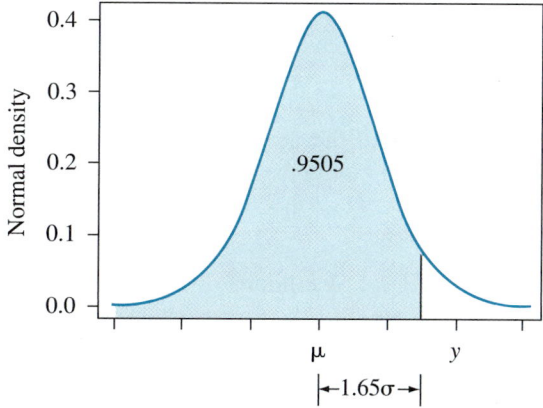

To determine the probability that a measurement will be less than some value y, we first calculate the number of standard deviations that y lies away from the mean by using the formula

$$z = \frac{y - \mu}{\sigma}$$

z score

The value of z computed using this formula is sometimes referred to as the **z score** associated with the y-value. Using the computed value of z, we determine the appropriate probability by using Table 1 in the Appendix. Note that we are merely coding the value y by subtracting μ and dividing by σ. (In other words, $y = z\sigma + \mu$.) Figure 4.12 illustrates the values of z corresponding to specific values

FIGURE 4.12

Relationship between specific values of y and $z = (y - \mu)/\sigma$

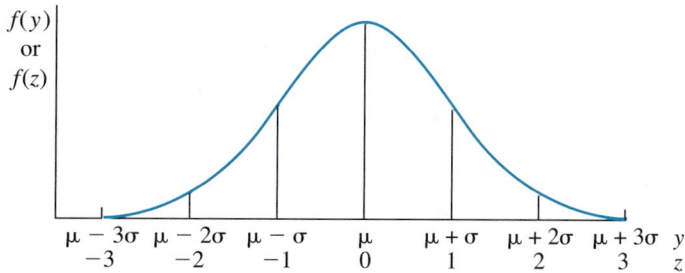

of y. Thus, a value of y that is 2 standard deviations below (to the left of) μ corresponds to $z = -2$.

EXAMPLE 4.12

Consider a normal distribution with $\mu = 20$ and $\sigma = 2$. Determine the probability that a measurement will be less than 23.

Solution   When first working problems of this type, it might be a good idea to draw a picture so that you can see the area in question, as we have in Figure 4.13.

FIGURE 4.13

Area less than $y = 23$ under normal curve, with $\mu = 20$, $\sigma = 2$

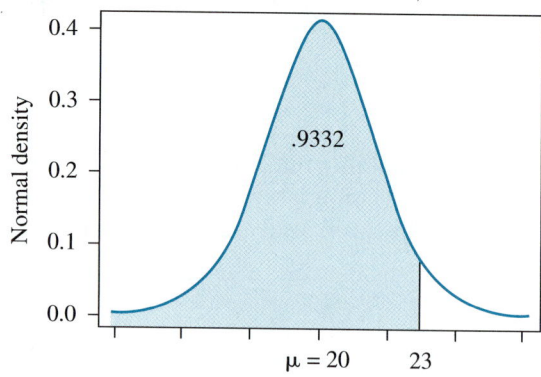

To determine the area under the curve to the left of the value $y = 23$, we first calculate the number of standard deviations $y = 23$ lies away from the mean.

$$z = \frac{y - \mu}{\sigma} = \frac{23 - 20}{2} = 1.5$$

Thus, $y = 23$ lies 1.5 standard deviations above $\mu = 20$. Referring to Table 1 in the Appendix, we find the area corresponding to $z = 1.5$ to be .9332. This is the probability that a measurement is less than 23.

EXAMPLE 4.13

For the normal distribution of Example 4.12 with $\mu = 20$ and $\sigma = 2$, find the probability that y will be less than 16.

Solution   In determining the area to the left of 16, we use

$$z = \frac{y - \mu}{\sigma} = \frac{16 - 20}{2} = -2$$

We find the appropriate area from Table 1 to be .0228; thus, .0228 is the probability that a measurement is less than 16. The area is shown in Figure 4.14.

FIGURE 4.14

Area less than $y = 16$ under normal curve, with $\mu = 20$, $\sigma = 2$

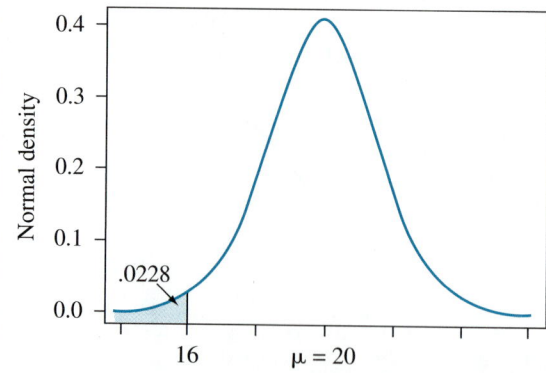

EXAMPLE 4.14

The mean daily milk production of a herd of Guernsey cows has a normal distribution with $\mu = 70$ pounds and $\sigma = 13$ pounds.

a. What is the probability that the milk production for a cow chosen at random will be less than 60 pounds?
b. What is the probability that the milk production for a cow chosen at random will be greater than 90 pounds?
c. What is the probability that the milk production for a cow chosen at random will be between 60 pounds and 90 pounds?

Solution   We begin by drawing pictures of the areas we are looking for (Figures 4.15 (a)–(c)). To answer part (a) we must compute the z values corresponding to the value of 60. The value $y = 60$ corresponds to a z score of

$$z = \frac{y - \mu}{\sigma} = \frac{60 - 70}{13} = -.77$$

From Table 1, the area to the left of 60 is .2206 (see Figure 4.15(a)).

FIGURE 4.15(a)

Area less than $y = 60$ under normal curve, with $\mu = 70$, $\sigma = 13$

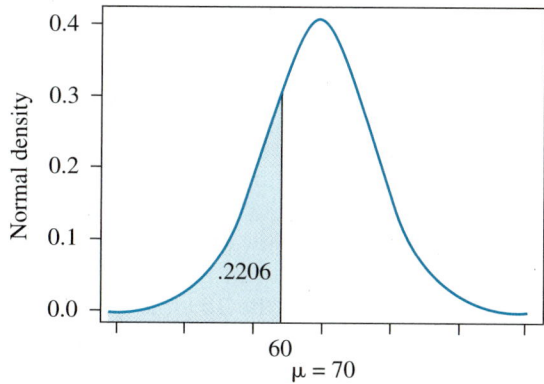

To answer part (b), the value $y = 90$ corresponds to a z score of

$$z = \frac{y - \mu}{\sigma} = \frac{90 - 70}{13} = 1.54$$

so from Table 1 we obtain .9382, the tabulated area less than 90. Thus, the area greater than 90 must be $1 - .9382 = .0618$, since the total area under the curve is 1 (see Figure 4.15(b)).

FIGURE 4.15(b)

Area greater than $y = 90$ under normal curve, with $\mu = 70$, $\sigma = 13$

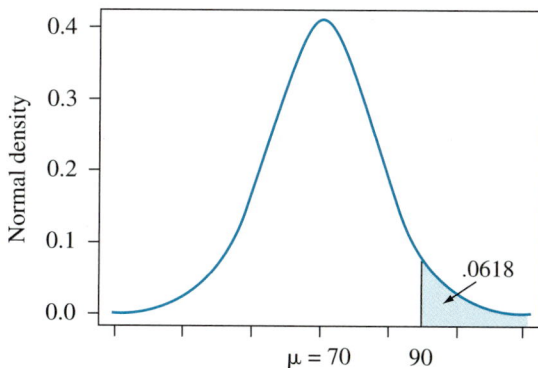

To answer part (c), we can use our results from (a) and (b). The area between two values y_1 and y_2 is determined by finding the difference between the areas to the left of the two values, (see Figure 4.15(c)). We have the area less than 60 is .2206, and the area less than 90 is .9382. Hence, the area between 60 and 90 is .9382 − .2206 = .7176. We can thus conclude that 22.06% of cow production is less than 60 pounds, 6.18% is greater than 90 pounds, and 71.76% is between 60 and 90 pounds.

FIGURE 4.15(c)
Area between 60 and 90 under normal curve, with $\mu = 70$, $\sigma = 13$

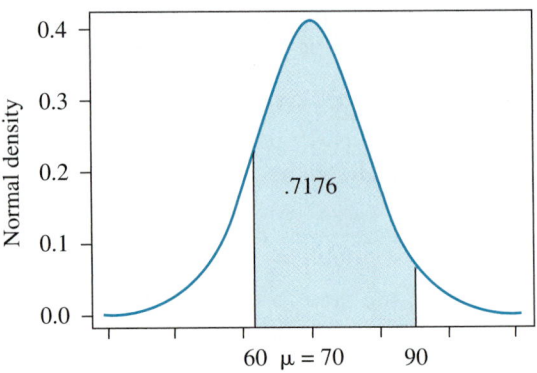

An important aspect of the normal distribution is that we can easily find the percentiles of the distribution. The **100pth percentile** of a distribution is that value, y_p, such that $100p$ % of the population values fall below y_p and $100\,(1 - p)$ % are above y_p. For example, the median of a population is the 50th percentile, $y_{.50}$, and the quartiles are the 25th and 75th percentiles. The normal distribution is symmetric, so the median and the mean are the same value, $y_{.50} = \mu$ (see Figure 4.16(a)).

100pth percentile

To find the percentiles of the standard normal distribution, we reverse our usage of Table 1. To find the 100pth percentile, z_p, we find the probability p in Table 1 and then read out its corresponding number, z_p, along the margins of the table. For example, to find the 80th percentile, $z_{.80}$, we locate the probability, $p = .8000$ in Table 1. The value nearest to .8000 is .7995, which corresponds to a z value of 0.84. Thus, $z_{.80} = 0.84$ (see Figure 4.16 (b)). Now, to find the 100pth percentile, y_p, of a normal distribution with mean μ and standard deviation σ, we need to apply the reverse of our standardization formula,

$$y_p = \mu + z_p \sigma$$

FIGURE 4.16

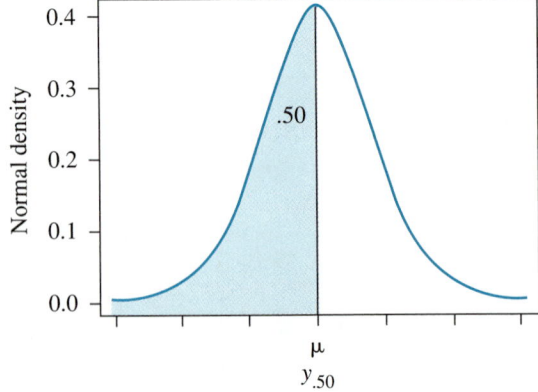

(a) For the normal curve, the mean and median agree

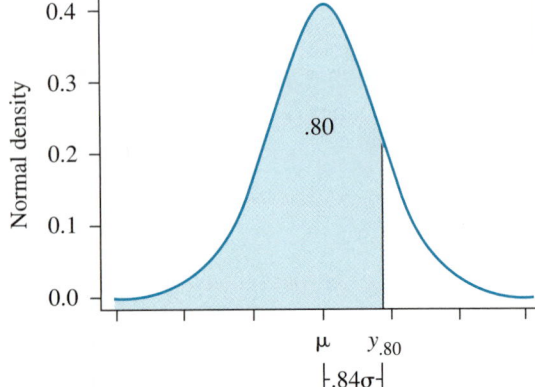

(b) The 80th percentile for the normal curve

Suppose we wanted to determine the 80th percentile of a population having a normal distribution with $\mu = 55$ and $\sigma = 3$. We have determined that $z_{.80} = 0.84$; thus, the 80th percentile for the population would be $y_{.80} = 55 + (.84)(3) = 57.52$.

EXAMPLE 4.15

The Scholastic Assessment Test (SAT) is an exam used to measure a person's readiness for college. The mathematics scores are scaled to have a normal distribution with mean 500 and standard deviation 100. What proportion of the people taking the SAT will score below 350? To identify a group of students needing remedial assistance, we want to determine the lower 10% of all scores; that is, we want to determine the tenth percentile, $y_{.10}$.

Solution   To find the proportion of scores below 350 (see Figure 4.17(a)), we need to find the area below 350:

$$z = \frac{y - \mu}{\sigma} = \frac{350 - 500}{100} = -1.5$$

FIGURE 4.17(a)

Area less than 350 under normal curve, with $\mu = 500$, $\sigma = 100$

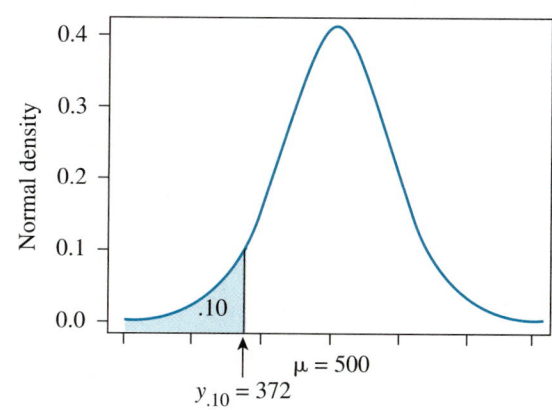

In a normal distribution, the area to the left of a value -1.5 standard deviations from the mean is, from Table 1, .0668. Hence, 6.68%, or approximately 7%, of the persons taking the exam scored below 350. The score of 350 would be approximately the seventh percentile, $y_{.07}$, of the population of all scores.

To find the tenth percentile (see Figure 4.17(b)), we first find $z_{.10}$ in Table 1. Since .1003 is the value nearest .1000 and its corresponding z value is -1.28,

FIGURE 4.17(b)

The tenth percentile for a normal curve, with $\mu = 500$, $\sigma = 100$

we take $z_{.10} = -1.28$. We then compute

$$y_{.10} = \mu + z_{.10}\sigma = 500 + (-1.28)(100) = 500 - 128 = 372$$

Thus, 10% of the scores on the SAT are less than 372.

EXAMPLE 4.16

An analysis of income tax returns from the previous year indicates that for a given income classification, the amount of money owed to the government over and above the amount paid in the estimated tax vouchers for the first three payments is approximately normally distributed with a mean of $530 and a standard deviation of $205. Find the 75th percentile for this distribution of measurements. The government wants to target that group of returns having the largest 25% of amounts owed.

Solution   We need to determine the 75th percentile, $y_{.75}$, (Figure 4.18). From Table 1, we find $z_{.75} = .67$ because the probability nearest .7500 is .7486, which corresponds to a z score of .67. We then compute

$$y_{.75} = \mu + z_{.75}\sigma = 530 + (.67)(205) = 667.35$$

FIGURE 4.18

The 75th percentile for a normal curve, with $\mu = 530$, $\sigma = 205$

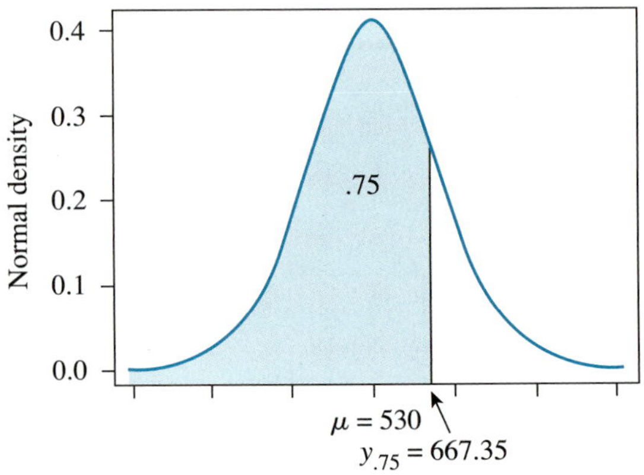

Thus, 25% of the tax returns in this classification exceed $667.35 in the amount owed the government.

EXERCISES   **Basic Techniques**

4.52   Use Table 1 of the Appendix to find the area under the normal curve between these values:
  a. $z = 0$ and $z = 1.3$
  b. $z = 0$ and $z = -1.9$

4.53   Repeat Exercise 4.52 for these values:
  a. $z = 0$ and $z = .7$
  b. $z = 0$ and $z = -1.2$

4.54   Repeat Exercise 4.52 for these values:
  a. $z = 0$ and $z = 1.29$
  b. $z = 0$ and $z = -.77$

4.55   Repeat Exercise 4.52 for these values:
  a. $z = -.21$ and $z = 1.35$
  b. $z = .37$ and $z = 1.20$

4.56   Repeat Exercise 4.52 for these values:
  a. $z = 1.43$ and $z = 2.01$
  b. $z = -1.74$ and $z = -.75$

4.57   Find the probability that z is greater than 1.75.

4.58   Find the probability that z is less than 1.14.

4.59   Find a value for z, say z_0, such that $P(z > z_0) = .5$.

4.60   Find a value for z, say z_0, such that $P(z > z_0) = .025$.

4.61   Find a value for z, say z_0, such that $P(z > z_0) = .0089$.

4.62   Find a value for z, say z_0, such that $P(z > z_0) = .05$.

4.63   Find a value for z, say z_0, such that $P(-z_0 < z < z_0) = .95$.

4.64   Let y be a normal random variable with mean equal to 100 and standard deviation equal to 8. Find the following probabilities:
  a. $P(y > 100)$
  b. $P(y > 110)$
  c. $P(y < 115)$
  d. $P(88 < y < 12)$
  e. $P(100 < y < 108)$

4.65   Let y be a normal random variable with $\mu = 500$ and $\sigma = 100$. Find the following probabilities:
  a. $P(500 < y < 696)$
  b. $P(y > 696)$
  c. $P(304 < y < 696)$
  d. k such that $P(500 - k < y < 500 + k) = .60$

4.66   Suppose that y is a normal random variable with $\mu = 100$ and $\sigma = 15$.
  a. Show that $y < 130$ is equivalent to $z < 2$.
  b. Convert $y > 82.5$ to the z-score equivalent.
  c. Find $P(y < 130)$ and $P(y > 82.5)$.
  d. Find $P(y > 106)$, $P(y < 94)$, and $P(94 < y < 106)$.
  e. Find $P(y < 70)$, $P(y > 130)$, and $P(70 < y < 130)$.

4.67   Use Table 1 in the Appendix to calculate the area under the curve between these values:
  a. $z = 0$ and $z = 1.5$
  b. $z = 0$ and $z = 1.8$

4.68   Repeat Exercise 4.67 for these values:
  a. $z = -1.96$ and $z = 1.96$
  b. $z = -2.33$ and $z = 2.33$

4.69   What is the value of z with an area of .05 to its right? To its left? (*Hint:* Use Table 2 in the Appendix.)

4.70   Find the value of z for these areas.
  a. an area .01 to the right of z
  b. an area .10 to the left of z

4.71   Find the probability of observing a value of z greater than these values.
  a. 1.96
  b. 2.21
  c. 2.86
  d. 0.73

4.72   Find the probability of observing a value of z less than these values.
  a. -1.20
  b. -2.62
  c. 1.84
  d. 2.17

Applications

Gov.   **4.73**   Records maintained by the office of budget in a particular state indicate that the amount of time elapsed between the submission of travel vouchers and the final reimbursement of funds has approximately a normal distribution with a mean of 39 days and a standard deviation of 6 days.

 a. What is the probability that the elapsed time between submission and reimbursement will exceed 50 days?

 b. If you had a travel voucher submitted more than 55 days ago, what might you conclude?

Edu.   **4.74**   The College Boards, which are administered each year to many thousands of high school students, are scored so as to yield a mean of 500 and a standard deviation of 100. These scores are close to being normally distributed. What percentage of the scores can be expected to satisfy each condition?

 a. greater than 600

 b. greater than 700

 c. less than 450

 d. between 450 and 600

Bus.   **4.75**   Monthly sales figures for a particular food industry tend to be normally distributed with mean of 150 (thousand dollars) and a standard deviation of 35 (thousand dollars). Compute the following probabilities:

 a. $P(y > 200)$

 b. $P(y > 220)$

 c. $P(y < 120)$

 d. $P(100 < y < 200)$

4.76   Refer to Exercise 4.74. An exclusive club wishes to invite those scoring in the top 10% on the College Boards to join.

 a. What score is required to be invited to join the club?

 b. What score separates the top 60% of the population from the bottom 40%? What do we call this value?

4.77   The mean for a normal distribution is 50 and the standard deviation is 10.

 a. What percentile is the value of 38? Choose the appropriate answer.

  88.49 38.49 49.99 0.01 11.51

 b. Which of the following is the z score corresponding to the 67th percentile?

  1.00 0.95 0.44 2.25 none of these

Soc.   **4.78**   The distribution of weights of a large group of high school boys is normal with $\mu = 120$ pounds and $\sigma = 10$ pounds. Which of the following is true?

 a. About 16% of the boys will be over 130 pounds.

 b. Probably fewer than 2.5% of the boys will be below 100 pounds.

 c. Half of the boys can be expected to weigh less than 120 pounds.

 d. All the above are true.

4.11   Random Sampling

Thus far in the text, we have discussed random samples and introduced various sampling schemes in Chapter 2. What is the importance of random sampling? We must know how the sample was selected so we can determine probabilities associated with various sample outcomes. The probabilities of samples selected *in a random manner* can be determined, and we can use these probabilities to make inferences about the population from which the sample was drawn.

Sample data selected in a nonrandom fashion are frequently distorted by a *selection bias*. A selection bias exists whenever there is a systematic tendency to overrepresent or underrepresent some part of the population. For example, a survey of households conducted during the week entirely between the hours of 9 A.M. and 5 P.M. would be severely biased toward households with at least one member at home. Hence, any inferences made from the sample data would be biased toward the attributes or opinions of those families with at least one member at home and may not be truly representative of the population of households in the region.

random sample   Now we turn to a definition of a **random sample** of n measurements selected from a population containing N measurements ($N > n$). (*Note*: This is a simple random sample as discussed in Chapter 2. Since most of the random samples discussed in this text will be simple random samples, we'll drop the adjective unless needed for clarification.)

DEFINITION 4.13

A sample of n measurements selected from a population is said to be a **random sample** if every different sample of size n from the population has an equal probability of being selected.

EXAMPLE 4.17

A study of crimes related to handguns is being planned for the ten largest cities in the United States. The study will randomly select two of the ten largest cities for an in-depth study following the preliminary findings. The population of interest is the ten largest cities $\{C_1, C_2, C_3, C_4, C_5, C_6, C_7, C_8, C_9, C_{10}\}$. List all possible different samples consisting of two cities that could be selected from the population of ten cities. Give the probability associated with each sample in a random sample of $n = 2$ cities selected from the population.

Solution   All possible samples are listed next.

Sample	Cities	Sample	Cities	Sample	Cities
1	C_1, C_2	16	C_2, C_9	31	C_5, C_6
2	C_1, C_3	17	C_2, C_{10}	32	C_5, C_7
3	C_1, C_4	18	C_3, C_4	33	C_5, C_8
4	C_1, C_5	19	C_3, C_5	34	C_5, C_9
5	C_1, C_6	20	C_3, C_6	35	C_5, C_{10}
6	C_1, C_7	21	C_3, C_7	36	C_6, C_7
7	C_1, C_8	22	C_3, C_8	37	C_6, C_8
8	C_1, C_9	23	C_3, C_9	38	C_6, C_9
9	C_1, C_{10}	24	C_3, C_{10}	39	C_6, C_{10}
10	C_2, C_3	25	C_4, C_5	40	C_7, C_8
11	C_2, C_4	26	C_4, C_6	41	C_7, C_9
12	C_2, C_5	27	C_4, C_7	42	C_7, C_{10}
13	C_2, C_6	28	C_4, C_8	43	C_8, C_9
14	C_2, C_7	29	C_4, C_9	44	C_8, C_{10}
15	C_2, C_8	30	C_4, C_{10}	45	C_9, C_{10}

Now, let us suppose that we select a random sample of $n = 2$ cities from the 45 possible samples. The sample selected is called a *random* sample if every sample has an equal probability, 1/45, of being selected.

random number table

One of the simplest and most reliable ways to select a random sample of n measurements from a population is to use a table of random numbers (see Table 13 in the Appendix). **Random number tables** are constructed in such a way that, no matter where you start in the table and no matter in which direction you move, the digits occur randomly and with equal probability. Thus, if we wished to choose a random sample of $n = 10$ measurements from a population containing 100 measurements, we could label the measurements in the population from 0 to 99 (or 1 to 100). Then by referring to Table 13 in the Appendix and choosing a random starting point, the next ten two-digit numbers going across the page would indicate the labels of the particular measurements to be included in the random sample. Similarly, by moving up or down the page, we would also obtain a random sample.

This listing of all possible samples is feasible only when both the sample size n and the population size N are small. We can determine the number, M, of distinct samples of size n that can be selected from a population of N measurements using the following formula:

$$M = \frac{N!}{n!(N - n)!}$$

In Example 4.17, we had $N = 10$ and $n = 2$. Thus,

$$M = \frac{10!}{2!(10 - 2)!} = \frac{10!}{2!8!} = 45$$

The value of M becomes very large even when N is fairly small. For example, if $N = 50$ and $n = 5$, then $M = 2,118,760$. Thus, it would be very impractical to list all 2,118,760 possible samples consisting of $n = 5$ measurements from a population of $N = 50$ measurements and then randomly select one of the samples. In practice, we construct a list of elements in the population by assigning a number from 1 to N to each element in the population, called the *sampling frame*. We then randomly select n integers from the integers $(1, 2, \ldots, N)$ by using a table of random numbers (see Table 13 in the Appendix) or by using a computer program. Most statistical software programs contain routines for randomly selecting n integers from the integers $(1, 2, \ldots, N)$, where $N > n$. Exercise 4.86 contains the necessary commands for using Minitab to generate the random sample.

EXAMPLE 4.18

A small community consists of 850 families. We wish to obtain a random sample of 20 families to ascertain public acceptance of a wage and price freeze. Refer to Table 13 in the Appendix to determine which families should be sampled.

Solution   Assuming that a list of all families in the community is available (such as a telephone directory), we can label the families from 0 to 849 (or, equivalently, from 1 to 850). Then, referring to Table 13 in the Appendix, we choose a starting point. Suppose we have decided to start at line 1, column 3. Going down the page we choose the first 20 three-digit numbers between 000 and 849. From Table 13, we have

015	110	482	333
255	564	526	463
225	054	710	337
062	636	518	224
818	533	524	055

These 20 numbers identify the 20 families that are to be included in our sample.

A telephone directory is not always the best source for names, especially in surveys related to economics or politics. In the 1936 presidential campaign, Franklin Roosevelt was running as the Democratic candidate against the Republican candidate, Governor Alfred Landon of Kansas. This was a difficult time for the nation; the country had not yet recovered from the Great Depression of the 1930s, and there were still 9 million people unemployed.

The *Literary Digest* set out to sample the voting public and predict the winner of the election. Using names and addresses taken from telephone books and club memberships, the *Literary Digest* sent out 10 million questionnaires and got 2.4 million back. Based on the responses to the questionnaire, the *Digest* predicted a Landon victory by 57% to 43%.

At this time, George Gallup was starting his survey business. He conducted two surveys. The first one, based on 3,000 people, predicted what the results of the *Digest* survey would be long before the *Digest* results were published; the second survey, based on 50,000, was used to forecast *correctly* the Roosevelt victory.

How did Gallup correctly predict what the *Literary Digest* survey would predict and then, with another survey, correctly predict the outcome of the election? Where did the *Literary Digest* go wrong? The first problem was a severe selection bias. By taking the names and addresses from telephone directories and club memberships, its survey systematically excluded the poor. Unfortunately for the *Digest,* the vote was split along economic lines; the poor gave Roosevelt a large majority, whereas the rich tended to vote for Landon. A second reason for the error could be due to a *nonresponse bias.* Because only 20% of the 10 million people returned their surveys, and approximately half of those responding favored Landon, one might suspect that maybe the nonrespondents had different preferences than did the respondents. This was, in fact, true.

How, then does one achieve a random sample? Careful planning and a certain amount of ingenuity are required to have even a decent chance to approximate random sampling. This is especially true when the universe of interest involves people. People can be difficult to work with; they have a tendency to discard mail questionnaires and refuse to participate in personal interviews. Unless we are very careful, the data we obtain may be full of biases having unknown effects on the inferences we are attempting to make.

We do not have sufficient time to explore the topic of random sampling further in this text; entire courses at the undergraduate and graduate levels can be devoted to sample survey research methodology. The important point to remember is that data from a random sample will provide the foundation for making statistical inferences in later chapters. Random samples are not easy to obtain, but with care we can avoid many potential biases that could affect the inferences we make.

EXERCISES   **Basic Techniques**

4.79 Define what is meant by a random sample. Is it possible to draw a truly random sample? Comment.

Gov.   **4.80** Suppose that we want to select a random sample of $n = 10$ persons from a population of 800. Use Table 13 in the Appendix to identify the persons to appear in the sample.

Pol. Sci.   **4.81** Refer to Exercise 4.80. Identify the elements of a population of $N = 1{,}000$ to be included in a random sample of $n = 15$.

Applications

Soc.   **4.82** City officials want to sample the opinions of the homeowners in a community regarding the desirability of increasing local taxes to improve the quality of the public schools. If a random number table is used to identify the homes to be sampled and a home is discarded if the homeowner is not home when visited by the interviewer, is it likely this process will approximate random sampling? Explain.

4.83 A local TV network wants to run an informal survey of individuals who exist from a local voting station to ascertain early results on a proposal to raise funds to move the city-owned historical museum to a new location. How might the network sample voters to approximate random sampling?

4.84 A psychologist was interested in studying women who are in the process of obtaining a divorce to determine whether the women experienced significant attitudinal changes after the divorce has been finalized. Existing records from the geographic area in question show that 798 couples have recently field for divorce. Assume that a sample of 25 women is needed for the study, and use Table 13 in the Appendix to determine which women should be asked to participate in the study. (*Hint:* Begin in column 2, row 1, and proceed down.)

4.85 Refer to Exercise 4.84. As is the case in most surveys, not all persons chosen for a study will agree to participate. Suppose that 5 of the 25 women selected refuse to participate. Determine 5 more women to be included in the study.

4.86 Suppose you have been asked to run a public opinion poll related to an upcoming election. There are 230 precincts in the city, and you need to randomly select 50 registered voters from each precinct. Suppose that each precinct has 1,000 registered voters and it is possible to obtain a list of these persons. You assign the numbers 1 to 1,000 to the 1,000 people on each list, with 1 to the first person on the list and 1,000 to the last person. You need to next obtain a random sample of 50 numbers from the numbers 1 to 1,000. The names on the sampling frame corresponding to these 50 numbers will be the 50 persons selected for the poll. A Minitab program is shown here for purposes of illustration. Note that you would need to run this program 230 separate times to obtain a new random sample for each of the 230 precincts.

Follow these steps:

Click on **Calc.**
Click on **Random Data.**
Click on **Integer.**
Type **5** in the **Generate rows of data** box.
Type **c1–c10** in the **Store in Column(s):** box.
Type **1** in the **Minimum value:** box.
Type **1000** in the **Maximum value:** box.
Click on **OK.**
Click on **File.**
Click on **Print Worksheet.**

	C1	C2	C3	C4	C5	C6	C7	C8	C9	C10
1	340	701	684	393	313	312	834	596	321	739
2	783	877	724	498	315	282	175	611	725	571
3	862	625	971	30	766	256	40	158	444	546
4	974	402	768	593	980	536	483	244	51	201
5	232	742	1	861	335	129	409	724	340	218

a. Using either a random number table or a computer program, generate a second random sample of 50 numbers from the numbers 1 to 1,000.

b. Give several reasons why you need to generate a different set of random numbers for each of the precincts. Why not use the same set of 50 numbers for all 230 precincts?

4.12 Sampling Distributions

We discussed several different measures of central tendency and variability in Chapter 3 and distinguished between numerical descriptive measures of a population (parameters) and numerical descriptive measures of a sample (statistics). Thus, μ and σ are parameters, whereas $\bar{y}$ and s are statistics.

The numerical value of a sample statistic cannot be predicted exactly in advance. Even if we knew that a population mean μ was \$216.37 and that the population standard deviation σ was \$32.90—even if we knew the complete population distribution—we could not say that the sample mean $\bar{y}$ would be exactly equal to \$216.37. A sample statistic is a random variable; it is subject to random variation because it is based on a random sample of measurements selected from the population of interest. Also, like any other random variable, a sample statistic has a probability distribution. We call the probability distribution of a sample statistic the *sampling distribution* of that statistic. Stated differently, the sampling distribution of a statistic is the population of all possible values for that statistic.

The actual mathematical derivation of sampling distributions is one of the basic problems of mathematical statistics. We will illustrate how the sampling distribution for $\bar{y}$ can be obtained for a simplified population. Later in the chapter, we will present several general results.

EXAMPLE 4.19

The sample $\bar{y}$ is to be calculated from a random sample of size 2 taken from a population consisting of ten values (2, 3, 4, 5, 6, 7, 8, 9, 10, 11). Find the sampling distribution of $\bar{y}$, based on a random sample of size 2.

Solution One way to find the sampling distribution is by counting. There are 45 possible samples of two items selected from the ten items. These are shown here:

Sample	Value of $\bar{y}$	Sample	Value of $\bar{y}$	Sample	Value of $\bar{y}$
2, 3	2.5	3, 10	6.5	6, 7	6.5
2, 4	3	3, 11	7	6, 8	7
2, 5	3.5	4, 5	4.5	6, 9	7.5
2, 6	4	4, 6	5	6, 10	8
2, 7	4.5	4, 7	5.5	6, 11	8.5
2, 8	5	4, 8	6	7, 8	7.5
2, 9	5.5	4, 9	6.5	7, 9	8
2, 10	6	4, 10	7	7, 10	8.5
2, 11	6.5	4, 11	7.5	7, 11	9
3, 4	3.5	5, 6	5.5	8, 9	8.5
3, 5	4	5, 7	6	8, 10	9
3, 6	4.5	5, 8	6.5	8, 11	9.5
3, 7	5	5, 9	7	9, 10	9.5
3, 8	5.5	5, 10	7.5	9, 11	10
3, 9	6	5, 11	8	10, 11	10.5

Assuming each sample of size 2 is equally likely, it follows that the sampling distribution for $\bar{y}$ based on $n = 2$ observations selected from the population {2, 3, 4, 5, 6, 7, 8, 9, 10, 11} is as indicated here.

$\bar{y}$	$P(\bar{y})$	$\bar{y}$	$P(\bar{y})$
2.5	1/45	7	4/45
3	1/45	7.5	4/45
3.5	2/45	8	3/45
4	2/45	8.5	3/45
4.5	3/45	9	2/45
5	3/45	9.5	2/45
5.5	4/45	10	1/45
6	4/45	10.5	1/45
6.5	5/45		

The sampling distribution is shown as a graph in Figure 4.19. Note that the distribution is symmetric, with a mean of 6.5 and a standard deviation of approximately 2.0 (the range divided by 4).

FIGURE 4.19

Sampling distribution for $\bar{y}$

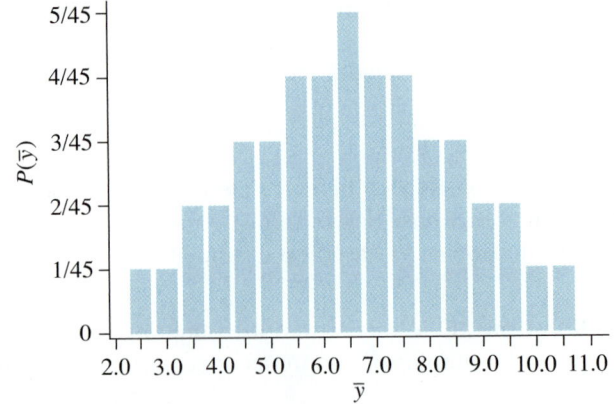

Example 4.19 illustrates for a very small population that we could in fact enumerate every possible sample of size 2 selected from the population and then compute all possible values of the sample mean. The next example will illustrate the properties of the sample mean, $\bar{y}$, when sampling from a larger population. This example will illustrate that the behavior of $\bar{y}$ as an estimator of μ depends on the sample size, n. Later in this chapter, we will illustrate the effect of the shape of the population distribution on the sampling distribution of $\bar{y}$.

EXAMPLE 4.20

In this example, the population values are known and, hence, we can compute the exact values of the population mean, μ, and population standard deviation, σ. We will then examine the behavior of $\bar{y}$ based on samples of size $n = 5$, 10, and 25 selected from the population. The population consists of 500 pennies from which we compute the age of each penny: Age = 2000 − Date on penny. The histogram of the 500 ages is displayed in Figure 4.20(a). The shape is skewed to the right with a very long right tail. The mean and standard deviation are computed to be $\mu = 13.468$ years and $\sigma = 11.164$ years. In order to generate the sampling distribution of $\bar{y}$ for $n = 5$, we would need to generate all possible samples of size $n = 5$ and then compute the $\bar{y}$ from each of these samples. This would be an enormous task since there are 255,244,687,600 possible samples of size 5 that could be selected from a population of 500 elements. The number of possible samples of size 10 or 25 is so large it makes even the national debt look small. Thus, we will use a computer program, *S-plus,* to select 25,000 samples of size 5 from the population of 500 pennies. For example, the first sample consists of pennies with ages 4, 12, 26, 16, and 9. The sample mean $\bar{y} = (4 + 12 + 26 + 16 + 9)/5 = 13.4$. We repeat 25,000 times the process of selecting 5 pennies, recording their ages, y_1, y_2, y_3, y_4, y_5, and then computing $\bar{y} = (y_1 + y_2 + y_3 + y_4 + y_5)/5$. The 25,000 values for $\bar{y}$ are then plotted in a frequency histogram, called the *sampling distribution* of $\bar{y}$ for $n = 5$. A similar procedure is followed for samples of size $n = 10$ and 25. The sampling distributions obtained are displayed in Figures 4.20(b)–(d).

Note that all three sampling distributions have nearly the same central value, approximately 13.5. (See Table 4.5.) The mean values of $\bar{y}$ for the three samples are nearly the same as the population mean, $\mu = 13.468$. In fact, if we had generated all possible samples for all three values of n, the mean of the possible values of $\bar{y}$ would agree exactly with μ.

The next characteristic to notice about the three histograms is their shape. All three are somewhat symmetric in shape, achieving a nearly normal distribution shape when $n = 25$. However, the histogram for $\bar{y}$ based on samples of size $n = 5$ is more spread out than the histogram based on $n = 10$, which, in turn, is more spread out than the histogram based on $n = 25$. When n is small, we are much more likely to obtain a value of $\bar{y}$ far from μ than when n is larger. What causes this increased dispersion in the values of $\bar{y}$? A single extreme y, either large or small relative to μ, in the sample has a greater influence on the size of $\bar{y}$ when n is small than when n is large. Thus, sample means based on small n are less accurate in their estimation of μ than their large-sample counterparts.

Table 4.5 contains summary statistics for the sampling distribution of $\bar{y}$. The sampling distribution of $\bar{y}$ has mean $\mu_{\bar{y}}$ and standard deviation $\sigma_{\bar{y}}$, which are related to the population mean, μ, and standard deviation, σ, by the following relationship:

$$\mu_{\bar{y}} = \mu \qquad \sigma_{\bar{y}} = \frac{\sigma}{\sqrt{n}}$$

FIGURE 4.20

(a) Histogram of ages for 500 pennies

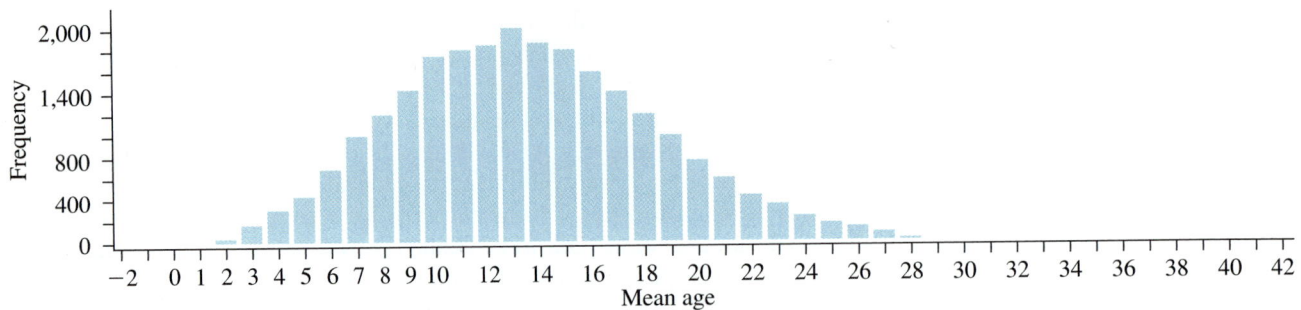

(b) Sampling distribution of $\bar{y}$ for $n = 5$

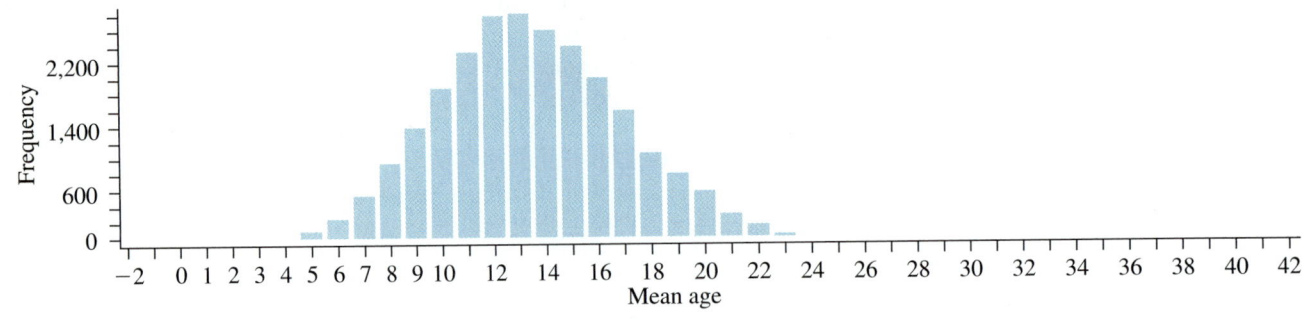

(c) Sampling distribution of $\bar{y}$ for $n = 10$

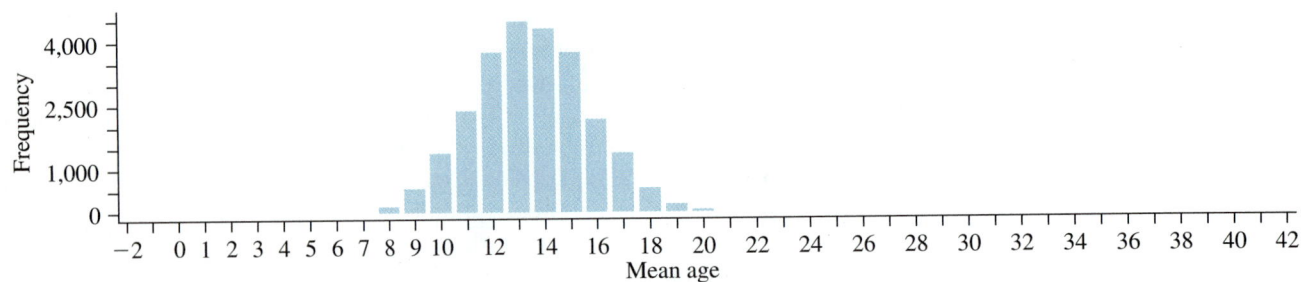

(d) Sampling distribution of $\bar{y}$ for $n = 25$

TABLE 4.5

Means and standard deviations for the sampling distributions of $\bar{y}$

Sample Size	Mean of $\bar{y}$	Standard Deviation of $\bar{y}$	$\dfrac{11.1638}{\sqrt{n}}$
1 (Population)	13.468 (μ)	11.1638 (σ)	11.1638
5	13.485	4.9608	4.9926
10	13.438	3.4926	3.5303
25	13.473	2.1766	2.2328

From Table 4.5, we note that the three sampling deviations have means that are approximately equal to the population mean. Also, the three sampling deviations have standard deviations that are approximately equal to $\sigma/\sqrt{n}$. If we had generated all possible values of $\bar{y}$, then the standard deviation of $\bar{y}$ would equal $\sigma/\sqrt{n}$ exactly. This quantity, $\sigma_{\bar{y}} = \sigma/\sqrt{n}$, is called the **standard error of $\bar{y}$.**

standard error of $\bar{y}$

Quite a few of the more common sample statistics, such as the sample median and the sample standard deviation, have sampling distributions that are nearly normal for moderately sized values of n. We can observe this behavior by computing the sample median and sample standard deviation from each of the three sets of 25,000 sample ($n = 5, 10, 25$) selected from the population of 500 pennies. The resulting sampling distributions are displayed in Figures 4.21(a)–(d), for the sample median, and Figures 4.22(a)–(d), for the sample standard deviation. The sampling distribution of both the median and the standard deviation are more highly skewed in comparison to the sampling distribution of the sample mean. In fact, the value of n at which the sampling distributions of the sample median and standard deviation have a nearly normal shape is much larger than the value required for the sample mean. A series of theorems in mathematical statistics called the **Central Limit Theorems** provide theoretical justification for our approximating the true sampling distribution of many sample statistics with the normal distribution. We will discuss one such theorem for the sample mean. Similar theorems exist for the sample median, sample standard deviation, and the sample proportion.

Central Limit Theorems

THEOREM 4.1

Central Limit Theorem for $\bar{y}$

Let $\bar{y}$ denote the sample mean computed from a random sample of n measurements from a population having a mean, μ, and finite standard deviation, σ. Let $\mu_{\bar{y}}$ and $\sigma_{\bar{y}}$ denote the mean and standard deviation of the sampling distribution of $\bar{y}$, respectively. Based on repeated random samples of size n from the population, we can conclude the following:

1. $\mu_{\bar{y}} = \mu$

2. $\sigma_{\bar{y}} = \sigma/\sqrt{n}$

3. When n is large, the sampling distribution of $\bar{y}$ will be approximately normal (with the approximation becoming more precise as n increases).

4. When the population distribution is normal, the sampling distribution of $\bar{y}$ is exactly normal for any sample size n.

FIGURE 4.21

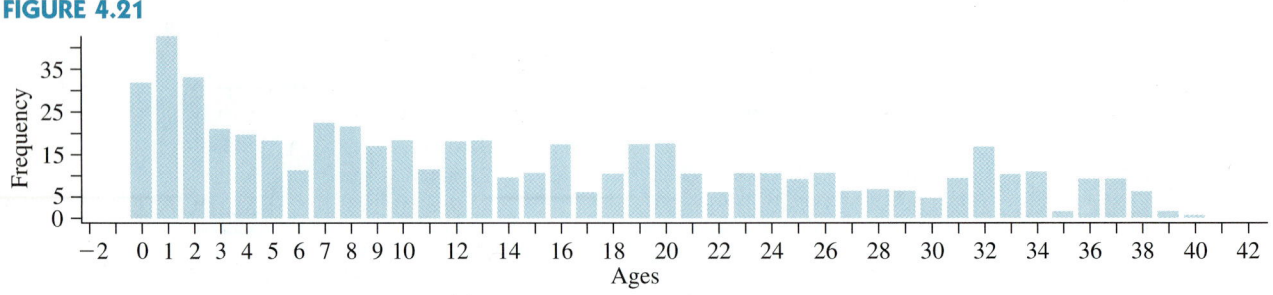

(a) Histogram of ages for 500 pennies

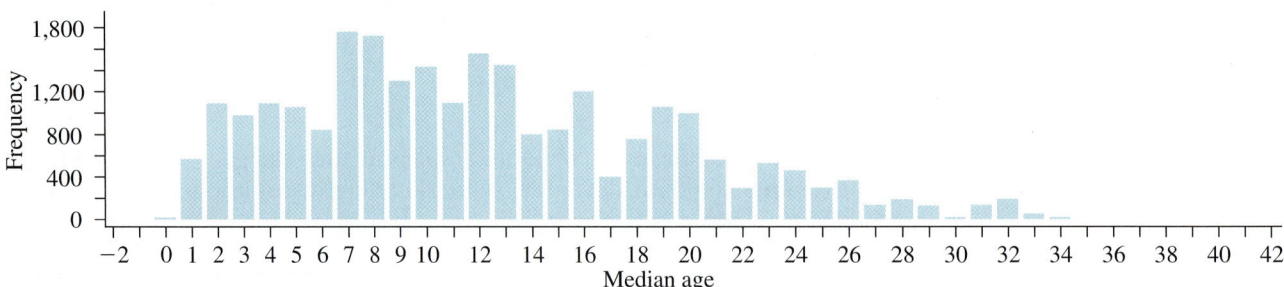

(b) Sampling distribution of median for $n = 5$

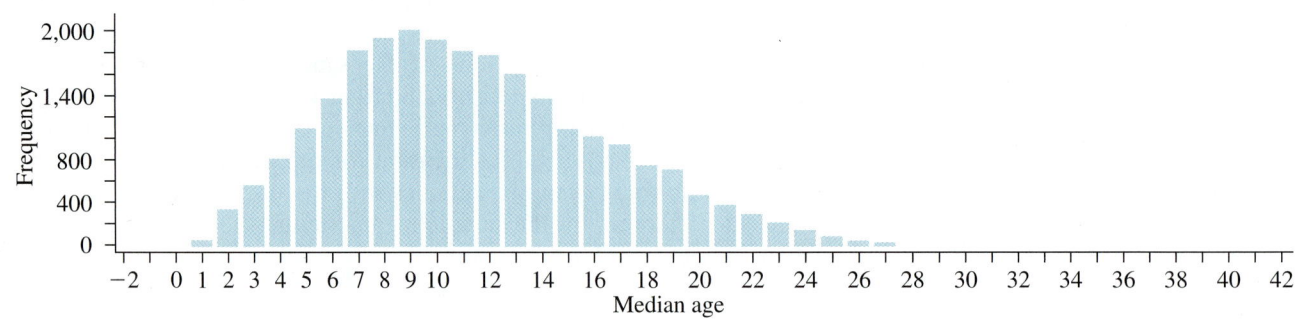

(c) Sampling distribution of median for $n = 10$

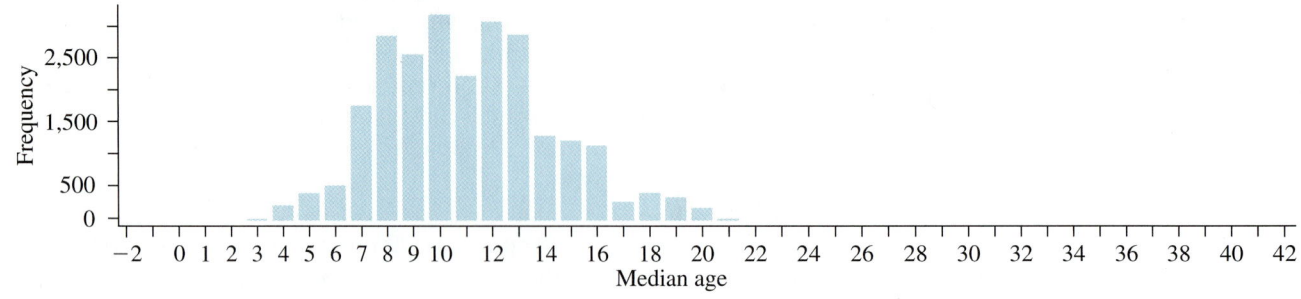

(d) Sampling distribution of median for $n = 25$

FIGURE 4.22

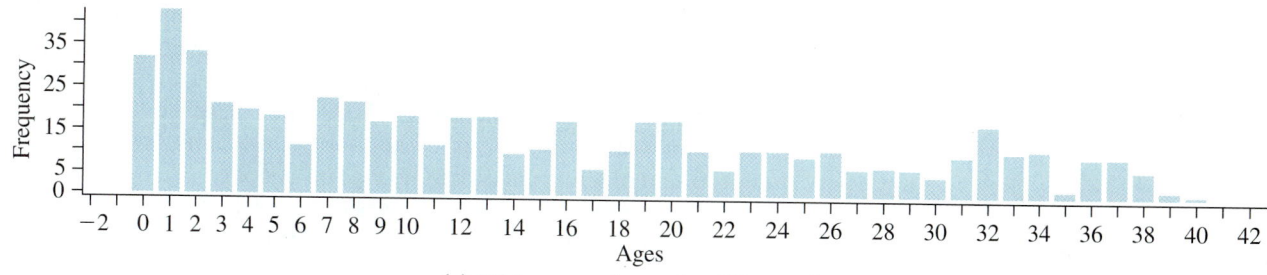

(a) Histogram of ages for 500 pennies

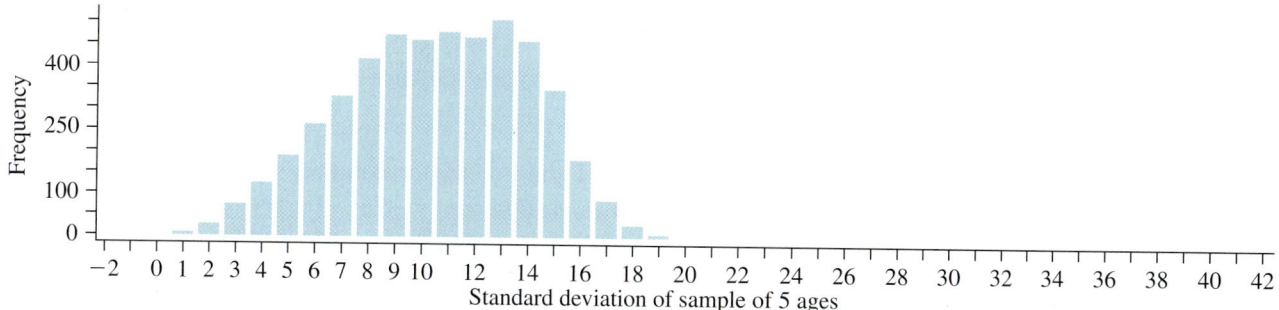

(b) Sampling distribution of standard deviation for $n = 5$

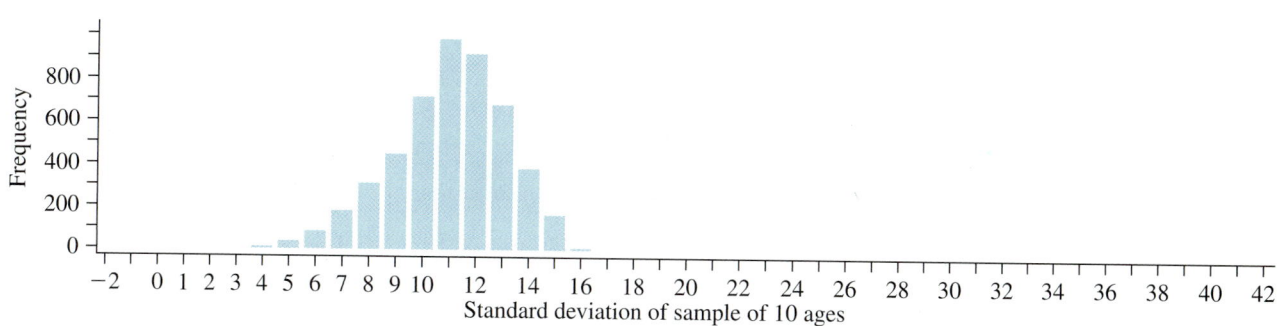

(c) Sampling distribution of standard deviation for $n = 10$

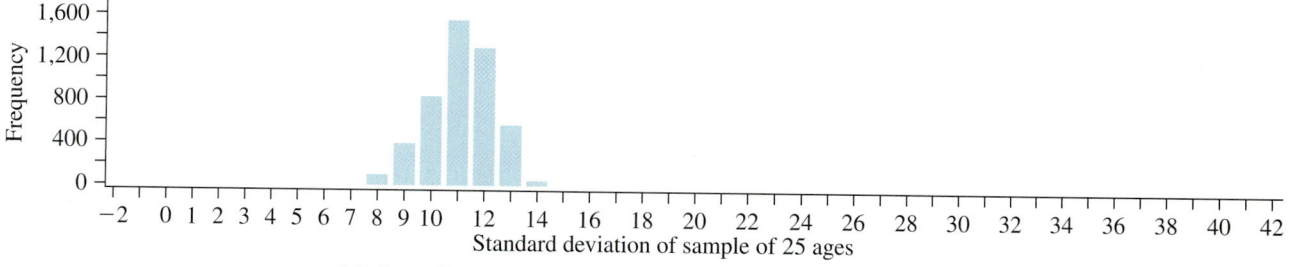

(d) Sampling distribution of standard deviation for $n = 25$

Figure 4.20 illustrates the Central Limit Theorem. Figure 4.20(a) displays the distribution of the measurements y in the population from which the samples are to be drawn. No specific shape was required for these measurements for the Central Limit Theorem to be validated. Figures 4.20(b)–(d) illustrate the sampling distribution for the sample mean $\bar{y}$ when n is 5, 10, and 25, respectively. We note that even for a very small sample size, $n = 10$, the shape of the sampling distribution of $\bar{y}$ is very similar to that of a normal distribution. This is not true in general. If the population distribution had many extreme values or several modes, the sampling distribution of $\bar{y}$ would require n to be considerably larger in order to achieve a symmetric bell shape.

We have seen that the sample size n has an effect on the shape of the sampling distribution of $\bar{y}$. The shape of the distribution of the population measurements also will affect the shape of the sampling distribution of $\bar{y}$. Figures 4.23 and 4.24 illustrate the effect of the population shape on the shape of the sampling distribution of $\bar{y}$. In Figure 4.23, the population measurements have a normal distribution. The sampling distribution of $\bar{y}$ is *exactly* a normal distribution for all values of n, as is illustrated for $n = 5$, 10, and 25 in Figure 4.23. When the population distribution is nonnormal, as depicted in Figure 4.24, the sampling distribution of $\bar{y}$ will not have a normal shape for small n (see Figure 4.24 with $n = 5$). However, for $n = 10$ and 25, the sampling distributions are nearly normal in shape, as can be seen in Figure 4.24.

FIGURE 4.23

Sampling distribution of $\bar{y}$ for $n = 5$, 10, 25 when sampling from a normal distribution

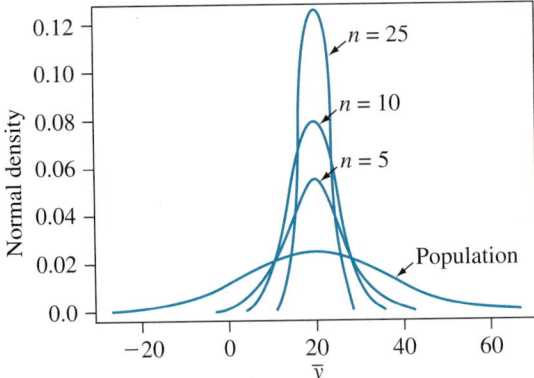

FIGURE 4.24

Sampling distribution of $\bar{y}$ for $n = 5$, 10, 25 when sampling from a skewed distribution

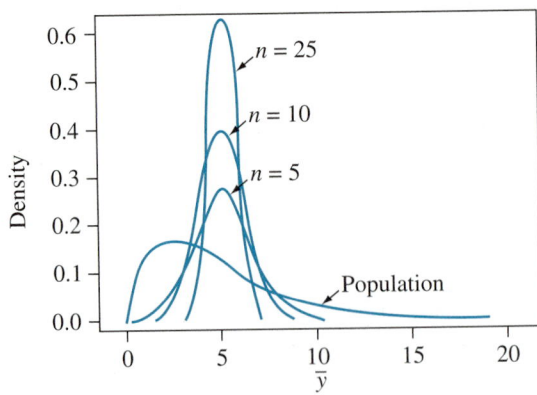

It is very unlikely that the exact shape of the population distribution will be known. Thus, the exact shape of the sampling distribution of $\bar{y}$ will not be

known either. The important point to remember is that the sampling distribution of $\bar{y}$ will be approximately normally distributed with a mean $\mu_{\bar{y}} = \mu$, the population mean, and a standard deviation $\sigma_{\bar{y}} = \sigma/\sqrt{n}$. The approximation will be more precise as n, the sample size for each sample, increases and as the shape of the population distribution becomes more like the shape of a normal distribution.

An obvious question is: How large should the sample size be for the Central Limit Theorem to hold? Numerous simulation studies have been conducted over the years and the results of these studies suggest that, in general, the Central Limit Theorem holds for $n > 30$. However, one should not apply this rule blindly. If the population is heavily skewed, the sampling distribution for $\bar{y}$ will still be skewed even for $n > 30$. On the other hand, if the population is symmetric, the Central Limit Theorem holds for $n < 30$.

Therefore, take a look at the data. If the sample histogram is clearly skewed, then the population will also probably be skewed. Consequently, a value of n much higher than 30 may be required to have the sampling distribution of $\bar{y}$ be approximately normal. Any inference based on the normality of $\bar{y}$ for $n = 30$ under this condition should be examined carefully.

As demonstrated in Figures 4.21 and 4.22, the Central Limit Theorem can be extended to many different sample statistics. The form of the Central Limit Theorem for the sample median and sample standard deviation is somewhat more complex than for the sample mean. Many of the statistics that we will encounter in later chapters will be either averages or sums of variables. The Central Limit Theorem for sums can be easily obtained from the Central Limit Theorem for the sample mean. Suppose we have a random sample of n measurements, $y_1, \ldots, y_n$, from a population and let $\Sigma y = y_1 + \cdots + y_n$.

THEOREM 4.2

Central Limit Theorem for Σy

Let Σy denote the sum of a random sample of n measurements from a population having a mean μ and finite standard deviation σ. Let $\mu_{\Sigma y}$ and $\sigma_{\Sigma y}$ denote the mean and standard deviation of the sampling distribution of Σy, respectively. Based on repeated random samples of size n from the population, we can conclude the following:

1. $\mu_{\Sigma y} = n\mu$
2. $\sigma_{\Sigma y} = \sqrt{n}\sigma$
3. When n is large, the sampling distribution of Σy will be approximately normal (with the approximation becoming more precise as n increases).
4. When the population distribution is normal, the sampling distribution of Σy is exactly normal for any sample size n.

Usually, a sample statistic is used as an estimate of a population parameter. For example, a sample mean $\bar{y}$ can be used to estimate the population mean μ from which the sample was selected. Similarly, a sample median and sample standard deviation estimate the corresponding population median and standard deviation. The sampling distribution of a sample statistic is then used to determine how accurate the estimate is likely to be. In Example 4.19, the population mean μ is known to be 6.5. Obviously, we do not know μ in any practical study or experiment. However, we can use the sampling distribution of $\bar{y}$ to determine the probability that the value of $\bar{y}$ for a random sample of $n = 2$ measurements from the population will be more than three units from μ. Using the data in Example 4.19, this probability is

$$P(2.5) + P(3) + P(10) + P(10.5) = \frac{4}{45}$$

In general, we would use the normal approximation from the Central Limit Theorem in making this calculation because the sampling distribution of a sample statistic is seldom known. This type of calculation will be developed in Chapter 5. Since a sample statistic is used to make inferences about a population parameter, the sampling distribution of the statistic is crucial in determining the accuracy of the inference.

interpretations of a sampling distribution   **Sampling distributions** can be **interpreted** in at least two ways. One way uses the long-run relative frequency approach. Imagine taking repeated samples of a fixed size from a given population and calculating the value of the sample statistic for each sample. In the long run, the relative frequencies for the possible values of the sample statistic will approach the corresponding sampling distribution probabilities. For example, if one took a large number of samples from the population distribution corresponding to the probabilities of Example 4.19 and, for each sample, computed the sample mean, approximately 9% would have $\bar{y} = 5.5$.

The other way to interpret a sampling distribution makes use of the classical interpretation of probability. Imagine listing all possible samples that could be drawn from a given population. The probability that a sample statistic will have a particular value (say, that $\bar{y} = 5.5$) is then the proportion of all possible samples that yield that value. In Example 4.19, $P(\bar{y} = 5.5) = 4/45$ corresponds to the fact that 4 of the 45 samples have a sample mean equal to 5.5. Both the repeated-sampling and the classical approach to finding probabilities for a sample statistic are legitimate.

In practice, though, a sample is taken only once, and only one value of the sample statistic is calculated. A sampling distribution is not something you can see in practice; it is not an empirically observed distribution. Rather, it is a theoretical concept, a set of probabilities derived from assumptions about the population and about the sampling method.

There's an unfortunate similarity between the phrase "sampling distribution," meaning the theoretically derived probability distribution of a statistic, and the phrase "sample distribution," which refers to the histogram of individual values actually observed in a particular sample. The two phrases mean very different things. To avoid confusion, we will refer to the distribution of sample values

sample histogram   as the **sample histogram** rather than as the sample distribution.

EXERCISES   **Basic Techniques**

4.87   A random sample of 16 measurements is drawn from a population with a mean of 60 and a standard deviation of 5. Describe the sampling distribution of $\bar{y}$, the sample mean. Within what interval would you expect $\bar{y}$ to lie approximately 95% of the time?

4.88   Refer to Exercise 4.87. Describe the sampling distribution for the sample sum $\Sigma\, y_i$. Is it unlikely (improbable) that $\Sigma\, y_i$ would be more than 70 units away from 960? Explain.

4.89   In Exercise 4.87, a random sample of 16 observations was to be selected from a population with $\mu = 60$ and $\sigma = 5$. Assume the original population of measurements is normally distributed. Use a computer program to simulate the sampling distribution of $\bar{y}$

based on 500 samples consisting of 16 observations each. A Minitab program that can be used in conducting the simulation is given in the last section of this chapter.

Applications

Psy.   **4.90** Psychomotor retardation scores for a large group of manic-depressive patients were approximately normal, with a mean of 930 and a standard deviation of 130.
  a. What fraction of the patients scored between 800 and 1,100?
  b. Less than 800?
  c. Greater than 1,200?

4.91 Refer to Exercise 4.90.
  a. Find the 90th percentile for the distribution of manic-depressive scores. (*Hint:* Solve for y in the expression $z = (y - \mu)/\sigma$, where z is the number of standard deviations the 90th percentile lies above the mean μ.)
  b. Find the interquartile range.

Soc.   **4.92** Federal resources have been tentatively approved for the construction of an out-patient clinic. In order to design a facility that will handle patient load requirements and stay within a limited budget, the designers studied patient demand. From studying a similar facility in the area, they found that the distribution of the number of patients requiring hospitalization during a week could be approximated by a normal distribution with a mean of 125 and a standard deviation of 32.
  a. Use the Empirical Rule to describe the distribution of y, the number of patients requesting service in a week.
  b. If the facility was built with a 160-patient capacity, what fraction of the weeks might the clinic be unable to handle the demand?

4.93 Refer to Exercise 4.92. What size facility should be built so the probability of the patient load's exceeding the clinic capacity is .05? .01?

Soc.   **4.94** Based on the 1990 census, the number of hours per day adults spend watching television is approximately normally distributed with a mean of 5 hours and a standard deviation of 1.3 hours.
  a. What proportion of the population spends more than 7 hours per day watching television?
  b. In a 1998 study of television viewing, a random sample of 500 adults reported that the average number of hours spent viewing television was greater than 5.5 hours per day. Do the results of this survey appear to be consistent with the 1990 census? (*Hint:* If the census results are still correct, what is the probability that the average viewing time would exceed 5.5 hours?)

Env.   **4.95** The level of a particular pollutant, nitrogen oxide, in the exhaust of a hypothetical model of car, the Polluter, when driven in city traffic has approximately a normal distribution with a mean level of 2.1 grams per mile (g/m) and a standard deviation of 0.3 g/m.
  a. If the EPA mandates that a nitrogen oxide level of 2.7 g/m cannot be exceeded, what proportion of Polluters would be in violation of the mandate?
  b. At most, 25% of Polluters exceed what nitrogen oxide level value (that is, find the 75th percentile)?
  c. The company producing the Polluter must reduce the nitrogen oxide level so that at most 5% of its cars exceed the EPA level of 2.7 g/m. If the standard deviation remains 0.3 g/m, to what value must the mean level be reduced so that at most 5% of Polluters would exceed 2.7 g/m?

4.96 Refer to Exercise 4.95. A company has a fleet of 150 Polluters used by its sales staff. Describe the distribution of the total amount, in g/m, of nitrogen oxide produced in the exhaust of this fleet. What are the mean and standard deviation of the total amount, in g/m, of nitrogen oxide in the exhaust for the fleet? (*Hint:* The total amount of nitrogen oxide can be represented as $\sum_{i=1}^{150} W_i$, where W_i is the amount of nitrogen oxide in the exhaust of the ith car. Thus, the Central Limit Theorem for sums is applicable.)

Soc.   **4.97**   The baggage limit for an airplane is set at 100 pounds per passenger. Thus, for an airplane with 200 passenger seats there would be a limit of 20,000 pounds. The weight of the baggage of an individual passenger is a random variable with a mean of 95 pounds and a standard deviation of 35 pounds. If all 200 seats are sold for a particular flight, what is the probability that the total weight of the passengers' baggage will exceed the 20,000-pound limit?

Med.   **4.98**   A patient visits her doctor with concerns about her blood pressure. If the systolic blood pressure exceeds 150, the patient is considered to have high blood pressure and medication may be prescribed. The problem is that there is a considerable variation in a patient's systolic blood pressure readings during a given day.

   a. If a patient's systolic readings during a given day have a normal distribution with a mean of 160 mm mercury and a standard deviation of 20 mm, what is the probability that a single measurement will fail to detect that the patient has high blood pressure?

   b. If five measurements are taken at various times during the day, what is the probability that the average blood pressure reading will be less than 150 and hence fail to indicate that the patient has a high blood pressure problem?

   c. How many measurements would be required so that the probability is at most 1% of failing to detect that the patient has high blood pressure?

4.13   Normal Approximation to the Binomial

A binomial random variable y was defined earlier to be the number of successes observed in n independent trials of a random experiment in which each trial resulted in either a success (S) or a failure (F) and $P(S) = \pi$ for all n trials. We will now demonstrate how the Central Limit Theorem for sums enables us to calculate probabilities for a binomial random variable by using an appropriate normal curve as an approximation to the binomial distribution. We said in Section 4.8 that probabilities associated with values of y can be computed for a binomial experiment for any values of n or π, but the task becomes more difficult when n gets large. For example, suppose a sample of 1,000 voters is polled to determine sentiment toward the consolidation of city and county government. What would be the probability of observing 460 or fewer favoring consolidation if we assume that 50% of the entire population favor the change? Here we have a binomial experiment with $n = 1,000$ and π, the probability of selecting a person favoring consolidation, equal to .5. To determine the probability of observing 460 or fewer favoring consolidation in the random sample of 1,000 voters, we could compute $P(y)$ using the binomial formula for $y = 460, 459, \ldots, 0$. The desired probability would then be

$$P(y = 460) + P(y = 459) + \cdots + P(y = 0)$$

There would be 461 probabilities to calculate with each one being somewhat difficult because of the factorials. For example, the probability of observing 460 favoring consolidation is

$$P(y = 460) = \frac{1000!}{460!540!}(.5)^{460}(.5)^{540}$$

A similar calculation would be needed for all other values of y.

   To justify the use of the Central Limit Theorem, we need to define n random variables, $I_1, \ldots, I_n$, by

$$I_i = \begin{cases} 1 & \text{if the } i\text{th trial results in a success} \\ 0 & \text{if the } i\text{th trial results in a failure} \end{cases}$$

The binomial random variable y is the number of successes in the n trials. Now, consider the sum of the random variable $I_1, \ldots, I_n$, $\sum_{i=1}^{n} I_i$. A 1 is placed in the sum for each S that occurs and a 0 for each F that occurs. Thus, $\sum_{i=1}^{n} I_i$ is the number of S's that occurred during the n trials. Hence, we conclude that $y = \sum_{i=1}^{n} I_i$. Because the binomial random variable y is the sum of independent random variables, each having the same distribution, we can apply the Central Limit Theorem for sums to y. Thus, the normal distribution can be used to approximate the binomial distribution when n is of an appropriate size. The normal distribution that will be used has a mean and standard deviation given by the following formula:

$$\mu = n\pi \qquad \sigma = \sqrt{n\pi(1 - \pi)}$$

These are the mean and standard deviation of the binomial random variable y.

EXAMPLE 4.21

Use the normal approximation to the binomial to compute the probability of observing 460 or fewer in a sample of 1,000 favoring consolidation if we assume that 50% of the entire population favor the change.

Solution The normal distribution used to approximate the binomial distribution will have

$$\mu = n\pi = 1,000(.5) = 500$$
$$\sigma = \sqrt{n\pi(1 - \pi)} = \sqrt{1,000(.5)(.5)} = 15.8$$

The desired probability is represented by the shaded area shown in Figure 4.25. We calculate the desired area by first computing

$$z = \frac{y - \mu}{\sigma} = \frac{460 - 500}{15.8} = -2.53$$

FIGURE 4.25

Approximating normal distribution for the binomial distribution, $\mu = 500$ and $\sigma = 15.8$

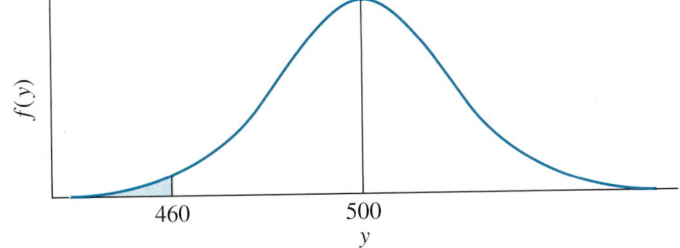

Referring to Table 1 in the Appendix, we find that the area under the normal curve to the left of 460 (for $z = -2.53$) is .0057. Thus, the probability of observing 460 or fewer favoring consolidation is approximately .0057.

The normal approximation to the binomial distribution can be unsatisfactory if $n\pi < 5$ or $n(1 - \pi) < 5$. If π, the probability of success, is small, and n, the sample size, is modest, the actual binomial distribution is seriously skewed to the right. In such a case, the symmetric normal curve will give an unsatisfactory approximation. If π is near 1, so $n(1 - \pi) < 5$, the actual binomial will be skewed to the left, and again the normal approximation will not be very accurate. The normal approximation, as described, is quite good when $n\pi$ and $n(1 - \pi)$ exceed about 20. In the middle zone, $n\pi$ or $n(1 - \pi)$ between 5 and 20, a modification

continuity correction   called a **continuity correction** makes a substantial contribution to the quality of the approximation.

    The point of the continuity correction is that we are using the continuous normal curve to approximate a discrete binomial distribution. A picture of the situation is shown in Figure 4.26.

FIGURE 4.26

Normal approximation to binomial

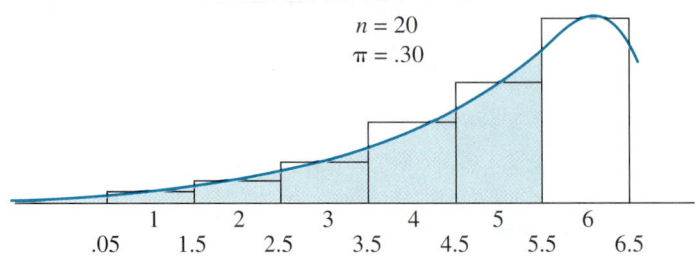

$$n = 20$$
$$\pi = .30$$

	1	2	3	4	5	6
.05	1.5	2.5	3.5	4.5	5.5	6.5

    The binomial probability that $y \leq 5$ is the sum of the areas of the rectangle above 5, 4, 3, 2, 1, and 0. This probability (area) is approximated by the area under the superimposed normal curve to the left of 5. Thus, the normal approximation ignores half of the rectangle above 5. The continuity correction simply includes the area between $y = 5$ and $y = 5.5$. For the binomial distribution with $n = 20$ and $\pi = .30$ (pictured in Figure 4.26), the correction is to take $P(y \leq 5)$ as $P(y \leq 5.5)$. Instead of

$$P(y \leq 5) = P[z \leq (5 - 20(.3))/\sqrt{20(.3)(.7)}] = P(z \leq -.49) = .3121$$

use

$$P(y \leq 5.5) = P[z \leq (5.5 - 20(.3))/\sqrt{20(.3)(.7)}] = P(z \leq -.24) = .4052$$

The actual binomial probability can be shown to be .4164. The general idea of the continuity correction is to add or subtract .5 from a binomial value before using normal probabilities. The best way to determine whether to add or subtract is to draw a picture like Figure 4.26.

Normal Approximation to the Binomial Probability Distribution

For large n and π not too near 0 or 1, the distribution of a binomial random variable y may be approximated by a normal distribution with $\mu = n\pi$ and $\sigma = \sqrt{n\pi (1 - \pi)}$. This approximation should be used only if $n\pi \geq 5$ and $n(1- \pi) \geq 5$. A continuity correction will improve the quality of the approximation in cases in which n is not overwhelmingly large.

EXAMPLE 4.22

A large drug company has 100 potential new prescription drugs under clinical test. About 20% of all drugs that reach this stage are eventually licensed for sale. What is the probability that at least 15 of the 100 drugs are eventually licensed? Assume that the binomial assumptions are satisfied, and use a normal approximation with continuity correction.

Solution   The mean of y is $\mu = 100(.2) = 20$; the standard deviation is $\sigma = \sqrt{100(.2)(.8)} = 4.0$. The desired probability is that 15 or more drugs are approved.

Because $y = 15$ is included, the continuity correction is to take the event as y greater than or equal to 14.5.

$$P(y \geq 14.5) = P\left(z \geq \frac{14.5 - 20}{4.0}\right) = P(z \geq -1.38) = 1 - P(z < -1.38)$$

$$= 1 - .0838 = .9162$$

4.14 Minitab Instructions

Generating Random Numbers

To generate 1,000 random numbers from the set $[0, 1, \ldots, 9]$,

1. Click on **Calc,** then **Random Data,** then **Integer.**
2. Type the number of rows of data: **Generate 20 rows of data.**
3. Type the columns in which the data are to be stored: **Store in column(s): c1—c50.**
4. Type the first number in the list: **Minimum value: 0.**
5. Type the last number in the list: **Maximum value: 9.**
6. Click on **OK.**

Note that we have generated (20) (50) = 1,000 random numbers.

Calculating Binomial Probabilities

To calculate binomial probabilities when $n = 10$ and $\pi = 0.6$,

1. Enter the values of x in column c1: **0, 1, 2, 3, 4, 5, 6, 7, 8, 9, 10.**
2. Click on **Calc,** then **Probability Distributions,** then **Binomial.**
3. Select either **Probability** [to compute $P(X = x)$] or **Cumulative probability** [to compute $P(X \leq x)$].
4. Type the value of n: **Number of trials: 10.**
5. Type the value of π: **Probability of success: 0.6.**
6. Click on **Input column.**
7. Type the column number where values of x are located: **C1.**
8. Click on **Optional storage.**
9. Type the column number to store probability: **C2.**
10. Click on **OK.**

Calculating Normal Probabilities

To calculate $P(X \leq 18)$ when X is normally distributed with $\mu = 23$ and $\sigma = 5$,

1. Click on **Calc,** then **Probability Distributions,** then **Normal.**
2. Click on **Cumulative probability.**
3. Type the value of μ: **Mean: 23.**
4. Type the value of σ: **Standard deviation: 5.**
5. Click on **Input constant.**
6. Type the value of x: **18.**
7. Click on **OK.**

Generating Sampling Distribution of $\bar{y}$

To create the sampling distribution of $\bar{y}$ based on 500 samples of size $n = 16$ from a normal distribution with $\mu = 60$ and $\sigma = 5$,

1. Click on **Calc,** then **Random Data,** then **Normal**.
2. Type the number of samples: **Generate 500 rows**.
3. Type the sample size n in terms of number of columns: **Store in column(s) c1-c16**.
4. Type in the value of μ: **Mean: 60**.
5. Type in the value of σ: **Standard deviation: 5**.
6. Click on **OK**. There are now 500 rows in columns c1–c16, 500 samples of 16 values each to generate 500 values of $\bar{y}$.
7. Click on **Calc,** then **Row Statistics,** then **mean**.
8. Type in the location of data: **Input Variables c1-c16**.
9. Type in the column in which the 500 means will be stored: **Store Results in c17**.
10. To obtain the mean of the 500 $\bar{y}$s, click on **Calc,** then **Column Statistics,** then **mean**.
11. Type in the location of the 500 means: **Input Variables c17**.
12. Click on **OK**.
13. To obtain the standard deviation of the 500 $\bar{y}$s, click on **Calc,** then **Column Statistics,** then **standard deviation**.
14. Type in the location of the 500 means: **Input Variables c17**.
15. Click on **OK**.
16. To obtain the sampling distribution of $\bar{y}$, click **Graph,** then **Histogram**.
17. Type **c17** in the Graph box.
18. Click on **OK**.

4.15 Summary

In this chapter, we presented an introduction to probability, probability distributions, and sampling distributions. Knowledge of the probabilities of sample outcomes is vital to a statistical inference. Three different interpretations of the probability of an outcome were given: the classical, relative frequency, and subjective interpretations. Although each has a place in statistics, the relative frequency approach has the most intuitive appeal because it can be checked.

Quantitative random variables are classified as either discrete or continuous random variables. The probability distribution for a discrete random variable y is a display of the probability $P(y)$ associated with each value of y. This display may be presented in the form of a histogram, table, or formula.

The binomial is a very important and useful discrete random variable. Many experiments that scientists conduct are similar to a coin-tossing experiment where dichotomous (yes–no) type data are accumulated. The binomial experiment frequently provides an excellent model for computing probabilities of various sample outcomes.

Probabilities associated with a continuous random variable correspond to areas under the probability distribution. Computations of such probabilities were illustrated for areas under the normal curve. The importance of this exercise is borne out by the Central Limit Theorem: Any random variable that is expressed as a sum or average of a random sample from a population having a finite standard deviation will have a normal distribution for a sufficiently large sample size. Direct application of the Central Limit Theorem gives the sampling distribution for the sample mean. Because many sample statistics are either sums or averages of random variables, application of the Central Limit Theorem provides us with information about probabilities of sample outcomes. These probabilities are vital for the statistical inferences we wish to make.

Key Formulas

1. Binomial probability distribution

$$P(y) = \frac{n!}{y!(n - y)!}\pi^y(1 - \pi)^{n-y}$$

2. Sampling distribution for $\bar{y}$

Mean: μ

Standard error: $\sigma_{\bar{y}} = \sigma/\sqrt{n}$

3. Normal approximation to the binomial

$$\mu = n\pi \qquad \sigma = \sqrt{n\pi(1 - \pi)}$$

provided that $n\pi$ and $n(1 - \pi)$ are greater than or equal to 5 or, equivalently, if

$$n \geq \frac{5}{\min(\pi, 1 - \pi)}$$

Supplementary Exercises

Bus. **4.99** One way to audit expense accounts for a large consulting firm is to sample all reports dated the last day of each month. Comment on whether such a sample constitutes a random sample.

Bus. **4.100** Critical key-entry errors in the data processing operation of a large district bank occur approximately .1% of the time. If a random sample of 10,000 entries is examined, determine the following:
   a. the expected number of errors
   b. the probability of observing fewer than five errors
   c. the probability of observing fewer than two errors

4.101 Use the binomial distribution with $n = 20$, $\pi = .5$ to compare accuracy of the normal approximation to the binomial.
   a. Compute the exact probabilities and corresponding normal approximations for $y < 5$.
   b. The normal approximation can be improved slightly by taking $P(y \leq 4.5)$. Why should this help? Compare your results.
   c. Compute the exact probabilities and corresponding normal approximations with the continuity correction for $P(8 < y < 14)$.

4.102 Let y be a binomial random variable with $n = 10$ and $\pi = .5$.
   a. Calculate $P(4 \leq y \leq 6)$.
   b. Use a normal approximation without the continuity correction to calculate the same probability. Compare your results. How well did the normal approximation work?

4.103 Refer to Exercise 4.102. Use the continuity correction to compute the probability $P(4 \leq y \leq 6)$. Does the continuity correction help?

Bus. **4.104** A marketing research firm believes that approximately 25% of all persons mailed a sweepstakes offer will respond if a preliminary mailing of 5,000 is conducted in a fixed region.
   a. What is the probability that 1,000 or fewer will respond?
   b. What is the probability that 3,000 or more will respond?

Engin.   **4.105**   The breaking strengths for 1-foot-square samples of a particular synthetic fabric are approximately normally distributed with a mean of 2,250 pounds per square inch (psi) and a standard deviation of 10.2 psi.

>   **a.** Find the probability of selecting a 1-foot-square sample of material at random that on testing would have a breaking strength in excess of 2,265 psi.
>   **b.** Describe the sampling distribution for $\bar{y}$ based on random samples of 15 1-foot sections.

4.106   Refer to Exercise 4.105. Suppose that a new synthetic fabric has been developed that may have a different mean breaking strength. A random sample of 15 one-foot sections is obtained and each section is tested for breaking strength. If we assume that the population standard deviation for the new fabric is identical to that for the old fabric, give the standard deviation for the sampling distribution of $\bar{y}$ using the new fabric.

4.107   Refer to Exercise 4.106. Suppose that the mean breaking strength for the sample of 15 one-foot sections of the new synthetic fabric is 2,268 psi. What is the probability of observing a value of $\bar{y}$ equal to or greater than 2,268, assuming that the mean breaking strength for the new fabric is 2,250, the same as that for the old?

4.108   Based on your answer in Exercise 4.107, do you believe the new fabric has the same mean breaking strength as the old? (Assume $\sigma = 10.2$.)

4.109   In Figure 4.19, we visually inspected the relative frequency histogram for sample means based on two measurements and noted its bell shape. Another way to determine whether a set of measurements is bell-shaped (normal) is to construct a **normal probability plot** of the sample data. If the plotted points are nearly a straight line, we say the measurements were selected from a normal population. We can generate a normal probability plot using the following Minitab code. If the plotted points fall within the curved dotted lines, we consider the data to be a random sample from a normal distribution.

Minitab code:

>   **1.** Enter the 45 measurements into C1 of the data spreadsheet.
>   **2.** Click on **Graph**, then **Probability Plot**.
>   **3.** Type **c1** in the box labeled **Variables:**.
>   **4.** Click on **OK**.

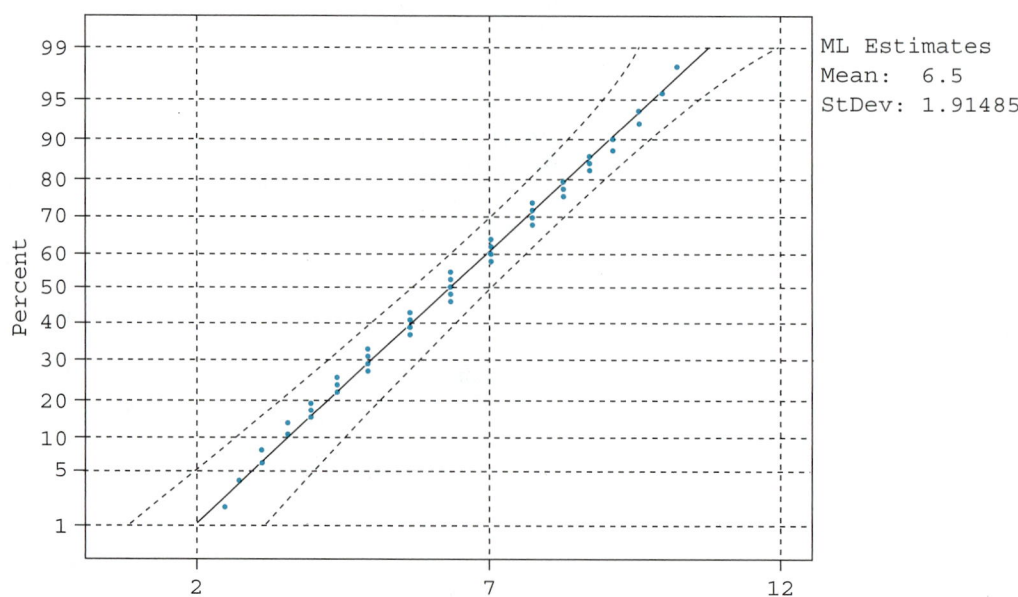

The 45 data values clearly fall within the two curved lines and fairly close to the straight line. Thus, we conclude there is a strong indication that the data values follow a normal distribution.

a. Suppose our population was the ten measurements (2, 3, 6, 8, 9, 12, 25, 29, 39, 50). Generate the 45 sample means based on $n = 2$ observations per sample and determine whether the sampling distribution of the sample mean is approximately normally distributed by constructing a histogram and normal probability plot of the 45 sample means.

b. Why do you think the plots for the means from the population (2, 3, 6, 8, 9, 12, 25, 29, 39, 50) differ greatly from the plots obtained for the means from the population (2, 3, 4, 5, 6, 7, 8, 9, 10, 11)?

H.R. **4.110** A labor union's examining board for the selection of apprentices has a record of admitting 70% of all applicants who satisfy a set of basic requirements. Five members of a minority group recently came before the board, and four of five were rejected. Find the probability that one or fewer would be accepted if the record is really .7. Did the board apply a lower probability of acceptance when reviewing the five members of the minority group?

Gov. **4.111** Suppose that you are a regional director of the IRS office and that you are charged with sampling 1% of the returns with gross income levels above $15,000. How might you go about this? Would you use random sampling? How?

Med. **4.112** Experts consider high serum cholesterol levels to be associated with an increased incidence of coronary heart disease. Suppose that the logarithm of cholesterol levels for males in a given age bracket are normally distributed with a mean of 2.35 and a standard deviation of 1.2.

a. What percentage of the males in this age bracket could be expected to have a serum cholesterol level greater than 250 mg/ml, the upper limit of the clinical normal range?

b. What percentage of the males could be expected to have serum cholesterol levels within the clinical normal range of 150–250 mg/ml?

c. If levels above 300 mg/ml are considered very risky, what percentage of the adult males in this age bracket could be expected to exceed 300?

Bus. **4.113** One of the major soft-drink companies changed the secret formula for its leading beverage to attract new customers. Recently, a marketing research firm interviewed 1,000 potential new customers and, after giving them a taste of the newly reformulated beverage, determined the number of these individuals planning to buy the reformulated beverage in the near future.

a. Identify the random variable for the population of y = values of interest.

b. Can you compute the mean and variance? Why or why not?

c. How would you calculate $P(y \leq 250)$?

Bus. **4.114** Many firms are using or exploring the possibility of using telemarketing techniques—that is, marketing their products via the telephone to supplement the more traditional marketing strategies. Assume a firm finds that approximately 1 in every 100 calls yields a sale.

a. Find the probability the first sale will occur somewhere in the first 5 calls.

b. Find the probability the first sale will occur sometime after 10 calls.

Bus. **4.115** Marketing analysts have determined that a particular advertising campaign should make at least 20% of the adult population aware of the advertised product. After a recent campaign, 25 of 400 adults sampled indicated that they had seen the ad and were aware of the new product.

a. Find the approximate probability of observing $y \leq 25$ given that 20% of the population is aware of the product through the campaign.

b. Based on your answer to part (a), does it appear the ad was successful? Explain.

Med. **4.116** One or more specific, minor birth defects occurs with probability .0001 (that is, 1 in 10,000 births). If 20,000 babies are born in a given geographic area in a given year, can we calculate the probability of observing at least one of the minor defects using the binomial or normal approximation to the binomial? Explain.

4.117 The sample mean to be calculated from a random sample of size $n = 4$ from a population consists of the eight measurements (2, 6, 9, 12, 25, 29, 39, 50). Find the sampling distribution of $\bar{y}$. (*Hint:* There are 70 samples of size 4 when sampling from a population of eight measurements.)

4.118 Plot the sampling distribution of $\bar{y}$ from Exercise 4.115.
   a. Does the sampling distribution appear to be approximately normal?
   b. Verify that the mean of the sampling distribution of $\bar{y}$ equals the mean of the eight population values.

4.119 Refer to Exercise 4.117. Use the same population to find the sampling distribution for the sample median based on samples of size $n = 4$.

4.120 Plot the sampling distribution of the sample median of Exercise 4.119.
   a. Does the sampling distribution appear to be approximately normal?
   b. Compute the mean of the sampling distribution of the sample median and compare this value to the population median.

4.121 Random samples of size 5, 20, and 80 are drawn from a population with mean $\mu = 100$ and standard deviation $\sigma = 15$.
   a. Give the mean of the sampling distribution of $\bar{y}$ for each of the sample sizes 5, 20, and 80.
   b. Give the standard deviation of the sampling distribution of $\bar{y}$ for each of the sample sizes 5, 20, and 80.
   c. Based on the results obtained in parts (a) and (b), what do you conclude about the accuracy of using the sample mean $\bar{y}$ as an estimate of population mean μ?

4.122 Refer to Exercise 4.121. To evaluate how accurately the sample mean $\bar{y}$ estimates the population mean μ, we need to know the chance of obtaining a value of $\bar{y}$ that is far from μ. Suppose it is important that the sample mean $\bar{y}$ is within 5 units of the population mean μ. Find the following probabilities for each of the three sample sizes and comment on the accuracy of using $\bar{y}$ to estimate μ.
   a. $P(\bar{y} \geq 105)$
   b. $P(\bar{y} \leq 95)$
   c. $P(95 \leq \bar{y} \geq 105)$

4.123 A random sample of $n = 36$ measurements is selected from a population with a mean equal to 40 and a standard deviation equal to 12.
   a. Describe the sampling distribution of $\bar{y}$.
   b. Find $P(\bar{y} > 36)$.
   c. Find $P(\bar{y} < 30)$.
   d. Find the value of $\bar{y}$ (say, k) such that $P(\bar{y} > k) = .05$.

4.124 Refer to Exercise 4.123.
   a. Describe the sampling distribution for the sample sum $\Sigma\, y_i$.
   b. Find $P(\Sigma\, y_i > 1{,}440)$.
   c. Find $P(\Sigma\, y_i > 1{,}540)$.
   d. Find the value of $\Sigma\, y_i$ (say, k) such that $P[k < \Sigma\, y_i < k_2] = .95$.

4.125 For each of the following situations, find the expected value and standard error of $\bar{y}$ based on a random sample of size n drawn from a population with mean μ and standard deviation σ.
   a. $n = 25$, $\mu = 10$, $\sigma = 10$
   b. $n = 100$, $\mu = 10$, $\sigma = 10$
   c. $n = 25$, $\mu = 10$, $\sigma = 20$
   d. $n = 100$, $\mu = 10$, $\sigma = 20$

4.126 Based on the results of Exercise 4.125, speculate on the effect of increasing the sample size and on the effect of an increase in σ on the standard error of $\bar{y}$.

PART 5

Analyzing Data: Central Values, Variances, and Proportions

5 Inferences about Population Central Values

6 Comparing Two Population Central Values

7 Inferences about Population Variances

8 Inferences about More than Two Population Central Values

9 Multiple Comparisons

10 Categorical Data Analysis

CHAPTER 5

Inferences about Population Central Values

5.1 Introduction and Case Study

5.2 Estimation of μ

5.3 Choosing the Sample Size for Estimating μ

5.4 A Statistical Test for μ

5.5 Choosing the Sample Size for Testing μ

5.6 The Level of Significance of a Statistical Test

5.7 Inferences about μ for Normal Population, σ Unknown

5.8 Inferences about Median

5.9 Summary

5.1 Introduction and Case Study

Inference, specifically decision making and prediction, is centuries old and plays a very important role in our lives. Each of us faces daily personal decisions and situations that require predictions concerning the future. The U.S. government is concerned with the balance of trade with countries in Europe and Asia. An investment advisor wants to know whether inflation will be rising in the next six months. A metallurgist would like to use the results of an experiment to determine whether a new light-weight alloy possesses the strength characteristics necessary for use in automobile manufacturing. A veterinarian investigates the effectiveness of a new chemical for treating heartworm in dogs. The inferences that these individuals make should be based on relevant facts, which we call observations, or data.

In many practical situations, the relevant facts are abundant, seemingly inconsistent, and, in many respects, overwhelming. As a result, a careful decision or prediction is often little better than an outright guess. You need only refer to the "Market Views" section of the *Wall Street Journal* or one of the financial news shows on cable TV to observe the diversity of expert opinion concerning future stock market behavior. Similarly, a visual analysis of data by scientists and engineers often yields conflicting opinions regarding conclusions to be drawn from an experiment.

Many individuals tend to feel that their own built-in inference-making equipment is quite good. However, experience suggests that most people are incapable of utilizing large amounts of data, mentally weighing each bit of relevant information, and arriving at a good inference. (You may test your own inference-making ability by using the exercises in Chapters 5 through 10. Scan the data and make an inference before you use the appropriate statistical procedure. Then compare the results.) The statistician, rather than relying upon his or her own intuition, uses statistical results to aid in making inferences. Although we touched on some of the notions involved in statistical inference in preceding chapters, we will now collect our ideas in a presentation of some of the basic ideas involved in statistical inference.

The objective of statistics is to make inferences about a population based on information contained in a sample. Populations are characterized by numerical descriptive measures called *parameters*. Typical population parameters are the mean μ, the median M, the standard deviation σ, and a proportion π. Most inferential problems can be formulated as an inference about one or more parameters of a population. For example, a study is conducted by the Wisconsin Education Department to assess the reading ability of children in the primary grades. The population consists of the scores on a standard reading test of all children in the primary grades in Wisconsin. We are interested in estimating the value of the population mean score μ and the proportion π of scores below a standard, which designates that a student needs remedial assistance.

Methods for making inferences about parameters fall into one of two categories. Either we will **estimate** (predict) the value of the population parameter of interest or we will **test a hypothesis** about the value of the parameter. These two methods of statistical inference—estimation and hypothesis testing—involve different procedures, and, more important, they answer two different questions about the parameter. In estimating a population parameter, we are answering the question, "What is the value of the population parameter?" In testing a hypothesis, we are answering the question, "Is the parameter value equal to this specific value?"

Consider a study in which an investigator wishes to examine the effectiveness of a drug product in reducing anxiety levels of anxious patients. The investigator uses a screening procedure to identify a group of anxious patients. After the patients are admitted into the study, each one's anxiety level is measured on a rating scale immediately before he or she receives the first dose of the drug and then at the end of one week of drug therapy. These sample data can be used to make inferences about the population from which the sample was drawn either by estimation or by a statistical test:

Estimation: Information from the sample can be used to estimate (or predict) the mean decrease in anxiety ratings for the set of all anxious patients who may conceivably be treated with the drug.

Statistical test: Information from the sample can be used to determine whether the population mean decrease in anxiety ratings is greater than zero.

Notice that the inference related to estimation is aimed at answering the question, "What is the mean decrease in anxiety ratings for the population?" In contrast, the statistical test attempts to answer the question, "Is the mean drop in anxiety ratings greater than zero?"

Case Study: Percentage of Calories from Fat

Many studies have proposed relationships between diet and many diseases. For example, the percentage of calories from fat in the diet may be related to the

(margin notes) estimation   hypothesis testing

incidence of certain types of cancer and heart disease. The assessment and quantification of a person's usual diet is crucial in evaluating the degree of relationship between diet and diseases. This is a very difficult task but is important in an effort to monitor dietary behavior among individuals. Rosner, Willett, and Spiegelman, in "Correction of logistic regression relative risk estimates and confidence intervals for systematic within-person measurement error," *Statistics in Medicine* (1989), 8: 1051–1070, describe Nurses' Health Study, which examined the diet of a large sample of women.

Designing the Data Collection   One of the objectives of the study was to determine the percent calories from fat in the diet of a population of women. There are many dietary assessment methodologies. The most commonly used method in large nutritional epidemiology studies is the food frequency questionnaire (FFQ), which uses a carefully designed series of questions to determine the dietary intakes of participants in the study. In the Nurses' Health Study, a sample of 168 women who represented a random sample from a population of female nurses completed a single FFQ. From the information gathered from the questionnaire, the percentage of calories from fat (PCF) was computed. The parameters of interest were the average value of PCF μ for the population of nurses, the standard deviation σ in PCF for the population of nurses, the proportion π of nurses having PCF greater than 50%, as well as other parameters. The complete data set, which contains the ages of the women and several other variables, may be found on the data disk.

The number of persons needed in the study was determined by specifying the degree of accuracy in the estimation of the parameters μ, σ, and π. In a later section, we will discuss several methods for determining the proper sample sizes. For this study, it was decided that a sample of 168 participants would be adequate.

Managing the Data   The researchers need to carefully examine the data from the questionnaires to determine whether the responses were recorded correctly. The data would then be transferred to computer files and prepared for analysis following the steps outlined in Section 2.5.

Analyzing the Data   The next step in the study is to summarize the data through plots and summary statistics. The PCF values for the 168 women are displayed in Figure 5.1 in a stem-and-leaf diagram, along with a table of summary statistics.

FIGURE 5.1

The percentage of calories from fat (PCF) for 168 women in a dietary study

```
1   5
2   0 0 4 4
2   5 5 6 6 6 6 7 7 8 8 8 9 9 9 9 9 9 9 9
3   0 0 0 0 0 1 1 1 1 1 1 1 2 2 2 2 2 2 2 3 3 3 3 3 3 3 3 3 3 3 4 4 4 4 4 4 4 4 4
3   5 5 5 5 5 5 5 5 5 5 5 5 5 5 6 6 6 6 6 6 6 7 7 7 7 7 7 7 7 7 8 8 8 8 8 8 8 8 8 8 8 8 9 9 9 9 9 9 9 9
4   0 0 0 0 0 0 0 1 1 1 1 1 1 1 1 1 1 1 1 1 1 1 2 2 2 2 2 2 3 3 3 4 4 4 4 4
4   5 5 5 5 5 6 6 6 7 7 8 9 9
5   0 3 4
5   5 7
```

```
         Descriptive Statistics for Percentage Calories from Fat Data

         Variable          N      Mean    Median   TrMean    StDev   SE Mean
         PCF             168    36.919   36.473   36.847    6.728    0.519

         Variable    Minimum   Maximum       Q1       Q3
         PCF          15.925    57.847   32.766   41.295
```

From the stem-and-leaf plot, it appears that the data are nearly normally distributed with PCF values ranging from 15% to 57%. The proportion of the women that have PCF greater than 50% is $\hat{\pi} = 4/168 = 2.4\%$. From the table of summary statistics, the sample mean is $\bar{y} = 36.919$ and sample standard deviation is $s = 6.728$. The researchers want to draw inferences from the random sample of 168 women to the population from which they were selected. Thus, we need to place bounds on our point estimates to reflect our degree of confidence in their estimation of the population values. Also, the researchers may be interested in testing hypotheses about the size of the population mean PCF μ or variance σ^2. For example, many nutritional experts recommend that one's daily diet have no more than 30% of total calories a day from fat. Thus, we would want to test the statistical hypotheses that μ is greater than 30 to determine whether the average value of PCF for this population exceeds the recommended value. In this chapter, we will develop interval estimators for the population mean μ and a statistical test about μ. The estimation and testing of the parameters σ and π will be examined in Chapters 7 and 10, respectively.

EXERCISES

Basic Techniques

Pol. Sci.

5.1 A researcher is interested in estimating the percentage of registered voters in her state who have voted in at least one election over the past 2 years.
   a. Identify the population of interest to the researcher.
   b. How might you select a sample of voters to gather this information?

5.2 In the case study on percentage of calories from fat,
   a. What is the population of interest?
   b. What dietary variables other than PCF might affect a person's health?
   c. What characteristics of the nurses other than dietary intake might be important in studying the nurses' health condition?
   d. Describe a method for randomly selecting which nurses participate in the study.
   e. State several hypotheses that may be of interest to the researchers.

Engin.

5.3 A manufacturer claims that the average lifetime of a particular fuse is 1,500 hours. Information from a sample of 35 fuses shows that the average lifetime is 1,380 hours. What can be said about the manufacturer's claim?
   a. Identify the population of interest to us.
   b. Would an answer to the question posed involve estimation or testing a hypothesis?

5.4 Refer to Exercise 5.3. How might you select a sample of fuses from the manufacturer to test the claim?

5.2   Estimation of μ

The first step in statistical inference is point estimation, in which we compute a single value (statistic) from the sample data to estimate a population parameter. Suppose that we are interested in estimating a population mean and that we are willing to assume the underlying population is normal. One natural statistic that could be used to estimate the population mean is the sample mean, but we also could use the median and the trimmed mean. Which sample statistic should we use?

A whole branch of mathematical statistics deals with problems related to developing point estimators (the formulas for calculating specific point estimates from sample data) of parameters from various underlying populations and determining whether a particular point estimator has certain desirable properties. Fortunately, we will not have to derive these point estimators—they'll be given to us for each parameter. When we know which point estimator (formula) to use for a given parameter, we can develop confidence intervals (interval estimates) for these same parameters.

In this section, we deal with point and interval estimation of a population mean μ. Tests of hypotheses about μ are covered in Section 5.4.

For most problems in this text, we will use sample mean $\bar{y}$ as a point estimate of μ; we also use it to form an interval estimate for the population mean μ. From the Central Limit Theorem for the sample mean given in Chapter 4, we know that for large n (crudely, $n \geq 30$), $\bar{y}$ will be approximately normally distributed, with a mean μ and a standard error $\sigma_{\bar{y}}$. Then from our knowledge of the Empirical Rule and areas under a normal curve, we know that the interval $\mu \pm 2\sigma_{\bar{y}}$, or more precisely, the interval $\mu \pm 1.96\sigma_{\bar{y}}$, includes 95% of the $\bar{y}$s in repeated sampling, as shown in Figure 5.2.

FIGURE 5.2
Sampling distribution for $\bar{y}$

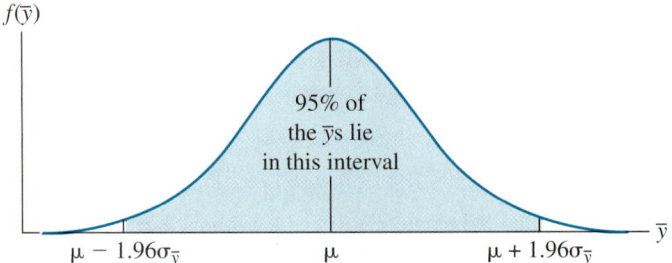

From Figure 5.2 we can observe that the sample mean $\bar{y}$ may not be very close to the population mean μ, the quantity it is suppose to estimate. Thus, when the value of $\bar{y}$ is reported, we should also provide an indication of how accurately $\bar{y}$ estimates μ. We will accomplish this by considering an interval of possible values for μ in place of using just a single value $\bar{y}$. Consider the interval $\bar{y} \pm 1.96\sigma_{\bar{y}}$. Any time $\bar{y}$ falls in the interval $\mu \pm 1.96\sigma_{\bar{y}}$, the interval $\bar{y} \pm 1.96\sigma_{\bar{y}}$ will contain the parameter μ (see Figure 5.3). The probability of $\bar{y}$ falling in the interval $\mu \pm 1.96\sigma_{\bar{y}}$ is .95, so we state that $\bar{y} \pm 1.96\sigma_{\bar{y}}$ is an **interval estimate** of μ with **level of confidence** .95.

interval estimate

level of confidence

FIGURE 5.3

When the observed value of $\bar{y}$ lies in the interval $\mu \pm 1.96\sigma_{\bar{y}}$, the interval $\bar{y} \pm 1.96\sigma_{\bar{y}}$ contains the parameter μ.

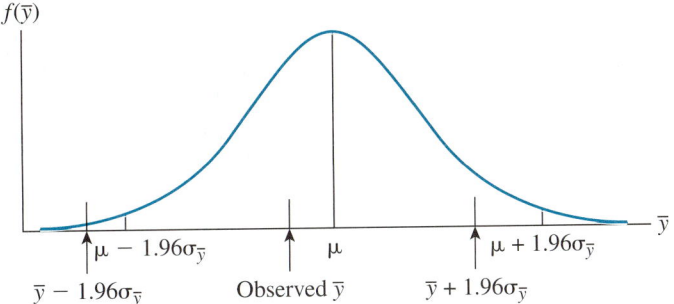

confidence coefficient

We evaluate the goodness of an interval estimation procedure by examining the fraction of times in repeated sampling that interval estimates would encompass the parameter to be estimated. This fraction, called the **confidence coefficient,** is .95 when using the formula $\bar{y} \pm 1.96\sigma_{\bar{y}}$; that is, 95% of the time in repeated sampling, intervals calculated using the formula $\bar{y} \pm 1.96\sigma_{\bar{y}}$ will contain the mean μ.

This idea is illustrated in Figure 5.4. Suppose we want to study a commercial process that produces shrimp for sale to restaurants. The shrimp are monitored for size by randomly selecting 40 shrimp from the tanks and measuring their length. We will consider a simulation of the shrimp monitoring. Suppose that the distribution of shrimp length in the tank had a normal distribution with a mean $\mu = 27$ cm and a standard deviation $\sigma = 10$ cm. Fifty samples of size $n = 40$ are drawn from the shrimp population. From each of these samples we compute the interval estimate $\bar{y} \pm 1.96\sigma_{\bar{y}} = \bar{y} \pm 1.96(10/\sqrt{40})$, because $\sigma_{\bar{y}} = \sigma/\sqrt{n}$. (See Table 5.1.) Note that although the intervals vary in location, only 2 of the 50 intervals failed to capture the population mean μ. The fact that two samples produced intervals that did not contain μ is not an indication that the procedure for producing intervals is faulty. Because our level of confidence is 95%, we would expect that, in a large collection of 95% confidence intervals, approximately 5% of the intervals would fail to include μ. Thus, in 50 intervals we would expect two or three intervals

FIGURE 5.4

Fifty interval estimates of the population mean (27)

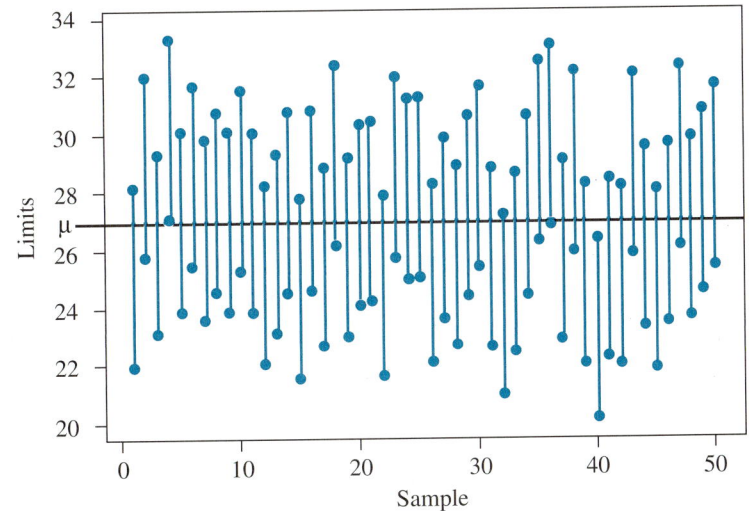

The interval from 35.90 to 37.94 forms a 95% confidence interval for μ. In other words, we are 95% certain that the average percent calories from fat is a value between 35.90 and 37.94. The researchers would next compare the relative size of these values to the mean PCF in other populations or to recommended values for PCF.

In Section 5.7, we present a procedure for obtaining a confidence interval for μ when σ is unknown. However, if the sample size is large—for example n larger than 30—we can estimate the population standard deviation σ with the sample standard deviation s in the confidence interval formula. Also, based on the results from the Central Limit Theorem, if the population distribution is not too nonnormal and the sample size is large, the level of confidence of our interval will be approximately the same as if we were sampling from a normal distribution.

99% confidence interval

There are many different confidence intervals for μ, depending on the confidence coefficient we choose. For example, the interval $\mu \pm 2.58\sigma_{\bar{y}}$ includes 99% of the values of $\bar{y}$ in repeated sampling, and the interval $\bar{y} \pm 2.58\sigma_{\bar{y}}$ forms a **99% confidence interval** for μ.

$(1 - \alpha)$ = confidence coefficient

We can state a general formula for a confidence interval for μ with a **confidence coefficient of $(1 - \alpha)$**, where α (Greek letter alpha) is between 0 and 1. For a specified value of $(1 - \alpha)$, a $100(1 - \alpha)\%$ confidence interval for μ is given by the following formula. Here we assume that σ is known or that the sample size is large enough to replace σ with s.

Confidence Interval for μ, σ Known

$$\bar{y} \pm z_{\alpha/2}\sigma_{\bar{y}}, \text{ where } \sigma_{\bar{y}} = \sigma/\sqrt{n}$$

$z_{\alpha/2}$

The quantity $z_{\alpha/2}$ is a value of z having a tail area of $\alpha/2$ to its right. In other words, at a distance of $z_{\alpha/2}$ standard deviations to the right of μ, there is an area of $\alpha/2$ under the normal curve. Values of $z_{\alpha/2}$ can be obtained from Table 1 in the Appendix by looking up the z-value corresponding to an area of $1 - (\alpha/2)$ (see Figure 5.6). Common values of the confidence coefficient $(1 - \alpha)$ and $z_{\alpha/2}$ are given in Table 5.2.

FIGURE 5.6

Interpretation of $z_{\alpha/2}$ in the confidence interval formula

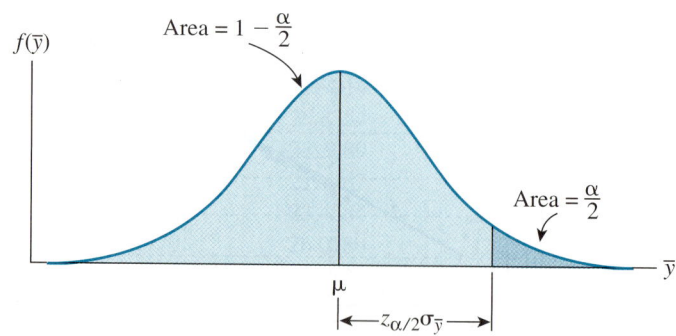

TABLE 5.2

Common values of the confidence coefficient $(1 - \alpha)$ and the corresponding z-value, $z_{\alpha/2}$

Confidence Coefficient $(1 - \alpha)$	Value of $\alpha/2$	Area in Table 1 $1 - \alpha/2$	Corresponding z-Value, $z_{\alpha/2}$
.90	.05	.95	1.645
.95	.025	.975	1.96
.98	.01	.99	2.33
.99	.005	.995	2.58

EXAMPLE 5.2

A forester wishes to estimate the average number of "count trees" per acre (trees larger than a specified size) on a 2,000-acre plantation. She can then use this information to determine the total timber volume for trees in the plantation. A random sample of $n = 50$ 1-acre plots is selected and examined. The average (mean) number of count trees per acre is found to be 27.3, with a standard deviation of 12.1. Use this information to construct a 99% confidence interval for μ, the mean number of count trees per acre for the entire plantation.

Solution   We use the general confidence interval with confidence coefficient equal to .99 and a $z_{\alpha/2}$-value equal to 2.58 (see Table 5.2). Substituting into the formula $\bar{y} \pm 2.58\sigma_{\bar{y}}$ and replacing σ with s in $\sigma_{\bar{y}} = \sigma/\sqrt{n}$, we have

$$27.3 \pm 2.58 \frac{12.1}{\sqrt{50}}$$

This corresponds to the confidence interval 27.3 ± 4.41—that is, the interval from 22.89 to 31.71. Thus, we are 99% sure that the average number of count trees per acre is between 22.89 and 31.71.

The discussion in this section has included one rather unrealistic assumption—namely, that the population standard deviation is known. In practice, it's difficult to find situations in which the population mean is unknown but the standard deviation is known. Usually both the mean and the standard deviation must be estimated from the sample. Because σ is estimated by the sample standard deviation s, the actual standard error of the mean, $\sigma/\sqrt{n}$, is naturally estimated by $s/\sqrt{n}$. This estimation introduces another source of random error (s will vary randomly, from sample to sample, around σ) and, strictly speaking, invalidates

substituting s for σ   our confidence interval formula. Fortunately, the formula is still a very good approximation for large sample sizes. As a very rough rule, we can use this formula when n is larger than 30; a better way to handle this issue is described in Section 5.7.

Statistical inference-making procedures differ from ordinary procedures in that we not only make an inference, but also provide a measure of how good that inference is. For interval estimation, the width of the confidence interval and the confidence coefficient measure the goodness of the inference. For a given value of the confidence coefficient, the smaller the width of the interval, the more precise the inference. The confidence coefficient, on the other hand, is set by the experimenter to express how much assurance he or she places in whether the interval estimate encompasses the parameter of interest. For a fixed sample size,

EXAMPLE 5.3

In the dietary intake example, the researchers wanted to estimate the mean PCF with a 95% confidence interval having a tolerable error of 3. From previous studies, the values of PCF ranged from 10% to 50%. How many nurses must the researchers include in the sample to achieve their specifications?

Solution   We want a 95% confidence interval with width 3, so $E = 3/2 = 1.5$ and $z_{\alpha/2} = z_{.025} = 1.96$. Our estimate of σ is range/4 $= (50 - 10)/4 = 10$. Substituting into the formula for n, we have

$$n = \frac{(z_{\alpha/2})^2 \sigma^2}{E^2} = \frac{(1.96)^2(10)^2}{(1.5)^2} = 170.7$$

Thus, a random sample of 171 nurses should give a 95% confidence interval for μ with the desired width of 3 provided 10 is a reasonable estimate of σ.

EXAMPLE 5.4

A federal agency has decided to investigate the advertised weight printed on cartons of a certain brand of cereal. The company in question periodically samples cartons of cereal coming off the production line to check their weight. A summary of 1,500 of the weights made available to the agency indicates a mean weight of 11.80 ounces per carton and a standard deviation of .75 ounce. Use this information to determine the number of cereal cartons the federal agency must examine to estimate the average weight of cartons being produced now, using a 99% confidence interval of width .50.

Solution   The federal agency has specified that the width of the confidence interval is to be .50, so $E = .25$. Assuming that the weights made available to the agency by the company are accurate, we can take $\sigma = .75$. The required sample size with $z_{\alpha/2} = 2.58$ is

$$n = \frac{(2.58)^2(.75)^2}{(.25)^2} = 59.91$$

Thus, the federal agency must obtain a random sample of 60 cereal cartons to estimate the mean weight to within $\pm.25$.

EXERCISES   **Basic Techniques**

5.18   Refer to Example 5.3. Suppose we continue to estimate σ with $\hat{\sigma} = 10$.
    a. If the level of confidence remains at 95% but the tolerable width is 2, how large a sample size is required?
    b. If the level of confidence increases to 99% but the specified width remains at 3, how large a sample size is required?
    c. If the level of confidence decreases to 90% but the specified width remains at 3, how large a sample size is required?

5.19   Based on your results in 5.18 (a), (b) and (c),
    a. What happens to the required sample size if we decrease the level of confidence with the width fixed?
    b. What happens to the required sample size if we increase the level of confidence with the width fixed?
    c. What happens to the required sample size if we decrease the width with the level of confidence fixed?

5.20 In general, if we keep the level of confidence fixed, how much would you need to increase the sample size to cut the width in half?

Applications

Bio. **5.21** A biologist wishes to estimate the effect of an antibiotic on the growth of a particular bacterium by examining the mean amount of bacteria present per plate of culture when a fixed amount of the antibiotic is applied. Previous experimentation with the antibiotic on this type of bacteria indicates that the standard deviation of the amount of bacteria present is approximately 13 cm^2. Use this information to determine the number of observations (cultures that must be developed and then tested) to estimate the mean amount of bacteria present, using a 99% confidence interval with a half-width of 3 cm^2.

Soc. **5.22** The city housing department wants to estimate the average rent for rent-controlled apartments. They need to determine the number of renters to include in the survey in order to estimate the average rent to within $50 using a 95% confidence interval. From past results, the rent for controlled apartments ranged from $200 to $1500 per month. How many renters are needed in the survey to meet the requirements?

5.23 Refer to Exercise 5.22. Suppose the mayor has reviewed the proposed survey and decides on the following changes:
   a. If the level of confidence is increased to 99% with the average rent estimated to within $25, what sample size is required?
   b. Suppose the budget for the project will not support both increasing the level of confidence and reducing the width of the interval. Explain to the mayor the impact on the estimation of the average rent of not raising the level of confidence from 95% to 99%.

Bus. **5.24** An insurance company is concerned about the number of worker compensation claims based on back injuries by baggers in grocery stores. They want to evaluate the fitness of baggers at the many grocery stores they insure. The workers selected for the study will be evaluated to determine the amount of weight that they can lift without undue back stress. From studies by other insurance companies, $\sigma \approx 25$ pounds.
   a. How many baggers must be included in the study to be 99% confident that the average weight lifted is estimated to within 8 pounds?
   b. If an estimate of σ is not known, suggest several ways in which the insurance company can obtain an idea of the size of σ.

5.4 A Statistical Test for μ

The second type of inference-making procedure is statistical testing (or hypothesis testing). As with estimation procedures, we will make an inference about a population parameter, but here the inference will be of a different sort. With point and interval estimates there was no supposition about the actual value of the parameter prior to collecting the data. Using sampled data from the population, we are simply attempting to determine the value of the parameter. In hypothesis testing, there is a preconceived idea about the value of the population parameter. For example, in studying the antipsychotic properties of an experimental compound, we might ask whether the average shock-avoidance response of rats treated with a specific dose of the compound is greater than 60, $\mu > 60$, the value that has been observed after extensive testing using a suitable standard drug. Thus, there are two theories or hypotheses involved in a statistical study. The first is the hypothesis being proposed by the person conducting the study, called the **research hypothesis**, $\mu > 60$ in our example. The second theory is the negation of this

research hypothesis

null hypothesis   hypothesis, called the **null hypothesis,** $\mu \le 60$ in our example. The goal of the study is to decide whether the data tend to support the research hypothesis.

statistical test   A **statistical test** is based on the concept of proof by contradiction and is composed of the five parts listed here.

> 1. Research hypothesis (also called the alternative hypothesis), denoted by H_a.
> 2. Null hypothesis, denoted by H_0.
> 3. Test statistics, denoted by T.S.
> 4. Rejection region, denoted by R.R.
> 5. Check assumptions and draw conclusions.

For example, the Texas A&M agricultural extension service wants to determine whether the mean yield per acre (in bushels) for a particular variety of soybeans has increased during the current year over the mean yield in the previous two years when μ was 520 bushels per acre. The first step in setting up a statistical test is determining the proper specification of H_0 and H_a. The following guidelines will be helpful:

> 1. The statement that μ equals a specific value will always be included in H_0. The particular value specified for μ is called its null value and is denoted μ_0.
> 2. The statement about μ that the researcher is attempting to support or detect with the data from the study is the research hypothesis, H_a.
> 3. The negation of H_a is the null hypothesis, H_0.
> 4. The null hypothesis is presumed correct unless there is overwhelming evidence in the data that the research hypothesis is supported.

In our example, μ_0 is 520. The research statement is that yield in the current year has increased above 520; that is, $H_a: \mu > 520$. (Note that we will include 520 in the null hypothesis.) Thus, the null hypothesis, the negation of H_a, is $H_0: \mu \le 520$.

To evaluate the research hypothesis, we take the information in the sample data and attempt to determine whether the data support the research hypothesis or the null hypothesis, but we will give the benefit of the doubt to the null hypothesis.

After stating the null and research hypotheses, we then obtain a random sample of 1-acre yields from farms throughout the state. The decision to state whether or not the data support the research hypothesis is based on a quantity

test statistic   computed from the sample data called the **test statistic.** If the population distribution is determined to be mound shaped, a logical choice as a test statistic for μ is $\bar{y}$ or some function of $\bar{y}$.

If we select $\bar{y}$ as the test statistic, we know that the sampling distribution of $\bar{y}$ is approximately normal with a mean μ and standard deviation $\sigma_{\bar{y}} = \sigma/\sqrt{n}$, provided the population distribution is normal or the sample size is fairly large. We are attempting to decide between $H_a: \mu > 520$ or $H_0: \mu \le 520$. The decision will be to either reject H_0 or fail to reject H_0. In developing our decision rule, we will assume that $\mu = 520$, the null value of μ. We will now determine the values

rejection region   of $\bar{y}$, called the **rejection region,** which we are very unlikely to observe if $\mu = 520$ (or if μ is any other value in H_0). The rejection region contains the values of $\bar{y}$ that support the research hypothesis and contradict the null hypothesis, hence the

FIGURE 5.7

Assuming that H_0 is true, contradictory values of $\bar{y}$ are in the upper tail.

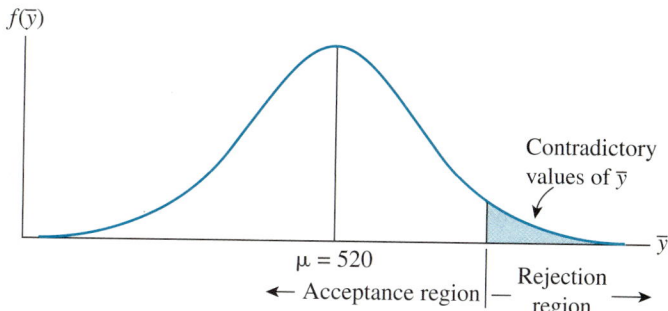

region of values for $\bar{y}$ that reject the null hypothesis. The rejection region will be the values of $\bar{y}$ in the upper tail of the null distribution ($\mu = 520$) of $\bar{y}$. See Figure 5.7.

As with any two-way decision process, we can make an error by falsely rejecting the null hypothesis or by falsely accepting the null hypothesis. We give these errors the special names **Type I error** and **Type II error.**

Type I error

Type II error

DEFINITION 5.1

A **Type I error** is committed if we reject the null hypothesis when it is true. The probability of a Type I error is denoted by the symbol α.

DEFINITION 5.2

A **Type II error** is committed if we accept the null hypothesis when it is false and the research hypothesis is true. The probability of a Type II error is denoted by the symbol β (Greek letter beta).

The two-way decision process is shown in Table 5.3 with corresponding probabilities associated with each situation.

TABLE 5.3

Two-way decision process

Decision	Null Hypothesis	
	True	**False**
Reject H_0	Type I error α	Correct $1 - \beta$
Accept H_0	Correct $1 - \alpha$	Type II error β

Although it is desirable to determine the acceptance and rejection regions to simultaneously minimize both α and β, this is not possible. The probabilities associated with Type I and Type II errors are inversely related. For a fixed sample size n, as we change the rejection region to increase α, then β decreases, and vice versa.

To alleviate what appears to be an impossible bind, the experimenter specifies a tolerable probability for a Type I error of the statistical test. Thus, the experimenter may choose α to be .01, .05, .10, and so on. Specification of a value for α then locates the rejection region. Determination of the associated probability of a Type II error is more complicated and will be delayed until later in the chapter.

Let us now see how the choice of α locates the rejection region. Returning to our soybean example, we will reject the null hypothesis for large values of the

Solution   The research hypothesis for this statistical test is H_a: $\mu \neq 2{,}600$ and the null hypothesis is H_0: $\mu = 2{,}600$. Using $\alpha = .05$, the two-tailed rejection region for this test is located as shown in Figure 5.11.

FIGURE 5.11

Rejection region for H_a: $\mu \neq 2{,}600$ when $\alpha = .05$

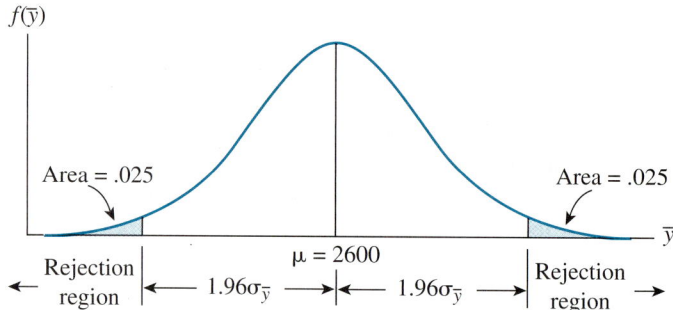

With a sample size of $n = 40$, the Central Limit Theorem should hold for $\bar{y}$. To determine how many standard errors our test statistic $\bar{y}$ lies away from $\mu = 2{,}600$, we compute

$$z = \frac{\bar{y} - \mu_0}{\sigma/\sqrt{n}} = \frac{2{,}752 - 2{,}600}{350/\sqrt{40}} = 2.75$$

The observed value for $\bar{y}$ lies more than 1.96 standard errors above the mean, so we reject the null hypothesis in favor of the alternative H_a: $\mu \neq 2{,}600$. We conclude that the mean number of miles driven is different from 2,600.

The mechanics of the statistical test for a population mean can be greatly simplified if we use z rather than $\bar{y}$ as a test statistic. Using

$$H_0: \quad \mu \leq \mu_0 \text{ (where } \mu_0 \text{ is some specified value)}$$
$$H_a: \quad \mu > \mu_0$$

and the test statistic

$$z = \frac{\bar{y} - \mu_0}{\sigma/\sqrt{n}}$$

then for $\alpha = .025$ we reject the null hypothesis if $z \geq 1.96$—that is, if $\bar{y}$ lies more than 1.96 standard errors above the mean. Similarly, for $\alpha = .05$ and H_a: $\mu \neq \mu_0$, we reject the null hypothesis if the computed value of $z \geq 1.96$ or the computed value of $z \leq -1.96$. This is equivalent to rejecting the null hypothesis if the computed value of $|z| \geq 1.96$.

test for a population mean   The statistical **test for a population mean** μ is summarized next. Three different sets of hypotheses are given with their corresponding rejection regions. In a given situation, you will choose only one of the three alternatives with its associated rejection region. The tests given are appropriate only when the population distribution is normal with known σ. The rejection region will be approximately the correct region even when the population distribution is nonnormal provided the sample size is large; in most cases, $n \geq 30$ is sufficient. We can then apply the results from the Central Limit Theorem with the sample standard deviation s replacing σ to conclude that the sampling distribution of $z = (\bar{y} - \mu_0)/(s/\sqrt{n})$ is approximately normal.

**Summary of a Statistical
Test for μ with a Normal
Population Distribution
(σ Known) or Large
Sample Size n**

Hypotheses:

Case 1. $H_0: \mu \leq \mu_0$ vs. $H_a: \mu > \mu_0$ (right-tailed test)
Case 2. $H_0: \mu \geq \mu_0$ vs. $H_a: \mu < \mu_0$ (left-tailed test)
Case 3. $H_0: \mu = \mu_0$ vs. $H_a: \mu \neq \mu_0$ (two-tailed test)

T.S.: $z = \dfrac{\bar{y} - \mu_0}{\sigma/\sqrt{n}}$

R.R.: For a probability α of a Type I error,

Case 1. Reject H_0 if $z \geq z_\alpha$.
Case 2. Reject H_0 if $z \leq -z_\alpha$.
Case 3. Reject H_0 if $|z| \geq z_{\alpha/2}$.

Note: These procedures are appropriate if the population distribution is normally distributed with σ known. In most situations, if $n \geq 30$, then the Central Limit Theorem allows us to use these procedures when the population distribution is nonnormal. Also, if $n \geq 30$, then we can replace σ with the sample standard deviation s. The situation in which $n < 30$ is presented later in this chapter.

EXAMPLE 5.7

As a part of her evaluation of municipal employees, the city manager audits the parking tickets issued by city parking officers to determine the number of tickets that were contested by the car owner and found to be improperly issued. In past years, the number of improperly issued tickets per officer had a normal distribution with mean $\mu = 380$ and $\sigma = 35.2$. Because there has recently been a change in the city's parking regulations, the city manager suspects that the mean number of improperly issued tickets has increased. An audit of 50 randomly selected officers is conducted to test whether there has been an increase in improper tickets. Use the sample data given here and $\alpha = .01$ to test the research hypothesis that the mean number of improperly issued tickets is greater than 380. The audit generates the following data: $n = 50$ and $\bar{y} = 390$.

Solution   Using the sample data with $\alpha = .01$, the five parts of a statistical test are as follows:

$H_0:$ $\mu \leq 380$
$H_a:$ $\mu > 380$

T.S.: $z = \dfrac{\bar{y} - \mu_0}{\sigma/\sqrt{n}} = \dfrac{390 - 380}{35.2/\sqrt{50}} = \dfrac{10}{35.2/7.07} = 2.01$

R.R.:   For $\alpha = .01$ and a right-tailed test, we reject H_0 if $z \geq z_{.01}$, where $z_{.01} = 2.33$.

Conclusion: Because the observed value of z, 2.01, does not exceed 2.33, we might be tempted to accept the null hypothesis that $\mu \leq 380$. The only problem with this conclusion is that we do not know β, the probability of incorrectly accepting the null hypothesis. To hedge somewhat in situations in which z does not fall in the rejection region and β has not been calculated, we recommend stating that there is insufficient evidence to reject the null hypothesis. To reach a conclusion about whether to accept H_0, the experimenter would have to compute β. If β is small for reasonable alternative values of μ, then H_0 is accepted. Otherwise, the experimenter should conclude that there is insufficient evidence to reject the null hypothesis.

computing β

We can illustrate the **computation of β,** the probability of a Type II error, using the data in Example 5.7. If the null hypothesis is H_0: $\mu \leq 380$, the probability of incorrectly accepting H_0 will depend on how close the actual mean is to 380. For example, if the actual mean number of improperly issued tickets is 400, we would expect β to be much smaller than if the actual mean is 387. The closer the actual mean is to μ_0 the more likely we are to obtain data having a value $\bar{y}$ in the acceptance region. The whole process of determining β for a test is a "what-if" type of process. In practice, we compute the value of β for a number of values of μ in the alternative hypothesis H_a and plot β versus μ in a graph called the

OC curve

OC curve. Alternatively, tests of hypotheses are evaluated by computing the probability that the test rejects false null hypotheses, called the **power** of the test.

power

We note that power $= 1 - \beta$. The plot of power versus the value of μ is called

power curve

the **power curve.** We attempt to design tests that have large values of power and hence small values for β.

Let us suppose that the actual mean number of improper tickets is 395 per officer. What is β? With the null and research hypotheses as before,

$$H_0: \quad \mu \leq 380$$
$$H_a: \quad \mu > 380$$

and with $\alpha = .01$, we use Figure 5.12(a) to display β. The shaded portion of Figure 5.12(a) represents β, as this is the probability of $\bar{y}$ falling in the acceptance region when the null hypothesis is false and the actual value of μ is 395. The power of the test for detecting that the actual value of μ is 395 is $1 - \beta$, the area in the rejection region.

Let us consider two other possible values for μ—namely, 387 and 400. The corresponding values of β are shown as the shaded portions of Figures 5.12(b) and (c), respectively; power is the unshaded portion in the rejection region of Figure 5.12(b) and (c). The three situations illustrated in Figure 5.12 confirm what we alluded to earlier; that is, the probability of a Type II error β decreases (and hence power increases) the further μ lies away from the hypothesized means under H_0.

The following notation will facilitate the calculation of β. Let μ_0 denote the null value of μ and let μ_a denote the actual value of the mean in H_a. Let $\beta(\mu_a)$ be the probability of a Type II error if the actual value of the mean is μ_a and let $\text{PWR}(\mu_a)$ be the power at μ_a. Note that $\text{PWR}(\mu_a)$ equals $1 - \beta(\mu_a)$. Although we never really know the actual mean, we select feasible values of μ and determine β for each of these values. This will allow us to determine the probability of a Type II error occurring if one of these feasible values happens to be the actual

FIGURE 5.12

The probability β of a
Type II error

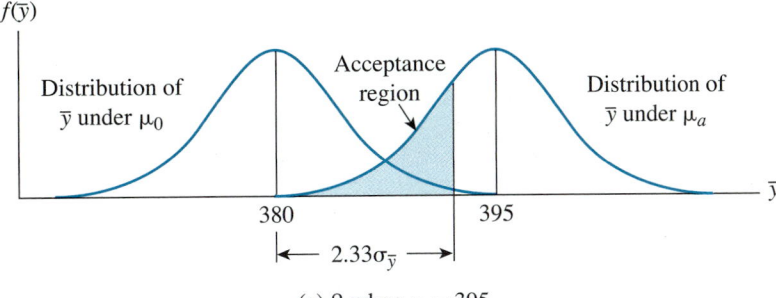

(a) β when $\mu_a = 395$

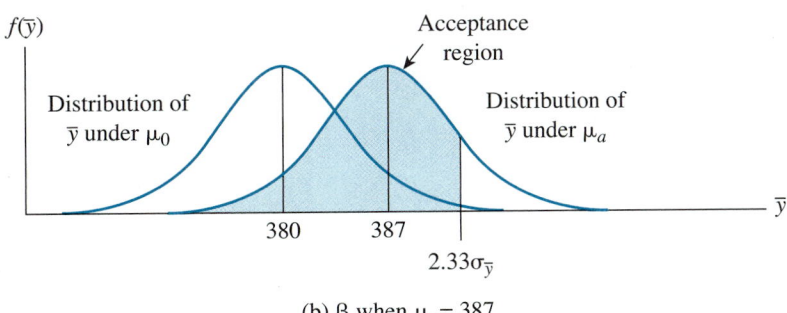

(b) β when $\mu_a = 387$

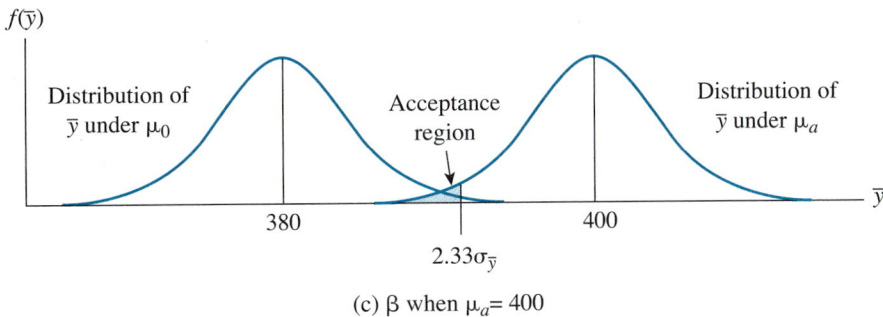

(c) β when $\mu_a = 400$

value of the mean. The decision whether or not to accept H_0 depends on the magnitude of β for one or more reasonable values for μ_a. Alternatively, researchers calculate the power curve for a test of hypotheses. Recall, that the power of the test at μ_a PWR(μ_a) is the probability the test will detect that H_0 is false when the actual value of μ is μ_a. Hence, we want tests of hypotheses in which PWR(μ_a) is large when μ_a is far from μ_0.

For a one-tailed test, $H_0: \mu \leq \mu_0$ or $H_0: \mu \geq \mu_0$, the value of β at μ_a is the probability that z is less than

$$z_\alpha - \frac{|\mu_0 - \mu_a|}{\sigma_{\bar{y}}}$$

This probability is written as

$$\beta(\mu_a) = P\left[z < z_\alpha - \frac{|\mu_0 - \mu_a|}{\sigma_{\bar{y}}}\right]$$

The value of $\beta(\mu_a)$ is found by looking up the probability corresponding to the number $z_\alpha - |\mu_0 - \mu_a|/\sigma_{\bar{y}}$ in Table 1 in the Appendix.

Formulas for β are given here for one- and two-tailed tests. Examples using these formulas follow.

Calculation of β for a One- or Two-Tailed Test about μ

1. One-tailed test:

$$\beta(\mu_a) = P\left(z \le z_\alpha - \frac{|\mu_0 - \mu_a|}{\sigma_{\bar{y}}}\right) \qquad \text{PWR}(\mu_a) = 1 - \beta(\mu_a)$$

2. Two-tailed test:

$$\beta(\mu_a) \approx P\left(z \le z_{\alpha/2} - \frac{|\mu_0 - \mu_a|}{\sigma_{\bar{y}}}\right) \qquad \text{PWR}(\mu_a) = 1 - \beta(\mu_a)$$

EXAMPLE 5.8

Compute β and power for the test in Example 5.7 if the actual mean number of improperly issued tickets is 395.

Solution The research hypothesis for Example 5.7 was $H_a: \mu > 380$. Using $\alpha = .01$ and the computing formula for β with $\mu_0 = 380$ and $\mu_a = 395$, we have

$$\beta(395) = P\left[z < z_{.01} - \frac{|\mu_0 - \mu_a|}{\sigma_{\bar{y}}}\right] = P\left[z < 2.33 - \frac{|380 - 395|}{35.2/\sqrt{50}}\right]$$

$$= P[z < 2.33 - 3.01] = P[z < -.68]$$

Referring to Table 1 in the Appendix, the area corresponding to $z = -.68$ is .2483. Hence, $\beta(395) = .2483$ and $\text{PWR}(395) = 1 - .2483 = .7517$.

Previously, when $\bar{y}$ did not fall in the rejection region, we concluded that there was insufficient evidence to reject H_0 because β was unknown. Now when $\bar{y}$ falls in the acceptance region, we can compute β corresponding to one (or more) alternative values for μ that appear reasonable in light of the experimental setting. Then provided we are willing to tolerate a probability of falsely accepting the null hypothesis equal to the computed value of β for the alternative value(s) of μ considered, our decision is to accept the null hypothesis. Thus, in Example 5.8, if the actual mean number of improperly issued tickets is 395, then there is about a .25 probability (1 in 4 chance) of accepting the hypothesis that μ is less than or equal to 380 when in fact μ equals 395. The city manager would have to analyze the consequence of making such a decision. If the risk was acceptable then she could state that the audit has determined that the mean number of improperly issued tickets has not increased. If the risk is too great, then the city manager would have to expand the audit by sampling more than 50 officers. In the next section, we will describe how to select the proper value for n.

EXAMPLE 5.9

Prospective salespeople for an encyclopedia company are now being offered a sales training program. Previous data indicate that the average number of sales per month for those who do not participate in the program is 33. To determine whether the training program is effective, a random sample of 35 new employees is given the sales training and then sent out into the field. One month later, the mean and standard deviation for the number of sets of encyclopedias sold are 35 and 8.4, respectively. Do the data present sufficient evidence to indicate that the training program enhances sales? Use $\alpha = .05$.

Solution The five parts to our statistical test are as follows:

$$H_0: \quad \mu \leq 33$$
$$H_a: \quad \mu > 33$$

T.S.: $\quad z = \dfrac{\bar{y} - \mu_0}{\sigma_{\bar{y}}} \approx \dfrac{35 - 33}{8.4/\sqrt{35}} = 1.41$

R.R.: For $\alpha = .05$ we will reject the null hypothesis if $z \geq z_{.05} = 1.645$.

Check assumptions and draw conclusions: With $n = 35$, the Central Limit Theorem should hold. Because the observed value of z does not fall into the rejection region, we reserve judgment on accepting H_0 until we calculate β. In other words, we conclude that there is insufficient evidence to reject the null hypothesis that persons in the sales program have the same or a smaller mean number of sales per month as those not in the program.

EXAMPLE 5.10

Refer to Example 5.9. Suppose that the encyclopedia company thinks that the cost of financing the sales program will be offset by increased sales if those in the program average 38 sales per month. Compute β for $\mu_a = 38$ and, based on the value of $\beta(38)$, indicate whether you would accept the null hypothesis.

Solution Using the computational formula for β with $\mu_0 = 33$, $\mu_a = 38$, and $\alpha = .05$, we have

$$\beta(38) = P\left[z \leq z_{.05} - \frac{|\mu_0 - \mu_a|}{\sigma_{\bar{y}}}\right] = P\left[z \leq 1.645 - \frac{|33 - 38|}{8.4/\sqrt{35}}\right]$$

$$= P[z \leq -1.88]$$

The area corresponding to $z = -1.88$ in Table 1 of the Appendix is .0301. Hence,

$$\beta(38) = .0301 \qquad \text{PWR}(38) = 1 - .0301 = .9699$$

Because β is relatively small, we accept the null hypothesis and conclude that the training program has not increased the average sales per month above the point at which increased sales would offset the cost of the training program.

The encyclopedia company wants to compute the chance of a Type II error for several other values of μ in H_a so they will have a reasonable idea of their chance of making a Type II error based on the data collected in the random sample of new employees. Repeating the calculations for obtaining $\beta(38)$, we obtain the values in Table 5.4.

TABLE 5.4

Probability of Type II error
and power for values
of μ in H_a

μ	33	34	35	36	37	38	39	40	41
$\beta(\mu)$	.9500	.8266	.5935	.3200	.1206	.0301	.0049	.0005	.0000
PWR(μ)	.0500	.1734	.4065	.6800	.8794	.9699	.9951	.9995	.9999

Figure 5.13 is a plot of the $\beta(\mu)$ values in Table 5.4 with a smooth curve through the points. Note that as the value of μ increases, the probability of Type II error decreases to 0 and the corresponding power value increases to 1.0. The company could examine this curve to determine whether the chances of Type II error are reasonable for values of μ in H_a that are important to the company. From Table 5.4 or Figure 5.13, we observe that $\beta(38) = .0301$, a relatively small number. Based on the results from Example 5.10, we find that the test statistic does not fall in the rejection region. Because $\beta(38)$ is small, we can now state that we accept the null hypothesis and conclude that the training program has not increased the average sales per month above the point at which increased sales would offset the cost of the training program.

FIGURE 5.13

Probability of Type II error

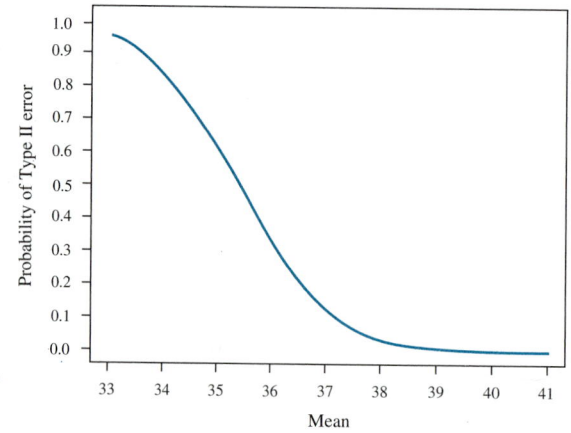

In Section 5.2, we discussed how we measure the effectiveness of interval estimates. The effectiveness of a statistical test can be measured by the magnitudes of the Type I and Type II errors, α and $\beta(\mu)$. When α is preset at a tolerable level by the experimenter, $\beta(\mu_a)$ is a function of the sample size for a fixed value of μ_a. The larger the sample size n, the more information we have concerning μ, the less likely we are to make a Type II error, and, hence, the smaller the value of $\beta(\mu_a)$. To illustrate this idea, suppose we are testing the hypotheses H_0: $\mu \leq 84$ against H_a: $\mu > 84$, where μ is the mean of a population having a normal distribution with $\sigma = 1.4$. If we take $\alpha = .05$, then the probability of Type II errors is plotted in Figure 5.14(a) for three possible sample sizes $n = 10, 18$, and 25. Note that $\beta(84.6)$ becomes smaller as we increase n from 10 to 25. Another relationship of interest is that between α and $\beta(\mu)$. For a fixed sample size n, if

FIGURE 5.14

(a) OC curve for $\alpha = .05$, $n = 10, 18, 25$. (b) OC curve for $n = 25$, $\alpha = .05, .01, .001$

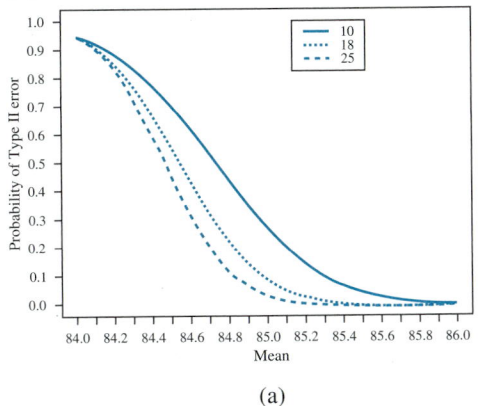
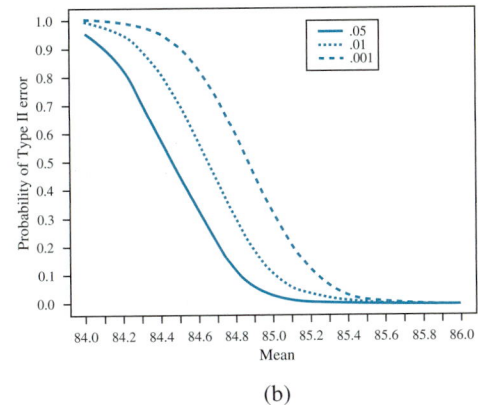

(a)   (b)

we change the rejection region to increase the value of α, the value of $\beta(\mu_a)$ will decrease. This relationship can be observed in Figure 5.14(b). Fix the sample size at 25 and plot $\beta(\mu)$ for three different values of $\alpha = .05, .01, .001$. We observe that $\beta(84.6)$ becomes smaller as α increases from .001 to .05. A similar set of graphs can be obtained for the power of the test by simply plotting $\text{PWR}(\mu) = 1 - \beta(\mu)$ vs. μ. The relationships described would be reversed; that is, for fixed α increasing the value of the sample size would increase the value of $\text{PWR}(\mu)$ and, for fixed sample size, increasing the value of α would increase the value of $\text{PWR}(\mu)$. We will consider now the problem of designing an experiment for testing hypotheses about μ when α is specified and $\beta(\mu_a)$ is preset for a fixed value μ_a. This problem reduces to determining the sample size needed to achieve the fixed values of α and $\beta(\mu_a)$. Note that in those cases in which the determined value of n is too large for the initially specified values of α and β, we can increase our specified value of α and achieve the desired value of $\beta(\mu_a)$ with a smaller sample size.

5.5 Choosing the Sample Size for μ

The quantity of information available for a statistical test about μ is measured by the magnitudes of the Type I and II error probabilities, α and $\beta(\mu)$ for various values of μ in the alternative hypothesis H_a. Suppose that we are interested in testing $H_0: \mu \leq \mu_0$ against the alternative $H_a: \mu > \mu_0$. First, we must specify the value of α. Next we must determine a value of μ in the alternative, μ_1, such that if the actual value of the mean is larger than μ_1, then the consequences of making a Type II error would be substantial. Finally, we must select a value for $\beta(\mu_1)$, β. Note that for any value of μ larger than μ_1, the probability of Type II error will be smaller than $\beta(\mu_1)$; that is,

$$\beta(\mu) < \beta(\mu_1), \text{ for all } \mu > \mu_1$$

Let $\Delta = \mu_1 - \mu_0$. The sample size necessary to meet these requirements is

$$n = \sigma^2 \frac{(z_\alpha + z_\beta)^2}{\Delta^2}$$

Note: If σ^2 is unknown, substitute an estimated value from previous studies or a pilot study to obtain an approximate sample size.

The same formula applies when testing $H_0: \mu \geq \mu_0$ against the alternative $H_a: \mu < \mu_0$, with the exception that we want the probability of a Type II error to be of magnitude β or less when the actual value of μ is *less* than μ_1, a value of the mean in H_a; that is,

$$\beta(\mu) < \beta, \text{ for all } \mu < \mu_1$$

with $\Delta = \mu_0 - \mu_1$.

EXAMPLE 5.11

A cereal manufacturer produces cereal in boxes having a labeled weight of 12 ounces. The boxes are filled by machines that are set to have a mean fill per box of 16.37 ounces. Because the actual weight of a box filled by these machines has a normal distribution with a standard deviation of approximately .225 ounces, the percentage of boxes having weight less than 16 ounces is 5% using this setting. The manufacturer is concerned that one of its machines is underfilling the boxes and wants to sample boxes from the machine's output to determine whether the mean weight μ is less than 16.37—that is, to test

$$
\begin{aligned}
H_0: &\quad \mu \geq 16.37 \\
H_a: &\quad \mu < 16.37
\end{aligned}
$$

with $\alpha = .05$. If the true mean weight is 16.27 or less, the manufacturer needs the probability of failing to detect this underfilling of the boxes with a probability of at most .01, or risk incurring a civil penalty from state regulators. Thus, we need to determine the sample size n such that our test of H_0 versus H_a has $\alpha = .05$ and $\beta(\mu)$ less than .01 whenever μ is less than 16.27 ounces.

Solution We have $\alpha = .05$, $\beta = .01$, $\Delta = 16.37 - 16.27 = .1$, and $\sigma = .225$. Using our formula with $z_{.05} = 1.645$ and $z_{.01} = 2.33$, we have

$$n = \frac{(.225)^2(1.645 + 2.33)^2}{(.1)^2} = 79.99 \approx 80$$

Thus, the manufacturer must obtain a random sample of $n = 80$ boxes to conduct this test under the specified conditions.

Suppose that after obtaining the sample, we compute $\bar{y} = 16.35$ ounces. The computed value of the test statistic is

$$z = \frac{\bar{y} - 16.37}{\sigma_{\bar{y}}} = \frac{16.35 - 16.37}{.225/\sqrt{80}} = -.795$$

Because the rejection region is $z < -1.645$, the computed value of z does not fall in the rejection region. What is our conclusion? In similar situations in previous sections, we would have concluded that there is insufficient evidence to reject H_0. Now, however, knowing that $\beta(\mu) \leq .01$ when $\mu \leq 16.27$, we would feel safe in our conclusion to accept $H_0: \mu \geq 16.37$. Thus, the manufacturer is somewhat secure in concluding that the mean fill from the examined machine is at least 16.37 ounces.

With a slight modification of the sample size formula for the one-tailed tests, we can test

$$H_0: \quad \mu = \mu_0$$
$$H_a: \quad \mu \neq \mu_0$$

for a specified α, β, and Δ, where

$$\beta(\mu) \leq \beta, \text{ whenever } |\mu - \mu_0| \geq \Delta$$

Thus, the probability of Type II error is at most β whenever the actual mean differs from μ_0 by at least Δ. A formula for an approximate sample size n when testing a two-sided hypothesis for μ is presented here.

Approximate Sample Size for a Two-Sided Test of H_0: $\mu = \mu_0$

$$n = \frac{\sigma^2}{\Delta^2}(z_{\alpha/2} + z_\beta)^2$$

Note: If σ^2 is unknown, substitute an estimated value to get an approximate sample size.

EXERCISES

Basic Techniques

5.25 Consider the data of Example 5.11. Suppose we want to test whether the mean is different from 16.37.

    a. Determine the sample size required for testing H_0: $\mu = 16.37$ against H_a: $\mu \neq 16.37$ in order to have $\alpha = .05$ and $\beta(\mu)$ less than .01 when the actual value of μ lies more than than .1 unit away from $\mu_0 = 16.37$.

    b. How does this sample size compare to the value in Example 5.11 for the one-sided test?

5.26 A researcher wanted to test the hypotheses H_0: $\mu \leq 38$ against H_a: $\mu > 38$ with $\alpha = .05$. A random sample of 50 measurements from a population yielded $\bar{y} = 40.1$ and $s = 5.6$.

    a. What conclusions can you make about the hypotheses based on the sample information?

    b. Could you have made a Type II error in this situation? Explain.

    c. Calculate the probability of a Type II error if the actual value of μ is at least 39.

5.27 For the data of Exercise 5.26, sketch the power curve for rejecting H_0: $\mu \leq 38$ by determining $\text{PWR}(\mu_a)$ for the following values of μ in the alternative hypothesis: 39, 40, 41, 42, 43, and 44.

    a. Interpret the values on your curve.

    b. Without actually recalculating the values for $\text{PWR}(\mu)$, sketch the power curve for $\alpha = .025$ and $n = 50$.

    c. Without actually recalculating the values for $\text{PWR}(\mu)$, sketch the power curve for $\alpha = .05$ and $n = 20$.

5.28 Using a computer software program, simulate 100 samples of size 16 from a normal distribution with $\mu = 40$ and $\sigma = 8$. We wish to test the hypotheses H_0: $\mu = 40$ vs. H_a: $\mu \neq 40$ with $\alpha = 0.10$. Assume that $\sigma = 8$ and that the population distribution is normal when performing the test of hypothesis using each of the 100 samples.

    a. How many of the 100 tests of hypotheses resulted in your making the incorrect decision of rejecting H_0?

    b. On the average, if you were to conduct 100 tests of hypotheses with $\alpha = 0.10$, how many times would you expect to make the wrong decision of rejecting H_0?

    c. What type of error are you making if you incorrectly reject H_0?

a. Is there sufficient evidence ($\alpha = .05$) in the data that the mean lead concentration exceeds 30 mg kg^{-1} dry weight?
b. What is the probability of a Type II error if the actual mean concentration is 50?
c. Do the data appear to have a normal distribution?
d. Based on your answer in (c), is the sample size large enough for the test procedures to be valid? Explain.

Med.   **5.39**   Tooth decay generally develops first on teeth that have irregular shapes (typically molars). The most susceptible surfaces on these teeth are the chewing surfaces. Usually the enamel on these surfaces contains tiny pockets that tend to hold food particles. Bacteria begins to eat the food particles to create an environment in which the tooth surface will decay.

Of particular importance in the decay rate of teeth, in addition to the natural hardness of the teeth, is the form of food eaten by the individual. Some forms of carbohydrates are particularly detrimental to dental health. Many studies have been conducted to verify these findings, and we can imagine how a study might have been run. A random sample of 60 adults was obtained from a given locale. Each person was examined and then maintained a diet supplemented with a sugar solution at all meals. At the end of a 1-year period, the average number of newly decayed teeth for the group was .70, and the standard deviation was .4.

a. Do these data present sufficient evidence ($\alpha = .05$) that the mean number of newly decayed teeth for the people whose diet includes a sugar solution is greater than .30? The value .30 was of interest as this is the rate that had been shown to apply to a person whose diet did not contain a sugar supplement.
b. Why would a two-tailed test be inappropriate?

5.6   The Level of Significance of a Statistical Test

In Section 5.4, we introduced hypothesis testing along rather traditional lines: we defined the parts of a statistical test along with the two types of errors and their associated probabilities α and $\beta(\mu_a)$. The problem with this approach is that if other researchers want to apply the results of your study using a different value for α then they must compute a new rejection region before reaching a decision concerning H_0 and H_a. An alternative approach to hypothesis testing follows the following steps: specify the null and alternative hypotheses, specify a value for α, collect the sample data, and determine the weight of evidence for rejecting the null hypothesis. This weight, given in terms of a probability, is called the **level of**

level of significance

p-value

significance (or **p-value**) of the statistical test. More formally, the level of significance is defined as follows: *the probability of obtaining a value of the test statistic that is as likely or more likely to reject H_0 as the actual observed value of the test statistic. This probability is computed assuming that the null hypothesis is true.* Thus, if the level of significance is a small value, then the sample data fail to support H_0 and our decision is to reject H_0. On the other hand, if the level of significance is a large value, then we fail to reject H_0. We must next decide what is a large or small value for the level of significance. The following decision rule yields results that will always agree with the testing procedures we introduced in Section 5.5.

Decision Rule for Hypothesis Testing Using the p-Value

1. If the p-value $\leq \alpha$, then reject H_0.
2. If the p-value $> \alpha$, then fail to reject H_0.

We illustrate the calculation of a level of significance with several examples.

EXAMPLE 5.12

Refer to Example 5.7.

a. Determine the level of significance (p-value) for the statistical test and reach a decision concerning the research hypothesis using $\alpha = .01$.

b. If the preset value of α is .05 instead of .01, does your decision concerning H_a change?

Solution

a. The null and alternative hypotheses are

$$H_0: \quad \mu \leq 380$$
$$H_a: \quad \mu > 380$$

From the sample data, the computed value of the test statistic is

$$z = \frac{\bar{y} - 380}{s/\sqrt{n}} = \frac{390 - 380}{35.2/\sqrt{50}} = 2.01$$

The level of significance for this test (i.e., the weight of evidence for rejecting H_0) is the probability of observing a value of $\bar{y}$ greater than or equal to 390 assuming that the null hypothesis is true; that is, $\mu = 380$. This value can be computed by using the z-value of the test statistic, 2.01, because

$$p\text{-value} = P(\bar{y} \geq 390, \text{ assuming } \mu = 380) = P(z \geq 2.01)$$

Referring to Table 1 in the Appendix, $P(z \geq 2.01) = 1 - P(z < 2.01) = 1 - .9778 = .0222$. This value is shown by the shaded area in Figure 5.15. Because the p-value is greater than α (.0222 > .01), we fail to reject H_0 and conclude that the data do not support the research hypothesis.

FIGURE 5.15

Level of significance for Example 5.12

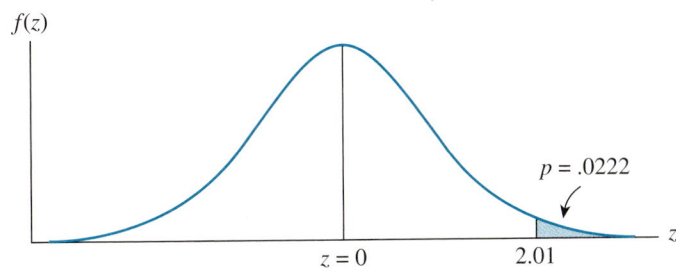

b. Another person examines the same data but with a preset value for $\alpha = .05$. This person is willing to support a higher risk of a Type I error, and hence the decision is to reject H_0 because the p-value is less than α (.0222 $\leq$ .05). It is important to emphasize that the value of α used in the decision rule is *preset* and not selected after calculating the p-value.

practical significance of your findings after drawing conclusions based on the statistical test.

EXERCISES   **Basic Techniques**

5.40   The sample data for a statistical test concerning μ yielded $n = 50$, $\bar{y} = 48.2$, $s = 12.57$. Determine the level of significance for testing $H_0: \mu \le 45$ versus $H_a: \mu > 45$. Is there significant evidence in the data to support the claim that μ is greater than 45 using $\alpha = .05$?

5.41   Refer to Exercise 5.40. If the researcher used $\alpha = .025$ in place of $\alpha = .05$, would the conclusion about μ change? Explain how the same data can reach a different conclusion about μ.

5.42   The sample data for a statistical test concerning μ yielded $n = 30$, $\bar{y} = 5.8$, $s = 4.11$. Determine the level of significance for testing $H_0: \mu = 4$ versus $H_a: \mu \ne 4$. Is there significant evidence in the data to support the research hypothesis that μ is different from 4 using $\alpha = .01$?

5.43   Refer to Exercise 5.42. If the research hypothesis was one-sided, $H_0: \mu \le 4$ versus $H_a: \mu > 4$, instead of two-sided, would your conclusion about μ change?

5.44   The researcher was interested in determining whether μ was less than 14. The sample data yielded $n = 40$, $\bar{y} = 13.5$, $s = 3.62$. Determine the level of significance for testing $H_0: \mu \ge 14$ versus $H_a: \mu < 14$. Is there significant evidence in the data to support the claim that μ is less than 14 using $\alpha = .05$?

5.45   Suppose the hypotheses in Exercise 5.44 were erroneously stated as $H_0: \mu \le 14$ versus $H_a: \mu > 14$. Determine the level of significance for this formulation of H_0 and H_a and show that the conclusion reached about μ has been reversed from the conclusion reached in Exercise 5.44.

Applications

Med.   **5.46**   A tobacco company advertises that the average nicotine content of its cigarettes is at most 14 milligrams. A consumer protection agency wants to determine whether the average nicotine content is in fact greater than 14. A random sample of 300 cigarettes of the company's brand yield an average nicotine content of 14.6 and a standard deviation of 3.8 milligrams. Determine the level of significance of the statistical test of the agency's claim that μ is greater than 14. If $\alpha = .01$, is there significant evidence that the agency's claim has been supported by the data?

Psy.   **5.47**   A psychological experiment was conducted to investigate the length of time (time delay) between the administration of a stimulus and the observation of a specified reaction. A random sample of 36 persons was subjected to the stimulus and the time delay was recorded. The sample mean and standard deviation were 2.2 and .57 seconds, respectively. Is there significant evidence that the mean time delay for the hypothetical population of all persons who may be subjected to the stimulus differs from 1.6 seconds? Use $\alpha = .05$. What is the level of significance of the test?

5.7   Inferences about μ for a Normal Population, σ Unknown

The estimation and test procedures about μ presented earlier in this chapter were based on the assumption that the population variance was known or that we had enough observations to allow s to be a reasonable estimate of σ. In this section, we present a test that can be applied when σ is unknown, no matter what the sample size, provided the population distribution is approximately normal. In Section 5.8, we will provide inference techniques for the situation where the population distribution is nonnormal. Consider the following example. Researchers would like to determine the average concentration of a drug in the bloodstream

1 hour after it is given to patients suffering from a rare disease. For this situation, it might be impossible to obtain a random sample of 30 or more observations at a given time. What test procedure could be used in order to make inferences about μ?

W. S. Gosset faced a similar problem around the turn of the century. As a chemist for Guinness Breweries, he was asked to make judgments on the mean quality of various brews, but was not supplied with large sample sizes to reach his conclusions.

Gosset thought that when he used the test statistic

$$z = \frac{\bar{y} - \mu_0}{\sigma/\sqrt{n}}$$

with σ replaced by s for small sample sizes, he was falsely rejecting the null hypothesis $H_0: \mu = \mu_0$ at a slightly higher rate than that specified by α. This problem intrigued him, and he set out to derive the distribution and percentage points of the test statistic

$$\frac{\bar{y} - \mu_0}{s/\sqrt{n}}$$

for $n < 30$.

For example, suppose an experimenter sets α at a nominal level—say, .05. Then he or she expects falsely to reject the null hypothesis approximately 1 time in 20. However, Gosset proved that the actual probability of a Type I error for this test was somewhat higher than the nominal level designated by α. He published the results of his study under the pen name Student, because at that time it was against company policy for him to publish his results in his own name. The quantity

$$\frac{\bar{y} - \mu_0}{s/\sqrt{n}}$$

Student's t   is called the t statistic and its distribution is called the *Student's t distribution* or, simply, **Student's t**. (See Figure 5.16.)

FIGURE 5.16

PDFs of two t distributions and a standard normal distribution

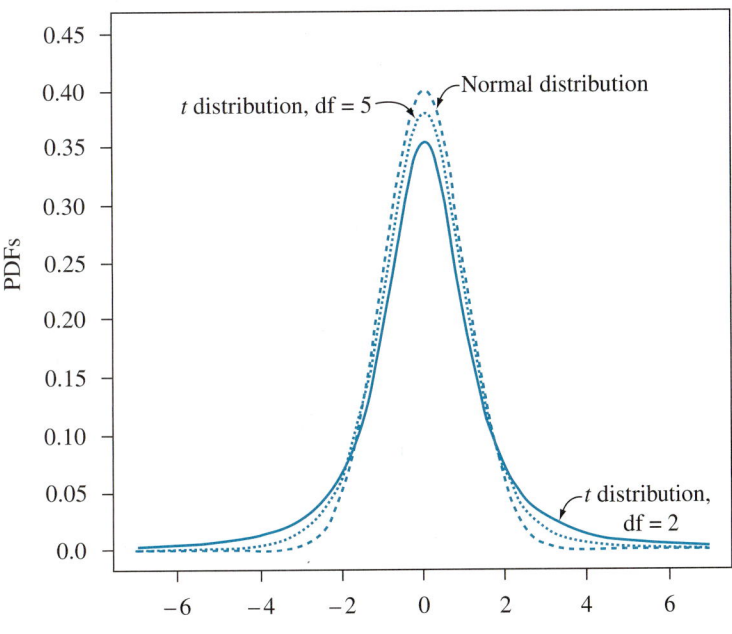

EXAMPLE 5.15

A massive multistate outbreak of food-borne illness was attributed to *Salmonella enteritidis*. Epidemiologists determined that the source of the illness was ice cream. They sampled nine production runs from the company that had produced the ice cream to determine the level of *Salmonella enteritidis* in the ice cream. These levels (MPN/g) are as follows:

.593   .142   .329   .691   .231   .793   .519   .392   .418

Use these data to determine whether the average level of *Salmonella enteritidis* in the ice cream is greater than .3 MPN/g, a level that is considered to be very dangerous. Set $\alpha = .01$.

Solution   The null and research hypotheses for this example are

$$H_0: \quad \mu \le .3$$
$$H_a: \quad \mu > .3$$

Because the sample size is small, we need to examine whether the data appear to have been randomly sampled from a normal distribution. Figure 5.18 is a normal probability plot of the data values. All 9 points fall nearly on the straight line. We conclude that the normality condition appears to be satisfied. Before setting up the rejection region and computing the value of the test statistic, we must first compute the sample mean and standard deviation. You can verify that

$$\bar{y} = .456 \quad \text{and} \quad s = .2128$$

FIGURE 5.18

Normal probability plot for *Salmonella* data

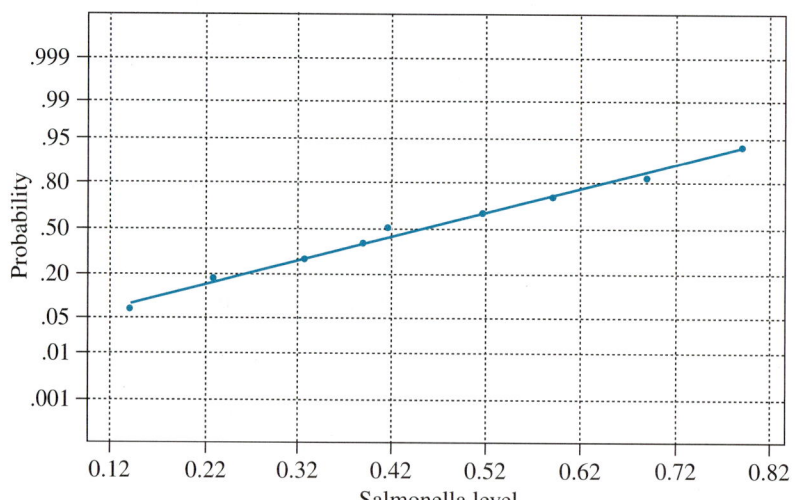

The rejection region with $\alpha = .01$ is

R.R.:   Reject H_0 if $t > 2.896$,

where from Table 2 in the Appendix, the value of $t_{.01}$ with df $= 9 - 1 = 8$ is 2.896. The computed value of t is

$$t = \frac{\bar{y} - \mu_0}{s/\sqrt{n}} = \frac{.456 - .3}{.2128/\sqrt{9}} = 2.21$$

The observed value of t is not greater than 2.896, so we have insufficient evidence to indicate that the average level of *Salmonella enteritidis* in the ice cream is greater than .3 MPN/g. The level of significance of the test is given by

$$p\text{-value} = P(t > \text{computed } t) = P(t > 2.21)$$

The t tables have only a few areas (a) for each value of df. The best we can do is bound the p-value. From Table 2 with df $= 8$, $t_{.05} = 1.860$ and $t_{.025} = 2.306$. Because computed $t = 2.21$, $.025 < p\text{-value} < .05$. However, with $\alpha = .01 < .025 < p\text{-value}$, we can still conclude that $p\text{-value} > \alpha$, and hence fail to reject H_0. The output from Minitab given here shows that the p-value $= .029$.

```
T-Test of the Mean

Test of mu ⇐ 0.3000 vs mu > 0.3000

Variable    N      Mean     StDev    SE Mean      T       P
Sal. Lev    9     0.4564    0.2128    0.0709    2.21    0.029


T Confidence Intervals

Variable    N      Mean     StDev    SE Mean        95.0 % CI
Sal. Lev    9     0.4564    0.2128    0.0709     (0.2928, 0.6201)
```

As we commented previously, in order to state that the level of *Salmonella enteritidis* is less than or equal to .3, we need to calculate the probability of Type II error for some crucial values of μ in H_a. These calculations are somewhat more complex than the calculations for the z test. We will use a set of graphs to determine $\beta(\mu_a)$. The value of $\beta(\mu_a)$ depends on three quantities, df $= n - 1$, α, and the distance d from μ_a to μ_0 in σ units,

$$d = \frac{|\mu_a - \mu_0|}{\sigma}$$

Thus, to determine $\beta(\mu_a)$, we must specify α, μ_a, and provide an estimate of σ. Then with the calculated d and df $= n - 1$, we locate $\beta(\mu_a)$ on the graph. Table 3 in the Appendix provides graphs of $\beta(\mu_a)$ for $\alpha = .01$ and $.05$ for both one-sided and two-sided hypotheses for a variety of values for d and df.

EXAMPLE 5.16

Refer to Example 5.15. We have $n = 9$, $\alpha = .01$, and a one-sided test. Thus, df $= 8$ and if we estimate $\sigma \approx .25$, we can compute the values of d corresponding to selected values of μ_a. The values of $\beta(\mu_a)$ can then be determined using the graphs in Table 3 in the Appendix. Figure 5.19 is the necessary graph for this example. To illustrate the calculations, let $\mu_a = .45$. Then

$$d = \frac{|\mu_a - \mu_0|}{\sigma} = \frac{|.45 - .3|}{.25} = .6$$

We draw a vertical line from $d = .6$ on the horizontal axis to the line labeled 8, our df. We then locate the value on the vertical axis at the height of the intersection,

skewed distributions

heavy-tailed distributions

There are two issues to consider when populations are assumed to be non-normal. First, what kind of nonnormality is assumed? Second, what possible effects do these specific forms of nonnormality have on the *t*-distribution procedures? The most important deviations from normality are **skewed distributions** and **heavy-tailed distributions.** Heavy-tailed distributions are roughly symmetric but have outliers relative to a normal distribution. Figure 5.21 displays four such distributions: Figure 5.21(a) is the standard normal distribution, Figure 5.21(b) is a heavy-tailed distribution (a *t* distribution with df = 3), Figure 5.21(c) is a distribution mildly skewed to the right, and Figure 5.21(d) is a distribution heavily skewed to the right.

To evaluate the effect of nonnormality as exhibited by skewness or heavy tails, we will consider whether the *t*-distribution procedures are still approximately correct for these forms of nonnormality and whether there are other more efficient procedures. For example, even if a test procedure for μ based on the *t* distribution gave nearly correct results for, say, a heavy-tailed population distribution, it might be possible to obtain a test procedure with a more accurate probability of Type I error and greater power if we test hypotheses about the population median in

FIGURE 5.21

(a) Density of the standard normal distribution. (b) Density of a heavy-tailed distribution. (c) Density of a lightly skewed distribution. (d) Density of a highly skewed distribution.

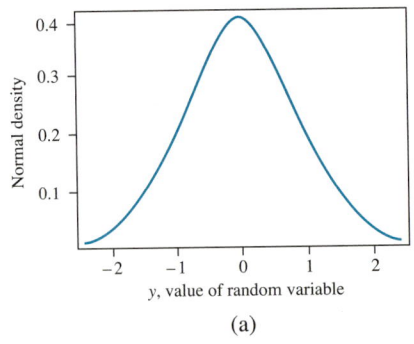

(a)

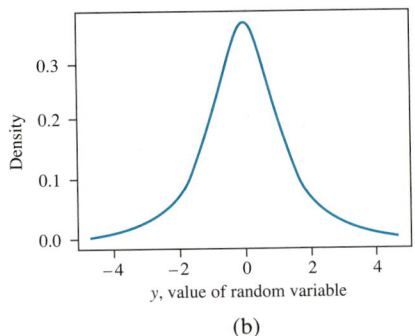

(b)

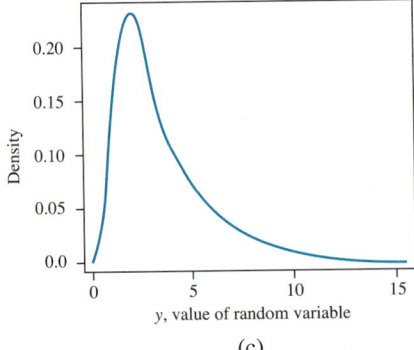

(c)

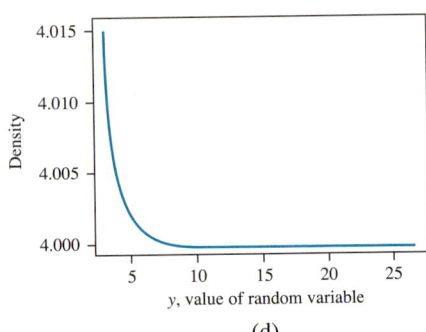

(d)

place of the population μ. Also, in the case of heavy-tailed or highly skewed population distributions, the median rather than μ is a more appropriate representation of the population center.

The question of approximate correctness of t procedures has been studied extensively. In general, probabilities specified by the t procedures, particularly the confidence level for confidence intervals and the Type I error for statistical tests, have been found to be fairly accurate, even when the population distribution is heavy-tailed. However, when the population is very heavy-tailed, as is the case in Figure 5.21(b), the tests of hypotheses tend to have probability of Type I errors smaller than the specified level, which leads to a test having much lower power and hence greater chances of committing Type II errors. Skewness, particularly with small sample sizes, can have an even greater effect on the probability of both Type I and Type II errors. When we are sampling from a population distribution that is normal, the sampling distribution of a t statistic is symmetric. However, when we are sampling from a population distribution that is highly skewed, the sampling distribution of a t statistic is skewed, not symmetric. Although the degree of skewness decreases as the sample size increases, there is no procedure for determining the sample size at which the sampling distribution of the t statistic becomes symmetric.

As a consequence, the level of a nominal $\alpha = .05$ test may actually have a level of .01 or less when the sample size is less than 20 and the population distribution looks like that of Figure 5.21(b), (c), or (d). Furthermore, the power of the test will be considerably less than when the population distribution is a normal distribution, thus causing an increase in the probability of Type II errors. A simulation study of the effect of skewness and heavy-tailedness on the level and power of the t test yielded the results given in Table 5.6. The values in the table are the power values for a level $\alpha = .05$ t test of $H_0: \mu \leq \mu_0$ versus $H_a: \mu > \mu_0$. The power values are calculated for shifts of size $d = |\mu_a - \mu_0|/\sigma$ for values of $d = 0, .2, .6, .8$. Three different sample sizes were used: $n = 10, 15$, and 20. When $d = 0$, the level of the test is given for each type of population distribution. We want to compare these values to .05. The values when $d > 0$ are compared to the corresponding values when sampling from a normal population. We observe that when sampling from the lightly skewed distribution and the heavy-tailed distribution, the levels are somewhat less than .05 with values nearly equal to .05 when using $n = 20$. However, when sampling from a heavily skewed distribution, even with $n = 20$ the level is only .011. The power values for the heavy-tailed and heavily skewed populations are considerably less than the corresponding values when sampling from a normal distribution. Thus, the test is much less likely

TABLE 5.6
Level and power values for t test

Population Distribution	$n = 10$ Shift d				$n = 15$ Shift d				$n = 20$ Shift d			
	0	.2	.6	.8	0	.2	.6	.8	0	.2	.6	.8
Normal	.05	.145	.543	.754	.05	.182	.714	.903	.05	.217	.827	.964
Heavy Tailed	.035	.104	.371	.510	.049	.115	.456	.648	.045	.163	.554	.736
Light Skewness	.025	.079	.437	.672	.037	.129	.614	.864	.041	.159	.762	.935
Heavy Skewness	.007	.055	.277	.463	.006	.078	.515	.733	.011	.104	.658	.873

to correctly detect that the alternative hypothesis H_a is true. This reduced power is present even when $n = 20$. When sampling from a lightly skewed population distribution, the power values are very nearly the same as the values for the normal distribution.

Because the t procedures have reduced power when sampling from skewed populations with small sample sizes, procedures have been developed that are **robust methods** not as affected by the skewness or extreme heavy-tailedness of the population distribution. These procedures are called **robust methods** of estimation and inference. Two robust procedures, the sign test and Wilcoxon signed rank test, will be considered in Section 5.8 and Chapter 6, respectively. They are both more efficient than the t test when the population distribution is very nonnormal in shape. Also, they maintain the selected α level of the test unlike the t test, which, when applied to very nonnormal data, has a true α value much different from the selected α value. The same comments can be made with respect to confidence intervals for the mean. When the population distribution is highly skewed, the coverage probability of a nominal $100(1 - \alpha)$ confidence interval is considerably less than $100(1 - \alpha)$.

So what is a nonexpert to do? First, examine the data through graphs. A boxplot or normal probability plot will reveal any gross skewness or extreme outliers. If the plots do not reveal extreme skewness or many outliers, the nominal t-distribution probabilities should be reasonably correct. Thus, the level and power calculations for tests of hypotheses and the coverage probability of confidence intervals should be reasonably accurate. If the plots reveal severe skewness or heavy-tailedness, the test procedures and confidence intervals based on the t-distribution will be highly suspect. In these situations, the median is a more appropriate measure of the center of the population than is the mean. In Section 5.8, we will develop test of hypotheses and confidence intervals for the median of a population. These procedures will maintain the nominal coverage probability of confidence intervals and the stated α level of tests of hypotheses when the population distribution is highly skewed or heavy-tailed.

We will now complete our analysis of the case study involving the percentage of calories from fat.

Analyzing Data for the Case Study One of the objectives of the study was to estimate the mean percentage of calories from fat in the diet of female nurses. Also, the researchers wanted to test whether the mean was greater than the recommended value of 30%. Recall that we had a random sample of 168 women and recorded the percentage of calories from fat (PCF) using the questionnaire. Before constructing confidence intervals or testing hypotheses, we must first check whether the data represent a random sample from normally distributed populations. From the stem-and-leaf plot displayed in Figure 5.1 and the normal probability plot in Figure 5.5, the data appear to follow a normal distribution. The mean and standard deviation of the PCF data were given by $\bar{y} = 36.92$ and $s = 6.73$. We can next construct a 95% confidence interval for the mean PCF for the population of nurses as follows:

$$36.92 \pm t_{.025,167} \frac{6.73}{\sqrt{168}}, \ 36.92 \pm 1.974 \frac{6.73}{\sqrt{168}}, \text{ or } 36.92 \pm 1.02$$

We are 95% confident that the mean PCF in the population of nurses is between 35.90 and 37.94. Thus, there is evidence that the mean PCF for the population of nurses exceeds the recommended value of 30. We will next formally test the hypotheses

$$H_0: \mu \leq 30 \text{ versus } H_0: \mu > 30$$

Because the data appear to be normally distributed and, in any case, the sample size is large, we can use the t test with rejection region as follows:

R.R.: For a one-tailed t test with $\alpha = .05$, we reject H_0 if

$$t = \frac{\bar{y} - 30}{s/\sqrt{168}} \geq t_{.05,167} = 1.654$$

Because $t = (36.92 - 30)/(6.73/\sqrt{168}) = 13.33$, we reject H_0. The p-value of the test is essentially 0, so we can conclude that the mean PCF value is significantly greater than 30. Thus, there is strong evidence that the population of nurses has an average PCF larger than the recommended value of 30. The experts in this field would have to determine the practical consequences of having a PCF value between 5.90 and 7.94 units higher than the recommended value.

Reporting Conclusions We need to write a report summarizing our findings from the study that would include the following items:

1. Statement of the objective for the study
2. Description of the study design and data collection procedures
3. Numerical and graphical summaries of data sets
4. Description of all inference methodologies:
   - t tests
   - t-based confidence interval on the population mean
   - Verification that all necessary conditions for using inference techniques were satisfied
5. Discussion of results and conclusions
6. Interpretation of findings relative to previous studies
7. Recommendations for future studies
8. Listing of the data set

EXERCISES

Basic Techniques

5.48 Why is the z test of Section 5.4 inappropriate for testing $H_0: \mu \geq \mu_0$ when $n < 30$ and σ is unknown?

5.49 Set up the rejection region based on t for the following conditions with $\alpha = .05$:
   a. $H_a: \mu < \mu_0$, $n = 15$
   b. $H_a: \mu \neq \mu_0$, $n = 23$
   c. $H_a: \mu > \mu_0$, $n = 6$

5.50 Repeat Exercise 5.49 with $\alpha = .01$.

5.51 The sample data for a t-test of $H_0: \mu \leq 15$ and $H_a: \mu > 15$ are $\bar{y} = 16.2$, $s = 3.1$, and $n = 18$. Use $\alpha = .05$ to draw your conclusions.

Applications

Edu.   **5.52** A new reading program was being evaluated in the fourth grade at an elementary school. A random sample of 20 students was thoroughly tested to determine reading speed and reading comprehension. Based on a fixed-length standardized test reading passage, the following speeds (in minutes) and comprehension scores (based on a 100-point scale) were obtained.

Student	1	2	3	4	5	6	7	8	9	10	11	12	13	14	15	16	17	18	19	20	n	$\bar{y}$	s
Speed	5	7	15	12	8	7	10	11	9	13	10	6	11	8	10	8	7	6	11	8	20	9.10	2.573
Comprehension	60	76	76	90	81	75	95	98	88	73	90	66	91	83	100	85	76	69	91	78	20	82.05	10.88

    a. Use the reading speed data to place a 95% confidence interval on μ, the average reading speed, for all fourth-grade students in the large school from which the sample was selected.
    b. Plot the reading speed data using a normal probability plot or boxplot to evaluate whether the data appear to be a random sample from a normal population distribution.
    c. Interpret the interval estimate in part (a).
    d. How would your inference change by using a 98% confidence interval in place of the 95% confidence interval?

5.53 Refer to Exercise 5.52. Using the reading comprehension data, is there significant evidence that the mean comprehension for all fourth graders is greater than 80, the statewide average for comparable students during the previous year? Give the level of significance for your test. Interpret your findings.

5.54 Refer to Exercise 5.53.
    a. Do you note any relationships in the data between the reading comprehension and reading speed of the individual students?
    b. What criticisms do you have of the study relative to evaluating the new reading program?

Bus.   **5.55** A consumer testing agency wants to evaluate the claim made by a manufacturer of discount tires. The manufacturer claims that their tires can be driven at least 35,000 miles before wearing out. To determine the average number of miles that can be obtained from the manufacturer's tires, the agency randomly selects 60 tires from the manufacturer's warehouse and places the tires on 15 cars driven by test drivers on a 2-mile oval track. The number of miles driven (in thousands of miles) until the tires are determined to be worn out is given in the following table.

Car	1	2	3	4	5	6	7	8	9	10	11	12	13	14	15	n	$\bar{y}$	s
Miles driven	25	27	35	42	28	37	40	31	29	33	30	26	31	28	30	15	31.47	5.04

    a. Place a 99% confidence interval on the average number of miles driven, μ, prior to the tires wearing out.
    b. Is there significant evidence ($\alpha = .01$) that the manufacturer's claim is false? What is the level of significance of your test? Interpret your findings.

5.56 Refer to Exercise 5.55. Using the Minitab output given, compare your results to the results given by the computer program.
    a. Does the normality assumption appear to be valid?
    b. How close to the true value were your bounds on the p-value?
    c. Is there a contradiction between the interval estimate of μ and the conclusion reached by your test of the hypotheses?

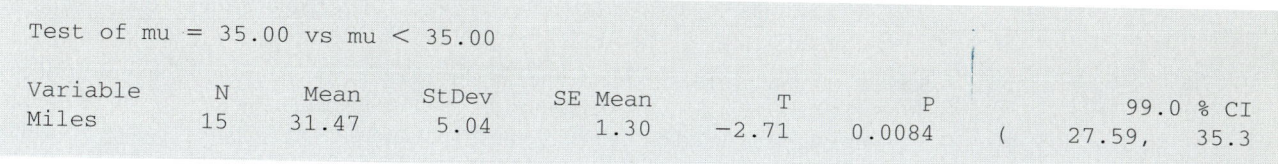

```
Test of mu = 35.00 vs mu < 35.00

Variable    N     Mean     StDev    SE Mean      T         P              99.0 % CI
Miles      15    31.47     5.04      1.30      -2.71    0.0084    (    27.59,    35.3
```

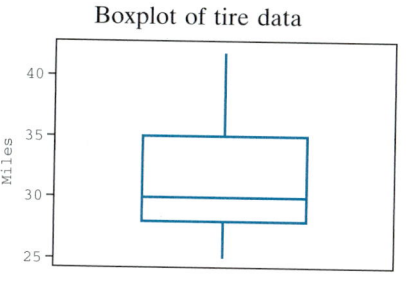

Boxplot of tire data

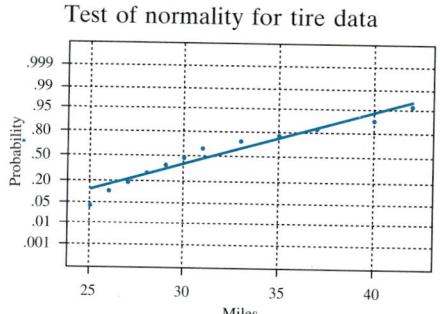

Test of normality for tire data

Env.   **5.57**   The amount of sewage and industrial pollutants dumped into a body of water affects the health of the water by reducing the amount of dissolved oxygen available for aquatic life. Over a 2-month period, 8 samples were taken from a river at a location 1 mile downstream from a sewage treatment plant. The amount of dissolved oxygen in the samples was determined and is reported in the following table. The current research asserts that the mean dissolved oxygen level must be at least 5.0 parts per million (ppm) for fish to survive.

Sample	1	2	3	4	5	6	7	8	n	$\bar{y}$	s
Oxygen (ppm)	5.1	4.9	5.6	4.2	4.8	4.5	5.3	5.2	8	4.95	.45

    a. Place a 95% confidence on the mean dissolved oxygen level during the 2-month period.

    b. Using the confidence interval from (a), does the mean oxygen level appear to be less than 5 ppm?

    c. Test the research hypothesis that the mean oxygen level is less than 5 ppm. What is the level of significance of your test? Interpret your findings.

Env.   **5.58**   A dealer in recycled paper places empty trailers at various sites. The trailers are gradually filled by individuals who bring in old newpapers and magazines, and are picked up on several schedules. One such schedule involves pickup every second week. This schedule is desirable if the average amount of recycled paper is more than 1,600 cubic feet per 2-week period. The dealer's records for eighteen 2-week periods show the following volumes (in cubic feet) at a particular site:

1,660   1,820   1,590   1,440   1,730   1,680   1,750   1,720   1,900
1,570   1,700   1,900   1,800   1,770   2,010   1,580   1,620   1,690
$\bar{y} = 1,718.3$ and $s = 137.8$

    a. Assuming the eighteen 2-week periods are fairly typical of the volumes throughout the year, is there significant evidence that the average volume μ is greater than 1,600 cubic feet?

    b. Place a 95% confidence interval on μ.

c. Compute the p-value for the test statistic. Is there strong evidence that μ is greater than 1,600?

Ag.   **5.59** Commercial growers of ornamental shrubs often want to retard the growth of the shrubs so that they do not become too large prior to being sold. A growth retardant, dikegulac, was evaluated on Kalanchoe, an ornamental shrub. The paper "Dikegulac alters growth and flowering of kalanchoe" (*HortScience* (1985), 20: 722–724) describes the results of these experiments. Ten shrubs were treated with dikegulac and another ten shrubs were untreated so the effect of the dikegulac on plant growth could be determined. The heights (cm) of the 20 plants were measured 13 weeks after the treatment date and the summary statistics are given in the following table.

	n	$\bar{y}$	s
Untreated	10	43.6	5.7
Treated	10	36.1	4.9

a. Construct 90% confidence intervals on the average height of the treated and untreated shrubs. Interpret these intervals.

b. Do the two confidence intervals overlap? What conclusions can you make about the effectiveness of dikegulac as a growth retardant?

Gov.   **5.60** A federal regulatory agency is investigating an advertised claim that a certain device can increase the gasoline mileage of cars (mpg). Ten such devices are purchased and installed in cars belonging to the agency. Gasoline mileage for each of the cars is recorded both before and after installation. The data are recorded here.

					Car								
	1	**2**	**3**	**4**	**5**	**6**	**7**	**8**	**9**	**10**	n	$\bar{x}$	s
Before (mpg)	19.1	29.9	17.6	20.2	23.5	26.8	21.7	25.7	19.5	28.2	10	23.22	4.25
After (mpg)	25.8	23.7	28.7	25.4	32.8	19.2	29.6	22.3	25.7	20.1	10	25.33	4.25
Change (mpg)	6.7	−6.2	11.1	5.2	9.3	−7.6	7.9	−3.4	6.2	−8.1	10	2.11	7.54

Place 90% confidence intervals on the average mpg for both the before and after phases of the study. Interpret these intervals. Does it appear that the device will significantly increase the average mileage of cars?

5.61 Refer to Exercise 5.60.
a. The cars in the study appear to have grossly different mileages before the devices were installed. Use the change data to test whether there has been a significant gain in mileage after the devices were installed. Use $\alpha = .05$.
b. Construct a 90% confidence interval for the mean change in mileage. On the basis of this interval, can one reject the hypothesis that the mean change is either zero or negative? (Note that the two-sided 90% confidence interval corresponds to a one-tailed $\alpha = .05$ test by using the decision rule: reject $H_0: \mu \le \mu_0$ if μ_0 is greater than the upper limit of the confidence interval.)

5.62 Refer to Exercise 5.60.
a. Calculate the probability of a Type II error for several values of μ_C, the average change in mileage. How do these values affect the conclusion you reached in Exercise 5.61?
b. Suggest some changes in the way in which this study in Exercise 5.60 was conducted.

5.8 Inferences about the Median

When the population distribution is highly skewed or very heavily tailed, the median is more appropriate than the mean as a representation of the center of the population. Furthermore, as was demonstrated in Section 5.7, the t procedures for constructing confidence intervals and for tests of hypotheses for the mean are not appropriate when applied to random samples from such populations with small sample sizes. In this section, we will develop a test of hypotheses and a confidence interval for the population median that will be appropriate for all types of population distributions.

The estimator of the population median M is based on the order statistics that were discussed in Chapter 3. Recall that if the measurements from a random sample of size n are given by $y_1, y_2, \ldots, y_n$, then the order statistics are these values ordered from smallest to largest. Let $y_{(1)} \leq y_{(2)} \leq \ldots \leq y_{(n)}$ represent the data in ordered fashion. Thus, $y_{(1)}$ is the smallest data value and $y_{(n)}$ is the largest data value. The estimator of the population median is the sample median $\hat{M}$. Recall that $\hat{M}$ is computed as follows.

If n is an odd number, then $\hat{M} = y_{(m)}$, where $m = (n + 1)/2$.

If n is an even number, then $\hat{M} = (y_{(m)} + y_{(m + 1)})/2$, where $m = n/2$.

To take into account the variability of $\hat{M}$ as an estimator of M, we next construct a confidence interval for M. A confidence interval for the population median M may be obtained by using the binomial distribution with $\pi = 0.5$.

100(1 − α)% Confidence Interval for the Median

A confidence interval for M with level of confidence at least $100(1 - \alpha)\%$ is given by

$$(M_L, M_U) = (y_{(L_{\alpha/2})}, y_{(U_{\alpha/2})})$$

where

$$L_{\alpha/2} = C_{\alpha(2),n}$$
$$U_{\alpha/2} = n - C_{\alpha(2),n} + 1.$$

Table 4 in the Appendix contains values for $C_{\alpha(2),n}$, which are percentiles from a binomial distribution with $\pi = 0.5$.

Because the confidence limits are computed using the binomial distribution, which is a discrete distribution, the level of confidence of (M_L, M_U) will generally be somewhat larger than the specified $100(1 - \alpha)\%$. The exact level of confidence is given by

$$\text{Level} = Pr(Bin(n, .5) \leq U_{\alpha/2} - 2) - Pr(Bin(n, .5) \leq L_{\alpha/2})$$

The following example will demonstrate the construction of the interval.

EXAMPLE 5.18

The sanitation department of a large city wants to investigate ways to reduce the amount of recyclable materials that are placed in the city's landfill. By separating the recyclable material from the remaining garbage, the city could prolong the life of the landfill site. More important, the number of trees needed to be harvested for paper products and aluminum needed for cans could be greatly reduced. From an analysis of recycling records from other cities, it is determined that if the average weekly amount of recyclable material is more than five pounds per household a commercial recycling firm could make a profit collecting the material. To determine the feasibility of the recycling plan, a random sample of 25 households is selected. The weekly weight of recyclable material (in pounds/week) for each household is given here.

| 14.2 | 5.3 | 2.9 | 4.2 | 1.2 | 4.3 | 1.1 | 2.6 | 6.7 | 7.8 | 25.9 | 43.8 | 2.7 |
| 5.6 | 7.8 | 3.9 | 4.7 | 6.5 | 29.5 | 2.1 | 34.8 | 3.6 | 5.8 | 4.5 | 6.7 | |

Determine an appropriate measure of the amount of recyclable waste from a typical household in the city.

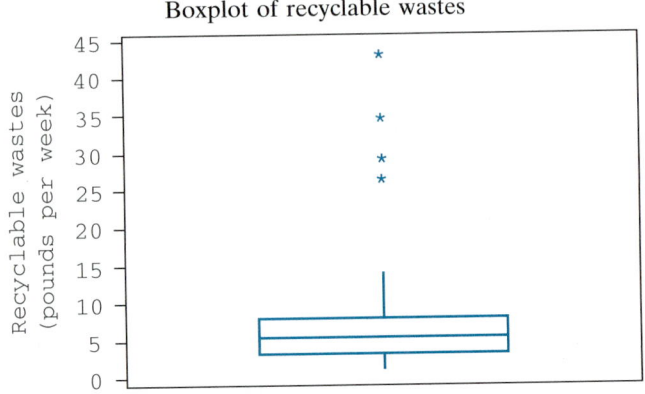

Boxplot of recyclable wastes

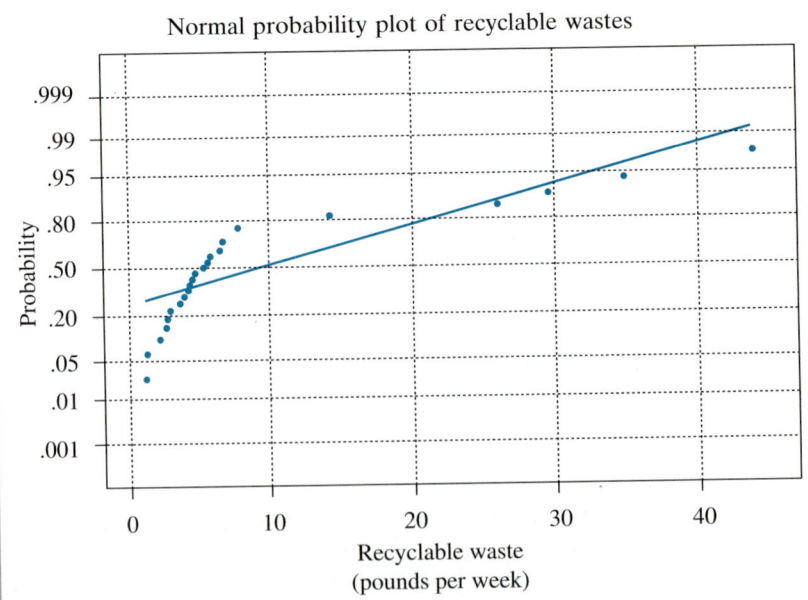

Normal probability plot of recyclable wastes

Solution A boxplot and normal probability of the recyclable waste data reveal the extreme right skewness of the data. Thus, the mean is not an appropriate representation of the typical household's potential recyclable material. The sample median and a confidence interval on the population are given by the following computations. First, we order the data from smallest value to largest value:

1.1	1.2	2.1	2.6	2.7	2.9	3.6	3.9	4.2	4.3	4.5	4.7	5.3
5.6	5.8	6.5	6.7	6.7	7.8	7.8	14.2	25.9	29.5	34.8	43.8	

The number of values in the data set is an odd number, so the sample median is given by

$$\hat{M} = y_{((25+1)/2)} = y_{(13)} = 5.3$$

The sample mean is calculated to be $\bar{y} = 9.53$. Thus, we have that 20 of the 25 households' weekly recyclable wastes are less than the sample mean. Note that 12 of the 25 waste values are less and 12 of the 25 are greater than the sample median. Thus, the sample median is more representative of the typical household's recyclable waste than is the sample mean. Next we will construct a 95% confidence interval for the population median.

From Table 4, we find

$$C_{\alpha(2),n} = C_{.05,25} = 7$$

Thus,

$$L_{.025} = C_{.05,25} = 7$$

$$U_{.025} = n - C_{.05,n} + 1 = 25 - 7 + 1 = 19$$

The 95% confidence interval for the population median is given by

$$(M_L, M_U) = (y_{(7)}, y_{(19)}) = (3.6, 7.8)$$

Using the binomial distribution, the exact level of coverage is given by $Pr(Bin(25, .5) \leq 19 - 2) - Pr(Bin(25, .5) \leq 7) = .9784 - .0216 = .9568$, which is slightly larger than the desired level 95%. Thus, we are at least 95% confident that the median amount of recyclable waste per household is between 3.6 and 7.8 pounds per week.

Large-Sample Approximation

When the sample size n is large, we can apply the normal approximation to the binomial distribution to obtain approximations to $C_{\alpha(2),n}$. The approximate value is given by

$$C_{\alpha(2),n} \approx \frac{n}{2} - z_{\alpha/2} \sqrt{\frac{n}{4}}$$

Because this approximate value for $C_{\alpha(2),n}$ is generally not an integer, we set $C_{\alpha(2),n}$ to be the largest integer that is less than or equal to the approximate value.

EXAMPLE 5.19

Using the data in Example 5.18, find a 95% confidence interval for the median using the approximation to $C_{\alpha(2),n}$.

Solution   We have $n = 25$ and $\alpha = .05$. Thus, $z_{.05/2} = 1.96$, and

$$C_{\alpha(2),n} \approx \frac{n}{2} - z_{\alpha/2}\sqrt{\frac{n}{4}} = \frac{25}{2} - 1.96\sqrt{\frac{25}{4}} = 7.6$$

Thus, we set $C_{\alpha(2),n} = 7$, and our confidence interval is identical to the interval constructed in Example 5.18. If n is larger than 30, the approximate and the exact value of $C_{\alpha(2),n}$ will often be the same integer.

In Example 5.18, the city wanted to determine whether the median amount of recyclable material was more than 5 pounds per household per week. We constructed a confidence interval for the median but we still have not answered the question of whether the median is greater than 5. Thus, we need to develop a test of hypotheses for the median.

We will use the ideas developed for constructing a confidence interval for the median in our development of the testing procedures for hypotheses concerning a population median. In fact, a $100(1 - \alpha)\%$ confidence interval for the population median M can be used to test two-sided hypotheses about M. If we want to test $H_0: M = M_0$ vs. $H_1: M \neq M_0$ at level α, then we construct a $100(1 - \alpha)\%$ confidence interval for M. If M_0 is contained in the confidence interval, then we fail to reject H_0. If M_0 is outside the confidence interval, then we reject H_0.

For testing one-sided hypotheses about M, we will use the binomial distribution to determine the rejection region. The testing procedure is called the **sign test** and is constructed as follows. Let $y_1, \ldots, y_n$ be a random sample from a population having median M. Let the null value of M be M_0 and define $W_i = y_i - M_0$. The sign test statistic B is the number of positive W_is. Note that B is simply the number of y_is that are greater than M_0. Because M is the population median, 50% of the data values are greater than M and 50% are less than M. Now, if $M = M_0$, then there is a 50% chance that y_i is greater than M_0 and hence a 50% chance that W_i is positive. Because the W_is are independent, each W_i has a 50% chance of being positive whenever $M = M_0$, and B counts the number of positive W_is under H_0, B is a binomial random variable with $\pi = .5$ and the percentiles from the binomial distribution with $\pi = .5$ given in Table 4 in the Appendix can be used to construct the rejection region for the test of hypothesis. The statistical **test for a population median M** is summarized next. Three different sets of hypotheses are given with their corresponding rejection regions. The tests given are appropriate for any population distribution.

sign test

test for a population median M

Summary of a Statistical Test for the Median M

Hypotheses:

Case 1. H_0: $M \leq M_0$ vs. H_a: $M > M_0$ (right-tailed test)
Case 2. H_0: $M \geq M_0$ vs. H_a: $M < M_0$ (left-tailed test)
Case 3. H_0: $M = M_0$ vs. H_a: $M \neq M_0$ (two-tailed test)

T.S.: Let $W_i = y_i - M_0$ and $B = $ number of positive W_is.

R.R.: For a probability α of a Type I error,

Case 1. Reject H_0 if $B \geq n - C_{\alpha(1),n}$.
Case 2. Reject H_0 if $B \leq C_{\alpha(1),n}$.
Case 3. Reject H_0 if $B \leq C_{\alpha(2),n}$ or $B \geq n - C_{\alpha(2),n}$.

The following example will illustrate the test of hypotheses for the population median.

EXAMPLE 5.20

Refer to Example 5.18. The Sanitation Department wanted to determine whether the median household recyclable wastes was greater than 5 pounds per week. Test this research hypothesis at level $\alpha = .05$ using the data from Exercise 5.18.

Solution The set of hypotheses are

$$H_0: \quad M \leq 5 \text{ versus } M > 5$$

The data set consisted of a random sample of $n = 25$ households. From Table 4 in the Appendix, we find $C_{\alpha(1),n} = C_{.05,25} = 7$. Thus, we will reject $H_0: M \leq 5$ if $B \geq n - C_{\alpha(1),n} = 25 - 7 = 18$. Let $W_i = y_i - M_0 = y_i - 5$, which yields

−3.9	−3.8	−2.9	−2.4	−2.3	−2.1	−1.4	−1.1	−0.8
−0.7	−0.5	−0.3	0.3	0.6	0.8	1.5	1.7	1.7
2.8	2.8	9.2	20.9	24.5	29.8	38.8		

The 25 values of W_i contain 13 positive values. Thus, $B = 13$, which is not greater than 18. We conclude the data set fails to demonstrate that the median household level of recyclable waste is greater than 5 pounds.

Large-Sample Approximation

When the sample size n is larger than the values given in Table 4 in the Appendix, we can use the normal approximation to the binomial distribution to set the rejection region. The standardized version of the sign test is given by

$$B_{ST} = \frac{B - (n/2)}{\sqrt{n/4}}$$

When M equals M_0, B_{ST} has approximately a standard normal distribution. Thus, we have the following decision rules for the three different research hypotheses.

Case 1. Reject H_0: $M \leq M_0$ if $B_{ST} \geq z_\alpha$, with p-value $= Pr(Z \geq B_{ST})$
Case 2. Reject H_0: $M \geq M_0$ if $B_{ST} \leq -z_\alpha$, with p-value $= Pr(Z \leq B_{ST})$
Case 3. Reject H_0: $M = M_0$ if $|B_{ST}| \geq z_{\alpha/2}$, with p-value $= 2Pr(Z \geq |B_{ST}|)$

where z_α is the standard normal percentile.

EXAMPLE 5.21

Using the information in Example 5.20, construct the large-sample approximation to the sign test, and compare your results to those obtained using the exact sign test.

Solution   Refer to Example 5.20, where we had $n = 25$ and $B = 13$. We conduct the large-sample approximation to the sign test as follows. We will reject H_0: $M \leq 5$ in favor of H_a: $M > 5$ if $B_{ST} \geq z_{.05} = 1.96$.

$$B_{ST} = \frac{B - (n/2)}{\sqrt{n/4}} = \frac{13 - (25/2)}{\sqrt{25/4}} = 0.2$$

Because B_{ST} is not greater than 1.96, we fail to reject H_0. The p-value $= Pr(Z \geq 0.2) = 1 - Pr(Z < 0.2) = 1 - .5793 = .4207$ using Table 1 in the Appendix. Thus, we reach the same conclusion as was obtained using the exact sign test.

In Section 5.7, we observed that the performance of the t test deteriorated when the population distribution was either very heavily tailed or highly skewed. In Table 5.7, we compute the level and power of the sign test and compare these values to the comparable values for the t test for the four population distributions depicted in Figure 5.21 in Section 5.7. Ideally, the level of the test should remain the same for all population distributions. Also, we want tests having the largest possible power values because the power of a test is its ability to detect false null hypotheses. When the population distribution is either heavy tailed or highly skewed, the level of the t test changes from its stated value of .05. In these situations, the level of the sign test stays the same because the level of the sign test is the same for all distributions. The power of the t test is greater than the power of the sign test when sampling from a population having a normal distribution. However, the power of the sign test is greater than the power of the t test when sampling from very heavily tailed distributions or highly skewed distributions.

TABLE 5.7

Level and power values of the t test versus the sign test

Population Distribution	Test Statistic	$n = 10$ $(M_a - M_0)/\sigma$				$n = 15$ $(M_a - M_0)/\sigma$				$n = 20$ $(M_a - M_0)/\sigma$			
		Level	.2	.6	.8	Level	.2	.6	.8	Level	.2	.6	.8
Normal	t	.05	.145	.543	.754	.05	.182	.714	.903	.05	.217	.827	.964
	Sign	.055	.136	.454	.642	.059	.172	.604	.804	.058	.194	.704	.889
Heavy Tailed	t	.035	.104	.371	.510	.049	.115	.456	.648	.045	.163	.554	.736
	Sign	.055	.209	.715	.869	.059	.278	.866	.964	.058	.325	.935	.990
Lightly Skewed	t	.055	.140	.454	.631	.059	.178	.604	.794	.058	.201	.704	.881
	Sign	.025	.079	.437	.672	.037	.129	.614	.864	.041	.159	.762	.935
Highly Skewed	t	.007	.055	.277	.463	.006	.078	.515	.733	.011	.104	.658	.873
	Sign	.055	.196	.613	.778	.059	.258	.777	.912	.058	.301	.867	.964

EXERCISES

Basic Techniques

5.63 Suppose we have a random sample of n measurements from a population having median M. We want to place a 90% confidence interval on M.
   a. If $n = 20$, find $L_{\alpha/2}$ and $U_{\alpha/2}$ using Table 4 in the Appendix.
   b. Use the large-sample approximation to find $L_{\alpha/2}$ and $U_{\alpha/2}$ and compare these values to the values found in part (a).

5.64 Suppose we have a random sample of n measurements from a population having median M. We want to place a 90% confidence interval on M.
   a. If $n = 40$, find $L_{\alpha/2}$ and $U_{\alpha/2}$ using Table 4 in the Appendix.
   b. Use the large-sample approximation to find $L_{\alpha/2}$ and $U_{\alpha/2}$ and compare these values to the values found in part (a).

5.65 Suppose we have a random sample of 30 measurements from a population having median M. We want to test H_0: $M \leq M_0$ versus H_a: $M > M_0$ at level $\alpha = .05$. Set up the rejection region for testing these hypotheses using the values in Table 4 of the Appendix.

5.66 Refer to Exercise 5.65. Use the large-sample approximation to set up the rejection region and compare your results to the rejection region obtained in Exercise 5.65.

5.67 Suppose we have a random sample of 50 measurements from a population having median M. We want to test H_0: $M = M_0$ versus H_a: $M \neq M_0$ at level $\alpha = .05$. Set up the rejection region for testing these hypotheses using the values in Table 4 of the Appendix.

5.68 Refer to Exercise 5.67. Use the large-sample approximation to set up the rejection region and compare your results to the rejection region obtained in Exercise 5.67.

Applications

Bus.

5.69 The amount of money spent on health care is an important issue for workers because many companies provide health insurance that only partial covers many medical procedures. The director of employee benefits at a midsize company wants to determine the amount spent on health care by the typical hourly worker in the company. A random sample of 35 workers is selected and the amount they spent on their families' health care needs during the past year is given here.

400	345	248	1,290	398	218	197	342	208	223	531	172	4,321
143	254	201	3,142	219	276	326	207	225	123	211	108	

    a. Graph the data using a boxplot or normal probability plot and determine whether the population has a normal distribution.

    b. Based on your answer to part (a), is the mean or the median cost per household a more appropriate measure of what the typical worker spends on health care needs?

    c. Place a 95% confidence interval on the amount spent on health care by the typical worker. Explain what the confidence interval is telling us about the amount spent on health care needs.

    d. Does the typical worker spend more than $400 per year on health care needs? Use $\alpha = .05$.

Gov.

5.70 Many states have attempted to reduce the blood-alcohol level at which a driver is declared to be legally drunk. There has been resistance to this change in the law by certain business groups who have argued that the current limit is adequate. A study was conducted to demonstrate the effect on reaction time of a blood-alcohol level of .1%, the current limit in many states. A random sample of 25 persons of legal driving age had their reaction time recorded in a standard laboratory test procedure before and after drinking a sufficient amount of alcohol to raise their blood alcohol to a .1% level. The difference (After–Before) in their reaction times in seconds was recorded as follows:

.01	.02	.04	.05	.07	.09	.11	.26	.27	.27	.28	.28	.29
.29	.30	.31	.31	.32	.33	.35	.36	.38	.39	.39	.40	

    a. Place a 99% confidence interval on both the mean and median difference in reaction times of drivers who have a blood-alcohol level of .1%.

    b. Is there sufficient evidence that a blood-alcohol level of .1% causes any increase in the mean reaction time?

    c. Is there sufficient evidence that a blood-alcohol level of .1% causes any increase in the median reaction time?

    d. Which summary of reaction time differences seems more appropriate, the mean or median? Justify your answer.

5.71 Refer to Exercise 5.70. The lobbyist for the business group has their expert examine the experimental equipment and determines that there may be measurement errors in recording the reaction times. Unless the difference in reaction time is at least .25 seconds, the expert claims that the two times are essentially equivalent.

    a. Is there sufficient evidence that the median difference in reaction time is greater than .25 seconds?

    b. What other factors about the drivers are important in attempting to decide whether moderate consumption of alcohol affects reaction time?

Soc.   **5.72** In an attempt to increase the amount of money people would receive at retirement from Social Security, the U.S. Congress during its 1999 session debated whether a portion of Social Security funds should be invested in the stock market. Advocates of mutual stock funds reassure the public by stating that most mutual funds would provide a larger retirement income than the income currently provided by Social Security. The annual rates of return of two highly recommended mutual funds for the years 1989 through 1998 are given here. (The annual rate of return is defined as $(P_1-P_0)/P_0$, where P_0 and P_1 are the prices of the fund at the beginning and end of the year, respectively.)

Year	1989	1990	1991	1992	1993	1994	1995	1996	1997	1998
Fund A	25.4	17.1	−8.9	26.7	3.6	−8.5	−1.3	32.9	22.9	26.6
Fund B	31.9	−8.4	41.8	6.2	17.4	−2.1	30.5	15.8	26.8	5.7

a. For both Fund A and Fund B, estimate the mean and median annual rate of return and construct a 95% confidence interval for each.

b. Which of the parameters, the mean or median, do you think best represents the annual rate of return for Fund A and for Fund B during the years 1989 through 1998? Justify your answer.

5.73 Refer to Exercise 5.72.

a. Is there sufficient evidence that the median annual rate of return for the two mutual funds is greater than 10%?

b. Is there sufficient evidence that the mean annual rate of return for the two mutual funds is greater than 10%?

5.74 What other summaries of the mutual fund's rate of return are of importance to a person selecting a retirement plan?

5.75 Using the information in Table 5.7, answer the following questions.

a. If the population has a normal distribution, then the population mean and median are identical. Thus, either the mean or median could be used to represent the center of the population. In this situation, why is the *t* test more appropriate than the sign test for testing hypotheses about the center of the distribution?

b. Suppose the population has a distribution that is highly skewed to the right. The researcher uses an $\alpha = .05$ *t* test to test hypotheses about the population mean. If the sample size $n = 10$, will the probability of a Type I error for the test be .05? Justify your answer.

c. When testing hypotheses about the mean or median of a highly skewed population, the difference in power between the sign and *t* test decreases as the size of $(M_a - M_0)$ increases. Verify this statement using the values in Table 5.7. Why do think this occurs?

d. When testing hypotheses about the mean or median of a lightly skewed population, the difference in power between the sign and *t* test is much less than that for a highly skewed population distribution. Verify this statement using the values in Table 5.7. Why do you think this occurs?

5.9   Summary

A population mean or median can be estimated using point or interval estimation. The selection of the median in place of the mean as a representation of the center of a population depends on the shape of the population distribution. The performance of an interval estimate is determined by the width of the interval and the confidence coefficient. The formulas for a $100(1 - \alpha)\%$ confidence interval for the mean μ and median M were given. A formula was provided for determining the necessary sample size in a study so that a confidence interval for μ would have a predetermined width and level of confidence.

Following the traditional approach to hypothesis testing, a statistical test consists of five parts: research hypothesis, null hypothesis, test statistic, rejection region, and checking assumptions and drawing conclusions. A statistical test employs the technique of proof by contradiction. We conduct experiments and studies to gather data to verify the research hypothesis through the contradiction of the null hypothesis H_0. As with any two-decision process based on variable data, there are two types of errors that can be committed. A Type I error is the rejection of H_0 when H_0 is true and a Type II error is the acceptance of H_0 when the alternative hypothesis H_a is true. The probability for a Type I error is denoted by α. For a given value of the mean μ_a in H_a, the probability of a Type II error is denoted by $\beta(\mu_a)$. The value of $\beta(\mu_a)$ decreases as the distance from μ_a to μ_0 increases. The power of a test of hypothesis is the probability that the test will reject H_0 when the value of μ resides in H_a. Thus, the power at μ_a equals $1 - \beta(\mu_a)$.

We also demonstrated that for a given sample size and value of the mean μ_a, α and $\beta(\mu_a)$ are inversely related; as α is increased, $\beta(\mu_a)$ decreases, and vice versa. If we specify the sample size n and α for a given test procedure, we can compute $\beta(\mu_a)$ for values of the mean μ_a in the alternative hypothesis. In many studies, we need to determine the necessary sample size n to achieve a testing procedure having a specified value for α and a bound on $\beta(\mu_a)$. A formula is provided to determine n such that a level α test has $\beta(\mu_a) \le \beta$ whenever μ_a is a specified distance beyond μ_0.

We developed an alternative to the traditional decision-based approach for a statistical test of hypotheses. Rather than relying on a preset level of α, we compute the weight of evidence in the data for rejecting the null hypothesis. This weight, expressed in terms of a probability, is called the level of significance for the test. Most professional journals summarize the results of a statistical test using the level of significance. We discussed how the level of significance can be used to obtain the same results as the traditional approach.

We also considered inferences about μ when σ is unknown (which is the usual situation). Through the use of the t distribution, we can construct both confidence intervals and a statistical test for μ. The t-based tests and confidence intervals do not have the stated levels or power when the population distribution is highly skewed or very heavy tailed and the sample size is small. In these situations, we may use the median in place of the mean to represent the center of the population. Procedures were provided to construct confidence intervals and tests of hypothesis for the population median.

Key Formulas

Estimation and tests for μ and the median

1. $100(1 - \alpha)\%$ confidence interval for μ (σ known) when sampling from a normal population or n large

$$\bar{y} \pm z_{\alpha/2}\sigma_{\bar{y}}, \text{ where } \sigma_{\bar{y}} = \sigma/\sqrt{n}$$

2. $100(1 - \alpha)\%$ confidence interval for μ (σ unknown) when sampling from a normal population or n large

$$\bar{y} \pm t_{\alpha/2}s/\sqrt{n}, \qquad df = n - 1$$

3. Sample size for estimating μ with a $100(1 - \alpha)\%$ confidence interval, $\bar{y} \pm E$

$$n = \frac{(z_{\alpha/2})^2\sigma^2}{E^2}$$

4. Statistical test for μ (σ known) when sampling from a normal population or n large

$$\text{Test statistic: } z = \frac{\bar{y} - \mu_0}{\sigma/\sqrt{n}}$$

5. Statistical test for μ (σ unknown) when sampling from a normal population or n large

$$\text{Test statistic: } t = \frac{\bar{y} - \mu_0}{s/\sqrt{n}}, \qquad df = n - 1$$

6. Calculation of $\beta(\mu_a)$ (and equivalent power) for a test on μ (σ known) when sampling from a normal population or n large
   a. One-tailed level α test

$$\beta(\mu_a) = P\left(z < z_\alpha - \frac{|\mu_0 - \mu_a|}{\sigma_{\bar{y}}}\right)$$

   where $\sigma_{\bar{y}} = \sigma/\sqrt{n}$

   b. Two-tailed level α test

$$\beta(\mu_a) \approx P\left(z < z_{\alpha/2} - \frac{|\mu_0 - \mu_a|}{\sigma_{\bar{y}}}\right)$$

   where $\sigma_{\bar{y}} = \sigma/\sqrt{n}$

7. Calculation of $\beta(\mu_a)$ (and equivalent power) for a test on μ (σ unknown) when sampling from a normal population or n large: Use Table 3 in the Appendix.

8. Sample size n for a statistical test on μ (σ known) when sampling from a normal population
   a. One-tailed level α test

$$n = \frac{\sigma^2}{\Delta^2}(z_{\alpha/2} + z_\beta)^2$$

   b. One-tailed level α test

$$n \approx \frac{\sigma^2}{\Delta^2}(z_{\alpha/2} + z_{\beta/2})^2$$

9. $100(1 - \alpha)\%$ confidence interval for the population median M

$$(y_{(L_{\alpha/2})}, y_{(U_{\alpha/2})}), \qquad \text{where } L_{\alpha/2} = C_{\alpha(2),n} \quad \text{and} \quad U_{\alpha/2} = n - C_{\alpha(2),n} + 1$$

10. Statistical test for median

Test statistic:

Let $W_i = y_i - M_0$ and B = number of positive W_is

Supplementary Exercises

Bus.   **5.76**   A paint manufacturer wishes to validate its advertisement statement that a gallon of its paint covers on the average more than 400 square feet. An independent testing laboratory is hired to evaluate the advertisement statement based on fifty 1-gallon cans of paint randomly selected from the manufacturer's warehouse.
  a. In words, what is the parameter of interest?
  b. What is the research hypothesis and the corresponding rejection region if the manufacturer wants the probability of a Type I error to be at most .05?
  c. If the random sample of 50 cans produces an average coverage of 412 square feet with a standard deviation of 38 square feet, is the manufacturer's statement supported by the data?
  d. Construct a 95% confidence interval on the average coverage of 1 gallon of the manufacturer's paint.
  e. Determine the p-value of the test. Does your conclusion about the manufacturer's statement change if α is reduced to .01?

Engin.   **5.77**   The transportation department of a large city remodeled one of its parking garages and increased the hourly parking rates. From the city's records, the average parking time over the past 5 years was 220 minutes. The department wants to know whether the remodeling and rate increases have changed the mean parking time. Over a 3-month period after the changes were made, a random sample of 100 cars had an average parking time of 208 minutes with a standard deviation of 55 minutes.
  a. What is the research hypothesis for the study?
  b. Do the data support the research hypothesis if $\alpha = .05$?
  c. What is the significance level (p-value) of the test?
  d. Construct a 95% confidence interval for the average parking time after the changes were made to the garage.

H.R.   **5.78**   An office manager has implemented an incentive plan that she thinks will reduce the mean time required to handle a customer complaint. The mean time for handling a complaint was 30 minutes prior to implementing the incentive plan. After the plan was in place for several months, a random sample of the records of 38 customers who had complaints revealed a mean time of 28.7 minutes with a standard deviation of 3.8 minutes.
  a. Give a point estimate of the mean time required to handle a customer complaint.
  b. What is the standard deviation of the point estimate given in (a)?
  c. Construct a 95% confidence on the mean time to handle a complaint after implementing the plan. Interpret the confidence interval for the office manager.
  d. Is there sufficient evidence that the incentive plan has reduced the mean time to handle a complaint?

Env.   **5.79**   The concentration of mercury in a lake has been monitored for a number of years. Measurements taken on a weekly basis yielded an average of 1.20 mg/m³ (milligrams per cubic meter) with a standard deviation of .32 mg/m³. Following an accident at a smelter on the shore of the lake, 15 measurements produced the following mercury concentrations.

1.60	1.77	1.61	1.08	1.07	1.79	1.34	1.07
1.45	1.59	1.43	2.07	1.16	0.85	2.11	

Bus. **5.89** Statistics has become a valuable tool for auditors, especially where large inventories are involved. It would be costly and time consuming for an auditor to inventory each item in a large operation. Thus, the auditor frequently resorts to obtaining a random sample of items and using the sample results to check the validity of a company's financial statement. For example, a hospital financial statement claims an inventory that averages $300 per item. An auditor's random sample of 20 items yielded a mean and standard deviation of $160 and $90, respectively. Do the data contradict the hospital's claimed mean value per inventoried item and indicate that the average is less than $300? Use $\alpha = .05$.

Bus. **5.90** Over the past 5 years, the mean time for a warehouse to fill a buyer's order has been 25 minutes. Officials of the company believe that the length of time has increased recently, either due to a change in the workforce or due to a change in customer purchasing policies. The processing time (in minutes) was recorded for a random sample of 15 orders processed over the past month.

$$
\begin{array}{ccccc}
28 & 25 & 27 & 31 & 10 \\
26 & 30 & 15 & 55 & 12 \\
24 & 32 & 28 & 42 & 38 \\
\end{array}
$$

Do the data present sufficient evidence to indicate that the mean time to fill an order has increased?

Engin. **5.91** If a new process for mining copper is to be put into full-time operation, it must produce an average of more than 50 tons of ore per day. A 15-day trial period gave the results shown in the accompanying table.

Day	1	2	3	4	5	6	7	8	9	10	11	12	13	14	15
Yield (tons)	57.8	58.3	50.3	38.5	47.9	157.0	38.6	140.2	39.3	138.7	49.2	139.7	48.3	59.2	49.7

a. Estimate the typical amount of ore produce by the mine using both a point estimate and a 95% confidence interval.
b. Is there significant evidence that on a typical day the mine produces more than 50 tons of ore? Test by using $\alpha = .05$.

Edu. **5.92** A test was conducted to determine the length of time required for a student to read a specified amount of material. The students were instructed to read at the maximum speed at which they could still comprehend the material because a comprehension test would be given on the material. Sixteen third-grade students were randomly selected from a large school district, and the results of their test are as follows (in minutes):

$$
\begin{array}{cccccccc}
25 & 18 & 27 & 29 & 20 & 19 & 25 & 24 \\
32 & 21 & 24 & 19 & 23 & 28 & 31 & 22 \\
\end{array}
$$

a. Estimate the mean length of time required for all third-grade students to read the material, using a 95% confidence interval.
b. What is the population for which the confidence interval in (a) is applicable?
c. Provide an interpretation of the confidence interval computed in (a).

Med. **5.93** A drug manufacturer produces an antibiotic in large fermentation vats. To determine the average potency for the batch of antibiotic being prepared, the vat is sampled at 12 different locations. The potency readings of the antibiotic are recorded as follows:

$$
\begin{array}{cccccc}
8.9 & 9.0 & 9.1 & 8.3 & 9.2 & 9.0 \\
8.4 & 9.2 & 9.0 & 8.7 & 9.3 & 9.1 \\
\end{array}
$$

a. Estimate the mean potency for the batch based on a 95% confidence interval. Interpret the interval.
b. How would you select the 12 samples from the vat?

c. If the potency of the antibiotic is stated to be 9.0, is there significant evidence that the mean potency differs from the stated value?

5.94 In a statistical test about μ, the null hypothesis was rejected. Based on this conclusion, which of the following statements are true?
- **a.** A Type I error was committed.
- **b.** A Type II error was committed.
- **c.** A Type I error could have been committed.
- **d.** A Type II error could have been committed.
- **e.** It is impossible to have committed both Type I and Type II errors.
- **f.** It is impossible that neither a Type I nor a Type II error was committed.
- **g.** Whether any error was committed is not known, but if an error was made, it was Type I.
- **h.** Whether any error was committed is not known, but if an error was made, it was Type II.

5.95 Answer "true" or "false" for each statement.
- **a.** In a level $\alpha = .05$ test of hypothesis, increasing the sample size will not affect the level of the test.
- **b.** In a level $\alpha = .05$ test of hypothesis, increasing the sample size will not affect the power of the test.
- **c.** The sample size n plays an important role in testing hypotheses because it measures the amount of data (and hence information) upon which we base a decision. If the data are quite variable and n is small, it is unlikely that we will reject the null hypothesis even when the null hypothesis is false.
- **d.** Suppose we are testing the following hypothesis about the population mean μ, $H_0: \mu \leq \mu_0$ versus $H_a: \mu > \mu_0$. If the sample size n is very large and the data are not highly variable, it is very likely that we will reject the null hypothesis even when the true value of μ is only trivially larger than μ_0.
- **e.** When making inferences about a population that has a distribution highly skewed to the right, an $\alpha = .05$ t test is less likely to make a Type II error than is an $\alpha = .05$ sign test.
- **f.** When making inferences about a population that has a distribution highly skewed to the right, an $\alpha = .05$ t test is less likely to make a Type I error than is an $\alpha = .05$ sign test.

5.96 Complete the following statements (more than one word may be needed).
- **a.** If we take all possible samples (of a given sample size) from a population, then the distribution of sample means tends to be _____ and the mean of these sample means is equal _____.
- **b.** The larger the sample size, other things remaining equal, the _____ the confidence interval.
- **c.** The larger the confidence coefficient, other things remaining equal, the _____ the confidence interval.
- **d.** The statement "If random samples of a fixed size are drawn from any population (regardless of the form of the population distribution), as n becomes larger, the distribution of sample means approaches normality," is known as the _____.
- **e.** By failing to reject a null hypothesis that is false, one makes a _____ error.

Med. **5.97** Suppose that the tar content of cigarettes is normally distributed with a mean of 10 and a standard deviation of 2.4 mg. A new manufacturing process is developed for decreasing the tar content. A sample of 16 cigarettes produced by the new process yields a mean of 8.8 mg. Use $\alpha = .05$.
- **a.** Do a test of hypothesis to determine whether the new process has significantly *decreased* the tar content. Use the following outline.

  Null hypothesis
  Alternative hypothesis

Assumptions
Rejection region(s)
Test statistic and computations
Conclusion in statistical terms
Conclusion in plain English

b. Based on your conclusion, could you have made a Type I error? A Type II error? Neither error? Both Type I and Type II errors?

Env. **5.98** The board of health of a particular state was called to investigate claims that raw pollutants were being released into the river flowing past a small residential community. By applying financial pressure, the state was able to get the violating company to make major concessions toward the installation of a new water purification system. In the interim, different production systems were to be initiated to help reduce the pollution level of water entering the stream. To monitor the effect of the interim system, a random sample of 50 water specimens was taken throughout the month at a location downstream from the plant. If $\bar{y} = 5.0$ and $s = .70$, use the sample data to determine whether the mean dissolved oxygen count of the water (in ppm) is less than 5.2, the average reading at this location over the past year.

a. List the five parts of the statistical test, using $\alpha = .05$.
b. Conduct the statistical test and state your conclusion.

Engin. **5.99** An automatic merge system has been installed at the entrance ramp to a major highway. Prior to the installation of the system, investigators found the average stress level of drivers to be 8.2 on a 10-point scale. After installation, a sample of 50 drivers showed $\bar{y} = 7.6$ and $s = 1.8$. Conduct a statistical test of the research hypothesis that the average stress at peak hours for drivers under the new system is less than 8.2, the average stress level prior to the installation of the automatic merge system. Determine the level of significance of the statistical test. Interpret your findings.

Env. **5.100** The search for alternatives to oil as a major source of fuel and energy will inevitably bring about many environmental challenges. These challenges will require solutions to problems in such areas as strip mining and many others. Let us focus on one. If coal is considered as a major source of fuel and energy, we will have to consider ways to keep large amounts of sulfur dioxide (SO_2) and particulates from getting into the air. This is especially important at large government and industrial operations. Here are some possibilities.

1. Build the smokestack extremely high.
2. Remove the SO_2 and particulates from the coal prior to combustion.
3. Remove the SO_2 from the gases after the coal is burned but before the gases are released into the atmosphere. This is accomplished by using a scrubber.

A new type of scrubber has been recently constructed and is set for testing at a power plant. Over a 15-day period, samples are obtained three times daily from gases emitted from the stack. The amounts of SO_2 emissions (in pounds per million BTU) are given here:

Time	Day														
	1	2	3	4	5	6	7	8	9	10	11	12	13	14	15
6 A.M.	.158	.129	.176	.082	.099	.151	.084	.155	.163	.077	.116	.132	.087	.134	.179
2 P.M.	.066	.135	.096	.174	.179	.149	.164	.122	.063	.111	.059	.118	.134	.066	.104
10 P.M.	.128	.172	.106	.165	.163	.200	.228	.129	.101	.068	.100	.119	.125	.182	.138

a. Estimate the average amount of SO_2 emissions during each of the three time periods using 95% confidence intervals.
b. Does there appear to be a significant difference in average SO_2 emissions over the three time periods?
c. Combining the data over the entire day, is the average SO_2 emissions using the new scrubber less than .145, the average daily value for the old scrubber?

Soc. **5.101** As part of an overall evaluation of training methods, an experiment was conducted to determine the average exercise capacity of healthy male army inductees. To do this, each male in a random sample of 35 healthy army inductees exercised on a bicycle ergometer (a device for measuring work done by the muscles) under a fixed work load until he tired. Blood pressure, pulse rates, and other indicators were carefully monitored to ensure that no one's health was in danger. The exercise capacities (mean time, in minutes) for the 35 inductees are listed here.

23	19	36	12	41	43	19
28	14	44	15	46	36	25
35	25	29	17	51	33	47
42	45	23	29	18	14	48
21	49	27	39	44	18	13

a. Use these data to construct a 95% confidence interval for μ, the average exercise capacity for healthy male inductees. Interpret your findings.
b. How would your interval change using a 99% confidence interval?

5.102 Using the data in Exercise 5.101, determine the number of sample observations that would be required to estimate μ to within 1 minute, using a 95% confidence interval. (*Hint:* Substitute $s = 12.36$ for σ in your calculations.)

Ag. **5.103** A study was conducted to examine the effect of a preparation of mosaic virus on tobacco leaves. In a random sample of $n = 32$ leaves, the mean number of lesions was 22, with a standard deviation of 3. Use these data and a 95% confidence interval to estimate the average number of lesions for leaves affected by a preparation of mosaic virus.

5.104 Refer to Exercise 5.103. Use the sample data to form a 99% confidence interval on μ, the average number of lesions for tobacco leaves affected by a preparation of mosaic virus.

Med. **5.105** We all remember being told, "Your fever has subsided, and your temperature has returned to normal." What do we mean by the word *normal*? Most people use the benchmark 98.6°F, but this does not apply to all people, only the "average" person. Without putting words into someone's mouth, we might define a person's normal temperature to be his or her average temperature when healthy, but even this definition is cloudy because a person's temperature varies throughout the day. To determine a person's normal temperature, we recorded it for a random sample of 30 days. On each day selected for inclusion in the sample, the temperature reading was made at 7 A.M. The sample mean and standard deviation for these 30 readings were, respectively, 98.4 and .15. Assuming the person was healthy on the days examined, use these data to estimate the person's 7 A.M. "normal" temperature using a 90% confidence interval.

5.106 Refer to the data of Exercise 5.101. Suppose that the random sample of 35 inductees was selected from a large group of new army personnel being subjected to a new (and hopefully improved) physical fitness program. Assume previous testing with several thousand personnel over the past several years has shown an average exercise capacity of 29 minutes. Run a statistical test for the research hypothesis that the average exercise capacity is improved for the new fitness program. Give the level of significance for the test. Interpret your findings.

5.107 Refer to Exercise 5.106.
  a. How would the research hypothesis change if we were interested in determining whether the new program is better or worse than the physical fitness program for inductees?
  b. What is the level of significance for your test?

5.108 In a random sample of 40 hospitals from a list of hospitals with over 100 semiprivate beds, a researcher collected information on the proportion of persons whose bills are covered by a group policy under a major medical insurance carrier. The sample proportions are given in the following chart.

.67	.74	.68	.63	.91	.81	.79	.73
.82	.93	.92	.59	.90	.75	.76	.88
.85	.90	.77	.51	.67	.67	.92	.72
.69	.73	.71	.76	.84	.74	.54	.79
.71	.75	.70	.82	.93	.83	.58	.84

Use the sample data to construct a 90% confidence interval on μ, the average proportion of patients per hospital with group medical insurance coverage. Interpret the interval.

H.R. **5.109** Faculty members in a state university system who resign within 10 years of initial employment are entitled to receive the money paid into a retirement system, plus 4% per year. Unfortunately, experience has shown that the state is extremely slow in returning this money. Concerned about such a practice, a local teachers' organization decides to investigate. From a random sample of 50 employees who resigned from the state university system over the past 5 years, the average time between the termination date and reimbursement was 75 days, with a standard deviation of 15 days. Use the data to estimate the mean time to reimbursement, using a 95% confidence interval.

5.110 Refer to Exercise 5.109. After a confrontation with the teachers' union, the state promised to make reimbursements within 60 days. Monitoring of the next 40 resignations yields an average of 58 days, with a standard deviation of 10 days. If we assume that these 40 resignations represent a random sample of the state's future performance, estimate the mean reimbursement time, using a 99% confidence interval.

Soc. **5.111** A random sample of birth rates from 40 inner-city areas shows an average of 35 per thousand, with a standard deviation of 6.3. Estimate the mean inner-city birth rate. Use a 95% confidence interval.

Soc. **5.112** A random sample of 30 standard metropolitan statistical areas (SMSAs) was selected and the ratio (per 1,000) of registered voters to the total number of persons 18 years and over was recorded in each area. Use the data given to test the research hypothesis that μ, the average ratio (per 1,000), is different from 675, last year's average ratio. Give the level of significance for your test.

802	497	653	600	729	812
751	730	635	605	760	681
807	747	728	561	696	710
641	848	672	740	818	725
694	854	674	683	695	803

Bus. **5.113** Improperly filled orders are a costly problem for mail-order houses. To estimate the mean loss per incorrectly filled order, a large firm plans to sample n incorrectly filled orders and to determine the added cost associated with each one. The firm estimates that the added cost is between $40 and $400. How many incorrectly filled orders must be sampled to estimate the mean additional cost using a 95% confidence interval of width $20?

5.114 Records from a particular hospital were examined to determine the average length of stay for patients being treated for lung cancer. Data from a sample of 100 records showed $\bar{y} = 2.1$ months and $s = 2.6$ months.

  a. Would a confidence interval for μ based on t be appropriate? Why or why not?
  b. Indicate an alternative procedure for estimating the center of the distribution.

Gov. **5.115** Investigators of food stamp fraud would like to estimate the average annual gross income of participants in the food stamp program to within $750 using a 95% confidence interval. If we assume the annual gross incomes for food stamp participants have a range of $20,000, determine the number of participants that should be included in the study.

Bus. **5.116** As indicated earlier, the stated weight on the new giant-sized laundry detergent package is 42 ounces. Also displayed on the box is the following statement: "Individual boxes of this product may weigh slightly more or less than the marked weight due to normal variations incurred with high-speed packaging machines, but each day's production of detergent average slightly above the marked weight." Discuss how you might attempt to test this claim. Would it be simpler to modify this claim slightly for testing purposes? State all parts of your test. Would there be any way to determine in advance the sample size required to pick up a specified alternative with power equal to .90, using $\alpha = .05$?

Med. **5.117** Congestive heart failure is known to be fatal in a high percentage of cases. A total of 182 patients with chronic left-ventricular failure who were symptomatic in spite of therapy were followed. The length of survival for these patients ranged from 1 to 41 months with a mean of 12 months and a standard deviation of 10. Would a confidence interval for the mean survival of these patients be appropriate? Why or why not?

Bus. **5.118** After a decade of steadily increasing popularity, the sales of automatic teller machines (ATMs) have been on the decline. In a recent month, a spot check of a random sample of 40 suppliers indicated that shipments averaged 20% lower than those for the corresponding period 1 year ago. Assume the sandard deviation is 6.2% and the percentage data appear mound-shaped. Use these data to construct a 99% confidence interval on the mean percentage decrease in shipments of ATMs.

5.119 Suppose the percentage change in shipments of ATMs from the 40 suppliers of Exercise 5.118 ranged from -40% (a 40% decrease) to $+16\%$ (a 16% increase) with a sample mean of -20%, a median decrease of -10%, and a 10% trimmed mean of -12%. Discuss the appropriateness of the t methods for examining the percentage change in shipments of ATMs.

Med. **5.120** Doctors have recommended that we try to keep our caffeine intake at 200 mg or less per day. With the following chart, a sample of 35 office workers was asked to record their caffeine intake for a 7-day period.

Coffee (6 oz)	100–150 mg
Tea (6 oz)	40–110 mg
Cola (12 oz)	30 mg
Chocolate cake	20–30 mg
Cocoa (6 oz)	5–20 mg
Milk chocolate (1 oz)	5–10 mg

After the 7-day period, the average daily intake was obtained for each worker. The sample mean and standard deviation of the daily averages were 560 mg and 160 mg, respectively. Use these data to estimate μ, the average daily intake, using a 90% confidence interval.

5.121 Refer to Exercise 5.120. How many additional observations would be needed to estimate μ to within ± 10 mg with 90% confidence?

Ag. **5.122** Investigators from the Ohio Department of Agriculture recently selected a junior high school in the area and took samples of the half-pint (8-ounce) milk cartons used for student lunches. Based on 25 containers, the investigators found that the cartons were .067 ounces short of a full half pint on the average, with a standard deviation of .02.

    a. Use these data to test the hypothesis that the average shortfall is zero against a one-sided alternative. Give the *p*-value for your test.

    b. Although .067 ounces is only a few drops, predict the annual savings (in pints) for the dairy if it sells 3 million 8-ounce cartons of milk each year with this shortweight.

5.123   Refer to the clinical trials database on the data disk to construct a 95% confidence interval for the mean HAM-D total score of treatment group C. How would this interval change for a 99% confidence interval?

5.124   Using the clinical trials database, give a 90% confidence interval for the Hopkins Obrist cluster score of treatment A.

Inferences Comparing Two Population Central Values

6.1 Introduction and Case Study

6.2 Inferences about $\mu_1 - \mu_2$: Independent Samples

6.3 A Nonparametric Alternative: The Wilcoxon Rank Sum Test

6.4 Inferences about $\mu_1 - \mu_2$: Paired Data

6.5 A Nonparametric Alternative: The Wilcoxon Signed-Rank Test

6.6 Choosing Sample Sizes for Inferences about $\mu_1 - \mu_2$

6.7 Summary

6.1 Introduction and Case Study

The inferences we have made so far have concerned a parameter from a single population. Quite often we are faced with an inference involving a comparison of parameters from different populations. We might wish to compare the mean corn crop yield for two different varieties of corn, the mean annual income for two ethnic groups, the mean nitrogen content of two different lakes, or the mean length of time between administration and eventual relief for two different antivertigo drugs.

In many sampling situations, we will select independent random samples from two populations to compare the populations' parameters. The statistics used to make these inferences will, in many cases, be the difference between the corresponding sample statistics. Suppose we select independent random samples of n_1 observations from one population and n_2 observations from a second population. We will use the difference between the sample means, $(\bar{y}_1 - \bar{y}_2)$, to make an inference about the difference between the population means, $(\mu_1 - \mu_2)$.

Case Study: Effects of Oil Spill on Plant Growth

On January 7, 1992, an underground oil pipeline ruptured and caused the contamination of a marsh along the Chiltipin Creek in San Patricio County, Texas. The cleanup process consisted of burning the contaminated regions in the marsh. To evaluate the influence of the oil spill on the flora, researchers designed a study of plant growth 1 year after the burning. In an unpublished Texas A&M University dissertation, Newman (1997) describes the researchers' findings with respect to *Distichlis spicata,* a flora of particular importance to the area of the spill.

Designing the Data Collection The researchers needed to determine the important characteristics of the flora that may be affected by the spill. Here are some of the questions to be answered before starting the study:

1. What are the factors that determine the viability of the flora?
2. How did the oil spill affect these factors?
3. Are there data on the important flora factors prior to the spill?
4. How should the researchers measure the flora factors in the oil spill region?
5. How many observations are necessary to confirm that the flora have undergone a change after the oil spill?
6. What type of experimental design or study is needed?
7. What statistical procedures are valid for making inferences about the change in flora parameters after the oil spill?
8. What types of information should be included in a final report to document any changes observed in the flora parameters?

After lengthy discussion, reading of the relevant literature, and searching many databases about similar sites and flora, the researchers found there was no specific information on the flora in this region prior to the oil spill. They determined that the flora parameters of interest were the average density μ of *Distichlis spicata* after burning the spill region, the variability σ in flora density, and the proportion π of the spill region in which the flora density was essentially zero. Because there was no relevant information on flora density in the spill region before the spill, it was necessary to evaluate the flora density in unaffected areas of the marsh to determine whether the plant density had changed after the oil spill. The researchers located several regions that had not been contaminated by the oil spill. The spill region and the unaffected regions were divided into tracts of nearly the same size. The number of tracts needed in the study was determined by specifying how accurately the parameters μ, σ, and π needed to be estimated to achieve a level of precision as specified by the width of 95% confidence intervals and by the power of tests of hypotheses. From these calculations and within budget and time limitations, it was decided that 40 tracts from both the spill and unaffected areas would be used in the study. Forty tracts of exactly the same size were randomly selected in these locations and the *Distichlis spicata* density was recorded. Similar measurements were taken within the spill area of the marsh.

Managing the Data The data consist of 40 measurements of flora density in the uncontaminated (control) sites and 40 density measurements in the spill (burned) sites. The researchers would next carefully examine the data from the field work to determine whether the measurements were recorded correctly. The data would then be transferred to computer files and prepared for analysis following the steps outlined in Section 2.5.

Analyzing the Data The next step in the study would be to summarize the data through plots and summary statistics. The data are displayed in Figure 6.1 with

FIGURE 6.1

Number of plants observed in tracts at oil spill and control sites. The data are displayed in stem-and-leaf plots

Control Tracts				Oil Spill Tracts	
Mean:	38.48	000	0	Mean:	26.93
Median:	41.50	7	0	59	Median:	26.00
St. Dev:	16.37	1	1	14	St. Dev:	9.88
n:	40	6	1	77799	n:	40
		4	2	2223444		
		9	2	555667779		
		0	3	11123444		
		55678	3	5788		
		000111222233	4	1		
		57	4			
		0112344	5	02		
		67789	5			

TABLE 6.1

Summary statistics for oil spill data

			Descriptive Statistics			
Variable	**Site Type**	**N**	**Mean**	**Median**	**Tr. Mean**	**St. Dev.**
No. plants	Control	40	38.48	41.50	39.50	16.37
	Oil spill	40	26.93	26.00	26.69	9.88
Variable	**Site Type**	**SE Mean**	**Minimum**	**Maximum**	**Q1**	**Q3**
No. plants	Control	2.59	0.00	59.00	35.00	51.00
	Oil spill	1.56	5.00	52.00	22.00	33.75

summary statistics given in Table 6.1. A boxplot of the data displayed in Figure 6.2 indicates that the control sites have a somewhat greater plant density than the oil spill sites. From the summary statistics, we have that the average flora density in the control sites is $\bar{y}_{Con} = 38.48$ with a standard deviation of $s_{Con} = 16.37$. The sites within the spill region have an average density of $\bar{y}_{Spill} = 26.93$ with a standard deviation of $s_{Spill} = 9.88$. Thus, the control sites have a larger average flora density and a greater variability in flora density than the sites within the spill region. Whether these observed differences in flora density reflect similar differences in

FIGURE 6.2

Number of plants observed in tracts at control sites (1) and oil spill sites (2)

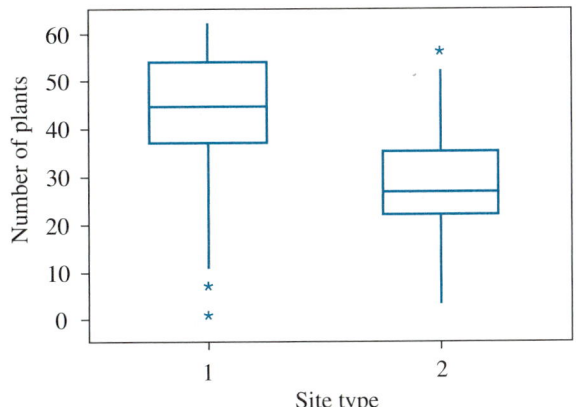

Confidence Interval for
$\mu_1 - \mu_2$, **Independent**
Samples

$$(\bar{y}_1 - \bar{y}_2) \pm t_{\alpha/2} s_p \sqrt{\frac{1}{n_1} + \frac{1}{n_2}}$$

where

$$s_p = \sqrt{\frac{(n_1 - 1)s_1^2 + (n_2 - 1)s_2^2}{n_1 + n_2 - 2}} \quad \text{and} \quad \text{df} = n_1 + n_2 - 2$$

The sampling distribution of $\bar{y}_1 - \bar{y}_2$ is a normal distribution, with standard deviation

$$\sigma_{\bar{y}_1 - \bar{y}_2} = \sqrt{\frac{\sigma_1^2}{n_1} + \frac{\sigma_2^2}{n_2}} = \sqrt{\frac{\sigma^2}{n_1} + \frac{\sigma^2}{n_2}} = \sigma \sqrt{\frac{1}{n_1} + \frac{1}{n_2}}$$

because we require that the two populations have the same standard deviation σ. If we knew the value of σ, then we would use $z_{\alpha/2}$ in the formula for the confidence interval. Because σ is unknown in most cases, we must estimate its value. This estimate is denoted by s_p and is formed by combining (pooling) the two independent estimates of σ, s_1, and s_2. In fact, s_p^2 is a **weighted average** of the sample variances s_1^2 and s_2^2. We have to estimate the standard deviation of the point estimate of $\mu_1 - \mu_2$, so we must use the percentile from the t-distribution $t_{\alpha/2}$ in place of the normal percentile, $z_{\alpha/2}$. The degrees of freedom for the t-percentile are df $= n_1 + n_2 - 2$, because we have a total of $n_1 + n_2$ data values and two parameters μ_1 and μ_2 that must be estimated prior to estimating the standard deviation σ. Remember that we use $\bar{y}_1$ and $\bar{y}_2$ in place of μ_1 and μ_2, respectively, in the formulas for s_1^2 and s_2^2.

s_p^2, **a weighted average**

Recall that we are assuming that the two populations from which we draw the samples have normal distributions with a common variance σ^2. If the confidence interval presented were valid only when these assumptions were met exactly, the estimation procedure would be of limited use. Fortunately, the confidence coefficient remains relatively stable if both distributions are mound-shaped and the sample sizes are approximately equal. For those situations in which these conditions do not hold, we will discuss alternative procedures in this section and in Section 6.3.

EXAMPLE 6.1

Company officials were concerned about the length of time a particular drug product retained its potency. A random sample of $n_1 = 10$ bottles of the product was drawn from the production line and analyzed for potency.

A second sample of $n_2 = 10$ bottles was obtained and stored in a regulated environment for a period of 1 year. The readings obtained from each sample are given in Table 6.2.

Suppose we let μ_1 denote the mean potency for all bottles that might be sampled coming off the production line and μ_2 denote the mean potency for all bottles that may be retained for a period of 1 year. Estimate $\mu_1 - \mu_2$ by using a 95% confidence interval.

TABLE 6.2

Potency reading for two
samples

Fresh		Stored	
10.2	10.6	9.8	9.7
10.5	10.7	9.6	9.5
10.3	10.2	10.1	9.6
10.8	10.0	10.2	9.8
9.8	10.6	10.1	9.9

Solution   The potency readings for the fresh and stored bottles are plotted in Figures 6.4(a) and (b) in normal probability plots to assess the normality assumption. We find that the plotted points in both plots fall very close to a straight line, and hence the normality condition appears to be satisfied for both types of bottles. The summary statistics for the two samples are presented next.

Fresh Bottles

$n_1 = 10$
$\bar{y}_1 = 10.37$
$s_1 = 0.3234$

Stored Bottles

$n_2 = 10$
$\bar{y}_2 = 9.83$
$s_2 = 0.2406$

FIGURE 6.4

(a) Normal probability plot: potency of fresh bottles;
(b) Normal probability plot: potency of stored bottles

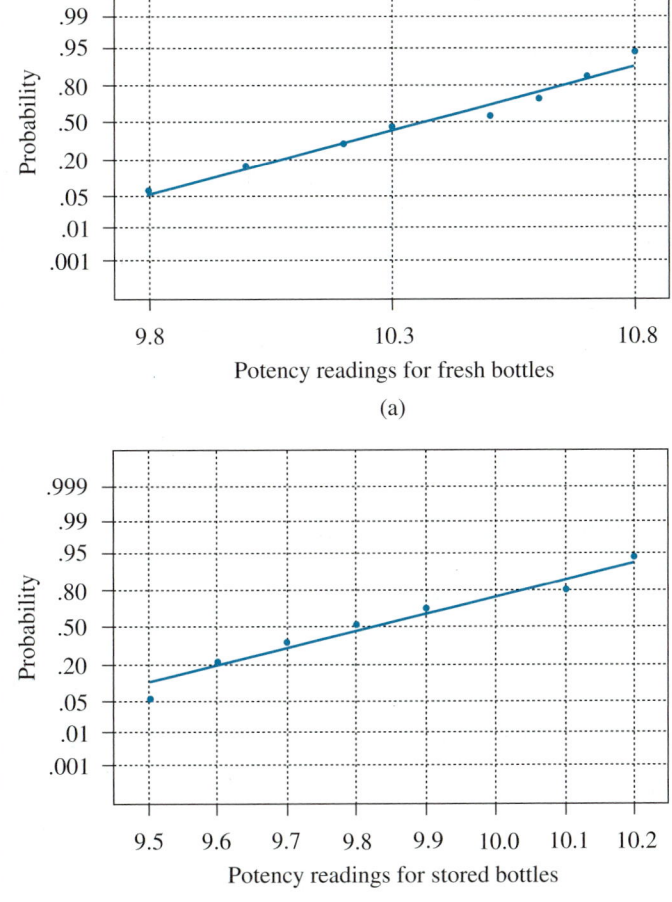

(a)

(b)

In Chapter 7, we will provide a test of equality for two population variances. However, for the above data, the computed sample standard deviations are approximately equal considering the small sample sizes. Thus, the required conditions necessary to construct a confidence interval on $\mu_1 - \mu_2$—that is, normality, equal variances, and independent random samples—appear to be satisfied. The estimate of the common standard deviation σ is

$$s_p = \sqrt{\frac{(n_1 - 1)s_1^2 + (n_2 - 1)s_2^2}{n_1 + n_2 - 2}} = \sqrt{\frac{9(.3234)^2 + 9(.2406)^2}{18}} = .285$$

The t-value based on df $= n_1 + n_2 - 2 = 18$ and $a = .025$ is 2.101. A 95% confidence interval for the difference in mean potencies is

$$(10.37 - 9.83) \pm 2.101(.285)\sqrt{1/10 + 1/10}$$

$$.54 \pm .268 \text{ or } (.272, .808)$$

We estimate that the difference in mean potencies for the bottles from the production line and those stored for 1 year, $\mu_1 - \mu_2$, lies in the interval .272 to .808. Company officials would then have to evaluate whether a decrease in mean potency of size between .272 and .808 would have a practical impact on the useful potency of the drug.

EXAMPLE 6.2

A school district decided that the number of students attending their high school was nearly unmanageable, so it was split into two districts, with District 1 students going to the old high school and District 2 students going to a newly constructed building. A group of parents became concerned with how the two districts were constructed relative to income levels. A study was thus conducted to determine whether persons in suburban District 1 have a different mean income from those in District 2. A random sample of 20 homeowners was taken in District 1. Although 20 homeowners were to be interviewed in District 2 also, one person refused to provide the information requested, even though the researcher promised to keep the interview confidential. Thus, only 19 observations were obtained from District 2. The data, recorded in thousands of dollars, produced sample means and variances as shown in Table 6.3. Use these data to construct a 95% confidence interval for $(\mu_1 - \mu_2)$.

TABLE 6.3

Income data for Example 6.2

	District 1	District 2
Sample Size	20	19
Sample Mean	18.27	16.78
Sample Variance	8.74	6.58

Solution A preliminary analysis using histograms plotted for the two samples suggests that the two populations are mound-shaped (near normal). Also, the sample variances are very similar. The difference in the sample means is

$$\bar{y}_1 - \bar{y}_2 = 18.27 - 16.78 = 1.49$$

The estimate of the common standard deviation σ is

$$s_p = \sqrt{\frac{(n_1 - 1)s_1^2 + (n_2 - 1)s_2^2}{n_1 + n_2 - 2}} = \sqrt{\frac{19(8.74) + 18(6.58)}{20 + 19 - 2}} = 2.77$$

The t-percentile for $a = \alpha/2 = .025$ and df $= 20 + 19 - 2 = 37$ is not listed in Table 2 of the Appendix, but taking the labeled value for the nearest df less than 37 (df $= 35$), we have $t_{.025} = 2.030$. A 95% confidence interval for the difference in mean incomes for the two districts is of the form

$$\bar{y}_1 - \bar{y}_2 \pm t_{\alpha/2} s_p \sqrt{\frac{1}{n_1} + \frac{1}{n_2}}$$

Substituting into the formula, we obtain

$$1.49 \pm 2.030(2.77) \sqrt{\frac{1}{20} + \frac{1}{19}} \quad \text{or} \quad 1.49 \pm 1.80$$

Thus, we estimate the difference in mean incomes to lie somewhere in the interval from $-.31$ to 3.29. If we multiply these limits by \$1,000, the confidence interval for the difference in mean incomes is $-\$310$ to \$3,290. This interval includes both positive and negative values for $\mu_1 - \mu_2$, so we are unable to determine whether the mean income for District 1 is larger or smaller than the mean income for District 2.

We can also test a hypothesis about the difference between two population means. As with any test procedure, we begin by specifying a research hypothesis for the difference in population means. Thus, we might, for example, specify that the difference $\mu_1 - \mu_2$ is greater than some value D_0. (*Note:* D_0 will often be 0.) The entire test procedure is summarized here.

A Statistical Test for $\mu_1 - \mu_2$, Independent Samples

The assumptions under which the test will be valid are the same as were required for constructing the confidence interval on $\mu_1 - \mu_2$: population distributions are normal with equal variances and the two random samples are independent.

H_0: 1. $\mu_1 - \mu_2 \leq D_0$ (D_0 is a specified value, often 0)
      2. $\mu_1 - \mu_2 \geq D_0$
      3. $\mu_1 - \mu_2 = D_0$

H_a: 1. $\mu_1 - \mu_2 > D_0$
      2. $\mu_1 - \mu_2 < D_0$
      3. $\mu_1 - \mu_2 \neq D_0$

T.S.: $t = \dfrac{(\bar{y}_1 - \bar{y}_2) - D_0}{s_p \sqrt{\dfrac{1}{n_1} + \dfrac{1}{n_2}}}$

R.R.: For a level α, Type I error rate and with df $= n_1 + n_2 - 2$,
      1. Reject H_0 if $t \geq t_\alpha$.
      2. Reject H_0 if $t \leq -t_\alpha$.
      3. Reject H_0 if $|t| \geq t_{\alpha/2}$.

Check assumptions and draw conclusions.

EXAMPLE 6.3

An experiment was conducted to evaluate the effectiveness of a treatment for tapeworm in the stomachs of sheep. A random sample of 24 worm-infected lambs of approximately the same age and health was randomly divided into two groups. Twelve of the lambs were injected with the drug and the remaining twelve were left untreated. After a 6-month period, the lambs were slaughtered and the following worm counts were recorded:

Drug-Treated Sheep	18	43	28	50	16	32	13	35	38	33	6	7
Untreated Sheep	40	54	26	63	21	37	39	23	48	58	28	39

a. Test whether the mean number of tapeworms in the stomachs of the treated lambs is less than the mean for untreated lambs. Use an $\alpha = .05$ test.
b. What is the level of significance for this test?
c. Place a 95% confidence interval on $\mu_1 - \mu_2$ to assess the size of the difference in the two means.

Solution

a. Boxplots of the worm counts for the treated and untreated lambs are displayed in Figure 6.5. From the plots, we can observe that the data for the untreated lambs are symmetric with no outliers and the data for the treated lambs are slightly skewed to the left with no outliers. Also, the widths of the two boxes are approximately equal. Thus, the condition that the population distributions are normal with equal variances appears to be satisfied. The condition of independence of the worm counts both between and within the two groups is evaluated by considering how the lambs were selected, assigned to the two groups, and cared for during the 6-month experiment. Because the 24 lambs were randomly selected from a representative herd of infected lambs, were randomly assigned to the treated and untreated groups, and were properly separated and cared for during the 6-month period of the experiment, the 24 worm counts are presumed to be independent random samples from the two populations. Finally, we can observe from the boxplots that the untreated lambs appear to have higher worm counts than the treated lambs because the median line is higher for the un-

FIGURE 6.5

Boxplots of worm counts for treated (1) and untreated (2) sheep

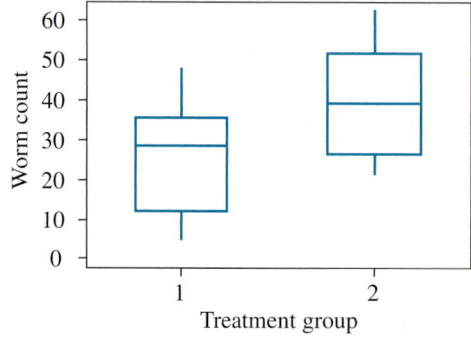

treated group. The following test confirms our observation. The data for the treated and untreated sheep are summarized next.

Drug-Treated Lambs

$n_1 = 12$
$\bar{y}_1 = 26.58$
$s_1 = 14.36$

Untreated Lambs

$n_2 = 12$
$\bar{y}_2 = 39.67$
$s_2 = 13.86$

The sample standard deviations are of a similar size, so from this and from our observation from the boxplot, the pooled estimate of the common population standard deviation σ is now computed:

$$s_p = \sqrt{\frac{(n_1 - 1)s_1^2 + (n_2 - 1)s_2^2}{n_1 + n_2 - 2}} = \sqrt{\frac{11(14.36)^2 + 11(13.86)^2}{22}} = 14.11$$

The test procedure for evaluating the research hypothesis that the treated lambs have mean tapeworm count (μ_1) less than the mean level (μ_2) for untreated lambs is as follows:

H_0: $\mu_1 - \mu_2 \geq 0$ (that is, the drug does not reduce mean worm count)

H_a: $\mu_1 - \mu_2 < 0$ (that is, the drug reduces mean worm count)

T.S.: $t = \dfrac{(\bar{y}_1 - \bar{y}_2) - D_0}{s_p \sqrt{\dfrac{1}{n_1} + \dfrac{1}{n_2}}} = \dfrac{(26.58 - 39.67) - 0}{14.11 \sqrt{\dfrac{1}{12} + \dfrac{1}{12}}} = -2.272$

R.R.:   For $\alpha = .05$, the critical t-value for a one-tailed test with df $= n_1 + n_2 - 2 = 22$ is obtained from Table 2 in the Appendix, using $\alpha = .05$. We will reject H_0 if $t \leq -1.717$.

Conclusion: Because the observed value of $t = -2.272$ is less than -1.717 and hence is in the rejection region, we have sufficient evidence to conclude that the drug treatment does reduce the mean worm count.

b. Using Table 2 in the Appendix with $t = -2.272$ and df $= 22$, we can bound the level of significance in the range $.01 < p\text{-value} < .025$. From the following computed output, we can observe that the exact level of significance is p-value $= .017$.

```
Two-Sample T-Test and Confidence Interval


Two-sample T for Treated vs Untreated


              N    Mean    StDev    SE Mean
Treated      12    26.6    14.4       4.1
Untreated    12    39.7    13.9       4.0


95% CI for mu Treated − mu Untreated:   ( −25.0,  −1.1)
T-Test mu Treated = mu Untreated (vs <): T = −2.27  P = 0.017  DF = 22
Both use Pooled StDev = 14.1
```

R.R.: For a level α, Type I error rate and with df $= n_1 + n_2 - 2$,
1. reject H_0 if $t' \geq t_\alpha$
2. reject H_0 if $t' \leq -t_\alpha$
3. reject H_0 if $|t'| \geq t_{\alpha/2}$

where

$$\text{df} = \frac{(n_1 - 1)(n_2 - 1)}{(1 - c)^2(n_1 - 1) + c^2(n_2 - 1)}, \quad \text{with } c = \frac{s_1^2/n_1}{\dfrac{s_1^2}{n_1} + \dfrac{s_2^2}{n_2}}$$

Note: If the computed value of df is not an integer, *round down* to the nearest integer.

The test based on the t' statistic is sometimes referred to as the *separate-variance t test* because we use the separate sample variances s_1^2 and s_2^2 rather than a pooled sample variance.

When there is a large difference between σ_1 and σ_2, we must also modify the confidence interval for $\mu_1 - \mu_2$. The following formula is developed from the separate-variance t test.

Approximate Confidence Interval for $\mu_1 - \mu_2$, Independent Samples with $\sigma_1 \neq \sigma_2$

$$(\bar{y}_1 - \bar{y}_2) \pm t_{\alpha/2} \sqrt{\frac{s_1^2}{n_1} + \frac{s_2^2}{n_2}}$$

where the t percentile has

$$\text{df} = \frac{(n_1 - 1)(n_2 - 1)}{(1 - c)^2(n_1 - 1) + c^2(n_2 - 1)}, \quad \text{with } c = \frac{s_1^2/n_1}{\dfrac{s_1^2}{n_1} + \dfrac{s_2^2}{n_2}}$$

We will now continue our analysis of the data in our oil spill example.

Analyzing Data for Oil Spill Case Study   The researchers hypothesized that the oil spill sites would have a lower plant density than the control sites. Thus, we will construct confidence intervals on the mean plant density in the control plots μ_1 and in the oil spill plots μ_2 to assess their average plant density. Also, we can construct confidence intervals on the difference $\mu_1 - \mu_2$ and test the research hypothesis that μ_1 is greater than μ_2. From Figure 6.1, the data from the oil spill area appear to have a normal distribution, whereas the data from the control area appear to be skewed to the left. The normal probability plots are given in Figure 6.6a and b to assess further whether the population distributions are in fact normal in shape. We observe that the data from the oil spill tracts appear to follow a normal distribution, but the data from the control tracts do not because their plotted points do not fall close to the straight line. Also, the variability in plant density is higher in control sites than in the oil spill sites. Thus, the approximate

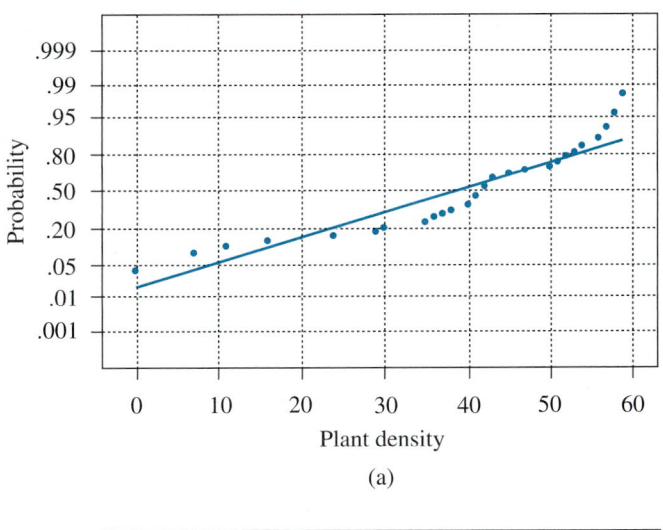

FIGURE 6.6

(a) Normal probability plot for control sites. (b) Normal probability plot oil for spill sites

(a)

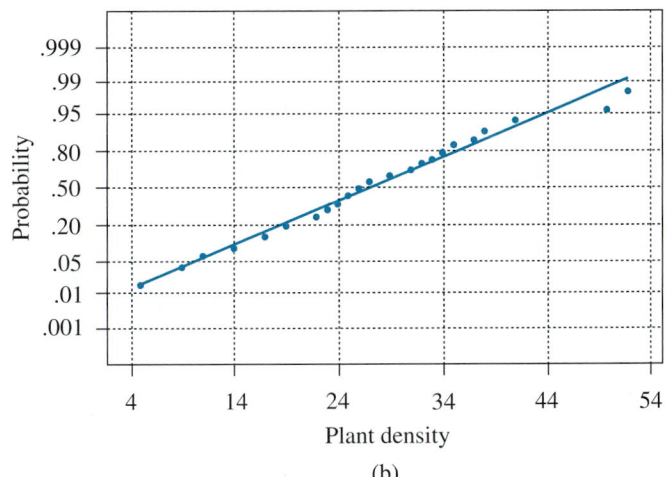

(b)

t procedures will be the most appropriate inference procedures. The sample data yielded the following values:

Control Plots   **Oil Spill Plots**

$n_1 = 40$ $n_2 = 40$

$\bar{y}_1 = 38.48$ $\bar{y}_2 = 26.93$

$s_1 = 16.37$ $s_2 = 9.88$

The research hypothesis is that the mean plant density for the control plots exceeds that for the oil spill plots. Thus, our approximate t test is set up as follows:

$$H_0: \mu_1 \le \mu_2 \text{ versus } H_a: \mu_1 > \mu_2$$

That is,

$$H_0: \mu_1 - \mu_2 \le 0$$

$$H_a: \mu_1 - \mu_2 > 0$$

$$\text{T.S.: } t' = \frac{(\bar{y}_1 - \bar{y}_2) - D_0}{\sqrt{\dfrac{s_1^2}{n_1} + \dfrac{s_2^2}{n_2}}} = \frac{(38.48 - 26.93) - 0}{\sqrt{\dfrac{(16.37)^2}{40} + \dfrac{(9.88)^2}{40}}} = 3.82$$

To compute the rejection region and p-value, we need to compute the approximate df for t':

$$c = \frac{s_1^2/n_1}{\dfrac{s_1^2}{n_1} + \dfrac{s_2^2}{n_2}} = \frac{(16.37)^2/40}{(16.37)^2/40 + (9.88)^2/40} = .73$$

$$df = \frac{(n_1 - 1)(n_2 - 1)}{(1 - c)^2(n_1 - 1) + c^2(n_2 - 1)} = \frac{(39)(39)}{(1 - .73)^2(39) + (.73)^2(39)}$$

$$= 64.38, \text{ which is rounded to } 64$$

Table 2 in the Appendix does not have df = 64, so we will use df = 60. In fact, the difference is very small when df becomes large: $t_{.05} = 1.671$ and 1.669 for df = 60 and 64, respectively.

R.R.: For $\alpha = .05$ and df = 60, reject H_0 if $t' > 1.671$.

Since $t' = 3.82$ is greater than 1.671, we reject H_0. We can bound the p-value using Table 2 in the Appendix with df = 60. With $t' = 3.82$, the level of significance is p-value $< .001$. Thus we can conclude that there is significant (p-value $< .001$) evidence that μ_1 is greater than μ_2. Although we have determined that there is a statistically significant difference between the mean plant densities at the control and oil spill sites, the question remains whether these differences have *practical* significance. We can estimate the size of the difference in the means by placing a 95% confidence interval on $\mu_1 - \mu_2$.

The appropriate 95% confidence interval for $\mu_1 - \mu_2$ is computed by using the following formula with df = 64, the same as was used for the R.R.

$$(\bar{y}_1 - \bar{y}_2) \pm t_{\alpha/2}\sqrt{\frac{s_1^2}{n_1} + \frac{s_2^2}{n_2}} \quad \text{or} \quad (38.48 - 26.93) \pm 2.0\sqrt{\frac{(16.37)^2}{40} + \frac{(9.88)^2}{40}}$$

$$\text{or} \quad 11.55 \pm 6.05$$

Thus, we are 95% confident that the mean plant densities differ by an amount between 5.5 and 17.6. The plant scientists would then evaluate whether a difference in this range is of practical importance.

Reporting Conclusions   We would need to write a report summarizing our findings from the study. The report would include the following items:

1. Statement of the objective for the study
2. Description of the study design and data collection procedures
3. Numerical and graphical summaries of the data sets

   - Table of means, medians, standard deviations, quartiles, range
   - Boxplots
   - Stem-and-leaf plots

4. Description of all inference methodologies:

   - Approximate t tests of differences in means
   - Approximate t-based confidence interval on population means
   - Verification that all necessary conditions for using inference techniques were satisfied using boxplots, normal probability plots

5. Discussion of results and conclusions
6. Interpretation of findings relative to previous studies
7. Recommendations for future studies
8. Listing of the data set

To illustrate that the separate-variance t test is less affected by unequal variances than the pooled t test, the data from the computer simulation reported in Table 6.4 was analyzed using the separate-variance t test. The proportion of the 1,000 tests that incorrectly rejected H_0 is presented in Table 6.5. If the separate-variance t test were unaffected by the unequal variances, we would expect the proportions to be close to .05, the intended level, in all cases.

TABLE 6.5

The effect of unequal variances on the Type I error rates of the separate-variance t test

| | | $\sigma_1 = k\sigma_2$ | | | | |
n_1	n_2	$k = .25$	.50	1	2	4
10	10	.055	.040	.056	.038	.052
10	20	.055	.044	.049	.059	.051
10	40	.049	.047	.043	.041	.055
15	15	.044	.041	.054	.055	.057
15	30	.052	.039	.051	.043	.052
15	45	.058	.042	.055	.050	.058

From the results in Table 6.5, we can observe that the separate-variance t test has a Type I error rate that is consistently very close to .05 in all the cases considered. On the other hand, the pooled t test had Type I error rates very different from .05 when the sample sizes were unequal and we sampled from populations having very different variances.

In this section, we developed pooled-variance t methods based on the requirement of independent random samples from normal populations with equal population variances. For situations when the variances are not equal, we introduced the separate-variance t' statistic. Confidence intervals and hypothesis tests based on these procedures (t or t') need not give identical results. Standard computer packages often report the results of both t and t' tests. Which of these results should you use in your report?

If the sample sizes are equal and the population variances are equal, the separate-variance t test and the pooled t test give algebraically identical results; that is, the computed t equals the computed t'. Thus, why not always use t' in place of t when $n_1 = n_2$? The reason we would select t over t' is that the df for t are nearly always larger than the df for t', and hence the power of the t test is greater than the power of the t' test when the variances are equal. When the sample sizes and variances are very unequal, the results of the t and t' procedures may differ greatly. The evidence in such cases indicates that the separate-variance methods are somewhat more reliable and more conservative than the results of the pooled t methods. However, if the populations have both different means and different variances, an examination of just the size of the difference in their means $\mu_1 - \mu_2$ would be an inadequate description of how the populations differ. We should always examine the size of the differences in both the means and the standard deviations of the populations being compared. In Chapter 7, we will discuss procedures for examining the difference in the standard deviations of two populations.

EXERCISES   **Basic Techniques**

6.1 Set up the rejection regions for testing $H_0: \mu_1 - \mu_2 = 0$ for the following conditions:
   a. $H_a: \mu_1 - \mu_2 \neq 0$, $n_1 = 12$, $n_2 = 14$, and $\alpha = .05$
   b. $H_a: \mu_1 - \mu_2 > 0$, $n_1 = n_2 = 8$, and $\alpha = .01$
   c. $H_a: \mu_1 - \mu_2 < 0$, $n_1 = 6$, $n_2 = 4$, and $\alpha = .05$

What assumptions must be made prior to applying a two-sample t test?

6.2 Conduct a test of $H_0: \mu_1 - \mu_2 \geq 0$ against the alternative hypothesis $H_a: \mu_1 - \mu_2 < 0$ for the sample data shown here. Use $\alpha = .05$.

	Population	
	1	**2**
Sample size	16	13
Sample mean	71.5	79.8
Sample variance	68.35	70.26

6.3 Refer to the data of Exercise 6.2. Give the level of significance for your test.

Applications

Med.   **6.4** In an effort to link cold environments with hypertension in humans, a preliminary experiment was conducted to investigate the effect of cold on hypertension in rats. Two random samples of 6 rats each were exposed to different environments. One sample of rats was held in a normal environment at 26°C. The other sample was held in a cold 5°C environment. Blood pressures and heart rates were measured for rats for both groups. The blood pressures for the 12 rats are shown in the accompanying table.
   a. Do the data provide sufficient evidence that rats exposed to a 5°C environment have a higher mean blood pressure than rats exposed to a 26°C environment? Use $\alpha = .05$.
   b. Evaluate the three conditions required for the test used in part (a).
   c. Provide a 95% confidence interval on the difference in the two population means.

26°C		**5°C**	
Rat	**Blood Pressure**	**Rat**	**Blood Pressure**
1	152	7	384
2	157	8	369
3	179	9	354
4	182	10	375
5	176	11	366
6	149	12	423

Env.   **6.5** A pollution-control inspector suspected that a riverside community was releasing semitreated sewage into a river and this, as a consequence, was changing the level of dissolved oxygen of the river. To check this, he drew 15 randomly selected specimens of river water at a location above the town and another 15 specimens below. The dissolved oxygen readings, in parts per million, are given in the accompanying table.

| **Above Town** | 5.2 | 4.8 | 5.1 | 5.0 | 4.9 | 4.8 | 5.0 | 4.7 | 4.7 | 5.0 | 4.7 | 5.1 | 5.0 | 4.9 | 4.9 |
| **Below Town** | 4.2 | 4.4 | 4.7 | 4.9 | 4.6 | 4.8 | 4.9 | 4.6 | 5.1 | 4.3 | 5.5 | 4.7 | 4.9 | 4.8 | 4.7 |

Use the computer output shown here to answer the following questions.

```
Two-Sample T-Test and Confidence Interval

Two-sample T for Above Town vs Below Town

              N      Mean    StDev   SE Mean
Above To   15       4.92    0.157    0.042
Below To   15       4.74    0.320    0.084


95% CI for mu Above To — mu Below To: ( −0.013,   0.378)
T-Test mu Above To = mu Below To (vs not =): T = 1.95   P = 0.065   DF = 20
```

Boxplots of above- and below-town specimens (means are indicated by solid circles)

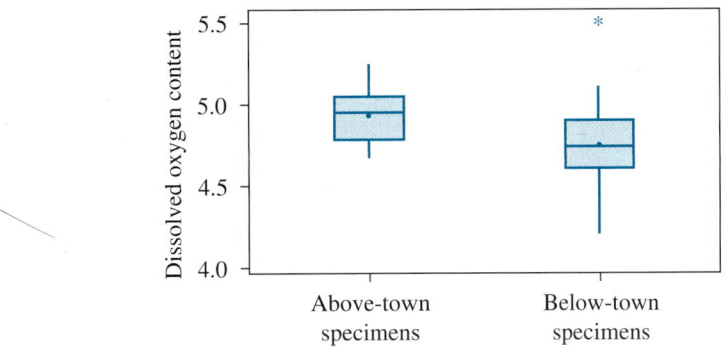

a. Do the data provide sufficient evidence to indicate a difference in mean oxygen content between locations above and below the town? Use $\alpha = .05$.

b. Was the pooled t test or the separate-variance t test used in the computer output?

c. Do the required conditions to use the test in (a) appear to be valid for this study? Justify your answer.

d. How large is the difference between the mean oxygen content above and below the town?

Engin.   **6.6**   An industrial engineer conjectures that a major difference between successful and unsuccessful companies is the percentage of their manufactured products returned because of defectives. In a study to evaluate this conjecture, the engineer surveyed the quality control departments of 50 successful companies (identified by the annual profit statement) and 50 unsuccessful companies. The companies in the study all produced products of a similar nature and cost. The percentage of the total output returned by customers in the previous year is summarized in the following graphs and tables.

```
Two-Sample T-Test and Confidence Interval

Two-sample T for Unsuccessful vs Successful

                N       Mean     StDev   Se Mean
Unsuccessful 50         9.08     1.97      0.28
Successful   50         5.40     2.88      0.41

95% CI for mu Unsuccessful − mu Successful: ( 2.70,  4.66)
T-Test mu Unsuccessful = mu Successful (vs >): T = 7.46  P = 0.0000  DF = 86
```

Boxplots of unsuccessful and successful businesses
(means are indicated by solid circles)

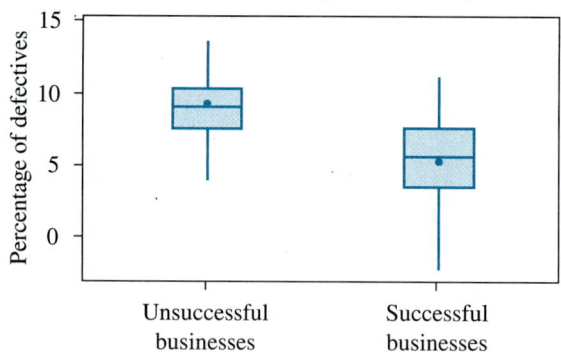

a. Do the data provide sufficient evidence that successful companies have a lower percentage of their products returned by customers? Use $\alpha = .05$.
b. Was the pooled t test or the separate-variance t test used in the computer output?
c. Do the required conditions to use the test in (a) appear to be valid for this study? Justify your answer.
d. How large is the difference between the percentage of returns for successful and unsuccessful companies?

Soc. 6.7 The number of households currently receiving a daily newspaper has decreased over the last 10 years, and many people state they obtain information about current events through television news and the Internet. To test whether people who receive a daily newspaper have a greater knowledge of current events than people who don't, a sociologist gave a current events test to 25 randomly selected people who subscribe to a daily newspaper and to 30 randomly selected persons who do not receive a daily newspaper. The following stem-and-leaf graphs give the scores for the two groups. Does it appear that people who receive a daily newspaper have a greater knowledge of current events? Be sure to evaluate all necessary conditions for your procedures to be valid.

```
Character Stem-and-Leaf Display

Stem-and-leaf of No Newspaper Deliver          Stem-and-leaf of Newspaper Subscribers
N=30                                           N=25
Leaf Unit = 1.0                                Leaf Unit = 1.0


      0 000
      0
      1 3
      1 59
      2 334                                        2 2
      2 57                                         2 99
      3 00234                                      3 2
      3 5589                                       3 66889
      4 00124                                      4 000112333
      4 5                                          4 55666

      5 0                                          5 2
      5 55                                         5 9
      6 2
```

```
Two-Sample T-Test and Confidence Interval

Two-sample T for No Newspaper vs Newspaper

                 N      Mean     StDev    SE Mean
No Newspaper    30      32.0     16.0       2.9
Newspaper       25     40.91     7.48       1.5

95% CI for mu No Newspaper  -  mu Newspaper: ( -15.5,   -2.2)
T-Test mu No Newspaper = mu Newspaper (vs <) : T = -2.70  P = 0.0049  DF = 42
```

Env.   **6.8**   Two different emission-control devices were tested to determine the average amount of nitric oxide emitted by an automobile over a 1-hour period of time. Twenty cars of the same model and year were selected for the study. Ten cars were randomly selected and equipped with a Type I emission-control device, and the remaining cars were equipped with Type II devices. Each of the 20 cars was then monitored for a 1-hour period to determine the amount of nitric oxide emitted.

Use the following data to test the research hypothesis that the mean level of emission for Type I devices (μ_1) is greater than the mean emission level for Type II devices (μ_2). Use $\alpha = .01$.

Type I Device		Type II Device	
1.35	1.28	1.01	0.96
1.16	1.21	0.98	0.99
1.23	1.25	0.95	0.98
1.20	1.17	1.02	1.01
1.32	1.19	1.05	1.02

Med.   **6.9** It is estimated that lead poisoning resulting from an unnatural craving (pica) for substances such as paint may affect as many as a quarter of a million children each year, causing them to suffer from severe, irreversible retardation. Explanations for why children voluntarily consume lead range from "improper parental supervision" to "a child's need to mouth objects." Some researchers, however, have investigated whether the habit of eating such substances has some nutritional explanation. One such study involved a comparison of a regular diet and a calcium-deficient diet on the ingestion of a lead-acetate solution in rats. Each rat in a group of 20 rats was randomly assigned to either an experimental or a control group. Those in the control group received a normal diet, while the experimental group received a calcium-deficient diet. Each of the rats occupied a separate cage and was monitored to observe the quantity of a .15% lead-acetate solution consumed during the study period. The sample results are summarized here.

Control group	5.4	6.2	3.1	3.8	6.5	5.8	6.4	4.5	4.9	4.0
Experimental group	8.8	9.5	10.6	9.6	7.5	6.9	7.4	6.5	10.5	8.3

  a. Plot the data for the 2 samples separately. Is there reason to think the assumptions for a t test have been violated?
  b. Run a test of the research hypothesis that the mean quantity of lead acetate consumed in the experimental group is greater than that consumed in the control group. Use $\alpha = .05$.

Med.   **6.10** The results of a 3-year study to examine the effect of a variety of ready-to-eat breakfast cereals on dental caries (tooth decay) in adolescent children were reported by Rowe, Anderson, and Wanninger (1974). A sample of 375 adolescent children of both genders from the Ann Arbor, Michigan, public schools was enrolled (after parental consent) in the study. Each child was provided with toothpaste and boxes of different varieties of ready-to-eat cereals. Although these were brand-name cereals, each type of cereal was packaged in plain white 7-ounce boxes and labeled as wheat flakes, corn cereal, oat cereal, fruit-flavored corn puffs, corn puffs, cocoa-flavored cereal, and sugared oat cereal. Note that the last four varieties of cereal had been presweetened and the others had not.

  Each child received a dental examination at the beginning of the study, twice during the study, and once at the end. The response of interest was the incremental decayed, missing, and filled (DMF) surfaces—that is, the difference between the final (poststudy) and initial (prestudy) number of DMF tooth surfaces. Careful records for each participant were maintained throughout the 3 years, and at the end of the study, a person was classified as "noneater" if he or she had eaten fewer than 28 boxes of cereal throughout the study. All others were classified as "eaters." The incremental DMF surface readings for each group are summarized here. Use these data to test the research hypothesis that the mean incremental DMF surface for noneaters is larger than the corresponding mean for eaters. Give the level of significance for your test. Interpret your findings.

	Sample Size	Sample Mean	Sample Standard Deviation
Noneaters	73	6.41	5.62
Eaters	302	5.20	4.67

6.11 Refer to Exercise 6.10. Although complete details of the original study have not been disclosed, critique the procedure that has been discussed.

Env.   **6.12** The study of concentrations of atmospheric trace metals in isolated areas of the world has received considerable attention because of the concern that humans might somehow alter the climate of the earth by changing the amount and distribution of trace metals in the atmosphere. Consider a study at the south pole, where at 10 different sampling periods throughout a 2-month period, 10,000 standard cubic meters (scm) of air were obtained and analyzed for metal concentrations. The results associated with magnesium and europium are listed here. (*Note:* Magnesium results are in units of 10^{-9} g/scm; europium results are in units of 10^{-15} g/scm.) Note that $s > \bar{y}$ for the magnesium data. Would you expect the data to be normally distributed? Explain.

	Sample Size	Sample Mean	Sample Standard Deviation
Magnesium	10	1.0	2.21
Europium	10	17.0	12.65

6.13   Refer to Exercise 6.12. Could we run a *t* test comparing the mean metal concentrations for magnesium and europium? Why or why not?

Env.   **6.14**   PCBs have been in use since 1929, mainly in the electrical industry, but it was not until the 1960s that they were found to be a major environmental contaminant. In the paper, "The ratio of DDE to PCB concentrations in Great Lakes herring gull eggs and its use in interpreting contaminants data" [appearing in the *Journal of Great Lakes Research* 24 (1):12–31, 1998], researchers report on the following study. Thirteen study sites from the five Great Lakes were selected. At each site, 9 to 13 herring gull eggs were collected randomly each year for several years. Following collection, the PCB content was determined. The mean PCB content at each site is reported in the following table for the years 1982 and 1996.

	Site												
Year	1	2	3	4	5	6	7	8	9	10	11	12	13
1982	61.48	64.47	45.50	59.70	58.81	75.86	71.57	38.06	30.51	39.70	29.78	66.89	63.93
1996	13.99	18.26	11.28	10.02	21.00	17.36	28.20	7.30	12.80	9.41	12.63	16.83	22.74

a. Legislation was passed in the 1970s restricting the production and use of PCBs. Thus, the active input of PCBs from current local sources has been severely curtailed. Do the data provide evidence that there has been a significant decrease in the mean PCB content of herring gull eggs?

b. Estimate the size of the decrease in mean PCB content from 1982 to 1996, using a 95% confidence interval.

c. Evaluate the conditions necessary to validly test hypotheses and construct confidence intervals using the collected data.

d. Does the independence condition appear to be violated?

6.15   Refer to Exercise 6.14. There appears to be a large variation in the mean PCB content across the 13 sites. How could we reduce the effect of variation in PCB content due to site differences on the evaluation of the difference in the mean PCB content between the two years?

H.R.   **6.16**   A firm has a generous but rather complicated policy concerning end-of-year bonuses for its lower-level managerial personnel. The policy's key factor is a subjective judgment of "contribution to corporate goals." A personnel officer took samples of 24 female and 36 male managers to see whether there was any difference in bonuses, expressed as a percentage of yearly salary. The data are listed here:

Gender	Bonus Percentage								
F	9.2	7.7	11.9	6.2	9.0	8.4	6.9	7.6	7.4
	8.0	9.9	6.7	8.4	9.3	9.1	8.7	9.2	9.1
	8.4	9.6	7.7	9.0	9.0	8.4			
M	10.4	8.9	11.7	12.0	8.7	9.4	9.8	9.0	9.2
	9.7	9.1	8.8	7.9	9.9	10.0	10.1	9.0	11.4
	8.7	9.6	9.2	9.7	8.9	9.2	9.4	9.7	8.9
	9.3	10.4	11.9	9.0	12.0	9.6	9.2	9.9	9.0

```
Two-Sample T-Test and Confidence Interval

Two-sample T for Female vs Male

          N     Mean    StDev   SE Mean
Female   24     8.53     1.19      0.24
Male     36     9.68     1.00      0.17

95% CI for mu Female − mu Male: ( −1.74,  −0.56)
T-Test mu Female = mu Male (vs <): T = −3.90   P = 0.0002   DF = 43

95% CI for mu Female − mu Male: ( −1.72,  −0.58)
T-Test mu Female = mu Male (vs <): T = −4.04   P = 0.0001   DF = 58
Both use Pooled StDev = 1.08
```

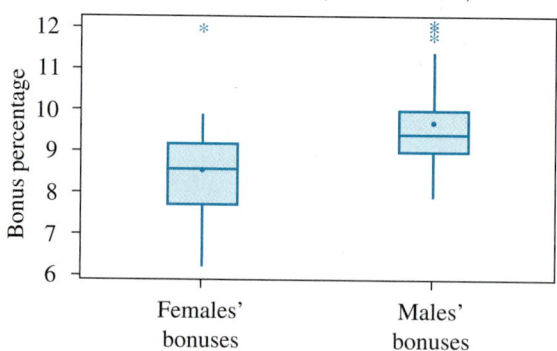

Boxplots of females' and males' bonuses
(means are indicated by solid circles)

a. Identify the value of the pooled-variance t statistic (the usual t test based on the equal variance assumption).

b. Identify the value of the t' statistic.

c. Use both statistics to test the research hypothesis of unequal means at $\alpha = .05$ and at $\alpha = .01$. Does the conclusion depend on which statistic is used?

Bus. **6.17** The costs of major surgery vary substantially from one state to another due to differences in hospital fees, malpractice insurance cost, doctors' fees, and rent. A study of hysterectomy costs was done in California and Montana. Based on a random sample of 200 patient records from each state, the sample statistics shown here were obtained.

State	n	Sample Mean	Sample Standard Deviation
Montana	200	$ 6,458	$250
California	200	$12,690	$890

a. Is there significant evidence that California has a higher mean hysterectomy cost than Montana?

b. Estimate the difference in the mean costs of the two states using a 95% confidence interval.

c. Justify your choice between using the pooled t test and the separate-variance t test in part (a).

Edu.

6.18 A national educational organization monitors reading proficiency for American students on a regular basis using a scale that ranges from 0 to 500. Sample results based on 500 students per category are shown here. Use these data to make the inferences listed next. Assume that the pooled standard deviation for any comparison is 100.

Age	Gender	Sample Mean*
9	Male	210
	Female	216
13	Male	253
	Female	262
17	Male	283
	Female	293

* Scale: 150—rudimentary reading skills; can follow basic directions. 200—basic skills; can identify facts from simple paragraphs. 250—intermediate skills; can organize information in lengthy passages. 300—adapt skills; can understand and explain complicated information. 350— advanced skills; can understand and explain specialized materials.

a. Construct a meaningful graph that shows age, gender, and mean proficiency scores.

b. Use the sample data to place a 95% confidence interval on the difference in mean proficiencies for females and males age 17 years.

c. Compare the mean scores for females, age 13 and 17 years, using a 90% confidence interval. Does the interval include 0? Why might these means be different?

6.3 A Nonparametric Alternative: The Wilcoxon Rank Sum Test

The two-sample t test of the previous section was based on several conditions: independent samples, normality, and equal variances. When the conditions of normality and equal variances are not valid but the sample sizes are large, the results using a t (or t') test are approximately correct. There is, however, an alternative test procedure that requires less stringent conditions. This procedure, called the **Wilcoxon rank sum test,** is discussed here.

Wilcoxon rank sum test

The assumptions for this test are that we have independent random samples taken from two populations whose distributions are identical except that one distribution may be shifted to the right of the other distribution, as shown in Figure 6.7. The Wilcoxon rank sum test does not require that populations have normal distributions. Thus, we have removed one of the three conditions that were required of the t-based procedures. The other conditions, equal variances and independence of the random samples, are still required for the Wilcoxon rank sum test. Because the two population distributions are assumed to be identical under the null hypothesis, independent random samples from the two populations should be similar if the null hypothesis is true. Because we are now allowing the population distributions to be nonnormal, the rank sum procedure must deal with the possibility of extreme observations in the data. One way to handle samples containing extreme values is to replace each data value with its rank (from lowest to highest) in the combined sample—that is, the sample consisting of the data from both populations. The smallest value in the combined sample is assigned the rank of 1 and the largest value is assigned the rank of $N = n_1 + n_2$. The ranks are not affected by how far the smallest (largest) data value is from next smallest (largest) data value. Thus, extreme values in data sets do not have a strong effect on the rank sum statistic as they did in the t-based procedures.

FIGURE 6.7

Skewed population distributions identical in shape but shifted

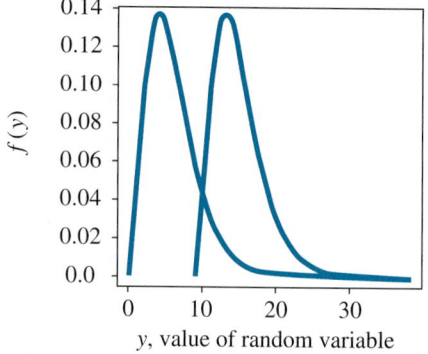

The calculation of the rank sum statistic consists of the following steps:

1. List the data values for both samples from smallest to largest.
2. In the next column, assign the numbers 1 to N to the data values with 1 to the smallest value and N to the largest value. These are the **ranks** of the observations.

ranks

3. If there are ties—that is, duplicated values—in the combined data set, the ranks for the observations in a tie are taken to be the average of the ranks for those observations.
4. Let T denote the sum of the ranks for the observations from population 1.

If the null hypothesis of identical population distributions is true, the n_1 ranks from population 1 are just a random sample from the N integers $1, \ldots, N$. Thus, under the null hypothesis, the distribution of the sum of the ranks T depends only on the sample sizes, n_1 and n_2, and does not depend on the shape of the population distributions. Under the null hypothesis, the sampling distribution of T has mean and variance given by

$$\mu_T = \frac{n_1(n_1 + n_2 + 1)}{2} \qquad \sigma_T^2 = \frac{n_1 n_2}{12}(n_1 + n_2 + 1)$$

Intuitively, if T is much smaller (or larger) than μ_T, we have evidence that the null hypothesis is false and in fact the population distributions are not equal. The rejection region for the rank sum test specifies the size of the difference between T and μ_T for the null hypothesis to be rejected. Because the distribution of T under the null hypothesis does not depend on the shape of the population distributions, Table 5 provides the critical values for the test regardless of the shape of the population distribution. The Wilcoxon rank sum test is summarized here.

Wilcoxon Rank Sum Test*

H_0: The two populations are identical.

H_a: 1. Population 1 is shifted to the right of population 2.
2. Population 1 is shifted to the left of population 2.
3. Populations 1 and 2 are shifted from each other.

$(n_1 \leq 10, n_2 \leq 10)$

T.S.: T, the sum of the ranks in sample 1

R.R.: For $\alpha = .05$, use Table 5 in the Appendix to find critical values for T_U and T_L;

1. Reject H_0 if $T > T_U$.
2. Reject H_0 if $T < T_L$.
3. Reject H_0 if $T > T_U$ or $T < T_L$.

Check assumptions and draw conclusions.

EXAMPLE 6.4

Many states are considering lowering the blood-alcohol level at which a driver is designated as driving under the influence (DUI) of alcohol. An investigator for a legislative committee designed the following test to study the effect of alcohol on reaction time. Ten participants consumed a specified amount of alcohol. Another group of ten participants consumed the same amount of a nonalcoholic drink, a placebo. The two groups did not know whether they were receiving alcohol or the placebo. The twenty participants' average reaction times (in seconds) to a series of simulated driving situations are reported in the following table. Does it appear that alcohol consumption increases reaction time?

---

*This test is equivalent to the Mann–Whitney U test, Conover (1998).

Placebo	0.90	0.37	1.63	0.83	0.95	0.78	0.86	0.61	0.38	1.97
Alcohol	1.46	1.45	1.76	1.44	1.11	3.07	0.98	1.27	2.56	1.32

a. Why is the t test inappropriate for analyzing the data in this study?

b. Use the Wilcoxon rank sum test to test the hypotheses:

H_0: The distributions of reaction times for the placebo and alcohol populations are identical.

H_a: The distribution of reaction times for the placebo consumption population is shifted to the left of the distribution for the alcohol population. (Larger reaction times are associated with the consumption of alcohol.)

c. Place 95% confidence intervals on the median reaction times for the two groups.

d. Compare the results from (b) to the results from Minitab.

Solution

a. A boxplot of the two samples is given here. The plots indicate that the population distributions are skewed to the right, because 10% of the data values are large outliers and the upper whiskers are longer than the lower whiskers. The sample sizes are both small, and hence the t test may be inappropriate for analyzing this study.

Boxplots of placebo and alcohol populations (means are indicated by solid circles)

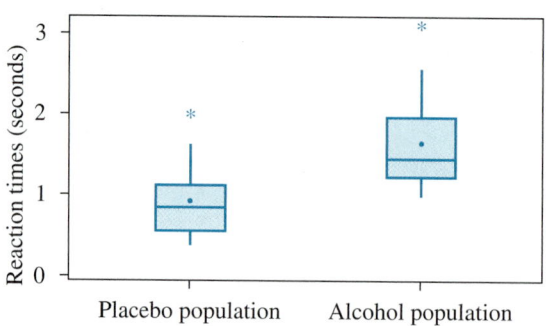

b. The Wilcoxon rank sum test will be conducted to evaluate whether alcohol consumption increases reaction time. Table 6.6 contains the ordered data for the combined samples, along with their associated ranks. We will designate observations from the placebo group as 1 and from the alcohol group as 2.

For $\alpha = .05$, reject H_0 if $T < 83$, using Table 5 in the Appendix with $\alpha = .05$, one-tailed, and $n_1 = n_2 = 10$. The value of T is computed by summing the ranks from group 1: $T = 1 + 2 + 3 + 4 + 5 + 6 + 7 + 8 + 16 + 18 = 70$. Because 70 is less than 83, we reject H_0 and conclude there is significant evidence that the placebo population has smaller reaction times than the population of alcohol consumers.

c. Because we have small sample sizes and the population distributions appear to be skewed to the right, we will construct confidence intervals

TABLE 6.6

Ordered reaction times and ranks

	Ordered Data	Group	Ranks		Ordered Data	Group	Ranks
1	0.37	1	1	11	1.27	2	11
2	0.38	1	2	12	1.32	2	12
3	0.61	1	3	13	1.44	2	13
4	0.78	1	4	14	1.45	2	14
5	0.83	1	5	15	1.46	2	15
6	0.86	1	6	16	1.63	1	16
7	0.90	1	7	17	1.76	2	17
8	0.95	1	8	18	1.97	1	18
9	0.98	2	9	19	2.56	2	19
10	1.11	2	10	20	3.07	2	20

on the median reaction times in place of confidence intervals on the mean reaction times. Using the methodology from Section 5.8, from Table 4 in the Appendix, we find

$$C_{\alpha(2),n} = C_{.05,10} = 1$$

Thus,

$$L_{.025} = C_{.05,10} + 1 = 2$$

and

$$U_{.025} = n - C_{.05,10} = 10 - 1 = 9$$

The 95% confidence intervals for the population medians are given by

$$(M_L, M_U) = (y_{(2)}, y_{(9)})$$

Thus, a 95% confidence interval is (.38, 1.63) for the placebo population median and (1.11, 2.56) for the alcohol population median. Because the sample sizes are very small, the confidence intervals are not very informative.

d. The output from Minitab is given here.

```
Mann-Whitney Confidence Interval and Test

PLACEBO   N =   10       Median =        0.845
ALCOHOL   N =   10       Median =        1.445
Point estimate for ETA1-ETA2 is    -0.600
95.5 Percent CI for ETA1-ETA2 is (-1.080,-0.250)
W = 70.0
Test of ETA1 = ETA2 vs ETA1 < ETA2 is significant at 0.0046
```

Minitab refers to the test statistic as the Mann–Whitney test. This test is equivalent to the Wilcoxon test statistic. In fact, the value of the test statistic $W = 70$ is identical to the Wilcoxon $T = 70$. The output provides the p-value $= 0.0046$ and a 95.5% confidence interval on the difference in the population medians, $(-1.08, -0.25)$.

When both sample sizes are more than 10, the sampling distribution of T is approximately normal; this allows us to use a z statistic in place of T when using the Wilcoxon rank sum test:

$$z = \frac{T - \mu_T}{\sigma_T}$$

The theory behind the Wilcoxon rank sum test requires that the population distributions be continuous, so the probability that any two data values are equal is zero. Because in most studies we only record data values to a few decimal places, we will often have ties—that is, observations with the same value. For these situations, each observation in a set of tied values receives a rank score equal to the average of the ranks for the set of values. When there are ties, the variance of T must be adjusted. The adjusted value of σ_T^2 is shown here.

$$\sigma_T^2 = \frac{n_1 n_2}{12}\left((n_1 + n_2 + 1) - \frac{\sum_{j=1}^{k} t_j(t_j^2 - 1)}{(n_1 + n_2)(n_1 + n_2 - 1)}\right)$$

where k is the number of tied groups and t_j denotes the number of tied observations in the jth group. Note that when there are no tied observations, $t_j = 1$ for all j, which results in

$$\sigma_T^2 = \frac{n_1 n_2}{12}(n_1 + n_2 + 1)$$

From a practical standpoint, unless there are many ties, the adjustment will result in very little change to σ_T^2. The normal approximation to the Wilcoxon rank sum test is summarized here.

Wilcoxon Rank Sum Test: Normal Approximation

$n_1 > 10$ and $n_2 > 10$

H_0: The two populations are identical.

H_a: 1. Population 1 is shifted to the right of population 2.
2. Population 1 is shifted to the left of population 2.
3. Population 1 and 2 are shifted from each other.

T.S.: $z = \dfrac{T - \mu_T}{\sigma_T}$, where T denotes the sum of the ranks in sample 1.

R.R.: For a specified value of α,

1. Reject H_0 if $z \geq z_\alpha$.
2. Reject H_0 if $z \leq -z_\alpha$.
3. Reject H_0 if $|z| \geq z_{\alpha/2}$.

Check assumptions and draw conclusions.

EXAMPLE 6.5

Environmental engineers were interested in determining whether a cleanup project on a nearby lake was effective. Prior to initiation of the project, they obtained 12 water samples at random from the lake and analyzed the samples for the amount of dissolved oxygen (in ppm). Due to diurnal fluctuations in the dissolved oxygen, all measurements were obtained at the 2 P.M. peak period. The before and after data are presented in Table 6.7.

TABLE 6.7
Dissolved oxygen
measurements (in ppm)

Before Cleanup		After Cleanup	
11.0	11.6	10.2	10.8
11.2	11.7	10.3	10.8
11.2	11.8	10.4	10.9
11.2	11.9	10.6	11.1
11.4	11.9	10.6	11.1
11.5	12.1	10.7	11.3

a. Use $\alpha = .05$ to test the following hypotheses:

H_0: The distributions of measurements for before cleanup and 6 months after the cleanup project began are identical.

H_a: The distribution of dissolved oxygen measurements before the cleanup project is shifted to the right of the corresponding distribution of measurements for 6 months after initiating the cleanup project. (Note that a cleanup project has been effective in one sense if the dissolved oxygen level drops over a period of time.)

For convenience, the data are arranged in ascending order in Table 6.7.
b. Has the correction for ties made much of a difference?

Solution

a. First we must jointly rank the combined sample of 24 observations by assigning the rank of 1 to the smallest observation, the rank of 2 to the next smallest, and so on. When two or more measurements are the same, we assign all of them a rank equal to the average of the ranks they occupy. The sample measurements and associated ranks (shown in parentheses) are listed in Table 6.8.

TABLE 6.8
Dissolved oxygen
measurements and ranks

Before Cleanup		After Cleanup	
11.0	(10)	10.2	(1)
11.2	(14)	10.3	(2)
11.2	(14)	10.4	(3)
11.2	(14)	10.6	(4.5)
11.4	(17)	10.6	(4.5)
11.5	(18)	10.7	(6)
11.6	(19)	10.8	(7.5)
11.7	(20)	10.8	(7.5)
11.8	(21)	10.9	(9)
11.9	(22.5)	11.1	(11.5)
11.9	(22.5)	11.1	(11.5)
12.1	(24)	11.3	(16)
	$T = 216$		

Because n_1 and n_2 are both greater than 10, we will use the test statistic z. If we are trying to detect a shift to the left in the distribution after the cleanup, we expect the sum of the ranks for the observations in sample 1 to be large. Thus, we will reject H_0 for large values of $z = (T - \mu_T)/\sigma_T$.

Grouping the measurements with tied ranks, we have 18 groups. These groups are listed next with the corresponding values of t_j, the number of tied ranks in the group.

Rank(s)	Group	t_j	Rank(s)	Group	t_j
1	1	1	14, 14, 14	10	3
2	2	1	16	11	1
3	3	1	17	12	1
4.5, 4.5	4	2	18	13	1
6	5	1	19	14	1
7.5, 7.5	6	2	20	15	1
9	7	1	21	16	1
10	8	1	22.5, 22.5	17	2
11.5, 11.5	9	2	24	18	1

For all groups with $t_j = 1$, there is no contribution for

$$\frac{\sum_j t_j(t_j^2 - 1)}{(n_1 + n_2)(n_1 + n_2 - 1)}$$

in σ_T^2, because $t_j^2 - 1 = 0$. Thus, we will need only $t_j = 2, 3$.

Substituting our data in the formulas, we obtain

$$\mu_T = \frac{n_1(n_1 + n_2 + 1)}{2} = \frac{12(12 + 12 + 1)}{2} = 150$$

$$\sigma_T^2 = \frac{n_1 n_2}{12}\left[(n_1 + n_2 + 1) - \frac{\sum t_j(t_j^2 - 1)}{(n_1 + n_2)(n_1 + n_2 - 1)} \right]$$

$$= \frac{12(12)}{12}\left[25 - \frac{6 + 6 + 6 + 24 + 6}{24(23)} \right]$$

$$= 12(25 - .0870) = 298.956$$

$$\sigma_T = 17.29$$

The computed value of z is

$$z = \frac{T - \mu_T}{\sigma_T} = \frac{216 - 150}{17.29} = 3.82$$

This value exceeds 1.645, so we reject H_0 and conclude that the distribution of before-cleanup measurements is shifted to the right of the corresponding distribution of after-cleanup measurements; that is, the after-cleanup measurements on dissolved oxygen tend to be smaller than the corresponding before-cleanup measurements.

b. The value of σ_T^2 without correcting for ties is

$$\sigma_T^2 = \frac{12(12)(25)}{12} = 300 \qquad \text{and} \qquad \sigma_T = 17.32$$

For this value of σ_T, $z = 3.81$ rather than 3.82 which was found by applying the correction. This should help you understand how little effect the correction has on the final result unless there are a large number of ties.

The Wilcoxon rank sum test is an alternative to the two-sample t test, with the rank sum test requiring fewer conditions than the t test. In particular, Wilcoxon's test does not require the two populations to have normal distributions; it only requires that the distributions are identical except possibly that one distribution is shifted from the other distribution. When both distributions are normal, the t test is more likely to detect an existing difference; that is, the t test has greater power than the rank sum test. This is logical, because the t test uses the magnitudes of the observations rather than just their relative magnitudes (ranks) as is done in the rank sum test. However, when the two distributions are non-normal, the Wilcoxon rank sum test has greater power; that is, it is more likely to detect a shift in the population distributions. Also, the level or probability of a Type I error for the Wilcoxon rank sum test will be equal to the stated level for all population distributions. The t test's *actual* level will deviate from its stated value when the population distributions are nonnormal. This is particularly true when nonnormality of the population distributions is present in the form of severe skewness or extreme outliers.

Randles and Wolfe (1979) investigated the effect of skewed and heavy-tailed distributions on the power of the t test and the Wilcoxon rank sum test. Table 6.9 contains a portion of the results of their simulation study. For each set of distributions, sample sizes and shifts in the populations, 5,000 samples were drawn and the proportion of times a level $\alpha = .05$ t test or Wilcoxon rank sum test rejected H_0 was recorded. The distributions considered were normal, double exponential (symmetric, heavy-tailed), Cauchy (symmetric, extremely heavy-tailed), and Weibull (skewed to the right). Shifts of size 0, $.6\sigma$, and 1.2σ were considered, where σ denotes the standard deviation of the distribution, with the exception of the Cauchy distribution, where σ is a general scale parameter.

TABLE 6.9
Power of t test (t) and Wilcoxon rank sum test (T) with $\alpha = .05$

Distribution		Normal			Double Exponential			Cauchy			Weibull		
Shift		**0**	**.6**	**1.2**	**0**	**.6**	**1.2**	**0**	**.6**	**1.2**	**0**	**.6**	**1.2**
n_1, n_2	**Test**												
5, 5	t	.044	.213	.523	.045	.255	.588	.024	.132	.288	.049	.221	.545
	T	.046	.208	.503	.049	.269	.589	.051	.218	.408	.049	.219	.537
5, 15	t	.047	.303	.724	.046	.304	.733	.056	.137	.282	.041	.289	.723
	T	.048	.287	.694	.047	.351	.768	.046	.284	.576	.049	.290	.688
15, 15	t	.052	.497	.947	.046	.507	.928	.030	.153	.333	.046	.488	.935
	T	.054	.479	.933	.046	.594	.962	.046	.484	.839	.046	.488	.927

When the distribution is normal, the *t* test is only slightly better—has greater power values—than the Wilcoxon rank sum test. For the double exponential, the Wilcoxon test has greater power than the *t* test. For the Cauchy distribution, the level of the *t* test deviates significantly from .05 and its power is much lower than for the Wilcoxon test. When the distribution was somewhat skewed, as in the Weibull distribution, the tests had similar performance. Furthermore, the level and power of the *t* test were nearly identical to the values when the distribution was normal. The *t* test is quite robust to skewness, except when there are numerous outliers.

EXERCISES   **Applications**

Bus.   **6.19**   A plumbing contractor was interested in making her operation more efficient by cutting down on the average distance between service calls while still maintaining at least the same level of business activity. One plumber (plumber 1) was assigned a dispatcher who monitored all his incoming requests for service and outlined a service strategy for that day. Plumber 2 was to continue as she had in the past, by providing service in roughly sequential order for stacks of service calls received. The total daily mileages for these two plumbers are recorded here for a total of 18 days (3 work weeks).

Plumber 1	88.2	94.7	101.8	102.6	89.3	95.7
	78.2	80.1	83.9	86.1	89.4	71.4
	92.4	85.3	87.5	94.6	92.7	84.6
Plumber 2	105.8	117.6	119.5	126.8	108.2	114.7
	90.2	95.6	110.1	115.3	109.6	112.4
	104.6	107.2	109.7	102.9	99.1	115.5

   a. Plot the sample data for each plumber and compute $\bar{y}$ and *s*.
   b. Based on your findings in part (a), which procedure appears more appropriate for comparing the distributions?

Med.   **6.20**   The paper, "Serum beta-2-microglobulin (SB2M) in patients with multiple myeloma treated with alpha interferon" [*Journal of Medicine* (1997) 28: 311–318] reports on the influence of alpha interferon administration in the treatment of patients with multiple myeloma (MM). Twenty newly diagnosed patients with MM were entered into the study. The researchers randomly assigned the 20 patients into the two groups. Ten patients were treated with both intermittent melphalan and sumiferon (treatment group), whereas the remaining ten patients were treated only with intermittent melphalan. The SB2M levels were measured before and at days 3, 8, 15, and months 1, 3, and 6 from the start of therapy. The measurement of SB2M was performed using a radioimmunoassay method. The measurements before treatment are given here.

Treatment Group	2.9	2.7	3.9	2.7	2.1	2.6	2.2	4.2	5.0	0.7
Control Group	3.5	2.5	3.8	8.1	3.6	2.2	5.0	2.9	2.3	2.9

   a. Plot the sample data for both groups using boxplots or normal probability plots.
   b. Based on your findings in part (a), which procedure appears more appropriate for comparing the distributions of SB2M?

    c. Is there significant evidence that there is a difference in the distribution of SB2M for the two groups?

    d. Discuss the implications of your findings in part (c) on the evaluation of the influence of alpha interferon.

Ag.   **6.21**   An experiment was conducted to compare the weights of the combs of roosters fed two different vitamin-supplemented diets. Twenty-eight healthy roosters were randomly divided into two groups, with one group receiving diet I and the other receiving diet II. After the study period, the comb weight (in milligrams) was recorded for each rooster. The data are given here.

Diet I	73	130	115	144	127	126	112	76	68	101	126	49	110	123
Diet II	80	72	73	60	55	74	67	89	75	66	93	75	68	76

    a. Plot the data for both diet I and diet II and evaluate whether the conditions for using the pooled t test appear to be satisfied.

    b. Use the appropriate t procedure to determine whether there is a difference in the distributions of comb weights for the two groups. Use $\alpha = .05$.

    c. Use the Wilcoxon rank sum test to determine whether there is a difference in the distributions of comb weights for the two groups. Use $\alpha = .05$.

    d. Which procedure, Wilcoxon or t, seems most appropriate for evaluating the results of this experiment? Justify your answer.

Bus.   **6.22**   Refer to Exercise 6.21. A second study was done the following year. A treatment group of 18 plumbers was assigned a dispatcher who monitored all the plumbers' incoming requests for service and outlined a service strategy for that day's activities. A control group of 18 plumbers was to conduct their activities as they did in the past, by providing service in roughly sequential order for stacks of service calls received. The average daily mileages for these 36 plumbers were computed over a 30-day period and are recorded here.

Treatment Group	62.2	79.3	83.2	82.2	84.1	89.3
	95.8	87.9	91.5	96.6	90.1	98.6
	85.2	87.9	86.7	99.7	101.1	88.6
Control Group	87.1	70.2	94.6	182.9	85.6	89.5
	109.5	101.7	99.7	193.2	105.3	92.9
	63.9	88.2	99.1	95.1	92.4	87.3

    a. The sample data are plotted next. Based on these plots, which procedure appears more appropriate for comparing the distributions of the two groups?

    b. Computer output is shown here for a t test and a Wilcoxon rank sum test (which is equivalent to the Mann–Whitney test shown in the output). Compare the results for these two tests and draw a conclusion about the effectiveness of the dispatcher program.

    c. Comment on the appropriateness or inappropriateness of the t test based on the plots of the data and the computer output.

    d. Does it matter which test is used here? Might it be reasonable to run both tests in certain situations? Why or why not?

```
Two-Sample T-Test and Confidence Interval

Two-sample T for Treatment vs Control

                N        Mean       StDev    SE Mean
Treatment      18       88.33        9.06        2.1
Control        18       102.1        33.2        7.8

95% CI for mu Treatment − mu Control: ( −30.3,  2.7)
T-Test mu Treatment = mu Control (vs <): T = −1.70   P = 0.049   DF = 34
Both use Pooled StDev = 24.3
```

```
Two-Sample T-Test and Confidence Interval

Two-sample T for Treatment vs Control

                N        Mean       StDev    SE Mean
Treatment      18       88.33        9.06        2.1
Control        18       102.1        33.2        7.8

95% CI for mu Treatment − mu Control: ( −30.8,  3.2)
T-Test mu Treatment = mu Control (vs <): T = −1.70   P = 0.053   DF = 19
```

```
Mann-Whitney Confidence Interval and Test

Treatment   N =   18        Median =         88.25
Control     N =   18        Median =         93.75
Point estimate for ETA1-ETA2 is         −5.20
95.2 Percent CI for ETA1-ETA2 is (−12.89,0.81)
W = 278.5
Test of ETA1 = ETA2   vs   ETA1 < ETA2 is significant at 0.0438
The test is significant at 0.0438 (adjusted for ties)
```

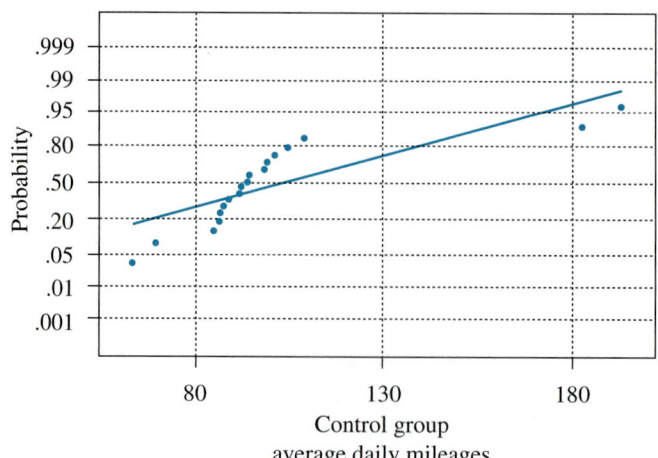

Normal probability plot for control group

Control group average daily mileages

Normal probability plot for
treatment group

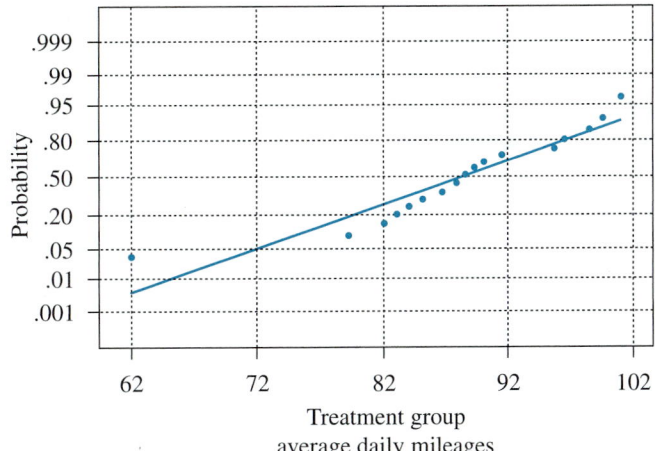

Control group
average daily mileages

Treatment group
average daily mileages

Boxplots of treatment and
control groups (means are
indicated by solid circles)

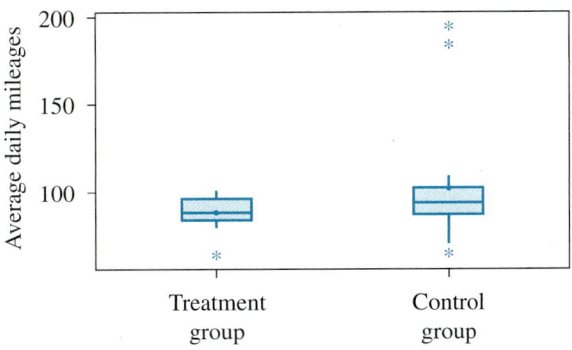

6.23 The results of the simulation study on the effect of heavy-tailed and skewed distributions on the performance of the t test and the Wilcoxon rank sum test were given in Table 6.9.

    a. For what distributions, if any, do the probabilities of a Type I error for the Wilcoxon rank sum test deviate grossly from the nominal value of $\alpha = .05$?

    b. For what distributions, if any, do the probabilities of a Type I error for the t test deviate grossly from the nominal value of $\alpha = .05$?

6.24 Refer to Exercise 6.23.

    a. Compare the power of the t test under the three nonnormal distributions to its power when the distributions are normal. Does skewness or heavy-tailedness seem to have the greatest effect?

    b. Is there much difference in the power of the Wilcoxon rank sum test across the four types of distributions? Explain your answer.

    c. For what type of distribution would you recommend using the Wilcoxon rank sum test? Explain your answer.

    d. For what type of distribution would you recommend using the t test? Explain your answer.

6.4   Inferences about $\mu_1 - \mu_2$: Paired Data

The methods we presented in the preceding four sections were appropriate for situations in which independent random samples are obtained from two populations. These methods are not appropriate for studies or experiments in which each measurement in one sample is *matched* or *paired* with a particular measurement in the other sample. In this section, we will deal with methods for analyzing "paired" data. We begin with an example.

EXAMPLE 6.6

Insurance adjusters are concerned about the high estimates they are receiving for auto repairs from garage I compared to garage II. To verify their suspicions, each of 15 cars recently involved in an accident was taken to both garages for separate estimates of repair costs. The estimates from the two garages are given in Table 6.10.

TABLE 6.10

Repair estimates (in hundreds of dollars)

Car	Garage I	Garage II
1	17.6	17.3
2	20.2	19.1
3	19.5	18.4
4	11.3	11.5
5	13.0	12.7
6	16.3	15.8
7	15.3	14.9
8	16.2	15.3
9	12.2	12.0
10	14.8	14.2
11	21.3	21.0
12	22.1	21.0
13	16.9	16.1
14	17.6	16.7
15	18.4	17.5
Totals	$\bar{y}_1 = 16.85$	$\bar{y}_2 = 16.23$
	$s_1 = 3.20$	$s_1 = 2.94$

A preliminary analysis of the data used a two-sample *t* test.

Solution   Computer output for these data is shown here.

```
Two-Sample T-Test and Confidence Interval

Two-sample T for Garage I vs Garage II

             N      Mean     StDev    SE Mean
Garage I    15     16.85      3.20      0.83
Garage II   15     16.23      2.94      0.76

95% CI for mu Garage I — mu Garage II: (−1.69, 2.92)
T-Test mu Garage I = mu Garage II (vs not =): T = 0.55   P = 0.59   DF = 27
```

From the output, we see there is a consistent difference in the sample means $(\bar{y}_1 - \bar{y}_2 = .62)$. However, this difference is rather small considering the variability of the measurements ($s_1 = 3.20$, $s_2 = 2.94$). In fact, the computed t-value (.55) has a p-value of .59, indicating very little evidence of a difference in the average claim estimates for the two garages.

A closer glance at the data in Table 6.10 indicates that something about the conclusion in Example 6.6 is inconsistent with our intuition. For all but one of the 15 cars, the estimate from garage I was higher than that from garage II. From our knowledge of the binomial distribution, the probability of observing garage I estimates higher in $y = 14$ or more of the $n = 15$ trials, assuming no difference ($\pi = .5$) for garages I and II, is

$$P(y = 14 \text{ or } 15) = P(y = 14) + P(y = 15)$$

$$= \binom{15}{14}(.5)^{14}(.5) + \binom{15}{15}(.5)^{15} = .000488$$

Thus, if the two garages in fact have the same distribution of estimates, there is approximately a 5 in 10,000 chance of having 14 or more estimates from garage I higher than those from garage II. Using this probability, we would argue that the observed estimates are highly contradictory to the null hypothesis of equality of distribution of estimates for the two garages. Why are there such conflicting results from the t test and the binomial calculation?

The explanation of the difference in the conclusions from the two procedures is that one of the required conditions for the t test, two samples being independent of each other, has been violated by the manner in which the study was conducted. The adjusters obtained a measurement from both garages for each car. For the two samples to be independent, the adjusters would have to take a random sample of 15 cars to garage I and a *different* random sample of 15 to garage II.

As can be observed in Figure 6.8, the repair estimates for a given car are about the same value, but there is a large variability in the estimates from each garage. The large variability *among* the 15 estimates from each garage diminishes the relative size of any difference *between* the two garages. When designing the

FIGURE 6.8

Repair estimates from two garages

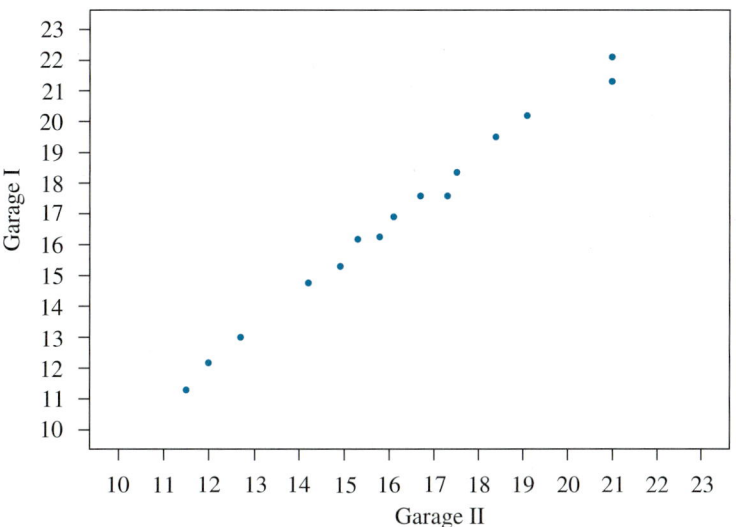

study, the adjusters recognized that the large differences in the amount of damage suffered by the cars would result in a large variability in the 15 estimates at both garages. By having both garages give an estimate on each car, the adjusters could calculate the difference between the estimates from the garages and hence reduce the large car-to-car variability.

This example illustrates a general design principle. In many situations, the available experimental units may be considerably different prior to their random assignment to the treatments with respect to characteristics that may affect the experimental responses. These differences will often then mask true treatment differences. In the previous example, the cars had large differences in the amount of damage suffered during the accident and hence would be expected to have large differences in their repair estimates no matter what garage gave the repair estimate. When comparing two treatments or groups in which the available experimental units have important differences prior to their assignment to the treatments or groups, the samples should be paired. There are many ways to design experiments to yield paired data. One method involves having the same group of experimental units receive both treatments, as was done in the repair estimates example. A second method involves having measurements taken before and after the treatment is applied to the experimental units. For example, suppose we want to study the effect of a new medicine proposed to reduce blood pressure. We would record the blood pressure of participants before they received the medicine and then after receiving the medicine. A third design procedure uses naturally occurring pairs such as twins or husbands and wives. A final method pairs the experimental units with respect to factors that may mask differences in the treatments. For example, a study is proposed to evaluate two methods for teaching remedial reading. The participants could be paired based on a pretest of their reading ability. After pairing the participants, the two methods are randomly assigned to the participants within each pair. Design principles are presented in Chapter 14.

A proper analysis of paired data needs to take into account the lack of independence between the two samples. The sampling distribution for the difference in the sample means, $(\bar{y}_1 - \bar{y}_2)$, will have mean and standard error

$$\mu_{\bar{y}_1-\bar{y}_2} = \mu_1 - \mu_2 \qquad \text{and} \qquad \sigma_{\bar{y}_1-\bar{y}_2} = \sqrt{\frac{\sigma_1^2 + \sigma_2^2 - 2\sigma_1\sigma_2\rho}{n}}$$

where ρ measures the amount of dependence between the two samples. When the two samples produce similar measurements, ρ is positive and the standard error of $\bar{y}_1 - \bar{y}_2$ is smaller than what would be obtained using two independent samples. This was the case in the repair estimates data. The size and sign of ρ can be determined by examining the plot of the paired data values. The magnitude of ρ is large when the plotted points are close to a straight line. The sign of ρ is positive when the plotted points follow an increasing line and negative when plotted points follow a decreasing line. From Figure 6.8, we observe that the estimates are close to an increasing line and thus ρ will be positive. The use of paired data in the repair estimate study will reduce the variability in the standard error of the difference in the sample means in comparison to using independent samples.

The actual analysis of paired data requires us to compute the differences in the n pairs of measurements, $d_i = y_{1i} - y_{2i}$, and obtain $\bar{d}$, s_d, the mean and standard deviations in the d_is. Also, we must formulate the hypotheses about μ_1 and μ_2 into hypotheses about the mean of the differences, $\mu_d = \mu_1 - \mu_2$. The conditions required to develop a t procedure for testing hypotheses and constructing confidence intervals for μ_d are

1. The sampling distribution of the d_is is a normal distribution.
2. The d_is are independent; that is, the pairs of observations are independent.

A summary of the test procedure is given here.

Paired _t_ test

H_0: 1. $\mu_d \leq D_0$ (D_0 is a specified value, often 0)
2. $\mu_d \geq D_0$
3. $\mu_d = D_0$

H_a: 1. $\mu_d > D_0$
2. $\mu_d < D_0$
3. $\mu_d \neq D_0$

T.S.: $t = \dfrac{\overline{d} - D_0}{s_d/\sqrt{n}}$

R.R.: For a level α Type I error rate and with df $= n - 1$
1. Reject H_0 if $t \geq t_\alpha$.
2. Reject H_0 if $t \leq -t_\alpha$.
3. Reject H_0 if $|t| \geq t_{\alpha/2}$.

Check assumptions and draw conclusions.

The corresponding $100(1 - \alpha)\%$ confidence interval on $\mu_d = \mu_1 - \mu_2$ based on the paired data is shown here.

$100(1 - \alpha)\%$ Confidence Interval for μ_d Based on Paired Data

$$\overline{d} \pm t_{\alpha/2}\frac{s_d}{\sqrt{n}}$$

where n is the number of pairs of observations (and hence the number of differences) and df $= n - 1$.

EXAMPLE 6.7

Refer to the data of Example 6.6 and perform a paired t test. Draw a conclusion based on $\alpha = .05$.

Solution For these data, the parts of the statistical test are

H_0: $\mu_d = \mu_1 - \mu_2 \leq 0$

H_a: $\mu_d > 0$

T.S.: $t = \dfrac{\overline{d}}{s_d/\sqrt{n}}$

R.R.: For df $= n - 1 = 14$, reject H_0 if $t \geq t_{.05}$.

Before computing t, we must first calculate $\bar{d}$ and s_d. For the data of Table 6.10, we have the differences d_i = garage I estimate − garage II estimate.

Car	1	2	3	4	5	6	7	8	9	10	11	12	13	14	15
d_i	.3	1.1	1.1	−.2	.3	.5	.4	.9	.2	.6	.3	1.1	.8	.9	.9

The mean and standard deviation are given here.

$$\bar{d} = .61 \quad \text{and} \quad s_d = .394$$

Substituting into the test statistic t, we have

$$t = \frac{\bar{d} - 0}{s_d/\sqrt{n}} = \frac{.61}{.394/\sqrt{15}} = 6.00$$

Indeed, $t = 6.00$ is far beyond all tabulated t values for df = 14, so the p-value is less than .005; in fact, the p-value is .000016. We conclude that the mean repair estimate for garage I is greater than that for garage II. This conclusion agrees with our intuitive finding based on the binomial distribution.

The point of all this discussion is not to suggest that we typically have two or more analyses that may give *very* conflicting results for a given situation. Rather, the point is that the analysis must fit the experimental situation; and for this experiment, the samples are dependent, demanding we use an analysis appropriate for dependent (paired) data.

After determining that there is a *statistically significant* difference in the means, we should estimate the size of the difference. A 95% confidence interval for $\mu_1 - \mu_2 = \mu_d$ will provide an estimate of the size of the difference in the average repair estimate between the two garages.

$$\bar{d} \pm t_{\alpha/2}\frac{s_d}{\sqrt{n}}$$

$$.61 \pm 2.145\frac{.394}{\sqrt{15}} \quad \text{or} \quad .61 \pm .22$$

Thus, we are 95% confident that the mean repair estimates differ by a value between $390 and $830. The insurance adjusters determined that a difference of this size is of practical significance.

The reduction in standard error of $\bar{y}_1 - \bar{y}_2$ by using the differences d_is in place of the observed values y_{1i}s and y_{2i}s will produce a t test having greater power and confidence intervals having smaller width. Is there any loss in using paired data experiments? Yes, the t procedures using the d_is have df = $n - 1$, whereas the t procedures using the individual measurements have df = $n_1 + n_2 - 2 = 2(n - 1)$. Thus, when designing a study or experiment, the choice between using an independent samples experiment and a paired data experiment will depend on how much difference exists in the experimental units prior to their assignment to the treatments. If there are only small differences, then the independent samples design is more efficient. If the differences in the experimental units are extreme, then the paired data design is more efficient.

EXERCISES   **Basic Techniques**

6.25 Consider the paired data shown here.

Pair	y_1	y_2
1	21	29
2	28	30
3	17	21
4	24	25
5	27	33

a. Run a paired t test and give the p-value for the test.
b. What would your conclusion be using an argument related to the binomial distribution? Does it agree with part (a)? When might these two approaches not agree?

Applications

Engin.   **6.26** Researchers are studying two existing coatings used to prevent corrosion in pipes that transport natural gas. The study involves examining sections of pipe that had been in the ground at least 5 years. The effectiveness of the coating depends on the pH of the soil, so the researchers recorded the pH of the soil at all 20 sites at which the pipe was buried prior to measuring the amount of corrosion on the pipes. The pH readings are given here. Describe how the researchers could conduct the study to reduce the effect of the differences in the pH readings on the evaluation of the difference in the two coatings' corrosion protection.

pH Readings at Twenty Research Sites									
Coating A	3.2	4.9	5.1	6.3	7.1	3.8	8.1	7.3	5.9	8.9
Coating B	3.7	8.2	7.4	5.8	8.8	3.4	4.7	5.3	6.8	7.2

Med.   **6.27** Suppose you are a participant in a project to study the effectiveness of a new treatment for high cholesterol. The new treatment will be compared to a current treatment by recording the change in cholesterol readings over a 10-week treatment period. The effectiveness of the treatment may depend on the participant's age, body fat percentage, diet, and general health. The study will involve at most 30 participants because of cost considerations.

a. Describe how you would conduct the study using independent samples.
b. Describe how you would conduct the study using paired samples.
c. How would you decide which method, paired or independent samples, would be more efficient in evaluating the change in cholesterol readings?

Ag.   **6.28** An agricultural experiment station was interested in comparing the yields for two new varieties of corn. Because the investigators thought that there might be a great deal of variability in yield from one farm to another, each variety was randomly assigned to a different 1-acre plot on each of seven farms. The 1-acre plots were planted; the corn was harvested at maturity. The results of the experiment (in bushels of corn) are listed here.

Farm	1	2	3	4	5	6	7
Variety A	48.2	44.6	49.7	40.5	54.6	47.1	51.4
Variety B	41.5	40.1	44.0	41.2	49.8	41.7	46.8

a. Use these data to test whether there is a difference in mean yields for the two varieties of corn. Use $\alpha = .05$.

b. Estimate the size of the difference in the mean yields of the two varieties.

Med.

6.29 The paper "Effect of long-term blood pressure control on salt sensitivity" [*Journal of Medicine* (1997) 28: 147–156]. The objective of the study was to evaluate salt sensitivity (SENS) after a period of antihypertensive treatment. Ten hypertensive patients (diastolic blood pressure between 90 and 115 mmHg) were studied after at least 18 months on antihypertensive treatment. SENS readings, which were obtained before and after the patients were placed on a antihypertensive treatment, are given here.

Patient	1	2	3	4	5	6	7	8	9	10
Before treatment	22.86	7.74	15.49	9.97	1.44	9.39	11.40	1.86	−6.71	6.42
After treatment	6.11	−4.02	8.04	3.29	−0.77	6.99	10.19	2.09	11.40	10.70

a. Is there significant evidence that the mean SENS value decreased after the patient received antihypertensive treatment?

b. Estimate the size of the change in the mean SENS value.

c. Do the conditions required for using the *t* procedures appear to be valid for these data? Justify your answer.

H.R.

6.30 Suppose we wish to estimate the difference between the mean monthly salaries of male and female sales representatives. Because there is a great deal of salary variability from company to company, we decided to filter out the variability due to companies by making male–female comparisons within each company. One male and one female with the required background and work experience will be selected from each company. If the range of differences in salaries (between males and females) within a company is approximately $300 per month, determine the number of companies that must be examined to estimate the difference in mean monthly salary for males and females. Use a 95% confidence interval with a half width of $5. (*Hint:* Refer to Section 5.3.)

6.31 Refer to Exercise 6.30. If $n = 35$, $\bar{d} = 120$, and $s_d = 250$, construct a 90% confidence interval for μ_d, the mean difference in salaries for male and female sales representatives.

Edu.

6.32 A study was designed to measure the effect of home environment on academic achievement of 12-year-old students. Because genetic differences may also contribute to academic achievement, the researcher wanted to control for this factor. Thirty sets of identical twins were identified who had been adopted prior to their first birthday, with one twin placed in a home in which academics were emphasized (Academic) and the other twin placed in a home in which academics were not emphasized (Nonacademic). The final grades (based on 100 points) for the 60 students are given here.

Set of Twins	Academic	Nonacademic	Set of Twins	Academic	Nonacademic
1	78	71	8	80	75
2	75	70	9	98	92
3	68	66	10	52	56
4	92	85	11	67	63
5	55	60	12	55	52
6	74	72	13	49	48
7	65	57	14	66	67

Set of Twins	Academic	Nonacademic	Set of Twins	Academic	Nonacademic
15	75	70	23	82	78
16	90	88	24	70	62
17	89	80	25	68	73
18	73	65	26	74	73
19	61	60	27	85	75
20	76	74	28	97	88
21	81	76	29	95	94
22	89	78	30	78	75

a. Use the following computer output to evaluate whether there is a difference in the mean final grade between the students in an academically oriented home environment and those in a nonacademic home environment.

b. Estimate the size of the difference in the mean final grades of the students in academic and nonacademic home environments.

c. Do the conditions for using the t procedures appear to be satisfied for these data?

d. Does it appear that using twins in this study to control for variation in final scores was effective as compared to taking a random sample of 30 students in both types of home environments? Justify your answer.

```
Paired T-Test and Confidence Interval

Paired T for Academic — Nonacademic

                 N        Mean      StDev    SE Mean
Academic        30       75.23      13.29      2.43
Nonacademic     30       71.43      11.42      2.09
Difference      30       3.800      4.205      0.768

95% CI for mean difference: (2.230, 5.370)
T-Test of mean difference = 0 (vs not = 0): T-value = 4.95   P-Value = 0.000
```

```
Two-Sample T-Test and Confidence Interval

Two-sample T for Academic vs Nonacademic

                N       Mean     StDev    SE Mean
Academic       30       75.2     13.3       2.4
Nonacademic    30       71.4     11.4       2.1

95% CI for mu Academic — mu Nonacademic: ( −2.6,  10.2)
T-Test mu Academic = mu Nonacademic (vs not =): T = 1.19   P = 0.24   DF = 56
```

Boxplot of differences
(with H_0 and 95% t confidence interval for the mean)

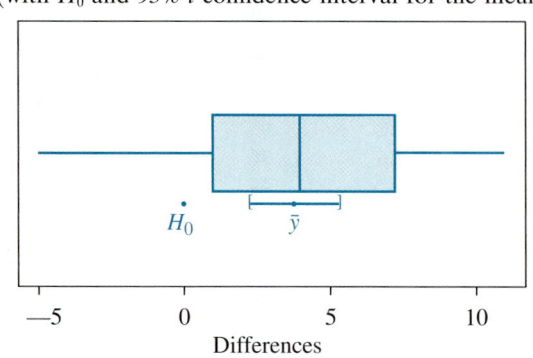

Normal probability plot of differences

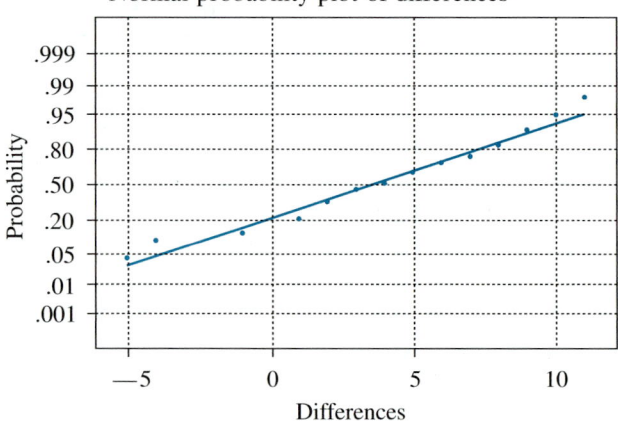

6.5 A Nonparametric Alternative: The Wilcoxon Signed-Rank Test

The Wilcoxon signed-rank test, which makes use of the sign and the magnitude of the rank of the differences between pairs of measurements, provides an alternative to the paired t test when the population distribution of the differences is nonnormal. The Wilcoxon signed-rank test requires that the population distribution of differences be symmetric about the unknown median M. Let D_0 be a specified hypothesized value of M. The test evaluates shifts in the distribution of differences to the right or left of D_0; in most cases, D_0 is 0. The computation of the signed-rank test involves the following steps:

1. Calculate the differences in the n pairs of observations.
2. Subtract D_0 from all the differences.
3. Delete all zero values. Let n be the number of nonzero values.
4. List the *absolute values* of the differences in increasing order, and assign them the ranks $1, \ldots, n$ (or the average of the ranks for ties).

We define the following notation before describing the Wilcoxon signed-rank test:

n = the number of pairs of observations with a nonzero difference
T_+ = the sum of the positive ranks; if there are no positive ranks, $T_+ = 0$
T_- = the sum of the negative ranks; if there are no negative ranks, $T_- = 0$
T = the smaller of T_+ and T_-

μ_T

$$\mu_T = \frac{n(n+1)}{4}$$

σ_T

$$\sigma_T = \sqrt{\frac{n(n+1)(2n+1)}{24}}$$

g groups

If we group together all differences assigned the same rank, and there are g such groups, the variance of T is

$$\sigma_T^2 = \frac{1}{24}\left[n(n+1)(2n+1) - \frac{1}{2}\sum_j t_j(t_j - 1)(t_j + 1)\right]$$

t_j where t_j is the number of tied ranks in the jth group. Note that if there are no tied ranks, $g = n$, and $t_j = 1$ for all groups. The formula then reduces to

$$\sigma_T^2 = \frac{n(n + 1)(2n + 1)}{24}$$

The Wilcoxon signed-rank test is presented here.

Wilcoxon Signed-Rank Test

H_0: The distribution of differences is symmetrical around D_0. (D_0 is specified; usually D_0 is 0.)

H_a: 1. The differences tend to be larger than D_0.
2. The differences tend to be smaller than D_0.
3. Either 1 or 2 is true (two-sided H_a).

$(n \leq 50)$

T.S.: 1. $T = T_-$
2. $T = T_+$
3. $T =$ smaller of T_+ and T_-

R.R.: For a specified value of α (one-tailed .05, .025, .01, or .005; two-tailed .10, .05, .02, .01) and fixed number of nonzero differences n, reject H_0 if the value of T is less than or equal to the appropriate entry in Table 6 in the Appendix.

$(n > 50)$

T.S.: Compute the test statistic

$$z = \frac{T - \dfrac{n(n + 1)}{4}}{\sqrt{\dfrac{n(n + 1)(2n + 1)}{24}}}$$

R.R.: For cases 1 and 2, reject H_0 if $z < -z_\alpha$; for case 3, reject H_0 if $z < -z_{\alpha/2}$.

Check assumptions and draw conclusions.

EXAMPLE 6.8

A city park department compared a new formulation of a fertilizer, brand A, to the previously used fertilizer, brand B, on each of 20 different softball fields. Each field was divided in half, with brand A randomly assigned to one half of the field and brand B to the other. Sixty pounds of fertilizers per acre were then applied to the fields. The effect of the fertilizer on the grass grown at each field was measured by the weight (in pounds) of grass clippings produced by mowing the grass at the fields over a 1-month period. Evaluate whether brand A tends to produce more grass than brand B. The data are given here.

Field	Brand A	Brand B	Difference	Field	Brand A	Brand B	Difference
1	211.4	186.3	25.1	11	208.9	183.6	25.3
2	204.4	205.7	−1.3	12	208.7	188.7	20.0
3	202.0	184.4	17.6	13	213.8	188.6	25.2
4	201.9	203.6	−1.7	14	201.6	204.2	−2.6
5	202.4	180.4	22.0	15	201.8	181.6	20.1
6	202.0	202.0	0	16	200.3	208.7	−8.4
7	202.4	181.5	20.9	17	201.8	181.5	20.3
8	207.1	186.6	20.4	18	201.5	208.7	−7.2
9	203.6	205.7	−2.1	19	212.1	186.8	25.3
10	216.0	189.1	26.9	20	203.4	182.9	20.5

Solution Evaluate whether brand A tends to produce more grass than brand B. Plots of the differences in grass yields for the 20 fields are given in Figure 6.9 (a) and (b). The differences appear to not follow a normal distribution and appear to form two distinct clusters. Thus, we will apply the Wilcoxon signed-rank test to evaluate the differences in grass yields from brand A and brand B. The null hypothesis is that the distribution of differences is symmetrical about 0 against the alternative that the differences tend to be greater than 0. First we must rank (from smallest to largest) the absolute values of the $n = 20 − 1 = 19$ nonzero differences. These ranks appear in Table 6.11.

FIGURE 6.9 (a)

Boxplot of differences (with H_0 and 95% t confidence interval for the mean)

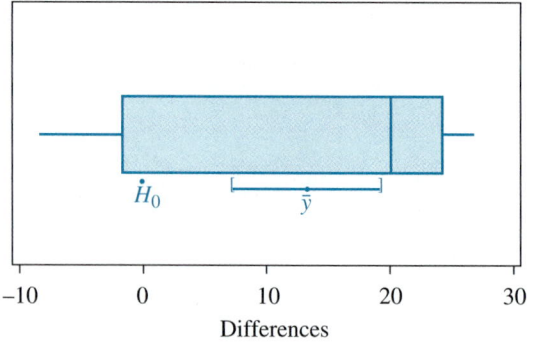

FIGURE 6.9 (b)

Normal probability plot of differences

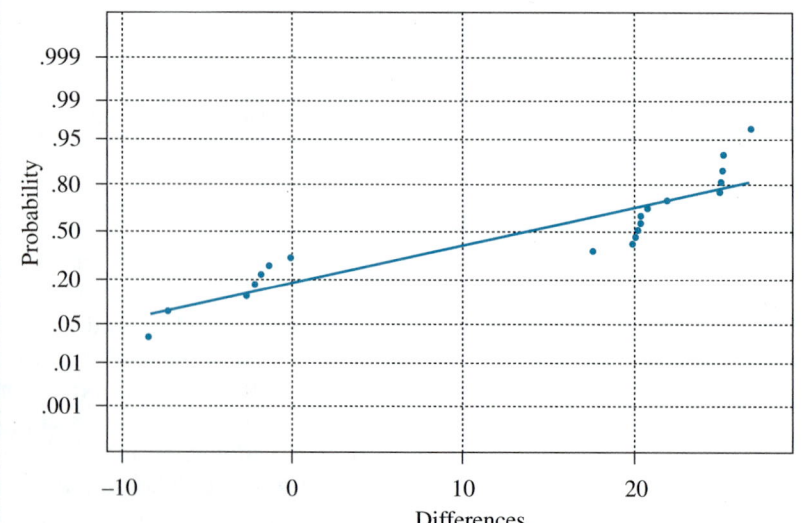

TABLE 6.11

Rankings of
grass yield data

Field	Difference	Rank of Absolute Difference	Sign of Difference	Field	Difference	Rank of Absolute Difference	Sign of Difference
1	25.1	15	Positive	11	25.3	17.5	Positive
2	−1.3	1	Negative	12	20.0	8	Positive
3	17.6	7	Positive	13	25.2	16	Positive
4	−1.7	2	Negative	14	−2.6	4	Negative
5	22.0	14	Positive	15	20.1	9	Positive
6	0	None	Positive	16	−8.4	6	Negative
7	20.9	13	Positive	17	20.3	10	Positive
8	20.4	11	Positive	18	−7.2	5	Negative
9	−2.1	3	Negative	19	25.3	17.5	Positive
10	26.9	19	Positive	20	20.5	12	Positive

The sum of the positive and negative ranks are

$$T_- = 1 + 2 + 3 + 4 + 5 + 6 = 21$$

and

$$T_+ = 7 + 8 + 9 + 10 + 11 + 12 + 13 + 14 + 15 + 16 + 17.5 + 17.5 + 19$$
$$= 169$$

Thus, T, the smaller of T_+ and T_-, is 21. For a one-sided test with $n = 19$ and $\alpha = .05$, we see from Table 6 in the Appendix that we will reject H_0 if T is less than or equal to 53. Thus, we reject H_0 and conclude that brand A fertilizer tends to produce more grass than brand B.

The choice of an appropriate paired-sample test depends on examining different types of deviations from normality. Because the level of the Wilcoxon signed-rank does not depend on the population distribution, its level is the same as the stated value for all symmetric distributions. The level of the paired t test may be different from its stated value when the population distribution is very nonnormal. Also, we need to examine which test has greater power. We will report a portion of a simulation study contained in Randles and Wolfe (1979). The population distributions considered were normal, uniform (short-tailed), double exponential (moderately heavy-tailed), and Cauchy (very heavy-tailed). Table 6.12 displays the proportion of times in 5,000 replications that the tests rejected H_0. The two populations were shifted by amounts 0, $.4\sigma$, and $.8\sigma$, where σ denotes the standard deviation of the distribution. (When the population distribution is Cauchy, σ denotes a scale parameter.)

TABLE 6.12

Empirical power of paired
t (t) and signed-rank (T)
tests with $\alpha = .05$

Distribution		Normal			Double Exponential			Cauchy			Uniform		
	Shift:	0	$.4\sigma$	$.8\sigma$	0	$.4\sigma$	$.8\sigma$	0	$.4\sigma$	$.8\sigma$	0	$.4\sigma$	$.8\sigma$
$n = 10$	t	.049	.330	.758	.047	.374	.781	.028	.197	.414	.051	.294	.746
	T	.050	.315	.741	.048	.412	.804	.049	.332	.623	.049	.277	.681
$n = 15$	t	.048	.424	.906	.049	.473	.898	.025	.210	.418	.051	.408	.914
	T	.047	.418	.893	.050	.532	.926	.050	.423	.750	.051	.383	.852
$n = 20$	t	.048	.546	.967	.044	.571	.955	.026	.214	.433	.049	.522	.971
	T	.049	.531	.962	.049	.652	.975	.049	.514	.849	.050	.479	.935

From Table 6.12, we can make the following observations. The level of the paired *t* test remains nearly equal to .05 for uniform and double exponential distributions, but is much less than .05 for the very heavy-tailed Cauchy distribution. The Wilcoxon signed-rank test's level is nearly .05 for all four distributions, as expected, because the level of the Wilcoxon test only requires that the population distribution be symmetric. When the distribution is normal, the *t* test has only slightly greater power values than the Wilcoxon signed-rank test. When the population distribution is short-tailed and uniform, the paired *t* test has slightly greater power than the signed-rank test. Note also that the power values for the *t* test are slightly less than the *t* power values when the population distribution is normal. For the double exponential, the Wilcoxon test has slightly greater power than the *t* test. For the Cauchy distribution, the level of the *t* test deviates significantly from .05 and its power is much lower than the Wilcoxon test. From other studies, if the distribution of differences is grossly skewed, the nominal *t* probabilities may be misleading. The skewness has less of an effect on the level of the Wilcoxon test.

Even with this discussion, you might still be confused as to which statistical test or confidence interval to apply in a given situation. First, plot the data and attempt to determine whether the population distribution is very heavy-tailed or very skewed. In such cases, use a Wilcoxon rank-based test. When the plots are not definitive in their detection of nonnormality, perform both tests. If the results from the different tests yield different conclusions, carefully examine the data to identify any peculiarities to understand why the results differ. If the conclusions agree and there are no blatant violations of the required conditions, you should be very confident in your conclusions. This particular "hedging" strategy is appropriate not only for paired data but also for many situations in which there are several alternative analyses.

EXERCISES   **Basic Techniques**

6.33   Refer to Exercise 6.29.
  a. Using the data in the table, run a Wilcoxon signed-rank test. Give the *p*-value and draw a conclusion.
  b. Compare your conclusions here to those in Exercise 6.29. Does it matter which test (*t* or signed rank) is used?

Applications

6.34   Use the information in Table 6.12 in forming your answers to the following questions.
  a. Does the sample size *n* affect how close the actual level of the *t* test is to the nominal level of .05?
  b. Does the shape of the population distribution affect how close the actual level of the *t* test is to the nominal level of .05?
  c. Does the shape of the population distribution affect how close the actual level of the Wilcoxon test is to the nominal level of .05?
  d. Suppose a level .05 test is to be applied to a data set that is highly skewed to the right. Will the Wilcoxon signed-rank test's "actual" level or the paired *t* test's actual level be closer to .05? Justify your answer.

Soc.   **6.35**   A study was conducted to determine whether automobile repair charges are higher for female customers than for male customers. Ten auto repair shops were randomly selected from the telephone book. Two cars of the same age, brand, and engine problem were used in the study. For each repair shop, the two cars were randomly assigned to a man and woman participant and then taken to the shop for an estimate of repair cost. The repair costs (in dollars) are given here.

Repair Shop	1	2	3	4	5	6	7	8	9	10	11	12	13	14	15	16	17	18	19	20
Female customers	871	684	795	838	1,033	917	1,047	723	1,179	707	817	846	975	868	1,323	791	1,157	932	1,089	770
Male customers	792	765	511	520	618	447	548	720	899	788	927	657	851	702	918	528	884	702	839	878

a. Which procedure, t or Wilcoxon, is more appropriate in this situation? Why?

b. Are repair costs generally higher for female customers than for male customers? Use $\alpha = .05$.

Boxplot of differences (with H_0 and 95% t confidence interval for the mean)

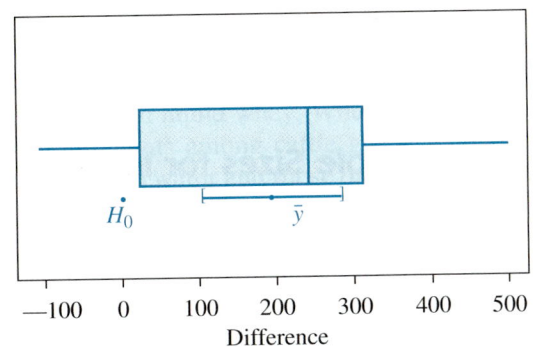

Normal probability plot of differences in cost

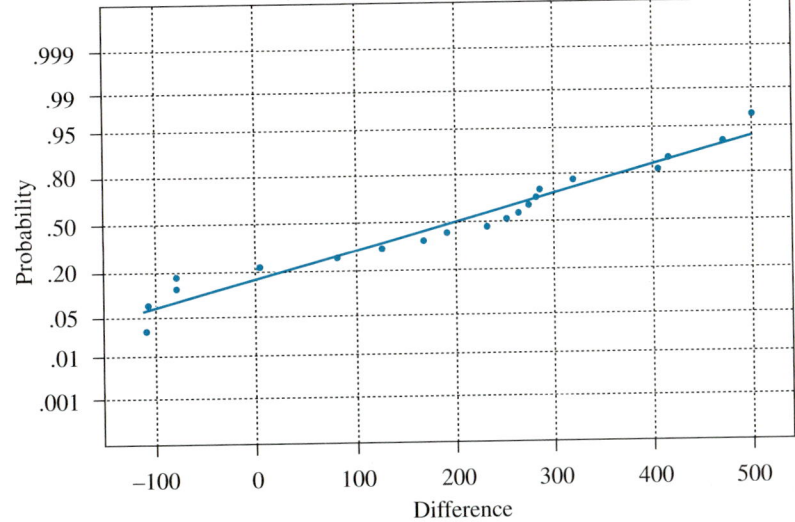

Bio.   **6.36**   The effect of Benzedrine on the heart rate of dogs (in beats per minute) was examined in an experiment on 14 dogs chosen for the study. Each dog was to serve as its own control, with half of the dogs assigned to receive Benzedrine during the first study period and the other half assigned to receive a placebo (saline solution). All dogs were examined to determine the heart rates after 2 hours on the medication. After 2 weeks in which no medication was given, the regimens for the dogs were switched for the second study period. The dogs previously on Benzedrine were given the placebo and the others received Benzedrine. Again heart rates were measured after 2 hours.

The following sample data are not arranged in the order in which they were taken but have been summarized by regimen. Use these data to test the research hypothesis that

where T denotes the sum of the ranks in sample 1

$$\mu_T = \frac{n_1(n_1 + n_2 + 1)}{2} \quad \text{and} \quad \sigma_T = \sqrt{\frac{n_1 n_2}{12}(n_1 + n_2 + 1)}$$

provided there are no tied ranks.

6. Paired t test; difference approximately normal

$$\text{T.S.:} \quad t = \frac{\overline{d} - D_0}{s_d/\sqrt{n}} \quad \text{df} = n - 1$$

where n is the number of differences.

7. $100(1 - \alpha)\%$ confidence interval for μ_d, paired data; differences approximately normal

$$\overline{d} \pm t_{\alpha/2} s_d/\sqrt{n}$$

8. Wilcoxon's signed-rank test, paired data

H_0: The distribution of differences is symmetrical about D_0.
T.S.: $(n \le 50)$ T_-, or T_+ or smaller of T_+ and T_- depending on the form of H_a.
T.S.: $n > 50$

$$z = \frac{T - \mu_T}{\sigma_T}$$

where

$$\mu_T = \frac{n(n + 1)}{4} \quad \text{and} \quad \sigma_T = \sqrt{\frac{n(n + 1)(2n + 1)}{24}}$$

provided there are no tied ranks.

9. Independent samples: sample sizes for estimating $\mu_1 - \mu_2$ with a $100(1 - \alpha)\%$ confidence interval, of the form $\overline{y}_1 - \overline{y}_2 \pm E$

$$n = \frac{2z_{\alpha/2}^2 \sigma^2}{E^2}$$

10. Independent samples: sample sizes for a test of $\mu_1 - \mu_2$
   a. One-sided test:

$$n = \frac{2\sigma^2(z_\alpha + z_\beta)^2}{\Delta^2}$$

   b. Two-sided test:

$$n = \frac{2\sigma^2(z_{\alpha/2} + z_\beta)^2}{\Delta^2}$$

11. Paired samples: sample size for estimating $\mu_1 - \mu_2$ with $100(1 - \alpha)\%$ confidence interval, of the form $\overline{d} \pm E$,

$$n = \frac{z_{\alpha/2}^2 \sigma_d^2}{E^2}$$

12. Paired samples: sample size for a test of $\mu_1 - \mu_2$
   a. One-sided test:

$$n = \frac{2\sigma_d^2 (z_\alpha + z_\beta)^2}{\Delta^2}$$

   b. Two-sided test:

$$n = \frac{2\sigma_d^2 (z_{\alpha/2} + z_\beta)^2}{\Delta^2}$$

Supplementary Exercises

Engin.   **6.37**   A new alloy is proposed for the manufacture of steel beams. The company's metallurgist designs a study to compare the strength of the new alloy to the currently used alloy. Ten beams of each type of alloy are manufactured. The load capacities (in tons) for the 20 beams are determined and are given here.

Beam	1	2	3	4	5	6	7	8	9	10
Old alloy	23.7	24.0	21.1	23.1	22.8	25.0	25.3	22.6	23.3	22.8
New alloy	26.6	32.3	28.0	31.1	29.6	28.5	31.2	23.6	29.1	28.5

   a. What are the populations of interest?
   b. Place 99% confidence intervals on the mean load capacity of beams produced with both the new and old alloy.
   c. Is the mean load capacity of the new alloy significantly greater than the mean load capacity of the currently used alloy? Use $\alpha = .01$. Report the p-value of your test.
   d. Do the conditions required for the statistical techniques used in (b) and (c) appear to be satisfied?
   e. The beams produced from the new alloy are more expensive than the beams produced from the currently used alloy. Thus, the new alloy will be used only if the mean load capacity is at least 5 tons greater than the mean load capacity of the currently used alloy. Based on this information, would you recommend that the company use the new alloy?

Med.   **6.38**   Long-distance runners have contended that moderate exposure to ozone increases lung capacity. To investigate this possibility, a researcher exposed 12 rats to ozone at the rate of 2 parts per million for a period of 30 days. The lung capacity of the rats was determined at the beginning of the study and again after the 30 days of ozone exposure. The lung capacities (in mL) are given here.

Rat	1	2	3	4	5	6	7	8	9	10	11	12
Before exposure	8.7	7.9	8.3	8.4	9.2	9.1	8.2	8.1	8.9	8.2	8.9	7.5
After exposure	9.4	9.8	9.9	10.3	8.9	8.8	9.8	8.2	9.4	9.9	12.2	9.3

   a. Is there sufficient evidence to support the conjecture that ozone exposure increases lung capacity? Use $\alpha = .05$. Report the p-value of your test.

b. Estimate the size of the increase in lung capacity after exposure to ozone using a 95% confidence interval.

c. After completion of the study, the researcher claimed that ozone causes increased lung capacity. Is this statement supported by this experiment?

Env.   **6.39**   In an environmental impact study for a new airport, the noise level of various jets was measured just seconds after their wheels left the ground. The jets were either wide-bodied or narrow-bodied. The noise levels in decibels (dB) are recorded here for 15 wide-bodied jets and 12 narrow-bodied jets.

Wide-Bodied Jet	109.5	107.3	105.0	117.3	105.4	113.7	121.7	109.2	108.1	106.4	104.6	110.5	110.9	111.0	112.4
Narrow-Bodied Jet	131.4	126.8	114.1	126.9	108.2	122.0	106.9	116.3	115.5	111.6	124.5	116.2			

a. Do the two types of jets have different mean noise levels? Report the level of significance of the test.

b. Estimate the size of the difference in mean noise level between the two types of jets using a 95% confidence interval.

c. How would you select the jets for inclusion in this study?

Ag.   **6.40**   An entomologist is investigating which of two fumigants, F_1 or F_2, is more effective in controlling parasities in tobacco plants. To compare the fumigants, nine fields of differing soil characteristics, drainage, and amount of wind shield were planted with tobacco. Each field was then divided into two plots of equal area. Fumigant F_1 was randomly assigned to one plot in each field and F_2 to the other plot. Fifty plants were randomly selected from each field, 25 from each plot, and the number of parasites were counted. The data are in the following table.

Field	1	2	3	4	5	6	7	8	9
Fumigant F_1	77	40	11	31	28	50	53	26	33
Fumigant F_2	76	38	10	29	27	48	51	24	32

a. What are the populations of interest?

b. Do the data provide sufficient evidence to indicate a difference in the mean level of parasites for the two fumigants? Use $\alpha = .10$. Report the p-value for the experimental data.

c. Estimate the size of the difference in the mean number of parasites between the two fumigants using a 90% confidence interval.

6.41   Refer to Exercise 6.40. An alternative design of the experiment would involve randomly assigning fumigant F_1 to nine of the plots and F_2 to the other nine plots, ignoring which fields the plots were from. What are some of the problems that may occur in using the alternative design?

Env.   **6.42**   Following the March 24, 1989 grounding of the tanker *Exxon Valdez* in Alaska, approximately 35,500 tons of crude oil were released into Prince William Sound. The paper, "The deep benthos of Prince William Sound, Alaska, 16 months after the *Exxon Valdez* oil spill" [*Marine Pollution Bulletin* (1998), 36: 118–130] reports on an evaluation of deep benthic infauna after the spill. Thirteen sites were selected for study. Seven of the sites were within the oil trajectory and six were outside the oil trajectory. Collection of environmental and biological data at two depths, 40 m and 100 m, occurred in the period July 1–23, 1990. One of the variables measured was population abundance (individuals per square meter). The values are given in the following table.

Site	Within Oil Trajectory							Outside Oil Trajectory					
	1	2	3	4	5	6	7	1	2	3	4	5	6
Depth 40 m	5,124	2,904	3,600	2,880	2,578	4,146	1,048	1,336	394	7,370	6,762	744	1,874
Depth 100 m	3,228	2,032	3,256	3,816	2,438	4,897	1,346	1,676	2,008	2,224	1,234	1,598	2,182

a. After combining the data from the two depths, does there appear to be a difference in population abundance between the sites within and outside the oil trajectory? Use $\alpha = .05$.

b. Estimate the size of the difference in the mean population abundance at the two types of sites using a 95% confidence interval.

c. What are the required conditions for the techniques used in parts (a) and (b)?

d. Check to see whether the required conditions are satisfied.

6.43 Refer to Exercise 6.42. Answer the following questions using the combined data for both depths.

a. Use the Wilcoxon rank sum test to assess whether there is a difference in population abundance between the sites within and outside the oil trajectory. Use $\alpha = .05$.

b. What are the required conditions for the techniques used in part (a)?

c. Are the required conditions satisfied?

d. Discuss any differences in the conclusions obtained using the t-procedures and the Wilcoxon rank sum test.

6.44 Refer to Exercise 6.42. The researchers also examined the effect of depth on population abundance.

a. Plot the four data sets using side-by-side boxplots to demonstrate the effect of depth on population abundance.

b. Separately for each depth, evaluate differences between the sites within and outside the oil trajectory. Use $\alpha = .05$.

c. Are your conclusions at 40 m consistent with your conclusions at 100 m?

6.45 Refer to Exercises 6.42–6.44.

a. Discuss the veracity of the statement, "The oil spill did not adversely affect the population abundance; in fact, it appears to have increased the population abundance."

b. A possible criticism of the study is that the six sites outside the oil trajectory were not comparable in many aspects to the seven sites within the oil trajectory. Suppose that the researchers had data on population abundance at the seven within sites prior to the oil spill. What type of analysis could be used on these data to evaluate the effect of the oil spill on population abundance? What are some advantages to using this data rather than the data in Exercise 6.43?

c. What are some possible problems with using the before and after oil spill data in assessing the effect of the spill on population abundance?

6.46 Refer to the data in Exercise 3.13.

a. Does the new therapy appear to have a longer mean survival time than the standard therapy? Use $\alpha = .05$.

b. Estimate the mean survival time using the new therapy using a 95% confidence interval.

c. Estimate the difference in mean survival time between the standard and new therapies using a 95% confidence interval. Is the confidence interval consistent with the results from (a)?

d. Are the required conditions for the procedures used to answer (a)–(c) satisfied?

Edu. **6.47** A research team at a leading university college of education designed a study to evaluate the conjecture that teachers' expectations of their students can have an effect on student performance. Two hundred students of comparable educational achievement were randomly assigned to either an experimental or a control group. The teachers of the experimental group were told that the students in their class were high achievers and would most likely show a large increase in their standardized test scores. The teachers in the control group were not told anything about their students. The students were given a standardized test at the beginning and end of the semester. The change in test scores are summarized in the following table and graph.

	Sample Size	Mean	Standard Deviation
Experimental group	100	26.5	24.2
Control group	100	17.0	9.1

a. Does it appear that the experimental group has a larger change in mean test score?
b. Report the *p*-value of the test.
c. Estimate the size of the difference in mean change in test score between the two groups.
d. Are the required conditions for the *t*-procedures used in (a)–(c) satisfied?
e. In nontechnical terms—that is, not using statistical terms—state your conclusions about the differences in the results for the two groups.

Boxplots of experimental and control groups (means are indicated by solid circles)

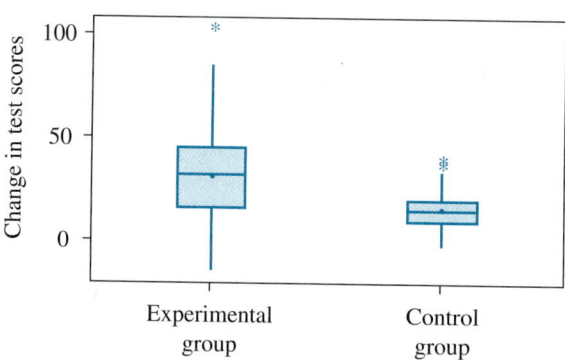

6.48 Refer to Exercise 6.47. Critics of the study state that widely varying teaching approaches and student abilities can affect the outcome of the study. Thus, because there were different teachers and students in the two groups, it is not possible to know whether the observed differences in test scores are due to the teachers' expectations, inherent differences in the students' abilities, or differences in individual teaching styles. Describe an alternative method of running the study to address some of these problems.

Med.   **6.49** An experiment was conducted to compare the mean lengths of time required for the bodily absorption of two drugs used in the treatment of epilepsy, D_1 and D_2. A group of 20 epileptic persons were randomly selected for inclusion in the study. Ten persons were randomly assigned to receive an oral dosage of drug D_1, with the other ten receiving drug D_2. The length of time (in minutes) for the drug to reach a specified level in the blood was recorded. The data are given in the following table.

Patient	1	2	3	4	5	6	7	8	9	10
Drug D_1	19.8	45.4	32.0	24.5	47.2	18.1	50.2	47.2	16.8	41.2
Drug D_2	23.0	37.3	11.4	60.6	72.1	41.4	42.6	43.8	42.8	65.3

a. Is there sufficient evidence that the two drugs have different mean absorption times?

b. Report the level of significance of your findings.

c. Estimate the size of the difference between the mean absorption times of the two drugs.

d. Do the conditions required for the statistical techniques used in (a)–(c) appear to be satisfied?

Engin. **6.50** A construction firm frequently estimates the amount of work accomplished on a construction site by a visual estimate of the amount of material used per day. The firm wanted to determine whether the accuracy of the estimates depended on the type of site. Two types of sites were considered: a high-rise building and a large one-story building. The firm employed hundreds of supervisors and randomly selected nine to participate in the study. Each supervisor approximated the number of bricks used in a given day at both sites. The difference (in thousands of bricks) between the approximation and the actual number of bricks used is recorded in the following table.

Supervisor	1	2	3	4	5	6	7	8	9
High-rise building	0.9	1.1	0.7	0.3	1.3	1.6	−0.8	1.4	1.7
One-story building	−1.6	2.5	1.1	−1.0	1.7	1.4	1.9	1.8	1.9

a. Is there significant evidence that the mean accuracy of the supervisors' approximations for the two types of sites? Use $\alpha = .05$. Report the p-value of the test results.

b. Estimate the size of the difference in the mean accuracy between the two types of sites.

c. Do the conditions required for the statistical techniques used in (a) and (b) appear to be satisfied? Justify your answer.

Bio. **6.51** A study was conducted to evaluate the effectiveness of an antihypertensive product. Three groups of 20 rats each were randomly selected from a strain of hypertensive rats. The 20 rats in the first group were treated with a low dose of an antihypertensive product, the second group with a higher dose of the same product, and the third group with an inert control. Note that negative values represent increases in blood pressure. The accompanying computer output can be used to answer the following questions.

Boxplot of blood pressure data

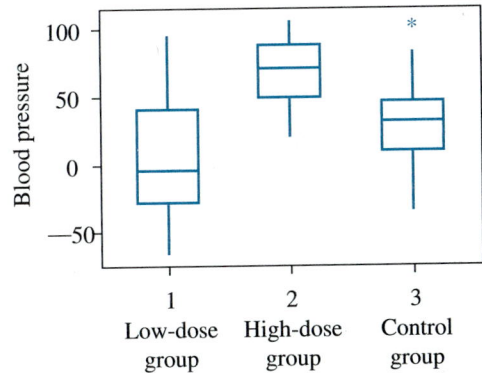

```
Row   Low Dose   High Dose   Control
 1      -45.1       54.2       18.2
 2      -59.8       89.1       17.2
 3       58.1       89.6       34.8
 4      -23.7       98.8        3.2
 5       64.9      107.3       42.9
 6       12.1       65.1      -27.2
 7       10.5       75.6       42.6
 8       42.5       52.0       10.0
 9       48.5       50.2      102.3
10       -1.7       80.9       61.0
11      -65.4       92.6      -33.1
12      -17.5       55.3       55.1
13       22.1      103.2       84.6
14      -15.4       45.4       40.3
15       96.5       70.9       30.5
16      -27.7       29.7       18.5
17      -16.7       40.3       29.3
18       39.5       73.3      -19.7
19       -4.2       21.0       37.2
20      -41.3       73.2       48.8
```

Two-sample T for Low Dose vs Control

            N   Mean   StDev   SE Mean
Low Dose   20   3.8   44.0   9.8
Control   20   29.8   34.0   7.6

95% CI for mu Low Dose − mu Control: (−51.3, −0.8)
T-Test mu Low Dose = mu Control (vs not =): T = −2.09 P = 0.044 DF = 35

Two-sample T for High Dose vs Control

            N   Mean   StDev   SE Mean
High Dose 20   68.4   24.5   5.5
Control   20   29.8   34.0   7.6

95% CI for mu High Dose − mu Control: (19.5, 57.6)
T-Test mu High Dose = mu Control (vs not =): T = 4.12 P = 0.0002 DF = 34

Two-sample T for Low Dose vs High Dose

            N   Mean   StDev   SE Mean
Low Dose   20   3.8   44.0   9.8
High Dose 20   68.4   24.5   5.5

95% CI for mu Low Dose − mu High Dose: (−87.6, −41.5)
T-Test mu Low Dose = mu High Dose (vs not =): T = −5.73 P = 0.0000 DF = 29
```
```

a. Compare the mean drop in blood pressure for the high-dose group and the control group. Use $\alpha = .05$ and report the level of significance.

b. Estimate the size of the difference in the mean drop for the high-dose and control groups using a 95% confidence interval.

c. Do the conditions required for the statistical techniques used in (a) and (b) appear to be satisfied? Justify your answer.

6.52 Refer to Exercise 6.51.

a. Compare the mean drop in blood pressure for the low-dose group and the control group. Use $\alpha = .05$ and report the level of significance.

b. Estimate the size of the difference in the mean drop for the low-dose and control groups using a 95% confidence interval.

c. Do the conditions required for the statistical techniques used in (a) and (b) appear to be satisfied? Justify your answer.

6.53 Refer to Exercise 6.51.

a. Compare the mean drop in blood pressure for the low-dose group and the high-dose group. Use $\alpha = .05$ and report the level of significance.

b. Estimate the size of the difference in the mean drop for the low-dose and high-dose groups using a 95% confidence interval.

c. Do the conditions required for the statistical techniques used in (a) and (b) appear to be satisfied? Justify your answer.

6.54 In Exercises 6.51–6.53, we tested three sets of hypotheses using portions of the same data sets in each of the sets of hypotheses. Let the experiment-wide Type I error rate be defined as the probability of making at least one Type I error in testing any set of hypotheses using the data from the experiment.

a. If we tested each of the three sets of hypotheses at the .05 level, estimate the experiment-wide Type I error rate.

b. Suggest a procedure by which we could be ensured that the experiment-wide Type I error rate would be at most .05.

Engin.   **6.55** A processor of recycled aluminum cans is concerned about the levels of impurities (principally other metals) contained in lots from two sources. Laboratory analysis of sample lots yields the following data (in kilograms of impurities per hundred kilograms of product):

Source I: 3.8 3.5 4.1 2.5 3.6 4.3 2.1 2.9 3.2 3.7 2.8 2.7
(mean = 3.267, standard deviation = .676)

Source II: 1.8 2.2 1.3 5.1 4.0 4.7 3.3 4.3 4.2 2.5 5.4 4.6
(mean = 3.617, standard deviation = 1.365)

a. Calculate the pooled variance and standard deviation.

b. Calculate a 95% confidence interval for the difference in mean impurity levels.

c. Can the processor conclude, using $\alpha = .05$, that there is a nonzero difference in means?

I.S.   **6.56** To compare the performance of microcomputer spreadsheet programs, teams of three students each choose whatever spreadsheet program they wish. Each team is given the same set of standard accounting and finance problems to solve. The time (in minutes) required for each team to solve the set of problems is recorded. The data shown here were obtained for the two most widely used programs; also displayed are the sample means, sample standard deviations, and sample sizes.

Program					Time						$\bar{y}$	s	n
A	39	57	42	53	41	44	71	56	49	63	51.50	10.46	10
B	43	38	35	45	40	28	50	54	37	29			
	36	27	52	33	31	30					38.00	8.67	16

a. Calculate the pooled variance.
b. Use this variance to find a 99% confidence interval for the difference of population means.
c. According to this interval, can the null hypothesis of equal means be rejected at $\alpha = .01$?

6.57 Redo parts (b) and (c) of Exercise 6.56 using a separate-variance (t') method. Which method is more appropriate in this case? How critical is it to use one or the other?

6.58 Refer to Exercise 3.32. Is there a significant difference in the mean DDE to PCB ratio for terrestrial and aquatic feeders? Use $\alpha = .05$ and report the level of significance of your test.

6.59 To assess whether degreed nurses received a more comprehensive training than registered nurses, a study was designed to compare the two groups. The state nursing licensing board randomly selected 50 nurses from each group for evaluation. They were given the state licensing board examination and their scores are summarized in the following tables and graphs.

Boxplots of degreed and registered nurses (means are indicated by solid circles)

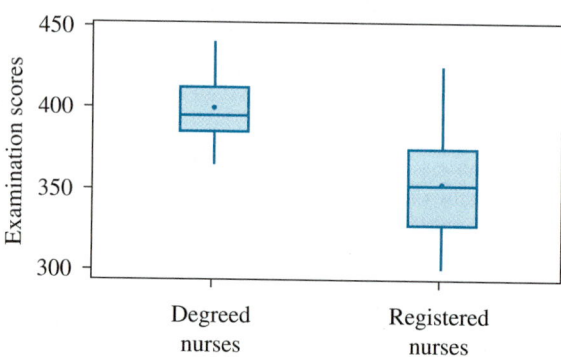

```
Two-Sample T-Test and Confidence Interval

Two-sample T for Degreed vs RN

              N    Mean    StDev    SE Mean
Degreed      50   399.9    17.2       2.4
RN           50   354.7    30.9       4.4

95% CI for mu Degreed - mu RN: (35.3, 55.2)
T-Test mu Degreed = mu RN (vs >): T = 9.04   P = 0.0000   DF = 76
```

a. Can the licensing board conclude that the mean score of nurses who receive a BS in nursing is higher than the mean score of registered nurses? Use $\alpha = .05$.
b. Report the approximated p-value for your test.
c. Estimate the size of the difference in the mean scores of the two groups of nurses using a 95% confidence interval.
d. The mean test scores are considered to have a meaningful difference only if they differ by more than 40 points. Is the observed difference in the mean scores a meaningful one?

Soc. **6.60** A study was to be designed to determine whether food prices charged in the inner city are higher than the prices charged in suburban areas. A market basket of goods was comprised, with the total cost of the goods obtained at n inner city stores and n suburban stores. Assume that cost of the goods has a normal distribution with a range of $115 to $135 for the inner city stores and $100 to $120 for the suburban stores. Determine the sample size needed so that we would be 95% confident that the estimated difference in the mean costs is within $4 of the true difference.

6.61 Refer to Exercise 6.60. The study was conducted using 13 randomly selected stores from the inner city and 14 randomly selected suburban stores. The costs of the market basket at each of the stores are given in the following table.

Store	1	2	3	4	5	6	7	8	9	10	11	12	13	14
Suburban	109.11	106.94	118.37	111.09	117.74	109.48	113.59	117.49	117.74	108.68	113.00	122.97	111.87	113.00
Inner City	131.76	121.39	122.37	130.47	125.81	134.47	122.15	135.47	128.26	129.90	129.24	134.72	126.17	

a. Estimate the difference in the mean cost of the market basket for the two areas.
b. Did the study achieve the goals stated in Exercise 6.60?
c. Is there substantial evidence that food costs are higher in the inner city? State your conclusions in *nonstatistical terms*.
d. What type of error (Type I or II) could you possibly have made?

6.62 Consider the oil spill case study in Section 6.1.
a. What are the populations of interest?
b. What factors other than flora density may indicate that the oil spill has affected the marsh?
c. Describe a method for randomly selecting the tracts to be included in the study.
d. Suppose the researchers had taken flora density readings 5 years prior to the oil spill on 20 tracts within the oil spill region. Describe an alternative analysis that makes use of this information.

Pol. Sci.   **6.63** All persons running for public office must report the amount of money spent during their campaign. Political scientists have contended that female candidates generally find it difficult to raise money and therefore spend less in their campaign than male candidates. Suppose the accompanying data represent the campaign expenditures of a randomly selected group of male and female candidates for the state legislature. Do the data support the claim that female candidates generally spend less in their campaigns for public office than male candidates?

Campaign Expenditures (in thousands of dollars)

Candidate	1	2	3	4	5	6	7	8	9	10	11	12	13	14	15	16	17	18	19	20
Female	169	206	257	294	252	283	240	207	230	183	298	269	256	277	300	126	318	184	252	305
Male	289	334	278	268	336	438	388	388	394	394	425	386	356	342	305	365	355	312	209	458

a. State the null and alternative hypotheses in
  i. plain English
  ii. statistical terms or symbols
b. Estimate the size of the difference in campaign expenditures for female and male candidates.
c. Is the difference statistically significant at the .05 level?
d. Is the difference of practical significance?

6.64 Refer to Exercise 6.63. What conditions must be satisfied in order to use the *t* procedures to analyze the data? Use the accompanying summary data and plot to determine whether these conditions have been satisfied for the data in Exercise 6.63.

```
Two-Sample T-Test and Confidence Interval

Two-sample T for Female vs Male

              N    Mean   StDev   SE Mean
Female   20   245.4   52.1      12
Male     20   350.9   61.9      14

95% CI for mu Female - mu Male: (-142, -69)
T-Test mu Female = mu Male (vs not =): T = -5.83   P = 0.0000   DF = 38
Both use Pooled StDev = 57.2
```

Boxplots of female and male candidates (means are indicated by solid circles)

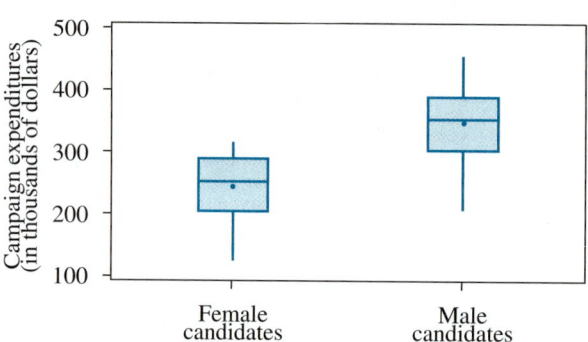

6.65 After strip mining for coal, the mining company is required to restore the land to its condition prior to mining. One of many factors that is considered is the pH of the soil, which is an important factor in determining what types of plants will survive in a given location. The area was divided into grids before the mining took place. Fifteen grids were randomly selected and the soil pH was measured before mining. When the mining was completed, the land was restored and another set of pH readings were taken on the same 15 grids; see the accompanying table.

Env.

Location	1	2	3	4	5	6	7	8	9	10	11	12	13	14	15
Before	10.02	10.16	9.96	10.01	9.87	10.05	10.07	10.08	10.05	10.04	10.09	10.09	9.92	10.05	10.13
After	10.21	10.16	10.11	10.10	10.07	10.13	10.08	10.30	10.17	10.10	10.06	10.37	10.24	10.19	10.13

a. The state land office analyzed the data and their computer output is given here. What is the level of significance of the test for a change in mean pH after reclamation of the land?

b. What is the research hypothesis that the land office was testing?

c. Estimate the change in mean soil pH after strip mining using a 99% confidence interval.

d. The land office assessed a fine on the mining company because the t test indicated a significant difference in mean pH after the reclaimation of the land. Do you think their findings are supported by the data? Justify your answer using the results from parts (a) and (c).

6.66 Refer to Exercise 6.65. Based on the land office's decision in the test of hypotheses, could they have made (select one of the following)

a. A Type I error?
b. A Type II error?
c. Both a Type I and a Type II error?
d. Neither a Type I nor a Type II error?

6.67 Refer to Exercise 3.13. The survival times for the standard and new therapy were analyzed with the results given in the following table and graph.

```
Two-Sample T-Test and Confidence Interval

Two-sample T for Standard vs New

             N     Mean    StDev   SE Mean
Standard    28    15.68     9.63      1.8
New         28    20.71     9.81      1.9

95% CI for mu Standard - mu New: ( -10.2,   0.2)
T-Test mu Standard = mu New (vs <): T = -1.94   P = 0.029   DF = 54
Both use Pooled StDev = 9.72
```

Boxplots of standard and new therapy (means are indicated by solid circles)

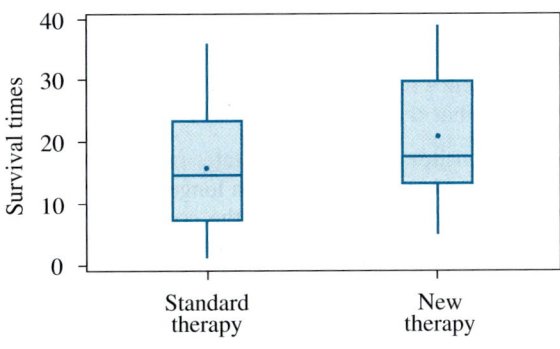

a. Do the data support the research hypothesis that the new therapy has a longer mean survival time than the standard therapy?
b. Estimate the amount of increase in the mean survival time using the new therapy with a 95% confidence interval.
c. What conditions must be satisfied in order to use the t procedures to analyze the data? Have these conditions been satisfied?

Soc.   **6.68** A study was carried out to determine whether nonworking wives from middle-class families have more voluntary association memberships than nonworking wives from working-class families. A random sample of housewives was obtained, and each was asked for information about her husband's occupation and her own memberships in voluntary associations. On the basis of their husbands' occupations, the women were divided into middle-class and working-class groups, and the mean number of voluntary association memberships was computed for each group.

For the 15 middle-class women, the mean number of memberships per woman was $\bar{y}_1 = 3.4$ with $s_1 = 2.5$. For the 15 working-class wives, $\bar{y}_2 = 2.2$ with $s_2 = 2.8$.
a. Use these data to construct a 95% confidence interval for the difference in the mean number of memberships per woman.
b. Do the data support the research hypothesis that women from middle-class families have more voluntary association memberships than women from working-class families?

  a. What is the research hypothesis?

  b. What are the values of the t and t' statistics? Why are they not equal for this data set?

  c. What are the p-values for the t and t' statistics?

  d. Are the conclusions concerning the research hypothesis the same for the two tests if we use $\alpha = .05$?

  e. Which test, t or t', is more appropriate for this data set?

Ag.   **6.75** A study was conducted on 16 dairy cattle. Eight cows were randomly assigned to a liquid regimen of water only (group 1); the others received liquid whey only (group 2). In addition, each animal was given 7.5 kg of grain per day and allowed to graze on hay at will. Although no significant differences were observed between the groups in the dairy-milk-production gauges, such as milk production and fat content of the milk, the following data on daily hay consumption (in kilograms/cow) were of interest:

Group 1	15.1	14.9	14.8	14.2	13.1	12.8	15.5	15.9
Group 2	6.8	7.5	8.6	8.4	8.9	8.1	9.2	9.5

  a. Use these data to test the research hypothesis that there is a difference in mean hay consumption for the two diets. Use $\alpha = .05$.

  b. Provide an estimate of the amount of difference in the mean hay consumption of the two groups.

6.76 Refer to Exercise 6.75. The weights (in hundreds of pounds) of the 16 cattle prior to the start of the study are given in the following table.

Cattle	1	2	3	4	5	6	7	8
Sample 1	12.3	16.0	15.6	12.4	14.2	18.6	16.6	15.8
Sample 2	22.7	20.1	17.9	21.0	20.5	19.5	20.6	21.4

  a. Do the data indicate a significant difference in the mean weights of the cattle in the two samples prior to the beginning of the study?

  b. Suppose we were to redesign the study using the information on the 16 cattle. How would you assign the two diets to the cattle to minimize the differences in the weights of the cattle?

Engin.   **6.77** An industrial concern has experimented with several different mixtures of the four components—magnesium, sodium nitrate, strontium nitrate, and a binder—that comprise a rocket propellant. The company has found that two mixtures in particular give higher flare-illumination values than the others. Mixture 1 consists of a blend composed of the proportions .40, .10, .42, and .08, respectively, for the four components of the mixture; mixture 2 consists of a blend using the proportions .60, .27, .10, and .05. Twenty different blends (10 of each mixture) are prepared and tested to obtain the flare-illumination values. These data appear here (in units of 1,000 candles).

Mixture 1	185	192	201	215	170	190	175	172	198	202
Mixture 2	221	210	215	202	204	196	225	230	214	217

  a. Plot the sample data. Which test(s) could be used to compare the mean illumination values for the two mixtures?

  b. Give the level of significance of the test and interpret your findings.

6.78 Refer to Exercise 6.77. Instead of conducting a statistical test, use the sample data to answer the question, What is the difference in mean flare illumination for the two mixtures?

6.79 Refer to Exercise 6.77. Suppose we wish to test the research hypothesis that $\mu_1 < \mu_2$ for the two mixtures. Assume that the population distributions are normally distributed with a common $\sigma = 12$. Determine the sample size required to obtain a test having $\alpha = .05$ and $\beta(\mu_d) < .10$ when $\mu_2 - \mu_1 \geq 15$.

6.80 Refer to the epilepsy study data on the data disk. An analysis of the data produced the following computer output. The measured variable is the number of seizures after 8 weeks in the study for patients on the placebo and for those treated with the drug progabide.

```
Two-Sample T-Test and Confidence Interval

Two-sample T for Placebo vs Progabide

               N       Mean      StDev     SE Mean
Placebo       28       7.96      7.63        1.4
Progabide     31        6.7      11.3        2.0

95% CI for mu Placebo − mu Progabide:  ( −3.8,   6.3)
T-Test mu Placebo = mu Progabide (vs >): T = 0.50   P = 0.31   DF = 57
Both use Pooled StDev = 9.71
```

Boxplots of placebo and progabide (means are indicated by solid circles)

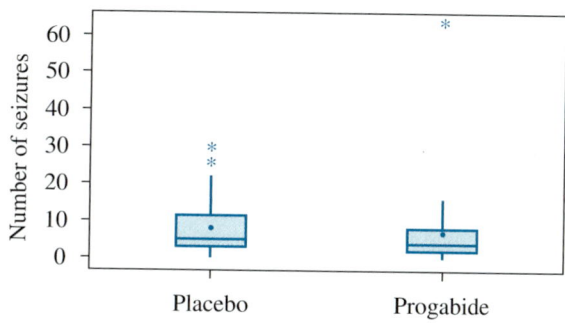

a. Do the data support the conjecture that progabide reduces the mean number of seizures for epileptics? Use both a *t* test and the Wilcoxon test with $\alpha = .05$.
b. Which test appears to be most appropriate for this study? Why?
c. Estimate the size of the differences in the mean number of seizures between the two groups.

Soc. **6.81** A study of anxiety was conducted among residents of a southeastern metropolitan area. Each person selected for the study was asked to check a "yes" or a "no" for the presence of each of 12 anxiety symptoms. Anxiety scores ranged from 0 to 12, with higher scores related to higher perceived presence of any anxiety symptoms. The results for a random sample of 50 residents, categorized by gender, are summarized in the table. Use these data to test the research hypothesis that the mean perceived anxiety score is different for males and females. Give the level of significance for your test.

	Sample Size	Mean	Standard Deviation
Female	26	5.26	3.2
Male	24	7.02	3.9

Med. **6.82** A clinical trial was conducted to determine the effectiveness of drug A in the treatment of symptoms associated with alcohol withdrawal. A total of 30 patients were treated (under blinded conditions) with the drug and another 30 with an identical-appearing placebo. The average symptom score for the two groups after 1 week of therapy was 1.5 and 6.3, respectively. (*Note:* Higher symptom scores indicate more withdrawal "problems.") The corresponding standard deviations were 3.1 and 4.2.

a. Compare the mean total symptom scores for the two groups. Give the *p*-value for a two-sample *t* test of $H_0: \mu_1 - \mu_2 = 0$ versus $H_a: \mu_1 - \mu_2 < 0$. Draw conclusions.

b. Suppose the average total symptom scores were 6.8 and 12.2 prior to therapy. How would this affect your conclusions? How could you guard against possible baseline (pretreatment) differences?

Env.   **6.83**   Two analysts, supposedly of identical abilities, each measure the parts per million of a certain type of chemical impurity in drinking water. It is claimed that analyst 1 tends to give higher readings than analyst 2. To test this theory, each of six water samples is divided and then analyzed by both analysts separately. The data are shown in the accompanying table (readings in ppm).

Water Sample	Analyst 1	Analyst 2
1	31.4	28.1
2	37.0	37.1
3	44.0	40.6
4	28.8	27.3
5	59.9	58.4
6	37.6	38.9

a. Is there evidence to indicate that analyst 1 reads higher on the average than analyst 2? Give the level of significance for your test.

b. What would be the conclusion using a Wilcoxon test? Compare your results to part (a).

Ag.   **6.84**   A single leaf was taken from each of 11 different tobacco plants. Each was divided in half; one half was chosen at random and treated with preparation I and the other half received preparation II. The object of the experiment was to compare the effects of the two preparations of mosaic virus on the number of lesions on the half leaves after a fixed period of time. These data are recorded in the table. For $\alpha = .05$, use Wilcoxon's signed-rank test to examine the research hypothesis that the distributions of lesions are different for the two populations.

Tobacco Plant	Number of Lesions on the Half Leaf	
	Preparation I	Preparation II
1	18	14
2	20	15
3	9	6
4	14	12
5	38	32
6	26	30
7	15	9
8	10	2
9	25	18
10	7	3
11	13	6

Env.   **6.85** An investigator plans to compare the mean number of particles of effluent in water collected at two different locations in a water treatment plant. If the standard deviation for particle counts is expected to be approximately 6 for the counts in samples taken at each of the locations, determine the sample sizes required to estimate the mean difference in particles of effluent using a 99% confidence interval of width 1 (particle).

Bus.   **6.86** Many people purchase sports utility vehicles (SUVs) because they think they are sturdier and hence safer than regular cars. However, preliminary data have indicated that the costs for repairs of SUVs are higher than for midsize cars when both vehicles are in an accident. A random sample of 8 new SUVs and 8 midsize cars are tested for front impact resistance. The amounts of damage (in hundreds of dollars) to the vehicles when crashed at 20 mph head on into a stationary barrier are recorded in the following table.

Car	1	2	3	4	5	6	7	8
SUV	14.23	12.47	14.00	13.17	27.48	12.42	32.59	12.98
Midsize	11.97	11.42	13.27	9.87	10.12	10.36	12.65	25.23

a. Plot the data to determine whether the conditions required for the t procedures are valid.
b. Do the data support the conjecture that the mean damage is greater for SUVs than for midsize vehicles? Use $\alpha = .05$ with both the t test and Wilcoxon test.
c. Which test appears to be the more appropriate procedure for this data set?
d. Do you reach the same conclusions from both procedures? Why or why not?

Boxplots of midsize and SUV damage amounts (means are indicated by solid circles)

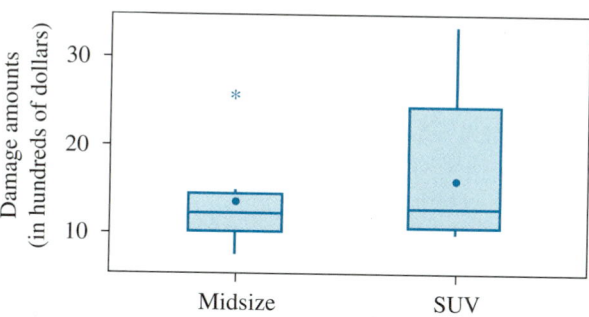

```
Two-Sample T-Test and Confidence Interval

Two-sample T for Midsize vs SUV

              N      Mean     StDev    SE Mean
Midsize   8        13.11      5.05       1.8
SUV       8        17.42      7.93       2.8

95% CI for mu Midsize — mu SUV: ( −11.4,   2.8)
T-Test mu Midsize = mu SUV (vs <): T = −1.30   P = 0.11   DF = 14
Both use Pooled StDev = 6.65
```

```
Mann-Whitney Confidence Interval and Test

Midsize     N =   8      Median =        11.69
SUV         N =   8      Median =        13.59
Point estimate for ETA1-ETA2 is         -2.32
95.9 Percent CI for ETA1-ETA2 is (-14.83,-0.33)
W = 48.0
Test of ETA1 = ETA2 vs ETA1 < ETA2 is significant at 0.0203
```

6.87 Refer to Exercise 6.86. The small number of vehicles in the study has led to criticism of the results. A new study is to be conducted with a larger sample size. Assume that the populations of damages are both normally distributed with a common $\sigma = \$700$.

a. Determine the sample size so that we are 95% confident that the estimate of the difference in mean repair cost is within $500 of the true difference.

b. For the research hypothesis H_a: $\mu_{SUV} > \mu_{MID}$, determine the sample size required to obtain a test having $\alpha = .05$ and $\beta(\mu_d) < .05$ when $\mu_{SUV} - \mu_{MID} \geq \$500$.

Law **6.88** The following memorandum opinion on statistical significance was issued by the judge in a trial involving many scientific issues. The opinion has been stripped of some legal jargon and has been taken out of context. Still, it can give us an understanding of how others deal with the problem of ascertaining the meaning of statistical significance. Read this memorandum and comment on the issues raised regarding statistical significance.

Memorandum Opinion

This matter is before the Court upon two evidentiary issues that were raised in anticipation of trial. First, it is essential to determine the appropriate level of statistical significance for the admission of scientific evidence.

With respect to statistical significance, no statistical evidence will be admitted during the course of the trial unless it meets a confidence level of 95%.

Every relevant study before the court has employed a confidence level of at least 95%. In addition, plaintiffs concede that social scientists routinely utilize a 95% confidence level. Finally, all legal authorities agree that statistical evidence is inadmissable unless it meets the 95% confidence level required by statisticians. Therefore, because plaintiffs advance no reasonable basis to alter the accepted approach of mathematicians to the test of statistical significance, no statistical evidence will be admitted at trial unless it satisfies the 95% confidence level.

Med. **6.89** Certain baseline determinations were made on 182 patients entered in a study of survival in males suffering from congestive heart failure. At the time these data were summarized, 88 deaths had been observed. This table summarizes the baseline data for the survivors and nonsurvivors. The variables listed below "heart rate" are measures of the severity of the heart failure. The arrows to the left of each variable indicate the direction of improvement.

a. Discuss these baseline findings.

b. What assumptions have the authors made when doing these t tests?

**Baseline Characteristics of Patients with
Severe Chronic Left-Ventricular Failure Due to Cardiomyopathy**

Variable	Nonsurvivors (*n* = 88)	Survivors (*n* = 94)	*t* Test *p*-Value
Age (y)	57 ± 10	56 ± 8	NS
Duration of symptoms (mo)	45 ± 43	39 ± 27	NS
Heart rate (beats/min)	87 ± 15	83 ± 16	NS
↓ Mean arterial pressure (mm Hg)	87 ± 13	94 ± 13	<0.001
↓ Left-ventricular filling pressure (mm Hg)	29 ± 7	24 ± 9	<0.001
↑ Cardiac index (L/min/m²)	2.0 ± 0.7	2.5 ± 0.8	<0.001
↑ Stroke volume (mL/beat)	45 ± 16	59 ± 5	<0.001
↓ Systemic vascular resistance (units)	25 ± 10	21 ± 8	<0.01
↑ Stroke work (g-m)	35 ± 19	56 ± 33	<0.001

Values are listed as mean ± standard deviation.

Bus.   **6.90**   Hospital administrators studied the patterns of length of hospital stays with particular attention paid to those patients having health-maintenance organization (HMO) payment sources versus those with non-HMO payment sources. The graph shown here summarizes the sample data.

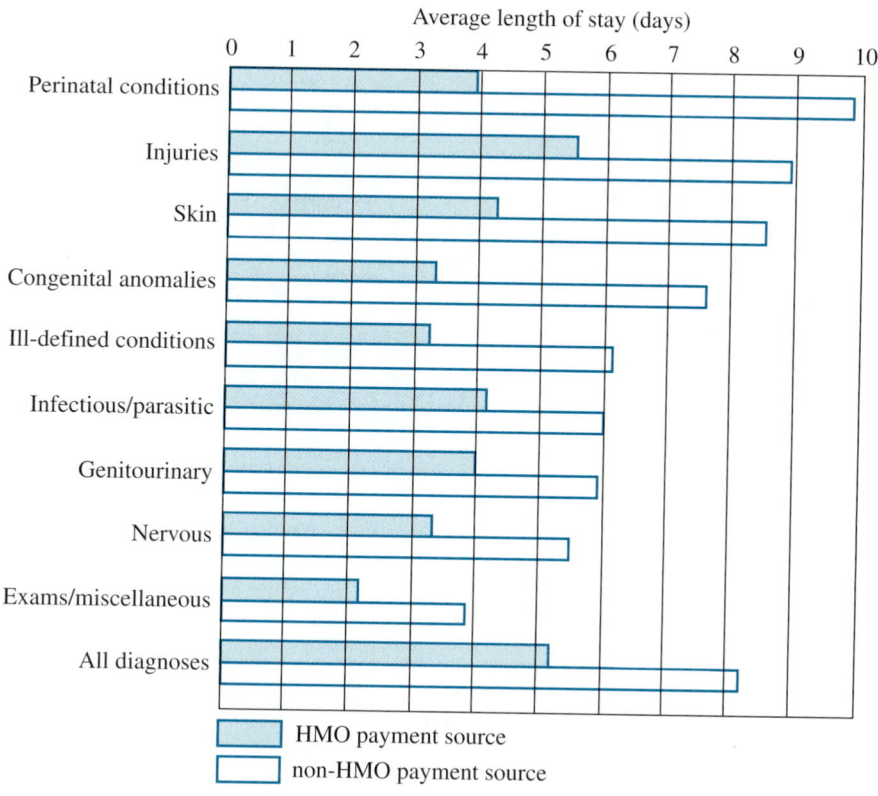

Sources: American Hospital Assn.; Twin Cities Metropolitan Health Board.

a. What general observations would you draw from the graph? What additional information would you need to make more definitive statements regarding these results?

b. Suppose that across all diagnoses, the sample statistics were as shown in the table. Use these data to test $H_0: \mu_1 - \mu_2 = 0$ versus $H_a: \mu_1 - \mu_2 \neq 0$. Give the p-value for your test.

	Sample Mean (days)	Sample Size	Sample Standard Deviation
HMO	5.0	120	1.3
Non-HMO	8.1	130	1.9

6.91   Refer to Exercise 6.90. Also run the t' test and compare your results. Which (if any) test is better for these data?

Med.   **6.92**   An abstract for the results of a study of ten congestive heart-failure patients is given here. Read the abstract and try to interpret the results.

Abstract

An experimental compound was studied in ten patients suffering from congestive heart failure. Certain variables were measured at baseline and then 4 hours after intravenous treatment with the compound. The compound was shown to increase cardiac index from 11.1% to 34.3% from a baseline average of 2.41 ± 0.49 L/min/m² ($p < .01$), heart rate by 6–10% from 72 ± 12 beats/min ($p < .02$), and decreased pulmonary capillary wedge pressure by 15.3–24.2% from 18.7 ($p < .001$).

Med.   **6.93**   Several antidepressant drugs have been studied in the treatment of cocaine abusers. One recent study showed that 20 cocaine abusers who were treated with an antidepressant in an outpatient setting experienced decreases in cravings after 2 weeks and some reduction in their actual use of cocaine. Comment on these results. Are they compelling? Why or why not?

Edu.   **6.94**   In April 1986, the *Australian Journal of Statistics* (30: 1, 23–44) published the results of a study of S. R. Butler and H. W. Marsh on reading and arithmetic achievement for students from non–English-speaking families. All kindergarten students from seven public schools in Sydney, Australia were included in the original sample of 392 children. Reading and arithmetic achievement tests were administered at the start of the study during kindergarten and then at years 1, 2, 3, and 6 of primary school.

The table shown here gives the characteristics of the 286 of the original 392 students who were available for testing at year 6 ($n = 226$ students from English-speaking families, and $n = 60$ from non–English-speaking families.)

Characteristics	Group	
	English-Speaking Family ($n = 226$) $\bar{y}$	**Non–English-Speaking Family** ($n = 60$) $\bar{y}$
Age (in months)	67.17	67.15
Gender (1 = male, 2 = female)	1.50	1.55
Number of children in family	2.54	2.62
Ordinal position in family (1 = oldest child, etc.)	1.89	1.82
Father's occupation (1 = most skilled, 17 = least skilled)	8.26*	11.50
Peabody Picture Vocabulary IQ	99.26*	74.45

*Statistically significant, $p < .01$

> **a.** Can you suggest better ways to summarize these baseline characteristics?
> **b.** What test(s) may have been used to compare these characteristics?
> **c.** What other characteristics could or should have been examined to make a direct comparison of reading and arithmetic achievement?
> **d.** What effect (if any) might the attrition rate have on the study results? Recall that 106 (27%) of the original 392 students were not available for testing at year 6.

6.95 Refer to the clinical trials database on the data disk. Use the HAM-D total score data to conduct a statistical test of H_0: $\mu_D - \mu_A = 0$ vs. H_a: $\mu_D - \mu_A > 0$; that is, we want to know whether the placebo group (D) has a higher (worse) mean total depression score at the end of the study than the group receiving treatment A. Use $\alpha = .05$. What are your conclusions?

6.96 Refer to Exercise 6.95 and repeat this same comparison with the placebo group for treatment B, and then for treatment C. Give the p-value for each of these tests. Which of the three treatment groups (A, B, or C) appears to have the lowest mean HAM-D total score?

6.97 Use the clinical trials database to construct a 95% confidence interval for $\mu_D - \mu_A$ based on the HAM-D anxiety score data. What can you conclude about $\mu_D - \mu_A$ based on this interval?

6.98 Refer to the clinical trials database on the data disk. Compare the mean ages for treatment groups B and D using a two-sided statistical test. Set up all parts of the test using $\alpha = .05$; draw a conclusion. Why might it be important to have patients with similar ages in the different treatment groups when studying the effects of several drug products on the treatment of depression?

6.99 Refer to Exercise 6.98. What other variables should be comparable among the treatment groups in order to draw conclusions about the effectiveness of the drug products for treating depression?

CHAPTER 7

Inferences about Population Variances

7.1 Introduction and Case Study

7.2 Estimation and Tests for a Population Variance

7.3 Estimation and Tests for Comparing Two Population Variances

7.4 Tests for Comparing $t > 2$ Population Variances

7.5 Summary

7.1 Introduction and Case Study

When people think of statistical inference, they usually think of inferences concerning population means. However, the population parameter that answers an experimenter's practical questions will vary from one situation to another. In many situations, the variability of a population's values is as important as the population mean. In the case of problems involving product improvement, product quality is defined as a product having mean value at the target value with low variability about the mean. For example, the producer of a drug product is certainly concerned with controlling the mean potency of tablets, but he or she must also worry about the variation in potency from one tablet to another. Excessive potency or an underdose could be very harmful to a patient. Hence, the manufacturer would like to produce tablets with the desired mean potency and with as little variation in potency (as measured by σ or σ^2) as possible. Another example is from the area of investment strategies. Investors search for a portfolio of stocks, bonds, real estate, and other investments having low risk. A measure used by investors to determine the uncertainty inherent in a particular portfolio is the variance in the value of the investments over a set period. At times, a portfolio with a high average value and a large standard deviation will have a value that is much lower than the average value. Investors thus need to examine the variability in the value of a portfolio along with its average value when determining its degree of risk.

Case Study: Evaluation of Method for Detecting *E. coli*

The outbreaks of bacterial disease in recent years due to the consumption of contaminated meat products have created a demand for new rapid pathogen-detecting methods that can be used in a meat surveillance program. The paper "Repeatability of the petrifilm HEC test and agreement with a hydrophobic grid membrane filtration method for the enumeration of *Escherichia coli* 0157:H7 on beef carcasses," [*Journal of Food Protection* (1998) 61:402–408] describes a formal comparison between a new microbial method for the detection of *E. coli,* the petrifilm HEC test, with an elaborate laboratory-based procedure, hydrophobic grid membrane filtration (HGMF). The HEC test is easier to inoculate, more compact to incubate, and safer to handle than conventional procedures. However, researchers had to compare the performances of the HEC test and the HGMF procedure to determine whether HEC may be a viable method for detecting *E. coli*.

Designing the Data Collection   The developers of the HEC method sought answers to the following questions:

1. What parameters associated with the HEC and HGMF readings must be compared?
2. How many observations are necessary for a valid comparison of HEC and HGMF?
3. What type of experimental design would produce the most efficient comparison of HEC and HGMF?
4. What are the valid statistical procedures for making the comparisons?
5. What types of information should be included in a final report to document the evaluation of HEC and HGMF?

What aspects of the *E. coli* counts should be of interest to the researchers? A comparison of only the mean concentration would indicate whether or not the two procedures were in agreement with respect to the average readings over a large number of determinations. However, we would not know whether HEC was more variable in its determination of *E. coli* than HGMF. For example, consider the two distributions in Figure 7.1. Suppose the distributions represent the population of *E. coli* concentration determinations from HEC and HGMF for a situation in which the true *E. coli* concentration is 7 $\log_{10}$ CFU/ml. The distributions would indicate that the HEC evaluation of a given meat sample may yield a reading very different from the true *E. coli* concentration, whereas the individual readings from HGMF are more likely to be near the true concentration. In this type of situation, it is crucial to compare both the means and standard deviations of the two procedures. In fact, we need to examine other aspects of the relationship

FIGURE 7.1

Hypothetical distribution of *E. coli* concentrations from HEC and HGMF

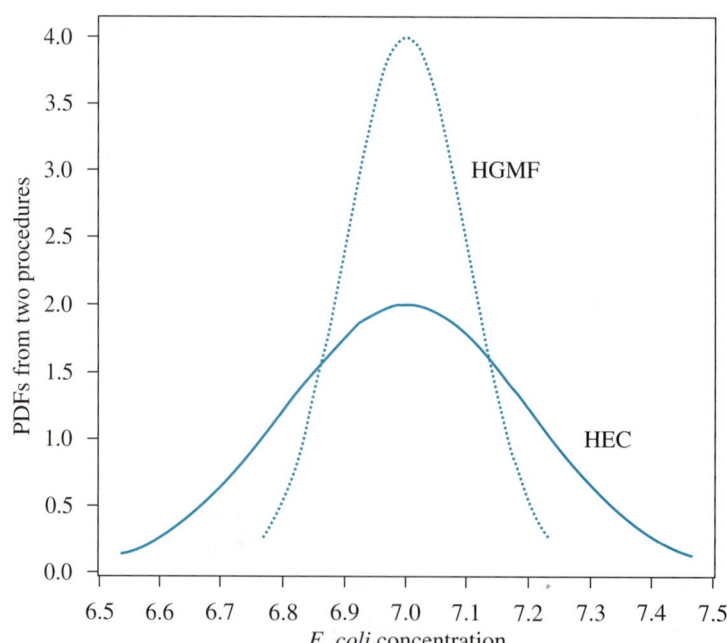

between HEC and HGMF determinations to evaluate the comparability of the two procedures. We will examine these ideas in the chapters on model building and analysis of variance.

The experiment was designed to have two phases. Phase one of the study applied both procedures to pure cultures of *E. coli* representing 10^7 CFU/ml of strain E318N. Based on the specified degree of precision in estimating the *E. coli* level (see Exercise 7.4), it was determined that the HEC and HGMF procedures would be applied to 24 pure cultures each. Thus, there were two independent samples of size 24 each. The determinations yielded the *E. coli* concentrations in transformed metric ($\log_{10}$ CFU/ml) given here in Table 7.1. (The values in Table 7.1 were simulated using the summary statistics given in the paper.)

TABLE 7.1

E. coli readings ($\log_{10}$ (CFU/ml)) from HGMF and HEC procedures

Sample	HGMF	HEC	Sample	HGMF	HEC	Sample	HGMF	HEC
1	6.65	6.67	9	6.89	7.08	17	7.07	7.25
2	6.62	6.75	10	6.90	7.09	18	7.09	7.28
3	6.68	6.83	11	6.92	7.09	19	7.11	7.34
4	6.71	6.87	12	6.93	7.11	20	7.12	7.37
5	6.77	6.95	13	6.94	7.11	21	7.16	7.39
6	6.79	6.98	14	7.03	7.14	22	7.28	7.45
7	6.79	7.03	15	7.05	7.14	23	7.29	7.58
8	6.81	7.05	16	7.06	7.23	24	7.30	7.54

Phase two of the study applied both procedures to artificially contaminated beef. Portions of beef trim were obtained from three Holstein cows that had tested negatively for *E. coli*. Eighteen portions of beef trim were obtained from the cows and then contaminated with *E. coli*. The HEC and HGMF procedures were applied to a portion of each of the 18 samples. The two procedures yielded *E. coli* concentrations in transformed metric ($\log^{10}$ CFU/ml). The data in this case would be 18 paired samples. The researchers were interested in determining a model to relate the two procedures' determinations of *E. coli* concentrations. We will only consider phase one in this chapter; phase two will be discussed in Chapter 11.

Managing the Data   The researchers next prepared the data for a statistical analysis following the steps described in Section 2.5.

Analyzing the Data   The researchers were interested in determining whether the two procedures yielded equivalent measures of *E. coli* concentrations. The box-plots of the experimental data are given in Figure 7.2. The two procedures appear to be very similar with respect to the width of box and length of whiskers, but HEC has a larger median than HGMF. The sample summary statistics are given here.

```
Descriptive Statistics

Variable          N      Mean    Median   TrMean     StDev   SE Mean
HEC              24    7.1346    7.1100   7.1373    0.2291    0.0468
HGMF             24    6.9529    6.9350   6.9550    0.2096    0.0428


Variable    Minimum   Maximum        Q1        Q3
HEC          6.6700    7.5400    6.9925    7.3250
HGMF         6.5600    7.3000    6.7900    7.1050
```

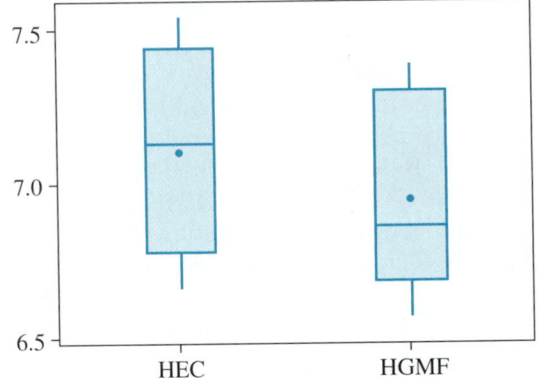

From the summary statistics we note that HEC yields a larger mean concentration HGMF. Also, the variability in concentration readings for HEC appears to be slightly greater than that for HGMF. Our initial conclusion might be that the two procedures yield different distributions of readings for their determination of *E. coli* concentrations. However, we need to determine whether the differences in their sample means and standard deviations infer a difference in the corresponding population values. We will summarize our findings after developing the appropriate procedures for comparing population variances.

Inferential problems about population variances are similar to the problems addressed in making inferences about the population mean. We must construct point estimators, confidence intervals, and the test statistics from the randomly sampled data to make inferences about the variability in the population values. We then can state our degree of certainty that observed differences in the sample data convey differences in the population parameters.

7.2 Estimation and Tests for a Population Variance

The sample variance

$$s^2 = \frac{\sum (y - \bar{y})^2}{n - 1}$$

unbiased estimator

can be used for inferences concerning a population variance σ^2. For a random sample of n measurements drawn from a population with mean μ and variance σ^2, s^2 is an **unbiased estimator** of σ^2. If the population distribution is normal, then the sampling distribution of s^2 can be specified as follows. From repeated samples of size n from a normal population whose variance is σ^2, calculate the statistic $(n - 1)s^2/\sigma^2$ and plot the histogram for these values. The shape of the histogram is similar to those depicted in Figure 7.3, because it can be shown that the statistic **chi-square distribution with df = n − 1** $(n - 1)s^2/\sigma^2$ follows a **chi-square distribution with df = n − 1**. The mathematical formula for the chi-square (χ^2, where χ is the Greek letter chi) probability distribution is very complex so we will not display it. However, some of the properties of the distribution are as follows:

1. The chi-square distribution is positively skewed with values between 0 and ∞ (see Figure 7.3).

2. There are many chi-square distributions and they are labeled by the parameter degrees of freedom (df). Three such chi-square distributions are shown in Figure 7.3 with df = 5, 15, and 30, respectively.
3. The mean and variance of the chi-square distribution are given by $\mu =$ df and $\sigma^2 = (2)$ df. For example, if the chi-square distribution has df = 30, then the mean and variance of that distribution are $\mu = 30$ and $\sigma^2 = 60$.

FIGURE 7.3

Densities of the chi-square (df = 5, 15, 30) distribution

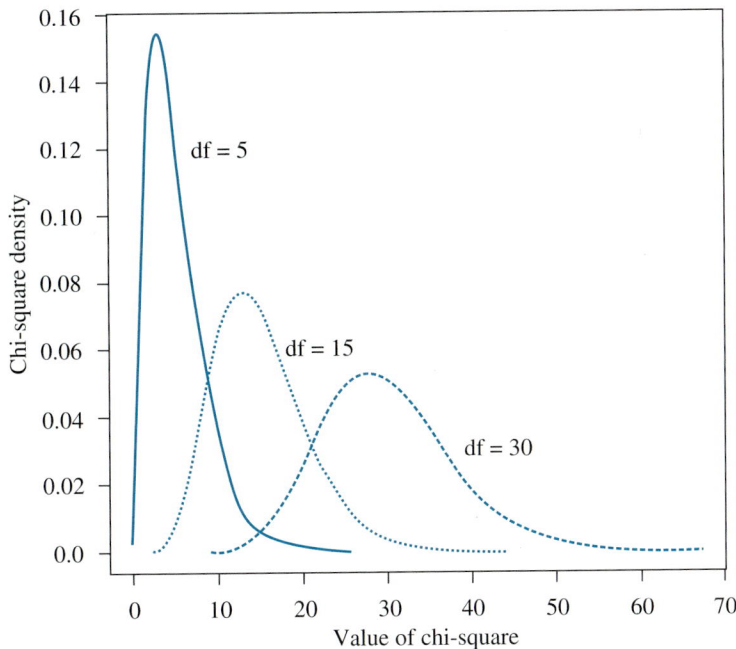

Upper-tail values of the chi-square distribution can be found in Table 7 in the Appendix. Entries in the table are values of χ^2 that have an area a to the right under the curve. The degrees of freedom are specified in the left column of the table, and values of a are listed across the top of the table. Thus, for df = 14, the value of chi-square with an area $a = .025$ to its right under the curve is 26.12 (see Figure 7.4). To determine the value of chi-square with an area .025 to its left

FIGURE 7.4

Critical values of the chi-square distribution with df = 14

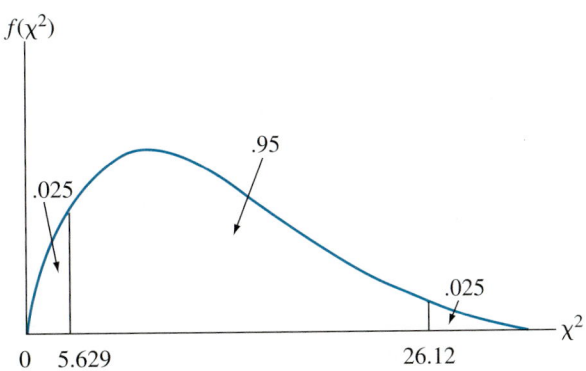

In addition to estimating a population variance, we can construct a statistical test of the null hypothesis that σ^2 equals a specified value, σ_0^2. This test procedure is summarized here.

Statistical Test for σ^2 (or σ)

H_0: 1. $\sigma^2 \leq \sigma_0^2$ H_a: 1. $\sigma^2 > \sigma_0^2$
　　　 2. $\sigma^2 \geq \sigma_0^2$ 　　　 2. $\sigma^2 < \sigma_0^2$
　　　 3. $\sigma^2 = \sigma_0^2$ 　　　 3. $\sigma^2 \neq \sigma_0^2$

T.S.: $\chi^2 = \dfrac{(n-1)s^2}{\sigma_0^2}$

R.R: For a specified value of α,
　1. Reject H_0 if χ^2 is greater than χ_U^2, the upper-tail value for $a = \alpha$ and df $= n - 1$.
　2. Reject H_0 if χ^2 is less than χ_L^2, the lower-tail value for $a = 1 - \alpha$ and df $= n - 1$.
　3. Reject H_0 if χ^2 is greater than χ_U^2, based on $a = \alpha/2$ and df $= n - 1$, or less than χ_L^2, based on $a = 1 - \alpha/2$ and df $= n - 1$.

Check assumptions and draw conclusions.

EXAMPLE 7.2

A manufacturer of a specific pesticide useful in the control of household bugs claims that its product retains most of its potency for a period of at least 6 months. More specifically, it claims that the drop in potency from 0 to 6 months will vary in the interval from 0% to 8%. To test the manufacturer's claim, a consumer group obtained a random sample of 20 containers of pesticide from the manufacturer. Each can was tested for potency and then stored for a period of 6 months at room temperature. After the storage period, each can was again tested for potency. The percentage drop in potency was recorded for each can and is given here.

| 0.5 | 3.5 | 4.4 | 6.0 | 6.6 | 5.4 | 7.9 | 4.6 | 5.4 | 5.7 |
| 2.5 | 1.1 | 5.9 | 2.7 | 2.3 | 1.4 | 1.8 | 5.8 | 0.2 | 7.1 |

Use these data to determine whether there is sufficient evidence to indicate that the population of potency drops has more variability than claimed by the manufacturer. Use $\alpha = .05$.

Solution   The manufacturer claimed that the population of potency reductions has a range of 8%. Dividing the range by 4, we obtain an approximate population standard deviation of $\sigma = 2\%$ (or $\sigma^2 = 4$).

The appropriate null and alternative hypotheses are

H_0: $\sigma^2 \leq 4$ (i.e., we assume the manufacturer's claim is correct)

H_a: $\sigma^2 > 4$ (i.e., there is more variability than claimed by the manufacturer)

To use our inference techniques for variances, we must first check the normality of the data. From Figure 7.7 we observe that the plotted points fall nearly on the

FIGURE 7.7

Normal probability plot for
potency data

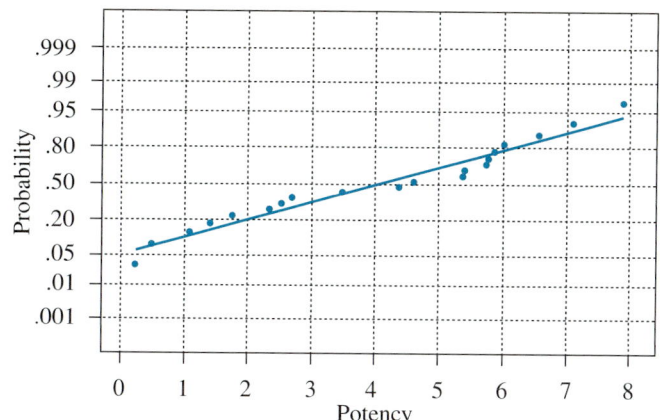

straight line. Thus, the normality condition appears to be satisfied. From the 20 data values, we compute $s^2 = 5.45$. The test statistic and rejection region are as follows:

$$\text{T.S.:} \quad \chi^2 = \frac{(n-1)s^2}{\sigma_0^2} = \frac{19(5.45)}{4} = 25.88$$

R.R.: For $\alpha = .05$, we will reject H_0 if the computed value of chi-square is greater than 30.14, obtained from Table 7 in the Appendix for $a = .05$ and df $= 19$.

Conclusion: Since the computed value of chi-square, 25.88, is less than the critical value, 30.14, there is insufficient evidence to reject the manufacturer's claim, based on $\alpha = .05$. The p-value $= P(\chi_{19}^2 > 25.88) = .14$ can be found using a computer program. Using Table 7 in the Appendix, we can only conclude that p-value $> .10$, because the p-value $= P(\chi_{19}^2 > 25.88) > P(\chi_{19}^2 > 27.20) = .10$. The sample size is relatively small and the p-value is only moderately large, so the consumer group is not prepared to accept H_0: $\sigma^2 \leq 4$. Rather, it would be wise to do additional testing with a larger sample size before reaching a definite conclusion.

The inference methods about σ are based on the condition that the random sample is selected from a population having a normal distribution similar to the requirements for using t distribution-based inference procedures. However, when sample sizes are moderate to large ($n \geq 30$), the t distribution-based procedures can be used to make inferences about μ even when the normality condition does not hold, because for moderate to large sample sizes the Central Limit Theorem provides that the sampling distribution of the sample mean is approximately normal. Unfortunately, the same type of result does not hold for the chi-square-based procedures for making inferences about σ; that is, if the population distribution is distinctly nonnormal, then these procedures for σ are not appropriate even if the sample size is large. Population nonnormality, in the form of skewness or heavy tails, can have serious effects on the nominal significance and confidence probabilities for σ. If a boxplot or normal probability plot of the sample data shows substantial skewness or a substantial number of outliers, the chi-square-based inference procedures should not be applied. There are some alternative approaches that involve computationally elaborate inference procedures. One such procedure is the bootstrap. Bootstrapping is a technique that provides a simple and practical way to estimate the uncertainty in sample statistics like

FIGURE 7.8

Normal probability plot of juice data

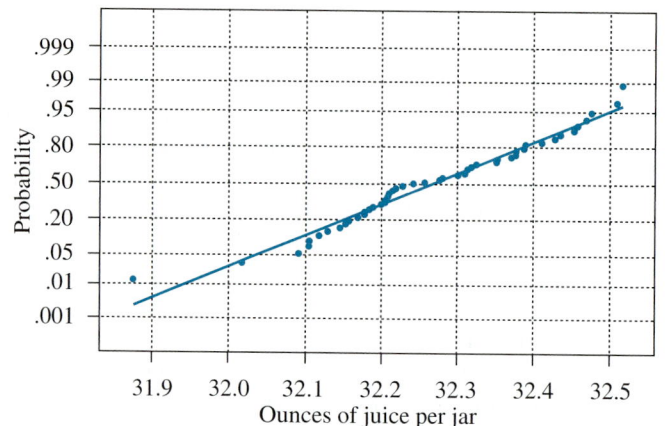

Ounces of juice per jar

```
Descriptive Statistics for Juice Data

Variable        N      Mean    Median    TrMean    StDev    SE Mean
Juice Jars      50    32.267   32.248    32.270    0.135     0.019

Variable     Minimum   Maximum        Q1        Q3
Juice Jars    31.874    32.515    32.177    32.376
```

a. If the process yields jars having a normal distribution with a mean of 32.30 ounces and a standard deviation of .15 ounces, what proportion of the jars filled on the packaging line will be underfilled?

b. Does the plot in Figure 7.8 suggest any violation of the conditions necessary to use the chi-square procedures for generating a confidence interval and a test of hypotheses about σ?

c. Construct a 95% confidence interval on the process standard deviation.

d. Do the data indicate that the process standard deviation is greater than .15? Use $\alpha = .05$.

e. Place bounds on the *p*-value of the test.

Engin.   **7.6**   A leading researcher in the study of interstate highway accidents proposes that a major cause of many collisions on the interstates is not the speed of the vehicles but rather the *difference* in speeds of the vehicles. When some vehicles are traveling slowly while other vehicles are traveling at speeds greatly in excess of the speed limit, the faster-moving vehicles may have to change lanes quickly, which can increase the chance of an accident. Thus, when there is a large variation in the speeds of the vehicles in a given location on the interstate, there may be a larger number of accidents than when the traffic is moving at a more uniform speed. The researcher believes that when the standard deviation in speed of vehicles exceeds 10 mph, the rate of accidents is greatly increased. During a 1-hour period of time, a random sample of 100 vehicles is selected from a section of an interstate known to have a high rate of accidents, and their speeds are recorded using a radar gun. The data are summarized here and in a boxplot.

```
Descriptive Statistics for Vehicle Speeds

Variable          N      Mean    Median    TrMean    StDev    SE Mean
Speed (mph)      100    64.48    64.20     64.46     11.35     1.13

Variable      Minimum   Maximum        Q1        Q3
Speed (mph)     37.85     96.51    57.42    71.05
```

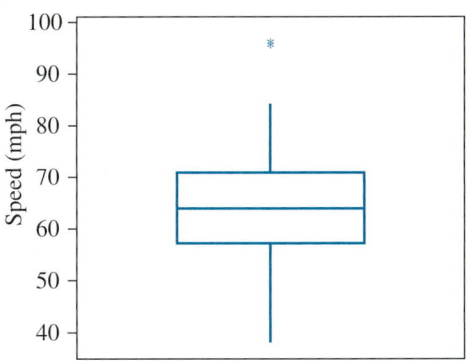

a. Does the boxplot suggest any violation of the conditions necessary to use the chi-square procedures for generating a confidence interval and a test of hypotheses about σ?
b. Estimate the standard deviation in the speeds of the vehicles on the interstate using a 95% confidence interval.
c. Do the data indicate at the 5% level that the standard deviation in vehicle speeds exceeds 10 mph?

Engin.   **7.7**   A certain part for a small assembly should have a diameter of 4.000 mm, and specifications allow a maximum standard deviation of .011 mm. A random sample of 26 parts shows the following diameters:

3.952	3.978	3.979	3.984	3.987	3.991	3.995	3.997	3.999	3.999	3.999
4.000	4.000	4.000	4.001	4.001	4.002	4.002	4.003	4.004	4.006	4.009
4.010	4.012	4.023	4.041							

a. Calculate the sample mean and standard deviation.
b. Can the research hypothesis that $\sigma > .011$ be supported (at $\alpha = .05$) by these data? State all parts of a statistical hypothesis test.

7.8   Calculate 90% confidence intervals for the true variance and for the true standard deviation for the data of Exercise 7.7.

7.9   Plot the data of Exercise 7.7. Does the plot suggest any violation of the assumptions underlying your answers to Exercises 7.7 and 7.8? Would such a violation have a serious effect on the validity of your answers?

Edu.   **7.10**   A large public school system was evaluating its elementary school reading program. In particular, educators were interested in the performance of students on a standardized reading test given to all third graders in the state. The mean score on the test was compared to the state average to determine the school system's rating. Also, the educators were concerned with the variation in scores. If the mean scores were at an acceptable level but the variation was high, this would indicate that a large proportion of the students still needed remedial reading programs. Also, a large variation in scores might indicate a need for programs for those students at the gifted level. Without accelerated reading programs, these students lose interest during reading classes. To obtain information about students early in the school year (the statewide test is given during the last month of the school year), a random sample of 150 third-grade students was given the exam used in the previous year. The possible scores on the reading test range from 0 to 100. The data are summarized here.

```
Descriptive Statistics for Reading Scores

Variable          N       Mean    Median   TrMean    StDev    SE Mean
Reading         150     70.571    71.226   70.514    9.537     0.779

Variable   Minimum    Maximum        Q1        Q3
Reading     44.509     94.570    65.085    76.144
```

a. Does the plot of the data suggest any violation of the conditions necessary to use the chi-square procedures for generating a confidence interval and a test of hypotheses about σ?

b. Estimate the variation in reading scores using a 99% confidence interval.

c. Do the data indicate that the variation in reading scores is greater than 90, the variation for all students taking the exam the previous year?

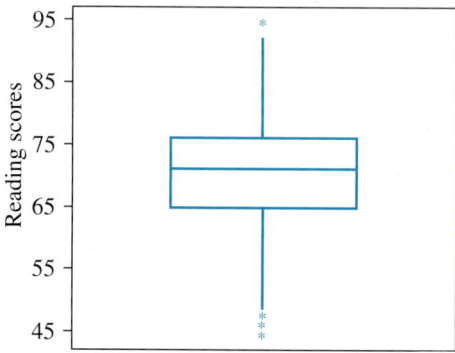

7.11   Place bounds on the *p*-value of the test in Exercise 7.10.

Engin.   **7.12**   Baseballs vary somewhat in their rebounding coefficient. A baseball that has a large rebound coefficient will travel further when the same force is applied to it than a ball with a smaller coefficient. To achieve a game in which each batter has an equal opportunity to hit a home run, the balls should have nearly the same rebound coefficient. A standard test has been developed to measure the rebound coefficient of baseballs. A purchaser of large quantities of baseballs requires that the mean coefficient value be 85 units and the standard deviation be less than 2 units. A random sample of 81 baseballs is selected from a large batch of balls and tested. The data are summarized here.

```
Descriptive Statistics for Rebound Coefficient Data

Variable          N       Mean    Median   TrMean    StDev    SE Mean
Rebound          81     85.296    85.387   85.285    1.771     0.197

Variable   Minimum    Maximum        Q1        Q3
Rebound     80.934     89.687    84.174    86.352
```

a. Does the plot indicate any violation of the conditions underlying the use of the chi-square procedures for constructing confidence intervals or testing hypotheses about σ?

b. Is there sufficient evidence that the standard deviation in rebound coefficient for the batch of balls is less than 2?

c. Estimate the standard deviation of the rebound coefficients using a 95% confidence interval.

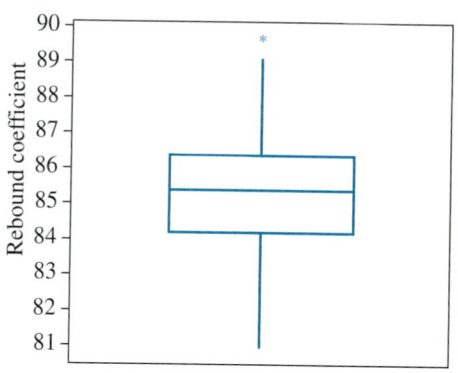

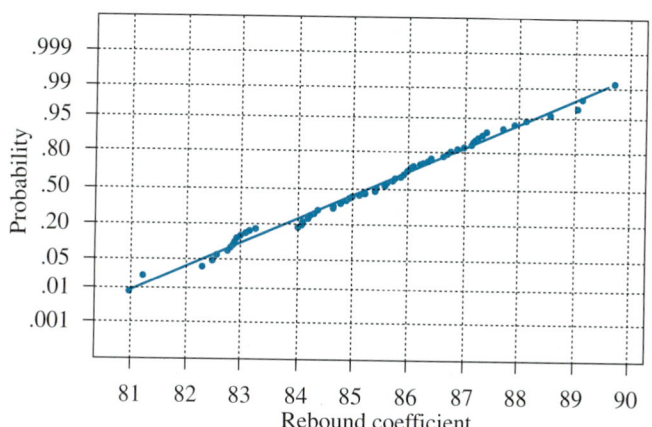

Estimation and Tests for Comparing Two Population Variances

evaluating

equal variance condition

In the case study about *E. coli* detection methods, we were concerned about comparing the standard deviations of the two procedures. In many situations in which we are comparing two processes or two suppliers of a product, we need to compare the standard deviations of the populations associated with process measurements. Another major application of a test for the equality of two population variances is for **evaluating** the validity of the **equal variance condition** (that is, $\sigma_1^2 = \sigma_2^2$) for a two-sample t test. The test developed in this section requires that the two population distributions both have normal distributions. We are interested in comparing the variance of population 1, σ_1^2, to the variance of population 2, σ_2^2.

When random samples of sizes n_1 and n_2 have been independently drawn from two normally distributed populations, the ratio

$$\frac{s_1^2/\sigma_1^2}{s_2^2/\sigma_2^2} = \frac{s_1^2/s_2^2}{\sigma_1^2/\sigma_2^2}$$

F distribution   possesses a probability distribution in repeated sampling referred to as an **F distribution.** The formula for the probability distribution is omitted here, but we will specify its properties.

Properties of the F Distribution

1. Unlike t or z but like χ^2, F can assume only positive values.
2. The F distribution, unlike the normal distribution or the t distribution but like the χ^2 distribution, is nonsymmetrical. (See Figure 7.9.)
3. There are many F distributions, and each one has a different shape. We specify a particular one by designating the degrees of freedom associated with s_1^2 and s_2^2. We denote these quantities by df_1 and df_2, respectively. (See Figure 7.9.)
4. Tail values for the F distribution are tabulated and appear in Table 8 in the Appendix.

FIGURE 7.9
Densities of two F distributions

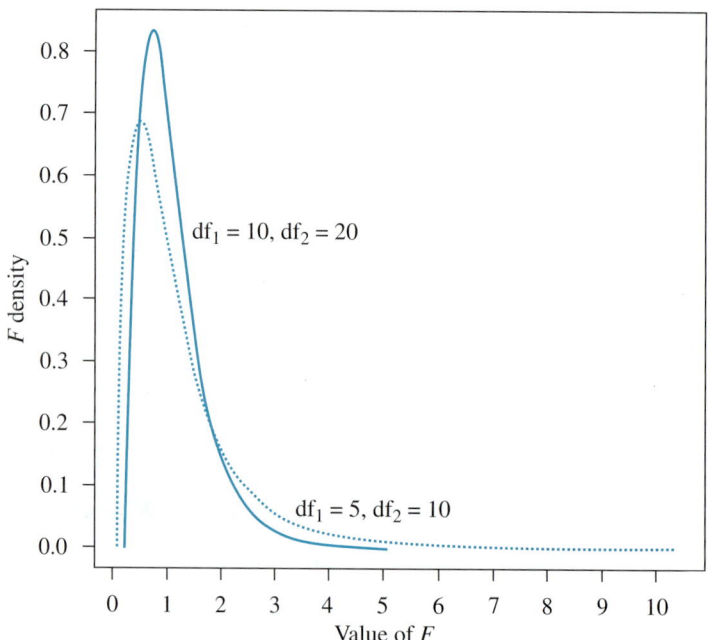

Table 8 in the Appendix records upper-tail values of F corresponding to areas $a = .25, .10, .05, .025, .01, .005,$ and $.001$. The degrees of freedom for s_1^2, designated by df_1, are indicated across the top of the table; df_2, the degrees of freedom for s_2^2, appear in the first column to the left. Values of a are given in the next column. Thus, for $df_1 = 5$ and $df_2 = 10$, the critical values of F corresponding to $a = .25, .10, .05, .025, .01, .005,$ and $.001$ are, respectively, 1.59, 2.52, 3.33, 4.24,

5.64, 6.78, and 10.48. It follows that only 5% of the measurements from an F distribution with $df_1 = 5$ and $df_2 = 10$ would exceed 3.33 in repeated sampling. (See Figure 7.10.) Similarly, for $df_1 = 24$ and $df_2 = 10$, the critical values of F corresponding to tail areas of $a = .01$ and $.001$ are, respectively, 4.33 and 7.64.

FIGURE 7.10
Critical value for the F distributions ($df_1 = 5$, $df_2 = 10$)

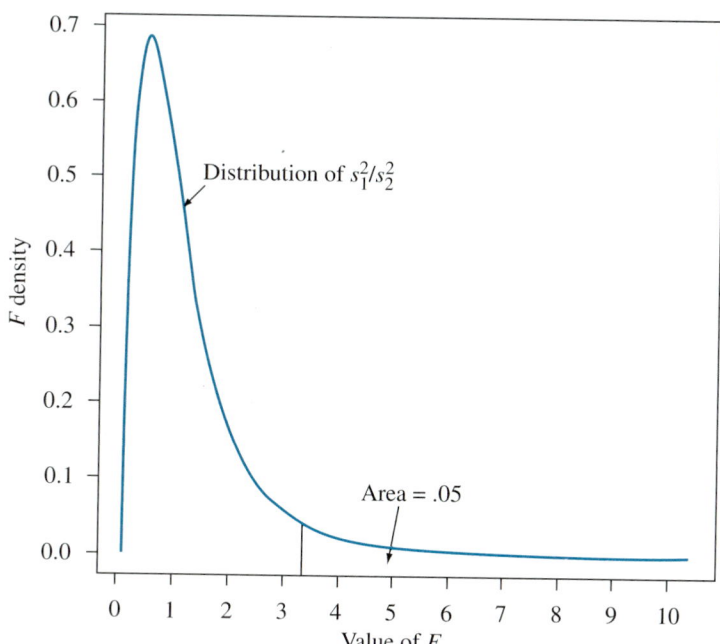

A statistical test comparing σ_1^2 and σ_2^2 utilizes the test statistic s_1^2/s_2^2. When $\sigma_1^2 = \sigma_2^2$, $\sigma_1^2/\sigma_2^2 = 1$ and s_1^2/s_2^2 follows an F distribution with $df_1 = n_1 - 1$ and $df_2 = n_2 - 1$. Table 8 in the Appendix provides the upper-tail values of the F distribution. The lower-tail values are obtained from the following relationship: Let F_{α,df_1,df_2} be the upper α percentile and $F_{1-\alpha,df_1,df_2}$ be the lower α percentile of an F distribution with df_1 and df_2. Then,

$$F_{1-\alpha,df_1,df_2} = \frac{1}{F_{\alpha,df_2,df_1}}$$

Note that the degrees of freedom have been reversed for the F percentile on the right-hand side of the equation.

EXAMPLE 7.4

Determine the lower .025 percentile for an F distribution with $df_1 = 4$ and $df_2 = 9$.

Solution   From Table 8 in the Appendix, the upper .025 percentile for the F distribution with $df_1 = 9$ and $df_2 = 4$ is 8.90. Thus,

$$F_{.975,4,9} = \frac{1}{F_{.025,9,4}} \quad \text{or} \quad F_{.975,4,9} = \frac{1}{8.90} = 0.11$$

For a one-tailed alternative hypothesis, the populations are designated 1 and 2 so that H_a is of the form $\sigma_1^2 > \sigma_2^2$. Then the rejection region is located in the upper-tail of the F distribution.

Voltage	Sample Size	Mean	Standard Deviation
110	10	20.04	.474
220	16	9.99	.233

The researchers wanted to estimate the relative size of the variation in life length under 110 and 220 volts. Use the data to construct a 90% confidence interval for σ_1/σ_2, the ratio of the standard deviations in life lengths for the components under the two operating voltages.

Solution   Before constructing the confidence interval, it is necessary to check whether the two populations of life lengths were both normally distributed. From the normal probability plots, it would appear that both samples of life lengths are from normal distribution. Next, we need to find the upper and lower $\alpha/2 = .10/2 = .05$ percentiles for the F distribution with $df_1 = 10 - 1 = 9$ and $df_2 = 16 - 1 = 15$. From Table 8 in the Appendix, we find

$$F_U = F_{.05,15,9} = 3.01 \quad \text{and} \quad F_L = F_{.95,15,9} = 1/F_{.05,9,15} = 1/2.59 = .386$$

Substituting into the confidence interval formula, we have a 90% confidence interval for σ_1^2/σ_2^2

$$\frac{(.474)^2}{(.233)^2}.386 \leq \frac{\sigma_1^2}{\sigma_2^2} \leq \frac{(.474)^2}{(.233)^2}3.01$$

$$1.5975 \leq \frac{\sigma_1^2}{\sigma_2^2} \leq 12.4569$$

It follows that our 90% confidence interval for σ_1/σ_2 is given by

$$\sqrt{1.5975} \leq \frac{\sigma_1}{\sigma_2} \leq \sqrt{12.4569} \quad \text{or} \quad 1.26 \leq \frac{\sigma_1}{\sigma_2} \leq 3.53$$

Thus, we are 90% confident that σ_1 is between 1.26 and 3.53 times as large as σ_2.

EXAMPLE 7.7

We will now complete our analysis of phase 1 of the *E. coli* case study.

Analyzing Data for Case Study   Because the objective of the study was to evaluate the HEC procedure for its performance in detecting *E. coli,* it is necessary to evaluate its repeatability and its agreement with an accepted method for *E. coli*—namely, the HGMF procedure. Thus, we need to compare both the level and variability in the two methods to determine *E. coli* concentrations; that is, we need to test hypotheses about both the means and standard deviations of HEC and HGMF *E. coli* concentrations. Recall that we had 24 independent observations from the HEC and HGMF procedures on pure cultures of *E. coli* having a specified level of 7 $\log_{10}$ CFU/ml. Prior to constructing confidence intervals or testing hypotheses, we must first check whether the data represent random samples from normally distributed populations. From the boxplots displayed in Figure 7.2 and the normal probability plots in Figure 7.11, the data from both procedures appear to follow a normal distribution.

FIGURE 7.11(a)

Normal probability plots for
HGMF and HEC

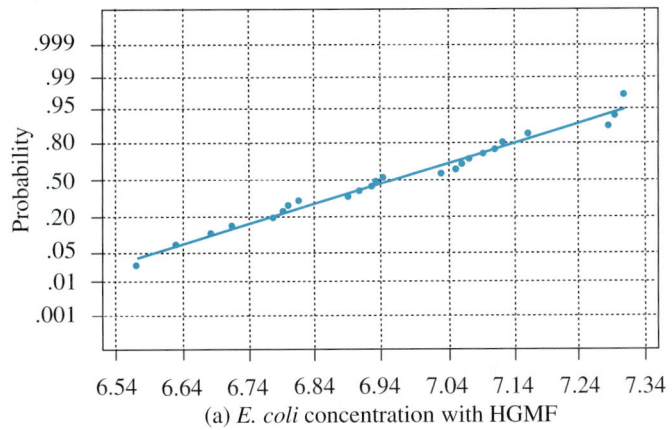

(a) *E. coli* concentration with HGMF

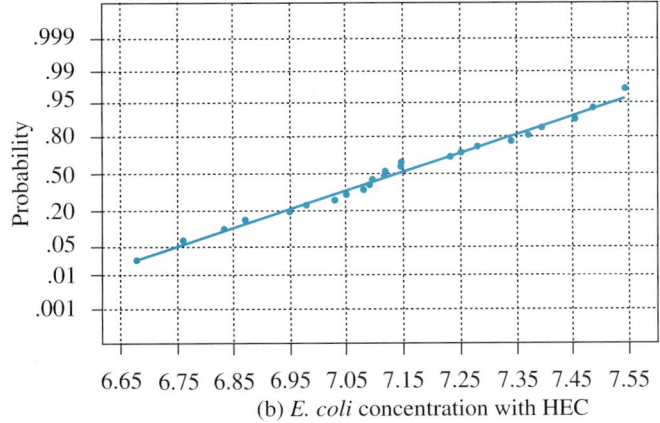

(b) *E. coli* concentration with HEC

We next will test the hypotheses

$$H_0: \sigma_1^2 = \sigma_2^2 \text{ vs. } H_a: \sigma_1^2 \neq \sigma_2^2$$

where we designate HEC as population 1 and HGMF as population 2. The summary statistics are given here.

Procedure	Sample Size	Mean	Standard Deviation
HEC	24	7.1346	.2291
HGMF	24	6.9529	.2096

R.R.: For a two-tailed test with $\alpha = .05$, we will reject H_0 if

$$F_0 = s_1^2/s_2^2 \leq F_{.975,23,23} = 1/F_{.025,23,23} = 1/2.31 = .43 \quad \text{or} \quad F \geq F_{.025,23,23} = 2.31$$

Because $F_0 = (.2291)^2/(.2096)^2 = 1.19$ is not less than .43 or greater than 2.31, we fail to reject H_0. Using a computer software program, we determine that the p-value of the test statistic is 0.672. Thus, we can conclude that HEC appears to have a similar degree of variability as HGMF in its determination of *E. coli* concentration. To obtain estimates of the variability in the HEC and HGMF readings, 95% confidence intervals on their standard deviations are given by (0.17, .23) for σ_{HEC} and (.16, .21) for σ_{HGMF}.

Both the HEC and HGMF *E. coli* concentration readings appear to be independent random samples from normal populations with a common standard deviation, so we can use a pooled *t* test to evaluate

$H_0: \mu_1 = \mu_2$ vs. $H_a: \mu_1 \neq \mu_2$

R.R.: For a two-tailed test with $\alpha = .05$, we will reject H_0 if

$$|t| = \frac{|\bar{y}_1 - \bar{y}_2|}{s_p \sqrt{\dfrac{1}{n_1} + \dfrac{1}{n_2}}} \geq t_{.025,46} = 2.01$$

Because $t = 2.87$ is greater than 2.01, we reject H_0. The *p*-value is .0062. Thus, there is significant evidence that the average HEC *E. coli* concentration readings differ from the average HGMF readings. To estimate the average readings, 95% confidence intervals are given by (7.04, 7.23) for μ_{HEC} and (6.86, 7.04) for μ_{HGMF}. The HEC readings are on the average somewhat higher than the HGMF readings. These findings then prepare us for the second phase of the study. In this phase, HEC and HGMF will be applied to the same sample of meats in a field study similar to what would be encountered in a meat quality monitoring setting. The two procedures have similar levels of variability, but HEC produced *E. coli* concentration readings higher than those of HGMF. Thus, the goal of phase two is to calibrate the HEC readings to the HGMF readings. We will discuss this phase of the study in a later chapter.

Reporting Conclusions   We need to write a report summarizing our findings concerning phase one of the study. The report should include

1. Statement of objective for study
2. Description of study design and data collection procedures
3. Numerical and graphical summaries of data sets
4. Description of all inference methodologies:

   - *t* and *F* tests
   - *t*-based confidence intervals on means
   - Chi-square-based confidence intervals on standard deviations
   - Verification that all necessary conditions for using inference techniques were satisfied

5. Discussion of results and conclusions
6. Interpretation of findings relative to previous studies
7. Recommendations for future studies
8. Listing of data sets

A simulation study was conducted to investigate the effect on the level of the *F* test of sampling from heavy-tailed and skewed distributions rather than the required normal distribution. The five distributions were described in Example 7.3.

For each pair of sample sizes $(n_1, n_2) = (10, 10), (10, 20)$, or $(20, 20)$, random samples of the specified sizes were selected from one of the five distributions. A test of $H_0: \sigma_1^2 = \sigma_2^2$ vs. $H_a: \sigma_1^2 \neq \sigma_2^2$ was conducted using an *F* test with $\alpha = .05$. This process was repeated 2,500 times for each of the five distributions and three sets of sample sizes. The results are given in Table 7.4.

TABLE 7.4

Proportion of times $H_0: \sigma_1^2 = \sigma_2^2$ was rejected $(\alpha = .05)$

Sample Sizes	Distribution				
	Normal	**Uniform**	**t (df = 5)**	**Gamma (shape = 1)**	**Gamma (shape = .1)**
(10, 10)	.054	.010	.121	.225	.693
(10, 20)	.056	.0068	.140	.236	.671
(20, 20)	.050	.0044	.150	.264	.673

The values given in Table 7.4 are estimates of the probability of Type I errors, α, for the F test of equality of two population variances. When the samples are from a normally distributed population, the value of α is nearly equal to the nominal level of .05 for all three pairs of sample sizes. This is to be expected, because the F test was constructed to test hypotheses when the population distributions have normal distributions. However, when the population distribution is a symmetric short-tailed distribution like the uniform distribution, the value of α is much smaller than the specified value of .05. Thus, the probability of Type II errors of the F test would most likely be much larger than what would occur when sampling from normally distributed populations. When we have population distributions that are symmetric and heavy-tailed, the t with df = 5, the values of α are two to three times larger than the specified value of .05. Thus, the F test commits many more Type I errors than would be expected when the population distributions are of this type. A similar problem occurs when we sample with skewed population distributions such as the two gamma distributions. In fact, the Type I error rates are extremely large in these situations, thus rendering the F test invalid for these types of distributions.

EXERCISES

Basic Techniques

7.13 Find the value of F that locates an area a in the upper tail of the F distribution for these conditions:
 a. $a = .05$, $df_1 = 7$, $df_2 = 12$
 b. $a = .05$, $df_1 = 3$, $df_2 = 10$
 c. $a = .05$, $df_1 = 10$, $df_2 = 20$
 d. $a = .01$, $df_1 = 8$, $df_2 = 15$
 e. $a = .01$, $df_1 = 12$, $df_2 = 25$

(Note: Your answers may not agree with those in the back of the book. As long as your answer is close to the recorded answer, it is satisfactory.)

7.14 Find approximate values for F_a for these conditions:
 a. $a = .05$, $df_1 = 11$, $df_2 = 24$
 b. $a = .05$, $df_1 = 14$, $df_2 = 14$
 c. $a = .05$, $df_1 = 35$, $df_2 = 22$
 d. $a = .01$, $df_1 = 22$, $df_2 = 24$
 e. $a = .01$, $df_1 = 17$, $df_2 = 25$

7.15 Random samples of $n_1 = 8$ and $n_2 = 10$ observations were selected from populations 1 and 2, respectively. The corresponding sample variances were $s_1^2 = 7.4$ and $s_2^2 = 12.7$. Do the data provide sufficient evidence to indicate a difference between σ_1^2 and σ_2^2? Test by using $\alpha = .10$. What assumptions have you made?

7.16 An experiment was conducted to determine whether there was sufficient evidence to indicate that data variation within one population, say population A, exceeded the variation within a second population, population B. Random samples of $n_A = n_B = 8$

measurements were selected from the two populations and the sample variances were calculated to be

$$s_A^2 = 2.87 \qquad s_B^2 = .91$$

Do the data provide sufficient evidence to indicate that σ_A^2 is larger than σ_B^2? Test by using $\alpha = .05$.

Applications

Engin.

7.17 A soft-drink firm is evaluating an investment in a new type of canning machine. The company has already determined that it will be able to fill more cans per day for the same cost if the new machines are installed. However, it must determine the variability of fills using the new machines, and wants the variability from the new machines to be equal to or smaller than that currently obtained using the old machines. A study is designed in which random samples of 61 cans are selected from the output of both types of machines and the amount of fill (in ounces) is determined. The data are summarized in the following table and boxplots.

Summary Data for Canning Experiment

Machine Type	Sample Size	Mean	Standard Deviation
Old	61	12.284	.231
New	61	12.197	.162

Boxplots of old machine and new machine (means are indicated by solid circles)

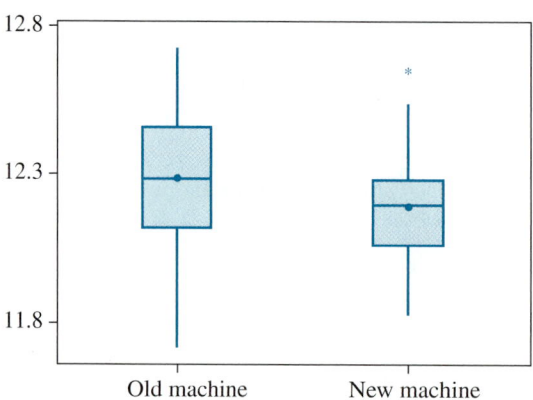

a. Estimate the standard deviations in fill for types of machines using 95% confidence intervals.

b. Do these data present sufficient evidence to indicate that the new type of machine has less variability of fills than the new machine?

c. Do the necessary conditions for conducting the inference procedures in parts (a) and (b) appear to be satisfied? Justify your answer.

Edu.

7.18 The Scholastic Assessment Test (SAT) is an exam taken by most high school students as part of their college admission requirements. A proposal has been made to alter the exam by having the students take the exam on a computer. The exam questions would be selected for the student in the following fashion. For a given section of questions, if the student answers the initial questions posed correctly, then the following questions become increasingly difficult. If the student provides incorrect answers for the initial questions asked in a given section, then the level of difficulty of latter questions does not increase.

The final score on the exams will be standardized to take into account the overall difficulty of the questions on each exam. The testing agency wants to compare the scores obtained using the new method of administering the exam to the scores using the current method. A group of 182 high school students is randomly selected to participate in the study with 91 students randomly assigned to each of the two methods of administering the exam. The data are summarized in the following table and boxplots for the math portion of the exam.

Summary Data for SAT Math Exams			
Testing Method	**Sample Size**	**Mean**	**Standard Deviation**
Computer	91	484.45	53.77
Conventional	91	487.38	36.94

Boxplots of conventional and computer methods (means are indicated by solid circles)

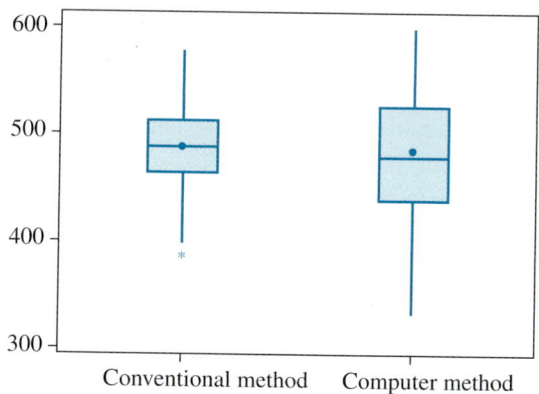

Evaluate the two methods of administering the SAT exam. Provide tests of hypotheses and confidence intervals. Are the means and standard deviations of scores for the two methods equivalent? Justify your answer using $\alpha = .05$.

7.19 Use the information in Tables 7.3 and 7.4 to answer the following questions.
   a. Are the effects of skewness and heavy-tailedness similar for both the chi-square and F tests?
   b. For a given population distribution, does increasing the sample size yield α values more nearly equal to the nominal value of .05? Justify your answer and provide reasons why this occurs.
   c. For the short-tailed distribution, the actual probability of a Type I error is actually smaller than the specified value of .05. What is the negative impact on the F test of the decrease in α, if any?

7.4 Tests for Comparing $t > 2$ Population Variances

In the previous section, we discussed a method for comparing variances from two normally distributed populations based on taking independent random samples from the populations. In many situations, we will need to compare more than two populations. For example, we may want to compare the variability in the level of nutrients of five different suppliers of a feed supplement or the variability in scores of the students using SAT preparatory materials from the three major publishers of those materials. Thus, we need to develop a statistical test that will allow us to compare $t > 2$ population variances. We will consider two procedures. The first procedure, Hartley's test, is very simple to apply but has the restriction that the population distributions must be normally distributed and the sample sizes equal. The second procedure, Levine's test, is more complex in its computations but does not restrict the population distributions or the sample sizes. Levine's test can be obtained from many of the statistical software packages. For example, SAS and Minitab both use Levine's test for comparing population variances.

Hartley F_{max} test   H. O. Hartley (1950) developed the **Hartley F_{max} test** for evaluating the hypotheses

$$H_0: \sigma_1^2 = \sigma_2^2 = \ldots = \sigma_t^2 \text{ vs. } H_a: \sigma_i^2 \text{s not all equal}$$

The Hartley F_{max} requires that we have independent random samples of the same size n from t normally distributed populations. With the exception that we require $n_1 = n_2 = \ldots = n_t = n$, the Hartley test is a logical extension of the F test from the previous section for testing $t = 2$ variances. With s_i^2 denoting the sample variance computed from the ith sample, let $s_{min}^2 = $ the smallest of the s_i^2s and $s_{max}^2 = $ the largest of the s_i^2s. The Hartley F_{max} test statistic is

$$F_{max} = \frac{s_{max}^2}{s_{min}^2}$$

The test procedure is summarized here.

Hartley's F_{max} Test for Homogeneity of Population Variances

H_0: $\sigma_1^2 = \sigma_2^2 = \ldots = \sigma_t^2$ homogeneity of variances

H_a: Population variances are not all equal

T.S.: $F_{max} = \dfrac{s_{max}^2}{s_{min}^2}$

R.R.:   For a specified value of α, reject H_0 if F_{max} exceeds the tabulated F value (Table 12) for $a = \alpha, t$, and $df_2 = n - 1$, where n is the common sample size for the t random samples.

Check assumptions and draw conclusions.

We will illustrate the application of the Hartley test with the following example.

EXAMPLE 7.8

Wludyka and Nelson [*Technometrics* (1997), 39: 274–285] describe the following experiment. In the manufacture of soft contact lenses, a monomer is injected into a plastic frame, the monomer is subjected to ultraviolet light and heated (the time, temperature, and light intensity are varied), the frame is removed, and the lens is hydrated. It is thought that temperature can be manipulated to target the power (the strength of the lens), so interest is in comparing the variability in power. The data are coded deviations from target power using monomers from three different suppliers. We wish to test $H_0: \sigma_1^2 = \sigma_2^2 = \sigma_3^2$.

					Sample						
Deviations from Target Power for Three Suppliers											
Supplier	**1**	**2**	**3**	**4**	**5**	**6**	**7**	**8**	**9**	n	s_i^2
1	191.9	189.1	190.9	183.8	185.5	190.9	192.8	188.4	189.0	9	8.69
2	178.2	174.1	170.3	171.6	171.7	174.7	176.0	176.6	172.8	9	6.89
3	218.6	208.4	187.1	199.5	202.0	211.1	197.6	204.4	206.8	9	80.22

Solution Before conducting the Hartley test, we must check the normality condition. The data are evaluated for normality using a boxplot given in Figure 7.12.

FIGURE 7.12

Boxplot of deviations from target power for three suppliers

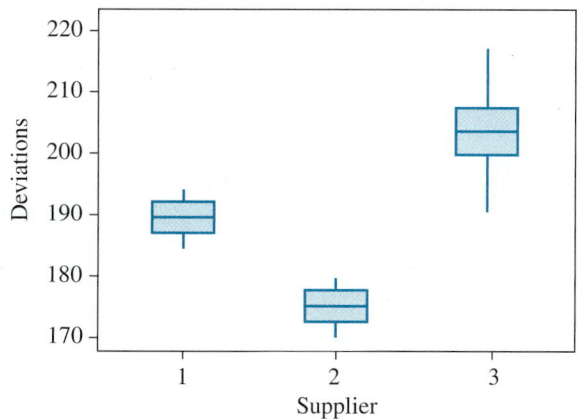

All three data sets appear to be from normally distributed populations. Thus, we will apply the Hartley F_{max} test to the data sets. From Table 12, with $a = .05$, $t = 3$, and $df_2 = 9 - 1 = 8$, we have $F_{max,.05} = 6.00$. Thus, our rejection region will be

R.R.: Reject H_0 if $F_{max} \geq F_{max,.05} = 6.00$

$s_{min}^2 = min(8.69, 6.89, 80.22) = 6.89$

and

$s_{max}^2 = max(8.69, 6.89, 80.22) = 80.22$

Thus,

$$F_{max} = \frac{s_{max}^2}{s_{min}^2} = \frac{80.22}{6.89} = 11.64 > 6.00$$

Thus, we reject H_0 and conclude that the variances are not all equal.

If the sample sizes are not all equal, we can take $n = n_{max}$, where n_{max} is the largest sample size. F_{max} no longer has an exact level α. In fact, the test is liberal in the sense that the probability of Type I error is slightly more than the nominal value α. Thus, the test is more likely to falsely reject H_0 than the test having all n_is equal when sampling from normal populations with the variances all equal.

The Hartley F_{max} test is quite sensitive to departures from normality. Thus, if the population distributions we are sampling from have a somewhat nonnormal distribution but the variances are equal, the F_{max} will reject H_0 and declare the variances to be unequal. The test is detecting the nonnormality of the population distributions, not the unequal variances. Thus, when the population distributions are nonnormal, the F_{max} is not recommended as a test of homongeneity of variances. An alternative approach that does not require the populations to have normal distributions is the Levine test. However, the Levine test involves considerably more calculation than the Hartley test. Also, when the populations have a normal distribution, the Hartley test is more powerful than the Levine test. Conover, Johnson, and Johnson [*Technometrics*, (1981), 23: 351–361], conducted a simulation study of a variety of tests of homongeneity of variance, including the Hartley and Levine test. They demonstrated the inflated α levels of the Hartley test when the populations have highly skewed distributions and recommended the Levine test as one of several alternative procedures.

    The Levine test involves replacing the jth observation from sample i, y_{ij}, with the random variable $z_{ij} = |y_{ij} - \tilde{y}|$, where $\tilde{y}$ is the sample median of the ith sample. We then compute the Levine test statistic on the z_{ij}s.

Levine's Test for Homogeneity of Population Variances	H_0: $\sigma_1^2 = \sigma_2^2 = \ldots = \sigma_t^2$ homogeneity of variances H_a: Population variances are not all equal T.S.: $L = \dfrac{\sum_{i=1}^{t} n_i(\overline{z}_{i.} - \overline{z}..)^2/(t-1)}{\sum_{i=1}^{t} \sum_{j=1}^{n_i} (z_{ij} - \overline{z}_{i.})^2/(N-t)}$ R.R.: For a specified value of α, reject H_0 if $L \geq F_{\alpha,df_1,df_2}$, where $df_1 = t - 1$, $df_2 = N - t$, $N = \sum_{i=1}^{t} n_i$, and F_{α,df_1,df_2} is the upper α percentile from the F distribution, Table 8. Check assumptions and draw conclusions.

We will illustrate the computations for the Levine test in the following example. However, in most cases, we would recommend using a computer software package such as SAS or Minitab for conducting the test.

EXAMPLE 7.9

Three different additives that are marketed for increasing the miles per gallon (mpg) for automobiles were evaluated by a consumer testing agency. Past studies have shown an average increase of 8% in mpg for economy automobiles after using the product for 250 miles. The testing agency wants to evaluate the variability in the increase in mileage over a variety of brands of cars within the economy class. The agency randomly selected 30 economy cars of similar age, number of miles on their odometer, and overall condition of the power train to be used in the study. It then randomly assigned 10 cars to each additive. The percentage increase in mpg obtained by each car was recorded for a 250-mile test drive. The testing agency wanted to evaluate whether there was a difference between the three additives with respect to their variability in the increase in mpg. The data are give here along with the intermediate calculations needed to compute the Levine's test statistic.

Solution   Using the plots in Figures 7.13 (a)–(d), we can observe that the samples from additive 1 and additive 2 do not appear to be samples from normally distrib-

FIGURE 7.13(a)
Boxplots of additive 1,
additive 2, and additive 3
(means are indicated by
solid circles)

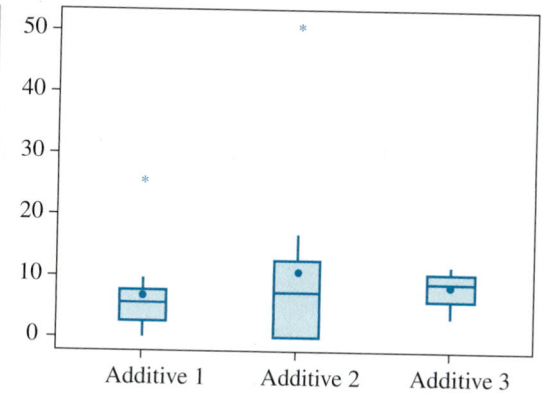

FIGURE 7.13(a)
Boxplots of additive 1,
additive 2, and additive 3
(means are indicated by
solid circles)

FIGURE 7.13(b)–(d)
Normal probability plots for
additives 1, 2, and 3

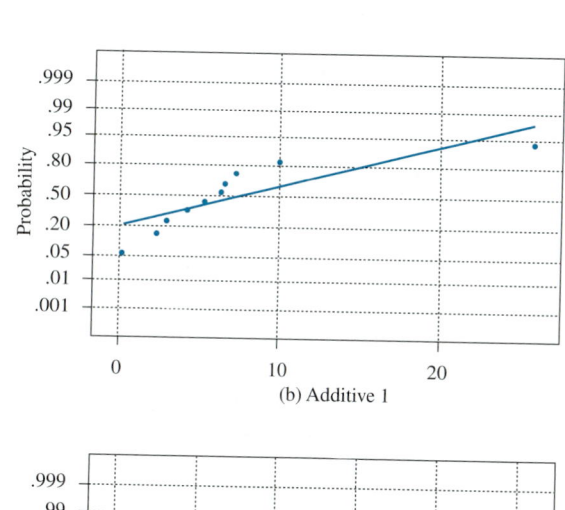

(b) Additive 1

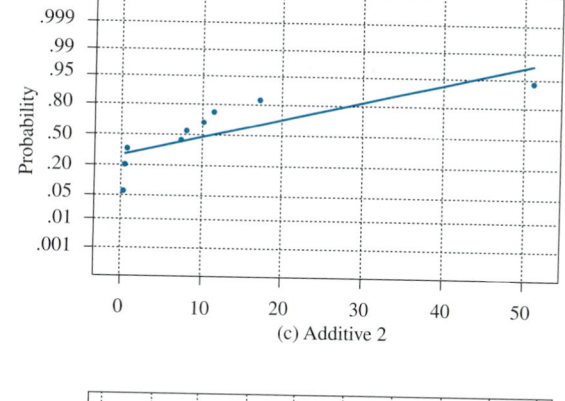

(c) Additive 2

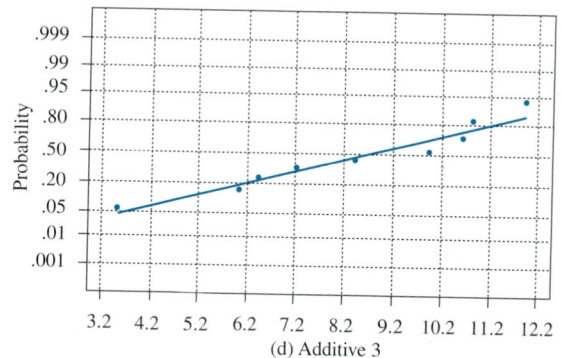

(d) Additive 3

uted populations. Hence, we should not use Hartley's F_{max} test for evaluating differences in the variances in this example. The information in Table 7.5 will assist us in calculating the value of the Levine test statistic. The medians of the percentage increase in mileage, y_{ij}s, for the three additives are 5.80, 7.55, and 9.15. We then calculate the absolute deviations of the data values about their respective medians—namely, $z_{1j} = |y_{1j} - 5.80|$, $z_{2j} = |y_{2j} - 7.55|$, and $z_{3j} = |y_{3j} - 9.15|$ for $j = 1, \ldots, 10$. These values are given in column 3 of the table. Next, we calculate the three means of these values, $\bar{z}_{1.} = 4.07$, $\bar{z}_{2.} = 8.88$, and $\bar{z}_{3.} = 2.23$. Next, we calculate the squared deviations of the z_{ij}s about their respective means, $(z_{ij} - \bar{z}_{i})^2$; that is, $(z_{1j} - 4.07)^2$, $(z_{2j} - 8.88)^2$, and $(z_{3j} - 2.23)^2$. These values are contained in column 6 of the table. Then we calculate the squared deviations of the z_{ij}s

TABLE 7.5

Percentage increase in mpg from cars driven using three additives

Additive	y_{1j}	$\tilde{y}_1$	$z_{1j} = \|y_{1j} - 5.80\|$	$\bar{z}_{1.}$	$(z_{1j} - 4.07)^2$	$(z_{ij} - 5.06)^2$
1	4.2	5.80	1.60	4.07	6.1009	11.9716
1	2.9		2.90		1.3689	4.6656
1	0.2		5.60		2.3409	0.2916
1	25.7		19.90		250.5889	220.2256
1	6.3		0.50		12.7449	20.7936
1	7.2		1.40		7.1289	13.3956
1	2.3		3.50		0.3249	2.4336
1	9.9		4.10		0.0009	0.9216
1	5.3		0.50		12.7449	20.7936
1	6.5		0.70		11.3569	19.0096

Additive	y_{2j}	$\tilde{y}_2$	$z_{2j} = \|y_{2j} - 7.55\|$	$\bar{z}_{2.}$	$(z_{1j} - 8.88)^2$	$(z_{2j} - 5.06)^2$
2	0.2	7.55	7.35	8.88	2.3409	5.2441
2	11.3		3.75		26.3169	1.7161
2	0.3		7.25		2.6569	4.7961
2	17.1		9.55		0.4489	20.1601
2	51.0		43.45		1,195.0849	1,473.7921
2	10.1		2.55		40.0689	6.3001
2	0.3		7.25		2.6569	4.7961
2	0.6		6.95		3.7249	3.5721
2	7.9		0.35		72.7609	22.1841
2	7.2		0.35		72.7609	22.1841

Additive	y_{3j}	$\tilde{y}_3$	$z_{3j} = \|y_{3j} - 9.15\|$	$\bar{z}_{3.}$	$(z_{1j} - 2.33)^2$	$(z_{3j} - 5.06)^2$
3	7.2	9.15	1.95	2.23	0.0784	9.6721
3	6.4		2.75		0.2704	5.3361
3	9.9		0.75		2.1904	18.5761
3	3.5		5.65		11.6964	0.3481
3	10.6		1.45		0.6084	13.0321
3	10.8		1.65		0.3364	11.6281
3	10.6		1.45		0.6084	13.0321
3	8.4		0.75		2.1904	18.5761
3	6.0		3.15		0.8464	3.6481
3	11.9		2.75		0.2704	5.3361
Total				5.06	1,742.6	1,978.4

about the overall mean, $\bar{z}_{..} = 5.06$—that is, $(z_{ij} - \bar{z}_{i.})^2 = (z_{ij} - 5.06)^2$. The last column in the table contains these values. The final step is to sum columns 6 and 7, yielding

$$T_1 = \sum_{i=1}^{3} \sum_{j=1}^{n_i} (z_{ij} - \bar{z}_{i.})^2 = 1742.6 \quad \text{and} \quad T_2 = \sum_{i=1}^{3} \sum_{j=1}^{n_i} (z_{ij} - \bar{z}_{..})^2 = 1{,}978.4$$

The value of Levine's test statistics, in an alternative form, is given by

$$L = \frac{(T_2 - T_1)/(t-1)}{T_1/(N-t)} = \frac{(1{,}978.4 - 1{,}742.6)/(3-1)}{1{,}742.6/(30-3)} = 1.827$$

The rejection region for Levine's test is to reject H_0 if $L \geq F_{\alpha,t-1,N-t} = F_{.05,3-1,30-3} = 3.35$. Because $L = 1.827$, we fail to reject H_0: $\sigma_1^2 = \sigma_2^2 = \sigma_3^2$ and conclude that there is insufficient evidence of a difference in the population variances of the percentage increase in mpg for the three additives.

EXERCISES   **Applications**

7.20 In Example 7.9 we stated that the Hartley test was not appropriate because there was evidence that two of the population distributions were nonnormal. The Levine test was then applied to the data and it was determined that the data did not support a difference in the population variances at an $\alpha = .05$ level. The data yielded the following summary statistics:

Additive	Sample Size	Mean	Median	Standard Deviation
1	10	7.05	5.80	7.11
2	10	10.60	7.55	15.33
3	10	8.53	9.15	2.69

a. Using the plots in Example 7.9, justify that the population distributions are not normal.
b. Use the Hartley test to test for differences in the population variances.
c. Are the results of the Hartley test consistent with those of the Levine test?
d. Which test is more appropriate for this data set? Justify your answer.
e. Which supplier of monomer would you recommend to the manufacturer of soft lenses? Justify your answer.

7.21 Refer to Example 7.8. Use the Levine test to test for differences in the population variances.

a. In Example 7.8, we stated that the population distributions appeared to be normally distributed. Justify this statement.
b. Are the results of the Levine test consistent with the conclusions obtained using the Hartley test?
c. Which test is more appropriate for testing differences in variances in this situation? Justify your answer.
d. Which of the additives appears to be a better product? Provide an explanation for your choice.

Bio.   **7.22** A wildlife biologist was interested in determining the effect of raising deer in captivity on the size of the deer. She decided to consider three populations: deer raised in the wild, deer raised on large hunting ranches, and deer raised in zoos. She randomly selected eight deer in each of the three environments and weighed the deer at age 1 year. The weights (in pounds) are given in the following table.

Environment	Weight (in pounds) of Deer							
Wild	114.7	128.9	111.5	116.4	134.5	126.7	120.6	129.59
Ranch	120.4	91.0	119.6	119.4	150.0	169.7	100.9	76.1
Zoo	103.1	90.7	129.5	75.8	182.5	76.8	87.3	77.3

a. The biologist hypothesized that the weights of deer from captive environments would have a larger level of variability than the weights from deer raised in the wild. Do the data support her contention?

b. Are the requisite conditions for the test you used in (a) satisfied in this situation? Provide plots to support your answer.

7.23 Describe an experiment in which you were involved or a study from the literature in your field in which a comparison of the variances of the measured responses was just as important in determining treatment differences as a difference in the treatment means.

7.24 Why do you think that Levine's statistic is more appropriate than Hartley's test for testing differences in population variances when the population distributions are highly skewed? (*Hint*: Which measure of population location is more highly affected by skewed distributions, the mean or median?)

Edu. **7.25** Many school districts are attempting to both reduce costs and motivate students by using computers as instructional aides. A study was designed to evaluate the use of computers in the classroom. A group of students enrolled in an alternative school were randomly assigned to one of four methods for teaching adding and multiplying fractions. The four methods were lectures only (L), lectures with remedial textbook assistance (L/R), lectures with computer assistance (L/C), and computer instruction only (C). After a 15-week instructional period, an exam was given. The students had taken an exam at the beginning of the 15-week period and the difference in the scores of the two exams is given in the following table. The school administrator wants to determine which method yields the largest increase in test scores and provides the most consistent gains in scores.

Method	Student									
	1	**2**	**3**	**4**	**5**	**6**	**7**	**8**	**9**	**10**
L	7	2	2	6	16	11	9	0	4	2
L/R	5	2	3	11	16	11	3			
L/C	9	12	2	17	12	20	20	31	21	
C	17	19	26	1	47	27	−8	10	20	

Boxplots of L, L/R, L/C, and C (means are indicated by solid circles)

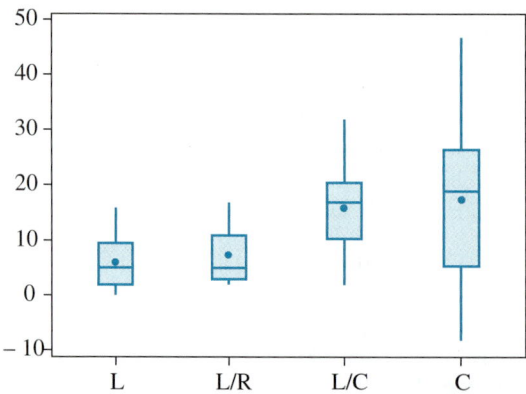

Which method of instruction appears to be the most successful? Provide all relevant tests, confidence intervals, and plots to justify your conclusion.

7.5   Summary

In this chapter, we discussed procedures for making inferences concerning population variances or, equivalently, population standard deviations. Estimation and statistical tests concerning σ make use of the chi-square distribution with df $= n - 1$. Inferences concerning the ratio of two population variances or standard deviations utilize the F distribution with $df_1 = n_1 - 1$ and $df_2 = n_1 - 2$. Finally, when we developed tests concerning differences in $t > 2$ population variances, we used the Hartley or Levine test statistic.

The need for inferences concerning one or more population variances can be traced to our discussion of numerical descriptive measures of a population in Chapter 3. To describe or make inferences about a population of measurements, we cannot always rely on the mean, a measure of central tendency. Many times in evaluating or comparing the performance of individuals on a psychological test, the consistency of manufactured products emerging from a production line, or the yields of a particular variety of corn, we gain important information by studying the population variance.

Key Formulas

1. $100(1 - \alpha)\%$ confidence interval for σ^2 (or σ)

$$\frac{(n-1)s^2}{\chi_U^2} < \sigma^2 < \frac{(n-1)s^2}{\chi_L^2} \quad \text{or} \quad \sqrt{\frac{(n-1)s^2}{\chi_U^2}} < \sigma < \sqrt{\frac{(n-1)s^2}{\chi_L^2}}$$

2. Statistical test for σ^2 (σ_0^2 specified)

$$\text{T.S.:} \quad \chi^2 = \frac{(n-1)s^2}{\sigma_0^2}$$

3. Statistical test for σ_1^2/σ_2^2

$$\text{T.S.:} \quad F = \frac{s_1^2}{s_2^2}$$

4. $100(1 - \alpha)\%$ confidence interval for σ_1^2/σ_2^2 (or σ_1/σ_2)

$$\frac{s_1^2}{s_2^2} F_L < \frac{\sigma_1^2}{\sigma_2^2} < \frac{s_1^2}{s_2^2} F_U$$

where

$$F_L = \frac{1}{F_{\alpha/2, df_1, df_2}} \quad \text{and} \quad F_U = F_{\alpha/2, df_2, df_1}$$

or

$$\sqrt{\frac{s_1^2}{s_2^2}} F_L < \frac{\sigma_1}{\sigma_2} < \sqrt{\frac{s_1^2}{s_2^2}} F_U$$

5. Statistical test for $H_0: \sigma_1^2 = \sigma_2^2 = \ldots = \sigma_t^2$

   a. When population distributions are normally distributed, the Hartley test should be used.

   $$\text{T.S.:} \quad F_{max} = \frac{s_{max}^2}{s_{min}^2}$$

   b. When population distributions are nonnormally distributed, the Levine test should be used.

   $$\text{T.S.:} \quad L = \frac{\sum_{i=1}^{t} n_i (\bar{z}_{i.} - \bar{z}_{..})^2 / (t - 1)}{\sum_{i=1}^{t} \sum_{j=1}^{n_i} (z_{ij} - \bar{z}_{i.})^2 / (N - t)}$$

   where $z_{ij} = |y_{ij} - \tilde{y}_{i.}|$, $\tilde{y}_{i.} = \text{median} \ (y_{i1}, \ldots, y_{in_i})$, $\bar{z}_{i.} = \text{mean} \ (z_{i1}, \ldots, z_{in_i})$, and $\bar{z}_{..} = \text{mean} \ (z_{11}, \ldots, z_{tn_t})$

Supplementary Exercises

7.26   Refer to Exercise 6.21, in which we were interested in comparing the weights of the combs of roosters fed one of two vitamin-supplemented diets. The Wilcoxon rank sum test was suggested as a test of the hypothesis that the two populations were identical. Would it have been appropriate to run a t test comparing the two population means? Explain.

Bus.   **7.27**   A consumer-protection magazine was interested in comparing tires purchased from two different companies that each claimed their tires would last 40,000 miles. A random sample of 10 tires of each brand was obtained and tested under simulated road conditions. The number of miles until the tread thickness reached a specified depth was recorded for all tires. The data are given next (in 1,000 miles).

Brand I	38.9	39.7	42.3	39.5	39.6	35.6	36.0	39.2	37.6	39.5
Brand II	44.6	46.9	48.7	41.5	37.5	33.1	43.4	36.5	32.5	42.0

   a. Plot the data and compare the distributions of longevity for the two brands.
   b. Construct 95% confidence intervals on the means and standard deviations for the number of miles until tread wearout occurred for both brands.
   c. Does there appear to be a difference in wear characteristics for the two brands? Justify your statement with appropriate plots of the data, tests of hypotheses, and confidence intervals.

Med.   **7.28**   A random sample of 20 patients, each of whom has suffered from depression, was selected from a mental hospital, and each patient was administered the Brief Psychiatric Rating Scale. The scale consists of a series of adjectives that the patient scores according to his or her mood. Extensive testing in the past has shown that ratings in certain mood adjectives tend to be similar and hence are grouped together as jointly measuring one or more components of a person's mood. For example, a group consisting of certain adjectives seems to be measuring depression. If a patient population has a standard deviation higher than 4, then the population will generally be split into at least two more consistent groups of patients for treatment purposes. The data for the 20 patients in the group are given next.

10	16	15	13	19	12	16	16	14	11
21	21	18	18	24	16	13	16	24	9

   a. Place a 95% confidence interval on the σ, the standard deviation of the population of patients' scores.

b. Does it appear that the population of patients needs to be split into several groups?

c. What are the conditions underlying the above inferences? Do they appear to be satisfied for this data set? Justify your answer with appropriate plots of the data.

Med.   **7.29** A pharmaceutical company manufactures a particular brand of antihistamine tablets. In the quality control division, certain tests are routinely performed to determine whether the product being manufactured meets specific performance criteria prior to release of the product onto the market. In particular, the company requires that the potencies of the tablets lie in the range of 90% to 110% of the labeled drug amount.

   a. If the company is manufacturing 25-mg tablets, within what limits must tablet potencies lie?

   b. A random sample of 30 tablets is obtained from a recent batch of antihistamine tablets. The data for the potencies of the tablets are given next. Is the assumption of normality warranted for inferences about the population variances?

   c. Translate the company's 90% to 110% specifications on the range of the product potency into a statistical test concerning the population variance for potencies. Draw conclusions based on $\alpha = .05$.

24.1	27.2	26.7	23.6	26.4	25.2
25.8	27.3	23.2	26.9	27.1	26.7
22.7	26.9	24.8	24.0	23.4	25.0
24.5	26.1	25.9	25.4	22.9	24.9
26.4	25.4	23.3	23.0	24.3	23.8

Engin.   **7.30** A study was conducted to compare the variabilities in strengths of 1-inch-square sections of a synthetic fiber produced under two different procedures. A random sample of nine squares from each process was obtained and tested.

   a. Plot the data for each sample separately.

   b. Is the assumption of normality warranted?

   c. If permissible from part (b), use the following data (psi) to test the research hypothesis that the population variances corresponding to the two procedures are different. Use $\alpha = .10$.

Procedure 1	74	90	103	86	75	102	97	85	69
Procedure 2	59	66	73	68	70	71	82	69	74

7.31 Refer to Example 7.2. Construct a 95% confidence interval for σ^2, and use this interval to help interpret the findings of the consumer group. Does it appear that the test of Example 7.2 had much power to detect an increase in σ^2 of 25% over the claimed value? Explain.

Bus.   **7.32** The risk of an investment is measured in terms of the variance in the return that could be observed. Random samples of 10 yearly returns were obtained from two different portfolios. The data are given next (in thousands of dollars).

Portfolio 1	130	135	135	131	129	135	126	136	127	132
Portfolio 2	154	144	147	150	155	153	149	139	140	141

   a. Does portfolio 2 appear to have a higher risk than portfolio 1?

   b. Give a p-value for your test and place a confidence interval on the ratio of the standard deviations of the two portfolios.

   c. Provide a justification that the required conditions have been met for the inference procedures used in parts (a) and (b).

7.33 Refer to Exercise 7.32. Are there any differences in the average returns for the two portfolios? Indicate the method you used in arriving at a conclusion, and explain why you used it.

Bus. **7.34** Two different modeling techniques for assessing the resale value of houses were considered. A random sample of 12 existing listings was taken and each house was valued using the two techniques. The data are shown here.

	Assessed Value Listing (000) Technique	
Listing	1	2
1	155	138
2	137	128
3	248	230
4	136	146
5	102	95
6	87	82
7	63	67
8	129	134
9	144	149
10	270	292
11	157	150
12	51	48

a. Plot the data. Does it appear that the two modeling techniques give similar results?
b. Give an estimate of the mean and standard error of the difference between estimates for the two methods.

7.35 Refer to Exercise 7.34. Place a 90% confidence interval on the variance of the difference in estimates. Give the corresponding interval for σ.

7.36 Refer to Exercises 7.34 and 7.35. What is the critical assumption concerning the sample data? How would you check this assumption? Do the data suggest that the assumption holds? Do you have any cautions about the inferences in Exercise 7.35?

Med. **7.37** An important consideration in examining the potency of a pharmaceutical product is the amount of drop in potency for a specific shelf life (time on a pharmacist's shelf). In particular, the variability of these drops in potency is very important. Researchers studied the drops in potency for two different drug products over a 6-month period. Suppose that drug 1 is an experimental drug product and drug 2 is a marketed product. The data are given here.

Experimental Drug	59.4	52.5	69.0	63.9	62.6	55.8	52.7	50.8	53.1	76.9	62.1	53.6	51.2	88.3	51.4
Marketed Drug	51.9	52.5	50.0	55.2	51.3	52.7	59.0	64.8	50.1	52.7	50.6	54.8	53.1	55.8	50.0

The researchers wanted to determine whether the experimental drug and marketed drug differ with respect to either the mean or standard deviation in the potency drop. In determining whether the two drugs differ, make sure to include the p-value of tests, confidence intervals on all relevant parameters, and plots to assist in determining whether the necessary conditions are satisfied with these data in order to use the various inference procedures.

Med. **7.38** Blood cholesterol levels for randomly selected patients with similar histories were compared for two diets, one a low-fat-content diet and the other a normal diet. The summary data appear here.

	Low-Fat Content	Normal
Sample size	19	24
Sample mean	170	196
Sample variance	198	435

    a. Do these data present sufficient evidence to indicate a difference in cholesterol level variabilities for the two diets? Use $\alpha = .10$.

    b. What other test might be of interest in comparing the two diets?

Med. **7.39** Sales from weight-reducing agents marketed in the United States represent sizable amounts of income for many of the companies that manufacture these products. Psychological as well as physical effects often contribute to how well a person responds to the recommended therapy. Consider a comparison of two weight-reducing agents, A and B. In particular, consider the length of time people remain on the therapy. A total of 26 overweight males, matched as closely as possible physically, were randomly divided into two groups. Those in group 1 received preparation A and those assigned to group 2 received preparation B. The data are given here (in days).

Preparation A	42	47	12	17	26	27	28	26	34	19	20	27	34
Preparation B	35	38	35	36	37	35	29	37	31	31	30	33	44

    Compare the lengths of times that people remain on the two therapies. Make sure to include all relevant plots, tests, confidence intervals, and a written conclusion concerning the two therapies.

7.40 Refer to Exercise 7.39. How would your inference procedures change if preparation A was an old product that had been on the market a number of years and preparation B was a new product, and we wanted to determine whether people would continue to use B a longer time in comparison to preparation A?

Engin. **7.41** A chemist at an iron ore mine suspects that the variance in the amount (weight, in ounces) of iron oxide per pound of ore tends to increase as the mean amount of iron oxide per pound increases. To test this theory, ten 1-pound specimens of iron ore are selected at each of two locations, one, location 1, containing a much higher mean content of iron oxide than the other, location 2. The amounts of iron oxide contained in the ore specimens are shown in the accompanying table.

Location 1	8.1	7.4	9.3	7.5	7.1	8.7	9.1	7.9	8.4	8.8
Location 2	3.9	4.4	4.7	3.6	4.1	3.9	4.6	3.5	4.0	4.2

    Do the data provide sufficient evidence to indicate that the amount of iron oxide per pound of ore is more variable at location 1 than at location 2? Use $\alpha = .05$. Include a confidence interval to demonstrate the size of the difference in the two variances.

H.R. **7.42** A personnel officer was planning to use a t test to compare the mean number of monthly unexcused absences for two divisions of a multinational company but then noticed a possible difficulty. The variation in the number of unexcused absences per month seemed to differ for the two groups. As a check, a random sample of 5 months was selected at each division, and for each month, the number of unexcused absences was obtained.

Category A	20	14	19	22	25
Category B	37	29	51	40	26

a. What assumption seemed to bother the personnel officer?

b. Do the data provide sufficient evidence to indicate that the variances differ for the populations of absences for the two employee categories? Use $\alpha = .05$.

Env.   **7.43**   A researcher was interested in weather patterns in Phoenix and Seattle. As part of the investigation, the researcher took a random sample of 20 days in (June) and observed the daily average temperatures. The data were collected over several years to ensure independence of daily temperatures. The data are given here (in degrees fahrenheit).

Phoenix	96	95	85	98	94	93	98	98	96	96	93	102	99	85	102	83	93	107	104	87
Seattle	60	66	65	62	67	69	64	65	61	57	55	58	75	67	63	60	71	59	50	63

Do the data suggest that there is a difference in the variability of average daily temperatures during June for the two cities? Is there a difference in mean temperatures for the two cities during June? Use $\alpha = .05$ for both tests.

7.44   Refer to the clinical trial database on the data disk to calculate the sample variances for the anxiety scores within each treatment group. Use these data to run separate tests comparing each of the treatments A, B, and C to the placebo group D. Use two-sided tests with $\alpha = .05$.

7.45   Do any of these tests in Exercise 7.44 negate the possibility of comparing treatment means for groups A, B, and C to the treatment mean for the placebo group using t tests? Explain.

7.46   Use the sleep disturbance scores from the clinical trial database to give a 98% confidence interval for σ_B^2/σ_C^2. Do the same for σ_B^2/σ_A^2.

Inferences about More Than Two Population Central Values

8.1 Introduction and Case Study

8.2 A Statistical Test about More Than Two Population Means: An Analysis of Variance

8.3 The Model for Observations in a Completely Randomized Design

8.4 Checking on the AOV Conditions

8.5 Alternative Analysis: Transformations of the Data

8.6 A Nonparametric Alternative: The Kruskal–Wallis Test

8.7 Summary

8.1 Introduction and Case Study

In Chapter 6, we presented methods for comparing two population means, based on independent random samples. Very often the two-sample problem is a simplification of what we encounter in practical situations. For example, suppose we wish to compare the mean hourly wage for nonunion farm laborers from three different ethnic groups (African American, Anglo-American, and Hispanic) employed by a large produce company. Independent random samples of farm laborers would be selected from each of the three ethnic groups (populations). Then, using the information from the three sample means, we would try to make an inference about the corresponding population mean hourly wages. Most likely, the sample means would differ, but this does not necessarily imply a difference among the population means for the three ethnic groups. How do you decide whether the differences among the sample means are large enough to imply that the corresponding population means are different? We will answer this question using a statistical testing procedure called an *analysis of variance*.

Case Study: Effect of Timing of the Treatment of Port-Wine Stains with Lasers

Port-wine stains are congenital vascular malformations that occur in an estimated 3 children per 1,000 births. The stigma of a disfiguring birthmark may have a substantial effect on a child's social and psychosocial adjustment. In 1985, the flash-pumped, pulsed-dye laser was advocated for the treatment of port-wine stains in children. Treatment with this type of laser was hypothesized to be more effective in children than in adults because the skin in children is thinner and the size of the port-wine stain is smaller; fewer treatments would therefore be necessary to achieve optimal clearance. These are all arguments for initiating treatment at an early age.

In a prospective study described in the paper, "Effect of the timing of treatment of port-wine stains with the flash-lamp-pumped pulsed-dye laser" (1998), *The New England Journal of Medicine*, 338: 1028–1033, the researchers investigated whether treatment at a young age would yield better results than treatment at an older age.

Designing the Data Collection   The researchers considered the following issues relative to the most effective treatment:

1. What objective measurements should be used to assess the effectiveness of the treatment in reducing the visibility of the port-wine stains?
2. How many different age groups should be considered for evaluating the treatment?
3. What type of experimental design would produce the most efficient comparison of the different treatments?
4. What are the valid statistical procedures for making the comparisons?
5. What types of information should be included in a final report to document for which age groups the laser treatment was most effective?

One hundred patients, 31 years of age or younger, with a previously untreated port-wine stain were selected for inclusion in the study. During the first consultation, the extent and location of the port-wine stain was recorded. Four age groups of 25 patients each were determined for evaluating whether the laser treatment was more effective for younger patients. Enrollment in an age group ended as soon as 25 consecutive patients had entered the group. A series of treatments was required to achieve optimal clearance of the stain. Before the first treatment, color slides were taken of each patient by a professional photographer in a studio under standardized conditions. Color of the skin was measured using a chromometer. The reproducibility of the color measurements was analyzed by measuring the same location twice in a single session before treatment. For each patient, subsequent color measurements were made at the same location. Treatment was discontinued if either the port-wine stain had disappeared or the three previous treatments had not resulted in any further lightening of the stain. The outcome measure of each patient was the reduction in the difference in color between the skin with the port-wine stain and the contralateral healthy skin.

Eleven of the 100 patients were not included in the final analysis due to a variety of circumstances that occurred during the study period. A variety of baseline characteristics were recorded for the 89 patients: sex, surface area and location of the port-wine stain, and any other medical conditions that might have implications of the effectiveness of the treatment. Also included were treatment characteristics such as average number of visits, level of radiation exposure, number of laser pulses per visit, and the occurrence of headaches after treatment. For all variables there were no significant differences between the four age groups with respect to these characteristics.

The two variables of main interest to the researchers were the difference in color between port-wine stain and contralateral healthy skin before treatment and the improvement in this difference in color after a series of treatments. The before-treatment differences in color are presented in Figure 8.1. The boxplots demonstrate that there were not sizable differences in the color differences between the four groups. This is important, because if the groups differed prior to treatment, then the effect of age group on the effectiveness of the treatment may have been masked by preexisting differences. (The values in Table 8.1 were simulated using the summary statistics given in the paper.)

FIGURE 8.1

Boxplots of stain color by age group (means are indicated by solid circles)

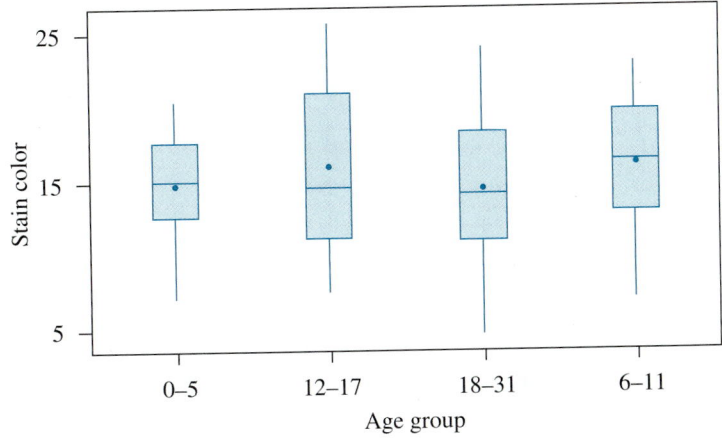

TABLE 8.1

Improvement in color of port-wine stains by age group

Patient	0–5 Years	6–11 Years	12–17 Years	18–31 Years
1	9.6938	13.4081	10.9110	1.4352
2	7.0027	8.2520	10.3844	10.7740
3	10.3249	12.0098	6.4080	8.4292
4	2.7491	7.4514	13.5611	4.4898
5	0.5637	6.9131	3.4523	13.6303
6	8.0739	5.6594	9.5427	4.1640
7	0.1440	8.7352	10.4976	5.4684
8	8.4572	0.2510	4.6775	4.8650
9	2.0162	8.9991	24.7156	3.0733
10	6.1097	6.6154	4.8656	12.3574
11	9.9310	6.8661	0.5023	7.9067
12	9.3404	5.5808	7.3156	9.8787
13	1.1779	6.6772	10.7833	2.3238
14	1.3520	8.2279	9.7764	6.7331
15	0.3795	0.1883	3.6031	14.0360
16	6.9325	1.9060	9.5543	0.6678
17	1.2866	7.7309	5.3193	2.7218
18	8.3438	7.9143	3.0053	2.3195
19	9.2469	1.8724	11.0496	1.6824
20	0.7416	12.5082	2.8697	1.8150
21	1.1072	6.2382	0.1082	5.9665
22		11.2425		0.5041
23		6.8404		5.4484
24		11.2774		

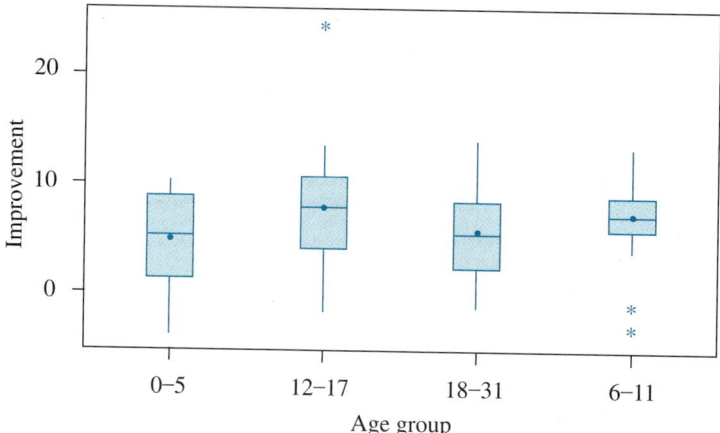

FIGURE 8.2

Boxplots of improvement by age group (means are indicated by solid circles)

Managing the Data Next, the researchers would prepare the data for a statistical analysis following the steps described in Section 2.5. The researchers need to verify that the stain colors were properly recorded and that all computer files were consistent with the field data.

Analyzing the Data The summary statistics are given in the following table along with boxplots for the four age groups (Figure 8.2). The 12–17 years group showed the greatest improvement, but the 6–11 years group had only a slightly smaller improvement. The other two groups had values at least 2 units less than the 12–17 years group. However, from the boxplots we can observe that the four groups do not appear to have that great a difference in improvement. We will now develop the analysis of variance procedure to confirm whether a statistically significant difference exists between the four age groups.

Descriptive Statistics for Port-Wine Stain Case Study

Variable		N	Mean	Median	TrMean	StDev	SE Mean
0-5	Years	21	4.999	6.110	4.974	3.916	0.855
6-11	Years	24	7.224	7.182	7.262	3.564	0.727
12-17	Years	21	7.757	7.316	7.270	5.456	1.191
18-31	Years	23	5.682	4.865	5.531	4.147	0.865

Variable		Minimum	Maximum	Q1	Q3
0-5	Years	0.144	10.325	1.143	8.852
6-11	Years	0.188	13.408	5.804	8.933
12-17	Years	0.108	24.716	3.528	10.640
18-31	Years	0.504	14.036	2.320	8.429

The reason we call the testing procedure an analysis of variance can be seen by using the example cited at the beginning of the section. Assume that we wish to compare the three ethnic mean hourly wages based on samples of five workers selected from each of the ethnic groups. We will use a sample of size five from each of the populations to illustrate the basic ideas, although this sample size is unreasonably small.

Suppose the sample data (hourly wages, in dollars) are as shown in Table 8.2. Do these data present sufficient evidence to indicate differences among the three population means? A brief visual inspection of the data indicates very little variation within a sample, whereas the variability among the sample means is much larger. Because the variability among the sample means is large *in comparison to the* **within-sample variation,** we might conclude intuitively that the corresponding population means are different.

within-sample variation

TABLE 8.2

A comparison of three sample means (small amount of within-sample variation)

Sample from Population		
1	**2**	**3**
5.90	5.51	5.01
5.92	5.50	5.00
5.91	5.50	4.99
5.89	5.49	4.98
5.88	5.50	5.02
$\bar{y}_1 = 5.90$	$\bar{y}_2 = 5.50$	$\bar{y}_3 = 5.00$

Table 8.3 illustrates a situation in which the sample means are the same as given in Table 8.2, but the variability within a sample is much larger, and the **between-sample variation** is small relative to the within-sample variability. We would be less likely to conclude that the corresponding population means differ based on these data.

between-sample variation

TABLE 8.3

A comparison of three sample means (large amount of within-sample variation)

Sample from Population		
1	**2**	**3**
5.90	6.31	4.52
4.42	3.54	6.93
7.51	4.73	4.48
7.89	7.20	5.55
3.78	5.72	3.52
$\bar{y}_1 = 5.90$	$\bar{y}_2 = 5.50$	$\bar{y}_3 = 5.00$

The variations in the two sets of data, Tables 8.2 and 8.3, are shown graphically in Figure 8.3. The strong evidence to indicate a difference in population

FIGURE 8.3

Dot diagrams for the data of Table 8.2 and Table 8.3: ○, measurement from sample 1; ●, measurement from sample 2; □, measurement from sample 3

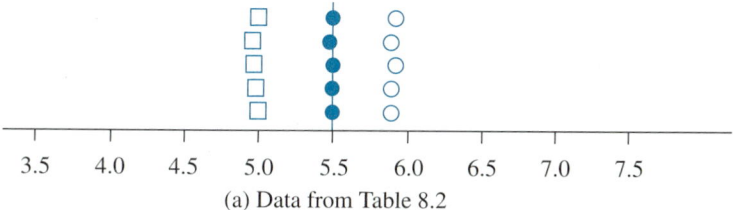

(a) Data from Table 8.2

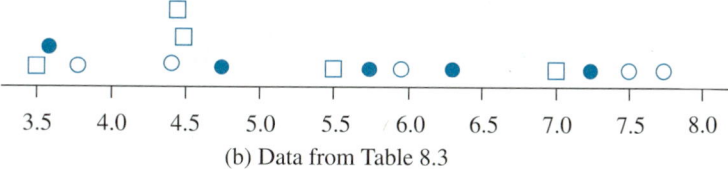

(b) Data from Table 8.3

means for the data of Table 8.2 is apparent in Figure 8.3(a). The lack of evidence to indicate a difference in population means for the data of Table 8.3 is indicated by the overlapping of data points for the samples in Figure 8.3(b).

analysis of variance

The preceding discussion, with the aid of Figure 8.3, should indicate what we mean by an **analysis of variance.** All differences in sample means are judged statistically significant (or not) by comparing them to the variation within samples. The details of the testing procedure will be presented next.

8.2 A Statistical Test about More Than Two Population Means: An Analysis of Variance

In Chapter 6, we presented a method for testing the equality of two population means. We hypothesized two normal populations (1 and 2) with means denoted by μ_1 and μ_2, respectively, and a common variance σ^2. To test the null hypothesis that $\mu_1 = \mu_2$, independent random samples of sizes n_1 and n_2 were drawn from the two populations. The sample data were then used to compute the value of the test statistic

$$t = \frac{\bar{y}_1 - \bar{y}_2}{s_p \sqrt{(1/n_1) + (1/n_2)}}$$

where

$$s_p^2 = \frac{(n_1 - 1)s_1^2 + (n_2 - 1)s_2^2}{(n_1 - 1) + (n_2 - 1)} = \frac{(n_1 - 1)s_1^2 + (n_2 - 1)s_2^2}{n_1 + n_2 - 2}$$

pooled estimate of σ^2

is a pooled estimate of the common population variance σ^2. The rejection region for a specified value of α, the probability of a Type I error, was then found using Table 2 in the Appendix.

Now suppose that we wish to extend this method to test the equality of more than two population means. The test procedure described here applies to only two means and therefore is inappropriate. Hence, we will employ a more general method of data analysis, the analysis of variance. We illustrate its use with the following example.

Students from five different campuses throughout the country were surveyed to determine their attitudes toward industrial pollution. Each student sampled was asked a specific number of questions and then given a total score for the interview. Suppose that 25 students are surveyed at each of the five campuses and we wish to examine the average student score for each of the five campuses.

We label the set of all test scores that could have been obtained from campus I as population I, and we will assume that this population possesses a mean μ_1. A random sample of $n_1 = 25$ measurements (scores) is obtained from this population to monitor student attitudes toward pollution. The set of all scores that could have been obtained from students on campus II is labeled population II (which has a mean μ_2). The data from a random sample of $n_2 = 25$ scores are obtained from this population. Similarly μ_3, μ_4, and μ_5 represent the means of the populations for scores from campuses III, IV, and V, respectively. We also obtain random samples of 25 student scores from each of these populations.

From each of these five samples, we calculate a sample mean and variance. The sample results can then be summarized as shown in Table 8.4.

TABLE 8.4

Summary of the sample results for five populations

	Population				
	I	**II**	**III**	**IV**	**V**
Sample mean	$\bar{y}_1$	$\bar{y}_2$	$\bar{y}_3$	$\bar{y}_4$	$\bar{y}_5$
Sample variance	s_1^2	s_2^2	s_3^2	s_4^2	s_5^2
Sample size	25	25	25	25	25

If we are interested in testing the equality of the population means (i.e., $\mu_1 = \mu_2 = \mu_3 = \mu_4 = \mu_5$), we might be tempted to run all possible pairwise comparisons of two population means. Hence, if we confirm that the five distributions are approximately normal with the same variance σ^2, we could run 10 t tests comparing all pairs of means, as listed here (see Section 6.2).

multiple *t* tests

Null Hypotheses

$\mu_1 = \mu_2$	$\mu_1 = \mu_4$	$\mu_2 = \mu_3$	$\mu_2 = \mu_5$	$\mu_3 = \mu_5$
$\mu_1 = \mu_3$	$\mu_1 = \mu_5$	$\mu_2 = \mu_4$	$\mu_3 = \mu_4$	$\mu_4 = \mu_5$

One obvious disadvantage to this test procedure is that it is tedious and time consuming. However, a more important and less apparent disadvantage of running multiple t tests to compare means is that the probability of falsely rejecting at least one of the hypotheses increases as the number of t tests increases. Thus, although we may have the probability of a Type I error fixed at $\alpha = .05$ for each individual test, the probability of falsely rejecting *at least one* of those tests is larger than .05. In other words, the combined probability of a Type I error for the set of 10 hypotheses would be larger than the value .05 set for each individual test. Indeed, it can be proved that the combined probability could be as large as .40.

What we need is a single test of the hypothesis "all five population means are equal" that will be less tedious than the individual t tests and can be performed with a specified probability of a Type I error (say, .05). This test is the analysis of variance.

The analysis of variance procedures are developed under the following conditions:

1. Each of the five populations has a normal distribution.
2. The variances of the five populations are equal; that is, $\sigma_1^2 = \sigma_2^2 = \sigma_3^2 = \sigma_4^2 = \sigma_5^2 = \sigma^2$.
3. The five sets of measurements are independent random samples from their respective populations.

From condition 2, we now consider the quantity

s_W^2

$$s_W^2 = \frac{(n_1 - 1)s_1^2 + (n_2 - 1)s_2^2 + (n_3 - 1)s_3^2 + (n_4 - 1)s_4^2 + (n_5 - 1)s_5^2}{(n_1 - 1) + (n_2 - 1) + (n_3 - 1) + (n_4 - 1) + (n_5 - 1)}$$

$$= \frac{(n_1 - 1)s_1^2 + (n_2 - 1)s_2^2 + (n_3 - 1)s_3^2 + (n_4 - 1)s_4^2 + (n_5 - 1)s_5^2}{n_1 + n_2 + n_3 + n_4 + n_5 - 5}$$

Note that this quantity is merely an extension of

$$s_p^2 = \frac{(n_1 - 1)s_1^2 + (n_2 - 1)s_2^2}{n_1 + n_2 - 2}$$

which is used as an estimate of the common variance for two populations for a test of the hypothesis $\mu_1 = \mu_2$ (Section 6.2). Thus, s_W^2 represents a combined estimate of the common variance σ^2, and it measures the variability of the observations within the five populations. (The subscript W refers to the within-sample variability.)

Next we consider a quantity that measures the variability between or among the population means. If the null hypothesis $\mu_1 = \mu_2 = \mu_3 = \mu_4 = \mu_5$ is true, then the populations are identical, with mean μ and variance σ^2. Drawing single samples from the five populations is then equivalent to drawing five different samples from the same population. What kind of variation might we expect for these sample means? If the variation is too great, we would reject the hypothesis that $\mu_1 = \mu_2 = \mu_3 = \mu_4 = \mu_5$.

To evaluate the variation in the five sample means, we need to know the sampling distribution of the sample mean computed from a random sample of 25 observations from a normal population. From our discussion in Chapter 4, we recall that the sampling distribution for $\bar{y}$ based on $n = 25$ measurements will have the same mean as the population μ but the variance of $\bar{y}$ will be $\sigma^2/25$. We have five random samples of 25 observations each, so we can estimate the variance of the distribution of sample means, $\sigma^2/25$, using the formula

$$\text{sample variance of five sample means} = \frac{\sum_{i=1}^{5}(\bar{y}_i - \bar{y}_{.})^2}{5 - 1}$$

where $\bar{y}_{.} = \sum_{i=1}^{5}\bar{y}_i/5$ is the average of the five $\bar{y}_i$s.

Note that we merely consider the $\bar{y}$s to be a sample of five observations and calculate the "sample variance." This quantity estimates $\sigma^2/25$, and hence $25 \times$ (sample variance of the means) estimates σ^2. We designate this quantity as s_B^2; the subscript B denotes a measure of the variability among the sample means for the five populations. For this problem $s_B^2 = $ (25 times the sample variance of the means).

s_B^2

Under the null hypothesis that all five population means are identical, we have two estimates of σ^2—namely, s_W^2 and s_B^2. Suppose the ratio

$$\frac{s_B^2}{s_W^2}$$

is used as the test statistic to test the hypothesis that $\mu_1 = \mu_2 = \mu_3 = \mu_4 = \mu_5$. What is the distribution of this quantity if we repeat the experiment over and over again, each time calculating s_B^2 and s_W^2?

For our example, s_B^2/s_W^2 follows an F distribution, with degrees of freedom that can be shown to be $\text{df}_1 = 4$ for s_B^2 and $\text{df}_2 = 120$ for s_W^2. The proof of these remarks is beyond the scope of this text. However, we will make use of this result for testing the null hypothesis $\mu_1 = \mu_2 = \mu_3 = \mu_4 = \mu_5$.

The test statistic used to test equality of the population means is

test statistic $$F = \frac{s_B^2}{s_W^2}$$

When the null hypothesis is true, both s_B^2 and s_W^2 estimate σ^2, and we expect F to assume a value near $F = 1$. When the hypothesis of equality is false, s_B^2 will tend to be larger than s_W^2 due to the differences among the population means. Hence, we will reject the null hypothesis in the upper tail of the distribution of $F = s_B^2/s_W^2$; for $\alpha = .05$, the critical value of $F = s_B^2/s_W^2$ is 2.45. (See Figure 8.4.) If the calculated value of F falls in the rejection region, we conclude that not all five population means are identical.

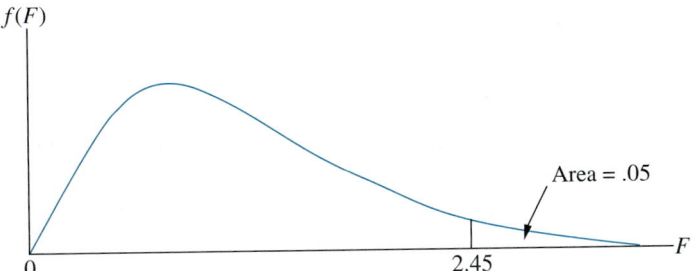

This procedure can be generalized (and simplified) with only slight modifications in the formulas to test the equality of t (where t is an integer equal to or greater than 2) population means from normal populations with a common variance σ^2. Random samples of sizes $n_1, n_2, \ldots, n_t$ are drawn from the respective populations. We then compute the sample means and variances. The null hypothesis $\mu_1 = \mu_2 = \cdots = \mu_t$ is tested against the alternative that at least one of the population means is different from the others.

Before presenting the generalized test procedure, we introduce the notation to be used in the formulas for s_B^2 and s_W^2.

The experimental setting in which a random sample of observations is taken from each of t different populations is called a **completely randomized design.** Consider a completely randomized design in which four observations are obtained from each of the five populations. If we let y_{ij} denote the jth observation from population i, we could display the sample data for this completely randomized design as shown in Table 8.5. Using Table 8.5, we can introduce notation that is helpful when performing an **analysis of variance (AOV)** for a completely randomized design.

completely randomized design

analysis of variance

Notation Needed for the AOV of a Completely Randomized Design

y_{ij}: The jth sample observation selected from population i. For example, y_{23} denotes the third sample observation drawn from population 2.

n_i: The number of sample observations selected from population i. In our data set, n_1, the number of observations obtained from population 1, is 4. Similarly, $n_2 = n_3 = n_4 = n_5 = 4$. However, it should be noted that the sample sizes need not be the same. Thus, we might have $n_1 = 12$, $n_2 = 3$, $n_3 = 6$, $n_4 = 10$, and so forth.

n_T: The total sample size; $n_T = \Sigma \, n_i$. For the data given in Table 8.5, $n_T = n_1 + n_2 + n_3 + n_4 + n_5 = 20$.

$\bar{y}_{i.}$: The average of the n_i sample observations drawn from population i, $\bar{y}_{i.} = \Sigma_j \, y_{ij}/n_i$.

$\bar{y}_{..}$: The average of all sample observations; $\Sigma_i \Sigma_j \, y_{ij}/n_T$.

TABLE 8.5

Summary of sample data for a completely randomized design

Population	Data				Mean
1	y_{11}	y_{12}	y_{13}	y_{14}	$\bar{y}_{1.}$
2	y_{21}	y_{22}	y_{23}	y_{24}	$\bar{y}_{2.}$
3	y_{31}	y_{32}	y_{33}	y_{34}	$\bar{y}_{3.}$
4	y_{41}	y_{42}	y_{43}	y_{44}	$\bar{y}_{4.}$
5	y_{51}	y_{52}	y_{53}	y_{54}	$\bar{y}_{5.}$

With this notation, it is possible to establish the following algebraic identities. (Although we will use these results in later calculations for s_W^2 and s_B^2, the proofs of these identities are beyond the scope of this text.) Let s_T^2 be the sample variance of the n_T measurements y_{ij}. We can measure the variability of the n_T sample measurements y_{ij} about the overall mean $\bar{y}_{..}$ using the quantity

$$\text{TSS} = \sum_{i=1}^{t} \sum_{j=1}^{n_i} (y_{ij} - \bar{y}_{..})^2 = (n_T - 1)s_T^2$$

total sum of squares

This quantity is called the **total sum of squares** (TSS) of the measurements about the overall mean. The double summation in TSS means that we must sum the squared deviations for all rows (i) and columns (j) of the one-way classification.

It is possible to partition the total sum of squares as follows:

$$\sum_{i,j} (y_{ij} - \bar{y}_{..})^2 = \sum_{i,j} (y_{ij} - \bar{y}_{i.})^2 + \sum_{i} n_i(\bar{y}_{i.} - \bar{y}_{..})^2$$

The first quantity on the right side of the equation measures the variability of an observation y_{ij} about its sample mean $\bar{y}_{i.}$. Thus,

$$\text{SSW} = \sum_{i,j} (y_{ij} - \bar{y}_{i.})^2 = (n_1 - 1)s_1^2 + (n_2 - 1)s_2^2 + \cdots + (n_t - 1)s_t^2$$

within-sample sum of squares

is a measure of the *within-sample* variability. SSW is referred to as the **within-sample sum of squares** and is used to compute s_W^2.

The second expression in the total sum of squares equation measures the variability of the sample means $\bar{y}_{i.}$ about the overall mean $\bar{y}_{..}$. This quantity, which measures the variability *between* (or among) the sample means, is referred to as the **sum of squares between samples** (SSB) and is used to compute s_B^2.

between-sample sum of squares

$$\text{SSB} = \sum_{i} n_i(\bar{y}_{i.} - \bar{y}_{..})^2$$

Although the formulas for TSS, SSW, and SSB are easily interpreted, they are not easy to use for calculations. Instead, we recommend using a computer software program.

An analysis of variance for a completely randomized design with t populations has the following null and alternative hypotheses:

H_0: $\mu_1 = \mu_2 = \mu_3 = \cdots = \mu_t$ (i.e., the t population means are equal)

H_a: At least one of the t population means differs from the rest.

The quantities s_B^2 and s_W^2 can be computed using the shortcut formulas

$$s_B^2 = \frac{\text{SSB}}{t-1} \qquad s_W^2 = \frac{\text{SSW}}{n_T - t}$$

where $t - 1$ and $n_T - t$ are the degrees of freedom for s_B^2 and s_W^2, respectively.

mean square

Historically, people have referred to a sum of squares divided by its degrees of freedom as a **mean square.** Hence, s_B^2 is often called the *mean square between samples* and s_W^2, the *mean square within samples*. The quantities are the mean squares because they both are averages of squared deviations. There are only $n_T - t$ linearly independent deviations ($y_{ij} - \bar{y}_{i.}$) in SSW because $\sum_j (y_{ij} - \bar{y}_{i.}) = 0$ for each of the t samples. Hence, we divide SSW by $n_T - t$ and not n_T. Similarly, there are only $t - 1$ linearly independent deviations ($\bar{y}_{i.} - \bar{y}_{..}$) in SSB, because $\sum_i n_i(\bar{y}_{i.} - \bar{y}_{..}) = 0$. Hence, we divide SSB by $t - 1$.

The null hypothesis of equality of the t population means is rejected if

$$F = \frac{s_B^2}{s_W^2}$$

exceeds the tabulated value of F for $a = \alpha$, $df_1 = t - 1$, and $df_2 = n_T - t$.

AOV table

After we complete the F test, we then summarize the results of a study in an *analysis of variance table*. The format of an **AOV table** is shown in Table 8.6. The AOV table lists the sources of variability in the first column. The second column lists the sums of squares associated with each source of variability. We showed that the total sum of squares (TSS) can be partitioned into two parts, so SSB and SSW must add up to TSS in the AOV table. The third column of the table gives the degrees of freedom associated with the sources of variability. Again, we have a check; $(t - 1) + (n_T - t)$ must add up to $n_T - 1$. The mean squares are found in the fourth column of Table 8.6, and the F test for the equality of the t population means is given in the fifth column.

TABLE 8.6

An example of an AOV table for a completely randomized design

Source	Sum of Squares	Degrees of Freedom	Mean Square	F Test
Between samples	SSB	$t - 1$	$s_B^2 = \text{SSB}/(t - 1)$	s_B^2/s_W^2
Within samples	SSW	$n_T - t$	$s_W^2 = \text{SSW}/(n_T - t)$	
Totals	TSS	$n_T - 1$		

EXAMPLE 8.1

A horticulturist was investigating the phosphorus content of tree leaves from three different varieties of apple trees (1, 2, and 3). Random samples of five leaves from each of the three varieties were analyzed for phosphorus content. The data are given in Table 8.7. Use these data to test the hypothesis of equality of the mean phosphorus levels for the three varieties. Use $\alpha = .05$.

TABLE 8.7

Phosphorus content of leaves from three different trees

Variety	Phosphorus Content					Sample Sizes	Means	Variances
1	.35	.40	.58	.50	.47	5	0.460	.00795
2	.65	.70	.90	.84	.79	5	0.776	.01033
3	.60	.80	.75	.73	.66	5	0.708	.00617
Total						15	0.648	

Solution The null and alternative hypotheses for this example are

H_0: $\mu_1 = \mu_2 = \mu_3$

H_a: At least one of the population means differs from the rest.

The sample sizes are $n_1 = n_2 = n_3 = 5$, which yields $n_T = 15$. Using the sample means and sample variances, the sum of squares within and between are

$$\text{SSB} = \sum_{i=1}^{3} n_i(\bar{y}_{i.} - \bar{y}_{..})^2 = 5(.46 - .648)^2 + 5(.776 - .648)^2 + 5(.708 - .648)^2$$

$$= .277$$

and

$$SSW = \sum_{i=1}^{3} (n_i - 1)s_i^2 = (5 - 1)(.00795) + (5 - 1)(.01033) + (5 - 1)(.00617)$$

$$= .0978$$

Finally, TSS = SSB + SSW = .277 + .0978 = .3748.

The AOV table for these data is shown in Table 8.8. The critical value of $F = s_B^2/s_W^2$ is 3.89, which is obtained from Table 8 in the Appendix for $a = .05$, $df_1 = 2$, and $df_2 = 12$. Because the computed value of F, 17.25, exceeds 3.89, we reject the null hypothesis of equality of the mean phosphorus content for the three varieties. It appears from the data that the mean for variety 1 is smaller than the means for varieties 2 and 3. We will develop techniques to confirm this observation in Chapter 9.

TABLE 8.8

AOV table for the data for Example 8.1

Source	Sum of Squares	Degrees of Freedom	Mean Square	F Test
Between samples	.277	2	.277/2 = .138	.138/.008 = 17.25
Within samples	.0978	12	.0978/12 = .008	
Totals	.3748	14		

EXAMPLE 8.2

A clinical psychologist wished to compare three methods for reducing hostility levels in university students, and used a certain test (HLT) to measure the degree of hostility. A high score on the test indicated great hostility. The psychologist used 24 students who obtained high and nearly equal scores in the experiment. Eight were selected at random from among the 24 problem cases and were treated with method 1. Seven of the remaining 16 students were selected at random and treated with method 2. The remaining nine students were treated with method 3. All treatments were continued for a one-semester period. Each student was given the HLT test at the end of the semester, with the results shown in Table 8.9. Use these data to perform an analysis of variance to determine whether there are differences among mean scores for the three methods. Use $\alpha = .05$.

TABLE 8.9

HLT test scores

Method	Test Scores								Mean	Standard Deviation	Sample Size
1	96	79	91	85	83	91	82	87	86.750	5.625	8
2	77	76	74	73	78	71	80		75.571	3.101	7
3	66	73	69	66	77	73	71	70	74	71.000	3.674	9

Solution The null and alternative hypothesis are

$$H_0: \quad \mu_1 = \mu_2 = \mu_3$$

$$H_a: \quad \text{At least one of the population means differs from the rest.}$$

For $n_1 = 8$, $n_2 = 7$, and $n_3 = 9$, we have a total sample size of $n_T = 24$. Using the sample means given in the table, we compute the overall mean of the 24 data values:

$$\bar{y}_{..} = \sum_{i=1}^{3} n_i\bar{y}_{i.}/n_T = (8(86.750) + 7(75.571) + 9(71.000))/24 = 1{,}861.997/24$$

$$= 77.5832$$

Using this value along with the means and standard deviations in Table 8.9, we can compute the three sums of squares as follows:

$$SSB = \sum_{i=1}^{3} n_i(\bar{y}_{i.} - \bar{y}_{..})^2 = 8(86.750 - 77.5832)^2 + 7(75.571 - 77.5832)^2$$

$$+ 9(71 - 77.5832)^2 = 1{,}090.6311$$

and

$$SSW = \sum_{i=1}^{3} (n_i - 1)s_i^2 = (8 - 1)(5.625)^2 + (7 - 1)(3.101)^2 + (9 - 1)(3.674)^2$$

$$= 387.1678$$

Finally, TSS = SSB + SSW = 1,090.6311 + 387.1678 = 1,477.80. The AOV table for these data is given in Table 8.10.

TABLE 8.10
AOV table for data
of Example 8.2

Source	SS	df	MS	F	p-value
Between samples	1,090.6311	2	545.316	545.316/18.4366 = 29.58	<.001
Within samples	387.1678	21	18.4366		
Totals	1,477.80	23			

The critical value of F is obtained from Table 8 in the Appendix for $\alpha = .05$, $df_1 = 2$, and $df_2 = 21$; this value is 3.47. Because the computed value of F is 29.57, which exceeds the critical value 3.47, we reject the null hypothesis of equality of the mean scores for the three methods of treatment. We can only place an upper bound on the p-value because the largest value in Table 8 for $df_1 = 2$ and $df_2 = 21$ is 9.77, which corresponds to a probability of .001. Thus, there is a very strong rejection of the null hypothesis. From the three sample means, we observe that the mean for method 1 is considerably larger than the means for methods 2 and 3. The researcher would need to determine whether all three population means differ or the means for methods 2 and 3 are equal. Also, we may want to place confidence intervals on the three method means and on their differences; this would provide the researcher with information concerning the degree of differences in the three methods. In the next chapter, we will develop techniques to construct these types of inferences. Computer output shown here is consistent with the results we obtained. In the computer printout, note that the names for the sum of squares are not given as between and within. The between sum of squares is labeled by the population name, in this example, Method. The within sum of squares is often labeled as the error sum of squares.

```
General Linear Models Procedure

Class Level Information

Class     Levels   Values
METHOD        3    1 2 3

Number of observations in data set = 24

Dependent Variable: SCORE

Source            DF   Sum of Squares   F Value   Pr > F
Model              2   1090.61904762     29.57    0.0001
Error             21    387.21428571
Corrected Total   23   1477.833333333
```

EXERCISES   **Applications**

Ag.   **8.1** A large laboratory has four types devices used to determine the pH of soil samples. The laboratory wants to determine whether there are differences in the average readings given by these devices. The lab uses 24 soil samples having known pH in the study, and randomly assigns six of the samples to each device. The soil samples are tested and the response recorded is the difference between the pH reading of the device and the known pH of the soil. These values, along with summary statistics, are given in the following table.

Device	Sample						Sample Size	Mean	Standard Deviation
	1	2	3	4	5	6			
A	−.307	−.294	.079	.019	−.136	−.324	6	−.1605	.1767
B	−.176	.125	−.013	.082	.091	.459	6	.0947	.2091
C	.137	−.063	.240	−.050	.318	.154	6	.1227	.1532
D	−.042	.690	.201	.166	.219	.407	6	.2735	.2492

a. Based on your intuition, is there evidence to indicate any difference among the mean differences in pH readings for the four devices?

b. Run an analysis of variance to confirm or reject your conclusion of part (a). Use $\alpha = .05$.

c. Compute the p-value of the F test in part (b).

d. What conditions must be satisfied for your analysis in parts (b) and (c) to be valid?

e. Suppose the 24 soil samples have widely different pH values. What problems may occur by simply randomly assigning the soil samples to the different devices?

Bus.   **8.2** A cigarette manufacturer has advertised that it has developed a new brand of cigarette, LowTar, that has a lower average tar content than the major brands. To evaluate this claim, a consumer testing agency randomly selected 100 cigarettes from each of the four leading brands of cigarettes and 100 from the new brand. The tar content (milligrams) of the cigarettes gave the following results:

Brand	$\bar{y}_i$	s_i	n_i
LowTar	9.64	.291	100
A	10.22	.478	100
B	10.77	.372	100
C	11.57	.352	100
D	13.59	.469	100

A boxplot of the data used to produce the table is given here.

Boxplots of tar content by brand for Exercise 8.2 (means are indicated by solid circles)

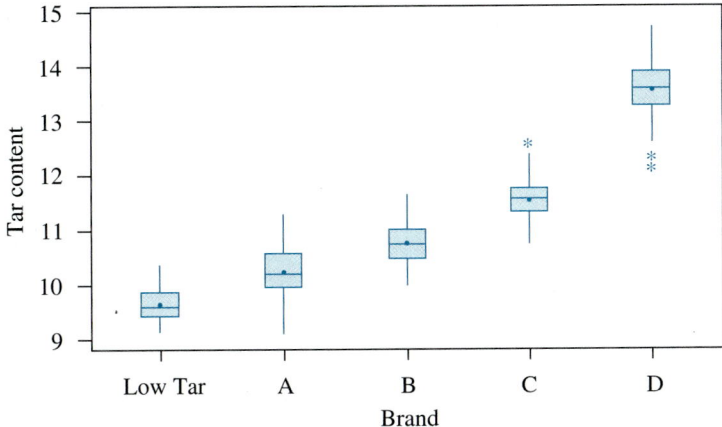

a. Based on the information contained in the boxplot, does the LowTar brand appear to have a lower average tar content than the other brands?

b. Using the computer output shown here, is there a significant ($\alpha = .01$) difference in the average tar content of the five brands of cigarettes?

c. What is the p-value of the test statistic in (b)?

d. What are the practical consequences of making a Type I error with respect to your test in (b)?

```
One-Way Analysis of Variance for Exercise 8.2

Analysis of Variance for Tar Cont
Source      DF        SS         MS         F       P
Brand        4    941.193    235.298   1478.39   0.000
Error      495     78.784     0.159
Total      499   1019.976
                                       Individual 95% CIs for Mean
                                       Based on Pooled StDev
Level     N       Mean      StDev   -+---------+---------+---------+-----
1       100      9.644      0.291    *)
2       100     10.221      0.478         *)
3       100     10.775      0.372               (*
4       100     11.570      0.352                      *)
5       100     13.592      0.469                                    *)

                                     -+---------+---------+---------+-----
Pooled StDev =    0.399               9.6      10.8      12.0      13.2
```

8.3 The Model for Observations in a Completely Randomized Design

In this section, we will consider a model for the completely randomized design (sometimes referred to as a one-way classification). This model will demonstrate the types of settings for which AOV testing procedures are appropriate. We can think of a model as a mathematical description of a physical setting. A model also enables us to computer-simulate the data that the physical process generates.

We will impose the following conditions concerning the sample measurements and the populations from which they are drawn:

1. The samples are independent random samples. Results from one sample in no way affect the measurements observed in another sample.
2. Each sample is selected from a normal population.
3. The mean and variance for population i are, respectively, μ_i and $\sigma^2 (i = 1, 2, \ldots, t)$.

Figure 8.5 depicts a setting in which these three conditions are satisfied. The population distributions are normal with the same standard deviation. Note that populations III and IV have the same mean, which differs from the means of populations I and II. To summarize, we assume that the t populations are independently normally distributed with different means but a common variance σ^2.

FIGURE 8.5

Distributions of four populations that satisfy AOV assumptions

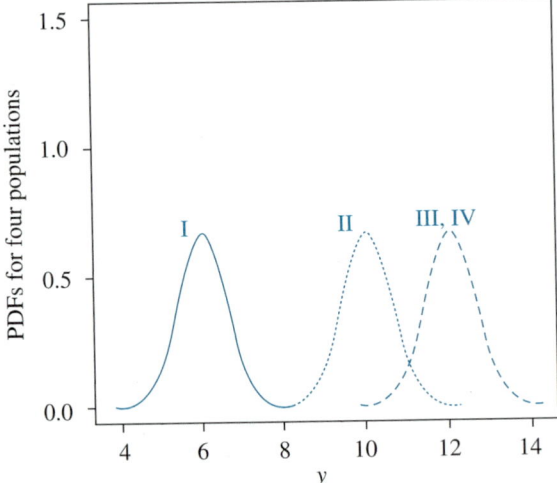

We can now formulate a model (equation) that encompasses these three assumptions. Recall that we previously let y_{ij} denote the jth sample observation from population i.

model
$$y_{ij} = \mu + \alpha_i + \varepsilon_{ij}$$

terms
This model states that y_{ij}, the jth sample measurement selected from population i, is the sum of three **terms.** The term μ denotes an overall mean that is an unknown constant. The term α_i denotes an effect due to population i; α_i is an unknown constant. The term μ denotes the overall mean across all t populations— that is, the mean of the population consisting of the observations from all t popula-

tions. The term α_i denotes the effect of population i on the overall variation in the observations. The terms μ and α_i are unknown constants, which will be estimated from the data obtained during the study or experiment. The term ε_{ij} represents the random deviation of y_{ij} about the ith population mean, μ_i. The ε_{ij}s are often referred to as *error terms*. The expression *error* is not to be interpreted as a mistake made in the experiment. Instead, the ε_{ij}s model the random variation of the y_{ij}s about their mean μ_i. The term *error* simply refers to the fact that the observations from the t populations differ by more than just their means. We assume that ε_{ij}s are independently normally distributed with a mean of 0 and a standard deviation of σ_ε. The independence condition can be interpreted as follows: The ε_{ij}s are independent if the size of the deviation of the y_{ij} observation from μ_i in no way affects the size of the deviation associated with any other observation.

Since y_{ij} is an observation from the ith population, it has mean μ_i. However, since the ε_{ij}s are distributed with mean 0, the mean or expected value of y_{ij}, denoted by $E(y_{ij})$, is

$$\mu_i = E(y_{ij}) = E(\mu + \alpha_i + \varepsilon_{ij}) = \mu + \alpha_i + E(\varepsilon_{ij}) = \mu + \alpha_i$$

that is, y_{ij} is a randomly selected observation from a population having mean $\mu_i = \mu + \alpha_i$. The effect α_i thus represents the deviation of the ith population mean μ_i from the overall mean μ. Thus, the α_is may assume a positive, zero, or negative value. Hence, the mean for population i can be greater than, equal to, or less than μ, the overall mean. The variance for each of the t populations can be shown to be σ_ε^2. Finally, because the εs are normally distributed, each of the t populations is normal. A summary of the assumptions for a one-way classification is shown in Table 8.11.

TABLE 8.11

Summary of some of the assumptions for a completely randomized design

Population	Population Mean	Population Variance	Sample Measurements
1	$\mu + \alpha_1$	σ_ε^2	$y_{11}, y_{12}, \ldots, y_{1n_1}$
2	$\mu + \alpha_2$	σ_ε^2	$y_{21}, y_{22}, \ldots, y_{2n_2}$
$\vdots$	$\vdots$	$\vdots$	$\vdots$
t	$\mu + \alpha_t$	σ_ε^2	$y_{t1}, y_{t2}, \ldots, y_{tn_t}$

The null hypothesis for a one-way analysis of variance is that $\mu_1 = \mu_2 = \cdots = \mu_t$. Using our model, this would be equivalent to the null hypothesis

$$H_0: \quad \alpha_1 = \alpha_2 = \cdots = \alpha_t = 0$$

If H_0 is true, then all populations have the same unknown mean μ. Indeed, many textbooks use this latter null hypothesis for the analysis of variance in a completely randomized design. The corresponding alternative hypothesis is

$$H_a: \quad \text{At least one of the } \alpha_i\text{s differs from 0.}$$

In this section, we have presented a brief description of the model associated with the analysis of variance for a completely randomized design. Although some authors bypass an examination of the model, we believe it is a necessary part of an analysis of variance discussion.

We have imposed several conditions on the populations from which the data are selected or, equivalently, on the experiments in which the data are generated, so we need to verify that these conditions are satisfied prior to making inferences from the AOV table. In Chapter 7, we discussed how to test the "equality of variances" condition using Hartley's F_{max} test or Levine's test. The normality condition is not as critical as the equal variance assumption when we have large sample sizes unless the populations are severely skewed or have very heavy tails. When we have small sample sizes, the normality condition and the equal variance condition become more critical. This situation presents a problem, because there generally will not be enough observations from the individual population to test validly whether the normality or equal variance condition is satisfied. In the next section, we will discuss a technique that can at least partially overcome this problem. Also, some alternatives to the AOV will be presented in later sections of this chapter that can be used when the populations have unequal variances or have nonnormal distributions. As we discussed in Chapter 6, the most critical of the three conditions is that the data values are independent. This condition can be met by carefully conducting the studies or experiments so as to not obtain data values that are dependent. In studies involving randomly selecting data from the t populations, we need to take care that the samples are truly random and that the samples from one population are not dependent on the values obtained from another population. In experiments in which t treatments are randomly assigned to experimental units, we need to make sure that the treatments are truly **randomly assigned.** Also, the experiments must be conducted so the experimental units do not interact with each other in a manner that could affect their responses.

8.4 Checking on the AOV Conditions

The assumption of equal population variances and the assumption of normality of the populations have been made in several places in the text, such as for the t test when comparing two population means and now for the analysis of variance F test in a completely randomized design.

Let us consider first an experiment in which we wish to compare t population means based on independent random samples from each of the populations. Recall that we assume we are dealing with normal populations with a common variance σ_ε^2 and possibly different means. We could verify the assumption of equality of the population variances using Hartley's test or Levine's test of Chapter 7.

Several comments should be made here. Most practitioners do not routinely run Hartley's test. One reason is that the test is extremely sensitive to departures from normality. Thus, in checking one assumption (constant variance), the practitioner would have to be very careful about departures from another analysis of variance assumption (normality of the populations). Fortunately, as we mentioned in Chapter 6, the assumption of homogeneity (equality) of population variances is less critical when the sample sizes are nearly equal, where the variances can be markedly different and the p-values for an analysis of variance will still be only mildly distorted. Thus, we recommend that Hartley's test or Levine's test be used only for the more extreme cases. In these extreme situations where homogeneity of the population variances is a problem, a transformation of the data may help to stabilize the variances. Then inferences can be made from an analysis of variance.

The normality of the population distributions can be checked using normal probability plots or boxplots, as we discussed in Chapters 5 and 6, when the sample sizes are relatively large. However, in many experiments, the sample sizes may be as small as 5 to 10 observations from each population. In this case, the plots will not be a very reliable indication of whether the population distributions are normal. By taking into consideration the model we introduced in the previous section, the evaluation of the normal condition will be evaluated using a **residuals analysis.**

residuals analysis

From the model, we have $y_{ij} = \mu + \alpha_i + \varepsilon_{ij} = \mu_i + \varepsilon_{ij}$. Thus, we can write $\varepsilon_{ij} = y_{ij} - \mu_i$. Then if the condition of equal variances is valid, the ε_{ij}s are a random sample from a normal population. However, μ_i is an unknown constant, but if we estimate μ_i with $\bar{y}_{i\cdot}$, and let

$$e_{ij} = y_{ij} - \bar{y}_{i\cdot}.$$

then we can use the e_{ij}s to evaluate the normality assumption. Even when the individual n_is are small, we would have n_T residuals, which would provide a sufficient number of values to evaluate the normality condition. We can plot the e_{ij}s in a boxplot or a normality plot to evaluate whether the data appear to have been generated from normal populations.

EXAMPLE 8.3

Because many HMOs either do not cover mental health costs or provide only minimal coverage, ministers and priests often need to provide counseling to persons suffering from mental illness. An interdenominational organization wanted to determine whether the clerics from different religions have different levels of awareness with respect to the causes of mental illness. Three random samples were drawn, one containing ten Methodist ministers, a second containing ten Catholic priests, and a third containing ten Pentecostal ministers. Each of the 30 clerics was then examined, using a standard written test, to measure his or her knowledge about causes of mental illness. The test scores are listed in Table 8.12. Does there appear to be a significant difference in the mean test scores for the three religions?

TABLE 8.12

Scores for clerics' knowledge of mental illness

Cleric	Methodist	Catholic	Pentecostal
1	62	62	37
2	60	62	31
3	60	24	15
4	25	24	15
5	24	22	14
6	23	20	14
7	20	19	14
8	13	10	5
9	12	8	3
10	6	8	2
$\bar{y}_{i\cdot}$	30.50	25.90	15.00
s_i	21.66	20.01	11.33
n_i	10	10	10
Median($\tilde{y}_i$)	23.5	21	14

Solution Prior to conducting an AOV test of the three means, we need to evaluate whether the conditions required for AOV are satisfied. Figure 8.6 is a boxplot of the mental illness scores by religion. There is an indication that the data may be somewhat skewed to the right. Thus, we will evaluate the normal-

FIGURE 8.6

Boxplots of score by religion (means are indicated by solid circles)

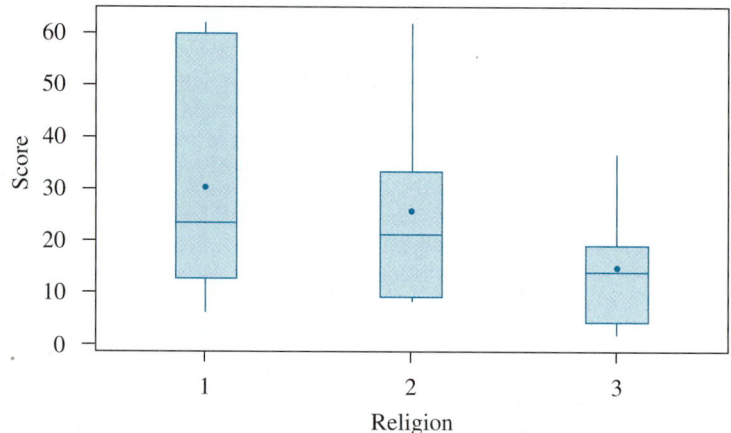

FIGURE 8.6

Boxplots of score by religion (means are indicated by solid circles)

ity condition. We need to obtain the residuals $e_{ij} = y_{ij} - \bar{y}_{i.}$. For example, $e_{11} = y_{11} - \bar{y}_{1.} = 62 - 30.50 = 31.50$. The remaining e_{ij}s are given in Table 8.13.

TABLE 8.13

Residuals e_{ij} for clerics' knowledge of mental illness

Cleric	Methodist	Catholic	Pentecostal
1	31.5	36.1	22.0
2	29.5	36.1	16.0
3	29.5	−1.9	0.0
4	−5.5	−1.9	0.0
5	−6.5	−3.9	−1.0
6	−7.5	−5.9	−1.0
7	−10.5	−6.9	−1.0
8	−17.5	−15.9	−10.0
9	−18.5	−17.9	−12.0
10	−24.5	−17.9	−13.0

The residuals are then plotted in Figures 8.7 and 8.8. The boxplot in Figure 8.8 displays three outliers out of 30 residuals. It is very unlikely that 10% of the

FIGURE 8.7

Normal probability plot for residuals

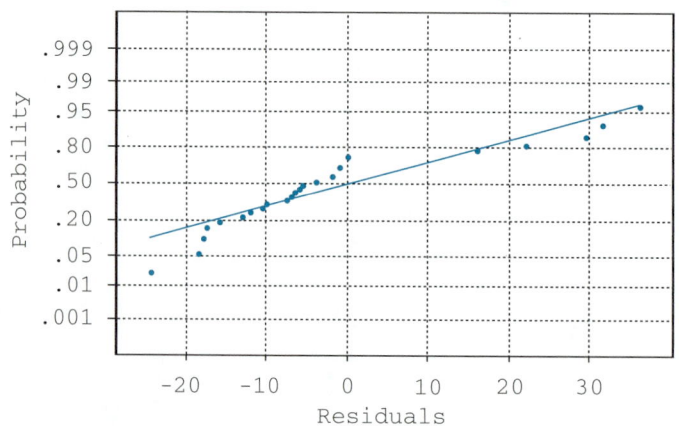

Average: −0.0000000
StDev: 17.5984
N: 30

Anderson-Darling Normality Test
A-Squared: 1.714
P-Value: 0.000

FIGURE 8.8

Boxplot of residuals

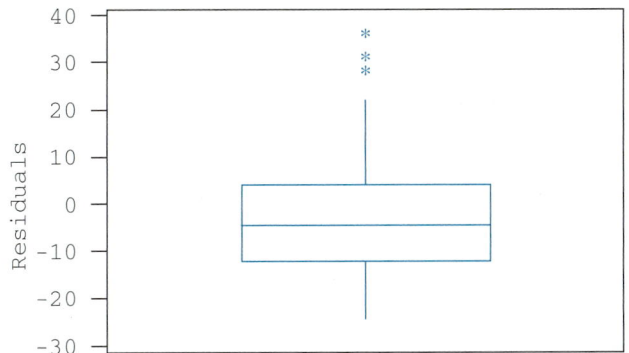

data values are outliers if the residuals are in fact a random sample from a normal distribution. This is confirmed in the normal probability plot displayed in Figure 8.7, which shows a lack of concentration of the residuals about the straight line. Furthermore, the test of normality has a p-value less than .001, which indicates a strong departure from normality. Thus, we conclude that the data have nonnormal characteristics. In Section 8.6, we will provide an alternative to the F test from the AOV table, which would be appropriate for this situation.

Because the data may be nonnormal, it would be inappropriate to test for equal variances using Hartley's F_{max} test. Thus, we will use Levine's test. An examination of the formula for Levine's test reveals that once we make the conversion of the data from y_{ij} to $z_{ij} = |y_{ij} - \tilde{y}_i|$, where $\tilde{y}_i$ is the sample median of the ith data set, Levine's test is equivalent to the F test from AOV applied to the z_{ij}s. Thus, we can simply use the formulas from AOV to compute Levine's test. The z_{ij}s are given in Table 8.14 using the medians from Table 8.12.

TABLE 8.14

Transformed data set,
$z_{ij} = |y_{ij} - \tilde{y}_i|$

Cleric	Methodist	Catholic	Pentecostal
1	38.5	41	23
2	36.5	41	17
3	36.5	3	1
4	1.5	3	1
5	0.5	1	0
6	0.5	1	0
7	3.5	2	0
8	10.5	11	9
9	11.5	13	11
10	17.5	13	12
$\bar{z}_{i.}$	15.70	12.90	7.40
s_i	15.80	15.57	8.29

Using the sample means given in the table, we compute the overall mean of the 30 data values:

$$\bar{z}_{..} = \sum_{i=1}^{3} n_i \bar{z}_{i.}/n_T = [10(15.70) + 10(12.90) + 10(7.40)]/30 = 360/30 = 12$$

Using this value along with the means and standard deviations in Table 8.14, we can compute the sum of squares as follows:

$$\text{SSB} = \sum_{i=1}^{3} n_i (\bar{z}_{i.} - \bar{z}_{..})^2 = 10(15.70 - 12)^2 + 10(12.90 - 12)^2 + 10(7.40 - 12)^2$$

$$= 356.6$$

and

$$\text{SSW} = \sum_{i=1}^{3} (n_i - 1)s_i^2 = (10 - 1)(15.80)^2 + (10 - 1)(15.57)^2$$

$$+ (10 - 1)(8.29)^2 = 5,047.10$$

The mean squares are MSB = SSB/$(t - 1)$ = 356.6/(3 − 1) = 178.3 and MSW = SSW/$(n_T - t)$ = 5,047.10/(30 − 3) = 186.9. Finally, we can next obtain the value of the Levine's test statistic from L = MSB/MSW = 178.3/186.9 = .95. The critical value of L, using α = .05, is obtained from the F tables with df_1 = 2 and df_2 = 27. This value is 3.35, and thus we fail to reject the null hypothesis that the standard deviations are equal. The p-value is greater than .25, because the smallest value in the F table with df_1 = 2 and df_2 = 27 is 1.46, which corresponds to a probability of 0.25. Thus, we have a high degree of confidence that the three populations have the same variance.

In Section 8.6, we will present the Kruskal–Wallis test, which can be used when the populations are nonnormal but have identical distributions under the null hypothesis. This test requires, as a minimum, that the populations have the same variance. Thus, the Kruskal–Wallis test would not be appropriate for the situation in which the populations have very different variances. The next section will provide procedures for testing for differences in population means when the population variances are unequal.

Analyzing the Data for the Case Study   The objective of the study was to evaluate whether the treatment of port-wine stains was more effective for younger children than for older ones. A summary of the data is given here.

Descriptive Statistics for Port-Wine Stain Case Study

Variable	N	Mean	Median	TrMean	StDev	SE Mean
0-5 Years	21	4.999	6.110	4.974	3.916	0.855
6-11 Years	24	7.224	7.182	7.262	3.564	0.727
12-17 Years	21	7.757	7.316	7.270	5.456	1.191
18-31 Years	23	5.682	4.865	5.531	4.147	0.865

Variable	Minimum	Maximum	Q1	Q3
0-5 Years	0.144	10.325	1.143	8.852
6-11 Years	0.188	13.408	5.804	8.933
12-17 Years	0.108	24.716	3.528	10.640
18-31 Years	0.504	14.036	2.320	8.429

We observed in Figure 8.1 that the boxplots were nearly of the same width with no outliers and whiskers of the same length. The means and medians were of a similar size for each of the four age groups. Thus, the assumptions of AOV would appear to be satisfied. To confirm this observation, we computed the residuals and plotted them in a normal probability plot (see Figure 8.9). From this plot we can observe that, with the exception of one data value, the points fall nearly on a straight line. Thus, there is a strong confirmation that the four populations of improvements in skin color have normal distributions.

FIGURE 8.9

Normal probability plot of the residuals for the case study

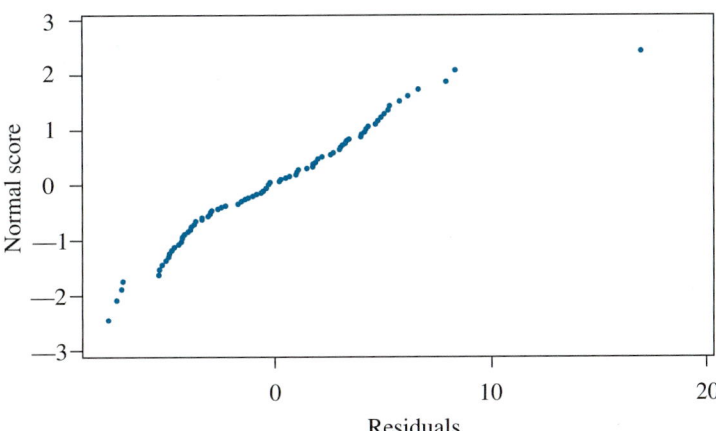

Next, we can check on the equal variance assumption by using Hartley's test or Levine's test. For Hartley's test, we obtain

$$F_{max} = \frac{(5.46)^2}{(3.564)^2} = 2.35$$

The critical value of F_{max} for $\alpha = .05$, $t = 4$, and $df_2 = 20$ is 3.29. This test is only approximate because the sample sizes are unequal. However, the sample sizes are very nearly the same: 21, 21, 23, and 24. Because F_{max} is not greater than 3.29, we conclude that there is not significant evidence that the four population variances differ. Levine's test yields a value of $L = 1.050$ with a p-value of .375 and thus agrees with the findings from Hartley's test. We feel comfortable that the normality and equal variance conditions of the AOV procedure are satisfied. The condition of independence of the data would be checked by discussing with the researchers the manner in which the study was conducted. The sequencing of treatment and the evaluation of the color of the stains should have been performed such that the determination of improvement in color of one patient would not in any way affect the determination of improvement in color of any other patient.

The problems that may arise in this type of experiment, which can cause dependencies in the data, would be due to equipment problems, technician biases, any relationships between patients, and other similar factors.

The research hypothesis is that the mean improvement in stain color after treatment is different for the four age groups:

$H_0: \quad \mu_1 = \mu_2 = \mu_3 = \mu_4$

$H_a: \quad$ At least one of the means differs from the rest.

The computer output for the AOV table is given here:

```
   One-Way Analysis of Variance for Improvement in Stain Color

Source       DF      SS      MS      F      P
Age Group     3   108.0    36.0   1.95  0.128
Error        85  1572.5    18.5
Total        88  1680.5
                                 Individual 95% CIs for Mean
                                 Based on Pooled StDev
Level      N    Mean   StDev   -----+---------+---------+---------+-
0-5       21   4.999   3.916   (--------*--------)
06-11     24   7.224   3.564                 (--------*--------)
12-17     21   7.757   5.456                  (---------*--------)
18-31     23   5.682   4.147        (---------*--------)
                                 -----+---------+---------+---------+-
Pooled StDev = 4.301                 4.0       6.0       8.0      10.0
```

From the output, the p-value for the F test is .128. Thus, there is not a significant difference in the mean improvement for the four groups. We can also compute 95% confidence intervals for the mean improvements. The four intervals are provided in the computer output. They are computed using the pooled standard deviation, $\hat{\sigma} = \sqrt{\text{MSW}} = \sqrt{18.5} = 4.30$ with df $= 85$. Thus, the intervals are of the form

$$\bar{y}_{i.} \pm \frac{t_{.025,85}\hat{\sigma}}{\sqrt{n_i}} = \bar{y}_{i.} \pm \frac{(1.99)(4.30)}{\sqrt{n_i}}$$

The four intervals are presented here:

Age Group	$\bar{y}_{i.}$	95% Confidence Interval
0–5	4.999	(3.13, 6.87)
6–11	7.224	(5.48, 8.97)
12–17	7.757	(5.89, 9.62)
18–31	5.682	(3.90, 7.47)

From the confidence intervals, we can observe the overall effect in the estimation of the mean improvement in stain color for the four groups. The youngest group has the smallest improvement but its upper bound is greater than the lower bound for the age group having the greatest improvement. The problem with this type of decision making is that the confidence intervals are not simultaneous confidence intervals, and hence we cannot attribute a level of certainty to our conclusions. In the next chapter, we will present simultaneous confidence intervals for the difference in treatment means, and hence will be able to decide which pairs of treatments in fact are significantly different. In our case study, however, we can safely conclude that all pairs of treatment means are not significantly different, since the AOV F test failed to reject the null hypothesis.

The researchers did not confirm the hypothesis that treatment of port-wine stains at an early age is more effective than treatment at a later age. The researchers did conclude that their results had implications for the timing of therapy in children.

Although facial port-wine stains can be treated effectively and safely early in life, treatment at a later age leads to similar results. Therefore, the age at which therapy is initiated should be based on a careful weighing of the anticipated benefit and the discomfort of treatment.

Reporting Conclusions We would need to write a report summarizing our findings of this prospective study of the treatment of port-wine stains. The report should include

1. Statement of objective for study
2. Description of study design and data collection procedures
3. Discussion of why the results from 11 of the 100 patients were not included in the data analysis
4. Numerical and graphical summaries of data sets
5. Description of all inference methodologies:
   - AOV table and F test
   - t-based confidence intervals on means
   - Verification that all necessary conditions for using inference techniques were satisfied
6. Discussion of results and conclusions
7. Interpretation of findings relative to previous studies
8. Recommendations for future studies
9. Listing of data sets

8.5 An Alternative Analysis: Transformations of the Data

transformation of data

A **transformation of the sample data** is defined to be a process in which the measurements on the original scale are systematically converted to a new scale of measurement. For example, if the original variable is y and the variances associated with the variable across the treatments are not equal (heterogeneous), it may be necessary to work with a new variable such as $\sqrt{y}$, log y, or some other transformed variable.

How can we select the appropriate transformation? This is no easy task and often takes a great deal of experience in the experimenter's area of application. In spite of these difficulties, we can consider several guidelines for choosing an appropriate transformation.

guidelines for selecting y_T

Many times the variances across the populations of interest are heterogeneous and seem to vary with the magnitude of the population mean. For example, it may be that the larger the population mean, the larger is the population variance. When we are able to identify how the variance varies with the population mean, we can define a suitable transformation from the variable y to a new variable y_T. Three specific situations are presented in Table 8.15.

The first row of Table 8.15 suggests that, if y is a Poisson* random variable, the variance of y is equal to the mean of y. Thus, if the different populations

---

* The Poisson random variable is a useful discrete random variable with applications as an approximation for the binomial (when n is large but $n\pi$ is small) and as a model for events occurring randomly in time.

TABLE 8.15

Transformation to achieve uniform variance

Relationship between μ and σ^2	y_T	Variance of y_T (for a given k)
$\sigma^2 = k\mu$ (when $k = 1$, y is a Poisson variable)	$y_T = \sqrt{y}$ or $\sqrt{y + .375}$	$1/4$; ($k = 1$)
$\sigma^2 = k\mu^2$	$y_T = \log y$ or $\log (y + 1)$	1; ($k = 1$)
$\sigma^2 = k\pi(1 - \pi)$ (when $k = 1/n$, y is a binomial variable)	$y_T = \arcsin \sqrt{y}$	$1/4n$; ($k = 1/n$)

correspond to different Poisson populations, the variances will be heterogeneous provided the means are different. The transformation that will stabilize the variances is $y_T = \sqrt{y}$; or, if the Poisson means are small (under 5), the transformation $y_T = \sqrt{y + .375}$ is better.

EXAMPLE 8.4

Marine biologists are studying a major reduction in the number of shrimp and commercial fish in the Gulf of Mexico. The area in which the Mississippi River enters the gulf is one of the areas of greatest concern. The biologists hypothesize that nutrient-rich water, including mainly nitrogens from the farmlands of the Midwest, flows into the gulf, which results in rapid growth in algae that feeds zooplankton. Bacteria then feed on the zooplankton pellets and dead algae, resulting in a depletion of the oxygen in the water. The more mobile marine life flees these regions while the less mobile marine life dies from hypoxia. To monitor this condition, the mean dissolved oxygen contents (in ppm) of four areas at increasing distance from the mouth of the Mississippi were determined. A random sample of 10 water samples were taken at a depth of 12 meters in each of the four areas. The sample data are given in Table 8.16. The biologists want to test whether the mean oxygen content is lower in those areas closer to the mouth of the Mississippi.

TABLE 8.16

Mean dissolved oxygen contents (in ppm) at four distances from mouth

Sample	1 KM	5 KM	10 KM	20 KM
1	1	4	20	37
2	5	8	26	30
3	2	2	24	26
4	1	3	11	24
5	2	8	28	41
6	2	5	20	25
7	4	6	19	36
8	3	4	19	31
9	0	3	21	31
10	2	3	24	33
Mean	$\bar{y}_{1.} = 2.2$	$\bar{y}_{2.} = 4.6$	$\bar{y}_{3.} = 21.2$	$\bar{y}_{4.} = 31.4$
Standard Deviation	$s_1 = 1.476$	$s_2 = 2.119$	$s_3 = 4.733$	$s_4 = 5.5220$

Distance to Mouth

a. Run a test of the equality of the population variances with $\alpha = .05$.
b. Transform the data if necessary to obtain a new data set in which the observations have equal variances.

Solution

a. Figure 8.10 depicts the data in a set of boxplots. The data do not appear noticeably skewed or heavy tailed. Thus, we will use Hartley's F_{max} test with $\alpha = .05$.

$$F_{max} = \frac{(5.522)^2}{(1.476)^2} = 14.0$$

The critical value of F_{max} for $\alpha = .05$, $t = 4$, and $df_2 = 10 - 1 = 9$ is 6.31. Since F_{max} is greater than 6.31, we reject the hypothesis of homogeneity of the population variances.

FIGURE 8.10

Boxplots of 1–20 KM (means are indicated by solid circles)

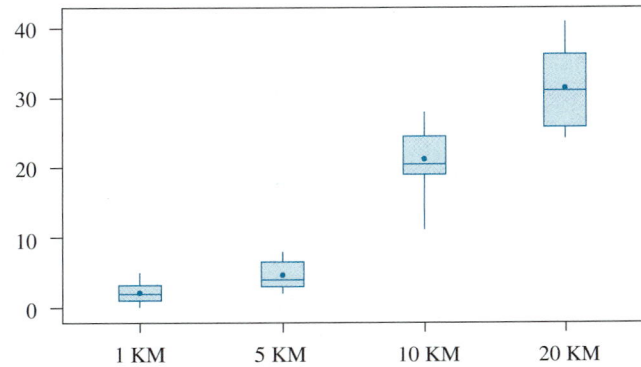

b. We next examine the relationship between the sample means $\bar{y}_{i.}$ and sample variances s_i^2.

$$\frac{s_1^2}{\bar{y}_{1.}} = .99 \qquad \frac{s_2^2}{\bar{y}_{2.}} = .97 \qquad \frac{s_3^2}{\bar{y}_{3.}} = 1.06 \qquad \frac{s_4^2}{\bar{y}_{4.}} = .97$$

Thus, it would appear that $\sigma_i^2 = k\mu_i$, with $k \approx 1$. From Table 8.15, the suggested transformation is $y_T = \sqrt{y + .375}$. The values of y_T appear in Table 8.17 along with their means and standard deviations. Although the original data had heterogeneous variances, the sample variances are all approximately .25, as indicated in Table 8.17.

TABLE 8.17

Transformation of data in Table 8.16: $y_T = \sqrt{y + .375}$

	Distance to Mouth			
Sample	**1 KM**	**5 KM**	**10 KM**	**20 KM**
1	1.173	2.092	4.514	6.114
2	2.318	2.894	5.136	5.511
3	1.541	1.541	4.937	5.136
4	1.173	1.837	3.373	4.937
5	1.541	2.894	5.327	6.432
6	1.541	2.318	4.514	5.037
7	2.092	2.525	4.402	6.031

(continues)

TABLE 8.17

Transformation of data
in Table 8.16:
$y_T = \sqrt{y + .375}$
(*continued*)

Sample	Distance to Mouth			
	1 KM	**5 KM**	**10 KM**	**20 KM**
8	1.837	2.092	4.402	5.601
9	0.612	1.837	4.623	5.601
10	1.541	1.837	4.937	5.777
Mean	1.54	2.19	4.62	5.62
Variances	.24	.22	.29	.24

$y_T = \log y$

coefficient of variation

The second transformation indicated in Table 8.15 is for an experimental situation in which the population variance is approximately equal to the square of the population mean, or equivalently, where $\sigma = \mu$. Actually, the logarithmic transformation is appropriate any time the **coefficient of variation** σ_i/μ_i is constant across the populations of interest.

EXAMPLE 8.5

Irritable bowel syndrome (IBS) is a nonspecific intestinal disorder characterized by abdominal pain and irregular bowel habits. Each person in a random sample of 24 patients having periodic attacks of IBS was randomly assigned to one of three treatment groups, A, B, and C. The number of hours of relief while on therapy is recorded in Table 8.18 for each patient.

a. Test for differences among the population variances. Use $\alpha = .05$.
b. There are no 0 y values, so use the transformation $y_T = \ln y$ ("ln" denotes logarithms to the base e) to try to stabilize the variances.
c. Compute the sample means and the sample standard deviations for the transformed data.

Solution

a. The Hartley F_{max} test for a test of the null hypothesis $H_0: \sigma_1^2 = \sigma_2^2 = \sigma_3^2$ is

$$F_{max} = \frac{(15.66)^2}{(3.22)^2} = \frac{245.24}{10.37} = 23.65.$$

TABLE 8.18

Data for hours of relief
while on therapy

Treatment		
A	**B**	**C**
4.2	4.1	38.7
2.3	10.7	26.3
6.6	14.3	5.4
6.1	10.4	10.3
10.2	15.3	16.9
11.7	11.5	43.1
7.0	19.8	48.6
3.6	12.6	29.5
$\bar{y} = 6.46$	$\bar{y} = 12.34$	$\bar{y} = 27.35$
$s = 3.22$	$s = 4.53$	$s = 15.66$

The computed value of F_{max} exceeds 6.94, the tabulated value (Table 12) for $a = .05$, $t = 3$, and $df_2 = 7$, so we reject H_0 and conclude that the population variances are different.

b. The transformed data are shown in Table 8.19. Note: Natural logs can be computed using a calculator or computer spreadsheet.

TABLE 8.19
Natural logarithms of the data in Table 8.18

Treatment		
A	**B**	**C**
1.435	1.411	3.656
.833	2.370	3.270
1.887	2.660	1.686
1.808	2.342	2.332
2.322	2.728	2.827
2.460	2.442	3.764
1.946	2.986	3.884
1.281	2.534	3.384

c. The sample means and standard deviations for the transformed data are given in Table 8.20. Hartley's test for the homogeneity of variances for the transformed data is

$$F_{max} = \frac{(.77)^2}{(.46)^2} = 2.80$$

TABLE 8.20
Sample means and standard deviations for the data of Table 8.19

	Treatment		
	A	**B**	**C**
Sample mean	1.75	2.43	3.10
Sample standard deviation	.54	.46	.77

The computed value of F_{max} is 2.80, which is less than 6.94, the tabulated value, so we fail to reject H_0 and conclude that there is insufficient evidence of a difference in the population variances. Thus, the transformation has produced data in which the three variances are approximately equal.

$y_T = \text{arcsin } \sqrt{y}$

The third transformation listed in Table 8.15 is particularly appropriate for data recorded as percentages or proportions. Recall that in Chapter 4 we introduced the binomial distribution, where y designates the number of successes in n identical trials and $\hat{\pi} = y/n$ provides an estimate of π, the proportion of experimental units in the population possessing the characteristic. Although we may not have mentioned this while studying the binomial, the variance of $\hat{\pi}$ is given by $\pi(1 - \pi)/n$. Thus, if the response variable is $\hat{\pi}$, the proportion of successes in a random sample of n observations, then the variance of $\hat{\pi}$ will vary, depending on the values of π for the populations from which the samples were drawn. See Table 8.21.

TABLE 8.21

Variance of $\hat{\pi}$, the sample proportion, for several values of π and $n = 20$

Values of π	$\pi(1 - \pi)/n$
.01	.0005
.05	.0024
.1	.0045
.2	.0080
.3	.0105
.4	.0120
.5	.0125

Because the variance of $\hat{\pi}$ is symmetrical about $\pi = .5$, the variance of $\hat{\pi}$ for $\pi = .7$ and $n = 20$ is .0105, the same value as for $\pi = .3$. Similarly, we can determine $\pi(1 - \pi)/n$ for other values of $\pi > .5$. The important thing to note is that if the populations have values of π in the vicinity of approximately .3 to .5, there is very little difference in the variances for $\hat{\pi}$. However, the variance of $\hat{\pi}$ is quite variable for either large or small values of π, and for these situations we should consider the possibility of transforming the sample proportions to stabilize the variances.

The transformation we recommend is arcsin $\sqrt{\hat{\pi}}$ (sometimes written as $\sin^{-1} \sqrt{\hat{\pi}}$); that is, we are transforming the sample proportion into the angle whose sine is $\sqrt{\hat{\pi}}$. Some experimenters express these angles in degrees, others in radians. For consistency, we will always express our angles in radians. Table 9* of the Appendix provides arcsin computations for various values of $\hat{\pi}$.

EXAMPLE 8.6

A national opinion poll was hired to evaluate the voting public's opinion concerning whether the FBI director's term of office should be of a fixed length of time (such as 10 years). Also, there may be differences in opinion depending on geographical location. For this poll, the country was divided into four regions (NE, SE, NW, SW). A random sample of 100 registered voters was obtained from each of six standard metropolitan statistical areas (SMSAs) located in each of the four regions. The following data are the sample proportions for the 24 SMSAs. Transform the data by using $y_T = 2$ arcsin $\sqrt{\hat{\pi}}$.

Region	SMSA 1	2	3	4	5	6	Mean	Standard Deviation
NE	.13	.20	.23	.05	.14	.31	.177	.0903
SE	.57	.47	.47	.51	.53	.20	.458	.1321
NW	.30	.10	.07	.13	.17	.23	.167	.0860
SW	.53	.72	.70	.63	.79	.87	.707	.1191

Solution Using a calculator, computer spreadsheet, or Table 9 in the Appendix, the transformed data are as follows:

---

* Table 9 in the Appendix gives 2 arcsin $\sqrt{\hat{\pi}}$.

| | | | SMSA | | | | | Standard |
Region	1	2	3	4	5	6	Mean	Deviation
NE	.74	.93	1.00	.45	.77	1.18	.845	.2515
SE	1.71	1.51	1.51	1.59	1.63	.93	1.480	.2799
NW	1.16	.64	.54	.74	.85	1.00	.822	.2307
SW	1.63	2.03	1.98	1.83	2.19	2.40	2.010	.2693

The four regions can now be compared with respect to their opinion using an AOV procedure.

when $\pi = 0, 1$ One comment should be made concerning the situation in which a **sample proportion of 0 or 1** is observed. For these cases, we recommend substituting $1/4n$ and $1 - (1/4n)$, respectively, as the corresponding sample proportions to be used in the calculations.

In this section, we have discussed how transformations of data can alleviate the problem of nonconstant variances prior to conducting an analysis of variance. As an added benefit, the transformations presented in this section also (sometimes) decrease the nonnormality of the data. Still, there will be times when the presence of severe skewness or outliers causes nonnormality that could not be eliminated by these transformations. Wilcoxon's rank sum test (Chapter 6) can be used for comparing two populations in the presence of nonnormality when working with two independent samples. For data based on more than two independent samples, we can address nonnormality using the Kruskal–Wallis test (Section 8.6). Note that these tests are also based on a transformation (the rank transformation) of the sample data.

EXERCISES   **8.3**   Refer to Example 8.6. Analyze the sample data using the arcsin transformation to determine whether there are differences among the four geographic locations. Use $\alpha = .05$.

8.4   Refer to Example 8.4. Analyze the sample data after performing the transformation to determine whether the oxygen content is related to the distance to the mouth of the Mississippi River.

8.5   Refer to Example 8.5. In many situations in which the difference in variances is not too great, the results from the AOV comparisons of the population means of the transformed data are very similar to those from the results that would have been obtained using the original data. In these situations, the researcher is inclined to ignore the transformations because the scale of the transformed data is not relevant to the researcher. Thus, confidence intervals constructed for the means using the transformed data may not be very relevant. One possible remedy for this problem is to construct confidence intervals using the transformed data, and then perform an inverse transformation of the endpoints of the intervals. Then we would obtain a confidence interval with values having the same scale units of measurements as the original data.

   a. Test the hypothesis that the mean hours of relief for patients on the three treatments differs using $\alpha = .05$. Use the original data.
   b. Place 95% confidence intervals on the mean hours of relief for the three treatments.
   c. Repeat the analysis in (a) and (b) using the transformed data.
   d. Comment on any differences in the results of the test of hypotheses.
   e. Perform an inverse transformation on the endpoints of the intervals constructed in (c). Compare these intervals to the ones constructed in (b).

Because there are groups of tied ranks, we will use H' and compare its value to H. To do this we form the g groups composed of identical ranks, shown in the accompanying table.

Rank	Group	t_i
1	1	1
2	2	1
3	3	1
4	4	1
5.5, 5.5	5	1
7	6	1
8	7	1
9	8	1
11, 11, 11	9	1
13.5, 13.5	10	1
15	11	1
16.5, 16.5	12	1
18	13	1
19	14	1
21, 21, 21	15	1
23	16	1
24	17	1
25	18	1
26.5, 26.5	19	1
29, 29, 29	20	1

From this information, we calculate the quantity

$$\sum_i \frac{(t_i^3 - t_i)}{n_T^3 - n_T}$$

$$= \frac{(2^3 - 2) + (3^3 - 3) + (2^3 - 2) + (2^3 - 2) + (3^3 - 3) + (2^3 - 2) + (3^3 - 3)}{30^3 - 30}$$

$$= .0036$$

Substituting this value into the formula for H', we have

$$H' = \frac{H}{1 - .0036} = \frac{3.24}{.9964} = 3.25$$

Thus, even with more than half of the measurements involved in ties, H' and H are nearly the same value. The critical value of the chi-square with $\alpha = .05$ and df $= k - 1 = 2$ can be found using Table 7 in the Appendix. This value is 5.991; we fail to reject the null hypothesis and conclude that there is no significant difference in the test scores of the three groups of clerics. It is interesting to note that the p-value for the Kruskal–Wallis test is .198, whereas the p-value from AOV F test applied to the original test scores was .168. Thus, even though the data did not have a normal distribution, the F test from AOV is robust against departures from normality. Only when the data are extremely skewed or very heavy tailed do the Kruskal–Wallis test and the F test from AOV differ.

EXERCISES

Hort.

8.6 A team of researchers wants to compare the yields (in pounds) of five different varieties (A, B, C, D, E) of 4-year-old orange trees in one orchard. They obtain a random sample of seven trees of each variety from the orchard. The yields for these trees are presented here.

A	B	C	D	E
13	27	40	17	36
19	31	44	28	32
39	36	41	41	34
38	29	37	45	29
22	45	36	15	25
25	32	38	13	31
10	44	35	20	30

a. Using tests and plots of the data, determine whether the conditions for using the AOV are satisfied.

b. Conduct an AOV test of the null hypothesis that the five varieties have the same mean yield. Use $\alpha = .01$.

c. Use the Kruskal–Wallis test to test the null hypothesis that the five varieties have the same yield distributions. Use $\alpha = .01$.

d. Are the conclusions you reached in (b) and (c) consistent?

8.7 How do the research hypotheses tested by the AOV test and the Kruskal–Wallis test differ?

Engin.

8.8 In the manufacture of soft contact lenses, the actual strength (power) of the lens needs to be very close to the target value for the lenses to properly fit the customer's needs. In the paper, "An ANOM-type test for variances from normal populations," *Technometrics* (1997), 39: 274–283, a comparison of several suppliers is made relative to the consistency of the power of the lenses. The following table contains the deviations from the target power of lenses produced using materials from three different suppliers:

| Supplier | Lens | | | | | | | | |
	1	2	3	4	5	6	7	8	9
A	189.9	191.9	190.9	183.8	185.5	190.9	192.8	188.4	189.0
B	156.6	158.4	157.7	154.1	152.3	161.5	158.1	150.9	156.9
C	218.6	208.4	187.1	199.5	202.0	211.1	197.6	204.4	206.8

a. Is there a significant difference in the distributions of deviations for the three suppliers? Use $\alpha = .01$.

b. Using the appropriate tests and plots given here, assess whether the data meet the necessary conditions to use an AOV to determine whether there is a significant difference in the mean deviations for the three suppliers.

c. Conduct an AOV with $\alpha = .05$ and compare your results with the conclusions from (a).

d. Suppose that a difference in mean deviation of 20 units would have commercial consequences for the manufacture of the lenses. Does there appear to be a *practical* difference in the three suppliers?

Supplementary Exercises

Mfr.   **8.9**   Company researchers conducted an experiment to compare the number of major defectives observed along each of five production lines in which changes were being instituted. They monitored production continuously during the period of changes, and recorded the number of major defectives per day for each line. The data are shown here.

Production Line				
1	2	3	4	5
34	54	75	44	80
44	41	62	43	52
32	38	45	30	41
36	32	10	32	35
51	56	68	55	58

  a. Compute $\bar{y}$ and s^2 for each sample. Does there appear to be a problem with nonconstant variances? Use Hartley's test based on $\alpha = .05$.
  b. Use a square root transformation on the data and conduct an analysis on the transformed data.
  c. Draw your conclusions concerning differences among production lines.

8.10   Do a Kruskal–Wallis test on the data represented in Exercise 8.9. Does this test confirm the conclusions drawn in Exercise 8.9? If the results differ, which analysis do you believe? Use $\alpha = .05$.

Ag.   **8.11**   The Agricultural Experiment Station of a university tested two different herbicides and their effects on crop yield. From 90 acres set aside for the experiment, the station used herbicide 1 on a random sample of 30 acres, herbicide 2 on a second random sample of 30 acres, and they used the remaining 30 acres as a control. At the end of the growing season, the yields (in bushels per acre) were

	Sample Mean	Sample Standard Deviation	Sample Sizes
Herbicide 1	90.2	6.5	30
Herbicide 2	89.3	7.8	30
Control 3	85.0	7.4	30

  a. Use these data to conduct a one-way analysis of variance to test whether there is a difference in the mean yields. Use $\alpha = .05$.
  b. Construct 95% confidence intervals on the mean yields μ_i.
  c. Which of the mean yields appear to be different?

Hort.   **8.12**   Researchers from the Department of Fruit Crops at a university compared four different preservatives to be used in freezing strawberries. The researchers prepared the yield from a strawberry patch for freezing and randomly divided it into four equal groups. Within each group they treated the strawberries with the appropriate preservative and packaged them into eight small plastic bags for freezing at 0°C. The bags in group I served as a control group, while those in groups II, III, and IV were assigned one of three newly developed preservatives. After all 32 bags of strawberries were prepared, they were stored at 0°C for a period of 6 months. At the end of this time, the contents of each bag were allowed to thaw and then rated on a scale of 1 to 10 points for discoloration. (Note that a low score indicates little discoloration.) The ratings are given here.

Group I	10	8	7.5	8	9.5	9	7.5	7
Group II	6	7.5	8	7	6.5	6	5	5.5
Group III	3	5.5	4	4.5	3	3.5	4	4.5
Group IV	2	1	2.5	3	4	3.5	2	2

a. Use the following plots of the residuals and a test of the homogeneity of variances to assess whether the conditions needed to use AOV techniques are satisfied with this data set.

b. Test whether there is a difference in the mean ratings using $\alpha = .05$.

c. Place 95% confidence intervals on the mean ratings for each of the groups.

d. Confirm your results with the computer output given here.

```
               One-Way Analysis of Variance for Exercise 8.12

Analysis of Variance for Ratings
Source      DF        SS         MS         F         P
Group        3    159.187     53.062     55.67     0.000
Error       28     26.687      0.953
Total       31    185.875
                                       Individual 95% CIs for Mean
                                       Based on Pooled StDev
Group        N       Mean      StDev   --+---------+---------+---------+----
I            8     8.3125     1.0670                              (---*--)
II           8     6.4375     1.0155                       (--*---)
III          8     4.0000     0.8452             (---*---)
IV           8     2.5000     0.9636    (---*--)
                                       --+---------+---------+---------+----
Pooled StDev =     0.9763             2.0       4.0       6.0       8.0
```

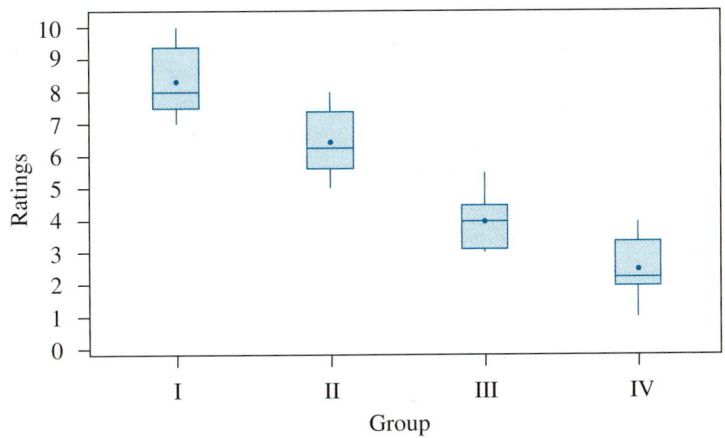

Boxplots of ratings by group for Exercise 8.12 (means are indicated by solid circles)

Normal probability plot of
residuals for Exercise 8.12

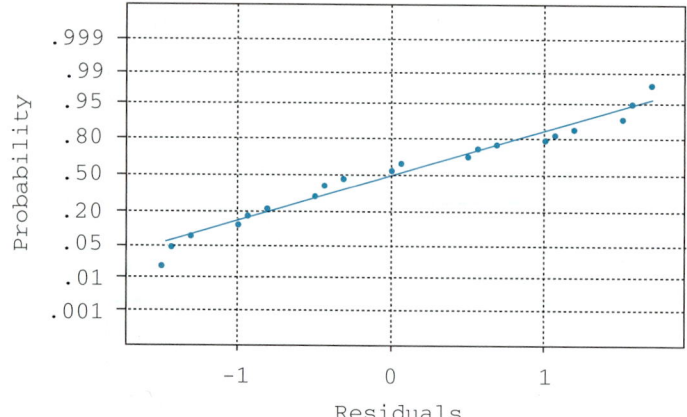

Average: 0
StDev: 0.927840
N: 32

Anderson-Darling Normality Test
            A-Squared: 0.503
            P-Value: 0.191

8.13 Refer to Exercise 8.12. In many situations in which the response is a rating rather than an actual measurement, it is recommended that the Kruskal–Wallis test be used.
  a. Apply the Kruskal–Wallis test to determine whether there is a shift in the distribution of ratings for the four groups.
  b. Is the conclusion reached using the Kruskal–Wallis test consistent with the conclusion reached in Exercise 8.12 using AOV?

Ag.   **8.14** Researchers conducted an experiment to compare the starch content of tomato plants grown in sandy soil supplemented by one of three different nutrients, A, B, or C. The researchers selected 18 tomato seedlings of one particular variety for the study, with six assigned to each of the nutrient groups. They planted all seedlings in a sand culture and maintained them in a controlled environment. Those seedlings assigned to nutrient A served as the control group (receiving distilled water only). Plants assigned to nutrient B were fed a weak concentration of Hoagland nutrient, while those assigned to nutrient C received the Hoagland nutrient at full strength. The researchers determined the stem starch contents 25 days after planting; the contents are recorded here, in micrograms per milligram.

Nutrient A	22	20	21	18	16	14
Nutrient B	12	14	15	10	9	6
Nutrient C	7	9	7	6	5	3

  a. Run an analysis of variance to test for differences in starch content for the three nutrient groups. Use $\alpha = .05$.
  b. Draw your conclusions.

8.15 Although we often have well-planned experiments with equal numbers of observations per treatment, we still end up with unequal numbers at the end of a study. Suppose that although six plants were allocated to each of the nutrient groups of Exercise 8.14, only five survived in group B and four in group C. The data for the stem starch contents are given here.

Nutrient A	22	20	21	18	16	14
Nutrient B	12	14	15	10	9	
Nutrient C	7	9	7	6		

a. Write an appropriate model for this experimental situation. Define all terms.
b. Assuming that nutrients B and C did not cause the plants to die, perform an analysis of variance to compare the treatment means. Use $\alpha = .05$.

H.R.   **8.16**   Salary disputes and their eventual resolutions often leave both employers and employees embittered by the entire ordeal. To assess employee reactions to a recently devised salary and fringe benefits plan, the personnel department obtained random samples of 15 employees from each of three divisions in the company: manufacturing, marketing, and research. The personnel staff asked each employee sampled to respond (in confidence) to a series of questions. Several employees refused to cooperate, as reflected in the unequal sample sizes. The data are given here.

	Manufacturing	Marketing	Research
Sample size	12	14	11
Sample mean	25.2	32.6	28.1
Sample variance	3.6	4.8	5.3

a. Write a model for this experimental situation.
b. Use the summary of the scored responses to compare the means for the three divisions (the higher a score, the higher the employee acceptance). Use $\alpha = .01$.

Ag.   **8.17**   Researchers record the yields of corn, in bushels per plot, for four different varieties of corn, A, B, C, and D. In a controlled greenhouse experiment, the researchers randomly assign each variety to eight of 32 plots available for the study. The yields are listed here.

A	2.5	3.6	2.8	2.7	3.1	3.4	2.9	3.5
B	3.6	3.9	4.1	4.3	2.9	3.5	3.8	3.7
C	4.3	4.4	4.5	4.1	3.5	3.4	3.2	4.6
D	2.8	2.9	3.1	2.4	3.2	2.5	3.6	2.7

a. Write an appropriate statistical model.
b. Perform an analysis of variance on these data and draw your conclusions. Use $\alpha = .05$.

8.18   Refer to Exercise 8.17. Perform a Kruskal–Wallis analysis of variance by ranks (with $\alpha = .05$) and compare your results to those in Exercise 8.17.

Bus.   **8.19**   Many corporations make use of the Wide Area Telephone System (WATS), where, for a fixed rent per month, the corporation can make as many long distance calls as it likes. Depending on the area of the country in which the corporation is located, it can rent a WATS line for certain geographic bands. For example, in Ohio, these bands might include the following states:

Band I:	Ohio	
Band II:	Indiana	Pennsylvania
	Kentucky	Tennessee
	Maryland	Virginia
	Michigan	West Virginia
	North Carolina	Washington, D.C.
Band III:	32 Eastern and Midwestern states, plus Washington, D.C.	

To monitor the use of the WATS lines, a corporation selected a random sample of 12 calls from each of the following areas in a given month, and recorded the length of the conversation (in minutes) for each call. (Band III excludes states in Band II and Ohio.)

Ohio	2	3	5	8	4	6	18	19	9	6	7	5
Band II	6	8	10	15	19	21	10	12	13	2	5	7
Band III	12	14	13	20	25	30	5	6	21	22	28	11

Perform an analysis of variance to compare the mean lengths of calls for the three areas. Use $\alpha = .05$.

Edu.

8.20   Doing homework is a nightly routine for most school-age children. The article, "Family involvement with middle-grades homework: effects of differential prompting," *Journal of Experimental Education,* 66: 31–48, examines the question of whether parents' involvement with their children's homework is associated with improved academic performance. Seventy-four sixth graders and their families participated in the study. Researchers assigned the students, similar in student academic ability and background, in one of three mathematics classes taught by the same teacher, and randomly assigned the classes to one of the three treatment groups.

Group I, student/family prompt: Students were prompted to seek assistance from a family member and the family was encouraged to provide assistance to students.
Group II, student prompt: Students were prompted to seek assistance from a family member but there was no specific encouragement of family members to provide assistance to students.
Group III, no prompts: Students were not prompted to seek assistance from a family member nor were family members encouraged to provide assistance to students.

Thus, one class was assigned to each of the three treatment groups. The researchers gave the students a posttest, with the results given here.

Treatment Group	Number of Students	Mean Posttest Score
Student/family prompt	22	68%
Student prompt	22	66%
No prompt	25	67%

The researchers concluded that higher levels of family involvement were not associated with higher student achievement in this study.
   a. What is the population of interest in this study?
   b. Based on the data collected, to what population can the results of this study be inferred?
   c. What is the effective sample for each of the treatment groups; that is, how many experimental units were randomly assigned to each of the treatment groups?
   d. What criticisms would you have for the design of this study?
   e. Suggest an improved design for addressing the research hypothesis that family involvement improves student performance in mathematics classes.

Gov.

8.21   In a 1994 Senate subcommittee hearing, an executive of a major tobacco company testified that the accusation that nicotine was added to cigarettes was false. Tobacco company scientists stated that the amount of nicotine in cigarettes was completely determined by the size of tobacco leaf, with smaller leaves having greater nicotine content. Thus, the variation in nicotine content in cigarettes occurred due to a variation in the size of the tobacco leaves and was not due to any additives placed in the cigarettes by the company. Furthermore, the company argued that the size of the leaves varied depending on the weather conditions during the growing season, for which they had no control. To study whether smaller tobacco leaves had a higher nicotine content, a consumer health organization conducted the following experiment. The major factors controlling leaf size are temperature and the amount of water received by the plants during the growing season. The experimenters created four types of growing conditions for tobacco plants. Condition A was average temperature and rainfall amounts. Condition B was lower than average temperature and rainfall conditions. Condition C was higher temperatures with lower rainfall. Finally, Condition D was higher than normal temperatures and rainfall. The scientists then planted 10 tobacco plants under each of the four conditions in a greenhouse where temperature and amount of moisture were carefully controlled. After growing the plants, the scientists recorded the leaf size and nicotine content, which are given here.

Plant	A Leaf Size	B Leaf Size	C Leaf Size	D Leaf Size
1	27.7619	4.2460	15.5070	33.0101
2	27.8523	14.1577	5.0473	44.9680
3	21.3495	7.0279	18.3020	34.2074
4	31.9616	7.0698	16.0436	28.9766
5	19.4623	0.8091	10.2601	42.9229
6	12.2804	13.9385	19.0571	36.6827
7	21.0508	11.0130	17.1826	32.7229
8	19.5074	10.9680	16.6510	34.5668
9	26.2808	6.9112	18.8472	28.7695
10	26.1466	9.6041	12.4234	36.6952

Plant	A Nicotine	B Nicotine	C Nicotine	D Nicotine
1	10.0655	8.5977	6.7865	9.9553
2	9.4712	8.1299	10.9249	5.8495
3	9.1246	11.3401	11.3878	10.3005
4	11.3652	9.3470	9.7022	9.7140
5	11.3976	9.3049	8.0371	10.7543
6	11.2936	10.0193	10.7187	8.0262
7	10.6805	9.5843	11.2352	13.1326
8	8.1280	6.4603	7.7079	11.8559
9	10.5066	8.2589	7.5653	11.3345
10	10.6579	5.0106	9.0922	10.4763

a. Perform a one-way analysis of variance to test whether there is a significant difference in the average leaf size under the four growing conditions. Use $\alpha = .05$.

b. What conclusions can you reach concerning the effect of growing conditions on the average leaf size?

c. Perform a one-way analysis of variance to test whether there is a significant difference in the average nicotine content under the four growing conditions. Use $\alpha = .05$.

d. What conclusions can you reach concerning the effect of growing conditions on the average nicotine content?

e. Based on the conclusions you reached in (b) and (d), do you think the testimony of the tobacco companies' scientists is supported by this experiment? Justify your conclusions.

8.22   Using the plots given here, do the nicotine content data in Exercise 8.21 suggest violations of the AOV conditions? If you determine that the conditions are not met, perform an alternative analysis and compare your results to those of Exercise 8.21.

Boxplots of leaf size by group for Exercise 8.22 (means are indicated by solid circles)

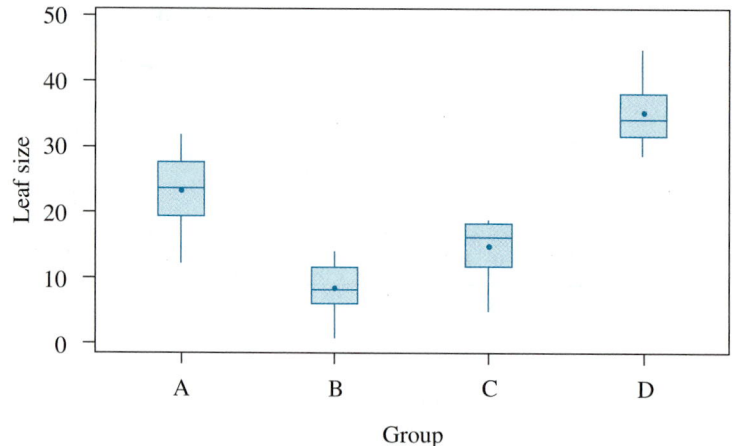

Probability plot of residuals leaf size for Exercise 8.22

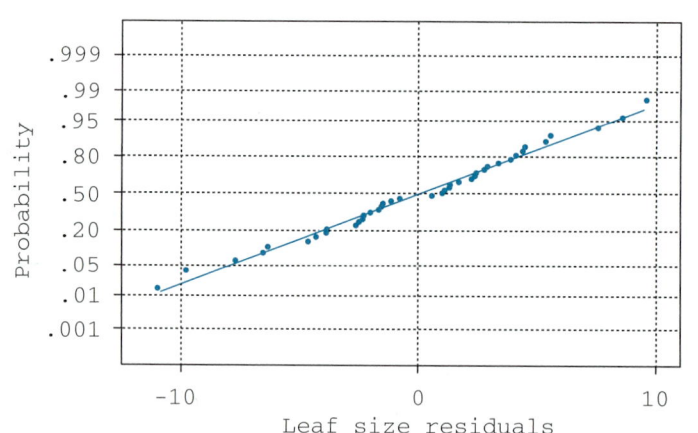

Average: 0.0000000   Anderson-Darling Normality Test
StDev: 4.75535   A-Squared: 0.205
N: 40   P-Value: 0.864

Boxplots of nicotine by group for Exercise 8.22 (means are indicated by solid circles)

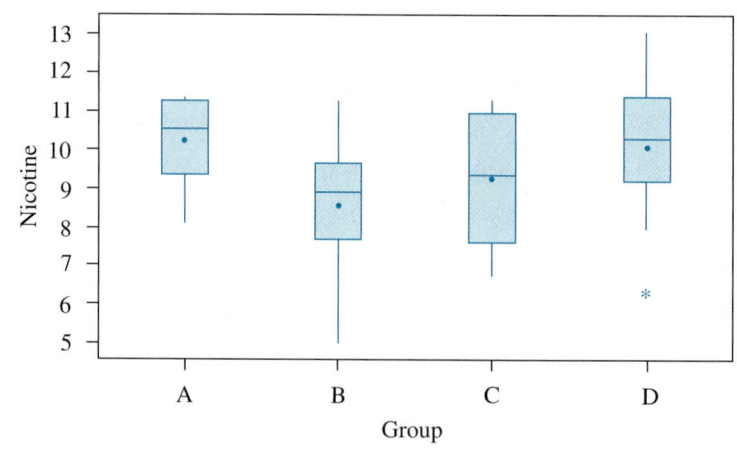

Probability plot of residuals
nicotine content for
Exercise 8.22

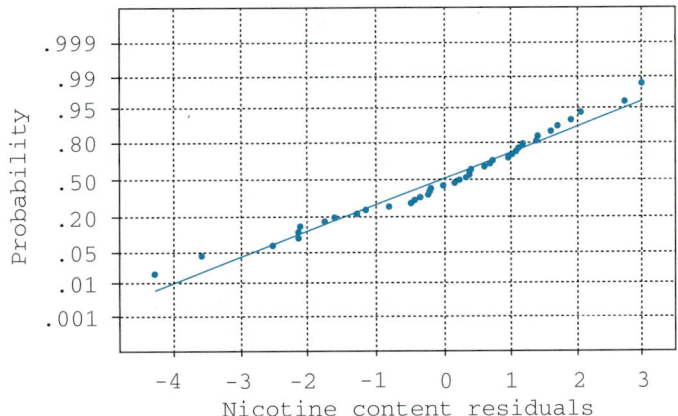

Average: 0.0000000   Anderson-Darling Normality Test
StDev: 1.62647   A-Squared: 0.443
N: 40   P-Value: 0.273

8.23 Select an article from a journal in your field of study that contains a one-way analysis of variance. Discuss any problems with the design of the experiment and the type of conclusions obtained by the researchers.

Ag. **8.24** Scientists conducted an experiment to test the effects of five different diets in turkeys. They randomly assigned six turkeys to each of the five diet groups and fed them for a fixed period of time.

Group	Weight Gained (pounds)
Control diet	4.1, 3.3, 3.1, 4.2, 3.6, 4.4
Control diet + level 1 of additive A	5.2, 4.8, 4.5, 6.8, 5.5, 6.2
Control diet + level 2 of additive A	6.3, 6.5, 7.2, 7.4, 7.8, 6.7
Control diet + level 1 of additive B	6.5, 6.8, 7.3, 7.5, 6.9, 7.0
Control diet + level 2 of additive B	9.5, 9.6, 9.2, 9.1, 9.8, 9.1

    a. Plot the data separately for each sample.
    b. Compute $\bar{y}$ and s^2 for each sample.
    c. Is there any evidence of unequal variances or nonnormality? Explain.
    d. Assuming that the five groups were comparable with respect to initial weights of the turkeys, use the weight-gained data to draw conclusions concerning the different diets. Use $\alpha = .05$.

8.25 Run a Kruskal–Wallis test for the data of Exercise 8.24. Do these results confirm what you concluded from an analysis of variance? What overall conclusions can be drawn? Use $\alpha = .05$.

Hort. **8.26** Some researchers have conjectured that stem-pitting disease in peach tree seedlings might be related to the presence or absence of nematodes in the soil. Hence, weed and soil treatment using herbicides might be effective in promoting seedling growth. Researchers conducted an experiment to compare peach tree seedling growth with soil and weeds treated with one of three herbicides:

    A: control (no herbicide)
    B: herbicide with Nemagone
    C: herbicide without Nemagone

The researchers randomly assigned six of the 18 seedlings chosen for the study to each treatment group. They treated soil and weeds in the growing areas for the three groups with the appropriate herbicide. At the end of the study period, they recorded the height (in centimeters) for each seedling. Use the following sample data to run an analysis of variance for detecting differences among the seedling heights for the three groups. Use $\alpha = .05$. Draw your conclusions.

Herbicide A	66	67	74	73	75	64
Herbicide B	85	84	76	82	79	86
Herbicide C	91	93	88	87	90	86

8.27   Refer to the data of Exercise 8.24. To illustrate the effect that an extreme value can have on conclusions from an analysis of variance, suppose that the weight gained by the fifth turkey in the level 2, additive B group was 15.8 rather than 9.8.

    a. What effect does this have on the assumptions for an analysis of variance?

    b. With 9.8 replaced by 15.8, if someone unknowingly ran an analysis of variance, what conclusions would he or she draw?

8.28   Refer to Exercise 8.27. What happens to the Kruskal–Wallis test if you replace the value 9.8 by 15.8? Might there be a reason to run both a Kruskal–Wallis test and an analysis of variance? Why?

8.29   Is the Kruskal–Wallis test more powerful than an analysis of variance, in certain situations, for detecting differences among treatment means? Explain.

Engin.   **8.30**   A small corporation makes insulation shields for electrical wires using three different types of machines. The corporation wants to evaluate the variation in the inside diameter dimension of the shields produced by the machines. A quality engineer at the corporation randomly selects shields produced by each of the machines and records the inside diameters of each shield (in millimeters). She wants to determine whether the means and standard deviations of the three machines differ.

Shield	Machine A	Machine B	Machine C
1	18.1	8.7	29.7
2	2.4	56.8	18.7
3	2.7	4.4	16.5
4	7.5	8.3	63.7
5	11.0	5.8	18.9
6			107.2
7			19.7
8			93.4
9			21.6
10			17.8

    a. Conduct a test for the homogeneity of the population variances. Use $\alpha = .05$.

    b. Would it be appropriate to proceed with an analysis of variance based on the results of this test? Explain.

    c. If the variances of the diameters are different, suggest a transformation that may alleviate their differences and then conduct an analysis of variance to determine whether the mean diameters differ. Use $\alpha = .05$.

    d. Compare the results of your analysis in (c) to the computer output given here, which was an analysis of variance on the original diameters.

e. How could the engineer have designed her experiment differently if she knew that the variance of machine B and machine C were so much larger than that of machine A?

```
              One-Way Analysis of Variance for Exercise 8.30

Analysis of Variance
Source      DF        SS        MS        F        P
Factor       2      4141      2071     2.73    0.094
Error       17     12907       759
Total       19     17048             Individual 95% CIs for Mean
                                     Based on Pooled StDev
Level        N      Mean     StDev   --------+---------+---------+--------
Machine      5      8.32      6.52   (---------*----------)
Machine      5     16.78     22.43     (----------*---------)
Machine     10     40.70     34.52                 (------*-------)
                                     --------+---------+---------+--------
Pooled StDev =     27.55                   0        25        50
```

8.31 The Kruskal–Wallis test is not as highly affected as the AOV test when the variances are unequal. Demonstrate this result by applying the Kruskal–Wallis test to both the original and transformed data and comparing the conclusions reached in this analysis for the data of Exercise 8.30.

CHAPTER 9

Multiple Comparisons

9.1 Introduction and Case Study

9.2 Linear Contrasts

9.3 Which Error Rate Is Controlled?

9.4 Fisher's Least Significant Difference

9.5 Tukey's W Procedure

9.6 Student–Newman–Keuls Procedure

9.7 Dunnett's Procedure: Comparison of Treatments to a Control

9.8 Scheffé's S Method

9.9 Summary

9.1 Introduction and Case Study

In Chapter 8, we introduced a procedure for testing the equality of t population means. We used the test statistic $F = s_B^2/s_W^2$ to determine whether the between-sample variability was large relative to the within-sample variability. If the computed value of F for the sample data exceeded the critical value obtained from Table 8 in the Appendix, we rejected the null hypothesis $H_0: \mu_1 = \mu_2 = \cdots = \mu_t$ in favor of the alternative hypothesis

H_a: At least one of the t population means differs from the rest.

Although rejection of the null hypothesis does give us some information concerning the population means, we do not know which means differ from each other. For example, does μ_1 differ from μ_2 or μ_3? Does μ_3 differ from the average of μ_2, μ_4, and μ_5? Is there an increasing trend in the treatment means $\mu_1, \ldots, \mu_t$? **Multiple-comparison procedures** and contrasts have been developed to answer questions such as these. Although many multiple-comparison procedures have been proposed, we will focus on just a few of the more commonly used methods. After studying these few procedures, you should be able to evaluate the results of most published material using multiple comparisons or to suggest an appropriate multiple-comparison procedure in an experimental situation.

A word of caution: It is tempting to analyze only those comparisons that appear to be interesting after seeing the sample data. This practice has sometimes **data dredging** been called **data dredging** or **data snooping,** and the confidence coefficient for a **data snooping** single comparison does not reflect the after-the-fact nature of the comparison. For example, we know from previous work that the interval estimate for the difference between two population means using the formula

$$(\bar{y}_1 - \bar{y}_2) \pm t_{\alpha/2} s_p \sqrt{\frac{1}{n_1} + \frac{1}{n_2}}$$

427

has a confidence coefficient of $1 - \alpha$. Suppose we had run an analysis of variance to test the hypothesis

$$H_0: \quad \mu_1 = \mu_2 = \mu_3 = \mu_4 = \mu_5 = \mu_6$$

for six populations, but decided to compute a confidence interval for μ_1 and μ_2 only after we saw that the largest sample mean was $\bar{y}_1.$ and the smallest was $\bar{y}_2..$ In this situation, the confidence coefficient would not be $1 - \alpha$ as originally thought; that value applies only to a preplanned comparison, one planned before looking at the sample data.

One way to allow for data snooping after observing the sample data is to use a multiple-comparison procedure that has a confidence coefficient to cover all comparisons that could be done after observing the sample data. Some of these procedures are discussed in this chapter.

exploratory hypothesis generation

confirmation

The other possibility is to use data-snooping comparisons as a basis for generating hypotheses that must be confirmed in future experiments or studies. Here, the data-snooping comparisons serve an exploratory, or hypothesis-generating, role, and inferences would not be made based on the data snoop. Further experimentation would be done to confirm (or not) the hypothesis generated in the data snoop.

Case Study: Are Interviewers' Decisions Affected by Different Handicap Types?

There are approximately 50 million people in the United States who report having a physical handicap. Furthermore, it is estimated that the unemployment rate of the noninstitutionalized handicapped people between ages of 18 and 64 is nearly double the unemployment rate of people with no impairment. Thus, it appears that people with disabilities have a more difficult time obtaining employment. One of the problems confronting people having a handicap may be a bias by employers during the employment interview.

The paper, "Interviewers' decisions related to applicant handicap type and rater empathy," *Human Performance,* 1990, 3: 157–171, describes a study that examines these issues. The purpose of the study was to investigate whether different types of physical handicaps produce different empathy by raters and to examine whether interviewers' evaluations are affected by the type of handicap of the person being interviewed.

Designing the Data Collection   The researchers videotaped five simulated employment interviews. To minimize bias across videotapes, the same male actors (job applicant and interviewer) and the same interview script, consisting of nine questions, were used in all five videotapes. The script was directed toward average qualifications of the applicant, since this type of applicant is the most likely to be susceptible to interview biases. The videotapes differed with respect to type of applicant disability, but all were depicted as permanent disabilities. The five conditions were labeled wheelchair, Canadian crutches, hard of hearing, leg amputee, and nonhandicapped (control).

Each participant in the study was asked to rate the applicant's qualifications for a computer sales position based on the questions asked during the videotaped interview. Prior to viewing the videotape, each participant completed the Hogan Empathy Scale. The researchers decided to have each participant view only one of the five videotapes. Based on the variability in scores of raters in previous studies, the researchers decided they would require 14 raters for each videotape in order to obtain a precise estimate of the mean rating for each of the five handicap conditions. Seventy undergraduate students were selected to participate in the study. For each of the five videotapes, 14 students were randomly assigned to view the videotapes. After viewing each videotape, the participant rated the applicant on two scales: an 11-item scale assessing the rater's liking of the applicant, and a 10-item scale that assessed the rater's evaluation of the applicant's job qualifications. For each scale, the average of the individual items form an overall assessment of the applicant. The researchers used these two variables to determine whether different types of physical handicaps are reacted to differently by raters and to determine the effect of rater empathy on evaluations of handicapped applicants.

These are some of the questions that the researchers were interested in:

1. Is there a difference in the average empathy scores of the 70 raters?
2. Do the raters' average qualification scores differ across the five handicap conditions?
3. Which pairs of handicap conditions produced different average qualification scores?
4. Is the average rating for the control group (no handicap) greater than the average ratings for all types of handicapped applicants?
5. Is the average qualification rating for the hard-of-hearing applicants different from the average ratings for those applicants who had a mobility handicap?
6. Is the average qualification rating for the crutches applicants different from the average ratings of the applicants who were either amputees or in wheelchairs?
7. Is the average rating for the amputee applicants different from the average rating for the wheelchair applicants?

The researchers conducted the experiments and obtained the following data from the 70 raters of the applicants. The data in Table 9.1 are a summary of the empathy values.

TABLE 9.1

Empathy values across the five handicap conditions

Condition	Control (None)	Hard of Hearing	Canadian Crutches	One-Leg Amputee	Wheelchair
Mean	21.43	22.71	20.43	20.86	19.86
Standard Deviation	3.032	3.268	3.589	3.035	3.348

The data in Table 9.2 are the applicant qualification scores of the 70 raters for the five handicap conditions along with their summary statistics. (These data were simulated using the summary statistics of the ratings given in the paper.)

TABLE 9.2
Ratings of applicant
qualification across the
five handicap conditions

Control	Hard of Hearing	Amputee	Crutches	Wheelchair
6.1	2.1	4.1	6.7	3.0
4.6	4.8	6.1	6.7	3.9
7.7	3.7	5.9	6.5	7.9
4.2	3.5	5.0	4.6	3.0
6.1	2.2	6.1	7.2	3.5
2.9	3.4	5.7	2.9	8.1
4.6	5.5	1.1	5.2	6.4
5.4	5.2	4.0	3.5	6.4
4.1	6.8	4.7	5.2	5.8
6.4	0.4	3.0	6.6	4.6
4.0	5.8	6.6	6.9	5.8
7.2	4.5	3.2	6.1	5.5
2.4	7.0	4.5	5.9	5.0
2.9	1.8	2.1	8.8	6.2

```
                    Descriptive Statistics for Case Study

Variable               N         Mean       Median      TrMean       StDev      SE Mean
Control               14        4.900        4.600       4.875        1.638       0.438
Hard of Hearing   .   14        4.050        4.100       4.108        1.961       0.524
Amputee               14        4.436        4.600       4.533        1.637       0.437
Crutches              14        5.914        6.300       5.925        1.537       0.411
Wheelchair            14        5.364        5.650       5.333        1.633       0.436


Variable         Minimum      Maximum           Q1           Q3
Control            2.400        7.700        3.725        6.175
Hard of Hearing    0.400        7.000        2.175        5.575
Amputee            1.100        6.600        3.150        5.950
Crutches           2.900        8.800        5.050        6.750
Wheelchair         3.000        8.100        3.800        6.400
```

The qualification scores are plotted in Figure 9.1. The boxplots display somewhat higher qualification scores from the raters viewing the crutches condition. The mean qualification scores for the hard of hearing and amputee conditions were somewhat smaller than those of the control and wheelchair conditions. The variability of the qualification scores was nearly the same for all five conditions. Also, the conditions show distributions that do not present any problem with extreme skewness or outliers; therefore, it seems reasonable to assume normality of the distributions.

Managing the Data   The researchers would next prepare the data for a statistical analysis following the steps described in Section 2.5.

In the following sections of Chapter 9, we will develop the various methodologies needed to answer the questions we have posed here.

FIGURE 9.1

Boxplots of ratings by handicap (means are indicated by solid circles)

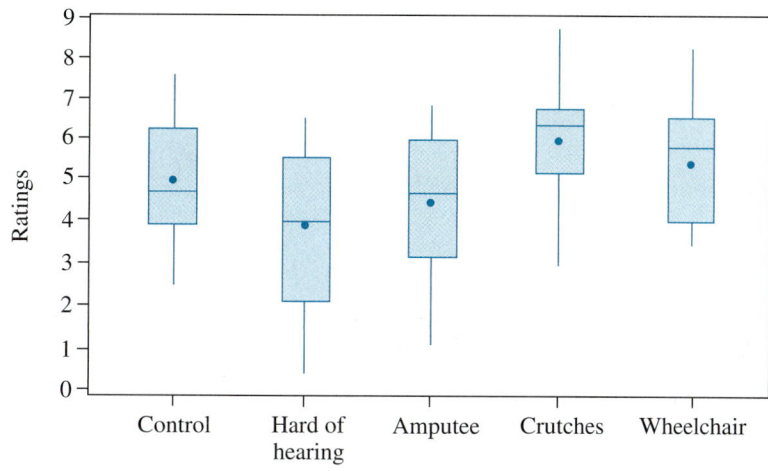

9.2 Linear Contrasts

Before developing several different multiple-comparison procedures, we need the following notation and definitions. Consider a one-way classification where we wish to make comparisons among the t population means $\mu_1, \mu_2, \ldots, \mu_t$. These comparisons among t population means can be written in the form

$$l = a_1\mu_1 + a_2\mu_2 + \cdots + a_t\mu_t = \sum_{i=1}^{t} a_i\mu_i$$

where the a_is are constants satisfying the property that $\Sigma \, a_i = 0$. For example, if we wanted to compare μ_1 to μ_2, we would write the linear form

$$l = \mu_1 - \mu_2$$

Note here that $a_1 = 1$, $a_2 = -1$, $a_3 = a_4 = \cdots = a_t = 0$, and $\Sigma_i \, a_i = 0$. Similarly, we could compare the mean for population 1 to the average of the means for populations 2 and 3. Then l would be of the form

$$l = \mu_1 - \frac{(\mu_2 + \mu_3)}{2}$$

where $a_1 = 1$, $a_2 = a_3 = -\frac{1}{2}$, $a_4 = a_5 = \cdots = a_t = 0$, and $\Sigma_i \, a_i = 0$.

We often write the contrasts with all the a_is as integer values. We accomplish this by rewriting the a_is with a common denominator and then multiplying the a_is by this common denominator. Suppose we have the following contrast in four treatment means:

$$a_1 = \frac{1}{4} \qquad a_2 = \frac{-1}{6} \qquad a_3 = \frac{-1}{3} \qquad a_4 = \frac{1}{4}$$

The common denominator is 12, which we multiply by each of the a_is, yielding

$$a_1 = 3 \qquad a_2 = -2 \qquad a_3 = -4 \qquad a_4 = 3$$

The two contrasts yield equivalent comparisons concerning the differences in the μs, but the integer form is somewhat easier to work with in many of our calculations.

An estimate of the linear form l, designated by $\hat{l}$, is formed by replacing the μ_is in l with their corresponding sample means $\bar{y}_i$. The estimate $\hat{l}$ is called a **linear contrast.**

$\hat{l}$

linear contrast

DEFINITION 9.1

$\hat{l} = a_1\bar{y}_1. + a_2\bar{y}_2. + \cdots + a_t\bar{y}_t. = \Sigma_i\, a_i\bar{y}_i.$ is called a **linear contrast** among the t sample means and can be used to estimate $l = \Sigma_i\, a_i\mu_i$. The a_is are constants satisfying the constraint $\Sigma_i\, a_i = 0$.

The variance of the linear contrast $\hat{l}$ can be estimated as follows:

$\hat{V}(\hat{l})$

$$\hat{V}(\hat{l}) = s_W^2 \left[\frac{a_1^2}{n_1} + \frac{a_2^2}{n_2} + \cdots + \frac{a_t^2}{n_t} \right] = s_W^2 \sum_i \frac{a_i^2}{n_i}$$

where n_i is the number of sample observations selected from population i and s_W^2 is the mean square within samples obtained from the analysis of variance table for the one-way classification. If all sample sizes are the same (i.e., all $n_i = n$), then

$$\hat{V}(\hat{l}) = \frac{s_W^2}{n} \sum_i a_i^2$$

Many different contrasts can be formed among the t sample means. A special set of contrasts are defined as follows:

DEFINITION 9.2

Two contrasts l_1 and l_2, where

$$\hat{l}_1 = \sum_i a_i\bar{y}_i. \qquad \text{and} \qquad \hat{l}_2 = \sum_i b_i\bar{y}_i.$$

are said to be **orthogonal** if

$$\frac{a_1 b_1}{n_1} + \frac{a_2 b_2}{n_2} + \cdots + \frac{a_t b_t}{n_t} = \sum_{i=1}^{t} \frac{a_i b_i}{n_i} = 0$$

Note: If the sample sizes are the same, then the condition becomes

$$a_1 b_1 + a_2 b_2 + \cdots + a_t b_t = \sum_{i=1}^{t} a_i b_i = 0$$

mutually orthogonal

A set of contrasts is said to be **mutually orthogonal** if all pairs of contrasts in the set are orthogonal.

EXAMPLE 9.1

Consider a one-way classification for comparing $t = 4$ population means. Are the following contrasts orthogonal?

$$\hat{l}_1 = \bar{y}_1. - \bar{y}_2. \qquad \hat{l}_2 = \bar{y}_3. - \bar{y}_4.$$

Solution We can rewrite the contrasts in the following form:

$$\hat{l}_1 = \bar{y}_{1.} - \bar{y}_{2.} + 0(\bar{y}_{3.}) + 0(\bar{y}_{4.})$$
$$\hat{l}_2 = 0(\bar{y}_{1.}) + 0(\bar{y}_{2.}) + \bar{y}_{3.} - \bar{y}_{4.}$$

where we see that $a_1 = 1$, $a_2 = -1$, $a_3 = 0$, $a_4 = 0$, and $b_1 = 0$, $b_2 = 0$, $b_3 = 1$, $b_4 = -1$. It is then apparent that

$$\sum_i a_i b_i = a_1 b_1 + a_2 b_2 + a_3 b_3 + a_4 b_4 = 0$$

and hence the contrasts are orthogonal.

EXAMPLE 9.2

Refer to Example 9.1. Are the given contrasts orthogonal?

$$\hat{l}_1 = \bar{y}_{1.} - \bar{y}_{2.} \qquad \text{and} \qquad \hat{l}_2 = \bar{y}_{1.} - \bar{y}_{3.}$$

Solution Rewriting the contrasts as

$$\hat{l}_1 = \bar{y}_{1.} - \bar{y}_{2.} + 0(\bar{y}_{3.}) + 0(\bar{y}_{4.})$$
$$\hat{l}_2 = \bar{y}_{1.} + 0(\bar{y}_{2.}) - \bar{y}_{3.} + (\bar{y}_{4.})$$

we see that

$$\sum_i a_i b_i = (1)(1) + (-1)(0) + (0)(-1) + (0)(0) = 1$$

which indicates that the two contrasts are not orthogonal.

The concept of orthogonality between linear contrasts is important because, if two contrasts are orthogonal, then one contrast conveys no information about the other contrast. We will demonstrate that $t - 1$ orthogonal contrasts can be formed using the t sample means, $\bar{y}_{i.}$s. These $t - 1$ **contrasts** form a set of mutually orthogonal contrasts. (An easy way to remember $t - 1$ is to refer to the number of degrees of freedom associated with the treatment (between-sample) source of variability in the AOV table.) In addition, it can be shown that the sums of squares for the $t - 1$ contrasts will add up to the treatment (between-sample) sum of squares. Mutual orthogonality is desirable because it leads to the independence of the $t - 1$ sums of squares associated with the $t - 1$ orthogonal contrasts. Thus, we can take the $t - 1$ degrees of freedom associated with the treatment sum of squares that describe any differences in the treatment means and break them into $t - 1$ independent explanations of how the treatment means may differ. We will now further develop these ideas and illustrate the concepts with an example.

A sum of squares associated with a treatment contrast is calculated to indicate the amount of variation in the treatment means that can be *explained* by that particular contrast. For each contrast $\hat{l} = \sum_{i=1}^{t} a_i \bar{y}_{i.}$, we can calculate a sum of squares associated with that contrast (SSC):

$$\text{SSC} = \frac{\left(\sum_{i=1}^{t} a_i \bar{y}_{i.}\right)^2}{\sum_{i=1}^{t} (a_i^2/n_i)} = \frac{(\hat{l})^2}{\sum_{i=1}^{t} (a_i^2/n_i)}$$

When we have equal sample sizes, this formula simplifies to

$$SSC = \frac{n(\hat{l})^2}{\sum_{i=1}^{t} a_i^2}$$

Associated with each such sum of squares is 1 degree of freedom. Thus, we can obtain $t - 1$ orthogonal contrasts such that the sum of squares treatment, which has $t - 1$ degrees of freedom, equals the total of the $t - 1$ sum of squares associated with each of the contrasts. The following example illustrates these calculations.

EXAMPLE 9.3

Various agents are used to control weeds in crops. Of particular concern is the overusage of chemical agents. Although effective in controlling weeds, these agents may also drain into the underground water system and cause health problems. Thus, several new biological weed agents have been proposed to eliminate the contamination problem present in chemical agents. Researchers conducted a study of biological agents to assess their effectiveness in comparison to the chemical weed agents. The study consisted of a control (no agent), two biological agents (Bio1 and Bio2), and two chemical agents (Chm1 and Chm2). Thirty 1-acre plots of land were planted with hay. Six plots were randomly assigned to receive one of the five treatments. The hay was harvested and the total yield in tons per acre was recorded. The data are given here.

Agent	1	2	3	4	5
Type	None	Bio1	Bio2	Chm1	Chm2
$\bar{y}_{i\cdot}$	1.175	1.293	1.328	1.415	1.500
s_i	.1204	.1269	.1196	.1249	.1265
n_i	6	6	6	6	6

Determine four orthogonal contrasts and demonstrate that the total of the four sums of squares associated with the four contrasts equals the sum of squares for treatment (between-samples).

Solution An analysis of variance was conducted on these data yielding the results summarized in the AOV table given here.

Source	df	SS	MS	F	p-value
Treatment	4	.3648	.0912	5.96	.0016
Error	25	.3825	.0153		
Totals	29	.7472			

From the AOV table, we have that $SS_{Trt} = .3648$. We will now construct four orthogonal contrasts in the five treatment means and demonstrate that SS_{Trt} can be partitioned into four terms, each representing a 1-degree of freedom sum of square associated with a particular contrast. Table 9.3 contains the coefficients and sum of squares for each of the four contrasts.

TABLE 9.3 Sum of squares computations for weed control experiment

| | Treatment | | | | | | | |
Contrast	1(Cntrl) a_1	2(Bio1) a_2	3(Bio2) a_3	4(Chm1) a_4	5(Chm2) a_5	$\sum_{i=1}^{5} a_i^2$	$\hat{l}$	SSC$_i$
Control vs. Agents	4	−1	−1	−1	−1	20	−.836	.2097
Biological vs. Chemical	0	1	1	−1	−1	4	−.294	.1297
Biol vs. Bio2	0	1	−1	0	0	2	−.035	.0037
Chm1 vs. Chm2	0	0	0	1	−1	2	−.085	.0217
$\bar{y}_{i \cdot}$	1.175	1.293	1.328	1.415	1.500			.3648

To illustrate the calculations involved in Table 9.3, we will compute the sum of squares associated with the first contrast, control vs. agents. First, note that the contrast represents a comparison of the yield for the control treatment versus the average yield of the four active agents. We initially would have written this contrast as

$$l = \mu_1 - \frac{(\mu_2 + \mu_3 + \mu_4 + \mu_5)}{4}$$

$$= (1)\mu_1 + \left(\frac{-1}{4}\right)\mu_2 + \left(\frac{-1}{4}\right)\mu_3 + \left(\frac{-1}{4}\right)\mu_4 + \left(\frac{-1}{4}\right)\mu_5$$

However, we can multiply each coefficient by 4 and change the coefficients from

$$a_1 = 1 \qquad a_2 = \frac{-1}{4} \qquad a_3 = \frac{-1}{4} \qquad a_4 = \frac{-1}{4} \qquad a_5 = \frac{-1}{4}$$

to

$$a_1 = 4 \qquad a_2 = -1 \qquad a_3 = -1 \qquad a_4 = -1 \qquad a_5 = -1$$

Next, we calculate

$$\sum_{i=1}^{5} a_i^2 = (4)^2 + (-1)^2 + (-1)^2 + (-1)^2 + (-1)^2 = 20$$

and

$$\hat{l} = (4)(1.175) + (-1)(1.293) + (-1)(1.328) + (-1)(1.415) + (-1)(1.500)$$

$$= -.836$$

Finally, we can obtain the sum of squares associated with the contrast from

$$SSC_1 = \frac{(\hat{l})^2}{\sum_{i=1}^{5}(a_i^2/n_i)} = \frac{n(\hat{l})^2}{\sum_{i=1}^{5} a_i^2} = \frac{6(-.836)^2}{20} = .2097$$

The remaining three sums of squares are calculated in a similar fashion. From Table 9.3 we thus obtain

$$SSC_1 + SSC_2 + SSC_3 + SSC_4 = .2097 + .1297 + .0037 + .0217 = .3648 = SS_{Trt}$$

You will verify that the four contrasts are indeed mutually orthogonal in an exercise.

Example 9.3 illustrated how we can decompose differences in the treatment means into individual contrasts that represent various comparisons of the treatment means. After defining the contrasts and obtaining their estimates and sum of squares, we need to determine which of the contrasts are significantly different from zero. A value of zero for a contrast would indicate that the difference in the means represented by the contrast does not exist. For example, if our contrast l_1 (control versus agents) was determined to be zero, then we would conclude that the average yield on plots assigned no agent (control) was equal to the average yield across all plots having one of the four agents. We will now present a test of the hypothesis that a contrast $l = \sum_{i=1}^{t} a_i \mu_i$ is different from zero. Our test will be a variation of the F test from AOV. Because the sum of squares associated with a contrast has 1 degree of freedom, its mean square is the same as its sum of squares. The test statistic is simply

$$F = \frac{\text{SSC}}{\text{MS}_{\text{Error}}} = \frac{\text{SSC}}{s_W^2}$$

The test procedure is summarized here.

F Test for Contrasts

H_0: $l = a_1\mu_1 + a_2\mu_2 + \cdots + a_t\mu_t = 0$

H_a: $l = a_1\mu_1 + a_2\mu_2 + \cdots + a_t\mu_t \neq 0$

T.S.: $F = \dfrac{\text{SSC}}{\text{MS}_{\text{Error}}}$

R.R.:   For a specified value of α, reject H_0 if F exceeds the tabled F value (Table 8) for $a = \alpha$, $df_1 = 1$, and $df_2 = n_T - t$.

Check assumptions and draw conclusions.

EXAMPLE 9.4

Refer to Example 9.3. The researchers were very interested in determining whether the biological agents would perform as well as the chemical agents. Is there a significant difference between the control treatment and the four active agents for weed control with respect to their effect on average hay production? Test each of the four contrasts for significance.

Solution   From the table of summary statistics in Example 9.3, the sample standard deviations are nearly equal. In fact, $F_{\text{max}} = (.1269)^2/(.1196)^2 = 1.13$. From Table 12, $F_{\text{max},.05,5,5} = 16.3$. Thus, we have very little reason to suspect that the five population variances are unequal. The AOV table in Example 9.3 has a p-value of .0016. Thus, we have a very strong rejection of H_0: $\mu_1 = \mu_2 = \mu_3 = \mu_4 = \mu_5$. We thus conclude that there are significant (p-value = .0016) differences in the five treatment means. We can investigate the types of differences in these means using the four contrasts that we constructed in Example 9.3. The four test statistics are computed here with $F_i = \text{SSC}_i/\text{MS}_{\text{Error}}$.

$$F_1 = \frac{.2097}{.0153} = 13.71 \qquad F_2 = \frac{.1297}{.0153} = 8.48 \qquad F_3 = \frac{.0037}{.0153} = 0.24$$

$$F_4 = \frac{.0217}{.0153} = 1.42$$

From Table 8, with $a = .05$, $df_1 = 1$, $df_2 = 30 - 5 = 25$, we obtain $F_{.05,1,25} = 4.24$. Thus, we conclude that contrasts l_1 and l_2 were significantly different from zero but contrasts l_3 and l_4 were not significantly different from zero. Using contrast l_1, we could thus conclude that the mean yields from plots using a weed control agent produced significantly higher yields than plots on which no agent was used. From contrast l_2, we infer that the mean yields from fields using biological agents for weed control would tend to have lower yields than those with chemical agents. However, we would need to investigate the size of the differences in the mean yields to determine whether the differences were of economical importance rather than just statistically significantly different. If the differences were economically significant, the ecological gains from using the biological agents may justify their use in place of chemical agents.

When we select contrasts for a study, the goal is not to obtain a set of orthogonal contrasts that yield a decomposition of the sum of squares treatment into $t - 1$ components. Rather, the goal is to obtain contrasts of the treatment means that will elicit a clear explanation of the pattern of differences in the treatment means of most benefit to the researcher. The mutual orthogonality of the contrasts is somewhat of a fringe benefit of the selection process. For example, in the analysis of the weed agents, we may have also been interested in comparing the control treatment to the average of the two biological agents. This contrast would not have been orthogonal to several of the contrasts we had already designed. We could have still used this contrast and tested its significance using the experimental data. The choice of which contrasts to evaluate should be determined by the overall goals of the experimenter and not by orthogonality.

One problem we do encounter when testing a number of contrasts is referred to as multiple comparisons. When we have tested several contrasts, each with a Type I error rate of α, the chance of at least one Type I error occurring during the several tests becomes somewhat larger than α. In the next section, we will address this difficulty.

EXERCISES

Basic Techniques

9.1 Consider the expressions

$$\hat{l}_1 = \bar{y}_1. + \bar{y}_2. - 2\bar{y}_3.$$

$$\hat{l}_2 = \bar{y}_1. + \bar{y}_2. - 2\bar{y}_4.$$

a. Are $\hat{l}_1$ and $\hat{l}_2$ linear contrasts?

b. Are $\hat{l}_1$ and $\hat{l}_2$ orthogonal?

9.2 In Example 9.3, we defined four contrasts in the five treatment means. Verify that the four contrasts are mutually orthogonal.

9.3 In the case study described earlier in this chapter, the researchers were interested in answering several questions concerning the difference in how the raters reacted to various handicaps. For each of the following questions, write a contrast in the five condition mean ratings that would attempt to answer the researchers' question.

a. Question 1: Is the average rating for the control group (no handicap) greater than the average rating for all types of handicapped applicants?

b. Question 2: Is the average qualification rating for the hard-of-hearing applicant different from the average rating for those applicants that had a mobility handicap?

c. Question 3: Is the average qualification rating for the crutches applicants different from the average rating of the applicants who were either amputees or in a wheelchairs?

d. Question 4: Is the average rating for the amputee applicants different from the average rating of the wheelchair applicants?

9.4 Refer to Exercise 9.3. For each of the following pairs of contrasts, determine whether they are orthogonal.
   a. Question 1 and question 2
   b. Question 1 and question 3
   c. Question 1 and question 4
   d. Question 2 and question 3
   e. Question 2 and question 4
   f. Question 3 and question 4
   g. Are the four contrasts mutually orthogonal?

9.5 Refer to Example 8.4. The researchers were interested in determining whether the mean oxygen content was lower for samples taken near the mouth of the Mississippi than for samples taken further away from the mouth. Write a contrast to answer each of the following questions and test whether the contrast is different from zero using $\alpha = .05$. Clearly summarize your results.
   a. Is the mean oxygen content at 20 KM different than the average of the mean oxygen content at 1 KM, 5 KM, and 10 KM?
   b. Is the mean oxygen content at 10 KM different than the average of the mean oxygen content at 1 KM and 5 KM?
   c. Is the mean oxygen content at 5 KM different than the average of the mean oxygen content at 1 KM?
   d. Are the three contrasts defined in (a), (b), and (c) mutually orthogonal?
   e. Do the three contrast sums of squares total to SS_{Trt}?

9.3 Which Error Rate Is Controlled?

An experimenter wishes to compare t population (treatment) means using m contrasts. Each of the m contrasts can be tested using the F test we introduced in the previous section. Suppose each of the contrasts is tested with the same value of α, which we will denote as α_I, called the **individual comparisons** Type I error rate. Thus, we have an α_I chance of making a Type I error on each of the m tests. We need to also consider the probability of falsely rejecting at least one of the m null hypotheses, called the **experimentwise** Type I error rate and denoted by α_E. The value of α_E takes into account that we are conducting m tests, each having an α_I chance of making a Type I error. Now, if MS_{Error} has an infinite number of degrees of freedom (so the tests are independent), then when all m null hypotheses are true, the probability of falsely rejecting at least one of the m null hypotheses can be shown to be $\alpha_E = 1 - (1 - \alpha_I)^m$. Table 9.4 contains values

individual comparisons

experimentwise error rate

TABLE 9.4
A comparison of the experimentwise error rate α_E for m independent contrasts among $t(t > m)$ sample means

m, Number of Contrasts	α_I Probability of a Type I Error on an Individual Test		
	.10	.05	.01
1	.100	.050	.010
2	.190	.097	.020
3	.271	.143	.030
4	.344	.185	.039
5	.410	.226	.049
⋮	⋮	⋮	⋮
10	.651	.401	.096

of α_E for various values of m and α_I. We can observe from Table 9.4 that as the number of tests m increases for a given value of α_I, the probability of falsely rejecting H_0 on at least one of the m tests α_E becomes quite large. For example, if an experimenter wanted to compare $t = 20$ population means by using $m = 10$ orthogonal contrasts, the probability of falsely rejecting H_0 on at least one of the t tests could be as high as .401 when each individual test was performed with $\alpha_I = .05$.

In any practical problem, the degrees of freedom for MS_{Error} will not be infinite and hence the tests will not be independent. Thus, the relationship between α_E and α_I is not generally as described in Table 9.4. It is difficult to obtain an expression equivalent to $\alpha_E = 1 - (1 - \alpha_I)^m$ for comparisons made with tests that are not independent. However, it can be shown that for most of the types of comparisons we will be making among the population means, the following upper bound exists for the experimentwise error rate:

$$\alpha_E \leq 1 - (1 - \alpha_I)^m$$

Thus, we know the largest possible value for α_E when we set the value of α_I for each of the individual tests. Suppose, for example, that we wish the experimentwise error rate for $m = 8$ contrasts among $t = 20$ population means to be at most .05. What value of α_I must we use on the m tests to achieve an overall error rate of $\alpha_E = .05$? We can use the previous upper bound to determine that if we select

$$\alpha_I = 1 - (1 - \alpha_E)^{1/m} = 1 - (1 - .05)^{1/8} = .0064$$

then we will have $\alpha_E \leq .05$. The only problem is that this procedure may be very conservative with respect to the experimentwise error rate, and hence an inflated probability of Type II error may result.

We will now consider a method that will work for any set of m tests and is much easier to apply in obtaining an upper bound on α_E. The results of Table 9.4 are disturbing when we are conducting a number of tests. The chance of making at least one Type I error may be considerably larger than the selected individual error rates. This could lead us to question significant results when they appear in our analysis of experimental results. The problem can be alleviated somewhat by *controlling the experimentwise error rate* α_E rather than the *individual error rate* α_I. We need to select a value of α_I that will provide us with an acceptable value

Bonferroni inequality

for α_E. The **Bonferroni inequality** provides us with a method for selecting α_I so that α_E is bounded below a specified value. This inequality states that the overall Type I error rate α_E is less than or equal to the sum of the individual error rates for the m tests. Thus, when each of the m tests have the same individual error rates α_I, the Bonferroni inequality yields

$$\alpha_E \leq m\alpha_I$$

If we wanted to guarantee that the chance of a Type I error was at most α, we could select

$$\alpha_I = \frac{\alpha}{m}$$

for each of the m tests. Then,

$$\alpha_E \leq m\alpha_I = m \left(\frac{\alpha}{m} \right) = \alpha$$

The experimentwise error rate is thus less than or equal to our specified value. Just as we mentioned earlier, this procedure may be very conservative with respect to the experimentwise error rate, and hence an inflated probability of Type II error may result.

EXAMPLE 9.5

Refer to Example 9.4, where we constructed $m = 4$ contrasts (comparisons) among the $t = 5$ treatment means. If we wanted to control the experimentwise error rate at a level of $\alpha_E = .05$, then we would take

$$\alpha_I = \frac{0.5}{4} = .0125$$

The critical value for the F tests would then be $F_{.0125,1,25} = 7.24$ as opposed to $F_{.05,1,25} = 4.24$ if we ignore the fact that we are conducting multiple tests on the treatment means. We would then reject H_0 if $SSC_i/MS_{Error} \geq 7.24$. From Example 9.4, the four F ratios were

$$F_1 = 13.71 \qquad F_2 = 8.48 \qquad F_3 = 0.24 \qquad F_3 = 1.42$$

Using the Bonferroni procedure, we would declare contrast l_1 and l_2 significantly different from 0 because their F ratios are greater than 7.24. Using the Bonferroni test procedure, we are assured that the chance of making at least one Type I error during the four tests is at most .05. Using $\alpha = .05$ for each of the four procedures would not have allowed us to assess the exact probability of making a Type I error among the four comparisons. However, this value would have been considerably larger than .05, possibly as large as .20.

The Bonferroni procedure gives us a method for evaluating a small number of contrasts that were selected prior to observing the data while preserving a selected experimentwise Type I error rate. In many experiments, the researcher will want to compare all pairs of treatments or compare all treatments to a control. In these situations, there are many methods for testing these types of contrasts among the treatment means. A major difference among these multiple-comparison procedures is the type of error rate that each procedure controls. We will discuss several of these procedures in the next sections.

9.4   Fisher's Least Significant Difference

Recall that we are interested in determining which population means differ after we have rejected the hypothesis of equality of t population means in an analysis of variance. R. A. Fisher (1949) developed a procedure for making pairwise comparisons among a set of t population means. The procedure is called Fisher's least significant difference (LSD).

The α-level of Fisher's LSD is valid for a given comparison only if the LSD is used for independent (orthogonal) comparisons or for preplanned comparisons. However, since many people find Fisher's LSD easy to compute and hence use it for making all possible pairwise comparisons (particularly those that look "interesting" following the completion of the experiment), researchers recommend applying Fisher's LSD only after the F test for treatments has been shown to be

Fisher's protected LSD   significant. This revised approach is sometimes referred to as **Fisher's protected**

LSD. Simulation studies [Cramer and Swanson (1973)] suggest that the error rate for the protected LSD is controlled on an experimentwise basis at a level approximately equal to the α-level for the F test.

We will illustrate Fisher's protected procedure, but will continue to call it Fisher's LSD. This procedure is summarized here.

Fisher's Least Significant Difference Procedure

1. Perform an analysis of variance to test $H_0: \mu_1 = \mu_2 = \cdots = \mu_t$ against the alternative hypothesis that at least one of the means differs from the rest.
2. If there is insufficient evidence to reject H_0 using $F = $ MSB/MSW, proceed no further.
3. If H_0 is rejected, define the **least significant difference (LSD)** to be the observed difference between two sample means necessary to declare the corresponding population means different.
4. For a specified value of α, the least significant difference for comparing μ_i to μ_j is

$$LSD_{ij} = t_{\alpha/2}\sqrt{s_W^2\left(\frac{1}{n_i} + \frac{1}{n_j}\right)}$$

where n_i and n_j are the respective sample sizes from population i and j and t is the critical t value (Table 2 of the Appendix) for $a = \alpha/2$ and df denoting the degrees of freedom for s_W^2. Note that for $n_1 = n_2 = \cdots = n_t = n$,

$$LSD_{ij} \equiv LSD = t_{\alpha/2}\sqrt{\frac{2s_W^2}{n}}$$

5. Then compare all pairs of sample means. If $|\bar{y}_{i\cdot} - \bar{y}_{j\cdot}| \geq LSD_{ij}$, declare the corresponding population means μ_i and μ_j different.
6. For each pairwise comparison of population means, the probability of a Type I error is fixed at a specified value of α.

Note: The LSD procedure is analogous to a two-sample t-test for any two population means μ_i and μ_j, except that we use s_W^2, the pooled estimator of the population variance σ^2 from all t samples, rather than the pooled sample variance from samples i and j. Also, the degrees of freedom for the t value is $df_W = n_T - t$ from the analysis of variance, rather than $n_i + n_j - 2$.

EXAMPLE 9.6

Refer to Example 9.3, where we had five different weed agents and $n = 6$ plots of land assigned to each of the agents. The analysis of variance is given in Table 9.5.

TABLE 9.5
AOV table for the data of Example 9.3

Source	df	SS	MS	F	p-value
Treatment	4	.3648	.0912	5.96	.0016
Error	25	.3825	.0153		
Totals	29	.7472			

steps for LSD procedure

Solution We can solve this problem by following the five steps listed for the LSD procedure.

Step 1. We use the AOV table in Table 9.5. The F test of H_0: $\mu_1 = \mu_2 = \cdots = \mu_5$ is based on

$$F = \frac{\text{MSB}}{\text{MSW}} = \frac{\text{MS}_\text{Trt}}{\text{MS}_\text{Error}} = 5.96$$

For $\alpha = .05$ with $df_1 = 4$ and $df_2 = 25$, we reject H_0 if F exceeds 2.76 (see Table 8 in the Appendix).

Steps 2, 3. Since 5.96 is greater than 2.76, we reject H_0 and conclude that at least one of the population means differs from the rest ($p = .0016$).

Step 4. The least significant difference for comparing two means based on samples of size 6 is then

$$\text{LSD} = t_{\alpha/2} \sqrt{\frac{2\,\text{MS}_\text{Error}}{6}} = 2.060 \sqrt{\frac{2(.0153)}{6}} = .1471$$

Note that the appropriate t value (2.060) was obtained from Table 2 with $a = \alpha/2 = .025$ and $df = 25$.

Step 5. When we have equal sample sizes, it is convenient to use the following procedures rather than make all pairwise comparisons among the sample means, because the same LSD is to be used for all comparisons.

a. We rank the sample means from lowest to highest.

Agent	1	2	3	4	5
$\bar{y}_{i\cdot}$	1.175	1.293	1.328	1.415	1.500

b. We compute the sample difference

$$\bar{y}_\text{largest} - \bar{y}_\text{smallest}$$

If this difference is greater than the LSD, we declare the corresponding population means significantly different from each other. Next we compute the sample difference

$$\bar{y}_\text{2nd largest} - \bar{y}_\text{smallest}$$

and compare the result to the LSD. We continue to make comparisons with $\bar{y}_\text{smallest}$:

$$\bar{y}_\text{3rd largest} - \bar{y}_\text{smallest}$$

and so on, until we find either that all sample differences involving $\bar{y}_\text{smallest}$ exceed the LSD (and hence the corresponding population means are different) or that a sample difference involving $\bar{y}_\text{smallest}$ is less than the LSD. In the latter case, we stop and make no further comparisons with $\bar{y}_\text{smallest}$. For our data, comparisons with $\bar{y}_\text{smallest}$, $\bar{y}_{1\cdot}$ give the following results:

Comparison		Conclusion
$\bar{y}_\text{largest} - \bar{y}_\text{smallest} = \bar{y}_{5\cdot} - \bar{y}_{1\cdot} = .325$		>LSD; proceed
$\bar{y}_\text{2nd largest} - \bar{y}_\text{smallest} = \bar{y}_{4\cdot} - \bar{y}_{1\cdot} = .240$		>LSD; proceed
$\bar{y}_\text{3rd largest} - \bar{y}_\text{smallest} = \bar{y}_{3\cdot} - \bar{y}_{1\cdot} = .153$		>LSD; proceed
$\bar{y}_\text{4th largest} - \bar{y}_\text{smallest} = \bar{y}_{2\cdot} - \bar{y}_{1\cdot} = .118$		<LSD; stop

To summarize our results, we make the following diagram:

$$\text{Agent} \quad \underline{1 \quad 2} \quad 3 \quad 4 \quad 5$$

Those populations joined by the underline have means that are not significantly different from $\bar{y}_{1.}$. Note that agents 3, 4, and 5 have sample differences with agent 1 that exceed LSD and hence are not underlined.

c. We now make similar comparisons with $\bar{y}_{\text{2nd smallest}}$, $\bar{y}_{2.}$. In this case, we use the procedure of part (b).

Comparison	Conclusion
$\bar{y}_{5.} - \bar{y}_{2.} = .207$	>LSD; proceed
$\bar{y}_{4.} - \bar{y}_{2.} = .122$	<LSD; stop

$$\text{Agent} \quad 1 \quad \underline{2 \quad 3 \quad 4} \quad 5$$

d. Continue the comparisons with $\bar{y}_{\text{3rd smallest}}$, $\bar{y}_{3.}$ in this example.

Comparison	Conclusion
$\bar{y}_{5.} - \bar{y}_{3.} = .172$	>LSD; proceed
$\bar{y}_{4.} - \bar{y}_{3.} = .087$	<LSD; stop

$$\text{Agent} \quad 1 \quad 2 \quad \underline{3 \quad 4} \quad 5$$

e. Continue the comparisons with $\bar{y}_{\text{4th smallest}}$, $\bar{y}_{4.}$ in this example.

Comparison	Conclusion
$\bar{y}_{5.} - \bar{y}_{4.} = .085$	<LSD; stop

$$\text{Agent} \quad 1 \quad 2 \quad 3 \quad \underline{4 \quad 5}$$

f. We can summarize steps (a) through (e) as follows:

$$\text{Agent} \quad \underline{1 \quad 2} \quad 3 \quad 4 \quad 5$$

Those populations not underlined by a common line are declared to have means that are significantly different according to the LSD criterion. Note that we can eliminate the third line from the top of part (f) since it is contained in the second line from the top. The revised summary of significant and nonsignificant results is

$$\text{Agent} \quad \underline{1 \quad 2} \quad 3 \quad 4 \quad 5$$

In conclusion, we have μ_1, μ_2, and μ_3 significantly less than μ_5. Also, μ_3 and μ_4 are significantly greater than μ_1.

Although the LSD procedure described in Example 9.6 may seem quite laborious, its application is quite simple. First, we run an analysis of variance. If we reject the null hypothesis of equality of the population means, we compute the LSD for all pairs of sample means. When the sample sizes are the same, this difference is a single number for all pairs. We can use the stepwise procedure described in steps 5(a) through 5(f) of Example 9.6. We need not write down all those steps, only the summary lines. The final summary, as given in step 5(f), gives a handy visual display of the pairwise comparisons using Fisher's LSD.

Several remarks should be made concerning the LSD method for pairwise comparisons. First, there is the possibility that the overall F test in our analysis of variance is significant but that no pairwise differences are significant using the LSD procedure. This apparent anomaly can occur because the null hypothesis $H_0: \mu_1 = \mu_2 = \cdots = \mu_t$ for the F test is equivalent to the hypothesis that all possible comparisons (paired or otherwise) among the population means are zero. For a given set of data, the comparisons that are significant might not be of the form $\mu_i - \mu_j$, the form we are using in our paired comparisons.

Fisher's confidence interval

Second, Fisher's LSD procedure can also be used to form a confidence interval for $\mu_i - \mu_j$. A $100(1 - \alpha)\%$ confidence interval has the form

$$(\bar{y}_{i\cdot} - \bar{y}_{j\cdot}) \pm \text{LSD}_{ij}$$

LSD for equal sample sizes

Third, when all sample sizes are the same, the LSD for all pairs is

$$t_{\alpha/2} \sqrt{\frac{2s_W^2}{n}}$$

9.5 Tukey's *W* Procedure

We are aware of the major drawback of a multiple-comparison procedure with a controlled per-comparison error rate. Even when $\mu_1 = \mu_2 = \cdots = \mu_t$, unless α, the per-comparison error rate (such as with Fisher's unprotected LSD) is quite small, there is a high probability of declaring at least one pair of means significantly different when running multiple comparisons. To avoid this, other multiple-comparison procedures have been developed that control different error rates.

Studentized range distribution

Tukey (1953) proposed a procedure that makes use of the **Studentized range distribution.** When more than two sample means are being compared, to test the largest and smallest sample means, we could use the test statistic

$$\frac{\bar{y}_{\text{largest}} - \bar{y}_{\text{smallest}}}{s_p \sqrt{1/n}}$$

where n is the number of observations in each sample and s_p is a pooled estimate of the common population standard deviation σ. This test statistic is very similar to that for comparing two means, but it does not possess a t distribution. One reason it does not is that we have waited to determine which two sample means (and hence population means) we would compare until we observed the largest and smallest sample means. This procedure is quite different from that of specifying $H_0: \mu_1 - \mu_2 = 0$, observing $\bar{y}_{1\cdot}$ and $\bar{y}_{2\cdot}$, and forming a t statistic.

The quantity

$$\frac{\bar{y}_{\text{largest}} - \bar{y}_{\text{smallest}}}{s_p \sqrt{1/n}}$$

follows a Studentized range distribution. We will not discuss the properties of this distribution, but will illustrate its use in Tukey's multiple-comparison procedure.

Tukey's W Procedure

W

$q_\alpha(t, v)$

1. Rank the t sample means.
2. Two population means μ_i and μ_j are declared different if

   $$|\bar{y}_{i\cdot} - \bar{y}_{j\cdot}| \geq W$$

   where

   $$W = q_\alpha(t, v) \sqrt{\frac{s_W^2}{n}}$$

   s_W^2 is the mean square within samples based on v degrees of freedom, $q_\alpha(t, v)$ is the **upper-tail critical value of the Studentized range** for comparing t different populations, and n is the number of observations in each sample. A discussion follows showing how to obtain values of $q_\alpha(t, v)$ from Table 10 in the Appendix.

experimentwise error rate

3. The error rate that is controlled is an **experimentwise error rate.** Thus, the probability of observing an experiment with one or more pairwise comparisons falsely declared to be significant is specified at α.

We can obtain values of $q_\alpha(t, v)$ from Table 10 in the Appendix. Values of v are listed along the left column of the table with values of t across the top row. Upper-tail values for the Studentized range are then presented for $a = .05$ and $.01$. For example, in comparing 10 population means based on 9 degrees of freedom for s_W^2, the .05 upper-tail critical value of the Studentized range is $q_{.05}(10, 9) = 5.74$.

EXAMPLE 9.7

Refer to the data of Example 9.3. Use Tukey's W procedure with $\alpha = .05$ to make pairwise comparisons among the five population means.

Solution   Step 1 is to rank the sample means from smallest to largest, to produce

Agent	1	2	3	4	5
$\bar{y}_{i\cdot}$	1.175	1.293	1.328	1.415	1.500

For the experiment described in Example 9.6, we have

$t = 5$ (we are making pairwise comparisons among five means)
$v = 25$ (s_W^2 had degrees of freedom equal to df_{Error} in the AOV)
$\alpha = .05$ (we specified α_E, the experimentwise error rate at .05)
$n = 6$ (there were six plots randomly assigned to each of the agents)

We find in Table 10 of the Appendix that

$$q_\alpha(t, v) = q_{.05}(5, 25) \approx 4.158$$

The absolute value of each difference in the sample means $|\bar{y}_i - \bar{y}_i|$ must then be compared to

$$W = q_\alpha(t, v) \sqrt{\frac{s_W^2}{n}} = 4.158 \sqrt{\frac{.0153}{6}} = .2100$$

By substituting W for LSD, we can use the same stepwise procedure for comparing sample means that we used in step 5 of the solution to Example 9.6. Having ranked the sample means from low to high, we compare against $\bar{y}_{smallest}$, which is $\bar{y}_1$, as follows:

Comparison		Conclusion
$\bar{y}_{largest} - \bar{y}_{smallest} = \bar{y}_{5.} - \bar{y}_{1.} = .325$		$>W$; proceed
$\bar{y}_{2nd\ largest} - \bar{y}_{smallest} = \bar{y}_{4.} - \bar{y}_{1.} = .240$		$>W$; proceed
$\bar{y}_{3rd\ largest} - \bar{y}_{smallest} = \bar{y}_{3.} - \bar{y}_{1.} = .153$		$<W$; stop

To summarize our results we make the following diagram:

Agent   <u>1   2   3</u>   4   5

Comparison with $\bar{y}_{2nd\ smallest}$, which is $\bar{y}_{2.}$, yields

Comparison	Conclusion
$\bar{y}_{5.} - \bar{y}_{2.} = .207$	$<W$; stop

Agent   1   <u>2   3   4   5</u>

Similarly, comparisons of $\bar{y}_{5.}$ with $\bar{y}_{3.}$ and $\bar{y}_{4.}$ yield

Agent   1   2   <u>3   4   5</u>

Combining our results, we obtain

Agent   <u>1   2   3</u>   4   5

which simplifies to

Agent   <u>1   2   3</u>   4   5

All populations not underlined by a common line have population means that are significantly different from each other; that is, μ_4 and μ_5 are significantly larger than μ_1. No other pairs of means are significantly different.

By examining the multiple-comparison summaries using the least significant difference (Example 9.6) and Tukey's W procedure (Example 9.7), we see that Tukey's procedure is more conservative (declares fewer significant differences) than the LSD procedure. For example, when applying Tukey's procedure to the data in Example 9.6, we found that μ_3 is no longer significantly larger than μ_1. Similarly, μ_2 and μ_3 are no longer significantly less than μ_5. The explanation for this is that although both procedures have an experimentwise error rate, we have shown the per-comparison error rate of the protected LSD method to be larger than that for Tukey's W procedure.

A limitation of Tukey's procedure is the requirement that all the sample means are based on the same number of data values. There are no available tables for the case in which the sample sizes are unequal. If the sample sizes are only somewhat different, then Miller (1981) recommends replacing the value of n in the formula for W with the harmonic mean of the n_is:

$$n = \frac{t}{\dfrac{1}{n_1} + \dfrac{1}{n_2} + \cdots + \dfrac{1}{n_t}}$$

For large departures from equal sample sizes, the experimenter should consider using either the LSD procedure or a Bonferroni test procedure.

Tukey's confidence interval   Tukey's procedure can also be used to construct confidence intervals for comparing two means. However, unlike the confidence intervals that we can form from Fisher's LSD, Tukey's procedure enables us to construct simultaneous confidence intervals for all pairs of treatment differences. For a specified α level from which we compute W, the overall probability is $1 - \alpha$ that all differences $\mu_i - \mu_j$ will be included in an interval of the form

$$(\bar{y}_{i\cdot} - \bar{y}_{j\cdot}) \pm W$$

that is, the probability is $1 - \alpha$ that all the intervals $(\bar{y}_{i\cdot} - \bar{y}_{j\cdot}) \pm W$ include the corresponding population differences $\mu_i - \mu_j$.

9.6   Student–Newman–Keuls Procedure

The Student–Newman–Keuls (SNK) procedure provides a modification of the Tukey W procedure. Although the SNK procedure also makes use of the Studentized range statistic, different critical values are used depending on the number of steps separating the means being tested. To compare the two procedures, let's refer to Example 9.3. Ranked in order from lowest to highest, the sample means are

Agent	1	2	3	4	5
$\bar{y}_{i\cdot}$	1.175	1.293	1.328	1.415	1.500

and the critical value of the Studentized range for Tukey's W procedure is

$$q_\alpha(t, v) = q_{.05}(5, 25) \approx 4.158$$

This same value of q is used for all pairwise comparisons of the five treatment means.

The SNK procedure makes use of a critical value

$$W_r = q_\alpha(r, v) \sqrt{\frac{s_W^2}{n}}$$

for means that are r steps apart when the t sample means are ranked from lowest to highest. For our example, $\bar{y}_{largest}$ and $\bar{y}_{smallest}$ are five steps apart, and they would be compared using

$$W_5 = q_\alpha(5, v) \sqrt{\frac{s_W^2}{n}}$$

$$= q_{.05}(5, 25) \sqrt{\frac{.0153}{6}} = 4.158 \sqrt{\frac{.0153}{6}} = .2100$$

(*Note:* This is W for Tukey's W procedure.) However, $\bar{y}_{largest}$ and $\bar{y}_{2nd\ smallest}$ are four steps apart, and they would be compared to

$$W_4 = q_\alpha(4, v) \sqrt{\frac{s_W^2}{n}}$$

$$= q_{.05}(4, 25) \sqrt{\frac{.0153}{6}} = 3.892 \sqrt{\frac{.0153}{6}} = .1965$$

The complete set of critical values W_r needed for the data of Example 9.3 is shown in Table 9.6. Values of $q_\alpha(r, v)$ are obtained from Table 10 in the Appendix by replacing t with r and approximating the value, because $v = 25$ is not given in the table.

TABLE 9.6

Values of r, $q_\alpha(r, v)$ and W_r for Example 9.3

r	2	3	4	5
$q_\alpha(r, v)$	2.915	3.523	3.892	4.158
W_r	.1472	.1779	.1965	.2100

The Student–Newman–Keuls procedure, which relies on the number of ordered steps between two sample means when determining the significance of an observed sample difference, has neither an experimentwise nor a per-comparison error rate. Rather, the error rate is defined for means the same number of ordered steps apart. Since the critical value W_r decreases as the number of steps between the means being compared decreases, the SNK procedure is less conservative and hence will generally declare more significant differences than will Tukey's W procedure, which utilizes the largest value for W no matter how many steps separate the means being compared. In fact, the critical value for Tukey's W is W_t, and $W_r < W_t$ for all $r < t$. Also, when the n_is are equal, the critical value for Fisher's LSD is W_2, and hence LSD $< W_r$ for all $r > 2$. Thus, SNK will generally declare fewer pairs significantly different than will Fisher's LSD procedure.

The SNK procedure is summarized here.

SNK Procedure

1. Rank the t sample means from lowest to highest.
2. For two means $\bar{y}_{i\cdot}$ and $\bar{y}_{j\cdot}$ that are r steps apart, we declare μ_i and μ_j different if

$$|\bar{y}_{i\cdot} - \bar{y}_{j\cdot}| \geq W_r$$

where $W_r = q_\alpha(r, v) \sqrt{s_W^2/n}$, n is the number of observations per sample, s_W^2 is the mean square within samples from the AOV table, v is the degrees of freedom for s_W^2, and $q_\alpha(r, v)$ is the critical value of the Studentized range. Values of $q_\alpha(r, v)$ are given in Table 10 in the Appendix for $\alpha = .05$ and .01.

(*Note:* Use the column labeled t to locate the desired value for r.)

EXAMPLE 9.8

Refer to the data of Example 9.3. Run the SNK procedure to make all pairwise comparisons based on $\alpha = .05$.

Solution   The critical values of W_r are given in Table 9.6.

1. Beginning with $\bar{y}_{largest}$, every sample mean is compared to $\bar{y}_{smallest}$, using the appropriate value of W_r. The results are summarized here.

Comparison	W_r	Conclusion
$\bar{y}_{5.} - \bar{y}_{1.} = .325$	.2100	$> W_5$; proceed
$\bar{y}_{4.} - \bar{y}_{1.} = .240$	.1965	$> W_4$; proceed
$\bar{y}_{3.} - \bar{y}_{1.} = .153$	.1779	$< W_3$; stop

2. Similarly, we can make comparisons with $\bar{y}_{2nd\ smallest}$.

Comparison	W_r	Conclusion
$\bar{y}_{5.} - \bar{y}_{2.} = .207$	.1965	$> W_4$; proceed
$\bar{y}_{4.} - \bar{y}_{2.} = .122$	.1779	$< W_3$; stop

3. Next, we can make comparisons with $\bar{y}_{3rd\ smallest}$.

Comparison	W_r	Conclusion
$\bar{y}_{5.} - \bar{y}_{3.} = .172$	.1779	$< W_3$; stop

4. Finally, we can make comparisons with $\bar{y}_{4th\ smallest}$.

Comparison	W_r	Conclusion
$\bar{y}_{5.} - \bar{y}_{4.} = .085$	.1471	$< W_2$; stop

The results of these multiple comparisons using the SNK procedure are shown here:

Agent   1   2   3   4   5

All populations not underlined by a common line have population means that are significantly different from each other; that is, μ_4 and μ_5 are significantly larger than μ_1. Also, μ_5 is significantly larger than μ_2 but no other pairs of means are significantly different using the SNK procedure. This example illustrates the fact that the SNK procedure tends to declare more differences (and hence be less conservative) than Tukey's W procedure.

The SNK procedure can be modified to account for different sample sizes. When $n_i \neq n_j$, the value of W_r for means r steps apart is modified in the same way as Tukey's procedure. Here we would use the harmonic mean of the ns in place of n in the formula for W_r:

$$n = \frac{t}{\dfrac{1}{n_1} + \dfrac{1}{n_2} + \cdots + \dfrac{1}{n_t}}$$

if the n_is were not too different. For large departures from equal sample sizes, the experimenter should consider using either the LSD procedure or a Bonferroni test procedure.

9.7

Dunnett's Procedure: Comparison of Treatments to a Control

In many studies and experiments, the researchers will include a control treatment for comparison purposes. There are many types of controls, but generally the control serves as a standard to which the other treatments may be compared. For example, in many situations the conditions under which the experiment is run may have such a strong effect on the response variable that generally effective treatments will not produce a favorable response in the experiment. For example, if the insect population is too dense, most insecticides used at a reasonable level would not provide a noticeable reduction in the insect population. Thus, a control spray with no active ingredient would reveal the level of insects in the sprayed region. A second situation in which a control is useful is when the experimental participants generate a favorable response whenever any reasonable treatment is applied; this is referred to as the **placebo effect.** In this type of study or experiment, the participants randomly assigned to the control treatment are handled exactly in the same manner as the participants receiving active treatments. In most clinical trials and experiments used to evaluate new drugs or medical treatments, a placebo treatment is included so as to determine the size of the placebo effect. Finally, a control may represent the current method or standard procedure to which any new procedures would be compared.

In experiments in which a control is included, the researchers would want to determine whether the mean responses from the active treatments differ from the mean for the control. Dunnett (1955) developed a procedure for comparisons to a control that controls the experimentwise Type I error rate. This procedure compares each treatment mean to the mean for the control $\bar{y}_c$ by comparing the difference in the sample means, $\bar{y}_i - \bar{y}_c$, to the critical difference

$$D = d_\alpha(k, v) \sqrt{\frac{s_W^2}{n}}$$

where $n_c = n_1 = \cdots n_{t-1} = n$. The Dunnett procedure requires equal sample sizes, $n_i = n_c$. The values for $d_\alpha(k, v)$ are given in Table 11 (Appendix). Dunnett (1964) describes adjustments to the values in Table 11 for the case of unequal n_i. The comparison can be either one-sided or two-sided, as is summarized here.

Dunnett's Procedure

1. For a specified value of α_E, Dunnett's D value for comparing μ_i to μ_c, the control mean is

$$D = d_\alpha(k, v) \sqrt{\frac{2s_W^2}{n}}$$

where n is the common sample size from treatment i and the control, $k = t - 1$, the number of noncontrol treatments; α is the desired experimentwise error rate; s_W^2 is the mean square within; v is the degrees of freedom associated with s_W^2; and $d_\alpha(k, v)$ is the critical Dunnett value (Table 11 of the Appendix).

2. For the two-sided alternative $H_a: \mu_i \neq \mu_c$, we declare μ_i different from μ_c if

$$|\bar{y}_i - \bar{y}_c| \geq D$$

where the value of $d_\alpha(k, v)$ is the two-sided value in Table 11.

3. For the one-sided alternative $H_a: \mu_i > \mu_c$, we declare μ_i greater than μ_c if

$$(\bar{y}_i - \bar{y}_c) \geq D$$

where the value of $d_\alpha(k, v)$ is the one-sided value in Table 11.

4. For the one-sided alternative $H_a: \mu_i > \mu_c$, we declare μ_i less than μ_c if

$$(\bar{y}_i - \bar{y}_c) \leq -D$$

where the value of $d_\alpha(k, v)$ is the one-sided value in Table 11.

5. The Type I error rate that is controlled is an *experimentwise error rate*. Thus, the probability of observing an experiment with one or more comparisons with the control falsely declared to be significant is specified at α.

EXAMPLE 9.9

Refer to the data of Example 9.3. Compare the two biological treatments and two chemical treatments to the control treatment using $\alpha = .05$.

Solution   We want to determine whether the biological and chemical treatments have increased hay production, so we will conduct one-sided comparisons with the control.

1. From Example 9.3, we had $s_W^2 = .0153$ with df $= 25$ and $t = 5$ treatments including the control treatment. The critical value of the Dunnett procedure is found in the one-sided portion of Table 11 with

$$\alpha = .05 \qquad k = 5 - 1 = 4 \qquad v = 25$$

yielding $d_{.05}(4, 25) = 2.28$. Since $n_c = n_2 = n_3 = n_4 = n_5 = 6$, we have

$$D = d_\alpha(k, v) \sqrt{\frac{2s_W^2}{n_c}} = 2.28 \sqrt{\frac{2(.0153)}{6}} = .163$$

2. We declare treatment mean μ_i greater than the control mean μ_c if $(\bar{y}_i - \bar{y}_c) \geq .163$. We can summarize the comparisons as follows:

Treatment	$(\bar{y}_i - \bar{y}_c)$	Comparison	Conclusion
Bio1	$(1.293 - 1.175) = .118$	$< D$	Not greater than control
Bio2	$(1.328 - 1.175) = .153$	$< D$	Not greater than control
Chm1	$(1.415 - 1.175) = .240$	$> D$	Greater than control
Chm2	$(1.500 - 1.175) = .325$	$> D$	Greater than control

3. We conclude that using either of the biological agents would result in an average hay production not greater than the production obtained

using no agent on the fields. Thus, at the $\alpha = .05$ level, the biological agents are not effective in controlling weeds in the hay fields. However, the average hay production using the chemical agents appears to be greater than the hay production on fields with no weed agents.

When the sample sizes are not equal, the Dunnett procedure does not produce an experimentwise error rate equal to α. Dunnett (1964) provided adjustments to the values given in Table 11 for the unequal sample sizes.

9.8   Scheffé's S Method

The five multiple-comparison procedures discussed so far have been developed for pairwise comparisons among t population means. A more general procedure, proposed by Scheffé (1953), can be used to make all possible comparisons among the t population means. Although Scheffé's procedure can be applied to pairwise comparisons among the t population means, it is more conservative (less sensitive) than any of the other three multiple comparison procedures for detecting significant differences among pairs of population means because the "family" of comparisons it is trying to protect is larger than that for pairwise comparisons.

Scheffé's S Method for Multiple Comparisons

1. Consider any linear comparison among the t population means of the form

$$l = a_1\mu_1 + a_2\mu_2 + \cdots + a_t\mu_t$$

We wish to test the null hypothesis

$$H_0: \quad l = 0$$

against the alternative

$$H_a: \quad l \neq 0$$

2. The test statistic is

$$\hat{l} = a_1\bar{y}_1. + a_2\bar{y}_2. + \cdots + a_t\bar{y}_t.$$

3. Let

$$S = \sqrt{\hat{V}(\hat{l})}\,\sqrt{(t-1)F_{\alpha,df_1,df_2}}$$

where, from Section 9.2,

$$\hat{V}(\hat{l}) = s_W^2 \sum_i \frac{a_i^2}{n_i}$$

t is the total number of population means, F_{α,df_1,df_2} is the upper-tail critical value of the F distribution with $a = \alpha$, $df_1 = t - 1$, and df_2 is the degrees of freedom for s_W^2.

4. For a specified value of α, we reject H_0 if $|\hat{l}| > S$.

5. The error rate that is controlled is an *experimentwise error rate*. If we consider all imaginable contrasts, the probability of observing an experiment with one or more contrasts falsely declared to be significant is designated by α.

EXAMPLE 9.10

Refer to Example 9.3. We defined four contrasts in the $t = 5$ treatment means in an attempt to investigate the differences in the average hay production on fields treated with either the control or one of the four weed agents. Use the sample data and Scheffé's procedure to determine which if any of the four contrasts are significantly different from zero. Use $\alpha = .05$.

Solution The four contrasts of interest are given in Table 9.7 along with their estimates. To illustrate the calculations involved in Table 9.7, we will compute the value of S for the first contrast, control vs. agents. To compute

$$S = \sqrt{\hat{V}(\hat{l})} \sqrt{(t - 1)F_{\alpha, df_1, df_2}}$$

we must first calculate $\hat{V}(\hat{l})$. Using the formula

$$\hat{V}(\hat{l}) = s_W^2 \sum_{i=1}^{t} \frac{a_i^2}{n_i}$$

with all samples sizes equal to 6 and $s_W^2 = .0153$, we have

$$\hat{V}(\hat{l}) = .0153 \left(\frac{(4)^2}{6} + \frac{1}{6} + \frac{1}{6} + \frac{1}{6} + \frac{1}{6} \right) = .0153 \frac{20}{6} = .0510$$

TABLE 9.7 Computations for Scheffé procedure in weed control experiment

Contrast	Control a_1	Bio1 a_2	Bio2 a_3	Chm1 a_4	Chm2 a_5	$\sum a_i^2/n_i$	$\hat{l}$	$\hat{V}(\hat{l})$	S	Conclusion
Control vs. agents	4	−1	−1	−1	−1	20/6	−.836	.0510	.750	Significant
Biological vs. chemical	0	1	1	−1	−1	4/6	−.294	.0102	.336	Not significant
Bio1 vs. Bio2	0	1	−1	0	0	2/6	−.035	.0051	.237	Not significant
Chm1 vs. Chm2	0	0	0	1	−1	2/6	−.085	.0051	.237	Not significant

From Table 8 for $\alpha = .05$, $df_1 = t - 1 = 4$, $df_2 = 25$ (the degrees of freedom for s_W^2), $F_{.05, 4, 25} = 2.76$. The computed value of S is then

$$S = \sqrt{.0510} \sqrt{4(2.76)} = (.2258)(3.323) = .750$$

Because the absolute value of $\hat{l}$ is $|-.836| = .836$ exceeds .750, we have significant evidence ($\alpha = .05$) to indicate that the average hay production from the fields treated with a weed agent exceeds the average yield in the fields having no treatment for weeds. The calculations for the other three contrasts are summarized in Table 9.7. Note that the value of S changes for the different contrasts. In our example, the only contrast significantly different from zero was the first contrast. The remaining three contrasts were not significant at the $\alpha = .05$ level. These conclusions are different from the conclusions we reached in Example 9.4, where we found that the second contrast was also significantly different from zero. The reason for the differences in conclusions is that the Scheffé procedure controls the experimentwise Type I error rate at level .05, whereas in Example 9.4 we only control the individual comparison rate at level .05.

Scheffé's confidence interval

Scheffé's method can also be used for constructing a simultaneous confidence interval for all possible (not necessarily pairwise) contrasts using the t treatment means. In particular, there is a probability equal to $1 - \alpha$ that all possible comparisons of the form $l = \Sigma a_i\mu_i$, where $\Sigma a_i = 0$, will be encompassed by intervals of the form

$$\hat{l} - S < l < \hat{l} + S$$

Now let us return to the case study introduced at the beginning of this chapter.

Analyzing Data for the Case Study   The objective of the study was to investigate whether an interviewer's evaluation of a job applicant is affected by the physical handicap of the person being interviewed. Prior to testing hypotheses and making comparisons amongst the five treatments, we need to verify that the conditions under which the tests and multiple comparison procedures are valid have been satisfied in this study.

We observed in Figure 9.1 that the boxplots were nearly of the same width, with no outliers and whiskers of nearly the same length. The means and medians were of a similar size for each of the five groups of applicants. Thus, the assumptions of AOV would appear to be satisfied. To confirm this observation, we computed the residuals and plotted them in a normal probability plot. (See Figure 9.2.) From this plot we can observe that, with the exception of two data values, the points fall nearly on a straight line. Thus, there is a strong confirmation that the five populations of ratings of applicants qualifications have normal distributions.

FIGURE 9.2

Normal probability plot of residuals

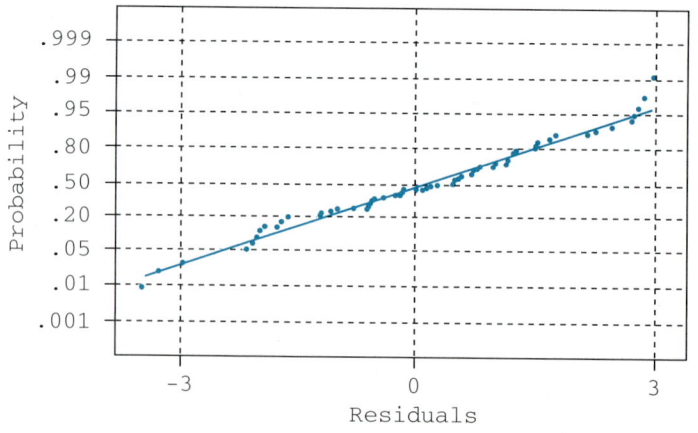

Average: -0.0000000
StDev: 1.63767
N: 70

Anderson-Darling Normality Test
    A-Squared: 0.384
    P-Value: 0.387

Next, we can check on the equal variance assumption. Based on the boxplot, the spreads were similar. From the summary statistics given in Table 9.2, we note that the standard deviations ranged from 1.537 to 1.961. Thus, there is very little difference in the sample standard deviation. To confirm this observation, we conduct a test of homogeneity of variance using Levine's test. We are testing

$$H_0: \sigma_1^2 = \sigma_2^2 = \sigma_3^2 = \sigma_4^2 = \sigma_5^2$$

H_a: Variances are not all equal.

We compute a value of $L = .405$. The critical value is $F_{.05,4,25} = 2.76$. Thus, we fail to reject H_0. Furthermore, we compute the p-value to be $P(F_{4,25} \geq .405) = .803$;

we are confident that the condition of homogeneity of variance has not been violated in this study.

We would check the condition of independence of the data by discussing with the researchers the manner in which the study was conducted. It would be important to make sure that the conditions in the room in which the interview tape was viewed remained constant throughout the study, so as not to introduce any distractions that could affect the raters' evaluations. Also, the initial check that the empathy scores were evenly distributed over the five groups of raters assures us that a difference in empathy levels did not exist in the five groups of raters prior to their evaluation of the applicants' qualifications. The research hypothesis is that the mean qualification ratings, μ_is, differ over the five handicap conditions:

$$H_0: \mu_1 = \mu_2 = \mu_3 = \mu_4 = \mu_5$$

H_a: At least one of the means differs from the rest.

The computer output for the AOV table is given here. The following notation is used in the output: Control (C), Hard of Hearing (H), Amputee (A), Crutches (R), and Wheelchair (W).

```
General Linear Models Procedure

Dependent Variable: Rating

Source            DF     Sum of Squares       F value     Pr > F

Model              4        30.47800000         2.68       0.0394
Error             65       185.05642857
Corrected Total   69       215.53442857
----------------------------------------------------------------------


Dunnett's One-tailed T Tests for Variable: Rating

Note: This test controls the type I experimentwise error for
      comparisons of all treatments against a control.

Alpha= 0.05  Confidence= 0.95  df= 65  MSE= 2.847022
Critical Value of Dunnett's T= 2.203
Minimum Significant Difference= 1.4049

Comparisons significant at the 0.05 level are indicated by '***'.

                   Simultaneous                 Simultaneous
                     Lower       Difference        Upper
          HC        Confidence    Between        Confidence
       Comparison     Limit        Means           Limit

  R   - C            -0.3907        1.0143          2.4192
  W   - C            -0.9407        0.4643          1.8692
  A   - C            -1.8692       -0.4643          0.9407
  H   - C            -2.2549       -0.8500          0.5549
----------------------------------------------------------------------
```

```
Tukey's Studentized Range (HSD) Test for Variable: Rating

NOTE: This test controls the type I experimentwise error rate, but
      generally has a higher type II error rate than REGWQ.

Alpha= 0.05  df= 65  MSE= 2.847022
Critical Value of Studentized Range= 3.968
Minimum Significant Difference= 1.7894

Means with the same letter are not significantly different.

Tukey Grouping                      Mean      N    HC

                 A                 5.9143     14    R
            B    A                 5.3643     14    W
            B    A                 4.9000     14    C
            B    A                 4.4357     14    A
            B                      4.0500     14    H
------------------------------------------------------------------
```

From the output, we see that the p-value for the F test is .0394. Thus, there is a significant difference in the mean ratings across the five types of handicaps. We next investigate what types of differences exist in the ratings for the groups. We make a comparison of the control (C) group to the four groups having handicaps—crutches (R), wheelchair (W), amputee (A), and hard of hearing (H)—using the Dunnett procedure at the $\alpha_E = .05$ level. We use a one-sided test of whether any of the four handicap groups had a lower mean rating than the control:

$$H_0: \mu_i \geq \mu_C$$

$$H_a: \mu_i < \mu_C$$

We reach the conclusion that the mean rating for the control (no handicap) group is not significantly greater than the mean rating for any of the handicap groups. Next, we run a multiple procedure to determine which group pairs produced different mean ratings. The analysis uses the Tukey procedure with $\alpha = .05$, with the results displayed in the computer output. All handicap types with the same Tukey grouping letter have mean ratings that are not significantly different from each other. Thus, the mean rating from the applicant using crutches was significantly higher than the mean rating for the applicant who was hard of hearing. No other pairs were found to be significantly different. To investigate the size of the differences in the pairs of rating means for the five handicap conditions, we computed simultaneous 95% confidence intervals for the ten pairs of mean differences using the Tukey procedure. The intervals are provided in the following computer output.

```
Tukey's Studentized Range (HSD) Test for Variable: R

NOTE: This test controls the type I experimentwise error rate.
```

```
Alpha= 0.05  Confidence= 0.95  df= 65  MSE= 2.847022
Critical Value of Studentized Range= 3.968
Minimum Significant Difference= 1.7894

Comparisons significant at the 0.05 level are indicated by '***'.

                Simultaneous           Simultaneous
                  Lower    Difference     Upper
        HC      Confidence   Between   Confidence
    Comparison    Limit      Means       Limit

      R - W      -1.2394     0.5500      2.3394
      R - C      -0.7751     1.0143      2.8037
      R - A      -0.3108     1.4786      3.2680
      R - H       0.0749     1.8643      3.6537     ***
      W - C      -1.3251     0.4643      2.2537
      W - A      -0.8608     0.9286      2.7180
      W - H      -0.4751     1.3143      3.1037
      C - A      -1.3251     0.4643      2.2537
      C - H      -0.9394     0.8500      2.6394
      A - H      -1.4037     0.3857      2.1751

    ----------------------------------------------------------------

    Contrast                DF     Contrast SS   F Value    PR > F

    Control vs Handicap     1       0.01889286    0.01      0.9353
    Hearing vs Mobility     1      14.82148810    5.21      0.0258
    Crutches vs Amp.&Wheel  1       9.60190476    3.37      0.0709
```

Finally, we constructed several contrasts to evaluate the remaining questions posed by researchers. The questions along with the corresponding contrasts are given here.

Question	Contrast
Control ratings vs. handicap ratings	$4\mu_C - \mu_R - \mu_W - \mu_A - \mu_H$
Hearing ratings vs. mobility handicap ratings	$0\mu_C - \mu_R - \mu_W - \mu_A + 3\mu_H$
Crutches ratings vs. amputee and wheelchair ratings	$0\mu_C + 2\mu_R - \mu_W - \mu_A + 0\mu_H$

The three contrasts can be tested using a null and alternative hypothesis of the following form:

$$H_0: l = 0$$

$$H_a: l \neq 0$$

We can use Bonferroni procedure with $\alpha_E = .05$ to test the three hypotheses. The individual comparison rate is set at $\alpha_I = \alpha_E/3 = .05/3 = .0167$. Thus, if the p-value for any of the three F tests for contrasts is less than .0167, we will declare that contrast to be significantly different from zero. From the computer output, the three *p*-values were .9353, .0258, and .0709. Thus, none of the three contrasts is significantly different from zero.

The only significant difference found in the five mean ratings was between the applicant with a hearing handicap and the applicant using crutches. The researchers discussed in detail in the article why this difference may have occurred.

Reporting Conclusions The report summarizing our findings of this study would need to include the following:

1. Statement of the objective for the study
2. Description of the study design, how raters were selected, and how the interviews were conducted
3. Discussion of the generalizability of results from the study
4. Numerical and graphical summaries of data sets
5. Description of all inference methodologies:
   AOV table and *F* test
   Multiple-comparison procedures, contrasts, and confidence intervals
   Verification that all necessary conditions for using inference techniques were satisfied
6. Discussion of results and conclusions
7. Interpretation of findings relative to previous studies
8. Recommendations for future studies
9. Listing of the data sets

9.9 Summary

We presented three different multiple-comparison procedures (Fisher's, Tukey's, and SNK) for making pairwise comparisons of *t* population means. Another procedure, Scheffé's, can be applied to any linear combination (including pairwise comparisons) of the means. For each procedure, we have tried to indicate which error rate is controlled and how conservative the procedure is relative to the others presented. Because all pairwise, multiple-comparison procedures compute the magnitude of the difference $|\bar{y}_{i.} - \bar{y}_{j.}|$ that is needed to declare μ_i and μ_j different, we can get some feel for how conservative one procedure is relative to another by comparing the magnitudes of the differences required for significance using the data of Example 9.3. This information is shown in Table 9.8.

TABLE 9.8
Critical difference for sample means *r* steps apart

Procedure	Number of Steps Separating Means			
	2	**3**	**4**	**5**
LSD	.1471	.1471	.1471	.1471
SNK	.1471	.1783	.1969	.2100
Tukey's	.2100	.2100	.2100	.2100
Scheffés	.2373	.2373	.2373	.2373

As you can see in Table 9.8, Scheffé's procedure is *very* conservative. The critical difference is 13% larger than Tukey's *W* and 61% larger than the LSD. Thus, we would never recommend using Scheffé's procedure for conducting a pairwise comparison of treatment means. Also, note that the value for the LSD equals the SNK value for $r = 2$ and Tukey's W equals SNK for $r = t = 5$. Thus, SNK is a compromise between the liberal LSD and the conservative W.

Which procedure should you use? We generally prefer the SNK procedure for efficacy (effectiveness) comparisons and Dunnett's procedure for comparisons to a control. However, our reasons for these choices have a great deal to do with our work setting and the regulations surrounding our decision. Because our environment may be entirely different from yours, the decision regarding which procedure to use, and when to use it, is up to the individual. For a given problem, determine whether your decisions regarding differences should, in general, be more (or less) conservative. Then choose a procedure that exhibits the desired characteristic.

Key Formulas

1. Fisher's LSD procedure

$$\text{LSD} = t_{\alpha/2} \sqrt{s_W^2 \left(\frac{1}{n_i} + \frac{1}{n_j} \right)}$$

2. Tukey's W procedure

$$W = q_\alpha(t, v) \sqrt{\frac{s_W^2}{n}}$$

3. SNK procedure

$$W_r = q_\alpha(r, v) \sqrt{\frac{s_W^2}{n}}$$

4. Approximate value for n in Tukey's and SNK procedures when sample sizes are modestly different

$$n = \frac{t}{\dfrac{1}{n_1} + \dfrac{1}{n_2} + \cdots + \dfrac{1}{n_t}}$$

5. Dunnett's procedure

$$D = d_\alpha(k, v) \sqrt{\frac{s_W^2}{n}}$$

6. Scheffé's method

$$S = \sqrt{\hat{V}(\hat{l})} \sqrt{(t - 1) F_{\alpha, df_1, df_2}}$$

where

$$\hat{V}(\hat{l}) = s_W^2 \sum \frac{a_i^2}{n_i}$$

Supplementary Exercises

Ag. **9.6** Refer to the data of Example 8.1. There a horticulturist was investigating the phosphorus content of the leaves from three different varieties of apple trees.
   a. Perform an analysis of variance.
   b. Use the SNK procedure to run all pairwise comparisons. Use $\alpha = .05$.
   c. Compare your conclusions in part (b) to those in the SAS computer output shown here.

```
General Linear Models Procedure
Class Level Information

Class        Levels      Values
VARIETY         3         1 2 3
```

```
Number of observations in data set = 15

Dependent Variable: P    PHOSPHORUS

Source                   DF    Sum of Squares    F Value    Pr > F
Model                     2        0.27664000     16.97    0.0003
Error                    12        0.09780000
Corrected Total          14        0.37444000

              R-Square                 C.V.              P Mean
              0.738810             13.93169          0.64800000

Source                   DF    Type III SS       F Value    Pr > F
VARIETY                   2        0.27664000     16.97    0.0003

Student-Newman-Keuls test for variable: PHOSPHORUS

NOTE: This test controls the type I experimentwise error rate un-
      der the complete null hypothesis but not under partial null
      hypotheses.

Alpha= 0.05  df= 12  MSE= 0.00815

Number of Means           2          3
Critical Range  0.1244028 0.1523193

Means with the same letter are not significantly different.

          SNK Grouping          Mean      N  VARIETY

                   A          0.77600     5   2
                   A          0.70800     5   3
                   B          0.46000     5   1
```

Med.   **9.7**   Researchers conducted an experiment to compare the effectiveness of four new weight-reducing agents to that of an existing agent. The researchers randomly divided a random sample of 50 males into five equal groups, with preparation A assigned to the first group, B to the second group, and so on. They then gave a prestudy physical to each person in the experiment and told him how many pounds overweight he was. A comparison of the mean number of pounds overweight for the groups showed no significant differences. The researchers then began the study program, and each group took the prescribed preparation for a fixed period of time. The weight losses recorded at the end of the study period are given here.

$A1$	12.4	10.7	11.9	11.0	12.4	12.3	13.0	12.5	11.2	13.1
$A2$	9.1	11.5	11.3	9.7	13.2	10.7	10.6	11.3	11.1	11.7
$A3$	8.5	11.6	10.2	10.9	9.0	9.6	9.9	11.3	10.5	11.2
$A4$	12.7	13.2	11.8	11.9	12.2	11.2	13.7	11.8	11.5	11.7
S	8.7	9.3	8.2	8.3	9.0	9.4	9.2	12.2	8.5	9.9

The standard agent is labeled agent S, and the four new agents are labeled A_1, A_2, A_3, and A_4. The data and a computer printout of an analysis are given here.

```
General Linear Models Procedure
Class Level Information

Class      Levels     Values

AGENT         5       1 2 3 4 S

Number of observations in data set = 50

Dependent Variable: L   WEIGHTLOSS

SOURCE                     DF     Sum of Squares    F Value    Pr > F
Model                       4        61.61800000      15.68    0.0001
Error                      45        44.20700000
Corrected Total            49       105.82500000

                  R-Square              C.V.              L Mean
                  0.582263            9.035093         10.9700000

Source                     DF      Type III SS    F Value    Pr > F
AGENT                       4       61.61800000     15.68     0.0001
-------------------------------------------------------------------
Level of           -------------L---------------
A          N         Mean                SD

1          10       12.0500000         0.82898867
2          10       11.0200000         1.12130876
3          10       10.2700000         1.02637442
4          10       12.2400000         0.75601293
S          10        9.2700000         1.15859110
-------------------------------------------------------------------


FISHER'S LSD  for variable: WEIGHTLOSS

Alpha= 0.05  df= 45  MSE= 0.982378
Critical Value of T= 2.01
Least Significant Difference= 0.8928

Means with the same letter are not significantly different.

T Grouping              Mean       N  A
                  A     12.2400     10  4
                  A     12.0500     10  1
                  B     11.0200     10  2
                  B     10.2700     10  3
                  C      9.2700     10  S
-------------------------------------------------------------------


Student-Newman-Keuls test for variable: L

Alpha= 0.05  df= 45  MSE= 0.982378
```

```
Number of Means          2        3        4        5
Critical Range   0.8927774 1.0742812 1.1824729 1.2594897

Means with the same letter are not significantly different.

SNK Grouping              Mean      N  A


                    A         12.2400     10  4
                    A         12.0500     10  1
                    B         11.0200     10  2
                    B         10.2700     10  3
                    C          9.2700     10  S
---------------------------------------------------------------------

Tukey's Studentized Range (HSD) Test for variable: L

Alpha= 0.05  df= 45  MSE= 0.982378
Critical Value of Studentized Range= 4.018
Minimum Significant Difference= 1.2595

Means with the same letter are not significantly different.

Tukey Grouping                Mean      N  A


                    A         12.2400     10  4
                    A         12.0500     10  1
              B     A         11.0200     10  2
              B     C         10.2700     10  3
                    C          9.2700     10  S
---------------------------------------------------------------------

Dunnett's One-tailed T tests for variable: L

Alpha= 0.05  Confidence= 0.95  df= 45   MSE= 0.982378
Critical Value of Dunnett's T= 2.222
Minimum Significant Difference= 0.9851

Comparisons significant at the 0.05 level are indicated by '***'.

                  Simultaneous          Simultaneous
                    Lower    Difference    Upper
        A         Confidence   Between   Confidence
     Comparison     Limit      Means       Limit

4    - s           1.9849     2.9700      3.9551     ***
1    - s           1.7949     2.7800      3.7651     ***
2    - s           0.7649     1.7500      2.7351     ***
3    - s           0.0149     1.0000      1.9851     ***
---------------------------------------------------------------------

Univariate Procedure

Variable=RESIDUAL
```

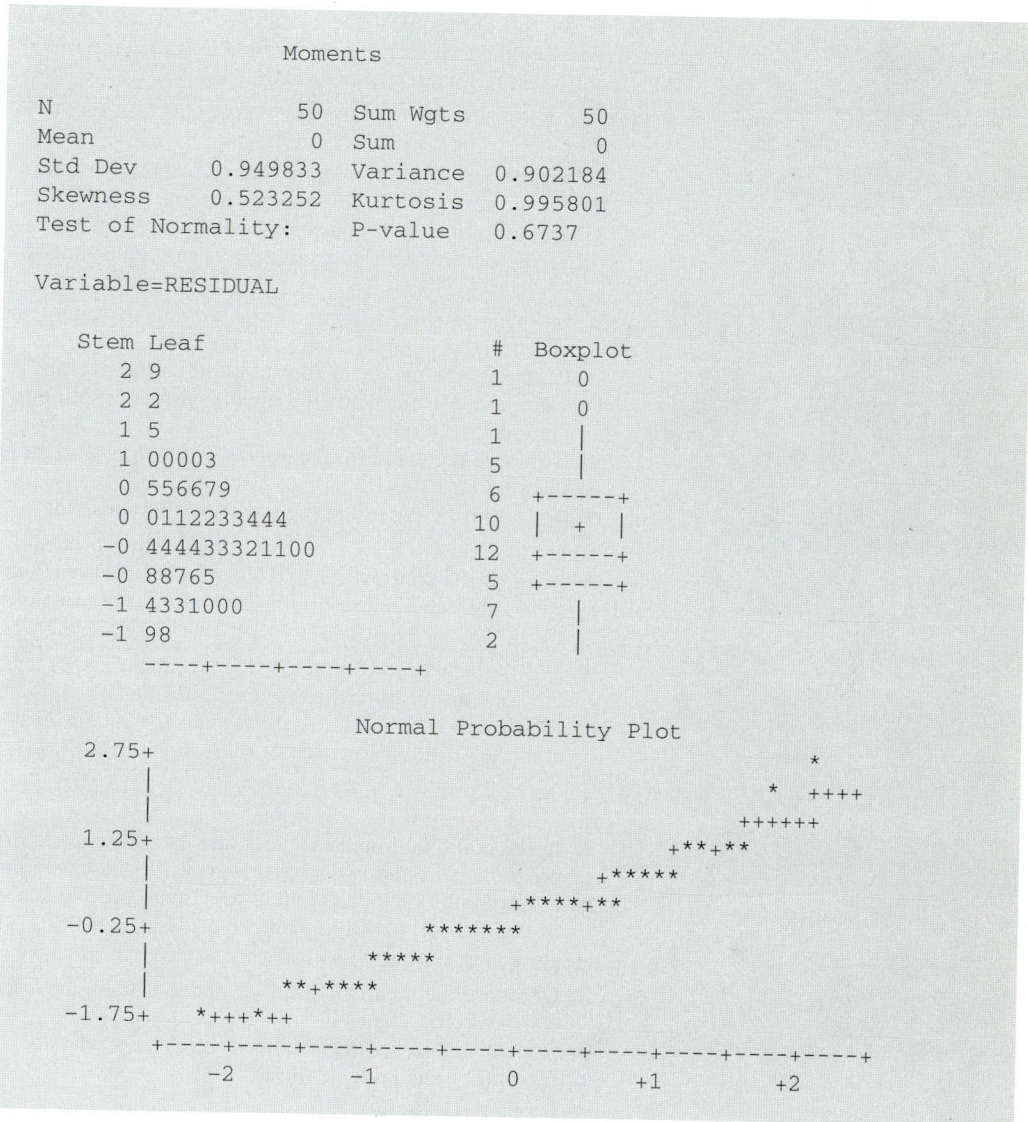

```
                    Moments

N                    50  Sum Wgts           50
Mean                  0  Sum                 0
Std Dev       0.949833  Variance      0.902184
Skewness      0.523252  Kurtosis      0.995801
Test of Normality:      P-value       0.6737


Variable=RESIDUAL

     Stem Leaf                    #   Boxplot
       2  9                       1      0
       2  2                       1      0
       1  5                       1      |
       1  00003                   5      |
       0  556679                  6   +-----+
       0  0112233444             10   |  +  |
      -0  444433321100           12   +-----+
      -0  88765                   5   +-----+
      -1  4331000                 7      |
      -1  98                      2      |
          ----+----+----+----+


                Normal Probability Plot
     2.75+                                          *
         |                               *    ++++
         |                                ++++++
     1.25+                             +**+**
         |                          +*****
         |                       +****+**
    -0.25+                  *******
         |               *****
         |           **+****
    -1.75+      *+++*++
          +----+----+----+----+----+----+----+----+----+----+
              -2        -1         0        +1        +2
```

Run an analysis of variance to determine whether there are any significant differences among the five weight-reducing agents. Use $\alpha = .05$. Do any of the AOV assumptions appear to be violated? What conclusions do you reach concerning the mean weight loss achieved using the five different agents?

9.8 Refer to Exercise 9.7. Using the computer output included there, determine the significantly different pairs of means using the following procedures.
   a. Fisher's LSD, $\alpha = .05$
   b. Tukey's W, $\alpha = .05$
   c. SNK procedure, $\alpha = .05$

9.9 Refer to Exercise 9.8. For each of the following situations, decide which of the multiple-comparison procedures would be most appropriate.
   a. The researcher is very concerned about falsely declaring any pair of agents significantly different.
   b. The researcher is very concerned about failing to declare a pair of agents significantly different when the population means are different.

9.10 Refer to Exercise 9.7. The researcher wants to determine which of the new agents produced a significantly larger mean weight loss in comparison to the standard agent. Use $\alpha = .05$ in making this determination.

9.11 Refer to Exercise 9.7. Suppose the weight-loss agents were of the following form:

A_1: Drug therapy with exercise and counseling
A_2: Drug therapy with exercise but no counseling
A_3: Drug therapy with counseling but no exercise
A_4: Drug therapy with no exercise and no counseling

Construct contrasts to make comparisons among the agent means that will address the following questions:
  a. Compare the mean for the standard to the average of the four agent means.
  b. Compare the mean for the agents with counseling to those without counseling. (Ignore the standard.)
  c. Compare the mean for the agents with exercise to those without exercise. (Ignore the standard.)
  d. Compare the mean for the agents with counseling to the standard.

9.12 Refer to Exercise 9.11. Use a multiple testing procedure to determine at the $\alpha = .05$ level which of the contrasts is significantly different from zero. Interpret your findings relative to the researcher's question about finding the most effective weight-loss method.

9.13 Refer to Exercise 8.2.
  a. Did the new brand LowTar have a reduced mean tar content when compared to the four brands of cigarettes currently on the marker? Use $\alpha = .05$.
  b. How large is the difference between the mean tar content for LowTar and the mean tar content for each of the four brands? Use a 95% confidence interval.

9.14 Refer to Exercise 8.6. Varieties A and B were planted in Texas, varieties C and D in Florida, and variety E in California.
  a. Is the combined mean yield of the Texas varieties different from the combined mean yield of the two varieties grown in Florida? Use $\alpha = .05$.
  b. Is the combined mean yield of the Texas and Florida varieties different from that of the variety grown in California? Use $\alpha = .05$.

9.15 Refer to Exercise 8.11.
  a. Compare the mean yields of herbicide 1 and herbicide 2 to the control treatment. Use $\alpha = .05$.
  b. Should the procedure you used in (a) be a one-sided or a two-sided procedure?
  c. Interpret your findings in (a).

9.16 Refer to Exercise 8.16.
  a. Compare the mean scores for the three divisions using an appropriate multiple-comparison procedure. Use $\alpha = .05$.
  b. What can you conclude about the differences in mean scores and the nature of the divisions from which any differences arise?

Ag. **9.17** The nitrogen contents of red clover plants inoculated with three strains of *Rhizobium* are given here.

3DOK1	3DOK5	3DOK7
19.4	18.2	20.7
32.6	24.6	21.0
27.0	25.5	20.5
32.1	19.4	18.8
33.0	21.7	18.6
	20.8	20.1
		21.3

a. Is there evidence of a difference in the effects of the three treatments on the mean nitrogen content? Analyze the data completely and draw conclusions based on your analysis. Use $\alpha = .01$.

b. Was there any evidence of a violation in the required conditions needed to conduct your analysis in (a)?

Ag.   **9.18** Scientists conducted an experiment to compare three methods of grass-seed preparation: mechanical scarification (ms), acid dip (ad), and hot water dip (hw). The scientists placed 100 grass seeds in each of 150 petri dishes. Among the 150 dishes, 50 were randomly assigned ms, 50 ad, and 50 hw. After a period of 2 weeks, the scientists checked the germination rates for each dish.

Method	Mean Germination	Standard Deviation
ms	65.3	7.2
ad	82.1	5.4
hw	73.8	6.5

Analyze these data using a one-way analysis and draw conclusions. Use $\alpha = .05$.

9.19 Refer to Exercise 9.18. Use the SNK procedure to identify method differences. Summarize your results.

Env.   **9.20** To assess the relative merits of four different gasoline blends, researchers conducted an experiment using 20 automobiles of the same type, model, and engine size with five randomly assigned to each of the blends. Summary test data for the blends are shown here.

Blend	Mean (mi/gal)	Standard Deviation
1 (control)	26.2	4.3
2 (control − additive x)	28.1	5.6
3 (control − additive y)	29.6	5.1
4 (control − additives x and y)	38.2	7.3

Run an analysis and draw conclusions. Use $\alpha = .05$.

9.21 Refer to Exercise 9.20 and consider the following linear contrasts. Describe what each contrast is measuring.
   a. $\hat{l}_1 = \bar{y}_{1.} + \bar{y}_{2.} - \bar{y}_{3.} - \bar{y}_{4.}$
   b. $\hat{l}_2 = \bar{y}_{1.} + \bar{y}_{3.} - \bar{y}_{2.} - \bar{y}_{4.}$
   c. $\hat{l}_3 = \bar{y}_{1.} - \bar{y}_{2.} - \bar{y}_{3.} + \bar{y}_{4.}$

9.22 Use Scheffé's method to conduct a test of significance on $\hat{l}_3$ of Exercise 9.21, based on $\alpha = .05$. What do you conclude?

9.23 Construct a confidence interval for l_1 and l_2 of Exercise 9.21 using Scheffé's method. Draw conclusions.

Vet.   **9.24** Researchers conducted a study of the effects of three drugs on the fat content of the shoulder muscles in labrador retrievers. They divided 80 dogs at random into four treatment groups. The dogs in group A were the untreated controls, while groups B, C, and D received one of three new heartworm medications in their diets. Ten dogs randomly selected from each of the four groups received varying lengths of treatment from 4 months to 2 years. The percentage fat content of the shoulder muscles was determined and is given here.

Examination Time	Treatment Group			
	A	**B**	**C**	**D**
4 months	2.84	2.43	1.95	3.21
	2.49	1.85	2.67	2.20
	2.50	2.42	2.23	2.32
	2.42	2.73	2.31	2.79
	2.61	2.07	2.53	2.94
8 months	2.23	2.83	2.32	2.45
	2.48	2.59	2.36	2.49
	2.48	2.53	2.46	2.95
	2.23	2.73	2.04	2.05
	2.65	2.26	2.30	2.31
1 year	2.30	2.70	2.85	2.53
	2.30	2.54	2.75	2.73
	2.38	2.70	2.62	2.65
	2.05	2.81	2.50	2.84
	2.13	2.70	2.69	2.92
2 years	2.64	3.24	2.90	2.91
	2.56	3.71	3.02	2.89
	3.30	2.95	3.78	3.21
	2.19	3.01	2.96	2.89
	2.45	3.08	2.87	2.68
Mean	2.411	2.694	2.605	2.698

Under the assumptions that conditions for an AOV were met, the researchers then computed an AOV to evaluate the difference in mean percentage fat content for dogs under the four treatments. The AOV computations did not takes into account the length of time on the medication. The AOV is given here.

Source	df	SS	MS	F ratio	p-value
Treatments	3	1.0796	.3599	3.03	.0345
Error	76	9.0372	.1189		
Totals	79	10.1168			

a. Is there a significant difference in the mean fat content in the four treatment groups? Use $\alpha = .05$.

b. Do any of the three treatments for heartworm appear to have increased the mean fat content over the level in the control group?

9.25 Refer to Exercise 9.24. Suppose the researchers conjectured that the new medications caused an increase in fat content and that this increase accumulated as the medication was continued in the dogs. How could we examine this question using the data given?

Med.　**9.26** The article "The Ames Salmonell/microsome mutagenicity assay: Issues of inference and validation" (1989, *Journal of American Statistical Association*, 84: 651–661) discusses the importance of chemically induced mutation for human health and the biological basis for the primary in vitro assay for mutagenicity, the Ames Salmonell/microsome assay. In an Ames test, the response obtained from a single sample is the number of visible colonies that result from plating approximately 10^8 microbes. A common protocol for an Ames test includes multiple samples at a control dose and four or five logarithmically spaced doses of a test compound. The following data are from one such experiment with 20 samples per dose level. The dose levels were μg/sample.

Dose	Number of Visible Colonies																				$\bar{y}_{i \cdot}$	s_i^2
Control	11	13	14	14	15	15	15	15	16	17	17	18	18	19	20	21	22	23	25	27	17.8	17.5
.3	39	39	42	43	44	45	46	50	50	50	51	52	52	52	55	61	62	63	67	70	51.7	81.0
1.0	88	90	92	92	102	104	104	106	109	113	117	117	119	119	120	120	121	122	130	133	110.9	175.4
3.0	222	233	251	251	253	255	259	275	276	283	284	294	299	301	306	312	315	323	337	340	283.5	1131.5
10.0	562	587	595	604	623	666	689	692	701	702	703	706	710	714	733	739	763	782	786	789	692.3	4584.4

We want to determine whether there is an increasing trend in the mean number of colonies as the dose level increases. One method of obtaining such a determination is to use a contrast with constants a_i determined in the following fashion. Suppose the treatment levels are t values of a continuous variable x: $x_1, x_2, \ldots, x_t$. Let $a_i = x_i - \bar{x}$ and $\hat{l} = \sum a_i \bar{y}_i$. If $\hat{l}$ is significantly different from zero and positive, then we state there is a positive trend in the μ_is. If $\hat{l}$ is significantly different from zero and negative, then we state there is a negative trend in the μ_is. In this experiment, the dose levels are the treatments $x_1 = 0$, $x_2 = .3$, $x_3 = 1.0$, $x_4 = 3.0$, $x_5 = 10.0$, with $\bar{x} = 2.86$. Thus, the coefficients for the contrasts are $a_1 = 0 - 2.86 = -2.86$, $a_2 = 0.3 - 2.86 = -2.56$, $a_3 = 1.0 - 2.86 = -1.86$, $a_4 = 3.0 - 2.86 = +.14$, $a_5 = 10.0 - 2.86 = +7.14$. We thus need to evaluate the significance of the following contrast in the treatment means given by $-2.86\bar{y}_C -2.56\bar{y}_{.3} -1.86\bar{y}_{1.0} +0.14\bar{y}_{3.0} +7.14\bar{y}_{10.0}$. If the contrast is significantly different from zero and is positive, we conclude that there is an increasing trend in the dose means.

a. Test whether there is an increasing trend in the dose mean. Use $\alpha = .05$.

b. Do there appear to be any violations in the conditions necessary to conduct the test in (a)? If there are violations, suggest a method that would enable us to validly test whether the positive trend exists.

9.27 In the case study concerning the evaluation of interviewers' decisions related to applicant handicap type, the raters were 70 undergraduate students, and the same male actors, both job applicant and interviewer, were used in all the videotapes of the job interview.

a. Discuss the limitations of this study in regard to using the undergraduate students, as the raters of the applicant's qualifications for the computer sales position.

b. Discuss the positive and negative points of using the same two actors for all five interview videotapes.

c. Discuss the limitations of not varying the type of job being sought by the applicant.

Med.   **9.28** The paper "The effect of an endothelin-receptor antagonist, bosentan, on blood pressure in patients with essential hypertension" (1998, *The New England Journal of Medicine*, 338: 784–790) discussed the contribution of bosentan to blood pressure regulation in patients with essential hypertension. The study involved 243 patients with mild-to-moderate essential hypertension. After a placebo run-in period, patients were randomly assigned to receive one of four oral doses of bosentan (100, 500, or 1,000 mg once daily, or 1,000 mg twice daily) or a placebo. The blood pressure was measured before treatment began and after a 4-week treatment period. The primary end point of the study was the change in blood pressure from the base line obtained prior to treatment to the blood pressure at the conclusion of the 4-week treatment period. A summary of the data is given in the following table.

	Blood Pressure Change				
	Placebo	**100 mg**	**500 mg**	**1,000 mg**	**2,000 mg**
Diastolic Pressure					
Mean	−1.8	−2.5	−5.7	−3.9	−5.7
Standard Deviation	6.71	7.30	6.71	7.21	7.30
Systolic Pressure					
Mean	−0.9	−2.5	−8.4	−10.3	−10.3
Standard Deviation	11.40	11.94	11.40	11.80	11.94
Sample Size	45	44	45	43	44

a. Which of the dose levels were associated with a significantly greater reduction in the diastolic pressure in comparison to the placebo? Use $\alpha = .05$.

b. Why was it important to include a placebo treatment in the study?

c. Using just the four treatments (ignore the placebo), construct a contrast to test for an increasing linear trend in the size of the systolic pressure reductions as the dose levels are increased. See Exercise 9.26 for the method for creating such a contrast.

d. Use the SNK procedure to test for pairwise differences in the mean systolic blood pressure reduction for the four treatment doses. Use $\alpha = .05$.

e. The researchers referred to their study as a *double-blind* study. Explain the meaning of this terminology.

CHAPTER 10

Categorical Data

10.1 Introduction and Case Study

10.2 Inferences about a Population Proportion π

10.3 Inferences about the Difference between Two Population Proportions, $\pi_1 - \pi_2$

10.4 Inferences about Several Proportions: Chi-Square Goodness-of-Fit Test

10.5 The Poisson Distribution

10.6 Contingency Tables: Tests for Independence and Homogeneity

10.7 Measuring Strength of Relation

10.8 Odds and Odds Ratios

10.9 Summary

10.1 Introduction and Case Study

Up to this point, we have been concerned primarily with sample data measured on a quantitative scale. However, we sometimes encounter situations in which levels of the variable of interest are identified by name or rank only and we are interested in the number of observations occurring at each level of the variable. Data obtained from these types of variables are called **categorical** or **count data.** For example, an item coming off an assembly line may be classified into one of three quality classes: acceptable, second, or reject. Similarly, a traffic study might require a count and classification of the type of transportation used by commuters along a major access road into a city. A pollution study might be concerned with the number of different alga species identified in samples from a lake and the number of times each species is identified. A consumer protection group might be interested in the results of a prescription fee survey to compare prices on a checklist of medications in different sections of a large city.

In this chapter, we will examine specific inferences that can be made from experiments involving categorical data.

469

Case Study: Franchise Expansion and Location Selection

The president of a motel chain is preparing to make a decision among four ownership groups competing for a franchise to open a new motel in a fast-growing "edge city" (a suburban area that has a large concentration of offices and shopping areas). The president of the chain has requested that a comparison of guest satisfaction be made for guests at motels currently operated by the four groups.

Designing the Data Collection   All four groups have operated other motels in the geographic area, so the chain has requested addresses of former guests so that information about consumer satisfaction could be obtained by using a mailed customer-satisfaction questionnaire. Initially, an analysis of the most recent guest survey results obtained by the four groups from their customers was to form the basis for the analysis. However, it was determined that the type and phrasing of questions was considerably different for the four groups. This would make the analysis of the data impossible because the data would not be consistent from all four groups. A questionnaire was designed to determine customers' satisfaction with their stay at each of the four groups' motels. The two key areas of interest, as far as the president is concerned, are guests' ratings of building quality (room, restaurant, exercise facility, etc.) and service quality. Both are rated on a 1 to 5 scale, with 1 denoting poor, 3 denoting average, and 5 denoting excellent, in the guests' opinions. Based on the size of necessary power of tests of hypotheses and the precision of confidence interval estimators, it was determined that a sample of at least 100 guests from each of the four ownership groups would be necessary. Because only about a 20% response rate could be anticipated from the mailed questionnaires, 500 recent guests of each group were randomly selected and sent a questionnaire.

Managing and Analyzing the Data   Of course, the president would prefer to grant the franchise to the ownership group that had achieved the best ratings. The financial arrangements negotiated with the groups were very similar, so much of the decision hinged on the ratings. The president realized that the survey covered only a small fraction of the people who had stayed at the various groups' motels in the recent past. However, the survey was the most unbiased information available about how the chain's customers perceived the four contending groups.

The president is not sure whether to emphasize the building ratings or the service ratings. In addition, there is a question of whether to compare the entire range of ratings or to concentrate on the proportion of guests giving an above average (4 or 5) rating.

The data from the questionnaires was summarized into the following two tables, one for building quality ratings and one for service quality ratings. The president asks you to analyze the ratings and provide inferences concerning the relative standings of the four ownership groups. He is concerned about how a survey of such a small percentage of the guests can accurately estimate the quality ratings of *all* guests of the motels. The methods of this chapter will enable you to answer the questions posed by the president.

Frequencies of building
ratings for case study

Ratings	Ownership Group				Totals
	G1	**G2**	**G3**	**G4**	
1	11	8	15	6	40
2	10	6	18	5	39
3	51	50	42	38	181
4	30	41	26	40	137
5	22	29	16	40	107
Totals	124	134	117	129	504

Frequencies of service
ratings for case study

Ratings	Ownership Group				Totals
	G1	**G2**	**G3**	**G4**	
1	15	16	23	11	65
2	18	21	17	18	74
3	36	31	33	24	124
4	29	35	21	33	118
5	26	31	23	43	123
Totals	124	134	117	129	504

10.2   Inferences about a Population Proportion π

In the binomial experiment discussed in Chapter 4, each trial results in one of two outcomes, which we labeled as either a success or a failure. We designated π as the probability of a success and $(1 - \pi)$ as the probability of a failure. Then the probability distribution for y, the number of successes in n identical trials, is

$$P(y) = \frac{n!}{y!(n - y)!}\,\pi^y(1 - \pi)^{n-y}$$

The point estimate of the binomial parameter π is one that we would choose intuitively. In a random sample of n from a population in which the proportion of elements classified as successes is π, the best estimate of the parameter π is the sample proportion of successes. Letting y denote the number of successes in the n sample trials, the sample proportion is

$$\hat{\pi} = \frac{y}{n}$$

We observed in Section 4.12 that y possesses a mound-shaped probability distribution that can be approximated by using a normal curve when

$$n \geq \frac{5}{\min(\pi, 1 - \pi)} \quad \text{(or equivalently, } n\pi \geq 5 \text{ and } n(1 - \pi) \geq 5\text{)}$$

In a similar way, the distribution of $\hat{\pi} = y/n$ can be approximated by a normal distribution with a mean and a standard error as given here.

Mean and Standard Error of $\hat{\pi}$

$$\mu_{\hat{\pi}} = \pi$$

$$\sigma_{\hat{\pi}} = \sqrt{\frac{\pi(1 - \pi)}{n}}$$

The normal approximation to the distribution of $\hat{\pi}$ can be applied under the same condition as that for approximating y by using a normal distribution. In fact, the approximation for both y and $\hat{\pi}$ becomes more precise for large n. Henceforth, in this text, we will assume that $\hat{\pi}$ can be adequately approximated by using a normal distribution, and we will base all our inferences on results from our previous study of the normal distribution.

A confidence interval can be obtained for π using the methods of Chapter 5 for μ, by replacing $\bar{y}$ with $\hat{\pi}$ and $\sigma_{\bar{y}}$ with $\sigma_{\hat{\pi}}$. A general $100(1 - \alpha)\%$ confidence interval for the binomial parameter is given here.

Confidence Interval for π, with Confidence Coefficient of $(1 - \alpha)$

$$\hat{\pi} \pm z_{\alpha/2}\hat{\sigma}_{\hat{\pi}} \quad \text{or} \quad (\hat{\pi} - z_{\alpha/2}\hat{\sigma}_{\hat{\pi}}, \hat{\pi} + z_{\alpha/2}\hat{\sigma}_{\hat{\pi}})$$

where

$$\hat{\pi} = \frac{y}{n} \quad \text{and} \quad \hat{\sigma}_{\hat{\pi}} = \sqrt{\frac{\hat{\pi}(1 - \hat{\pi})}{n}}$$

EXAMPLE 10.1

Researchers in the development of new treatments for cancer patients often evaluate the effectiveness of new therapies by reporting the proportion of patients who survive for a specified period of time after completion of the treatment. A new genetic treatment of 870 patients with a particular type of cancer resulted in 330 patients surviving at least 5 years after treatment. Estimate the proportion of all patients with the specified type of cancer who would survive at least 5 years after being administered this treatment. Use a 90% confidence interval.

Solution   For these data,

$$\hat{\pi} = \frac{330}{870} = .38$$

$$\hat{\sigma}_{\hat{\pi}} = \sqrt{\frac{(.38)(.62)}{870}} = .016$$

The confidence coefficient for our example is .90. Recall from Chapter 5 that we can obtain $z_{\alpha/2}$ by looking up the z-value in Table 1 in the Appendix corresponding to an area of $(\alpha/2)$. For a confidence coefficient of .90, the z-value corresponding to an area of .05 is 1.645. Hence, the 90% confidence interval on the proportion of cancer patients who will survive at least 5 years after receiving the new genetic treatment is

$$.38 \pm 1.645(.016) \quad \text{or} \quad .38 \pm .026$$

The confidence interval for π is based on a normal approximation to a binomial, which is appropriate provided n is sufficiently large. The rule we have specified is that both $n\pi$ and $n(1 - \pi)$ should be at least 5, but since π is the unknown parameter, we'll require that $n\hat{\pi}$ and $n(1 - \hat{\pi})$ be at least 5. When the sample size is too small and violates this rule, the confidence interval usually will be too wide to be of any use. For example, with $n = 20$ and $\hat{\pi} = .2$, the rule is not satisfied, since $n\hat{\pi} = 4$. The 95% confidence interval based on these data would be $.025 < \pi < .375$, which is practically useless. Very few product managers would be willing to launch a new product if the expected increase in market share was between .025 and .375.

Another problem that arises in the estimation of π occurs when π is very close to zero or one. In these situations, the population proportion would often be estimated to be 0 or 1, respectively, unless the sample size is extremely large. These estimates are not realistic since they would suggest that either no successes or no failures exist in the population. Rather than estimate π using the formula $\hat{\pi}$ given previously, adjustments are provided to prevent the estimates from being so extreme. One of the proposed adjustments is to use

$$\hat{\pi}_{\text{Adj.}} = \frac{\frac{3}{8}}{(n + \frac{3}{4})} \qquad \text{when } y = 0, \qquad \text{and}$$

$$\hat{\pi}_{\text{Adj.}} = \frac{(n + \frac{3}{8})}{(n + \frac{3}{4})} \qquad \text{when } y = n$$

When computing the confidence interval for π in those situations where $y = 0$ or $y = 1$, the confidence intervals using the normal approximation would not be valid. We can use the following confidence intervals, which are derived from using the binomial distribution.

100(1 − α)% Confidence Interval for π, when $y = 0$ or $y = 1$

When $y = 0$, the confidence interval is $(0, 1 - (\alpha/2)^{1/n})$.
When $y = n$, the confidence interval is $((\alpha/2)^{1/n}, 1)$.

EXAMPLE 10.2

A new PC operating system is being developed. The designer claims the new system will be compatible with nearly all computer programs currently being run on Microsoft's Windows operating system. A sample of 50 programs are run and all 50 programs perform without error. Estimate π, the proportion of all Microsoft's Windows-compatible programs that would run without change on the new operating system. Compute a 95% confidence interval for π.

Solution   If we used the standard estimator of π, we would obtain

$$\hat{\pi} = \frac{50}{50} = 1.0$$

Thus, we would conclude that 100% of all programs that are Microsoft's Windows-compatible programs would run without alteration on the new operating system. Would this conclusion be valid? Probably not, since we have only investigated a tiny fraction of all Microsoft's Windows-compatible programs. Thus, we will use the alternative estimators and confidence interval procedures. The point estimator would be given by

$$\hat{\pi}_{Adj.} = \frac{(n + \frac{3}{8})}{(n + \frac{3}{4})} = \frac{(50 + \frac{3}{8})}{(50 + \frac{3}{4})} = .993$$

A 95% confidence interval for π would be

$$((\alpha/2)^{1/n}, 1) = ((.05/2)^{1/50}, 1) = ((.025)^{.02}, 1) = (.929, 1.0)$$

We would now conclude that we are reasonably confident (95%) a high proportion (between 92.9% and 100%) of all programs that are Microsoft's Windows-compatible would run without alteration on the new operating system.

Keep in mind, however, that a sample size that is sufficiently large to satisfy the rule *does not* guarantee that the interval will be informative. It only judges the adequacy of the normal approximation to the binomial—the basis for the confidence level.

Sample size calculations for estimating π follow very closely the procedures we developed for inferences about μ. The required sample size for a $100(1 - \alpha)\%$ confidence interval for π of the form $\hat{\pi} \pm E$ (where E is specified) is found by solving the expression

$$z_{\alpha/2}\sigma_{\hat{\pi}} = E$$

for n. The result is shown here.

Sample Size Required for a 100(1 − α)% Confidence Interval for π of the Form $\hat{\pi} \pm E$

$$n = \frac{z_{\alpha/2}^2 \pi(1 - \pi)}{E^2}$$

Note: Since π is not known, either substitute an educated guess or use $\pi = .5$. Use of $\pi = .5$ will generate the largest possible sample size for the specified confidence interval width, $2E$, and thus will give a conservative answer to the required sample size.

EXAMPLE 10.3

A large public opinion polling agency plans to conduct a national survey to determine the proportion of employed adults who fear losing their job within the next year. How many workers must the agency poll to estimate to within .02 using a 95% confidence interval?

Solution By design, the agency wants the interval of the form $\hat{\pi} \pm .02$. The sample size necessary to achieve this accuracy is given by

$$n = \frac{z_{\alpha/2}^2 \pi(1 - \pi)}{E^2}$$

where $z_{\alpha/2} = 1.96$ and $E = .02$. If a previous survey was run recently, we could use the sample proportion from that survey to substitute for π; otherwise, we could use $\pi = .5$. Using $\pi = .5$, the required sample size is

$$n = \frac{(1.96)^2(.5)(.5)}{(.02)^2} = 2,401$$

that is, 2,401 workers would have to be surveyed to estimate π to within .02.

A statistical test about a binomial parameter π is very similar to the large-sample test concerning a population mean presented in Chapter 5. These results are summarized next, with three different alternative hypotheses along with their corresponding rejection regions. Recall that only one alternative is chosen for a particular problem.

Summary of a Statistical Test for π, π_0 Is Specified

H_0: 1. $\pi \leq \pi_0$ H_a: 1. $\pi > \pi_0$
 2. $\pi \geq \pi_0$ 2. $\pi < \pi_0$
 3. $\pi = \pi_0$ 3. $\pi \neq \pi_0$

T.S.: $z = \dfrac{\hat{\pi} - \pi_0}{\sigma_{\hat{\pi}}}$

R.R.: For a probability α of a Type I error

1. Reject H_0 if $z > z_\alpha$.
2. Reject H_0 if $z < -z_\alpha$.
3. Reject H_0 if $|z| > z_{\alpha/2}$.

Note: Under H_0,

$$\sigma_{\hat{\pi}} = \sqrt{\frac{\pi_0(1 - \pi_0)}{n}}$$

Also, n must satisfy both $n\pi_0 \geq 5$ and $n(1 - \pi_0) \geq 5$.

Check assumptions and draw conclusions.

EXAMPLE 10.4

Sports car owners in a town complain that the state vehicle inspection station judges their cars differently from family-style cars. Previous records indicate that 30% of all passenger cars fail the inspection on the first time through. In a random sample of 150 sports cars, 60 failed the inspection on the first time through. Is there sufficient evidence to indicate that the percentage of first failures for sports cars is higher than the percentage for all passenger cars? Use $\alpha = .05$.

Solution   The appropriate statistical test is as follows:

H_0: $\pi \leq .30$

H_a: $\pi > .30$

T.S.: $z = \dfrac{\hat{\pi} - \pi_0}{\sigma_{\hat{\pi}}}$

R.R.:   For $\alpha = .05$, we will reject H_0 if $z > 1.645$.

Using the sample data,

$$\hat{\pi} = \frac{60}{150} = .4 \quad \text{and} \quad \sigma_{\hat{\pi}} = \sqrt{\frac{(.3)(.7)}{150}} = .037$$

Also,

$$n\pi_0 = 150(.3) = 45 \geq 5$$

and

$$n(1 - \pi_0) = 150(.7) = 105 \geq 5$$

The test statistic is then

$$z = \frac{.4 - .3}{.037} = 2.70$$

Since the observed value of z exceeds 1.645, we conclude that sports cars at the vehicle inspection station have a first-failure rate greater than .3 (p-value = .0035). However, we must be careful not to attribute this difference to a difference in standards for sports cars and family-style cars. Parallel testing of sports cars versus other cars would have to be conducted to eliminate other sources of variability that would perhaps account for the higher first-failure rate for sports cars.

Most computer packages do not include this test. However, by coding successes as 1s and failures as 0s, we can approximate this test. This trick works exactly if the package includes a z test. Specify σ as $\sqrt{(\pi_0)(1-\pi_0)}$. If the package includes a one-sample t test, the result will be slightly different, but often close enough. For example, suppose we want to compare two versions of a product. We ask a random sample of 100 potential customers to choose between a new version of the product and the existing version. The new product will be adopted only if there is clear evidence that more than half of the customers prefer it. Thus we take the null hypothesis as H_0: $\pi \leq .5$ and the research hypothesis as H_a: $\pi >$ .5. If 68 of the 100 potential customers prefer the new product, we can run a z test and an approximate t test using Minitab. To do so, we input as data 68 1s and 32 0s in a column labeled "Yes_No." We obtained the following Minitab results; note that we specified σ as $\sqrt{(\pi_0)(1-\pi_0)} = \sqrt{(.5)(1-.5)} = .5$:

```
MTB > ZTest .5 .5 'Yes_No';
SUBC> Alternative 0.


Z-Test


Test of mu = 0.5000 vs mu not = 0.5000
The assumed sigma = 0.500


Variable       N       Mean      StDev     SE Mean        Z     P-Value
Yes_No        100     0.6800     0.4688     0.0500      3.60     0.0003

MTB > TTest 0.5'Yes_No';
SUBC> Alternative 0.


T-Test of the Mean


Test of mu = 0.500 vs mu not = 0.5000


Variable       N       Mean      StDev     SE Mean        T     P-Value
Yes_No        100     0.6800     0.4688     0.0469      3.84     0.0002
```

The results of the z-test procedure are exactly what we obtained by hand. The results of the t-test procedure are not quite the same, basically because that procedure used a sample standard deviation instead of $\sqrt{(\pi_0)(1-\pi_0)}$. The conclusion is the same, however: The research hypothesis of a difference in preferences is strongly supported, with a very low p-value. Though the t test based on 1s and 0s is not exactly the same as the z test, usually the results are so similar that we needn't worry too much about the difference.

sample-size requirement

We said that the z test for π is approximate and works best if n is large and π_0 is not too near 0 or 1. A natural next question is: When can we use it? There are several rules to answer the question; none of them should be considered sacred. Our sense of the many studies that have been done is this: If either $n\pi_0$ or $n(1-\pi_0)$ is less than about 2, treat the results of a z test very skeptically. If $n\pi_0$ and $n(1-\pi_0)$ are at least 5, the z test should be reasonably accurate. For the same sample size, tests based on extreme values of π_0 (for example, .001) are less accurate than tests for values of π_0, such as .05 or .10. For example, a test of H_0: $\pi = .0001$ with $n\pi_0 = 1.2$ is much more suspect than one for H_0: $\pi = .10$ with $n\pi_0 = 50$. If the issue becomes crucial, it's best to interpret the results skeptically or use exact tests (see Conover, 1999).

EXERCISES

Basic Techniques

10.1 Hypothetical sample results from a binomial experiment with $n = 150$ yielded $\hat{\pi} = .2$.
   a. Does this experiment satisfy the sample-test requirement for a confidence interval for π based on z? What sample sizes would be suspect, given the same sample proportion?
   b. Construct a 90% confidence interval for π.

10.2 Under what conditions can the formula $\hat{\pi} \pm z_{\alpha/2}\sigma_{\hat{\pi}}$ be used to express a confidence interval for π?

10.3 A random sample of 1,500 is drawn from a binomial population. If there are $y = 1,200$ successes,
   a. Construct a 95% confidence interval for π.
   b. Construct a 90% confidence interval for π.

10.4 Refer to Exercise 10.3. Explain the difference in the interpretation of the two confidence intervals.

Applications

Soc.

10.5 Experts have predicted that approximately 1 in 12 tractor-trailer units will be involved in an accident this year. One reason for this is that 1 in 3 tractor-trailer units has an imminently hazardous mechanical condition, probably related to the braking systems on the vehicle. A survey of 50 tractor-trailer units passing through a weighing station confirmed that 19 had a potentially serious braking system problem.
   a. Do the binomial assumptions hold?
   b. Can a normal approximation to the binomial be applied here to get a confidence interval for π?
   c. Give a 95% confidence interval for π using these data. Is the interval informative? What could be done to decrease the width of the interval, assuming $\hat{\pi}$ remained the same?

Psy.

10.6 In a study of self-medication practices, a random sample of 1,230 adults completed a survey. Some of the medical conditions that were self-treated are shown here. Summarize the results of this part of the survey using a 95% confidence interval for each medical condition.

Medical Condition	Home Remedy	% Responding
Sore throat—not related to a cold	Salt water or baking soda mouthwash	30
Burns—other than sunburn	Cold water/butter	28
Overindulgence in alcohol	Homebrew	25
Overweight	Diet	22
Pain associated with injury	Hot or cold compress	21

Psy. **10.7** In the survey discussed in Exercise 10.6, 441 of the adults reported they had a cough or cold recently and 260 of the respondents said they had treated the condition with an over-the-counter (OTC) remedy. The data are summarized here.

Survey respondents reporting problem	441
Number of patients using any OTC remedy	260
Patients using specific classes of OTC remedies:	
Adult pain relievers	110
Adult cold caps/tabs	57
Cough remedies	44
Allergy/hay fever remedies	9
Liquid cold remedies	35
Sprays/inhalers	4
Children's pain reliever	22
Cough drops	13
Sore-throat lozenges/gum	9
Children's cold caps/tabs	13
Nose drops	9
Chest rubs/ointments	9
Anesthetic throat lozenges	4
Room vaporizers	4
Other product	4

a. How might you organize and summarize these data? Would percentages help? Do the percentages add to 100%? Why or why not?

b. Based on these data, which classes of OTC remedies could you summarize using a 95% confidence interval for π?

Med. **10.8** Many individuals over the age of 40 develop an intolerance for milk and milk-based products. A dairy has developed a line of lactose-free products that are more tolerable to such individuals. To assess the potential market for these products, the dairy commissioned a market research study of individuals over 40 in its sales area. A random sample of 250 individuals showed that 86 of them suffer from milk intolerance. Based on the sample results, calculate a 90% confidence interval for the population proportion that suffers milk intolerance. Interpret this confidence interval.

Soc. **10.9** Shortly before April 15 of the previous year, a team of sociologists conducted a survey to study their theory that tax cheaters tend to allay their guilt by holding certain beliefs. The team interviewed a total of 500 adults and asked them under what situations they think cheating on an income tax return is justified. The responses include:

56% agree that "other people don't report all their income."
50% agree that "the government is often careless with tax dollars."
46% agree that "cheating can be overlooked if one is generally law abiding."

Assuming that the data are a simple random sample of the population of taxpayers (or tax-nonpayers), calculate 95% confidence intervals for the population proportion that agrees with each statement.

Pol. Sci.   **10.10**   A national columnist recently reported the results of a survey on marriage and the family. Part of the column is paraphrased here.

The Ingredients of Marriage

The Gallup people offered respondents a list of well-known ingredients. Here in the United States, such elements as faithfulness, mutual respect, and understanding ranked at the top. These were followed by enough money, same background, good housing and agreement in politics. Seventy-five percent of the respondents voted for "a good sex life," 59% for children, 52% for common interests, 48% for "living away from in-laws," and 43% for "sharing household chores." (In West Germany, by contrast, only 52% voted for a good sex life and only 19% for sharing household chores.)

a. How could you display the results of this survey in a graph or table?
b. Would you use a confidence interval to convey more information about the "true" percentages expressing an opinion on the various ingredients of a good marriage? Why or why not?
c. What qualms might you have about the way this survey has been reported?

Edu.   **10.11**   A substantial part of the U.S. population is "technologically illiterate," according to experts at a National Technological Literacy Conference organized by the National Science Foundation and Pennsylvania State University. At this conference, the results of a national survey of 2,000 adults showed that:

- 70% do not understand radiation.
- 40% think space-rocket launchings change the weather and that some unidentified flying objects are actually visitors from other planets.
- More than 80% do not understand how telephones work.
- 75% do not have a clear understanding of what computer software is.
- 72% do not understand the gross national product.

a. How might you display these data in a graph or table? Construct the display.
b. Many newspaper articles reporting the results of a survey give conclusions without sufficient details about the study for the reader to assess the data and reach a separate conclusion. What details are missing here for you to reach your own conclusion?

Bus.   **10.12**   The sales manager for a hardware wholesaler finds that 229 of the previous 500 calls to hardware store owners resulted in new product placements. Assuming that the 500 calls represent a random sample, find a 95% confidence interval for the long-run proportion of new product placements.

10.13   Give a careful verbal interpretation of the confidence interval found in Exercise 10.12.

Bus.   **10.14**   As part of a market research study, in a sample of 125, 84 individuals are aware of a certain product. Calculate a 90% confidence interval for the proportion of individuals in the population who are aware of the product. Interpret the interval.

Therefore, the 95% confidence interval is

$$(.784 - .645) - 1.96(.0264) \le \pi_1 - \pi_2 \le (.784 - .645) + 1.96(.0264)$$

or

$$.087 \le \pi_1 - \pi_2 \le .191$$

which indicates that somewhere between 8.7% and 19.1% more Wichita consumers than Grand Rapids consumers are aware of the product.

rule for sample sizes

This confidence interval method is based on the normal approximation to the binomial distribution. In Chapter 4, we indicated as a general rule that $n\hat{\pi}$ and $n(1 - \hat{\pi})$ should both be at least 5 to use this normal approximation. For this confidence interval to be used, the rule should hold for each sample. In practice, sample sizes that come even close to violating this rule aren't very useful because they lead to excessively wide confidence intervals. For instance, even though $n\hat{\pi}$ and $n(1 - \hat{\pi})$ are greater than 5 for both samples when $n_1 = 30$, $\hat{\pi}_1 = .20$ and $n_2 = 60$, $\hat{\pi}_2 = .10$, the 95% confidence interval is $-.06 \le \pi_1 - \pi_2 < .26$; π_1 could be anything from 6 percentage points lower than π_2 to 26 percentage points higher.

The reason for confidence intervals that seem very wide and unhelpful is that each measurement conveys very little information. In effect, each measurement conveys only one "bit": a 1 for a success or a 0 for a failure. For example, surveys of the compensation of chief executive officers of companies often give a manager's age in years. If we replaced the actual age by a category such as "over 55 years old" versus "under 55," we definitely would have far less information. When there is little information per item, we need a large number of items to get an adequate total amount of information. Wherever possible, it is better to have a genuinely numerical measure of a result rather than mere categories. When numerical measurement isn't possible, relatively large sample sizes will be needed.

Hypothesis testing about the difference between two population proportions is based on the z statistic from a normal approximation. The typical null hypothesis is that there is no difference between the population proportions, though any specified value for $\pi_1 - \pi_2$ may be hypothesized. The procedure is very much like a t test of the difference of means, and is summarized next.

Statistical Test for the Difference between Two Population Proportions

H_0: 1. $\pi_1 - \pi_2 \le 0$
2. $\pi_1 - \pi_2 \ge 0$
3. $\pi_1 - \pi_2 = 0$

H_a: 1. $\pi_1 - \pi_2 > 0$
2. $\pi_1 - \pi_2 < 0$
3. $\pi_1 - \pi_2 \ne 0$

$$\text{T.S.: } z = \frac{(\hat{\pi}_1 - \hat{\pi}_2)}{\sqrt{\dfrac{\hat{\pi}_1(1 - \hat{\pi}_1)}{n_1} + \dfrac{\hat{\pi}_2(1 - \hat{\pi}_2)}{n_2}}}$$

R.R.: 1. $z > z_\alpha$
2. $z < -z_\alpha$
3. $|z| > z_{\alpha/2}$

Check assumptions and draw conclusions.

Note: This test should be used only if $n_1\hat{\pi}_1$, $n_1(1 - \hat{\pi}_1)$, $n_2\hat{\pi}_2$, and $n_2(1 - \hat{\pi}_2)$ are all at least 5.

EXAMPLE 10.6

An educational researcher designs a study to compare the effectiveness of teaching English to non-English-speaking people by a computer software program and by the traditional classroom system. The researcher randomly assigns 125 students from a class of 300 to instruction using the computer. The remaining 175 students are instructed using the traditional method. At the end of a 6-month instructional period, all 300 students are given an examination with the results reported in the following table.

Exam Results	Computer Instruction	Traditional Instruction
Pass	94	113
Fail	31	62
Total	125	175

Does instruction using the computer software program appear to increase the proportion of students passing the examination in comparison to the pass rate using the traditional method of instruction? Use $\alpha = .05$.

Solution   Denote the proportion of all students passing the examination using the computer method of instruction and the traditional method of instruction by π_1 and π_2, respectively. We will test the hypotheses

$$H_0: \pi_1 - \pi_2 \le 0$$
$$H_a: \pi_1 - \pi_2 > 0$$

We will reject H_0 if the test statistic z is greater than $z_{.05} = 1.645$. From the data we compute the estimates

$$\hat{\pi}_1 = \frac{94}{125} = .752 \quad \text{and} \quad \hat{\pi}_2 = \frac{113}{175} = .646$$

From these we compute the test statistic to be

$$z = \frac{\hat{\pi}_1 - \hat{\pi}_2}{\sqrt{\dfrac{\hat{\pi}_1(1 - \hat{\pi}_1)}{n_1} + \dfrac{\hat{\pi}_2(1 - \hat{\pi}_2)}{n_2}}} = \frac{.752 - .646}{\sqrt{\dfrac{.752(1 - .752)}{125} + \dfrac{.646(1 - .646)}{175}}} = 2.00$$

Since $z = 2.00$ is greater than 1.645, we reject H_0 and conclude that the observations support the hypothesis that the computer instruction has a higher pass rate than the traditional approach. The p-value of the observed data is given by p-value $= P(z \ge 2.00) = .0228$, using the standard normal tables. A 95% confidence interval on the effect size $\pi_1 - \pi_2$ is given by

$$.752 - .646 \pm 1.96 \sqrt{\frac{.752(1 - .752)}{125} + \frac{.646(1 - .646)}{175}} \quad \text{or} \quad .106 \pm .104$$

We are 95% confident that the proportion passing the examination is between .2% and 21% higher for students using computer instruction than those using the traditional approach. For our conclusions to have a degree of validity, we need to check whether the sample sizes were large enough. Now, $n_1\hat{\pi}_1 = 94$, $n_1(1 - \hat{\pi}_1) = 31$, $n_2\hat{\pi}_2 = 113$, and $n_2(1 - \hat{\pi}_2) = 62$, thus all four quantities are greater than 5. Hence, the large sample criterion would appear to be satisfied.

EXERCISES   **Basic Techniques**

10.24   A random sample of $n_1 = 1,000$ observations was obtained from a binomial population with $\pi_1 = .4$. Another random sample, independent of the first sample, was selected from a binomial population with $\pi_2 = .2$. Does the normal approximation hold? Describe the sampling distribution for $\hat{\pi}_1 - \hat{\pi}_2$.

10.25   Refer to Exercise 10.24. How large a sample should we take from each population to have a 90% confidence interval of the form $\hat{\pi}_1 - \hat{\pi}_2 \pm .01$? (Hint: Assuming that equal sample sizes will be taken from the two populations, solve the expression

$$z_{\alpha/2}\sigma_{\hat{\pi}_1 - \hat{\pi}_2} = .01$$

for n, the common sample size. Use $\hat{\pi}_1 = .4$ and $\hat{\pi}_2 = .2$ from Exercise 10.24.)

Applications

Pol. Sci.   **10.26**   A law student believes that the proportion of registered Republicans in favor of additional tax incentives is greater than the proportion of registered Democrats in favor of such incentives. The student acquired independent random samples of 200 Republicans and 200 Democrats and found 109 Republicans and 86 Democrats in favor of additional tax incentives. Use these data to test $H_0: \pi_1 - \pi_2 \leq 0$ versus $H_a: \pi_1 - \pi_2 > 0$. Give the level of significance for your test.

Bus.   **10.27**   A retail computer dealer is trying to decide between two methods for servicing customers equipment. The first method emphasizes preventive maintenance; the second emphasizes quick response to problems. The dealer serves samples of customers by one of the two methods. After six months, it finds that 171 of 200 customers served by the first method are very satisfied with the service, as compared to 153 of 200 customers served by the second method. Execustat output, based on 1s and 0s for data, follows.

```
              TWO SAMPLE ANALYSIS FOR SATISFIED BY METHOD

                                1                        2

Sample size          200                 200
Mean                      0.855               0.765        diff. = 0.09
Variance                  0.124598            0.180678     ratio = 0.689612

Std. deviation            0.352984            0.425063

95% confidence intervals
     mu1 − mu2: (0.0131926, 0.166807) assuming equal variances
     mu1 − mu2: (0.0131847, 0.166815) not assuming equal variances

             HYPOTHESIS TEST—DIFFERENCE OF MEANS

Null hypothesis: difference of means = 0
Alternative: not equal
Equal variances assumed: no

Computed t statistic = 2.30362
              P value = 0.0218
```

Test the research hypothesis that the population proportions are different. Use $\alpha = .05$. State your conclusion carefully.

10.28   Locate a confidence interval for the difference of proportions in Exercise 10.27. Show that it reaches the same conclusion as the formal test about the research hypothesis.

Bus.   **10.29**   The media selection manager for an advertising agency inserts the same advertisement for a client bank in two magazines, similarly placed in each. One month later, a market research study finds that 226 of 473 readers of the first magazine are aware of the banking services offered in the ad, as are 165 of 439 readers of the second magazine (readers of both magazines are excluded). The following Minitab output was based on the appropriate number of 1s and 0s as data:

```
                    MTB > TWOT 'AWARE?' 'MAGAZINE'

             Two
  Sample T-Test and Confidence Interval

  Twosample T for Aware?
  Magazine    N     Mean    StDev    SE Mean
  1          473   0.478    0.500     0.023
  2          439   0.376    0.485     0.023

  95% C.I. for mu 1 - mu 2: (0.038, 0.166)
  T-Test mu 1 = mu 2 (vs not =): T = 3.13 P = 0.0018 DF = 908
```

    a. Calculate by hand a 95% confidence interval for the difference of proportions of readers who are aware of the advertised services. Compare your answer to the interval given by Minitab.

    b. Are the sample sizes adequate to use the normal approximation?

    c. Does the confidence interval indicate that there is a statistically significant difference using $\alpha = .05$?

10.30   Using the output of Exercise 10.29, perform a formal test of the null hypothesis of equal populations. Use $\alpha = .05$.

10.31   To test reliability, we subject samples of 30 electric motors for dot matrix printers from two suppliers to severe testing. Of the motors from supplier 1, 22 pass the test; of the motors from supplier 2, only 16 pass.

    a. Show that the difference is not statistically significant at $\alpha = .05$ (two-tailed).

    b. Can we claim to have shown that the two suppliers provide equally reliable motors?

10.32   Use the data of Exercise 10.31 to calculate a 95% confidence interval for the difference of proportions. Interpret the result carefully in terms of the relative reliability of the two suppliers.

Bio.

10.33   In a comparison of the incidence of tumor potential in two strains of rats, 100 rats (50 males, 50 females) were selected from each of two strains and were examined for a period of 1 year. All the rats were approximately the same age and were housed and fed under comparable conditions. Use the accompanying 1-year sample data to construct a 95% confidence interval for the difference in the proportions of rats exhibiting tumor potential for the two strains.

	Strain A	Strain B
Sample size	100	100
Number exhibiting tumor potential	25	15

Med.

10.34   There is a remedy for male pattern baldness—at least that's what millions of males hope since the FDA approved Upjohn's minoxidil for such a use. Minoxidil was investigated in a large, 27-center study where patients were randomly assigned to receive topical minoxidil or an identical-appearing placebo. Ignoring the center-to-center variation, suppose the preliminary results were as follows:

	Sample Size	% with New Hair Growth
Minoxidil group	310	32
Placebo	309	20

 a. Use these data to test $H_0: \pi_1 - \pi_2 = 0$ versus $H_a: \pi_1 - \pi_2 \neq 0$. Give the p-value for your test.

 b. If you were working for the FDA, what additional information might you want to examine in this study?

Bio.   **10.35** Is cocaine deadlier than heroin? A study reported in the *Journal of the American Medical Association* found that rats with unlimited access to cocaine had poorer health, had more behavior disturbances, and died at a higher rate than did a corresponding group of rats given unlimited access to heroin. The death rates after 30 days on the study were as follows:

	% Dead at 30 Days
Cocaine group	90
Heroin group	36

 a. Suppose that 100 rats were used in each group. Conduct a test of H_0: $\pi_1 - \pi_2 \leq 0$ versus $H_a: \pi_1 - \pi_2 > 0$. Give the p-value for your test.

 b. What implications are there for human use of the two drugs?

10.4   Inferences about Several Proportions: Chi-Square Goodness-of-Fit Test

We can extend the binomial sampling scheme of Chapter 4 to situations in which each trial results in one of k possible outcomes ($k > 2$). For example, a random sample of registered voters is classified according to political party (Republican, Democrat, Socialist, Green, Independent, etc.) or patients in a clinical trial are evaluated with respect to the degree of improvement in their medical condition (substantially improved, improved, no change, worse). This type of experiment or study is called a multinomial experiment with the characteristics listed here.

The Multinomial Experiment

1. The experiment consists of n identical trials.
2. Each trial results in one of k outcomes.
3. The probability that a single trial will result in outcome i is π_i for $i = 1, 2, \ldots, k$, and remains constant from trial to trial. (*Note:* $\Sigma\pi_i = 1$).
4. The trials are independent.
5. We are interested in n_i, the number of trials resulting in outcome i. (*Note:* $\Sigma n_i = n$).

  The probability distribution for the number of observations resulting in each of the k outcomes, called the **multinomial distribution**, is given by the formula

$$P(n_1, n_2, \ldots, n_k) = \frac{n!}{n_1! n_2! \ldots n_k!} \pi_1^{n_1} \pi_2^{n_2} \cdots \pi_k^{n_k}$$

Recall from Chapter 4, where we discussed the binomial probability distribution, that

$$n! = n(n - 1) \cdots 1$$

and

$$0! = 1$$

We can use the formula for the multinomial distribution to compute the probability of particular events.

EXAMPLE 10.7

Previous experience with the breeding of a particular herd of cattle suggests that the probability of obtaining one healthy calf from a mating is .83. Similarly, the probabilities of obtaining zero or two healthy calves are, respectively, .15 and .02. A farmer breeds three dams from the herd; find the probability of obtaining exactly three healthy calves.

Solution   Assuming the three dams are chosen at random, this experiment can be viewed as a multinomial experiment with $n = 3$ trials and $k = 3$ outcomes. These outcomes are listed with the corresponding probabilities.

Outcome	Number of Progeny	Probability, π_i
1	0	.15
2	1	.83
3	2	.02

Note that outcomes 1, 2, and 3 refer to the events that a dam produces zero, one, or two healthy calves, respectively. Similarly, n_1, n_2, and n_3 refer to the number of dams producing zero, one, or two healthy progeny, respectively. To obtain exactly three healthy progeny, we must observe one of the following possible events.

$$A: \begin{cases} \text{1 dam gives birth to no healthy progeny: } n_1 = 1 \\ \text{1 dam gives birth to 1 healthy progeny: } n_2 = 1 \\ \text{1 dam gives birth to 2 healthy progeny: } n_3 = 1 \end{cases}$$

$$B: \quad \text{3 dams give birth to 1 healthy progeny: } \begin{cases} n_1 = 0 \\ n_2 = 3 \\ n_3 = 0 \end{cases}$$

For event A with $n = 3$ and $k = 3$,

$$P(n_1 = 1, n_2 = 1, n_3 = 1) = \frac{3!}{1!1!1!} (.15)^1 (.83)^1 (.02)^1 \cong .015$$

Similarly, for event B,

$$P(n_1 = 0, n_2 = 3, n_3 = 0) = \frac{3!}{0!3!0!} (.15)^0 (.83)^3 (.02)^0 = (.83)^3 \cong .572$$

Thus, the probability of obtaining exactly three healthy progeny from three dams is the sum of the probabilities for events A and B; namely, $.015 + .572 \approx .59$.

Our primary interest in the multinomial distribution is as a probability model underlying statistical tests about the probabilities $\pi_1, \pi_2 \ldots, \pi_k$. We will hypothesize specific values for the πs and then determine whether the sample data agree with the hypothesized values. One way to test such a hypothesis is to examine the observed number of trials resulting in each outcome and to compare this to the number we would *expect* to result in each outcome. For instance, in our previous example, we gave the probabilities associated with zero, one, and two progeny as .15, .83, and .02. If we were to examine a sample of 100 mated dams, we would **expect to observe** 15 dams that produce no healthy progeny. Similarly, we would expect to observe 83 dams that produce one healthy calf and two dams that produce two healthy calves.

expected number of outcomes

DEFINITION 10.1

> In a multinomial experiment in which each trial can result in one of k outcomes, the **expected number of outcomes** of type i in n trials is $n\pi_i$, where π_i is the probability that a single trial results in outcome i.

In 1900, Karl Pearson proposed the following test statistic to test the specified probabilities:

$$\chi^2 = \sum_i \left[\frac{(n_i - E_i)^2}{E_i} \right]$$

where n_i represents the number of trials resulting in outcome i and E_i represents the number of trials we would expect to result in outcome i when the hypothesized probabilities represent the actual probabilities assigned to each outcome. Frequently, we will refer to the probabilities $\pi_1, \pi_2, \ldots, \pi_k$ as **cell probabilities**, one cell corresponding to each of the k outcomes. The observed numbers $n_1, n_2, \ldots, n_k$ corresponding to the k outcomes will be called **observed cell counts**, and the expected numbers $E_1, E_2, \ldots, E_k$ will be referred to as **expected cell counts**.

cell probabilities

observed cell counts

expected cell counts

Suppose that we hypothesize values for the cell probabilities $\pi_1, \pi_2, \ldots, \pi_k$. We can then calculate the expected cell counts by using Definition 10.1 to examine how well the observed data fit, or agree, with what we would expect to observe. Certainly, if the hypothesized π-values are correct, the observed cell counts n_i should not deviate greatly from the expected cell counts E_i, and the computed value of χ^2 should be small. Similarly, when one or more of the hypothesized cell probabilities are incorrect, the observed and expected cell counts will differ substantially, making χ^2 large.

chi-square distribution

The distribution of the quantity χ^2 can be approximated by a **chi-square distribution** provided that the expected cell counts E_i are fairly large.

The chi-square goodness-of-fit test based on k specified cell probabilities will have $k - 1$ degrees of freedom. We will explain why we have $k - 1$ degrees of freedom at the end of this section. Upper-tail values of the test statistic

$$\chi^2 = \sum_i \left[\frac{(n_i - E_i)^2}{E_i} \right]$$

can be found in Table 7 in the Appendix. See Figure 10.1 for a chi-square distribution with df = 4.

We can now summarize the chi-square goodness-of-fit test concerning k specified cell probabilities.

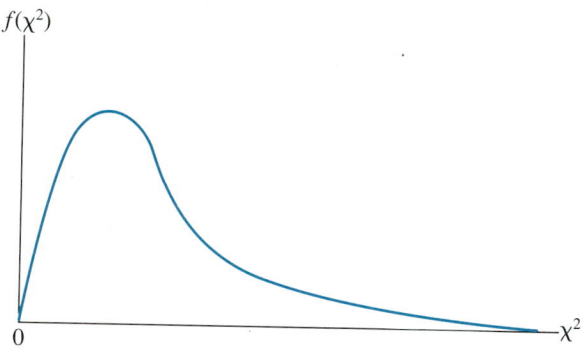

FIGURE 10.1
Chi-square probability
distribution for df = 4

Chi-Square Goodness-of-Fit Test

Null hypothesis: $\pi_i = \pi_{i0}$ for categories $i = 1, \ldots, k$, π_{i0} are specified probabilities or proportions.

Alternative hypothesis: At least one of the cell probabilities differs from the hypothesized value.

Test statistic: $\chi^2 = \sum \left[\dfrac{(n_i - E_i)^2}{E_i} \right]$, where n_i is the observed number in category i and $E_i = n\pi_{i0}$ is the expected number under H_0.

Rejection region: Reject H_0 if χ^2 exceeds the tabulated critical value for $a = \alpha$ and df $= k - 1$.

Check assumptions and draw conclusions.

The approximation of the sampling distribution of the chi-square goodness-of-fit test statistic by a chi-square distribution improves as the sample size n becomes larger. The accuracy of the approximation depends on both the sample size n and the number of cells k. Cochran (1954) indicates that the approximation should be adequate if no E_i is less than 1 and no more than 20% of the E_is are less than 5. The values of n/k that provide adequate approximations for the chi-square goodness-of-fit test statistic tends to decrease as k increases. Agresti (1990) discusses situations in which the chi-squared approximation tends to be poor for studies having small observed cell counts even if the expected cell counts are moderately large. Agresti concludes that it is hopeless to determine a single rule concerning the appropriate sample size to cover all cases. However, we recommend applying Cochran's guidelines for determining whether the chi-square goodness-of-fit test statistic can be adequately approximated with a chi-square distribution. When some of the E_is are too small, there are several alternatives. Researchers combine levels of the categorical variable to increase the observed cell counts. However, combining categories should not be done unless there is a natural way to redefine the levels of the categorical variable that does not change the nature of the hypothesis to be tested. When it is not possible to obtain observed cell counts large enough to permit the chi-squared approximation, Agresti (1990) discusses *exact* methods to test the hypotheses.

EXAMPLE 10.8

A laboratory is comparing a test drug to a standard drug preparation that is useful in the maintenance of patients suffering from high blood pressure. Over many clinical trials at many different locations, the standard therapy was administered to patients with comparable hypertension (as measured by the New York Heart Association (NYHA) Classification). The lab then classified the responses to therapy for this large patient group into one of four response categories. Table 10.1 lists the categories and percentages of patients treated on the standard preparation who have been classified in each category.

TABLE 10.1
Results of clinical trials using
the standard preparation

Category	Percentage
Marked decrease in blood pressure	50
Moderate decrease in blood pressure	25
Slight decrease in blood pressure	10
Stationary or slight increase in blood pressure	15

The lab then conducted a clinical trial with a random sample of 200 patients with high blood pressure. All patients were required to be listed according to the same hypertensive categories of the NYHA Classification as those studied under the standard preparation. Use the sample data in Table 10.2 to test the hypothesis that the cell probabilities associated with the test preparation are identical to those for the standard. Use $\alpha = .05$.

TABLE 10.2
Sample data for
example

Category	Observed Cell Counts
1	120
2	60
3	10
4	10

Solution   This experiment possesses the characteristics of a multinomial experiment, with $n = 200$ and $k = 4$ outcomes.

Outcome 1: A person's blood pressure will decrease markedly after treatment with the test drug.

Outcome 2: A person's blood pressure will decrease moderately after treatment with the test drug.

Outcome 3: A person's blood pressure will decrease slightly after treatment with the test drug.

Outcome 4: A person's blood pressure will remain stationary or increase slightly after treatment with the test drug.

The null and alternative hypotheses are then

$$H_0: \quad \pi_1 = .50, \pi_2 = .25, \pi_3 = .10, \pi_4 = .15$$

and

H_a: At least one of the cell probabilities is different from the hypothesized value.

Before computing the test statistic, we must determine the expected cell numbers. These data are given in Table 10.3.

TABLE 10.3

Observed and expected cell numbers for example

Category	Observed Cell Number, n_i	Expected Cell Number, E_i
1	120	$200(.50) = 100$
2	60	$200(.25) = 50$
3	10	$200(.10) = 20$
4	10	$200(.15) = 30$

Because all the expected cell numbers are relatively large, we may calculate the chi-square statistic and compare it to a tabulated value of the chi-square distribution.

$$\chi^2 = \sum_i \left[\frac{(n_i - E_i)^2}{E_i} \right]$$

$$= \frac{(120 - 100)^2}{100} + \frac{(60 - 50)^2}{50} + \frac{(10 - 20)^2}{20} + \frac{(10 - 30)^2}{30}$$

$$= 4 + 2 + 5 + 13.33 = 24.33$$

For the probability of a Type I error set at $\alpha = .05$, we look up the value of the chi-square statistic for $a = .05$ and df $= k - 1 = 3$. The critical value from Table 7 in the Appendix is 7.815.

R.R.: Reject H_0 if $\chi^2 > 7.815$.

Conclusion: The computed value of χ^2 is greater than 7.815, so we reject the null hypothesis and conclude that at least one of the cell probabilities differs from that specified under H_0. Practically, it appears that a much higher proportion of patients treated with the test preparation falls into the moderate and marked improvement categories. The p-value for this test is $p < .001$. (See Table 7 in the Appendix.)

The assumptions needed for running a chi-square goodness-of-fit test are those associated with a multinomial experiment, of which the key ones are independence of the trials and constant cell probabilities. Independence of the trials would be violated if, for example, several patients from the same family were included in the sample because hypertension has a strong hereditary component. The assumption of constant cell probabilities would be violated if the study were conducted over a period of time during which the standards of medical practice shifted, allowing for other "standard" therapies.

The test statistic for the chi-square goodness-of-fit test is the sum of k terms, which is the reason the degrees of freedom depend on k, the number of categories, rather than on n, the total sample size. However, there are only $k - 1$ degrees of freedom, rather than k, because the sum of the $n_i - E_i$ terms must be equal to $n - n = 0$; $k - 1$ of the observed minus expected differences are free to vary, but the last one (kth) is determined by the condition that the sum of the $n_i - E_i$ equals zero.

This goodness-of-fit test has been used extensively over the years to test various scientific theories. Unlike previous statistical tests, however, the hypothesis of interest is the null hypothesis, not the research (or alternative) hypothesis. Unfortunately, the logic behind running a statistical test does not hold. In the standard situation in which the research (alternative) hypothesis is the one of interest to the scientist, we formulate a suitable null hypothesis and gather data to reject H_0 in favor of H_a. Thus, we "prove" H_a by contradicting H_0.

We cannot do the same with the chi-square goodness-of-fit test. If a scientist has a set theory and wants to show that sample data conform to or "fit" that theory, she wants to accept H_0. From our previous work, there is the potential for committing a Type II error in accepting H_0. Here, as with other tests, the calculation of β probabilities is difficult. In general, for a goodness-of-fit test, the potential for committing a Type II error is high if n is small or if k, the number of categories, is large. Even if the expected cell counts E_i conform to our recommendations, the probability of a Type II error could be large. Therefore, the results of a chi-square goodness-of-fit test should be viewed suspiciously. Don't automatically accept the null hypothesis as fact given that H_0 was not rejected.

EXERCISES   **Basic Techniques**

10.36   List the characteristics of a multinomial experiment.

10.37   How does a binomial experiment relate to a multinomial experiment?

10.38   Under what conditions is it appropriate to use the chi-square goodness-of-fit test for a multinomial experiment? What qualification(s) might one have to make if the sample data do not yield rejection of the null hypothesis?

10.39   What restrictions are placed on the sample size n to use the chi-square goodness-of-fit test?

Applications

10.40   Hypothetical data are presented here. Use these data to run a chi-square goodness-of-fit test with H_0: $\pi_1 = .2$, $\pi_2 = .15$, $\pi_3 = .40$, $\pi_4 = .15$, and $\pi_5 = .10$. Use $\alpha = .05$. Do the data fit the hypothesized probabilities?

Category	Observed Cell Number, n_i
1	60
2	50
3	130
4	40
5	20
Total	300

10.41   Use the data of Exercise 10.40 to run a chi-square goodness-of-fit test with this new null hypothesis—H_0: $\pi_1 = .15$, $\pi_2 = .20$, $\pi_3 = .45$, $\pi_4 = .15$, and $\pi_5 = .05$. Again use $\alpha = .05$. Compare your results to those of Exercise 10.40. How sensitive does this test appear to be for the cell probabilities specified under H_0? What conclusion can you draw if you do *not* reject H_0?

Bus.   **10.42**   Over the past 5 years, an insurance company has had a mix of 40% whole life policies, 20% universal life policies, 25% annual renewable-term (ART) policies, and 15% other types of policies. A change in this mix over the long haul could require a change in the commission structure, reserves, and possibly investments. A sample of 1,000 policies issued over the last few months gave the results shown here. Use these data to assess whether there has been a shift from the historical percentages. Give the p-value for your test. Which policies (if any) seem to be more popular?

Category	Observed Cell Number, n_i
Whole life	320
Universal life	280
ART	240
Other	160
Total	1,000

Soc.

10.43 A university and several industries developed a work-study program in the surrounding community. Students were to work with industrial sociologists during a 3-month internship. Equal numbers of students from the university worked in a chemical, a textile, and a pharmaceutical industry. Students completing the program were classified according to the industry in which they interned. Consider the following data as a random sample of the many students who could have completed the program. Test the null hypothesis that the probability that a finishing student interned in a pharmaceutical, chemical, or textile industry is 1/3. Use $\alpha = .01$ with n_i the number of students in group i finishing the program.

Group	n_i
Pharmaceutical	20
Chemical	13
Textile	30

Soc.

10.44 Researchers conducted an experiment to determine whether the proportion of mentally ill patients of each social class housed in a county facility agrees with the social class distribution of the county. The observed cell numbers for the 400 patients classified are given here.

Lower: 215   Upper-middle: 60
Lower-middle: 100   Upper: 25

Use these data to test the null hypothesis

$\pi_1 = .25$ $\pi_3 = .20$
$\pi_2 = .48$ $\pi_4 = .07$

where the πs are the hypothesized proportions of persons in the respective social-class categories in the county. Use $\alpha = .05$ and draw conclusions.

Pol. Sci.

10.45 In previous presidential elections in a given locality, 50% of the registered voters were Republicans, 40% were Democrats, and 10% were registered as independents. Prior to the upcoming election, a random sample of 200 registered voters showed that 90 were registered as Republicans, 80 as Democrats, and 30 as independents. Test the research hypothesis that the distribution of registered voters is different from that in previous election years. Give the p-value for your test. Draw conclusions.

Med.

10.46 A local doctor suspects that there is a seasonal trend in the occurrence of the common cold. He estimates that 40% of the cases each year occur in the winter, 40% in the spring, 10% in the summer, and 10% in the fall. The doctor collected the following information from a random sample of 1,000 cases of patients with the common cold over the past year. Would you agree with the doctor's estimates, based on the sample information? Perform a statistical test using $\alpha = .05$. Draw conclusions.

Season	Frequency
Winter	374
Spring	292
Summer	169
Fall	165

10.47   Refer to Exercise 10.46. What would the null hypothesis be if the doctor claimed that there are no differences in the percentages of cases over the seasons? Test the hypothesis that there is no seasonal trend in the occurrence of the common cold. Give the level of significance of your test. Do you have any reservations about your conclusion?

Med.   **10.48**   Previous experimentation with a drug product developed for the relief of depression was conducted with normal adults with no signs of depression. We will assume a large data bank is available from studies conducted with normals and, for all practical purposes, the data bank can represent the population of responses for normals. Each of the adults participating in one of these studies was asked to rate the drug as ineffective, mildly effective, or effective. The percentages of respondents in these categories were 60%, 30%, and 10%, respectively. In a new study of depressed adults, a random sample of 85 adults responded as follows:

Ineffective: 30
Mildly effective: 35
Effective: 20

Is there evidence to indicate a different percentage distribution of responses for depressed adults than for nondepressed? Give the level of significance for your test and draw conclusions.

Pol. Sci.   **10.49**   In random sampling, an interviewer asked 40 newspaper editors their opinions on the degree of future suppression of freedom of the press brought about by recent court decisions. The editors' opinions are summarized here. Use these data to test the null hypothesis that each category is equally preferred. Use $\alpha = .05$. Draw conclusions from these data. What reservation(s), if any, do you have concerning your conclusions?

Degree of Suppression	Frequency
None	8
Very little	8
Moderate	10
Severe	14

Bus.   **10.50**   A researcher obtained a sample of 125 securities analysts and asked each analyst to select four stocks on the New York Stock Exchange that were expected to outperform the Standard and Poor's Index over a 3-month period. One theory suggests that the securities analysts would be expected to do no better than chance and that the number of correct guesses from the four selected had a multinomial distribution as shown here.

Number correct	0	1	2	3	4
Multinomial probabilities (π_i)	.0625	.2500	.3750	.2500	.0625

If the number of correct guesses from the sample of 125 analysts had a frequency distribution as shown here, use these data to conduct a chi-square goodness-of-fit test. Use $\alpha = .05$. Draw conclusions.

Number correct	0	1	2	3	4
Frequency	3	23	51	39	9

10.5 The Poisson Distribution

In Chapter 4 (and again in this chapter), we indicated that the normal distribution provides a good approximation to the binomial distribution provided $n \geq 5/\min (\pi, 1 - \pi)$. This requirement was needed to ensure that the binomial distribution was reasonably symmetric. However, there are many instances when the binomial probability distribution is sufficiently skewed so as to render the normal approximation inappropriate. For example, in observing patients administered a new drug product in a properly conducted clinical trial, the number of persons experiencing a particular side effect might be quite small. If π (the probability of observing a person with the side effect) is .001, $\min(\pi, 1 - \pi) = .001$. For this example, in order to approximate the binomial with a normal distribution, the sample size would have to be equal to or greater than $5/.001 = 5,000$.

Poisson distribution   In 1837, S. D. Poisson developed a discrete probability distribution, suitably called the **Poisson distribution**, which provides a good approximation to the binomial when π is small and n is large but $n\pi$ is less than 5. The probability of observing y successes in the n trials is given by the formula

$$P(y) = \frac{\mu^y e^{-\mu}}{y!}$$

e = 2.71828   where e is a constant approximately equal to 2.71828, and μ is the average value of y. Table 15 in the Appendix gives Poisson probabilities for various values of the parameter μ. For approximating binomial probabilities using the Poisson distribution, take

$$\mu = n\pi$$

EXAMPLE 10.9

Refer to the clinical trial mentioned at the beginning of this section, where $n = 1,000$ patients were treated with a new drug. Compute the probability that none of a sample of $n = 1,000$ patients experiences a particular side effect (such as nausea) when $\pi = .001$.

Solution   The mean of the binomial distribution is $\mu = n\pi = 1,000(.001) = 1$. Substituting into the Poisson probability distribution with $\mu = 1$, we have

$$P(y = 0) = \frac{(1)^0 e^{-1}}{0!} = e^{-1} = \frac{1}{2.71828} = .367879$$

(Note also from Table 15 in the Appendix that the entry corresponding to $y = 0$ and $\mu = 1$ is 0.3679.)

EXAMPLE 10.10

Suppose that after a clinical trial of a new medication involving 1,000 patients, no patient experienced nausea. Would it be reasonable to infer that less than .001 of the entire population would experience this side effect while taking the drug?

Solution   Certainly not. We computed the probability of observing $y = 0$ in $n = 1,000$ trials assuming $\pi = .001$ (i.e., assuming .1% of the population would experience nausea) to be .368. Since this probability is quite large, it would not be wise to infer that $\pi < .001$.

Although the Poisson distribution provides a useful approximation to the binomial under certain conditions, the application of the Poisson distribution is not limited to these situations. In particular, the Poisson distribution has been useful in finding the probability of y occurrences of an event that occurs randomly over an interval of time, volume, space, and so on, provided certain assumptions are met.

1. Events occur one at a time; two or more events do not occur precisely at the same time.
2. The occurrence of an event in a given period is independent of the occurrence of the event in a nonoverlapping period; that is, the occurrence (or nonoccurrence) of an event during one period does not change the probability of an event occurring in some later period.

In many discussions of this topic, a third assumption is added:

3. The expected number of events during any one period is the same as that during any other period.

Although these assumptions seem to be somewhat restrictive, many situations appear to satisfy these conditions. For example, the number of arrivals of customers at a checkout counter, parking lot toll booth, inspection station, or garage repair shop during a specified time interval (such as 1 minute) could be approximated with a Poisson probability distribution. Similarly, the number of clumps of algae of a particular species observed in a unit volume of lake water visible under a microscope could be approximated by a Poisson probability distribution.

Confronted with a set of measurements, we may now wish to check the assumption that the data follow a Poisson probability distribution. To do this we make use of the goodness-of-fit test of Section 10.4, using the test statistic

tests using Poisson distribution

$$\chi^2 = \sum_i \left[\frac{(n_i - E_i)^2}{E_i} \right]$$

There are two types of null hypotheses. The first hypothesis is that the data arise from a Poisson distribution with $\mu = \mu_0$; that is, we wish to test $H_0: \mu = \mu_0$ (μ_0 is specified) against the alternative hypothesis $H_a: \mu \neq \mu_0$. The quantity n_i denotes the number of observations in cell i and E_i is the expected number of observations in cell i obtained from the probabilities for a Poisson distribution with mean μ_0. The computed value of the test statistic is then compared to the tabulated chi-square value in Table 7 in the Appendix with $a = \alpha$ and df $= k - 1$, where k is the number of cells.

The second null hypothesis we might be interested in is less specific. We test

H_0: The observed cell counts all come from a common Poisson distribution with mean μ (unspecified).

The alternative is that not all cell counts come from a common Poisson distribution. The test statistic is

$$\chi^2 = \sum_i \left[\frac{(n_i - E_i)^2}{E_i} \right]$$

where for all cells E_i is the expected number of observations in cell i obtained from the probabilities for a Poisson distribution with a mean estimated from the sample data. The rejection region is then located for $a = \alpha$ and df $= k - 2$. Note the difference in the degrees of freedom for the two null hypotheses. In the latter test, we lose one degree of freedom because we must estimate the Poisson parameter μ.

EXAMPLE 10.11

Environmental engineers often utilize information contained in the number of different alga species and the number of cell clumps per species to measure the health of a lake. Those lakes exhibiting only a few species but many cell clumps are classified as oligotrophic. In one such investigation, a lake sample was analyzed under a microscope to determine the number of clumps of cells per microscope field. These data are summarized here for 150 fields examined under a microscope. Here y_i denotes the number of cell clumps per field and n_i denotes the number of fields with y_i cell clumps.

y_i	0	1	2	3	4	5	6	≥ 7
n_i	6	23	29	31	27	13	8	13

Use $\alpha = .05$ to test the null hypothesis that the sample data were drawn from a Poisson probability distribution.

Solution Before we can compute the value of χ^2, first we must estimate the Poisson parameter μ and then compute the expected cell counts. The Poisson mean μ is estimated by using the sample mean $\bar{y}$. For these data,

$$\bar{y} = \frac{\sum_i n_i y_i}{n} = \frac{486}{150} \approx 3.3$$

Note that the sample mean was computed to be 3.3 by using all the sample data before the 13 largest values were collapsed into the final cell. This is why the sample mean computed here was rounded up to 3.3.

The Poisson probabilities for $y = 0, 1, \ldots, 7$ or more can be found in Table 15 in the Appendix with $\mu = 3.3$. These probabilities are shown here.

y_i	0	1	2	3	4	5	6	≥ 7
$P(y_i)$ for $\mu = 3.3$	.037	.122	.201	.221	.182	.120	.066	.051

The expected cell count E_i can be computed for any cell using the formula $E_i = nP(y_i)$. Hence, for our data (with $n = 150$), the expected cell counts are as shown here.

y_i	0	1	2	3	4	5	6	≥ 7
E_i	5.55	18.30	30.15	33.15	27.30	18.00	9.90	7.65

Substituting these values into the test statistic, we have

$$\chi^2 = \sum_i \left[\frac{(n_i - E_i)^2}{E_i} \right]$$

$$= \frac{(6 - 5.55)^2}{5.55} + \frac{(23 - 18.30)^2}{18.30} + \cdots + \frac{(13 - 7.65)^2}{7.65} = 6.93$$

The tabulated value of chi-square for $a = .05$ and df $= k - 2 = 6$ is 12.59. Since the computed value of chi-square does not exceed 12.59, we have insufficient evidence to reject the null hypothesis that the data were collected from a Poisson distribution.

A word of caution is given here for situations in which we are considering this test procedure. As we mentioned previously, when using a chi-square statistic, we should have all expected cell counts fairly large. In particular, we want all $E_i > 1$ and not more than 20% less than 5. In Example 10.11, if values of $y \geq 7$ had been considered individually, the Es would not have satisfied the criteria for the use of χ^2. That is why we combined all values of $y \geq 7$ into one category.

EXERCISES   **Basic Techniques**

10.51 Compute the following Poisson probabilities using Table 15 in the Appendix.
  a. $P(y = 1)$ given $\mu = .5$, $\mu = 1.0$, and $\mu = 3.0$
  b. $P(y > 1)$ given $\mu = 1.7$, $\mu = 2.5$, and $\mu = 4.2$
  c. $P(y < 5)$ given $\mu = 0.2$, $\mu = 1.0$, and $\mu = 2.0$

Applications

Engin.   **10.52** Cars arrive at the exit gate for airport long-term parking at a rate of six per minute during rush hour. Find the following probabilities using Table 15 in the Appendix. (y is the number of cars arriving during any given minute in rush hour.)
  a. $P(y = 0)$
  b. $P(y > 1)$
  c. $P(y > .3)$

Bus.   **10.53** A firm is considering using telemarketing techniques to supplement traditional marketing methods. It is estimated that one of every 100 calls results in a sale. Suppose that 250 calls are made in a single day:
  a. Write an expression for the probability that there are less than six sales—do not do the mathematics.
  b. What assumptions are you making in part (a)?
  c. Use a normal approximation to compute $P(y < 6)$.
  d. Compute $P(y < 6)$ using the Poisson distribution.
  e. Which approximation [part (c) or part (d)] appears better? Why?

Med.   **10.54** A certain birth defect occurs with probability .0001; that is, one of every 10,000 babies has this defect. If 5,000 babies are born at a particular hospital in a given year, what approximation should be used? What is the approximate probability that there is at least one baby with the defect?

Gov.   **10.55** One portion of a government study to determine the effectiveness of an exclusive bus lane was directed at examining the number of conflicts (driving situations that could result in an accident) at a major intersection during a specified period of time. A previous study prior to the installation of the exclusive bus lane indicated that the number of conflicts per 5 minutes during the 7:00 to 9:00 A.M. peak period could be adequately approximated by a Poisson distribution with $\mu = 2$. The following data were based on a sample of 40 days; y_i denotes the number of conflicts and n_i denotes the number of 5-minute periods during which y was observed.

y_i	0	1	2	3	4	5	≥ 6
n_i	90	230	240	130	68	30	12

a. Does the Poisson assumption appear to hold?
b. Use these data to test the research hypothesis that the mean number of conflicts per 5 minutes differs from 2. (*Hint:* Use a chi-square test based on Poisson probabilities.)

Engin.   **10.56**   The number of shutdowns per day caused by a breaking of the thread was noted for a nylon spinning process over a period of 90 days. Use the sample data below to determine if the number of shutdowns per day follows a Poisson distribution. Use $\alpha = .05$. In the listing of the data, y_i denotes the number of shutdowns per day and n_i denotes the number of days with y_i shutdowns.

y_i	0	1	2	3	4	≥ 5
n_i	20	28	15	8	7	12

Bio.   **10.57**   Entomologists study the distribution of insects across agricultural fields. A study of fire ant hills across pasture lands is conducted by dividing pastures into 50-meter squares and counting the number of fire ant hills in each square. The null hypothesis of a Poisson distribution for the counts is equivalent to a random distribution of the fire ant hills over the pasture. Rejection of the hypothesis of randomness may occur due to one of two possible alternatives. The distribution of fire ant hills may be uniform; that is, the same number of hills per 50-meter square or the distribution of fire ants may be clustered across the pasture. A random distribution would have the variance in counts equal to the mean count, $\sigma^2 = \mu$. If the distribution is more uniform than random, then the distribution is said to be underdispersed, $\sigma^2 < \mu$. If the distribution is more clustered than random, then the distribution is said to be overdispersed, $\sigma^2 > \mu$. The number of fire ant hills was recorded on one hundred 50-meter squares. In the data set, y_i is the number of fire ant hills per square and n_i denotes the number of 50-meter squares with y_i ant hills.

y_i	0	1	2	3	4	5	6	7	8	9	12	15	20
n_i	2	6	8	10	12	15	13	12	10	6	3	2	1

a. Estimate the mean and variance of the number of fire ant hills per 50-meter square; that is, compute $\bar{y}$ and s^2 using the formulas from Chapter 3.
b. Do the fire ant hills appear to be randomly distributed across the pastures? Use a chi-square test of the adequacy of the Poisson distribution to fit the data using $\alpha = .05$.
c. If you reject the Poisson distribution as a model for the distribution of fire ant hills, does it appear that fire ant hills are more clustered or uniformly distributed across the pastures?

10.6   Contingency Tables: Tests for Independence and Homogeneity

In Section 10.3, we showed a test for comparing two proportions. The data were simply counts of how many times we got a particular result in two samples. In this section, we extend that test. First, we present a single test statistic for testing whether several deviations of sample data from theoretical proportions could plausibly have occurred by chance.

When we first introduced probability ideas in Chapter 4, we started by using tables of frequencies (counts). At the time, we treated these counts as if they represented the whole population. In practice, we'll hardly ever know the complete population data; we'll usually have only a sample. When we have counts from a sample, they're usually arranged in **cross tabulations** or **contingency tables**. In this section, we'll describe one particular test that is often used for such tables, a chi-square test of independence.

In Chapter 4, we introduced the idea of independence. In particular, we discussed the idea that **dependence** of variables means that one variable has some value for predicting the other. With sample data, there usually appears to be some degree of dependence. In this section, we develop a χ^2 test that assesses whether the perceived dependence in sample data may be a fluke—the result of random variability rather than real dependence.

First, the frequency data are to be arranged in a cross tabulation with r rows and c columns. The possible values of one variable determine the rows of the table, and the possible values of the other determine the columns. We denote the population proportion (or probability) falling in row i, column j as π_{ij}. The total proportion for row i is $\pi_{i \cdot}$ and the total proportion for column j is $\pi_{\cdot j}$. If the row and column proportions (probabilities) are independent, then $\pi_{ij} = \pi_{i \cdot} \pi_{\cdot j}$. For instance, suppose that a personnel manager for a large firm wants to assess the popularity of three alternative flexible time-scheduling (flextime) plans among clerical workers in four different offices. The following indicates a set of proportions (π_{ij}) that exhibit independence. The proportion of all clerical workers who favor plan 2 and work in office 1 is $\pi_{21} = .03$, the proportion of all workers favoring plan 2 is $\pi_{2 \cdot} = .30$, and the proportion working in office 1 is $\pi_{\cdot 1} = .10$. Independence holds for that cell because $\pi_{21} = .03 = (\pi_{2 \cdot})(\pi_{\cdot 1}) = (.30)(.10)$. Independence also holds for all other cells.

| | Office | | | | |
Favored Plan	1	2	3	4	Total
1	.05	.20	.15	.10	.50
2	.03	.12	.09	.06	.30
3	.02	.08	.06	.04	.20
Total	.10	.40	.30	.20	

The null hypothesis for this χ^2 test is independence. The research hypothesis specifies only that there is some form of dependence—that is, that it is not true that $\pi_{ij} = \pi_{i \cdot} \pi_{\cdot j}$ in every cell of the table. The test statistic is once again the sum over all cells of

(observed value − expected values)^2/expected value

The computation of expected values E_{ij} under the null hypothesis is different for the independence test than for the goodness-of-fit test. The null hypothesis of independence does not specify numerical values for the row probabilities $\pi_{i \cdot}$ and column probabilities $\pi_{\cdot j}$, so these probabilities must be estimated by the row and column relative frequencies. If n_i is the actual frequency in row i, estimate π_i by $\hat{\pi}_{i\cdot} = n_{i\cdot}/n$; similarly $\hat{\pi}_{\cdot j} = n_{\cdot j}/n$. Assuming the null hypothesis of independence is true, it follows that $\hat{\pi}_{ij} = \hat{\pi}_{i\cdot}\hat{\pi}_{\cdot j} = (n_{i\cdot}/n)(n_{\cdot j}/n)$.

DEFINITION 10.2

Under the hypothesis of independence, the **estimated expected value** in row i, column j is

$$\hat{E}_{ij} = n\hat{\pi}_{ij} = n\frac{(n_{i\cdot})}{n}\frac{(n_{\cdot j})}{n} = \frac{(n_{i\cdot})(n_{\cdot j})}{n}$$

the row total multiplied by the column total divided by the grand total.

EXAMPLE 10.12

Suppose that in the flexible time-scheduling illustration, a random sample of 216 workers yields the following frequencies:

		Office			
Favored Plan	**1**	**2**	**3**	**4**	**Total**
1	15	32	18	5	70
2	8	29	23	18	78
3	1	20	25	22	68
Total	24	81	66	45	216

Calculate a table of $\hat{E}_{ij}$ values.

Solution For row 1, column 1 the estimated expected number is

$$\hat{E}_{11} = \frac{(\text{row 1 total})(\text{column 1 total})}{\text{grand total}} = \frac{(70)(24)}{216} = 7.78$$

Similar calculations for all cells yield the following table.

		Office			
Plan	**1**	**2**	**3**	**4**	**Total**
1	7.78	26.25	21.39	14.58	70.00
2	8.67	29.25	23.83	16.25	78.00
3	7.56	25.50	20.78	14.17	68.01
Totals	24.01	81.00	66.00	45.00	216.01

Note that the row and column totals in the $\hat{E}_{ij}$ table equal (except for round-off error) the corresponding totals in the observed (n_{ij}) table.

χ^2 Test of Independence

H_0: The row and column variables are independent.

H_a: The row and column variables are dependent (associated).

T.S.: $\chi^2 = \sum_{i,j} (n_{ij} - \hat{E}_{ij})^2 / \hat{E}_{ij}$

R.R.: Reject H_0 if $\chi^2 > \chi^2_\alpha$, where χ^2_α cuts off area α in a χ^2 distribution with $(r - 1)(c - 1)$ df; r = number of rows, c = number of columns.

Check assumptions and draw conclusions.
The test statistic is sometimes called the Pearson χ^2 statistic.

df for table

The degrees of freedom for the χ^2 test of independence relate to the number of cells in the two-way table that are free to vary while the marginal totals remain fixed. For example, in a 2 × 2 table (2 rows, 2 columns), only one cell entry is free to vary. Once that entry is fixed, we can determine the remaining cell entries by subtracting from the corresponding row or column total. In Table 10.4(a), we have indicated some (arbitrary) totals. The cell indicated by * could take any value (within the limits implied by the totals), but then all remaining cells would be determined by the totals. Similarly, with a 2 × 3 table (2 rows, 3 columns), two of the cell entries, as indicated by *, are free to vary. Once these entries are set, we determine the remaining cell entries by subtracting from the appropriate row or column total [see Table 10.4(b)]. In general, for a table with r rows and c columns, $(r - 1)(c - 1)$ of the cell entries are free to vary. This number represents the degrees of freedom for the χ^2 test of independence.

TABLE 10.4
(a) One df in a 2 × 2 table;
(b) two df in a 2 × 3 table

Category A	Category B		Total
	*		16
			34
Totals	21	29	50

(a)

Category A	Category B			Total
	*	*		51
				40
Totals	28	41	22	91

(b)

This χ^2 test of independence is also based on an approximation. A conservative rule is that each $\hat{E}_{ij}$ must be at least 5 to use the approximation comfortably. Standard practice if some $\hat{E}_{ij}$s are too small is to lump together those rows (or columns) with small totals until the rule is satisfied.

EXAMPLE 10.13

Carry out the χ^2 test of independence for the data of Example 10.8. First use $\alpha = .05$; then obtain a bound for the p-value.

Solution The null and alternative hypotheses are:

H_0: The popularity of the scheduling plan is independent of the office location.

H_a: The popularity of the scheduling plan depends on the office location.

The test statistic can be computed using the n_{ij} and E_{ij} from Example 10.12:

T.S.: $\chi^2 = \sum (n_{ij} - E_{ij})^2/E_{ij}$

$$= (15 - 7.78)^2/7.78 + (32 - 26.25)^2/26.25$$
$$+ \cdots + (22 - 14.17)^2/14.17$$
$$= 6.70 + 1.26 + \cdots + 4.33 = 27.13$$

R.R.: For df $= (3 - 1)(4 - 1) = 6$ and $\alpha = .05$, the critical value from Table 7 in the Appendix is 12.59. Since $\chi^2 = 27.13$ exceeds 12.59, H_0 is rejected. In fact, since 27.13 is larger than the value in Table 7 for $a = .001$, the p-value is $<.001$.

Check assumptions and draw conclusions: Since each of the expected values exceeds 5, the χ^2 approximation should be good. There is strong evidence ($p < .001$) that the popularity of the scheduling plan depends on the office location.

likelihood ratio statistic

There is an alternative χ^2 statistic called the **likelihood ratio statistic** that is often shown in computer outputs. It is defined as

likelihood ratio $\chi^2 = \sum_{ij} n_{ij} \ln(n_{ij}/(n_i n_j))$

where $n_{i.}$ is the total frequency in row i, $n_{.j}$ is the total in column j, and ln is the natural logarithm (base $e = 2.71828$). Its value should also be compared to the χ^2 distribution with the same $(r - 1)(c - 1)$ df. Although it isn't at all obvious, this form of the χ^2 independence test is approximately equal to the Pearson form. There is some reason to believe that the Pearson χ^2 yields a better approximation to table values, so we prefer to rely on it rather than on the likelihood ratio form.

The only function of a χ^2 test of independence is to determine whether apparent dependence in sample data may be a fluke, plausibly a result of random variation. Rejection of the null hypothesis indicates only that the apparent association is not reasonably attributable to chance. It does not indicate anything about the **strength** or type **of association**.

strength of association

The same χ^2 test statistic applies to a slightly different sampling procedure. An implicit assumption of our discussion surrounding the χ^2 test of independence is that the data result from a single random sample from the whole population. Often, separate random samples are taken from the subpopulations defined by the column (or row) variable. In the flextime example (Example 10.12), the data might have resulted from separate samples (of respective sizes 24, 81, 66, and 45) from the four offices rather than from a single random sample of 216 workers.

In general, suppose the column categories represent c distinct subpopulations. Random samples of size $n_1, n_2, \ldots, n_c$ are selected from these subpopulations. The observations from each subpopulation are then classified into the r values of a categorical variable represented by the r rows in the contingency table. The research hypothesis is that there is a difference in the distribution of subpopulation units into the r levels of the categorical variable. The null hypothesis is that the set of r proportions for each subpopulation $(\pi_{1j}, \pi_{2j}, \ldots, \pi_{rj})$ is the same for all c subpopulations. Thus, the null hypothesis is given by

$$H_0: \quad (\pi_{11}, \pi_{21}, \ldots, \pi_{r1}) = (\pi_{12}, \pi_{22}, \ldots, \pi_{r2}) = \cdots = (\pi_{1c}, \pi_{2c}, \ldots, \pi_{rc})$$

test of homogeneity

The test is called a **test of homogeneity** of distributions. The mechanics of the test of homogeneity and the test of independence are identical. However, note that the sampling scheme and conclusions are different. With the test of indepen-

10.62   The data of Exercise 10.60 are combined as follows:

	Age		
	Up to 39	**40 and Over**	**Total**
Promoted	38	42	80
Not promoted	82	88	170
Total	120	130	

The Minitab results are as follows:

```
MTB > Table 'promoted' 'combined';
SUBC>    Counts;
SUBC>    ColPercents;
SUBC>    ChiSquare.
 ROWS:    promoted      COLUMNS: combined
              1           2        ALL
    1        38          42        80
           31.67       32.31     32.00
            38          42        80

    2        82          88        170
           68.33       67.69     68.00
            82          88        170

 ALL        120         130       250
          100.00      100.00    100.00
            120         130       250

 CHI-SQUARE =          0.012   WITH D.F. =   1
```

a. Can the hypothesis of independence be rejected using a reasonable α?
b. What is the effect of combining age categories? Compare the answers to those for Exercise 10.60.

10.7   Measuring Strength of Relation

The χ^2 test we discussed in Section 10.5 has a built-in limitation. By design, the test only answers the question of whether there is a statistically detectable (significant) relation among the categories. It cannot answer the question of whether the relation is strong, interesting, or relevant. This is not a criticism of the test; no hypothesis test can answer these questions. In this section, we discuss methods for assessing the strength of relation shown in cross-tabulated data.

The simplest (and often the best) method for assessing the strength of a relation is simple percentage analysis. If there is no relation (that is, if complete independence holds), then percentages by row or by column show no relation. For example, suppose that a direct-mail company tests two different offers to see whether the response rates differ. Their results are as shown here:

Offer	Response		
	Yes	**No**	**Total**
A	40	160	200
B	80	320	400
Totals	120	480	600

To check the relation, if any, we calculate percentages of response for each offer. We see that $(40/200) = .20$ (that is, 20%) respond to offer A and $(80/400) = .20$ respond to offer B. Because the percentages are exactly the same, there is no indication of relation. Alternatively, we note that one-third of the "yes" respondents and one-third of the "no" respondents were given offer A. Because these fractions are exactly the same, there is no indication of a statistical relation.

Of course, it is rare to have data that show absolutely no relation in the sample. More commonly, the percentages by row or by column differ, which suggest some relation. For example, a firm planning to market a cleaning product commissions a market research study of the leading current product. The variables of interest are the frequency of use and the rating of the leading product. The data are shown here:

Use	Rating			Total
	Fair	**Good**	**Excellent**	
Rare	64	123	137	324
Occasional	131	256	129	516
Frequent	209	171	45	425
Totals	404	550	311	1265

One natural analysis of the data takes the frequencies of use as givens and looks at the ratings as functions of use. The analysis essentially looks at conditional probabilities of the rating factor, given the use factor, but it recognizes that the data are only a sample, not the population. When use is rare, the best estimate is that $64/324 = .1975$ (or 19.75%) will rate the product as fair, that $123/324 = .3796$ will rate it good, and that $137/324 = .4228$ will rate it excellent. The corresponding proportions for occasional users are $131/516 = .2539$, $256/516 = .4961$, and $129/516 = .2500$. For frequent users, the proportions are .4918, .4024, and .1059. The proportions (or percentages, if one multiplies by 100) are quite different for the three use categories, which indicates that rating is related to use. Alternatively, we may calculate the use categories as percentages of the rating categories. In either case, there appears to be a relation. Because the proportions of ratings differ quite a bit as one varies use (or the proportions of use differ quite a bit as one varies rating), there is a suggestion that there is a fairly strong relation between use and rating.

Another way to analyze relations in data is to consider predictability. The stronger the relation exhibited in data, the better one can predict the value of one variable from the value of the other. We can imagine a situation in which every rare user rated the product as excellent, every occasional user rated the product as good, and every frequent user rated the product fair. In such a case, there would be a perfect statistical relation; in terms of predictability, given the use, one could predict the rating exactly. Of course, in practice, prediction and relation are not perfect; we need a measure of strength of relation defined as degree of predictability.

dependent variable   We need to distinguish between a **dependent variable**—the variable one is
independent variable   trying to predict—and an **independent variable**—the variable one is using to make the prediction. If one is trying to predict rating given use, use serves as the independent variable and rating as the dependent variable. No cause-and-effect connotations are intended; the choice of independent and dependent variables is entirely up to the person who is analyzing the data.

The simplest prediction rule is to predict the most common value (the mode) of the dependent variable; this rule is the basis of the λ (lambda) predictability measure. In our use–rating example, when use is rare, the most common rating is excellent, with 137 responses; if one predicts excellent for every rare case, one makes $187 = 64 + 123 = 324 - 137$ prediction errors. Similarly for occasional use, a prediction of rating as good gives $260 = 131 + 129 = 516 - 256$ errors; given frequent use, a prediction of rating as fair gives $216 = 171 + 45 = 425 - 209$ errors. The total number of errors is $187 + 260 + 216 = 663$. By comparison, if use is not known, we would have to predict the most common rating—namely, good—and we would make $715 = 404 + 311 = 1265 - 550$ errors. Reasonably enough, we do better in predicting rating when we have information about use than when we have no information.

The λ measure indicates how much better we predict. When use is the independent variable and rating is the dependent variable, the difference in prediction errors is $715 - 663 = 52$; we take this difference as a fraction of the errors made not knowing the independent variable—namely, 715. If use is the independent variable, then

$$\lambda = \frac{\text{errors with unknown independent variable} - \text{errors with known independent variable}}{\text{errors with unknown independent variable}}$$

$$= \frac{715 - 663}{715} = .073$$

The value of λ ranges between 0 and 1; $\lambda = 1$ indicates that, at least in the sample, the dependent variable is predicted perfectly given the independent variable. A value of $\lambda = 0$ occurs if there is independence in the data, or if the same value of the dependent variable is always predicted. To interpret other values of λ, note that it is a proportionate reduction in error (PRE) measure. The value $\lambda = .073$ that we found means that we make 7.3% fewer errors predicting rating given use than we would predicting rating without information about use. Values of λ above about .30 are rare in real data; thus, $\lambda = .073$ indicates a modest relation between rating and use. Note that the value of λ depends on which variable is taken as the dependent variable. For rating predicting use, $\lambda = .115$.

DEFINITION 10.3

Calculation of λ

For every level of the independent variable (every row or every column), find the modal value of the dependent variable (if there are two or more modal values with equal frequency, choose one arbitrarily). Add the frequencies in all nonmodal cells to find K = number of prediction errors with known independent variable value.

Refer to the marginal (total) frequencies of the dependent variables. Add the frequencies in all nonmodal categories to find U = number of prediction errors with unknown independent variable values.

$$\lambda = (U - K)/U$$

EXAMPLE 10.15

An internal survey of samples of clerical workers, supervisory personnel, and junior managers includes their opinions on a proposed flextime schedule. The following output (SAS) is obtained:

```
                    TABLE OF OPINION BY LEVEL

OPINION            LEVEL
FREQUENCY          |
  COL PCT          |          |         | junior|
                   |clerical|  superv |manager|   TOTAL
-----------------+--------+---------+-------+
strongly oppose  |     5  |     8   |    9  |    22
                 | 12.50  | 26.67   | 45.00 |
-----------------+--------+---------+-------+
         oppose  |     8  |    10   |    5  |    23
                 | 20.00  | 33.33   | 25.00 |
-----------------+--------+---------+-------+
          favor  |    14  |     7   |    4  |    25
                 | 35.00  | 23.33   | 20.00 |
-----------------+--------+---------+-------+
 strongly favor  |    13  |     5   |    2  |    20
                 | 32.50  | 16.67   | 10.00 |
-----------------+--------+---------+-------+
TOTAL                 40       30       20

         STATISTICS FOR TABLE OF OPINION BY LEVEL
STATISTIC                       DF     VALUE    PROB
--------------------------------------------------------
CHI-SQUARE                      6     12.110    0.060

STATISTIC                              VALUE    ASE
--------------------------------------------------------
LAMBDA ASYMMETRIC C|R                  0.120    0.106
LAMBDA ASYMMETRIC R|C                  0.123    0.079
SAMPLE SIZE = 90
ASE IS THE ASYMPTOTIC STANDARD ERROR.
R|C MEANS ROW VARIABLE DEPENDENT ON COLUMN VARIABLE.
```

a. What does the χ^2 test indicate about the relation shown in the data?
b. What can you see by examining the column percentages?
c. Locate the value for λ with *opinion* taken as the dependent variable. Interpret the number.
d. Verify the calculation of λ.

Solution

a. The χ^2 statistic is shown as CHI-SQUARE = 12.110 with 6 df. The *p*-value is shown as the PROB = .060. We cannot reject the null hypothesis of independence at $\alpha = .05$, but we can reject it at $\alpha = .10$. There is only limited evidence that the apparent relation is nonrandom; therefore, we should not place too much reliance on any apparent dependence.

b. The clerical staff seems to be largely in favor (35%) or strongly in favor (32.5%). The junior managers are largely strongly opposed (45%) or opposed (25%), and the supervisors are more or less evenly spread out over various opinions. Again, the result of the χ^2 test indicates that we can't rely too heavily on this apparent relation; it could conceivably have arisen by sheer random variation.

c. Here the row variable is the dependent variable. The statement that R|C MEANS ROW VARIABLE DEPENDENT ON COLUMN VARIABLE indicates that we want LAMBDA ASYMMETRIC R|C, which is .123.

d. The predicted cells, with the column listed first, are (CLERICAL, FAVOR), (SUPERV, OPPOSE), (JUNIOR MANAGER, STRONGLY OPPOSE). Adding the frequencies in all other cells, we find $K = 5 + 8 + 13 + 8 + 7 + 5 + 5 + 4 + 2 = 57$. In the TOTAL column for OPINION, we find that the most frequent opinion is FAVOR. Adding the frequencies for the other categories, we find $U = 22 + 23 + 20 = 65$. Therefore,

$$\lambda = \frac{65 - 57}{65} = .123$$

Percentage analyses and values of λ play a fundamentally different role than does the χ^2 test. The point of a χ^2 test is to see how much evidence there is that there *is* a relation, whatever the size may be. The point of percentage analyses and λ is to see *how strong* the relation appears to be, taking the data at face value. The two types of analyses are complementary.

Here are some final ideas about count data and relations:

1. A χ^2 goodness-of-fit test compares counts to theoretical probabilities that are specified outside the data. In contrast, a χ^2 independence test compares counts in one subset (one row, for example) to counts in other rows within the data. One way to decide which test is needed is to ask whether there is an externally stated set of theoretical probabilities. If so, the goodness-of-fit test is in order.

2. As is true of any significance test, the only purpose of a χ^2 test is to see whether differences in sample data might reasonably have arisen by chance alone. A test cannot tell you directly how large or important the difference is.

3. In particular, a statistically detectable (significant) χ^2 independence test does not necessarily mean a strong relation, nor does a nonsignificant goodness-of-fit test necessarily mean that the sample fractions are very close to the theoretical probabilities.
4. Looking thoughtfully at percentages is crucial in deciding whether the results show practical importance.

EXAMPLE 10.16

A total of 9,035 granted mortgage loans were categorized by the institution granting the loan (banks, bank-owned mortgage companies, or independent mortgage companies), as well as by income level of the borrower. Analysts wanted to know whether there was a statistical relation between loan-granting institution and income level of borrowers who were granted loans. The data are as follows:

	A	B	C	D	E
1		Banks	Bank	Mortgage	Total
2			Owned	Company	
3					
4					
5	<$21,000	943	166	151	1260
6	$21,000 to <$33,000	2186	503	711	3400
7	$33,000 to <$42,000	708	211	589	1508
8	$42,000 to <$50,000	424	150	397	971
9	$50,000 +	948	239	709	1896
10					
11	Total	5209	1269	2557	9035

Based on the Excel results shown in Figure 10.2, do we see a clear indication of some degree of statistical relation? If so, does the relation appear to be a strong one?

Solution   First, note that the type of mortgage lender is a qualitative variable. Also, the amount of the mortgage has been categorized into five groups, so we should treat it as qualitative as well.

To test for the presence of a statistical relation, we use the χ^2 statistic shown in the Excel results. The p-value is tiny; remember that the "E-110" notation means "move the decimal point 110 places to the left." Thus the p-value is a point followed by 109 zeros followed by 3!! This is slightly more than adequate evidence that there is a relation.

The small p-value does *not* necessarily imply a strong relation; it could also be the result of a weak relation and large sample size. To assess the strength of the relation, look at the percentages for each loan size in the lower left corner of the output. There is a clear pattern. As the size of the loan increases, an increasing percentage of the loans is held by mortgage companies and a decreasing percentage by banks. This trend reverses itself slightly for amounts over $50,000, but it generally holds fairly well. The change in percentages is not enormous, however, so we would not say that the relation was extremely strong.

Rather, it's a clear but modest relation.

FIGURE 10.2
Excel results for mortgage data

	A	B	C	D	E	F	G	H	I	J	K
1		Banks	Bank	Mortgage	Total			Banks	Bank	Mortgage	Total
2			Owned	Company					Owned	Company	
3											
4											
5	<$21,000	943	166	151	1260		<$21,000	726.435	176.972	356.593	1260
6	$21,000 to <$33,000	2186	503	711	3400		$21,000 to $33,000	1960.22	477.543	962.236	3400
7	$33,000 to <$42,000	708	211	589	1508		$33,000 to <$42,000	869.416	211.804	426.78	1508
8	$42,000 to <$50,000	424	150	397	971		$42,000 to <$50,000	559.816	136.381	274.803	971
9	$50,000+	948	239	709	1896		$50,000+	1093.11	266.3	536.588	1896
10											
11	Total	5209	1269	2557	9035						
12											
13	Chi-square statistic	534.48									
14	p-value	3E-110									
15											
16		Banks	Bank	Mortgage	Total						
17			Owned	Company							
18	<$21,000	78.84	13.17	11.98	100						
19	$21,000 to <$33,000	64.29	14.79	20.91	100						
20	$33,000 to <$42,000	46.95	13.99	39.06	100						
21	$42,000 to <$50,000	43.67	15.45	40.89	100						
22	$59,000+	50.00	12.61	37.39	100						
23											
24	Total	57.65	14.05	28.3	100						

10.8   Odds and Odds Ratios

Another way to analyze count data on qualitative variables is to use the concept of odds. This approach is widely used in biomedical studies and could be useful in some market research contexts as well. The basic definition of odds is the ratio of the probability that an event happens to the probability that it does not happen.

DEFINITION 10.4

$$\textbf{Odds of an event } A = \frac{P(A)}{1 - P(A)}$$

If an event has probability 2/3 of happening, the odds are $\frac{2/3}{1/3} = 2$. Usually this is reported as "the odds of the event happening are 2 to 1." Odds are used in horse racing and other betting establishments. The horse racing odds are given as the odds against the horse winning. Therefore odds of 4 to 1 means that it is 4 times more likely the horse will lose (not win) than not. Based on the odds, a horse with 4 to 1 odds is a better "bet" than, say, a horse with 20 to 1 odds. What about a horse with 1 to 2 odds (or equivalently, .5 to 1) against winning? This horse is highly favored because it is twice as likely (2 to 1) that the horse will win as not.

In working with odds, just make certain what the event of interest is. Also it is easy to convert the odds of an event back to the probability of the event. For event A,

$$P(A) = \frac{\text{odds of event } A}{1 + \text{odds of event } A}$$

Thus, if the odds of a horse (not winning) are stated as 9 to 1, then the probability of the horse not winning is

$$\text{Probability (not winning)} = \frac{9}{1 + 9} = .9$$

Similarly, the probability of winning is .1.

Odds are a convenient way to see how the occurrence of a condition changes the probability of an event. Recall from Chapter 4 that the conditional probability of an event A given another event B is

$$P(A|B) = P(A \text{ and } B)/P(B)$$

The odds favoring an event A given another event B turn out after a little algebra to be

$$\frac{P(A|B)}{P(\text{not } A|B)} = \frac{P(A)}{P(\text{not } A)} \frac{P(B|A)}{P(B|\text{not } A)}$$

The initial odds are multiplied by the *likelihood ratio*, the ratio of the probability of the conditioning event given A to its probability given not A. If B is more likely to happen when A is true than when it is not, the occurrence of B makes the odds favoring A go up.

EXAMPLE 10.17

Consider both a population in which 1 of every 1,000 people carried the HIV virus and a test that yielded positive results for 95% of those who carry the virus and (false) positive results for 2% of those who do not carry it. If a randomly chosen person obtains a positive test result, should the odds that that person carries the HIV virus go up or go down? By how much?

Solution   We certainly would think that a positive test result would increase the odds of carrying the virus. It would be a strange test indeed if a positive result decreased the chance of having the disease! Take the event A to be "carries HIV" and the event B to be "positive test result."

Before the test is made, the odds of a randomly chosen person carrying HIV are

$$\frac{.001}{.999} \approx .001$$

The occurrence of a positive test result causes the odds to change to

$$\frac{P(\text{HIV}|\text{positive})}{P(\text{not HIV}|\text{positive})} = \frac{P(\text{HIV})}{P(\text{not HIV})} \frac{P(\text{positive}|\text{HIV})}{P(\text{positive}|\text{not HIV})} = \frac{.001}{.999} \frac{.95}{.02} = .0475$$

The odds of carrying HIV do go up given a positive test result, from about .001 (to 1) to about .0475 (to 1).

odds ratio   A closely related idea, widely used in biomedical studies, is the **odds ratio**. As the name indicates, it is the ratio of the odds of an event (for example, contracting a certain form of cancer) for one group (for example, men) to the odds of the same event for another group (for example, women). The odds ratio is usually defined using conditional probabilities but can be stated equally well in terms of joint probabilities.

DEFINITION 10.5

Odds Ratio of an Event for Two Groups

If A is any event with probabilities $P(A|\text{group 1})$ and $P(A|\text{group 2})$, the odds ratio is

$$\frac{P(A|\text{group 1})/[1 - P(A|\text{group 1})]}{P(A|\text{group 2})/[1 - P(A|\text{group 2})]}$$

$$= \frac{P(A \text{ and group 1})/[1 - P(A \text{ and group 1})]}{P(A \text{ and group 2})/[1 - P(A \text{ and group 2})]}$$

The odds ratio equals 1 if the event A is statistically independent of group.

For example, suppose we have the following table of frequencies of purchase of a special suspension package for two brands of sports sedans:

	Yes	No	Total
Brand 1	250	750	1,000
Brand 2	400	1600	2,000
Total	650	2350	3,000

We would estimate the conditional probabilities of purchasing or not purchasing the package, given the brand, as

	Yes	No	Total
Brand 1	.250	.750	1.000
Brand 2	.200	.800	1.000

The odds ratio is $(.250/.750)/(.200/.800) = 1.333$, indicating that the odds of purchasing the package are 33.3% higher for purchasers of brand 1 than for purchasers of brand 2. It does not appear that purchasing the package is independent of brand (although we will have to allow for the fact that we have limited sample data). We could equally well have calculated the odds as

$$\frac{(250/3,000)(750/3,000)}{(400/3,000)/(1,600/3,000)} = \frac{250/750}{400/1,600} = 1.333$$

Inference about the odds ratio is usually done by way of the natural logarithm of the odds ratio. Recall that "ln" is the usual notation for the natural logarithm (base $e = 2.71828$) and that $\ln(1) = 0$. When the natural logarithm of the odds ratio is estimated from count data, it has approximately a normal distribution, with expected value the natural logarithm of the population odds ratio. Its standard error can be estimated by summing up 1/frequency for all four counts in a table, and then taking the square root. For the suspension package purchasing data, ln(odds ratio) = ln(1.333) = 0.2874; the estimated standard error is

$$\sqrt{1/250 + 1/750 + 1/400 + 1/1,600} = 0.0920$$

With these results, we can compute a 95% confidence interval as

$$0.2874 - 1.96(0.0920) \leq \ln(\text{odds ratio}) \leq 0.2874 + 1.96(0.0920)$$

or

$$0.1071 \leq \ln(\text{odds ratio}) \leq 0.4677$$

We may convert this to a statement about the odds ratio by "unlogging"—exponentiating—the interval.

$$e^{0.1071} \leq \text{odds ratio} \leq e^{0.4677}$$

or

$$1.113 \leq \text{odds ratio} \leq 1.596$$

Because the interval does not include an odds ratio of 1.000 (or, equivalently, the interval for ln(odds ratio) does not include 0.000), we may conclude that there is a statistically detectable relation in the data.

The odds ratio is a useful way to compare two population proportions π_1 and π_2 and may be more meaningful than their difference $(\pi_1 - \pi_2)$ when π_1 and π_2 are small. For example, suppose the rate of reinfarction for a sample of 5,000 coronary bypass patients treated with compound 1 is $\hat{\pi}_1 = .05$ and the corresponding rate for another sample of 5,000 coronary bypass patients treated with compound 2 is $\hat{\pi}_2 = .02$. Then their difference $\hat{\pi}_1 - \hat{\pi}_2 = .03$ may be less important and less informative than the odds ratio. See Table 10.5.

TABLE 10.5

	Reinfarction?		
	Yes	**No**	
Compound 1	250 (5%)	4,750	$n_1 = 5,000$
Compound 2	100 (2%)	4,900	$n_2 = 5,200$

The reinfarction odds for compounds 1 and 2 are as follows:

$$\text{Compound 1 odds} = \frac{250/5,000}{4,750/5,000} = \frac{250}{4,750} = .053$$

$$\text{Compound 2 odds} = \frac{100/5,000}{4,900/5,000} = \frac{100}{4,900} = .020$$

The corresponding odds ratio is .053/.020 = 2.65. Note that although the difference in reinfarction rates is only .03, the odds of having a reinfarction after treatment with compound 1 are 2.65 as likely as a reinfarction following treatment with compound 2.

10.9 Summary

In this chapter, we dealt with categorical data. Categorical data on a single variable arise in a number of situations. We first examined estimation and test procedures for a population proportion π and for two population proportions $(\pi_1 - \pi_2)$ based on independent samples. The extension of these procedures to comparing several population proportions (more than two) gave rise to the chi-square goodness-of-fit test.

Two-variable categorical data problems were discussed using the chi-square tests for independence and for homogeneity based on data displayed in an $r \times c$ contingency table. A measure of the strength of a relation between two categorical variables was presented using λ.

Finally, we discussed odds and odds ratios, which are especially useful in biomedical trials involving binomial proportions.

Key Formulas

1. Confidence interval for π

$$\hat{\pi} \pm z_{\alpha/2}\sigma_{\hat{\pi}}$$

where

$$\hat{\pi} = \frac{y}{n}$$

and

$$\sigma_{\hat{\pi}} = \sqrt{\frac{\hat{\pi}(1 - \hat{\pi})}{n}}$$

2. Sample size required for a $100(1 - \alpha)\%$ confidence interval of the form $\hat{\pi} \pm E$

$$n = \frac{z_{\alpha/2}^2\pi(1 - \pi)}{E^2}$$

(*Hint:* Use $\pi = .5$ if no estimate is available.)

3. Statistical test for π

$$\text{T.S.: } z = \frac{\hat{\pi} - \pi_0}{\sigma_{\hat{\pi}}}$$

where

$$\sigma_{\hat{\pi}} = \sqrt{\frac{\pi_0(1 - \pi_0)}{n}}$$

4. Confidence interval for $\pi_1 - \pi_2$

$$\hat{\pi}_1 - \hat{\pi}_2 \pm z_{\alpha/2}\sigma_{\hat{\pi}_1 - \hat{\pi}_2}$$

where

$$\sigma_{\hat{\pi}_1 - \hat{\pi}_2} = \sqrt{\frac{\hat{\pi}_1(1 - \hat{\pi}_1)}{n_1} + \frac{\hat{\pi}_2(1 - \hat{\pi}_2)}{n_2}}$$

5. Statistical test for $\pi_1 - \pi_2$

$$\text{T.S.: } z = \frac{\hat{\pi}_1 - \hat{\pi}_2}{\sigma_{\hat{\pi}_1 - \hat{\pi}_2}}$$

where

$$\sigma_{\hat{\pi}_1 - \hat{\pi}_2} = \sqrt{\frac{\hat{\pi}_1(1 - \hat{\pi}_1)}{n_1} + \frac{\hat{\pi}_2(1 - \hat{\pi}_2)}{n_2}}$$

6. Multinomial distribution

$$P(n_1, n_2, \ldots, n_k) = \frac{n!}{n_1!n_2!\ldots n_k!}\pi_1^{n_1}\pi_2^{n_2}\cdots\pi_k^{n_k}$$

7. Chi-square goodness-of-fit test

$$\text{T.S.: } \chi^2 = \sum_i \left[\frac{(n_i - E_i)^2}{E_i}\right]$$

where

$$E_i = n\pi_{i0}$$

8. Chi-square test of independence

$$\chi^2 = \sum_{i,j} \left[\frac{(n_{ij} - E_{ij})^2}{E_{ij}} \right]$$

where

$$E_{ij} = \frac{(\text{row } i \text{ total})(\text{column } j \text{ total})}{n}$$

9.
$$\lambda = \frac{\begin{array}{l}\text{errors with unknown}\\ \text{independent variable}\\ - \text{ errors with known}\\ \text{independent variable}\end{array}}{\begin{array}{l}\text{errors with unknown}\\ \text{independent variable}\end{array}}$$

10. Odds of event $A = \dfrac{P(A)}{1 - P(A)}$

$$\left(\text{in a binomial situation, odds of} \right.$$

$$\left. \text{a success} = \frac{\pi}{1 - \pi} \right)$$

11. Odds ratio for binomial situation, two groups

$$\frac{\text{odds for group 1}}{\text{odds for group 2}} = \frac{\pi_1/(1 - \pi_1)}{\pi_2/(1 - \pi_2)}$$

Supplementary Exercises

Soc.

10.63 A speaker who advises managers on how to avoid being unionized claims that only 25% of industrial workers favor union membership, 40% are indifferent, and 35% are opposed. In addition, the adviser claims that these opinions are independent of actual union membership. A random sample of 600 industrial workers yields the following data:

	Favor	Indifferent	Opposed	Total
Members	140	42	18	200
Nonmembers	70	198	132	400
Total	210	240	150	600

a. What part of the data is relevant to the 25%, 40%, 35% claim?
b. Test this hypothesis using $\alpha = .01$.

10.64 What can be said about the p-value in Exercise 10.63?

10.65 Test the hypothesis of independence in the data of Exercise 10.63. How conclusively is it rejected?

10.66 Calculate (for the data of Exercise 10.63) percentages of workers in favor of unionization, indifferent to it, and opposed to it; do so separately for members and for nonmembers. Do the percentages suggest there is a strong relation between membership and opinion?

Pol. Sci.

10.67 Three different television commercials are advertising an established product. The commercials are shown separately to theater panels of consumers; each consumer views only one of the possible commercials and then states an opinion of the product. Opinions range from 1 (very favorable) to 5 (very unfavorable). The data are as follows.

	Age					
Commercial	1	2	3	4	5	Total
A	32	87	91	46	44	300
B	53	141	76	20	10	300
C	41	93	67	36	63	300
Total	126	321	234	102	117	900

H.R.

10.76 The benefits manager for a major bank surveyed a sample of 353 employees (out of several thousand) to obtain their opinions of two alternative medical benefits plans. The variables of interest were: age (five categories, with 1 being a code for the youngest employees, 5 for the oldest); opinion (five categories, with 1 being most in favor of a health maintenance organization option, 5 being most in favor of a traditional fee-for-service option, and 3 being neutral); and a code for whether the employee has dependents covered by the plan (0 if not, 1 if so). The responses are listed (in order stated for columns 1–3) in the EX1076.DAT file in the Web site data sets. Load that file into your software package.

  a. Obtain a table of frequencies for each combination of age and opinion codes. If the software package will do so, obtain percentages in each opinion code for each age category. Are the opinion percentages similar for the various age codes?

  b. Have the software package carry out a formal test of the null hypothesis that age and opinion are independent. Can the null hypothesis be rejected at $\alpha = .05$?

10.77 The benefits manager in Exercise 10.76 suspected that there might be an indirect relation between age and opinion: age might be related to whether dependents are covered, and whether dependents are covered might be related to opinion.

  a. Have the software package test for dependence between age and dependents. Is the relation conclusively established?

  b. Do the same analysis for dependents and opinion.

  c. Have the software package test for dependence between age and opinion. Have the software package test separately for those employees with dependents covered and for those employees without dependents covered. In these tests, is there any evidence of a relation?

Bio.

10.78 A carcinogenicity study was conducted to examine the tumor potential of a drug product scheduled for initial testing in humans. A total of 300 rats (150 males and 150 females) were studied for a 6-month period. At the beginning of the study, 100 rats (50 males, 50 females) were randomly assigned to the control group, 100 to the low-dose group, and the remaining 100 (50 males, 50 females) to the high-dose group. On each day of the 6-month period, the rats in the control group received an injection of an inert solution, whereas those in the drug groups received an injection of the solution plus drug. The sample data are shown in the accompanying table.

	Number of Tumors	
Rat Group	**One or More**	**None**
Control	10	90
Low dose	14	86
High dose	19	81

  a. Give the percentage of rats with one or more tumors for each of the three treatment groups.

  b. Conduct a test of whether there is a significant difference in the proportion of rats having one or more tumors for the three treatment groups with $\alpha = .05$.

  c. Does there appear to be a drug-related problem regarding tumors for this drug product? That is, as the dose is increased, does there appear to be an increase in the proportion of rats with tumors?

10.79 SAS computer output for the data of Exercise 10.78 is shown here. Compare the results from the output with your results in Exercise 10.78.

```
RAT_GRP                 N_TUMORS

Frequency      |
Expected       |
Cell Chi-Square|
Percent        |
Row Pct        |
Col Pct        |NONE    |ONE-MORE|   Total
---------------+--------+--------+
CONTROL        |    90 |     10 |    100
               | 85.667 | 14.333 |
               | 0.2192 | 1.3101 |
               |  30.00 |   3.33 |   33.33
               |  90.00 |  10.00 |
               |  35.02 |  23.26 |
---------------+--------+--------+
HIGHDOSE       |    81 |     19 |    100
               | 85.667 | 14.333 |
               | 0.2542 | 1.5194 |
               |  27.00 |   6.33 |   33.33
               |  81.00 |  19.00 |
               |  31.52 |  44.19 |
---------------+--------+--------+
LOWDOSE        |    86 |     14 |    100
               | 85.667 | 14.333 |
               | 0.0013 | 0.0078 |
               |  28.67 |   4.67 |   33.33
               |  86.00 |  14.00 |
               |  33.46 |  32.56 |
---------------+--------+--------+
Total               257       43        300
                  85.67    14.33     100.00
STATISTICS FOR TABLE OF RAT_GRP BY N_TUMORS

Statistic                          DF     Value     Prob
-------------------------------------------------------
Chi-Square                          2     3.312     0.191
Likelihood Ratio Chi-Square         2     3.327     0.189
Mantel-Haenszel Chi-Square          1     0.649     0.420
Phi Coefficient                           0.105
Contingency Coefficient                   0.104
Cramer's V                                0.105

Sample Size = 300
```

Soc. **10.80** A sociological study was conducted to determine whether there is a relationship between the length of time blue-collar workers remain in their first job and the amount of their education. From union membership records, a random sample of persons was classified. The data are shown here.

Years on First Job	Years of Education			
	0–4.5	4.5–9	9–13.5	13.5
0–2.5	5	21	30	33
2.5–5	15	35	40	30
5–7.5	22	16	15	30
7.5	28	10	8	10

a. Use the computer output that follows to identify the expected cell numbers.
b. Test the research hypothesis that the variable "length of time on first job" is related to the variable "amount of education."
c. Give the level of significance for the test.
d. Draw your conclusions using $\alpha = .05$.

```
YRS_JOB                YRS_EDU

Frequency       |
Expected        |
Cell Chi-Square |
Percent         |
Row Pct         |
Col Pct         |0-4.5   |13.5    |4.5-9   |9-13.5 |  Total
----------------+--------+--------+--------+-------+
0-2.5           |      5 |     33 |     21 |     30|     89
                | 17.902 | 26.342 | 20.971 | 23.784|
                | 9.2988 | 1.6829 | 394E-7 | 1.6243|
                |   1.44 |   9.48 |   6.03 |   8.62|  25.57
                |   5.62 |  37.08 |  23.60 |  33.71|
                |   7.14 |  32.04 |  25.61 |  32.26|
----------------+--------+--------+--------+-------+
2.5-5           |     15 |     30 |     35 |     40|    120
                | 24.138 | 35.517 | 28.276 | 32.069|
                | 3.4594 | 0.857  | 1.599  | 1.9614|
                |   4.31 |   8.62 |  10.06 |  11.49|  34.48
                |  12.50 |  25.00 |  29.17 |  33.33|
                |  21.43 |  29.13 |  42.68 |  43.01|
----------------+--------+--------+--------+-------+
5-7.5           |     22 |     30 |     16 |     15|     83
                | 16.695 | 24.566 | 19.557 | 22.181|
                | 1.6854 | 1.202  | 0.6471 | 2.3248|
                |   6.32 |   8.62 |   4.60 |   4.31|  23.85
                |  26.51 |  36.14 |  19.28 |  18.07|
                |  31.43 |  29.13 |  19.51 |  16.13|
----------------+--------+--------+--------+-------+
7.5             |     28 |     10 |     10 |      8|     56
                | 11.264 | 16.575 | 13.195 | 14.966|
                | 24.864 | 2.608  | 0.7738 | 3.242 |
                |   8.05 |   2.87 |   2.87 |   2.30|  16.09
                |  50.00 |  17.86 |  17.86 |  14.29|
                |  40.00 |   9.71 |  12.20 |   8.60|
----------------+--------+--------+--------+-------+
Total                70      103       82       93      348
                    20.11    29.60    23.56    26.72   100.00


STATISTICS FOR TABLE OF YRS_JOB BY YRS_EDU

Statistic                     DF      Value      Prob
-----------------------------------------------------
Chi-Square                     9      57.830     0.001
Likelihood Ratio Chi-Square    9      55.605     0.001
Mantel-Haenszel Chi-Square     1      31.376     0.001
Phi Coefficient                       0.408
Contingency Coefficient               0.377
Cramer's V                            0.235

Sample Size = 348
```

Psy. **10.81**   Two researchers at Johns Hopkins University studied the use of drug products in the elderly. Patients in a recent study were asked the extent to which physicians counseled them with regard to their drug therapies. The researchers found the following:

- 25.4% of the patients said their physicians did not explain what the drug was supposed to do.
- 91.6% indicated they were not told how the drug might "bother" them.
- 47.1% indicated their physicians did not ask how the drug "helped" or "bothered" them after therapy was started.
- 87.7% indicated the drug was not changed after discussion on how the therapy was helping or bothering them.

a. Assume that 500 patients were interviewed in this study. Summarize each of these results using a 95% confidence interval.
b. Do you have any comments about the validity of any of these results?

Med.   **10.82** People over the age of 40 years tend to notice changes in their digestive systems that alter what and how much they can eat. A study was conducted to see whether this observation applies across different ethnic segments of our society. Random samples of Anglo-Saxons, Germans, Latin Americans, Italians, Spaniards, and African Americans were obtained. The data from this survey are summarized here:

Ethnic Group	Sample Size Responding (60 of Each Group Were Contacted)	Number Reporting Altered Digestive System
Anglo-Saxon	55	7
German	58	6
Latin American	52	34
Italian	54	38
Spanish	30	20
African American	49	31

a. Does it appear that there may be a bias due to the response rates?
b. Compare the rates (π_is) for the Anglo-Saxon and German groups using a 95% confidence interval.

10.83 Refer to Exercise 10.82. There seem to be two distinct rates—those around 12% and those around 70%. Combine the sample data for the first two groups and for the last four groups. Use these data to test the hypothesis $H_0: \pi_1 - \pi_2 \geq 0$ versus $H_a: \pi_1 - \pi_2 < 0$. Here, π_1 corresponds to the population rate for the first combined group, and π_2 is the corresponding proportion for the second combined group. Give the p-value for your test.

Bus.   **10.84** The following data give the observed frequencies of errors per page of unread galley proof for a sample of 40 pages from a certain journal publisher.

Errors/Page	Observed Frequencies
0	5
1	9
2	5
3	7
4	4
5	2
6	3
7	2
8	1
9	0
10	2

Conduct a test to determine whether the errors per page follow a Poisson distribution with a mean rate of 3.2. Use $\alpha = .10$.

Hort.   **10.85** An entomologist was interested in studying the infestation of adult European red mites on apple trees in a Michigan orchard. She randomly selected 50 leaves from each of 10 similar apple trees in the orchard, examined the leaves, and recorded the number of mites on each of the 500 leaves. As a part of a larger study, she wanted to simulate the distribution of mites on the trees in the orchard. Thus, the Poisson distribution was suggested as a possible model. Based on the data given here, does the Poisson distribution appear to be a plausible model for the concentration of European red mites on apple trees?

Mites per Leaf	0	1	2	3	4	5	6	7
Frequency	233	127	57	33	30	10	7	3

10.86 Refer to the case study. Analyze the data to answer the president's questions. Be sure to include confidence intervals, test of hypotheses, and any pertinent graphs. Computer output is given here.

ANALYSIS OF CASE STUDY

TABLE OF BUILDING RATINGS BY GROUP

BUILDING RATINGS   GROUP

```
Frequency       |
Expected        |
Cell Chi-Square |
Percent         |
Row Pct         |
Col Pct         |G1       G2        G3         G4       Total
----------------+-------+--------+---------+-------+
1               |    11 |      8 |      15 |     6 |    40
                | 9.8413| 10.635 |  9.2857 |10.238 |
                | 0.1364| 0.6528 |  3.5165 |1.7544 |
                |  2.18 |   1.59 |    2.98 |  1.19 |  7.94
                | 27.50 |  20.00 |   37.50 | 15.00 |
                |  8.87 |   5.97 |   12.82 |  4.65 |
----------------+-------+--------+---------+-------+
2               |    10 |      6 |      18 |     5 |    39
                | 9.5952| 10.369 |  9.0536 |9.9821 |
                | 0.0171| 1.8409 |  8.8406 |2.4866 |
                |  1.98 |   1.19 |    3.57 |  9.99 |  7.74
                | 25.64 |  15.38 |   46.15 | 12.82 |
                |  8.06 |   4.48 |   15.38 |  3.88 |
----------------+-------+--------+---------+-------+
3               |    51 |     50 |      42 |    38 |   181
                | 44.532| 48.123 |  42.018 |46.327 |
                | 0.9395| 0.0732 |  759E-8 |1.4969 |
                | 10.12 |   9.92 |    8.33 |  7.54 | 35.91
                | 28.18 |  27.62 |   23.20 | 20.99 |
                | 41.13 |  37.31 |   35.90 | 29.46 |
----------------+-------+--------+---------+-------+
4               |    30 |     41 |      26 |    40 |   137
                | 33.706| 36.425 |  31.804 |35.065 |
                | 0.4076| 0.5747 |  1.059  |0.6944 |
                |  5.95 |   8.13 |    5.16 |  7.94 | 27.18
                | 21.90 |  29.93 |   18.98 | 29.20 |
                | 24.19 |  30.60 |   22.22 | 31.01 |
----------------+-------+--------+---------+-------+
5               |    22 |     29 |      16 |    40 |   107
                | 26.325| 28.448 |  24.839 |27.387 |
                | 0.7107| 0.0107 |  3.1455 |5.809  |
                |  4.37 |   5.75 |    3.17 |  7.94 | 21.23
                | 20.56 |  27.10 |   14.95 | 37.38 |
                | 17.74 |  21.64 |   13.68 | 31.01 |
----------------+-------+--------+---------+-------+
Total               124      134       117     129      504
                  24.60    26.59     23.21   25.60  100.00
```

STATISTICS FOR TABLE OF BUILDING RATINGS BY GROUP

Statistic	DF	Value	Prob
Chi-Square	12	34.167	0.001
Likelihood Ratio Chi-Square	12	32.737	0.001
Mantel-Haenszel Chi-Square	1	4.139	0.042
Phi Coefficient		0.260	
Contingency Coefficient		0.252	
Cramer's V		0.150	
Sample Size = 504			

ANALYSIS OF CASE STUDY

TABLE OF SERVICE RATINGS BY GROUP

SERVICE RATINGS   GROUP

```
Frequency       |
Expected        |
Cell Chi-Square |
Percent         |
Row Pct         |
Col Pct         |G1       G2        G3         G4       Tota
----------------+-------+--------+---------+-------+
1               |    11 |      8 |      15 |     6 |    4
                | 9.8413| 10.635 |  9.2857 |10.238 |
                | 0.1364| 0.6528 |  3.5165 |1.7544 |  7.9
                |  2.18 |   1.59 |    2.98 |  1.19 |
                | 27.50 |  20.00 |   37.50 | 15.00 |
                |  8.87 |   5.97 |   12.82 |  4.65 |
----------------+-------+--------+---------+-------+
2               |    10 |      6 |      18 |     5 |    3
                | 9.5952| 10.369 |  9.0536 |9.9821 |
                | 0.0171| 1.8409 |  8.8406 |2.4866 |  7.7
                |  1.98 |   1.19 |    3.57 |  0.99 |
                | 25.64 |  15.38 |   46.15 | 12.82 |
                |  8.06 |   4.48 |   15.38 |  3.88 |
----------------+-------+--------+---------+-------+
3               |    51 |     50 |      42 |    38 |   18
                | 44.532| 48.123 |  42.018 |46.327 |
                | 0.9395| 0.0732 |  759E-8 |1.4969 | 35.9
                | 10.12 |   9.92 |    8.33 |  7.54 |
                | 28.18 |  27.62 |   23.20 | 20.99 |
                | 41.13 |  37.31 |   35.90 | 29.46 |
----------------+-------+--------+---------+-------+
4               |    30 |     41 |      26 |    40 |   13
                | 33.706| 36.425 |  31.804 |35.065 |
                | 0.4076| 0.5747 |  1.059  |0.6944 | 27.1
                |  5.95 |   8.13 |    5.16 |  7.94 |
                | 21.90 |  29.93 |   18.98 | 29.20 |
                | 24.19 |  30.60 |   22.22 | 31.01 |
----------------+-------+--------+---------+-------+
5               |    22 |     29 |      16 |    40 |   10
                | 26.325| 28.448 |  24.839 |27.387 |
                | 0.7107| 0.0107 |  3.1455 |5.809  | 21.2
                |  4.37 |   5.75 |    3.17 |  7.94 |
                | 20.56 |  27.10 |   14.95 | 37.38 |
                | 17.74 |  21.64 |   13.68 | 31.01 |
----------------+-------+--------+---------+-------+
Total               124      134       117     129      50
                  24.60    26.59     23.21   25.60  100.0
```

STATISTICS FOR TABLE OF BUILDING RATINGS BY GROUP

Statistic	DF	Value	Prob
Chi-Square	12	34.167	0.001
Likelihood Ratio Chi-Square	12	32.737	0.001
Mantel-Haenszel Chi-Square	1	4.139	0.042
Phi Coefficient		0.260	
Contingency Coefficient		0.252	
Cramer's V		0.150	
Sample Size = 504			

10.87 Refer to the case study. Write a nontechnical explanation of what your analysis reveals.

PART

6

Analyzing

Data:

Regression

Methods and

Model

Building

11 Linear Regression
      and Correlation

12 Multiple Regression and the
      General Linear Model

13 More on Multiple Regression

CHAPTER 11

Linear Regression and Correlation

11.1 Introduction and Case Study

11.2 Estimating Model Parameters

11.3 Inferences about Regression Parameters

11.4 Predicting New y Values Using Regression

11.5 Examining Lack of Fit in Linear Regression

11.6 The Inverse Regression Problem (Calibration)

11.7 Correlation

11.8 Summary

11.1 Introduction and Case Study

Predicting future values of a variable is a crucial management activity. Financial officers must predict future cash flows, production managers must predict needs for raw materials, and human resource managers must predict future personnel needs. Explanation of past variation is also important. Explaining the past variation in number of clients of a social service agency can help a manager understand demand for the agency's services. Finding the variables that explain deviations from an automobile component's specifications can help to improve the quality of that component. The basic idea of regression analysis is to use data on a *quantitative* independent variable to predict or explain variation in a *quantitative* dependent variable.

prediction vs. explanation   We can distinguish between prediction (reference to future values) and explanation (reference to current or past values). Because of the virtues of hindsight, explanation is easier than prediction. However, it is often clearer to use the term *prediction* to include both cases. Therefore, in this book, we sometimes blur the distinction between prediction and explanation.

For prediction (or explanation) to make much sense, there must be some connection between the variable we're predicting (the dependent variable) and the variable we're using to make the prediction (the independent variable). No doubt, if you tried long enough, you could find 28 common stocks whose price changes over a year have been accurately predicted by the won–lost percentage of the 28 major league baseball teams on the fourth of July. However, such a prediction is absurd because there is no connection between the two variables.

unit of association

Prediction requires a **unit of association**; there should be an entity that relates the two variables. With time-series data, the unit of association may simply be time. The variables may be measured at the same time period or, for genuine prediction, the independent variable may be measured at a time period before the dependent variable. For cross-sectional data, an economic or physical entity should connect the variables. If we are trying to predict the change in market share of various soft drinks, we should consider the promotional activity for those drinks, not the advertising for various brands of spaghetti sauce. The need for a unit of association seems obvious, but many predictions are made for situations in which no such unit is evident.

simple regression

In this chapter, we consider simple linear regression analysis, in which there is a single independent variable and the equation for predicting a dependent variable y is a linear function of a given independent variable x. Suppose, for example, that the director of a county highway department wants to predict the cost of a resurfacing contract that is up for bids. We could reasonably predict the costs to be a function of the road miles to be resurfaced. A reasonable first attempt is to use a linear production function. Let y = total cost of a project in thousands of dollars, x = number of miles to be resurfaced, and $\hat{y}$ = the predicted cost, also in thousands of dollars. A prediction equation $\hat{y} = 2.0 + 3.0x$ (for example) is a

intercept

linear equation. The constant term, such as the 2.0, is the **intercept** term and is interpreted as the predicted value of y when $x = 0$. In the road resurfacing example, we may interpret the intercept as the fixed cost of beginning the project. The

slope

coefficient of x, such as the 3.0, is the **slope** of the line, the predicted change in y when there is a one-unit change in x. In the road resurfacing example, if two projects differed by 1 mile in length, we would predict that the longer project cost 3 (thousand dollars) more than the shorter one. In general, we write the prediction equation as

$$\hat{y} = \hat{\beta}_0 + \hat{\beta}_1 x$$

where $\hat{\beta}_0$ is the intercept and $\hat{\beta}_1$ is the slope. See Figure 11.1.

FIGURE 11.1
Linear prediction function

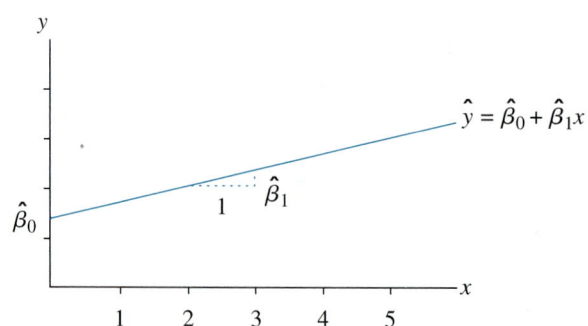

The basic idea of simple linear regression is to use data to fit a prediction line that relates a dependent variable y and a single independent variable x. The first assumption in simple regression is that the relation is, in fact, linear. According

assumption of linearity

to the **assumption of linearity**, the slope of the equation does not change as x changes. In the road resurfacing example, we would assume that there were no (substantial) economies or diseconomies from projects of longer mileage. There is little point in using simple linear regression unless the linearity assumption makes sense (at least roughly).

Linearity is not always a reasonable assumption, on its face. For example, if we tried to predict y = number of drivers that are aware of a car dealer's midsummer sale using x = number of repetitions of the dealer's radio commercial, the assumption of linearity means that the first broadcast of the commercial leads to no greater an increase in aware drivers than the thousand-and-first. (You've heard commercials like that.) We strongly doubt that such an assumption is valid over a wide range of x values. It makes far more sense to us that the effect of repetition would diminish as the number of repetitions got larger, so a straight-line prediction wouldn't work well.

Assuming linearity, we would like to write y as a linear function of x: $y = \beta_0 + \beta_1 x$. However, according to such an equation, y is an exact linear function of x; no room is left for the inevitable errors (deviation of actual y values from their predicted values). Therefore, corresponding to each y we introduce a **random error term** ε_i and assume the model

random error term

$$y = \beta_0 + \beta_1 x + \varepsilon$$

We assume the random variable y to be made up of a predictable part (a linear function of x) and an unpredictable part (the random error ε_i). The coefficients β_0 and β_1 are interpreted as the true, underlying intercept and slope. The error term ε includes the effects of all other factors, known or unknown. In the road resurfacing project, unpredictable factors such as strikes, weather conditions, and equipment breakdowns would contribute to ε, as would factors such as hilliness or prerepair condition of the road—factors that might have been used in prediction but were not. The combined effects of unpredictable and ignored factors yield the random error terms ε.

For example, one way to predict the gas mileage of various new cars (the dependent variable) based on their curb weight (the independent variable) would be to assign each car to a different driver, say, for a 1-month period. What unpredictable and ignored factors might contribute to prediction error? Unpredictable (random) factors in this study would include the driving habits and skills of the drivers, the type of driving done (city versus highway), and the number of stoplights encountered. Factors that would be ignored in a regression analysis of mileage and weight would include engine size and type of transmission (manual versus automatic).

In regression studies, the values of the independent variable (the x_i values) are usually taken as predetermined constants, so the only source of randomness is the ε_i terms. Although most economic and business applications have fixed x_i values, this is not always the case. For example, suppose that x_i is the score of an applicant on an aptitude test and y_i is the productivity of the applicant. If the data are based on a random sample of applicants, x_i (as well as y_i) is a random variable. The question of fixed versus random in regard to x is not crucial for regression studies. If the x_is are random, we can simply regard all probability statements as conditional on the observed x_is.

When we assume that the x_is are constants, the only random portion of the model for y_i is the random error term ε_i. We make the following formal assumptions.

DEFINITION 11.1

Formal assumptions of regression analysis:

1. The relation is, in fact, linear, so that the errors all have expected value zero: $E(\varepsilon_i) = 0$ for all i.
2. The errors all have the same variance: $\mathrm{Var}(\varepsilon_i) = \sigma_\varepsilon^2$ for all i.
3. The errors are independent of each other.
4. The errors are all normally distributed; ε_i is normally distributed for all i.

These assumptions are illustrated in Figure 11.2. The actual values of the dependent variable are distributed normally, with mean values falling on the regression line and the same standard deviation at all values of the independent variable. The only assumption not shown in the figure is independence from one measurement to another.

FIGURE 11.2

Theoretical distribution of y in regression

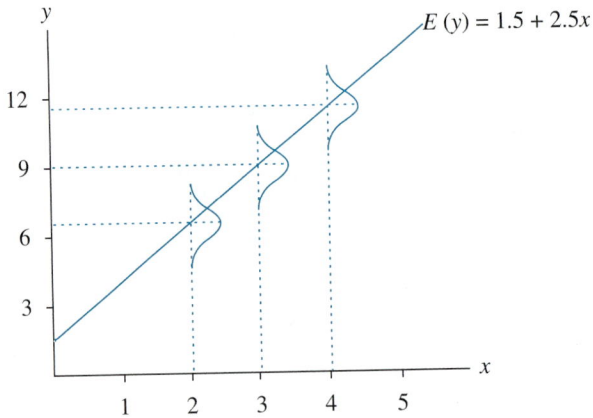

These are the formal assumptions, made in order to derive the significance tests and prediction methods that follow. We can begin to check these assumptions by looking at a **scatterplot** of the data. This is simply a plot of each (x, y) point, with the independent variable value on the horizontal axis, and the dependent variable value measured on the vertical axis. Look to see whether the points basically fall around a straight line or whether there is a definite curve in the pattern. Also look to see whether there are any evident outliers falling far from the general pattern of the data. A scatterplot is shown in the top part of Figure 11.3.

smoothers   Recently, **smoothers** have been developed to sketch a curve through data without necessarily assuming any particular model. If such a smoother yields something close to a straight line, then linear regression is reasonable. One such method is called LOWESS (locally weighted scatterplot smoother). Roughly, a smoother takes a relatively narrow "slice" of data along the x axis, calculates a line that fits the data in that slice, moves the slice slightly along the x axis, recalculates the line, and so on. Then all the little lines are connected in a smooth curve. The width of the slice is called the *bandwidth*; this may often be controlled in the computer program that does the smoothing. The plain scatterplot (Figure 11.3a) is shown again (Figure 11.3b) with a LOWESS curve through it. The scatterplot shows a curved relation; the LOWESS curve confirms that impression.

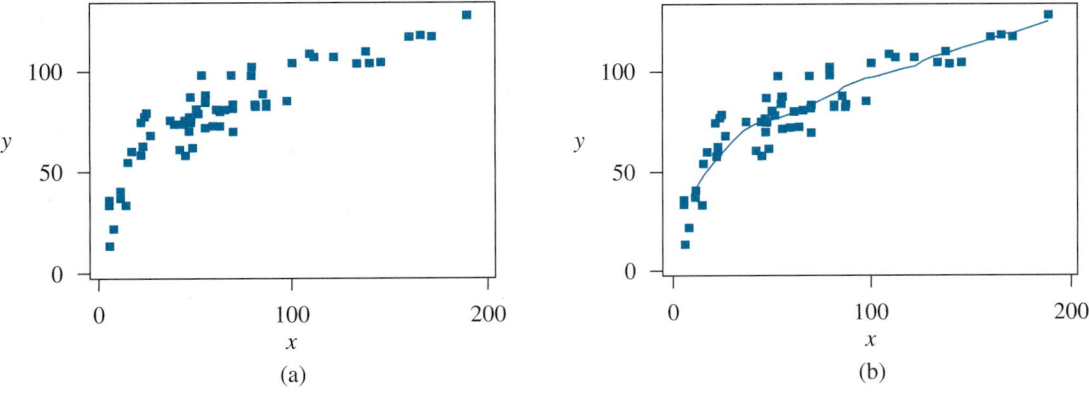

FIGURE 11.3 (a) Scatterplot and (b) LOWESS curve

spline fit

Another type of scatterplot smoother is the **spline fit**. It can be understood as taking a narrow slice of data, fitting a curve (often a cubic equation) to the slice, moving to the next slice, fitting another curve, and so on. The curves are calculated in such a way as to form a connected, continuous curve.

Many economic relations are not linear. For example, any diminishing returns pattern will tend to yield a relation that increases, but at a decreasing rate. If the scatterplot does not appear linear, by itself or when fitted with a LOWESS curve,

transformation

it can often be "straightened out" by a **transformation** of either the independent variable or the dependent variable. A good statistical computer package or a spreadsheet program will compute such functions as the square root of each value of a variable. The transformed variable should be thought of as simply another variable.

For example, a large city dispatches crews each spring to patch potholes in its streets. Records are kept of the number of crews dispatched each day and the number of potholes filled that day. A scatterplot of the number of potholes patched and the number of crews and the same scatterplot with a LOWESS curve through it are shown in Figure 11.4. The relation is not linear. Even without the LOWESS curve, the decreasing slope is obvious. That's not surprising; as the city sends out more crews, they will be using less effective workers, the crews will have to travel

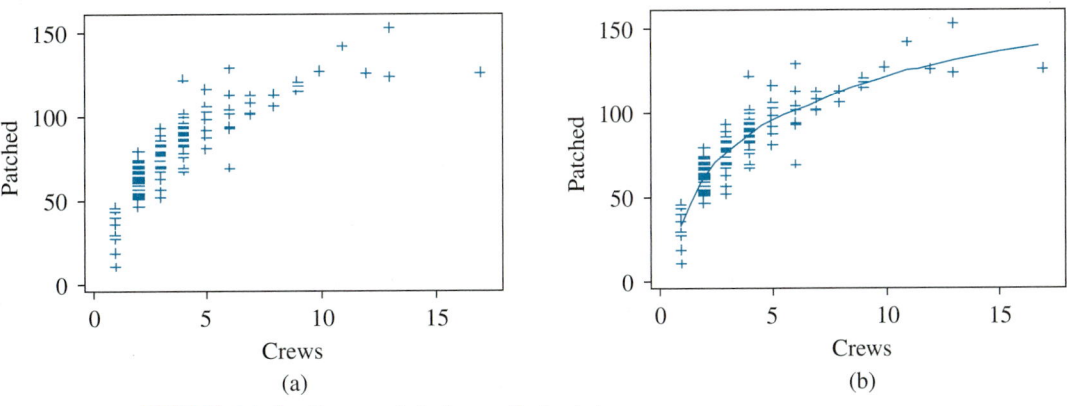

FIGURE 11.4 Scatterplots for pothole data

farther to find holes, and so on. All these reasons suggest that diminishing returns will occur.

We can try several transformations of the independent variable to find a more linear scatterplot. Three common transformations are square root, natural logarithm, and inverse (one divided by the variable). We applied each of these transformations to the pothole repair data. The results are shown in Figure 11.5a–c, with LOWESS curves. The square root (a) and inverse transformations (c) didn't really give us a straight line. The natural logarithm (b) worked very well, however. Therefore, we would use LnCrew as our independent variable.

Finding a good transformation often requires trial and error. Following are some suggestions to try for transformations. Note that there are *two* key features to look for in a scatterplot. First, is the relation nonlinear? Second, is there a pattern

FIGURE 11.5

Scatterplots with transformed predictor

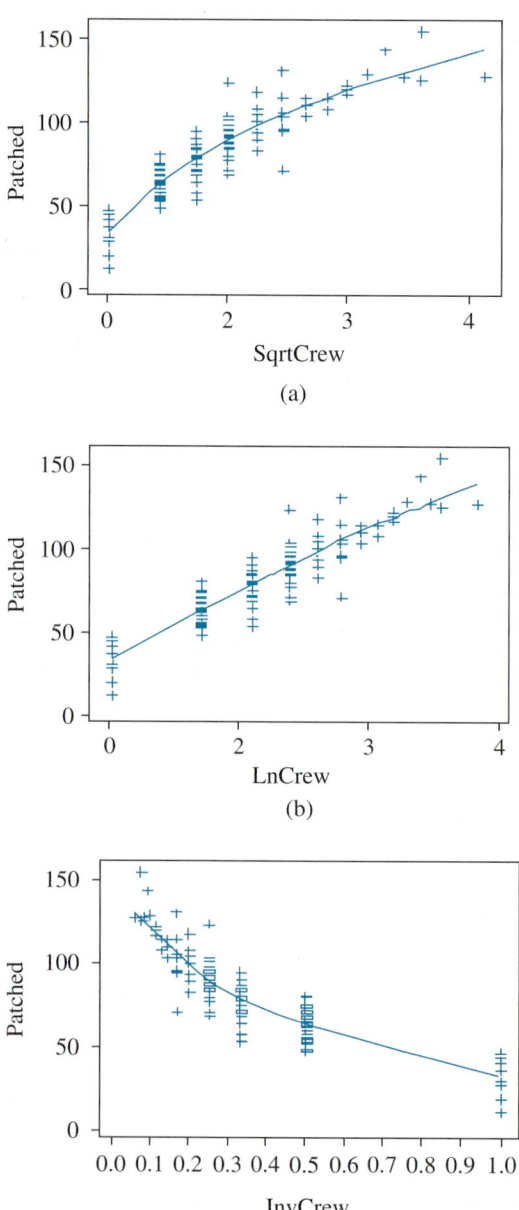

(a)

(b)

(c)

of increasing variability along the y (vertical) axis? If there is, the assumption of constant variance is questionable. These suggestions don't cover all the possibilities, but do include the most common problems.

DEFINITION 11.2

Follow these steps in choosing a transformation:

1. If the plot indicates a relation that is increasing but at a decreasing rate, and if variability around the curve is roughly constant, transform x using square root, logarithm, or inverse transformations.

2. If the plot indicates a relation that is increasing at an increasing rate, and if variability is roughly constant, try using both x and x^2 as predictors. Because this method uses two variables, the multiple regression methods of the next two chapters are needed.

3. If the plot indicates a relation that increases to a maximum and then decreases, and if variability around the curve is roughly constant, again try using both x and x^2 as predictors.

4. If the plot indicates a relation that is increasing at a decreasing rate, and if variability around the curve increases as the predicted y value increases, try using y^2 as the dependent variable.

5. If the plot indicates a relation that is increasing at an increasing rate, and if variability around the curve increases as the predicted y value increases, try using $\ln(y)$ as the dependent variable. It sometimes may also be helpful to use $\ln(x)$ as the independent variable. Note that a change in a natural logarithm corresponds quite closely to a percentage change in the original variable. Thus, the slope of a transformed variable can be interpreted quite well as a percentage change.

EXAMPLE 11.1

An airline has seen a very large increase in the number of free flights used by participants in its frequent flyer program. To try to predict the trend in these flights in the near future, the director of the program assembled data for the last 72 months. The dependent variable y is the number of thousands of free flights; the independent variable x is month number. A scatterplot with a LOWESS smoother, done using Minitab, is shown in Figure 11.6. What transformation is suggested?

FIGURE 11.6
Frequent flyer free flights by month

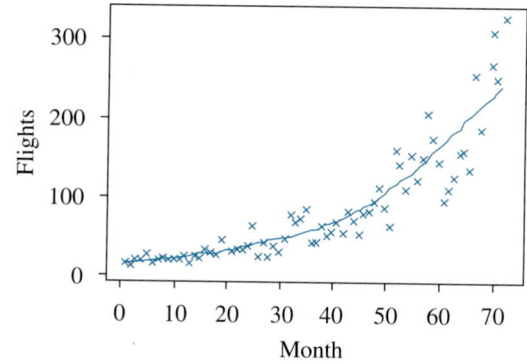

Solution The pattern shows flights increasing at an increasing rate. The LOWESS curve is definitely turning upward. In addition, variation (up and down) around the curve is increasing. The points around the high end of the curve (on the right, in this case) scatter much more than the ones around the low end of the curve. The increasing variability suggests transforming the *y* variable. A natural logarithm (ln) transformation often works well. Minitab computed the logarithms and replotted the data, as shown in Figure 11.7. The pattern is much closer to a straight line, and the scatter around the line is much closer to constant.

FIGURE 11.7
Result of logarithm
transformation

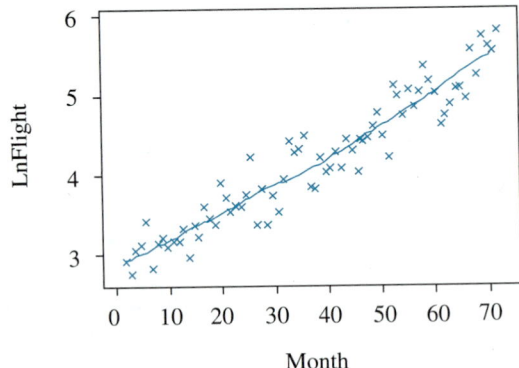

We will have more to say about checking assumptions in Chapter 12. For a simple regression with a single predictor, careful checking of a scatterplot, ideally with a smooth curve fit through it, will help avoid serious blunders.

Once we have decided on any mathematical transformations, we must estimate the actual equation of the regression line. In practice, only sample data are available. The population intercept, slope, and error variance all have to be estimated from limited sample data. The assumptions we made in this section allow us to make inferences about the true parameter values from the sample data.

Case Study: Comparison of Two Methods for Detecting *E. coli*

The case study in Chapter 7 described a new microbial method for the detection of *E. coli*, Petrifilm HEC test. The researcher wanted to evaluate the agreement of the results obtained using the HEC test with results obtained from an elaborate laboratory-based procedure, hydrophobic grid membrane filtration (HGMF). The HEC test is easier to inoculate, more compact to incubate, and safer to handle than conventional procedures. However, prior to using the HEC procedure it was necessary to compare the readings from the HEC test to readings from the HGMF procedure obtained on the same meat sample to determine whether the two procedures were yielding the same readings. If the readings differed but an equation could be obtained that could closely relate the HEC reading to the HGMF reading, then the researchers could calibrate the HEC readings to predict what readings would have been obtained using the HGMF test procedure. If the HEC test results were unrelated to the HGMF test procedure results, then the HEC test could not be used in the field in detecting *E. coli*.

Designing the Data Collection In Chapter 7 we described phase one of the experiment. Phase two of the study was to apply both procedures to artificially contaminated beef. Portions of beef trim were obtained from three Holstein cows who had tested negatively for *E. coli*. Eighteen portions of beef trim were obtained from the cows and then contaminated with *E. coli*. The HEC and HGMF procedures were applied to a portion of each of the 18 samples. The two procedures yielded *E. coli* concentrations in transformed metric ($\log_{10}$ CFU/ml). The data in this case would be 18 paired samples and are given here.

RUN	HEC	HGMF
1	0.50	0.42
2	0.06	0.20
3	0.20	0.42
4	0.61	0.33
5	0.20	0.42
6	0.56	0.64
7	−0.82	−0.82
8	0.67	1.06
9	1.02	1.21
10	1.20	1.25
11	0.93	0.83
12	2.27	2.37
13	2.02	2.21
14	2.32	2.44
15	2.14	2.28
16	2.09	2.69
17	2.30	2.43
18	−0.10	1.07

Managing the Data   Next the researchers would prepare the data for a statistical analysis following the steps described in Section 2.5. They would carefully review experimental procedures to make sure that each pair of meat samples were nearly identical so as not to introduce any differences in the HEC and HGMF readings that were not part of the differences in the two procedures. During such a review, procedural problems during run 18 were discovered and this pair of observations was excluded from the analysis.

Analyzing the Data   The researchers were interested in determining whether the two procedures yielded measures of *E. coli* concentrations that were strongly related. The scatterplot of the experimental data is given on the next page.

A 45° line was placed in the scatterplot to display the relative agreement between the readings from the two procedures. If the plotted points fell on this line, then the two procedures would be in complete agreement in their determination of *E. coli* concentrations. The 17 points fall close to the line but have some variation about it. Thus, the researchers would like to determine the degree of agreement and then obtain an equation that would relate the readings from the two procedures. If the readings from the two procedures could be accurately related using a regression equation, the researchers would want to predict the reading of the HGMF procedure given the HEC reading on a meat sample. This would enable them to compare *E. coli* concentrations obtained from meat samples in the field using the HEC procedure to the readings obtained in the laboratory using the HGMF procedure. We will obtain a detail analysis of the data following our discussion of calibration in Section 11.6.

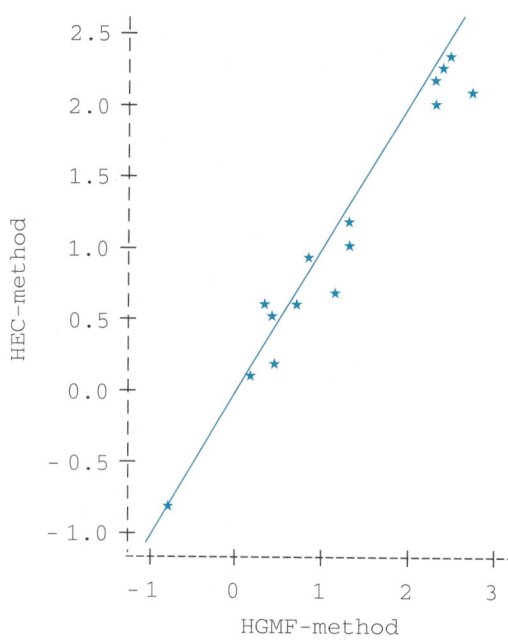

Plot of HEC-method versus HGMF-method

NOTE: 2 obs hidden.

11.2 Estimating Model Parameters

The intercept β_0 and slope β_1 in the regression model

$$y = \beta_0 + \beta_1 x + \varepsilon$$

are population quantities. We must estimate these values from sample data. The error variance σ_ε^2 is another population parameter that must be estimated. The first regression problem is to obtain estimates of the slope, intercept, and variance: we discuss how to do so in this section.

The road resurfacing example of Section 11.1 is a convenient illustration. Suppose the following data for similar resurfacing projects in the recent past are available. Note that we do have a unit of association: The connection between a particular cost and mileage is that they're based on the same project.

Cost y_i (in thousands of dollars):	6.0	14.0	10.0	14.0	26.0
Mileage x_i (in miles):	1.0	3.0	4.0	5.0	7.0

A first step in examining the relation between y and x is to plot the data as a scatterplot. Remember that each point in such a plot represents the (x, y) coordinates of one data entry, as in Figure 11.8. The plot makes it clear that there is an imperfect but generally increasing relation between x and y. A straight-line relation appears plausible; there is no evident transformation with such limited data.

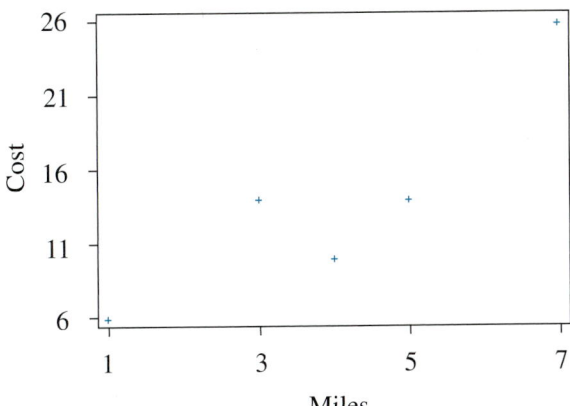

FIGURE 11.8

Scatterplot of cost versus mileage

The regression analysis problem is to find the best straight-line prediction. The most common criterion for "best" is based on squared prediction error. We find the equation of the prediction line—that is, the slope $\hat{\beta}_1$ and intercept $\hat{\beta}_0$ that minimize the total squared prediction error. The method that accomplishes this goal is called the **least-squares method** because it chooses $\hat{\beta}_0$ and $\hat{\beta}_1$ to minimize the quantity.

least-squares method

$$\sum_i (y_i - \hat{y}_i)^2 = \sum_i [y_i - (\hat{\beta}_0 + \hat{\beta}_1 x_i)]^2$$

The prediction errors are shown on the plot of Figure 11.9 as vertical deviations from the line. The deviations are taken as vertical distances because we're trying to predict y values, and errors should be taken in the y direction. For these data, the least-squares line can be shown to be $\hat{y} = 2.0 + 3.0x$; one of the deviations from it is indicated by the smaller brace. For comparison, the mean $\bar{y} = 14.0$ is also shown; deviation from the mean is indicated by the larger brace. The least-squares principle leads to some fairly long computations for the slope and intercept. Usually, these computations are done by computer.

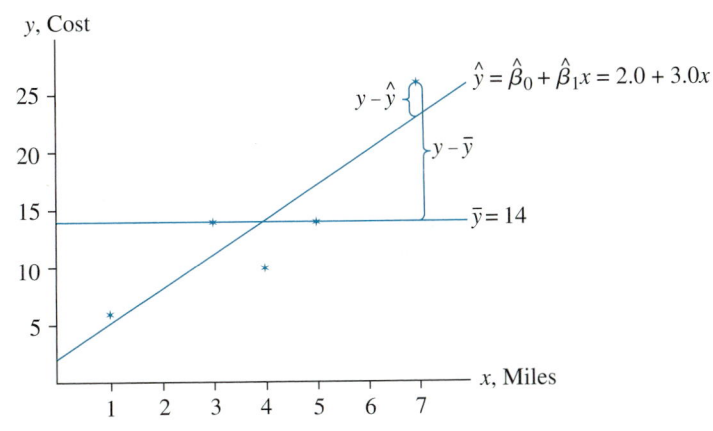

FIGURE 11.9

Deviations from the least-squares line from the mean

DEFINITION 11.3

The **least-squares estimates of slope and intercept** are obtained as follows:

$$\hat{\beta}_1 = \frac{S_{xy}}{S_{xx}} \quad \text{and} \quad \hat{\beta}_0 = \bar{y} - \hat{\beta}_1 \bar{x}$$

where

$$S_{xy} = \sum_i (x_i - \bar{x})(y_i - \bar{y}) \quad \text{and} \quad S_{xx} = \sum_i (x_i - \bar{x})^2$$

Thus, S_{xy} is the sum of x deviations times y deviations and S_{xx} is the sum of x deviations squared.

For the road resurfacing data, $n = 5$ and

$$\sum x_i = 1.0 + \cdots + 7.0 = 20.0$$

so $\bar{x} = \dfrac{20.0}{5} = 4.0$. Similarly,

$$\sum y_i = 70.0, \bar{y} = \frac{70.0}{5} = 14.0$$

Also,

$$
\begin{aligned}
S_{xx} &= \sum (x_i - \bar{x})^2 \\
&= (1.0 - 4.0)^2 + \cdots + (7.0 - 4.0)^2 \\
&= 20.00
\end{aligned}
$$

and

$$
\begin{aligned}
S_{xy} &= \sum (x_i - \bar{x})(y_i - \bar{y}) \\
&= (1.0 - 4.0)(6.0 - 14.0) + \cdots + (7.0 - 4.0)(26.0 - 14.0) \\
&= 60.0
\end{aligned}
$$

Thus,

$$\hat{\beta}_1 = \frac{60.0}{20.0} = 3.0 \quad \text{and} \quad \hat{\beta}_0 = 14.0 - (3.0)(4.0) = 2.0$$

From the value $\hat{\beta}_1 = 3$, we can conclude that the estimated average increase in cost for each additional mile is $3,000.

EXAMPLE 11.2

Data from a sample of 10 pharmacies are used to examine the relation between prescription sales volume and the percentage of prescription ingredients purchased directly from the supplier. The sample data are shown here:

Pharmacy	Sales Volume, *y* (in $1,000)	% of Ingredients Purchased Directly, *x*
1	25	10
2	55	18
3	50	25
4	75	40
5	110	50
6	138	63
7	90	42
8	60	30
9	10	5
10	100	55

a. Find the least-squares estimates for the regression line $\hat{y} = \hat{\beta}_0 + \hat{\beta}_1 x$.
b. Predict sales volume for a pharmacy that purchases 15% of its prescription ingredients directly from the supplier.
c. Plot the (x, y) data and the prediction equation $\hat{y} = \hat{\beta}_0 + \hat{\beta}_1 x$.
d. Interpret the value of $\hat{\beta}_1$ in the context of the problem.

Solution

a. The equation can be calculated by virtually any statistical computer package; for example, here is abbreviated Minitab output:

```
MTB > Regress 'Sales' on 1 variable 'Directly'

The regression equation is
Sales = 4.70 + 1.97 Directly

Predictor      Coef      Stdev     t-ratio        p
Constant      4.698      5.952        0.79    0.453
Directly     1.9705     0.1545       12.75    0.000
```

To see how the computer does the calculations, you can obtain the least-squares estimates from the following table:

	y	*x*	$y - \bar{y}$	$x - \bar{x}$	$(x - \bar{x})(y - \bar{y})$	$(x - \bar{x})^2$
	25	10	−46.3	−22.8	1,101.94	566.44
	55	18	−16.3	−15.8	257.54	249.64
	50	25	−21.3	−8.8	187.44	77.44
	75	40	3.7	6.2	22.94	38.44
	110	50	38.7	16.2	626.94	262.44
	138	63	66.7	29.2	1,947.64	852.64
	90	42	18.7	8.2	153.34	67.24
	60	30	−11.3	−3.8	42.94	14.44
	10	5	−61.3	−28.8	1,765.44	829.44
	100	55	28.7	21.2	608.44	449.44
Totals	713	338	0	0	6,714.60	3,407.60
Means	71.3	33.8				

$$S_{xx} = \sum (x - \bar{x})^2 = 3{,}407.6$$

$$S_{xy} = \sum (x - \bar{x})(y - \bar{y}) = 6{,}714.6$$

Substituting into the formulas for $\hat{\beta}_0$ and $\hat{\beta}_1$,

$$\hat{\beta}_1 = \frac{S_{xy}}{S_{xx}} = \frac{6{,}714.6}{3{,}407.6} = 1.9704778 \qquad \text{rounded to } 1.97$$

$$\hat{\beta}_0 = \bar{y} - \hat{\beta}_1 \bar{x} = 71.3 - 1.9704778(33.8) = 4.6978519 \qquad \text{rounded to } 4.70$$

b. When $x = 15\%$, the predicted sales volume is $\hat{y} = 4.70 + 1.97(15) = 34.25$ (that is, \$34,250).

c. The (x, y) data and prediction equation are shown in Figure 11.10.

FIGURE 11.10
Sample data and least-squares prediction equation

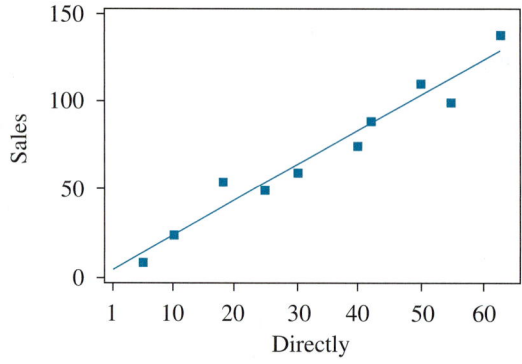

d. From $\hat{\beta}_1 = 1.97$, we conclude that if a pharmacy would increase by 1% the percentage of ingredients purchased directly, then the estimated increase in average sales volume would be \$1,970.

EXAMPLE 11.3

Use the following Statistix output to identify the least-squares estimates for the road resurfacing data:

```
PREDICTOR
VARIABLES    COEFFICIENT   STD ERROR    STUDENT'S T    P

CONSTANT       2.00000      3.82970        0.52       0.6376
MILES          3.00000      0.85634        3.50       0.0394

R-SQUARED               0.8036   RESID. MEAN SQUARE (MSE)   14.6666
ADJUSTED R-SQUARED      0.7381   STANDARD DEVIATION          3.82970

SOURCE          DF    SS        MS         F        P

REGRESSION      1   180.000   180.000    12.27    0.0394
RESIDUAL        3   44.0000   14.6666
TOTAL           4   224.000
```

Solution   The intercept is shown in the COEFFICIENT column as $\hat{\beta}_0 = 2.00000$. The slope (coefficient of $x =$ miles) is $\hat{\beta}_1 = 3.00000$.

The estimate of the regression slope can potentially be greatly affected by **high leverage point** **high leverage points**. These are points that have very high or very low values of the independent variable—outliers in the x direction. They carry great weight in the estimate of the slope. A high leverage point that also happens to correspond **high influence point** to a y outlier is a **high influence point**. It will alter the slope and twist the line badly.

A point has high influence if omitting it from the data will cause the regression line to change substantially. To have high influence, a point must first have high leverage and, in addition, must fall outside the pattern of the remaining points. Consider the two scatterplots in Figure 11.11. In plot (a), the point in the upper left corner is far to the left of the other points; it has a much lower x value and therefore has high leverage. If we drew a line through the other points, the line would fall far below this point, so the point is an outlier in the y direction as well. Therefore, it also has high influence. Including this point would change the slope of the line greatly. In contrast, in plot (b), the y outlier point corresponds to an x value very near the mean, having low leverage. Including this point would pull the line upward, increasing the intercept, but it wouldn't increase or decrease the slope much at all. Therefore, it does not have great influence.

FIGURE 11.11
(a) High influence and (b) low influence points

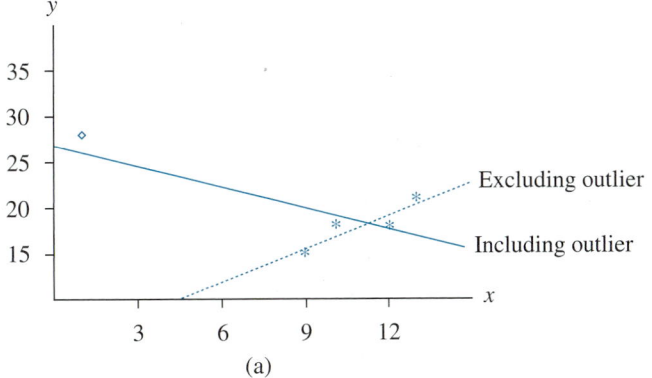

(a)

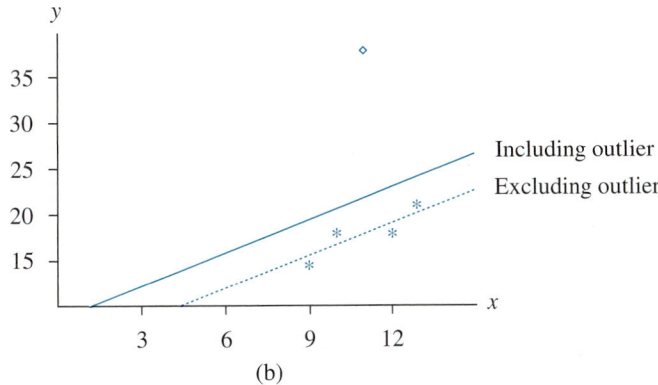

(b)

A high leverage point indicates only a *potential* distortion of the equation. Whether or not including the point will "twist" the equation depends on its influence (whether or not the point falls near the line through the remaining points). A point must have *both* high leverage and an outlying y value to qualify as a high influence point.

Mathematically, the effect of a point's leverage can be seen in the S_{xy} term that enters into the slope calculation. One of the many ways this term can be written is

$$S_{xy} = \sum (x_i - \bar{x}) y_i$$

We can think of this equation as a weighted sum of y values. The weights are large positive or negative numbers when the x value is far from its mean and has high leverage. The weight is almost 0 when x is very close to its mean and has low leverage.

diagnostic measures   Most computer programs that perform regression analyses will calculate one or another of several **diagnostic measures** of leverage and influence. We won't try to summarize all of these measures. We only note that very large values of any of these measures correspond to very high leverage or influence points. The distinction between high leverage (x outlier) and high influence (x outlier and y outlier) points is not universally agreed upon yet. Check the program's documentation to see what definition is being used.

The standard error of the slope $\hat{\beta}_1$ is calculated by all statistical packages. Typically, it is shown in output in a column to the right of the coefficient column. Like any standard error, it indicates how accurately one can estimate the correct population or process value. The quality of estimation of $\hat{\beta}_1$ is influenced by two quantities: the error variance σ_ε^2 and the amount of variation in the independent variable S_{xx}:

$$\sigma_{\hat{\beta}_1} = \frac{\sigma_\varepsilon}{\sqrt{S_{xx}}}$$

The greater the variability σ_ε of the y value for a given value of x, the larger $\sigma_{\hat{\beta}_1}$ is. Sensibly, if there is high variability around the regression line, it is difficult to estimate that line. Also, the smaller the variation in x values (as measured by S_{xx}), the larger $\sigma_{\hat{\beta}_1}$ is. The slope is the predicted change in y per unit change in x; if x changes very little in the data, so that S_{xx} is small, it is difficult to estimate the rate of change in y accurately. If the price of a brand of diet soda has not changed for years, it is obviously hard to estimate the change in quantity demanded when price changes.

$$\sigma_{\hat{\beta}_0} = \sigma_\varepsilon \sqrt{\frac{1}{n} + \frac{\bar{x}^2}{S_{xx}}}$$

The standard error of the estimated intercept $\hat{\beta}_0$ is influenced by n, naturally, and also by the size of the square of the sample mean, $\bar{x}^2$, relative to S_{xx}. The intercept is the predicted y value when $x = 0$; if all the x_i are, for instance, large positive numbers, predicting y at $x = 0$ is a huge extrapolation from the actual data. Such extrapolation magnifies small errors, and the standard error of $\hat{\beta}_0$ is large. The ideal situation for estimating $\hat{\beta}_0$ is when $\bar{x} = 0$.

To this point, we have considered only the estimates of intercept and slope. We also have to estimate the true error variance σ_ε^2. We can think of this quantity as "variance around the line," or as the mean squared prediction error. The
residuals   estimate of σ_ε^2 is based on the **residuals** $y_i - \hat{y}_i$, which are the prediction errors in the sample. The estimate of σ_ε^2 based on the sample data is the sum of squared residuals divided by $n - 2$, the degrees of freedom. The estimated variance is often shown in computer output as MS(Error) or MS(Residual). Recall that MS stands for "mean square" and is always a sum of squares divided by the appropriate degrees of freedom:

$$s_\varepsilon^2 = \frac{\sum_i (y_i - \hat{y}_i)^2}{n - 2} = \frac{\text{SS(Residual)}}{n - 2}$$

In the computer output for Example 11.3, SS(Residual) is shown to be 44.0.

Just as we divide by $n - 1$ rather than by n in the ordinary sample variance s^2 (in Chapter 3), we divide by $n - 2$ in s_ε^2, the estimated variance around the line. The reduction from n to $n - 2$ occurs because in order to estimate the variability around the regression line, we must first estimate the two parameters β_0 and β_1 to obtain the estimated line. The effective sample size for estimating σ_ε^2 is thus $n - 2$. In our definition, s_ε^2 is undefined for $n = 2$, as it should be. Another argument is that dividing by $n - 2$ makes s_ε^2 an unbiased estimator of σ_ε^2. In the computer output of Example 11.3, $n - 2 = 5 - 2 = 3$ is shown as DF (degrees of freedom) for RESIDUAL and $s_\varepsilon^2 = 14.6666$ is shown as MS for RESIDUAL.

residual standard deviation

The square root s_ε of the sample variance is called the **sample standard deviation around the regression line**, the **standard error of estimate**, or the **residual standard deviation**. Because s_ε estimates σ_ε, the standard deviation of y_i, σ_ε estimates the standard deviation of the population of y values associated with a given value of the independent variable x. The Statistix output in Example 11.3 labels s_ε as STANDARD DEVIATION; it shows that s_ε, rounded off, is 3.830.

Like any other standard deviation, the residual standard deviation may be interpreted by the Empirical Rule. About 95% of the prediction errors will fall within ±2 standard deviations of the mean error; the mean error is always 0 in the least-squares regression model. Therefore, a residual standard deviation of 3.830 means that about 95% of prediction errors will be less than ±2(3.830) = 7.660.

The estimates $\hat{\beta}_0$, $\hat{\beta}_1$, and s_ε are basic in regression analysis. They specify the regression line and the probable degree of error associated with y values for a given value of x. The next step is to use these sample estimates to make inferences about the true parameters.

EXAMPLE 11.4

The human resources director of a chain of fast-food restaurants studied the absentee rate of employees. Whenever employees called in sick, or simply didn't appear, the restaurant manager had to find replacements in a hurry, or else work short-handed. The director had data on the number of absences per 100 employees per week (y) and the average number of months' experience at the restaurant (x) for 10 restaurants in the chain. The director expected that longer-term employees would be more reliable and absent less often.

For the following data and Minitab output, do the following:

a. Examine the scatterplot and decide whether a straight line is a reasonable model.

b. Identify the least-squares estimates for β_0 and β_1 in the model $y = \beta_0 + \beta_1 x + \varepsilon$.

c. Predict y for $x = 19.5$.

d. Identify s_ε, the sample standard deviation about the regression line.

e. Interpret the value of $\hat{\beta}_1$.

| y: | 31.5 | 33.1 | 27.4 | 24.5 | 27.0 | 27.8 | 23.3 | 24.7 | 16.9 | 18.1 |
| x: | 18.1 | 20.0 | 20.8 | 21.5 | 22.0 | 22.4 | 22.9 | 24.0 | 25.4 | 27.3 |

```
MTB > Regress 'y' on 1 predictor 'x'.

The regression equation is
y = 64.7 - 1.75 x
```

```
Predictor        Coef      Stdev     t-ratio         p
Constant       64.672      6.762        9.56     0.000
x             -1.7487     0.2995       -5.84     0.000

s = 2.388      R-sq = 81.0%      R-sq(adj) = 78.6%

Analysis of Variance

SOURCE          DF        SS         MS         F         P
Regression       1     194.45     194.45     34.10     0.000
Error            8      45.61       5.70
Total            9     240.06
```

Solution

a. A scatterplot drawn by the Statistix package is shown in Figure 11.12; the data appear to fall approximately along a downward-sloping line. There is no reason to use a more complicated model.

FIGURE 11.12

Scatterplot of absences (y) versus average length of employment (x)

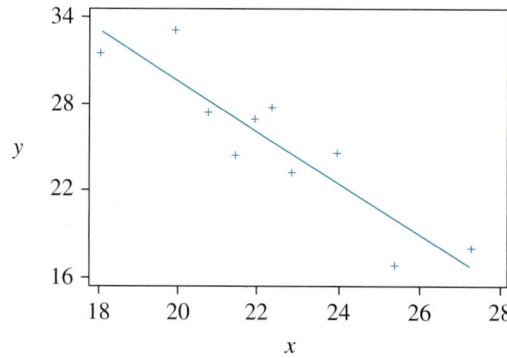

b. The output shows the coefficients twice, with differing numbers of digits. The intercept (constant) is 64.672 and the slope (coefficient of x) is -1.7487. Note that the negative slope corresponds to a downward-sloping line.

c. The least-squares prediction value when $x = 19.5$ is

$$\hat{y} = 64.672 - 1.7487(19.5) = 30.57$$

d. The standard deviation around the line (the residual standard deviation) is shown as $s = 2.388$. Therefore, about 95% of the prediction errors should be less than $\pm 2(2.388) = 4.776$.

e. From $\hat{\beta}_1 = -1.7487 \approx 1.75$, we conclude that for a 1-month increase in the average experience at a restaurant, there is an estimated decrease of 1.75 in the average number of absences per 100 employees per week.

EXERCISES   **Basic Techniques**

11.1 Plot the data shown here in a scatter diagram and sketch a line through the points.

x	5	10	12	15	18	24
y	10	19	21	28	34	40

11.2 Use the equation $y = 1.8 + 2.0x$.
  a. Predict y when $x = 3$.
  b. Plot the equation on a graph with the horizontal axis scaled from 0 to 5 and the vertical axis scaled from 0 to 12.

11.3 Use the accompanying data to determine the least-squares prediction equation.

x	1	2	3	4	5
y	2	4	6	7	9

11.4 Use the accompanying data to answer (a) and (b).

x	1	3	5	7	9
y	1	4	8	9	12

  a. Determine the least-squares prediction equation.
  b. Use the least-squares prediction equation to predict y when $x = 6$.

11.5 Refer to the data of Exercise 11.1. Find the least-squares prediction equation and compare it to the freehand regression line you sketched through the points.

11.6 A computer solution using SAS for the least-squares prediction equation to the data is shown here.

```
SAS CODE:

option ls =70 ps = 55 nocenter nodate;
title 'EXERCISE 11.6';
data linreg;
     input X Y;
     CARDS;
10     25
18     55
25     50
40     75
50     110
63     138
42     90
30     60
 5     10
55     100
RUN;
PROC PLOT; PLOT Y*X='*';
PROC REG; MODEL Y = X;
OUTPUT OUT=NEW P=PRED R=RESID;
LABEL PRED='PREDICTED VALUE' RESID='RESIDUALS';
PROC PRINT; VAR Y X PRED RESID;
RUN;
```

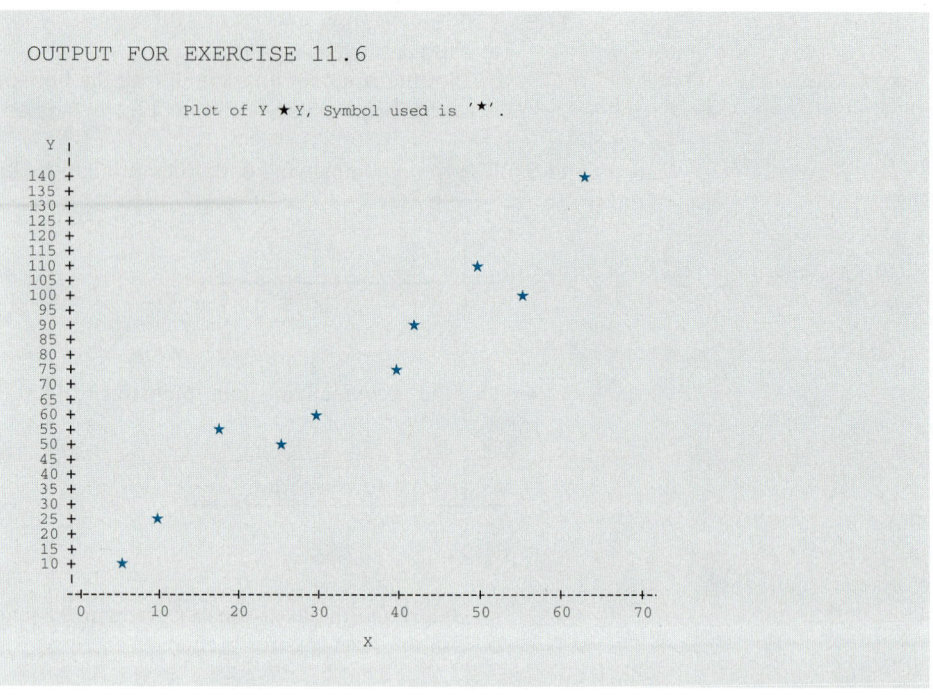

```
OUTPUT FOR EXERCISE 11.6

                    Plot of Y ★ Y, Symbol used is '★'.

    Y │
      │
  140 +                                                    ★
  135 +
  130 +
  125 +
  120 +
  115 +
  110 +                                      ★
  105 +
  100 +                                          ★
   95 +
   90 +                                 ★
   85 +
   80 +
   75 +                              ★
   70 +
   65 +
   60 +                        ★
   55 +              ★
   50 +                 ★
   45 +
   40 +
   35 +
   30 +
   25 +        ★
   20 +
   15 +
   10 +   ★
      │
      └─+─────────+─────────+─────────+─────────+─────────+─────────+─────────+─
        0        10        20        30        40        50        60        70
                                      X
```

```
Dependent Variable: Y

Analysis of Variance

                           Sum of          Mean
Source            DF       Squares        Square      F Value      Prob>F

Model              1    13230.96994    13230.96994    162.560      0.0001
Error              8      651.13006       81.39126
C Total            9    13882.10000


       Root MSE        9.02171      R-square      0.9531
       Dep Mean       71.30000      Adj R-sq      0.9472
       C.V.           12.65317

Parameter Estimates

                    Parameter      Standard     T for HO:
Variable   DF       Estimate         Error    Parameter=0     Prob > |T|

INTERCEP    1       4.697852      5.95202071        0.789        0.4527
X           1       1.970478      0.15454842       12.750        0.0001


OBS      Y      X      PREDICTED    RESIDUALS
                        VALUES
  1     25     10       24.403        0.5974
  2     55     18       40.166       14.8335
  3     50     25       53.960       -3.9598
```

4	75	40	83.517	-8.5170
5	110	50	103.222	6.7783
6	138	63	128.838	9.1620
7	90	42	87.458	2.5421
8	60	30	63.812	-3.8122
9	10	5	14.550	-4.5502
10	100	55	113.074	-13.0741

a. Determine the least-squares prediction equation from the output here and draw the regression line in the data plot.
b. Does the prediction equation seem to represent the data adequately?
c. Predict y for $x = 35$.

APPLICATIONS

Ag.   **11.7**   A food processor conducted an experiment to examine the effect of different concentrations of pectin on the firmness of canned sweet potatoes, using three concentrations: 0%, 1.5%, and 3% pectin by weight. The processor packed six number 303 × 406 cans with sweet potatoes in a 25% (by weight) sugar solution. Two cans were randomly assigned to each of the pectin concentrations with the appropriate percentage of pectin added to the sugar syrup. The cans were then sealed and placed in a 25°C environment for 30 days. At the end of the storage time, the cans were opened and a firmness determination made for the contents of each can. These data appear here:

Pectin concentration	0%, 0%	1.5%, 1.5%	3.0%, 3.0%
Firmness reading	50.5, 46.8	62.3, 67.7	80.1, 79.2

a. Let x denote the pectin concentration of a can and y denote the firmness reading following the 30 days of storage at 25°C. Plot the sample data in a scatter diagram.
b. Obtain least-squares estimates for the parameters in the model $y = \beta_0 + \beta_1 x + \varepsilon$.

11.8   Refer to Exercise 11.7. Predict the firmness for a can of sweet potatoes treated with a 1% concentration of pectin (by weight) after 30 days of storage at 25°C.

Env.   **11.9**   In a study conducted to examine the quality of fish after 7 days in ice storage, ten raw fish of the same kind and approximately the same size were caught and prepared for ice storage. Two of the fish were placed in storage immediately after being caught, two were placed in storage 3 hours after being caught, and two each were placed in storage at 6, 9, and 12 hours after being caught. Let y denote a measurement of fish quality (on a 10-point scale) after the 7 days of storage, and let x denote the time after being caught that the fish were placed in ice packing. The sample data appear here:

y	8.5	8.4	7.9	8.1	7.8	7.6	7.3	7.0	6.8	6.7
x	0	0	3	3	6	6	9	9	12	12

a. Plot the sample data in a scatter diagram.
b. Use the method of least squares to obtain estimates of the parameters in the model $y = \beta_0 + \beta_1 x + \varepsilon$.
c. Interpret the value of $\hat{\beta}_1$ in the context of this problem.

11.10   Refer to Exercise 11.9. Predict the 7-day quality score of a fish placed in ice storage 10 hours after being caught. Would you be willing to predict a quality score for fish placed in storage 18 hours after being caught?

11.11   A regression study yielded the following data and Statistix plots.

	x:	1	1	1	3	3	3
$x' = \log_{10}$	x':	.000	.000	.000	.477	.477	.477
	y:	13.5	15.4	16.1	18.3	19.9	20.9

	x:	5	5	5	7	7	7
$x' = \log_{10}$	x':	.699	.699	.699	.845	.845	.845
	y:	20.8	23.1	22.1	22.8	24.9	24.5

a. In the plot of y versus x, approximate the slope as the difference between predicted values for $x = 7$ and for $x = 1$ divided by the difference in x values—namely, 6.

b. In the plot of y versus x', approximate the slope of the prediction line.

c. Which plot appears more nearly linear to you?

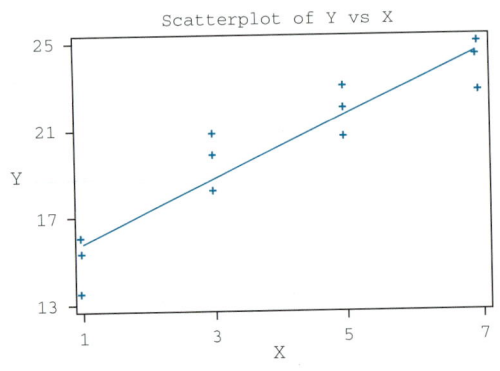

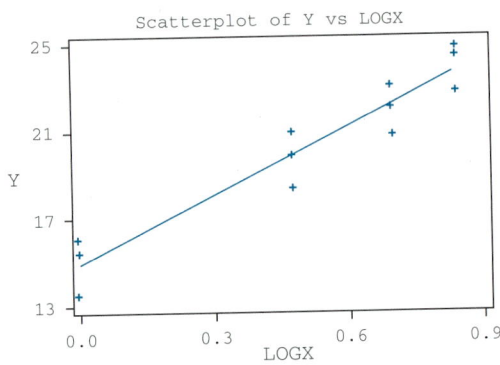

```
Analysis of Variance

SOURCE        DF          SS          MS          F           p
Regression    1        130.54      130.54      72.07       0.000
Error        10         18.11        1.81
Total        11        148.65


MTB > Regress 'y' 1 'logx'.


The regression equation is
y = 14.9 + 10.5 logx


Predictor          Coef        Stdev      t-ratio          p
Constant        14.8755       0.6106       24.36       0.000
logx            10.522        1.021        10.30       0.000


s = 1.131        R-sq = 91.4%      R-sq(adj) = 90.5%


Analysis of Variance

SOURCE        DF          SS          MS          F           p
Regression    1        135.85      135.85      106.17      0.000
Error        10         12.80        1.28
Total        11        148.65
```

11.12 Refer to the data of Exercise 11.11 and the following Minitab output:

```
MTB > Regress 'y' 1 'x'.

The regression equation is
y = 14.3 + 1.48 x

Predictor        Coef      Stdev     t-ratio         p
Constant      14.2917     0.7962       17.95     0.000
x              1.4750     0.1737        8.49     0.000

s = 1.346        R-sq = 87.8%        R-sq(adj) = 86.6%
```

    a. Locate the least-squares equation $\hat{y} = \hat{\beta}_0 + \hat{\beta}_1 x$.
    b. Locate the residual standard deviation.

11.13 Refer to the output for the data in Exercise 11.11.
    a. Find the least-squares equation $\hat{y} = \hat{\beta}_0 + \hat{\beta}_1 \log x$.
    b. What is the residual standard deviation?

11.14 Compare the residual standard deviations s_ε in the two preceding exercises. Which is smaller? Does this confirm your opinion about the choice of model based on the plots of Exercise 11.11?

Bus. **11.15** A mail-order retailer spends considerable effort in "picking" orders—selecting the ordered items and assembling them for shipment. A small study took a sample of 100 orders. An experienced picker carried out the entire process. The time in minutes needed was recorded for each order. A scatterplot and spline fit, created using JMP, are shown. What sort of transformation is suggested by the plot?

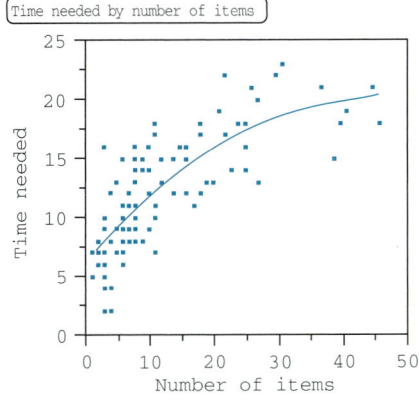

Time needed by number of items
Fitting — Smoothing spline fit, lambda = 10000

11.16 The order-picking time data in Exercise 11.15 were transformed by taking the square root of the number of items. A scatterplot of the result and regression results follow.
    a. Does the transformed scatterplot appear reasonably linear?
    b. Write out the prediction equation based on the transformed data.

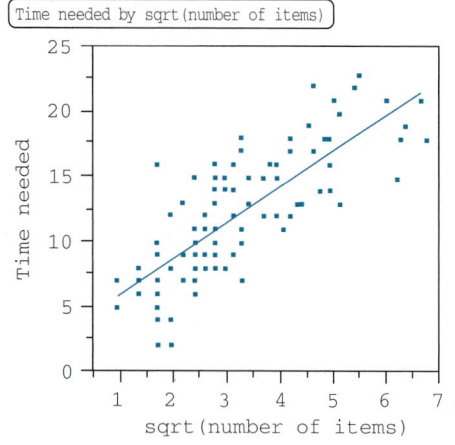

Summary of Fit

RSquare	0.624567
RSquare Adj	0.620736
Root Mean Square Error	2.923232
Mean of Response	12.29
Observations (or Sum Wgts)	100

Analysis of Variance

Source	DF	Sum of Squares	Mean Square	F Ratio
Model	1	1393.1522	1393.15	163.0317
Error	98	837.4378	8.55	Prob>F
C Total	99	2230.5900		0.0000

Parameter Estimates

Term	Estimate	Std Error	t Ratio	Prob>|t|
Intercept	3.097869	0.776999	3.99	0.0001
sqrt (Number of items)	2.7633138	0.216418	12.77	0.0000

11.17 In the JMP output of Exercise 11.16, the residual standard deviation is called "Root Mean Square Error." Locate and interpret this number.

11.18 In the preceding exercises, why can the residual standard deviation for the transformed data be compared to the residual standard deviation for the original data?

Bus. **11.19** As one part of a study of commercial bank branches, data are obtained on the number of independent businesses (x) located in sample zip code areas and the number of bank branches (y) located in these areas. The commercial centers of cities are excluded.

x:	92	116	124	210	216	267	306	378	415	502	615	703
y:	3	2	3	5	4	5	5	6	7	7	9	9

Output (StataQuest) for the analysis of the data is as follows:

```
. regress Branches Business

      Source |       SS       df       MS              Number of obs =      12
-------------+------------------------------           F(  1,    10) = 172.60
       Model | 53.7996874    1   53.7996874            Prob > F      = 0.0000
    Residual | 3.11697922   10   .311697922            R-square      = 0.9452
-------------+------------------------------           Adj R-square  = 0.9398
       Total | 56.9166667   11   5.17424242            Root MSE      = .5583
```

```
------------------------------------------------------------------------------
    Branches |      Coef.   Std. Err.       t      P>|t|     [95% Conf. Interval]
-------------+----------------------------------------------------------------
    Business |   .0111049   .0008453    13.138    0.000     .0092216    .0129883
       _cons |   1.766846   .3211751     5.501    0.000     1.051223    2.482469
------------------------------------------------------------------------------
```

a. Plot the data. Does a linear equation relating y to x appear plausible?
b. Locate the regression equation (with y as the dependent variable).
c. Interpret the value of $\hat{\beta}_1$ in the context of this problem.
d. Locate the sample residual standard deviation s_ε.

11.20 Does it appear that variability of y increases with x in the data plot of Exercise 11.19? (This would violate the assumption of constant variance.)

Engin.

11.21 A manufacturer of cases for sound equipment requires drilling holes for metal screws. The drill bits wear out and must be replaced; there is expense not only in the cost of the bits but also for lost production. Engineers varied the rotation speed of the drill and measured the lifetime y (thousands of holes drilled) of four bits at each of five speeds x. The data were:

x:	60	60	60	60	80	80	80	80	100	100
y:	4.6	3.8	4.9	4.5	4.7	5.8	5.5	5.4	5.0	4.5

x:	100	100	120	120	120	120	140	140	140	140
y:	3.2	4.8	4.1	4.5	4.0	3.8	3.6	3.0	3.5	3.4

a. Create a scatterplot of the data. Does there appear to be a relation? Does it appear to be linear?

b. Is there any evident outlier? If so, does it have high influence?

11.22 The data of Exercise 11.21 were analyzed using Excel's regression function. The following output was obtained:

	A	B	C	D	E	F
1	SUMMARY OUTPUT					
2						
3	Regression Statistics					
4	Multiple R	0.6254				
5	R Square	0.3911				
6	Adjusted R Square	0.3573				
7	Standard Error	0.6324				
8	Observations	20				
9						
10						
11	ANOVA					
12		df	SS	MS	F	Significance F
13	Regression	1	4.624	4.624	11.563	0.0032
14	Residual	18	7.198	0.400		
15	Total	19	11.822			
16						
17						
18		Coefficient	Standard Error	t Stat	P-value	
19	Intercept	6.03	0.5195	11.606	8.617E-10	
20	Speed	-0.017	0.005	-3.400	3.188E-03	

a. Locate the intercept and slope of the least-squares regression line.

b. What does the sign of the slope indicate about the relation between speed and bit lifetime?

c. Locate the residual standard deviation. Interpret the resulting number.

11.23 Again refer to Exercise 11.21.

a. Use the regression line of Exercise 11.22 to calculate predicted values for $x =$ 60, 80, 100, 120, and 140.

b. For which x values are most of the actual y values larger than the predicted values? For which x values are most y values lower than predicted? What does this pattern indicate about whether there is a linear relation?

Bus. **11.24** A realtor studied the relation between y = yearly income (in thousands of dollars per year) of home purchasers and y = sale price of the house (in thousands of dollars). The realtor gathered data from mortgage applications for 24 sales in the realtor's basic sales area in one season. Stata output was obtained, as shown after the data.

x:	25.0	28.5	29.2	30.0	31.0	31.5	31.9	32.0	33.0
y:	84.9	94.0	96.5	93.5	102.9	99.5	101.0	105.0	99.9

x:	33.5	34.0	35.9	36.0	39.0	39.0	40.5	40.9	42.5
y:	110.0	100.0	116.0	110.0	125.0	119.9	130.6	120.8	129.9

x:	44.0	45.0	50.0	54.6	65.0	70.0
y:	135.5	140.0	150.7	170.0	110.0	185.0

```
. regress Price Income

    Source |       SS           df       MS              Number of obs =      24
-----------+----------------------------------           F( 1, 22)     =   45.20
     Model | 9432.58336          1    9432.58336         Prob > F      =  0.0000
  Residual |  4590.6746         22    208.667027         R-square      =  0.6726
-----------+----------------------------------           Adj R-square  =  0.6578
     Total |  14023.258         23    609.706868         Root MSE      =  14.445

-----------------------------------------------------------------------------
     Price |      Coef.   Std. Err.       t      P>|t|     [95% Conf. Interval]
-----------+-----------------------------------------------------------------
    Income |    1.80264   .2681147      6.723    0.000     1.246604    2.358676
     _cons |   47.15048  10.93417       4.312    0.000      24.4744   69.82657
-----------------------------------------------------------------------------

.drop in 23
(1 observation deleted)
```

```
. regress Price Income

    Source |       SS           df       MS              Number of obs =      23
-----------+----------------------------------           F( 1, 21)     =  512.02
     Model | 13407.5437          1    13407.5437         Prob > F      =  0.0000
  Residual | 549.902031         21    26.185811          R-square      =  0.9606
-----------+----------------------------------           Adj R-square  =  0.9587
     Total | 13957.4457         22    634.429351         Root MSE      =  5.1172

-----------------------------------------------------------------------------
     Price |      Coef.   Std. Err.       t      P>|t|     [95% Conf. Interval]
-----------+-----------------------------------------------------------------
    Income |   2.461967    .108803     22.628    0.000     2.235699    2.688236
     _cons |   24.35755   4.286011      5.683    0.000      15.4443   33.27079
-----------------------------------------------------------------------------
```

a. A scatterplot with a LOWESS smooth, drawn using Minitab, follows. Does the relation appear to be basically linear?

b. Are there any high leverage points? If so, which ones seem to have high influence?

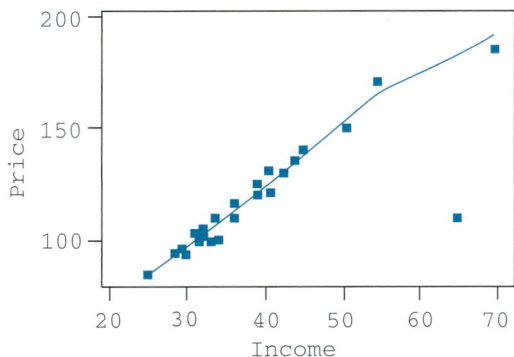

11.25   For Exercise 11.24,
   a. Locate the least-squares regression equation for the data.
   b. Interpret the slope coefficient. Is the intercept meaningful?
   c. Find the residual standard deviation.

11.26   The output of Exercise 11.24 also contains a regression line when we omit the point with $x = 65.0$ and $y = 110.0$. Does the slope change substantially? Why?

11.3   Inferences about Regression Parameters

The slope, intercept, and residual standard deviation in a simple regression model are all estimates based on limited data. As with all other statistical quantities, they are affected by random error. In this section, we consider how to allow for that random error. The concepts of hypothesis tests and confidence intervals that we have applied to means and proportions apply equally well to regression summary figures.

The t distribution can be used to make significance tests and confidence intervals for the true slope and intercept. One natural null hypothesis is that the

t test for β_1 true slope β_1 equals 0. If this H_0 is true, a change in x yields no predicted change in y, and it follows that x has no value in predicting y. We know from the previous section that the sample slope $\hat{\beta}_1$ has the expected value β_1 and standard error

$$\sigma_{\hat{\beta}_1} = \sigma_\varepsilon \sqrt{\frac{1}{S_{xx}}}$$

In practice, σ_ε is not known and must be estimated by s_ε, the residual standard deviation. In almost all regression analysis computer outputs, the estimated standard error is shown next to the coefficient. A test of this null hypothesis is given by the t statistic

$$t = \frac{\hat{\beta}_1 - \beta_1}{\text{estimated standard error } (\hat{\beta}_1)} = \frac{\hat{\beta}_1 - \beta_1}{s_\varepsilon \sqrt{1/S_{xx}}}$$

The most common use of this statistic is shown in the following summary.

Summary of a Statistical Test for β_1

Hypotheses:

Case 1 $H_0: \beta_1 \leq 0$ vs. $H_a: \beta_1 > 0$

Case 2 $H_0: \beta_1 \geq 0$ vs. $H_a: \beta_1 < 0$

Case 3 $H_0: \beta_1 = 0$ vs. $H_a: \beta_1 \neq 0$

T.S.: $t = \dfrac{\hat{\beta}_1 - 0}{s_\varepsilon / \sqrt{S_{xx}}}$

R.R.: For df $= n - 2$ and Type I error α,
1. Reject H_0 if $t > t_\alpha$.
2. Reject H_0 if $t < -t_\alpha$.
3. Reject H_0 if $|t| > t_{\alpha/2}$.

Check assumptions and draw conclusions.

All regression analysis outputs show this t value.

In most computer outputs, this test is indicated after the standard error and labeled as T TEST or T STATISTIC. Often, a p-value is also given, which eliminates the need for looking up the t value in a table.

EXAMPLE 11.5

Use the computer output of Example 11.3 (reproduced here) to locate the value of the t statistic for testing $H_0: \beta_1 = 0$ in the road resurfacing example. Give the observed level of significance for the test.

```
PREDICTOR
VARIABLES      COEFFICIENT     STD ERROR     STUDENT'S T        P

CONSTANT          2.00000        3.82970          0.52        0.6376
MILES             3.00000        0.85634          3.50        0.0394


R-SQUARED               0.8036    RESID. MEAN SQUARE (MSE)     14.6666
ADJUSTED R-SQUARED      0.7381    STANDARD DEVIATION           3.82970


SOURCE          DF        SS            MS          F        P

REGRESSION       1     180.000       180.000     12.27    0.0394
RESIDUAL         3     44.0000       14.6666
TOTAL            4     224.000
```

Solution It is clear from the output that the value of the test statistic in the column labeled STUDENT'S T is $t = 3.50$. The p-value for the two-tailed alternative H_a: $\beta_1 \neq 0$, labeled as P, is .0394. Because this value is fairly small, we can reject the hypothesis that mileage has no effect on predicting cost.

EXAMPLE 11.6

The following data show mean ages of executives of 15 firms in the food industry and the previous year's percentage increase in earnings per share of the firms. Use the Systat output shown to test the hypothesis that executive age has no predictive value for change in earnings. Should a one-sided or two-sided alternative be used?

Mean age	x:	38.2	40.0	42.5	43.4	44.6	44.9	45.0	45.4
Change, earnings per share	y:	8.9	13.0	4.7	−2.4	12.5	18.4	6.6	13.5

	x:	46.0	47.3	47.3	48.0	49.1	50.5	51.6
	y:	8.5	15.3	18.9	6.0	10.4	15.9	17.1

```
DEP VAR: CHGEPS  N: 15  MULTIPLE R: 0.383  SQUARED MULTIPLE R: 0.147
STANDARD ERROR OF ESTIMATE: 5.634

VARIABLE     COEFFICIENT    STD ERROR    STD COEF      T      P(2 TAIL)
CONSTANT       -16.991        18.866       0.000      0.901     0.384
MEANAGE          0.617         0.413       0.383      1.496     0.158


                        ANALYSIS OF VARIANCE

SOURCE          SUM-OF-SQUARES    DF    MEAN-SQUARE    F-RATIO     P
REGRESSION            71.055       1         71.055     2.239    0.158
RESIDUAL            412.602       13         31.739
```

Solution In the model $y = \beta_0 + \beta_1 x + \epsilon$, the null hypothesis is $H_0: \beta_1 = 0$. The myth in American business is that younger managers tend to be more aggressive and harder driving, but it is also possible that the greater experience of the older executives leads to better decisions. Therefore, there is a good reason to choose a two-sided research hypothesis, $H_a: \beta_1 \neq 0$. The t statistic is shown in the output column marked T, reasonably enough. It shows $t = 1.496$, with a (two-sided) p-value of 0.158. There is not enough evidence to conclude that there is any relation between age and change in earnings.

In passing, note that the interpretation of $\hat{\beta}_0$ is rather interesting in this example; it would be the predicted change in earnings of a firm with mean age of its managers equal to 0. Hmm.

It is also possible to calculate a confidence interval for the true slope. This is an excellent way to communicate the likely degree of inaccuracy in the estimate of that slope. The confidence interval once again is simply the estimate plus or minus a t table value times the standard error.

Confidence Interval for Slope β_1

$$\hat{\beta}_1 - t_{\alpha/2} s_\epsilon \sqrt{\frac{1}{S_{xx}}} \leq \beta_1 \leq \hat{\beta}_1 + t_{\alpha/2} s_\epsilon \sqrt{\frac{1}{S_{xx}}}$$

The required degrees of freedom for the table value $t_{\alpha/2}$ is $n - 2$, the error df.

EXAMPLE 11.7

Compute a 95% confidence interval for the slope β_1 using the output from Example 11.3.

Solution   In the output, $\hat{\beta}_1 = 3.000$ and the estimated standard error of $\hat{\beta}_1$ is shown as .856, rounded off. Because n is 5, there are $5 - 2 = 3$ df for error. The required table value for $\alpha/2 = .05/2 = .025$ is 3.182. The corresponding confidence interval for the true value of β_1 is then

$$3.00 \pm 3.182(.856) \qquad \text{or} \qquad .276 \text{ to } 5.724$$

The predicted cost per additional mile of resurfacing could be anywhere from $276 to $5,724. The enormous width of this interval results largely from the small sample size.

There is an alternative test, an F test, for the null hypothesis of no predictive value. It was designed to test the null hypothesis that *all* predictors have no value in predicting y. This test gives the same result as a two-sided t test of $H_0\colon \beta_1 = 0$ in simple linear regression; to say that all predictors have no value is to say that the (only) slope is 0. The F test is summarized next.

F Test for H_0; $\beta_1 = 0$

$H_0\colon \quad \beta_1 = 0$

$H_a\colon \quad \beta_1 \neq 0$

T.S.: $F = \dfrac{\text{SS(Regression)}/1}{\text{SS(Residual)}/(n - 2)} = \dfrac{\text{MS(Regression)}}{\text{MS(Residual)}}$

R.R.:   With $\text{df}_1 = 1$ and $\text{df}_2 = n - 2$, reject H_0 if $F > F_\alpha$.

Check assumptions and draw conclusions.

SS(Regression) is the sum of squared deviations of predicted y values from the y mean. SS(Regression) $= \Sigma(\hat{y}_i - \bar{y})^2$. SS(Residual) is the sum of squared deviations of actual y values from predicted y values. SS(Residual) $= \Sigma(y_i - \hat{y}_i)^2$.

Virtually all computer packages calculate this F statistic. In the road resurfacing example, the output shows $F = 12.27$ with a p-value of .0394. Again, the hypothesis of no predictive value can be rejected. It is always true for simple linear regression problems that $F = t^2$; in the example, $12.27 = (3.50)^2$, to within round-off error. The F and two-sided t tests are equivalent in simple linear regression; they serve different purposes in multiple regression.

EXAMPLE 11.8

For the output of Example 11.4, reproduced here, use the F test for $H_0\colon \beta_1 = 0$. Show that $t^2 = F$.

```
The regression equation is
y = 64.7 - 1.75 x

Predictor          Coef        Stdev        t-ratio          p
Constant         64.672        6.762           9.56      0.000
x               -1.7487       0.2995          -5.84      0.000

S = 2.388          R-sq = 81.0%        R-sq(adj) = 78.6%

Analysis of Variance

SOURCE           DF          SS          MS          F          P
Regression        1      194.45      194.45      34.10      0.000
Error             8       45.61        5.70
Total             9      240.06
```

Solution The F statistic is shown in the output as 34.10, with a p-value of 0.000 (indicating that the actual p-value is something smaller than 0.0005). Note that the t statistic is -5.84, and that $t^2 = (-5.84)^2 = 34.11$, equal to F, to within round-off error.

You should be able to work out comparable hypothesis testing and confidence interval formulas for the intercept β_0 using the estimated standard error of $\hat{\beta}_0$ as

$$\sigma_{\hat{\beta}_0} = s_\varepsilon \sqrt{\frac{1}{n} + \frac{\bar{x}^2}{S_{xx}}}$$

In practice, this parameter is of less interest than the slope. In particular, there is often no reason to hypothesize that the true intercept is zero (or any other particular value). Computer packages almost always test the null hypothesis of zero slope, but some don't bother with a test on the intercept term.

EXERCISES

11.27 Refer to the data of Exercise 11.15.
  a. Calculate a 90% confidence interval for β_1.
  b. What is the interpretation of $H_0: \beta_1 = 0$ in Exercise 11.15?
  c. What is the natural research hypothesis H_a for that problem?
  d. Do the data support H_a at $\alpha = .05$? Clearly state your assumptions.

11.28 Find the p-value of the test of $H_0: \beta_1 = 0$ for the previous exercise.

Ag. **11.29** Suppose a researcher is interested in examining the relationship between different concentrations of pectin (0%, 1.5%, and 3% by weight) on the firmness of canned sweet potatoes after storage in a controlled 25°C environment. The sample data for six cans are shown here.

y (firmness)	50.5	46.8	62.3	67.7	80.1	79.2
x (concentration of pectin)	0	0	1.5	1.5	3.0	3.0

  a. Obtain the least-squares estimates for the parameters in the model $y = \beta_0 + \beta_1 x + \varepsilon$.
  b. Obtain an estimate of σ_ε^2.
  c. Give the standard error of $\hat{\beta}_1$.

11.30 Refer to Exercise 11.29. Perform a statistical test of the null hypothesis that there is no linear relationship between the concentration of pectin and the firmness of canned sweet potatoes after 30 days of storage at 25°C. Give the p-value for this test and draw conclusions.

Bio. **11.31** The extent of disease transmission can be affected greatly by the viability of infectious organisms suspended in the air. Because of the infectious nature of the disease under study, the viability of these organisms must be studied in an airtight chamber. One way to do this is to disperse an aerosol cloud, prepared from a solution containing the organisms, into the chamber. The biological recovery at any particular time is the percentage of the total number of organisms suspended in the aerosol that are viable. The data in the accompanying table are the biological recovery percentages computed from 13 different aerosol clouds. For each of the clouds, recovery percentages were determined at different times.

  a. Plot the data.
  b. Since there is some curvature, try to linearize the data using the log of the biological recovery.

Cloud	Time, x (in minutes)	Biological Recovery (%)
1	0	70.6
2	5	52.0
3	10	33.4
4	15	22.0
5	20	18.3
6	25	15.1
7	30	13.0
8	35	10.0
9	40	9.1
10	45	8.3
11	50	7.9
12	55	7.7
13	60	7.7

11.32 Refer to Exercise 11.31.
  a. Fit the linear regression model $y = \beta_0 + \beta_1 x + \varepsilon$, where y is the log biological recovery.
  b. Compute an estimate of σ_ε.
  c. Identify the standard errors of $\hat{\beta}_0$ and $\hat{\beta}_1$.

11.33 Refer to Exercise 11.31. Conduct a test of the null hypothesis that $\beta_1 = 0$. Use $\alpha = .05$.

11.34 Refer to Exercise 11.31. Place a 95% confidence interval on β_0, the mean log biological recovery percentage at time zero. Interpret your findings. (*Note:* $E(y) = \beta_0$ when $x = 0$.)

Ag. **11.35** A researcher conducts an experiment to examine the relationship between the weight gain of chickens whose diets had been supplemented by different amounts of amino acid lysine and the amount of lysine ingested. Since the percentage of lysine is known, and we can monitor the amount of feed consumed, we can determine the amount of lysine eaten. A random sample of twelve 2-week-old chickens was selected for the study. Each was caged separately and was allowed to eat at will from feed composed of a base supplemented with lysine. The sample data summarizing weight gains and amounts of lysine eaten over the test period are given here. (In the data, y represents weight gain in grams, and x represents the amount of lysine ingested in grams.)

a. Refer to the output. Does a linear model seem appropriate?
b. From the output, obtain the estimated linear regression model $\hat{y} = \hat{\beta}_0 + \hat{\beta}_1 x$.

Plot of Y★X. Symbol used is '★'.

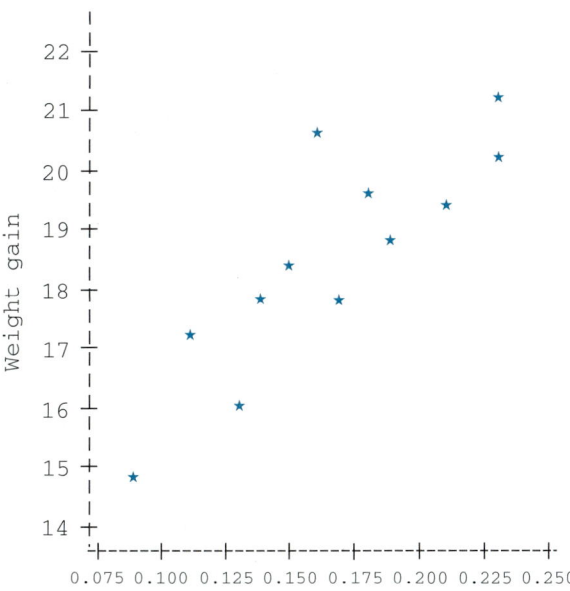

Chick	y	x	Chick	y	x
1	14.7	.09	7	17.2	.11
2	17.8	.14	8	18.7	.19
3	19.6	.18	9	20.2	.23
4	18.4	.15	10	16.0	.13
5	20.5	.16	11	17.8	.17
6	21.1	.23	12	19.4	.21

OUTPUT FOR EXERCISE 11.35

Dependent Variable: Y   WEIGHT GAIN

Analysis of Variance

Source	DF	Sum of Squares	Mean Square	F Value	Prob > F
Model	1	28.35785	28.35785	26.522	0.0004
Error	10	10.69215	1.06921		
C Total	11	39.05000			

Root MSE	1.03403	R-square	0.7262
Dep Mean	18.45000	Adj R-sq	0.6988
C.V.	5.60449		

```
Parameter Estimates

                  Parameter      Standard       T for H0:
Variable   DF     Estimate         Error      Parameter = 0     Prob > |T|

INTERCEP    1    12.508525     1.19168259         10.497          0.0001
X           1    35.827989     6.95693918          5.150          0.0004


                 Variable
Variable   DF    Label

INTERCEP    1     Intercept
X           1     LYSINE INGESTED


OBS      Y       X      PREDICTED     RESIDUALS
                        VALUES


  1    14.7    0.09    15.7330       -1.03304
  2    17.8    0.14    17.5244        0.27556
  3    19.6    0.18    18.9576        0.64244
  4    18.4    0.15    17.8827        0.51728
  5    20.5    0.16    18.2410        2.25900
  6    21.1    0.23    20.7490        0.35104
  7    17.2    0.11    16.4496        0.75040
  8    18.7    0.19    19.3158       -0.61584
  9    20.2    0.23    20.7490       -0.54896
 10    16.0    0.13    17.1662       -1.16616
 11    17.8    0.17    18.5993       -0.79928
 12    19.4    0.21    20.0324       -0.63240
```

11.36 Refer to the output of Exercise 11.35.
   a. Estimate σ^2_ε.
   b. Identify the standard error of $\hat{\beta}_1$.
   c. Conduct a statistical test of the research hypothesis that for this diet preparation and length of study, there is a direct (positive) linear relationship between weight gain and the amount of lysine eaten.

11.37 Refer to Exercise 11.35.
   a. For this example, would it make sense to give any physical interpretation to β_0? (Hint: The lysine was mixed in the feed.)
   b. Consider an alternative model relating weight gain to amount of lysine ingested:

   $$y = \beta_1 x + \varepsilon$$

   Distinguish between this model and the model $y = \beta_0 + \beta_1 x + \varepsilon$.

11.38 **a.** Refer to part (b) of Exercise 11.37. From the output on the next page, obtain $\hat{\beta}_1$ for the model $y = \beta_1 x + \varepsilon$, where

   $$\hat{\beta}_1 = \frac{\sum xy}{\sum x^2}$$

   b. Which of the two models, $y = \beta_0 + \beta_1 x + \varepsilon$ or $y = \beta_1 x + \varepsilon$, appears to give a better fit to the sample data? (*Hint:* Examine the two prediction equations on a graph of the sample observations.)

```
                          OUTPUT FOR EXERCISE 11.38

NOTE: No Intercept in model. R-square is redefined.

Dependent Variable: Y            WEIGHT GAIN


Analysis of Variance

                        Sum of           Mean
   Source     DF        Squares          Square     F Value    Prob > F

   Model       1      3995.38497      3995.38497    342.031     0.0001
   Error      11       128.49503        11.68137
   U Total    12      4123.88000


       Root MSE      3.41780      R-square    0.9688
       Dep Mean     18.45000      Adj R-sq    0.9660
       C.V.         18.52467

Parameter Estimates

                      Parameter      Standard       T for H0:
   Variable   DF      Estimate         Error     Parameter = 0    Prob > |T|

   X           1     106.523715     5.75988490        18.494        0.0001

                       Variable
   Variable   DF        Label

   X           1     LYSINE INGESTED

   OBS     Y       X      PREDICTED     RESIDUALS
                           VALUES

     1    14.7    0.09     9.5871        5.11287
     2    17.8    0.14    14.9133        2.88668
     3    19.6    0.18    19.1743        0.42573
     4    18.4    0.15    15.9786        2.42144
     5    20.5    0.16    17.0438        3.45621
     6    21.1    0.23    24.5005       -3.40045
     7    17.2    0.11    11.7176        5.48239
     8    18.7    0.19    20.2395       -1.53951
     9    20.2    0.23    24.5005       -4.30045
    10    16.0    0.13    13.8481        2.15192
    11    17.8    0.17    18.1090       -0.30903
    12    19.4    0.21    22.3700       -2.96998
```

Bus.   **11.39** A firm that prints automobile bumper stickers investigates the relation between the total direct cost of a lot of stickers and the number produced in the printing run. The data are analyzed by the Execustat computer package. The relevant output is as follows:

```
                    Simple Regression Analysis

      Linear model: TotalCost = 99.777 + 5.19179*Runsize

                        Table of Estimates

                              Standard      t         P
                  Estimate     Error     Value      Value

      Intercept    99.777      2.8273     35.29     0.0000
      Slope         5.19179    0.0586455  88.53     0.0000


      R-squared = 99.64%
      Correlation coeff. = 0.998
      Standard error of estimation = 12.2065
      Durbin-Watson statistic = 2.67999

                        Analysis of Variance

                      Sum of                                          P
      Source          Squares      D.F.     Mean Square   F-Ratio   Value

      Model        1.16775e+006      1     1.16775e+006   7837.26   0.0000
      Error        4171.98          28      148.999

      Total (corr.)  1.17192e+006   29
```

a. Plot the data. Do you detect any difficulties with using a linear regression model? Can you see any blatant violations of assumptions? The raw data are as follows:

Runsize:	2.6	5.0	10.0	2.0	.8	4.0	2.5	.6	.8	1.0	2.0	
Total cost:	230	341	629	187	159	327	206	124	155	147	209	
Runsize:	3.0	.4	.5	5.0	20.0	5.0	2.0	1.0	1.5	.5	1.0	1.0
Total cost:	247	135	125	366	1146	339	208	150	179	128	155	143
Runsize:	.6	2.0	1.5	3.0	6.5	2.2	1.0					
Total cost:	131	219	171	258	415	226	159					

b. Write the estimated regression equation indicated in the output. Find the residual standard deviation.

c. Calculate a 95% confidence interval for the true slope. What are the interpretations of the intercept and slope in this problem?

11.40 Refer to the computer output of Exercise 11.39.
   a. Locate the value of the t statistic for testing $H_0: \beta_1 = 0$.
   b. Locate the p-value for this test. Is the p-value one-tailed or two-tailed? If necessary, calculate the p-value for the appropriate number of tails.

11.41 Refer to the computer output of Exercise 11.39.
   a. Locate the value of the F statistic and the associated p-value.
   b. How do the p-values for this F test and the t test of Exercise 11.40 compare? Why should this relation hold?

11.4   Predicting New *y* Values Using Regression

In all the regression analyses we have done so far, we have been summarizing and making inferences about relations in data that have already been observed. Thus, we have been predicting the past. One of the most important uses of regression is trying to forecast the future. In the road resurfacing example, the county highway director wants to predict the cost of a new contract that is up for bids. In a regression predicting quantity sold given price, a manager will want to predict the demand at a new price. In this section, we discuss how to make such regression forecasts and how to determine the plus or minus probable error factor.

There are two possible interpretations of a *y* prediction based on a given *x*. Suppose that the highway director substitutes $x = 6$ miles in the regression equation $\hat{y} = 2.0 + 3.0x$ and gets $\hat{y} = 20$. This can be interpreted as either

"The average cost $E(y)$ of *all* resurfacing contracts for 6 miles of road will be $20,000."

or

"The cost *y* of *this specific* resurfacing contract for 6 miles of road will be $20,000."

The best-guess prediction in either case is 20, but the plus or minus factor differs. It is easier to predict an average value $E(y)$ than an individual *y* value, so the plus or minus factor should be less for predicting an average. We discuss the plus or minus range for predicting an average first, with the understanding that this is an intermediate step toward solving the specific-value problem.

In the mean-value forecasting problem, suppose that the value of the predictor *x* is known. Because the previous values of *x* have been designated x_1, . . . , x_n, call the new value x_{n+1}. Then $\hat{y}_{n+1} = \beta_0 + \beta_1 x_{n+1}$ is used to predict $E(y_{n+1})$. Because $\hat{\beta}_0$ and $\hat{\beta}_1$ are unbiased, $\hat{y}_{n+1}$ is an unbiased predictor of $E(y_{n+1})$. The standard error of $\hat{y}_{n+1}$ can be shown to be

$$\sigma_\varepsilon \sqrt{\frac{1}{n} + \frac{(x_{n+1} - \bar{x})^2}{S_{xx}}}$$

Here S_{xx} is the sum of squared deviations of the original *n* values of x_i; it can be calculated from most computer outputs as

$$\left(\frac{s_\varepsilon}{\text{standard error}(\hat{\beta}_1)}\right)^2$$

Again, *t* tables with $n - 2$ df (the error df) must be used. The usual approach to forming a confidence interval—namely, estimate plus or minus *t* (standard error)—yields a confidence interval for $E(y_{n+1})$. Some of the better statistical computer packages will calculate this confidence interval if a new *x* value is specified without specifying a corresponding *y*.

Confidence Interval for $E(y_{n+1})$

$$\hat{y}_{n+1} - t_{\alpha/2}s_\varepsilon \sqrt{\frac{1}{n} + \frac{(x_{n+1} - \bar{x})^2}{S_{xx}}} \le E(y_{n+1})$$

$$\le \hat{y}_{n+1} + t_{\alpha/2}s_\varepsilon \sqrt{\frac{1}{n} + \frac{(x_{n+1} - \bar{x})^2}{S_{xx}}}$$

where t_α cuts off area a in the right tail of the t distribution with $n - 2$ df.

For the resurfacing example, the computer output displayed here shows the estimated value of $E(y_{n+1})$ to be 20 when $x = 6$. The corresponding 95% confidence interval on $E(y_{n+1})$ is 12.29 to 27.71.

```
Resurfacing Data


PREDICTOR
VARIABLES     COEFFICIENT    STD ERROR    STUDENT'S T       P
CONSTANT        2.00000       3.82970        0.52         0.6376
MILES           3.00000       0.85634        3.50         0.0394


R-SQUARED               0.8036    RESID. MEAN SQUARE (MSE)    14.6666
ADJUSTED R-SQUARED      0.7381    STANDARD DEVIATION           3.82970


SOURCE         DF       SS          MS          F          P


REGRESSION      1     180.000     180.000     12.27      0.0394
RESIDUAL        3     44.0000     14.6666
TOTAL           4     224.000


PREDICTED/FITTED VALUES OF COST


LOWER PREDICTED BOUND     5.5791    LOWER FITTED BOUND    12.291
PREDICTED VALUE          20.000     FITTED VALUE          20.000
UPPER PREDICTED BOUND    34.420     UPPER FITTED BOUND    27.708
SE (PREDICTED VALUE)      4.5313    SE (FITTED VALUE)      2.4221


PREDICTOR VALUES: MILES = 6.0000
```

The forecasting plus or minus term in the confidence interval for $E(y_{n+1})$ depends on the sample size n and the standard deviation around the regression line, as one might expect. It also depends on the squared distance of x_{n+1} from $\bar{x}$ (the mean of the previous x_i values) relative to S_{xx}. As x_{n+1} gets farther from $\bar{x}$, the term

$$\frac{(x_{n+1} - \bar{x})^2}{S_{xx}}$$

gets larger. When x_{n+1} is far away from the other x values, so that this term is large, the prediction is a considerable extrapolation from the data. Small errors in estimating the regression line are magnified by the extrapolation. The term **extrapolation penalty** $(x_{n+1} - \bar{x})^2/S_{xx}$ could be called an **extrapolation penalty** because it increases with the degree of extrapolation.

Extrapolation—predicting the results at independent variable values far from the data—is often tempting and always dangerous. Using it requires an assumption that the relation will continue to be linear, far beyond the data. By definition, you have no data to check this assumption. For example, a firm might find a negative correlation between the number of employees (ranging between 1,200 and 1,400) in a quarter and the profitability in that quarter; the fewer the employees, the greater the profit. It would be spectacularly risky to conclude from this fact that cutting the number of employees to 600 would vastly improve profitability. (Do you suppose we could have a negative number of employees?) Sooner or later, the declining number of employees must adversely affect the business so that profitability turns downward. The extrapolation penalty term actually understates the risk of extrapolation. It is based on the assumption of a linear relation, and that assumption gets very shaky for large extrapolations.

The confidence and prediction intervals also depend heavily on the assumption of constant variance. In some regression situations, the variability around the line increases as the predicted value increases, violating this assumption. In such a case, the confidence and prediction intervals will be too wide where there is relatively little variability and too narrow where there is relatively large variability. A scatterplot that shows a "fan" shape indicates nonconstant variance. In such a case, the confidence and prediction intervals are not very accurate.

EXAMPLE 11.9

For the data of Example 11.4, and the following Minitab output from that data, obtain a 95% confidence interval for $E(y_{n+1})$ based on an assumed x_{n+1} of 22.4. Compare the width of the interval to one based on an assumed x_{n+1} of 30.4.

```
MTB > regress 'y' on 1 variable 'x';
SUBC> predict at 22.4;
SUBC> predict at 30.4.

The regression equation is
y = 64.7 - 1.75 x

Predictor        Coef      Stdev     t-ratio         p
Constant       64.672      6.762        9.56     0.000
x             -1.7487      0.2995       -5.84     0.000

s = 2.388     R-sq = 81.0%     R-sq(adj) = 78.6%


Analysis of Variance

SOURCE          DF         SS         MS         F         p
Regression       1     194.45     194.45     34.10     0.000
Error            8      45.61       5.70
Total            9     240.06

   Fit     Stdev.Fit           95% C.I.                 95% P.I.
25.500        0.755    ( 23.758, 27.242)      ( 19.723, 31.277)
11.510        2.500    (  5.742, 17.278)      (  3.535, 19.485) XX

X denotes a row with X values away from the center
XX denotes a row with very extreme X values
```

Solution For $x_{n+1} = 22.4$, the first of the two Fit entries shows a predicted value equal to 25.5. The confidence interval is shown as 23.758 to 27.242. For $x_{n+1} = 30.4$, the predicted value is 11.51, with a confidence interval of 5.742 to 17.278. The second interval has a width of about 11.5, much larger than the first interval's width of about 3.5. The value $x_{n+1} = 30.4$ is far outside the range of x data; the extrapolation penalty makes the interval very wide.

Usually, the more relevant forecasting problem is that of predicting an individual y_{n+1} value rather than $E(y_{n+1})$. In most computer packages, the interval for predicting an individual value is called a **prediction interval.** The same best guess $\hat{y}_{n+1}$ is used, but the forecasting plus or minus term is larger when predicting y_{n+1} than $E(y_{n+1})$. In fact, it can be shown that the plus or minus forecasting error using $\hat{y}_{n+1}$ to predict y_{n+1} is as follows.

prediction interval

Prediction Interval for y_{n+1}

$$\hat{y}_{n+1} - t_{\alpha/2}s_\varepsilon \sqrt{1 + \frac{1}{n} + \frac{(x_{n+1} - \overline{x})^2}{S_{xx}}} \leq y_{n+1}$$

$$\leq \hat{y}_{n+1} + t_{\alpha/2}s_\varepsilon \sqrt{1 + \frac{1}{n} + \frac{(x_{n+1} - \overline{x})^2}{S_{xx}}}$$

where $t_{\alpha/2}$ cuts off area $\alpha/2$ in the right tail of the t distribution with $n - 2$ df.

In the road resurfacing example, the corresponding 95% prediction limits for y_{n+1} when $x = 6$ are 5.58 to 34.42, (see output on p. 568). The 95% intervals for $E(y_{n+1})$ and for y_{n+1} are plotted in Figure 11.13; the inner curves are for $E(y_{n+1})$ and the outer ones for y_{n+1}.

FIGURE 11.13
Predicted versus observed values with 95% limits

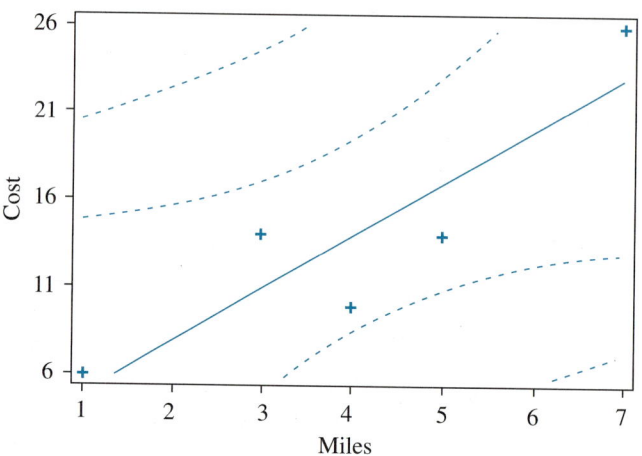

Cost = 2.0000 + 3.0000 * Miles   95% conf. and pred. intervals

The only difference between prediction of a mean $E(y_{n+1})$ and prediction of an individual y_{n+1} is the term +1 in the standard error formula. The presence of this extra term indicates that predictions of individual values are less accurate than predictions of means. The extrapolation penalty term still applies, as does the warning that it understates the risk of extrapolation. If n is large and the extrapolation term is small, the +1 term dominates the square root factor in the prediction interval. In such cases, the interval becomes approximately $\hat{y}_{n+1} - t_{\alpha/2}s_\varepsilon \leq y_{n+1} \leq \hat{y}_{n+1} + t_{\alpha/2}s_\varepsilon$. Thus, for large n, roughly 68% of the residuals (forecast errors) are less than $\pm 1s_\varepsilon$ and 95% less than $\pm 2s_\varepsilon$. There is not much point in devising rules for when to ignore the other terms in the square root factor. They are normally calculated in computer outputs and it does no harm to include them.

EXAMPLE 11.10

Using the output of Example 11.9 (reproduced here), find a 95% prediction interval for y_{n+1} with $x_{n+1} = 22.4$, and find the interval with $x_{n+1} = 30.4$. Compare these to widths estimated by the $\pm 2s_\varepsilon$ rules just discussed.

```
MTB > regress 'y' on 1 variable 'x';
SUBC> predict at 22.4;
SUBC> predict at 30.4.


The regression equation is
y = 64.7 - 1.75 x


Predictor       Coef      Stdev      t-ratio        p
Constant      64.672      6.762         9.56    0.000
x            -1.7487     0.2995        -5.84    0.000


s = 2.388    R-sq = 81.0%    R-sq(adj) = 78.6%


Analysis of Variance


SOURCE          DF         SS         MS         F        p
Regression       1     194.45     194.45     34.10    0.000
Error            8      45.61       5.70
Total            9     240.06


   Fit    Stdev.Fit           95% C.I.              95% P.I.
25.500        0.755    ( 23.758, 27.242)     ( 19.723, 31.277)
11.510        2.500    (  5.742, 17.278)     (  3.535, 19.485) XX


X denotes a row with X values away from the center
XX denotes a row with very extreme X values
```

Solution As in Example 11.9, $\hat{y}_{n+1} = 25.5$ if $x_{n+1} = 22.4$. The prediction interval is shown as

$$19.72 \leq y_{n+1} \leq 31.28$$

The $\pm 2s_\varepsilon$ range is

$$25.5 - (2)(2.388) \leq y_{n+1} \leq 25.5 + (2)(2.388) \quad \text{or} \quad 20.72 \leq y_{n+1} \leq 30.28$$

The latter interval is a bit too narrow, mostly because the tabled t value with only 8 df is quite a bit larger than 2.

For $x_{n+1} = 30.4$, $\hat{y}_{n+1} = 11.51$, the 95% prediction interval is

$$3.54 \leq y_{n+1} \leq 19.48$$

The $\pm 2s_\varepsilon$ range is

$$11.5 - (2)(2.388) \leq y_{n+1} \leq 11.5 + (2)(2.388) \qquad \text{or} \qquad 6.72 \leq y_{n+1} \leq 16.28$$

The latter is much too narrow. Not only is the tabled t value larger than 2, but also the large extrapolation penalty is not reflected. The output labels this prediction XX and notes that the x value used is far from the data. Be warned.

EXERCISES

Basic Techniques

11.42   Refer to Exercise 11.31. For the least-squares equation

$$\hat{y} = \hat{\beta}_0 + \hat{\beta}_1 x$$

estimate the mean log biological recovery percentage at 30 minutes, using a 95% confidence interval.

11.43   Using the data of Exercise 11.42, construct a 95% prediction interval for the log biological recovery percentage at 30 minutes. Compare your result to the confidence interval on $E(y)$ of Exercise 11.42.

Applications

Engin.   **11.44**   A chemist is interested in determining the weight loss y of a particular compound as a function of the amount of time the compound is exposed to the air. The data in the following table give the weight losses associated with $n = 12$ settings of the independent variable, exposure time.

**Weight Loss and Exposure
Time Data**

Weight Loss, y (in pounds)	Exposure Time (in hours)
4.3	4
5.5	5
6.8	6
8.0	7
4.0	4
5.2	5
6.6	6
7.5	7
2.0	4
4.0	5
5.7	6
6.5	7

a. Find the least-squares prediction equation for the model

$$y = \beta_0 + \beta_1 x + \varepsilon$$

b. Test $H_0: \beta_1 \leq 0$; give the p-value for $H_a: \beta_1 > 0$ and draw conclusions.

11.45   Refer to Exercise 11.44 and the SAS computer output shown here.
   a. Identify the 95% confidence bands for $E(y)$ when $4 \le x \le 7$.
   b. Identify the 95% prediction bands for y, $4 \le x \le 7$.
   c. Distinguish between the meaning of the confidence bands and prediction bands in parts (a) and (b).

```
Dependent Variable: Y    WEIGHT LOSS

Analysis of Variance

                   Sum of        Mean
Source     DF      Squares       Square     F Value    Prob>F

Model       1     26.00417     26.00417      40.223    0.0001
Error      10      6.46500      0.64650
C Total    11     32.46917

        Root MSE      0.80405     R-square    0.8009
        Dep Mean      5.50833     Adj R-sq    0.7810
        C.V.         14.59701

Parameter Estimates

                   Parameter      Standard      T for H0:
Variable    DF     Estimate         Error      Parameter=0    Prob > |T|

INTERCEP     1    -1.733333      1.16518239      -1.488        0.1677
X            1     1.316667      0.20760539       6.342        0.0001


          Predict   Std Err   Lower95%   Upper95%   Lower95%   Upper95%
X    Y    Value     Predict   Mean       Mean       Predict    Predict     Residual

4   4.3   3.5333    0.388     2.6679     4.3987     1.5437     5.5229      0.7667
5   5.5   4.8500    0.254     4.2835     5.4165     2.9710     6.7290      0.6500
6   6.8   6.1667    0.254     5.6001     6.7332     4.2877     8.0456      0.6333
7   8.0   7.4833    0.388     6.6179     8.3487     5.4937     9.4729      0.5167
4   4.0   3.5333    0.388     2.6679     4.3987     1.5437     5.5229      0.4667
5   5.2   4.8500    0.254     4.2835     5.4165     2.9710     6.7290      0.3500
6   6.6   6.1667    0.254     5.6001     6.7332     4.2877     8.0456      0.4333
7   7.5   7.4833    0.388     6.6179     8.3487     5.4937     9.4729      0.0167
4   2.0   3.5333    0.388     2.6679     4.3987     1.5437     5.5229     -1.5333
5   4.0   4.8500    0.254     4.2835     5.4165     2.9710     6.7290     -0.8500
6   5.7   6.1667    0.254     5.6001     6.7332     4.2877     8.0456     -0.4667
7   6.5   7.4833    0.388     6.6179     8.3487     5.4937     9.4729     -0.9833

Sum of Residuals                       0
Sum of Squared Residuals          6.4650
Predicted Resid SS (Press)       10.0309
```

11.46 Another part of the output of Exercise 11.39 is shown here.

Table of Predicted Values

Row	Runsize	Predicted TotalCost	95.00% Prediction Limits Lower	Upper	95.00% Confidence Limits Lower	Upper
1	2	203.613	178.169	229.057	198.902	208.323

a. Predict the mean total direct cost for all bumper sticker orders with a print run of 2,000 stickers (that is, with Runsize = 2.0).
b. Locate a 95% confidence interval for this mean.

11.47 Does the prediction in Exercise 11.46 represent a major extrapolation?

11.48 Refer to Exercise 11.46.
a. Predict the total direct cost for a particular bumper sticker order with a print run of 2,000 stickers. Obtain a 95% prediction interval.
b. Would an actual total direct cost of $250 be surprising for this order?

11.49 A heating contractor sends a repair person to homes in response to calls about heating problems. The contractor would like to have a way to estimate how long the

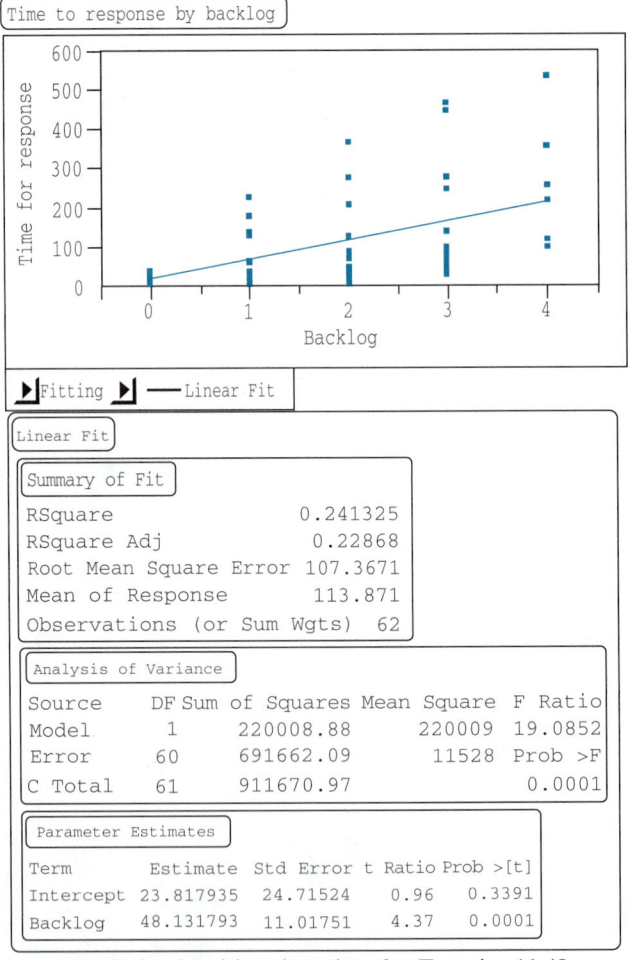

Analysis of waiting time data for Exercise 11.49

customer will have to wait before the repair person can begin work. Data on the number of minutes of wait and the backlog of previous calls waiting for service were obtained. A scatterplot and regression analysis of the data, obtained from JMP are shown on the previous page.

a. Calculate the predicted value and an approximate 95% prediction interval for the time to response of a call when the backlog is 6. Neglect the extrapolation penalty.

b. If we had calculated the extrapolation penalty, would it most likely be very small?

11.50 In the prediction interval of the previous exercise, is the calculated interval likely to be too narrow or too wide?

11.51 Here is some of the output of Exercise 11.11.

```
MTB > Regress 'y' 1 'x';
SUBC> Predict 1.301.

The regression equation is
y = 14.3 + 1.48 x

Predictor        Coef      Stdev     t-ratio        p
Constant      14.2917     0.7962       17.95    0.000
x              1.4750     0.1737        8.49    0.000

s = 1.346     R-sq = 87.8%     R-sq(adj) = 86.6%

   Fit     Stdev.Fit         95% C.I.              95% P.I.
43.792        2.807    ( 37.536, 50.047)    ( 36.854, 50.729) XX

X denotes a row with X values away from the center
XX denotes a row with very extreme X values

MTB > Regress 'y' 1 'logx';
SUBC> Predict 20.
```

```
The regression equation is
y = 14.9 + 10.5 logx

Predictor        Coef      Stdev     t-ratio        p
Constant      14.8755     0.6106       24.36    0.000
logx          10.522      1.021       10.30    0.000

s = 1.131     R-sq = 91.4%     R-sq(adj) = 90.5%

   Fit     Stdev.Fit         95% C.I.              95% P.I.
28.565        0.8756    ( 26.614, 30.516)    ( 25.378, 31.752) XX

X denotes a row with X values away from the center
XX denotes a row with very extreme X values
```

a. If the model with *y* a linear function of *x* is adopted, what is the predicted *y* when $x = 20$?

b. If the model with *y* a linear function of $x' = \log_{10} x$ is adopted, what is the predicted value of *y* when $x = 20$?

c. Which of the two predictions seems more reasonable (or perhaps less unreasonable)?

11.52 Give a 95% prediction interval for the prediction you selected in the preceding exercise.

11.5 Examining Lack of Fit in Linear Regression

In our study of linear regression, we have been concerned with how well a linear regression model $y = \beta_0 + \beta_1 x + \varepsilon$ fits, but only from an intuitive standpoint. We could examine a scatterplot of the data to see whether it looked linear and we could test whether the slope differed from 0; however, we had no way of testing to see whether a higher-order model would be a more appropriate model for the relationship between y and x. This section will outline situations in which we can test for the validity of a linear regression model.

Pictures (or graphs) are always a good starting point for examining lack of fit. First, use a scatterplot of y versus x. Second, a plot of residuals $y_i - \hat{y}_i$ versus predicted values $\hat{y}_i$ may give an indication of the following problems:

1. Outliers or erroneous observations. In examining the residual plot, your eye will naturally be drawn to data points with unusually high (in absolute value) residuals.
2. Violation of the assumptions. For the model $y = \beta_0 + \beta_1 x + \varepsilon$, we have assumed a linear relation between y and the dependent variable x, and independent, normally distributed errors with a constant variance.

The residual plot for a model and data set that has none of these apparent problems would look much like the plot in Figure 11.14. Note from this plot that there are no extremely large residuals (and hence no apparent outliers) and there is no trend in the residuals to indicate that the linear model is inappropriate. When a higher-order model is more appropriate, a residual plot more like that shown in Figure 11.15 would be observed.

FIGURE 11.14

Residual plot with no apparent pattern

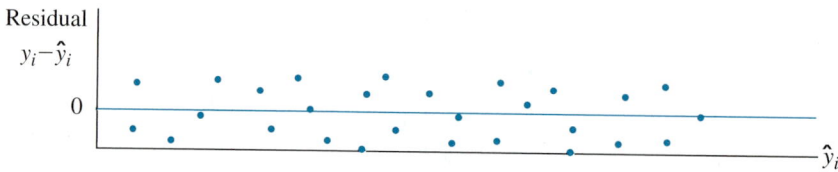

FIGURE 11.15

Residual plot showing the need for a higher-order model

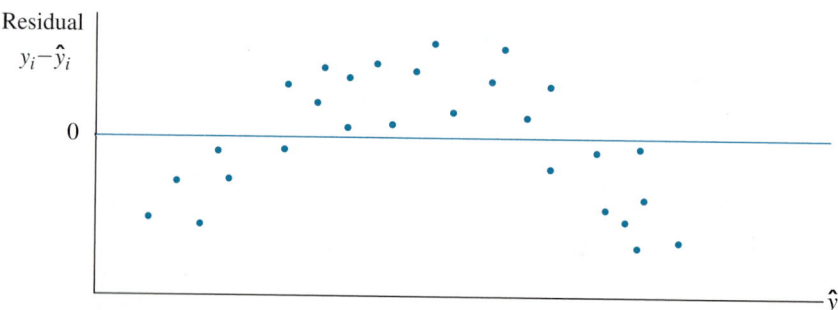

A check of the constant variance assumption can be addressed in the y versus x scatterplot or with a plot of the residuals ($y_i - \hat{y}_i$) versus x_i. For example, a pattern of residuals as shown in Figure 11.16 indicates homogeneous error variances across values of x; Figure 11.17 indicates that the error variances increase with increasing values of x.

FIGURE 11.16

Residual plot showing homogeneous error variances

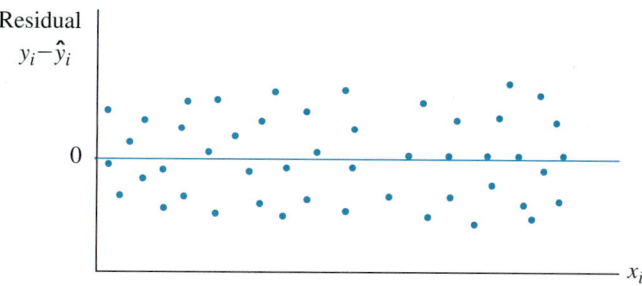

FIGURE 11.17

Residual plot showing error variances increasing with x

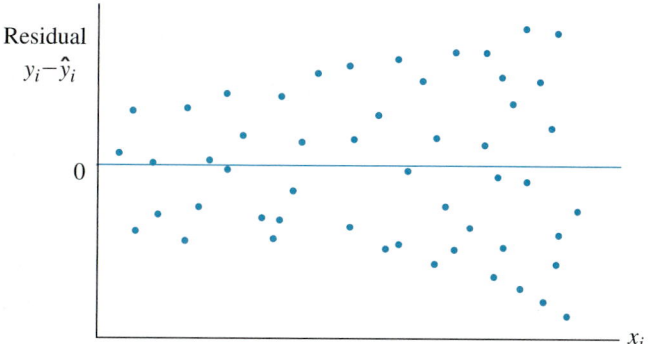

The question of independence of the errors and normality of the errors is addressed later in Chapter 13. We illustrate some of the points we have learned so far about residuals by way of an example.

EXAMPLE 11.11

The manufacturer of a new brand of thermal panes examined the amount of heat loss by random assignment of three different panes to each of the three outdoor temperature settings being considered. For each trial, the window temperature was controlled at 68°F and 50% relative humidity.

Outdoor Temperature (°F)	Heat Loss
20	86, 80, 77
40	78, 84, 75
60	33, 38, 43

a. Plot the data.
b. Fit the linear regression model $y = \beta_0 + \beta_1 x + \varepsilon$ and test $H_0: \beta_1 = 0$ (give the p-value for your test).
c. Compute $\hat{y}_i$ and $y_i - \hat{y}_i$ for the nine observations. Plot $y_i - \hat{y}_i$ versus $\hat{y}_i$.
d. Does the constant variance assumption seem reasonable?

Solution The computer output shown here can be used to address the four parts of this example.

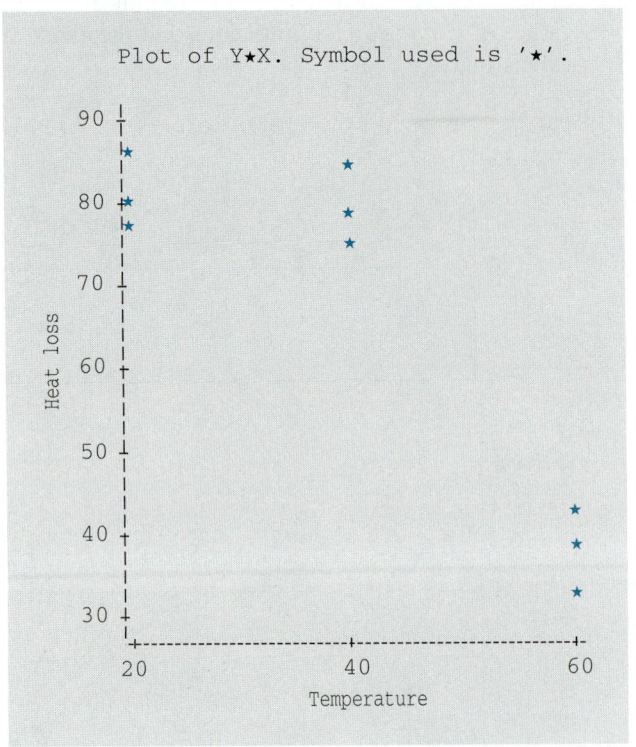

Plot of Y*X. Symbol used is '*'.

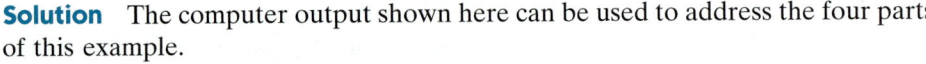

Dependent Variable: Y   HEAT LOSS

Analysis of Variance

Source	DF	Sum of Squares	Mean Square	F Value	Prob>F
Model	1	2773.50000	2773.50000	21.704	0.0023
Error	7	894.50000	127.78571		
C Total	8	3668.00000			

Root SE	11.30423	R-square	0.7561
Dep Mean	66.00000	Adj R-sq	0.7213
C.V.	17.12763		

Parameter Estimates

Variable	DF	Parameter Estimate	Standard Error	T for H0: Parameter=0	Prob > |T|
INTERCEP	1	109.000000	9.96939762	10.933	0.0001
X	1	-1.075000	0.23074672	-4.659	0.0023

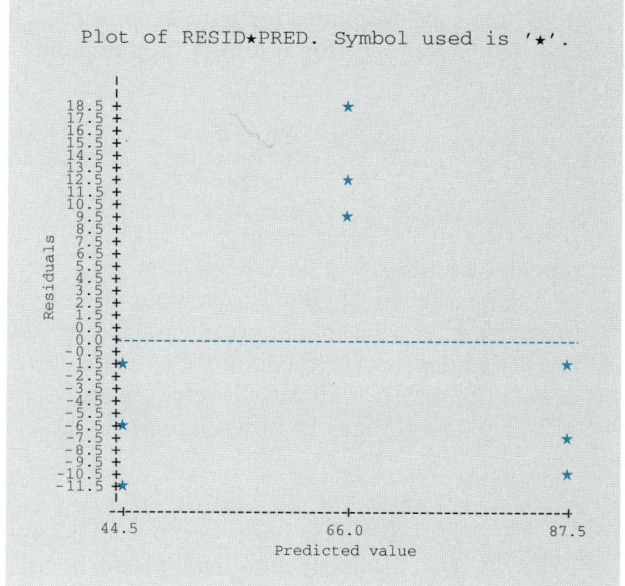

```
OBS      X      Y      PRED      RESID

 1      20     86      87.5      -1.5
 2      20     80      87.5      -7.5
 3      20     77      87.5     -10.5
 4      40     78      66.0      12.0
 5      40     84      66.0      18.0
 6      40     75      66.0       9.0
 7      60     33      44.5     -11.5
 8      60     38      44.5      -6.5
 9      60     43      44.5      -1.5
```

Plot of RESID*PRED. Symbol used is '*'.

a. The scatterplot of y versus x certainly shows a downward linear trend, and there may be evidence of curvature as well.

b. The linear regression model seems to fit the data well, and the test of $H_0: \beta_1 = 0$ is significant at the $p = .0023$ level. However, is this the best model for the data?

c. The plot of residuals $(y_i - \hat{y}_i)$ against the predicted values $\hat{y}_i$ is similar to Figure 11.15, suggesting that we may need additional terms in our model.

d. Because residuals associated with $x = 20$ (the first three), $x = 40$ (the second three), and $x = 60$ (the third three) are easily located, we really do not need a separate plot of residuals versus x to examine the constant variance assumption. It is clear from the original scatterplot and the residual plot shown that we do not have a problem.

How can we test for the apparent lack of fit of the linear regression model in Example 11.11? When there is more than one observation per level of the independent variable, we can conduct a test for lack of fit of the fitted model by partitioning SS (Residuals) into two parts, one **pure experimental error** and the **pure experimental error** other **lack of fit**. Let y_{ij} denote the response for the jth observation at the ith level

lack of fit

of the independent variable. Then, if there are n_i observations at the ith level of the independent variable, the quantity

$$\sum_j (y_{ij} - \bar{y}_i)^2$$

provides a measure of what we will call pure experimental error. This sum of squares has $n_i - 1$ degrees of freedom.

Similarly, for each of the other levels of x, we can compute a sum of squares due to pure experimental error. The pooled sum of squares

$$SSP_{exp} = \sum_{ij} (y_{ij} - \bar{y}_i)^2$$

called the sum of squares for pure experimental error, has $\sum_i (n_i - 1)$ degrees of freedom. With SS_{Lack} representing the remaining portion of SSE, we have

$$SS\,(\text{Residuals}) = \underset{\substack{\text{due to pure}\\\text{experimental}\\\text{error}}}{SSP_{exp}} + \underset{\substack{\text{due to lack}\\\text{to fit}}}{SS_{Lack}}$$

If SS (Residuals) is based on $n - 2$ degrees of freedom in the linear regression model, then SS_{Lack} will have df $= n - 2 - \sum_i (n_i - 1)$.

Under the null hypothesis that our model is correct, we can form independent estimates of σ_ε^2, the model error variance, by dividing SSP_{exp} and SS_{Lack} by their respective degrees of freedom; these estimates are called **mean squares** and are denoted by MSP_{exp} and MS_{Lack}, respectively.

mean squares

The test for lack of fit is summarized here.

A Test for Lack of Fit in Linear Regression

H_0: A linear regression model is appropriate.
H_a: A linear regression model is not appropriate.

T.S.: $F = \dfrac{MS_{Lack}}{MSP_{exp}}$,

where

$$MS_{exp} = \frac{SSP_{exp}}{\sum (n_i - 1)} = \frac{\sum_{ij} (y_{ij} - \bar{y}_i)^2}{\sum_i (n_i - 1)}$$

and

$$MS_{Lack} = \frac{SS\,(\text{Residuals}) - SSP_{exp}}{n - 2 - \sum (n_i - 1)}$$

R.R.: For specified value of α, reject H_0 (the adequacy of the model) if the computed value of F exceeds the table value for df$_1 = n - 2 - \sum_i (n_i - 1)$ and df$_2 = \sum_i (n_i - 1)$.

Conclusion: If the F test is significant, this indicates that the linear regression model is inadequate. A nonsignificant result indicates that there is insufficient evidence to suggest that the linear regression model is inappropriate.

EXAMPLE 11.12

Refer to the data of Example 11.11. Conduct a test for lack of fit of the linear regression model.

Solution It is easy to show that the contributions to experimental error for the differential levels of x are as shown here.

Level of x	$\bar{y}_i$	Contribution to Pure Experimental Error $\Sigma_i\,(y_{ij} - \bar{y}_i)^2$	$n_i - 1$
20	81	42	2
40	79	42	2
60	38	50	2
Total		134	6

Summarizing these results, we have

$$\text{SSP}_{\text{exp}} = \sum_{ij} (y_{ij} - \bar{y}_i)^2 = 134$$

The output shown for Example 11.11 gives SS (Residual) = 894.5; hence, by subtraction,

$$\text{SS}_{\text{Lack}} = \text{SS (Residual)} - \text{SSP}_{\text{exp}} = 894.5 - 134 = 760.5$$

The sum of squares due to pure experimental error has $\Sigma_i\,(n_i - 1) = 6$ degrees of freedom; it therefore follows that with $n = 9$, SS_{Lack} has $n - 2 - \Sigma_i\,(n_i - 1) = 1$ degree of freedom. We find that

$$\text{MSP}_{\text{exp}} = \frac{\text{SSP}_{\text{exp}}}{6} = \frac{134}{6} = 22.33$$

and

$$\text{MS}_{\text{Lack}} = \frac{\text{SS}_{\text{Lack}}}{1} = 760.5$$

The F statistic for the test of lack of fit is

$$F = \frac{\text{MS}_{\text{Lack}}}{\text{MSP}_{\text{exp}}} = \frac{760.5}{22.33} = 34.06$$

Using $\text{df}_1 = 1$, $\text{df}_2 = 6$, and $\alpha = .05$, we will reject H_0 if $F \geq 5.99$.

Because the computed value of F exceeds 5.99, we reject H_0 and conclude that there is significant lack of fit for a linear regression model. The scatterplot shown in Example 11.11 confirms this nonlinearity.

To summarize: In situations for which there is more than one y-value at one or more levels of x, it is possible to conduct a formal test for lack of fit of the linear regression model. This test should precede any inferences made using the fitted linear regression line. If the test for lack of fit is significant, some higher-order polynomial in x may be more appropriate. A scatterplot of the data and a residual plot from the linear regression line should help in selecting the appropriate model. More information on the selection of an appropriate model will be discussed along with multiple regression (Chapters 12 and 13).

If the F test for lack of fit is not significant, proceed with inferences based on the fitted linear regression line.

EXERCISES   **Applications**

Engin.   **11.53** A manufacturer of laundry detergent was interested in testing a new product prior to market release. One area of concern was the relationship between the height of the detergent suds in a washing machine as a function of the amount of detergent added in the wash cycle. For a standard size washing machine tub filled to the full level, the manufacturer made random assignments of amounts of detergent and tested them on the washing machine. The data appear next.

Height, y	Amount, x
28.1, 27.6	6
32.3, 33.2	7
34.8, 35.0	8
38.2, 39.4	9
43.5, 46.8	10

a. Plot the data.
b. Fit a linear regression model.
c. Use a residual plot to investigate possible lack of fit.

11.54 Refer to Exercise 11.53.
a. Conduct a test for lack of fit of the linear regression model.
b. If the model is appropriate, give a 95% prediction band for y.

11.6   The Inverse Regression Problem (Calibration)

In experimental situations, we are often interested in estimating the value of the independent variable corresponding to a measured value of the dependent variable. This problem will be illustrated for the case in which the dependent variable y is linearly related to an independent variable x.

Consider the calibration of an instrument that measures the flow rate of a chemical process. Let x denote the actual flow rate and y denote a reading on the calibrating instrument. In the calibration experiment, the flow rate is controlled at n levels x_i, and the corresponding instrument readings y_i are observed. Suppose we assume a model of the form

$$y_i = \beta_0 + \beta_1 x_i + \varepsilon_i$$

where the ε_is are independent, identically distributed normal random variables with mean zero and variance σ_ε^2. Then, using the n data points (x_i, y_i), we can obtain the least-squares estimates $\hat{\beta}_0$ and $\hat{\beta}_1$. Sometime in the future the experimenter will be interested in estimating the flow rate x from a particular instrument reading y.

The most commonly used estimate is found by replacing $\hat{y}$ by y and solving the least-squares equation $\hat{y} = \hat{\beta}_0 + \hat{\beta}_1 x$ for x:

$$\hat{x} = \frac{y - \hat{\beta}_0}{\hat{\beta}_1}$$

Two different inverse prediction problems will be discussed here. The first is for predicting x corresponding to an *observed* value of y; the second is for predicting x corresponding to the mean of $m > 1$ values of y that were obtained independent of the regression data. The solution to the first inverse problem is shown here.

Case 1: Predicting x Based on an Observed y-Value

Predictor of x: $\hat{x} = \dfrac{y - \hat{\beta}_0}{\hat{\beta}_1}$

$100(1 - \alpha)\%$ prediction limits for x:

$$\hat{x}_U = \bar{x} + \frac{1}{1 - c^2}[(\hat{x} - \bar{x}) + d]$$

$$\hat{x}_L = \bar{x} + \frac{1}{1 - c^2}[(\hat{x} - \bar{x}) - d]$$

where

$$d = \frac{t_{\alpha/2}s_\varepsilon}{\hat{\beta}_1}\sqrt{\frac{n+1}{n}(1 - c^2) + \frac{(\hat{x} - \bar{x})^2}{S_{xx}}}, \qquad s_\varepsilon^2 = \frac{SSE}{n - 2}, \qquad c^2 = \frac{t_{\alpha/2}^2 s_\varepsilon^2}{\hat{\beta}_1^2 S_{xx}}$$

and $t_{\alpha/2}$ is based on df $= n - 2$.

Note that because

$$t = \frac{\hat{\beta}_1}{s_\varepsilon/\sqrt{S_{xx}}}$$

is the test statistic for H_0: $\beta_1 = 0$, $c = t_{\alpha/2}/t$. We will require that $|t| > t_{\alpha/2}$; that is, β_1 must be significantly different from zero. Then $c^2 < 1$ and $0 < (1 - c^2) < 1$. The greater the strength of the linear relationship between x and y, the larger the quantity $(1 - c^2)$, making the width of the prediction interval narrower. Note also that we will get a better prediction of x when $\hat{x}$ is closer to the center of the experimental region, as measured by $\bar{x}$. Combining a prediction at an endpoint of the experimental region with a weak linear relationship between x and y ($t \approx t_{\alpha/2}$ and $c^2 < 1$) can create extremely wide limits for the prediction of x.

EXAMPLE 11.13

An engineer is interested in calibrating a flow meter to be used on a liquid-soap production line. For the test, 10 different flow rates are fixed and the corresponding meter readings observed. The data are shown here. Use these data to place a 95% prediction interval on x, the actual flow rate corresponding to an instrument reading of 4.0.

Solution   For these data, we find that $S_{xy} = 74.35$, $S_{xx} = 82.5$, and $S_{yy} = 67.065$. It follows that $\hat{\beta}_1 = 74.35/82.5 = .9012$, $\hat{\beta}_0 = \bar{y} - \hat{\beta}_1\bar{x} = 5.45 - (.9012)(5.5) = .4934$, and SS(Residual) $= S_{yy} - \hat{\beta}_1 S_{xy} = 67.065 - (.9012)(74.35) = .0608$. The estimate of σ_ε^2 is based on $n - 2 = 8$ degrees of freedom.

Data for the Calibration Problem	
Flow Rate, x	Instrument Reading, y
1	1.4
2	2.3
3	3.1
4	4.2
5	5.1
6	5.8
7	6.8
8	7.6
9	8.7
10	9.5

$$s_\varepsilon^2 = \frac{\text{SS(Residual)}}{n-2} = \frac{.0608}{8} = .0076$$

$$s_\varepsilon = .0872$$

For $\alpha = .05$, the t-value for df $= 8$ and $a = .025$ is 2.306.

$$c^2 = \frac{t_{\alpha/2}^2 s_\varepsilon^2}{\hat{\beta}_1^2 S_{xx}} = \frac{(2.306)^2(.0076)}{(.9012)^2(82.5)} = .0006$$

and $1 - c^2 = .9994$. Using $\hat{x} = 3.8910$, the upper and lower prediction limits for x when $y = 4.0$ are as follows:

$$\hat{x}_U = 5.5 + \frac{1}{.9994}\left[-1.6090 + \frac{2.306(.0872)}{.9012}\sqrt{\frac{11}{10}(.9994) + \frac{(-1.6090)^2}{82.5}}\right]$$

$$= 5.5 + \frac{1}{.9994}(-1.6090 + .2373) = 4.1274$$

$$\hat{x}_L = 5.5 + \frac{1}{.9994}(-1.6090 - .2373) = 3.6526$$

Thus, the 95% prediction limits for x are 3.65 to 4.13. These limits are shown in Figure 11.18.

FIGURE 11.18

95% prediction interval for x when $y = 4.0$

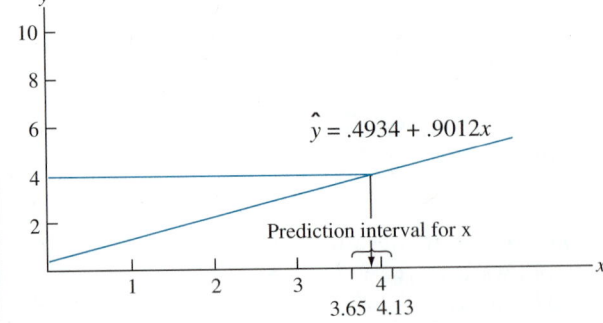

The solution to the second inverse prediction problem is summarized next.

Case 2: Predicting *x* Based on *m* *y*-Values

Predicting the value of x corresponding to $100P\%$ of the mean of m independent y values. For $0 \le P \le 1$,

$$\text{Predictor of } x: \hat{x} = \frac{P\bar{y}_m - \hat{\beta}_0}{\hat{\beta}_1}$$

$$\hat{x}_U = \bar{x} + \frac{1}{1 - c^2}[(\hat{x} - \bar{x}) + g]$$

$$\hat{x}_L = \bar{x} + \frac{1}{1 - c^2}[(\hat{x} - \bar{x}) - g]$$

where

$$g = \frac{t_{\alpha/2}}{\hat{\beta}_1} \sqrt{\left(s_{\bar{y}}^2 P^2 + \frac{s_\varepsilon^2}{n}\right)(1 - c^2) + \frac{(\hat{x} - \bar{x})^2 s_\varepsilon^2}{S_{xx}}}$$

and $\bar{y}_m$ and $s_{\bar{y}}$ are the mean and standard error, respectively, of m independent y-values.

We have now developed the techniques required to analyze the data from the case study that we introduced at the beginning of this chapter.

Analyzing Data from the *E. coli* Concentrations Case Study   The researchers were interested in assessing the degree to which the HEC and HGMF procedures agreed in determining the level of *E. coli* concentrations in meat samples. If there was a strong relationship between the two sets of readings, they would then also want to obtain the inverse regression equations so as to predict HGMF readings from the HEC readings obtained in the field. We will first obtain the regression relationship with HEC serving as the dependent variable and HGMF as the independent variable because the HGMF procedure has a known reliability in determining *E. coli* concentrations.

The computer output for analyzing the 17 pairs of *E. coli* concentrations is given here along with a plot of the residuals.

```
Dependent Variable: HEC                          HEC-METHOD

Analysis of Variance

                        Sum of        Mean
Source          DF      Squares       Square      F Value      Prob>F

Model            1      14.22159     14.22159     441.816      0.0001
Error           15       0.48283      0.03219
C Total         16      14.70442


        Root MSE       0.17941      R-square      0.9672
        Dep Mean       1.07471      Adj R-sq      0.9650
        C.V.          16.69413
```

Variable	DF	Parameter Estimate	Standard Error	T for H0: Parameter=0	Prob > |T|
INTERCEP	1	-0.023039	0.06797755	-0.339	0.7394
HGMF	1	0.915685	0.04356377	21.019	0.0001

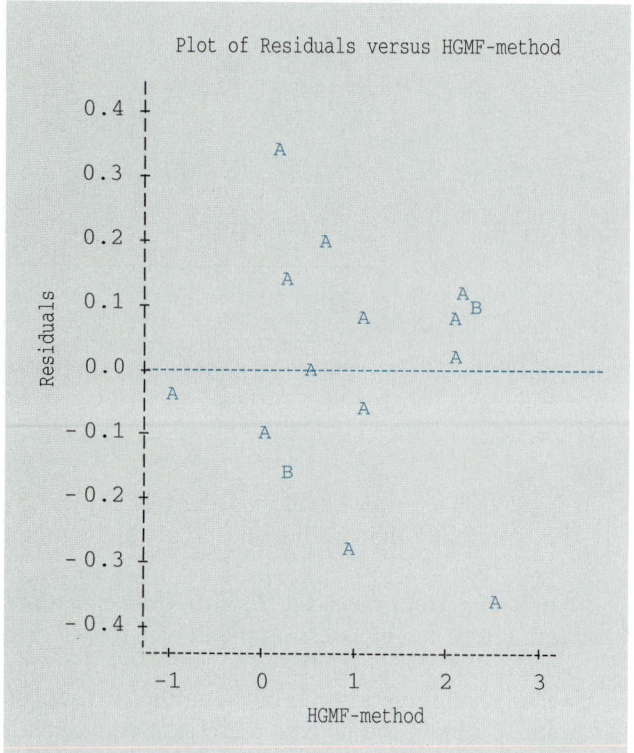

Plot of Residuals versus HGMF-method

The R^2 value of .9672 indicates a strong linear relationship between HEC and HGMF concentrations. An examination of the residual plots does not indicate the necessity for higher-order terms in the model or heterogeneity in the variances. The least-squares equation relating HEC to HGMF concentrations is given here.

$$\widehat{HEC} = -.023 + .9157 * HGMF$$

Thus, we can assess whether there is an exact relationship between the two methods of determining *E. coli* concentrations by testing the hypotheses:

$$H_0: \beta_0 = 0, \beta_1 = 1 \quad \text{versus} \quad H_0: \beta_0 \neq 0 \quad \text{or} \quad \beta_1 \neq 1$$

If H_0 were accepted, then we would have a strong indication that the relationship $\hat{H}EC = 0 + 1 * HGMF$ was valid. That is, HEC and HGMF were yielding essentially the same values for *E. coli* concentrations. From the output we have a *p*-value = .7394 for testing $H_0: \beta_0 = 0$ and we can test $H_0: \beta_1 = 1$ using the test statistic:

$$t = \frac{\hat{\beta}_1 - 1}{\widehat{SE}(\hat{\beta}_1)} = \frac{.915685 - 1}{.04356377} = -1.935$$

The *p*-value of the test statistic is $\Pr(|t_{15}| \geq 1.935) = .0721$. To obtain an overall α-value of .05, we evaluate the hypotheses of $H_0: \beta_1 = 1$ and $H_0: \beta_1 = 1$ individually using α = .025; that is, we reject either of the two hypotheses only if one of the p-values is less than .025. Because our p-values are .7394 and .0721, we fail to reject either null hypothesis and conclude that the data do not support the hypothesis that HEC and HGMF are yielding significantly different *E. coli* concentrations.

Because there were only 17 pairs of HEC and HGMF determinations, we will construct the calibration curves to determine the degree of accuracy to which HEC concentration readings would predict HGMF readings. Using the calibration equations, we obtain

$$\widehat{\text{HGMF}} = (\text{HEC} + .023)/.9157$$

with 95% prediction intervals

$$\widehat{\text{HGMF}}_L = 1.1988 + 1.0104 * (\widehat{\text{HGMF}} - 1.1988 - d)$$

$$\widehat{\text{HGMF}}_U = 1.1988 + 1.0104 * (\widehat{\text{HGMF}} - 1.1988 + d)$$

with $d = .4175 \sqrt{1.0479 + (\widehat{\text{HGMF}} - 1.1988)^2/16.9612}$.

We next plot $\widehat{\text{HGMF}}_L$ and $\widehat{\text{HGMF}}_U$ for HEC ranging from -1 to $+2$ to obtain an indication of the range of values that would be obtained in predicting HGMF readings from observed HEC readings.

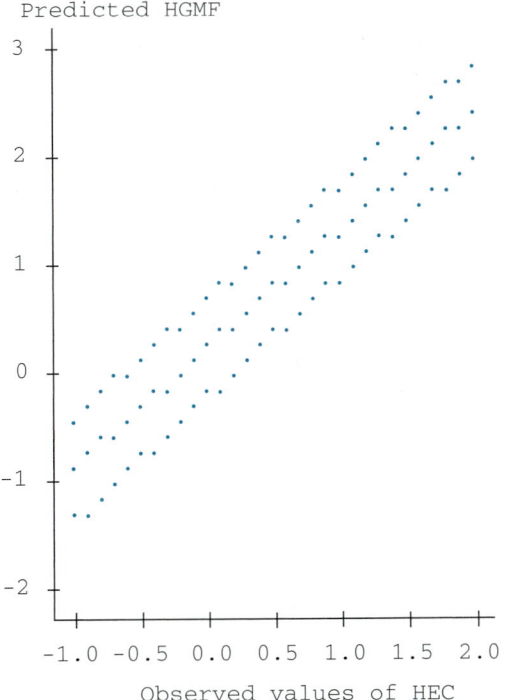

Plot of predicted HGMF for observed HEC with 95% prediction bounds

The width of the 95% prediction intervals was slightly less than one unit for most values of HEC. Thus, HEC determinations in the field of *E. coli* concentrations in the -1 to 2 range would result in 95% prediction intervals for the corresponding HGMF determinations. This degree of accuracy would not be acceptable. One way to reduce the width of the intervals would be to conduct an expanded study involving considerably more observations than the 17 obtained in this study, provided the same general degree of relationship held between HEC and HGMF in the new study.

Applications

Ag.   **11.55**   A forester has become adept at estimating the volume (in cubic feet) of trees on a particular site prior to a timber sale. Since his operation has now expanded, he would like to train another person to assist in estimating the cubic-foot volume of trees. He decides to calibrate his assistant's estimations of actual tree volume. The forester selects a random sample of trees soon to be felled. For each tree, the assistant is to guess the cubic-foot volume y. The forester also obtains the actual cubic-foot volume x after the tree has been chopped down. From these data, the forester obtains the calibration curve for the model

$$y = \beta_0 + \beta_1 x + \varepsilon$$

In the near future he can then use the calibration curve to correct the assistant's estimates of tree volumes. The sample data are summarized here.

Tree	1	2	3	4	5	6	7	8	9	10
Estimated volume, y	12	14	8	12	17	16	14	14	15	17
Actual volume, x	13	14	9	15	19	20	16	15	17	18

Fit the calibration curve using the method of least squares. Do the data indicate that the slope is significantly greater than 0? Use $\alpha = .05$.

11.56   Refer to Exercise 11.55.
   a. Predict the actual tree volume for a tree the assistant estimates to have a cubic-foot volume of 13.
   b. Place a 95% prediction interval on x, the actual tree volume in part (a).

Med.   **11.57**   A researcher obtains data from 24 patients to examine the relationship between dose (amount of drug) and cumulative urine volume (CUMVOL) for a drug product being studied as a diuretic. The data are shown here in the computer output. The initial fit of the data yielded a nonlinear relationship between dose and CUMVOL. The researcher decided on the transformations natural logarithm of dose and arcsine of the square root of CUMVOL/100, labeled as LOG (DOSE) and TRANS. CUMVOL on the output.
   a. Locate the linear regression equation. Identify the independent and dependent variables.
   b. Use the output to predict dose based on individual y values of 10, 14, and 19 cm³. What are the corresponding 99% prediction limits for each of those cases?

```
                    OUTPUT FOR EXERCISE 11.57
  OBS   DOSE    LOG (DOSE)   CUMVOL    TRANS. CUMVOL
   1    6.00    1.79176        7.1       0.26972
   2    6.00    1.79176       11.5       0.34598
   3    6.00    1.79176        8.4       0.29405
   4    6.00    1.79176        8.0       0.28676
   5    6.00    1.79176        9.4       0.31161
   6    6.00    1.79176       12.0       0.35374
   7    9.00    2.19722       13.2       0.37183
   8    9.00    2.19722       14.7       0.39348
   9    9.00    2.19722       12.7       0.36438
  10    9.00    2.19722       15.5       0.40465
  11    9.00    2.19722       18.4       0.44333
  12    9.00    2.19722       14.4       0.38923
```

13	13.50	2.60269	12.1	0.35528
14	13.50	2.60269	15.8	0.40878
15	13.50	2.60269	13.8	0.38061
16	13.50	2.60269	20.4	0.46863
17	13.50	2.60269	22.7	0.49661
18	13.50	2.60269	17.0	0.42499
19	20.25	3.00815	19.8	0.46114
20	20.25	3.00815	15.6	0.40603
21	20.25	3.00815	25.3	0.52706
22	20.25	3.00815	13.5	0.37624
23	20.25	3.00815	24.8	0.52129
24	20.25	3.00815	20.9	0.47481
25	10.00	2.30259	.	.
26	14.00	2.63906	.	.
27	19.00	2.94444	.	.

```
                       OUTPUT FOR EXERCISE 11.57

Dependent Variable: Y            TRANSFORMED CUMVOL

Analysis of Variance
```

		Sum of	Mean		
Source	DF	Squares	Square	F Value	Prob>F
Model	1	0.06922	0.06922	32.750	0.0001
Error	22	0.04650	0.00211		
C Total	23	0.11572			

Root MSE	0.04597	R-square	0.5982
Dep Mean	0.39709	Adj R-sq	0.5799
C.V.	11.57773		

Parameter Estimates

Variable	DF	Parameter Estimate	Standard Error	T for H0: Parameter=0	Prob > |T|
INTERCEP	1	0.112770	0.05056109	2.230	0.0362
X	1	0.118470	0.02070143	5.723	0.0001

OBS	X	Y	PRED	L95PRED	U95PRED	L95MEAN	U95MEAN
1	1.79176	0.26972	0.32504	0.22429	0.42579	0.29247	0.35761
2	1.79176	0.34598	0.32504	0.22429	0.42579	0.29247	0.35761
3	1.79176	0.29405	0.32504	0.22429	0.42579	0.29247	0.35761
4	1.79176	0.28676	0.32504	0.22429	0.42579	0.29247	0.35761
5	1.79176	0.31161	0.32504	0.22429	0.42579	0.29247	0.35761
6	1.79176	0.35374	0.32504	0.22429	0.42579	0.29247	0.35761
7	2.19722	0.37183	0.37307	0.27537	0.47077	0.35175	0.39439
8	2.19722	0.39348	0.37307	0.27537	0.47077	0.35175	0.39439
9	2.19722	0.36438	0.37307	0.27537	0.47077	0.35175	0.39439
10	2.19722	0.40465	0.37307	0.27537	0.47077	0.35175	0.39439
11	2.19722	0.44333	0.37307	0.27537	0.47077	0.35175	0.39439
12	2.19722	0.38923	0.37307	0.27537	0.47077	0.35175	0.39439

```
13  2.60269  0.35528  0.42111  0.32341  0.51881  0.39979  0.44243
14  2.60269  0.40878  0.42111  0.32341  0.51881  0.39979  0.44243
15  2.60269  0.38061  0.42111  0.32341  0.51881  0.39979  0.44243
16  2.60269  0.46863  0.42111  0.32341  0.51881  0.39979  0.44243
17  2.60269  0.49661  0.42111  0.32341  0.51881  0.39979  0.44243
18  2.60269  0.42499  0.42111  0.32341  0.51881  0.39979  0.44243
19  3.00815  0.46114  0.46914  0.36839  0.56990  0.43658  0.50171
20  3.00815  0.40603  0.46914  0.36839  0.56990  0.43658  0.50171
21  3.00815  0.52706  0.46914  0.36839  0.56990  0.43658  0.50171
22  3.00815  0.37624  0.46914  0.36839  0.56990  0.43658  0.50171
23  3.00815  0.52129  0.46914  0.36839  0.56990  0.43658  0.50171
24  3.00815  0.47481  0.46914  0.36839  0.56990  0.43658  0.50171
25  2.30259     .      0.38556  0.28816  0.48296  0.36565  0.40546
26  2.63906     .      0.42542  0.32757  0.52327  0.40341  0.44742
27  2.94444     .      0.46160  0.36152  0.56168  0.43118  0.49201

Sum of Residuals                          0
Sum of Squared Residuals             0.0465
Predicted Resid SS (Press)           0.0560
```

11.58 Refer to the output of Exercise 11.57. Suppose the investigator wanted to predict the dose of the diuretic that would produce a response equivalent to 50% (and 75%) of the response obtained from four patients treated with a known diuretic. Predict x and give appropriate limits for each of these situations.

11.7 Correlation

Once we have found the prediction line $\hat{y} = \hat{\beta}_0 + \hat{\beta}_1 x$, we need to measure how well it predicts actual values. One way to do so is to look at the size of the residual standard deviation in the context of the problem. About 95% of the prediction errors will be within $\pm 2s_\varepsilon$. For example, suppose we are trying to predict the yield of a chemical process, where yields range from 0.50 to 0.94. If a regression model had a residual standard deviation of 0.01, we could predict most yields within ± 0.02—fairly accurate in context. However, if the residual standard deviation were 0.08, we could predict most yields within ± 0.16, which is not very impressive given that the yield range is only $0.94 - 0.50 = 0.44$. This approach, though, requires that we know the context of the study well; an alternative, more general approach is based on the idea of correlation.

Suppose that we compare the squared prediction error for two prediction methods: one using the regression model, the other ignoring the model and always predicting the mean y value. In the road resurfacing example of previous sections, if we are given the mileage values x_i, we could use the prediction equation $\hat{y}_i = 2.0 + 3.0x_i$ to predict costs. The deviations of actual values from predicted values, the residuals, measure prediction errors. These errors are summarized by the sum of squared residuals, SS(Residual) $= \Sigma(y_i - \hat{y}_i)^2$, which is 44 for these data. For comparison, if we were not given the x_i values, the best squared error predictor of y would be the mean value $\bar{y} = 14$, and the sum of squared prediction errors would, in this case, be $\Sigma_i(y_i - \bar{y}_i)^2 =$ SS(Total) $= 224$. The proportionate reduction in error would be

$$\frac{\text{SS(Total)} - \text{SS(Residual)}}{\text{SS(Total)}} = \frac{224 - 44}{224} = .804$$

In words, use of the regression model reduces squared prediction error by 80.4%, which indicates a fairly strong relation between the mileage to be resurfaced and the cost of resurfacing.

correlation coefficient

This proportionate reduction in error is closely related to the **correlation coefficient** of x and y. A *correlation measures the strength of the linear relation between x and y.* The stronger the correlation, the better x predicts y.

Given n pairs of observations (x_i, y_i), we compute the sample correlation r as

$$r_{yx} = \frac{\Sigma (x_i - \bar{x})(y_i - \bar{y})}{\sqrt{S_{xx}S_{yy}}} = \frac{S_{xy}}{\sqrt{S_{xx}S_{yy}}}$$

where S_{xy} and S_{xx} are defined as before and

$$S_{yy} = \sum_i (y_i - \bar{y})^2 = SS(Total)$$

In the example,

$$r_{yx} = \frac{60}{\sqrt{(20)(224)}} = .896$$

Generally, the correlation r_{yx} is a positive number if y tends to increase as x increases; r_{yx} is negative if y tends to decrease as x increases; and r_{yx} is zero if there is either no relation between changes in x and changes in y, or there is a nonlinear relation such that patterns of increase and decrease in y (as x increases) cancel each other.

Figure 11.19 illustrates four possible situations for the values of r. In Figure 11.19 (d), there is a strong relationship between y and x but $r = 0$. This is a result of symmetric positive and negative nearly linear relationships canceling each other. When $r = 0$, there is not a "linear" relationship between y and x. However, higher-order (nonlinear) relationships may exist. This situation illustrates the importance of plotting the data in a scatterplot. In Chapter 12, we will develop techniques for modeling nonlinear relationships between y and x.

FIGURE 11.19
Interpretation of r

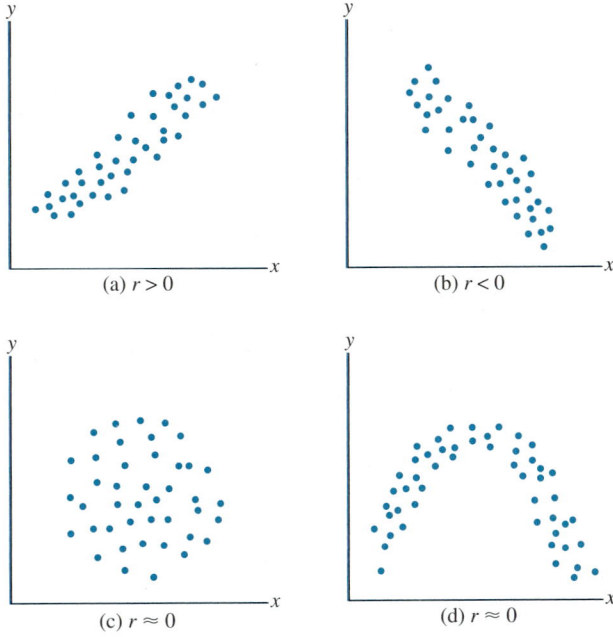

(a) $r > 0$ (b) $r < 0$ (c) $r \approx 0$ (d) $r \approx 0$

EXAMPLE 11.14

Consider the following data:

y:	25	41	47	59	54	56	49	43	30
x:	10	20	20	30	30	30	40	40	50

a. Should the correlation be positive or negative?
b. Calculate the correlation.

Solution

a. Note that as x increases from 10 to 50, y first increases and then de-creases. Therefore, the correlation should be small. The y values do not decrease quite back to where they started, so the correlation should be positive.

b. By easy calculation, the sample means are $\bar{x} = 30.0000$ and $\bar{y} = 44.8889$.

$$S_{xx} = (10 - 30.0000)^2 + \cdots + (50 - 30.0000)^2 = 1,200$$

$$S_{yy} = (25 - 44.8889)^2 + \cdots + (30 - 44.8889)^2 = 1,062.8889$$

$$S_{xy} = (10 - 30.0000)(25 - 44.8889) + \cdots + (50 - 30.0000)(30 - 44.8889)$$

$$= 140$$

$$r_{yx} = \frac{140}{\sqrt{(1,200)(1,062.8889)}} = .1240$$

The correlation is indeed a small positive number.

Correlation and regression predictability are closely related. The proportion-ate reduction in error for regression we defined earlier is called the **coefficient of determination**. The coefficient of determination is simply the square of the correlation coefficient,

coefficient of determination

$$r_{yx}^2 = \frac{SS(Total) - SS(Residual)}{SS(Total)}$$

which is the proportionate reduction in error. In the resurfacing example, $r_{yx} = .896$ and $r_{yx}^2 = .804$.

A correlation of zero indicates no predictive value in using the equation $y = \hat{\beta}_0 + \hat{\beta}_1 x$; that is, one can predict y as well without knowing x as one can knowing x. A correlation of 1 or -1 indicates perfect predictability—a 100% reduction in error attributable to knowledge of x. A correlation coefficient should routinely be interpreted in terms of its squared value, the coefficient of determina-tion. Thus, a correlation of $-.3$, say, indicates only a 9% reduction in squared prediction error. Many books and most computer programs use the equation

$$SS(Total) = SS(Residual) + SS(Regression)$$

where

$$SS(\text{Regression}) = \sum_i (\hat{y}_i - \bar{y})^2$$

Because the equation can be expressed as $SS(\text{Residual}) = (1 - r_{yx}^2)SS(\text{Total})$, it follows that $SS(\text{Regression}) = r_{yx}^2 SS(\text{Total})$, which again says that regression on x explains a proportion r_{yx}^2 of the total squared error of y.

EXAMPLE 11.15

Find SS(Total), SS(Regression), and SS(Residual) for the data of Example 11.14.

Solution $SS(\text{Total}) = S_{yy}$, which we computed to be 1,062.8889 in Example 11.14. We also found that $r_{yx} = .1240$, so $r_{yx}^2 = (.1240)^2 = .0154$. Using the fact that $SS(\text{Regression}) = r_{yx}^2 SS(\text{Total})$, we have $SS(\text{Regression}) = (.0154)(1,062.8889) = 16.3685$. Because $SS(\text{Residual}) = SS(\text{Total}) - SS(\text{Regression})$, $SS(\text{Residual}) = 1,062.8889 - 16.3685 = 1,046.5204$.

Note that SS(Regression) and r_{yx}^2 are very small. This suggests that x is not a good predictor of y. The reality, though, is that the relation between x and y is extremely nonlinear. A *linear* equation in x does not predict y very well, but a nonlinear equation would do far better.

What values of r_{yx} indicate a "strong" relationship between y and x? Figure 11.20 displays 15 scatterplots obtained by randomly selecting 1,000 pairs (x_i, y_i) from 15 populations having bivariate normal distributions with correlations ranging from -0.99 to 0.99. We can observe that unless $|r_{yx}|$ is greater than 0.6 there is very little trend in the plot.

The sample correlation r_{yx} is the basis for estimation and significance testing of the population correlation ρ_{yx}. Statistical inferences are always based on assumptions. The assumptions of regression analysis—linear relation between x and y and constant variance around the regression line, in particular—are also assumed in correlation inference. In regression analysis, we regard the x values as predetermined constants. In correlation analysis, we regard the x values as randomly selected (and the regression inferences are conditional on the sampled x values). If the xs are not drawn randomly, it is possible that the correlation estimates are biased. In some texts, the additional assumption is made that the x values are drawn from a normal population. The inferences we make do not depend crucially on this normality assumption.

assumptions for correlation inference

The most basic inference problem is potential bias in estimation of ρ_{yx}. A problem arises when the x values are predetermined, as often happens in regression analysis. The choice of x values can systematically increase or decrease the sample correlation. In general, a wide range of x values tends to increase the magnitude of the correlation coefficient and a small range to decrease it. This effect is shown in Figure 11.21. If all the points in this scatterplot are included, there is an obvious, strong correlation between x and y. Suppose, however, we consider only x values in the range between the dashed vertical lines. By eliminating the outside parts of the scatter diagram, the sample correlation coefficient (and the coefficient of determination) are much smaller. Correlation coefficients can be affected by systematic choices of x values; the residual standard deviation is *not* affected systematically, although it may change randomly if part of the x range changes. Thus, it is a good idea to consider the residual standard deviation s_ε and

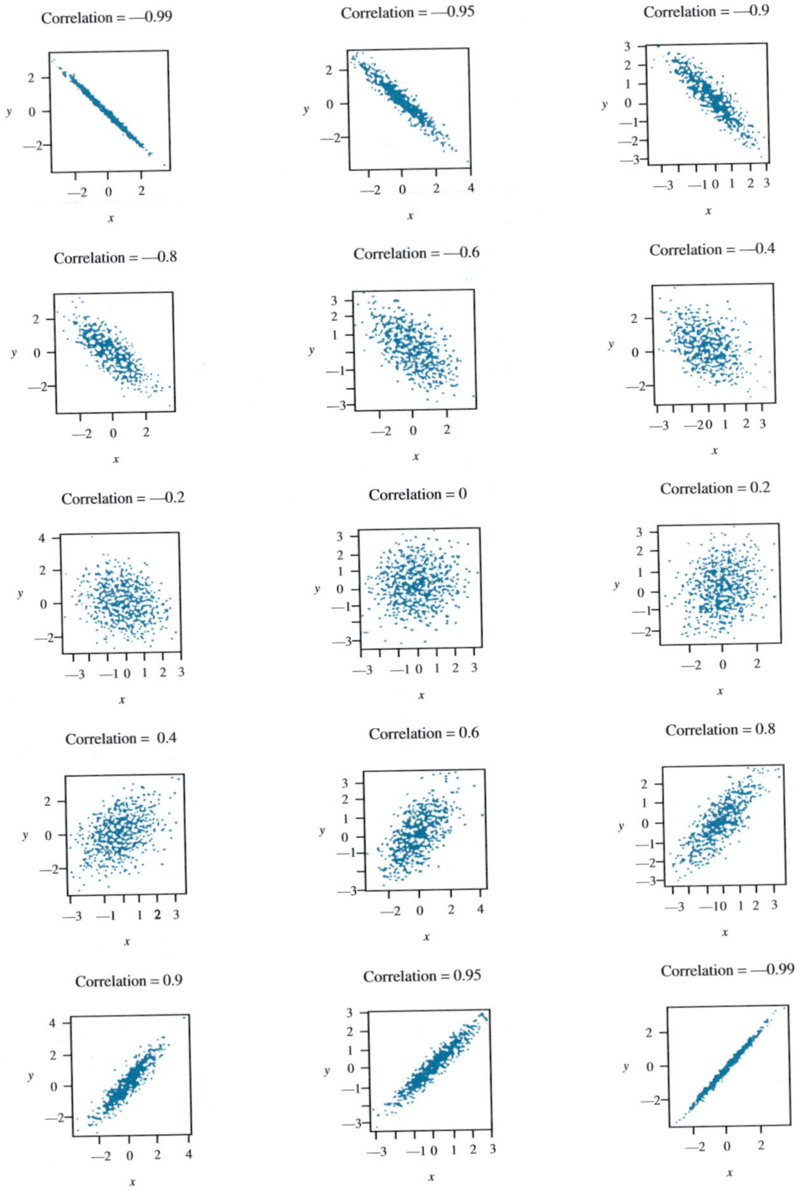

FIGURE 11.20

Samples of size 1,000 from
the bivariate normal
distribution

Correlation = —0.99

Correlation = —0.95

Correlation = —0.9

Correlation = —0.8

Correlation = —0.6

Correlation = —0.4

Correlation = —0.2

Correlation = 0

Correlation = 0.2

Correlation = 0.4

Correlation = 0.6

Correlation = 0.8

Correlation = 0.9

Correlation = 0.95

Correlation = —0.99

FIGURE 11.21

Effect of limited x range on
sample correlation coefficient

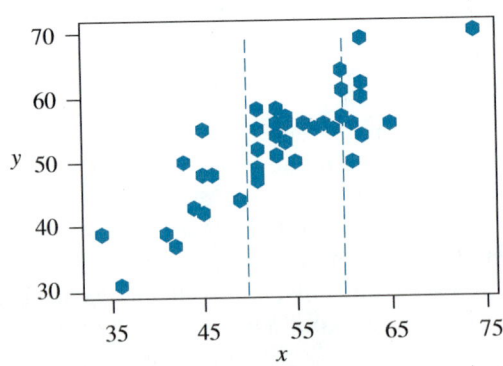

the magnitude of the slope when you decide how well a linear regression line predicts y.

EXAMPLE 11.16

Suppose that a company has the following data on productivity y and aptitude test score x for 12 data-entry operators:

```
y:  41  39  47  51  43  40  57  46  50  59  61  52
x:  24  30  33  35  36  36  37  37  38  40  43  49
```

Is the correlation larger or smaller if we consider only the last six values?

```
                    Simple Regression Analysis

    Linear model: y = 20.5394 + 0.775176*x

                      Table of Estimates

                               Standard      t          P
                   Estimate      Error     Value      Value
    Intercept      20.5394     10.7251     1.92      0.0845
    Slope          0.775176     0.289991    2.67      0.0234


    R-squared = 41.68%
    Correlation coeff. = 0.646
    Standard error of estimation = 5.99236


    File subset has been turned on, based on x>=37.

                    Simple Regression Analysis

    Linear model: y = 44.7439 + 0.231707*x

                      Table of Estimates

                               Standard      t          P
                   Estimate      Error     Value      Value
    Intercept      44.7439     24.8071     1.80      0.1456
    Slope          0.231707     0.606677    0.38      0.7219


    R-squared = 3.52%
    Correlation coeff. = 0.188
    Standard error of estimation = 6.34357
```

Solution For all 12 observations, the output shows a correlation coefficient of .646; the residual standard deviation is labeled as the standard error of estimation, 5.992. For the six highest x scores, shown as the subset having x greater than or equal to 37, the correlation is .188 and the residual standard deviation is 6.344. In going from all 12 observations to the six observations with the highest x values, the correlation has decreased drastically, but the residual standard deviation has hardly changed at all.

Just as we could run a statistical test for β_i, we can do it for ρ_{yx}.

Summary of a Statistical Test for ρ_{yx}

Hypotheses:
Case 1: $H_0: \rho_{yx} \leq 0$ vs. $H_0: \rho_{yx} > 0$
Case 2: $H_0: \rho_{yx} \geq 0$ vs. $H_a: \rho_{yx} < 0$
Case 3: $H_0: \rho_{yx} = 0$ vs. $H_a: \rho_{yx} \neq 0$

T.S.: $t = r_{yx} \dfrac{\sqrt{n-2}}{\sqrt{1 - r_{yx}^2}}$

R.R.: With $n - 2$ df and Type I error probability α,

1. $t > t_\alpha$
2. $t < -t_\alpha$
3. $|t| > t_{\alpha/2}$

Check assumptions and draw conclusions.

We tested the hypothesis that the true slope is zero (in predicting resurfacing cost from mileage) in Example 11.5; the resulting t statistic was 3.50. For those data, we can calculate r_{yx} as .896421 and r_{yx}^2 as .803571. Hence, the correlation t statistic is

$$\frac{.896\sqrt{3}}{\sqrt{1 - .803571}} = 3.50$$

An examination of the formulas for r and the slope $\hat{\beta}_1$ of the least-squares equation

$$\hat{y} = \hat{\beta}_0 + \hat{\beta}_1 x$$

yields the following relationship:

$$\hat{\beta}_1 = \frac{S_{xy}}{S_{xx}} = \frac{S_{xy}}{\sqrt{S_{xx}S_{yy}}} \sqrt{\frac{S_{yy}}{S_{xx}}} = r_{yx} \sqrt{\frac{S_{yy}}{S_{xx}}}$$

Thus, the t tests for a slope and for a correlation give identical results; it does not matter which form is used. It follows that the t test is valid for any choice of x values. The bias we mentioned previously does not affect the sign of the correlation.

EXAMPLE 11.17

Perform t tests for the null hypothesis of zero correlation and zero slope for the data of Example 11.18 (all observations). Use an appropriate one-sided alternative.

Solution   First, the appropriate H_a ought to be $\rho_{yx} > 0$ (and therefore $\beta_1 > 0$). It would be nice if an aptitude test had a positive correlation with the productivity score it was predicting! In Example 11.16, $n = 12$, $r_{yx} = .646$, and

$$t = \frac{.646\sqrt{12 - 2}}{\sqrt{1 - (.646)^2}} = 2.68$$

Because this value falls between the tabled t values for df $= 10$, $\alpha = .025(2.228)$ and for df $= 10$, $\alpha = .01(2.764)$, the p-value lies between .010 and .025. Hence, H_0 may be rejected.

The t statistic for testing the slope β_1 is shown in the output of Example 11.16 as 2.67, which equals (to within round-off error) the correlation t statistic, 2.68.

The test for a correlation provides a neat illustration of the difference between statistical significance and statistical importance. Suppose that a psychologist has devised a skills test for production-line workers and tests it on a huge sample of 40,000. If the sample correlation between test score and actual productivity is .02, then

$$t = \frac{.02\sqrt{39,998}}{\sqrt{1 - (.02)^2}} = 4.0$$

We would reject the null hypothesis at any reasonable α level, so the correlation is "statistically significant." However, the test accounts for only $(.02)^2 = .0004$ of the squared error in skill scores, so it is *almost* worthless as a predictor. Remember, the rejection of the null hypothesis in a statistical test is the conclusion that the sample results cannot plausibly have occurred by chance if the null hypothesis is true. The test itself does not address the practical significance of the result. Clearly, for a sample size of 40,000, even a trivial sample correlation like .02 is not likely to occur by mere luck of the draw. There is no practically meaningful relationship between these test scores and productivity scores in this example.

EXERCISES

11.59 The output of Exercise 11.19 is reproduced here. Calculate the correlation coefficient r_{yx} from the R-square (r_{yx}^2) value. Should its sign be positive or negative?

```
.regress Branches Business
   Source |      SS          df        MS              Number of obs =      12
----------+---------------------------------           F(1, 10)      =  172.60
    Model | 53.7996874        1     53.7996874         Prob > F       =  0.0000
 Residual | 3.11697922       10      .311697922        R-square       =  0.9452
----------+---------------------------------           Adj R-square   =  0.9398
    Total | 56.9166667       11     5.17424242         Root MSE       =   .5583
----------+---------------------------------------------------------------------

 Branches |   Coef.   Std. Err.        t    P>|t|      [95% Conf. Interval]
----------+---------------------------------------------------------------------
 Business | .0111049  .0008453    13.138  0.000       .0092216     .0129883
    _cons | 1.766846  .3211751     5.501  0.000       1.051223     2.482469
----------+---------------------------------------------------------------------
```

11.60   **a.** For the data in Exercise 11.59, test the hypothesis of no true correlation between x and y. Use a one-sided H_a and $\alpha = 0.05$.

   b. Compare the result of this test to the t test of the slope found in the output.

11.61 Refer to the computer output of Exercise 11.39 (reproduced here).

```
                    Simple Regression Analysis
Linear model: TotalCost = 99.777 + 5.19179+Runsize

                        Table of Estimates

                            Standard           t          P
              Estimate       Error           Value      Value

Slope         5.19179       0.0586455        88.53      0.0000
```

```
R-squared = 99.64%
Correlation coeff. = 0.998
Standard error of estimation = 12.2065

                          Analysis of Variance

                Sum of                                              P
Source          Squares        D.F.    Mean Square     F-Ratio     Value

Model         1.16775e+006       1     1.16775e+006    7837.26    0.0000
Error         4171.98           28     148.999

Total (corr.) 1.17192e+006      29
```

a. Locate r_{yx}^2. How is its very large value reflected in the Sum of Squares shown in the output?

b. The estimated slope $\hat{\beta}_1$ is positive; what must be the sign of the sample correlation coefficient?

c. Suppose that the study in Exercise 11.39 had been restricted to RUNSIZE values less than 1.8. Would you anticipate a larger or smaller r_{yx} value?

Bus. **11.62** Suppose that an advertising campaign for a new product is conducted in 10 test cities. The intensity of the advertising x, measured as the number of exposures per evening of prime-time television, is varied across cities; the awareness percentage y is found by survey after the ad campaign:

x:	4.0	4.5	5.0	5.5	6.0	6.5	7.0	7.5	8.0	8.5
y:	10.1	10.3	10.4	21.7	36.7	51.5	67.0	68.5	68.2	69.3

```
MTB > Correlation 'Intensty' 'Aware'.
Correlation of Intensty and Aware = 0.956
```

a. Interpret the correlation coefficient r_{yx}.

b. Plot the data. Does the relation appear linear to you? Does it appear to be generally increasing?

Edu. **11.63** A survey of recent M.B.A. graduates of a business school obtained data on first-year salary and years of prior work experience. The following results were obtained using the Systat package:

CASE	EXPER	SALARY	CASE	EXPER	SALARY
1	8.000	53.900	12	10.000	53.500
2	5.000	52.500	13	2.000	38.300
3	5.000	49.000	14	2.000	37.200
4	11.000	65.100	15	5.000	51.300
5	4.000	51.600	16	13.000	64.700
6	3.000	52.700	17	1.000	45.300
7	3.000	44.500	18	5.000	47.000
8	3.000	40.100	19	1.000	43.800
9	0.000	41.100	20	5.000	47.400
10	13.000	66.900	21	5.000	40.200
11	14.000	37.900	22	7.000	52.800

CASE	EXPER	SALARY	CASE	EXPER	SALARY
23	4.000	40.700	38	2.000	50.600
24	3.000	47.300	39	4.000	41.800
25	3.000	43.700	40	1.000	44.400
26	7.000	61.800	41	5.000	46.600
27	7.000	51.700	42	1.000	43.900
28	9.000	56.200	43	4.000	45.000
29	6.000	48.900	44	1.000	37.900
30	6.000	51.900	45	2.000	44.600
31	4.000	36.100	46	7.000	46.900
32	6.000	53.500	47	5.000	47.600
33	5.000	50.400	48	1.000	43.200
34	1.000	38.700	49	1.000	41.600
35	13.000	60.100	50	0.000	39.200
36	1.000	38.900	51	1.000	41.700
37	6.000	48.400			

a. By scanning the numbers, can you sense there is a relation? In particular, does it appear that those with less experience have smaller salaries?

b. Can you notice any cases that seem to fall outside the pattern?

11.64 The data in Exercise 11.63 were plotted by Systat's "influence plot." This plot is a scatterplot, with each point identified as to how much its removal would change the correlation. The larger the point, the more its removal would change the correlation. The plot is shown in the figure. Does there appear to be an increasing pattern in the plot? Do any points clearly fall outside the basic pattern?

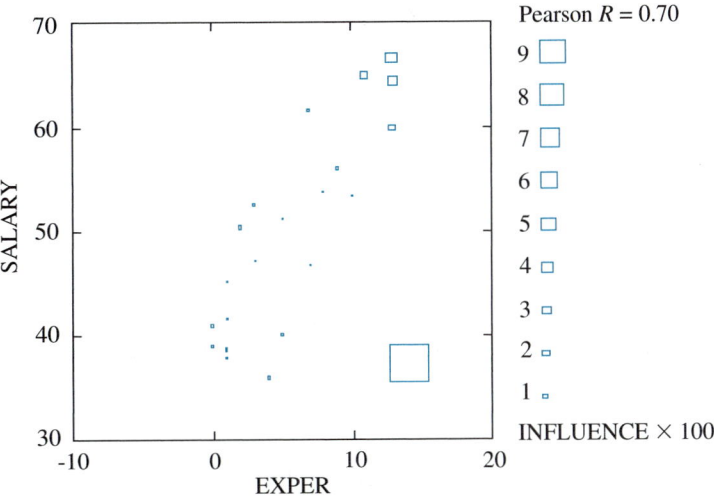

11.65 Systat computed a regression equation with salary from Exercise 11.64 as the dependent variable. A portion of the output is shown here:

```
DEP VAR: SALARY  N: 51  MULTIPLE R: 0.703 SQUARED MULTIPLE R: 0.494
ADJUSTED SQUARED MULTIPLE R:.484   STANDARD ERROR OF ESTIMATE: 5.402
VARIABLE     COEFFICIENT    STD ERROR     STD COEF    T    P(2 TAIL)
CONSTANT        40.507        1.257         0.000   32.219   0.000
EXPER            1.470        0.213         0.703    6.916   0.000
```

```
                     ANALYSIS OF VARIANCE

   SOURCE      SUM-OF-SQUARES   DF   MEAN-SQUARE    F-RATIO     P
   REGRESSION      1395.959     1     1395.959      47.838   0.000
   RESIDUAL        1429.868    49       29.181
```

a. Write out the prediction equation. Interpret the coefficients. Is the constant term (intercept) meaningful in this context?

b. Locate the residual standard deviation. What does the number mean?

c. Is the apparent relation statistically detectable (significant)?

d. How much of the variability in salaries is accounted for by variation in years of prior work experience?

11.66   The 11th person in the data of Exercise 11.63 went to work for a family business in return for a low salary but a large equity in the firm. This case (the high influence point in the influence plot) was removed from the data and the results reanalyzed using Systat. A portion of the output follows:

```
DEP VAR: SALARY  N: 50  MULTIPLE R: 0.842  SQUARED MULTIPLE R: 0.709
ADJUSTED SQUARED MULTIPLE R:.703    STANDARD ERROR OF ESTIMATE: 4.071
VARIABLE      COEFFICIENT    STD ERROR     STD COEF    T    P(2 TAIL)
CONSTANT         39.188        0.971        0.000   40.353   0.000
EXPER             1.863        0.172        0.842   10.812   0.000
```

a. Should removing the high influence point in the plot increase or decrease the slope? Did it?

b. In which direction (larger or smaller) should the removal of this point change the residual standard deviation? Did it? How large was the change?

c. How should the removal of this point change the correlation? How large was this change?

11.8   Summary

This chapter introduces regression analysis and is devoted to simple regression, using only one independent variable to predict a dependent variable. The basic questions involve the nature of the relation (linear or curved), the amount of variability around the predicted value, whether that variability is constant over the range of prediction, how useful the independent variable is in predicting the dependent variable, and how much to allow for sampling error. The key concepts of the chapter include the following:

1. The data should be plotted in a scatterplot. A smoother such as LOWESS or a spline curve is useful in deciding whether a relation is nearly linear or is clearly curved. Curved relations can often be made nearly linear by transforming either the independent variable or the dependent variable or both.

2. The coefficients of a linear regression are estimated by least squares, which minimizes the sum of squared residuals (actual values minus predicted). Because squared error is involved, this method is sensitive to outliers.

3. Observations that are extreme in the x (independent variable) direction have high leverage in fitting the line. If a high leverage point also falls well off the line, it has high influence, in that removing the observation substantially changes the fitted line. A high influence point should be omitted if it comes from a different population than the remainder. If it must be kept in the data, a method other than least squares should be used.

4. Variability around the line is measured by the standard deviation of the residuals. This standard deviation may be interpreted using the Empirical Rule. The standard deviation sometimes increases as the predicted value increases. In such a case, try transforming the dependent variable.

5. Hypothesis tests and confidence intervals for the slope of the line (and, less interestingly, the intercept) are based on the t distribution. If there is no relation, the slope is 0. The line is estimated most accurately if there is a wide range of variation in the x variable.

6. The fitted line may be used to forecast at a new x value, again using the t distribution. This forecasting is potentially inaccurate if the new x value is extrapolated far from the previous ones.

7. A standard method of measuring the strength of relation is the coefficient of determination, the square of the correlation. This measure is diminished by nonlinearity or by an artificially limited range of x variation.

One of the most important uses of statistics for managers is prediction. A manager may want to forecast the cost of a particular contracting job given the size of that job, to forecast the sales of a particular product given the current rate of growth of the gross national product, or to forecast the number of parts that will be produced given a certain size workforce. The statistical method most widely used in making predictions is *regression analysis.*

In the regression approach, past data on the relevant variables are used to develop and evaluate a prediction equation. The variable that is being predicted by this equation is the dependent variable. A variable that is being used to make the prediction is an independent variable. In this chapter, we discuss regression methods involving a single independent variable. In Chapter 12, we extend these methods to multiple regression, the case of several independent variables.

A number of tasks can be accomplished in a regression study:

1. The data can be used to obtain a prediction equation.
2. The data can be used to estimate the amount of variability or uncertainty around the equation.
3. The data can be used to identify unusual points far from the predicted value, which may represent unusual problems or opportunities.
4. Because the data are only a sample, inferences can be made about the true (population) values for the regression quantities.
5. The prediction equation can be used to predict a reasonable range of values for future values of the dependent variable.
6. The data can be used to estimate the degree of correlation between dependent and independent variables, a measure that indicates how strong the relation is.

Key Formulas

1. Least-squares estimates of slope and intercept

$$\hat{\beta}_1 = \frac{S_{xy}}{S_{xx}}$$

and

$$\hat{\beta}_0 = \bar{y} - \hat{\beta}_1\bar{x}$$

where

$$S_{xy} = \sum_i (x_i - \bar{x})(y_i - \bar{y})$$

and

$$S_{xx} = \sum_i (x_i - \bar{x})^2$$

2. Estimate of σ_ε^2

$$s_\varepsilon^2 = \frac{\sum_i (y_i - \hat{y}_i)^2}{n - 2}$$

$$= \frac{SS(\text{Residual})}{n - 2}$$

3. Statistical test for β_1

$$H_0: \beta_1 = 0 \text{ (two-tailed)}$$

$$\text{T.S.}: t = \frac{\hat{\beta}_1}{\dfrac{s_\varepsilon}{\sqrt{S_{xx}}}}$$

4. Confidence interval for β_1

$$\hat{\beta}_1 \pm t_{\alpha/2}\, s_\varepsilon \sqrt{\frac{1}{S_{xx}}}$$

5. F test for $H_0: \beta_1 = 0$ (two-tailed)

$$\text{T.S.}: F = \frac{MS(\text{Regression})}{MS(\text{Residual})}$$

6. Confidence interval for $E(y_{n+1})$

$$\hat{y}_{n+1} \pm t_{\alpha/2}s_\varepsilon \sqrt{\frac{1}{n} + \frac{(x_{n+1} - \bar{x})^2}{S_{xx}}}$$

7. Prediction interval for y_{n+1}

$$\hat{y}_{n+1} \pm t_{\alpha/2}s_\varepsilon \sqrt{1 + \frac{1}{n} + \frac{(x_{n+1} - \bar{x})^2}{S_{xx}}}$$

8. Test for lack of fit in linear regression

$$\text{T.S.}: = \frac{MS_{\text{Lack}}}{MS_{\text{exp}}}$$

where

$$MS_{\text{Lack}} = \frac{SSP_{\text{exp}}}{\sum_i (n_i - 1)}$$

$$= \frac{\sum_{ij} (y_{ij} - \bar{y}_i)^2}{\sum_i (n_i - 1)}$$

and

$$MS_{\text{Lack}} = \frac{SS(\text{Residual}) - SSP_{\text{exp}}}{(n - 2) - \sum_i (n_i - 1)}$$

9. Prediction limits for x based on a single y value

$$\hat{x} = \frac{y - \hat{\beta}_0}{\hat{\beta}_1}$$

$$\hat{x}_U = \bar{x} + \frac{1}{1 - c^2}[(\hat{x} - \bar{x}) + d]$$

$$\hat{x}_L = \bar{x} + \frac{1}{1 - c^2}[(\hat{x} - \bar{x}) - d]$$

where

$$c^2 = \frac{t_{\alpha/2}^2 s_\varepsilon^2}{\hat{\beta}_1^2 S_{xx}}$$

and

$$d = \frac{t_{\alpha/2}s_\varepsilon}{\hat{\beta}_1} \sqrt{\frac{n + 1}{n}(1 - c^2) + \frac{(\hat{x} - \bar{x})^2}{S_{xx}}}$$

10. Prediction interval for x based on m y-values

$$\hat{x}_U = \bar{x} + \frac{1}{1 - c^2}[(\hat{x} - \bar{x}) + g]$$

$$\hat{x}_L = \bar{x} + \frac{1}{1 - c^2}[(\hat{x} - \bar{x}) - g]$$

where

$$\hat{x} = \frac{P\bar{y}_m - \hat{\beta}_0}{\hat{\beta}_1}$$

and

$$g = \frac{t_{\alpha/2}}{\hat{\beta}_1} \sqrt{\left(s_{\hat{y}}^2 P^2 + \frac{s_\varepsilon^2}{n}\right)(1 - c^2) + \frac{(\hat{x} - \bar{x})^2 s_\varepsilon^2}{S_{xx}}}$$

11. Correlation coefficient

$$r_{yx} = \frac{S_{xy}}{\sqrt{S_{xx} S_{yy}}} = \hat{\beta}_1 \sqrt{\frac{S_{xx}}{S_{yy}}}$$

12. Coefficient of determination

$$r_{yx}^2 = \frac{\text{SS(Total)} - \text{SS(Residual)}}{\text{SS(Total)}}$$

13. Statistical test for ρ_{yx}

$$H_0: \rho_{yx} = 0 \text{ (two-tailed)}$$

$$\text{T.S.:} t = r_{yx} \frac{\sqrt{n-2}}{\sqrt{1 - r_{yx}^2}}$$

Supplementary Exercises

11.67 Consider the data shown here:

x:	10	12	14	15	18	19	23
y:	25	30	36	37	42	50	55

a. Plot the data.
b. Using the data, find the least-squares estimates for the model $y_i = \beta_0 + \beta_1 x_i + \varepsilon_i$.
c. Predict y when $x = 21$.

11.68 Refer to Exercise 11.67.
a. Calculate s_ε, the residual standard deviation.
b. Compute the residuals for these data. Do most lie within $\pm 2 s_\varepsilon$ of zero?

Gov. **11.69** A government agency responsible for awarding contracts for much of its research work is under careful scrutiny by a number of private companies. One company examines the relationship between the amount of the contract ($\times$ \$10,000) and the length of time between the submission of the contract proposal and contract approval:

Length (in months) y:	3	4	6	8	11	14	20
Size ($\times$\$10,000) x:	1	5	10	50	100	500	1000

A plot of y versus x and Stata output follow:

```
.regress Length Size

  Source |       SS       df       MS              Number of obs =       7
---------+------------------------------           F( 1,     5) =   33.78
   Model |  191.389193     1   191.389193          Prob > F      =  0.0021
Residual |  28.3250928     5   5.66501856          R-square      =  0.8711
---------+------------------------------           Adj R-square  =  0.8453
   Total |  219.714286     6   36.6190476          Root MSE      =  2.3801

------------------------------------------------------------------------------
  Length |      Coef.   Std. Err.       t     P>|t|     [95% Conf. Interval]
---------+--------------------------------------------------------------------
    Size |   .0148652   .0025575     5.812    0.002      .008291    .0214394
   _cons |   5.890659   1.086177     5.423    0.003     3.098553    8.682765
------------------------------------------------------------------------------
```

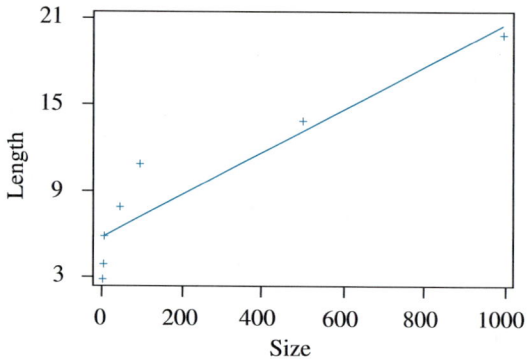

a. What is the least-squares line?

b. Conduct a test of the null hypothesis $H_0: \beta_1 \leq 0$. Give the p-value for your test, assuming $H_a: \beta_1 > 0$.

11.70 Refer to the data of Exercise 11.69. A plot of y versus the (natural) logarithm of x is shown and more Stata output is given here:

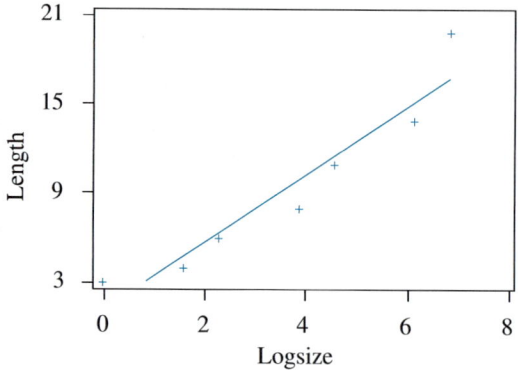

```
.regress Length lnSize

  Source |       SS          df          MS          Number of obs  =        7
---------+------------------------------          F( 1,    5) =      49.20
   Model | 199.443893          1   199.443893       Prob > F     =   0.0009
Residual |  20.2703932         5   4.05407863       R-square     =   0.9077
---------+------------------------------          Adj R-square =   0.8893
   Total | 219.714286          6   36.6190476       Root MSE     =   2.0135

-----------------------------------------------------------------------------
  Length |     Coef.    Std. Err.      t      P>|t|     [95% Conf. Interval]
---------+-------------------------------------------------------------------
  lnSize |   2.307015   .3289169     7.014    0.000     1.461508    3.152523
   _cons |   1.007445   1.421494     0.709    0.510    -2.646622    4.661511
-----------------------------------------------------------------------------
```

a. What is the regression line using log x as the independent variable?

b. Conduct a test of $H_0: \beta_1 \leq 0$, and give the level of significance for a one-sided alternative, $H_a: \beta_1 > 0$.

11.71 Use the results of Exercises 11.69 and 11.70 to determine which regression model provides the better fit. Give reasons for your choice.

11.72 Refer to the outputs of the previous two exercises.
a. Give a 95% confidence interval for β_1, the slope of the linear regression line.
b. Locate a 95% confidence interval for the slope in the logarithm model.

11.73 Use the model you prefer for the data of Exercise 11.70 to predict the length of time in months before approval of a $750,000 contract. Give a rough estimate of a 95% prediction interval.

Env. **11.74** An airline studying fuel usage by a certain type of aircraft obtains data on 100 flights. The air mileage x in hundreds of miles and the actual fuel use y in gallons are recorded. Statistix output follows and a plot is shown.

```
UNWEIGHTED LEAST SQUARES LINEAR REGRESSION OF GALLONS


PREDICTOR
VARIABLES      COEFFICIENT    STD ERROR      STUDENT'S T       P


CONSTANT          140.074       44.1293          3.17         0.0099
MILES             0.61896       0.04855         12.75         0.0000


R-SQUARED              0.9420    RESID. MEAN SQUARE (MSE)      1182.34
ADJUSTED R-SQUARED     0.9362    STANDARD DEVIATION            34.3852


SOURCE          DF        SS            MS          F         P


REGRESSION       1     1.921E+05    1.921E+05    162.48    0.0000
RESIDUAL        10       11823.4      1182.34
TOTAL           11     2.039E+05

PREDICTED/FITTED VALUES OF GALLONS


LOWER PREDICTED BOUND      678.33     LOWER FITTED BOUND     733.68
PREDICTED VALUE            759.03     FITTED VALUE           759.03
UPPER PREDICTED BOUND      839.73     UPPER FITTED BOUND     784.38
SE (PREDICTED VALUE)       36.218     SE (FITTED VALUE)      11.377


UNUSUALNESS (LEVERAGE)     0.1095
PERCENT COVERAGE            95.0
CORRESPONDING T             2.23


PREDICTOR VALUES: MILES = 1000.0
```

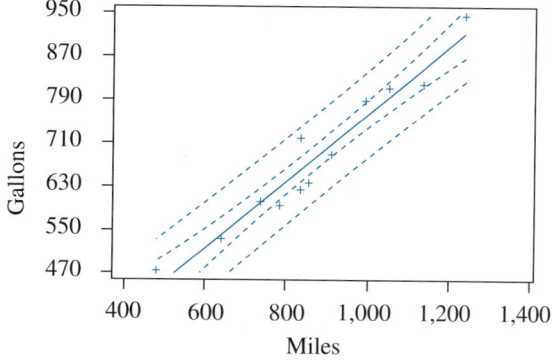

a. Locate the regression equation.

b. What are the sample correlation coefficient and coefficient of determination? Interpret these numbers.

c. Is there any point in testing $H_0: \beta_1 \leq 0$?

11.75 Refer to the data and output of Exercise 11.74.

a. Predict the mean fuel usage of all 1,000-mile flights. Give a 95% confidence interval.

b. Predict the fuel usage of a particular 1,000-mile flight. Would a usage of 628 gallons be considered exceptionally low?

11.76 What is the interpretation of $\hat{\beta}_1$ in the situation of Exercise 11.74? Is there a sensible interpretation of $\hat{\beta}_0$?

Bus. **11.77** A large suburban motel derives income from room rentals and purchases in its restaurant and lounge. It seems very likely that there should be a relation between room occupancy and restaurant/lounge sales, but the manager of the motel does not have a sense of how close that relation is. Data were collected for 36 nonholiday weekdays (Monday through Thursday nights) on the number of rooms occupied and the restaurant/lounge sales. A scatterplot of the data and regression results are shown.

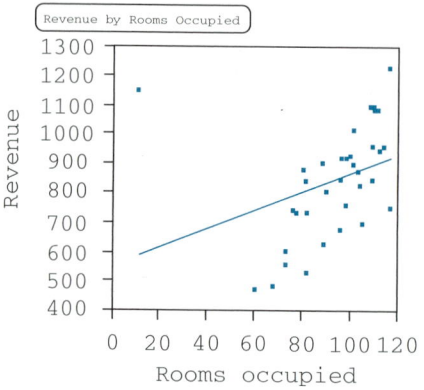

Linear Fit

Summary of Fit

RSquare	0.118716
RSquare Adj	0.092796
Root Mean Square Error	182.253
Mean of Response	854.1514
Observations (or Sum Wgts)	36

Analysis of Variance

Source	DF	Sum of Squares	Mean Square	F Ratio
Model	1	152132.2	152132	4.5801
Error	34	1129349.3	33216	Prob>F
C Total	35	1281481.6		0.0396

Parameter Estimates

Term	Estimate	Std Error	t Ratio	Prob>[t]
Intercept	557.72428	141.8019	3.93	0.0004
Rooms occupied	3.1760047	1.484039	2.14	0.0396

a. According to the output, is there a statistically significant relation between rooms occupied and revenue?

b. If the point at the upper left of the scatterplot is deleted, will the slope increase or decrease? Do you expect a substantial change?

11.78 One point in the hotel data was a data-entry error, with occupancy listed as 10 rather than 100. The error was corrected, leading to the output shown.

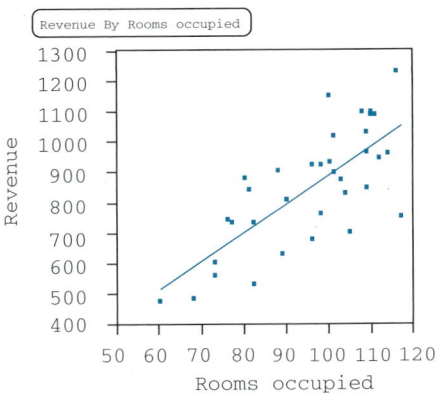

Linear Fit

Summary of Fit

RSquare	0.552922
RSquare Adj	0.539773
Root Mean Square Error	129.81
Mean of Response	854.1514
Observations (or Sum Wgts)	36

Analysis of Variance

Parameter Estimates

Term	Estimate	Std Error	t Ratio	Prob>|t|
Intercept	-50.18525	141.1283	-0.36	0.7243
Rooms occupied	9.4365563	1.455236	6.48	0.0000

a. How has the slope changed as a result of the correction?
b. How has the intercept changed?
c. Did the outlier make the residual standard deviation (root mean square error) larger or smaller?
d. Did the outlier make the r^2 value larger or smaller?

Engin.

11.79 The management science staff of a grocery products manufacturer is developing a linear programming model for the production and distribution of its cereal products. The model requires transportation costs for a very large number of origins and destinations. It is impractical to do the detailed tariff analysis for every possible combination, so a sample of 50 routes is selected. For each route, the mileage x and shipping rate y (in dollars per 100 pounds) are found. A regression analysis is performed, yielding the following scatterplot and Excel output:

	A	B	C	D	E	F	G
1	SUMMARY OUTPUT						
2							
3	Regression Statistics						
4	Multiple R	0.9929					
5	R Square	0.9859					
6	Adjusted R Square	0.9856					
7	Standard Error	2.2021					
8	Observations	48					

	A	B	C	D	E	F	G
9							
10							
11	ANOVA						
12		df	SS	MS	F	Significance F	
13	Regression	1	15558.63	15558.6	3208.47	0.00	
14	Residual	46	223.06	4.85			
15	Total	47	15781.7				
16							
17							
18		Coefficients	Standard Error	t Stat	P-value	Lower 95%	Upper 95%
19	Intercept	9.7709	0.4740	20.6122	0.0000	8.8167	10.7251
20	Mileage	0.0501	0.0009	56.6434	0.0000	0.0483	0.0519

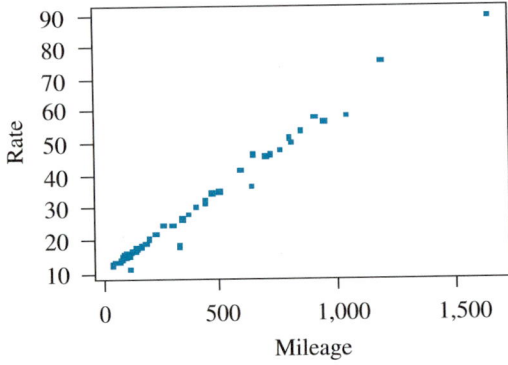

The data are as follows:

Mileage:	50	60	80	80	90	90	100	100	100	110	110	110
Rate:	12.7	13.0	13.7	14.1	14.6	14.1	15.6	14.9	14.5	15.3	15.5	15.9

Mileage:	120	120	120	120	130	130	140	150	170	190	200	230
Rate:	16.4	11.1	16.0	15.8	16.0	16.7	17.2	17.5	18.6	19.3	20.4	21.8

Mileage:	260	300	330	340	370	400	440	440	480	510	540	600
Rate:	24.7	24.7	18.0	27.1	28.2	30.6	31.8	32.4	34.5	35.0	36.3	41.4

Mileage:	650	700	720	760	800	810	850	920	960	1,050	1,200	1,650
Rate:	46.4	45.8	46.6	48.0	51.7	50.2	53.6	57.9	56.1	58.7	75.8	89.0

a. Write the regression equation and the residual standard deviation.

b. Calculate a 90% confidence interval for the true slope.

11.80   In the plot of Exercise 11.79, do you see any problems with the data?

11.81   For Exercise 11.79, predict the shipping rate for a 340-mile route. Obtain a 95% prediction interval. How serious is the extrapolation problem in this exercise?

Soc.

11.82   Suburban towns often spend a large fraction of their municipal budgets on public safety (police, fire, and ambulance) services. A taxpayers' group felt that very small towns were likely to spend large amounts per person because they have such small financial bases. The group obtained data on the per capita expenditure for public safety of 29 suburban towns in a metropolitan area, as well as the population of each town. The data were analyzed using the Minitab package. A regression model with dependent variable 'expendit' and independent variable 'townpopn' yields the following output:

```
MTB > regress 'expendit' 1 'townpopn'

The regression equation is
expendit = 119 +0.000532 townpopn

Predictor         Coef        Stdev      t-ratio          p
Constant        118.96        23.26         5.11      0.000
townpopn      0.0005324    0.0006181         0.86      0.397


s = 43.31         R-sq = 2.7%        R-sq(adj) = 0.0%

Analysis of Variance

SOURCE          DF           SS           MS         F         p
Regression       1         1392         1392      0.74     0.397
Error           27        50651         1876
Total           28        52043


Unusual Observations
Obs. townpopn    expendit        Fit   Stdev.Fit    Residual    St. Resid
  8      74151      334.00     158.43       25.32      175.57       5.00RX

R denotes an obs. with a large st. resid.
X denotes an obs. whose X value gives it large influence.
```

a. If the taxpayers' group is correct, what sign should the slope of the regression model have?

b. Does the slope in the output confirm the opinion of the group?

11.83 Minitab produced a scatterplot and LOWESS smoothing of the data in Exercise 11.82, shown here. Does this plot indicate that the regression line is misleading? Why?

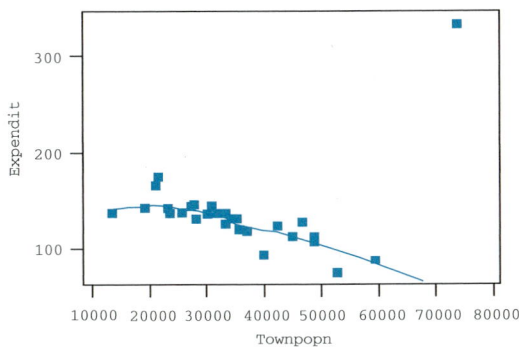

11.84 One town in the database of Exercise 11.82 is the home of an enormous regional shopping mall. A very large fraction of the town's expenditure on public safety is related to the mall; the mall management pays a yearly fee to the township that covers these expenditures. That town's data were removed from the database and the remaining data were reanalyzed by Minitab. A scatterplot is shown.

a. Explain why removing this one point from the data changed the regression line so substantially.

b. Does the revised regression line appear to conform to the opinion of the taxpayers' group in Exercise 11.82?

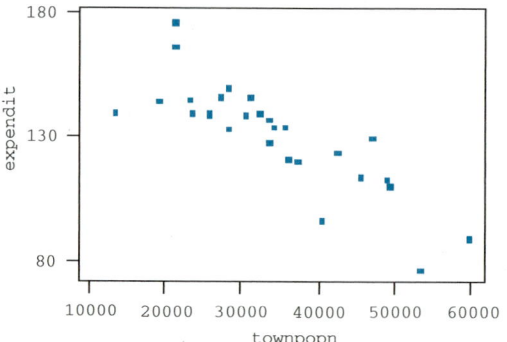

11.85   Regression output for the data of Exercise 11.82, excluding the one unusual town, is shown here. How has the slope changed from the one obtained previously?

```
MTB > regress 'expendit' 1 'townpopn'

The regression equation is
expendit = 184 - 0.00158 townpopn

Predictor          Coef        Stdev     t-ratio         p
Constant        184.240        7.481       24.63     0.000
townpopn     -0.0015766    0.0002099       -7.51     0.000

s = 12.14         R-sq = 68.5%        R-sq(adj) = 67.2%
```

```
Analysis of variance

SOURCE         DF              SS         MS         F        p
Regression      1          8322.7     8322.7     56.43    0.000
Error          26          3834.5      147.5
Total          27         12157.2

Unusual Observations
Obs. townpopn     expendit        Fit    Stdev.Fit    Residual    St.Resid
   5    40307        96.00     120.69         2.66      -24.69      -2.08R
   6    13457       139.00     163.02         4.87      -24.02      -2.16R
  13    59779        89.00      89.99         5.89       -0.99      -0.09 X
  22    21701       176.00     150.03         3.44       25.97       2.23R
  27    53322        76.00     100.17         4.67      -24.17      -2.16R

R denotes an obs. with a large st. resid.
X denotes an obs. whose X value gives it large influence.
```

Bio.

11.86   In screening for compounds useful in treating hypertension (high blood pressure), researchers assign six rats to each of three groups. The rats in group 1 receive .1 mg/kg of a test compound; those in groups 2 and 3 receive .2 and .4 mg/kg, respectively. The response of interest is the decrease in blood pressure 2 hours postdose, compared to the corresponding predose blood pressure. The data are shown here:

	Dose, x	Blood Pressure Drop (mm Hg), y					
Group 1	.1 mg/kg	10	12	15	16	13	11
Group 2	.2 mg/kg	25	22	26	19	18	24
Group 3	.4 mg/kg	30	32	35	27	26	29

a. Use a software package to fit the model

$$y = \beta_0 + \beta_1 \log_{10} x + \varepsilon$$

b. Use residual plots to examine the fit to the model in part(a).
c. Conduct a statistical test of $H_0: \beta_1 \leq 0$ versus $H_a: \beta_1 > 0$. Give the p-value for your test.

Ag.   **11.87**   A laboratory conducts a study to examine the effect of different levels of nitrogen on the yield of lettuce plants. Use the data shown here to fit a linear regression model. Test for possible lack of fit of the model.

Coded Nitrogen	Yield (Emergent Stalks per Plot)
1	21, 18, 17
2	24, 22, 26
3	34, 29, 32

Med.   **11.88**   Researchers measured the specific activity of the enzyme sucrase extracted from portions of the intestines of 24 patients who underwent an intestinal bypass. After the sections were extracted, they were homogenized and analyzed for enzyme activity [Carter (1981)]. Two different methods can be used to measure the activity of sucrase: the homogenate method and the pellet method. Data for the 24 patients are shown here for the two methods:

Sucrase Activity as Measured by the Homogenate and Pellet Methods		
Patient	Homogenate Method, y	Pellet Method, x
1	18.88	70.00
2	7.26	55.43
3	6.50	18.87
4	9.83	40.41
5	46.05	57.43
6	20.10	31.14
7	35.78	70.10
8	59.42	137.56
9	58.43	221.20
10	62.32	276.43
11	88.53	316.00
12	19.50	75.56
13	60.78	277.30
14	77.92	331.50
15	51.29	133.74
16	77.91	221.50

(*continues*)

(*continued*)

Sucrase Activity as Measured by the Homogenate and Pellet Methods

Patient	Homogenate Method, y	Pellet Method, x
17	36.65	132.93
18	31.17	85.38
19	66.09	142.34
20	115.15	294.63
21	95.88	262.52
22	64.61	183.56
23	37.71	86.12
24	100.82	226.55

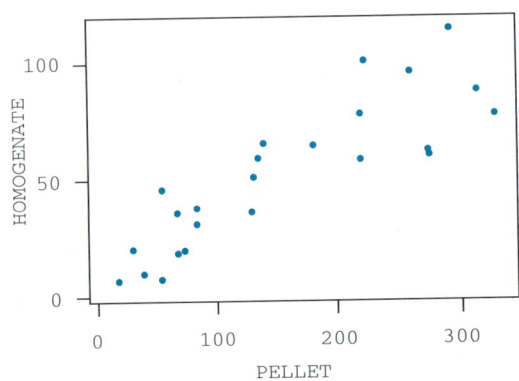

Relationship between Homogenate and Pellet

a. Examine the scatterplot of the data. Might a linear model adequately describe the relationship between the two methods?

```
Regression Analysis: HOMOGENATE versus PELLET


The regression equation is
HOMOGENATE = 10.3 + 0.267 PELLET

Predictor        Coef      SE Coef          T            P
Constant       10.335        5.995       1.72        0.099
PELLET        0.26694      0.03251       8.21        0.000

S = 15.62       R-Sq = 75.4%      R-Sq(adj) = 74.3%

Analysis of Variance

Source              DF           SS          MS          F          P
Regression           1        16440       16440      67.41      0.000
Residual Error      22         5366         244
Total               23        21806
```

Obs	PELLET	HOMOGENA	Fit	SE Fit	Residual	St Resid
1	70	18.88	29.02	4.24	-10.14	-0.67
2	55	7.26	25.13	4.57	-17.87	-1.20
3	19	6.50	15.37	5.49	-8.87	-0.61
4	40	9.83	21.12	4.93	-11.29	-0.76
5	57	46.05	25.67	4.52	20.38	1.36
6	31	20.10	18.65	5.17	1.45	0.10
7	70	35.78	29.05	4.24	6.73	0.45
8	138	59.42	47.06	3.24	12.36	0.81
9	221	58.43	69.38	3.83	-10.95	-0.72
10	276	62.32	84.13	5.04	-21.81	-1.48
11	316	88.53	94.69	6.10	-6.16	-0.43
12	76	19.50	30.50	4.13	-11.00	-0.73
13	277	60.78	84.36	5.07	-23.58	-1.60
14	332	77.92	98.83	6.53	-20.91	-1.47
15	134	51.29	46.04	3.27	5.25	0.34
16	222	77.91	69.46	3.83	8.45	0.56
17	133	36.65	45.82	3.28	-9.17	-0.60
18	85	31.17	33.13	3.93	-1.96	-0.13
19	142	66.09	48.33	3.22	17.76	1.16
20	295	115.15	88.98	5.52	26.17	1.79
21	263	95.88	80.41	4.70	15.47	1.04
22	184	64.61	59.33	3.31	5.28	0.35
23	86	37.71	33.32	3.92	4.39	0.29
24	227	100.82	70.81	3.92	30.01	1.99

```
Regression Line for Homogenate versus Pellet
    HOMOGENATE = 10.3348 + 0.266940 PELLET
 S = 15.6169  R-Sq = 75.4%  R-Sq(adj) = 74.3%
```

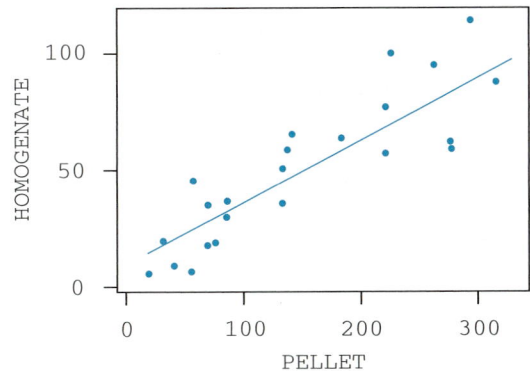

b. Examine the residual plot: are there any potential problems uncovered by the plot?

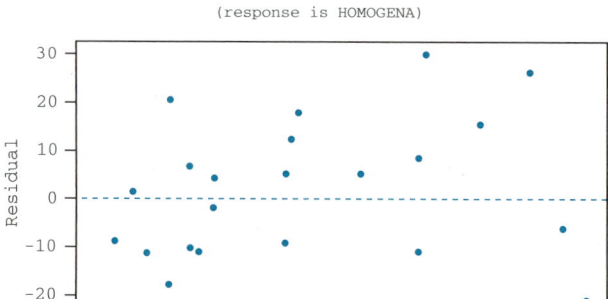

Residuals versus the Fitted Values
(response is HOMOGENA)

c. In general, the pellet method is more time-consuming than the homogenate method, yet it provides a more accurate measure of sucrase activity. How might you estimate the pellet reading based on a particular homogenate reading?

d. How would you develop a confidence (prediction) interval about your point estimate?

Bus.   **11.89**   A realtor in a suburban area attempted to predict house prices solely on the basis of size. From a multiple listing service, the realtor obtained size in thousands of square feet and asking price in thousands of dollars. The information is stored in the EX 1189.DAT file in the Web site data sets, with price in column 1 and size in column 2. Have your statistical software program read this file.

a. Obtain a plot of price against size. Does it appear there is an increasing relation?

b. Locate an apparent outlier in the data. Is it a high leverage point?

c. Obtain a regression equation and include the outlier in the data.

d. Delete the outlier and obtain a new regression equation. How much does the slope change without the outlier? Why?

e. Locate the residual standard deviations for the outlier-included and outlier-excluded models. Do they differ much? Why?

11.90   Obtain the outlier-excluded regression model for the data of Exercise 11.89.

a. Interpret the intercept (constant) term. How much meaning does this number have in this context?

b. What would it mean in this context if the slope were 0? Can the null hypothesis of zero slope be emphatically rejected?

c. Calculate a 95% confidence interval for the true population value of the slope. The computer output should give you the estimated slope and its standard error, but you will probably have to do the rest of the calculations by hand.

11.91   **a.** If possible, use your computer program to obtain a 95% prediction interval for the asking price of a home of 5,000 square feet, based on the outlier-excluded data of Exercise 11.89. If you must do the computations by hand, obtain the mean and standard deviation of the size data from the computer, and find $S_{xx} = (n - 1)s^2$ by hand. Would this be a wise prediction to make, based on the data?

b. Obtain a plot of the price against the size. Does the constant-variance assumption seem reasonable, or does variability increase as size increases?

c. What does your answer to part(b) say about the prediction interval obtained in part(a)?

Bus.   **11.92**   A lawn care company tried to predict the demand for its service by zip code, using the housing density in the zip code area as a predictor. The owners obtained the number of houses and the geographic size of each zip code and calculated their sales per thousand homes and number of homes per acre. The data are stored in the EX1192.DAT file in the Web site data sets. Sales data are in column 1 and density (homes/acre) are in column 2. Read the data into your computer package.

    a. Obtain the correlation between two variables. What does its sign mean?

    b. Obtain a prediction equation with sales as the dependent variable and density as the independent variable. Interpret the intercept (yes, we know the interpretation will be a bit strange) and the slope numbers.

    c. Obtain a value for the residual standard deviation. What does this number indicate about the accuracy of prediction?

11.93   **a.** Obtain a value of the t statistic for the regression model of Exercise 11.92. Is there conclusive evidence that density is a predictor of sales?

    b. Calculate a 95% confidence interval for the true value of the slope. The package should have calculated the standard error for you.

11.94   Obtain a plot of the data of Exercise 11.92, with sales plotted against density. Does it appear that straight-line prediction makes sense?

11.95   Refer to Exercise 11.92. Have your computer program calculate a new variable as 1/density.

    a. What is the interpretation of the new variable? In particular, if the new variable equals 0.50, what does that mean about the particular zip code area?

    b. Plot sales against the new variable. Does a straight-line prediction look reasonable here?

    c. Obtain the correlation of sales and the new variable. Compare its magnitude to the correlation obtained in Exercise 11.94 between sales and density. What explains the difference?

Engin.   **11.96**   A manufacturer of paint used for marking road surfaces developed a new formulation that needs to be tested for durability. One question concerns the concentration of pigment in the paint. If the concentration is too low, the paint will fade quickly; if the concentration is too high, the paint will not adhere well to the road surface. The manufacturer applies paint at various concentrations to sample road surfaces and obtains a durability measurement for each sample. The data are stored in the EX1196.DAT file in the Web site data sets, with durability in column 1 and concentration in column 2.

    a. Have your computer program calculate a regression equation with durability predicted by concentration. Interpret the slope coefficient.

    b. Find the coefficient of determination. What does it indicate about the predictive value of concentration?

11.97   In the regression model of Exercise 11.96, is the slope coefficient significantly different from 0 at $\alpha = .01$?

11.98   Obtain a plot of the data of Exercise 11.96, with durability on the vertical axis and concentration on the horizontal axis.

    a. What does this plot indicate about the wisdom of using straight-line prediction?

    b. What does this plot indicate about the correlation found in Exercise 11.96?

Bus.   **11.99**   Previously, we considered a group of builders who were considering a method for estimating the cost of constructing custom houses. They have come back to you for additional advice.

    Recall that the builders used the method to estimate the cost of 10 "spec" houses that were built without a commitment from a customer. The builders obtained the actual costs (exclusive of land costs) of completing each house, to compare with the estimated costs.

    "We went back to our accountant, who did a regression analysis of the data and gave us these results. The accountant says that the estimates are quite accurate, with an 80% correlation and a very low p-value. We're still pretty skeptical of whether this new method gives us decent estimates. We only clear a profit of about 10 percent, so a few bad estimates would hurt us. Can you explain to us what this output says about the estimating method?"

Write a brief, not-too-technical explanation for them. Focus on the builder's question about the accuracy of the estimates. A plot is shown here.

```
MTB > Regress 'Actual' on 1 variable 'Estimate'.
 The regression equation is
Actual = -34739 + 1.25 Estimate

Predictor       Coef        Stdev      t-ratio        p
Constant       -34739       60147        -0.58      0.579
Estimate        1.2474       0.3293       3.79      0.005

s = 19313     R-sq = 64.2%     R-sq(adj) = 59.7%

Analysis of Variance

SOURCE          DF          SS           MS          F          p
Regression       1  5350811136   5350811136      14.35      0.005
Error            8  2983948032    372993504
Total            9  8334758912

Unusual Observations
Obs. Estimate  Actual      Fit  Stdev.Fit    Residual   St.Resid   2    186200   152134
   197531       6286    -45397              -2.49R

R denotes an obs. with a large st. resid.

MTB > Correlation 'Estimate' 'Actual'.

Correlation of Estimate and Actual = 0.801
```

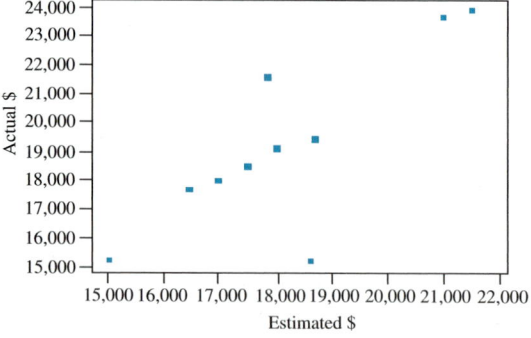

CHAPTER 12

Multiple Regression and the General Linear Model

12.1 Introduction and Case Study

12.2 The General Linear Model

12.3 Estimating Multiple Regression Coefficients

12.4 Inferences in Multiple Regression

12.5 Testing a Subset of Regression Coefficients

12.6 Forecasting Using Multiple Regression

12.7 Comparing the Slopes of Several Regression Lines

12.8 Logistic Regression

12.9 Some Multiple Regression Theory (Optional)

12.10 Summary

12.1 Introduction and Case Study

Case Study: Designing an Electric Drill That Will Not Heat Up under Strenuous Use

Engineers for a manufacturer of power tools for home use were trying to design an electric drill that did not heat up under strenuous use. The three key design factors were insulation thickness, quality of the wire used in the motor, and size of the vents in the body of the drill.

Designing the Data Collection The engineers had learned a little about off-line quality control, so they designed an experiment that varied these design factors. They created 10 drills using each combination of the three design factors, split them into two lots, and tested the lots under two (supposedly equivalent) "torture tests." The temperature of each drill was measured at the end of each test; for each lot, the mean temperature and the logarithm of the variance of temperatures were computed. The engineers wanted to minimize both the mean and the logarithm of the variance.

617

Managing the Data For this experiment, there are three key design factors:

IT is the insulating thickness of the drill (IT = 2, 3, 4, 5, or 6)

QW is the quality of wire used in the motor (QW = 6, 7, or 8)

and

VS is the size of the vent in the body of the drill (VS = 10, 11, or 12)

There are $5 \times 3 \times 3 = 45$ different combinations of these design factors. For each combination of factors, ten drills were made and divided into two lots of five drills. Each drill was subjected to a torture test and the temperature recorded at the end of the test. Avtem represents the average temperature for the five drills of a lot for a given combination of the design factors. The 90 measurements are given here ($5 \times 3 \times 3 \times 2$ lots = 90). Also included in the data set are

logv = logarithm of the variance of temperatures for a given combination of factors and lot

and I2, Q2, and V2, which are squared terms for the three design factors computed as

$$(\text{design factor} - \text{mean design factor})^2$$

avtem	logv	IT	QW	VS	I2	Q2	V2	Lot	avtem	logv	IT	QW	VS	I2	Q2	V2	Lot
185	3.6	2	6	10	4	1	1	1	168	3.4	4	7	11	0	0	0	2
176	3.7	2	6	10	4	1	1	2	160	2.9	4	7	12	0	0	1	1
177	3.6	2	6	11	4	1	0	1	154	3.1	4	7	12	0	0	1	2
184	3.7	2	6	11	4	1	0	2	169	2.8	4	8	10	0	1	1	1
178	3.6	2	6	12	4	1	1	1	156	2.9	4	8	10	0	1	1	2
169	3.4	2	6	12	4	1	1	2	168	2.7	4	8	11	0	1	0	1
185	3.2	2	7	10	4	0	1	1	161	2.7	4	8	11	0	1	0	2
184	3.2	2	7	10	4	0	1	2	156	2.6	4	8	12	0	1	1	1
180	3.2	2	7	11	4	0	0	1	158	2.7	4	8	12	0	1	1	2
184	3.5	2	7	11	4	0	0	2	164	3.7	5	6	10	1	1	1	1
179	3.0	2	7	12	4	0	1	1	163	3.7	5	6	10	1	1	1	2
173	3.2	2	7	12	4	0	1	2	161	3.7	5	6	11	1	1	0	1
179	2.9	2	8	10	4	1	1	1	158	3.4	5	6	11	1	1	0	2
185	2.7	2	8	10	4	1	1	2	154	3.4	5	6	12	1	1	1	1
180	2.8	2	8	11	4	1	0	1	162	3.7	5	6	12	1	1	1	2
180	2.7	2	8	11	4	1	0	2	163	2.8	5	7	10	1	0	1	1
169	2.9	2	8	12	4	1	1	1	166	3.0	5	7	10	1	0	1	2
177	2.8	2	8	12	4	1	1	2	159	3.3	5	7	11	1	0	0	1
172	3.6	3	6	10	1	1	1	1	156	3.3	5	7	11	1	0	0	2
171	3.9	3	6	10	1	1	1	2	152	3.3	5	7	12	1	0	1	1
172	3.8	3	6	11	1	1	0	1	150	3.3	5	7	12	1	0	1	2
167	3.6	3	6	11	1	1	0	2	165	2.9	5	8	10	1	1	1	1
165	3.3	3	6	12	1	1	1	1	156	2.7	5	8	10	1	1	1	2
159	3.4	3	6	12	1	1	1	2	155	2.8	5	8	11	1	1	0	1
169	3.0	3	7	10	1	0	1	1	155	3.2	5	8	11	1	1	0	2
174	3.3	3	7	10	1	0	1	2	149	2.6	5	8	12	1	1	1	1
163	3.3	3	7	11	1	0	0	1	152	2.9	5	8	12	1	1	1	2
170	3.3	3	7	11	1	0	0	2	165	3.4	6	6	10	4	1	1	1
169	3.2	3	7	12	1	0	1	1	160	3.7	6	6	10	4	1	1	2
163	3.2	3	7	12	1	0	1	2	157	3.7	6	6	11	4	1	0	1
178	2.7	3	8	10	1	1	1	1	149	3.7	6	6	11	4	1	0	2
165	2.7	3	8	10	1	1	1	2	149	3.8	6	6	12	4	1	1	1
167	2.8	3	8	11	1	1	0	1	145	3.7	6	6	12	4	1	1	2

(*continues*)

(continued)

avtem	logv	IT	QW	VS	I2	Q2	V2	Lot	avtem	logv	IT	QW	VS	I2	Q2	V2	Lot
171	2.8	3	8	11	1	1	0	2	154	3.4	6	7	10	4	0	1	1
166	2.9	3	8	12	1	1	1	1	153	3.2	6	7	10	4	0	1	2
166	2.7	3	8	12	1	1	1	2	150	3.0	6	7	11	4	0	0	1
161	3.7	4	6	10	0	1	1	1	156	3.1	6	7	11	4	0	0	2
162	3.7	4	6	10	0	1	1	2	146	3.2	6	7	12	4	0	1	1
169	3.4	4	6	11	0	1	0	1	153	3.3	6	7	12	4	0	1	2
162	3.7	4	6	11	0	1	0	2	161	2.8	6	8	10	4	1	1	1
159	3.5	4	6	12	0	1	1	1	160	2.9	6	8	10	4	1	1	2
168	3.4	4	6	12	0	1	1	2	156	2.9	6	8	11	4	1	0	1
169	3.1	4	7	10	0	0	1	1	150	2.7	6	8	11	4	1	0	2
165	3.2	4	7	10	0	0	1	2	149	2.9	6	8	12	4	1	1	1
163	3.2	4	7	11	0	0	0	1	151	2.8	6	8	12	4	1	1	2

Analyzing the Data The engineers want the data analyzed and they want to minimize both the mean and the logarithm of the variance for temperatures. They have asked you to try to figure out which of the design factors seem to affect the mean (and by how much), which affect the variance, which squared terms seem to matter, and, finally, whether the lot number (corresponding to the type of test) is relevant. These data will be analyzed using the techniques of this chapter in Exercises 12.65–12.67.

The simplest type of regression model relating the dependent variable y to a quantitative independent variable x is the one discussed in Chapter 11,

$$y = \beta_0 + \beta_1 x + \varepsilon$$

Under the assumption that the average value of ε (also called the **expected value of ε**) for a given value of x is $E(\varepsilon) = 0$, this model indicates that the expected value of y for a given value of x is described by the straight line

$$E(y) = \beta_0 + \beta_1 x$$

Not all data sets are adequately described by a model for which the expectation is a straight line. For example, consider the data of Table 12.1, which gives

TABLE 12.1

Yield of 14 equal-sized plots of tomato plantings for different amounts of fertilizer

Plot	Yield, y (in bushels)	Amount of Fertilizer, x (in pounds per plot)
1	24	12
2	18	5
3	31	15
4	33	17
5	26	20
6	30	14
7	20	6
8	25	23
9	25	11
10	27	13
11	21	8
12	29	18
13	29	22
14	26	25

FIGURE 12.1

Scatterplot of the yield
versus fertilizer data in
Table 12.1

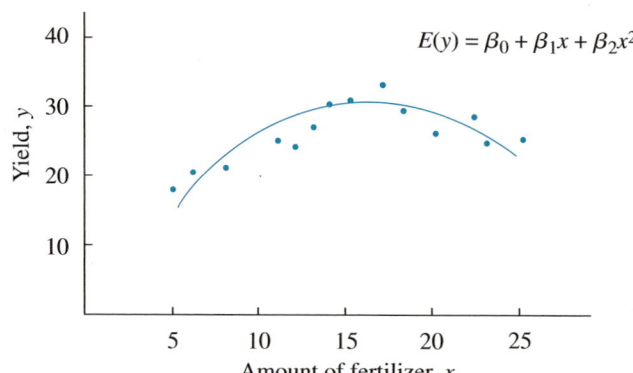

the yields (in bushels) for 14 equal-sized plots planted in tomatoes for different levels of fertilization. It is evident from the scatterplot in Figure 12.1 that a linear equation will not adequately represent the relationship between yield and the amount of fertilizer applied to the plot. The reason for this is that, whereas a modest amount of fertilizer may well enhance the crop yield, too much fertilizer can be destructive.

A model for this physical situation might be

$$y = \beta_0 + \beta_1 x + \beta_2 x^2 + \varepsilon$$

Again with the assumption that $E(\varepsilon) = 0$, the expected value of y for a given value of x is

$$E(y) = \beta_0 + \beta_1 x + \beta_2 x^2$$

One such line is plotted in Figure 12.1, superimposed on the data of Table 12.1.

A general polynomial regression model relating a dependent variable y to a single quantitative independent variable x is given by

$$y = \beta_0 + \beta_1 x + \beta_2 x^2 + \cdots + \beta_p x^p + \varepsilon$$

with

$$E(y) = \beta_0 + \beta_1 x + \beta_2 x^2 + \cdots + \beta_p x^p$$

The choice of p and hence the choice of an appropriate regression model will depend on the experimental situation.

multiple regression model   The **multiple regression model** that relates a dependent variable y to a set of quantitative independent variables is a direct extension of a polynomial regression model in one independent variable. We write the multiple regression model as

$$y = \beta_0 + \beta_1 x_1 + \beta_2 x_2 + \cdots + \beta_k x_k + \varepsilon$$

Any of the independent variables may be powers of other independent variables; for example, x_2 might be x_1^2. In fact, there are many other possibilities; x_3 might **cross-product term**   be a **cross-product term** equal to $x_1 x_2$, x_4 might be log x_1, and so on. The only restriction is that no x is a perfect linear function of any other x.

first-order model   The simplest type of multiple regression equation is a **first-order model**, in which each of the independent variables appears, but there are no cross-product terms or terms in powers of the independent variables. For example, when three quantitative independent variables are involved, the first-order multiple regression model is

$$y = \beta_0 + \beta_1 x_1 + \beta_2 x_2 + \beta_3 x_3 + \varepsilon$$

For these first-order models, we can attach some meaning to the βs. The parameter β_0 is the y-intercept, which represents the expected value of y when each x is zero. For cases in which it does not make sense to have each x be zero, β_0 (or its estimate) should be used only as part of the prediction equation, and not given an interpretation by itself.

partial slopes

The other parameters $(\beta_1, \beta_2, \ldots, \beta_k)$ in the multiple regression equation are sometimes called **partial slopes**. In linear regression, the parameter β_1 is the slope of the regression line and it represents the expected change in y for a unit increase in x. In a first-order multiple regression model, β_1 represents the expected change in y for a unit increase in x_1 *when all other xs are held constant*. In general then, $\beta_j (j \neq 0)$ represents the expected change in y for a unit increase in x_j while holding all other xs constant. The usual assumptions for a multiple regression model are shown here.

DEFINITION 12.1

The **assumptions for multiple regression** are as follows:

1. The mathematical form of the relation is correct, so $E(\varepsilon_i) = 0$ for all i.
2. $\text{Var}(\varepsilon_i) = \sigma_\varepsilon^2$ for all i.
3. The ε_is are independent.
4. ε_i is normally distributed.

additive effects

There is an additional assumption that is implied when we use a first-order multiple regression model. Because the expected change in y for a unit change in x_j is constant and does not depend on the value of any other x, we are in fact assuming that the effects of the independent variables are **additive.**

EXAMPLE 12.1

A brand manager for a new food product collected data on y = brand recognition (percent of potential consumers who can describe what the product is), x_1 = length in seconds of an introductory TV commercial, and x_2 = number of repetitions of the commercial over a 2-week period. What does the brand manager assume if a first-order model

$$\hat{y} = 0.31 + 0.042x_1 + 1.41x_2$$

is used to predict y?

Solution First, the manager assumes a straight-line, consistent rate of change. The manager assumes that a 1-second increase in length of the commercial will lead to a 0.042 percentage point increase in recognition, whether the increase is from, say, 10 to 11 seconds or from 59 to 60 seconds. Also, every additional repetition of the commercial is assumed to give a 1.41 percentage point increase in recognition, whether it is the second repetition or the twenty-second.

Second, there is a no-interaction assumption. The first-order model assumes that the effect of an additional repetition (that is, an increase in x_2) of a given length commercial (that is, holding x_1 constant) doesn't depend on *where* that length is held constant (at 10 seconds, 27 seconds, 60 seconds, whatever).

When might the additional assumption of additivity be warranted? Figure 12.2(a) shows a scatterplot of y versus x_1; Figure 12.2(b) shows the same plot with

FIGURE 12.2

(a) Scatterplot of y versus x_1;
(b) scatterplot of y versus x_1,
indicating additivity of
effects for x_1 and x_2

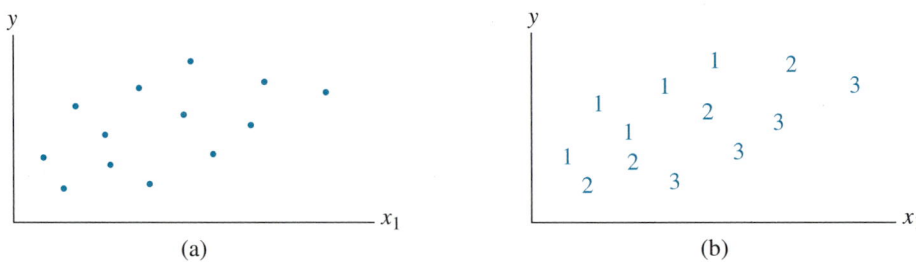

an ID attached to the different levels of a second independent variable x_2 (x_2 takes on the values of 1, 2, or 3). From Figure 12.2(a), we see that y is approximately linear in x_1. The parallel lines of Figure 12.2(b) corresponding to the three levels of the independent variable x_2 indicate that the expected change in y for a unit change in x_1 remains the same no matter which level of x_2 is used. These data suggest that the effects of x_1 and x_2 are additive; hence, a first-order model of the form $y = \beta_0 + \beta_1 x_1 + \beta_2 x_2 + \varepsilon$ is appropriate.

interaction   Figure 12.3 displays a situation in which **interaction** is present between the variables x_1 and x_2. Even though a scatterplot of y versus x_1 is as shown in Figure 12.2(a), the nonparallel lines of Figure 12.3 indicate that the expected change in y for a unit change in x_1 now depends on the level of x_2. When this occurs, the independent variables x_1 and x_2 are said to interact. A first-order model, which assumes additivity of the effects, would not be appropriate here. At the very least, we would include a cross-product term $(x_1 x_2)$ in the model.

FIGURE 12.3

Scatterplot of y versus x_1,
indicating nonadditivity
(interaction) of effects
between x_1 and x_2

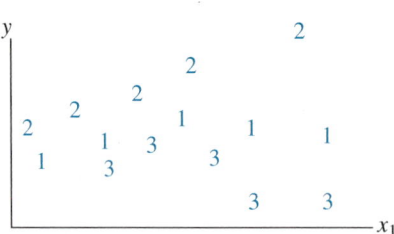

The simplest model allowing for interaction between x_1 and x_2 is

$$y = \beta_0 + \beta_1 x_1 + \beta_2 x_2 + \beta_3 x_1 x_2 + \varepsilon$$

Note that for a given value of x_2 (say, $x_2 = 2$), the expected value of y is

$$E(y) = \beta_0 + \beta_1 x_1 + \beta_2(2) + \beta_3 x_1(2)$$
$$= (\beta_0 + 2\beta_2) + (\beta_1 + 2\beta_3)x_1$$

Here the intercept and slope are $(\beta_0 + 2\beta_2)$ and $(\beta_1 + 2\beta_3)$, respectively. The corresponding intercept and slope for $x_2 = 3$ can be shown to be $(\beta_0 + 3\beta_2)$ and $(\beta_1 + 3\beta_3)$. Clearly, the slopes of the two regression lines are not the same, and hence we have nonparallel lines.

Not all experiments can be modeled using a first-order multiple regression model. For these situations, in which a higher-order multiple regression model may be appropriate, it will be more difficult to assign a literal interpretation to the βs because of the presence of terms that contain cross-products or powers of the independent variables. Our focus will be on finding a multiple regression model that provides a good fit to the sample data, not on interpreting individual βs, except as they relate to the overall model.

The models that we have described briefly have been for regression problems for which the experimenter is interested in developing a model to relate a response to one or more *quantitative* independent variables. The problem of modeling an experimental situation is not restricted to the quantitative independent-variable case.

Consider the problem of writing a model for an experimental situation in which a response y is related to a set of *qualitative* independent variables or to both quantitative and qualitative independent variables. For the first situation (relating y to one or more qualitative independent variables), let us suppose that we want to compare the average number of lightning discharges per minute for a storm, as measured from two different tracking posts located 30 miles apart. If we let y denote the number of discharges recorded on an oscilloscope during a 1-minute period, we could write the following two models:

For tracking post 1: $y = \mu_1 + \varepsilon$

For tracking post 2: $y = \mu_2 + \varepsilon$

Thus, we assume that observations at tracking post 1 randomly "bob" about a population mean μ_1. Similarly, at tracking post 2, observations differ from a population mean μ_2 by a random amount ε. These two models are not new and could have been used to describe observations when comparing two population means in Chapter 6. What is new is that we can combine these two models into a single model of the form

$$y = \beta_0 + \beta_1 x_1 + \varepsilon$$

dummy variable

where β_0 and β_1 are unknown parameters, ε is a random error term, and x_1 is a **dummy variable** with the following interpretation. We let

$x_1 = 1$ if an observation is obtained from tracking post 2

$x_1 = 0$ if an observation is obtained from tracking post 1

For observations obtained from tracking post 1, we substitute $x_1 = 0$ into our model to obtain

$$y = \beta_0 + \beta_1(0) + \varepsilon = \beta_0 + \varepsilon$$

Hence, $\beta_0 = \mu_1$, the population mean for observations from tracking post 1. Similarly, by substituting $x_1 = 1$ in our model, the equation for observations from tracking post 2 is

$$y = \beta_0 + \beta_1(1) + \varepsilon = \beta_0 + \beta_1 + \varepsilon$$

Because $\beta_0 = \mu_1$ and $\beta_0 + \beta_1$ must equal μ_2, we have $\beta_1 = \mu_2 - \mu_1$, the difference in means between observations from tracking posts 2 and 1.

This model, $y = \beta_0 + \beta_1 x_1 + \varepsilon$, which relates y to the qualitative independent variable tracking post, can be extended to a situation in which the qualitative variable has more than two levels. We do this by using more than one dummy variable. Consider an experiment in which we're interested in four levels of qualita-

treatments

tive variables. We call these levels **treatments.** We could write the model

$$y = \beta_0 + \beta_1 x_1 + \beta_2 x_2 + \beta_3 x_3 + \varepsilon$$

where

$x_1 = 1$ if treatment 2, $x_1 = 0$ otherwise

$x_2 = 1$ if treatment 3, $x_2 = 0$ otherwise

$x_3 = 1$ if treatment 4, $x_3 = 0$ otherwise

To interpret the βs in this equation, it is convenient to construct a table of the expected values. Because ε has expectation zero, the general expression for the expected value of y is

$$E(y) = \beta_0 + \beta_1 x_1 + \beta_2 x_2 + \beta_3 x_3$$

The expected value for observations on treatment 1 is found by substituting $x_1 = 0$, $x_2 = 0$, and $x_3 = 0$; after this substitution, we find $E(y) = \beta_0$. The expected value for observations on treatment 2 is found by substituting $x_1 = 1$, $x_2 = 0$, and $x_3 = 0$ into the $E(y)$ formula; this substitution yields $E(y) = \beta_0 + \beta_1$. Substitutions of $x_1 = 0$, $x_2 = 1$ $x_3 = 0$ and $x_1 = 0$, $x_2 = 0$, $x_3 = 1$ yield expected values for treatments 3 and 4, respectively. These expected values are summarized in Table 12.2.

TABLE 12.2

Expected values for an experiment with four treatments

Treatment			
1	2	3	4
$E(y) = \beta_0$	$E(y) = \beta_0 + \beta_1$	$E(y) = \beta_0 + \beta_2$	$E(y) = \beta_0 + \beta_3$

If we identify the mean of treatment 1 as μ_1, the mean of treatment 2 as μ_2, and so on, then from Table 12.2 we have

$$\mu_1 = \beta_0 \qquad \mu_2 = \beta_0 + \beta_1 \qquad \mu_3 = \beta_0 + \beta_2 \qquad \mu_4 = \beta_0 + \beta_3$$

Solving these equations for the βs, we have

$$\beta_0 = \mu_1 \qquad \beta_1 = \mu_2 - \mu_1 \qquad \beta_2 = \mu_3 - \mu_1 \qquad \beta_3 = \mu_4 - \mu_1$$

Any comparison among the treatment means can be phrased in terms of the βs. For example, the comparison $\mu_4 - \mu_3$ could be written as $\beta_3 - \beta_2$, and $\mu_3 - \mu_2$ could be written as $\beta_2 - \beta_1$.

EXAMPLE 12.2

Consider a hypothetical situation for an experiment with four treatments ($t = 4$) in which we know the means for the four treatments. If $\mu_1 = 7$, $\mu_2 = 9$, $\mu_3 = 6$, and $\mu_4 = 15$, determine values for β_0, β_1, β_2, and β_3 in the model

$$y = \beta_0 + \beta_1 x_1 + \beta_2 x_2 + \beta_3 x_3 + \varepsilon$$

where

$$x_1 = 1 \text{ if treatment 2}, \qquad x_1 = 0 \text{ otherwise}$$
$$x_2 = 1 \text{ if treatment 3}, \qquad x_2 = 0 \text{ otherwise}$$
$$x_3 = 1 \text{ if treatment 4}, \qquad x_3 = 0 \text{ otherwise}$$

Solution Based on what we saw in Table 12.2, we know that

$$\beta_0 = \mu_1 \qquad \beta_1 = \mu_2 - \mu_1 \qquad \beta_2 = \mu_3 - \mu_1 \qquad \beta_3 = \mu_4 - \mu_1$$

Using the known values for μ_1, μ_2, μ_3, and μ_4, it follows that

$$\beta_0 = 7 \qquad \beta_1 = 9 - 7 = 2 \qquad \beta_2 = 6 - 7 = -1 \qquad \beta_3 = 15 - 7 = 8$$

EXAMPLE 12.3

Refer to Example 12.2. Express $\mu_3 - \mu_2$ and $\mu_3 - \mu_4$ in terms of the βs. Check your findings by substituting values for the βs.

Solution   Using the relationship between the βs and the μs, we can see that

$$\beta_2 - \beta_1 = (\mu_3 - \mu_1) - (\mu_2 - \mu_1) = \mu_3 - \mu_2$$

and

$$\beta_2 - \beta_3 = (\mu_3 - \mu_1) - (\mu_4 - \mu_1) = \mu_3 - \mu_4$$

Substituting computed values for the βs, we have

$$\beta_2 - \beta_1 = -1 - (2) = -3$$

and

$$\beta_2 - \beta_3 = -1 - (8) = -9$$

These computed values are identical to the "known" differences for $\mu_3 - \mu_2$ and $\mu_3 - \mu_4$, respectively.

EXAMPLE 12.4

Use dummy variables to write the model for an experiment with t treatments. Identify the βs.

Solution   We can write the model in the form

$$y = \beta_0 + \beta_1 x_1 + \beta_2 x_2 + \cdots + \beta_{t-1} x_{t-1} + \varepsilon$$

where

$$x_1 = 1 \text{ if treatment 2,} \qquad x_1 = 0 \text{ otherwise}$$

$$x_2 = 1 \text{ if treatment 3,} \qquad x_2 = 0 \text{ otherwise}$$

$$\vdots \qquad\qquad\qquad \vdots$$

$$x_{t-1} = 1 \text{ if treatment } t, \qquad x_{t-1} = 0 \text{ otherwise}$$

The table of expected values would be

	Treatment		
1	**2**	$\cdots$	t
$E(y) = \beta_0$	$E(y) = \beta_0 + \beta_1$	$\cdots$	$E(y) = \beta_0 + \beta_{t-1}$

from which we obtain

$$\beta_0 = \mu_1$$

$$\beta_1 = \mu_2 - \mu_1$$

$$\vdots$$

$$\beta_{t-1} = \mu_t - \mu_1$$

In the procedure just described, we have a response related to the qualitative variable "treatments," and for t levels of the treatments, we enter $(t - 1)$ βs into our model, using dummy variables. More will be said about the use of the models for more than one qualitative independent variable in Chapters 15 and 16, where we consider the analysis of variance for several different experimental designs.

12.2   The General Linear Model

It is important at this point to recognize that a single general model can be used for multiple regression models in which a response is related to a set of quantitative independent variables, and for models that relate y to a set of qualitative indepen-

general linear model

dent variables. This model, called the **general linear model,** has the form

$$y = \beta_0 + \beta_1 x_1 + \beta_2 x_2 + \cdots + \beta_k x_k + \varepsilon$$

For multiple regression models, the xs represent quantitative independent variables (such as weight or amount of water), independent variables raised to powers, and cross-product terms involving the independent variables. We discussed a few regression models in Section 12.1; more about the use of the general linear model in regression will be discussed in the remainder of this chapter and in Chapter 13.

When y is related to a set of qualitative independent variables the xs of the general linear model represent dummy variables (coded 0 and 1) or products of dummy variables. We discussed how to use dummy variables for representing y in terms of a single qualitative variable in Section 12.1; the same approach can be used to relate y to more than one qualitative independent variable. This will be discussed in Chapter 15, where we present more analysis of variance techniques.

The general linear model can also be used for the case in which y is related to both qualitative and quantitative independent variables. A particular example of this is discussed in Section 12.7, and other applications are presented in Chapter 16.

Why is this model called the general *linear* model, especially as it can be used for polynomial models? The word "linear" in the general linear model refers to how the βs are entered in the model, not to how the independent variables appear in the model. A general linear model is linear (used in the usual algebraic sense) in the βs.

Why are we discussing the general linear model now? The techniques that we will develop in this chapter for making inferences about a single β, a set of βs, and $E(y)$ in multiple regression are those that apply to any general linear model. Thus, using general linear model techniques we have a common thread to inferences about multiple regression (Chapters 12 and 13) and the analysis of variance (Chapters 15 through 19). As you study these six chapters, try whenever possible to make the connection back to a general linear model; we'll help you with this connection. For Sections 12.3 through 12.10 of this chapter, we will concentrate on multiple regression, which is a special case of a general linear model.

EXERCISES

Basic Techniques

12.1 **a.** Write a first-order multiple regression model relating a response y to three qualitative independent variables.

b. Show how this model can be written as a general linear model.

12.2 Write a second-order multiple regression model relating a response y to three quantitative independent variables. Include all possible terms. (*Hint:* A first-order model contains terms in the x_j; a second-order model includes these terms as well as squares and cross-products.)

12.3 Refer to Exercise 12.2. Show that the model you wrote can be written in the form of a general linear model. Identify the terms.

12.4 Consider the model

$$y = \beta_0 + \beta_1 x_1 + \beta_2 x_2 + \varepsilon$$

where

$$x_1 = \begin{cases} 1 & \text{if treatment 2} \\ 0 & \text{otherwise} \end{cases} \qquad x_2 = \begin{cases} 1 & \text{if treatment 3} \\ 0 & \text{otherwise} \end{cases}$$

a. Interpret the βs is the model.

b. Identify the difference in mean responses for treatments 2 and 3 using the model.

12.5 (Optional) Refer to Exercise 12.4. Suppose that the model is expanded to include the term $\beta_3 x_3$, where x_3 is a dummy variable for the qualitative variable "location."

$$x_3 = \begin{cases} 1 & \text{if location 2} \\ 0 & \text{otherwise} \end{cases}$$

a. Interpret the βs for this model. (*Hint:* Consider all combinations of the three treatments and two locations.)
b. Write the difference in mean response for treatments 2 and 3 for location 2. Is it the same for location 1?
c. Identify an experimental situation in which this model might be a reasonable approximation.

12.6 (Optional) A study examined the effect of a quantitative independent variable (age) on reaction time (as measured by braking time). The experiment included males and females. Two models were proposed:

$$y = \beta_0 + \beta_1 x_1 + \beta_2 x_2 + \varepsilon \qquad \text{and} \qquad y = \beta_0 + \beta_1 x_1 + \beta_2 x_2 + \beta_3 x_1 x_2$$

where

$$x_1 = \text{age (in years)} \qquad \text{and} \qquad x_2 = \begin{cases} 1 & \text{female} \\ 0 & \text{if male} \end{cases}$$

Interpret the βs for the two models and explain a practical difference between the two models.

12.3 Estimating Multiple Regression Coefficients

The multiple regression model relates a response y to a set of quantitative independent variables. For a random sample of n measurements, we can write the ith observation as

$$y_i = \beta_0 + \beta_1 x_{i1} + \beta_2 x_{i2} + \cdots + \beta_k x_{ik} + \varepsilon_i \qquad (i = 1, 2, \ldots, n; n > k)$$

where $x_{i1}, x_{i2}, \ldots, x_{ik}$ are the settings of the quantitative independent variables corresponding to the observation y_i.

To find least-squares estimates for $\beta_0, \beta_1, \ldots,$ and β_k in a multiple regression model, we follow the same procedure that we did for a linear regression model in Chapter 11. We obtain a random sample of n observations; we find the least-squares prediction equation

$$\hat{y} = \hat{\beta}_0 + \hat{\beta}_1 x_1 + \cdots + \hat{\beta}_k x_k$$

by choosing $\hat{\beta}_0, \hat{\beta}_1, \ldots, \hat{\beta}_k$ to minimize SS(Residual) $= \Sigma_i (y_i - \hat{y}_i)^2$. However, although it was easy to write down the solutions to $\hat{\beta}_0$ and $\hat{\beta}_1$ for the linear regression model,

$$y = \beta_0 + \beta_1 x + \varepsilon$$

we must find the estimates for $\beta_0, \beta_1, \ldots, \beta_k$ by solving a set of simultaneous equations, called the *normal equations,* shown here.

	y_i	$\hat{\beta}_0$	$x_{i1}\hat{\beta}_1$	$\cdots$	$x_{ik}\hat{\beta}_k$
1	Σy_i $=$ $n\hat{\beta}_0$ $+$ $\Sigma x_{i1}\hat{\beta}_1$			$+ \cdots +$	$\Sigma x_{ik}\hat{\beta}_k$
x_{i1}	$\Sigma x_{i1} y_i = \Sigma x_{i1}\hat{\beta}_0 +$ $\Sigma x_{i1}^2 \hat{\beta}_1$			$+ \cdots +$	$\Sigma x_{i1} x_{ik}\hat{\beta}_k$
$\vdots$	$\vdots$				
x_{ik}	$\Sigma x_{ik} y_i = \Sigma x_{ik}\hat{\beta}_0 + \Sigma x_{ik} x_{i1}\hat{\beta}_1$			$+ \cdots +$	$\Sigma x_{ik}^2 \hat{\beta}_k$

Note the pattern associated with these equations. By labeling the rows and columns as we have done, we can obtain any term in the normal equations by multiplying the row and column elements and summing. For example, the last term in the second equation is found by multiplying the row element (x_{i1}) by the column element $(x_{ik}\hat{\beta}_k)$ and summing; the resulting term is $\Sigma\ x_{i1}x_{ik}\hat{\beta}_k$. Because all terms in the normal equations can be formed in this way, it is fairly simple to write down the equations to be solved to obtain the least-squares estimates $\hat{\beta}_0, \hat{\beta}_1, \ldots,$ $\hat{\beta}_k$. The solution to these equations is not necessarily trivial; that's why we'll enlist the help of various statistical software packages for their solution.

EXAMPLE 12.5

In Exercise 11.44, we presented data for the weight loss of a compound for different amounts of time the compound was exposed to the air. Additional information was also available on the humidity of the environment during exposure. The complete data are presented in Table 12.3.

TABLE 12.3

Weight loss, exposure time, and relative humidity data

Weight Loss, y (pounds)	Exposure Time, x_1 (hours)	Relative Humidity, x_2
4.3	4	.20
5.5	5	.20
6.8	6	.20
8.0	7	.20
4.0	4	.30
5.2	5	.30
6.6	6	.30
7.5	7	.30
2.0	4	.40
4.0	5	.40
5.7	6	.40
6.5	7	.40

a. Set up the normal equations for this regression problem if the assumed model is

$$y = \beta_0 + \beta_1 x_1 + \beta_2 x_2 + \varepsilon$$

where x_1 is exposure time and x_2 is relative humidity.

b. Use the computer output shown here to determine the least-squares estimates of β_0, β_1, and β_2. Predict weight loss for 6.5 hours of exposure and a relative humidity of .35.

```
OUTPUT FOR EXAMPLE 12.5

OBS    WT_LOSS    TIME    HUMID

 1      4.3       4.0     0.20
 2      5.5       5.0     0.20
 3      6.8       6.0     0.20
```

```
OBS   WT_LOSS   TIME   HUMID

 4      8.0      7.0    0.20
 5      4.0      4.0    0.30
 6      5.2      5.0    0.30
 7      6.6      6.0    0.30
 8      7.5      7.0    0.30
 9      2.0      4.0    0.40
10      4.0      5.0    0.40
11      5.7      6.0    0.40
12      6.5      7.0    0.40
13       .       6.5    0.35
```

Dependent Variable: WT_LOSS   WEIGHT LOSS

Analysis of Variance

Source	DF	Sum of Squares	Mean Square	F Value	Prob>F
Model	2	31.12417	15.56208	104.133	0.0001
Error	9	1.34500	0.14944		
C Total	11	32.46917			

Root MSE	0.38658	R-square	0.9586
Dep Mean	5.50833	Adj R-sq	0.9494
C.V.	7.01810		

Parameter Estimates

Variable	DF	Parameter Estimate	Standard Error	T for H0: Parameter=0	Prob > |T|
INTERCEP	1	0.666667	0.69423219	0.960	0.3620
TIME	1	1.316667	0.09981464	13.191	0.0001
HUMID	1	−8.000000	1.36676829	−5.853	0.0002

OBS	WT_LOSS	PRED	RESID	L95MEAN	U95MEAN
1	4.3	4.33333	−0.03333	3.80985	4.85682
2	5.5	5.65000	−0.15000	5.23519	6.06481
3	6.8	6.96667	−0.16667	6.55185	7.38148
4	8.0	8.28333	−0.28333	7.75985	8.80682
5	4.0	3.53333	0.46667	3.11091	3.95576
6	5.2	4.85000	0.35000	4.57346	5.12654
7	6.6	6.16667	0.43333	5.89012	6.44321
8	7.5	7.48333	0.01667	7.06091	7.90576
9	2.0	2.73333	−0.73333	2.20985	3.25682
10	4.0	4.05000	−0.05000	3.63519	4.46481
11	5.7	5.36667	0.33333	4.95185	5.78148
12	6.5	6.68333	−0.18333	6.15985	7.20682
13	.	6.42500	.	6.05269	6.79731

```
Sum of Residuals                    0
Sum of Squared Residuals        1.3450
Predicted Resid SS (Press)      2.6123
```

Solution

a. The three normal equations for this model are shown here.

	y_i	$\hat{\beta}_0$	$x_{i1}\hat{\beta}_1$	$x_{i2}\hat{\beta}_2$
1	$\Sigma\, y_i$ =	$n\hat{\beta}_0$	+ $\Sigma\, x_{i1}\hat{\beta}_1$	+ $\Sigma\, x_{i2}\hat{\beta}_2$
x_{i1}	$\Sigma\, x_{i1}y_i$ =	$\Sigma\, x_{i1}\hat{\beta}_0$ +	$\Sigma\, x_{i1}^2\hat{\beta}_1$	+ $\Sigma\, x_{i1}x_{i2}\hat{\beta}_2$
x_{i2}	$\Sigma\, x_{i2}y_i$ =	$\Sigma\, x_{i2}\hat{\beta}_0$ +	$\Sigma\, x_{i2}x_{i1}\hat{\beta}_1$ +	$\Sigma\, x_{i2}^2\hat{\beta}_2$

For these data, we have

$$\Sigma\, y_i = 66.10 \qquad \Sigma\, x_{i1} = 66 \qquad \Sigma\, x_{i2} = 3.60$$

$$\Sigma\, x_{i1}y_i = 383.3 \qquad \Sigma\, x_{i2}y_i = 19.19 \qquad \Sigma\, x_{i1}x_{i2} = 19.8$$

$$\Sigma\, x_{i1}^2 = 378 \qquad \Sigma\, x_{i2}^2 = 1.16$$

Substituting these values into the normal equation yields the result shown here:

$$66.1 = 12\hat{\beta}_0 + 66\hat{\beta}_1 + 3.6\hat{\beta}_2$$

$$383.3 = 66\hat{\beta}_0 + 378\hat{\beta}_1 + 19.8\hat{\beta}_2$$

$$19.19 = 3.6\hat{\beta}_0 + 19.8\hat{\beta}_1 + 1.16\hat{\beta}_2$$

b. The normal equations of part (a) could be solved to determine $\hat{\beta}_0$, $\hat{\beta}_1$, and $\hat{\beta}_2$. The solution would agree with that shown here in the output. The least-squares prediction equation is

$$\hat{y} = 0.667 + 1.317x_1 - 8.000x_2$$

where x_1 is exposure time and x_2 is relative humidity. Substituting $x_1 = 6.5$ and $x_2 = .35$, we have

$$\hat{y} = 0.667 + 1.317(6.5) - 8.000(.35) = 6.428$$

This value agrees with the predicted value shown as observation 13 in the output, except for rounding errors.

There are many software programs that provide the calculations to obtain least-squares estimates for parameters in the general linear model (and hence for multiple regression). The output of such programs typically has a list of variable names, together with the estimated partial slopes, labeled COEFFICIENTS (or ESTIMATES or PARAMETERS). The intercept term $\hat{\beta}_0$ is usually called INTERCEPT (or CONSTANT); sometimes it is shown along with the slopes but with no variable name.

EXAMPLE 12.6

The data for three variables (shown here) are analyzed with the Excel spreadsheet program. Identify the estimates of the partial slopes and the intercept.

y:	25	34	28	40	36	42	44	53	49
x_1:	-10	-10	-10	0	0	0	10	10	10
x_2:	-5	0	5	-5	0	5	-5	0	5

	Coefficients	Standard Error	t Stat	P-value
Intercept	39.0	1.256	31.055	7.4E-08
X1	0.983	0.154	6.393	0.0007
X2	0.333	0.308	1.084	0.3202

Solution   The intercept value 39.0 is labeled as such. The estimated partial slopes .983 and .333 are associated with x_1 and x_2, respectively. Most programs label the coefficients similarly, in a column.

The coefficient of an independent variable x_j in a multiple regression equation does not, in general, equal the coefficient that would apply to that variable in a simple linear regression. In multiple regression, the coefficient refers to the effect of changing that x_j variable while other independent variables stay constant. In simple regression, all other potential independent variables are ignored. If other independent variables are correlated with x_j (and therefore don't tend to stay constant while x_j changes), simple linear regression with x_j as the only independent variable captures not only the direct effect of changing x_j but also the indirect effect of the associated changes in other xs. In multiple regression, by holding the other xs constant, we eliminate that indirect effect.

EXAMPLE 12.7

Compare the coefficients of x_1 in the multiple regression model and in the simple (one-predictor) regression model shown in the following StataQuest output. Explain why the two coefficients differ.

```
. regress y  x1 x2

------------------------------------------------------------------------------
      y |   Coef.    Std. Err.      t      P>|t|      [95% Conf. Interval]
--------+---------------------------------------------------------------------
     x1 |     1       1.870829    0.535    0.646     -7.049526     9.049526
     x2 |     3       4.1833      0.717    0.548    -14.99929     20.99929
  _cons |    10       1.183216    8.452    0.014      4.909033    15.09097
------------------------------------------------------------------------------

. regress y x1

------------------------------------------------------------------------------
      y |   Coef.    Std. Err.      t      P>|t|      [95% Conf. Interval]
--------+---------------------------------------------------------------------
     x1 |    2.2      .7659417    2.872    0.064      -.2375683     4.637568
  _cons |    10       1.083205    9.232    0.003      6.552758     13.44724
------------------------------------------------------------------------------

. correlate  y x1 x2

         |      y        x1       x2
---------+---------------------------
       y |  1.0000
      x1 |  0.8563    1.0000
      x2 |  0.8704    0.8944   1.0000
```

Solution In the multiple regression model, the coefficient is shown as 1, but in the simple regression model, it's 2.2. The difference occurs because the two xs are correlated (correlation .8944 in the output). In the multiple regression model, we're thinking of varying x_1 while holding x_2 constant; in the simple regression model, we're thinking of varying x_1 and letting x_2 go wherever it goes.

residual standard deviation

In addition to estimating the intercept and partial slopes, it is important to estimate the **residual standard deviation** s_ϵ, sometimes called the *standard error of estimate*. The residuals are defined as before, as the difference between the observed value and the predicted value of y:

$$y_i - \hat{y}_i = y_i - (\hat{\beta}_0 + \hat{\beta}_1 x_{i1} + \hat{\beta}_2 x_{i2} + \cdots + \hat{\beta}_k x_{ik})$$

The sum of squared residuals, SS(Residual), also called SS(Error), is defined exactly as it sounds. Square the prediction errors and sum the squares:

$$\text{SS(Residual)} = \Sigma(y_i - \hat{y}_i)^2$$
$$= \Sigma[y_i - (\hat{\beta}_0 + \hat{\beta}_1 x_{i1} + \hat{\beta}_2 x_{i2} + \cdots + \hat{\beta}_i x_{ik})]^2$$

The df for this sum of squares is $n - (k + 1)$. One df is subtracted for the intercept and 1 df is subtracted for each of the k partial slopes. The mean square residual, MS(Residual), also called MS(Error), is the residual sum of squares divided by $n - (k + 1)$. Finally, the residual standard deviation s_ϵ is the square root of MS(Residual).

The residual standard deviation may be called "std dev," "standard error of estimate," or "root MSE." If the output is not clear, you can take the square root of MS(Residual) by hand. As always, interpret the standard deviation by the Empirical Rule. About 95% of the prediction errors will be within ±2 standard deviations of the mean (and the mean error is automatically zero):

$$s_\epsilon = \sqrt{\text{MS(Residual)}} = \sqrt{\frac{\text{SS(Residual)}}{n - (k + 1)}}$$

EXAMPLE 12.8

Identify SS(Residual) and s_ϵ in the output shown here for the data of Example 12.6.

	A	B	C	D	E	F
1						
2	The regression equation is y=39.0 + .983 x1 + .333 x2					
3						
4	Predictor	Coef	Stdev	t-ratio	p	
5						
6	Constant	39.000	1.256	31.05	0.000	
7	x1	0.9833	0.1538	6.39	0.001	
8	x2	0.3333	0.3076	1.08	0.320	
9						
10	s=3.768	R-sq=87.5%	R-sq(adj)= 83.3%			
11						
12						

	A	B	C	D	E	F
13	Analysis of Variance					
14						
15						
16	SOURCE	DF	SS	MS	F	P
17	Regression	2	596.83	298.420	21.02	0.002
18	Error	6	85.17	14.190		
19	Total	8	682.00			

Solution   In the section of the output labeled Analysis of Variance, SS(Residual) is shown as SS(Error) = 85.17, with 6 df. MS(Error) is 14.19. The residual standard deviation is indicated by $s = 3.768$. Note that $3.768 = \sqrt{14.19}$ to within round-off error.

The residual standard deviation is crucial in determining the probable error of a prediction using the regression equation. The precise standard error to be used in forecasting an individual y value is stated in Section 12.4. A rough approximation, ignoring extrapolation and df effects, is that the probable error is $\pm 2s_\epsilon$. This approximation can be used as a rough indicator of the forecasting quality of a regression model.

EXAMPLE 12.9

The admissions office of a business school develops a regression model that uses aptitude test scores and class rank to predict the grade average (4.00 = straight A; 2.00 = C average, the minimum graduation average; 0.00 = straight F). The residual standard deviation is $s_\epsilon = .46$. Does this value suggest highly accurate prediction?

Solution   A measure of the probable error of prediction is $2s_\epsilon = .92$. For example, if a predicted average is 2.80, then an individual's grade is roughly between $2.80 - .92 = 1.88$ (not good enough to graduate) and $2.80 + .92 = 3.72$ (good enough to graduate magna cum laude)! This is *not* an accurate forecast.

EXERCISES   **Applications**

Med.   **12.7**   A pharmaceutical firm would like to obtain information on the relationship between the dose level and potency of a drug product. To do this, each of 15 test tubes is inoculated with a virus culture and incubated for 5 days at 30°C. Three test tubes are randomly assigned to each of the five different dose levels to be investigated (2, 4, 8, 16, and 32 mg). Each tube is injected with only one dose level and the response of interest (a measure of the protective strength of the product against the virus culture) is obtained. The data are given here.

Dose Level	Response
2	5, 7, 3
4	10, 12, 14
8	15, 17, 18
16	20, 21, 19
32	23, 24, 29

a. Plot the data.
b. Fit a linear regression model to these data.
c. What other regression model might be appropriate?
d. SAS computer output is shown for both a linear and quadratic regression equation. Which regression equation appears to fit the data better? Why?

```
OUTPUT FOR EXERCISE 12.7

OBS    DOSE    RESPONSE
 1      2         5
 2      2         7
 3      2         3
 4      4        10
 5      4        12
 6      4        14
 7      8        15
 8      8        17
 9      8        18
10     16        20
11     16        21
12     16        19
13     32        23
14     32        24
15     32        29
```

Dependent Variable: RESPONSE   PROTECTIVE STRENGTH

Analysis of Variance

Source	DF	Sum of Squares	Mean Square	F Value	Prob>F
Model	1	590.91613	590.91613	44.280	0.0001
Error	13	173.48387	13.34491		
C Total	14	764.40000			

Root MSE	3.65307	R-square	0.7730
Dep Mean	15.80000	Adj R-sq	0.7556
C.V.	23.12069		

Parameter Estimates

Variable	DF	Parameter Estimate	Standard Error	T for H0: Parameter=0	Prob > |T|
INTERCEP	1	8.666667	1.42786770	6.070	0.0001
DOSE	1	0.575269	0.08645016	6.654	0.0001

OBS	DOSE	RESPONSE	PRED	RESID
1	2	5	9.8172	-4.81720
2	2	7	9.8172	-2.81720
3	2	3	9.8172	-6.81720
4	4	10	10.9677	-0.96774
5	4	12	10.9677	1.03226
6	4	14	10.9677	3.03226

OBS	DOSE	RESPONSE	PRED	RESID
7	8	15	13.2688	1.73118
8	8	17	13.2688	3.73118
9	8	18	13.2688	4.73118
10	16	20	17.8710	2.12903
11	16	21	17.8710	3.12903
12	16	19	17.8710	1.12903
13	32	23	27.0753	-4.07527
14	32	24	27.0753	-3.07527
15	32	29	27.0753	1.92473

Sum of Residuals   0
Sum of Squared Residuals   173.4839
Predicted Resid SS (Press)   238.0013

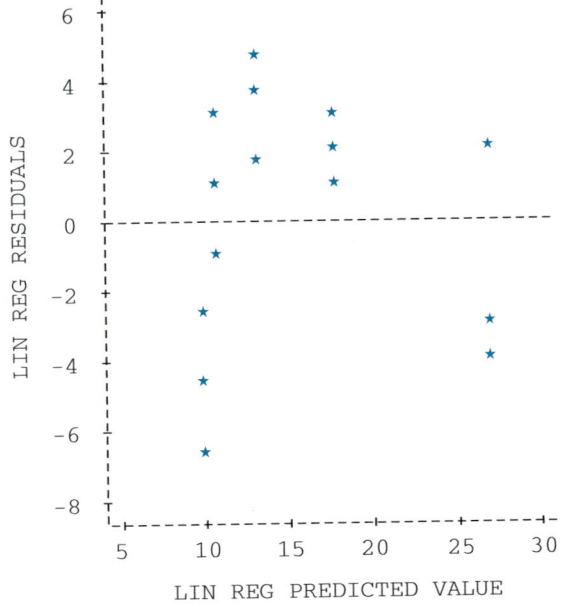

QUADRATIC REGRESSION ANALYSIS

Dependent Variable: RESPONSE   PROTECTIVE STRENGTH

Analysis of Variance

Source	DF	Sum of Squares	Mean Square	F Value	Prob>F
Model	2	673.82062	336.91031	44.634	0.0001
Error	12	90.57938	7.54828		
C Total	14	764.40000			

Root MSE	2.74741	R-square	0.8815
Dep Mean	15.80000	Adj R-sq	0.8618
C.V.	17.38869		

```
Parameter Estimates

                   Parameter      Standard     T for H0:
Variable    DF     Estimate         Error    Parameter=0    Prob > |T|

INTERCEP    1      4.483660     1.65720388       2.706        0.0191
DOSE        1      1.506325     0.28836373       5.224        0.0002
DOSE2       1     -0.026987     0.00814314      -3.314        0.0062

OBS    DOSE    RESPONSE    PREDICTED    RESIDUAL
 1      2         5          7.3884     -2.38836
 2      2         7          7.3884     -0.38836
 3      2         3          7.3884     -4.38836
 4      4        10         10.0772     -0.07717
 5      4        12         10.0772      1.92283
 6      4        14         10.0772      3.92283
 7      8        15         14.8071      0.19292
 8      8        17         14.8071      2.19292
 9      8        18         14.8071      3.19292
10     16        20         21.6762     -1.67615
11     16        21         21.6762     -0.67615
12     16        19         21.6762     -2.67615
13     32        23         25.0512     -2.05123
14     32        24         25.0512     -1.05123
15     32        29         25.0512      3.94877
```

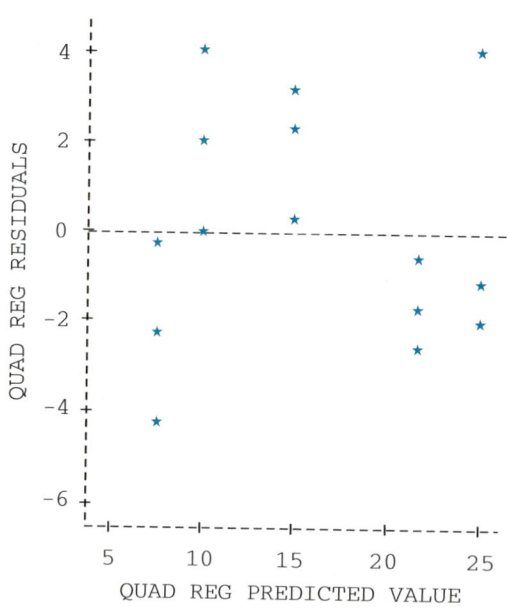

12.8   Refer to the data of Exercise 12.7. Often a logarithmic transformation can be used on the dose levels to linearize the response with respect to the independent variable.

   a. Refer to a set of log tables (see, for example, the Chemical Rubber Company tables) or use a calculator to obtain the logarithms of the five dose levels.

   b. If x_i denotes the log dose, fit the model

$$y = \beta_0 + \beta_1 x_1 + \varepsilon$$

A residual plot is shown here in the output.

c. Compare your results in part (b) to those for Exercise 12.7. Does the logarithmic transformation provide a better linear fit than that in Exercise 12.7?

REGRESSION ANALYSIS USING NATURAL LOGARITHM OF DOSE

Dependent Variable: RESPONSE   PROTECTIVE STRENGTH

Analysis of Variance

Source	DF	Sum of Squares	Mean Square	F Value	Prob>F
Model	1	710.53333	710.53333	171.478	0.0001
Error	13	53.86667	4.14359		
C Total	14	764.40000			

Root MSE	2.03558	R-square	0.9295	
Dep Mean	15.80000	Adj R-sq	0.9241	
C.V.	12.88342			

Parameter Estimates

Variable	DF	Parameter Estimate	Standard Error	T for H0: Parameter=0	Prob > |T|
INTERCEP	1	1.200000	1.23260547	0.974	0.3480
LOG_DOSE	1	7.021116	0.53616972	13.095	0.0001

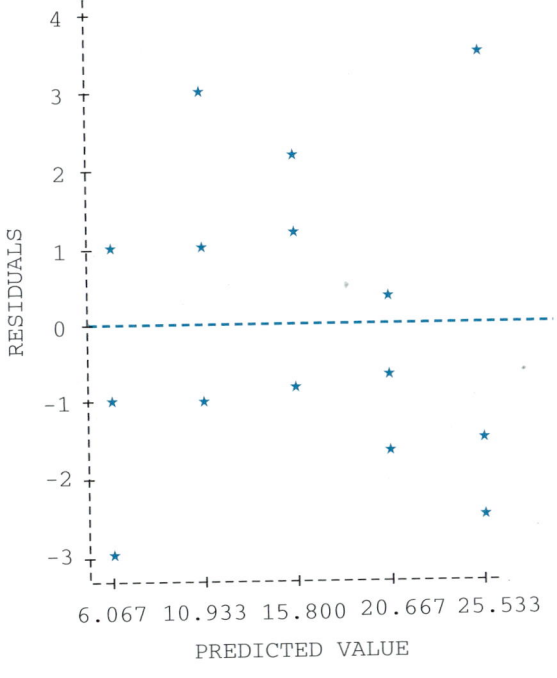

Bus. **12.9** A manufacturer of industrial chemicals investigates the effect on its sales of promotion activities (primarily direct contact and trade shows), direct development expenditures, and short-range research effort. Data are assembled for 24 quarters (6 years) and analyzed by the Stata multiple regression program, as shown here (in $100,000 per quarter):

```
. regress Sale Promo Devel Research

    Source |       SS           df       MS              Number of obs =      24
-----------+----------------------------------           F(  3,     20) =   22.28
     Model | 43901.7677       3   14633.9226            Prob > F       =  0.0000
  Residual | 13136.2323      20   656.811614            R-square       =  0.7697
-----------+----------------------------------           Adj R-square   =  0.7351
     Total |    57038.00      23   2479.91304            Root MSE       =  25.628


     Sales |      Coef.   Std. Err.       t     P>|T|    [95% Conf. Interval]
-----------+----------------------------------------------------------------
     Promo |   136.0983    28.10759     4.842   0.000     77.46689    194.7297
     Devel |  -61.17526    50.94102    -1.201   0.244    -167.4364    45.08585
  Research |  -43.69508    48.32298    -0.904   0.377    -144.495    57.10489
     _cons |   326.3893    241.6129     1.351   0.192    -177.6063    830.3849
```

a. Write the estimated regression equation.
b. Locate MS(Residual) and its square root, the residual standard deviation.
c. State the interpretation of $\hat{\beta}_i$, the coefficient of promotion expenses.

Bus.   **12.10**   A feeder airline transports passengers from small cities to a single larger hub airport. A regression study tried to predict the revenue generated by each of 22 small cities, based on the distance of each city (in miles) from the hub and on the population of the small cities. The correlations and scatterplots were obtained as shown.

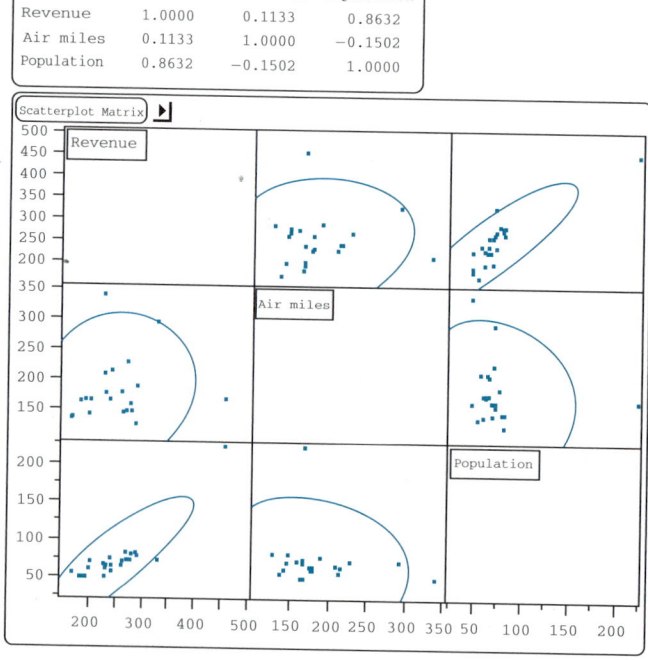

Correlations

Variable	Revenue	Air miles	Population
Revenue	1.0000	0.1133	0.8632
Air miles	0.1133	1.0000	-0.1502
Population	0.8632	-0.1502	1.0000

   a. Are the independent variables severely correlated?
   b. Do the scatterplots indicate that there may be a problem with high leverage points?

12.11 The feeder airline data were used in a multiple regression analysis using JMP. Some of the results are shown next.

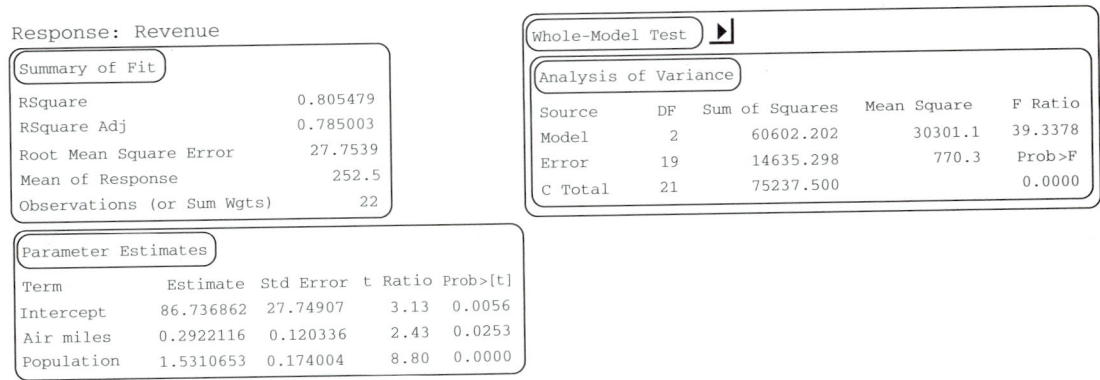

   a. Without considering the output, what sign (positive or negative) would you expect the slopes for air miles and population to have?
   b. Do the slopes in the output have the anticipated signs?
   c. State the meaning of the coefficient of air miles in the output.

Engin. **12.12** A manufacturer tested the abrasive effect of a wear tester for experimental fabrics on a particular fabric while run at six different machine speeds. Forty-eight identical 5-inch-square pieces of fabric were cut, with eight squares randomly assigned to each of the six machine speeds 100, 120, 140, 160, 180, and 200 revolutions per minute (rev/min). The order of assignment of the squares to the machine was random, with each square tested for a 3-minute period at the appropriate machine setting. The amount of wear was measured and recorded for each square. The data appear here.

Machine Speed (in rev/min)	Wear
100	23.0, 23.5, 24.4, 25.2, 25.6, 26.1, 24.8, 25.6
120	26.7, 26.1, 25.8, 26.3, 27.2, 27.9, 28.3, 27.4
140	28.0, 28.4, 27.0, 28.8, 29.8, 29.4, 28.7, 29.3
160	32.7, 32.1, 31.9, 33.0, 33.5, 33.7, 34.0, 32.5
180	43.1, 41.7, 42.4, 42.1, 43.5, 43.8, 44.2, 43.6
200	54.2, 43.7, 53.1, 53.8, 55.6, 55.9, 54.7, 54.5

   a. Generate a graph of the data. (The variability within a speed is about the same for all speeds, so you can save time while still maintaining the trend by plotting the sample mean for each speed.)
   b. What type of regression model appears appropriate?
   c. Output for linear, quadratic, and cubic regression models is shown on the following pages. Which regression equation gives a better fit? Why?
   d. Is there anything peculiar about the data? What might have happened?

```
            LINEAR REGRESSION ANALYSIS FOR WEAR TESTER DATA

Dependent Variable: FABRIC WEAR


Analysis of Variance

                        Sum of          Mean
Source          DF      Squares         Square      F Value      Prob>F

Model           1      4326.79207     4326.79207    291.474      0.0001
Error           46      682.84710       14.84450
C Total         47     5009.63917


        Root MSE         3.85286      R-square      0.8637
        Dep Mean        34.92917      Adj R-sq      0.8607
        C.V.            11.03048

Parameter Estimates

                   Parameter      Standard     T for H0:
Variable   DF      Estimate        Error      Parameter=0    Prob > |T|

INTERCEP   1      -6.765476      2.50470943     -2.701         0.0096
SPEED      1       0.277964      0.01628129     17.073         0.0001

                  Variable
Variable   DF     Label

INTERCEP   1      Intercept
SPEED      1      MACHINE SPEED


        OBS     SPEED     WEAR      PRED1       RESID1

          1      100      23.0     21.0310      1.96905
          2      100      23.5     21.0310      2.46905
          3      100      24.4     21.0310      3.36905
          4      100      25.2     21.0310      4.16905
          5      100      25.6     21.0310      4.56905
          6      100      26.1     21.0310      5.06905
          7      100      24.8     21.0310      3.76905
          8      100      25.6     21.0310      4.56905
          9      120      26.7     26.5902      0.10976
         10      120      26.1     26.5902     -0.49024
         11      120      25.8     26.5902     -0.79024
         12      120      26.3     26.5902     -0.29024
         13      120      27.2     26.5902      0.60976
         14      120      27.9     26.5902      1.30976
         15      120      28.3     26.5902      1.70976
         16      120      27.4     26.5902      0.80976
         17      140      28.0     32.1495     -4.14952
         18      140      28.4     32.1495     -3.74952
         19      140      27.0     32.1495     -5.14952
         20      140      28.8     32.1495     -3.34952
         21      140      29.8     32.1495     -2.34952
```

OBS	SPEED	WEAR	PRED1	RESID1
22	140	29.4	32.1495	−2.74952
23	140	28.7	32.1495	−3.44952
24	140	29.3	32.1495	−2.84952
25	160	32.7	37.7088	−5.00881
26	160	32.1	37.7088	−5.60881
27	160	31.9	37.7088	−5.80881
28	160	33.0	37.7088	−4.70881
29	160	33.5	37.7088	−4.20881
30	160	33.7	37.7088	−4.00881
31	160	34.0	37.7088	−3.70881
32	160	32.5	37.7088	−5.20881
33	180	43.1	43.2681	−0.16810
34	180	41.7	43.2681	−1.56810
35	180	42.4	43.2681	−0.86810
36	180	42.1	43.2681	−1.16810
37	180	43.5	43.2681	0.23190
38	180	43.8	43.2681	0.53190
39	180	44.2	43.2681	0.93190
40	180	43.6	43.2681	0.33190
41	200	54.2	48.8274	5.37262
42	200	43.7	48.8274	−5.12738
43	200	53.1	48.8274	4.27262
44	200	53.8	48.8274	4.97262
45	200	55.6	48.8274	6.77262
46	200	55.9	48.8274	7.07262
47	200	54.7	48.8274	5.87262
48	200	54.5	48.8274	5.67262

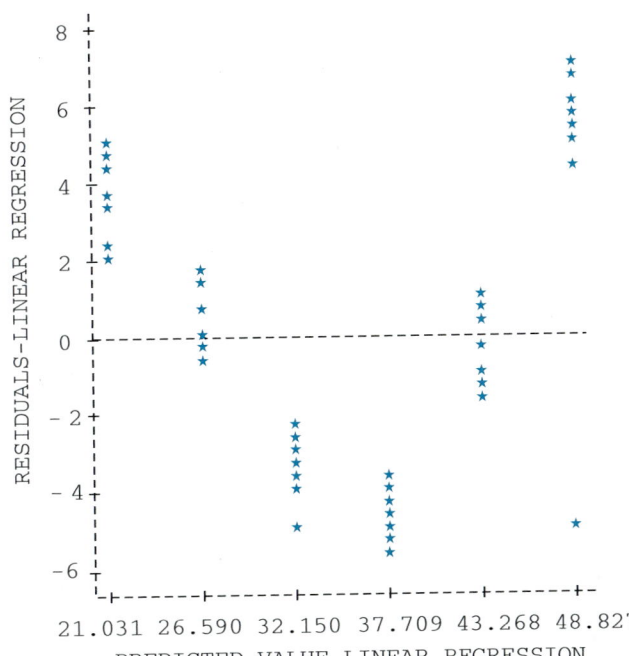

NOTE: 6 obs hidden

```
QUADRATIC REGRESSION ANALYSIS FOR WEAR TESTER DATA

Dependent Variable: FABRIC WEAR


Analysis of Variance

                        Sum of          Mean
Source          DF      Squares         Square       F Value      Prob>F

Model           2     4839.89302     2419.94651      641.532      0.0001
Error          45      169.74614        3.77214
C Total        47     5009.63917


        Root MSE        1.94220      R-square       0.9661
        Dep Mean       34.92917      Adj R-sq       0.9646
        C.V.            5.56039

Parameter Estimates

                    Parameter       Standard     T for H0:
Variable    DF       Estimate         Error      Parameter=0    Prob > |T|

INTERCEP    1       63.139286      6.12529888       10.308        0.0001
SPEED       1       -0.705071      0.08468583       -8.326        0.0001
SPEED2      1        0.003277      0.00028096       11.663        0.0001

                    Variable
Variable    DF       Label

INTERCEP    1     Intercept
SPEED       1     MACHINE SPEED
SPEED2      1     SPEED SQUARED


    OBS       SPEED      WEAR       PRED2       RESID2

      1        100       23.0      25.4000     -2.40000
      2        100       23.5      25.4000     -1.90000
      3        100       24.4      25.4000     -1.00000
      4        100       25.2      25.4000     -0.20000
      5        100       25.6      25.4000      0.20000
      6        100       26.1      25.4000      0.70000
      7        100       24.8      25.4000     -0.60000
      8        100       25.6      25.4000      0.20000
      9        120       26.7      25.7164      0.98357
     10        120       26.1      25.7164      0.38357
     11        120       25.8      25.7164      0.08357
     12        120       26.3      25.7164      0.58357
     13        120       27.2      25.7164      1.48357
     14        120       27.9      25.7164      2.18357
     15        120       28.3      25.7164      2.58357
     16        120       27.4      25.7164      1.68357
     17        140       28.0      28.6543     -0.65429
     18        140       28.4      28.6543     -0.25429
     19        140       27.0      28.6543     -1.65429
```

OBS	SPEED	WEAR	PRED2	RESID2
20	140	28.8	28.6543	0.14571
21	140	29.8	28.6543	1.14571
22	140	29.4	28.6543	0.74571
23	140	28.7	28.6543	0.04571
24	140	29.3	28.6543	0.64571
25	160	32.7	34.2136	-1.51357
26	160	32.1	34.2136	-2.11357
27	160	31.9	34.2136	-2.31357
28	160	33.0	34.2136	-1.21357
29	160	33.5	34.2136	-0.71357
30	160	33.7	34.2136	-0.51357
31	160	34.0	34.2136	-0.21357
32	160	32.5	34.2136	-1.71357
33	180	43.1	42.3943	0.70571
34	180	41.7	42.3943	-0.69429
35	180	42.4	42.3943	0.00571
36	180	42.1	42.3943	-0.29429
37	180	43.5	42.3943	1.10571
38	180	43.8	42.3943	1.40571
39	180	44.2	42.3943	1.80571
40	180	43.6	42.3943	1.20571
41	200	54.2	53.1964	1.00357
42	200	43.7	53.1964	-9.49643
43	200	53.1	53.1964	-0.09643
44	200	53.8	53.1964	0.60357
45	200	55.6	53.1964	2.40357
46	200	55.9	53.1964	2.70357
47	200	54.7	53.1964	1.50357
48	200	54.5	53.1964	1.30357

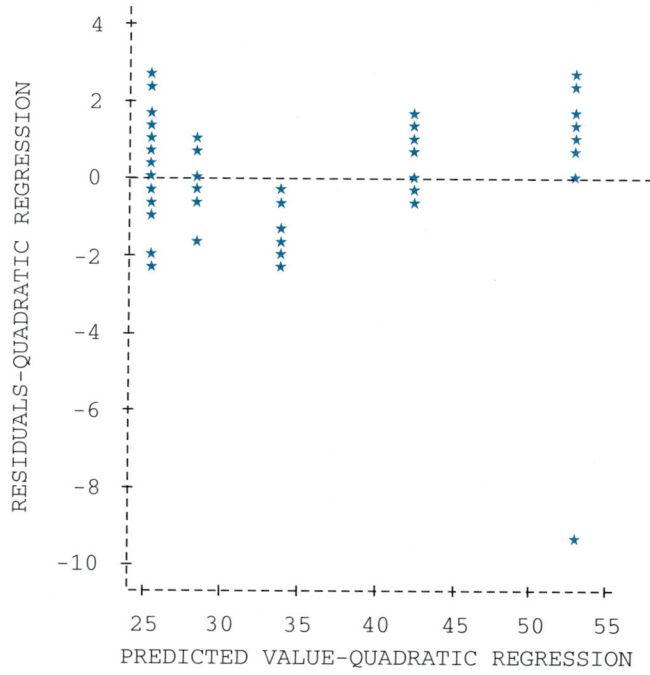

```
CUBIC REGRESSION ANALYSIS FOR WEAR TESTER DATA

Dependent Variable: FABRIC WEAR

Analysis of Variance

                       Sum of          Mean
Source        DF      Squares         Square      F Value     Prob>F

Model          3    4846.78202     1615.59401     436.494     0.0001
Error         44     162.85714        3.70130
C Total       47    5009.63917

        Root MSE        1.92388     R-square      0.9675
        Dep Mean       34.92917     Adj R-sq      0.9653
        C.V.            5.50794

Parameter Estimates

                     Parameter      Standard     T for H0:
Variable   DF         Estimate         Error    Parameter=0    Prob > |T|

INTERCEP    1       18.872619    33.00952220        0.572        0.5704
SPEED       1        0.238477     0.69668199        0.342        0.7338
SPEED2      1       -0.003208     0.00476113       -0.674        0.5040
SPEED2      1     0.000014410     0.00001056        1.364        0.1794

                     Variable
Variable   DF         Label

INTERCEP    1     Intercept
SPEED       1     MACHINE SPEED
SPEED2      1     SPEED SQUARED
SPEED3      1     SPEED CUBED


        OBS     SPEED     WEAR       PRED3        RESID3

          1      100      23.0      25.0542      -2.05417
          2      100      23.5      25.0542      -1.55417
          3      100      24.4      25.0542      -0.65417
          4      100      25.2      25.0542       0.14583
          5      100      25.6      25.0542       0.54583
          6      100      26.1      25.0542       1.04583
          7      100      24.8      25.0542      -0.25417
          8      100      25.6      25.0542       0.54583
          9      120      26.7      26.2006       0.49940
         10      120      26.1      26.2006      -0.10060
         11      120      25.8      26.2006      -0.40060
         12      120      26.3      26.2006       0.09940
         13      120      27.2      26.2006       0.99940
         14      120      27.9      26.2006       1.69940
         15      120      28.3      26.2006       2.09940
         16      120      27.4      26.2006       1.19940
         17      140      28.0      28.9310      -0.93095
         18      140      28.4      28.9310      -0.53095
```

OBS	SPEED	WEAR	PRED3	RESID3
19	140	27.0	28.9310	−1.93095
20	140	28.8	28.9310	−0.13095
21	140	29.8	28.9310	0.86905
22	140	29.4	28.9310	0.46905
23	140	28.7	28.9310	−0.23095
24	140	29.3	28.9310	0.36905
25	160	32.7	33.9369	−1.23690
26	160	32.1	33.9369	−1.83690
27	160	31.9	33.9369	−2.03690
28	160	33.0	33.9369	−0.93690
29	160	33.5	33.9369	−0.43690
30	160	33.7	33.9369	−0.23690
31	160	34.0	33.9369	0.06310
32	160	32.5	33.9369	−1.43690
33	180	43.1	41.9101	1.18988
34	180	41.7	41.9101	−0.21012
35	180	42.4	41.9101	0.48988
36	180	42.1	41.9101	0.18988
37	180	43.5	41.9101	1.58988
38	180	43.8	41.9101	1.88988
39	180	44.2	41.9101	2.28988
40	180	43.6	41.9101	1.68988
41	200	54.2	53.5423	0.65774
42	200	43.7	53.5423	−9.84226
43	200	53.1	53.5423	−0.44226
44	200	53.8	53.5423	0.25774
45	200	55.6	53.5423	2.05774
46	200	55.9	53.5423	2.35774
47	200	54.7	53.5423	1.15774
48	200	54.5	53.5423	0.95774

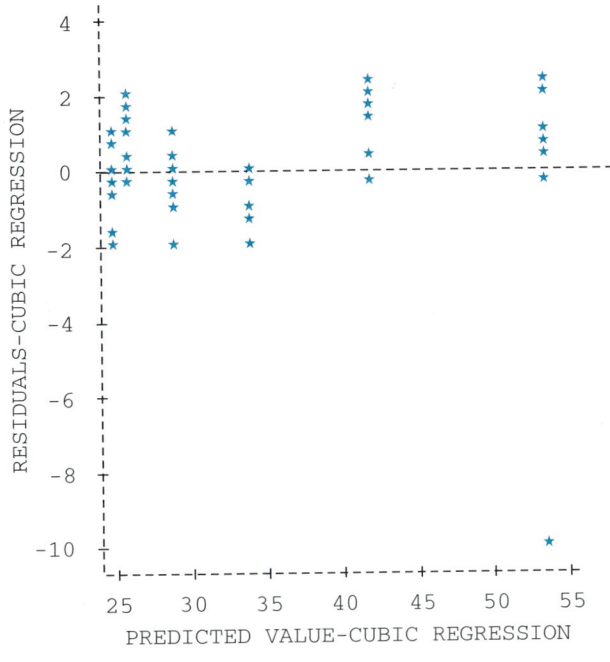

12.13   Refer to the data of Exercise 12.12. Suppose that another variable was controlled and that the first four squares at each speed were treated with a .2 concentration of protective coating, whereas the second four squares were treated with a .4 concentration of the same coating. x_1 denotes the machine speed and x_2 denotes the concentration of the protective coating. Fit these models using available statistical software. Which model seems to provide a better fit to the data? Why?

$$y = \beta_0 + \beta_1 x_1 + \beta_2 x_1^2 + \beta_3 x_2 + \varepsilon$$

$$y = \beta_0 + \beta_1 x_1 + \beta_2 x_1^2 + \beta_3 x_2 + \beta_4 x_1 x_2 + \beta_5 x_1^2 x_2 + \varepsilon$$

12.4   Inferences in Multiple Regression

We make inferences about any of the parameters in the general linear model (and hence in multiple regression) as we did for β_0 and β_1 in the linear regression model, $y = \beta_0 + \beta_1 x + \varepsilon$.

coefficient of determination

Before we do this, however, we must introduce the *coefficient of determination*, R^2. The **coefficient of determination,** R^2, is defined and interpreted very much like the r^2 value in Chapter 11. (The customary notation is R^2 for multiple regression and r^2 for simple linear regression.) As in Chapter 11, we define the coefficient of determination as the proportional reduction in the squared error of y, which we obtain by knowing the values of $x_1, \ldots, x_k$. For example, if we have the multiple regression model with three x values, and $R^2_{y \cdot x_1 x_2 x_3} = .736$, then we can account for 73.6% of the variability of the y values by variability in x_1, x_2, and x_3. Formally,

$$R^2_{y \cdot x_1 \cdots x_k} = \frac{\text{SS(Total)} - \text{SS(Residual)}}{\text{SS(Total)}}$$

where

$$\text{SS(Total)} = \Sigma (y_i - \bar{y})^2$$

EXAMPLE 12.10

Locate the value of $R^2_{y \cdot x_1 x_2}$ in the computer output of Example 12.8.

Solution   We want R-sq = 87.5%, not the one that is adj. Alternatively, SS(Total) = 682.00 and SS(Residual) = 85.17 are shown in the output, and we can compute $R^2_{y \cdot x_1 x_2} = (682.00 - 85.17)/682.00 = .875$.

There is no general relation between the multiple R^2 from a multiple regression equation and the individual coefficients of determination $r^2_{y x_1}, r^2_{y x_2}, \ldots, r^2_{y x_k}$ other than that multiple R^2 must be at least as big as any of the individual r^2 values. If all the independent variables are themselves perfectly uncorrelated with each other, then multiple R^2 is just the sum of the individual r^2 values. Equivalently, if all the xs are uncorrelated with each other, SS(Regression) for the all-predictors model is equal to the sum of SS(Regression) values for simple regressions using one x at a time. If the xs are correlated, it is much more difficult to break apart the overall predictive value of $x_1, x_2, \ldots, x_k$ as measured by $R^2_{y \cdot x_1 \cdots x}$ into separate pieces that can be attributable to x_1 alone, to x_2 alone, $\ldots$, to x_k alone.

collinearity

When the independent variables are themselves correlated, **collinearity** (sometimes called *multicollinearity*) is present. In multiple regression, we are trying to separate out the predictive value of several predictors. When the predictors are highly correlated, this task is very difficult. For example, suppose that we try to explain variation in regional housing sales over time, using gross domestic product (GDP) and national disposable income (DI) as two of the predictors. DI has been almost exactly a fraction of GDP, so the correlation of these two predictors will be extremely high. Now, is variation in housing sales attributable more to variation in GDP or to variation in DI? Good luck taking those two apart! It is very likely that either predictor alone will explain variation in housing sales almost as well as both together.

Collinearity is usually present to some degree in a multiple regression study. It is a small problem for slightly correlated xs but a more severe one for highly correlated xs. Thus, if collinearity occurs in a regression study—and it usually does to some degree—it is not easy to break apart the overall $R^2_{y \cdot x_1 x_2 \cdots x_k}$ into separate components associated with each x variable. The correlated xs often account for overlapping pieces of the variability in y, so that often, but not inevitably,

$$R^2_{y \cdot x_1 x_2 \cdots x_k} < r^2_{yx1} + r^2_{yx2} + \cdots + r^2_{yxk}$$

sequential sums of squares

Many statistical computer programs will report **sequential sums of squares.** These SS are *incremental* contributions to SS(Regression) when the independent variables enter the regression model in the order you specify to the program. Sequential sums of squares depend heavily on the particular order in which the independent variables enter the model. Again, the trouble is collinearity. For example, if all variables in a regression study are strongly and positively correlated (as often happens in economic data), whichever independent variable happens to be entered first typically accounts for most of the explainable variation in y and the remaining variables add little to the sequential SS. The explanatory power of any x given all the other xs (which is sometimes called the *unique predictive value* of that x) is small. When the data exhibit severe collinearity, separating out the predictive value of the various independent variables is very difficult indeed.

EXAMPLE 12.11

Interpret the sequential sums of squares in the following output for the data of Example 12.6. If x_2 and x_1 were used as predictors (in that order), would we obtain the same sequential sums of squares numbers?

```
MTB > Correlation 'y' 'x1' 'x2'.

              y        x1
x1        0.922
x2        0.156     0.000


MTB > Regress 'y' 2 'x1' 'x2'.

The regression equation is
y = 39.0 + 0.983 x1 + 0.333 x2

Predictor        Coef      Stdev     t-ratio         p
Constant       39.000      1.256       31.05     0.000
x1             0.9833     0.1538        6.39     0.001
x2             0.3333     0.3076        1.08     0.320

s = 3.768     R-sq = 87.5%     R-sq(adj) = 83.3%
```

```
Analysis of Variance

SOURCE          DF         SS          MS        F         p
Regression      2       596.83      298.42    21.02    0.002
Error           6        85.17       14.19
Total           8       682.00

SOURCE          DF      SEQ SS
x1              1       580.17
x2              1        16.67
```

Solution   The SEQ SS column shows that x_1 by itself accounts for 580.17 of the total variation in y and that adding x_2 after x_1 accounts for another 16.67 of the y variation. This example is a rarity in that the predictors are completely uncorrelated; in this unusual case, the order of adding predictors does not matter.

```
MTB > Regress 'y' 2 'x2' 'x1'.

The regression equation is
y = 39.0 + 0.333 x2 + 0.983 x1

Predictor       Coef       Stdev     t-ratio        p
Constant      39.000       1.256      31.05     0.000
x2            0.3333      0.3076       1.08     0.320
x1            0.9833      0.1538       6.39     0.001

s = 3.768     R-sq = 87.5%     R-sq(adj) = 83.3%

Analysis of Variance

SOURCE          DF         SS          MS        F         p
Regression      2       596.83      298.42    21.02    0.002
Error           6        85.17       14.19
Total           8       682.00

SOURCE          DF      SEQ SS
x2              1        16.67
x1              1       580.17
```

The ideas of Section 12.4 involve point (best guess) estimation of the regression coefficients and the standard deviation s_ϵ. Because these estimates are based on sample data, they will be in error to some extent, and a manager should allow for that error in interpreting the model. We now present tests about the partial slope parameters in a multiple regression model.

First, we examine a test of an overall null hypothesis about the partial slopes $(\beta_1, \beta_2, \ldots, \beta_k)$ in the multiple regression model. According to this hypothesis, $H_0: \beta_1 = \beta_2 = \cdots \beta_k = 0$, none of the variables included in the multiple regression has any predictive value at all. This is the "nullest" of null hypotheses; it says that all those carefully chosen predictors are absolutely useless. The research hypothesis is a very general one—namely, H_a: At least one $\beta_j \neq 0$. This merely says that there is some predictive value somewhere in the set of predictors.

The test statistic is the F statistic of Chapter 11. To state the test, we first define the sum of squares attributable to the regression of y on the variables x_1, $x_2, \ldots, x_k$. We designate this sum of squares as SS(Regression); it is also called SS(Model) or the explained sum of squares. It is the sum of squared differences between predicted values and the mean y value.

DEFINITION 12.2

$$SS(\text{Regression}) = \Sigma(\hat{y}_i - \bar{y})^2$$

$$SS(\text{Total}) = \Sigma(y_i - \bar{y})^2$$

$$= SS(\text{Regression}) + SS(\text{Residual})$$

Unlike SS(Total) and SS(Residual), we don't interpret SS(Regression) in terms of prediction error. Rather, it measures the extent to which the predictions $\hat{y}_i$ vary as the xs vary. If SS(Regression) $= 0$, the predicted y values ($\hat{y}$) are all the same. In such a case, information about the xs is useless in predicting y. If SS(Regression) is large relative to SS(Residual), the indication is that there is real predictive value in the independent variables $x_1, x_2, \ldots, x_k$. We state the test statistic in terms of mean squares rather than sums of squares. As always, a mean square is a sum of squares divided by the appropriate df.

F Test of H_0: $\beta_1 = \beta_2 =$ $\cdots = \beta_k = 0$

H_0: $\beta_1 = \beta_2 = \cdots = \beta_k = 0$

H_a: At least one $\beta \neq 0$.

T.S.: $F = \dfrac{SS(\text{Regression})/k}{SS(\text{Residual})/[n - (k + 1)]} = \dfrac{MS(\text{Regression})}{MS(\text{Residual})}$

R.R.:   With $df_1 = k$ and $df_2 = n - (k + 1)$, reject H_0 if $F > F_\alpha$.

Check assumptions and draw conclusions.

EXAMPLE 12.12

a. Locate SS(Regression) in the computer output of Example 12.11, reproduced here.

b. Locate the F statistic.

c. Can we safely conclude that the independent variables x_1 and x_2 together have at least some predictive power?

```
MTB > regress c1 on 2 vars c2 c3

The regression equation is
y = 39.0 + 0.983 x1 + 0.333 x2
```

```
Predictor         Coef      Stdev     t-ratio        p
Constant        39.000      1.256       31.05    0.000
x1              0.9833     0.1538        6.39    0.001
x2              0.3333     0.3076        1.08    0.320

s = 3.768     R-sq = 87.5%     R-sq(adj) = 83.3%

Analysis of Variance

SOURCE         DF         SS         MS         F        p
Regression      2     596.83     298.42     21.02    0.002
Error           6      85.17      14.19
Total           8     682.00
```

Solution

a. SS(Regression) is shown in the Analysis of Variance section of the output as 596.83.

b. The MS(Regression) and MS(Residual) values are also shown there. MS(Residual) is labeled as MS(Error), a common alternative name.

$$F = \frac{MS(\text{Regression})}{MS(\text{Residual})} = \frac{298.42}{14.19} = 21.02$$

c. For $df_1 = 2$, $df_2 = 6$, and $\alpha = .01$, the tabled F value is 10.92. Therefore, we have strong evidence (p-value well below .01, shown as .002) to reject the null hypothesis and conclude that the xs collectively have at least some predictive value.

This F test may also be stated in terms of R^2. Recall that $R^2_{y \cdot x_1 \cdots x_k}$ measures the reduction in squared error for y attributed to knowledge of all the x predictors. Because the regression of y on the xs accounts for a proportion $R^2_{y \cdot x_1 \cdots x_k}$ of the total squared error in y,

$$SS(\text{Regression}) = R^2_{y \cdot x_1 \cdots x_k} SS(\text{Total})$$

The remaining fraction, $1 - R^2$, is incorporated in the residual squared error:

$$SS(\text{Residual}) = (1 - R^2_{y \cdot x_1 \cdots x_k})SS(\text{Total})$$

F and R^2

The overall F test statistic can be rewritten as

$$F = \frac{MS(\text{Regression})}{MS(\text{Residual})} = \frac{R^2_{y \cdot x_1 \cdots x_k}/k}{(1 - R^2_{y \cdot x_1 \cdots x_k})/[n - (k + 1)]}$$

This statistic is to be compared with tabulated F values for $df_1 = k$ and $df_2 = n - (k + 1)$.

EXAMPLE 12.13

A large city bank studies the relation of average account size in each of its branches to per capita income in the corresponding zip code area, number of business accounts, and number of competitive bank branches. The data are analyzed by Statistix, as shown here:

```
CORRELATIONS (PEARSON)

          ACCTSIZE  BUSIN    COMPET
BUSIN     -0.6934
COMPET     0.8196  -0.6527
INCOME     0.4526   0.1492   0.5571


UNWEIGHTED LEAST SQUARES LINEAR REGRESSION OF ACCTSIZE


PREDICTOR
VARIABLES     COEFFICIENT    STD ERROR    STUDENT'S T      P      VIF

CONSTANT        0.15085       0.73776        0.20       0.8404
BUSIN          -0.00288       8.894E-04     -3.24       0.0048    5.2
COMPET         -0.00759       0.05810       -0.13       0.8975    7.4
INCOME          0.26528       0.10127        2.62       0.0179    4.3


R-SQUARED              0.7973    RESID. MEAN SQUARE (MSE)    0.03968
ADJUSTED R-SQUARED     0.7615    STANDARD DEVIATION          0.19920


SOURCE        DF      SS        MS        F        P

REGRESSION     3    2.65376   0.88458   22.29    0.0000
RESIDUAL      17    0.67461   0.03968
TOTAL         20    3.32838
```

a. Identify the multiple regression prediction equation.
b. Use the R^2 value shown to test $H_0: \beta_1 = \beta_2 = \beta_3 = 0$. (*Note: n = 21.*)

Solution
a. From the output, the multiple regression forecasting equation is

$$\hat{y} = 0.15085 - 0.00288x_1 - 0.00759x_2 + 0.26528x_3$$

b. The test procedure based on R^2 is

$$H_0: \quad \beta_1 = \beta_2 = \beta_3 = 0$$

$$H_a: \quad \text{At least one } \beta_j \text{ differs from zero.}$$

$$\text{T.S.:} \quad F = \frac{R^2_{y \cdot x_1 x_2 x_3}/3}{(1 - R^2_{y \cdot x_1 x_2 x_3})/(21 - 4)} = \frac{.7973/3}{.2027/17} = 22.29$$

R.R.: For $df_1 = 3$ and $df_2 = 17$, the critical .05 value of F is 3.20.

Because the computed F statistic, 22.29, is greater than 3.20, we reject H_0 and conclude that one or more of the x values has some predictive power. This also follows because the *p*-value, shown as .0000, is (much) less than .05. Note that the F value we compute is the same as that shown in the output.

Rejection of the null hypothesis of this F test is not an overwhelmingly impressive conclusion. This rejection merely indicates that there is good evidence of *some* degree of predictive value *somewhere* among the independent variables.

It does not give any direct indication of how strong the relation is, nor any indication of which individual independent variables are useful. The next task, therefore, is to make inferences about the individual partial slopes.

To make these inferences, we need the estimated standard error of each partial slope. As always, the standard error for any estimate based on sample data indicates how accurate that estimate should be. These standard errors are computed and shown by most regression computer programs. They depend on three things: the residual standard deviation, the amount of variation in the predictor variable, and the degree of correlation between that predictor and the others. The expression that we present for the standard error is useful in considering the effect of collinearity (correlated independent variables), but it is *not* a particularly good way to do the computation. Let a computer program do the arithmetic.

DEFINITION 12.3

Estimated standard error of $\hat{\beta}_j$ in a multiple regression:

$$s_{\hat{\beta}_j} = s_\epsilon \sqrt{\frac{1}{\sum (x_{ij} - \bar{x}_j)^2 (1 - R^2_{x_j \cdot x_1 \cdots x_{j-1} x_{j+1} \cdots x_k})}}$$

where $R^2_{x_j \cdot x_1 \cdots x_{j-1} x_{j+1} \cdots x_k}$ is the R^2 value obtained by letting x_j be the *dependent* variable in a multiple regression, with all other xs independent variables. Note that s_ϵ is the residual standard deviation for the multiple regression of y on $x_1, x_2, \ldots, x_k$.

As in simple regression, the larger the residual standard deviation, the larger the uncertainty in estimating coefficients. Also, the less variability there is in the predictor, the larger is the standard error of the coefficient. The most important **effect of collinearity**   use of the formula for estimated standard error is to illustrate the **effect of collinearity**. If the independent variable x_j is highly collinear with one or more other independent variables, $R^2_{x_j \cdot x_1 \cdots x_{j-1} x_{j+1} \cdots x_k}$ is by definition very large and $1 - R^2_{x_j \cdot x_1 \cdots x_{j-1} x_{j+1} \cdots x_k}$ is near zero. Division by a near-zero number yields a very large standard error. Thus, one important effect of severe collinearity is that it results in very large standard errors of partial slopes and therefore very inaccurate estimates of those slopes.

variance inflation factor   The term $1/(1 - R^2_{x_j \cdot x_1 \cdots x_{j-1} x_{j+1} \cdots x_k})$ is called the **variance inflation factor** (VIF). It measures how much the variance (square of the standard error) of a coefficient is increased because of collinearity. This factor is printed out by some computer packages and is helpful in assessing how serious the collinearity problem is. If the VIF is 1, there is no collinearity at all. If it is very large, such as 10 or more, collinearity is a serious problem.

A large standard error for any estimated partial slope indicates a large probable error for the estimate. The partial slope $\hat{\beta}_j$ of x_j estimates the effect of increasing x_j by one unit while all other xs remain constant. If x_j is highly collinear with other xs, when x_j increases, the other xs also vary rather than staying constant. Therefore, it is difficult to estimate β_j, and its probable error is large when x_j is severely collinear with other independent variables.

The standard error of each estimated partial slope $\hat{\beta}_j$ is used in a confidence interval and statistical test for $\hat{\beta}_j$. The confidence interval follows the familiar format of estimate plus or minus (table value) (estimated standard error). The table value is the t table with the error df, $n - (k + 1)$.

DEFINITION 12.4

The confidence interval for β_j is

$$\hat{\beta}_j - t_{\alpha/2} s_{\hat{\beta}_j} \leq \beta_j \leq \hat{\beta}_j + t_{\alpha/2} s_{\hat{\beta}_j}$$

where $t_{\alpha/2}$ cuts off area $\alpha/2$ in the tail of a t distribution with df = $n - (k + 1)$, the error df.

EXAMPLE 12.14

Calculate a 95% confidence interval for β_1 in the two-predictor model for the data of Example 12.7. Relevant output follows:

```
. regress y   x1 x2


-----------------------------------------------------------------------
      y |   Coef.   Std. Err.    t      P>|t|       [95% Conf. Interval]
--------+--------------------------------------------------------------
     x1 |      1    1.870829   0.535    0.646      -7.049526    9.049526
     x2 |      3    4.1833     0.717    0.548     -14.99929    20.99929
  _cons |     10    1.183216   8.452    0.014       4.909033   15.09097
-----------------------------------------------------------------------
```

Solution $\hat{\beta}_1$ is 1.00 and the standard error is shown as 1.870829. The t value that cuts off an area of .025 in a t distribution with df = $n - (k + 1) = 5 - (2 + 1) = 2$ is 4.303. The confidence interval is $1.00 - 4.303 (1.870829) \leq \beta_1 \leq 1.00 + 4.303(1.870829)$, or $-7.050 \leq \beta_1 \leq 9.050$. The output shows this interval to more decimal places.

EXAMPLE 12.15

Locate the estimated partial slope for x_2 and its standard error in the output of Example 12.12. Calculate a 90% confidence interval for β_2.

```
MTB > Regress 'y' on 2 vars 'x1' 'x2'

The regression equation is
y = 39.0 + 0.983 x1 + 0.333 x2

Predictor       Coef      Stdev    t-ratio        p
Constant      39.000      1.256      31.05    0.000
x1            0.9833     0.1538       6.39    0.001
x2            0.3333     0.3076       1.08    0.320
```

Solution $\hat{\beta}_2$ is .3333 with standard error (labeled Stdev) .3076. The tabled t value is 1.943 [tail area .05, $9 - (2 + 1) = 6$ df]. The desired interval is $.3333 - 1.943(.3076) \leq \beta_2 \leq .3333 + 1.943 (.3076)$, or $-.2644 \leq \beta_2 \leq .9310$.

interpretation of H_0:
$\beta_j = 0$

The usual null hypothesis for inference about β_j is $H_0: \beta_j = 0$. This hypothesis does not assert that x_j has no predictive value by itself. It asserts that it has no *additional* predictive value over and above that contributed by the other independent variables; that is, if all other xs had already been used in a regression model and then x_j was added last, the prediction would not improve. The test of H_0: $\beta_j = 0$ measures whether x_j has any additional (e.g., unique) predictive value. The t test of this H_0 is summarized next.

Summary for Testing β_j

H_0:   1. $\beta_j \leq 0$
          2. $\beta_j \geq 0$
          3. $\beta_j = 0$

H_a: 1. $\beta_j > 0$
          2. $\beta_j < 0$
          3. $\beta_j \neq 0$

T.S.: $t = \hat{\beta}_j / s_{\hat{\beta}_j}$

R.R.:   1. $t > t_\alpha$
          2. $t < t_\alpha$
          3. $|t| > t_{\alpha/2}$

where t_α cuts off a right-tail area a in the t distribution with df $= n - (k + 1)$.

Check assumptions and draw conclusions.

This test statistic is shown by virtually all multiple regression programs.

EXAMPLE 12.16

a. Use the information given in Example 12.14 to test H_0: $\beta_1 = 0$ at $\alpha = .05$. Use a two-sided alternative.
b. Is the conclusion of the test compatible with the confidence interval?

Solution

a. The test statistic for H_0: $\beta = 0$ versus H_a: $\beta_1 \neq 0$ is $t = \hat{\beta}_1 / s_{\hat{\beta}_1} = 1.00/1.871 = .535$. Because the .025 point for the t distribution with $5 - (2 + 1) = 2$ df is 4.303. H_0 must be retained; x_1 has not been shown to have any additional predictive power in the presence of the other independent variable x_2.
b. The 95% confidence interval includes zero, which also indicates that H_0: $\beta_1 = 0$ must be retained at $\alpha = .05$, two-tailed.

EXAMPLE 12.17

Locate the t statistic for testing H_0: $\beta_2 = 0$ in the output of Example 12.15. Can H_a: $\beta_2 > 0$ be supported at any of the usual α levels?

Solution   The t statistics are shown under the heading t-ratio. For x_2, the t statistic is 1.08. The t table value for 6 df and $\alpha = .10$ is 1.440, so H_0 cannot be rejected even at $\alpha = .10$. Alternatively, the p-value is .320, larger than $\alpha = .10$, so again H_0 cannot be rejected.

The multiple regression F and t tests that we discuss in this chapter test different null hypotheses. It sometimes happens that the F test results in the rejection of H_0: $\beta_1 = \beta_2 = \cdots = \beta_k = 0$, whereas no t test of H_0: $\beta_j = 0$ is significant. In such a case, we can conclude that there is predictive value in the equation as a whole, but we cannot identify the specific variables that have predictive value. Remember that each t test is testing the unique predictive value. Does this variable add predictive value, given all the other predictors? When two or more predictor variables are highly correlated among themselves, it often happens that no x_j can be shown to have significant, unique predictive value, even though the xs together have been shown to be useful. If we are trying to predict housing sales based on gross domestic product and disposable income, we probably cannot prove that GDP adds value given DI, or that DI adds value given GDP.

EXERCISES

12.14 Refer to the computer output of Exercise 12.9. Here it is again.

```
. regress Sale Promo Devel Research

    SOURCE |       SS           df        MS              Number of obs =       24
-----------+------------------------------              F(  3,     20) =    22.28
     MODEL |   43901.7677        3   14633.9226          Prob > F       =   0.0000
  Residual |   13136.2323       20   656.811614          R-square       =   0.7697
-----------+------------------------------              Adj R-square   =   0.7351
     Total |     57038.00       23   2479.91304          Root MSE       =   25.628


--------------------------------------------------------------------------------
    SALES |      Coef.   Std. Err.      t     P>|T|    [95% Conf. Interval]
----------+---------------------------------------------------------------------
    Promo |   136.0983   28.10759    4.842   0.000    77.46689     194.7297
    Devel |  -61.17526   50.94102   -1.201   0.244   -167.4364     45.08585
 Research |  -43.69508   48.32298   -0.904   0.377   -144.495      57.10489
    _cons |   326.3893   241.6129    1.351   0.192   -177.6063     830.3849
--------------------------------------------------------------------------------
```

a. Locate the F statistic.
b. Can the hypothesis of no overall predictive value be rejected at $\alpha = .01$?
c. Locate the t statistic for the coefficient of promotion $\hat{\beta}_1$.
d. Test the research hypothesis that $\beta_1 \neq 0$. Use $\alpha = 0.5$.
e. State the conclusion of the test in part (d).

12.15 Locate the p-value for the test of the previous exercise, part (d). Is it one-tailed or two-tailed?

12.16 Summarize the results of the t tests in Exercise 12.14. What null hypotheses are being tested?

12.17 The following artificial data are designed to illustrate the effect of correlated and uncorrelated independent variables:

y:	17	21	26	22	27	25	28	34	29	37	38	38
x:	1	1	1	1	2	2	2	2	3	3	3	3
w:	1	2	3	4	1	2	3	4	1	2	3	4
v:	1	1	2	2	3	3	4	4	5	5	6	6

The output for these data is shown here.

```
MTB > Regress 'Y' 3 'X' 'W' 'V'.

The regression equation is
y = 10.0 + 5.00 X + 2.00 W + 1.00 V

Predictor        Coef      Stdev     t-ratio        p
Constant       10.000      5.766        1.73    0.121
X               5.000      6.895        0.73    0.489
W               2.000      1.528        1.31    0.227
V               1.000      3.416        0.29    0.777

s = 2.646      R-sq = 89.5%    R-sq(adj) = 85.6%

Analysis of Variance

SOURCE          DF         SS         MS          F        p
Regression       3     479.00     159.67      22.81    0.000
Error            8      56.00       7.00
Total           11     535.00

SOURCE          DF     SEQ SS
X                1     392.00
W                1      86.40
V                1       0.60
```

a. Locate MS(Regression) and MS(Residual).
b. What is the value of the F statistic?
c. Determine the p-value for the F test.
d. What conclusion can be established from the F test?
e. Calculate a 95% confidence interval for the true coefficient of X.

12.18 In Exercise 12.11, we considered predicting revenue from each origin as a function of population and air miles to the hub airport. JMP output is shown again here. Is there a clear indication that the two independent variables together have at least some value in predicting revenue?

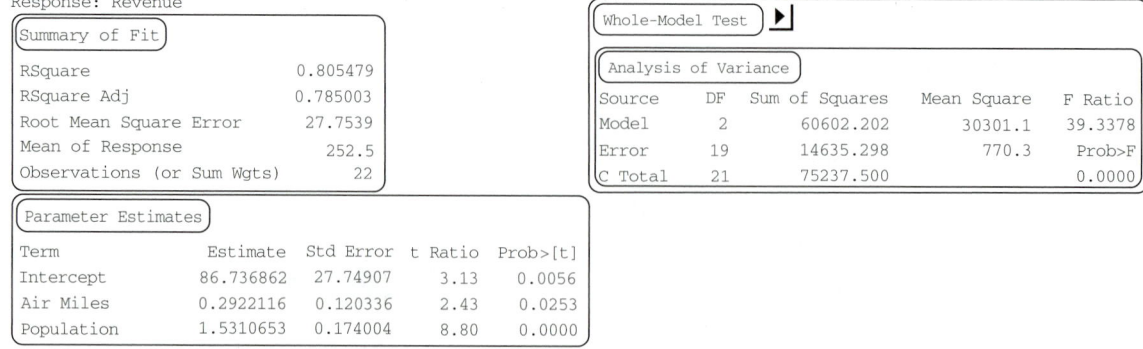

Response: Revenue

Summary of Fit

RSquare	0.805479
RSquare Adj	0.785003
Root Mean Square Error	27.7539
Mean of Response	252.5
Observations (or Sum Wgts)	22

Parameter Estimates

Term	Estimate	Std Error	t Ratio	Prob>|t|
Intercept	86.736862	27.74907	3.13	0.0056
Air Miles	0.2922116	0.120336	2.43	0.0253
Population	1.5310653	0.174004	8.80	0.0000

Whole-Model Test ▶

Analysis of Variance

Source	DF	Sum of Squares	Mean Square	F Ratio
Model	2	60602.202	30301.1	39.3378
Error	19	14635.298	770.3	Prob>F
C Total	21	75237.500		0.0000

12.19   In the feeder airline regression, is there strong evidence that each independent variable is adding predictive value, given the other?

12.20   Use the feeder airline output for Exercise 12.18 to calculate 90% confidence intervals for the two partial slopes. The relevant degrees of freedom are those shown for Error in the output.

12.21   A metalworking firm conducts an energy study using multiple regression methods. The dependent variable is y = energy consumption cost per day (in thousands of dollars), and the independent variables are x_1 = tons of metal processed in the day, x_2 = average external temperature $-60°F$ (a union contract requires cooling of the plant whenever outside temperatures reach 60°), x_3 = rated wattage for machinery in use, and $x_4 = x_1 x_2$. The data are analyzed by Statistix. Selected output is shown here:

```
CORRELATIONS (PEARSON)

              ENERGY    METAL    METXTEMP  TEMP
METAL         0.6128
METXTEMP      0.4929    0.1094
TEMP          0.4007   -0.0606   0.9831
WATTS         0.5775    0.2239   0.3630    0.3529


UNWEIGHTED LEAST SQUARES LINEAR REGRESSION OF ENERGY


PREDICTOR
VARIABLES    COEFFICIENT   STD ERROR   STUDENT'S T     P       VIF

CONSTANT       7.20439      17.5322       0.41       0.6855
METAL          1.36291       0.92438      1.47       0.1559    8.8
TEMP           0.30588       1.62104      0.19       0.8522    250.0
WATTS          0.01024       0.00473      2.16       0.0427    1.5
METXTEMP      -0.00277       0.07722     -0.04       0.9717    246.4


R-SQUARED              0.6636     RESID. MEAN SQUARE (MSE)    6.51555
ADJUSTED R-SQUARED     0.5963     STANDARD DEVIATION          2.55255


SOURCE        DF     SS        MS        F        P

REGRESSION     4   257.048   64.2622    9.86    0.0001
RESIDUAL      20   130.311    6.51555
TOTAL         24   387.360

CASES INCLUDED 25   MISSING CASES 0
```

   a. Write the estimated model.
   b. Summarize the results of the various t tests.
   c. Calculate a 95% confidence interval for the coefficient of METXTEMP.
   d. What does the VIF column of the output indicate about collinearity problems?

12.5   Testing a Subset of Regression Coefficients

F test for several β_js

In the last section, we presented an F test for testing *all* the coefficients in a regression model and a t test for testing *one* coefficient. Another F test of the null hypothesis tests that *several* of the true coefficients are zero—that is, that several

12.27 The credit rating data were reanalyzed, using only the monthly income variable as a predictor. JMP results are shown.

  a. By how much has the regression sum of squares been reduced by eliminating age and debt percentage as predictors?

  b. Do these variables add statistically significant (at normal α levels) predictive value, once income is given?

```
Response: Rating score
```

Summary of Fit

RSquare	0.895261
RSquare Adj	0.895051
Root Mean Square Error	4.571792
Mean of Response	65.044
Observations (or Sum Wgts)	500

Lack of Fit

Parameter Estimates

Term	Estimate	Std Error	t Ratio	Prob>|t|
Intercept	30.152827	0.572537	52.67	0.0000
Monthly income	0.0135544	0.000208	65.24	0.0000

Engin. **12.28** A chemical firm tests the yield that results from the presence of varying amounts of two catalysts. Yields are measured for five different amounts of catalyst 1 paired with four different amounts of catalyst 2. A second-order model is fit to approximate the anticipated nonlinear relation. The variables are y = yield, x_1 = amount of catalyst 1, x_2 = amount of catalyst 2, $x_3 = x_1^2$, $x_4 = x_1 x_2$, and $x_5 = x_2^2$. Selected output from the regression analysis is shown here.

```
                    Multiple Regression Analysis

      Dependent variable: Yield

                          Table of Estimates
```

	Estimate	Standard Error	t Value	P Value
Constant	50.0195	4.3905	11.39	0.0000
Cat1	6.64357	2.01212	3.30	0.0052
Cat2	7.3145	2.73977	2.67	0.0183
@Cat1Sq	-1.23143	0.301968	-4.08	0.0011
@Cat1Cat2	-0.7724	0.319573	-2.42	0.0299
@Cat2Sq	-1.1755	0.50529	-2.33	0.0355

```
R-squared = 86.24%
Adjusted R-squared = 81.33%
Standard error of estimation = 2.25973
```

```
                              Analysis of Variance

                  Sum of                                            P
Source            Squares      D.F.    Mean Square    F-Ratio    Value
Model             448.193       5        89.6386       17.55    0.0000
Error              71.489      14         5.10636


Total (corr.)     519.682      19

                          Conditional Sums of Squares

                  Sum of                                            P
Source            Squares      D.F.    Mean Square    F-Ratio    Value

Cat1              286.439       1       286.439        56.09    0.0000
Cat2               19.3688      1        19.3688        3.79    0.0718
@Cat1Sq            84.9193      1        84.9193       16.63    0.0011
@Cat1Cat2          29.8301      1        29.8301        5.84    0.0299
@Cat2Sq            27.636       1        27.636         5.41    0.0355

Model             448.193       5


                        Multiple Regression Analysis

Dependent variable: Yield


                                Table of Estimates

                           Standard           t            P
               Estimate      Error          Value        Value

Constant        70.31       2.57001         27.36       0.0000
Cat1            -2.676      0.560822        -4.77       0.0002
Cat2            -0.8802     0.70939         -1.24       0.2315


R-squared = 58.85%
Adjusted R-squared = 54.00%
Standard error of estimation = 3.54695

                              Analysis of Variance

                  Sum of                                            P
Source            Squares      D.F.    Mean Square    F-Ratio    Value

Model             305.808       2        152.904       12.15    0.0005
Error             213.874      17         12.5808


Total (corr.)     519.682      19
```

  a. Write the estimated complete model.
  b. Write the estimated reduced model.
  c. Locate the R^2 values for the complete and reduced models.
  d. Is there convincing evidence that the addition of the second-order terms improves the predictive ability of the model?

```
MTB > regress 'catch' on 4 variables 'residenc' 'size' 'access'
'structur';
SUBC> predict at 8 .7 1 55;
SUBC> predict at 55 1.0 1 40.

The regression equation is
catch = - 1.94 + 0.0193 residenc + 0.332 size + 0.836 access
        + 0.0477 structur

Predictor      Coef       Stdev      t-ratio      p
Constant    -1.9378      0.9081       -2.13     0.050
residenc     0.01929     0.01018       1.90     0.077
size         0.3323      0.2458        1.35     0.196
access       0.8355      0.2250        3.71     0.002
structur     0.047714    0.005056      9.44     0.000


s = 0.4336       R-sq = 88.2%      R-sq(adj) = 85.0%

    Fit   Stdev.Fit       95% C.I.            95% P.I.
 1.9090      0.6812   ( 0.4567, 3.3613)  ( 0.1874, 3.6306) XX
 2.1998      0.1850   ( 1.8054, 2.5941)  ( 1.1947, 3.2049)
X  denotes a row with X values away from the center
XX denotes a row with very extreme X values
```

Locate the 95% prediction intervals for the two lakes. Why is the first interval so much wider than the second?

Solution   The prediction intervals are the respective 95% P.I. values, 0.1874 to 3.6306 for the first lake and 1.1947 to 3.2049 for the second. The first interval carries a warning that the y values are "very extreme." If we check back with the data, we find that no lake was even close to eight residences per square mile. Thus, the prediction is a severe extrapolation, which makes the interval very wide. In this case, the problem is with one predictor; the remaining x values are well within the range of the data.

EXERCISES   **12.29**   A prediction was made based on the data of Exercise 12.17. Recall that x varied from 1 to 3, w from 1 to 4, and v from 1 to 6. Here is relevant Minitab output:

```
MTB > Correlation 'y' 'x' 'w' 'v'.

             y         x         w
x         0.856
w         0.402     0.000
v         0.928     0.956     0.262

MTB > Regress 'y' 3 'x' 'w' 'v';
SUBC>   Predict at x 3 w 1 v 6.
```

```
The regression equation is
y = 10.0 + 5.00 x + 2.00 w + 1.00 v

s = 2.646      R-sq = 89.5%     R-sq(adj) = 85.6%

    Fit   Stdev.Fit           95% C.I.                95% P.I.
  33.000        4.077   ( 23.595,  42.405)   ( 21.788,  44.212) XX

X  denotes a row with X values away from the center
XX denotes a row with very extreme X values
```

Locate the 95% prediction interval. Explain why Minitab gave the "very extreme X values" warning.

12.30 Refer to the chemical firm data of Exercise 12.28. Predicted yields for $x_1 = 3.5$ and $x_2 = 0.35$ (observation 21) and also for $x_1 = 3.5$ and $x_2 = 2.5$ (observation 22) are calculated based on models with and without second-order terms. Execustat output follows:

```
                    Multiple Regression Analysis

Dependent variable: Yield

                    Table of Estimates
```

	Estimate	Standard Error	t Value	P Value
Constant	50.0195	4.3905	11.39	0.0000
Cat1	6.64357	2.01212	3.30	0.0052
Cat2	7.3145	2.73977	2.67	0.0183
@Cat1Sq	-1.23143	0.301968	-4.08	0.0011
@Cat1Cat2	-0.7724	0.319573	-2.42	0.0299
@Cat2Sq	-1.1755	0.50529	-2.33	0.0355

```
R-squared = 86.24%
Adjusted R-squared = 81.33%
Standard error of estimation = 2.25973

            Table of Predicted Values (Missing Data Only)
```

Row	Predicted Yield	95.00% Prediction Limits Lower	Upper	95.00% Confidence Limits Lower	Upper
21	59.926	54.7081	65.1439	57.993	61.8589
22	62.3679	57.0829	67.6529	60.2605	64.4753

```
                    Multiple Regression Analysis

Dependent variable: Yield
```

```
                        Table of Estimates

                                  Standard         t           P
                    Estimate        Error        Value       Value

Constant             70.31         2.57001       27.36       0.0000
Cat1                 -2.676        0.560822      -4.77       0.0002
Cat2                 -0.8802       0.70939       -1.24       0.2315

R-squared = 58.85%
Adjusted R-squared = 54.00%
Standard error of estimation = 3.54695

            Table of Predicted Values (Missing Data Only)

                          95.00%                      95.00%
            Predicted   Prediction Limits       Confidence Limits
Row          Yield     Lower      Upper         Lower       Upper
21          57.8633   50.028     65.6986        55.5416     60.185
22          58.7435   51.0525    66.4345        56.9687     60.5183
```

a. Locate the 95% limits for individual prediction in the model $\hat{y} = 50.0195 + 6.6436x_1 + 7.3145x_2 - 1.2314x_1^2 - 0.7724x_1x_2 - 1.1755x_2^2$.

b. Locate the 95% limits for individual prediction in the model $\hat{y} = 70.3100 - 2.6760x_1 - 0.8802x_2$.

c. Are the limits for the model of part (a) much tighter than those for the model of part (b)?

12.7 Comparing the Slopes of Several Regression Lines

This topic represents a special case of the general problem of constructing a multiple regression equation for both qualitative and quantitative independent variables. The best way to illustrate this particular problem is by way of an example.

EXAMPLE 12.21

An investigator was interested in comparing the responses of rats to different doses of two drug products (A and B). The study called for a sample of 60 rats of a particular strain to be randomly allocated into two equal groups. The first group of rats was to receive drug A, with 10 rats randomly assigned to each of three doses (5, 10, and 20 mg). Similarly, the 30 rats in group 2 were to receive drug B, with 10 rats randomly assigned to the 5-, 10-, and 20-mg doses. In the study, each rat received its assigned dose, and after a 30-minute observation period, it was scored for signs of anxiety on a 0-to-30-point scale. Assume that a rat's anxiety score is a linear function of the dosage of the drug. Write a model relating a rat's scores to the two independent variables "drug product" and "drug dose." Interpret the βs.

Solution   For this experimental situation, we have one qualitative variable (drug product) and one quantitative variable (drug dose). Letting x_1 denote the drug dose, we have the model

$$y = \beta_0 + \beta_1 x_1 + \beta_2 x_2 + \beta_3 x_1 x_2 + \varepsilon$$

where

$x_1 =$ drug dose

$x_2 = 1$ if product B, $\qquad x_2 = 0$ otherwise

The expected value for y in our model is

$$E(y) = \beta_0 + \beta_1 x_1 + \beta_2 x_2 + \beta_3 x_1 x_2$$

Substituting $x_2 = 0$ and $x_2 = 1$, respectively, for drugs A and B, we obtain the expected rat anxiety score for a given dose:

drug A: $\quad E(y) = \beta_0 + \beta_1 x_1$

drug B: $\quad E(y) = \beta_0 + \beta_1 x_1 + \beta_2 + \beta_3 x_1 = (\beta_0 + \beta_2) + (\beta_1 + \beta_3) x_1$

linear regression lines

These two expected values represent **linear regression lines.** The parameters in the model can be interpreted in terms of the slopes and intercepts associated with these regression lines. In particular,

y-intercept

β_0: **y-intercept** for product A regression line

slope

β_1: **slope** of product A regression line

β_2: difference in y-intercepts of regression lines for products B and A

β_3: difference in slopes of regression lines for products B and A

intersecting lines
parallel lines

Figure 12.4(a) indicates a situation in which $\beta_3 \neq 0$ (that is, there is an interaction between the two variables "drug product" and "drug dose"). Thus, the regression lines are not parallel. Figure 12.4(b) indicates a case in which $\beta_3 = 0$ (no interaction), which results in parallel regression lines.

FIGURE 12.4

Comparing two regression lines

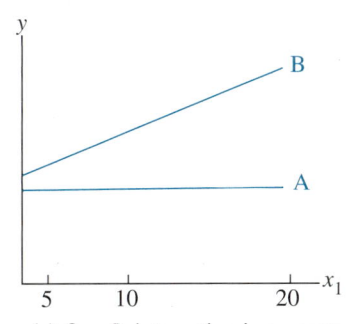 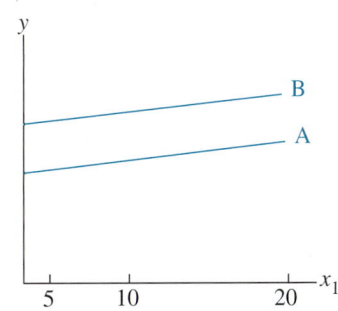

(a) $\beta_3 \neq 0$; interaction is present; intersecting lines

(b) $\beta_3 = 0$; interaction is not present; parallel lines

Indeed, many other experimental situations are possible, depending on the signs and magnitudes of the parameters β_0, β_1, β_2, and β_3.

EXAMPLE 12.22

Sample data for the experiment discussed in Example 12.21 are listed in Table 12.4. The response of interest is an anxiety score obtained from trained investigators. Use these data to fit the general linear model

$$y = \beta_0 + \beta_1 x_1 + \beta_2 x_2 + \beta_3 x_1 x_2 + \varepsilon$$

TABLE 12.4
Rat anxiety scores

Drug Product	Drug Dose (mg) 5		10		20	
A	15	16	18	16	20	17
	16	15	17	15	19	18
	18	16	18	19	21	21
	13	17	19	18	18	20
	19	15	20	16	19	17
	av = 16		av = 17.6		av = 19.0	
B	16	15	19	18	24	23
	17	15	21	20	25	24
	18	18	22	21	23	22
	17	17	23	22	25	26
	15	16	20	19	25	24
	av = 16.4		av = 20.5		av = 24.1	

Of particular interest to the experimenter is a comparison between the slopes of the regression lines. A difference in slopes would indicate that the drug products have different effects on the anxiety of the rats. Conduct a statistical test of the equality of the two slopes. Use $\alpha = .05$.

Solution Using the complete model

$$y = \beta_0 + \beta_1 x_1 + \beta_2 x_2 + \beta_3 x_1 x_2 + \varepsilon$$

we obtain a least-squares fit of

$$\hat{y} = 15.30 + .19x_1 - .70x_2 + .30x_1 x_2$$

with SS(Regression, complete) = 442.10 and SS(Residual, complete) = 133.63. (See the computer output that follows.)

The reduced model corresponding to the null hypothesis $H_0: \beta_3 = 0$ (that is, the slopes are the same) is

$$y = \beta_0 + \beta_1 x_1 + \beta_2 x_2 + \varepsilon$$

```
REGRESSION ANALYSIS OF ANXIETY TREATMENTS-COMPLETE MODEL

Model: MODEL1
Dependent Variable: SCORE


Analysis of Variance

                        Sum of         Mean
Source         DF      Squares        Square      F Value      Prob>F

Model           3     442.10476     147.36825      61.758      0.0001
Error          56     133.62857       2.38622
C Total        59     575.73333
```

```
     Root MSE        1.54474    R-square      0.7679
     Dep Mean       18.93333    Adj R-sq      0.7555
     C.V.            8.15884
```

Parameter Estimates

Variable	DF	Parameter Estimate	Standard Error	T for H0: Parameter=0	Prob > |T|
INTERCEP	1	15.300000	0.59827558	25.573	0.0001
DOSE	1	0.191429	0.04522538	4.233	0.0001
PRODUCT	1	−0.700000	0.84608944	−0.827	0.4116
PRD_DOSE	1	0.300000	0.06395835	4.691	0.0001

Variable	DF	Variable Label
INTERCEP	1	Intercept
DOSE	1	DRUG DOSE LEVEL
PRODUCT	1	DRUG PRODUCT
PRD_DOSE	1	PRODUCT TIMES DOSE

REGRESSION ANALYSIS OF ANXIETY TREATMENTS-REDUCED MODEL

Model: MODEL1
Dependent Variable: SCORE

Analysis of Variance

Source	DF	Sum of Squares	Mean Square	F Value	Prob>F
Model	2	389.60476	194.80238	59.656	0.0001
Error	57	186.12857	3.26541		
C Total	59	575.73333			

```
     Root MSE        1.80705    R-square      0.6767
     Dep Mean       18.93333    Adj R-sq      0.6654
     C.V.            9.54425
```

Parameter Estimates

Variable	DF	Parameter Estimate	Standard Error	T for H0: Parameter=0	Prob > |T|
INTERCEP	1	13.550000	0.54711020	24.766	0.0001
DOSE	1	0.341429	0.03740940	9.127	0.0001
PRODUCT	1	2.800000	0.46657715	6.001	0.0001

```
                    Variable
      Variable  DF    Label

      INTERCEP   1   Intercept
      DOSE       1   DRUG DOSE LEVEL
      PRODUCT    1   DRUG PRODUCT
```

for which we obtain

$$\hat{y} = 13.55 + .34x_1 - 2.80x_2$$

and SS(Regression, reduced) = 389.60. The reduction in the sum of squares for error attributed to x_1x_2 is

$$\text{SS}_{\text{drop}} = \text{SS(Regression, complete)} - \text{SS(Regression, reduced)}$$

$$= 442.10 - 389.60 = 52.50$$

It follows that

$$F = \frac{[\text{SS(Regression, complete)} - \text{SS(Regression, reduced)}]/k - g}{\text{SS(Residual, complete)}/[n - (k + 1)]}$$

$$= \frac{52.50/1}{133.63/56} = 22.00.$$

Because the observed value of F exceeds 4.00, the value for $\text{df}_1 = 1$, $\text{df}_2 = 56$ (actually, 60), and $a = .05$ in Appendix Table 8, we reject H_0 and conclude that the slopes for the two groups are different. Note that we could have obtained the same result by testing $H_0: \beta_3 = 0$ using a t test. From the computer output, the t statistic is 4.69, which is significant at the .0001 level. For this type of test, the t statistic and F statistic are related; namely, $t^2 = F$ (here $4.69^2 \approx 22$).

The results presented here for comparing the slope of two regression lines can be readily extended to the comparison of three or more regression lines by including additional dummy variables and all possible interaction terms between the quantitative variable x_1 and the dummy variables. Thus, for example, in comparing the slopes of three regression lines, the model would contain the quantitative variable x_1, two dummy variables x_2 and x_3, and two interaction terms x_1x_2 and x_1x_3.

EXERCISES   **Applications**

Med.   **12.31**   An experimenter wished to compare the potencies of three different drug products. To do this, 12 test tubes were inoculated with a culture of the virus under study and incubated for 2 days at 35°C. Four dosage levels (.2, .4, .8, and 1.6 μg per tube) were to be used from each of the three drug products, with only one dose–drug product combination for each of the 12 test-tube cultures. One means of comparing the drug products would be to examine their slopes (with respect to dose).

   a. Write a general linear model relating the response y to the independent variables "dose" and "drug product." Make the expected response a linear function of log dose (x_1). Identify the parameters in the model.

   b. It would seem reasonable to assume that the three separate response lines have a common intercept β_0, because this would correspond to a 0 dosage level of any of the drug products. Change the model of part (a) to reflect this change.

12.32 Refer to Exercise 12.31.

    a. Use the following data to make a comparison among the three slopes. Fit a complete and a reduced model for your test. Use $\alpha = .05$.

	Drug Product		
Dose	**A**	**B**	**C**
.2	2.0	1.8	1.3
.4	4.3	4.1	2.0
.8	6.5	4.9	2.8
1.6	8.9	5.7	3.4

    b. Is there evidence to indicate the slopes are equal?

    c. Suggest how you could test the null hypothesis that the intercepts are all equal to zero.

12.8 Logistic Regression

In many research studies, the response variable may be represented as one of two possible values. Thus, the response variable is a binary random variable taking on the values 0 and 1. For example, in a study of a suspected carcinogen, aflatoxin B_1, a number of levels of the compound were fed to test animals. After a period of time, the animals were sacrificed and the number of animals having liver tumors was recorded. The response variable is $y = 1$ if the animal has a tumor and $y = 0$ if the animal fails to have a tumor. Similarly, a bank wants to determine which customers are most likely to repay their loan. Thus, they want to record a number of independent variables that describe the customer's reliability and then determine whether these variables are related to the binary variable, $y = 1$ if the customer repays the loan and $y = 0$ if the customer fails to repay the loan. A model that relates a binary variable y to explanatory variables will be developed next.

When the response variable y is binary, the distribution of y reduces to a single value, the probability $p = \Pr(y = 1)$. We want to relate p to a linear combination of the independent variables. The difficulty is that p varies between zero and one, whereas linear combinations of the explanatory variables can vary between $-\infty$ and $+\infty$. In Chapter 10, we introduced the transformation of probabilities into an odds ratio. As the probabilities vary between zero and one, the odds ratio varies between zero and infinity. By taking the logarithm of the odds ratio, we will have a transformed variable that will vary between $-\infty$ and $+\infty$ when the probabilities vary between zero and one. The model often used to study the association between a binary response and a set of explanatory variables is given by **logistic regression analysis.** In this model, the natural logarithm of the odds ratio is related to the explanatory variables by a linear model. We will consider the situation where we have a single independent variable, but this model can be generalized to multiple independent variables. Let $p(x)$ be the probability that y equals 1 when the independent variable equals x. We model the log-odds ratio to a linear model in x, a **simple logistic regression model:**

logistic regression analysis

simple logistic regression model

$$\ln\left(\frac{p(x)}{1 - p(x)}\right) = \beta_0 + \beta_1 x$$

This transformation can be formulated directly in terms of $p(x)$ as

$$p(x) = \frac{e^{\beta_0 + \beta_1 x}}{1 + e^{\beta_0 + \beta_1 x}}$$

For example, the probability of a tumor being present in an animal exposed to x units of the aflatoxin B_1 would be given by $p(x)$ as expressed by the above equation. The values of β_0 and β_1 would be estimated from the observed data using maximum likelihood estimation.

We can interpret the parameters β_0 and β_1 in the logistic regression model in terms of $p(x)$. The intercept parameter β_0 permits the estimation of the probability of the event associated with $y = 1$ when the independent variable $x = 0$. For example, the probability of a tumor being present when the animal is not exposed to aflatoxin B_1 would correspond to the probability of $y = 1$ when $x = 0$—that is, $p(0)$. The logistic regression model would yield

$$p(0) = \frac{e^{\beta_0}}{1 + e^{\beta_0}}$$

The slope parameter β_1 measures the degree of association between the probability of the event occurring and the value of the independent variable x. When $\beta_1 = 0$, the probability of the event occurring is not associated with size of the value of x. In our example, the chance of an animal developing a liver tumor would remain constant no matter the amount of aflatoxin B_1 the animal was exposed to. Figure 12.5 displays two simple logistic regression functions. If $\beta_1 > 0$, the probability of the event occurring increases as the value of the independent variable increases. If $\beta_1 < 0$, the probability of the event occurring decreases as the value of the independent variable increases.

FIGURE 12.5
Logistic regression functions

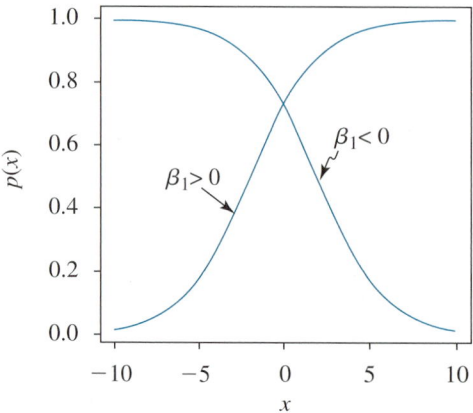

In the situation where both β_0 and β_1 are zero, the event is as likely to occur as not to occur because

$$p(x) = \frac{e^0}{1 + e^0} = \frac{1}{1 + 1} = \frac{1}{2}$$

This would indicate that the probability of the occurrence of the event indicated by $y = 1$ is not related to the independent variable x. Thus, the model is noninformative in determining the probability of the event's occurrence, hence an equal chance of occurrence or nonoccurence of the event no matter the value of the independent variable.

Whether we are using the simple logistic regression model or multiple logistic regression models, the computational techniques used to estimate the model parameters require the use of computer software. We will use an example to illustrate the use of logistic regression models.

EXAMPLE 12.23

A study reported by A. F. Smith (1967), *Lancet*, 2, 178, recorded the level of an enzyme, creatinine kinase (CK), for patients who were suspected of having a heart attack. The objective of the study was to assess whether measuring the amount of CK on admission to the hospital was a useful diagnostic indicator of whether patients admitted with a diagnosis of a heart attack had really had a heart attack. The enzyme CK was measured in 360 patients on admission to the hospital. After a period of time, a doctor reviewed the records of these patients to decide which of the 360 patients had actually had a heart attack. The data are given in the following table with the CK values given as the midpoint of the range of values in each of 13 classes of values.

CK Value	Number of Patients with Heart Attack	Number of Patients without Heart Attack
20	2	88
60	13	26
100	30	8
140	30	5
180	21	0
220	19	1
260	18	1
300	13	1
340	19	0
380	15	0
420	7	0
460	8	0
500	35	0

The computer output for obtaining the estimated logistic regression curve and 95% confidence intervals on the predicted probabilities of a heart attack are given here.

```
               LOGISTIC REGRESSION ANALYSIS EXAMPLE

The LOGISTIC Procedure

Data Set: WORK.LOGREG
Response Variable (Events): R
Response Variable (Trials): N
Number of Observations: 13
Link Function: Logit
```

```
             Response Profile

Ordered  Binary
  Value   Outcome      Count

      1   EVENT          230
      2   NO EVENT       130


Model Fitting Information and Testing Global Null Hypothesis BETA=0

                                   Intercept
                      Intercept       and
Criterion               Only       Covariates   Chi-Square for Covariates


AIC                    472.919      191.773         .
SC                     476.806      199.545         .
-2 LOG L               470.919      187.773      283.147 with 1 DF (p=0.0001)
Score                     .            .         159.142 with 1 DF (p=0.0001)


              Analysis of Maximum Likelihood Estimates


              Parameter Standard    Wald        Pr >     Standardized
Variable DF   Estimate   Error   Chi-Square  Chi-Square    Estimate


INTERCPT 1    -3.0284   0.3670    68.0948      0.0001
CK       1     0.0351   0.00408   73.9842      0.0001      3.100511


              LOGISTIC REGRESSION ANALYSIS EXAMPLE


         OBS     CK       PRED        LCL          UCL


          1      20     0.08897    0.05151      0.14937
          2      60     0.28453    0.21224      0.36988
          3     100     0.61824    0.51935      0.70821
          4     140     0.86833    0.78063      0.92436
          5     180     0.96410    0.91643      0.98502
          6     220     0.99094    0.97067      0.99724
          7     260     0.99776    0.99000      0.99950
          8     300     0.99945    0.99662      0.99991
          9     340     0.99986    0.99886      0.99998
         10     380     0.99997    0.99962      1.00000
         11     420     0.99999    0.99987      1.00000
         12     460     1.00000    0.99996      1.00000
         13     500     1.00000    0.99999      1.00000
```

a. Is CK level significantly related to the probability of a heart attack through the logistic regression model?

b. From the computer output, obtain the estimated coefficients β_0 and β_1.

c. Construct the estimated probability of a heart attack as a function of CK level. In particular, estimate this probability for a patient having a CK level of 140.

Solution

a. From the computer output, we obtain, p-value $= .0001$ for testing the hypothesis $H_0: \beta_1 = 0$ versus $H_0: \beta_1 \neq 0$ in the logistic regression model. Thus, CK is significantly related to the probability of a heart attack.

b. From the computer output, we obtain $\hat{\beta}_0 = -3.0284$ and $\hat{\beta}_1 = 0.0351$. Note that $\hat{\beta}_1$ is positive. This would indicate that patients having higher levels of CK are associated with a larger probability that a heart attack had occurred.

c. The estimated probability of a heart attack as a function of CK level in the patient is given by

$$p(\widehat{CK}) = \frac{e^{-3.0284+.0351*CK}}{1 + e^{-3.0284+.0351*CK}}$$

We can use this formula to calculate the probability that a patient had a heart attack when the CK level in the patient was 140. This value is given by

$$p(\widehat{CK}) = \frac{e^{-3.0284+.0351*140}}{1 + e^{-3.0284+.0351*140}} = \frac{e^{1.886}}{1 + e^{1.886}} = .868$$

From the computer printout, we obtain 95% confidence intervals for this probability as .781 to .924. Thus, we are 95% confident that between 78.1% and 92.4% of patients with a CK level of 140 would have a heart attack. The estimated probabilities of a heart attack along with 95% confidence intervals on these probabilities are plotted in Figure 12.6. We note that the estimated probability of a heart attack increases very rapidly with increasing CK levels in the patients. This would indicate that CK levels are a useful indicator of heart attack potential.

FIGURE 12.6

Estimated probability of heart attack with 95% confidence limits

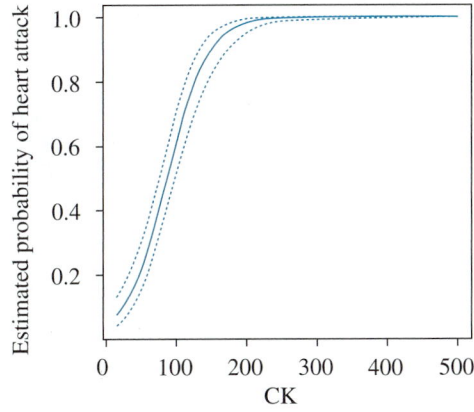

EXERCISES

Applications

Engin.

12.33 A quality control engineer studied the relationship between years of experience of a system control engineer on the capacity of the engineer to complete within a given time a complex control design including the debugging of all computer programs and control devices. A group of 25 engineers having a wide difference in experience (measured in months of experience) were given the same control design project. The results of the study are given in the following table with $y = 1$ if the project was successfully completed in the allocated time and $y = 0$ if the project was not successfully completed.

Months of Experience	Project Success	Months of Experience	Project Success
2	0	15	1
4	0	16	1
5	0	17	0
6	0	19	1
7	0	20	1
8	1	22	0
8	1	23	1
9	0	24	1
10	0	27	1
10	0	30	0
11	1	31	1
12	1	32	1
13	0		

a. Use the computer output given here to determine whether experience is associated with the probability of completing the task.

b. Compute the probability of successfully completing the task for an engineer having 24 months of experience. Place a 95% confidence interval on your estimate.

```
SAS Code for Logistic Regression
option ls=70 ps=55 nocenter nodate;
data logreg;
input x y @@;
label x='MONTHS EXPERIENCE' y='SUCCESS INDICATOR';cards;
2 0 4 0 5 0 6 0 7 0 8 1 8 1 9 0 10 0 10 0 11 1 12 1
13 0 15 1 16 1 17 0 19 1 20 1 22 0 23 1 24 1 27 1 30 0 31 1 32 1
run;
proc print;
proc logistic descending;
model y=x;
output out=new p=pred lower=lcl upper=ucl;
proc sort; by x;
proc print; var x pred lcl ucl;
run;

------------------------------------------------------------------


The LOGISTIC Procedure

Data Set: WORK.LOGREG
Response Variable: Y          SUCCESS INDICATOR
Response Levels: 2
Number of Observations: 25
Link Function: Logit
```

```
        Response Profile

Ordered
  Value      Y      Count

    1        1        13
    2        0        12

            Analysis of Maximum Likelihood Estimates

             Parameter Standard    Wald       Pr >     Standardized
  Variable DF Estimate   Error  Chi-Square Chi-Square   Estimate

INTERCPT 1    -1.6842   0.9451    3.1759     0.0747         .
X        1     0.1194   0.0589    4.1091     0.0427     0.585706
```

The LOGISTIC Procedure

Association of Predicted Probabilities and Observed Responses

```
Concordant = 77.6%        Somers' D = 0.551
Discordant = 22.4%        Gamma     = 0.551
Tied       = 0.0%         Tau-a     = 0.287
(156 pairs)               c         = 0.776
```

The LOGISTIC Procedure

```
                                 95% Lower  95% Upper
        OBS    X      PRED        Limit      Limit

          1    2    0.19070      0.04320    0.55155
          2    4    0.23029      0.06487    0.56339
          3    5    0.25213      0.07884    0.57042
          4    6    0.27530      0.09518    0.57839
          5    7    0.29974      0.11399    0.58749
          6    8    0.32538      0.13526    0.59794
          7    8    0.32538      0.13526    0.59794
          8    9    0.35211      0.15884    0.61001
          9   10    0.37980      0.18434    0.62397
         10   10    0.37980      0.18434    0.62397
         11   11    0.40830      0.21117    0.64011
         12   12    0.43742      0.23858    0.65863
         13   13    0.46698      0.26568    0.67964
         14   15    0.52660      0.31574    0.72839
         15   16    0.55623      0.33753    0.75512
         16   17    0.58547      0.35684    0.78239
         17   19    0.64199      0.38830    0.83514
         18   20    0.66894      0.40092    0.85917
         19   22    0.71954      0.42133    0.90040
         20   23    0.74299      0.42962    0.91732
         21   24    0.76512      0.43691    0.93186
         22   27    0.82333      0.45436    0.96307
         23   30    0.86958      0.46732    0.98065
```

OBS	X	PRED	95% Lower Limit	95% Upper Limit
24	31	0.88253	0.47097	0.98447
25	32	0.89435	0.47436	0.98756

12.34 An additive to interior house paint has been recently developed that may greatly increase the ability of the paint to resist staining. An investigation was conducted to determine whether the additive is safe when exposed to children. Various amounts of the additive were fed to test animals and the number of animals developing liver tumors was recorded. The data are given in the following table.

Amount (ppm)	0	10	25	50	100	200
Number of Test Animals	30	20	20	30	30	30
Number of Animals with Tumors	0	2	2	7	25	30

a. Use the computer output given here to determine whether the amount of additive given to the test animals is associated with the probability of a tumor developing in the liver of the animal.

b. Compute the probability of a tumor developing in the liver of a test animal exposed to 100 ppm of the additive. Place a 95% confidence interval on your estimate.

```
SAS Code for Exercise

option ls=70 ps=55 nocenter nodate;
TITLE 'OUTPUT FOR EXERCISE';
data logreg;
input x R N @@;
label x='AMOUNT (PPM)' ;cards;
0 0 30 10 2 20 25 2 20 50 7 30 100 25 30 200 30 30
run;
proc print;
proc logistic descending;
model R/N=x;
output out=new p=pred lower=lcl upper=ucl;
proc sort; by x;
proc print; var x pred lcl ucl;
run;

------------------------------------------------------------
OUTPUT FOR EXERCISE

OBS      X       R      N

 1        0       0      30
 2       10       2      20
 3       25       2      20
 4       50       7      30
 5      100      25      30
 6      200      30      30
```

```
The LOGISTIC Procedure

          Response Profile

Ordered  Binary
  Value  Outcome       Count

      1  EVENT            66
      2  NO EVENT         94


          Analysis of Maximum Likelihood Estimates

              Parameter Standard    Wald       Pr >    Standardized
  Variable DF Estimate   Error   Chi-Square Chi-Square   Estimate

INTERCPT 1    -3.6429    0.5530   43.3998    0.0001           .
X        1     0.0521    0.00824  39.9911    0.0001      2.044518

        Analysis of Maximum
        Likelihood Estimates

              Odds    Variable
  Variable    Ratio    Label

INTERCPT       .      Intercept
X            1.053    AMOUNT (PPM)


                     95% Lower  95% Upper
  OBS     X    PRED    Limit      Limit

    1     0   0.02551  0.00878   0.07182
    2    10   0.04221  0.01681   0.10203
    3    25   0.08783  0.04308   0.17077
    4    50   0.26156  0.16907   0.38142
    5   100   0.82738  0.66925   0.91905
    6   200   0.99886  0.98818   0.99989
```

12.9 Some Multiple Regression Theory (Optional)

In this section, we use matrix notation to sketch some of the mathematics underlying multiple regression. The focus is on how multiple regression calculations are actually done, whether by hand or by computer. We do not prove most of the results; proofs are available in many specialized texts, such as Draper and Smith (1998).

The starting point for the use of matrix notation is the multiple regression model itself. Recall that a model relating a response y to a set of independent variables of the form

$$y = \beta_0 + \beta_1 x_1 + \beta_2 x_2 + \cdots + \beta_k x_k + \varepsilon$$

is called the *general linear model*. The least-squares estimates $\hat{\beta}_0, \hat{\beta}_1, \ldots, \hat{\beta}_k$ of the intercept and partial slopes in the general linear model can be obtained using matrices.

Let the $n \times 1$ matrix $\mathbf{Y}$

$$\mathbf{Y} = \begin{bmatrix} y_1 \\ y_2 \\ \vdots \\ y_n \end{bmatrix}$$

be the matrix of observations, and let the $n \times (k + 1)$ matrix $\mathbf{X}$

$$\mathbf{X} = \begin{bmatrix} 1 & x_{11} & \cdots & x_{1k} \\ 1 & x_{21} & \cdots & x_{2k} \\ \vdots & \vdots & & \vdots \\ 1 & x_{n1} & \cdots & x_{nk} \end{bmatrix}$$

be a matrix of settings for the independent variables augmented with a column of 1s. The first row of $\mathbf{X}$ contains a 1 and the settings for the k independent variables for the first observation. Row 2 contains a 1 and corresponding settings on the independent variables for y_2. Similarly, the other rows contain settings for the remaining observations.

Next we turn to the least-squares estimates $\hat{\beta}_0, \hat{\beta}_1, \ldots, \hat{\beta}_k$ of the intercept and partial slopes in the multiple regression model. Recall that the least-squares principle involves choosing the estimates to minimize the sum of squared residuals. Those familiar with the calculus will see that the solution can be found by differentiating SS(Residual) with respect to $\hat{\beta}_j (j = 0, \ldots, k)$ and setting the result to zero. The resulting normal equations, in matrix notation, are

$$(\mathbf{X'X})\hat{\boldsymbol{\beta}} = \mathbf{X'Y}$$

where

$$\hat{\boldsymbol{\beta}} = \begin{bmatrix} \hat{\beta}_0 \\ \hat{\beta}_1 \\ \vdots \\ \hat{\beta}_k \end{bmatrix}$$

is the desired vector of estimated coefficients. Provided that the matrix $\mathbf{X'X}$ has an inverse (it does as long as no x_j is perfectly collinear with other xs), the solution is

$$\hat{\boldsymbol{\beta}} = (\mathbf{X'X})^{-1}\mathbf{X'Y}$$

EXAMPLE 12.24

Suppose that in a given experimental situation,

$$\mathbf{Y} = \begin{bmatrix} 25 \\ 19 \\ 33 \\ 23 \end{bmatrix} \quad \text{and} \quad \mathbf{X} = \begin{bmatrix} 1 & -2 & 5 \\ 1 & -2 & -5 \\ 1 & 2 & 5 \\ 1 & 2 & -5 \end{bmatrix}$$

Obtain the least-squares estimates for the prediction equation

$$\hat{y} = \hat{\beta}_0 + \hat{\beta}_1 x_1 + \hat{\beta}_2 x_2$$

Solution   For these data,

$$\mathbf{X'X} = \begin{bmatrix} 4 & 0 & 0 \\ 0 & 16 & 0 \\ 0 & 0 & 100 \end{bmatrix}$$

$$\mathbf{X'Y} = \begin{bmatrix} 100 \\ 24 \\ 80 \end{bmatrix}$$

The $\mathbf{X'X}$ matrix is a diagonal one, so inverting the matrix is easy. The solution is

$$\hat{\boldsymbol{\beta}} = (\mathbf{X'X})^{-1}\mathbf{X'Y}$$

$$= \begin{bmatrix} .25 & 0 & 0 \\ 0 & .0625 & 0 \\ 0 & 0 & .01 \end{bmatrix}\begin{bmatrix} 100 \\ 24 \\ 80 \end{bmatrix} = \begin{bmatrix} 25 \\ 1.5 \\ 0.8 \end{bmatrix}$$

and the prediction equation is

$$\hat{y} = 25 + 1.5x_1 + .8x_2$$

The hard part of the arithmetic in multiple regression is computing the inverse of $\mathbf{X'X}$. For the most realistic multiple regression problems, this task takes hours by hand and fractions of a second by computer. This is the major reason why most multiple regression problems are done with computer software.

Once the inverse of the $\mathbf{X'X}$ matrix is found and the $\hat{\boldsymbol{\beta}}$ vector is calculated, the next task is to compute the residual standard deviation. The hard work is to compute SS(Residual) $= \Sigma(y_i - \hat{y}_i)^2$, which can be written as SS(Residual) $= \mathbf{Y'Y} - \hat{\boldsymbol{\beta}}'(\mathbf{X'Y})$.

EXAMPLE 12.25

Compute SS(Residual) for the data of Example 12.24.

Solution $\hat{\boldsymbol{\beta}}$ and $\mathbf{X'Y}$ were calculated to be $\begin{bmatrix} 25 \\ 1.5 \\ 0.8 \end{bmatrix}$ and $\begin{bmatrix} 100 \\ 24 \\ 80 \end{bmatrix}$ respectively, and

$$\mathbf{Y'Y} = \begin{bmatrix} 25 & 19 & 33 & 23 \end{bmatrix}\begin{bmatrix} 25 \\ 19 \\ 33 \\ 23 \end{bmatrix} = 2,604$$

The shortcut formula yields

$$\text{SS(Residual)} = 2{,}604 - \begin{bmatrix} 25 & 1.5 & 0.8 \end{bmatrix} \begin{bmatrix} 100 \\ 24 \\ 80 \end{bmatrix} = 4$$

Similar calculations yield SS(Regression) and SS(Total). Although the formulas for these sums can be expressed artificially in pure matrix notation, they can be expressed more easily in mixed matrix and algebraic notation:

$$\text{SS(Regression)} = \hat{\boldsymbol{\beta}}'(\mathbf{X}'\mathbf{Y}) - \frac{(\sum y_i)^2}{n}$$

$$\text{SS(Total)} = \mathbf{Y}'\mathbf{Y} - \frac{(\sum y_i)^2}{n}$$

EXAMPLE 12.26

Calculate SS(Regression) and SS(Total) for the data of Example 12.23.

Solution $\sum y_i = 100$ and $n = 4$. The relevant matrix calculations were performed in the previous example.

$$\text{SS(Regression)} = 2{,}600 - \frac{(100)^2}{4} = 100$$

$$\text{SS(Total)} = 2{,}604 - \frac{(100)^2}{4} = 104$$

Note that SS(Total) = 104 = 100 + 4 = SS(Regression) + SS(Residual).

These sum-of-squares calculations are necessary for making inferences based on R^2 using F tests. For inferences about individual coefficients using t tests, the estimated standard errors of the coefficients are necessary. In Section 12.4, we presented a conceptually useful but computationally cumbersome formula for these estimated standard errors. A much easier way of computing them involves only the standard deviation s_ε and the main diagonal elements of the $(\mathbf{X}'\mathbf{X})^{-1}$ matrix.

DEFINITION 12.5

The estimated standard error of $\hat{\beta}_j$ is

$$S_{\hat{\beta}_j} = s_\varepsilon \sqrt{v_{jj}}$$

where s_ε is the standard deviation from the regression equation and v_{jj} is the entry in row $j + 1$, column $j + 1$ of $(\mathbf{X}'\mathbf{X})^{-1}$:

$$(\mathbf{X}'\mathbf{X})^{-1} = \begin{bmatrix} v_{00} & & & \\ & v_{11} & & \\ & & \ddots & \\ & & & v_{kk} \end{bmatrix}$$

Because the $(\mathbf{X}'\mathbf{X})^{-1}$ matrix must be computed to obtain the $\hat{\beta}_j$s, it is easy to get the estimated standard errors.

12.10 Summary

This chapter consolidates the material for expressing a response y as a function of one or more independent variables. Multiple regression models (where all the independent variables are quantitative) and models that incorporate information on qualitative variables were discussed and can be represented in the form of a general linear model

$$y = \beta_0 + \beta_1 x_1 + \beta_2 x_2 + \cdots + \beta_k x_k + \varepsilon$$

After discussing various models and the interpretation of βs in these models, we presented the normal equations used in obtaining the least-squares estimates $\hat{\beta}$.

A confidence interval and statistical test about an individual parameter β_j were developed using $\hat{\beta}_j$ and the standard error of $\hat{\beta}_j$. We also considered a statistical test about a set of βs, a confidence interval for $E(y)$ based on a set of xs, and a prediction interval for a given set of xs.

All of these inferences involve a fair to moderate amount of numerical calculation unless statistical software programs of packages are available. Sometimes these calculations can be done by hand if one is familiar with matrix operations (see Section 12.9). However, even these methods become unmanageable as the number of independent variables increases. Thus, the message should be very clear. Inferences about general linear models should be done using available computer software to facilitate the analysis and to minimize computational errors. Our job in these situations is to review and interpret the output.

Aside from a few exercises that will probe your understanding of the mechanics involved with these calculations, most of the exercises in the remainder of this chapter and in the regression problems of the next chapter will make extensive use of computer output.

Here are some reminders about multiple regression concepts:

1. Regression coefficients in a first-order model (one not containing transformed values, such as squares of a variable or product terms) should be interpreted as partial slopes—the predicted change in a dependent variable when an independent variable is increased by one unit, while other variables are held constant.

2. Correlations are important, not only between an independent variable and the dependent variable, but also between independent variables. Collinearity—correlation between independent variables—implies that regression coefficients will change as variables are added to or deleted from a regression model.

3. The effectiveness of a regression model can be indicated not only by the R^2 value but also by the residual standard deviation. It's often helpful to use that standard deviation to see roughly how much of a plus or minus must be allowed around a prediction.

4. As always, the various statistical tests in a regression model only indicate how strong the evidence is that the apparent pattern is more than random. They don't directly indicate how good a predictive model is. In particular, a large overall F statistic may merely indicate a weak prediction in a large sample.

5. A t test in a multiple regression assesses whether that independent variable adds unique, predictive value as a predictor in the model. It is quite possible that several variables may not add a statistically detect-

able amount of unique, predicted value, yet deleting all of them from the model causes a serious drop in predictive value. This is especially true when there is severe collinearity.

6. The variance inflation factor (VIF) is a useful indicator of the overall impact of collinearity in estimating the coefficient of an independent variable. The higher the VIF number, the more serious is the impact of collinearity on the accuracy of a slope estimate.

7. Extrapolation in multiple regression can be subtle. A new set of x values may not be unreasonable when considered one by one, but the combination of values may be far outside the range of previous data.

Key Formulas

1. $R^2_{y \cdot x_1 \cdots x_k} = \dfrac{\text{SS(Total)} - \text{SS(Residual)}}{\text{SS(Total)}}$

    where

    $$\text{SS(Total)} = \Sigma(y_i - \bar{y})^2$$

    and

    $$\text{SS(Residual)} = \Sigma(y_i - \hat{y}_i)^2$$

2. $\text{SS(Regression)} = \Sigma(\hat{y}_i - \bar{y})^2$

    and

    $$\text{SS(Total)} = \Sigma(y_i - \bar{y})^2 = \text{SS(Regression)} + \text{SS(Residual)}$$

3. F test for $H_0: \beta_1 = \beta_2 = \cdots = \beta_k = 0$

    $$F = \frac{\text{SS(Regression)}/k}{\text{SS(Residual)}/[n - (k + 1)]}$$

4. $s_{\hat{\beta}_j} = s_\varepsilon \sqrt{\dfrac{1}{\Sigma(x_{ij} - \bar{x})^2(1 - R^2_{x_j \cdot x_1 \cdots x_{j-1} x_{j+1} \cdots x_k})}}$

    and

    $$s_\varepsilon = \frac{\text{MS(Residual)}}{n - (k + 1)}$$

5. Confidence interval for β_j

    $$\hat{\beta}_j - t_{\alpha/2} s_{\hat{\beta}_j} \leq \beta_j \leq \hat{\beta}_j + t_{\alpha/2} s_{\hat{\beta}_j}$$

6. Statistical test for β_j

    T.S.: $t = \dfrac{\hat{\beta}_j}{s_{\hat{\beta}_j}}$

7. Testing a subset of predictors

    $$H_0: \quad \beta_{g+1} = \beta_{g+2} = \cdots = \beta_k = 0$$

    T.S.: $F = \dfrac{[\text{SS(Regression, complete)} - \text{SS(Regression, reduced)}]/(k - g)}{\text{SS(Residual, complete)}/[n - (k + 1)]}$

Supplementary Exercises

Bus.   **12.35**   A study of demand for imported subcompact cars incorporates data from 12 metropolitan areas. The variables are as follows:

Demand: Imported subcompact car sales as a percentage of total sales

Educ: Average number of years of schooling completed by adults

Income: Per capita income

Popn: Area population

Famsize: Average size of intact families

Minitab output is as follows:

```
MTB > Regress 'Demand' 4 'Educ' 'Income' 'Popn' 'Famsize'.

The regression equation is
Demand = - 1.3 + 5.55 Educ + 0.89 Income + 1.92 Popn - 11.4 Famsize

Predictor        Coef       Stdev       t-ratio          p
Constant        -1.32       57.98        -0.02        0.982
Educ            5.550        2.702         2.05        0.079
Income          0.885        1.308         0.68        0.520
Popn            1.925        1.371         1.40        0.203
Famsize       -11.389        6.669        -1.71        0.131

s = 2.686      R-sq = 96.2%   R-sq(adj) = 94.1%

Analysis of Variance

SOURCE         DF         SS          MS         F          p
Regression      4     1295.70      323.93      44.89     0.000
Error           7       50.51        7.22
Total          11     1346.22

SOURCE         DF      SEQ SS
Educ            1     1239.95
Income          1       32.85
Popn            1        1.86
Famsize         1       21.04

Unusual Observations
Obs.     Educ     Demand       Fit   Stdev.Fit    Residual    St.Resid
  9       9.3     13.100     9.760      2.149       3.340       2.07R

R denotes an obs. with a large st. resid.
```

a. Write the regression equation. Place the standard error of each coefficient below the coefficient, perhaps in parentheses.

b. Locate R^2 and the residual standard deviation.

c. The Unusual Observations entry in the output indicates that observation 9 had a value 2.07 standard deviations away from the predicted Fit value. Does this indicate that observation 9 is a very serious outlier?

12.36   Summarize the conclusions of the F test and the various t tests in the output of Exercise 12.35.

12.37   Another analysis of the data of Exercise 12.35 uses only Educ and Famsize to predict Demand. The output is as follows:

```
MTB > Regress 'Demand' 2 'Educ' 'Famsize'.

The regression equation is
Demand = - 19.2 + 7.79 Educ + 9.46 Famsize

Predictor        Coef       Stdev      t-ratio          p
Constant       -19.17       45.87        -0.42      0.686
Educ            7.793        2.490         3.13      0.012
Famsize        -9.464        5.207        -1.82      0.103

s = 2.939          R-sq = 94.2%    R-sq(adj) = 92.9%

Analysis of Variance

SOURCE          DF          SS          MS          F          p
Regression       2     1268.48      634.24      73.43      0.000
Error            9       77.73        8.64
Total           11     1346.22
```

a. Locate the R^2 value for this reduced model.
b. Test the null hypothesis that the true coefficients of Income and Popn are zero. Use $\alpha = .05$. What is the conclusion?

Bus.

12.38   One of the functions of bank branch offices is to arrange profitable loans to small businesses and individuals. As part of a study of the effectiveness of branch managers, a bank collected data from a sample of branches on current total loan volumes (the dependent variable), the total deposits held in accounts opened at that branch, the number of such accounts, the average number of daily transactions, and the number of employees at the branch. Correlations and a scatterplot matrix are shown in the figure.
a. Which independent variable is the best predictor of loan volume?
b. Is there a substantial collinearity problem?
c. Do any points seem extremely influential?

```
Correlations

Variable                 Loan volume (millions) Deposit volume (millions) Number of accounts  Transactions  Employees
Loan volume (millions)            1.0000                  0.9369               0.9403         0.8766      0.6810
Deposit volume (millions)         0.9369                  1.0000               0.9755         0.9144      0.7377
Number of accounts                0.9403                  0.9755               1.0000         0.9299      0.7487
Transactions                      0.8766                  0.9144               0.9299         1.0000      0.8463
Employees                         0.6810                  0.7377               0.7487         0.8463      1.0000
```

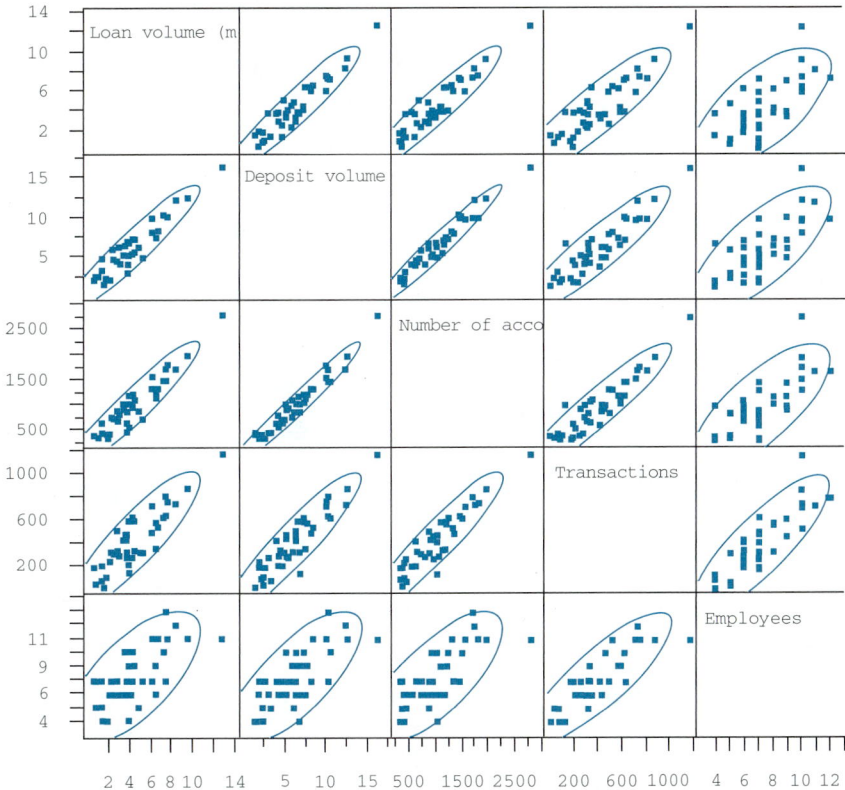

12.39 A regression model was created for the bank branch office data using JMP. Some of the results are shown here.

    a. Use the R^2 value shown to compute an overall F statistic. Is there clear evidence that there is predictive value in the model, using $\alpha = .01$?

    b. Which individual predictors have been shown to have unique, predictive value, again using $\alpha = .01$?

    c. Explain the apparent contradiction between your answers to the first two parts.

Response: Loan volume (millions)

Summary of Fit

RSquare	0.894477
RSquare Adj	0.883369
Root Mean Square Error	0.870612
Mean of Response	4.383395
Observations (or Sum Wgts)	43

Parameter Estimates

Term	Estimate	Std Error	t Ratio	Prob>[t]
Intercept	0.2284381	0.6752	0.34	0.7370
Deposit volume (millions)	0.3222099	0.191048	1.69	0.0999
Number of accounts	0.0025812	0.001314	1.96	0.0569
Transactions	0.0010058	0.001878	0.54	0.5954
Employees	-0.119898	0.130721	-0.92	0.3648

12.40 Another multiple regression model used only deposit volume and number of accounts as independent variables, with results as shown here.

a. Does omitting the transactions and employees variables seriously reduce R^2?

b. Use the R^2 values to test the null hypothesis that the coefficients of transactions and employees are zero. What is your conclusion?

Response: Loan volume (millions)

Summary of Fit

RSquare	0.892138
RSquare Adj	0.886744
Root Mean Square Error	0.857923
Mean of Response	4.383395
Observations (or Sum Wgts)	43

Parameter Estimates

Term	Estimate	Std Error	t Ratio	Prob > [t]
Intercept	-0.324812	0.290321	-1.12	0.2699
Deposit volume (millions)	0.3227636	0.187509	1.72	0.0929
Number of accounts	0.002684	0.001166	2.30	0.0266

Engin.

12.41 The manager of documentation for a computer software firm wants to forecast the time required to document moderate-size computer programs. Records are available for 26 programs. The variables are y = number of writer-days needed, x_1 = number of subprograms, x_2 = average number of lines per subprogram, $x_3 = x_1 x_2$, $x_4 = x_2^2$, and $x_5 = x_1 x_2^2$. A portion of the output from a regression analysis of the data is shown here:

```
                     Multiple Regression Analysis

Dependent variable: Y

                     Table of Estimates

                                  Standard        t          P
                   Estimate        Error       Value       Value

Constant          -16.8198        11.631       -1.45       0.1636
X1                  1.47019        0.365944      4.02       0.0007
X2                  0.994778       0.611441      1.63       0.1194
@X1X2              -0.0240071      0.0237565    -1.01       0.3243
@X2Sq             -0.01031         0.007374     -1.40       0.1774
@X1X2Sq           -0.000249574     0.000351779   0.71       0.4862


R-squared = 91.72%
Adjusted R-squared = 89.65%
Standard error of estimation = 3.39011
Durbin-Watson statistic = 2.12676
Mean absolute error = 2.4127
```

```
                          Analysis of Variance

                    Sum of                                              P
     Source          Squares      D.F.     Mean Square    F-Ratio    Value

     Model           2546.03        5        509.205       44.31     0.0000
     Error           229.857       20         11.4929

     Total (corr.)   2775.88       25
```

a. Write the multiple regression model and locate the residual standard deviation.
b. Does x_3 have a significant, unique predictive value?

12.42   The model $y = \beta_0 + \beta_1 x_1 + \varepsilon$ is fit to the data of Exercise 12.41. Selected output is shown here:

```
                  Multiple Regression Analysis

Dependent variable: Y

                        Table of Estimates

                                 Standard          t          P
                   Estimate        Error        Value       Value

Constant           0.840085       3.43375        0.24       0.8089
X1                 1.01583        0.0792925      12.81       0.0000
X2                 0.0558262      0.0515066       1.08       0.2897
R-squared = 90.64%
Adjusted R-squared = 89.83%
Standard error of estimation = 3.36066
Durbin-Watson statistic = 2.2053
Mean absolute error = 2.57584

                     Analysis of Variance

                    Sum of                                              P
     Source          Squares      D.F.     Mean Square    F-Ratio    Value

     Model           2516.12        2        1258.06       111.39    0.0000
     Error           259.763       23          11.294

     Total (corr.)   2775.88       25
```

a. Write the complete and reduced-form estimated models.
b. Is the improvement in R^2 obtained by adding x_3, x_4, and x_5 statistically significant at $\alpha = .05$? Approximately what is the p-value for this test?

Bus.   **12.43**   A chain of small convenience food stores performs a regression analysis to explain variation in sales volume among 16 stores. The variables in the study are as follows:

Sales: Average daily sales volume of a store, in thousands of dollars

Size: Floor space in thousands of square feet

Parking: Number of free parking spaces adjacent to the store

Income: Estimated per household income of the zip code area of the store

Output from a regression program (StataQuest) is shown here:

```
. regress Sale Size Parking Income

   Source |      SS         df       MS              Number of obs =      16
----------+-----------------------------             F( 3,    12) =   15.16
    Model | 27.1296056      3   9.04320188           Prob > F      = 0.0002
 Residual | 7.15923792     12    .59660316           R-square      = 0.7912
----------+-----------------------------             Adj R-square  = 0.7390
    Total | 34.2888436     15    2.2859229           Root MSE      =  .7724


----------------------------------------------------------------------------
    Sales |    Coef.   Std. Err.      t     P>|T|   [95% Conf. Interval]
----------+-----------------------------------------------------------------
     Size | 2.547936   1.200827    2.122   0.055   -.0684405    5.164313
  Parking | .2202793    .1553877   1.418   0.182   -.1182814     .5588401
   Income | .5893221    .1780576   3.310   0.006    .2013679     .9772763
    _cons | .872716    1.945615    0.449   0.662   -3.366415    5.111847
----------------------------------------------------------------------------

. correlate Sales Size Parking Income
(obs=16)

          |   Sales     Size    Parking    Income
----------+-------------------------------------------
    Sales | 1.0000
     Size | 0.7415    1.0000
  Parking | 0.6568    0.6565    1.0000
   Income | 0.7148    0.4033    0.3241    1.0000
```

a. Write the regression equation. Indicate the standard errors of the coefficients.
b. Carefully interpret each coefficient.
c. Locate R^2 and the residual standard deviation.
d. Is there a severe collinearity problem in this study?

12.44 Summarize the results of the F and t tests for the output of Exercise 12.43.

Ag. **12.45** A producer of various feed additives for cattle conducts a study of the number of days of feedlot time required to bring beef cattle to market weight. Eighteen steers of essentially identical age and weight are purchased and brought to a feedlot. Each steer is fed a diet with a specific combination of protein content, antibiotic concentration, and percentage of feed supplement. The data are as follows:

STEER:	1	2	3	4	5	6	7	8	9
PROTEIN:	10	10	10	10	10	10	15	15	15
ANTIBIO:	1	1	1	2	2	2	1	1	1
SUPPLEM:	3	5	7	3	5	7	3	5	7
TIME:	88	82	81	82	83	75	80	80	75

STEER:	10	11	12	13	14	15	16	17	18
PROTEIN:	15	15	15	20	20	20	20	20	20
ANTIBIO:	2	2	2	1	1	1	2	2	2
SUPPLEM:	3	5	7	3	5	7	3	5	7
TIME:	77	76	72	79	74	75	74	70	69

Computer output from a Systat regression analysis follows:

```
CORRELATIONS (PEARSON)

              TIME    . PROTEIN    ANTIBIO
PROTEIN    -0.7111
ANTIBIO    -0.4180     0.0000
SUPPLEM    -0.4693     0.0000     -0.0000


CASES INCLUDED 18  MISSING CASES 0

UNWEIGHTED LEAST SQUARES LINEAR REGRESSION OF TIME


PREDICTOR
VARIABLES      COEFFICIENT      STD ERROR      STUDENT'S T        P        VIF

CONSTANT         102.708         2.31037         44.46          0.0000
PROTEIN         -0.83333         0.09870         -8.44          0.0000     1.0
ANTIBIO         -4.00000         0.80589         -4.96          0.0002     1.0
SUPPLEM         -1.37500         0.24675         -5.57          0.0001     1.0


R-SQUARED                 0.9007    RESID. MEAN SQUARE (MSE)     2.92261
ADJUSTED R-SQUARED        0.8794    STANDARD DEVIATION           1.70956


SOURCE          DF        SS           MS          F         P

REGRESSION       3      371.083     123.694      42.32     0.0000
RESIDUAL        14      40.9166       2.92261
TOTAL           17     412.000


PREDICTED/FITTED VALUES OF TIME

LOWER PREDICTED BOUND     73.566    LOWER FITTED BOUND     76.469
PREDICTED VALUE           77.333    FITTED VALUE           77.333
UPPER PREDICTED BOUND     81.100    UPPER FITTED BOUND     78.197
SE (PREDICTED VALUE)       1.7564   SE (FITTED VALUE)       0.4029


UNUSUALNESS (LEVERAGE)     0.0556
PERCENT COVERAGE          95.0
CORRESPONDING T            2.14

PREDICTOR VALUES: PROTEIN = 15.000, ANTIBIO = 1.5000, SUPPLEM = 5.0000
```

a. Write the regression equation.
b. Find the standard deviation.
c. Find the R^2 value.
d. How much of a collinearity problem is there with these data?

12.46   Refer to Exercise 12.45.
  a. Predict the feedlot time required for a steer fed 15% protein, 1.5% antibiotic concentration, and 5% supplement.
  b. Do these values of the independent variables represent a major extrapolation from the data?
  c. Give a 95% confidence interval for the mean time predicted in part (a).

12.47   The data of Exercise 12.45 are also analyzed by a regression model using only protein content as an independent variable, with the following output:

```
UNWEIGHTED LEAST SQUARES LINEAR REGRESSION OF TIME


PREDICTOR
VARIABLES       COEFFICIENT      STD ERROR     STUDENT'S T        P


CONSTANT           89.8333        3.20219         28.05         0.0000
PROTEIN           -0.83333        0.20598         -4.05         0.0009


R-SQUARED                0.5057    RESID. MEAN SQUARE (MSE)     12.7291
ADJUSTED R-SQUARED       0.4748    STANDARD DEVIATION           3.56779


SOURCE         DF       SS         MS          F          P


REGRESSION      1    208.333    208.333      16.37      0.0009
RESIDUAL       16    203.666    12.7291
TOTAL          17    412.000
```

  a. Write the regression equation.
  b. Find the R^2 value.
  c. Test the null hypothesis that the coefficients of ANTIBIO and SUPPLEM are zero at $\alpha = .05$.

H.R.   **12.48**   A sex discrimination suit alleges that a small college discriminated against women faculty in terms of salaries. A regression study considers the following variables:

Salary: Base salary per year (thousands of dollars)

Senior: Seniority at the college (in years)

Sex: 1 for men, 0 for women

RankD1:   1 for full professors, 0 for others

RankD2:   1 for associate professors, 0 for others

RankD3:   1 for assistant professors, 0 for others

Doct: 1 for holders of doctorate, 0 for others

Note that lecturers and instructors have value 0 for all three RankD variables. Computer output (Excel) from the study is shown here.

	A	B	C	D	E	F
1						
2	Regression Statistics					
3						
4	Multiple R	0.9716				
5	R Square	0.944				
6	Adjusted R Square	0.9294				
7	Standard Error	2.3375				
8	Observations	30				
9						
10						
11	ANOVA					
12		df	SS	MS	F	Significance F
13	Regression	6	2119.347	353.225	64.646	0.000
14	Residual	23	125.672	5.464		
15	Total	29	2245.019			
16						
17						
18		Coefficients	Standard Error	t Stat	P-value	
19	Intercept	18.6784	1.3788	13.5470	0.0000	
20	Senior	0.5420	0.0762	7.1176	0.0000	
21	Sex	1.2074	0.0649	1.1339	0.2685	
22	RankD1	8.7779	1.9380	4.5293	0.0002	
23	RankdD2	4.4211	1.7797	2.4842	0.0207	
24	RankD3	2.7165	1.4239	1.9079	0.0690	
25	Doct	0.9225	1.2589	0.7328	0.4711	

a. Write the regression equation.

b. What is the interpretation of the coefficient of Sex?

c. What is the interpretation of the coefficient of RankD1?

12.49 Refer to Exercise 12.48.

a. Test the hypothesis that the true coefficient of Sex is positive. Use $\alpha = .05$.

b. What does the conclusion of this test indicate about allegations of discrimination?

12.50 a. Locate the value of the F statistic in Exercise 12.48.

b. What null hypothesis is being tested by this statistic?

c. Is this null hypothesis rejected at $\alpha = .01$? How plausible is this null hypothesis?

12.51 Another regression model of the data of Exercise 12.48 omits Sex and Doct from the list of independent variables. The output is as follows:

```
Regression Statistics
Multiple R              0.9697
R Square                0.9403
Adjusted R Square       0.9307
Standard Error          2.3160
Observations                30
```

```
ANOVA
                 df           SS          MS        F       Significance F
Regression        4       2110.925     527.731   98.389        0.0000
Residual         25        134.093       5.364
Total            29       2245.019


               Coefficients   Standard Error   t Stat   P-value
Intercept        19.7113         1.0776       18.2913   0.0000
Senior            0.5572         0.0744        7.4893   0.0000
RankD1            9.2414         1.8214        5.0738   0.0000
RankD2            5.1050         1.5875        3.2158   0.0036
RankD3            3.2243         1.3204        2.4418   0.0220
```

a. Locate R^2 for this reduced model.

b. Test the null hypothesis that the true coefficients of Sex and Doct are zero. Use $\alpha = .01$.

12.52 A survey of information systems managers was used to predict the yearly salary of beginning programmer/analysts in a metropolitan area. Managers specified their standard salary for a beginning programmer/analyst, the number of employees in the firm's information processing staff, the firm's gross profit margin in cents per dollar of sales, and the firm's information processing cost as a percentage of total administrative costs. The data are stored in the EX1252.DAT file in the Web site data sets, with salary in column 1, number of employees in column 2, profit margin in column 3, and information processing cost in column 4.

a. Obtain a multiple regression equation with salary as the dependent variable and the other three variables as predictors. Interpret each of the (partial) slope coefficients.

b. Is there conclusive evidence that the three predictors together have at least some value in predicting salary? Locate a p-value for the appropriate test.

c. Which of the independent variables, if any, have statistically detectable ($\alpha = .05$) predictive value as the last predictor in the equation?

12.53 a. Locate the coefficient of determination (R^2) for the regression model in Exercise 12.52.

b. Obtain another regression model with number of employees as the only independent variable. Find the coefficient of determination for this model.

c. By hand, test the null hypothesis that adding profit margin and information processing cost does not yield any additional predictive value, given the information about number of employees. Use $\alpha = .10$. What can you conclude from this test?

12.54 Obtain correlations for all pairs of predictor variables in Exercise 12.52. Does there seem to be a major collinearity problem in the data?

Gov.   **12.55** A government agency pays research contractors a fee to cover overhead costs, over and above the direct costs of a research project. Although the overhead cost varies considerably among contracts, it is usually a substantial share of the total contract cost. An agency task force obtained data on overhead cost as a fraction of direct costs, number of employees of the contractor, size of contract as a percentage of the contractor's yearly income, and personnel costs as a percentage of direct cost. These four variables are stored (in the order given) in the EX1255.DAT file in the Web site data sets.

a. Obtain correlations of all pairs of variables. Is there a severe collinearity problem with the data?

b. Plot overhead cost against each of the other variables. Locate a possible high influence outlier.

c. Obtain a regression equation (Overhead cost as dependent variable) using all the data including any potential outlier.

d. Delete the potential outlier and get a revised regression equation. How much did the slopes change?

12.56 Consider the outlier-deleted regression model of Exercise 12.55.

a. Locate the F statistic. What null hypothesis is being tested? What can we conclude based on the F statistic?

b. Locate the t statistic for each independent variable. What conclusions can we reach based on the t tests?

12.57 Use the outlier-deleted data of Exercise 12.55 to predict overhead cost of a contract when the contractor has 500 employees, the contract is 2.50% of the contractor's income, and personnel cost is 55% of the direct cost. Obtain a 95% prediction interval. Would an overhead cost equal to 88.9% of direct cost be unreasonable in this situation?

Bus.   **12.58** The owner of a rapidly growing computer store tried to explain the increase in biweekly sales of computer software, using four explanatory variables: Number of titles displayed, Display footage, Current customer base of IBM-compatible computers, and Current customer base of Apple-compatible computers. The data are stored in time-series order in the EX1258.DAT file in the Web site data sets, with sales in column 1, titles in 2, footage in 3, IBM base in 4, and Apple base in 5.

a. Before doing the calculations, consider the economics of the situation and state what sign you would expect for each of the partial slopes.

b. Obtain a multiple regression equation with sales as the dependent variable and all other variables as independent. Does each partial slope have the sign you expected in part (a)?

c. Calculate a 95% confidence interval for the coefficient of the Titles variable. The computer output should contain the calculated standard error for this coefficient. Does the interval include 0 as a plausible value?

12.59 **a.** In the regression model of Exercise 12.58, can the null hypothesis that none of the variables has predictive value be rejected at normal α levels?

b. According to t tests, which predictors, if any, add statistically detectable predictive value ($\alpha = .05$) given all the others?

12.60 Obtain correlation coefficients for all pairs of variables from the data of Exercise 12.58. How severe is the collinearity problem in the data?

12.61 Compare the coefficient of determination (R^2) for the regression model of Exercise 12.58 to the square of the correlation between sales and titles in Exercise 12.60. Compute the incremental F statistic for testing the null hypothesis that footage, IBM base, and Apple base add no predictive value given titles. Can this hypothesis be rejected at $\alpha = .01$?

Bus.   **12.62** The market research manager of a catalog clothing supplier has begun an investigation of what factors determine the typical order size the supplier receives from customers. From the sales records stored on the company's computer, the manager obtained average order size data for 180 zip code areas. A part-time intern looked up the latest census information on per capita income, average years of formal education, and median price of an existing house in each of these zip code areas. (The intern couldn't find house price data for two zip codes, and entered 0 for those areas.) The manager also was curious whether climate had any bearing on order size, and included data on the average daily high temperature in winter and in summer.

The market research manager has asked for your help in analyzing the data. The output provided is only intended as a first try. The manager would like to know whether there was any evidence that the temperature variables mattered much, and also which of the other variables seemed useful. There is some question about whether putting in 0 for the missing house price data was the right thing to do, or whether that might distort the results. Please provide a basic, not too technical explanation of the results in this output and any other analyses you choose to perform.

```
MTB > name c1 'AvgOrder' c2 'Income' c3 'Educn' &
CONT> c4 'HousePr' c5 'WintTemp' c6 'SummTemp'
MTB > correlations of c1-c6

            AvgOrder   Income    Educn   HousePr  WintTemp
Income       0.205
Educn        0.171     0.913
HousePr      0.269     0.616    0.561
WintTemp    -0.134    -0.098    0.014    0.066
SummTemp    -0.068    -0.115    0.005    0.018     0.481

MTB > regress c1 on 5 variables in c2-c6

The regression equation is
AvgOrder = 36.2 + 0.078 Income - 0.019 Educn
   + 0.0605 HousePr - 0.223 WintTemp + 0.006 SummTemp

Predictor        Coef        Stdev      t-ratio        p
Constant        36.18        12.37        2.92      0.004
Income         0.0780       0.4190        0.19      0.853
Educn         -0.0189       0.5180       -0.04      0.971
HousePr       0.06049      0.02161        2.80      0.006
WintTemp      -0.2231       0.1259       -1.77      0.078
SummTemp       0.0063       0.1646        0.04      0.969

s = 4.747        R-sq = 9.6%       R-sq(adj) = 7.0%

Analysis of Variance

SOURCE          DF          SS          MS         F         p
Regression       5       417.63       83.53      3.71     0.003
Error          174      3920.31       22.53
Total          179      4337.94

SOURCE          DF      SEQ SS
Income           1      182.94
Educn            1        7.18
HousePr          1      142.63
WintTemp         1       84.84
SummTemp         1        0.03

Unusual Observations
Obs.   Income AvgOrder       Fit   Stdev.Fit   Residual    St.Resid
 25      17.1   23.570    36.555       0.632    -12.985      -2.76R
 78      11.9   24.990    34.950       0.793     -9.960      -2.13R
 83      13.4   36.750 .  29.136       2.610      7.614       1.92X
 87      14.3   45.970    35.918       0.463     10.052       2.13R
111      11.1   21.720    33.570       0.802    -11.850      -2.53R
113      10.4   43.500    33.469       0.817     10.031       2.15R
143      16.1   20.350    27.915       3.000     -7.565      -2.06RX
149      13.2   44.970    35.369       0.604      9.601       2.04R
169      13.5   44.650    34.361       0.660     10.289       2.19R
180      13.7   23.050    34.929       0.469    -11.879      -2.51R
R denotes an obs. with a large st. resid.
X denotes an obs. whose X value gives it large influence.
```

12.63 The accompanying table gives demographic data for 12 male patients with congestive heart failure enrolled in a study of an experimental compound.

Demographic Data for Patients with Heart Failure
(NYHA Class III or IV)

Patient	Age (yrs)	Disease Duration	Height (cm)	Weight (kg)	Baseline Cardiac Index (L/min/m²)	Baseline Pulmonary Capillary Wedge Pressure (mm Hg)
01	67	5 yr	172.0	57.0	1.6	40
02	45	2 yr	170.0	67.0	2.4	25
03	59	8 yr	172.7	102.0	2.2	39
04	63	1 yr	175.3	74.9	1.7	39
05	55	1 yr	172.7	92.0	2.3	34
06	65	1 yr	178.0	90.0	1.6	36
07	62	2 yr	163.0	67.0	1.4	36
08	60	1 yr	182.5	72.0	2.2	17
09	72	2 yr	168.0	71.0	1.3	37
10	44	3 mo	163.0	68.0	2.4	28
11	63	5 yr	172.0	82.0	2.1	38
12	63	1 yr	163.0	64.0	1.1	36

a. Summarize these data using a boxplot for each variable.
b. Construct scatterplots to display (1) age by cardiac index (CI) and by pulmonary capillary wedge pressure (PCWP) and (2) disease duration by CI and by PCWP. Is there evidence of a correlation between age and CI or PCWP? What about correlation between duration of disease and CI or PCWP?

12.64 The data of Exercise 12.63 were used to fit several multiple regression models; y_1 = CI, y_2 = PCWP, x_1 = age, x_2 = disease duration.

  a. $y_1 = \beta_0 + \beta_1 x_1 + \beta_2 x_2 + \varepsilon$
  b. $y_1 = \beta_0 + \beta_1 x_1 + \beta_2 x_2 + \beta_3 x_1 x_2 + \varepsilon$
  c. $y_2 = \beta_0 + \beta_1 x_1 + \beta_2 x_2 + \varepsilon$
  d. $y_2 = \beta_0 + \beta_1 x_1 + \beta_2 x_2 + \beta_3 x_1 x_2 + \varepsilon$

```
              REGRESSION ANALYSIS,    MODEL I


Dependent Variable: CI

Analysis of Variance
                    Sum of         Mean
  Source    DF      Squares       Square      F Value      Prob>F
  Model      2      1.56955      0.78478       9.298       0.0065
  Error      9      0.75961      0.08440
  C Total   11      2.32917


      Root MSE     0.29052    R-square     0.6739
      Dep Mean     1.85833    Adj R-sq     0.6014
      C.V.        15.63333
```

Parameter Estimates

Variable	DF	Parameter Estimate	Standard Error	T for H0: Parameter=0	Prob > |T|
INTERCEP	1	4.475622	0.63976685	6.996	0.0001
AGE	1	-0.046203	0.01083529	-4.264	0.0021
DURATION	1	0.060395	0.03852829	1.568	0.1514

REGRESSION ANALYSIS, MODEL II

Dependent Variable: CI

Analysis of Variance

Source	DF	Sum of Squares	Mean Square	F Value	Prob>F
Model	3	1.57161	0.52387	5.532	0.0237
Error	8	0.75755	0.09469		
C Total	11	2.32917			

Root MSE	0.30772	R-square	0.6748	
Dep Mean	1.85833	Adj R-sq	0.5528	
C.V.	16.55915			

Parameter Estimates

Variable	DF	Parameter Estimate	Standard Error	T for H0: Parameter=0	Prob > |T|
INTERCEP	1	4.599307	1.07814691	4.266	0.0027
AGE	1	-0.048340	0.01848097	-2.616	0.0309
DURATION	1	-0.022410	0.56287924	-0.040	0.9692
AGE_DUR	1	0.001376	0.00932590	0.147	0.8864

Variable	DF	Variable Label
INTERCEP	1	Intercept
AGE	1	
DURATION	1	
AGE_DUR	1	AGE TIMES DURATION

REGRESSION ANALYSIS, MODEL III

Dependent Variable: PCWP

Analysis of Variance

Source	DF	Sum of Squares	Mean Square	F Value	Prob>F
Model	2	221.88101	110.94051	3.259	0.0862
Error	9	306.36899	34.04100		
C Total	11	528.25000			

Root MSE	5.83447	R-square	0.4200	
Dep Mean	33.75000	Adj R-sq	0.2911	
C.V.	17.28731			

```
Parameter Estimates
                   Parameter       Standard     T for H0:
Variable   DF      Estimate         Error      Parameter=0    Prob > |T|
INTERCEP   1       7.298786      12.84835977       0.568        0.5839
AGE        1       0.400475       0.21760372       1.840        0.0989
DURATION   1       1.021327       0.77375900       1.320        0.2194



              REGRESSION ANALYSIS,   MODEL IV


Dependent Variable: PCWP

Analysis of Variance
                      Sum of         Mean
Source       DF       Squares        Square     F Value     Prob>F
Model         3      228.56515      76.18838      2.034      0.1878
Error         8      299.68485      37.46061
C Total      11      528.25000


     Root MSE      6.12051    R-square      0.4327
     Dep Mean     33.75000    Adj R-sq      0.2199
     C.V.         18.13484


Parameter Estimates
                   Parameter       Standard     T for H0:
Variable   DF      Estimate         Error      Parameter=0    Prob > |T|
INTERCEP   1      14.344026      21.44389171       0.669        0.5224
AGE        1       0.278775       0.36757883       0.758        0.4700
DURATION   1      -3.695301      11.19543293      -0.330        0.7498
AGE_DUR    1       0.078352       0.18548824       0.422        0.6838


                     Variable
Variable   DF        Label
INTERCEP   1  Intercept
AGE        1
DURATION   1
AGE_DUR    1    AGE TIMES DURATION
```

Which model provides the best fit to the cardiac index data? To the pulmonary capillary wedge pressure data? Do these analyses confirm what you concluded in Exercise 12.63? Explain.

The following exercises all refer to the case study introduced at the beginning of the chapter.

12.65 The initial analysis of the data involves plotting the performance variables avtem and logv versus the design factors in order to obtain an idea of which design factors may be related to the performance variables. Also, these plots display whether the relationships between performance variables and design factors is linear or requires a higher-order model.

    a. Construct the six scatterplots of performance variables versus design factors; that is, plot avtem versus IT, QW, and VS. Then plot logv versus IT, QW, and VS.

    b. Examine the three scatterplots involving avtem and describe the relationships, if any, between avtem and the three design factors.

    c. Examine the three scatterplots involving logv and describe the relationships, if any, between logv and the three design factors.

12.66 After examining the scatterplots, we want to examine a number of models to determine which model provides the best overall fit to the avtem data without overfitting the model—that is, placing too many terms in the model. Fit the following models to the avtem data.

Model 1: $\text{avtem} = \beta_0 + \beta_1 IT + \beta_2 QW + \beta_3 VS + \varepsilon$

Model 2: $\text{avtem} = \beta_0 + \beta_1 IT + \beta_2 QW + \beta_3 VS + \beta_4 I2 + \beta_5 Q2 + \beta_6 V2 + \varepsilon$

Model 3: $\text{avtem} = \beta_0 + \beta_1 IT + \beta_2 QW + \beta_3 VS + \beta_4 IT * QW + \beta_5 IT * VS$
$+ \beta_6 QW * VS + \varepsilon$

Model 4: $\text{avtem} = \beta_0 + \beta_1 IT + \beta_2 QW + \beta_3 VS + \beta_4 I2 + \beta_5 Q2 + \beta_6 V2$

$+ \beta_7 IT * QW + \beta_8 IT * VS + \beta_9 QW * VS + \varepsilon$

a. Based on the values of R^2 for the four models, which model would you select as providing the "best" fit to the data?
b. Test the hypothesis that model 2 is not significantly different from model 1. Use $\alpha = .05$.
c. Test the hypothesis that model 3 is not significantly different from model 1. Use $\alpha = .05$.
d. Test the hypothesis that model 4 is not significantly different from model 3. Use $\alpha = .05$.
e. Test the hypothesis that model 4 is not significantly different from model 2. Use $\alpha = .05$.
f. Using the scatterplots and the results of (a)–(e), which model would you recommend to the engineers? Explain your reasons for selecting the model.

12.67 After examining the scatterplots, we want to examine a number of models to determine which model provides the best overall fit to the logv data without overfitting the model—that is, without placing too many terms in the model. Fit the following models to the logv data.

Model 1: $\text{logv} = \beta_0 + \beta_1 IT + \beta_2 QW + \beta_3 VS + \varepsilon$

Model 2: $\text{logv} = \beta_0 + \beta_1 IT + \beta_2 QW + \beta_3 VS + \beta_4 I2 + \beta_5 Q2 + \beta_6 V2 + \varepsilon$

Model 3: $\text{logv} = \beta_0 + \beta_1 IT + \beta_2 QW + \beta_3 VS + \beta_4 IT * QW + \beta_5 IT * VS$

$+ \beta_6 QW * VS + \varepsilon$

Model 4: $\text{logv} = \beta_0 + \beta_1 IT + \beta_2 QW + \beta_3 VS + \beta_4 I2 + \beta_5 Q2 + \beta_6 V2$

$+ \beta_7 IT * QW + \beta_8 IT * VS + \beta_9 QW * VS + \varepsilon$

a. Based on the values of R^2 for the four models, which model would you select as providing the "best" fit to the data?
b. Test the hypothesis that model 2 is not significantly different from model 1. Use $\alpha = .05$.
c. Test the hypothesis that model 3 is not significantly different from model 1. Use $\alpha = .05$.
d. Test the hypothesis that model 4 is not significantly different from model 3. Use $\alpha = .05$.
e. Test the hypothesis that model 4 is not significantly different from model 2. Use $\alpha = .05$.
f. Using the scatterplots and the results of (a)–(e), which model would you recommend to the engineers? Explain your reasons for selecting the model.

CHAPTER 13

More on Multiple Regression

13.1 Introduction and Case Study

13.2 Selecting the Variables (Step 1)

13.3 Formulating the Model (Step 2)

13.4 Checking Model Assumptions (Step 3)

13.5 Summary

13.1 Introduction and Case Study

In Chapter 12, we presented the background information needed to use multiple regression. We discussed the general linear model and its use in multiple regression and introduced the normal equations, a set of simultaneous equations used in obtaining least-squares estimates for the βs of a multiple regression equation. Next, we presented standard errors associated with the $\hat{\beta}_j$ and their use in inferences about a single parameter β_j, a set of βs, $E(y)$, and a future value of y. We also considered special situations—comparing the slopes of several regression lines and the logistic regression problem. Finally, we condensed all of these inferential techniques using matrices.

This chapter is devoted to putting multiple regression into practice. How does one begin to develop an appropriate multiple regression for a given problem? Although there are no hard and fast rules, we can offer a few hints.

First, for each problem you must decide on the dependent variable and candidate independent variables for the regression equation. This selection process will be discussed in Section 13.2. In Section 13.3, we consider how one selects the form of the multiple regression equation. The final step in the process of developing a multiple regression is to check for violation of the underlying assumptions. Tools for assessing the validity of the assumptions will be discussed in Section 13.4.

Following these steps *once* for a given problem will not ensure that you have an appropriate model. Rather, the regression equation seems to evolve as these steps are applied repeatedly, depending on the problem. For example, having considered candidate independent variables (step 1) and selected the form for a regression model involving some of these variables (step 2), we may find that certain assumptions have been violated (step 3). This will mean that we may have to return to either step 1 or step 2, but, hopefully, we have learned from our previous deliberations and can modify the variables under consideration and/or the model(s) selected for consideration. Eventually, a regression model will emerge that meets the needs of the experimenter. Then the analysis techniques of Chapter 12 can be used to draw inferences about model parameters $E(y)$ and y.

Case Study: Finding a Regression Model That Will Assist Marketing Managers of an Office Products Company in Evaluating the Performance of Field Sales Representatives

The marketing managers of an office products company have some difficulty in evaluating the field sales representatives' performance. The representatives travel among the outlets that carry the company's products, create displays, try to increase volume, introduce new products, and discover any problems that the outlets are having with the company's products. The job involves a great deal of travel time.

Designing the Data Collection   The marketing managers believe that one important factor in the representatives' performance is the degree of motivation to spend a great deal of time on the road. Other variables also have an effect. Some sales districts have more potential than others, either because of differences in population or differences in the number of retail outlets. Large districts are difficult because of the extra travel time.

One important variable is compensation. Some of the representatives are paid a salary plus a commission on sales; others work solely for a larger commission on sales. The marketing managers suspect there is a difference in effectiveness between the two groups, although some managers argue that the important factor is the combination of commission status and number of outlets. In particular, they suspect that commission-only representatives with many outlets to cover are highly productive. Also, the managers suspect that profit may be inflated for representatives with many outlets, they would prefer measuring profit per outlet.

Managing the Data   Data are collected on 51 representatives. The data include DISTRICT number, PROFIT (net profit margin for all orders placed through the representative—the dependent variable of interest), AREA (of the district in thousands of square miles), POPN (millions of people in the district), OUTLETS (number of outlets in the district), and COMMIS, which is 1 for full-commission representatives and 0 for partially salaried representatives. The data are shown here:

DIST	PROFIT	AREA	POPN	OUTLETS	COMMIS
1	1011	16.96	3.881	213	1
2	1318	7.31	3.141	158	1
3	1556	7.81	3.766	203	1
4	1521	7.31	4.587	170	1
5	979	19.84	3.648	142	1
6	1290	12.37	3.456	159	1
7	1596	6.15	3.695	178	1
8	1155	14.21	3.609	182	1
9	1412	7.45	3.801	181	1
10	1194	14.43	3.322	148	1
11	1054	6.12	5.124	227	0
12	1157	11.71	4.158	139	1
13	1001	9.36	3.887	179	0
14	831	19.14	2.230	124	1
15	857	11.75	4.468	205	0
16	188	40.34	.297	85	1

DIST	PROFIT	AREA	POPN	OUTLETS	COMMIS
17	1030	7.16	4.224	211	0
18	1331	9.37	3.427	145	1
19	643	7.62	4.031	205	1
20	992	27.54	2.370	166	1
21	795	15.97	3.903	149	1
22	1340	12.97	3.423	186	1
23	689	17.36	2.390	141	0
24	1726	6.24	4.947	223	1
25	1056	11.20	4.166	176	0
26	989	18.09	4.063	187	1
27	895	13.32	3.105	131	1
28	1028	14.97	4.116	170	0
29	771	21.92	1.510	144	1
30	484	34.91	.741	126	1
31	917	8.46	5.260	234	0
32	1786	7.52	5.744	210	0
33	1063	14.43	2.703	141	1
34	1001	15.37	3.583	158	0
35	1052	11.20	4.469	167	1
36	1610	7.20	4.951	174	1
37	1486	13.49	3.474	211	1
38	1576	6.56	4.637	172	1
39	1665	9.35	3.900	185	1
40	878	11.12	3.766	166	0
41	849	10.58	3.876	189	0
42	775	17.82	2.753	164	0
43	1012	10.03	4.449	193	0
44	1436	10.01	4.680	157	1
45	798	10.70	4.806	200	0
46	519	24.38	2.367	142	0
47	1701	6.57	5.563	199	0
48	1387	6.64	4.357	166	1
49	1717	9.24	4.670	221	1
50	1032	11.62	3.993	180	0
51	973	12.85	3.923	193	0

Analyzing the Data   Use the data to perform a multiple regression analysis. Find out whether the variables suspected by the managers as having an effect on PROFIT actually do have an effect; in particular, try to discover whether there is a combination effect of COMMIS and OUTLETS. Consider whether PROFIT itself or PROFIT divided by OUTLETS works better as a dependent variable. Omit variables that show little predictive value. Locate and, if possible, correct any serious violations of assumptions. Write a brief nontechnical report to the marketing managers and explain your findings. The data will be analyzed in Exercise 13.70 based on our discussions in this chapter.

13.2   Selecting the Variables (Step 1)

Perhaps the most critical decision in constructing a multiple regression model is the initial selection of independent variables. In later sections of this chapter, we consider many methods for refining a multiple regression analysis, but first we must make a decision about which independent (x) variables to consider for inclusion—and, hence, which data to gather. If we do not have useful data, we are unlikely to come up with a useful predictive model.

Although initially it may appear that an optimum strategy might be to construct a monstrous multiple regression model with very many variables, such models are difficult to interpret and are much more costly from a data-gathering and analysis time standpoint. How can a manager make a reasonable selection of initial variables to include in a regression analysis?

selection of the independent variables

Knowledge of the problem area is critically important in the initial selection of data. First, identify the dependent variable to be studied. Individuals who have had experience with this variable by observing it, trying to predict it, and trying to explain changes in it often have remarkably good insight as to what factors (independent variables) affect it. As a consequence, the first step involves consulting those who have the most experience with the dependent variable of interest. For example, suppose that the problem is to forecast the next quarter's sales volume of an inexpensive brand of computer printer for each of 40 districts. The dependent variable y is then district sales volume. Certain independent variables, such as the advertising budget in each district and the number of sales outlets, are obvious candidates. A good district sales manager undoubtedly could suggest others.

collinearity

A major consideration in selecting predictor variables is the problem of **collinearity**—that is, severely correlated independent variables. A partial slope in multiple regression estimates the predictive effect of changing one independent variable while holding all others constant. However, when some or all of the predictors vary together, it can be almost impossible to separate out the predictive effects of each one. A common result when predictors are highly correlated is that the overall F test is highly significant, but none of the individual t tests comes close to significance. The significant F result indicates only that there is detectable predictive value somewhere among the independent variables; the nonsignificant t values indicate that we cannot detect *additional* predictive value for any variable, given all the others. The reason is that highly correlated predictors are surrogates for each other; any of them individually may be useful, but adding others will not be. When seriously collinear independent variables are all used in a multiple regression model, it can be virtually impossible to decide which predictors are in fact related to the dependent variable.

correlation matrix

There are several ways to assess the amount of collinearity in a set of independent variables. The simplest method is to look at a (Pearson) **correlation matrix**, which can be produced by almost all computer packages. The higher these correlations, the more severe the collinearity problem is. In most situations, any correlation over .9 or so definitely indicates a serious problem.

scatterplot matrix

Some computer packages can produce a **scatterplot matrix**, a set of scatterplots for each pair of variables. Collinearity appears in such a matrix as a close linear relation between two of the *independent* variables. For example, a sample of automotive writers rated a new compact car on 0 to 100 scales for performance, comfort, appearance, and overall quality. The promotion manager doing the study wanted to know which variables best predicted the writers' rating of overall quality. A Minitab scatterplot matrix is shown in Figure 13.1. There are clear linear relations among the performance, comfort, and appearance ratings, indicating substantial collinearity. The following matrix of correlations confirms that fact:

```
MTB > correlations c1-c4
              Correlations (Pearson)
            overall  perform  comfort
perform    0.698
comfort    0.769    0.801
appear     0.630    0.479    0.693
```

FIGURE 13.1

Scatterplot matrix for auto writers data

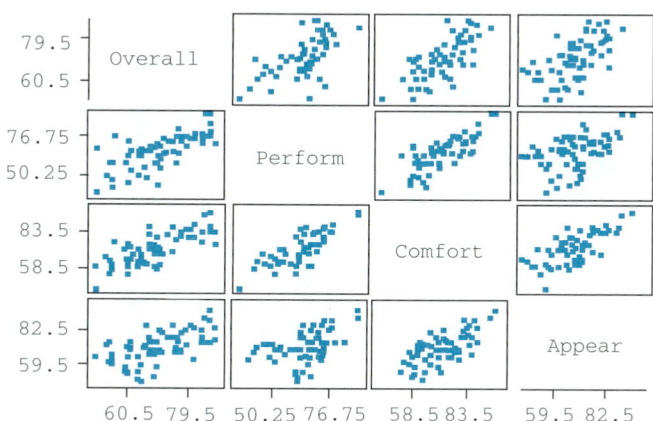

A scatterplot matrix can also be useful in detecting nonlinear relations or outliers. The matrix contains scatterplots of the dependent variable against each independent variable separately. Sometimes a curve or a serious outlier will be clear in the matrix. Other times, the effect of other independent variables may conceal a problem. The analysis of residuals, discussed later in this chapter, is another good way to look for assumption violations.

The correlation matrix and scatterplot matrix may not reveal the full extent of a collinearity problem. Sometimes two predictors together predict a third all too well, even though either of the two by itself shows a more modest correlation with the third one. (Direct labor hours and indirect labor hours together predict total labor hours remarkably well, even if either one predicts the total imperfectly.) A number of more sophisticated ways of diagnosing collinearity are built into various computer packages. One such diagnostic is the variance inflation factor (VIF) discussed in Chapter 12. It is $1/(1 - R^2)$, where this R^2 refers to how much of the variation in one *independent* variable is explained by the others. The VIF takes into account all relations among predictors, so it is more complete than simple correlations. The books by Cook and Weisberg (1982) and by Belsley, Kuh, and Welsch (1980) define several diagnostic measures for collinearity. The manuals for most statistical computer programs will indicate which of these can be computed and what the results indicate.

EXAMPLE 13.1

A supermarket chain staged a promotion for a superpremium brand of ice cream. Data on actual sales of the brand for the weekend of the promotion were obtained from scanner data gathered at the checkout stands. Three explanatory variables being considered were the size of the store (in thousands of square feet), the number of customers processed in the store (in hundreds), and the average size of purchase (also obtained from scanner data). A scatterplot matrix is shown in Figure 13.2. What does it indicate about collinearity? Does it show any other problems?

FIGURE 13.2

Scatterplot matrix for ice cream data

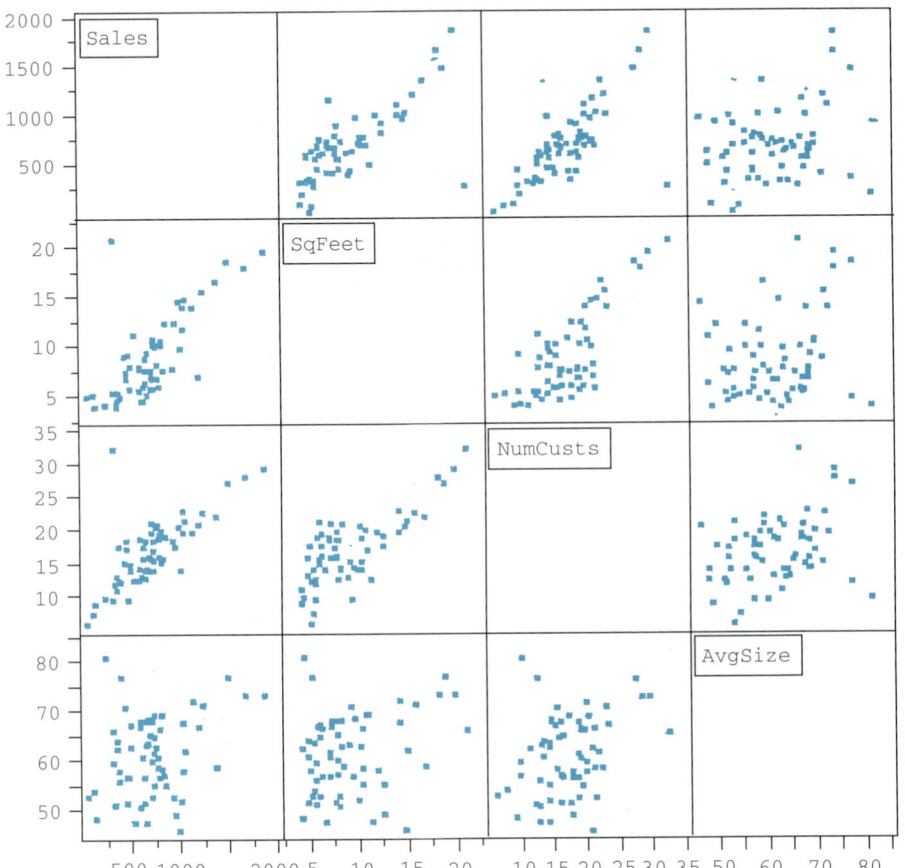

Solution   Look at the six scatterplots in the upper right of the matrix. There is a clear increasing relation, but not extremely strong, between SqFeet, the size of the store, and NumCusts, the number of customers. There is little correlation between either of these independent variables and AvgSize of purchase. Thus, there is a modest collinearity problem. There is no indication of a serious nonlinearity problem in the matrix. However, there is an outlier. In the plot of Sales against SqFeet, one store has the largest size but nearly the smallest sales. This outlier has fairly high leverage; it is extreme on two of the independent variables (size and number of customers). It may also have substantial influence, because it falls well off the line for predicting sales from either of these predictors. Further checking showed that the ice cream display case at store 41 had lost power during the weekend, forcing the store manager to remove all the ice cream from the case. The store was omitted from the regression analysis.

One of the best ways to avoid collinearity problems is to choose predictor variables intelligently, right at the beginning of a regression study. Try to find independent variables that should correlate decently with the dependent variable but do not have obvious correlations with each other. If possible, try to find independent variables that reflect various components of the dependent variable. For example, suppose we want to predict the sales of inexpensive printers for personal computers in each of 40 sales districts. Total sales are made up of several sectors of buyers. We might identify the important sectors as college students, home users, small businesses, and computer network workstations. Therefore, we might try number of college freshmen, household income, small business starts, and new network installations as independent variables. Each one makes sense as a predictor of printer sales, and there is no obvious correlation among the predictors. People who are knowledgeable about the variable you want to predict can often identify components and suggest reasonable predictors for the different components.

EXAMPLE 13.2

A firm that sells and services minicomputers is concerned about the volume of service calls. The firm maintains several district service branches within each sales region, and computer owners requiring service call the nearest branch. The branches are staffed by technicians trained at the main office. The key problem is whether technicians should be assigned to main office duty or to service branches; assignment decisions have to be made monthly. The required number of service branch technicians grows in almost exact proportion to the number of service calls. Discussion with the service manager indicates that the key variables in determining the volume of service calls seem to be the number of computers in use, the number of new installations, whether or not a model change has been introduced recently, and the average temperature. (High temperatures, or possibly the associated high humidity, lead to more frequent computer troubles, especially in imperfectly air conditioned offices.) Which of these variables can be expected to correlate with the others?

Solution   It is hard to imagine why temperature should be correlated with any of the other variables. There should be some correlation between number of computers in use and number of new installations, if only because every new installation is a computer in use. Unless the firm has been growing at an increasing rate, we would not expect a severe correlation (we would, however, like to see the data). The correlation of model change to number in use and new installations is not at all obvious; surely data should be collected and correlations analyzed.

A manager who begins a regression study may try to put too many independent variables into a regression model; hence, we need some sensible guidelines to help or select the independent variables to be included in the final regression model from potential candidates.

One way to sort out which independent variables should be included in a regression model from the list of variables generated from discussions with experts is to resort to any one of a number of selection procedures. We will consider several of these in this text; for further details, the reader can consult Neter, Kutner, Nachtsheim, Wasserman (1996).

The first selection procedure involves performing *all possible regressions* with the dependent variable and one or more of the independent variables from the list of candidate variables. Obviously, this approach should not be attempted unless the analyst has access to a computer with suitable software and sufficient core to run a large number of regression models relatively efficiently.

As an illustration, we will use hypothetical data on prescription sales data (volume per month) obtained for a random sample of 20 independent pharmacies. These data, along with data on the total floor space, percentage of floor space allocated to the prescription department, the number of parking spaces available for the store, whether the pharmacy is in a shopping center, and the per capita income for the surrounding community are recorded in Table 13.1.

TABLE 13.1

Data on 20 independent pharmacies

OBS	PHARMACY	VOLUME	FLOOR—SP	PRESC—RX	PARKING	SHOPCNTR	INCOME
1	1	22	4900	9	40	1	18
2	2	19	5800	10	50	1	20
3	3	24	5000	11	55	1	17
4	4	28	4400	12	30	0	19
5	5	18	3850	13	42	0	10
6	6	21	5300	15	20	1	22
7	7	29	4100	20	25	0	8
8	8	15	4700	22	60	1	15
9	9	12	5600	24	45	1	16
10	10	14	4900	27	82	1	14
11	11	18	3700	28	56	0	12
12	12	19	3800	31	38	0	8
13	13	15	2400	36	35	0	6
14	14	22	1800	37	28	0	4
15	15	13	3100	40	43	0	6
16	16	16	2300	41	20	0	5
17	17	8	4400	42	46	1	7
18	18	6	3300	42	15	0	4
19	19	7	2900	45	30	1	9
20	20	17	2400	46	16	0	3

N = 20

Before running all possible regressions for the data of Table 13.1, we need to consider what criterion should be used to select the best-fitting equation from all possible regressions. The first and perhaps simplest criterion for selecting the best regression equation from the set of all possible regression equations is to compute an estimate of the error variance σ_ε^2 using $s_\varepsilon^2 = $ MS(Residual) $=$ SS(Residual)$/[n - (k + 1)]$. Since this quantity is used in most inferences (statistical tests and confidence intervals) about model parameters and $E(y)$, it would seem reasonable to choose the model that has the smallest value of s_ε^2. A second criterion makes use of the *coefficient of determination R^2* for each model; by examining in detail the models that have the highest R^2 values, we can see whether there is some consistent pattern that suggests the number and identity of the variables to include in the model.

EXAMPLE 13.3

Refer to the data of Table 13.1. Use the R^2 criterion to determine the best-fitting regression equation for 1, 2, 3, and 4 independent variables.

Solution   SAS output is provided here, and the regression equations with the highest R^2 values are summarized in Table 13.2.

```
SAS OUTPUT FOR PROC RSQUARE
ALL POSSIBLE SUBSETS ANALYSIS
N = 20        Regression Models for Dependent Variable: VOLUME


NUMBER
IN MODEL R-square      C(p) Variables in Model


   1  0.4393318   10.1709 PRESC_RX
   1  0.1479899   23.7702 INCOME
   1  0.0410534   28.7618 SHOPCNTR
   1  0.0335317   29.1129 FLOOR_SP
   1  0.0048042   30.4539 PARKING
----------------------------------
   2  0.6656627    1.6062 FLOOR_SP PRESC_RX
   2  0.6470647    2.4744 PRESC_RX SHOPCNTR
   2  0.5474878    7.1224 PRESC_RX INCOME
   2  0.5314244    7.8722 PRESC_RX PARKING
   2  0.4957679    9.5366 SHOPCNTR INCOME
   2  0.2565364   20.7035 FLOOR_SP SHOPCNTR
   2  0.2348733   21.7147 FLOOR_SP INCOME
   2  0.2054310   23.0890 PARKING INCOME
   2  0.0685567   29.4780 FLOOR_SP PARKING
   2  0.0421078   30.7126 PARKING SHOPCNTR
-------------------------------------------
   3  0.6907243    2.4364 FLOOR_SP PRESC_RX SHOPCNTR
   3  0.6794331    2.9635 FLOOR_SP PRESC_RX PARKING
   3  0.6664115    3.5713 FLOOR_SP PRESC_RX INCOME
   3  0.6625912    3.7496 PRESC_RX PARKING SHOPCNTR
   3  0.6471156    4.4720 PRESC_RX SHOPCNTR INCOME
   3  0.6024323    6.5577 PRESC_RX PARKING INCOME
   3  0.5001258   11.3332 FLOOR_SP SHOPCNTR INCOME
   3  0.4982807   11.4193 PARKING SHOPCNTR INCOME
   3  0.2650711   22.3051 FLOOR_SP PARKING SHOPCNTR
   3  0.2556961   22.7427 FLOOR_SP PARKING INCOME
----------------------------------------------------
   4  0.6987395    4.0623 FLOOR_SP PRESC_RX PARKING SHOPCNTR
   4  0.6932666    4.3177 FLOOR_SP PRESC_RX SHOPCNTR INCOME
   4  0.6805857    4.9097 FLOOR_SP PRESC_RX PARKING INCOME
   4  0.6630086    5.7301 PRESC_RX PARKING SHOPCNTR INCOME
   4  0.5012890   13.2789 FLOOR_SP PARKING SHOPCNTR INCOME
-------------------------------------------------------------
   5  0.7000737    6.0000 FLOOR_SP PRESC_RX PARKING SHOPCNTR INCOME
-------------------------------------------------------------
```

TABLE 13.2

Best-fitting models, R^2 criterion, Example 13.2

Number of Independent Variables	R^2	Variables
1	.439	Prescription sales
2	.666	Floor space, prescription sales
3	.691	Floor space, prescription sales, shopping center
4	.699	All except per capita income

Although there is a good jump in R^2 going from one to two independent variables, very little improvement is seen thereafter. Hence, the best-fitting model based on the R^2 criterion involves the independent variables floor space and prescription sales.

One problem with using R^2 as a criterion for the best-fitting regression equation is that R^2 increases for each independent variable, even when the new x has very little predictive power. Other possible criteria for selecting the best regression that do not increase with the addition of each are presented here.

Keep in mind that the object of our search is to choose the subset of independent variables that generates the best prediction equation for *future* values of y; unfortunately, however, because we do not know these future values, we focus on criteria that choose the best-fitting regression equations to the known sample y-values. One possible bridge between this emphasis on the best fit to the known sample y-values and that on choosing the best predictor of future y-values is to split the sample data into two parts—one part used for fitting the various regression equations and the other part for validating how well the prediction equations can predict "future" values. Although there is no universally accepted rule for deciding how many of the data should be included in the "fitting" portion of the sample and how many go into the "validating" portion of the sample, it is reasonable to split the total sample in half, provided the total sample size n is greater than $2p + 20$, where p is the number of parameters in the largest potential regression model. A possible criterion for the best prediction equation would be to minimize $\Sigma (y_i - \hat{y}_i)^2$ for the validating portion of the total sample.

Once the regression model is selected from the data-splitting approach, the entire set of sample data is used to obtain the final prediction equation. Thus, even though it appears we would only use part of the data, the entire data set is used to obtain the final prediction equation.

Observations do cost money, however, and it may be impractical to obtain enough observations to apply the data-splitting approach for choosing the best-fitting regression equation. In these situations, a form of validation can be accomplished using the PRESS statistic. For a sample of y-values and a proposed regression model relating y to a set of xs, we first remove the first observation and fit the model using the remaining $n - 1$ observations. Based on the fitted equation, we estimate the first observation (denoted by $\hat{y}_1^*$) and compute the residual $y_1 - \hat{y}_1^*$. This process is repeated $n - 1$ times, successively removing the second, third, . . . , nth observation, each time computing the residual for the removed observation. The PRESS statistic is defined as

$$\text{PRESS} = \sum_{i=1}^{n} (y_i - \hat{y}_i^*)^2$$

The model that gives the smallest value for the PRESS statistic is chosen as the best-fitting model.

EXAMPLE 13.4

Compute the PRESS statistic for the data of Table 13.1 to determine the best-fitting regression equation.

Solution SAS output is provided here. The best-fitting model based on the lowest value of the PRESS statistic involves the independent variables floor space and prescription sales.

```
SAS OUTPUT FOR ALL POSSIBLE SUBSET ANALYSIS
           PRESS STATISTIC
N = 20    Regression Models for Dependent Variable: VOLUME

NUMBER IN    PRESS STATISTIC   VARIABLES IN MODEL
MODEL

    1         516.391          PRESC_RX
    1         772.163          INCOME
    1         869.668          FLOOR_SP
    1         887.636          SHOPCNTR
    1         907.636          PARKING
------------------------------------------------------
    2         347.007          FLOOR_SP PRESC_RX
    2         368.757          PRESC_RX SHOPCNTR
    2         479.976          PRESC_RX PARKING
    2         485.820          SHOPCNTR INCOME
    2         547.150          PRESC_RX INCOME
    2         762.507          FLOOR_SP SHOPCNTR
    2         787.578          PARKING INCOME
    2         797.404          FLOOR_SP INCOME
    2         916.644          FLOOR_SP PARKING
    2         975.912          PARKING SHOPCNTR
------------------------------------------------------
    3         370.843          FLOOR_SP PRESC_RX SHOPCNTR
    3         371.671          FLOOR_SP PRESC_RX PARKING
    3         378.166          PRESC_RX PARKING SHOPCNTR
    3         455.424          PRESC_RX SHOPCNTR INCOME
    3         482.387          FLOOR_SP PRESC_RX INCOME
    3         513.246          PRESC_RX PARKING INCOME
    3         523.006          PARKING SHOPCNTR INCOME
    3         602.214          FLOOR_SP SHOPCNTR INCOME
    3         819.792          FLOOR_SP PARKING SHOPCNTR
    3         890.550          FLOOR_SP PARKING INCOME
------------------------------------------------------
    4         405.832          FLOOR_SP PRESC_RX PARKING SHOPCNTR
    4         458.014          PRESC_RX PARKING SHOPCNTR INCOME
    4         471.086          FLOOR_SP PRESC_RX SHOPCNTR INCOME
    4         513.468          FLOOR_SP PRESC_RX PARKING INCOME
    4         684.190          FLOOR_SP PARKING SHOPCNTR INCOME
------------------------------------------------------
    5         513.915          FLOOR_SP PRESC_RX PARKING SHOPCNTR INCOME
------------------------------------------------------
```

To this point, we have considered criteria for selecting the best-fitting regression model from a subset of independent variables. In general, if we choose a model that leaves out one or more "important" predictor variables, our model is *underspecified* and the additional variability in the y-values that would be accounted for with these variables becomes part of the estimated error variance. At the other end of the spectrum, if we choose a model that contains one or more "extraneous" predictor variables, our model is *overspecified* and we stand the chance of having a *multicollinearity* problem. We will deal with this problem later.

The point is that a final criterion, based on the C_p statistic, seems to balance some pros and cons of previously presented selection criteria, along with the problems of over- and underspecification, to arrive at a choice of the best-fitting subset regression equation. The C_p statistic [see Mallows (1973)] is

$$C_p = \frac{\text{SS(Residual)}_p}{s_\varepsilon^2} - (n - 2p)$$

where SSR_p is the sum of squares for error from a model with p parameters (including β_0) and s_ε^2 is the mean square error from the regression equation with the largest number of independent variables. For a given selection problem, compute C_p for every regression equation that is fit. Theory suggests that the best-fitting model should have $C_p \approx p$.

EXAMPLE 13.5

Refer to the output of Example 13.3. Determine the value of C_p for all possible regressions with 1, 2, 3, 4, and 5 independent variables. Select the best-fitting equation for 1, 2, 3, and 4 independent variables. Which regression equation seems to give the best overall fit, based on the C_p statistic?

Solution   The best-fitting models are summarized in Table 13.3. Based on the C_p criterion, there would be very little difference between the best-fitting models for 2, 3, or 4 independent variables in the model. The most "important" predictive variables seem to be floor space and prescription sales because they appear in the best-fitting models for 2, 3, and 4 independent variables. Note that these are the same important independent variables found in Example 13.3.

TABLE 13.3
Best-fitting models, C_p criterion

Number of Independent Variables	p	C_p	Variables
1	2	10.17	Prescription sales
2	3	1.61	Floor space, prescription sales
		2.47	Prescription sales, shopping center
3	4	2.96	Floor space, prescription sales, parking space
4	5	4.06	Floor space, prescription sales, parking spaces, shopping center

Best subset regression provides another procedure for finding the best-fitting regression equation from a set of k candidate independent variables. This procedure uses an algorithm that avoids running all possible regressions. The computer program prints a listing of the best M (the user selects M) regression equations with one independent variable in the model, two independent variables in the model, three independent variables in the model, and so on, up to the model containing all K independent variables in the model. Some programs allow the user to specify the criterion for "best" (for example, C_p or maximum R^2), whereas other programs fix the criterion. For instance, the Minitab program uses maximum R^2 to select the M best subsets of each size. The program computes the M regressions having the largest R^2 for each value of $K = 1, 2, \ldots, k$ independent variables in the model. We will illustrate this procedure with the data of Table 13.1.

EXAMPLE 13.6

Use the Minitab output shown here to find the $M = 2$ best subset regression equations of size 1 to 5 based on the maximum R^2 criterion for the data of Table 13.1. From the various "best" regression equations, select the regression equation that has the "best" overall R^2.

Solution. The output is shown here. The program identified two best subsets of each size. The values of adjusted R^2, C_p, and $\sqrt{MS\,(Residual)} = s$ are given for each subset. Based on the maximum R^2, the subset with all independent variables will always be the best regression. However, based on adjusted R^2 or C_p our conclusion would differ from the best obtained from the maximum R^2. Minitab does not provide the least-squares regression line in this output. The subset of independent variables selected as best would next be run in the Minitab regression program to obtain the regression equation. Note that R^2 is expressed as a percentage in the Minitab output, $100R^2$.

```
Best Subsets Regression: VOLUME versus FLOOR_SP, PRESC_RX, PARKING,
SHOPCNT, INCOME


Response is VOLUME                          VARIABLES INCLUDED
                                               Indicated by X


                                            F P   S
                                            L R P H
                                            O E A O I
                                            O S R P N
                                            R C K C C
                                            - - I N O
No.                                         S R N T M
Vars                                        P X G R E
In
Model  R-Sq      R-Sq(adj)   C-p     S

  1    43.9        40.8      10.2  4.8351   X
  1    14.8        10.1      23.8  5.9604         X
  2    66.6        62.6       1.6  3.8420   X X
  2    64.7        60.6       2.5  3.9474   X     X
  3    69.1        63.3       2.4  3.8089   X X   X
  3    67.9        61.9       3.0  3.8778   X X X
  4    69.9        61.8       4.1  3.8825   X X X X
  4    69.3        61.1       4.3  3.9176   X X   X X
  5    70.0        59.3       6.0  4.0099   X X X X X
```

A number of other procedures can be used to select the best regression and, although we will not spend a great deal more time on this subject, we will mention briefly the **backward elimination** method and **stepwise regression** procedure.

backward elimination

stepwise regression

The backward elimination method begins with fitting the regression model, which contains all the candidate independent variables. For each independent variable x_j, we compute

$$F_j = \frac{\text{SSdrop}_j}{\text{MS(Residual)}} \qquad j = 1, 2, \ldots$$

where SSdrop_j is the drop in the sum of squares error obtained for the complete model, which contains all xs except x_j. MS(Residual) is the mean square error for the complete model. Let min F_j denote the smallest F_j value. If min $F_j < F_\alpha$, where α is the preselected significance level, remove the independent variable corresponding to min F_j from the regression equation. The backward elimination process then begins all over again with one variable removed from the list of candidate independent variables.

Backward elimination starts with the complete model with all independent variables entered and eliminates variables one at a time until a reasonable candidate regression model is found. This occurs when, in a particular step, min $F_j > F_\alpha$; the resulting complete model is the best-fitting regression equation. Stepwise regression, on the other hand, works in the other direction starting with the model $y = \beta_0 + \varepsilon$ and adding variables one at a time until a stopping criterion is satisfied. At the initial stage of the process, the first variable entered into the equation is the one with the largest F test for regression. At the second stage, the two variables to be included in the model are the variables with the largest F test for regression of two variables. Note that the variable entered in the first step might not be included in the second step; that is, the best single variable might not be one of the best two variables. Because of this, some people use a simplified stepwise regression (sometimes called *forward selection*) whereby, once a variable is entered, it cannot be eliminated from the regression equation at a later stage.

EXAMPLE 13.7

Use the data of Example 13.3 to find the variables to be included in a regression equation based on backward elimination. Comment on your findings.

Solution   SAS output is shown for a backward elimination procedure applied to the data of Table 13.1. As indicated, backward elimination begins with all (five) candidate variables in the regression equation. This is designated as step 0 in the backward elimination process. Then one by one, independent variables are eliminated until min $F_j > F_\alpha$. Note that in step 1, the variable income is removed and in step 2, the variable parking is removed from the regression equation. Step 3 is the final step in the process for this example; the variable shopping center is removed. As indicated in the output, the remaining variables comprise the best-fitting regression equation based on backward elimination. That equation is

$$\hat{y} = 48.291 - .004(\text{floor space}) - .582(\text{prescription sales})$$

which is identical to the result we obtained from the other variable selection procedures.

```
REGRESSION ANALYSIS, USING BACKWARD ELIMINATION

Backward Elimination Procedure for Dependent Variable VOLUME

Step 0   All Variables Entered    R-square = 0.70007369   C(p) = 6.00000000
```

	DF	Sum of Squares	Mean Square	F	Prob>F
Regression	5	525.44030541	105.08806108	6.54	0.0025
Error	14	225.10969459	16.07926390		
Total	19	750.55000000			

Variable	Parameter Estimate	Standard Error	Type II Sum of Squares	F	Prob>F
INTERCEP	42.08710826	10.43775070	261.42703544	16.26	0.0012
FLOOR_SP	-0.00241878	0.00183889	27.81923726	1.73	0.2095
PRESC_RX	-0.50046955	0.16429694	149.19783807	9.28	0.0087
PARKING	-0.03690284	0.06546687	5.10907792	0.32	0.5819
SHOPCNTR	-3.09957355	3.24983522	14.62673442	0.91	0.3564
INCOME	0.10666360	0.42742012	1.00135642	0.06	0.8066

Bounds on condition number: 7.823107,   117.1991

Step 1   Variable INCOME Removed   R-square = 0.69873952   C(p) = 4.06227626

	DF	Sum of Squares	Mean Square	F	Prob>F
Regression	4	524.43894899	131.10973725	8.70	0.0008
Error	15	226.11105101	15.07407007		
Total	19	750.55000000			

Variable	Parameter Estimate	Standard Error	Type II Sum of Squares	F	Prob>F
INTERCEP	43.46782063	8.56960161	387.83321233	25.73	0.0001
FLOOR_SP	-0.00228513	0.00170330	27.13112543	1.80	0.1997
PRESC_RX	-0.52910174	0.11386382	325.48983690	21.59	0.0003
PARKING	-0.03952477	0.06256589	6.01580808	0.40	0.5371
SHOPCNTR	-2.71387948	2.76799605	14.49041122	0.96	0.3424

Bounds on condition number: 5.071729,   46.98862

Step 2   Variable PARKING Removed   R-square = 0.69072432   C(p) = 2.43641080

	DF	Sum of Squares	Mean Square	F	Prob>F
Regression	3	518.42314091	172.80771364	11.91	0.0002
Error	16	232.12685909	14.50792869		
Total	19	750.55000000			

Variable	Parameter Estimate	Standard Error	Type II Sum of Squares	F	Prob>F
INTERCEP	42.82702645	8.34803435	381.83242065	26.32	0.0001
FLOOR_SP	-0.00247284	0.00164539	32.76871130	2.26	0.1523
PRESC_RX	-0.52941361	0.11170410	325.87978038	22.46	0.0002
SHOPCNTR	-3.03834296	2.66836223	18.81002755	1.30	0.2716

Bounds on condition number: 4.917388,   30.31995

Step 3   Variable SHOPCNTR Removed   R-square = 0.66566267   C(p) = 1.60624219

	DF	Sum of Squares	Mean Square	F	Prob>F
Regression	2	499.61311336	249.80655668	16.92	0.0001
Error	17	250.93688664	14.76099333		
Total	19	750.55000000			

```
                Parameter       Standard          Type II
Variable         Estimate          Error    Sum of Squares          F    Prob>F

INTERCEP       48.29085530     6.89043477      725.02357305      49.12    0.0001
FLOOR_SP       -0.00384228     0.00113262      169.87259933      11.51    0.0035
PRESC_RX       -0.58189034     0.10263739      474.44587802      32.14    0.0001


Bounds on condition number:      2.290122,        9.160487
----------------------------------------------------------------------------


All variables left in the model are significant at the 0.1000 level.

Summary of Backward Elimination Procedure for Dependent Variable VOLUME

          Variable   Number   Partial    Model
Step       Removed      In     R**2       R**2        C(p)          F    Prob>F

  1        INCOME        4    0.0013     0.6987      4.0623     0.0623    0.8066
  2        PARKING       3    0.0080     0.6907      2.4364     0.3991    0.5371
  3        SHOPCNTR      2    0.0251     0.6657      1.6062     1.2965    0.2716
```

EXAMPLE 13.8

Describe the results of stepwise regression applied to the data of Table 13.1.

Solution   The SAS output for the data of Table 13.1 is shown here. Stepwise regression begins with the model $y = \beta_0 + \varepsilon$ and adds variables one at a time. For these data, the variable prescription sales was entered in step 1 of the stepwise procedure, the variable floor space was added to the regression model in step 2, and the variable shopping center was added in step 3. No other variables met the entrance criterion of $p = .5$ for inclusion in the model. If the criterion was more selective, requiring a relatively small p-value (say, .15 or less) for each new independent variable, the stepwise regression procedure would not include the variable shopping center in step 3 (with a p-value of .2716) and we would arrive at the same best-fitting regression equation that we obtained previously with other methods.

```
REGRESSION ANALYSIS, USING FORWARD ELIMINATION

Forward Selection Procedure for Dependent Variable VOLUME

Step 1    Variable PRESC_RX Entered   R-square = 0.43933184    C(p) = 10.17094219

                DF          Sum of Squares          Mean Square          F    Prob>F

Regression       1            329.74051403          329.74051403      14.10    0.0014
Error           18            420.80948597           23.37830478
Total           19            750.55000000

                Parameter       Standard          Type II
Variable         Estimate          Error    Sum of Squares          F    Prob>F

INTERCEP       25.98133346     2.58814791     2355.90463660     100.77    0.0001
PRESC_RX       -0.32055657     0.08535423      329.74051403      14.10    0.0014


Bounds on condition number:            1,            1
----------------------------------------------------------------------------
```

```
Step 2    Variable FLOOR_SP Entered   R-square = 0.66566267    C(p) = 1.60624219

                  DF          Sum of Squares        Mean Square        F    Prob>F

Regression         2          499.61311336        249.80655668     16.92    0.0001
Error             17          250.93688664         14.76099333
Total             19          750.55000000

                Parameter       Standard          Type II
Variable         Estimate         Error      Sum of Squares        F    Prob>F

INTERCEP        48.29085530      6.89043477     725.02357305     49.12    0.0001
FLOOR_SP        -0.00384228      0.00113262     169.87259933     11.51    0.0035
PRESC_RX        -0.58189034      0.10263739     474.44587802     32.14    0.0001

Bounds on condition number:     2.290122,         9.160487
--------------------------------------------------------------------------------

Step 3   Variable SHOPCNTR Entered   R-square = 0.69072432     C(p) = 2.43641080

                  DF          Sum of Squares        Mean Square        F    Prob>F

Regression         3          518.42314091        172.80771364     11.91    0.0002
Error             16          232.12685909         14.50792869
Total             19          750.55000000

                Parameter       Standard          Type II
Variable         Estimate         Error      Sum of Squares        F    Prob>F

INTERCEP        42.82702645      8.34803435     381.83242065     26.32    0.0001
FLOOR_SP        -0.00247284      0.00164539      32.76871130      2.26    0.1523
PRESC_RX        -0.52941361      0.11170410     325.87978038     22.46    0.0002
SHOPCNTR        -3.03834296      2.66836223      18.81002755      1.30    0.2716

Bounds on condition number:     4.917388,         30.31995
--------------------------------------------------------------------------------


No other variable met the 0.5000 significance level for entry into the model.

Summary of Forward Selection Procedure for Dependent Variable VOLUME

          Variable   Number   Partial   Model
Step      Entered      In      R**2      R**2      C(p)         F    Prob>F

  1       PRESC_RX      1      0.4393    0.4393   10.1709   14.1046   0.0014
  2       FLOOR_SP      2      0.2263    0.6657    1.6062   11.5082   0.0035
  3       SHOPCNTR      3      0.0251    0.6907    2.4364    1.2965   0.2716
```

In a typical regression problem, you ascertain which variables are potential candidates for inclusion in a regression model (step 1) by discussions with experts and/or by using any one of a number of possible selection procedures. For example, we could run all possible regressions, apply a best-subset regression approach, or follow a stepwise regression (a backward elimination) procedure. This list is by no means exhaustive. Sometimes the various criteria do single out the same model as best (or near best, as seen with the data of Table 13.1). At other times you may get different models from the different criteria. Which approach is best? Which one should we believe and use?

The most important response to these questions is that with the availability and accessibility of a computer and applicable software systems, it is possible to work effectively with any of these selection procedures; no one procedure is universally accepted as better than the others. Hence, rather than attempting to use some or all of the procedures, you should begin to use one method (perhaps because of the availability of particular software in your computer facility) and learn as much as you can about it by continued use. Then you will be well equipped to solve almost any regression problem to which you are exposed.

EXERCISES

Applications

Edu.

13.1 Class Project The director of admissions at your college or university is interested in developing a regression model that will be useful in predicting a student's end-of-the-year grade point average (GPA) based on his or her high school record. Discuss this project among yourselves and seek out additional experts to develop a list of candidate independent variables for inclusion in the regression model. Should only one model be developed or should you consider more than one regression model? Might dummy variables be useful?

13.2 Class Project See Exercise 13.1. Obtain data from the admissions office and apply one of the selection procedures to identify a possible regression model.

Soc.

13.3 An interviewer asked a random sample of 45 students at a state university to decide whether each of the following acts should be considered a crime: aggravated assault, armed robbery, arson, atheism, auto theft, burglary, civil disobedience, communism, drug addiction, embezzlement, forcible rape, gambling, homosexuality, land fraud, nazism, pay-ola, price fixing, prostitution, sexual abuse of children, sexual discrimination, shoplifting, strikes, strip mining, treason, and vandalism. For each student, the interviewer determined the number of acts considered a crime and other information concerning the interviewee (years of college education, age, income of parents, and gender). The data are shown here. Refer to the output to identify potential collinearity issues. Use the output of a best-subset regression program to ascertain which variables should be included in the model. Can you suggest other variables that should have been addressed in the interview?

```
LISTING OF DATA

OBS     CRIME     AGE     COLLEGE     INCOME     SEX
 1       23       16        2           63        1
 2       25       18        2           72        1
 3       22       18        2           75        1
 4       16       18        2           61        0
 5       19       19        2           65        1
 6       19       19        2           70        1
 7       18       20        2           78        1
 8       16       19        2           76        0
 9       12       18        2           53        0
10       13       19        2           56        0
11       16       19        2           59        1
12       13       20        2           55        0
13       13       21        2           60        0
14       14       20        2           52        0
15       14       24        3           54        0
16       13       25        3           55        0
17       16       25        3           55        0
18       16       27        4           56        1
19       14       28        4           52        1
20       20       38        4           59        0
21       25       29        4           63        1
```

OBS	CRIME	AGE	COLLEGE	INCOME	SEX
22	19	30	4	55	1
23	23	31	4	59	0
24	25	32	4	52	1
25	22	32	4	55	1
26	25	31	4	57	0
27	17	30	4	46	1
28	14	29	4	35	0
29	12	29	4	32	0
30	10	28	4	30	0
31	8	27	4	29	0
32	7	26	4	28	0
33	5	25	4	25	0
34	9	24	3	33	0
35	7	23	3	26	0
36	9	23	3	28	1
37	10	22	3	38	0
38	4	22	3	24	0
39	6	22	3	28	0
40	8	21	3	29	1
41	11	21	2	35	1
42	10	20	2	33	0
43	6	19	2	27	0
44	7	21	3	24	0
45	15	21	2	53	1

Plot of CRIME versus AGE

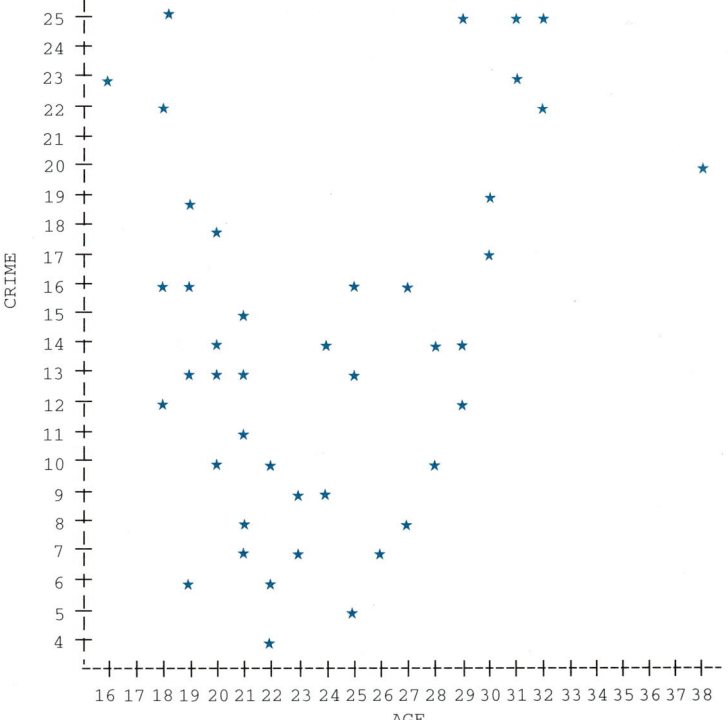

Plot of CRIME versus GENDER

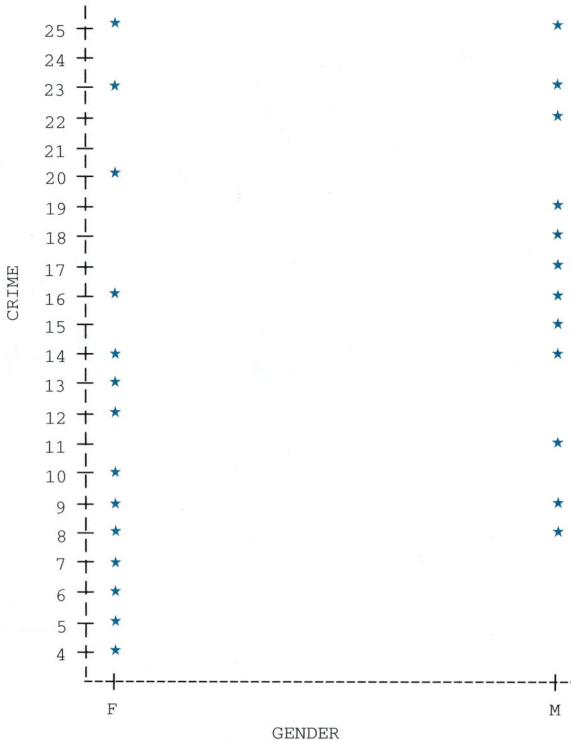

Plot of CRIME versus YEARS OF COLLEGE EDUCATION

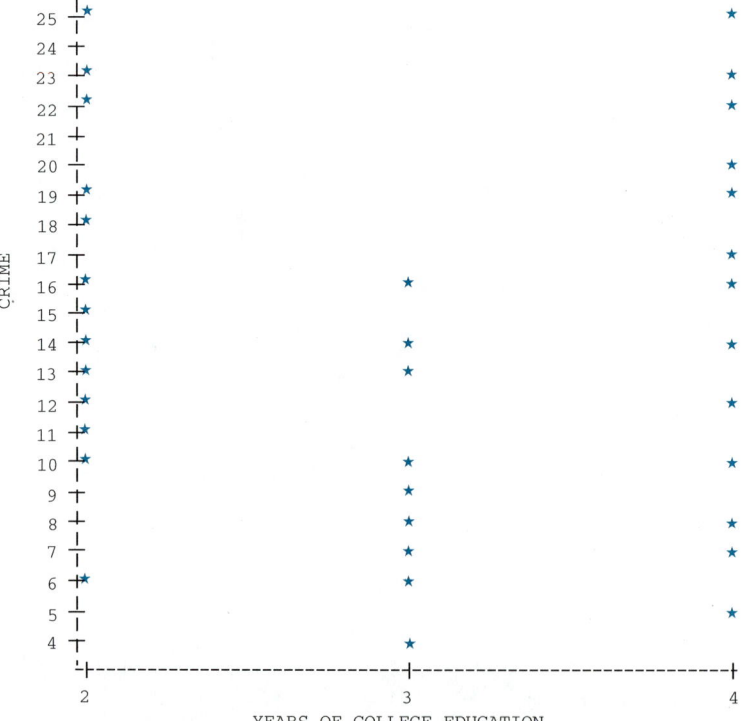

Plot of CRIME versus INCOME OF PARENTS

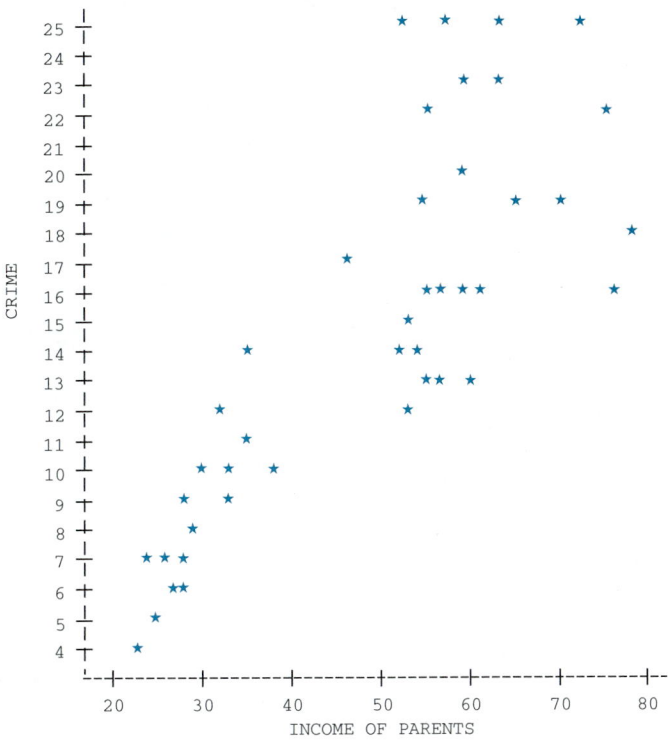

```
Backward Elimination Procedure for Dependent Variable CRIME

Step 0     All Variables Entered    R-square = 0.82783940    C(p) = 5.00000000

                    DF        Sum of Squares      Mean Square           F   Prob>F

Regression           4        1301.62108953      325.40527238       48.09   0.0001
Error               40         270.69002158        6.76725054
Total               44        1572.31111111

                 Parameter        Standard          Type II
Variable          Estimate           Error    Sum of Squares           F   Prob>F

INTERCEP       -10.82338752      2.39210442      138.54102767       20.47   0.0001
AGE              0.43238152      0.20236447       30.89427247        4.57   0.0388
COLLEGE         -0.02399594      1.22148794        0.00261162        0.00   0.9844
INCOME           0.29025487      0.03141812      577.57817022       85.35   0.0001
SEX              2.45416550      0.87466592       53.27648156        7.87   0.0077

Bounds on condition number:     7.476669,    68.21544
-------------------------------------------------------------------------------

Step 1     Variable COLLEGE Removed  R-square = 0.82783774   C(p) = 3.00038592

                    DF        Sum of Squares      Mean Square           F   Prob>F

Regression           3        1301.61847791      433.87282597       65.72   0.0001
Error               41         270.69263320        6.60225935
Total               44        1572.31111111
```

```
                 Parameter       Standard        Type II
Variable          Estimate          Error    Sum of Squares        F    Prob>F

INTERCEP       -10.82193315     2.36163189      138.63658941    21.00   0.0001
AGE              0.42872187     0.07806990      199.10244384    30.16   0.0001
INCOME           0.29058236     0.02630415      805.71727230   122.04   0.0001
SEX              2.45108843     0.84997169       54.90370062     8.32   0.0062

Bounds on condition number:      1.202437,       10.21103
---------------------------------------------------------------------------

All variables left in the model are significant at the 0.1000 level.

Summary of Backward Elimination Procedure for Dependent Variable CRIME

          Variable    Number    Partial      Model
Step      Removed        In     R**2         R**2       C(p)          F    Prob>F

   1      COLLEGE         3    0.0000       0.8278     3.0004     0.0004    0.9844
```

13.4 Refer to Exercise 13.3. Computer output from a stepwise regression program is shown here. Comment on the results of this analysis compared to that done in Exercise 13.3.

```
REGRESSION ANALYSIS, FORWARD SELECTION

Forward Selection Procedure for Dependent Variable CRIME

Step 1   Variable INCOME Entered     R-square = 0.66453936    C(p) = 36.94132731

                 DF        Sum of Squares       Mean Square         F    Prob>F

Regression        1        1044.86262180     1044.86262180     85.18    0.0001
Error            43         527.44848931       12.26624394
Total            44        1572.31111111

                 Parameter       Standard        Type II
Variable          Estimate          Error    Sum of Squares        F    Prob>F

INTERCEP        -0.19647505     1.66089569        0.17164917     0.01    0.9064
INCOME           0.30177022     0.03269660     1044.86262180    85.18    0.0001

Bounds on condition number:           1,            1
---------------------------------------------------------------------------

Step 2   Variable AGE Entered        R-square = 0.79291863    C(p) = 9.11353325

                 DF        Sum of Squares       Mean Square         F    Prob>F

Regression        2        1246.71477730      623.35738865     80.41    0.0001
Error            42         325.59633381        7.75229366
Total            44        1572.31111111

                 Parameter       Standard        Type II
Variable          Estimate          Error    Sum of Squares        F    Prob>F

INTERCEP       -11.33832496     2.55169607      153.06296650     19.74   0.0001
AGE              0.43163600     0.08458942      201.85215549     26.04   0.0001
INCOME           0.32018698     0.02624270     1154.03879316    148.86   0.0001

Bounds on condition number:      1.01928,        4.077119
---------------------------------------------------------------------------
```

```
Step 3    Variable SEX Entered       R-square = 0.82783774    C(p) = 3.00038592

                 DF          Sum of Squares        Mean Square          F    Prob>F

Regression        3          1301.61847791       433.87282597      65.72    0.0001
Error            41           270.69263320         6.60225935
Total            44          1572.31111111

               Parameter          Standard           Type II
Variable        Estimate             Error      Sum of Squares          F    Prob>F

INTERCEP      -10.82193315        2.36163189       138.63658941      21.00    0.0001
AGE             0.42872187        0.07806990       199.10244384      30.16    0.0001
INCOME          0.29058236        0.02630415       805.71727230     122.04    0.0001
SEX             2.45108843        0.84997169        54.90370062       8.32    0.0062

Bounds on condition number:      1.202437,      10.21103
-------------------------------------------------------------------------------

No other variable met the 0.5000 significance level for entry into the model.

Summary of Forward Selection Procedure for Dependent Variable CRIME

         Variable   Number   Partial     Model
Step     Entered      In      R**2        R**2        C(p)            F    Prob>F

  1      INCOME        1      0.6645      0.6645    36.9413      85.1820    0.0001
  2      AGE           2      0.1284      0.7929     9.1135      26.0377    0.0001
  3      SEX           3      0.0349      0.8278     3.0004       8.3159    0.0062
```

Ag. **13.5** A company is interested in the effects of various food additives (protein and antibiotics) on the amount of time it takes to bring cattle to a desired market weight. Discuss what variable should be examined in arriving at a multiple regression equation for predicting the time to market weight.

13.3 Formulating the Model (Step 2)

In Section 13.2, we suggested several ways to develop a list of candidate independent variables for a given regression problem. We can and should seek the advice of experts in the subject matter area to provide a starting point and we can employ any one of several selection procedures to come up with a possible regression model. This section involves refining the information gleaned from step 1 to develop a useful multiple regression model.

Having chosen a subset of k independent variables to be candidates for inclusion in the multiple regression and the dependent variable y, we still may not know the actual relationship between the dependent and independent variables. Suppose the assumed regression model is of a lower order than is the actual model relating y to $x_1, x_2, \ldots, x_k$. Then provided there is more than one observation per factor–level combination of the independent variables, we can conduct a test of the inadequacy of a fitted polynomial model using the equation $F = \text{MS}_{\text{Lack}}/\text{MS}(\text{Residual})$ as discussed in Chapter 11.

Another way to examine an assumed (fitted) model for lack of fit is to examine scatterplots of residuals $(y_i - \hat{y}_i)$ versus x_j. For example, suppose that step 1 has indicated that the variables x_1, x_2, and x_3 constitute a reasonable subset of independent variables to be related to a response y using a multiple regression equation. Not knowing which polynomial function of the independent variables to use, we could start by fitting the multiple linear regression model

$$y = \beta_0 + \beta_1 x_1 + \beta_2 x_2 + \beta_3 x_3 + \varepsilon$$

to obtain the least-squares prediction equation $\hat{y} = \hat{\beta}_0 + \hat{\beta}_1 x_1 + \hat{\beta}_2 x_2 + \hat{\beta}_3 x_3$. A plot of the residuals $(y_i - \hat{y}_i)$ versus each one of the xs would shed some light as to which higher-degree terms may be appropriate. We'll illustrate the concepts using residuals by way of a regression problem for one independent variable and then extend the concepts to a multiple regression situation.

EXAMPLE 13.9

In a radioimmunoassay, a hormone with a radioactive trace is added to a test tube containing an antibody that is specific to that hormone. The two will combine to form an antigen–antibody complex. To measure the extent of the reaction of the hormone with the antibody, we measure the amount of hormone that is bound to the antibody relative to the amount remaining free. Typically, experimenters measure the ratio of the bound/free radioactive count (y) for each dose of hormone (x) added to a test tube. Frequently, the relation between y and x is nearly linear. Data from 11 test tubes in a radioimmunoassay experiment are shown in Table 13.4.

TABLE 13.4
Radioimmunoassay data

Bound/Free Count	Dose (concentration)
9.900	0.00
10.465	0.25
10.312	0.50
13.633	0.75
20.784	1.00
36.164	1.25
62.045	1.50
78.327	1.75
90.307	2.00
97.348	2.25
102.686	2.50

a. Plot the sample data and fit the linear regression model

$$y = \beta_0 + \beta_1 x + \varepsilon$$

b. Plot the residuals versus count and versus $\hat{y}$. Does a linear model adequately fit the data?
c. Suggest an alternative (if appropriate).

Solution Computer output is shown here.

```
Data Display

Row   BOUND/FREE   COUNT   DOSE   DOSE_2
1                  9.900   0.00   0.0000
2                 10.465   0.25   0.0625
3                 10.312   0.50   0.2500
4                 13.633   0.75   0.5625
```

```
Row    BOUND/FREE    COUNT    DOSE    DOSE_2
  5                 20.784    1.00    1.0000
  6                 36.164    1.25    1.5625
  7                 62.045    1.50    2.2500
  8                 78.327    1.75    3.0625
  9                 90.307    2.00    4.0000
 10                 97.348    2.25    5.0625
 11                102.686    2.50    6.2500


Regression Analysis: BOUND/FREE COUNT versus DOSE

The regression equation is
BOUND/FREE COUNT = -7.19 + 44.4 DOSE

Predictor          Coef      SE Coef          T          P
Constant         -7.189        6.226      -1.15      0.278
DOSE             44.440        4.210       10.56      0.000


S = 11.04        R-Sq = 92.5%     R-Sq(adj) = 91.7%

Analysis of Variance

Source            DF          SS          MS         F         P
Regression         1       13577       13577    111.44     0.000
Residual Error     9        1097         122
Total             10       14674
```

Plot of BOUND/FREE COUNT versus DOSE

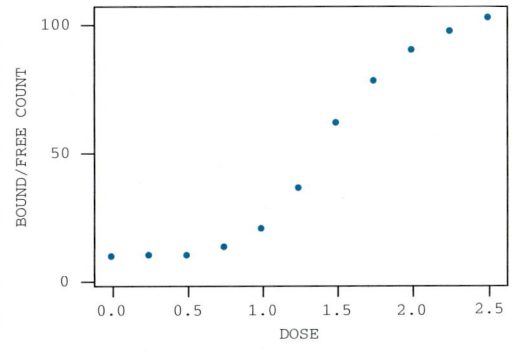

Residuals versus DOSE
(response is BOUND/FR)

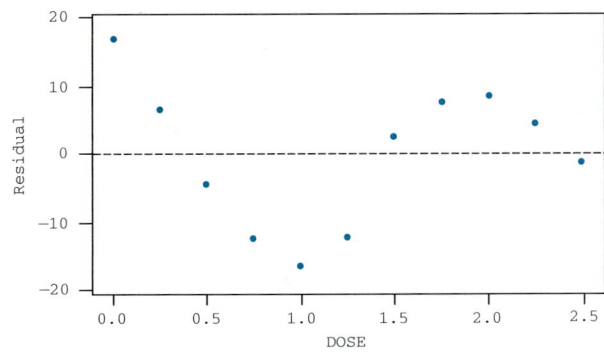

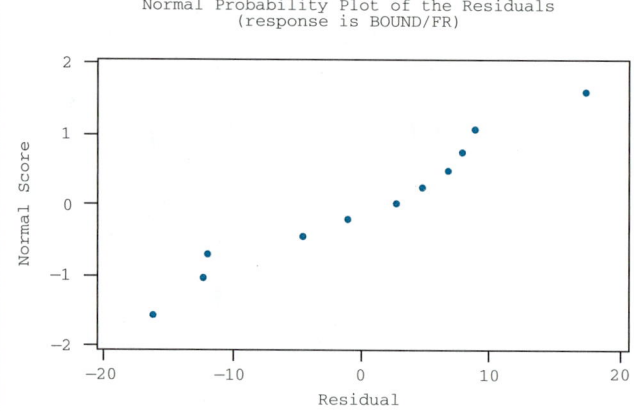

Normal Probability Plot of the Residuals
(response is BOUND/FR)

a, b. The linear fit is

$$\hat{y} = -7.189 + 44.440x$$

The plot of y (count) versus x (concentration) clearly shows a lack of fit of the linear regression model; the residual plots confirm this same lack of fit. The linear regression underestimates counts at the lower and upper ends of the concentration scale and overestimates at the middle concentrations.

c. A possible alternative model would be a quadratic model in concentration,

$$y = \beta_0 + \beta_1 x + \beta_2 x^2 + \varepsilon$$

More will be said about this later in the chapter.

Scatterplots are not very helpful in detecting interactions among the independent variables, other than for the two independent variable case. The reason is that there are too many variables for most practical problems and it is difficult to present the interrelationships among independent variables and their joint effects on the response y using two-dimensional scatterplots. Perhaps the most reasonable suggestion is to use one of the best subset regression methods of the previous section, some trial-and-error fitting of models using the candidate independent variables, and a bit of common sense to determine which interaction terms should be used in the multiple regression model.

The presence of dummy variables (for qualitative independent variables) presents no major problem for ascertaining the adequacy of the fit of a polynomial model. The important thing to remember is that when quantitative and dummy variables are included in the same regression model, for each setting of the dummy variables, we obtain a regression in the quantitative variables. Hence, plotting methods for detecting an inadequate fit should be applied separately for each setting of the dummy variables. By examining these plots carefully, we can also detect potential differences in the forms of the polynomial models for different settings of the dummy variables.

EXAMPLE 13.10

A company analyst is interested in developing a regression model for predicting automobile sales for standard and luxury models of a particular make in a given territory. Empirical discussions and some substantive research into previous sales patterns for the company in that territory tend to indicate that the prevailing interest rate for car loans and the price per gallon of gasoline are the key predictive variables. The number of cars sold per month (in 1,000s) for the previous 18 months is shown here for gasoline-powered standard and luxury models. Fit a linear regression model and use residual plots to determine what (if any) higher-order terms are required. Do the same conclusions hold for standard and luxury models? Make suggestions for additional terms in the multiple regression equation.

Solution   A multiple regression model of the form

$$y = \beta_0 + \beta_1 x_1 + \beta_2 x_2 + \beta_3 x_3 + \varepsilon$$

where

y = number of sales per month (in 1,000s)

x_1 = price per gallon

x_2 = interest rate

$x_3 = \begin{cases} 1 \text{ if standard} \\ 0 \text{ if luxury} \end{cases}$

was fit to the data. From the output, the regression equation is

$$\hat{y} = 56.074 - 16.144 x_1 - 2.332 x_2 + 14.422 x_3$$

Substituting $x_3 = 0$ and 1 into this equation, we obtain the separate regression equations for the luxury and standard cars, respectively:

$x_3 = 0$ (luxury cars)
$$\hat{y} = 56.074 - 16.144 x_1 - 2.332 x_2$$

$x_3 = 1$ (standard cars)
$$\hat{y} = 56.074 - 16.144 x_1 - 2.332 x_2 + 14.422$$
$$= 70.496 - 16.144 x_1 - 2.332 x_2$$

Plots of y versus x_1 and x_2 for the two model types show clear negative linear relationships between sales and price per gallon of gasoline or interest rates. However, the slopes appear to be greater for the standard model than for the luxury model. This is borne out in the residual plots for the two models.

```
MULTIPLE REGRESSION ANALYSIS

DATA LISTING

           MONTHLY    PRICE/
           SALES      GALLON     INTEREST
   MONTH   (1000)     GASOLINE   RATE (%)    TYPECAR

     1      22.1       1.89        6.1          1
     1       7.2       1.89        6.1          0
     2      15.4       1.94        6.2          1
     2       5.4       1.94        6.2          0
     3      11.7       1.95        6.3          1
     3       7.6       1.95        6.1          0
     4      10.3       1.82        8.2          1
```

MONTH	MONTHLY SALES (1000)	PRICE/ GALLON GASOLINE	INTEREST RATE (%)	TYPECAR
4	2.5	1.82	8.2	0
5	11.4	1.85	9.8	1
5	2.4	1.85	9.8	0
6	7.5	1.78	10.3	1
6	1.7	1.78	10.3	0
7	13.0	1.76	10.5	1
7	4.3	1.76	10.5	0
8	12.8	1.76	8.7	1
8	3.7	1.76	8.7	0
9	14.6	1.75	7.4	1
9	3.9	1.75	7.4	0
10	18.9	1.74	6.9	1
10	7.0	1.74	6.9	0
11	19.3	1.70	5.2	1
11	6.8	1.70	5.2	0
12	30.1	1.70	4.9	1
12	10.1	1.70	4.9	0
13	28.2	1.68	4.3	1
13	9.4	1.68	4.3	0
14	25.6	1.60	3.7	1
14	7.9	1.60	3.7	0
15	37.5	1.61	3.6	1
15	14.1	1.61	3.6	0
16	36.1	1.64	3.1	1
16	14.5	1.64	3.1	0
17	39.8	1.67	1.8	1
17	14.9	1.67	1.8	0
18	44.3	1.68	2.3	1
18	15.6	1.68	2.3	0

Dependent Variable: SALES

Analysis of Variance

Source	DF	Sum of Squares	Mean Square	F Value	Prob>F
Model	3	3716.34235	1238.78078	54.900	0.0001
Error	32	722.05765	22.56430		
C Total	35	4438.40000			

Root MSE	4.75019	R-square	0.8373
Dep Mean	14.93333	Adj R-sq	0.8221
C.V.	31.80931		

Parameter Estimates

Variable	DF	Parameter Estimate	Standard Error	T for H0: Parameter=0	Prob > |T|
INTERCEP	1	50.620740	15.18488648	3.334	0.0022
GASPRICE	1	-16.436642	9.26565678	-1.774	0.0856
INTEREST	1	-2.328968	0.36053082	-6.460	0.0001
TYPECAR	1	14.448100	1.58340161	9.125	0.0001

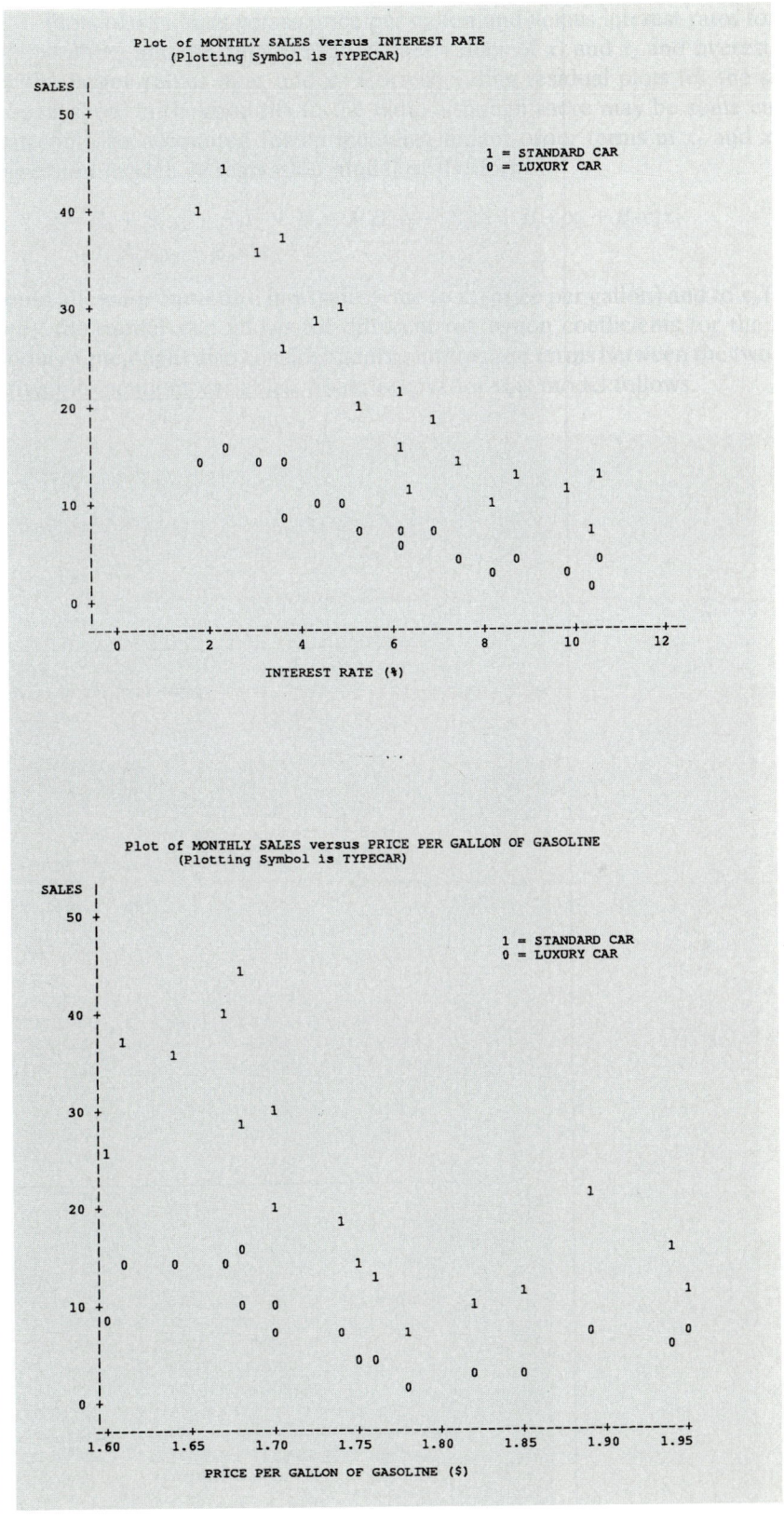

Plot of MONTHLY SALES versus INTEREST RATE
(Plotting Symbol is TYPECAR)

Plot of MONTHLY SALES versus PRICE PER GALLON OF GASOLINE
(Plotting Symbol is TYPECAR)

The problem is to find a method for obtaining estimates $\hat{\alpha}_1$, $\hat{\alpha}_2$, . . . that will minimize SS(Residual). The set of simultaneous equations used for finding these estimates is again called the set of normal equations, but unlike least squares for the general linear model, the form of the normal equations depends on the form of the nonlinear model being used. Also, because the normal equations involve nonlinear functions of the parameters, their solutions can be quite complicated. Because of this technical difficulty, a number of iterative methods have been developed for obtaining a solution to the normal equations.

For those of you with a background in calculus, the normal equations for a nonlinear model involve partial derivatives of the nonlinear function with respect to each of the parameters α_i. Fortunately, most of the computer software packages currently marketed (for example, SAS, NONLIN, Splus) approximate the derivative and do not require one to give the form of the normal equations; only the form of the nonlinear equation is needed. We will illustrate this with the data from a previous example.

Recall that in Example 13.9 we fit a linear regression model to the radioimmunoassay data; a residual plot for that model suggested that a quadratic model might be more appropriate:

$$y = \beta_0 + \beta_1 x + \beta_2 x^2 + \varepsilon$$

Computer output for the revised model is shown here. Note that the cyclical pattern is still apparent in the residual plot and hence the quadratic model is still inadequate.

```
Regression Analysis: BOUND/FREE COUNT versus DOSE, DOSE_2


The regression equation is
BOUND/FREE COUNT = 2.88 + 17.6 DOSE + 10.7 DOSE_2


Predictor          Coef       SE Coef            T          P
Constant          2.884         7.175         0.40      0.698
DOSE              17.58         13.35         1.32      0.225
DOSE_2           10.745         5.144         2.09      0.070


S = 9.418          R-Sq = 95.2%        R-Sq(adj) = 94.0%


Analysis of Variance

Source              DF            SS           MS         F        P
Regression           2       13964.4       6982.2     78.72    0.000
Residual Error       8         709.6         88.7
Total               10       14674.0


Source    DF       Seq SS
DOSE       1      13577.4
DOSE_2     1        386.9
```

Plot of BOUND/FREE COUNT versus DOSE

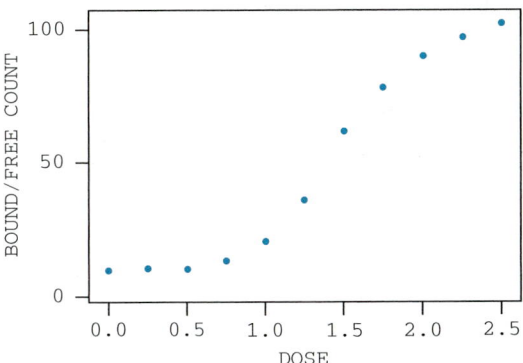

Residuals versus DOSE
(response is BOUND/FR)

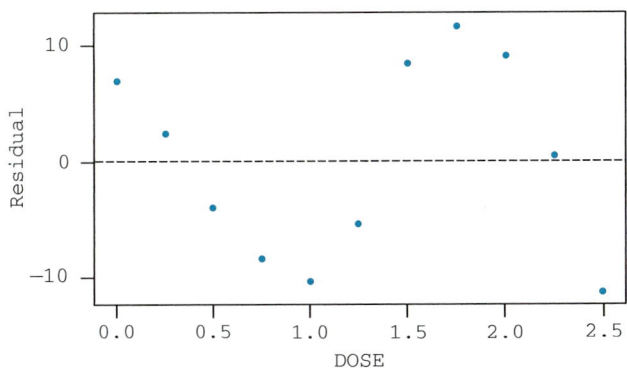

Residuals versus the Fitted Values
(response is BOUND/FR)

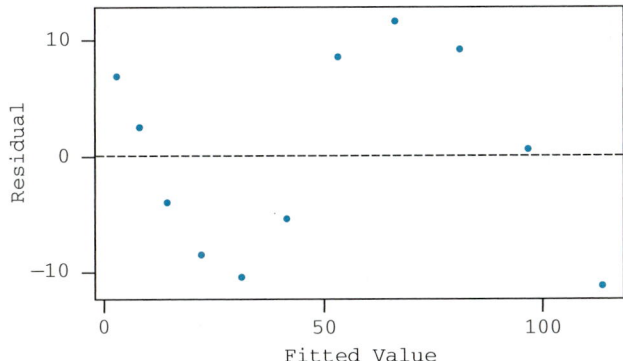

A nonlinear model that may help to flatten the S-shape of the data plot shown in the output has the following form:

$$y = \frac{\beta_0 - \beta_3}{1 + (x/\beta_2)^{\beta_1}} + \beta_3$$

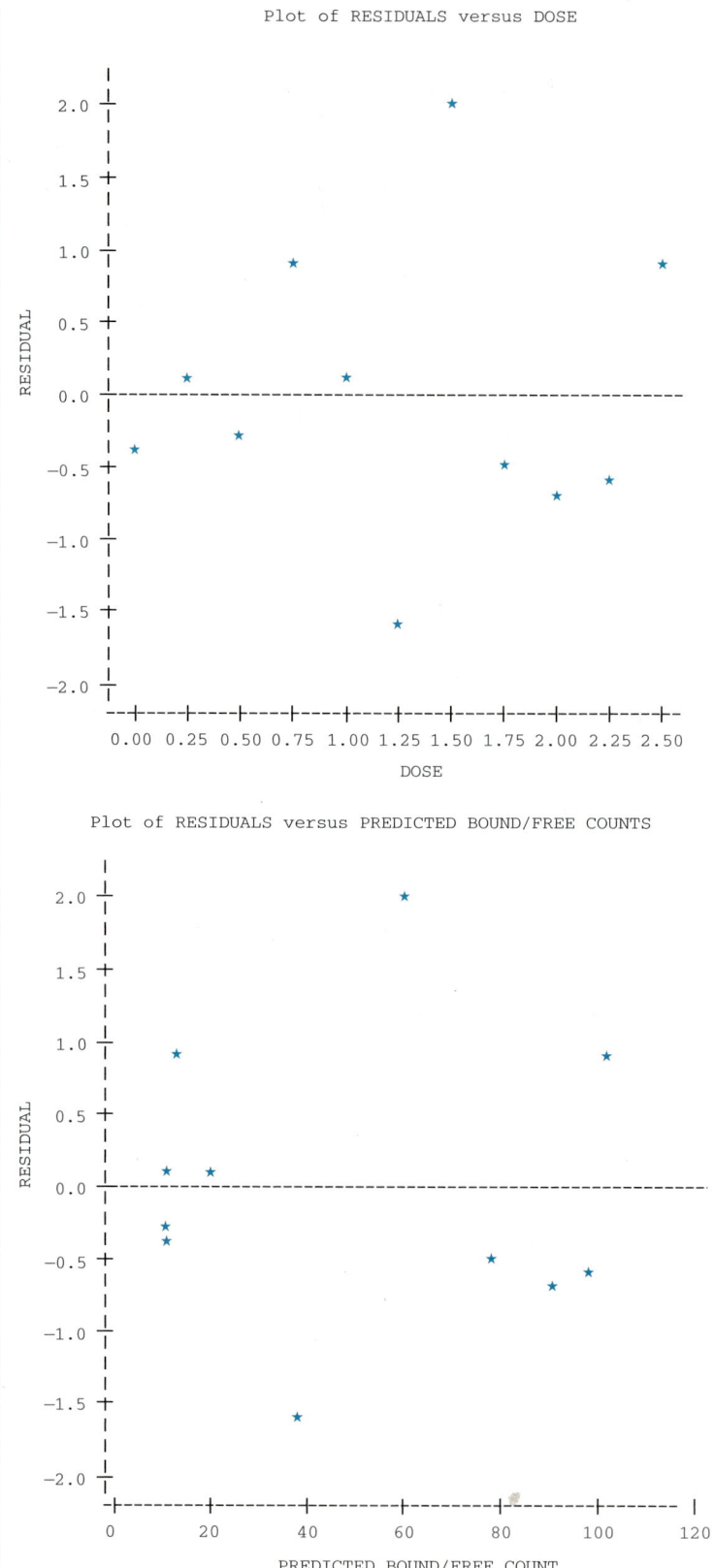

We can also use the fitted equation to predict y (count ratio) based on concentration.

EXERCISES **Applications**

Med. **13.6** Peak blood level data (in mg/ml) are available for 20 patients for a single dose of a drug product, along with the patient's weight (lb) and the amount of drug (mg). Use the output shown here to fit a linear regression line and use residual plots to identify possible additional terms to be included in the regression model.

```
SAS OUTPUT FOR EXERCISE 13.6

DATA LISTING

OBS     BLOOD     DOSE     WEIGHT
 1       300       1        120
 2       250       1        135
 3       210       1        150
 4       150       1        128
 5       210       2        150
 6       230       2        160
 7       350       2        135
 8       270       2        180
 9       380       4        132
10       330       4        148
11       270       4        190
12       240       4        195
13       340       8        150
14       330       8        160
15       180       8        200
16       320       8        140
17       270      16        195
18       290      16        170
19       315      16        161
20       350      16        145
```

Dependent Variable: PEAK BLOOD LEVEL

Analysis of Variance

Source	DF	Sum of Squares	Mean Square	F Value	Prob > F
Model	2	22290.44079	11145.22040	3.684	0.0468
Error	17	51423.30921	3024.90054		
C Total	19	73713.75000			

Root MSE	54.99910	R-square	0.3024
Dep Mean	279.25000	Adj R-sq	0.2203
C.V.	19.69529		

Parameter Estimates

Variable	DF	Parameter Estimate	Standard Error	T for H0: Parameter=0	Prob >|T|
INTERCEP	1	432.602294	84.69454320	5.108	0.0001
DOSE	1	5.546666	2.40278001	2.308	0.0338
WEIGHT	1	-1.194285	0.55838151	-2.139	0.0473

Variable	DF	Variable Label
INTERCEP	1	Intercept
DOSE	1	AMOUNT OF DRUG (mg)
WEIGHT	1	PATIENT'S WEIGHT (lb)

a. How much larger is the R^2 for this model than the R^2 for the first-order model of Exercise 13.17?

b. Use the F test for complete and reduced models to test the null hypothesis that the addition of the squared terms yields no additional predictive value. Use $\alpha = .05$.

c. Do the t statistics indicate that either squared term is a statistically significant ($\alpha = .05$) predictor as last predictor in?

13.19 A forward-selection stepwise regression is run using a first-order model for the data of Exercise 13.16. The following output is obtained:

```
PEARSON CORRELATION MATRIX

               CHILL       KNIFE        MELT        REPEL        SPEED
    CHILL      1.000
    KNIFE     -0.000       1.000
     MELT     -0.000       0.000       1.000
    REPEL      0.000       0.000       0.000       1.000
    SPEED      0.000      -0.000      -0.000      -0.000       1.000
    STIFF      0.138      -0.308       0.059      -0.886       0.030

MINIMUM TOLERANCE FOR ENTRY INTO MODEL =    .010000

STEP #    1 R=  .886 RSQUARE=    .786
TERM ENTERED: REPEL

VARIABLE        COEFFICIENT  STD ERROR  STD COEF   TOLERANCE     F      'P'
  1 CONSTANT
  4 REPEL            -1.838      -0.175    -0.886     .1E+01   109.988   0.000

STEP #    2 R=  .938 RSQUARE=    .880
TERM ENTERED: KNIFE

VARIABLE        COEFFICIENT  STD ERROR  STD COEF   TOLERANCE     F      'P'
  1 CONSTANT
  4 REPEL            -1.838      -0.133    -0.886     .1E+01   190.301   0.000
  6 KNIFE            -0.319      -0.067    -0.308     .1E+01    22.906   0.000

STEP #    3 R=  .948 RSQUARE=    .899
TERM ENTERED: CHILL

VARIABLE        COEFFICIENT  STD ERROR  STD COEF   TOLERANCE     F      'P'
  1 CONSTANT
  3 CHILL             0.128       0.056     0.138     .1E+01     5.253   0.030
  4 REPEL            -1.838      -0.124    -0.886     .1E+01   218.210   0.000
  6 KNIFE            -0.319      -0.062    -0.308     .1E+01    26.265   0.000

STEP #    4 R= .950 RSQUARE=    .903
TERM ENTERED: MELT

VARIABLE        COEFFICIENT  STD ERROR  STD COEF   TOLERANCE     F      'P'
  1 CONSTANT
  2 MELT              0.028       0.028     0.059     .1E+01     0.977   0.332
  3 CHILL             0.128       0.056     0.138     .1E+01     5.249   0.030
  4 REPEL            -1.838      -0.124    -0.886     .1E+01   218.028   0.000
  6 KNIFE            -0.319      -0.062    -0.308     .1E+01    26.243   0.000
```

a. List the order in which the independent variables enter the regression model.

b. List the independent variables from largest (in absolute value) to smallest correlation with STIFF.

c. Compare the ordering of the variables given by the two lists.

13.20 Refer to Exercise 13.19. Use the F test for complete and reduced models, described in Section 12.5, to test the hypothesis that the last two variables entered in the stepwise regression have no predictive value.

13.21 The consultant of Exercise 13.10 runs a regression model with CONTRIB/INCOME as the dependent variable.

```
MTB > regress c9 on 7 vars in 'Income'-'Matching'

The regression equation is
Cont/Inc = 0.0211 -0.000093 Income + 0.00153 Size + 0.00168 DPDummy
           + 0.00713 ELDummy + 0.00281 CHDummy - 0.0144 EdLevel
           + 0.00138 Matching

Predictor        Coef       Stdev     t-ratio        p
Constant      0.021085    0.003497       6.03    0.000
Income       -0.0000933   0.0001033     -0.90    0.372
Size          0.0015301   0.0006699      2.28    0.028
DPDummy       0.001684    0.004720       0.36    0.723
ELDummy       0.007132    0.006355       1.12    0.269
CHDummy       0.002808    0.003810       0.74    0.466
EdLevel      -0.01436     0.01597       -0.90    0.374
Matching      0.001381    0.002092       0.66    0.513

s = 0.006057    R-sq = 21.3%    R-sq(adj) = 6.8%

Analysis of Variance

SOURCE        DF         SS          MS         F        p
Regression     7    0.00037698   0.00005385    1.47    0.208
Error         38    0.00139418   0.00003669
Total         45    0.00177115
```

a. Can the hypothesis that none of the independent variables has predictive value be rejected (using reasonable α values)?

b. Which variables have been shown to have statistically significant (say, $\alpha = .05$) predictive value as last predictor in?

13.22 A simpler regression model than that of Exercise 13.21 is obtained by regressing the dependent variable on the independent variables DPDummy, ELDummy, EdLevel, and Matching. The following output is obtained:

```
MTB > regress c9 on 4 vars 'DPDummy' 'ELDummy' 'EdLevel' 'Matching'

The regression equation is
Cont/Inc = 0.0202 - 0.00378 DPDummy - 0.00099 ELDummy
           + 0.0097 EdLevel + 0.00197 Matching

Predictor        Coef       Stdev     t-ratio        p
Constant      0.020233    0.002304       8.78    0.000
DPDummy      -0.003775    0.003912      -0.97    0.340
ELDummy      -0.000995    0.005131      -0.19    0.847
EdLevel       0.00969     0.01276        0.76    0.452
Matching      0.001974    0.001995       0.99    0.328

s = 0.006343  R-sq = 6.9%  R-sq(adj) = 0.0%

Analysis of Variance

SOURCE        DF         SS          MS         F        p
Regression     4    0.00012167   0.00003042   0.76    0.560
Error         41    0.00164949   0.00004023
Total         45    0.00177115
```

a. What is the increment to R^2 for the model of Exercise 13.21, as opposed to the model considered here?
b. Is this increment statistically significant by an F test, at $\alpha = .05$?
c. Compute C_p for this model, treating the previous model as the "all coefficients" model. Which of the two models do you think is more sensible, given the information you have?

13.4   Checking Model Assumptions (Step 3)

Now that we have identified possible independent variables (step 1) and considered the form of the multiple regression model (step 2), we should check whether the assumptions underlying the chosen model are valid. Recall that in Chapter 11 we indicated that the basic assumptions for a regression model of the form

$$y_i = \beta_0 + \beta_1 x_{i1} + \beta_2 x_{i2} + \cdots + \beta_k x_{ik} + \varepsilon_i$$

are as follows:

1. Zero expectation: $E(\varepsilon_i) = 0$ for all i.
2. Constant variance: $V(\varepsilon_i) = \sigma_\varepsilon^2$ for all i.
3. Normality: ε_i is normally distributed.
4. Independence: The ε_i are independent.

Note that because the assumptions for multiple regression are written in terms of the random errors ε_i, it would seem reasonable to check the assumptions by using the residuals $y_i - \hat{y}_i$, which are *estimates* of the ε_i.

The first assumption, zero expectation, deals with model selection and whether additional independent variables need to be included in the model. If we have done our job in steps 1 and 2, assumption 1 should hold. The use of residual plots to check for inadequacy (lack of fit) of the model was discussed briefly in Chapter 11 and again in Section 13.3. If we have not done our job in steps 1 and 2, then a plot of the residuals should help detect this.

standardized residuals   Recall that residuals are differences between actual y values and predicted values using the regression model. In plotting residuals, we often use **standardized residuals.** A standardized residual is expressed in standard deviation units, so a standardized residual of -3.00 means that the point is 3 standard deviations from the regression line. Often, subtracting out the predictive part of the data reveals other structure more clearly. In particular, plotting the residuals from a first-order (linear terms only) model against each independent variable often reveals further structure in the data that can be used to improve the regression model.

One possibility is nonlinearity. We discussed nonlinearity and transformations earlier in the chapter. A noticeable curve in the residuals reflects a curved relation in the data, indicating that a different mathematical form for the regression equation would improve the predictive value of the model. A plot of residuals against each independent variable x often reveals this problem. A scatterplot smoother, such as LOWESS, can be useful in looking for curves in residual plots. For example, Figure 13.4 shows a scatterplot of y against x_2 and a residual plot against x_2. We think that the curved relation is more evident in the residual plot. The LOWESS curve helps considerably in both plots.

FIGURE 13.4

y and residual plots showing curvature

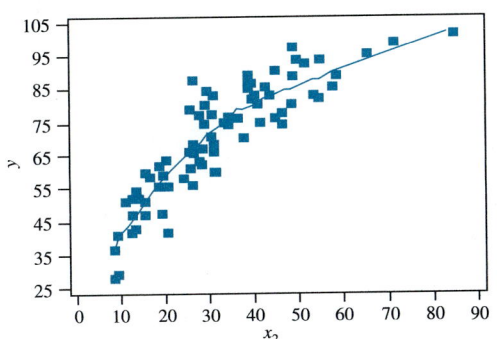 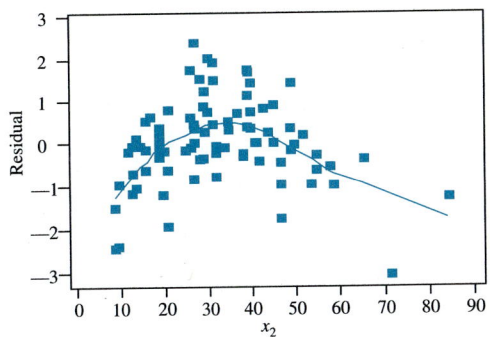

When nonlinearity is found, try transforming either independent or dependent variables. One standard method for doing this is to use (natural) logarithms of all variables except dummy variables. Such a model essentially estimates *percentage* changes in the dependent variable for a small percentage change in an independent variable, other independent variables held constant. Other useful transformations are logarithms of one or more independent variables only, square roots of independent variables, or inverses of the dependent variable or an independent variable. With a good computer package, a number of these transformations can be tested easily.

Assumption 2, the property of constant variance, can be examined using residual plots. One of the simplest residual plots for detecting nonconstant variance is a plot of the residuals versus the predicted values, $\hat{y}_i$. Most of the available statistical software systems can provide these plots as part of the regression analysis.

EXAMPLE 13.14

The data shown in Table 13.5 were fit to the model $y = \beta_0 + \beta_1 x + \beta_2 x^2 + \varepsilon$ using SAS. Examine the plot residuals versus $\hat{y}_i$ to detect possible nonconstant variance. Can you identify a pattern of nonconstant variance?

TABLE 13.5

Data for Example 13.14

y	11	10	2	14	22	10	20	19	32	23	40	37
x	.5	1	1.2	1.4	1.7	1.8	2	2.3	2.5	2.8	3	3.1
y	30	43	55	29	45	60	53	30	42	25	63	51
x	3.5	3.6	3.8	4.2	4.4	5.1	5.2	5.4	5.5	6	6.2	6.3

Solution   As we can see from the SAS residual plot, the magnitudes of the residuals are generally increasing with the magnitudes of the predicted values of *y*, suggesting possible nonconstant variance. Also, because y_i is directly related to *x* via the regression model (i.e., *y* increases with *x*), the residuals are increasing with the magnitude of the *x*s. This pattern in the residuals suggests that the variance of ε_i (and hence $V(y_i)$) is increasing with *x*. The accompanying plot of *y* versus *x* tends to bear this out.

```
DATA LISTING

OBS    Y      X       X2

  1    11    0.5     0.25
  2    10    1.0     1.00
  3     2    1.2     1.44
  4    14    1.4     1.96
  5    22    1.7     2.89
  6    10    1.8     3.24
  7    20    2.0     4.00
  8    19    2.3     5.29
  9    32    2.5     6.25
 10    23    2.8     7.84
 11    40    3.0     9.00
 12    37    3.1     9.61
 13    30    3.5    12.25
 14    43    3.6    12.96
 15    55    3.8    14.44
 16    29    4.2    17.64
 17    45    4.4    19.36
 18    60    5.1    26.01
 19    53    5.2    27.04
 20    30    5.4    29.16
 21    42    5.5    30.25
 22    25    6.0    36.00
 23    63    6.2    38.44
 24    51    6.3    39.69
```

Dependent Variable: Y

Analysis of Variance

Source	DF	Sum of Squares	Mean Square	F Value	Prob>F
Model	2	4458.40552	2229.20276	20.448	0.0001
Error	21	2289.42782	109.02037		
C Total	23	6747.83333			

Root MSE	10.44128	R-square	0.6607	
Dep Mean	31.91667	Adj R-sq	0.6284	
C.V.	32.71420			

Parameter Estimates

Variable	DF	Parameter Estimate	Standard Error	T for H0: Parameter=0	Prob > |T|
INTERCEP	1	-6.871747	8.87156858	-0.775	0.4472
X	1	17.105361	5.73414408	2.983	0.0071
X2	1	-1.349036	0.79186923	-1.704	0.1032

Variable	DF	Variable Label
INTERCEP	1	Intercept
X	1	
X2	1	X-SQUARED

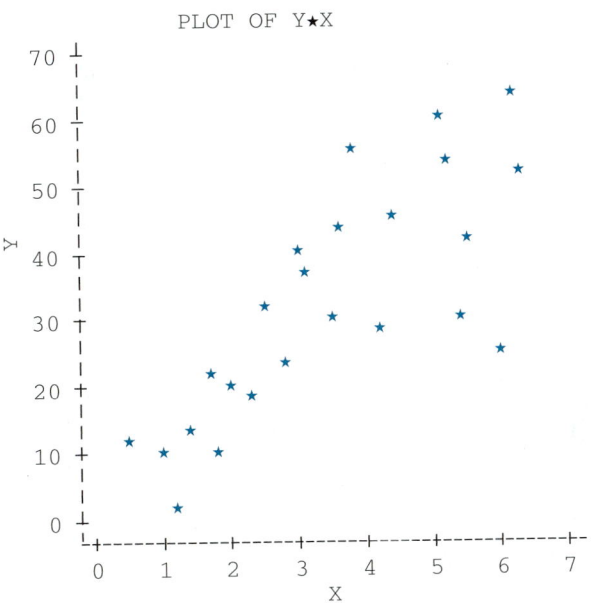

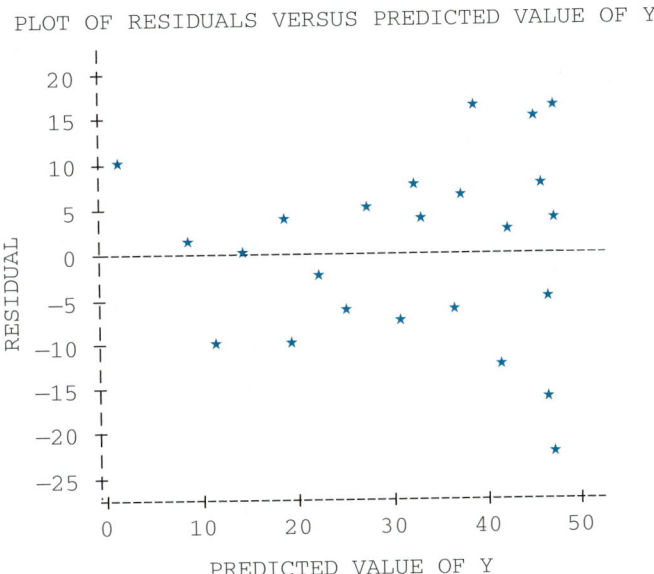

What are the consequences of having a nonconstant variance problem in a regression model? First, if the variance about the regression line is not constant, the least-squares estimates may not be as accurate as possible. A technique called **weighted least squares** [see Draper and Smith (1998)] will give more accuracy. Perhaps more important, however, the weighted least-squares technique improves the statistical tests (F and t tests) on model parameters and the interval estimates for parameter because they are, in general, based on smaller standard errors.

weighted least squares

The more serious pitfall involved with inferences in the presence of noncon-stant variance seems to be for estimates $E(y)$ and predictions of y. For these inferences, the point estimate y is sound but the width of the interval may be too large or too small depending on whether we're predicting in a low or high variance section of the experimental region.

The remedy for nonconstant variance seems to be weighted least squares. We will not cover this technique in the text. However, when the nonconstant variance possesses a pattern related to y, a reexpression (transformation) of y may resolve the problem, accomplishing the same end as weighted least squares. Several transformations for y were discussed in Chapter 11; ones that help to stabilize the variance when there is a pattern to the nonconstant variance were discussed in Chapter 8 for the analysis of variance. They can also be applied in certain regression situations.

EXAMPLE 13.15

Refer to the data of Example 13.14, where we detected a nonconstant variance problem. Because the variance about the regression line seemed to increase with x, a square root transformation on y was tried to stabilize the variance. Examine the computer output and residual plot shown here to determine whether the nonconstant variance problem has been eliminated. The SAS package produced the following analysis.

```
DATA LISTING

OBS     Y       X      SQRT_Y      X-SQUARED

 1     11     0.5     3.31662       0.25
 2     10     1.0     3.16228       1.00
 3      2     1.2     1.41421       1.44
 4     14     1.4     3.74166       1.96
 5     22     1.7     4.69042       2.89
 6     10     1.8     3.16228       3.24
 7     20     2.0     4.47214       4.00
 8     19     2.3     4.35890       5.29
 9     32     2.5     5.65685       6.25
10     23     2.8     4.79583       7.84
11     40     3.0     6.32456       9.00
12     37     3.1     6.08276       9.61
13     30     3.5     5.47723      12.25
14     43     3.6     6.55744      12.96
15     55     3.8     7.41620      14.44
16     29     4.2     5.38516      17.64
```

OBS	Y	X	SQRT_Y	X-SQUARED
17	45	4.4	6.70820	19.36
18	60	5.1	7.74597	26.01
19	53	5.2	7.28011	27.04
20	30	5.4	5.47723	29.16
21	42	5.5	6.48074	30.25
22	25	6.0	5.00000	36.00
23	63	6.2	7.93725	38.44
24	51	6.3	7.14143	39.69

Dependent Variable: SQRT_Y   SQUARE ROOT OF Y

Analysis of Variance

Source	DF	Sum of Squares	Mean Square	F Value	Prob>F
Model	2	45.00905	22.50453	24.683	0.0001
Error	21	19.14653	0.91174		
C Total	23	64.15558			

Root MSE	0.95485	R-square	0.7016
Dep Mean	5.40773	Adj R-sq	0.6731
C.V.	17.65715		

Parameter Estimates

Variable	DF	Parameter Estimate	Standard Error	T for H0: Parameter=0	Prob > |T|
INTERCEP	1	1.189795	0.81130083	1.467	0.1573
X	1	1.990218	0.52438482	3.795	0.0011
X2	1	-0.176856	0.07241607	-2.442	0.0235

Variable	DF	Variable Label
INTERCEP	1	Intercept
X	1	
X2	1	X-SQUARED

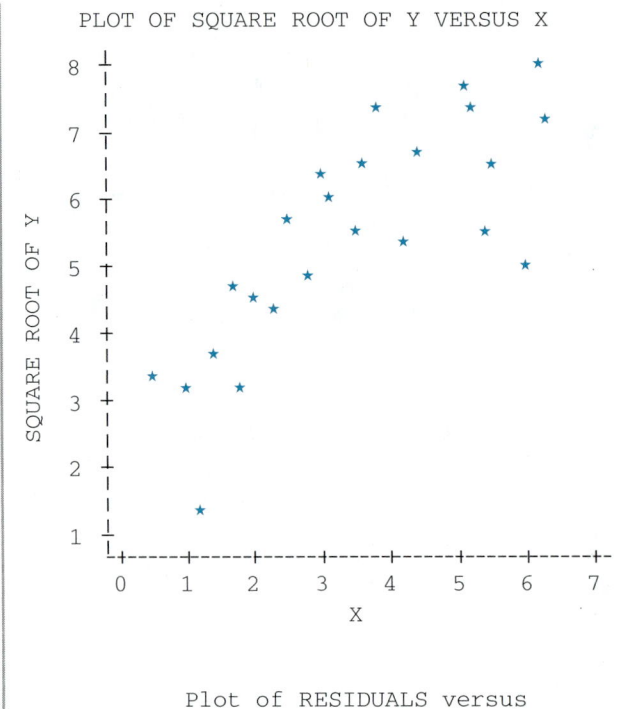

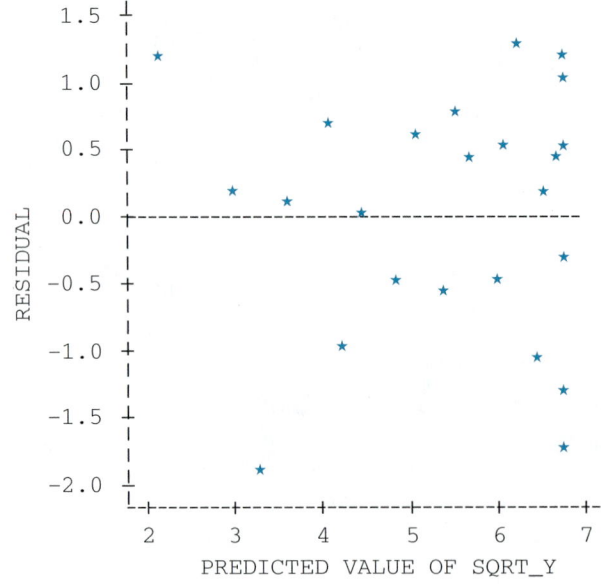

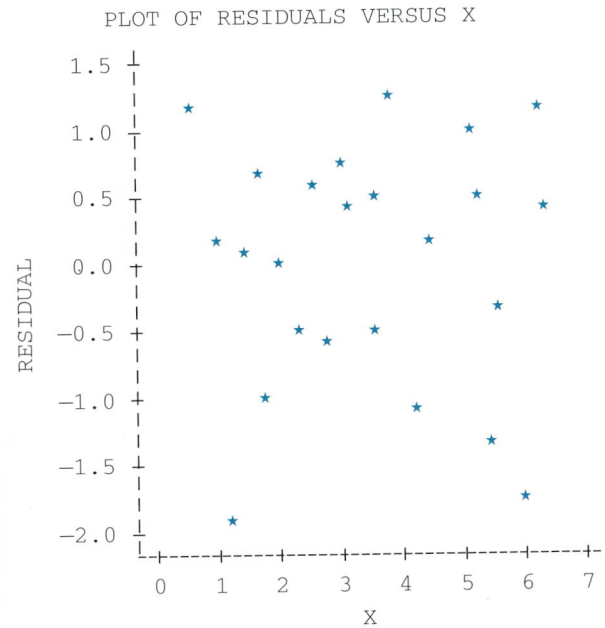

Solution   The output shown here documents that this model provides a much better fit to the sample data; note, especially, the residual plot.

The third assumption for multiple regression is that of normality of the ε_i Skewness and/or outliers are examples of forms of nonnormality that may be detected through the use of certain scatterplots and residual plots.

A plot of the residuals in the form of a histogram or a stem-and-leaf plot will help to detect skewness. By assumption, the ε_i are normally distributed with mean 0. If a histogram of the residuals is not symmetrical about 0, some skewness is present. For example, the residual plot in Figure 13.5 (a) is symmetrical on 0 and suggests no skewness. In contrast, the residual plot in Figure 13.5 (b) is skewed to the right.

FIGURE 13.5

Top: residuals centered on zero; bottom: residuals skewed to right

(a) Middle of interval   Number of observations

Middle of interval		
−2.0	3	× × ×
−1.5	10	× × × × × × × × × ×
−1.0	16	× × × × × × × × × × × × × × × ×
−0.5	15	× × × × × × × × × × × × × × ×
0.0	20	× × × × × × × × × × × × × × × × × × × ×
0.5	15	× × × × × × × × × × × × × × ×
1.0	11	× × × × × × × × × × ×
1.5	6	× × × × × ×
2.0	3	× × ×
2.5	0	
3.0	1	×

(b) Middle of
     interval   Number of observations

```
−2.0   3   × × ×
−1.5  10   × × × × × × × × × ×
−1.0  16   × × × × × × × × × × × × × × × ×
−0.5  15   × × × × × × × × × × × × × × ×
 0.0  18   × × × × × × × × × × × × × × × × × ×
 0.5  12   × × × × × × × × × × × ×
 1.0   7   × × × × × × ×
 1.5   5   × × × × ×
 2.0   3   × × ×
 2.5   3   × × ×
 3.0   2   × ×
 3.5   0
 4.0   1   ×
```

probability plot   Another way to detect nonnormality is through the use of a normal **probability plot** of the residuals. The idea behind the plot is that if the residuals are normally distributed, the normal probability plot will be approximately a straight line. Most computer packages in statistics offer an option to obtain normal probability plots. We'll use them when needed to do our plots.

EXAMPLE 13.16

Refer to Example 13.15. Use the computer output shown here to determine whether there is evidence of nonnormality for the residuals.

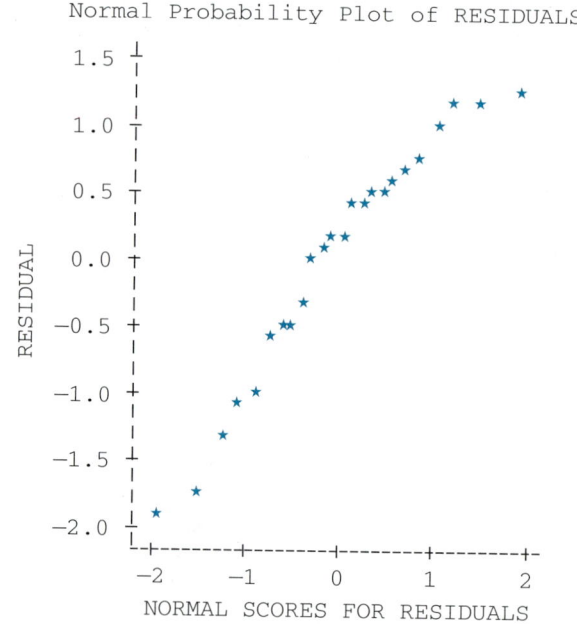

Solution   The normal probability plot is nearly linear. Thus, we can be reasonably assured that the residuals have a normal distribution.

The presence of one or more outliers is perhaps a more subtle form of nonnormality that may be detected by using a scatterplot and one or more residual plots. An outlier is a data point that falls away from the rest of the data. Recall from Chapter 11 that we must be concerned about the leverage (x outlier) and influence (both x and y outlier) properties of a point. A high influence point may seriously distort the regression equation. In addition, some outliers may signal a need for management action. If a regression analysis indicates that the price of a particular parcel of land is very much lower than predicted, that parcel may be an excellent purchase. A sales office that has far better results than a regression model predicts may have employees who are doing outstanding work that can be copied. Conversely, a sales office that has far poorer results than the model predicts may have problems. Sometimes it is possible to isolate the reason for the outlier; other times it is not. An outlier may arise because of an error in recording the data or in entering it into a computer, or because the observation is obtained under different conditions from the other observations. If such a reason can be found, the data entry can be corrected or the point omitted from the analysis. If there is no identifiable reason to correct or omit the point, run the regression both with and without it to see which results are sensitive to that point. No matter what the source or reason for outliers, if they go undetected they can cause serious distortions in a regression equation.

For the linear regression model $y = \beta_0 + \beta_1 x + \varepsilon$, a scatterplot of y versus x will help detect the presence of an outlier. This is shown in Table 13.6 and Figure 13.6. It certainly appears that the circled data point is an outlier. Computer output for a linear fit to the data of Table 13.6 is shown here, along with a residual plot and a normal probability plot. Again the data point corresponding to the suspected outlier (62, 125) is circled in each plot. The Minitab program produced the following analysis.

TABLE 13.6

Listing of data

Obs	x	y
1	10	120
2	20	115
3	21	250
4	27	210
5	29	300
6	33	330
7	40	295
8	44	400
9	52	380
10	56	460
11	62	125
12	68	510

$N = 12$

FIGURE 13.6

Scatterplot of the data in Table 13.6

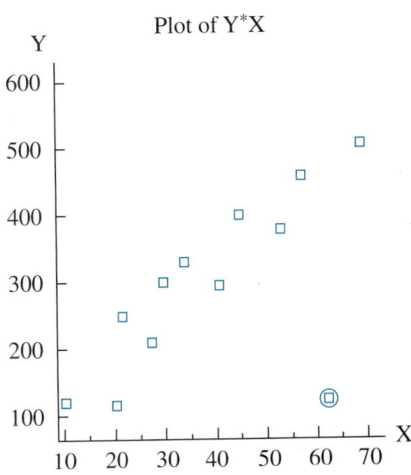

Plot of Y*X

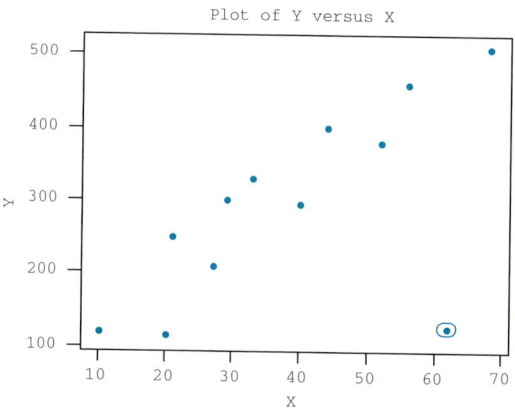

Plot of Y versus X

```
Regression Analysis: y versus x


The regression equation is
y = 114 + 4.59 x


Predictor       Coef      SE Coef         T          P
Constant        114.36       75.53      1.51      0.161
x                4.595        1.787      2.57      0.028


s = 108.1        R-Sq = 39.8%       R-Sq(adj) = 33.8%

Analysis of Variance


Source          DF          SS          MS        F         P
Regression       1        77201       77201      6.61     0.028
Residual Error  10       116755       11676
Total           11       193956


Obs        x           y           Fit        SE Fit      Residual    Standardized
                                                                      Residual
  1       10.0       120.0        160.3         59.7        -40.3        -0.45
  2       20.0       115.0        206.2         45.4        -91.2        -0.93
  3       21.0       250.0        210.8         44.2         39.2         0.40
  4       27.0       210.0        238.4         37.4        -28.4        -0.28
  5       29.0       300.0        247.6         35.5         52.4         0.51
  6       33.0       330.0        266.0         32.7         64.0         0.62
  7       40.0       295.0        298.1         31.3         -3.1        -0.03
  8       44.0       400.0        316.5         32.7         83.5         0.81
  9       52.0       380.0        353.3         39.4         26.7         0.27
 10       56.0       460.0        371.7         44.2         88.3         0.90
 11       62.0       125.0        399.2         52.3       -274.2        -2.90R
 12       68.0       510.0        426.8         61.2         83.2         0.93


R denotes an observation with a large standardized residual
```

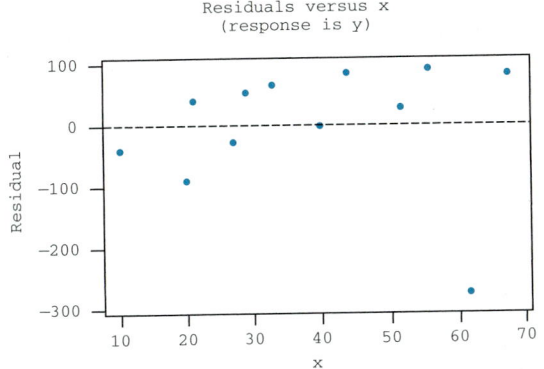

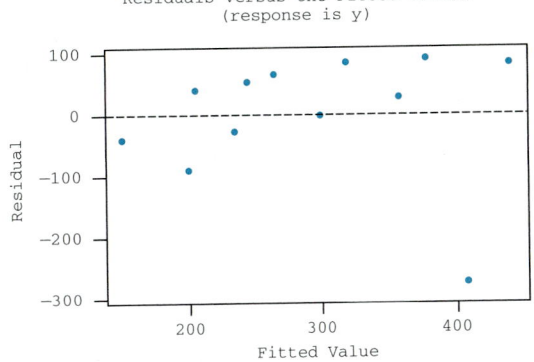

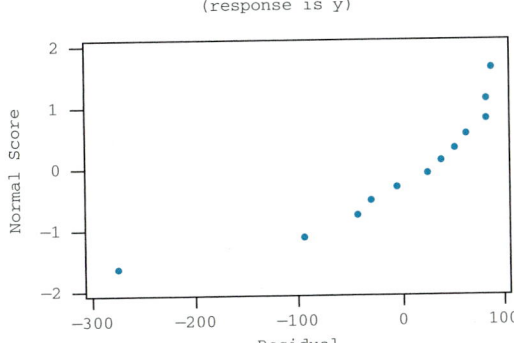

This data set helps to illustrate one of the problems in trying to identify outliers. Sometimes a single plot is not sufficient. For this example, the scatterplot and the probability plot clearly identify the outlier, whereas the residual plot is less conclusive because the outlier adversely affects the linear fit to the data by pulling the fitted line toward the outlier. This makes some of the other residuals larger than they should be. The message is clear: *Don't jump to conclusions without examining the data in several different ways.* The problem becomes even more difficult with multiple regression, where simple scatterplots are not possible.

Another approach to detecting outliers is the *jackknife method.* This involves calculating a succession of regression coefficients for the model, each time excluding one data point. When an outlier is excluded, the regression coefficients change substantially. This procedure could be extended in the case of multiple outliers to exclude, say, two or three points at a time, but the number of times the regression model is fit becomes prohibitively large. Thus, although the one-at-a-time jackknife method may not always catch multiple outliers, it is double and often picks up outliers.

In practice, it may be necessary to consider a combination of techniques for examining the sample data for outliers. First, simple x, y scatterplots may suggest that certain observations are outliers. An examination of the residuals may (or may not) confirm this suspicion. If neither the scatterplots nor residuals suggest the existence of one or more outliers, one can probably end the search. However, identification of possible outliers could require additional work with jackknife techniques to isolate specific outliers.

If you detect outliers, what should you do with them? Of course, recording or transcribing errors should simply be corrected. Sometimes an outlier obviously comes from a different population than the other data points. For example, a Fortune 500 conglomerate firm doesn't belong in a study of small manufacturers. In such situations, the outliers can reasonably be omitted from the data. Unless a compelling reason can be found, throwing out a data point is inappropriate.

EXAMPLE 13.17

Suppose the data for a regression study are as follows:

x:	10	13	16	18	20	22	24	27	30
y:	31	35	42	45	51	53	59	31	70

Draw a scatterplot of the data, identify the outlier, and fit a simple regression model with and without the outlier point.

Solution   A scatterplot of the data (Figure 13.7) shows that any line with slope about 2 and intercept about 10 fits all the data points fairly well, except for the $x = 27$, $y = 31$ point. If that point is included, the least-squares equation is

$$\hat{y} = 19.94 + 1.32x$$

FIGURE 13.7

Effect of an outlier

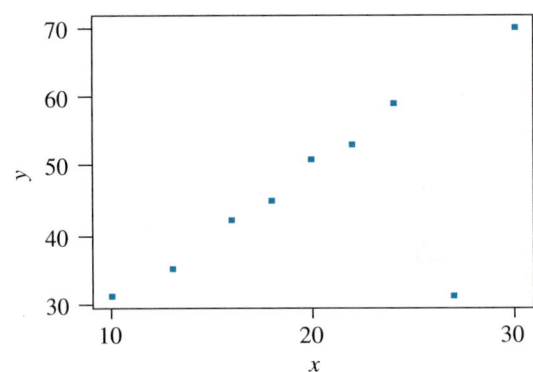

If it is excluded, the prediction equation is

$$\hat{y} = 9.93 + 2.00x$$

The scatterplot shows clearly that the observation (27, 31) is a high-influence outlier and that the regression equation is distorted by inclusion of the data point.

EXAMPLE 13.18

Apply a jackknife procedure, eliminating one data point at a time, to the data of Example 13.17. Examine the estimated slopes and intercepts to locate possible outliers.

Solution   We ran repeated regression analyses by computer, each time omitting one of the points. The estimated slopes and intercepts are listed next. Note that the last two data points appear to be outliers because omitting them caused a large change in the equation.

Data Point Excluded	Slope	Intercept
10, 31	1.21286	22.47672
13, 35	1.26116	21.42333
16, 42	1.33281	19.55234
18, 45	1.32834	19.60120
20, 51	1.31953	19.35947
22, 53	1.29235	19.97601
24, 59	1.21563	21.04531
27, 31	2.00354	9.93239
30, 70	.79712	28.42905

Thus, although the scatterplot of Figure 13.7 identified one potential outlier (the point 27, 31), an examination of the residuals as well as the jackknife procedure detects a second potential outlier (the point 30, 70). An examination of residuals from the regression omitting the point 27, 31 indicates that the point 30, 70 is not in fact an outlier.

```
Regression Analysis: Y versus X (With the point (27, 31) Removed)


The regression equation is
Y = 9.93 + 2.00 X

Predictor        Coef      SE Coef          T         P
Constant        9.932        1.190       8.34     0.000
X             2.00354      0.05944      33.71     0.000


S = 0.9997      R-Sq = 99.5%      R-Sq(adj) = 99.4%

Analysis of Variance

Source            DF           SS          MS          F         P
Regression         1       1135.5      1135.5    1136.17     0.000
Residual Error     6          6.0         1.0
Total              7       1141.5

Obs          X            Y          Fit      SE Fit    Residual    St Resid
  1       10.0       31.000       29.968       0.647       1.032        1.36
  2       13.0       35.000       35.978       0.507      -0.978       -1.14
  3       16.0       42.000       41.989       0.399       0.011        0.01
  4       18.0       45.000       45.996       0.360      -0.996       -1.07
  5       20.0       51.000       50.003       0.357       0.997        1.07
  6       22.0       53.000       54.010       0.393      -1.010       -1.10
  7       24.0       59.000       58.017       0.457       0.983        1.11
  8       30.0       70.000       70.038       0.737      -0.038       -0.06
```

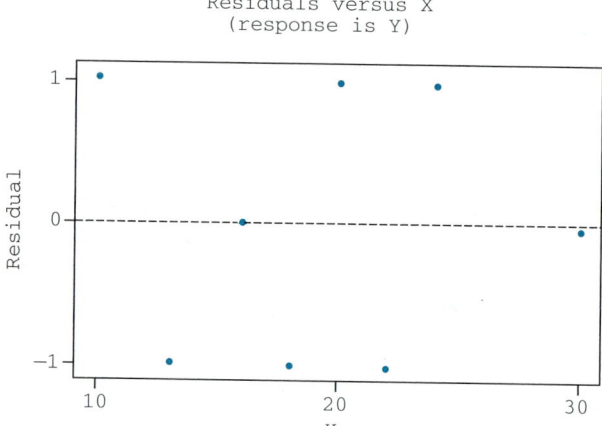

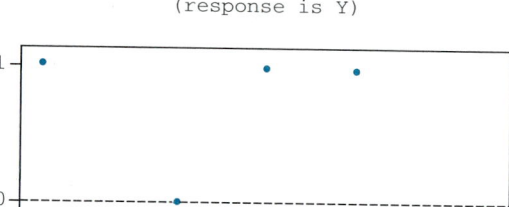

The final assumption is that the ε_i are statistically independent and hence uncorrelated. When the time sequence of the observations is known, as is the case with **time series** data, where observations are taken at successive points in time, it is possible to construct a plot of the residuals versus time to observe where the residuals are **serially correlated**. If, for example, there is a positive serial correlation, adjacent residuals (in time) tend to be similar; negative serial correlation implies that adjacent residuals are dissimilar. These patterns of positive and negative serial correlation are displayed in Figures 13.8(a) and 13.8(b), respectively. Figure 13.8(c) shows a residual plot with no apparent serial correlation.

time series

serial correlation

FIGURE 13.8

(a) Positive serial correlation; (b) negative serial correlation; (c) no apparent serial correlation

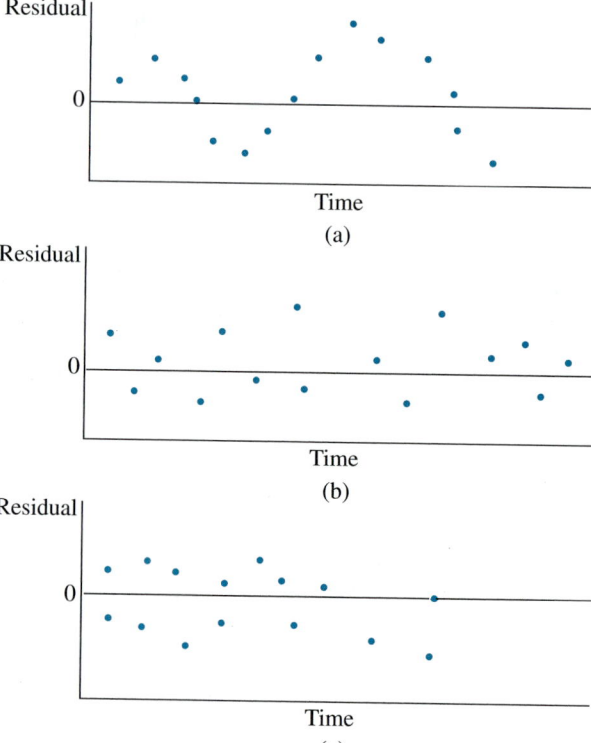

Durbin–Watson test statistic

A formal statistical test for serial correlation is based on the *Durbin–Watson statistic*. Let $\hat{\varepsilon}_t$ denote the residual at time t and n the total number of time points. Then the **Durbin–Watson test statistic** is

$$d = \frac{\sum_{t=1}^{n-1} (\hat{\varepsilon}_{t+1} - \hat{\varepsilon}_t)^2}{\sum_t \hat{\varepsilon}_t^2}$$

The logic behind this statistic is as follows: If there is a positive serial correlation, then successive residuals will be similar and their squared difference $(\hat{\varepsilon}_{t+1} - \hat{\varepsilon}_t)^2$ will tend to be smaller than it would be if the residuals were uncorrelated. Similarly, if there is a negative serial correlation among the residuals, the squared difference of successive residuals will tend to be larger than when no correlation exists.

positive and negative serial correlation

When there is no serial correlation, the expected value of the Durbin–Watson test statistic d is approximately 2.0; **positive serial correlation** makes $d < 2.0$ and **negative serial correlation** makes $d > 2.0$. Although critical values of d have been tabulated by J. Durbin and G. S. Watson (1951), values of d less than approximately 1.5 (or greater than approximately 2.5) lead one to suspect positive (or negative) serial correlation.

EXAMPLE 13.19

Sample data corresponding to retail sales for a particular line of personalized computers by month are shown here.

Month, x	Sales (millions of dollars), y
1	6.0
2	6.3
3	6.1
4	6.8
5	7.5
6	8.0
7	8.1
8	8.5
9	9.0
10	8.7
11	7.9
12	8.2
13	8.4
14	9.0

Plot the data. Also plot the residuals by time based on a linear regression equation. Does there appear to be serial correlation?

Solution   It is clear from the scatterplot of the sample data and from the residual plot of the linear regression that serial correlation is present in the data.

If serial correlation is suspected, then the proposed multiple regression model is inappropriate and some alternative must be sought. A study of the many approaches to analyzing time series data where the errors are not independent can consume many years; hence, we cannot expect to solve many of these problems within the confines of this text. We will, however, suggest a simplified regression approach, based on *first differences,* which may alleviate the problem.

Regression based on first differences is simple to use and, as might be expected, is only a crude approach to the problem of serial correlation. For a simple linear regression of y on x, we compute the differences $y_t - y_{t-1}$ and $x_t - x_{t-1}$. A regression of the $n - 1$ y differences on the corresponding $n - 1$ x differences may eliminate the serial correlation. If not, you should consult someone more familiar with analyzing time series data.

The residual plots that we have discussed can be useful in diagnosing problems in fitting regression models to data. Unfortunately however, they, too, can be misleading because the residuals are subject to random variation. Some researchers have suggested that it is better to use "standardized" residuals to detect problems with a fitted regression model. One particular type of standardized residual, called the Studentized residual, has become part of the output for some of the major software packages such as SAS.

If the software package you use works with standardized residuals, you can replace plots of the ordinary residuals with plots of the standardized residuals to perform the diagnostic evaluation of the fit of a regression model. In theory, these standardized residuals have a mean of 0 and a standard deviation of 1. Large residuals would be ones with an absolute value of, say, 3 or more.

EXERCISES   **Basic Techniques**

13.23   Several different patterns of residuals are shown in the following plots. Indicate whether the plot suggests a problem, and, if so, indicate the potential problem and a possible solution.

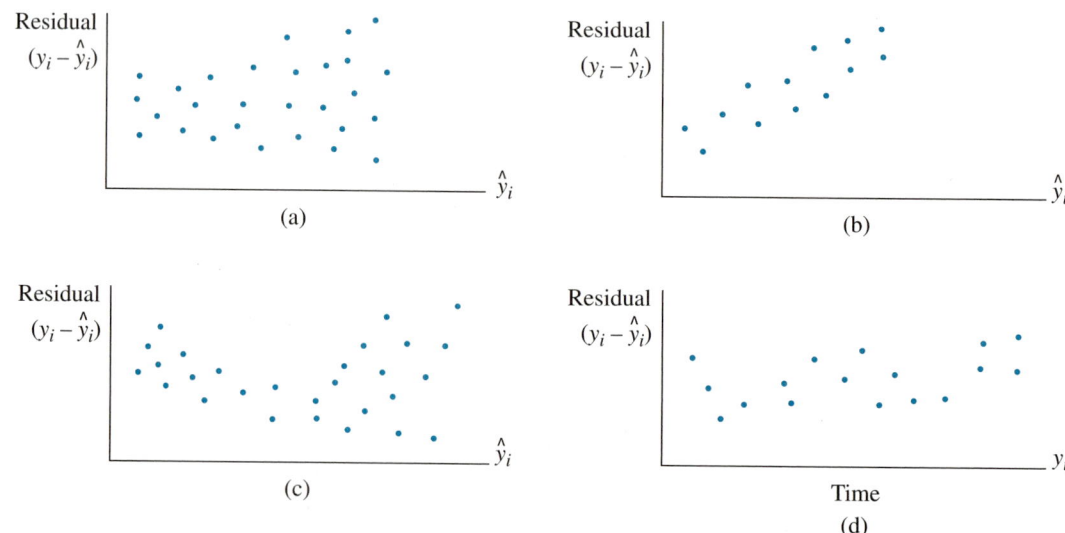

13.24   Refer to the data of Example 13.14. Form first differences and regress the y differences on the x differences. Is there evidence of serial correlation for the difference model? What plot(s) did you use to reach a conclusion?

13.25 What is the value of the Durbin–Watson statistic for the data of Exercise 13.24? Does it agree with your previous conclusion?

Applications

Soc. **13.26** A researcher in the social sciences examined the relationship between the rate (per 1,000) of nonviolent crimes y based on the rate 5 years ago x_1, the present unemployment rate x_2 for cities. Data from 20 different cities are shown here.

CITY	PRESENT RATE	RATE 5 YEARS AGO	PRESENT UNEMPLOYMENT RATE
1	13	14	5.1
2	8	10	2.7
3	14	16	4.0
4	10	10	3.4
5	12	16	3.1
6	11	12	4.3
7	7	8	3.8
8	6	7	3.2
9	10	12	3.2
10	16	20	4.1
11	16	14	5.9
12	9	10	4.0
13	11	10	4.1
14	18	20	5.0
15	9	13	3.1
16	10	6	6.3
17	15	10	5.7
18	14	14	5.2
19	17	16	4.9
20	6	8	3.0

Use the output shown here to:
  a. Determine the fit to the model

$$y = \beta_0 + \beta_1 x_1 + \beta_2 x_2 + \beta_3 x_1 x_2 + \varepsilon.$$

  b. Examine the assumptions underlying the regression model. Discuss whether the assumptions appear to hold. If they don't, suggest possible remedies.

```
SAS OUTPUT FOR EXERCISES 13.26


DATA LISTING


OBS    RATE    RATE_5    UNEMPLOY    RT5_UNEP
  1     13       14        5.1        71.4
  2      8       10        2.7        27.0
  3     14       16        4.0        64.0
  4     10       10        3.4        34.0
  5     12       16        3.1        49.6
  6     11       12        4.3        51.6
  7      7        8        3.8        30.4
```

```
OBS     RATE    RATE_5    UNEMPLOY    RT5_UNEP
 8       6        7         3.2         22.4
 9      10       12         3.2         38.4
10      16       20         4.1         82.0
11      16       14         5.9         82.6
12       9       10         4.0         40.0
13      11       10         4.1         41.0
14      18       20         5.0        100.0
15       9       13         3.1         40.3
16      10        6         6.3         37.8
17      15       10         5.7         57.0
18      14       14         5.2         72.8
19      17       16         4.9         78.4
20       6        8         3.0         24.0
```

```
MULTIPLE REGRESSION ANALYSIS

Dependent Variable: RATE        NONVIOLENT CRIME RATE PER 1000

Analysis of Variance

                     Sum of          Mean
Source       DF      Squares         Square      F Value     Prob>F

Model         3     234.27348       78.09116      67.442      0.0001
Error        16      18.52652        1.15791
C Total      19     252.80000


      Root MSE      1.07606     R-square      0.9267
      Dep Mean     11.60000     Adj R-sq      0.9130
      C.V.          9.27639
```

```
Parameter Estimates

                  Parmeter      Standard    T for H0:
Variable   DF     Estimate        Erro      Parameter=0    Prob > |T|

INTERCEP    1    -2.704052     3.37622689    -0.801          0.4349
RATE_5      1     0.517215     0.30264512     1.709          0.1068
UNEMPLOY    1     1.449811     0.74635173     1.943          0.0699
RT5_UNEP    1     0.035338     0.06631783     0.533          0.6015


                  Variable
Variable   DF     Label

INTERCEP    1     Intercept
RATE_5      1     CRIME RATE 5 YEARS AGO
UNEMPLOY    1     PRESENT UNEMPLOYMENT RATE
RT5_UNEP    1     RATE_5 TIMES UNEMPLOY

Durbin-Watson D                 2.403
(For Number of Obs.)               20
1st Order Autocorrelation      -0.269
```

Plot of PRESENT NONVIOLENT CRIME RATE versus CRIME RATE 5 YEARS AGO

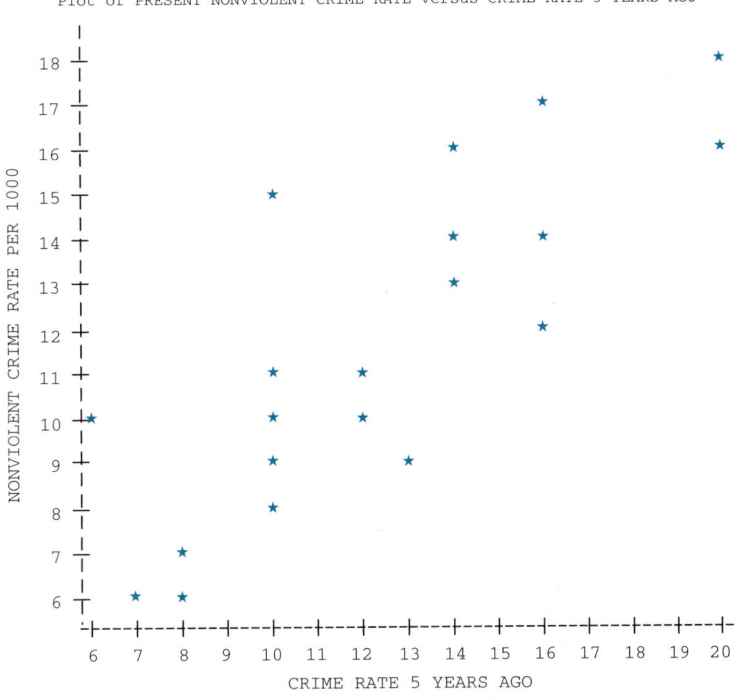

Plot of PRESENT NONVIOLENT CRIME RATE versus PRESENT UNEMPLOYMENT RATE

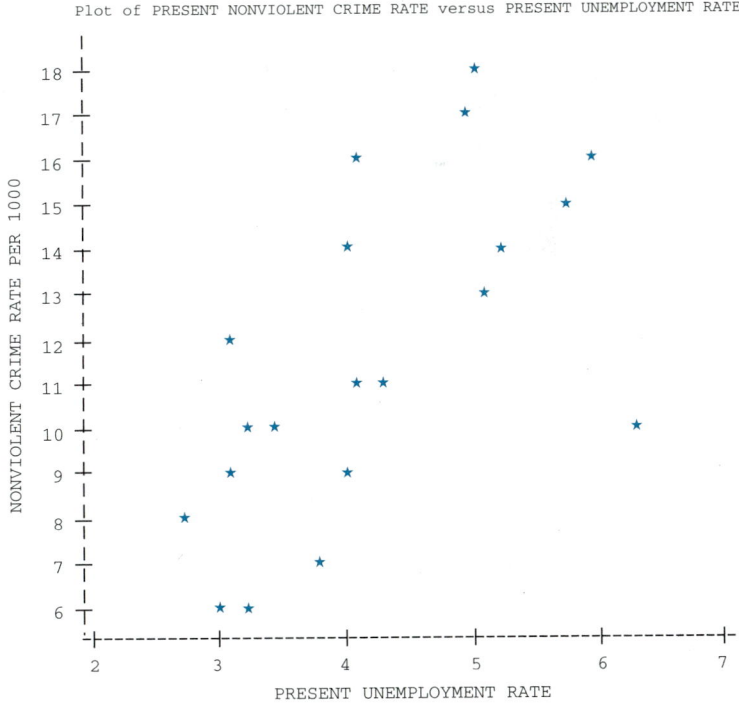

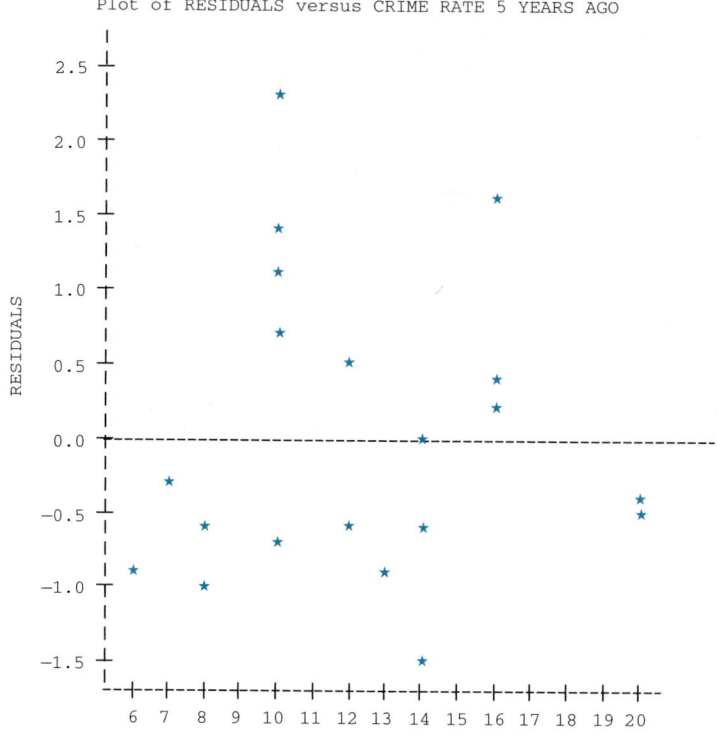

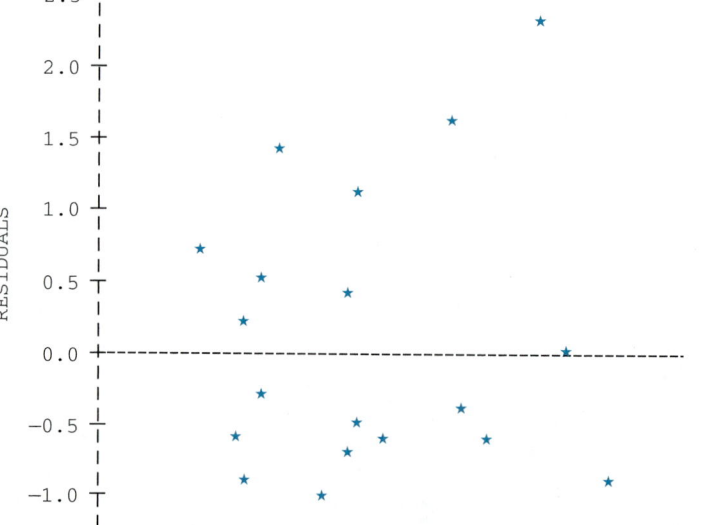

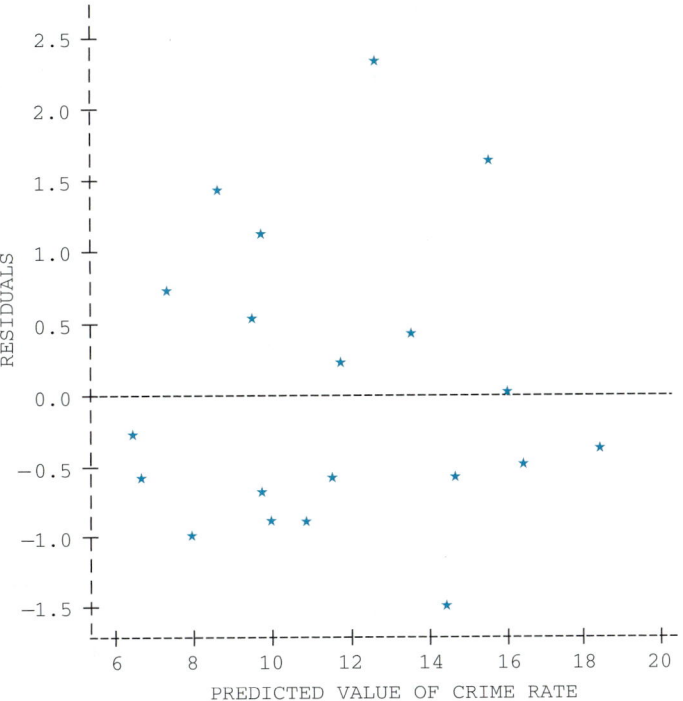

Plot of RESIDUAL versus PREDICTED CRIME RATE

SAS UNIVARIATE PROCEDURE FOR RESIDUAL ANALYSIS

Variable=Residual

```
Stem Leaf                          #  Boxplot
   2  3                            1    |
   1  6                            1    |
   1  14                           2    |
   0  57                           2  +------+
   0  24                           2  |  +   |
  -0  430                          3  *------*
  -0  9976665                      7  +------+
  -1  0                            1    |
  -1  5                            1    |
      ----+-----+-----+-----+
```

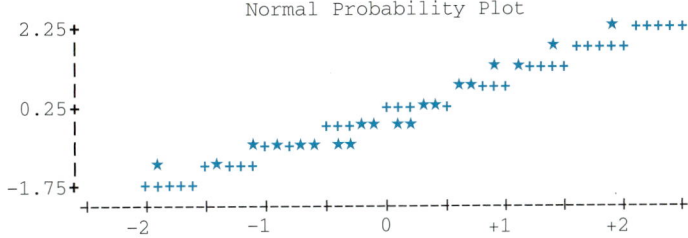

Normal Probability Plot

13.27   Refer to Exercise 13.26. Predict present crime rate for a city having a crime rate of 9 and an unemployment rate of 16% 5 years ago. Might there be a problem with this prediction? If so, why?

13.28   Estimates ($\hat{y}$s) and residuals from a securities firm's regression model for the prediction of earnings per share (per quarter) are shown here for 25 different high-technology companies. Is there any evidence that the assumptions have been violated? Are any additional tests or plots warranted?

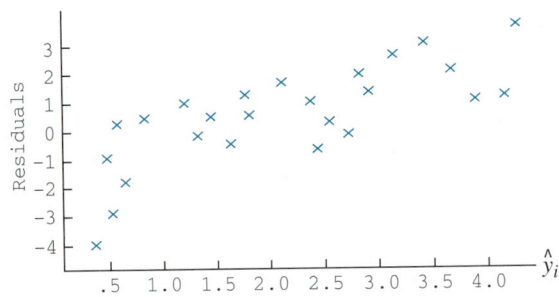

13.5   Summary

This key chapter presents some of the practical problems associated with multiple regression problems. Step 1 of the process is to decide on the dependent variable and a set of candidate independent variables for inclusion in the model. We discussed the invaluable nature of information from an expert in the subject matter field and the utility of some of the best subset regression techniques for choosing which variables to include in the model.

Step 2 involves the actual polynomial form of the particular multiple regression equation. In particular, attention should be paid to lack of fit of a proposed model to data collected on the dependent and independent variables of interest. A formal test for lack of fit of a polynomial model is possible where there are repetitions of observations at one or more settings of the independent variables. Lack of fit can also be examined using residual plots.

Following steps 1 and 2 as we've discussed them can sometimes be a problem, depending on the data that are available. For example, if data are available on many variables at the time that the multiple regression model is being formulated, then consultation with experts and application of one (or more) of the best subset regression techniques can be useful in culling the list of potential independent variables (step 1). The regression model is then modified in step 2 based on the discussions and analyses of step 1. Sometimes, however, data are not available on many possible independent variables. For these situations, step 1 consists of discussions with experts to determine which variables may be important predictors; data are then gathered on these variables. After the data are obtained on these candidate independent variables, the subset regression techniques and the model formulation techniques of step 2 can be applied to refine the model.

The final step of the multiple regression problem is to check the underlying assumptions of multiple regression: zero expectation, constant variance, normality, and independence. Although some formal tests were presented, violation of the assumption is checked best by closely examining the data using scatterplots and various and probability residual plots. The more experience one gains in examining and interpreting data with these plots, the better will be the resulting regression equations.

Key Formulas

1. C_P statistic

$$C_P = \frac{SS(\text{Residual})_P}{s_\varepsilon^2} - (n - 2_P)$$

2. Backward elimination

$$F_d = \frac{SS_{\text{drop}_j}}{MS(\text{Residual})}, \quad j = 1, 2, \ldots$$

3. Durbin–Watson statistic

$$d = \frac{\sum_{t=1}^{n-1} (\hat{\varepsilon}_{t+1} - \hat{\varepsilon}_t)^2}{\sum_t \hat{\varepsilon}_t^2}$$

Supplementary Exercises

13.29 Use the following data to fit a model. Plot the data and suggest a polynomial model.

y	7	8	6	12	15	13	7	10	11	14	16	17
x	10	10	10	15	15	15	20	20	20	25	25	25

13.30 Refer to the data of Exercise 13.29.
    a. Fit the model $y = \beta_0 + \beta_1 x + \beta_2 x^2 + \beta_3 x^3 + \varepsilon$.
    b. Test for lack of fit using $\alpha = .05$.
    c. Examine a residual plot for violation of the regression assumptions.

13.31 Refer to Exercise 13.29. Suppose that the third, fifth, sixth, and tenth observations are missing.
    a. Fit a cubic model.
    b. Examine the residuals and compare the fits for the models of Exercises 13.30 and 13.31.

Med.

13.32 A pharmaceutical firm wanted to obtain information on the relationship between the dose level of a drug product and its potency. To do this, each of 15 test tubes were inoculated with a virus culture and incubated for 5 days at 30°C. Three test tubes were randomly assigned to each of the five different dose levels to be investigated (2, 4, 8, 16, and 32 mg). Each tube was injected with only one dose level and the response of interest (a measure of the protective strength of the product against the virus culture) was obtained. The data are given here.

Dose Level	Response
2	5, 7, 3
4	10, 12, 14
8	15, 17, 18
16	20, 21, 19
32	23, 24, 29

    a. Plot the data.
    b. Fit both a linear and a quadratic model to these data.
    c. Which model seems more appropriate?
    d. Compare your results in part (b) to those obtained in the SAS computer output that follows.

```
SAS OUTPUT FOR EXERCISE 13.32

            DATA LISTING

   OBS        DOSE    RESPONSE    DOSE2
    1          2          5          4
    2          2          7          4
    3          2          3          4
    4          4         10         16
    5          4         12         16
    6          4         14         16
    7          8         15         64
    8          8         17         64
    9          8         18         64
   10         16         20        256
   11         16         21        256
   12         16         19        256
   13         32         23       1024
   14         32         24       1024
   15         32         29       1024
```

```
REGRESSION ANALYSIS WITH LINEAR DOSE TERM IN MODEL

Dependent Variable: RESPONSE    POTENCY OF DRUG

Analysis of Variance
                        Sum of         Mean
Source        DF       Squares        Square      F Value    Prob>F

Model          1     590.91613     590.91613      44.280     0.0001
Error         13     173.48387      13.34491
C Total       14     764.40000


        Root MSE      3.65307     R-square     0.7730
        Dep Mean     15.80000     Adj R-sq     0.7556
        C.V.         23.12069

Parameter Estimates


                  Parameter      Standard     T for HO:
Variable   DF      Estimate         Error    Parameter=0    Prob > |T|

INTERCEP    1      8.666667    1.42786770        6.070        0.0001
DOSE        1      0.575269    0.08645016        6.654        0.0001


                 Variable
Variable   DF      Label


INTERCEP    1    Intercept
DOSE        1    DOSE LEVEL OF DRUG
```

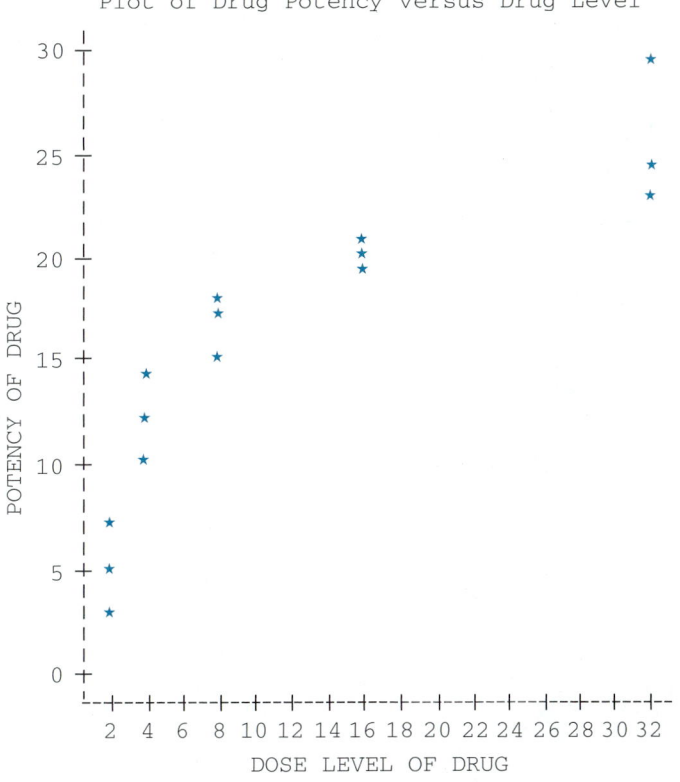

Plot of Drug Potency versus Drug Level

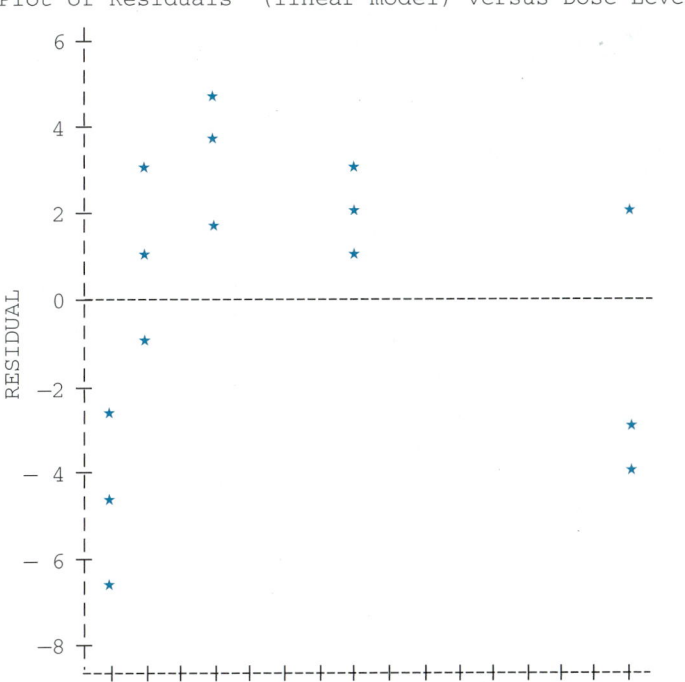

Plot of Residuals (linear model) versus Dose Level

```
REGRESSION ANALYSIS WITH QUADRATIC TERM IN DOSE

Dependent Variable: RESPONSE    POTENCY OF DRUG

Analysis of Variance

                         Sum of          Mean
Source          DF       Squares         Square        F Value     Prob>F

Model            2      673.82062      336.91031        44.634      0.0001
Error           12       90.57938        7.54828
C Total         14      764.40000

        Root MSE        2.74741    R-square        0.8815
        Dep Mean       15.80000    Adj R-sq        0.8618
        C.V.           17.38869

Parameter Estimates

                   Parameter      Standard      T for HO:
Variable DF        Estimate         Error      Parameter=0     Prob > |T|

INTERCEP  1        4.483660       1.65720388       2.706        0.0191
DOSE      1        1.506325       0.28836373       5.224        0.0002
DOSE2     1       -0.026987       0.00814314      -3.314        0.0062

                  Variable
Variable DF       Label

INTERCEP  1       Intercept
DOSE      1       DOSE LEVEL OF DRUG
DOSE2     1       DOSE SQUARED
```

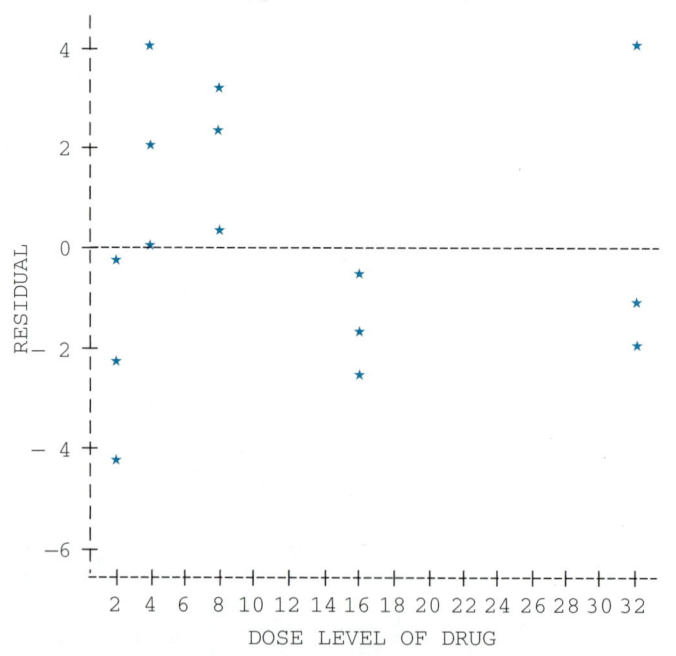

Plot of Residuals (quadratic model) versus Dose Level

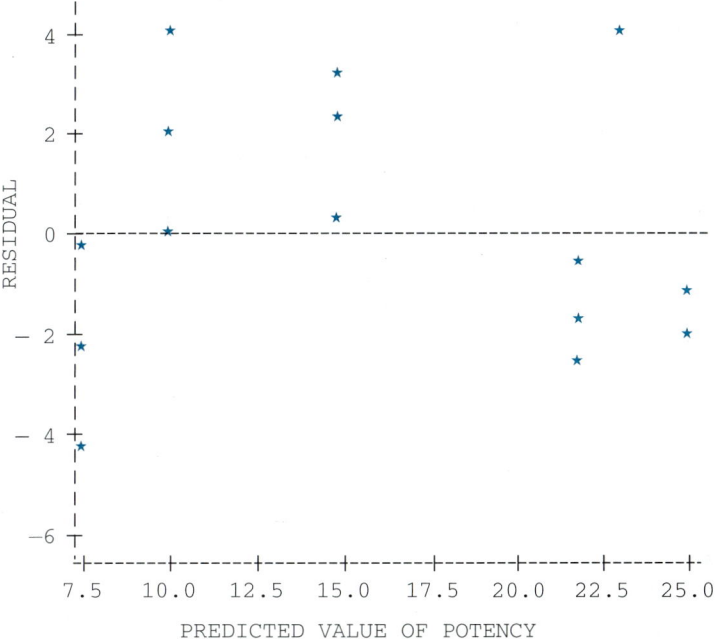

Plot of Residuals (quadratic model) versus Predicted Potency

13.33 Refer to the data of Exercise 13.32. Many times, a logarithmic transformation can be used on the dose levels to linearize the response with respect to the independent variable.

a. Refer to a set of log tables or an electronic calculator to obtain the logarithms of the five dose levels.

b. Where x_1 denotes the log dose, fit the model

$$y = \beta_0 + \beta_1 x_1 + \varepsilon$$

c. Compare your results in part (b) to those shown in the computer printout that follows.

d. Which of the three models seems more appropriate? Why?

```
SAS OUTPUT FOR EXERCISE 13.33
         DATA LISTING


   OBS    DOSE    RESPONSE    LOG_DOSE

    1      2         5         0.69315
    2      2         7.        0.69315
    3      2         3         0.69315
    4      4        10         1.38629
    5      4        12         1.38629
    6      4        14         1.38629
    7      8        15         2.07944
    8      8        17         2.07944
    9      8        18         2.07944
   10     16        20         2.77259
   11     16        21         2.77259
   12     16        19         2.77259
   13     32        23         3.46574
   14     32        24         3.46574
   15     32        29         3.46574
```

```
Dependent Variable: RESPONSE   POTENCY OF DRUG

Analysis of Variance

                        Sum  of       Mean
Source          DF      Squares       Square       F Value      Prob>F

Model            1      710.53333     710.53333    171.478      0.0001
Error           13       53.86667       4.14359
C Total         14      764.40000

         Root MSE         2.03558      R-square      0.9295
         Dep Mean        15.80000      Adj R-sq      0.9241
         C.V.            12.88342

Parameter Estimates

                   Parameter        Standard      T for HO:
Variable   DF      Estimate         Error         Parameter=0      Prob > |T|

INTERCEP    1      1.200000        1.23260547          0.974          0.3480
LOG_DOSE    1      7.021116        0.53616972         13.095          0.0001

                   Variable
Variable   DF      Label

INTERCEP    1      Intercept
LOG_DOSE    1      NATURAL LOGARITHM OF DOSE
```

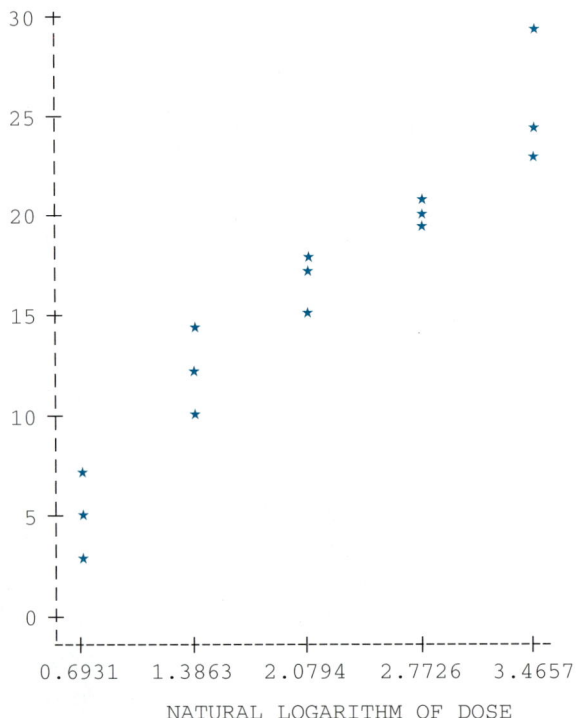

Plot of Drug Potency versus Natural Logarithm of Dose Level

Plot of Residuals versus Predicted Potency of Drug

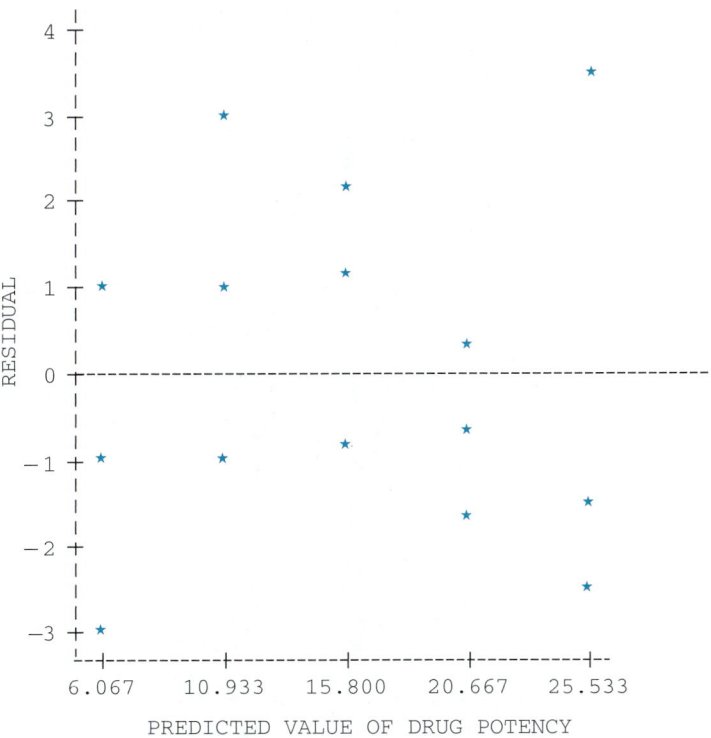

Plot of Residuals versus Natural Logarithm of Dose Level

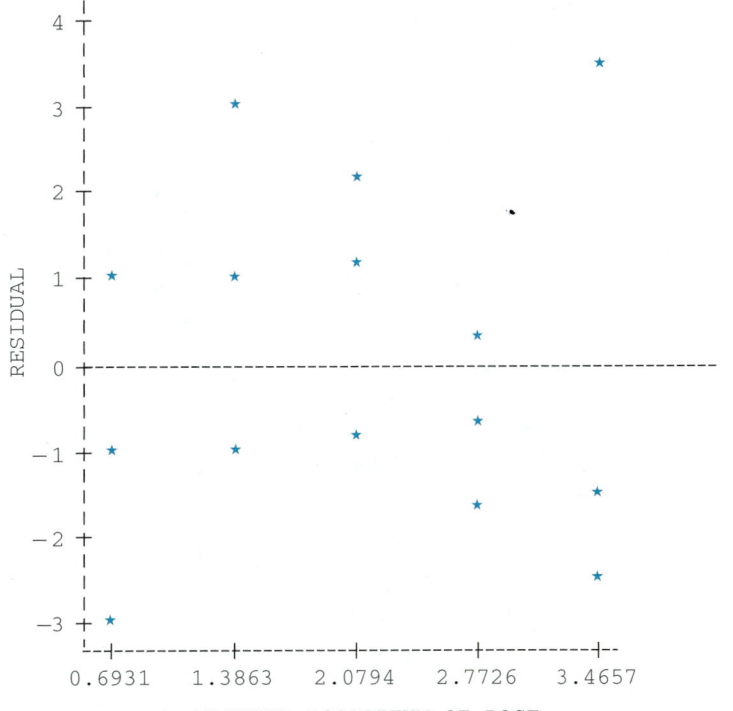

Engin.   **13.34**   An experiment was conducted to examine the weather resistance of a new commercial paint as a function of two independent variables, temperature x_1 and exposure time x_2. The sample data are listed here.

y	120	101	110	105	92	130
x_1 (°C)	-10	-10	0	0	10	10
x_2 (months)	1	3	2	2	1	3

a. Fit the model

$$y = \beta_0 + \beta_1 x_1 + \beta_2 x_2 + \beta_3 x_1 x_2 + \varepsilon$$

b. Examine the residuals and comment on your findings.

13.35   Refer to Exercise 13.34.
a. Could we fit the following model?

$$y = \beta_0 + \beta_1 x_1 + \beta_2 x_1^2 + \beta_3 x_2 + \beta_4 x_2^2 + \beta_5 x_1 x_2 + \beta_6 x_1 x_2^2 + \beta_7 x_1^2 x_2 + \beta_8 x_1^2 x_2^2 + \varepsilon$$

b. Test for lack of fit of the model in Exercise 13.34. Make a recommendation.

Engin.   **13.36**   The abrasive effect of a wear tester for experimental fabrics was tested on a particular fabric while run at six different machine speeds. Forty-eight identical 5-inch-square pieces of fabric were cut, with eight squares randomly assigned to each of the six machine speeds 100, 120, 140, 160, 180, and 200 revolutions per minute (rev/min). The order of assignment of the squares to the machine was random, with each square tested for a 3-minute period at the appropriate machine setting. The amount of wear was measured and recorded for each square. The data appear in the accompanying table.
a. Plot the mean data per revolutions per minute level and suggest a model.
b. Fit the suggested model to the data.
c. Suggest which residual plots might be useful in checking the assumptions underlying the model.

Machine Speed (rev/min)	Wear
100	23.0, 23.5, 24.4, 25.2, 25.6, 26.1, 24.8, 25.6
120	26.7, 26.1, 25.8, 26.3, 27.2, 27.9, 28.3, 27.4
140	28.0, 28.4, 27.0, 28.8, 29.8, 29.4, 28.7, 29.3
160	32.7, 32.1, 31.9, 33.0, 33.5, 33.7, 34.0, 32.5
180	43.1, 41.7, 42.4, 42.1, 43.5, 43.8, 44.2, 43.6
200	54.2, 43.7, 53.1, 53.8, 55.6, 55.9, 54.7, 54.5

13.37   Refer to the data of Exercise 13.36. Suppose that another variable was controlled and that the first four squares at each speed were treated with a .2 concentration of protective coating, and the second four squares were treated with a .4 concentration of the same coating. Given that x_1 denotes the machine speed and x_2 denotes the concentration of the protective coating, fit these models:

$$y = \beta_0 + \beta_1 x_1 + \beta_2 x_1^2 + \beta_3 x_2 + \varepsilon$$

$$y = \beta_0 + \beta_1 x_1 + \beta_2 x_1^2 + \beta_3 x_2 + \beta_4 x_1 x_2 + \beta_5 x_1^2 x_2 + \varepsilon$$

Engin. **13.38** A laundry detergent manufacturer wished to test a new product prior to market release. One area of concern was the relationship between the height of the detergent suds in a washing machine as a function of the amount of detergent added and the degree of agitation in the wash cycle. For a standard size washing machine tub filled to the full level, random assignments of different agitation levels (measured in minutes) and amounts of detergent were made and tested on the washing machine. The data are shown in the accompanying table.

    a. Plot the data and suggest a model.

    b. Does the assumption of normality appear to hold?

    c. Fit an appropriate model.

    d. Use residual plots to detect possible violations of the assumptions.

Height, y	Agitation, x_1	Amount, x_2
28.1	1	6
32.3	1	7
34.8	1	8
38.2	1	9
43.5	1	10
60.3	2	6
63.7	2	7
65.4	2	8
69.2	2	9
72.9	2	10
88.2	3	6
89.3	3	7
94.1	3	8
95.7	3	9
100.6	3	10

13.39 Refer to Exercise 13.38. Would the following model be more appropriate? Why or why not?

$$y = \beta_0 + \beta_1 x_1 + \beta_2 x_1^2 + \beta_3 x_2 + \beta_4 x_2^2 + \beta_5 x_1 x_2 + \beta_6 x_1 x_2^2 + \beta_7 x_1^2 x_2 + \beta_8 x_1^2 x_2^2 + \varepsilon$$

13.40 Refer to the data of Exercise 13.38.

    a. Can we test for lack of fit for the following model?

$$y = \beta_0 + \beta_1 x_1 + \beta_2 x_1^2 + \beta_3 x_2 + \beta_4 x_2^2 + \beta_5 x_1 x_2 + \beta_6 x_1 x_2^2 + \beta_7 x_1^2 x_2 + \beta_8 x_1^2 x_2^2 + \varepsilon$$

    b. Write the complete model for the sample data. Note that if there were replication at one or more design points, the number of degrees of freedom for SS_{Lack} would be identical to the difference between the number of parameters in the complete model and the number of parameters in the model of part (a).

13.41 Refer to Example 13.10.

    a. Identify the parameters in the model.

    b. Fit the "complete" model.

    c. Draw conclusions relative to the standard and luxury models.

Chem. **13.42** The solubility of a solution was examined for six different temperature settings, shown in the accompanying table.

y, Solubility by Weight	x, Temperature (°C)
43, 45, 42	0
32, 33, 37	25
21, 28, 29	50
15, 14, 9	75
12, 10, 8	100
7, 6, 2	125

   a. Plot the data, and fit as appropriate.
   b. Test for lack of fit if possible. Use $\alpha = .05$.
   c. Examine the residuals and draw conclusions.

13.43   Refer to Exercise 13.42. Suppose we are missing observations 5, 8, and 14.
   a. Fit the model $y = \beta_0 + \beta_1 x + \beta_2 x^2 + \varepsilon$.
   b. Test for lack of fit, using $\alpha = .05$.
   c. Again examine the residuals.

13.44   Refer to the data of Exercise 13.37.
   a. Test for lack of fit of the model

$$y = \beta_0 + \beta_1 x_1 + \beta_2 x_1^2 + \beta_3 x_2 + \beta_4 x_1 x_2 + \beta_5 x_1^2 x_2 + \varepsilon$$

   b. Write the complete model for this experimental situation.

13.45   Refer to the data of Exercise 13.32. Test for lack of fit of a quadratic model.

Psy.   **13.46**   A psychologist wants to examine the effects of sleep deprivation on a person's ability to perform simple arithmetic tasks. To do this, prospective subjects are screened to obtain individuals whose daily sleep patterns were closely matched. From this group, 20 subjects are chosen. Each individual selected is randomly assigned to one of five groups, four individuals per group.

   Group 1:   0 hours of sleep

   Group 2:   2 hours of sleep

   Group 3:   4 hours of sleep

   Group 4:   6 hours of sleep

   Group 5:   8 hours of sleep

All subjects are then placed on a standard routine for the next 24 hours.
   The following day after breakfast, each individual is tested to determine the number of arithmetic additions done correctly in a 10-minute period. That evening the amount of sleep each person is allowed depends on the group to which he or she had been assigned. The following morning after breakfast, each person is again tested using a different but equally difficult set of additions.
   Let the response of interest be the difference in the number of correct responses on the first test day minus the number correct on the second test day. The data are presented here.

Group	Response, y
1	39, 33, 41, 40
2	25, 29, 34, 26
3	10, 18, 14, 17
4	4, 6, −1, 9
5	−5, 0, −3, −8

  a. Plot the sample data and use the plot to suggest a model.
  b. Fit the suggested model.
  c. Examine the fitted model for possible violation of assumptions.

Engin.   **13.47**   An experiment was conducted to determine the relationship between the amount of warping y for a particular alloy and the temperature (in °C) under which the experiment was conducted. The sample data appear in the accompanying table. Note that three observations were taken at each temperature setting. Use the computer output that follows to complete parts (a) through (d).

Amount of Warping	Temperature (°C)
10, 13, 12	15
14, 12, 11	20
14, 12, 16	25
18, 19, 22	30
25, 21, 20	35
23, 25, 26	40
30, 31, 34	45
35, 33, 38	50

  a. Plot the data to determine whether a linear or quadratic model appears more appropriate.
  b. If a linear model is fit, indicate the prediction equation. Superimpose the prediction equation over the scatter diagram of y versus x.
  c. If a quadratic model is fit, identify the prediction equation. Superimpose the quadratic prediction equation on the scatter diagram. Which fit looks better, the linear or the quadratic?
  d. Predict the amount of warping at a temperature of 27°C, using both the linear and the quadratic prediction equations.

```
SAS OUTPUT FOR EXERCISE 13.47
        DATA LISTING

OBS     WARPING     TEMP     TEMP2

 1          10        15       225
 2          13        15       225
 3          12        15       225
 4          14        20       400
 5          12        20       400
 6          11        20       400
 7          14        25       625
 8          12        25       625
 9          16        25       625
10          18        30       900
11          19        30       900
12          22        30       900
13          25        35      1225
14          21        35      1225
15          20        35      1225
16          23        40      1600
17          25        40      1600
```

OBS	WARPING	TEMP	TEMP2
18	26	40	1600
19	30	45	2025
20	31	45	2025
21	34	45	2025
22	35	50	2500
23	33	50	2500
24	38	50	2500

```
LINEAR REGRESSION OF WARPING ON TEMPERATURE

Dependent Variable: AMOUNT OF WARPING

Analysis of Variance
```

Source	DF	Sum of Squares	Mean Square	F Value	Prob>F
Model	1	1571.62698	1571.62698	265.546	0.0001
Error	22	130.20635	5.91847		
C Total	23	1701.83333			

Root MSE	2.43279	R-square	0.9235
Dep Mean	21.41667	Adj R-sq	0.9200
C.V.	11.35933		

```
Parameter Estimates
```

Variable	DF	Parameter Estimate	Standard Error	T for H0: Parameter=0	Prob > |T|
INTERCEP	1	−1.539683	1.49370995	−1.031	0.3138
TEMP	1	0.706349	0.04334604	16.296	0.0001

Variable	DF	Variable Label
INTERCEP	1	Intercept
LOG_DOSE	1	TEMPERATURE (in C)

```
Durbin-Watson D              0.908
(For Number of Obs.)           24
1st Order Autocorrelation   0.474
```

Plot of AMOUNT OF WARPING versus TEMPERATURE

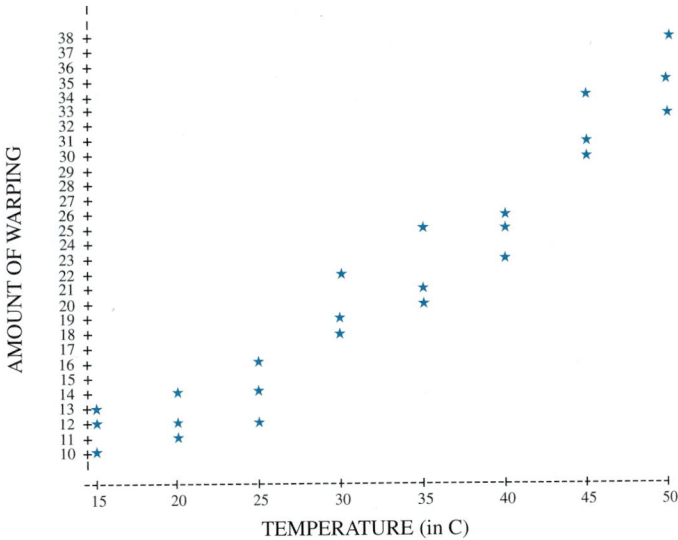

Plot of RESIDUAL versus TEMPERATURE

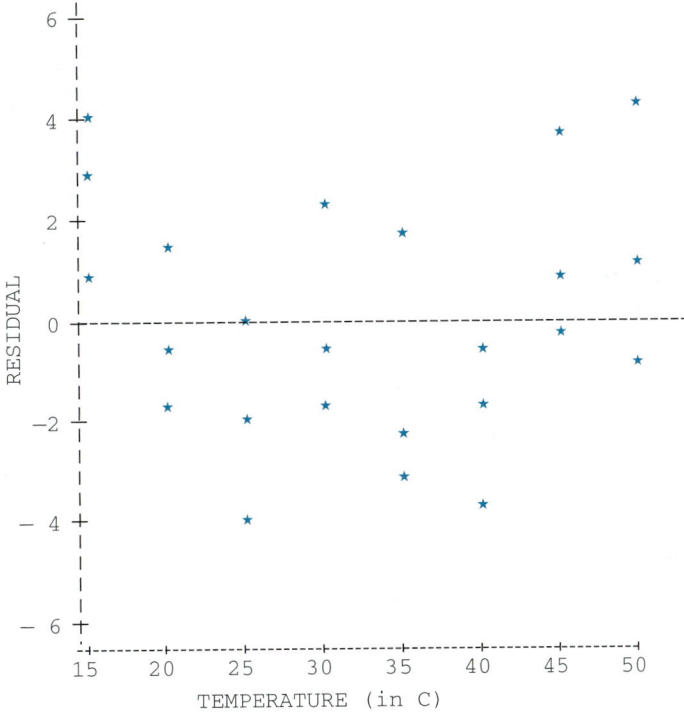

```
QUADRATIC REGRESSION OF WARPING ON TEMPERATURE
Dependent Variable: AMOUNT OF WARPING
Analysis of Variance

                          Sum of          Mean
Source          DF        Squares         Square      F Value      Prob>F
Model            2       1613.92063      806.96032    192.761      0.0001
Error           21         87.91270        4.18632
C Total         23       1701.83333

        Root MSE        2.04605      R-square      0.9483
        Dep Mean       21.41667      Adj R-sq      0.9434
        C.V.            9.55354

Parameter Estimates

                    Parameter       Standard      T for H0:
Variable    DF      Estimate          Error       Parameter=0      Prob > |T|

INTERCEP     1      9.178571        3.59852022        2.551         0.0186
TEMP         1     -0.046825        0.23974742       -0.195         0.8470
TEMP2        1      0.011587        0.00364553        3.178         0.0045
                    Variable
Variable    DF      Label

INTERCEP     1      Intercept
TEMP         1      TEMPERATURE (in C)
TEMP2        1      TEMPERATURE SQUARED

Durbin-Watson D                    1.451
(For Number of Obs.)                 24
1st Order Autocorrelation          0.240
```

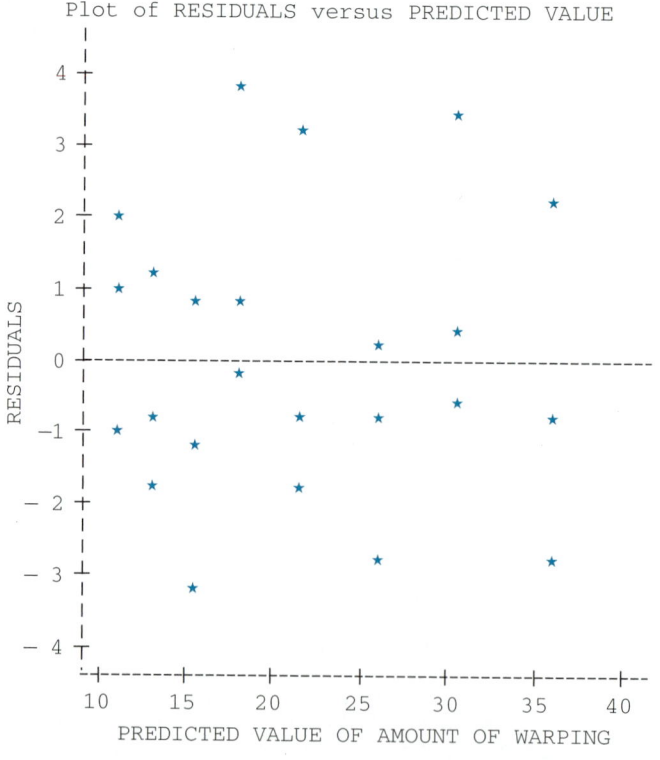

Plot of RESIDUALS versus PREDICTED VALUE

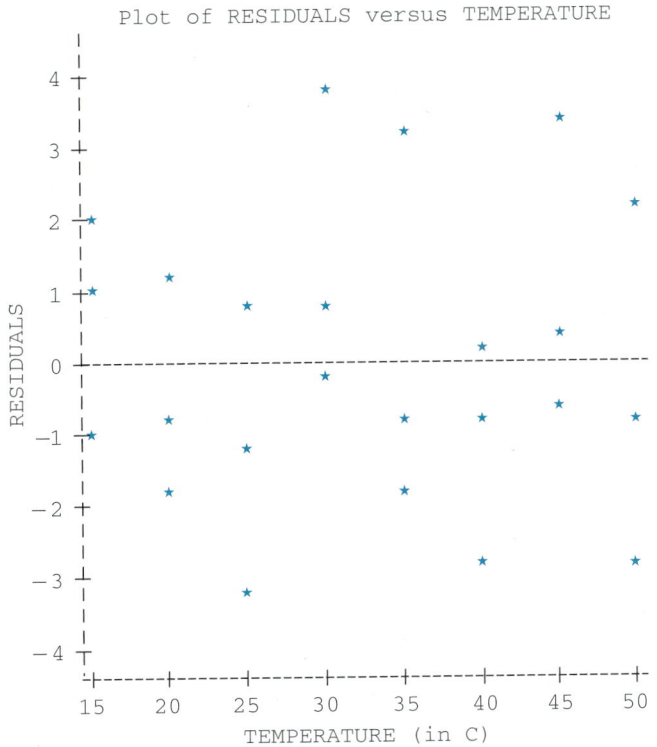

Plot of RESIDUALS versus TEMPERATURE

Bus.   **13.48**   One use of multiple regression is in the setting of performance standards. In other words, a regression equation can be used to predict how well an individual ought to perform when certain conditions are met. In a study of this type, designed to identify an equation that could be used to predict the sales of individual salespeople, data from a random sample of 50 sales territories from four sections of the country (northeast, southeast, midwest, and west) were collected. Data on individual sales performances, as well as on several potential predictor variables, were collected. The variables were as follows.

y = sales–territory performance measured by aggregate sales, in units credited to territory salesperson

x_1 = time with company (months)

x_2 = advertising, or company effort (dollar expenditures in ads in territory)

x_3 = market share (the weighted average of past market share magnitudes for four previous years)

x_4 = indicator variable for section of country (1 = northeast, 0 = otherwise)

x_5 = indicator variable for section of country (1 = southeast, 0 = otherwise)

x_6 = indicator variable for section of country (1 = midwest, 0 = otherwise)

x_7 = indicator variable (1 = male salesperson, 0 = female salesperson)

These data were analyzed using Minitab, with the following results:

```
MTB > DESCRIBE C1-C10

         N       MEAN      MEDIAN      TRMEAN       STDEV      SEMEAN
Y       50       3335        3396        3277        1579         223
X1      50      96.62       85.00       93.86       66.33        9.38
X2      50       5002        5069        4915        2370         335
X3      50      7.335       7.305       7.297       1.668       0.236
C5      50      2.460       2.000       2.455       1.129       0.160
X4      50     0.8200      1.0000      0.8636      0.3881      0.0549
X5      50     0.2600      0.0000      0.2273      0.4431      0.0627
X6      50     0.2600      0.0000      0.2273      0.4431      0.0627
X7      50     0.2400      0.0000      0.2045      0.4314      0.0610


          MIN        MAX          Q1          Q3
Y         131       7205        2033        4367
X1        000     237.00       40.00      144.25
X2        222      10832        3038        6564
X3      4.131     11.205       5.987       8.569
C5      1.000      4.000       1.000       3.250
X4     0.0000     1.0000      1.0000      1.0000
X5     0.0000     1.0000      0.0000      1.0000
X6     0.0000     1.0000      0.0000      1.0000
X7     0.0000     1.0000      0.0000      0.2500
MTB > REGRESS 'Y' ON 7 'X1' 'X2' 'X3' 'X4' 'X5' 'X6' 'X7'

The regression equation is
Y = 16.4 - 0.000546X1 + 0.667X2 + 0.0302X3 - 0.116X4 - 0.041X5 -33.3X6 - 33.6X7

Predictor        Coef        Stdev        t-ratio
Constant       16.3944       0.2931         55.94
    X1       -0.0005463     0.0007607       -0.72
    X2        0.666689       0.000047    14315.675
    X3        0.03024        0.06467         0.47
    X4       -0.1163         0.1128         -1.03
    X5       -0.0412         0.1201         -0.34
    X6      -33.3155         0.1204       -276.81
    X7      -33.6118         0.1185       -283.70
S = 0.2864    R-sq =100.0%      R-sq(adj) = 100.0%

Analysis of Variance
SOURCE         DF        SS        MS
Regression      7    122189056  17455576
Error          42          3         0
Total          49    122189056


SOURCE      DF      SEQ SS
    X1       1     33243924
    X2       1     88931584
    X3       1            1
    X4       1           80
    X5       1         4972
    X6       1         1880
    X7       1         6602
```

Obs.	X1	Y	Fit	Stdev. Fit	Residual	St. Resid.
1	62	3407.00	3406.54	0.09	0.46	1.68
2	70	131.00	131.17	0.14	-0.17	-0.69
3	186	4650.00	4649.93	0.09	0.07	0.27
4	13	1971.00	1970.91	0.11	0.09	0.35
5	20	4168.00	4167.94	0.11	0.06	0.21
6	0	3047.00	3047.28	0.10	-0.28	-1.03
7	31	1196.00	1195.91	0.13	0.09	0.36
8	61	2415.00	2414.91	0.10	0.09	0.34
9	48	1987.00	1987.12	0.09	-0.12	-0.46
10	101	2214.00	2213.84	0.10	0.16	0.61
11	145	4333.00	4333.14	0.27	-0.14	-1.36X
12	200	6253.00	6253.08	0.12	-0.08	-0.29
13	81	1714.00	1713.87	0.12	0.13	0.49
14	124	5146.00	5146.01	0.09	-0.01	-0.04
15	24	3469.00	3469.27	0.11	-0.27	-1.04
16	216	4124.00	4123.60	0.11	0.40	1.53
17	232	3851.00	3851.17	0.14	-0.17	-0.69
18	109	2172.00	2171.83	0.10	0.17	0.64
19	75	1743.00	1743.25	0.12	-0.25	-0.97
20	5	2269.00	2268.93	0.11	0.07	0.27
21	12	3429.00	3429.24	0.10	-0.24	-0.88
22	90	1986.00	1985.83	0.10	0.17	0.64
23	209	3623.00	3623.21	0.12	-0.21	-0.82
24	167	5429.00	5429.16	0.15	-0.16	-0.64
25	170	4511.00	4511.22	0.10	-0.22	-0.81
26	42	1478.00	1477.94	0.12	0.06	0.24
27	167	3385.00	3385.22	0.11	-0.22	-0.84
28	98	1660.00	1660.84	0.11	-0.84	-3.16R
29	144	1212.00	1211.69	0.12	0.31	1.20
30	78	4592.00	4592.00	0.09	0.00	0.00
31	116	2876.00	2875.85	0.09	0.15	0.55
32	89	4349.00	4349.02	0.09	-0.02	-0.06
33	37	2096.00	2095.80	0.09	0.20	0.72
34	34	5308.00	5308.07	0.11	-0.07	-0.26
35	165	5731.00	5730.01	0.10	0.99	3.70R
36	41	1121.00	1120.84	0.11	0.16	0.62
37	80	2356.00	2355.91	0.12	0.09	0.34
38	140	7205.00	7204.80	0.13	0.20	0.79
39	48	3562.00	3561.96	0.13	0.04	0.15
40	203	4133.00	4132.94	0.11	0.06	0.23
41	71	2049.00	2049.12	0.09	-0.12	-0.42
42	13	2512.00	2511.90	0.09	0.10	0.36
43	144	3722.00	3721.89	0.09	0.11	0.40
44	11	2806.00	2805.74	0.13	0.26	1.01
45	34	1477.00	1477.10	0.09	-0.10	-0.37
46	94	4040.00	4039.96	0.08	0.04	0.16
47	237	6633.00	6633.36	0.12	-0.36	-1.37
48	115	3203.00	3203.04	0.12	-0.04	-0.17
49	66	4423.00	4423.27	0.10	-0.27	-1.00
50	113	5563.00	5563.38	0.10	-0.38	-1.40

R denotes an obs. with a large st. resid.
X denotes an obs. whose X value gives it large influence.

Conduct a test to determine whether salespersons in the west make more (other things being equal) than salespersons in the northeast. Give the null and alternative hypotheses, the computed and the critical values of the test statistic, and your conclusion. Use $\alpha = .05$.

13.49 Refer to Exercise 13.48. What is the estimated average increase in sales territory performance of a salesperson when advertising in the territory increases by $1,000?

13.50 Refer to Exercise 13.48. Conduct a test to determine whether males sell (on average) 200 units more than females (other things being equal). Use $\alpha = .05$.

13.51 Refer to Exercise 13.48. A particular concern of one company sales manager is that different regional attitudes may well affect the performance of males and females unequally.
   a. Suggest a new regression model that allows for the possibility of an interaction effect between the four regions of the country and the gender of the salesperson.
   b. Interpret the "new" βs in this model.

Bus.   **13.52** A random sample of 22 residential properties was used in a regression of price on nine different independent variables. The variables used in this study were as follows:

PRICE = selling price (dollars)

BATHS = number of baths (powder room = 1/2 bath)

BEDA = dummy variable for number of bedrooms (1 = 2 bedrooms, 0 = otherwise)

BEDB = dummy variable for number of bedrooms (1 = 3 bedrooms, 0 = otherwise)

BEDC = dummy variable for number of bedrooms (1 = 4 bedrooms, 0 = otherwise)

CARA = dummy variable for type of garage (1 = no garage, 0 = otherwise)

CARB = dummy variable for type of garage (1 = one-car garage, 0 = otherwise)

AGE = age in years

LOT = lot size in square yards

DOM = days on the market

In this study, homes had two, three, four, or five bedrooms and either no garage or one- or two-car garages. Hence, we are using two dummy variables to code for the three categories of garage.

The data were analyzed using Minitab, with the results that follow. Using the full regression model (nine independent variables), estimate the average difference in selling price between
   a. Properties with no garage and properties with a one-car garage.
   b. Properties with a one-car garage and properties with a two-car garage.
   c. Properties with no garage and properties with a two-car garage.

```
MINITAB OUTPUT FOR EXERCISE 13.52

DATA DISPLAY
```

Row	PRICE	BATHS	BEDA	BEDB	BEDC	CARA	CARB	AGE	LOT	DOM
1	25750	1.0	1	0	0	1	0	23	9680	164
2	37950	1.0	0	1	0	0	1	7	1889	67
3	46450	2.5	0	1	0	0	0	9	1941	315
4	46550	2.5	0	0	1	1	0	18	1813	61
5	47950	1.5	1	0	0	0	1	2	1583	234
6	49950	1.5	0	1	0	0	0	10	1533	116
7	52450	2.5	0	0	1	0	0	4	1667	162
8	54050	2.0	0	1	0	0	1	5	3450	80

Row	PRICE	BATHS	BEDA	BEDB	BEDC	CARA	CARB	AGE	LOT	DOM
9	54850	2.0	0	1	0	0	0	5	1733	63
10	52050	2.5	0	1	0	0	0	5	3727	102
11	54392	2.5	0	1	0	0	0	7	1725	48
12	53450	2.5	0	1	0	0	0	3	2811	423
13	59510	2.5	0	1	0	0	1	11	5653	130
14	60102	2.5	0	1	0	0	0	7	2333	159
15	63850	2.5	0	0	1	0	0	6	2022	314
16	62050	2.5	0	0	0	0	0	5	2166	135
17	69450	2.0	0	1	0	0	0	15	1836	71
18	82304	2.5	0	0	1	0	0	8	5066	338
19	81850	2.0	0	1	0	0	0	0	2333	147
20	70050	2.0	0	1	0	0	0	4	2904	115
21	112450	2.5	0	0	1	0	0	1	2930	11
22	127050	3.0	0	0	1	0	0	9	2904	36

Descriptive Statistics: PRICE, BATHS, BEDA, BEDB, BEDC, CARA, CARB, AGE, LOT, DO

Variable	N	Mean	Median	TrMean	StDev	SE Mean
PRICE	22	62023	54621	60585	22749	4850
BATHS	22	2.182	2.500	2.200	0.524	0.112
BEDA	22	0.0909	0.0000	0.0500	0.2942	0.0627
BEDB	22	0.591	1.000	0.600	0.503	0.107
BEDC	22	0.2727	0.0000	0.2500	0.4558	0.0972
CARA	22	0.0909	0.0000	0.0500	0.2942	0.0627
CARB	22	0.1818	0.0000	0.1500	0.3948	0.0842
AGE	22	7.45	6.50	7.05	5.48	1.17
LOT	22	2895	2250	2624	1868	398
DOM	22	149.6	123.0	142.9	109.8	23.4

Variable	Minimum	Maximum	Q1	Q3
PRICE	25750	127050	49450	69600
BATHS	1.000	3.000	2.000	2.500
BEDA	0.0000	1.0000	0.0000	0.0000
BEDB	0.000	1.000	0.000	1.000
BEDC	0.0000	1.0000	0.0000	1.0000
CARA	0.0000	1.0000	0.0000	0.0000
CARB	0.0000	1.0000	0.0000	0.0000
AGE	0.00	23.00	4.00	9.25
LOT	1533	9680	1793	3060
DOM	11.0	423.0	66.0	181.5

Regression Analysis: PRICE versus BATHS, BEDA, BEDB, BEDC, CARA, CARB, AGE, LOT, DOM

The regression equation is
PRICE = 39617 + 11686 BATHS + 15128 BEDA + 2477 BEDB + 26114 BEDC − 44023 CARA
        − 12375 CARB − 506 AGE + 3.40 LOT − 86.0 DOM

Predictor	Coef	SE Coef	T	P
Constant	39617	30942	1.28	0.225
BATHS	11686	10428	1.12	0.284
BEDA	15128	26254	0.58	0.575

```
Predictor          Coef       SE Coef          T          P
BEDB               2477         17783       0.14      0.892
BEDC              26114         18118       1.44      0.175
CARA             -44023         22775      -1.93      0.077
CARB             -12375         10759      -1.15      0.272
AGE                -506          1111      -0.46      0.657
LOT               3.399         2.504       1.36      0.200
DOM              -86.05         35.72      -2.41      0.033


S = 16531       R-Sq = 69.8%      R-Sq(adj) = 47.2%

Analysis of Variance

Source            DF            SS           MS         F          P
Regression         9    7588195915    843132879      3.09      0.036
Residual Error    12    3279393939    273282828
Total             21   10867589854

Source         DF       Seq SS
BATHS           1    3352323167
BEDA            1      24291496
BEDB            1     668205893
BEDC            1     261898228
CARA            1    1261090278
CARB            1     133807628
AGE             1          5848
LOT             1     300736097
DOM             1    1585837280

Unusual Observations
Obs      BATHS       PRICE        Fit      SE Fit     Residual     St Resid
  7       2.50       52450      84651        7506       -32201       -2.19R
 16       2.50       62050      62050       16531           -0         * X
```

R denotes an observation with a large standardized residual
X denotes an observation whose X value gives it large influence.

Regression Analysis: PRICE versus BATHS, BEDA, BEDC, CARA, CARB, LOT, DOM

The regression equation is
PRICE = 39091 + 11712 BATHS + 14183 BEDA + 24531 BEDC - 50962 CARA - 12121 CARB
        + 3.08 LOT - 84.8 DOM

```
Predictor          Coef       SE Coef          T          P
Constant          39091         21445       1.82      0.090
BATHS             11712          9531       1.23      0.239
BEDA              14183         16759       0.85      0.412
BEDC              24531          9021       2.72      0.017
CARA             -50962         15878      -3.21      0.006
CARB             -12121         10010      -1.21      0.246
LOT               3.082         2.231       1.38      0.189
DOM              -84.81         33.24      -2.55      0.023


S = 15443       R-Sq = 69.3%      R-Sq(adj) = 53.9%
```

Analysis of Variance

Source	DF	SS	MS	F	P
Regression	7	7528777484	1075539641	4.51	0.008
Residual Error	14	3338812370	238486598		
Total	21	10867589854			

Source	DF	Seq SS
BATHS	1	3352323167
BEDA	1	24291496
BEDC	1	929454598
CARA	1	1261501483
CARB	1	133856231
LOT	1	274447991
DOM	1	1552902518

Unusual Observations

Obs	BATHS	PRICE	Fit	SE Fit	Residual	St Resid
7	2.50	52450	84299	6973	-31849	-2.31R

R denotes an observation with a large standardized residual

Regression Analysis: PRICE versus BATHS, BEDC, CARA, CARB, LOT, DOM

The regression equation is
PRICE = 44534 + 8336 BATHS + 24649 BEDC - 47007 CARA - 10588 CARB + 3.54 LOT
        - 76.7 DOM

Predictor	Coef	SE Coef	T	P
Constant	44534	20264	2.20	0.044
BATHS	8336	8574	0.97	0.346
BEDC	24649	8934	2.76	0.015
CARA	-47007	15030	-3.13	0.007
CARB	-10588	9751	-1.09	0.295
LOT	3.539	2.144	1.65	0.120
DOM	-76.67	31.51	-2.43	0.028

S = 15296   R-Sq = 67.7%   R-Sq(adj) = 54.8%

Analysis of Variance

Source	DF	SS	MS	F	P
Regression	6	7357974702	1226329117	5.24	0.004
Residual Error	15	3509615152	233974343		
Total	21	10867589854			

Source	DF	Seq SS
BATHS	1	3352323167
BEDC	1	883193335
CARA	1	1307168140
CARB	1	111305152
LOT	1	318872879
DOM	1	1385112029

```
Unusual Observations
Obs      BATHS        PRICE        FIT     SE Fit    Residual    St Resid
 7        2.50        52450       83502      6843      -31052       -2.27R
```

R denotes an observation with a large standardized residual

Regression Analysis: PRICE versus BEDC, CARA, CARB, LOT, DOM

The regression equation is
PRICE = 62606 + 28939 BEDC - 52659 CARA - 14153 CARB + 3.52 LOT - 75.6 DOM

```
Predictor       Coef     SE Coef         T        P
Constant       62606        8056      7.77    0.000
BEDC           28939        7755      3.73    0.002
CARA          -52659       13837     -3.81    0.002
CARB          -14153        9019     -1.57    0.136
LOT            3.523       2.140      1.65    0.119
DOM           -75.64       31.44     -2.41    0.029
```

S = 15270   R-Sq = 65.7%   R-Sq(adj) = 54.9%

Analysis of Variance

Source	DF	SS	MS	F	P
Regression	5	7136792581	1427358516	6.12	0.002
Residual Error	16	3730797273	233174830		
Total	21	10867589854			

Source	DF	Seq SS
BEDC	1	2901187555
CARA	1	2274636373
CARB	1	292810426
LOT	1	318495206
DOM	1	1349663021

Unusual Observations

Obs	BEDC	PRICE	Fit	SE Fit	Residual	St Resid
1	0.00	25750	31641	13849	-5891	-0.92 X
4	1.00	46550	40659	13849	5891	0.92 X
7	1.00	52450	85164	6614	-32714	-2.38R
22	1.00	127050	99052	7948	27998	2.15R

R denotes an observation with a large standardized residual
X denotes an observation whose X value gives it large influence.

Regression Analysis: PRICE versus BEDC, CARA, CARB, LOT, DOM

The regression equation is
PRICE = 59313 + 31921 BEDC - 48742 CARA + 3.02 LOT - 69.0 DOM

```
Predictor       Coef     SE Coef         T        P
Constant       59313        8105      7.32    0.000
BEDC           31921        7836      4.07    0.001
CARA          -48742       14183     -3.44    0.003
LOT            3.025       2.206      1.37    0.188
DOM           -69.00       32.46     -2.13    0.049
```

```
S = 15913        R-Sq = 60.4%      R-Sq(adj) = 51.1%
```

Analysis of Variance

Source	DF	SS	MS	F	P
Regression	4	6562672180	1640668045	6.48	0.002
Residual Error	17	4304917674	253230451		
Total	21	10867589854			

Source	DF	Seq SS
BEDC	1	2901187555
CARA	1	2274636373
LOT	1	242949284
DOM	1	1143898968

Unusual Observations

Obs	BEDC	PRICE	Fit	SE Fit	Residual	St Resid
1	0.00	25750	28533	14284	-2783	-0.40 X
4	1.00	46550	43767	14284	2783	0.40 X
7	1.00	52450	85098	6893	-32648	-2.28R
22	1.00	127050	97533	8221	29517	2.17R

R denotes an observation with a large standardized residual
X denotes an observation whose X value gives it large influence.

Regression Analysis: PRICE versus BEDC, CARA, DOM

The regression equation is
PRICE = 66338 + 30129 BEDC - 38457 CARA - 60.4 DOM

Predictor	Coef	SE Coef	T	P
Constant	66338	6433	10.31	0.000
BEDC	30129	7913	3.81	0.001
CARA	-38457	12329	-3.12	0.006
DOM	-60.41	32.62	-1.85	0.081

```
S = 16298        R-Sq = 56.0%      R-Sq(adj) = 48.7%
```

Analysis of Variance

Source	DF	SS	MS	F	P
Regression	3	6086432104	2028810701	7.64	0.002
Residual Error	18	4781157750	265619875		
Total	21	10867589854			

Source	DF	Seq SS
BEDC	1	2901187555
CARA	1	2274636373
DOM	1	910608176

```
Unusual Observations
Obs       BEDC        PRICE        Fit     SE Fit     Residual    St Resid
  1       0.00        25750       17975      12322         7775       0.73 X
  4       1.00        46550       54325      12322        -7775      -0.73 X
  7       1.00        52450       86682       6960       -34232      -2.32R
 22       1.00       127050       94293       8065        32757       2.31R
```

R denotes an observation with a large standardized residual
X denotes an observation whose X value gives it large influence.

Regression Analysis: PRICE versus BEDC, CARA

The regression equation is
PRICE = 57231 + 29518 BEDC - 35840 CARA

```
Predictor        Coef     SE Coef         T        P
Constant        57231        4403     13.00    0.000
BEDC            29518        8396      3.52    0.002
CARA           -35840       13006     -2.76    0.013
```

S = 17308   R-Sq = 47.6%   R-Sq(adj) = 42.1%

Analysis of Variance

```
Source          DF          SS          MS        F        P
Regression       2  5175823928  2587911964     8.64    0.002
Residual Error  19  5691765926   299566628
Total           21 10867589854
```

```
Source        DF       Seq SS
BEDC           1   2901187555
CARA           1   2274636373
```

Unusual Observations
```
Obs       BEDC        PRICE        Fit     SE Fit     Residual    St Resid
  1       0.00        25750       21391      12939         4359       0.38 X
  4       1.00        46550       50909      12939        -4359      -0.38 X
  7       1.00        52450       86749       7391       -34299      -2.19R
 22       1.00       127050       86749       7391        40301       2.58R
```

R denotes an observation with a large standardized residual
X denotes an observation whose X value gives it large influence.

Regression Analysis: PRICE versus BEDC

The regression equation is
PRICE = 54991 + 25785 BEDC

```
Predictor        Coef     SE Coef         T        P
Constant        54991        4989     11.02    0.000
BEDC            25785        9554      2.70    0.014
```

S = 19958   R-Sq = 26.7%   R-Sq(adj) = 23.0%

```
Analysis of Variance

Source            DF           SS           MS         F        P
Regression         1     2901187555   2901187555      7.28    0.014
Residual Error    20     7966402299    398320115
Total             21    10867589854

Unusual Observations
Obs       BEDC        PRICE        Fit     SE Fit      Residual      St Resid
 22       1.00       127050      80776       8148         46274         2.54R

R denotes an observation with a large standardized residual
```

13.53   Refer to Exercise 13.52. Conduct a test using the full regression model to determine whether the depreciation (decrease) in house price per year of age is less than $2,500. Give the null hypothesis for your test and the *p*-value. Draw a conclusion. Use $\alpha = .05$.

13.54   Refer to Exercise 13.52. Suppose that we wished to modify our nine-variable model to allow for the possibility that the relationship between "price" and "age" differs depending on the number of bedrooms.
   a. Formulate such a model.
   b. What combination of model parameters represents the difference between a five-bedroom, one-garage home and a two-bedroom, two-garage home?

13.55   Refer to Exercise 13.52. What is your choice of a "best" model from the original set of nine variables? Why did you choose this model?

13.56   Refer to Exercise 13.52. In another study involving the same 22 properties, PRICE was regressed on a single independent variable, LIST, which was the listing price of the property in thousands of dollars.

```
Best Subsets Regression: PRICE versus BATHS, BEDA, BEDB, BEDC,
                                      CARA, CARB, AGE, LOT, DOM

Response is PRICE

                                          B
                                        A B B B C C
                                        T E E E A A A L D
                                        H D D D R R G O O
Vars   R-Sq   R-Sq(adj)    C-p     S    S A B C A B E T M

  1    30.8      27.4       9.5   19385  X
  1    26.7      23.0      11.2   19958      X
  2    47.6      42.1       4.8   17308      X X
  2    39.4      33.1       8.1   18612      X       X
  3    56.0      48.7       3.5   16298      X X           X
  3    51.0      42.8       5.5   17200  X   X X
  4    60.4      51.1       3.8   15913      X X       X X
  4    60.2      50.8       3.8   15950  X   X X         X
  5    65.7      54.9       3.7   15270      X X X     X X
  5    65.2      54.3       3.9   15382  X   X X       X X
  6    67.7      54.8       4.8   15296  X   X X X     X X
  6    66.5      53.1       5.3   15576      X X X X X X
  7    69.3      53.9       6.2   15443  X X X X X     X X
  7    68.6      52.9       6.5   15611  X   X X X X X X
```

```
                                             B
                                             A B B B C C
                                             T E E E A A A L D
                                             H D D D R R G O O
Vars    R-Sq    R-Sq(adj)    C-p        S    S A B C A B E T M

  8     69.8        51.2     8.0    15896    X X   X X X X X X
  8     69.3        50.4     8.2    16019    X X X X X X   X X
  9     69.8        47.2    10.0    16531    X X X X X X X X X
```

Data Display

```
Row      PRICE       LIST
  1      25750      29900
  2      37950      39900
  3      46450      44900
  4      46550      47500
  5      47950      49900
  6      49950      49900
  7      52450      53000
  8      54050      54900
  9      54850      54900
 10      52050      55900
 11      54392      55900
 12      53450      56000
 13      59510      62000
 14      60102      62500
 15      63850      63900
 16      62050      66900
 17      69450      72500
 18      82304      82254
 19      81850      82900
 20      70050      99900
 21     112450     117000
 22     127050     139000
```

Descriptive Statistics: PRICE, LIST

Variable	N	Mean	Median	TrMean	StDev	SE Mean
PRICE	22	62023	54621	60585	22749	4850
LIST	22	65521	55950	63628	25551	5447

Variable	Minimum	Maximum	Q1	Q3
PRICE	25750	127050	49450	69600
LIST	29900	139000	49900	74939

Regression Analysis: PRICE versus LIST

The regression equation is
PRICE = 5406 + 0.864 LIST

Predictor	Coef	SE Coef	T	P
Constant	5406	3363	1.61	0.124
LIST	0.86411	0.04797	18.01	0.000

S = 5616   R-Sq = 94.2%   R-Sq(adj) = 93.9%

```
Analysis of Variance

Source              DF           SS            MS            F          P
Regression           1   10236690015   10236690015      324.51      0.000
Residual Error      20     630899838      31544992
Total               21   10867589854


Unusual Observations
Obs       LIST        PRICE          Fit       SE Fit       Residual       St Resid
 20      99900        70050        91731         2038         -21681         -4.14R
 22     139000       127050       125518         3723           1532          0.36 X


R denotes an observation with a large standardized residual
X denotes an observation whose X value gives it large influence.
```

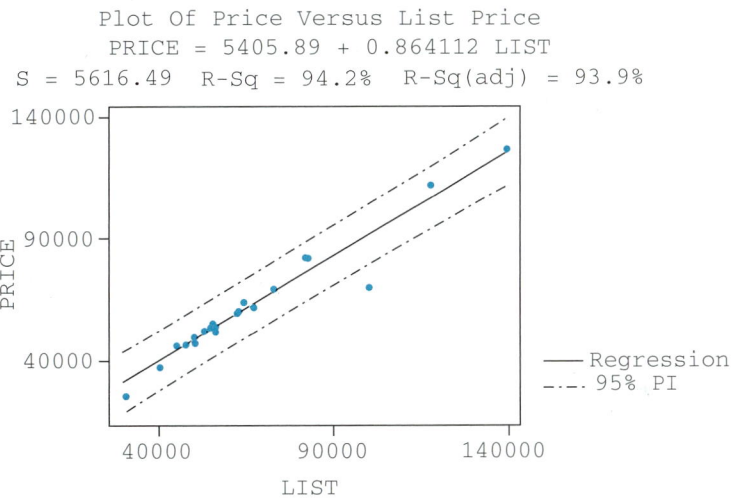

Plot Of Price Versus List Price
PRICE = 5405.89 + 0.864112 LIST
S = 5616.49   R-Sq = 94.2%   R-Sq(adj) = 93.9%

a. Using the regression results, predict the selling price of a home that is listed at $70,000.

b. What is the chance that your prediction is off by more than $3,000?

Soc.   **13.57** Using the selling price data of Exercise 13.52, examine the relationship between the selling price (in thousands of dollars) of a home and two independent variables, the number of rooms and the number of square feet. Use the following data.

Row	Price	Rooms	Square Feet
1	25.75	5	986
2	37.95	5	998
3	46.45	7	1,690
4	46.55	8	1,829
5	47.95	6	1,186
6	49.95	6	1,734
7	52.45	7	1,684
8	54.05	7	1,846
9	54.85	7	1,690

(*continues*)

(continued)

Row	Price	Rooms	Square Feet
10	52.05	7	1,910
11	54.39	7	1,784
12	53.45	6	1,690
13	59.51	7	1,590
14	60.10	8	1,855
15	63.85	8	2,212
16	62.05	10	2,784
17	69.45	7	2,190
18	82.30	8	2,259
19	81.85	7	1,919
20	70.05	7	1,685
21	112.45	10	2,654
22	127.05	10	2,756

Use the computer output shown here to address parts (a), (b), and (c).

```
MULTIPLE REGRESSION ANALYSIS

Dependent Variable: PRICE        SELLING PRICE (1000$)

Analysis of Variance

                          Sum of          Mean
Source        DF         Squares         Square       F Value      Prob>F

Model          2       6816.77693      3408.38847      15.987       0.0001
Error         19       4050.68890       213.19415
C Total       21      10867.46584


     Root MSE        14.60117     R-square        0.6273
     Dep Mean        62.02273     Adj R-sq        0.5880
     C.V.            23.54164

Parameter Estimates

                  Parameter        Standard      T for H0:
Variable  DF      Estimate          Error      Parameter=0     Prob > |T|

INTERCEP   1     -16.975979      18.94658431      -0.896         0.3815
ROOMS      1       4.336062       6.04912439       0.717         0.4822
SQFT       1       0.025511       0.01737891       1.468         0.1585


                  Variable
Variable  DF       Label

INTERCEP   1    Intercept
ROOMS      1    NUMBER OF ROOMS
SQFT       1    SQUARE FEET
```

a. Conduct a test to see whether the two variables, ROOMS and SQUARE FEET, taken together, contain information about PRICE. Use $\alpha = .05$.

b. Conduct a test to see whether the coefficient of ROOMS is equal to 0. Use $\alpha = .05$.

c. Conduct a test to see whether the coefficient of SQUARE FEET is equal to 0. Use $\alpha = .05$.

13.58 Refer to Exercise 13.57.

a. Explain the apparent inconsistency between the result of part (a) and the results of parts (b) and (c).

b. What do you think would happen to the t-value of SQUARE FEET if ROOMS were dropped from the model?

Med. **13.59** A study was conducted to determine whether infection surveillance and control programs have reduced the rates of hospital-acquired infection in U.S. hospitals. This data set consists of a random sample of 28 hospitals selected from 338 hospitals participating in a larger study. Each line of the data set provides information on variables for a single hospital. The variables are as follows:

RISK = output variable, average estimated probability of acquiring infection in hospital (in percent)

STAY = input variable, average length of stay of all patients in hospital (in days)

AGE = input variable, average age of patients (in years)

RCR = input variable, ratio of number of cultures performed to number of patients without signs or symptoms of hospital-acquired infection (times 100)

SCHOOL = dummy input variable for medical school affiliation, 1 = yes, 0 = no

DV_1 = dummy input variable for region of country, 1 = northeast, 0 = other

DV_2 = dummy input variable for region of country, 1 north central, 0 = other

DV_3 = dummy input variable for region of country, 1 = south, 0 = other

(Note that there are four geographic regions of the country—northeast, north central, south, and west. These four regions of the country require only three dummy variables to code for them.) The data were analyzed using SAS with the following results.

```
DATA LISTING
```

OBS	RISK	STAY	AGE	INS	SCHOOL	RC1	RC2	RC3
1	4.1	7.13	55.7	9.0	0	0	0	1
2	1.6	8.82	58.2	3.8	0	1	0	0
3	2.7	8.34	56.9	8.1	0	0	1	0
4	5.6	8.95	53.7	18.9	0	0	0	1
5	5.7	11.20	56.5	34.5	0	0	0	0
6	5.1	9.76	50.9	21.9	0	1	0	0
7	4.6	9.68	57.8	16.7	0	0	1	0
8	5.4	11.18	45.7	60.5	1	1	0	0
9	4.3	8.67	48.2	24.4	0	0	1	0

OBS	RISK	STAY	AGE	INS	SCHOOL	RC1	RC2	RC3
10	6.3	8.84	56.3	29.6	0	0	0	0
11	4.9	11.07	53.2	28.5	1	0	0	0
12	4.3	8.30	57.2	6.8	0	0	1	0
13	7.7	12.78	56.8	46.0	1	0	0	0
14	3.7	7.58	56.7	20.8	0	1	0	0
15	4.2	9.00	56.3	14.6	0	0	1	0
16	5.6	10.12	51.7	14.9	1	0	1	0
17	5.5	8.37	50.7	15.1	0	1	0	0
18	4.6	10.16	54.2	8.4	1	0	0	1
19	6.5	19.56	59.9	17.2	0	0	0	0
20	5.5	10.90	57.2	10.6	0	1	0	0
21	1.8	7.67	51.7	2.5	0	0	1	0
22	4.2	8.88	51.5	10.1	0	0	1	0
23	5.6	11.48	57.6	20.3	0	0	0	0
24	4.3	9.23	51.6	11.6	0	1	0	0
25	7.6	11.41	61.1	16.6	0	0	0	0
26	7.8	12.07	43.7	52.4	0	1	0	0
27	3.1	8.63	54.0	8.4	0	0	0	0
28	3.9	11.15	56.5	7.7	0	0	0	0

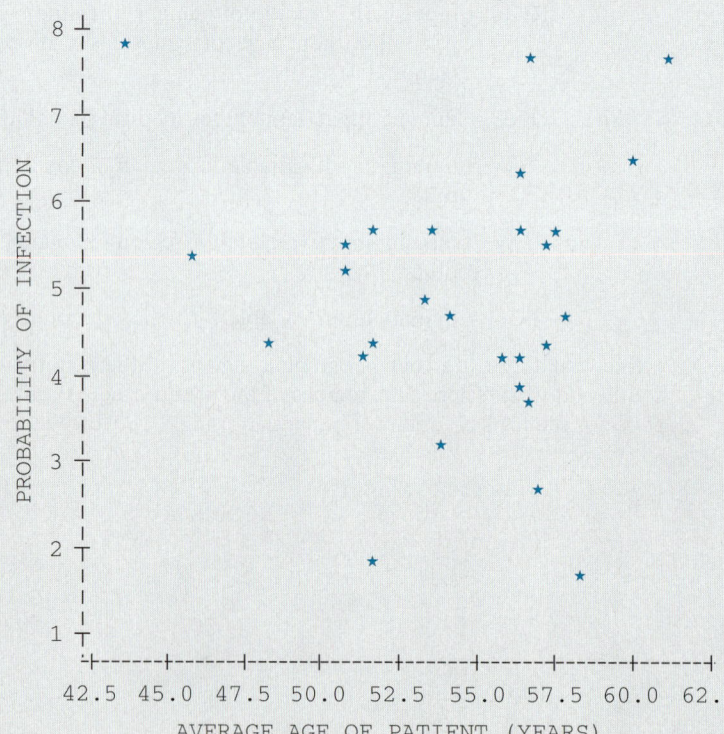

Plot of Risk versus Average Age of Patient

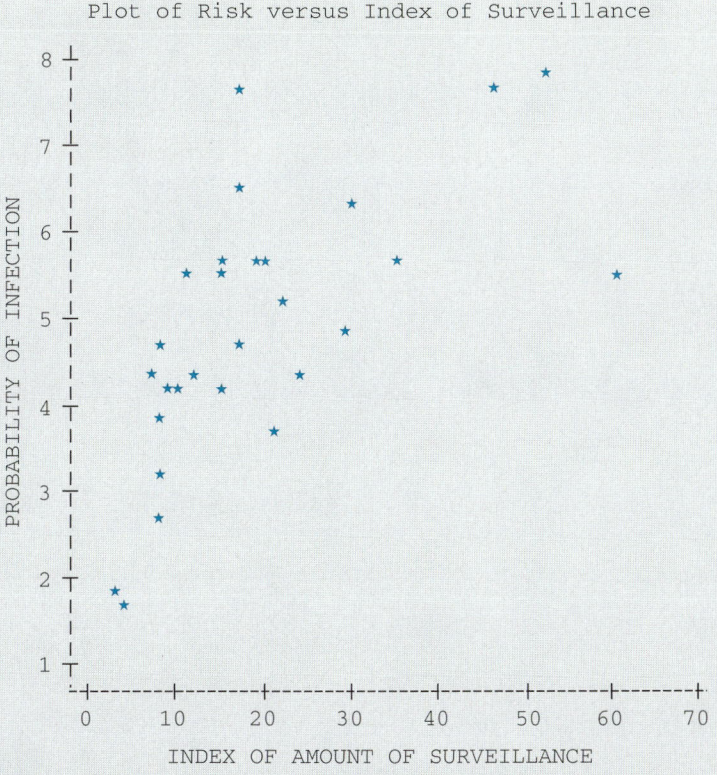

Plot of Risk versus Index of Surveillance

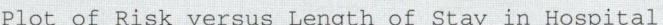

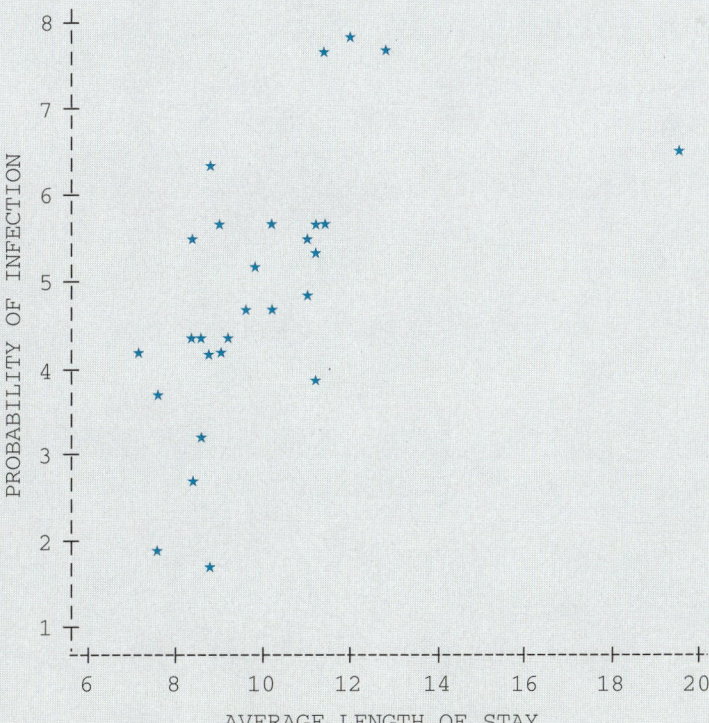

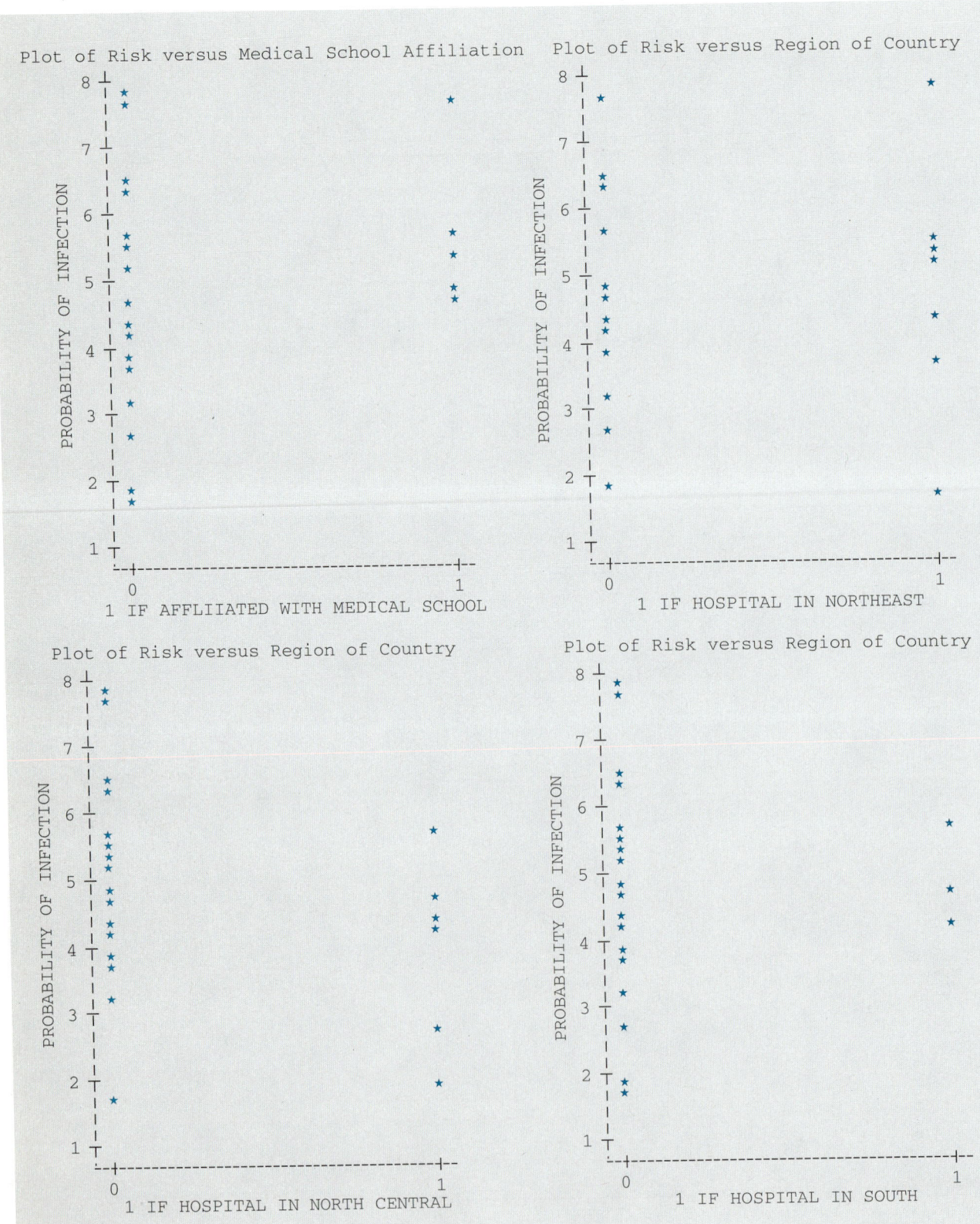

```
Correlation Analysis of the Independent Variables:

  7 'VAR' Variables:    STAY      AGE       INS       SCHOOL      RC1       RC2
                        RC3


                          Simple Statistics


Variable        N        Mean     Std Dev        Sum    Minimum     Maximum


STAY           28      10.0332     2.3729    280.9300     7.1300     19.5600
AGE            28      54.3393     4.0802        1522    43.7000     61.1000
INS            28      19.2821    14.3288    539.9000     2.5000     60.5000
SCHOOL         28       0.1786     0.3900      5.0000          0      1.0000
RC1            28       0.2857     0.4600      8.0000          0      1.0000
RC2            28       0.2857     0.4600      8.0000          0      1.0000
RC3            28       0.1071     0.3150      3.0000          0      1.0000


Pearson Correlation Coefficients / Prob > |R| under Ho: Rho=0 / N = 28


              STAY        AGE        INS      SCHOOL        RC1        RC2        RC3


STA        1.00000    0.18019    0.35014    0.20586   -0.07993   -0.32591   -0.19127
              0.0       0.3589     0.0678     0.2933     0.6860     0.0906     0.3296


AGE        0.18019    1.00000   -0.47243    0.23498   -0.39490   -0.06737    0.01678
            0.3589       0.0       0.0111     0.2287     0.0375     0.7334     0.9325


INS        0.35014   -0.47243    1.00000    0.41016    0.23847   -0.31552   -0.17682
            0.0678     0.0111       0.0       0.0302     0.2217     0.1019     0.3681


SCHOOL     0.20586   -0.23498    0.41016    1.00000   -0.08847   -0.08847    0.13998
            0.2933     0.2287     0.0302       0.0       0.6544     0.6544     0.4774


RC1       -0.07993   -0.39490    0.23847   -0.08847    1.00000   -0.40000   -0.21909
            0.6860     0.0375     0.2217     0.6544       0.0       0.0349     0.2627


RC2       -0.32591   -0.06737   -0.31552   -0.08847   -0.40000    1.00000   -0.21909
            0.0906     0.7334     0.1019     0.6544     0.0349       0.0       0.2627


RC3       -0.19127    0.01678   -0.17682    0.13998   -0.21909   -0.21909    1.00000
            0.3296     0.9325     0.3681     0.4774     0.2627     0.2627       0.0


Backward Elimination Procedure for Dependent Variable RISK

Step 0    All Variables Entered    R-square = 0.60724861    C(p) = 8.00000000


                  DF          Sum of Squares      Mean Square        F     Prob>F


Regression         7            39.49805177       5.64257882       4.42    0.0041
Error             20            25.54623394       1.27731170
Total             27            65.04428571
```

Variable	Parameter Estimate	Standard Error	Type II Sum of Squares	F	Prob>F
INTERCEP	-1.07800774	4.69134824	0.06744431	0.05	0.8206
STAY	0.23613428	0.11569116	5.32126218	4.17	0.0547
AGE	0.04359681	0.07810854	0.39793239	0.31	0.5829
INS	0.06923673	0.02278287	11.79650358	9.24	0.0065
SCHOOL	-0.41516871	0.64822732	0.52395194	0.41	0.5291
RC1	-0.26955673	0.68941266	0.19527144	0.15	0.6999
RC2	-0.19268071	0.71943459	0.09162010	0.07	0.7916
RC3	0.70243224	0.88962481	0.79632801	0.62	0.4390

Bounds on condition number: 2.315515,   94.11721

Step 1   Variable RC2 Removed   R-square = 0.60584002   C(p) = 6.07172885

	DF	Sum of Squares	Mean Square	F	Prob>F
Regression	6	39.40643167	6.56773861	5.38	0.0017
Error	21	25.63785404	1.22085019		
Total	27	65.04428571			

Variable	Parameter Estimate	Standard Error	Type II Sum of Squares	F	Prob>F
INTERCEP	-1.81224950	3.72184905	0.28945486	0.24	0.6314
STAY	0.24597088	0.10725430	6.42096620	5.26	0.0322
AGE	0.05262498	0.06888511	0.71251762	0.58	0.4534
INS	0.07154787	0.02061408	14.70713325	12.05	0.0023
SCHOOL	-0.42280540	0.63312506	0.54445805	0.45	0.5115
RC1	-0.15497958	0.52853481	0.10496975	0.09	0.7722
RC3	0.83288104	0.72780215	1.59882767	1.31	0.2653

Bounds on condition number: 1.929521,   53.56369

QUADRATIC REGRESSION OF WARPING ON TEMPERATURE   173

Step 2   Variable RC1 Removed   R-square = 0.60422621   C(p) = 4.15390906

	DF	Sum of Squares	Mean Square	F	Prob>F
Regression	5	39.30146193	7.86029239	6.72	0.0006
Error	22	25.74282379	1.17012835		
Total	27	65.04428571			

Variable	Parameter Estimate	Standard Error	Type II Sum of Squares	F	Prob>F
INTERCEP	-2.21637907	3.38468174	0.50174830	0.43	0.5194
STAY	0.24760767	0.10486035	6.52437780	5.58	0.0275
AGE	0.05898907	0.06400415	0.99394033	0.85	0.3667
INS	0.07087867	0.02005725	14.61240661	12.49	0.0019
SCHOOL	-0.38736862	0.60843670	0.47429829	0.41	0.5309
RC3	0.87192445	0.70049715	1.81291925	1.55	0.2263

Bounds on condition number: 1.905871,   36.65382

Step 3   Variable SCHOOL Removed   R-square = 0.59693428   C(p) = 2.52523447

	DF	Sum of Squares	Mean Square	F	Prob>F
Regression	4	38.82716364	9.70679091	8.52	0.0002
Error	23	26.21712207	1.13987487		
Total	27	65.04428571			

Variable	Parameter Estimate	Standard Error	Type II Sum of Squares	F	Prob>F
INTERCEP	-2.30479519	3.33782686	0.54349337	0.48	0.4968
STAY	0.23848508	0.10252510	6.16764346	5.41	0.0292
AGE	0.06257589	0.06292612	1.12722159	0.99	0.3304
INS	0.06713326	0.01892561	14.34276871	12.58	0.0017
RC3	0.76072793	0.66954727	1.47147677	1.29	0.2676

Bounds on condition number: 1.741914,   23.03492

Step 4   Variable AGE Removed   R-square = 0.57960421   C(p) = 1.40772979

	DF	Sum of Squares	Mean Square	F	Prob>F
Regression	3	37.69994205	12.56664735	11.03	0.0001
Error	24	27.34434367	1.13934765		
Total	27	65.04428571			

Variable	Parameter Estimate	Standard Error	Type II Sum of Squares	F	Prob>F
INTERCEP	0.88480344	0.92355510	1.04574126	0.92	0.3476

QUADRATIC REGRESSION OF WARPING ON TEMPERATURE   174

Variable	Parameter Estimate	Standard Error	Type II Sum of Squares	F	Prob>F
STAY	0.28060533	0.09334523	10.29588785	9.04	0.0061
INS	0.05622030	0.01541554	15.15391450	13.30	0.0013
RC3	0.74723631	0.66925498	1.42032908	1.25	0.2753

Bounds on condition number: 1.162616,   10.11556

Step 5   Variable RC3 Removed   R-square = 0.55776787   C(p) = 0.51969728

	DF	Sum of Squares	Mean Square	F	Prob>F
Regression	2	36.27961297	18.13980648	15.77	0.0001
Error	25	28.76467275	1.15058691		
Total	27	65.04428571			

Variable	Parameter Estimate	Standard Error	Type II Sum of Squares	F	Prob>F
INTERCEP	1.15123509	0.89658440	1.89699030	1.65	0.2109
STAY	0.26598212	0.09287658	9.43651980	8.20	0.0084
INS	0.05416385	0.01538042	14.26927648	12.40	0.0017

Bounds on condition number: 1.139728,   4.558912

All variables left in the model are significant at the 0.1000 level.

Summary of Backward Elimination Procedure for Dependent Variable RISK

Step	Variable Removed Label	Number In	Partial R**2	Model R**2	C(p)	F	Prob>F
1	RC2	6	0.0014	0.6058	6.0717	0.0717	0.7916
	1 IF HOSPITAL IN NORTH CENTRAL						
2	RC1	5	0.0016	0.6042	4.1539	0.0860	0.7722
	1 IF HOSPITAL IN NORTHEAST						
3	SCHOOL	4	0.0073	0.5969	2.5252	0.4053	0.5309
	1 IF AFFLIATED WITH MEDICAL SCHOOL, 0 IF						
4	AGE	3	0.0173	0.5796	1.4077	0.9889	0.3304
	AVERAGE AGE OF PATIENT (YEARS)						
5	RC3	2	0.0218	0.5578	0.5197	1.2466	0.2753
	1 IF HOSPITAL IN SOUTH						

Does the set of seven input variables contain information about the output variable, RISK? Give a p-value for your test.

Based on the full regression model (seven input variables), can we be at least 95% certain that hospitals in the south have at least .5% higher risk of infection than hospitals in the west, all other things being equal?

13.60 Refer to Exercise 13.59.

  a. Consider the following two statements:

  There is multicollinearity between region of the country and whether a hospital has a medical school.

  There is an interaction effect between region of the country and whether a hospital has a medical school.

  What is the difference between these two statements? What evidence is needed to ascertain the truth or falsity of the statements? Is this evidence present in the accompanying output? If it is, do you think the statements are true or false?

  b. Construct a model that allows for the possibility of an interaction effect between region of the country and medical school affiliation. For this model, what is the difference in intercept between a hospital in the northeast affiliated with a medical school and a hospital in the west not affiliated with one?

13.61 Refer to Exercise 13.59. Suppose that we decide to eliminate from the full model some variables that we think contribute little to explaining the output variable. What would your final choice of a model be? Why would you choose this model?

13.62 Refer to Exercise 13.59. Predict the infection risk of a patient in a medical school–affiliated hospital in the northeast, where the average stay of patients is 10 days, the average age is 64, and the routine culturing ratio is 20%. Is this prediction an interpolation or an extrapolation? How do you know?

Bio. **13.63** Thirty volunteers participated in the following experiment. The subjects took their own pulse rates (which is easiest to do by holding the thumb and forefinger of one hand on the pair of arteries on the side of the neck). They were then asked to flip a coin. If their coin came up heads, they ran in place for 1 minute. Then all subjects took their own pulse rates again. The difference in the before and after pulse rates was recorded, as well as other data on student characteristics. A regression was run to "explain" the pulse rate differences using the other variables as independent variables. The variables were:

PULSE = difference between the before and after pulse rates
RUN = dummy variable, 1 = did not run in place, 0 = ran in place
SMOKE = dummy variable, 1 = does not smoke, 0 = smokes
HEIGHT = height in inches
WEIGHT = weight in pounds
PHYS1 = dummy variable, 1 = a lot of physical exercise, 0 = otherwise
PHYS2 = dummy variable, 1 = moderate physical exercise, 0 = otherwise

a. Perform an appropriate test to determine whether the entire set of independent variables explains a significant amount of the variability of "pulse." Draw a conclusion based on $\alpha = .01$.

b. Does multicollinearity seem to be a problem here? What is your evidence? What effect does multicollinearity have on your ability to make predictions using regression?

c. Based on the full regression model (six dependent variables), compute a point estimate of the average increase in "pulse" for individuals who engaged in a lot of physical activity compared to those who engaged in little physical activity. Can we be 95% certain that the actual average increase is greater than 0?

```
LISTING OF DATA FOR EXERCISE 13.63
OBS      PULSE     RUN     SMOKE     HEIGHT     WEIGHT     PHYS1     PHYS2
 1        -29        0        1         66        140         0         1
 2        -17        0        1         72        145         0         1
 3        -14        0        0         73        160         1         0
 4        -22        0        0         73        190         0         0
 5        -21        0        1         69        155         0         1
 6        -25        0        1         73        165         0         0
 7         -5        0        1         72        150         1         0
 8         -9        0        1         74        190         0         1
 9        -18        0        1         72        195         0         1
10        -23        0        1         71        138         0         1
11        -14        0        0         74        160         0         0
12        -21        0        1         72        155         0         1
13          8        0        0         70        153         1         0
14        -13        0        1         67        145         0         1
15        -21        0        1         71        170         1         0
16         -1        0        1         72        175         1         0
17        -16        0        0         69        175         0         1
18        -15        1        1         68        145         0         0
19          4        1        0         75        190         0         1
20         -3        1        1         72        180         1         0
21          2        1        0         67        140         0         1
22         -5        1        1         70        150         0         1
23         -1        1        1         73        155         0         1
24         -5        1        1         74        148         1         0
25         -6        1        0         68        150         0         1
26         -6        1        0         73        155         0         1
27          8        1        0         66        130         0         1
28         -1        1        1         69        160         0         1
29         -5        1        1         66        135         1         0
30         -3        1        1         75        160         1         0

Correlation Analysis

   6 'VAR' Variables: RUN      SMOKE      HEIGHT      WEIGHT      PHYS1      PHYS2
```

```
                         Simple Statistics
Variable        N        Mean    Std Dev         Sum       Minimum    Maximum
RUN            30       0.4333    0.5040     13.0000             0     1.0000
SMOKE          30       0.6667    0.4795     20.0000             0     1.0000
HEIGHT         30      70.8667    2.7759         2126      66.0000    75.0000
WEIGHT         30     158.6333   17.5391         4759     130.0000   195.0000
PHYS1          30       0.3000    0.4661      9.0000             0     1.0000
PHYS2          30       0.5667    0.5040     17.0000             0     1.0000
```

Pearson Correlation Coefficients / Prob > |R| under HO: Rho = 0 / N = 30

```
                 RUN       SMOKE      HEIGHT      WEIGHT        PHYS1       PHYS2
RUN          1.00000    -0.09513    -0.12981    -0.25056      0.01468     0.08597
                 0.0      0.6170      0.4942      0.1817       0.9386      0.6515


SMOKE       -0.09513     1.00000     0.01727    -0.06834      0.15430    -0.04757
              0.6170        0.0       0.9278      0.7197       0.4156      0.8029


HEIGHT      -0.12981     0.01727     1.00000     0.59885      0.19189    -0.28919
              0.4942      0.9278        0.0       0.0005       0.3097      0.1211


WEIGHT      -0.25056    -0.06834     0.59885     1.00000      0.01392    -0.11221
              0.1817      0.7197      0.0005        0.0        0.9418      0.5549


PHYS1        0.01468     0.15430     0.19189     0.01392      1.00000    -0.74863
              0.9386      0.4156      0.3097      0.9418         0.0       0.0001


PHYS2        0.08597    -0.04757    -0.28919    -0.11221     -0.74863     1.00000
              0.6515      0.8029      0.1211      0.5549       0.0001        0.0
```

Backward Elimination Procedure for Dependent Variable PULSE

Step 0   All Variables Entered   R-square = 0.62973045   C(p) = 7.00000000

```
                 DF      Sum of Squares         Mean Square        F    Prob>F
Regression        6       1850.58887109       308.43147852      6.52    0.0004
Error            23       1088.11112891        47.30917952
Total            29       2938.70000000
```

```
               Parameter      Standard            Type II
Variable        Estimate         Error    Sum of Squares        F    Prob>F
INTERCEP     -31.68830679    36.42360015       35.80780871     0.76    0.3933
RUN           11.40166481     2.66171908      868.07553823    18.35    0.0003
SMOKE         -6.89029281     2.74454278      298.18154585     6.30    0.0195
HEIGHT         0.13169561     0.60021947        2.27754970     0.05    0.8283
WEIGHT         0.02303608     0.09440380        2.81697901     0.06    0.8094
PHYS1         13.43465041     4.25117641      472.47616161     9.99    0.0044
PHYS2          7.80635269     3.97815470      182.17065424     3.85    0.0619
```

Bounds on condition number: 2.464274,   62.50691

```
Step 1   Variable HEIGHT Removed   R-square = 0.62895543        C(p) = 5.04814181


                DF      Sum of Squares    Mean Square              F    Prob>F
Regression       5       1848.31132139    369.66226428          8.14    0.0001
Error           24       1090.38867861     45.43286161
Total           29       2938.70000000


             Parameter      Standard        Type II
Variable      Estimate         Error    Sum of Squares          F    Prob>F
INTERCEP   -24.25519127   13.11118684     155.48750216         3.42    0.0767
RUN         11.43076116    2.60516294     874.68284765        19.25    0.0002
SMOKE       -6.85327902    2.68448142     296.10525519         6.52    0.0175
WEIGHT       0.03529782    0.07456145      10.18209732         0.22    0.6402
PHYS1       13.44838310    4.16556957     473.54521380        10.42    0.0036
PHYS2        7.65315557    3.83795325     180.65576063         3.98    0.0576


Bounds on condition number:   2.406131,      40.22006
----------------------------------------------------------------------------
Step 2   Variable WEIGHT Removed   R-square = 0.62549060        C(p) = 3.26336637


                DF      Sum of Squares    Mean Square              F    Prob>F
Regression       4       1838.12922407    459.53230602         10.44    0.0001
Error           25       1100.57077593     44.02283104
Total           29       2938.70000000


             Parameter      Standard        Type II
Variable      Estimate         Error    Sum of Squares          F    Prob>F
INTERCEP   -18.30152045    3.64892257    1107.44716129        25.16    0.0001
RUN         11.13212935    2.48810400     881.24648295        20.02    0.0001
SMOKE       -6.96302377    2.63262467     307.96107626         7.00    0.0139
PHYS1       13.32514812    4.09240540     466.72897076        10.60    0.0032
PHYS2        7.45071026    3.75440264     173.37705597         3.94    0.0583


Bounds on condition number:   2.396734,          27.36375
All variables left in the model are significant at the 0.1000 level.


Summary of Backward Elimination Procedure for Dependent Variable PULSE


        Variable   Number   Partial    Model
Step    Removed      In      R**2      R**2       C(p)          F       Prob>F
        Label


  1     HEIGHT        5      0.0008    0.6290     5.0481       0.0481    0.8283
        HEIGHT (INCHES)
  2     WEIGHT        4      0.0035    0.6255     3.2634       0.2241    0.6402
        WEIGHT (POUNDS)
```

13.64 Refer to Exercise 13.63.
> **a.** Give the implied regression line of pulse-rate difference on height and weight for a smoker who did not run in place and who has engaged in little physical activity.

b. Consider the following two statements:
1. There is multicollinearity between the smoke variable and the physical activity dummy variables.
2. There is an interaction effect between the smoke variable and the physical activity dummy variables.

Is there any difference between these two statements? Explain the relationships that would exist in the data set if each of these two statements were correct.

13.65 Refer to Exercise 13.63.
a. What is your choice of a good predictive equation? Why did you choose that particular equation?
b. The model as constructed does not contain any interaction effects. Construct a model that allows for the possibility of an interaction effect between each pair of qualitative variables.

Chem.

13.66 The data for this exercise were taken from a chemical assay of calcium discussed in Brown, Healy, and Kearns (1981). A set of standard solutions is prepared and these and the unknowns are read on a spectrophotometer in arbitrary units (y). A linear regression model is fit to the standards, and the values of the unknowns (x) are read off from this. The preparation of the standard and unknown solutions involves a fair amount of laboratory manipulation, and the actual concentrations of the standards may differ slightly from their target values, the very precise instrumentation being capable of detecting this. The target values are 2.0, 2.0, 2.5, 3.0, 3.0 mmol per liter; the "duplicates" are made up independently. The sequence of reading the standards and unknowns is repeated four times. Two specimens of each unknown are included in each assay and the four sequences of readings are done twice, first with the flame conditions in the instrument optimized, and then with a slightly weaker flame. y is spectrophotometer reading and x is actual mmol per liter.

The data in the following table relate to assays on the above pattern of a set of six unknowns performed by four laboratories. The standards are identified as 2.0A, 2.0B, 2.5, 3.0A, 3.0B; the unknowns are identified as U1, U2, W1, W2, Y1, Y2.

Laboratory/Solution	Measurements				Laboratory/Solution	Measurements			
1 W1	1206	1202	1202	1201	3 W1	1090	1098	1090	1100
1 2.0A	1068	1071	1067	1066	3 2.0A	969	975	969	972
1 W2	1194	1193	1189	1185	3 U2	1088	1092	1087	1085
1 2.0B	1072	1068	1064	1067	3 2.0B	969	960	960	966
1 U1	1387	1387	1384	1380	3 U1	1270	1261	1261	1269
1 2.5	1333	1321	1326	1317	3 2.5	1196	1196	1209	1200
1 U2	1394	1390	1383	1376	3 W2	1261	1268	1270	1273
1 3.0A	1579	1576	1578	1572	3 3.0A	1451	1440	1439	1449
1 Y1	1478	1480	1473	1466	3 Y1	1352	1349	1353	1343
1 3.0B	1579	1571	1579	1567	3 3.0B	1439	1433	1433	1445
1 Y2	1483	1477	1482	1472	3 Y2	1349	1353	1349	1355
2 W1	1017	1017	1012	1020	4 2.0A	1122	1117	1119	1120
2 2.0A	910	916	915	915	4 W2	1256	1254	1256	1263
2 W2	1012	1018	1015	1023	4 W1	1260	1251	1252	1264
2 2.0B	913	923	914	921	4 2.0B	1122	1110	1111	1116
2 U1	1188	1199	1197	1202	4 U2	1453	1447	1451	1455
2 2.5	1129	1148	1136	1147	4 2.5	1386	1381	1381	1387
2 U2	1186	1196	1193	1199	4 U1	1450	1446	1448	1457
2 3.0A	1359	1378	1370	1373	4 3.0A	1656	1663	1659	1665
2 Y1	1263	1280	1280	1279	4 Y2	1543	1548	1543	1545
2 3.0B	1349	1361	1359	1363	4 3.0B	1658	1658	1661	1660
2 Y2	1259	1269	1259	1265	4 Y1	1545	1546	1548	1544

a. Plot y versus x for the standards, one graph for each laboratory.

b. Fit the linear regression equation $y = \beta_0 + \beta_1 x + \varepsilon$ for each laboratory and predict the value of x corresponding to the y for each of the unknowns. Compute the standard deviation of predicted values of x based on the four predicted x-values for each of the unknowns.

c. Which laboratory appears to make better predictions of x, mmol of calcium per liter? Why?

13.67   Refer to Exercise 13.66. Suppose you average the y-values for each of the unknowns and fit the ys in the linear regression model of Exercise 13.66.

a. Do your linear regression lines change for each of the laboratories?

b. Will predictions of x change based on these new regression lines for the four laboratories? Explain.

13.68   Refer to Exercise 13.66. Using the independent variable x, suggest a single general linear model that could be used to fit the data from all four laboratories. Identify the parameters in this general linear model.

13.69   Refer to Exercise 13.68.

a. Fit the data to the model of Exercise 13.68.

b. Give separate regression models for each of the laboratories.

c. How do these regression models compare to the previous regression equations for the laboratories?

d. What advantage(s) might there be to fitting a single model rather than separate models for the laboratories?

13.70   Refer to the case study introduced at the beginning of this chapter. Use the computer output given here to answer the marketing managers' questions. In particular, determine whether the variables suspected by the managers as having an effect on profit in fact do have an effect. Consider the individual variables and possible functions of these variables. For example, does the relationship between profit and the number of outlets differs for the two types of commissions? Omit any variables that show little predictive value. Examine the conditions required by the regression models and determine whether any of these conditions have been violated. If there are any violations, make the necessary corrections so that the results of the regression analysis are valid. Write a nontechnical explanation of what your analysis reveals. Include in the report your findings relevant to the market managers' questions.

```
Data Display


Row    DIST    PROFIT    AREA     POPN    OUTLETS    COMMIS    AREA_2     POPN_2
  1      1      1011     16.96    3.881     213         1       287.64    15.0622
  2      2      1318      7.31    3.141     158         1        53.44     9.8659
  3      3      1556      7.81    3.766     203         1        61.00    14.1828
  4      4      1521      7.31    4.587     170         1        53.44    21.0406
  5      5       979     19.84    3.648     142         1       393.63    13.3079
  6      6      1290     12.37    3.456     159         1       153.02    11.9439
  7      7      1596      6.15    3.695     178         1        37.82    13.6530
  8      8      1155     14.21    3.609     182         1       201.92    13.0249
  9      9      1412      7.45    3.801     181         1        55.50    14.4476
 10     10      1194     14.43    3.322     148         1       208.22    11.0357
 11     11      1054      6.12    5.124     227         0        37.45    26.2554
 12     12      1157     11.71    4.158     139         1       137.12    17.2890
 13     13      1001      9.36    3.887     179         0        87.61    15.1088
 14     14       831     19.41    2.230     124         1       376.75     4.9729
 15     15       857     11.75    4.468     205         0       138.06    19.9630
```

Row	DIST	PROFIT	AREA	POPN	OUTLETS	COMMIS	AREA_2	POPN_2
16	16	188	40.34	0.297	85	1	1627.32	0.0882
17	17	1030	7.16	4.224	211	0	51.27	17.8422
18	18	1331	9.37	3.427	145	1	87.80	11.7443
19	19	643	7.62	4.031	205	1	58.06	16.2490
20	20	992	27.54	2.370	166	1	758.45	5.6169
21	21	795	15.97	3.903	149	1	255.04	15.2334
22	22	1340	12.97	3.423	186	1	168.22	11.7169
23	23	689	17.36	2.390	141	0	301.37	5.7121
24	24	1726	6.24	4.947	223	1	38.94	24.4728
25	25	1056	11.20	4.166	176	0	125.44	17.3556
26	26	989	18.09	4.063	187	1	327.25	16.5080
27	27	895	13.32	3.105	131	1	177.42	9.6410
28	28	1028	14.97	4.116	170	0	224.10	16.9415
29	29	771	21.92	1.510	144	1	480.49	2.2801
30	30	484	34.91	0.741	126	1	1218.71	0.5491
31	31	917	8.46	5.260	234	0	71.57	27.6676
32	32	1786	7.52	5.744	210	0	56.55	32.9935
33	33	1063	14.43	2.703	141	1	208.22	7.3062
34	34	1001	15.37	3.583	158	0	236.24	12.8379
35	35	1052	11.20	4.469	167	1	125.44	19.9720
36	36	1610	7.20	4.951	174	1	51.84	24.5124
37	37	1486	13.49	3.474	211	1	181.98	12.0687
38	38	1576	6.56	4.637	172	1	43.03	21.5018
39	39	1665	9.35	3.900	185	1	87.42	15.2100
40	40	878	11.12	3.766	166	0	123.65	14.1828
41	41	849	10.58	3.876	189	0	111.94	15.0234
42	42	775	17.82	2.753	164	0	317.55	7.5790
43	43	1012	10.03	4.449	193	0	100.60	19.7936
44	44	1436	10.01	4.680	157	1	100.20	21.9024

Scatterplot Matrix for Case Study

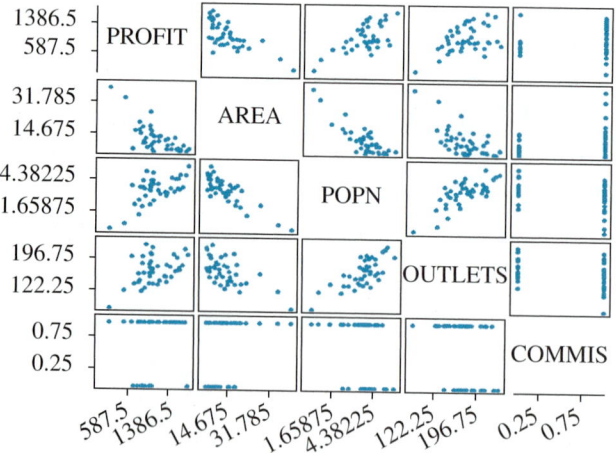

CORRELATION ANALYSIS FOR CASE STUDY
    5 'VAR' VARIABLES: PROFIT   AREA   POPN   OUTLETS   COMMIS

```
                        SIMPLE STATISTICS

VARIABLE          N        MEAN      STD DEV         SUM       MINIMUM      MAXIMUM
PROFIT           44        1114     346.6525       48995     188.0000         1786
AREA             44     13.2798       7.2163      584.3100      6.1200      40.3400
POPN             44      3.6757       1.0910      161.7310      0.2970       5.7440
OUTLETS          44    172.1364      31.0795        7574      85.0000     234.0000
COMMIS           44      0.6818       0.4712      30.0000           0       1.0000
```

PEARSON CORRELATION COEFFICIENTS / PROB > |R| UNDER HO: RHO = 0 / N = 44

```
                 PROFIT         AREA          POPN        OUTLETS        COMMIS
PROFIT          1.00000      -0.68852       0.60822       0.43273       0.23584
                    0.0        0.0001        0.0001        0.0033        0.1233


AREA           -0.68852       1.00000      -0.83383      -0.63678       0.18534
                 0.0001           0.0        0.0001        0.0001        0.2284


POPN            0.60822      -0.83383       1.00000       0.73194      -0.28712
                 0.0001        0.0001           0.0        0.0001        0.0588


OUTLETS         0.43273      -0.63678       0.73194       1.00000      -0.33842
                 0.0033        0.0001        0.0001           0.0        0.0246


COMMIS          0.23584       0.18534      -0.28712      -0.33842       1.00000
                 0.1233        0.2284        0.0588        0.0246           0.0
```

Best Subsets Regression: PROFIT versus AREA, POPN, . . .
Response is PROFIT

```
                                                    C C C
                                      O             O O O
                                      U C A P M M M
                                      T O R O M M M
                                      A P L M E P _ _ _
                                      R O E M A N A P D
                                      E P T I _ _ R O I
Vars   R-Sq   R-Sq(adj)    C-p     S  A N S S 2 2 E P S

  1    47.4      46.2     23.4  254.37  X
  1    40.9      39.4     31.3  269.75          X
  2    62.8      61.0      6.9  216.62  X                   X
  2    61.4      59.5      8.5  220.52  X               X
  3    65.7      63.1      5.4  210.61  X           X       X
  3    65.0      62.4      6.2  212.67      X       X X
  4    66.5      63.1      6.4  210.60      X       X X X
  4    66.1      62.7      6.8  211.79  X X         X       X
  5    70.9      67.1      3.0  198.78      X       X X X X
  5    70.0      66.1      4.1  201.86  X X         X X X
  6    71.5      66.8      4.4  199.64  X X         X X X X
  6    71.3      66.6      4.6  200.21      X   X X X X X
  7    71.7      66.2      6.1  201.58  X X X       X X X X
  7    71.5      66.0      6.3  202.14    X X X     X X X X
  8    71.8      65.3      8.0  204.19  X X X X     X X X X
  8    71.8      65.3      8.1  204.22  X X X   X X X X X X
  9    71.8      64.3     10.1  207.05  X X X X X X X X X
```

Regression Analysis: PROFIT versus AREA, POPN, . . .

CHAPTER 14

Design Concepts for Experiments and Studies

14.1 Introduction

14.2 Types of Studies

14.3 Designed Experiments: Terminology

14.4 Controlling Experimental Error

14.5 Randomization of Treatments to Experimental Units

14.6 Determining the Number of Replications

14.7 Summary

14.1 Introduction

The design of an experiment is the process of establishing a framework through which the comparison of treatments or groups can be made in terms of a recorded response. A scientific study may be conducted in many different ways. In some studies the researcher is interested in collecting information from an undisturbed natural process or setting—for example, a study of the differences in reading scores of second-grade students in public, religious, and private schools. In other studies, the scientist is working within a highly controlled laboratory, which may be a completely artificial setting for the study. The study of the affect of humidity and temperature on the length of life cycles of ticks would be conducted in a laboratory because it would be impossible to control the humidity or temperature in the tick's natural environment. This control of the factors under study allows the entomologist to obtain results that can then be more easily attributed to differences in the levels of the temperature and humidity because nearly all other conditions remain constant throughout the experiment. In a natural setting, however, many other factors vary that may also result in changes in the life cycles of the ticks. Thus, the greater the control is in these artificial settings, the less likely the experiment is to portray the true state of nature. A careful balance between control of conditions and depiction of reality must be maintained for the experiments to be useful. In this chapter, we will present some standard designs of experiments and methods for analyzing the data obtained from the experiment.

Section 14.2 describes several types of studies and the terminology associated with a designed experiment. In Section 14.3, we will discuss the important elements in a study or experiment that need to be considered prior to actually running the experiment. In Sections 14.4 and 14.5, we will discuss the role randomization plays in a designed experiment with respect to both controlling experimental error and providing a valid setting for conducting inferences. Section 14.6 will provide techniques for determining the number of replications needed to obtain proper inferences from the data to the populations and treatments under consideration.

14.2   Types of Studies

In Chapter 2, we discussed the differences between observational studies and designed experiments. Observational studies in the form of polls, surveys, epidemiological studies, and so on are used in many different settings to address questions of interest. Surveys are used to measure the changing opinion of the nation with respect to issues such as gun control, interest rates, taxes, welfare, Social Security, and the national debt. Similarly, we are informed on a daily basis through newspapers, magazines, television, radio, and the Internet of the results of public opinion polls concerning other relevant (and sometimes irrelevant) political, social, educational, and health issues.

In observational studies, the factors (or treatments) of interest are not manipulated while making the measurements. For example, a public health question of considerable interest is the relationship between heart disease and the amount of fat in one's diet. It would be unethical and far too time-consuming and costly to randomly assign volunteers to one of several fat-content diets and then monitor the people over time to observe whether heart disease develops. Without being able to manipulate the factor of interest (fat content of the diet), the scientist must use an observational study to address the issue. This could be done by comparing the diets for a sample of people with heart disease with those for a sample of people without heart disease. Great care would have to be taken to record other relevant factors such as family history of heart disease, smoking habits, exercise routine, age and gender for each person, and so on, so that differences between the two groups could be adjusted to eliminate all factors except fat content of the diet. Even with these adjustments, it would be difficult to assign a cause-and-effect relationship between high fat content of a diet and heart disease. In fact, if the diet fat content for the heart disease group tended to be higher than that for the group free of heart disease after adjusting for relevant factors, the study results would be reported as an *association* between high diet fat content and heart disease, not a *causal* relationship.

Stated slightly differently, in observational studies we are sampling from populations where the factors (or treatments) are already present and we compare the samples with respect to the factor (treatments) of interest. In contrast, in the controlled environment of an experimental study, we are able to randomly assign the people as objects under study to the factors (or treatments) and then observe the response of interest. For our heart disease example, the distinction is shown here:

Observational study: We sample from the heart disease population and the heart disease-free population and compare the fat content of the diets for the two groups.

Experimental study: Ignoring the ethical situation, we would assign volunteers to one of several different fat content diets (the treatments) and compare the different treatments with respect to the response of interest (heart disease) after a period of time.

In this chapter, we focus on designed experiments. In this type of study, the researcher controls the crucial factors by one of two methods.

Method 1: The subjects in the experiment are randomly assigned to the treatments. For example, ten rats are randomly assigned to one of the four dose levels of an experimental drug under investigation.

Method 2: Subjects are randomly selected from different populations of interest. For example, 50 male and 50 female dogs are randomly selected from animal shelters in large and small cities, and tested for the presence of heartworm.

In method 1, the researcher selects experimental units from a homogeneous population of experimental units and then has complete control over the random assignment of the units to the various treatments. In method 2, the reseacher has control over the random sampling from the treatment populations but not over the assignment of the experimental units to the treatments.

In designed experiments, it is crucial that the scientist follows a systematic plan that was established prior to running the experiment. The plan includes how all randomization is conducted, either the assignment of experimental units to treatments or the selection of units from the treatment populations. Extraneous factors may be present that may affect the experimental units. These factors may appear as subtle differences in the experimental units or slight differences in the surrounding environment during the conducting of the experiment. The randomization process ensures that, on the average, any large differences observed in the responses of the experimental units in different treatment groups can be attributed to the differences in the factor of interest and not to factors that were not controlled during the experiment. The plan should also include many other aspects on how to conduct the experiment. A list of some of the items that should be included in such a plan are listed here:

1. The research objectives of the experiment
2. The selection of the factors that will be varied (the treatments)
3. The identification of extraneous factors that may be present in the experimental units or in the environment of the experimental setting (the blocking factors)
4. The characteristics to be measured on the experimental units (response variable)
5. The method of randomization, either random selection from treatment populations or the random assignment of experimental units to treatments
6. The procedures to use in recording the responses from the experimental units
7. The selection of the number of experimental units assigned to each treatment, which may require designating the level of significance and power of tests or the precision and reliability of confidence intervals
8. A complete listing of available resources and materials

The procedures described in Chapters 15 through 19 of this book have been developed for the analysis of designed experiments.

14.3 Designed Experiments: Terminology

We will now distinguish between the common usage of words and the meaning that we will give these terms and concepts within our designed experiment structure. Stated more succinctly, a **designed experiment** is an investigation in which a specified framework is provided to observe, measure, and evaluate groups with

designed experiment

respect to a designated response. The researcher controls the elements of the framework during the experiment to obtain data from which statistical inferences can provide valid comparisons of the groups of interest.

We use the following example to illustrate the concepts and terminology that will be defined here.

EXAMPLE 14.1

A researcher is studying the conditions under which commercially raised shrimp reach maximum weight gain. Three water temperatures (25°, 30°, 35°) and four water salinity levels (10%, 20%, 30%, 40%) are selected for study. Shrimp are raised in containers with specified water temperatures and salinity levels. The weight gain of the shrimp in each container is recorded after a 6-week study period. There are many other factors that may affect weight gain, such as density of shrimp in the containers, variety of shrimp, size of shrimp, and type of feeding. The experiment is conducted as follows: Twenty-four containers are available for the study. A specific variety and size of shrimp is selected for study. The density of shrimp in the container is fixed at a given amount. One of the three water temperatures and one of the four salinity levels is randomly assigned to each of the 24 containers. All other identifiable conditions are specified to be maintained at the same level for all 24 containers for the duration of the study. In reality there will be some variation in the levels of these variables. After 6 weeks in the tanks, the shrimp are harvested and weighed.

factors

measurements

observations

There are two types of variables in a designed experiment. Controlled variables called **factors** are selected by the researchers for comparison. Response variables are **measurements** or **observations** that are recorded but not controlled by the reseacher. The controlled variables form the comparison groups that are defined by the research hypothesis. Water temperature and salinity level are the control variables or factors in our example, and shrimp weight is the response variable.

treatments

one-way classification

treatment design

factorial treatment design

The **treatments** in a designed experiment are the conditions constructed from the factors. The factors are selected by examining the questions raised by the research hypothesis. In some experiments, there may only be a single factor and hence the treatments and levels of the factor would be the same. This type of treatment design is referred to as a **one-way classification.** In most cases we will have several factors, and the treatments are formed by combining levels of the factors. This type of **treatment design** is called a **factorial treatment design.** For the shrimp study described in Example 14.1, there are two factors: water temperature at three levels (25°, 30°, and 35°) and water salinity at four levels (10%, 20%, 30%, and 40%). We can thus create 3(4) = 12 treatments from the combination of levels of the two factors. These factor–level combinations representing the 12 treatments are shown here:

(25°, 10%)   (25°, 20%)   (25°, 30%)   (25°, 40%)
(30°, 10%)   (30°, 20%)   (30°, 30%)   (30°, 40%)
(35°, 10%)   (35°, 20%)   (35°, 30%)   (35°, 40%)

Twenty-four containers were used in Example 14.1, and containers were randomly assigned to the 12 treatments so that there were two containers for each of the 12 treatments.

In other circumstances, there may be a large number of factors and hence the number of treatments may be so large that only a subset of all possible treatments would be examined in the experiment. For example, suppose we were investigating the effect of the following factors on the yield per acre of soybeans: factor 1—five varieties of soybeans; factor 2—three planting densities; factor 3—four levels of fertilization; factor 4—six locations within Texas; and factor 5—three irrigation rates. From the five factors, we can form $(5)(3)(4)(6)(3) = 1,080$ distinct treatments. This would make for a very large and expensive experiment. In this type of situation, a subset of the 1,080 possible treatments would be selected for studying the relationship between the five factors and the yield of soybeans. This type of experiment is called a **fractional factorial experiment,** because only a fraction of the possible treatments are actually used in the experiment. A great deal of care must be taken in selecting which treatments should be used in the experiment so as to be able to answer as many of the researcher's questions as possible.

fractional factorial experiment

A special type of treatment is called the **control treatment.** This treatment is the benchmark to which the effectiveness of the remaining treatments are compared. There are three situations in which a control treatment is particularly necessary. First, the conditions under which the experiments are conducted may prevent generally effective treatments from demonstrating their effectiveness. In this case, the control treatment consisting of no treatment may help to demonstrate that the experimental conditions are not allowing the treatments to demonstrate the differences in their effectiveness. For example, an experiment is conducted to determine the most effective level of nitrogen in a garden that is growing tomatoes. If the soil used in the study already has a high level of fertility prior to adding nitrogen to the soil, all levels of nitrogen will appear to be equally effective. However, if a treatment consisting of adding no nitrogen, the control, is used in the study, the high fertility of the soil will be revealed because the control treatment will be as effective as the nitrogen-added treatments.

control treatment

A second type of control is the standard method treatment to which all other treatments are compared. In this situation, several new procedures are proposed to replace an already existing, well-established procedure. Finally, the placebo control is used when a response may be obtained from the subject simply by the manipulation of the subject during the experiment. For example, a person may demonstrate a temporary reduction in pain level simply by visiting with the physician and having a treatment prescribed. Thus, in evaluating several different methods of reducing pain level in patients, a treatment with no active ingredients, the placebo, is given to a set of patients without the patients' knowledge. The treatments with active ingredients are then compared to the placebo to determine their true effectiveness.

experimental unit

The **experimental unit** is the physical entity to which the treatment is randomly assigned or the subject that is randomly selected from one of the treatment populations. For the shrimp study of Example 14.1, the experimental unit is the container.

Consider another experiment in which a researcher is testing various dose levels (treatments) of a new drug on laboratory rats. If the researcher randomly assigned a single dose of the drug to each rat, then the experimental unit would be the individual rat. Once the treatment is assigned to an experimental unit, a single **replication** of the treatment has occurred. In general, we will randomly assign several experimental units to each treatment. We will thus obtain several

replication

measurement unit

independent observations on any particular treatment and hence will have several replications of the treatments. In Example 14.1, we had two replications of each treatment.

Distinct from the experimental unit is the **measurement unit.** This is the physical entity on which a measurement is taken. In many experiments, the experimental and measurement unit are identical. In Example 14.1, the measurement unit is the container, the same as the experimental unit. If however, the individual shrimp were measured in each container, the experimental unit would be the container, because the treatments were applied to the containers, but the measurement unit would be the shrimp.

EXAMPLE 14.2

Consider the following experiment. Four types of protective coatings for frying pans are to be evaluated. Five frying pans are randomly assigned to each of the four coatings. A measure of the abrasive resistance of the coatings is taken at three locations on each of the 20 pans. Identify the following items for this study: experimental design, treatments, replications, experimental unit, measurement, and total number of measurements.

Solution

Experimental design: One-way classification
Treatments: Four types of protective coatings
Replication: There are five frying pans (replications) for each treatment
Experimental unit: Frying pan, because coatings (treatments) are randomly assigned to the frying pans
Measurement unit: Location in the frying pan

Total number of measurements: $4(5)(3) = 60$

experimental error

The term **experimental error** is used to describe the variation in the responses among experimental units, which are assigned the same treatment and are observed under the "same" experimental conditions. The reasons that the experimental error is not zero include (a) the natural differences in the experimental units prior to their receiving the treatment, (b) the variation in the devices that record the measurements, (c) the variation in setting the treatment conditions, (d) the effect on the response variable of all extraneous factors other than the treatment factors.

EXAMPLE 14.3

Refer to the previously discussed laboratory experiment in which the researcher randomly assigns a single dose of the drug to each rat and measures the level of drug in the rat's bloodstream after 2 hours. For this experiment, the experimental unit and measurement unit are the same: the rat. Identify the four possible sources of experimental error for this study. [See (a) to (d) above.]

Solution   We can address these sources as follows:

(a) Natural differences in experimental units prior to receiving the treatment. There will be slight physiological differences among rats, so two rats receiving the exact same dose level (treatment) will have slightly different blood levels after 2 hours.

(b) Variation in the devices used to record the measurement. There will be differences in the responses due to the method by which the quantity of the drug in the rat is determined by the laboratory technician. If several determinations of drug level were made in the blood of the same rat, there may be differences in the amount of drug found due to equipment variation, technician variation, or conditions in the laboratory variation.

(c) Variation in setting the treatment conditions. If there is more than one replication per treatment, the treatment may not be the same from one rat to another. Suppose, for example, that we had ten replications of each dose (treatment); it is highly unlikely that each of the ten rats receives exactly the same dose of drug specified by the treatment. There could be slightly different amounts of drug in the syringes and slightly different amounts could be injected and enter the bloodstream.

(d) The effect on the response (blood level) of all extraneous factors other than the treatment factors. Presumably, the rats are all placed in cages and given the same amount of food and water prior to determining the amount of drug in their blood. However, the temperature, humidity, external stimulation, and other conditions may be somewhat different in the cages. This may have an effect on the responses of the rats.

Thus, these differences and variation in the external conditions within the laboratory during the experiment all contribute to the size of the experimental error in the experiment.

variance of experimental error

To test research hypotheses and to construct confidence intervals on functions of the treatment population means, we need to obtain an estimate of the **variance of experimental error.** We will develop procedures to obtain such an estimate in the following chapters. The experimental design, treatment design, and the number of replications will all affect this determination. In the most basic experiment, in which we have a single factor having t levels and with the experimental unit and measurement unit identical, the estimate of the variance of experimental error is the pooled variance of responses from experimental units receiving the same treatment. In our laboratory study of the drug levels in rats, we had the rats randomly assigned to the four treatments, ten rats per treatment. An estimate of the variance of experimental ever would be obtained by measuring the variance in the ten blood-level responses at each dose level and pooling the four separate estimates of the variance into a single estimate.

EXAMPLE 14.4

Refer to Example 14.1. Suppose that each treatment is assigned to two containers and that 40 shrimp are placed in the containers. After 6 weeks, the individual shrimp are weighed. Identify the experimental units, measurement units, factors, treatments, number of replications, and possible sources of experimental error.

Solution This was a factorial treatment design with two factors: temperature and salinity level. The treatments are constructed by selecting a temperature and salinity level to be assigned to a particular container. We have a total of $(3)(4) = 12$ possible treatments for this experiment. The 12 treatments are as follows:

(25°, 10%)	(25°, 20%)	(25°, 30%)	(25°, 40%)
(30°, 10%)	(30°, 20%)	(30°, 30%)	(30°, 40%)
(35°, 10%)	(35°, 20%)	(35°, 30%)	(35°, 40%)

We next randomly assigned two containers to each of the 12 treatments. This resulted in two replications of each treatment. The experimental unit is the container because the treatments were randomly assigned to individual containers. Forty shrimp were placed in the containers, and after 6 weeks, the weights of the individual shrimp were recorded. The measurement unit is the individual shrimp because this is the physical entity on which an observation was made. Thus, in this experiment, the experimental and measurement units are different. Several possible sources of experimental error are the difference in the weights of the shrimp prior to being placed in the container, the accuracy at which the temperature and salinity levels are maintained over the 6-week study period, the accuracy with which the shrimp are weighed at the conclusion of the study, the consistency of the amount of food fed to the shrimp (was each shrimp given exactly the same quantity of food over the 6 weeks?), and the variation in any other conditions that may affect shrimp growth.

14.4 Controlling Experimental Error

As we observed in Examples 14.3 and 14.4, there are many potential sources of experimental error in an experiment. When the variance of experimental errors is large, the precision of our inferences will be greatly compromised. The estimates of treatment means will have a large standard deviation, which results in imprecise (wide) confidence intervals as well as tests of hypotheses having large values for their probability of Type II error. Thus, any techniques that can be implemented to reduce experimental error will lead to a much improved experiment and more precise inferences.

The researcher may be able to control many of the potential sources of experimental errors. Some of these sources are: (1) the procedures under which the experiment is conducted, (2) the choice of experimental units and measurement units, (3) the procedure by which measurements are taken and recorded, (4) blocking of the experimental units, (5) type of experimental design, and (6) the use of covariates. We will now address how each of these sources may affect experimental error and how the researcher may minimize the effect of these sources on the size of the variance of experimental error.

Experimental Procedures

When the individual procedures required to conduct an experiment are not done in a careful precise manner, the result is an increase in the variance of the response variable. This involves not only the personnel used to conduct the experiments and to measure the response variable but also the equipment used in their procedures. Personnel must be trained properly in constructing the treatments and carrying out the experiments. The consequences of their performance on the success of the experiment should be emphasized. The researcher needs to provide the technicians with equipment that will produce the most precise measurements within budget constraints. It is crucial that equipment be maintained and calibrated at frequent intervals throughout the experiment. The conditions under which the experiments are run must be as nearly constant as possible during the duration of the experiment. Otherwise, differences in the responses may be due to changes in the experimental conditions and not due to treatment differences.

When experimental procedures are not of high quality, the variance of the response variable may be inflated. Improper techniques used when taking measurements, improper calibration of instruments, or uncontrolled conditions within a laboratory may result in extreme observations that are not truly reflective of the effect of the treatment on the response variable. Extreme observations may also occur due to recording errors by the laboratory technician or the data manager. In either case, the researcher must investigate the circumstances surrounding extreme observations and then decide whether to delete the observations from the analysis. If an observation is deleted, an explanation of why the data value was not included should be given in the appendix of the final report.

When experimental procedures are not uniformly conducted throughout the study period, two possible outcomes are an inflation in the variance of the response variable and a bias in the estimation of the treatment mean. For example, suppose we are measuring the amount of drug in the blood of rats injected with one of four possible doses of a drug. The equipment used to measure the precise amount of drug to be injected is not working properly. For a given dosage of the drug, the first rats injected were given a dose that was less than the prescribed dose, whereas the last rats injected were given more than the prescribed amount. Thus, when the amount of drug in the blood is measured, there will be an increase in the variance in these measurements but the treatment mean may be estimated without bias because the overdose and underdose may cancel each other. On the other hand, if all the rats receiving the lowest dose level are given too much of the drug and all the rats receiving the highest dose level are not given enough of the drug, then the estimation of the treatment means will be biased. The treatment mean for the low dose will be overestimated, whereas the high dose will have an underestimated treatment mean. Thus, it is crucial to the success of the study that experimental procedures are conducted uniformly across all experimental units. The same is true concerning the environmental conditions within a laboratory or in a field study. Extraneous factors such as temperature, humidity, amount of sunlight, exposure to pollutants in the air, and other uncontrolled factors when not uniformly applied to the experimental units may result in a study with both an inflated variance and a biased estimation of treatment means.

Selecting Experimental and Measurement Units

When the experimental units used in an experiment are not similar with respect to those characteristics that may affect the response variable, the experimental error variance will be inflated. One of the goals of a study is to determine whether there is a difference in the mean responses of experimental units receiving different treatments. The researcher must determine the population of experimental units that are of interest. The experimental units are randomly selected from that population and then randomly assigned to the treatments. This is of course the idealized situation. In practice, the researcher is somewhat limited in the selection of experimental units by cost, availability, and ethical considerations. Thus, the inferences that can be drawn from the experimental data may be somewhat restricted. When examining the pool of potential experimental units, sets of units that are more similar in characteristics will yield more precise comparisons of the treatment means. However, if the experimental units are overly uniform, then the population to which inferences may be properly made will be greatly restricted. Consider the following example.

EXAMPLE 14.5

A sales campaign to market children's products will use television commercials as its central marketing technique. A marketing firm hired to determine whether the attention span of children is different depending on the type of product being advertised decided to examine four types of products: sporting equipment, healthy snacks, shoes, and video games. The firm selected 100 fourth-grade students from a New York City public school to participate in the study. Twenty-five students were randomly assigned to view a commercial for each of the four types of products. The attention spans of the 100 children were then recorded. The marketing firm thought that by selecting participants of the same grade level and from the same school system it would achieve a homogeneous group of subjects. What problems exist with this selection procedure?

Solution   The marketing firm was probably correct in assuming that by selecting the students from the same grade level and school system it would achieve a more homogeneous set of experimental units than by using a more general selection procedure. However, this procedure has severely limited the inferences that can be made from the study. The results may be relevant only to students in the fourth grade and residing in a very large city. A selection procedure involving other grade levels and children from smaller cities would provide a more realistic study.

Reducing the Variance of Experimental Error through Blocking

When we are concerned that the pool of available experimental units has large differences with respect to important characteristics, the use of blocking may prove to be highly effective in reducing the experimental error variance. The experimental units are placed into groups based on their similarity with respect to characteristics that may affect the response variable. This results in sets or blocks of experimental units that are homogeneous within the block, but there is a broad coverage of important characteristics when considering the entire unit. The treatments are randomly assigned separately within each block. The comparison of the treatments is within the groups of homogeneous units and hence yield a comparison of the treatments that is not masked by the large differences in the original set of experimental units. The blocking design will enable us to separate the variability associated with the characteristics used to block the units from the experimental error.

There are many criteria used to group experimental units into blocks; they include the following:

1. Physical characteristics such as age, weight, sex, health, and education of the subjects

2. Units that are related such as twins or animals from the same litter
3. Spatial location of experimental units such as neighboring plots of land or position of plants on a laboratory table
4. Time at which experiment is conducted such as the day of the week, because the environmental conditions may change from day to day
5. Person conducting the experiment, because if several operators or technicians are involved in the experiment they may have some differences in how they make measurements or manipulate the experimental units

In all of these examples, we are attempting to observe all the treatments at each of the levels of the blocking criterion. Thus, if we were studying the number of cars with a major defect coming off each of three assembly lines, we might want to use day of the week as a blocking variable and be certain to compare each of the assembly lines on all 5 days of the work week.

Using Covariates to Reduce Variability

A covariate is a variable that is related to the response variable. Physical characteristics of the experimental units are used to create blocks of homogeneous units. For example, in a study to compare the effectiveness of a new diet to a control diet in reducing the weight of dogs, suppose the pool of dogs available for the study varied in age from 1 year to 12 years. We could group the dogs into three blocks: B_1—under 3 years, B_2—3 years to 8 years, B_3—over 8 years. A more exacting methodology records the age of the dog and then incorporates the age directly into the model when attempting to assess the effectiveness of the diet. The response variable would be adjusted for the age of the dog prior to comparing the new diet to the control diet. Thus, we have a more exact comparison of the diets. Instead of using a range of ages as is done in blocking, we are using the exact age of the dog, which reduces the variance of the experimental error.

Candidates for covariates in a given experiment depend on the particular experiment. The covariate needs to have a relationship to the response variable, it must be measurable, and it cannot be affected by the treatment. In most cases the covariate is measured on the experimental unit before the treatment is given to the unit. Examples of covariates are soil fertility, amount of impurity in a raw material, weight of an experimental unit, SAT score of student, cholesterol level of subject, and insect density in the field. The following example will illustrate the use of a covariate.

EXAMPLE 14.6

In this study the effect of two treatments, supplemental lighting (SL) and partial shading (PS), on the yield of soybean plants were compared with normal lighting (NL). Normal lighting will serve as a control. Each type of lighting was randomly assigned to 15 soybean plants and the plants were grown in a greenhouse study. When setting up the experiment, the researcher recognized that the plants were of differing size and maturity. Consequently, the height of the plant, a measurable characteristic of plant vigor, was determined at the start of the experiment and will serve as a covariate. This will allow the researcher to adjust the yields of the individual soybeans plants depending on the initial size of the plant. On each plant we record two variables, (x, y) where x is the height of the plant at the beginning of the study and y is the yield of soybeans at the conclusion of the study. To determine whether the covariate has an effect on the response variable, we plot the two variables to assess any possible relationship. If no relationship exists, then the covariate need not be used in the analysis. If the two variables are related, then we must use the techniques of analysis of covariance to properly adjust the response variable prior to comparing the mean yields of the three treatments. An initial assessment of the viability of the relationship is simply to plot the response variable versus the covariate with a separate plotting characteristic for each treatment. Figure 14.1 contains this plot for the soybean data.

FIGURE 14.1

Plot of plant height versus yield: S = Supplemental Lighting, C = Normal Lighting, P = Partial Shading

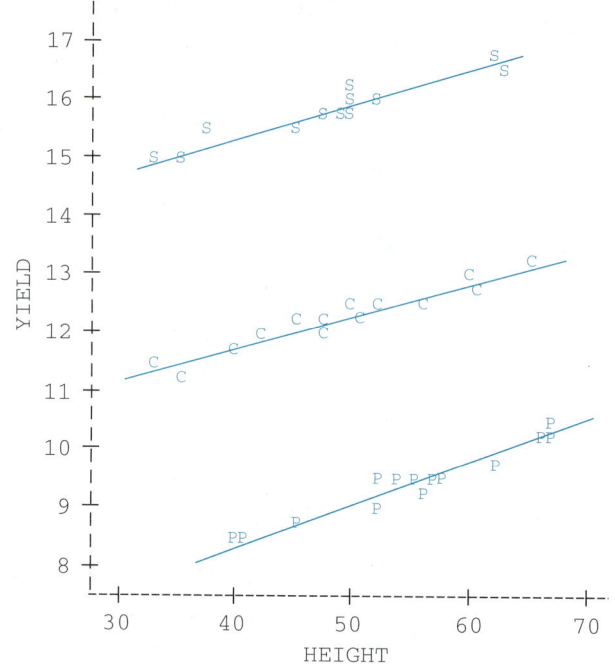

From Figure 14.1, we observe that there appears to be an increasing relationship between the covariate, initial plant height, and the response variable, yield. Also, the three treatments appear to have differing yields; some of the variation in the response variable is related to the initial height as well as to the difference in the amount of lighting the plant received. Thus, we must identify the amount of variation associated with initial height prior to testing for differences in the average yields of the three treatment. We can accomplish this using the techniques of analysis of variance. The analysis of covariance procedures will be discussed in detail in Chapter 16.

14.5   Randomization of Treatments to Experimental Units

As discussed in previous chapters, the statistical procedures are based on the condition that the data from the experiment are equivalent to a random sample from a normally distributed population of responses. When the experiment consists of randomly selecting experimental units from established treatment populations, we can in fact verify whether or not this condition is valid. However, in those experiments in which we have selected the experimental units to meet specific criteria or the experimental units are available plots of land in an agricultural research farm, the idea that the responses form a random sample from a specific population is somewhat questionable.

When we are dealing with the situation in which we are randomly assigning treatments to the experimental units and then observing the responses, it is a requirement that these observations are independent. The statistical theory that is used in this textbook to estimate parameters, construct confidence intervals, and test hypotheses is based on the condition that observations are independent of one another. In more advanced books, methods are available for dealing with dependent data such as time series data or spatially correlated data. To obtain valid results we need to be assured that the observations are independent. The data values would not be independent if there were physical relationships between the experimental units. Consider the following example.

EXAMPLE 14.7

Suppose we are comparing four root stimulators (S_1, S_2, S_3, S_4) with respect to their effect on production of root density in spruce seedlings. Thirty-two seedlings are placed in pots on a greenhouse table as depicted in Figure 14.2. In greenhouse experiments, if there are slight temperature gradients or some variations in the amount of sunlight in the greenhouse, then plants near each other are more likely to respond similarly than plants that are far apart. The responses are positively correlated. In other settings we may obtain the opposite result; that is, experimental units in close proximity tend to respond in opposite directions from the overall mean due to competition for resources. For example, if animals are fed from the same container, there is competition for the available food. If a few animals consume more than their share and grow faster, the remaining animals will tend to be smaller.

FIGURE 14.2

Arrangement of pots on table

II	I
S_2 S_2 S_2 S_2	S_1 S_1 S_1 S_1
S_2 S_2 S_2 S_2	S_1 S_1 S_1 S_1
III	IV
S_3 S_3 S_3 S_3	S_4 S_4 S_4 S_4
S_3 S_3 S_3 S_3	S_4 S_4 S_4 S_4

To conduct the experiment in the most efficient manner, the technician assigns the stimulants as follows:

Stimulant S_1 is applied to all eight seedlings in quadrant I.
Stimulant S_2 is applied to all eight seedlings in quadrant II.
Stimulant S_3 is applied to all eight seedlings in quadrant III.
Stimulant S_4 is applied to all eight seedlings in quadrant IV.

Suppose the air conditioning flows throughout the greenhouse such that there are subtle differences in the temperature on the tables. In particular, the pots in quadrant I receive slightly cooler air and the pots in quadrant III receive slightly warmer air. These differences were not identified when the experiment was initially designed. Under this arrangement, the plants receiving S_1 may be at a competitive disadvantage to those plants receiving the other stimulants since they would be growing at lower temperatures.

To avoid situations where the physical proximity of experimental units or the order in which a technician conducts the experiment produces responses from the experimental units that are correlated, it is necessary to randomly assign the treatments to the experimental units. Random assignment provides us with a justification for the model condition that the data values are independent observations. However, the random assignment of treatments to experimental units does not completely eliminate the problem of correlated data values. Correlation can also result from other circumstances that may occur during the experiment. Thus, the researcher always must be aware of any physical mechanisms that can enter the experimental setting and result in the responses from some of experimental units having an effect on the responses of other experimental units.

Suppose we have N homogeneous experimental units and t treatments. We want to randomly assign r_i experimental units to the ith treatment, where $r_1 + r_2 + \cdots + r_t = N$. The random assignment involves the following steps:

1. Number the experimental units from 1 to N.
2. Use a random number table or a computer software program to obtain a list of numbers that is a random permutation of the numbers 1 to N.
3. Give treatment #1 to the experimental units having the first r_1 numbers in the list. Treatment #2 goes to the experimental units having the next r_2 numbers in the list. We continue the assignment of experimental units to treatments until treatment #t goes to r_t experimental units.

We will illustrate this procedure in the following example.

EXAMPLE 14.8

Suppose we examine the study described in Example 14.5, in which we wanted to compare four root stimulators (S_1, S_2, S_3, S_4) with respect to their effect on production of root density in spruce seedlings. Thirty-two seedlings were available for the study and each type of stimulator was to be applied to eight seedlings. Following the procedure outlined above, we would number the 32 seedlings from 1 to 32. Next, we would obtain a random permutation of the numbers 1 to 32. Using a software package, we obtain the following random permutation of the numbers:

| 31 | 21 | 4 | 7 | 28 | 3 | 1 | 24 | 17 | 15 | 13 | 29 | 9 | 22 | 20 | 27 |
| 25 | 12 | 14 | 5 | 18 | 26 | 8 | 30 | 23 | 32 | 11 | 19 | 16 | 6 | 10 | 2 |

We thus obtain the following assignment of the seedlings to the stimulators.

Seedling	31	21	4	7	28	3	1	24	17	15	13	29	9	22	20	27
Stimulator	S_1	S_1	S_1	S_1	S_1	S_1	S_1	S_1	S_2	S_2	S_2	S_2	S_2	S_2	S_2	S_2
Seedling	25	12	14	5	18	26	8	30	23	32	11	19	16	6	10	2
Stimulator	S_3	S_3	S_3	S_3	S_3	S_3	S_3	S_3	S_4	S_4	S_4	S_4	S_4	S_4	S_4	S_4

FIGURE 14.3
Random arrangement of treatments on table

II				I			
S_2	S_4	S_4	S_3	S_1	S_4	S_1	S_1
S_2	S_3	S_2	S_4	S_3	S_4	S_1	S_3
III				**IV**			
S_2	S_3	S_4	S_2	S_3	S_3	S_2	S_1
S_1	S_2	S_4	S_1	S_2	S_3	S_1	S_4

Figure 14.3 shows the arrangement of stimulators to seedlings on the laboratory table. The seedlings numbered 1–8 were in quadrant I, seedlings numbered 9–16 in quadrant II, seedlings numbered 17–24 in quadrant III, and seedlings numbered 25–32 in quadrant IV.

Suppose this arrangement of the 32 pots was on four different tables and not simply on different quadrants of the same table. The tables are a considerable distance from each other and the researcher determines that the tables are exposed to distinct temperature and sunlight exposures. She examines the arrangement in Figure 14.3 and notes that 4 of the 8 seedlings on table I have been randomly assigned to received stimulator 1, which may result in a bias in the experiment if the conditions on table I are highly favorable to root growth. An alternative method of assigning treatments to the experimental units is called a **randomized complete block design.** In this type of design, the experimental units are not homogeneous. The researcher must first examine the experimental units and place them into groups such that the experimental units within a group are more homogeneous than are experimental units in different groups. If there are sufficient experimental units within each group, the treatments are then randomized within each of the groups. This assignment procedure is comparable to running separate **completely randomized designs** in each of the groups.

randomized complete block design

completely randomized designs

The procedure for assigning the treatments to the experimental units is conducted in the following manner. Suppose we have t treatments and N experimental units. Further assume that the experimental units can be grouped into b groups each containing t experimental units. This requires that the number of experimental units satisfies $N = bt$.

Randomization Procedure for a Randomized Complete Block Design

1. Group the experimental units into b groups, each containing t homogeneous experimental units.
2. In group 1, number the experimental unit from 1 to t.
3. Use a random number table or computer software to obtain a random permutation of the numbers from 1 to t.
4. In group 1, the experimental unit corresponding to the first number in the permutation receives treatment 1, the experimental unit corresponding to the second number in the permutation receives treatment 2, and so on.
5. Repeat steps 2–4 for each of the remaining groups (blocks).

The randomized complete block design, which has one replication of each treatment per block, can be generalized to have r replications of each treatment per block. The randomization procedure can be modified for a generalized randomized complete block design with t treatments, b blocks, and r replications. These experimental units would be grouped into b groups, each containing rt homogeneous experimental units. (Note that $N = brt$.)

Randomization Procedure for a Generalized Randomized Complete Block Design

1. Group the experimental units into b groups, each containing rt homogeneous experimental units.
2. In group 1, number the experimental units from 1 to rt.
3. Use a random number table or a computer software program to obtain a list of numbers that is a random permutation of the numbers 1 to rt.
4. In group 1, treatment #1 receives the experimental units having numbers given by the first r numbers in the list. Treatment #2 receives the experimental units having the next r numbers in the list. Continue the assignment of experimental units to treatments until treatment #t receives r experimental units.
5. Repeat steps 2–4 for the remaining groups of experimental units.

We will illustrate this procedure in the following example.

EXAMPLE 14.9

Suppose the four tables have considerably different temperatures and sunlight exposures. The 32 pots of seedlings are placed eight pots to a table. We have the situation where two pots on the same table are exposed to a more uniform set of conditions than two pots placed on different tables. Thus, we want to conduct a randomized complete block design with the four stimulators randomly assigned to two different pots within each table. Note that we have a generalized randomized complete block design with $N = 32$, $t = 4$, $b = 4$, and $r = 2$.

Solution

1. The 32 pots are randomly placed eight pots to a table.
2. The pots in each table are numbered 1 to 8.
3. We obtain the following random permutation of the numbers 1 to 8: 6, 7, 1, 4, 3, 8, 5, 2.
4. On table I, pots 6 and 7 are assigned to S_1, pots 1 and 4 are assigned to S_2, pots 3 and 8 are assigned to S_3, and pots 5 and 2 are assigned to S_4.
5. Permutation for table II: 1, 8, 2, 5, 4, 3, 7, 6.
   Permutation for table III: 5, 7, 3, 8, 2, 4, 1, 6.
   Permutation for table IV: 1, 4, 5, 7, 3, 2, 8, 6.
6. Table II: S_1 to pots 1, 8; S_2 to pots 2, 5; S_3 to pots 4, 3; S_4 to pots 7, 6.
   Table III: S_1 to pots 5, 7; S_2 to Pots 3, 8; S_3 to pots 2, 4; S_4 to pots 1, 6.
   Table IV: S_1 to pots 1, 4; S_2 to Pots 5, 7; S_3 to pots 3, 2; S_4 to pots 8, 6.

We obtain the random assignment given in Figure 14.4.

In this arrangement, each of the four treatments has been randomly assigned to two pots on each of the tables in the greenhouse. Now, if there are large differences between the conditions the pots are exposed to on the four tables,

FIGURE 14.4
Random block arrangement of treatments on four tables

II	I
S_3 S_3 S_2 S_1	S_2 S_3 S_4 S_2
S_2 S_4 S_4 S_1	S_4 S_1 S_1 S_3
III	**IV**
S_3 S_2 S_3 S_4	S_1 S_3 S_3 S_1
S_1 S_4 S_1 S_2	S_2 S_4 S_2 S_4

this arrangement provides for a more equitable distribution of the effect due to table differences across the four treatments. Furthermore, the randomization of the treatments on each table will assist in distributing any differences that may exist in the seedlings prior to applying the treatment and any differences in the conditions between pots on each of the four tables.

14.6 Determining the Number of Replications

The number of replications in an experiment is the crucial element in determining the accuracy of estimators of the treatment means and the power of tests of hypotheses concerning differences between the treatment means. In most situations, the greater the number of replications, the greater will be the accuracy of the estimators, the more precise will be confidence intervals on treatment means, and the greater will be the power of the test of hypotheses. The conditions that constrain the researcher from using very large numbers of replications are the cost of running the experiment, the time needed to handle a large number of experimental units, and the availability of experimental units. Thus, the researcher must determine the minimum number of replications required to meet reasonable specifications on the accuracy of estimators or on the power of tests of hypotheses.

Accuracy of Estimator Specifications to Determine the Number of Replications

We can determine the number of replications by specifying the desired width of a $100(1 - \alpha)\%$ confidence interval on the treatment mean. In Chapter 5, we provided a formula for determining the sample size needed so that we were $100(1 - \alpha)\%$ confident that the sample estimate was within E units of the true treatment mean. If we let r be the number of replications, σ be the experimental standard deviation, and E be the desired accuracy of the estimator, then we can approximate the value of r using the following formula.

Sample Size r Required to Be $100(1 - \alpha)\%$ Confident That the Estimator Is Within E Units of the Treatment Mean μ

$$r = \frac{(z_{\alpha/2})^2 \hat{\sigma}^2}{E^2}$$

In using this formula, the experimenter must specify

1. The desired level of confidence, $100(1 - \alpha)\%$.
2. The level of precision, E.
3. An estimate of σ. The estimate of σ may be obtained from a pilot study, similar past experiments, literature on similar experiments, or the use of a rough estimator $\hat{\sigma} = $ (largest value $-$ smallest value)/4. The following example will illustrate these calculations.

EXAMPLE 14.10

A researcher is designing a project to study the yield of pecans under four rates of nitrogen applications. The researcher wants to obtain estimates of the treatment means μ_1, μ_2, μ_3, and μ_4 such that she will be 95% confident that the estimates are within 4 pounds of the true mean yield. She wants to determine the necessary number of replications to achieve these goals.

Solution From previous experiments, the yields have ranged from 40 pounds to 60 pounds. Thus, an estimate of σ is given by

$$\hat{\sigma} = \frac{70 - 40}{4} = 7.5$$

From the normal tables, $z_{.025} = 1.96$. The value of E is 4 pounds as specified by the researcher. Thus, we determine that the number of replications is

$$r = \frac{(z_{\alpha/2})^2 \hat{\sigma}^2}{E^2} = \frac{(1.96)^2(7.5)^2}{(4)^2} = 13.51$$

Thus, the researcher should use 14 replications on each of the treatments to obtain the desired precision.

Using this technique to determine the number of replications does not take into account the power of the F test discussed in Chapter 8 to detect specified differences in the treatment means. Thus, the following method of determining the number of replications is preferred in most studies.

Specifying the Power of the *F* test to Determine Number of Replications

In a study involving t treatments, one of the goals is to test the hypotheses

$H_0: \mu_1 = \mu_2 = \cdots = \mu_t$

$H_a:$ Not all μs are equal.

Recall from Chapter 8 that our test procedure was given by

$$F = \frac{\text{MST}}{\text{MSE}}$$

Reject H_0 if $F \geq F_{\alpha, t-1, N-t}$

where MST and MSE are the mean squares from the AOV table. The number of replications, with $r_1 = r_2 = \cdots = r_t$, will be determined by specifying the following parameters with respect to the test statistic:

1. The significance level, α
2. The size of the difference $D = |\mu_i - \mu_j|$ in two treatment means, which is of practical significance
3. The probability of a Type II error if any pair of treatments have means greater than $D = |\mu_i - \mu_j|$
4. The variance σ^2

The probability of a Type II error, $\beta(\lambda)$, is determine by using the *noncentral F* distribution with degrees of freedom ν_1, ν_2, and *noncentrality parameter*

$$\lambda = \frac{r \sum_{i=1}^{t} (\mu_i - \mu.)^2}{\sigma^2}$$

The minimum value of λ for the situation in which at least one pair of treatments has means differing by D units or more is given by

$$\lambda = \frac{rD^2}{2\sigma^2}$$

Table 14 in the Appendix contains the power of the *F test*, which is the same as $1 - \beta(\lambda)$. The table uses the parameter $\phi = \sqrt{\lambda/t}$ to specify the alternative values of the μ_is. Using this table, we can determine the necessary number of replications to meet the given specifications. The following example will illustrate the requisite calculations.

EXAMPLE 14.11

Refer to Example 14.10, in which a researcher is designing a project to study the yield of pecans under four rates of nitrogen applications. The researcher knows that if the average pecan yields differ by more than 15 pounds, there is an economical advantage in using the treatment providing the higher yield. Thus, the researcher wants to determine the necessary replications to be 90% certain that the F test will reject H_0 and hence detect a difference in the average yields whenever any pair of nitrogen rates produce pecans having average yields differing by more than 15 pounds. The test must have $\alpha = .05$.

Solution From previous experiments, the yields have ranged from 40 pounds to 60 pounds. Thus, an estimate of σ is given by

$$\hat{\sigma} = \frac{70 - 40}{4} = 7.5$$

We have $\alpha = .05$, $t = 4$, $\nu_1 = t - 1 = 4 - 1 = 3$, and $\nu_2 = N - t = rt - t = t(r - 1) = 4(r - 1)$, where r is the required number of replications. Furthermore, the value of $D = 15$, and hence

$$\phi = \sqrt{\frac{rD^2}{2t\hat{\sigma}^2}} = \sqrt{\frac{r(15)^2}{2(4)(7.5)^2}} = .707\sqrt{r}$$

Figure 14.5 contains the power curves needed to solve this problem. Note that $\nu_1 = 3$, $\alpha = .05$, and the curves are labeled ν_2. We will determine the value of r such that the power is at least .90 when $\phi = .707\sqrt{r}$. We will accomplish this by selecting values of r until we reach the necessary threshold.

The method of determining the proper value for r is by trial and error. First, we guess $r = 6$. Next, we compute $\nu_2 = 4(6 - 1) = 20$ and $\phi = .707\sqrt{6} = 1.73$. In Figure 14.5, we locate $\phi = 1.73$ on the axis labeled ϕ and draw a vertical line

FIGURE 14.5

Power of the analysis of
variance test ($\alpha = .05$, $t = 4$)

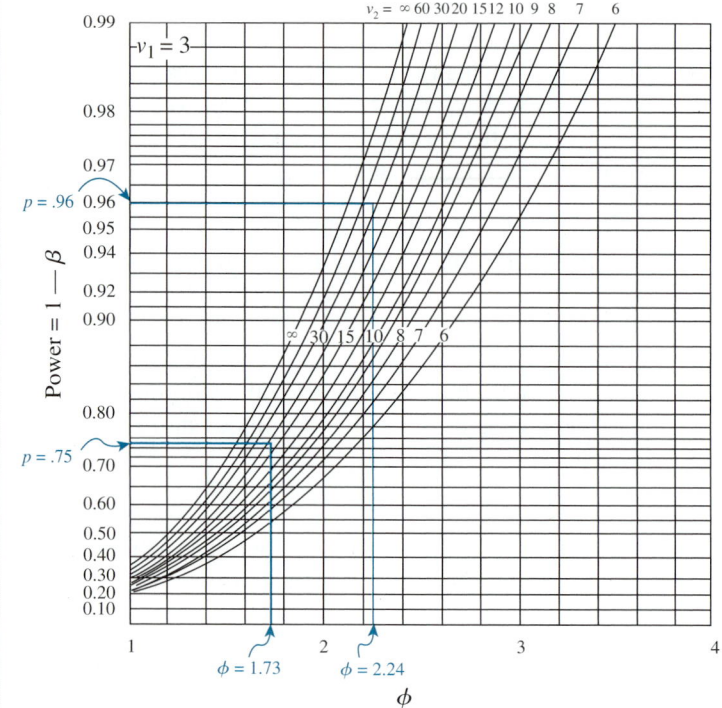

from 1.73 to the curve labeled 20. We then draw a horizontal line to the axis labeled power $= 1 - \beta$ and read the value .75. Thus, if we used six replications in the experiment, our power would only be .75 when $D = 15$, which is too small. We next try $r = 10$ and find that the power is .96. This value would be acceptable; however, a smaller value of r may achieve our goal. Thus, we try $r = 8$ and find that the power equals .89. This value is just slightly too small. Finally, we find that the power is .93 when $r = 9$. Thus, the experiment requires nine replications to meet its specifications. The calculations are summarized next.

TABLE 14.1

Determining the number
of replications

r	$v_2 = 4(r - 1)$	$\phi = .707\sqrt{r}$	Power
6	20	1.73	.75
10	36	2.24	.96
8	28	2.00	.89
9	32	2.12	.93

After determining the number of replications needed, the number of experimental units may be such that it is physically impossible to conduct the complete experiment at the same time or in the same location. In this type of situation, we can use the ideas of randomized complete block designs with the blocks being either time or location. In Example 14.11, we determined that nine replications of the four treatments or 36 experimental units were needed. Suppose that we only had 12 experimental units at a given location within an agricultural research center. However, there were three such locations, each containing 12 experimental plots. We could thus run three replications of each treatment at each of the three locations. The locations would serve as blocks for the experimental design.

14.7 Summary

In this chapter, we examined various types of studies that provide information to researchers. In particular, we described several designs that are used in controlled experiments. The types of conditions and variables that may affect the variation in the responses from experimental units were discussed. This is an extremely important issue because it affects the precision of our estimates of treatment means and the power of test statistics. Through the use of the proper experimental design, the size of the variance of experimental error can be greatly reduced. An important issue in designed experiments is the randomization process. It is crucial that the experimental units are randomly assigned to the treatments. Without randomization, many of the statistical properties of our estimators, confidence intervals, and test statistics would not be valid. The last section described how to determine the number of replications in a designed experiment.

EXERCISES
Engin.

14.1 Researchers ran a quality control study to evaluate the quality of plastic irrigation pipes. The study design involved a total of 24 pipes, with 12 pipes randomly selected from each of two manufacturing plants. The compressive strength was recorded at five locations on each of the pipes. The pipes were manufactured under one of two water temperatures and one of three types of hardeners. The experimental conditions are as follows:

Pipe No.	Plant	Temperature (°F)	Hardener	Pipe No.	Plant	Temperature (°F)	Hardener
1	1	200	H_1	13	1	200	H_3
2	1	175	H_2	14	1	175	H_3
3	2	200	H_1	15	2	200	H_3
4	2	175	H_2	16	2	175	H_3
5	1	200	H_1	17	1	200	H_2
6	1	175	H_2	18	1	175	H_1
7	2	200	H_1	19	2	200	H_2
8	2	175	H_2	20	2	175	H_1
9	1	200	H_3	21	1	200	H_2
10	1	175	H_3	22	1	175	H_1
11	2	200	H_3	23	2	200	H_2
12	2	175	H_3	24	2	175	H_1

Identify each of the following components of the experimental design.
  a. factors
  b. factor levels
  c. blocks
  d. experimental units
  e. measurement units
  f. replications
  g. covariates
  h. treatments

14.2 Consider a research study or experiment that you might run to answer a research question in your field of study. State the research question and why it is important. Describe how you would run the experiment to obtain data to answer your question. In your description be sure to include the following components:

- research question
- description of experimental units
- description of measurement units
- treatment design

- design structure
- method of randomization
- number of replications
- possible blocking

14.3   In each of the following descriptions of experiments, identify the important features of each design. Include as many of the components from Exercise 14.1 as needed to adequately describe the design.

a. A horticulturalist is measuring the vitamin C concentration in oranges in an orchard on a research farm in south Texas. He is interested in the variation in vitamin C concentration across the orchard, across the productive months, and within each tree. He divides the orchard into eight sections and randomly selects a tree from each section during October through May, the months in which the trees are in production. During each month, from eight trees he selects 10 oranges near the top of the tree, 10 oranges near the middle of the tree and 10 oranges near the bottom of the tree. The horticulturalist wants to monitor the vitamin C concentration across the productive season and determine whether there is a substantial difference in vitamin C concentration in oranges at various locations in the tree.

b. A medical specialist wants to compare two different treatments (T_1, T_2) for treating a particular illness. She will use eight hospitals for the study. She believes there may be differences in the response between hospitals. Each hospital has four wards of patients. She will randomly select four patients in each ward to participate in the study. Within each hospital, two wards are randomly assigned to get T_1, the other two wards will received T_2. All patients in a ward will get the same treatment. A single response variable is measured on each patient.

c. In place of the design described in (b), make the following change. Within each hospital, the two treatments will be randomly assigned to the patients, with two patients in each ward receiving T_1 and two patients receiving T_2.

d. An experiment is planned to compare three types of schools—public, private-nonparochial, parochial—all with respect to the reading abilities of students in sixth-grade classes. The researcher selected two large cities in each of five geographical regions of the United States for the study. In each city, she randomly selected one school of each of the three types and randomly selected a single sixth-grade class within each school. The scores on a standardized test were recorded for each of 20 students in each classroom. The researcher was concerned about differences in family income levels among the 30 schools, so she obtained the family income for each of the students that participated in the study.

Vet.   **14.4**   An experiment is designed to evaluate the effect of different levels of exercise on the health of dogs. The two levels are L_1—daily 2-mile walk and L_2—1-mile walk every other day. At the end of a 3-month study period, each dog will undergo measurements of respiratory and cariovascular fitness from which a fitness index will be computed. There are 16 dogs available for the study. They are all in good health and are of the same general size, which is within the normal range for their breed. The following table provides information about the sex and age of the 16 dogs.

Dog	Sex	Age	Dog	Sex	Age
1	F	5	9	F	8
2	F	3	10	F	9
3	M	4	11	F	6
4	M	7	12	M	8
5	M	2	13	F	2
6	M	3	14	F	1
7	F	5	15	M	6
8	M	9	16	M	3

a. How would you group the dogs prior to assigning the treatments to obtain a study having as small an experimental error variance as possible? List the dogs in each of your groups.
b. Use a random number generator or a random number table to obtain random numbers. Describe your procedure for assigning the treatments to the individual dogs.

Bus. **14.5** A computer magazine wants to rate four software programs used to prepare annual federal income tax forms based on the amount of time needed to complete the form. The study will select individuals who have incomes less than $100,000 and who itemize their deductions. Determine how many individuals would be needed for each software program to declare a difference in the average completion times at the $\alpha = .05$ level of significance with a power of .90 if the difference between any pair of means is greater than 30 minutes. From previous studies using similar software, the standard deviation in completion time is thought to be about 12.25 minutes.

14.6 A future experiment will analyze the difference in six treatments. Determine how many experimental units will be needed for each treatment to declare a difference in the treatment means at the .05 level of significance with a power of .80 if the difference between any pair of treatment means is greater than 20 units. From previous studies, the standard deviation in the responses was approximately 9 units.

Bio. **14.7** A research specialist for a large seafood company plans to investigate bacterial growth on oysters and mussels subjected to three different storage temperatures. Nine cold storage units are available. She plans to use three storage units for each of the three temperatures. One package of oysters and one package of mussels will be stored in each of the storage units for 2 weeks. At the end of the storage period, the packages will be removed and the bacterial count made for two samples from each package. The "treatment" factors of interest are temperature (levels: 0, 5, 10°C) and seafood (levels: oysters, mussels). She will also record the bacterial count for each package prior to placing seafood in the cooler. Identify each of the following components of the experimental design.
  a. factors
  b. factor levels
  c. blocks
  d. experimental units
  e. measurement units
  f. replications
  g. covariates
  h. treatments

14.8 Refer to Exercise 14.7. The experimenter could obtain the same number of observations using only three storage units, one per temperature. She could have three samples of each seafood for each temperature in a single storage unit per temperature. What are the potential difficulties with conducting the experiment in this manner?

Bus. **14.9** Four cake recipes are to be compared for moistness. The researcher will conduct the experiment by preparing and then baking the cake. Each preparation of a recipe makes only one cake. All recipes require the same cooking temperature and the same length of cooking time. The oven is large enough that four cakes may be baked during any one baking period, in positions P_1 through P_4, as shown here.

| P_1 | P_2 |
| P_3 | P_4 |

a. Discuss an appropriate experimental design and randomization procedure if there are to be r cakes for each recipe.

b. Suppose the experimenter is concerned that significant differences could exist due to the four baking positions in the oven (front vs. back, left side vs. right side). Is your design still appropriate? If not, describe an appropriate design.

c. For the design or designs described in (b), suggest modifications if there are five recipes to be tested but only four cakes may be cooked at any one time.

14.10   In each of the following situations, identify whether the design is a completely randomized design, randomized complete block design, or Latin square. If there is a factorial structure of treatments, specify whether it has a two-factor or three-factor structure. If the experiment measurement units different from the experimental units, identify both.

a. The 48 treatments were comprised of 3, 4, and 4 levels of fertilizers N, P, and K, respectively, in all possible combinations. Five peanut farms were randomly selected and the 48 treatments assigned at random at each farm to 48 plots of peanut plants.

b. Ten different software packages were randomly assigned to 30 graduate students. The time to complete a specified task was determined.

c. Four different glazes are applied to clay pots at two different thicknesses. The kiln used in the glazing can hold eight pots at a time, and it takes 1 day to apply the glazes. The experimenter wanted eight replications of the experiment. Because the conditions in the kiln vary somewhat from day to day, the experiment is conducted over an 8-day period. Each combination of a thickness and type of glaze is randomly assigned to one pot in the kiln each day.

14.11   A colleague has approached you for help with an experiment she is conducting. The experiment consists of asking a sample of consumers to taste five different recipes for meat loaf. When a consumer tastes a sample he or she will give scores to several characteristics and these scores will be combined into a single overall score. Hence, there will be one value for each recipe for a consumer. The literature indicates that in this kind of study some consumers tend to give low scores to all samples, others tend to give high scores to all samples.

a. There are two possible experimental designs. Design A would use a random sample of 100 consumers. From this group, 20 would be randomly assigned to each of the five recipes, so that each consumer tastes only one recipe. Design B would use a random sample of 100 consumers, with each consumer tasting all five recipes, the recipes being presented in a random order for each consumer. Which design would you recommend? Justify your answer.

b. When asked how the experiment is going, the researcher replies that one recipe smelled so bad that she eliminated it from the analysis. Is this a problem for the analysis if design B was used? Why or why not? Would it be a problem if design A was used? Why or why not?

CHAPTER 15

Analysis of Variance for Standard Designs

15.1 Introduction and Case Study

15.2 Completely Randomized Design with Single Factor

15.3 Randomized Complete Block Design

15.4 Latin Square Design

15.5 Factorial Treatment Structure in a Completely Randomized Design

15.6 Factorial Treatment Structure in a Randomized Complete Block Design

15.7 Estimation of Treatment Differences and Comparisons of Treatment Means

15.8 Summary

15.1 Introduction and Case Study

In Chapter 14, we introduced the concepts involved in designing an experiment. These concepts are fundamental to the scientific process, in which hypotheses are formulated, experiments (studies) are planned, data are collected and analyzed, and conclusions are reached, which in turn leads to the formulation of new hypotheses. To obtain logical conclusions from the experiments (studies), it is mandatory that the hypotheses are precisely and clearly stated and that experiments have been carefully designed, appropriately conducted, and properly analyzed. The analysis of a designed experiment requires the development of a model of the physical setting and a clear statement of the conditions under which this model is appropriate. Finally, a scientific report of the results of the experiment should contain graphical representations of the data, a verification of model conditions, a summary of the statistical analysis, and conclusions concerning the research hypotheses. In this chapter, we will discuss some standard experimental designs and their analyses.

Section 15.2 reviews the analysis of variance for a completely randomized design discussed in Chapter 8. Here the focus of interest is the comparison of treatment means. Sections 15.3 and 15.4 deal with extensions of the completely randomized design, where the focus remains the same—namely, treatment mean comparisons—but where other "nuisance" variables must be controlled. For each of these designs, we will consider the arrangement of treatments, the advantages and disadvantages of the design, a model, and an analysis of variance for data from such a design. Section 15.5 introduces factorial experiments that focus on the evaluation of the effects of two or more independent variables (factors) on a response rather than on comparisons of treatment means as in the designs of Sections 15.2 through 15.4. Particular attention is given to measuring the effects of each factor alone or in combination with the other factors. Not all designs focus on either comparison of treatment means or examination of the effects of factors on a response. In Section 15.6, we discuss designs that combine the attributes of the "block" designs of Sections 15.3 and 15.4 with those of factorial experiments in Section 15.5. The remaining sections of the chapter deal with estimation and comparisons of the treatment means for the different experimental designs, procedures to check the validity of model conditions, and alternative procedures when the standard model conditions are not satisfied.

Case Study: Texture of Low-Fat Bologna

Dietary health concerns and consumer demand for low-fat products have prompted meat companies to develop a variety of low-fat meat products. Numerous ingredients have been evaluated as fat replacements with the goal of maintaining product yields and minimizing formulation costs while retaining acceptable palatability. The paper "Utilization of soy protein isolate and konjac blends in a low-fat bologna (model system)" (1999), *Meat Science,* 53: 45–57, describes an experiment that examines several of these issues. The researchers determined that lowering the cost of production without affecting the quality of the low-fat meat product required the substitution of a portion of the meat block with non-meat ingredients such as soy protein isolates (SPI). Previous experiments have demonstrated SPI's effect on the characteristics of comminuted meats, but studies evaluating SPI's effect in low-fat meat applications are limited. Konjac flour has been incorporated into processed meat products to improve gelling properties and water-holding capacity while reducing fat content. Thus, when replacing meat with SPI, it is necessary to incorporate Konjac flour into the product to maintain the high-fat characteristics of the product.

Designing the Data Collection   The three factors identified for study were type of Konjac blend, amount of Konjac blend, and percentage of SPI substitution in the meat product. There were many other possible factors of interest, including cooking time, temperature, type of meat product, and length of curing. However, the researchers selected the commonly used levels of these factors in a commercial preparation of bologna and narrowed the study to the three most important factors. This resulted in an experiment having 12 treatments as displayed in Table 15.1.

The objective of this study was to evaluate various types of Konjac blends as a partial lean-meat replacement, and to characterize their effects in a very low-fat bologna model system. Two types of Konjac blends (KSS = Konjac flour/ starch and KNC = Konjac flour/carrageenan/starch), at levels .5% and 1%, and three meat protein replacement levels with SPI (1.1, 2.2, and 4.4%, DWB) were selected for evaluation.

The experiment was conducted as a completely randomized design with a $2 \times 2 \times 3$ three-factor factorial treatment structure and three replications of the 12 treatments. There were a number of response variables measured on the 36 runs of the experiment, but we will discuss the results for the texture of the final product as measured by an Instron universal testing machine. The mean responses are given in Table 15.2.

TABLE 15.1
Treatment design for low-fat bologna study

Treatment	Level of Blend (%)	Konjac Blend	SPI (%)
1	.5	KSS	1.1
2	.5	KSS	2.2
3	.5	KSS	4.4
4	.5	KNC	1.1
5	.5	KNC	2.2
6	.5	KNC	4.4
7	1	KSS	1.1
8	1	KSS	2.2
9	1	KSS	4.4
10	1	KNC	1.1
11	1	KNC	2.2
12	1	KNC	4.4

Managing the Data The researchers next prepared the data for a statistical analysis following the steps described in Section 2.5. The researchers needed to verify that the texture readings were properly recorded and that all computer files were consistent with the field data. (The values in Table 15.2 were simulated using the summary statistics given in the paper.)

TABLE 15.2
Mean values for meat texture in low-fat bologna study

Konjac Level (%)	Konjac Blend	SPI (%)	Texture Readings	Mean Texture
.5	KSS	1.1	107.3, 110.1, 112.6	110.0
.5	KSS	2.2	97.9, 100.1, 102.0	100.0
.5	KSS	4.4	86.8, 88.1, 89.1	88.0
.5	KNC	1.1	108.1, 110.1, 111.8	110.0
.5	KNC	2.2	108.6, 110.2, 111.2	110.0
.5	KNC	4.4	95.0, 95.4, 95.5	95.3
1	KSS	1.1	97.3, 99.1, 100.6	99.0
1	KSS	2.2	92.8, 94.6, 96.7	94.7
1	KSS	4.4	86.8, 88.1, 89.1	88.0
1	KNC	1.1	94.1, 96.1, 97.8	96.0
1	KNC	2.2	95.7, 97.6, 99.8	97.7
1	KNC	4.4	90.2, 92.1, 93.7	92.0

Analyzing the Data The researchers were interested in evaluating the relationship between mean texture of low-fat bologna as the percentage of SPI was increased and in comparing this relationship for the two types of Konjac blend at the set two levels. We will discuss the analysis of the data for this case study at the end of Section 15.5.

15.2 Completely Randomized Design with Single Factor

Recall that the completely randomized design is concerned with the comparison of t population (treatment) means $\mu_1, \mu_2, \ldots, \mu_t$. We assume that there are t different populations from which we are to draw independent random samples of sizes $n_1, n_2, \ldots, n_t$, respectively. In the terminology of the design of experiments, we assume that there are $n_1 + n_2 + \cdots + n_t$ homogeneous *experimental units* (people or objects on which a measurement is made). The treatments are randomly allocated to the experimental units in such a way that n_1 units receive treatment 1, n_2 receive treatment 2, and so on. The objective of the experiment is to make inferences about the corresponding treatment (population) means.

Consider the data for a completely randomized design as arranged in Table 15.3.

TABLE 15.3

A completely randomized design

Treatment					Mean
1	y_{11}	y_{12}	$\cdots$	y_{1n_1}	$\bar{y}_{1.}$
2	y_{21}	y_{22}	$\cdots$	y_{2n_2}	$\bar{y}_{2.}$
$\cdots$	$\cdots$	$\cdots$	$\cdots$	$\cdots$	$\cdots$
t	y_{t1}	y_{t2}	$\cdots$	y_{tn_t}	$\bar{y}_{t.}$

The model for a completely randomized design with t treatments and n_i observations per treatment can be written in the form

$$y_{ij} = \mu + \alpha_i + \varepsilon_{ij}$$

where the terms of the model are defined as follows:

y_{ij}: Observation on jth experimental unit receiving treatment i.

μ: Overall treatment mean, an unknown constant.

α_i: An effect due to treatment i, an unknown constant.

ε_{ij}: A random error associated with the response from the jth experimental unit receiving treatment i. We require that the ε_{ij}s have a normal distribution with mean 0 and common variance σ_ε^2. In addition, the errors must be independent.

The conditions given above for our model can be shown to imply that the jth recorded response from the ith treatment y_{ij} is normally distributed with mean $\mu + \alpha_i$ and variance σ_ε^2. The treatment means differ by an amount α_i, the treatment effect. Thus, a test of

$$H_0: \quad \mu_1 = \mu_2 = \cdots = \mu_t \text{ versus } H_a: \text{ Not all } \mu_i\text{'s are equal}$$

is equivalent to testing

$$H_0: \quad \alpha_1 = \alpha_2 = \cdots = \alpha_t = 0 \text{ versus } H_a: \text{ Not all } \alpha_i\text{'s are 0}$$

total sum of squares   Our test statistic is developed using the idea of a partition of the **total sum of squares** of the measurements about their mean $\bar{y}_{..} = \Sigma_{ij}\, y_{ij}$, which we defined in Chapter 8 as

$$\text{TSS} = \sum_{ij} (y_{ij} - \bar{y}_{..})^2$$

The total sum of squares is partitioned into two separate sources of variability: one due to variability among treatments and one due to the variability among the y_{ij}'s within each treatment. The second source of variability is called "error" because it accounts for the variability that is not explained by treatment differences.

partition of TSS

The **partition of TSS** can be shown to take the following form:

$$\sum_{ij} (y_{ij} - \bar{y}_{..})^2 = \sum_{i} n_i (\bar{y}_{i.} - \bar{y}_{..})^2 + \sum_{ij} (y_{ij} - \bar{y}_{i.})^2$$

When the number of replications is the same for all treatments—that is, $n_1 = n_2 = \cdots = n_t = n$—the partition becomes

$$\sum_{ij} (y_{ij} - \bar{y}_{..})^2 = n \sum_{i} (\bar{y}_{i.} - \bar{y}_{..})^2 + \sum_{ij} (y_{ij} - \bar{y}_{i.})^2$$

between-treatment sum of squares

The first term on the right side of the equal sign measures the variability of the treatment means $\bar{y}_{i.}$ about the overall mean $\bar{y}_{..}$. Thus, it is called the **between-treatment sum of squares** (SST) and is a measure of the variability in the y_{ij}'s due to differences between the treatment means, μ_i's. It is given by

$$SST = n \sum_{i} (\bar{y}_{i.} - \bar{y}_{..})^2$$

sum of squares for error

The second quantity is referred to as the **sum of squares for error** (SSE) and it represents the variability in the y_{ij}'s not explained by differences in the treatment means. This variability represents the differences in the experimental units prior to applying the treatments and the differences in the conditions that each experimental unit is exposed to during the experiment. It is given by

$$SSE = \sum_{ij} (y_{ij} - \bar{y}_{i.})^2$$

Recall from Chapter 8 that we summarized this information in an analysis of variance (AOV) table, as represented in Table 15.4.

TABLE 15.4

Analysis of variance table for a completely randomized design

Source	SS	df	MS	F
Treatments	SST	$t - 1$	$MST = SST/(t - 1)$	MST/MSE
Error	SSE	$N - t$	$MSE = SSE/(N - t)$	
Total	TSS	$N - 1$		

unbiased estimates

expected mean squares

When $H_0: \alpha_1 = \alpha_2 = \cdots = \alpha_t = 0$ is true, both MST and MSE are **unbiased estimates** of σ_ε^2, the variance of the experimental error. That is, when H_0 is true, both MST and MSE have a mean value in repeated sampling, called the **expected mean squares,** equal to σ_ε^2. We express these terms as

$$E(MST) = \sigma_\varepsilon^2 \quad \text{and} \quad E(MSE) = \sigma_\varepsilon^2$$

Thus, we would expect $F = MST/MSE$ to be near 1 when H_0 is true. When H_a is true and there is a difference in the treatment means, the mean of MSE is still an unbiased estimate of σ_ε^2,

$$E(MSE) = \sigma_\varepsilon^2$$

However, MST is no longer unbiased for σ_ε^2. In fact, the expected mean square for treatments can be shown to be

$$E(MST) = \sigma_\varepsilon^2 + n\theta_T$$

where $\theta_T = 1/(t - 1) \sum_i \alpha_i^2$. When H_a is true, some of the α_i's are not zero, and θ_T is positive. Thus, MST will tend to overestimate σ_ε^2. Hence, under H_a, the ratio

$F = $ MST/MSE will tend to be greater than 1, and we will reject H_0 in the upper tail of the distribution of F.

In particular, for selected values of the probability of Type I error α, we will reject H_0: $\alpha_1 = \alpha_2 = \cdots = \alpha_t = 0$ if the computed value of F exceeds $F_{\alpha, t-1, N-t}$, the critical value of F found in Table 8 in the Appendix with $a = \alpha$, $df_1 = t - 1$, $df_2 = N - t$. Note that df_1 and df_2 correspond to the degrees of freedom for MST and MSE, respectively, in the AOV table.

The completely randomized design has several advantages and disadvantages when used as an experimental design for comparing t treatment means.

Advantages and Disadvantages of a Completely Randomized Design

Advantages

1. The design is extremely easy to construct.
2. The design is easy to analyze even though the sample sizes might not be the same for each treatment.
3. The design can be used for any number of treatments.

Disadvantages

1. Although the completely randomized design can be used for any number of treatments, it is best suited for situations in which there are relatively few treatments.
2. The experimental units to which treatments are applied must be as homogeneous as possible. Any extraneous sources of variability will tend to inflate the error term, making it more difficult to detect differences among the treatment means.

EXAMPLE 15.1

An important factor in road safety on rural roads is the use of reflective paint to mark the lanes on highways. This provides lane references for drivers on roads with little or no evening lighting. A problem with the currently used paint is that it does not maintain its reflectivity over long periods of time. A researcher will be conducting a study to compare three new paints (P_2, P_3, P_4) to the currently used paint (P_1). The paints will be applied to sections of highway 6 feet in length. The response variable will be the percentage decrease in reflectivity of the markings 6 months after application. There are 16 sections of highway, and each type of paint is randomly applied to four sections of highway. The reflective coating is applied to the highway, and 6 months later the decrease in reflectivity is computed at each section. The resulting measurements are given in Table 15.5.

TABLE 15.5
Reflectivity measurements

Section	1	2	3	4	Mean
Paint P_1	28	35	27	21	27.75
P_2	21	36	25	18	25
P_3	26	38	27	17	27
P_4	16	25	22	18	20.25

It appears that paint P_4 is able to maintain its reflectivity longer than the other three paints, because it has the smallest decrease in reflectivity. We will now attempt to confirm this observation by testing the hypotheses

$$H_0: \quad \mu_1 = \mu_2 = \mu_3 = \mu_4 \qquad H_a: \quad \text{Not all } \mu_i'\text{s are equal.}$$

We will construct the AOV table by computing the sum of squares using the formulas given previously:

$$\bar{y}_{..} = \frac{y_{..}}{N} = \frac{400}{16} = 25$$

$$\text{TSS} = \sum_{ij} (y_{ij} - \bar{y}_{..})^2$$

$$= (28 - 25)^2 + (35 - 25)^2 + \cdots + (22 - 25)^2 + (18 - 25)^2 = 692$$

$$\text{SST} = n \sum_i (\bar{y}_{i.} - \bar{y}_{..})^2$$

$$= 4[(27.75 - 25)^2 + (25 - 25)^2 + (27 - 25)^2 + (20.25 - 25)^2]$$

$$= 136.5$$

$$\text{SSE} = \text{TSS} - \text{SST} = 692 - 136.5 = 555.5$$

We can now complete the AOV table as follows:

Source	SS	df	MS	F	p-value
Treatments	136.5	3	45.5	0.98	0.4346
Error	555.5	12	46.292		
Total	692	15			

Because p-value $= 0.4346 > .05 = \alpha$, we fail to reject H_0. There is not a significant difference in the mean decrease in reflectivity for the four types of paints.

The researcher is somewhat concerned about the results of the study described in Example 15.1, because he was certain that at least one of the paints would show some improvement over the currently used paint. He examines the road conditions and amount of traffic flow on the 16 sections used in the study and finds that the roadways had a very low traffic volume during the study period. He decides to redesign the study to improve the generalization of the results, and will include four different locations having different amounts of traffic volumes in the new study. Section 15.3 will describe how to conduct this experiment, in which we may have a second source of variability, location of the sections.

15.3 Randomized Complete Block Design

We will now modify the reflective paint study to incorporate four different locations into the design. The researcher identifies four sections of roadway of length 6 feet at each of the four locations. If we randomly assigned the four paints to the 16 sections, we might end up with a randomization scheme like the one listed in Table 15.6.

TABLE 15.6
Random assignment of
the four paints to
the 16 sections

Location			
1	2	3	4
P_1	P_2	P_3	P_4
P_1	P_2	P_3	P_4
P_1	P_2	P_3	P_4
P_1	P_2	P_3	P_4

Even though we still have four observations for each treatment in this design, any differences that we may observe among the reflectivity of the road markings for the four types of paints may be due entirely to the differences in the road conditions and traffic volumes among the four locations. Because the factors **confounded** location and type of paint are **confounded,** we cannot determine whether any observed differences in the decrease in reflectivity of the road markings are due to differences in the locations of the markings or due to differences in the type of paint used in creating the markings. This example illustrates a situation in which the 16 road markings are affected by an extraneous source of variability: the location of road marking. If the four locations present different environmental conditions or different traffic volumes, the 16 experimental units would not be a homogeneous set of units on which we could base an evaluation of the effects of the four treatments, the four types of paint.

The completely randomized design just described is not appropriate for this experimental setting. We need to use a randomized complete block design in order to take into account the differences that exist in the experimental units prior to assigning the treatments. In Chapter 14, we described how we *restrict* our randomization of treatments to experimental units to ensure that each location has a section of roadway painted with each of the four types of paint. One such randomization is listed in Table 15.7. Note that each location contains four sections of roadway, one section treated with each of the four paints. Hence, the variability in the reflectivity of paints due to differences in roadway conditions at the four locations can now be addressed and controlled. This will allow pairwise comparisons among the four paints that utilize the sample means to be free of the variability among locations. For example, if we ran the test

$$H_0: \quad \mu_{P_1} - \mu_{P_2} = 0$$

$$H_a: \quad \mu_{P_1} - \mu_{P_2} \neq 0$$

and rejected H_0, the differences between μ_{P_1} and μ_{P_2} would be due to a difference between the reflectivity properties of the two paints and not due to a difference

TABLE 15.7
Randomized complete block
assignment of the four paints
to the 16 sections

Location			
1	2	3	4
P_2	P_2	P_1	P_1
P_1	P_4	P_3	P_2
P_3	P_1	P_4	P_4
P_4	P_3	P_2	P_3

among the locations, since both paint P_1 and P_2 were applied to a section of roadway at each of the four locations.

In Chapter 14, we discussed how the random assignment of the treatments to the experimental units is conducted separately within each block, the location of the roadways in this example. The four sections within a given location would tend to be more alike with respect to environmental conditions and traffic volume than sections of roadway in two different locations. Thus, we are in essence conducting four independent completely randomized designs, one for each of the four locations. By using the randomized complete block design, we have effectively filtered out the variability among the locations, enabling us to make more precise comparisons among the treatment means μ_{P_1}, μ_{P_2}, μ_{P_3}, and μ_{P_4}.

In general, we can use a randomized complete block design to compare t treatment means when an extraneous source of variability (blocks) is present. If there are b different blocks, we would randomly assign each of the t treatments to an experimental unit in each block in order to filter out the block-to-block variability. In our example, we had $t = 4$ treatments (types of paint) and $b = 4$ blocks (locations).

We can formerly define a randomized complete block design as follows.

DEFINITION 15.1

A **randomized complete block design** is an experimental design for comparing t treatments in b blocks. The blocks consist of t homogeneous experimental units. Treatments are randomly assigned to experimental units within a block, with each treatment appearing exactly once in every block.

The randomized complete block design has certain advantages and disadvantages, as shown here.

Advantages and Disadvantages of the Randomized Complete Block Design

Advantages

1. The design is useful for comparing t treatment means in the presence of a single extraneous source of variability.
2. The statistical analysis is simple.
3. The design is easy to construct.
4. It can be used to accommodate any number of treatments in any number of blocks.

Disadvantages

1. Because the experimental units within a block must be homogeneous, the design is best suited for a relatively small number of treatments.
2. This design controls for only one extraneous source of variability (due to blocks). Additional extraneous sources of variability tend to increase the error term, making it more difficult to detect treatment differences.
3. The effect of each treatment on the response must be approximately the same from block to block.

Consider the data for a randomized complete block design as arranged in Table 15.8. Note that although these data look similar to the data presentation for a completely randomized design (see Table 15.3), there is a difference in the way treatments were assigned to the experimental units.

TABLE 15.8

Data for a randomized complete block design

Treatment	Block				Mean
	1	2	$\cdots$	b	
1	y_{11}	y_{12}	$\cdots$	y_{1b}	$\bar{y}_{1.}$
2	y_{21}	y_{22}	$\cdots$	y_{2b}	$\bar{y}_{2.}$
$\cdots$	$\cdots$	$\cdots$	$\cdots$	$\cdots$	$\cdots$
t	y_{t1}	y_{t2}	$\cdots$	y_{tb}	$\bar{y}_{t.}$
Mean	$\bar{y}_{.1}$	$\bar{y}_{.2}$	$\cdots$	$\bar{y}_{.b}$	$\bar{y}_{..}$

The model for an observation in a randomized complete block design can be written in the form

$$y_{ij} = \mu + \alpha_i + \beta_j + \varepsilon_{ij}$$

where the terms of the model are defined as follows:

y_{ij}:   Observation on experimental unit in jth block receiving treatment i.

μ: Overall mean, an unknown constant.

α_i: An effect due to treatment i, an unknown constant.

β_j: An effect due to block j, an unknown constant.

ε_{ij}:   A random error associated with the response from an experimental unit in block j receiving treatment i. We require that the ε_{ij}s have a normal distribution with mean 0 and common variance σ_ε^2. In addition, the errors must be independent.

The conditions given above for our model can be shown to imply that the recorded response from the ith treatment in the jth block, y_{ij}, is normally distributed with mean

$$E(y_{ij}) = \mu + \alpha_i + \beta_j$$

and variance σ_ε^2. Table 15.9 gives the population means (expected values) for the data of Table 15.8.

TABLE 15.9

Expected values for the y_{ij}s in a randomized complete block design

Treatment	Block			
	1	2	$\cdots$	b
1	$E(y_{11}) = \mu + \alpha_1 + \beta_1$	$E(y_{12}) = \mu + \alpha_1 + \beta_2$	$\cdots$	$E(y_{1b}) = \mu + \alpha_1 +$
2	$E(y_{21}) = \mu + \alpha_2 + \beta_1$	$E(y_{22}) = \mu + \alpha_2 + \beta_2$	$\cdots$	$E(y_{2b}) = \mu + \alpha_2 +$
$\cdots$	$\cdots$	$\cdots$	$\cdots$	$\cdots$
t	$E(y_{t1}) = \mu + \alpha_t + \beta_1$	$E(y_{t2}) = \mu + \alpha_t + \beta_2$	$\cdots$	$E(y_{tb}) = \mu + \alpha_t +$

Several comments should be made concerning the table of expected values. First, any pair of observations that receive the same treatment (appear in the same row of Table 15.9) have population means that differ only by their block effects (β_j's). For example, the expected values associated with y_{11} and y_{12} (two observations receiving treatment 1) are

$$E(y_{11}) = \mu + \alpha_1 + \beta_1 \qquad E(y_{12}) = \mu + \alpha_1 + \beta_2$$

Thus, the difference in their means is

$$E(y_{11}) - E(y_{12}) = (\mu + \alpha_1 + \beta_1) - (\mu + \alpha_1 + \beta_2) = \beta_1 - \beta_2$$

which accounts for the fact that y_{11} was recorded in block 1 and y_{12} was recorded in block 2 but both were responses from experimental units receiving treatment 1. Thus, there is no treatment effect, but a possible block effect may be present. Second, two observations appearing in the same block (in the same column of Table 15.9) have means that differ by a treatment effect only. For example, y_{11} and y_{21} both appear in block 1. The difference in their means, from Table 15.9, is

$$E(y_{11}) - E(y_{21}) = (\mu + \alpha_1 + \beta_1) - (\mu + \alpha_2 + \beta_1) = \alpha_1 - \alpha_2$$

which accounts for the fact that the experimental units received different treatments but were observed in the same block. Hence, there is a possible treatment effect but no block effect. Finally, when two experimental units receive different treatments and are observed in different blocks, their expected values differ by effects due to both treatment differences and block differences. Thus, observations y_{11} and y_{22} have expectations that differ by

$$E(y_{11}) - E(y_{22}) = (\mu + \alpha_1 + \beta_1) - (\mu + \alpha_2 + \beta_2) = (\alpha_1 - \alpha_2) + (\beta_1 - \beta_2)$$

filtering

Using the information we have learned concerning the model for a randomized block design, we can illustrate the concept of **filtering** and show how the randomized block design filters out the variability due to blocks. Consider a randomized block design with $t = 3$ treatments (1, 2, and 3) laid out in $b = 3$ blocks as shown in Table 15.10.

TABLE 15.10

Randomized complete block design with $t = 3$ treatments and $b = 3$ blocks

Block	Treatment		
1	1	2	3
2	1	3	2
3	3	1	2

The model for this randomized block design is

$$y_{ij} = \mu + \alpha_i + \beta_j + \varepsilon_{ij} \qquad (i = 1, 2, 3; j = 1, 2, 3)$$

Suppose we wish to estimate the difference in mean response for treatments 2 and 1—namely, $\alpha_2 - \alpha_1$. The difference in sample means, $\bar{y}_{2.} - \bar{y}_{1.}$, would represent a point estimate of $\alpha_2 - \alpha_1$. By substituting into our model, we have

$$\bar{y}_{1.} = \frac{1}{3} \sum_j y_{1j}$$

$$= \frac{1}{3}[(\mu + \alpha_1 + \beta_1 + \varepsilon_{11}) + (\mu + \alpha_1 + \beta_2 + \varepsilon_{12}) + (\mu + \alpha_1 + \beta_3 + \varepsilon_{13})]$$

$$= \mu + \alpha_1 + \bar{\beta} + \bar{\varepsilon}_1$$

where $\bar{\beta}$ represents the mean of the three block effects β_1, β_2, and β_3, and $\bar{\varepsilon}_1$ represents the mean of the three random errors ε_{11}, ε_{12}, and ε_{13}. Similarly, it is easy to show that

$$\bar{y}_2. = \mu + \alpha_2 + \bar{\beta} + \bar{\varepsilon}_2$$

and hence

$$\bar{y}_2. - \bar{y}_1. = (\alpha_2 - \alpha_1) + (\bar{\varepsilon}_2 - \bar{\varepsilon}_1)$$

Note how the block effects cancel, leaving the quantity $(\bar{\varepsilon}_2 - \bar{\varepsilon}_1)$ as the error of estimation using $\bar{y}_2. - \bar{y}_1.$ to estimate $\alpha_2 - \alpha_1$.

If a completely randomized design had been employed instead of a randomized block design, treatments would have been assigned to experimental units at random and it is quite unlikely that each treatment would have appeared in each block. When the same treatment appears more than once in a block and we calculate an estimate of $\alpha_2 - \alpha_1$ using $\bar{y}_2. - \bar{y}_1.$, all block effects would not cancel out as they did previously. Then the error of estimation would include not only $\bar{\varepsilon}_2 - \bar{\varepsilon}_1$ but also the block effects that do not cancel; that is,

$$\bar{y}_2. - \bar{y}_1. = \alpha_2 - \alpha_1 + [(\bar{\varepsilon}_2 - \bar{\varepsilon}_1) + (\text{block effects that do not cancel})]$$

Hence, the randomized block design filters out variability due to blocks by decreasing the error of estimation for a comparison of treatment means.

A plot of the expected values, μ_{ij} in Figure 15.1, demonstrates that the size of the difference between the means of observations receiving the same treatment but in different blocks (say, j and j') is the same for all treatments. That is,

$$\mu_{ij} - \mu_{ij'} = \beta_j - \beta_{j'}, \qquad \text{for all } i = 1, \ldots t$$

A consequence of this condition is that the lines connecting the means having the same treatment form a set of parallel lines.

FIGURE 15.1

Treatment means in a randomized block design

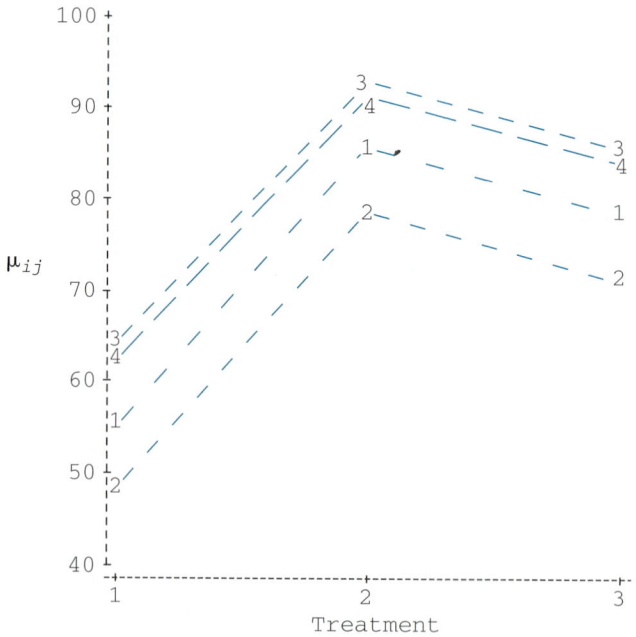

Plot of Treatment Mean by Treatment
(symbol is value of block)

The main goal in using the randomized complete block design was to examine differences in the t treatment means $\mu_{1.}, \mu_{2.}, \ldots, \mu_{t.}$, where $\mu_{i.}$ is the mean response of treatment i. The null hypothesis is *no difference among treatment means* versus the research hypothesis *treatment means differ*. That is,

$$H_0: \quad \mu_{1.} = \mu_{2.} = \cdots = \mu_{t.} \qquad H_a: \quad \text{At least one } \mu_{i.} \text{ differs from the rest.}$$

This set of hypothesis is equivalent to testing

$$H_0: \quad \alpha_1 = \alpha_2 = \cdots = \alpha_t = 0 \qquad H_a: \quad \text{At least one } \alpha_i \text{ different from 0.}$$

The two sets of hypotheses are equivalent because, as we observed in Table 15.9, when comparing the mean response of two treatments (say, i and i') observed in the same block, the difference in their mean response is

$$\mu_{i.} - \mu_{i.'} = \alpha_i - \alpha_{i'}$$

Thus, under H_0, we are assuming that treatments have the same mean response with a given block. Our test statistic will be obtained by examining the model for a randomized block design and partitioning the total sum of squares to include terms for treatment effects, block effects, and random error effects. Using Table 15.8 we can introduce notation that is needed in the partitioning of the total sum of squares. This notation is presented here.

y_{ij}: Observation for treatment i in block j

t: Number of treatments

b: Number of blocks

$\bar{y}_{i.}$: Sample mean for treatment i, $\bar{y}_{i.} = \dfrac{1}{b} \sum_{j=1}^{b} y_{ij}$

$\bar{y}_{.j}$: Sample mean for block j, $\bar{y}_{.j} = \dfrac{1}{t} \sum_{i=1}^{t} y_{ij}$

$\bar{y}_{..}$: Overall sample mean, $\bar{y}_{..} = \dfrac{1}{tb} \sum_{ij} y_{ij}$

total sum of squares
The **total sum of squares** of the measurements about their mean $\bar{y}_{..}$ is defined as before:

$$\text{TSS} = \sum_{ij} (y_{ij} - \bar{y}_{..})^2$$

This sum of squares will be partitioned into three separate sources of variability: one due to the variability among treatments, one due to the variability among blocks, and one due to the variability from all sources not accounted for by either treatment differences or block differences. We call this source of variability **error.**

error
partition of TSS
The **partition of TSS** follows from an examination of the randomized complete block model:

$$y_{ij} = \mu + \alpha_i + \beta_j + \varepsilon_{ij}$$

The parameters in the model have sample estimates:

$$\hat{\mu} = \bar{y}_{..} \qquad \hat{\alpha}_i = \bar{y}_{i.} - \bar{y}_{..} \qquad \text{and} \qquad \hat{\beta}_j = \bar{y}_{.j} - \bar{y}_{..}$$

It can be shown algebraically that TSS takes the following form:

$$\sum_{ij} (y_{ij} - \bar{y}_{..})^2 = b \sum_{i} (\bar{y}_{i.} - \bar{y}_{..})^2 + t \sum_{j} (\bar{y}_{.j} - \bar{y}_{..})^2 + \sum_{ij} (y_{ij} - \bar{y}_{i.} - \bar{y}_{.j} + \bar{y}_{..})^2$$

We will interpret the terms in the partition using the parameter estimates. The first quantity on the right-hand side of the equal sign measures the variability of the treatment means $\bar{y}_{i.}$ from the overall mean $\bar{y}_{..}$. Thus,

$$\text{SST} = b \sum_i (\bar{y}_{i.} - \bar{y}_{..})^2 = b \sum_i (\hat{\alpha}_i)^2$$

called the **between-treatment sum of squares,** is a measure of the variability in the y'_{ij}s due to differences in the treatment means. Similarly, the second quantity,

$$\text{SSB} = t \sum_j (\bar{y}_{.j} - \bar{y}_{..})^2 = t \sum_j (\hat{\beta}_j)^2$$

measures the variability between the block means $\bar{y}_{.j}$ and the overall mean. It is called the **between-block sum of squares.** The third source of variability, referred to as the **sum of squares for error,** SSE, represents the variability in the y'_{ij}s not accounted for by the block and treatment differences. There are several forms for this term:

$$\text{SSE} = \sum_{ij} (e_{ij})^2 = \sum_{ij} (y_{ij} - \hat{\mu} - \hat{\alpha}_i - \hat{\beta}_j)^2 = \text{TSS} - \text{SST} - \text{SSB}$$

where $e_{ij} = y_{ij} - \hat{\mu} - \hat{\alpha}_i - \hat{\beta}_j$ are the residuals used to check model conditions. We can summarize our calculations in an AOV table as given in Table 15.11.

TABLE 15.11

Analysis of variance table for a randomized complete block design

Source	SS	df	MS	F
Treatments	SST	$t - 1$	$\text{MST} = \text{SST}/(t - 1)$	MST/MSE
Blocks	SSB	$b - 1$	$\text{MSB} = \text{SSB}/(b - 1)$	MSB/MSE
Error	SSE	$(b - 1)(t - 1)$	$\text{MSE} = \text{SSE}/(b - 1)(t - 1)$	
Total	TSS	$bt - 1$		

The test statistic for testing

$$H_0: \quad \alpha_1 = \alpha_2 = \cdots = \alpha_t = 0 \qquad H_a: \quad \text{At least one } \alpha_i \text{ is different from } 0$$

is the ratio

$$F = \frac{\text{MST}}{\text{MSE}}$$

When $H_0: \alpha_1 = \alpha_2 = \cdots = \alpha_t = 0$ is true, both MST and MSE are **unbiased estimates** of σ_ε^2, the variance of the experimental error. That is, when H_0 is true, both MST and MSE have a mean value in repeated sampling, called the **expected mean squares,** equal to σ_ε^2. We express these terms as

$$E(\text{MST}) = \sigma_\varepsilon^2 \qquad E(\text{MSE}) = \sigma_\varepsilon^2$$

We would thus expect $F = \text{MST}/\text{MSE}$ to have a value near 1.

When H_a is true, the expected value of MSE is still σ_ε^2. However, MST is no longer unbiased for σ_ε^2. In fact, the expected mean square for treatments can be shown to be

$$E(\text{MST}) = \sigma_\varepsilon^2 + b\theta_T, \qquad \text{where } \theta_T = \frac{1}{t - 1} \sum_i \alpha_i^2$$

Thus, a large difference in the treatment means will result in a large value for θ_T. The expected value of MST will then be larger than the expected value of MSE and we would expect $F = \text{MST/MSE}$ to be larger than 1. Thus, our test statistic F rejects H_0 when we observe a value of F larger than a value in the upper tail of the F distribution.

The above discussion leads to the following decision rule for a specified probability of a Type I error:

Reject H_0: $\alpha_1 = \alpha_2 = \cdots = \alpha_t = 0$ when $F = \text{MST/MSE}$ exceeds $F_{a,\text{df}_1,\text{df}_2}$

where $F_{a,\text{df}_1,\text{df}_2}$ is from the F tables in Appendix Table 8 with $a = $ specified value of probability Type I error, $\text{df}_1 = \text{df}_{\text{MST}} = t - 1$, and $\text{df}_2 = \text{df}_{\text{MSE}} = (b - 1)(t - 1)$. Alternatively, we can compute the p-value for the observed value of the test statistic F_{obs} by computing

$$p\text{-value} = P(F_{\text{df}_1,\text{df}_2} > F_{\text{obs}})$$

where the F-distribution with $\text{df}_1 = t - 1$ and $\text{df}_2 = (b - 1)(t - 1)$ is used to compute the probability. We would then compare the p-value to a selected value for the probability of Type I error, with small p-values supporting the research hypothesis and large p-values failing to reject H_0.

The block effects are generally assessed only to determine whether or not the blocking was efficient in reducing the variability in the experimental units. Thus, hypotheses about the block effects are not tested. However, we might still ask whether blocking has increased our precision for comparing treatment means in a given experiment. Let MSE_{RCB} and MSE_{CR} denote the mean square errors for a randomized complete block design and a completely randomized design, respectively. One measure of precision for the two designs is the variance of the estimate of the ith treatment mean, $\hat{\mu}_i = \bar{y}_{i.}$ $(i = 1, 2, \ldots, t)$. For a randomized complete block design, the estimated variance of $\bar{y}_{i.}$ is $\text{MSE}_{\text{RCB}}/b$. For a completely randomized design, the estimated variance of $\bar{y}_{i.}$ is MSE_{CR}/r, where r is the number of observations (replications) of each treatment required to satisfy the relationship

$$\frac{\text{MSE}_{\text{CR}}}{r} = \frac{\text{MSE}_{\text{RCB}}}{b} \qquad \text{or} \qquad \frac{\text{MSE}_{\text{CR}}}{\text{MSE}_{\text{RCB}}} = \frac{r}{b}$$

relative efficiency
RE(RCB, CR)

The quantity r/b is called the **relative efficiency** of the randomized complete block design compared to a completely randomized design **RE(RCB, CR).** The larger the value of MSE_{CR} compared to MSE_{RCB}, the larger r must be to obtain the same level of precision for estimating a treatment mean in a completely randomized design as obtained using the randomized complete block design. Thus, if the blocking is effective, we would expect the variability in the experimental units to be smaller in the randomized complete block design than would be obtained in a completely randomized design. The ratio $\text{MSE}_{\text{CR}}/\text{MSE}_{\text{RCB}}$ should be large, which would result in r being much larger than b. Thus, the amount of data needed to obtain the same level of precision in estimating μ_i would be larger in the completely randomized design than in the randomized complete block design. When the blocking is not effective, then the ratio $\text{MSE}_{\text{CR}}/\text{MSE}_{\text{RCB}}$ would be nearly 1 and r and b would be equal.

In practice, evaluating the efficiency of the randomized complete block design relative to a completely randomized design cannot be accomplished because the completely randomized design was not conducted. However, we can use the mean squares from the randomized complete block design, MSB and MSE, to obtain the relative efficiency RE(RCB, CR) by using the formula

$$RE(RCB, CR) = \frac{MSE_{CR}}{MSE_{RCB}} = \frac{(b-1)MSB + b(t-1)MSE}{(bt-1)MSE}$$

When RE(RCB, CR) is much larger than 1, then r is greater than b and we would conclude that the blocking was efficient, because many more observations would be required in a completely randomized design than would be required in the randomized complete block design.

EXAMPLE 15.2

A researcher conducted an experiment to compare the effects of three different insecticides on a variety of string beans. To obtain a sufficient amount of data, it was necessary to use four different plots of land. Since the plots had somewhat different soil fertility, drainage characteristics, and sheltering from winds, the researcher decided to conduct a randomized complete block design with the plots serving as the blocks. Each plot was subdivided into three rows. A suitable distance was maintained between rows within a plot so that the insecticides could be confined to a particular row. Each row was planted with 100 seeds and then maintained under the insecticide assigned to the row. The insecticides were randomly assigned to the rows within a plot so that each insecticide appeared in one row within all four plots. The response y_{ij} of interest was the number of seedlings that emerged per row. The data and means are given in Table 15.12.

TABLE 15.12

Number of seedlings by insecticide and plot for Example 15.2

Insecticide	Plot 1	Plot 2	Plot 3	Plot 4	Insecticide Mean
1	56	48	66	62	58
2	83	78	94	93	87
3	80	72	83	85	80
Plot Mean	73	66	81	80	75

a. Write an appropriate statistical model for this experimental situation.
b. Run an analysis of variance to compare the effectiveness of the three insecticides. Use $\alpha = .05$.
c. Summarize your results in an AOV table.
d. Compute the relative efficiency of the randomized block design relative to a completely randomized design.

Solution   We recognize this experimental design as a randomized complete block design with $b = 4$ blocks (plots) and $t = 3$ treatments (insecticides) per block. The appropriate statistical model is

$$y_{ij} = \mu + \alpha_i + \beta_j + \varepsilon_{ij} \qquad i = 1, 2, 3 \qquad j = 1, 2, 3, 4$$

From the information in Table 15.12, we can estimate the treatment means $\mu_{i.}$ by $\hat{\mu}_{i.} = \bar{y}_{i.}$, which yields

$$\hat{\mu}_{1.} = 58 \qquad \hat{\mu}_{2.} = 87 \qquad \hat{\mu}_{3.} = 80$$

It would appear that the rows treated with insecticide 1 yielded many fewer plants than the other two insecticides. We will next estimate the model parameters and construct the AOV table. Recall that $\hat{\mu} = \bar{y}_{..}$, $\hat{\alpha}_i = \bar{y}_{i.} - \bar{y}_{..}$, and $\hat{\beta}_j = \bar{y}_{.j} - \bar{y}_{..}$. Thus, with $\hat{\mu} = \bar{y}_{..} = 75$, we obtain

Insecticide Effects
$\hat{\alpha}_1 = 58 - 75 = -17$
$\hat{\alpha}_2 = 87 - 75 = 12$
$\hat{\alpha}_3 = 80 - 75 = 5$

Plot Effects
$\hat{\beta}_1 = 73 - 75 = -2$
$\hat{\beta}_2 = 66 - 75 = -9$
$\hat{\beta}_3 = 81 - 75 = 6$
$\hat{\beta}_4 = 80 - 75 = 5$

Substituting into the formulas for the sum of squares, we have

$$\text{TSS} = \sum_{ij} (y_{ij} - \bar{y}_{..})^2 = (56 - 75)^2 + (48 - 75)^2 + \cdots + (85 - 75)^2 = 2{,}296$$

$$\text{SST} = b \sum_i (\hat{\alpha}_i)^2 = 4[(-17)^2 + (12)^2 + (5)^2] = 1{,}832$$

$$\text{SSB} = t \sum_j (\hat{\beta}_j)^2 = 3[(-2)^2 + (-9)^2 + (6)^2 + (5)^2] = 438$$

By subtraction, we have

$$\text{SSE} = \text{TSS} - \text{SST} - \text{SSB} = 2{,}296 - 1{,}832 - 438 = 26$$

The analysis of variance table in Table 15.13 summarizes our results. Note that the mean square for a source in the AOV table is computed by dividing the sum of squares for that source by its degrees of freedom.

TABLE 15.13
AOV table for the data of Example 15.2

Source	SS	df	MS	F	p-value
Treatments	1,832	2	916	211.38	.0001
Blocks	438	3	146	33.69	.0004
Error	26	6	4.3333		
Total	2,296	11			

The F test for differences in the treatment means—namely,

$$H_0: \quad \alpha_1 = \alpha_2 = \cdots = \alpha_t = 0 \text{ versus } H_a: \text{ at least one } \alpha_i \text{ is different from } 0$$

makes use of the F statistic MST/MSE. Since the computed value of F, 211.38, is greater than the tabulated F-value, 5.14, based on $\text{df}_1 = 2$, $\text{df}_2 = 6$, and $a = .05$, we reject H_0 and conclude that there are significant (p-value $< .0001$) differences in the mean number of seedlings among the three insecticides.

We will next assess whether the blocking was effective in increasing the precision of the analysis relative to a completely randomized design. From the AOV table, we have MSB = 146 and MSE = 4.3333. Hence, the relative efficiency of this randomized block design relative to a completely randomized design is

$$\text{RE(RCB, CR)} = \frac{(b - 1)\text{MSB} + b(t - 1)\text{MSE}}{(bt - 1)\text{MSE}}$$
$$= \frac{(4 - 1)(146) + 4(3 - 1)(4.3333)}{[(4)(3) - 1](4.3333)} = 9.92$$

That is, approximately ten times as many observations of each treatment would be required in a completely randomized design to obtain the same precision for estimating the treatment means as with this randomized complete block design. The plots were considerably different in their physical characteristics and hence it was crucial that blocking be used in this experiment.

The results in Example 15.2 are valid only if we can be assured that the conditions placed on the model are consistent with the observed data. Thus, we use the residuals $e_{ij} = y_{ij} - \hat{\mu} - \hat{\alpha}_i - \hat{\beta}_j$ to assess whether the conditions of normality, equal variance, and independence appear to be satisfied for the observed data. The following example includes the computer output for such an analysis.

EXAMPLE 15.3

The computer output for the experiment described in Example 15.2 is displayed here. Compare the results to those obtained using the definition of the sum of squares and assess whether the model conditions appear to be valid.

```
Dependent Variable: NUMBER OF SEEDLINGS


                                Sum of         Mean
Source                  DF     Squares       Square    F Value   Pr > F
Model                    5   2270.0000     454.0000    104.77    0.0001
Error                    6     26.0000       4.3333
Corrected Total         11   2296.0000


Source                  DF   Type I SS   Mean Square   F Value   Pr > F
INSECTICIDES             2   1832.0000      916.0000    211.38   0.0001
PLOTS                    3    438.000       146.0000     33.69   0.0004


RESIDUAL ANALYSIS

Variable=RESIDUALS


                              Moments


         N                  12   Sum Wgts          12
         Mean                0   Sum                0
         Std Dev      1.537412   Variance    2.363636
         Skewness     -0.54037   Kurtosis    -0.25385
                                                            Test of Normality
        W:Normal       0.942499   Pr<W          0.4938


    Stem Leaf                        #           Boxplot
     2 00                            2              |
     1 000                           3           +-----+
     0 000                           3           *--+--*
    -0                                              |     |
    -1 00                            2           +-----+
    -2 0                             1              |
    -3 0                             1              |
        ----+----+----+----+
```

Variable=RESIDUALS

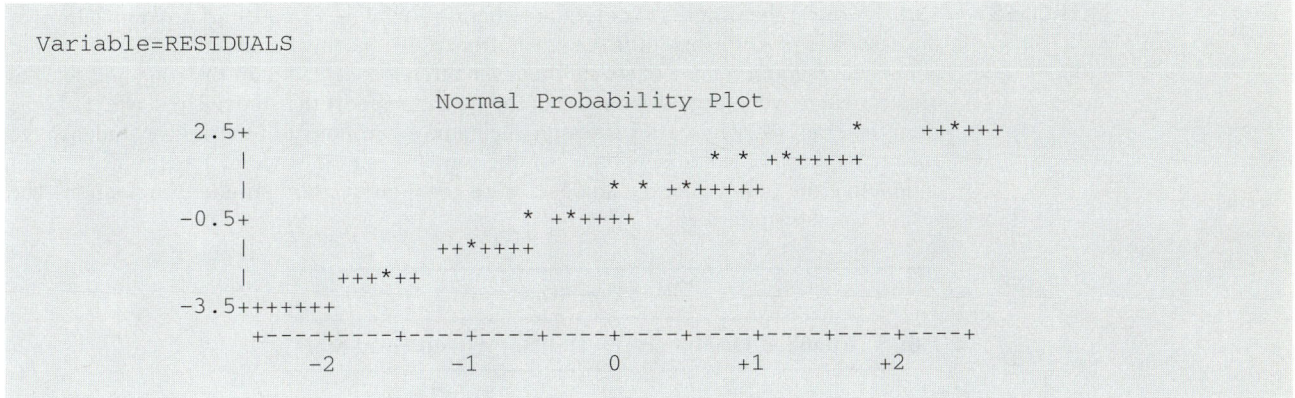

Solution Note that our hand calculations yielded the same values as are given in the computer output. Generally, there will be some rounding errors in our hand calculations, which can lead to values that will differ from these given in the computer output. It is strongly recommended that a computer software program be used in the analysis of variance calculations because of the potential for rounding errors. In assessing whether the model conditions have been met, we first note that in regard to the normality condition the test of H_0: residuals have normal distribution, the p-value from the Shapiro–Wilks test is p-value = .4938. Thus, we would not reject H_0 and the normality condition appears to be satisfied. Also, the stem and leaf plot, boxplot, and normal probability plot are also consistent with the condition that the residuals have a normal distribution. Figure 15.2 is a plot of the residuals versus the estimated treatment means. From this plot it would appear that the variability in the residuals is somewhat constant across the treatments.

FIGURE 15.2

Residuals versus treatment means from Example 15.2

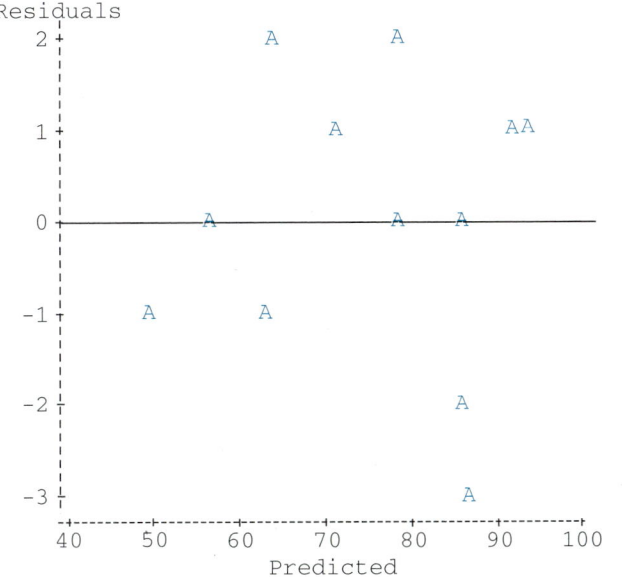

15.1   A researcher wanted to determine whether attending a Head Start program improves the academic performance of first graders from a low-income community. The researcher obtained a random sample of six children who attended a Head Start program and six who did not. There were large differences in the type of support the students received at home from their parents. Thus, after assessing the home environment of the twelve students, the researcher paired the students based on the similarities in their home environment. After completing the first grade, the students were given an overall aptitude examination. The results are shown here.

	Exam Score	
Pair	**Attended Head Start**	**Did Not Attend Head Start**
1	58	47
2	73	67
3	85	69
4	76	62
5	88	77
6	90	77

a. Do the students who attended a Head Start program appear to have higher mean aptitude scores than the students who did not attend such a program? Use $\alpha = .05$.
b. Give the efficiency of the randomized complete block design relative to a completely randomized design. Interpret your findings.

15.2   Refer to Exercise 15.1. Analyze these same data using the paired t test. Compare your results to the results from Exercise 15.1. (The F test for testing treatment differences in a randomized complete block design when there are only two treatments is equivalent to the paired t test of Chapter 6. This can be shown by noting that $t^2 = F$, where F is the F test for treatment differences.)

15.3   An experiment compares four different mixtures of the components oxidizer, binder, and fuel used in the manufacturing of rocket propellant. The four mixtures under test, corresponding to settings of the mixture proportions for oxide, are shown here.

Mixture	**Oxidizer**	**Binder**	**Fuel**
1	.4	.4	.2
2	.4	.2	.4
3	.6	.2	.2
4	.5	.3	.2

To compare the four mixtures, five different samples of propellant are prepared from each mixture and readied for testing. Each of five investigators is randomly assigned one sample of each of the four mixtures and asked to measure the propellant thrust. These data are summarized next.

	Investigator				
Mixture	**1**	**2**	**3**	**4**	**5**
1	2,340	2,355	2,362	2,350	2,348
2	2,658	2,650	2,665	2,640	2,653
3	2,449	2,458	2,432	2,437	2,445
4	2,403	2,410	2,418	2,397	2,405

a. Identify the blocks and treatments for this experimental design.
b. Indicate the method of randomization.
c. Why would this design be preferable to a completely randomized design?

15.4 Refer to Exercise 15.3.
a. Write a model for this experimental setting.
b. Estimate the parameters in the model.
c. Use the computer output shown here to conduct an analysis of variance. Use $\alpha = .05$.
d. What conclusions can you draw concerning the best mixture from the four tested? (*Note*: The higher the response value, the better is the rocket propellant thrust.)
e. Compute the relative efficiency of the randomized block design relative to a completely randomized design. Interpret this value. Were the blocks effective in reducing the variability in experimental units? Explain.

```
General Linear Models Procedure For Data in Exercise 15.3

Dependent Variable: THRUST

                                  Sum of          Mean
Source                   DF       Squares         Square    F Value    Pr > F
Model                     7      261713.45       37387.64    542.96    0.0001
Error                    12         826.30          68.86
Corrected Total          19      262539.75

                R-Square               C.V.       Root MSE            Y Mean
                0.996853           0.336807         8.2981            2463.8

Source                   DF    Type I SS   Mean Square    F Value    Pr > F
M                         3    261260.95      87086.98    1264.73    0.0001
I                         4       452.50        113.12       1.64    0.2273
```

Psy.   **15.5** An industrial psychologist working for a large corporation designs a study to evaluate the effect of background music on the typing efficiency of secretaries. The psychologist selects a random sample of seven secretaries from the secretarial pool. Each subject is exposed to three types of background music: no music, classical music, and hard rock music. The subject is given a standard typing test that combines an assessment of speed with a penalty for typing errors. The particular order of the three experiments is randomized for each of the seven subjects. The results are given here with a high score indicating a superior performance. This is a special type of randomized complete block design in which a single experimental unit serves as a block and receives all treatments.

Type of Music	Subject						
	1	2	3	4	5	6	7
No Music	20	17	24	20	22	25	18
Hard Rock	20	18	23	18	21	22	19
Classical	24	20	27	22	24	28	16

a. Write a statistical model for this experiment and estimate the parameters in your model.
b. Are there differences in the mean typing efficiency for the three types of music? Use $\alpha = .05$.

c. Does the additive model for a randomized complete block design appear to be appropriate? (*Hint:* Plot the data as was done in Figure 15.1.)
d. Compute the relative efficiency of the randomized block design relative to a completely randomized design. Interpret this value. Were the blocks effective in reducing the variability in experimental units? Explain.

15.6 Refer to Exercise 15.5. The computer output for the data in Exercise 15.5 follows. Compare your results with the results given here. Do the model conditions appear to be satisfied?

```
General Linear Models Procedure for Exercise 15.5


Dependent Variable: TYPING EFFICIENCY


                                Sum of          Mean
Source                  DF      Squares         Square      F Value    Pr > F
Model                    8      180.28571       22.53571       9.53    0.0004
Error                   12       28.38095        2.36508
Corrected Total         20      208.66667


                R-Square            C.V.        Root MSE               Y Mean
                0.863989         7.208819         1.5379               21.333


Source                  DF      Type I SS    Mean Square    F Value    Pr > F
M                        2       30.95238       15.47619       6.54    0.0120
S                        6      149.33333       24.88889      10.52    0.0003


RESIDUAL ANALYSIS:


Variable=RESIDUAL


                              Moments


            N                21   Sum Wgts          21
            Mean              0   Sum                0
            Std Dev    1.191238   Variance    1.419048
            Skewness   -0.77527   Kurtosis    2.587721


            W:Normal   0.936418   Pr< W        0.1813


        Stem Leaf                      #          Boxplot
          2 5                          1             0
          1 03                         2             |
          0 001355789                  9          +--+--+
         -0 9985211                    7          +-----+
         -1 8                          1             |
         -2
         -3 3                          1             0
            ----+----+----+----+
```

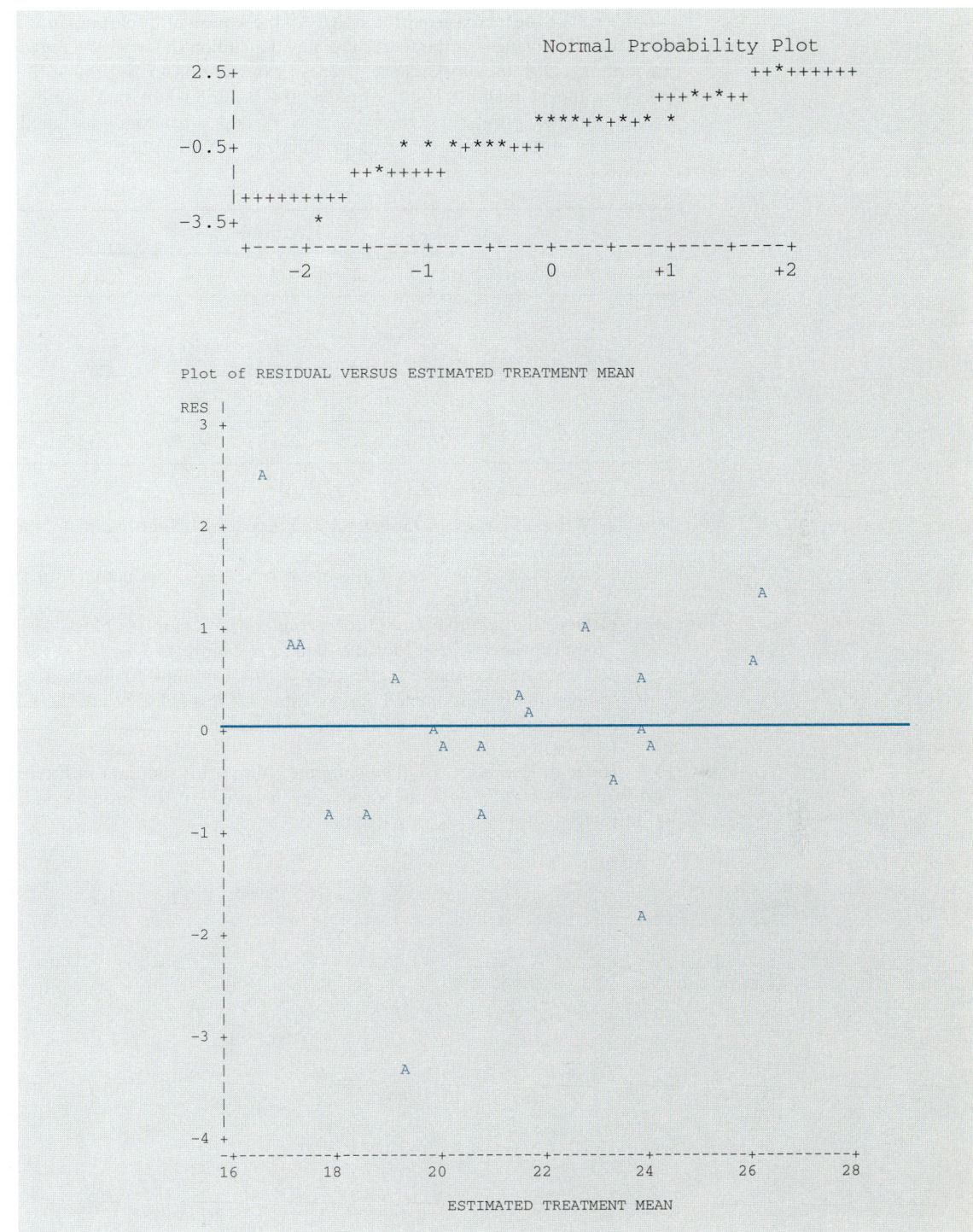

Psy. **15.7** A quality control engineer is considering implementing a workshop to instruct workers on the principles of total quality management (TQM). The program would be quite
expensive to implement across the whole corporation; hence the engineer has designed a
study to evaluate which of four types of workshops would be most effective. The response

variable will be the increase in productivity of the worker after participating in the workshop. Since the effectiveness of the workshop may depend on the worker's preconceived attitude concerning TQM, the workers are given an examination to determine their attitude prior to taking the workshop. Their attitudes are classified into five groups. There are four workers in each group, and the type of workshop is randomly assigned to the workers within each group. The increases in productivity are given here.

| | **Attitude** | | | | | |
Type of Workshop	**1**	**2**	**3**	**4**	**5**	**Mean**
A	33	38	39	42	62	42.8
B	35	37	43	47	71	46.6
C	40	42	45	52	74	50.6
D	54	50	55	62	84	61.0
Mean	40.5	41.75	45.5	50.75	72.75	50.25

a. Write a statistical model for this experiment and estimate the parameters in your model.
b. Are there differences in the mean increase in productivity for the four types of workshops? Use $\alpha = .05$.
c. Does the additive model for a randomized complete block design appear to be appropriate? (*Hint*: Plot the data as in Figure 15.1.)
d. Compute the relative efficiency of the randomized block design relative to a completely randomized design. Interpret this value. Were the blocks effective in reducing the variability in experimental units? Explain.

15.8 Refer to Exercise 15.7. The computer output for the data in Exercise 15.7 follows. Compare your results with the results given here. Do the model conditions appear to be satisfied?

```
General Linear Models Procedure for Exercise 15.7


Dependent Variable: INCREASE IN PRODUCTIVITY


                            Sum of          Mean
Source            DF        Squares         Square     F Value   Pr > F
Model              7       3708.0500       529.7214    114.12    0.0001
Error             12         55.7000         4.6417
Corrected Total   19       3763.7500


                R-Square          C.V.         Root MSE          Y Mean
                0.985201       4.287468         2.1545          50.250


Source            DF        Type I SS     Mean Square   F Value   Pr > F
W                  3        922.5500       307.5167      66.25    0.0001
A                  4       2785.5000       696.3750     150.03    0.0001
```

```
RESIDUAL ANALYSIS:

                              Moments
          N              20  Sum Wgts          20
          Mean            0  Sum               0
          Std Dev   1.712185 Variance    2.931579
          Skewness  0.207266 Kurtosis    0.149301


          W:Normal  0.985267 Pr<W          0.9725

     Stem Leaf                    #          Boxplot
        3 7                       1             |
        2 8                       1             |
        1 029                     3             |
        0 5599                    4          +--+--+
       -0 88110                   5          *-----*
       -1 8321                    4          +-----+
       -2 5                       1             |
       -3 3                       1             |
          ----+---+----+----+
                   Normal Probability Plot
      3.5+                                      +*++++
         |                                   ++*+++
         |                               +++*+*
         |                          ++***+  *
         |                       +***+  *
         |                  *+*+**
         |             +++*++
     -3.5+   ++++*+
         +----+----+----+----+----+----+----+----+----+----+
             -2        -1         0        -1        +2
```

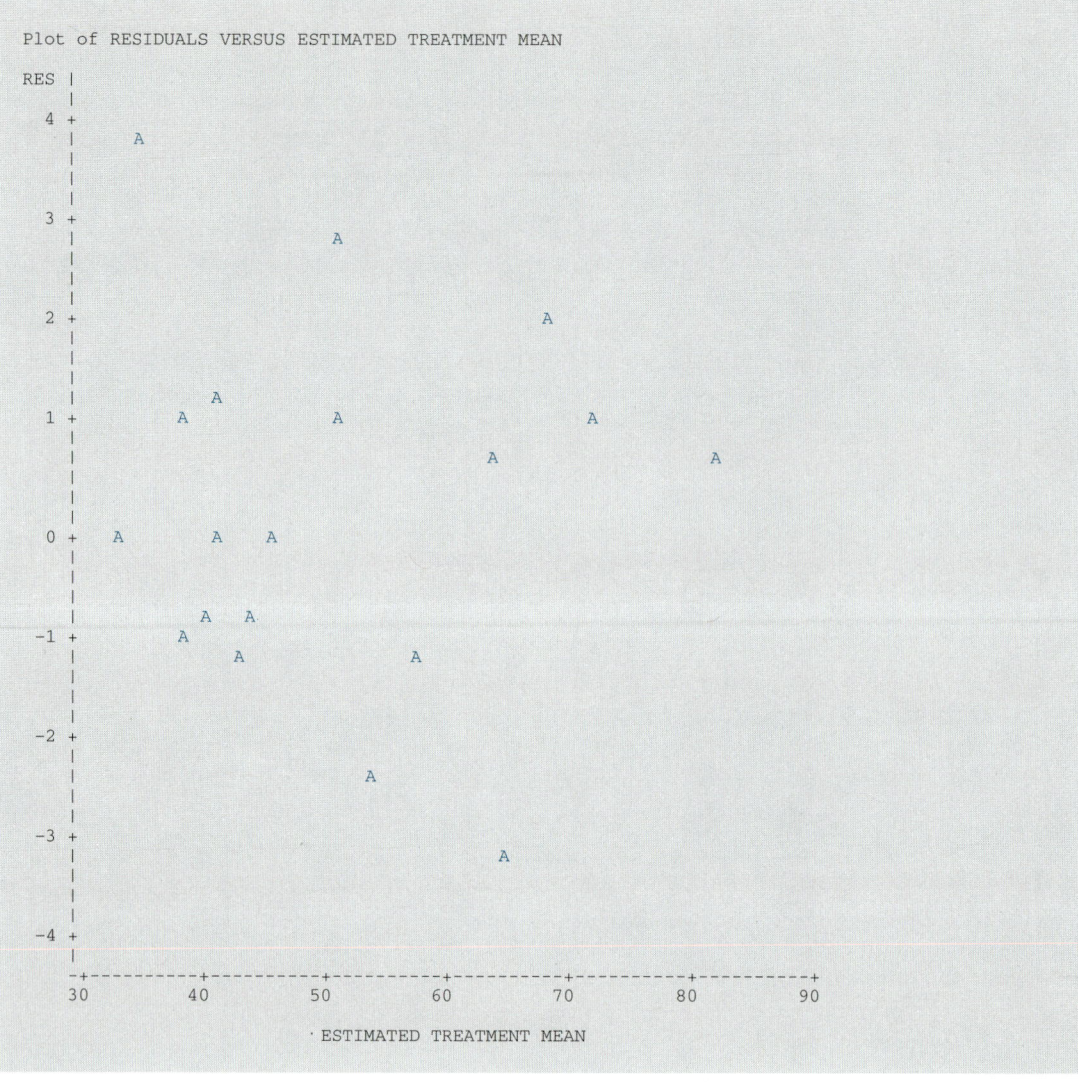

15.4 Latin Square Design

The randomized complete block design is used when there is one factor of interest and the experimenter wants to control a single source of extraneous variation. When there are two possible sources of extraneous variation, a **Latin square design** is the appropriate design for the experiment. Consider the following example.

EXAMPLE 15.4

A large law firm is studying which of four spreadsheets would be most appropriate for their secretarial pool. All four spreadsheets are from nationally known software companies and hence are acceptable to the company with respect to quality of their output. The final choice will thus be determined by which of the four is the easiest to learn. The software consultant for the firm notes that the time to learn various subroutines in a spreadsheet depends on the individual secretary and the type of problem being analyzed. Since the secretaries involved in the study would be unavailable for normal work, the law firm decides to use only four secretaries from the pool and restricts the time that they can be away from their regular assignments. Thus the consultant decides to have each secretary complete a task involving one of four types of problems. The factors to be considered in the study are

Spreadsheet: A, B, C, D

Secretary: 1, 2, 3, 4

Problem: I (accounting), II (data tables), III (summary statistics), IV (graphics)

The factors secretary and problem type are extraneous sources of variation that must be taken into account but are not of central importance to the consultant. The response variable will be the length of time required for the secretary to complete the assigned task. Each secretary will be assigned four problems. The consultant at first considers using the randomized complete block design displayed in Table 15.14.

TABLE 15.14

A randomized complete block design for the spreadsheet study

	Secretary			
Problem	**1**	**2**	**3**	**4**
I	A	A	C	A
II	B	D	A	D
III	D	B	D	B
IV	C	C	B	C

In this design, the type of spreadsheet used is randomly assigned to the problems separately for each secretary. Suppose the type of problem has a strong influence on the amount of time needed to complete the task. In particular, problem type I is by far the most time-consuming of the four problems, whereas problem type IV is the least time-consuming. This design would then produce a strong negative bias for spreadsheet A since it was applied three times using a type I problem and a positive bias to spreadsheet C since it was used three times on type IV problems. Thus, if it is found that spreadsheet C produces the shortest mean completion time for the four tasks, we could not be certain whether spreadsheet C was the better program or whether the results were due to three of its four tasks being type IV problems.

This example illustrates a situation in which the experimental units (problem) are affected by two sources of extraneous variation, the type of problem and the secretary solving the problem. We can modify the randomized complete block design to filter out this second source of variability, the variability among problem type, in addition to filtering out the first source, variability among secretaries. To do this, we restrict our randomization to ensure that each treatment appears in each row (problem type) and in each column (secretary). One such randomization is shown in Table 15.15. Note that the spreadsheets have been assigned to problem types and to secretaries so that each spreadsheet is applied once to each of the problem types and once to each of the secretaries. Hence, pairwise comparisons among spreadsheets that involve the sample means have been adjusted for the variability among problem types and secretaries.

TABLE 15.15

A Latin square design for the spreadsheet study

	Secretary			
Problem	**1**	**2**	**3**	**4**
I	A	B	C	D
II	B	C	D	A
III	C	D	A	B
IV	D	A	B	C

Latin square design

This experimental design is called a **Latin square design.** In general, a Latin square design can be used to compare t treatment means in the presence of two extraneous sources of variability, which we block off into t rows and t columns. The t treatments are then randomly assigned to the rows and columns so that each treatment appears in every row and every column of the design (see Table 15.15). The advantages and disadvantages of the Latin square design are listed here.

Advantages and Disadvantages of the Latin Square Design

Advantages

1. The design is particularly appropriate for comparing t treatment means in the presence of two sources of extraneous variation, each measured at t levels.
2. The analysis is quite simple.

Disadvantages

1. Although a Latin square can be constructed for any value of t, it is best suited for comparing t treatments when $5 \leq t \leq 10$.
2. Any additional extraneous sources of variability tend to inflate the error term, making it more difficult to detect differences among the treatment means.
3. The effect of each treatment on the response must be approximately the same across rows and columns.

The definition of a Latin square design is given here.

DEFINITION 15.2

> A $t \times t$ **Latin square design** contains t rows and t columns. The t treatments are randomly assigned to experimental units within the rows and columns so that each treatment appears in every row and in every column.

The model for a response in a Latin square design can be written in the form

$$y_{ij} = \mu + \alpha_k + \beta_i + \gamma_j + \varepsilon_{ij}$$

where the terms of the model are defined as follows:

y_{ij}: Observation on experimental unit in the ith row and jth column.

μ: Overall mean, an unknown constant.

α_k: An effect due to treatment k, an unknown constant.

β_i: An effect due to row i, an unknown constant.

γ_j: An effect due to column j, an unknown constant.

ε_{ij}: A random error associated with the response from an experimental unit in row i and column j. We require that the ε_{ij}s have a normal distribution with mean 0 and common variance σ_ε^2. In addition, the errors must be independent.

The conditions given above for our model can be shown to imply that the recorded response in the ith row and jth column, y_{ij}, is normally distributed with mean

$$E(y_{ij}) = \mu + \alpha_k + \beta_i + \gamma_j$$

and variance σ_ε^2. Note that we need not specify which treatment the observation is receiving, since once we know in which row and column the experimental unit is located the treatment is then specified.

filtering
We can use the model to illustrate how a Latin square design **filters** out extraneous variability due to row and column sources of variability. To illustrate, we will consider a Latin square design with $t = 4$ treatments (I, II, III, IV) and two sources of extraneous variability, each with $t = 4$ levels. This design is displayed in Table 15.16.

TABLE 15.16
A 4 × 4 Latin square design

Row	Column 1	2	3	4
1	I	II	III	IV
2	II	III	IV	I
3	III	IV	I	II
4	IV	I	II	III

If we wish to estimate $\alpha_3 - \alpha_1$, the difference in the mean response for treatments III and I, using the difference in sample means $\bar{y}_3 - \bar{y}_1$, we can substitute into our model to obtain expressions for $\bar{y}_3$ and $\bar{y}_1$, carefully noting in which rows and columns the treatments appear. With y_{ij} denoting the observation in row i and column j, we have, from Table 15.16,

$$\bar{y}_1 = \frac{1}{4}(y_{11} + y_{24} + y_{33} + y_{42})$$

$$= \mu + \alpha_1 + \frac{1}{4}(\beta_1 + \beta_2 + \beta_3 + \beta_4) + \frac{1}{4}(\gamma_1 + \gamma_2 + \gamma_3 + \gamma_4) + \bar{\varepsilon}_1$$

where $\bar{\varepsilon}_1$ is the mean of the random errors for the four observations on treatment I. Similarly,

$$\bar{y}_3 = \frac{1}{4}(y_{13} + y_{22} + y_{31} + y_{44})$$

$$= \mu + \alpha_3 + \frac{1}{4}(\beta_1 + \beta_2 + \beta_3 + \beta_4) + \frac{1}{4}(\gamma_1 + \gamma_2 + \gamma_3 + \gamma_4) + \bar{\varepsilon}_3$$

Then the sample difference is

$$\bar{y}_3 - \bar{y}_1 = \alpha_3 - \alpha_1 + (\bar{\varepsilon}_3 - \bar{\varepsilon}_1)$$

and the error of estimation for $\alpha_3 - \alpha_1$ is $\bar{\varepsilon}_3 - \bar{\varepsilon}_1$.

If a randomized block design had been used with blocks representing rows, treatments would be randomized within the rows only. It is quite possible for the same treatment to appear more than once in the same column. Then the sample difference would be

$$\bar{y}_3 - \bar{y}_1 = \alpha_3 - \alpha_1 + [(\bar{\varepsilon}_3 - \bar{\varepsilon}_1) + (\text{column effects that do not cancel})]$$

Thus, the error of estimation would be inflated by the column effects that do not cancel out. Following the same reasoning, if a completely randomized design was used when a Latin square design was appropriate, the error of estimation would be inflated by both row and column effects that do not cancel out.

test for treatment effects   We can **test specific hypotheses concerning the parameters in our model.** In particular, we may wish to test the hypothesis of no difference among the t treatment means. This hypothesis can be stated in the form

$$H_0: \quad \alpha_1 = \alpha_2 = \cdots = \alpha_t = 0 \text{ (i.e., the } t \text{ treatment means are identical)}$$

The alternative hypothesis would be

$$H_a: \quad \text{At least one of the } \alpha_i\text{s is not equal to zero (i.e., at least one treatment mean is different from the others)}$$

Our test statistic will be obtained by examining the model for a Latin square design and partitioning the total sum of squares to include terms for treatment effects, row effects, column effects, and random error effects.

total sum of squares   The **total sum of squares** of the measurements about their mean $\bar{y}_{..}$ is defined as before:

$$\text{TSS} = \sum_{ij}(y_{ij} - \bar{y}_{..})^2$$

This sum of squares will be partitioned into four separate sources of variability: one due to the variability among treatments, one due to the variability among rows, one due to the variability among columns, and one due to the variability from all sources not accounted for by either treatment differences or block differ-

error   ences. We call this source of variability **error.** The **partition of TSS** follows from

partition of TSS   an examination of the Latin square model:

$$y_{ij} = \mu + \alpha_k + \beta_i + \gamma_j + \varepsilon_{ij}$$

The parameters in the model have sample estimates:

$$\hat{\mu} = \bar{y}_{..} \qquad \hat{\alpha}_k = \bar{y}_k - \bar{y}_{..} \qquad \hat{\beta}_i = \bar{y}_{i.} - \bar{y}_{..} \qquad \hat{\gamma}_j = \bar{y}_{.j} - \bar{y}_{..}$$

$$\text{TSS} = t \sum_k (\bar{y}_k - \bar{y}_{..})^2 + t \sum_i (\bar{y}_{i.} - \bar{y}_{..})^2 + t \sum_j (\bar{y}_{.j} - \bar{y}_{..})^2 + \text{SSE}$$

We will interpret the terms in the partition using the parameter estimates. The first quantity on the right-hand side of the equal sign measures the variability of the treatment means $\bar{y}_k$ from the overall mean $\bar{y}_{..}$. Thus,

$$\text{SST} = t \sum_k (\bar{y}_k - \bar{y}_{..})^2 = t \sum_k (\hat{\alpha}_k)^2$$

called the **between-treatment sum of squares,** is a measure of the variability in the y'_{ij}s due to differences in the treatment means. The second quantity,

$$\text{SSR} = t \sum_i (\bar{y}_{i.} - \bar{y}_{..})^2 = t \sum_i (\hat{\beta}_i)^2$$

measures the variability between the row means $\bar{y}_{i.}$ and the overall mean. It is called the **between-rows sum of squares.** The third source of variability, referred to as the **between-columns sum of squares,** measures the variability between the column means $\bar{y}_{.j}$ and the overall mean. It is given by

$$\text{SSC} = t \sum_j (\bar{y}_{.j} - \bar{y}_{..})^2 = t \sum_j (\hat{\gamma}_j)^2$$

The final source of variability, designated as the **sum of squares for error,** SSE, represents the variability in the y_{ij}s not accounted for by the row, column, and treatment differences. It is given by

$$\text{SSE} = \text{TSS} - \text{SST} - \text{SSR} - \text{SSC}$$

We can summarize our calculations in an AOV table as given in Table 15.17.

TABLE 15.17

Analysis of variance table for a $t \times t$ Latin square design

Source	SS	df	MS	F
Treatments	SST	$t - 1$	MST = SST/$(t - 1)$	MST/MSE
Rows	SSR	$t - 1$	MSR = SSR/$(t - 1)$	MSR/MSE
Columns	SSC	$t - 1$	MSC = SSC/$(t - 1)$	MSC/MSE
Error	SSE	$(t - 1)(t - 2)$	MSE = SSE/$(t - 1)(t - 2)$	
Total	TSS	$t^2 - 1$		

The test statistic for testing

$$H_0: \quad \alpha_1 = \alpha_2 = \cdots = \alpha_t = 0 \qquad H_a: \quad \text{At least one } \alpha_k \text{ is different from 0}$$

is the ratio

$$F = \frac{\text{MST}}{\text{MSE}}$$

For our model,

$$E(\text{MSE}) = \sigma_\varepsilon^2 \qquad \text{and} \qquad E(\text{MST}) = \sigma_\varepsilon^2 + t\theta_T$$

where $\theta_T = 1/(t - 1)\sum_k \alpha_k^2$. When H_0 is true, α_k equals 0 for all $k = 1, \ldots, t$, and hence $\theta_T = 0$. Thus, when H_0 is true we would expect MST/MSE to be close to 1. However, under the research hypothesis, H_a, θ_T would be positive since at least one α_k is not 0. Thus, a large difference in the treatment means will result in a large value for θ_T. The expected value of MST will then be larger than the expected value of MSE and we would expect $F = $ MST/MSE to be larger than 1. Thus, our test statistic F rejects H_0 when we observe a value of F larger than a value in the upper tail of the F distribution.

The above discussion leads to the following decision rule for a specified probability of a Type I error:

Reject H_0: $\alpha_1 = \alpha_2 = \cdots = \alpha_t = 0$ when $F = $ MST/MSE exceeds $F_{a, \text{df}_1, \text{df}_2}$

where $F_{a, \text{df}_1, \text{df}_2}$ is from the F tables of Appendix Table 8 with $a = $ specified value of probability Type I error, $\text{df}_1 = \text{df}_{\text{MST}} = t - 1$, and $\text{df}_2 = \text{df}_{\text{MSE}} = (t - 1)(t - 2)$. Alternatively, we can compute the p-value for the observed value of the test statistic F_{obs} by computing

$$p\text{-value} = P(F_{\text{df}_1, \text{df}_2} > F_{\text{obs}})$$

where the F-distribution with $\text{df}_1 = t - 1$ and $\text{df}_2 = (t - 1)(t - 2)$ is used to compute the probability. We would then compare the p-value to a selected value for the probability of Type I error, with small p-values supporting the research hypothesis and large values of the p-value failing to reject H_0.

The row and column effects are generally assessed only to determine whether or not accounting for the two extraneous sources of variability was efficient in reducing the variability in the experimental units. Thus, hypotheses about the row and column effects are not generally tested. As with the randomized block design, we can compare the efficiency to that of the completely randomized design. We want to determine whether accounting for the row and column sources of variability has increased our precision for comparing treatment means in a given experiment. Let MSE_{LS} and MSE_{CR} denote the mean square errors for a Latin square design and a completely randomized design, respectively. The **relative efficiency** of the Latin square design compared to a completely randomized design is denoted **RE(LS, CR)**. We can use the mean squares from the Latin square design, MSR, MSC, and MSE, to obtain the relative efficiency RE(LS, CR) by using the formula

relative efficiency

RE(LS, CR)

$$\text{RE(LS, CR)} = \frac{\text{MSE}_{\text{LS}}}{\text{MSE}_{\text{CR}}} = \frac{\text{MSR} + \text{MSC} + (t - 1)\text{MSE}}{(t + 1)\text{MSE}}$$

When RE(LS, CR) is much larger than 1, we conclude that accounting for the row and/or column sources of variability was efficient, since many more observations would be required in a completely randomized design than would be required in Latin square design to obtain the same degree of precision in estimating the treatment means.

EXAMPLE 15.5

The law firm conducted a study of which spreadsheet to implement, and the data are shown in Table 15.18. Use these data to answer the following questions.

a. Write an appropriate statistical model for this experimental situation.
b. Run an analysis of variance to compare the mean time to completion for the spreadsheets. Use $\alpha = .05$.
c. Summarize your results in an AOV table.
d. Compute the relative efficiency of the Latin square design relative to a completely randomized design.

TABLE 15.18

Time required to complete task (in hours), a 4 × 4 Latin square design

Problem	Secretary 1	Secretary 2	Secretary 3	Secretary 4	Row Mean	Spreadsheet Mean
I	A(0.3)	B(1.8)	C(0.7)	D(1.2)	1.0	A: .45
II	B(1.4)	C(1.4)	D(1.1)	A(0.5)	1.1	B: 1.5
III	C(0.5)	D(1.5)	A(0.5)	B(1.1)	0.9	C: 1.05
IV	D(1.0)	A(0.5)	B(1.7)	C(1.6)	1.2	D: 1.2
Column Mean	0.8	1.3	1.0	1.1	1.05	

Solution   We recognize this experimental design as a Latin square design with $t = 4$ rows (problems), $t = 4$ columns (secretaries), and $t = 4$ treatments (spreadsheets). The appropriate statistical model is

$$y_{ij} = \mu + \alpha_k + \beta_i + \gamma_j + \varepsilon_{ij} \qquad i, j, k = 1, 2, 3, 4$$

From the information in Table 15.18, we can estimate the treatment means μ_k by $\hat{\mu}_k = \bar{y}_k$ yielding

$$\hat{\mu}_1 = 0.45 \qquad \hat{\mu}_2 = 1.5 \qquad \hat{\mu}_3 = 1.05 \qquad \hat{\mu}_4 = 1.2$$

It would appear that spreadsheet A has a somewhat shorter mean time to completion than the other three spreadsheets. We will next estimate the model parameters and construct the AOV table. Recall that $\hat{\mu} = \bar{y}_{..}$, $\hat{\alpha}_k = \bar{y}_k - \bar{y}_{..}$, $\hat{\beta}_i = \bar{y}_{i.} - \bar{y}_{..}$, and $\hat{\gamma}_j = \bar{y}_{.j} - \bar{y}_{..}$. Thus, with $\hat{\mu} = \bar{y}_{..} = 1.05$, we obtain

Spreadsheet Effects	Problem Effects	Secretary Effects
$\hat{\alpha}_1 = .45 - 1.05 = -.6$	$\hat{\beta}_1 = 1 - 1.05 = -.05$	$\hat{\gamma}_1 = .8 - 1.05 = -.25$
$\hat{\alpha}_2 = 1.5 - 1.05 = .45$	$\hat{\beta}_2 = 1.1 - 1.05 = .05$	$\hat{\gamma}_2 = 1.3 - 1.05 = .25$
$\hat{\alpha}_3 = 1.05 - 1.05 = 0$	$\hat{\beta}_3 = .9 - 1.05 = -.15$	$\hat{\gamma}_3 = 1 - 1.05 = -.05$
$\hat{\alpha}_4 = 1.2 - 1.05 = .15$	$\hat{\beta}_4 = 1.2 - 1.05 = .15$	$\hat{\gamma}_4 = 1.1 - 1.05 = .05$

Substituting into the formulas for the sum of squares, we have

$$\text{TSS} = \sum_{ij} (y_{ij} - \bar{y}_{..})^2 = (.3 - 1.05)^2 + (1.8 - 1.05)^2 + \cdots + (1.6 - 1.05)^2$$

$$= 3.66$$

$$\text{SST} = t \sum_k (\hat{\alpha}_k)^2 = 4[(-.6)^2 + (.45)^2 + (0)^2 + (.15)^2] = 2.34$$

$$\text{SSR} = t \sum_i (\hat{\beta}_i)^2 = 4[(-.05)^2 + (.05)^2 + (-.15)^2 + (.15)^2] = .2$$

$$\text{SSC} = t \sum_j (\hat{\gamma}_j)^2 = 4[(-.25)^2 + (.25)^2 + (-.05)^2 + (.05)^2] = .52$$

By subtraction, we have

$$\text{SSE} = \text{TSS} - \text{SST} - \text{SSB} = 3.66 - 2.34 - .2 - .52 = .6$$

The analysis of variance table in Table 15.19 summarizes our results. Note that the mean squares for a source in the AOV table is computed by dividing the sum of squares for that source by its degrees of freedom.

The F test for differences in the treatment means—namely,

$$H_0: \quad \alpha_1 = \alpha_2 = \cdots = \alpha_t = 0 \text{ versus } H_a: \quad \text{At least one } \alpha_k \text{ is different from 0}$$

TABLE 15.19
AOV table for the
spreadsheet study
of Example 15.5

Source	SS	df	MS	F	p-value
Spreadsheets	2.34	3	.78	7.8	.0171
Problem	0.2	3	.067	0.67	.6025
Secretary	0.52	3	.173	1.73	.2592
Error	0.6	6	.1		
Total	3.66	15			

makes use of the F statistic MST/MSE. Since the computed value of F, 7.8, is greater than the tabulated F-value, based on $df_1 = 3$, $df_2 = 6$, and $a = .05$, we reject H_0 and conclude that there are significant (p-value $= .0171$) differences in the mean completion times of the four spreadsheets. It would appear that spreadsheet A has a considerably shorter mean completion time than the other three spreadsheets.

Next we will assess whether taking into account the two extraneous sources of variation was effective in increasing the precision of the analysis relative to a completely randomized design. From the AOV table, we have MSR $= .067$, MSC $= .173$, and MSE $= 0.1$. Hence, the relative efficiency of this Latin square design relative to a completely randomized design is

$$RE(LS, CR) = \frac{MSR + MSC + (t - 1)MSE}{(t + 1)MSE}$$

$$= \frac{0.067 + 0.173 + (4 - 1)(0.1)}{(4 + 1)(0.1)} = 1.08$$

That is, approximately 8% more observations per each treatment would be required in a completely randomized design to obtain the same precision for estimating the treatment means as with this Latin square design. The Latin square did not improve the precision of estimation that much because there was very little difference in the completion times for the different types of problems. There was a somewhat larger difference in the mean completion times for the secretaries but these differences did not contribute a large variability to the overall study results.

The results in Example 15.5 are valid only if we can be assured the conditions placed on the model are consistent with the observed data. Thus, we use the residuals $e_{ij} = y_{ij} - \hat{\mu} - \hat{\alpha}_k - \hat{\beta}_i - \hat{\gamma}_j$ to assess whether the conditions of normality, equal variance, and independence appear to be satisfied for the observed data. The following example will display the computer output for such an analysis.

EXAMPLE 15.6

The computer output for the experiment described in Example 15.5 is displayed here. Compare the results to those obtained using the definition of the sum of squares and assess whether the model conditions appear to be valid.

Solution   Note that our hand calculations yielded nearly the same values as are given in the computer output. The differences are due to round-off errors. There can be large rounding errors in our hand calculations, which can lead to results that will differ from the values given in the computer output. It is strongly recommended that a computer software program be used in the analysis of variance calculations because of the potential for rounding errors. In assessing whether the model conditions have been met, we first note that in regard to the normality condition the test of H_0: residuals have normal distribution, the p-value from the Shapiro–Wilks test is p-value = .9098. Thus, we would not reject H_0 and the normality condition appears to be satisfied. Also, the stem and leaf plot, boxplot, and normal probability plot are also consistent with the condition that the residuals have a normal distribution. Figure 15.3 is a plot of the residuals versus the estimated treatment means. From this plot it would appear that the variability in the residuals is somewhat constant across the treatments.

FIGURE 15.3   Plot of Residuals Versus Estimated Treatment Means

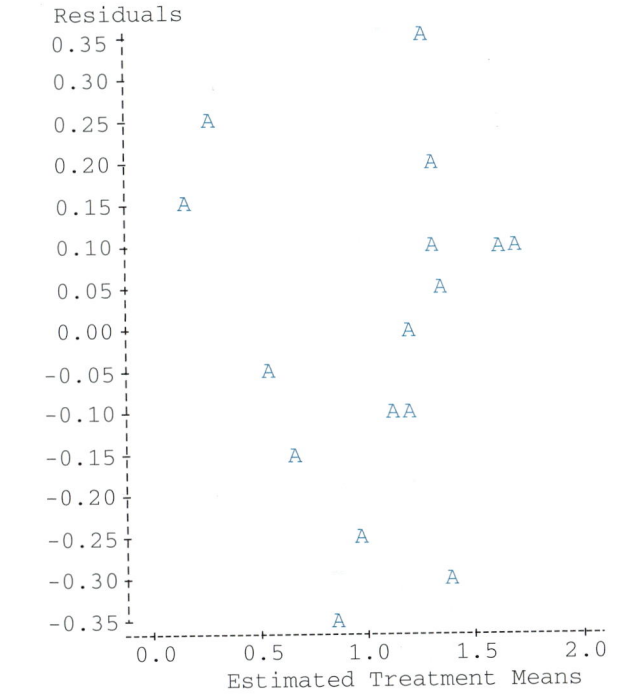

General Linear Models Procedure for Example 15.5

Dependent Variable: TIME TO COMPLETION

Source	DF	Sum of Squares	Mean Square	F Value	Pr > F
Model	9	3.0600000	0.3400000	3.40	0.0751
Error	6	0.6000000	0.1000000		
Corrected Total	15	3.6600000			

R-Square	C.V.	Root MSE	Y Mean
0.836066	30.11693	0.3162	1.0500

Source	DF	Type I SS	Mean Square	F Value	Pr > F
R	3	0.2000000	0.0666667	0.67	0.6025
C	3	0.5200000	0.1733333	1.73	0.2592
T	3	2.3400000	0.7800000	7.80	0.0171

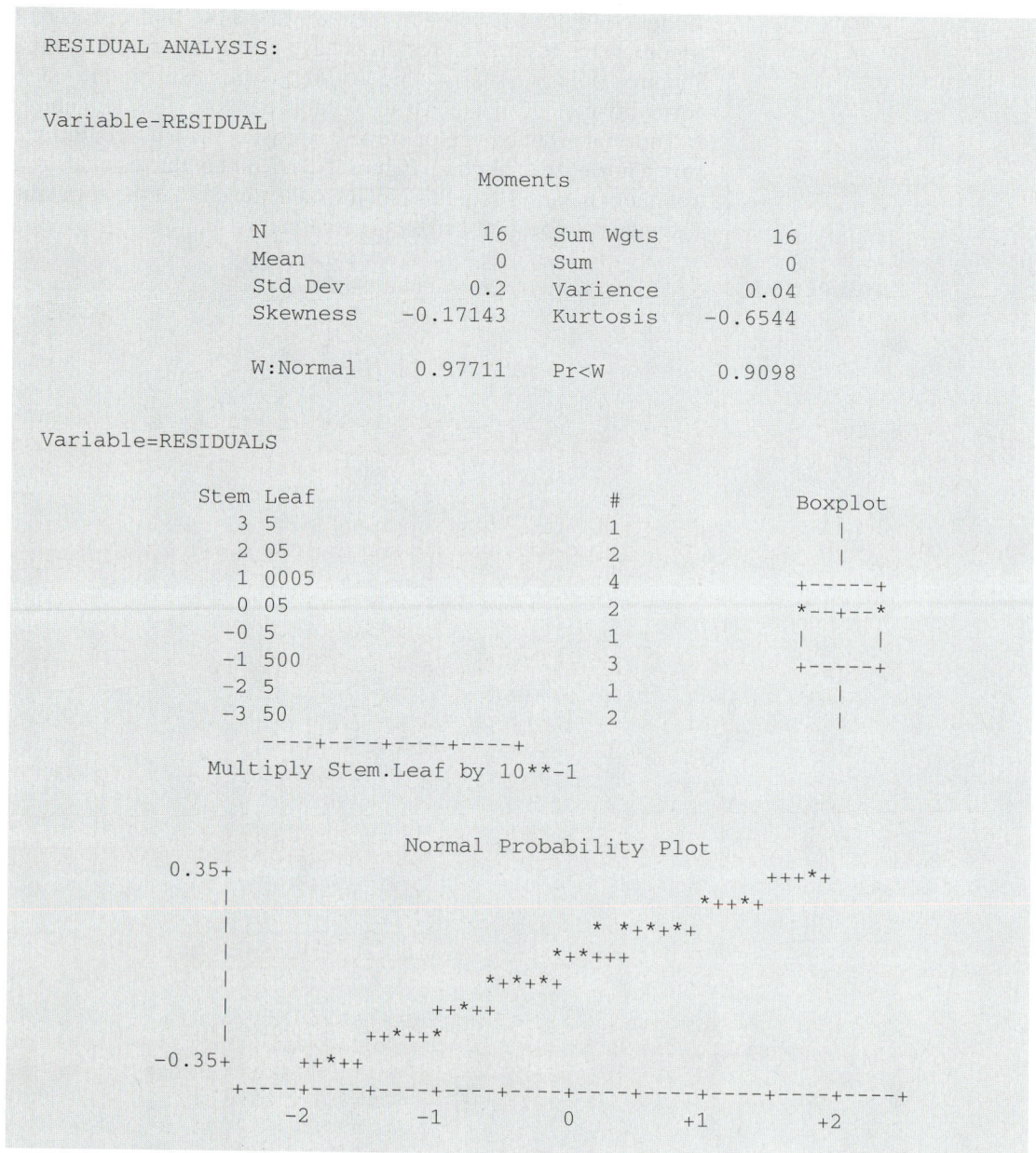

```
RESIDUAL ANALYSIS:


Variable-RESIDUAL


                              Moments

         N                 16   Sum Wgts        16
         Mean               0   Sum              0
         Std Dev          0.2   Varience      0.04
         Skewness    -0.17143   Kurtosis   -0.6544

         W:Normal     0.97711   Pr<W         0.9098


Variable=RESIDUALS


         Stem Leaf                    #         Boxplot
          3  5                        1           |
          2  05                       2           |
          1  0005                     4         +-----+
          0  05                       2         *--+--*
         -0  5                        1         |     |
         -1  500                      3         +-----+
         -2  5                        1           |
         -3  50                       2           |
            ----+----+----+----+
         Multiply Stem.Leaf by 10**-1
```

```
                 Normal Probability Plot
     0.35+                                   +++*+
         |                                 *++*+
         |                             *  *+*+*+
         |                           *+*+++
         |                        *+*+*+
         |                     ++*++
         |                 ++*++*
    -0.35+          ++*++
         +----+----+----+----+----+----+----+----+----+----+
             -2        -1         0        +1        +2
```

EXERCISES

Applications

Ag.

15.9 An experiment compared two different fertilizer placements (broadcast, band) and two different rates of fertilizer flow on watermelon yields. Recent research has shown that broadcast application (scattering over the outer area) of fertilizer is superior to bands of fertilizer applied near the seed for watermelon yields. For this experiment the investigators wished to compare two nitrogen–phosphorus–potassium (broadcast and band) fertilizers applied at a rate of 160–70–135 pounds per acre and two brands of micronutrients (A and B). These four combinations were to be studied in a Latin square field plot.

The treatments were randomly assigned according to a Latin square design conducted over a large farm plot, which was divided into rows and columns. A watermelon plant dry weight was obtained for each row–column combination 30 days after the emergence of the plants. The data are shown next.

Row	Column							
	1		2		3		4	
1	1	1.75	3	1.43	4	1.28	2	1.66
2	2	1.70	1	1.78	3	1.40	4	1.31
3	4	1.35	2	1.73	1	1.69	3	1.41
4	3	1.45	4	1.36	2	1.65	1	1.73

Treatment 1 broadcast, A   Treatment 3 band, A
Treatment 2 broadcast, B   Treatment 4 band, B

a. Write an appropriate statistical model for this experiment.
b. Use the data to run an analysis of variance. Give the p-value for each test and draw conclusions.

Engin.

15.10 A petroleum company was interested in comparing the miles per gallon achieved by four different gasoline blends (A, B, C, and D). Because there can be considerable variability due to differences in driving characteristics and car models, these two extraneous sources of variability were included as "blocking" variables in the study. The researcher selected four different brands of cars and four different drivers. The drivers and brands of cars were assigned to blends in the manner displayed in the following table. The mileage (in mpg) obtained over each test run was recorded as follows.

Driver	Car Model			
	1	2	3	4
1	A(15.5)	B(33.8)	C(13.7)	D(29.2)
2	B(16.3)	C(26.4)	D(19.1)	A(22.5)
3	C(10.5)	D(31.5)	A(17.5)	B(30.1)
4	D(14.0)	A(34.5)	B(19.7)	C(21.6)

a. Write a model for this experimental setting.
b. Estimate the parameters in the model.
c. Conduct an analysis of variance. Use $\alpha = .05$.
d. What conclusions can you draw concerning the best gasoline blend?
e. Compute the relative efficiency of the Latin square design relative to a completely randomized design. Interpret this value. Were the blocking variables effective in reducing the variability in experimental units? Explain.
f. If future studies were to be conducted, would you recommend using both car model and driver as blocking variables? Explain.

15.11 Refer to the computer output given as follows for the data of Exercise 15.10.
a. Compare your results to those of Exercise 15.10.
b. Do the model conditions appear to be satisfied for this set of data? Explain.

```
                    OBS    R    C    T      Y
                     1     1    1    1    15.5
                     2     1    2    2    33.8
                     3     1    3    3    13.7
                     4     1    4    4    29.2
                     5     2    1    2    16.3
                     6     2    2    3    26.4
                     7     2    3    4    19.1
                     8     2    4    1    22.5
                     9     3    1    3    10.5
                    10     3    2    4    31.5
                    11     3    3    1    17.5
                    12     3    4    2    30.1
                    13     4    1    4    14.0
                    14     4    2    1    34.5
                    15     4    3    2    19.7
                    16     4    4    3    21.6
```

General Linear Models Procedure
Dependent Variable: MILES PER GALLON

Source	DF	Sum of Squares	Mean Square	F Value	Pr > F
Model	9	869.97563	96.66396	22.42	0.0006
Error	6	25.86375	4.31062		
Corrected Total	15	895.83937			

R-Square	C.V.	Root MSE	Y Mean
0.971129	9.333878	2.0762	22.244

Source	DF	Type I SS	Mean Square	F Value	Pr > F
R	3	8.33187	2.77729	0.64	0.6143
C	3	755.37188	251.79063	58.41	0.0001
T	3	106.27188	35.42396	8.22	0.0151

RESIDUALS ANALYSIS:
Variable = RESIDUALS

Moments

N	16	Sum Wgts	16
Mean	0	Sum	0
Std Dev	1.313107	Variance	1.72425
Skewness	0.043408	Kurtosis	-0.41
W:Normal	0.985867	Pr<W	0.9840

```
Stem Leaf                         #           Boxplot
   2 5                            1              |
   1 346                          3           +-----+
   0 2457                         4           |  +  |
  -0 7443                         4           *-----*
  -1 543                          3           +-----+
  -2 4                            1              |
     ----+----+----+----+
```

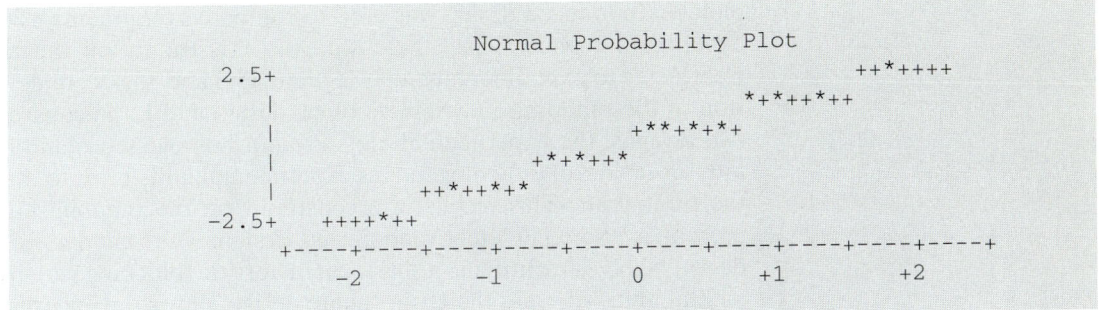

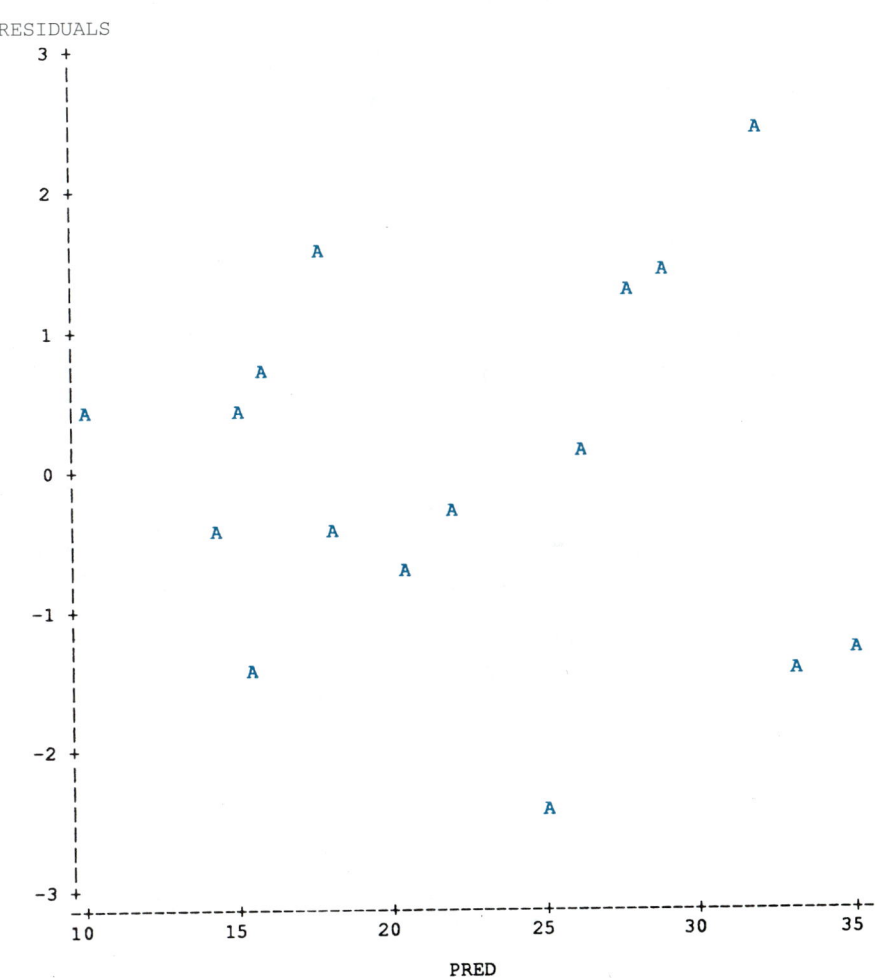

15.5 Factorial Treatment Structure in a Completely Randomized Design

In Chapter 14, we introduced the two components in an experimental design—the randomization method and the treatment structure. In Sections 15.2, 15.3, and 15.4, we were concerned with the randomization technique and the controlling of extraneous sources of variation through the use of blocking variables. The completely randomized design was used to compare t treatments using homogeneous experimental units. Thus, there were no constraints on the randomization of treatments to experimental units. Sections 15.3 and 15.4 were devoted to a discussion of the randomized complete block design and Latin square design. In these two designs, the experimental units are not homogeneous and must be grouped into sets (blocks) of homogeneous experimental units prior to randomly assigning the treatments. This results in a constraint on the randomization that was not present in the completely randomized design. The randomized complete block design involves identifying a single characteristic (blocking variable) of the experimental units, whereas the Latin square design allows the experimenter to use two different characteristics of the experimental units in creating the sets of homogeneous experimental units.

In this section, we will discuss the treatment design; that is, how treatments are constructed from several factors rather than just being t levels of a single factor. These types of experiments are involved with examining the effect of two or more independent variables on a response variable y. For example, suppose a company has developed a new adhesive for use in the home and wants to examine the effects of temperature and humidity on the bonding strength of the adhesive. Several treatment design questions arise in any study. First, we must consider what factors (independent variables) are of greatest interest. Second, the number of levels and the actual settings of these levels for each of the factors must be determined for each factor. Third, having separately selected the levels for each factor, we must choose the factor–level combinations (treatments) that will be applied to the experimental units.

The ability to choose the factors and the appropriate settings for each of the factors depends on budget, time to complete the study, and most important, the experimenter's knowledge of the physical situation under study. In many cases, this will involve conducting a detailed literature review to determine the current state of knowledge in the area of interest. Then, assuming that the experimenter has chosen the levels of each independent variable, he or she must decide which factor–level combinations are of greatest interest and are viable. In some situations, certain of the factor–level combinations will not produce an experimental setting that can elicit a reasonable response from the experimental unit. Certain combinations may not be feasible due to toxicity or practicality issues.

one-at-a-time approach   As discussed in Chapter 2, one approach for examining the effects of two or more factors on a response is the **one-at-a-time approach.** To examine the effect of a single variable, an experimenter changes the levels of this variable while holding the levels of the other independent variables fixed. This process is continued for each variable while holding the other independent variables constant. Suppose that an experimenter is interested in examining the effects of two independent variables, nitrogen and phosphorus, on the yield of a crop. For simplicity we will assume two levels of each variable have been selected for the study: 40 and 60 pounds per plot for nitrogen, 10 and 20 pounds per plot for phosphorus. For this study the experimental units are small, relatively homogeneous plots that have been partitioned from the acreage of a farm. For our experiment the factor–level combinations chosen might be as shown in Table 15.20. These factor–level combinations are illustrated in Figure 15.4.

From the graph in Figure 15.4, we see that there is one difference that can be used to measure the effects of nitrogen and phosphorus separately. The difference in response for combinations 1 and 2 would estimate the effect of nitrogen;

TABLE 15.20

Factor–level combinations for a one-at-a-time approach

Combination	Nitrogen	Phosphorus
1	60	10
2	40	10
3	40	20

FIGURE 15.4

Factor–level combinations for a one-at-a-time approach

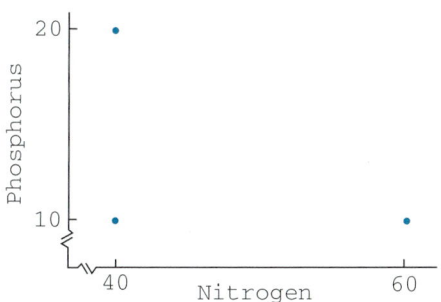

the difference in response for combinations 2 and 3 would estimate the effect of phosphorus.

Hypothetical yields corresponding to the three factor–level combinations of our experiment are given in Table 15.21. Suppose the experimenter is interested in using the sample information to determine the factor–level combination that will give the maximum yield. From the table, we see that crop yield increases when the nitrogen application is increased from 40 to 60 (holding phosphorus at 10). Yield also increases when the phosphorus setting is changed from 10 to 20 (at a fixed nitrogen setting of 40). Thus, it might seem logical to predict that increasing both the nitrogen and phosphorus applications to the soil will result in a larger crop yield. The fallacy in this argument is that our prediction is based on the assumption that the effect of one factor is the same for both levels of the other factor.

TABLE 15.21

Yields for the three factor–level combinations

Observation (yield)	Nitrogen	Phosphorus
145	60	10
125	40	10
160	40	20
?	60	20

We know from our investigation what happens to yield when the nitrogen application is increased from 40 to 60 for a phosphorus setting of 10. But will the yield also increase by approximately 20 units when the nitrogen application is changed from 40 to 60 at a setting of 20 for phosphorus?

To answer this question, we could apply the factor–level combination of 60 nitrogen–20 phosphorus to another experimental plot and observe the crop yield. If the yield is 180, then the information obtained from the three factor–level combinations would be correct and would have been useful in predicting the factor–level combination that produces the greatest yield. However, suppose the

interaction

yield obtained from the high settings of nitrogen and phosphorus turns out to be 110. If this happens, the two factors nitrogen and phosphorus are said to **interact.** That is, the effect of one factor on the response does not remain the same for different levels of the second factor, and the information obtained from the one-at-a-time approach would lead to a faulty prediction.

The two outcomes just discussed for the crop yield at the 60–20 setting are displayed in Figure 15.5, along with the yields at the three initial design points. Figure 15.5(a) illustrates a situation with no interaction between the two factors. The effect of nitrogen on yield is the same for both levels of phosphorus. In contrast, Figure 15.5(b) illustrates a case in which the two factors nitrogen and phosphorus do interact.

FIGURE 15.5

Yields of the three design points and possible yield at a fourth design point

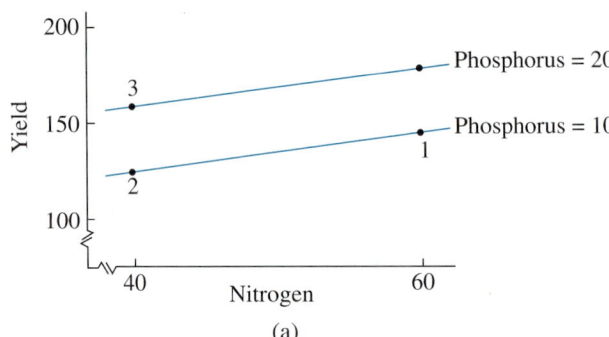

(a)

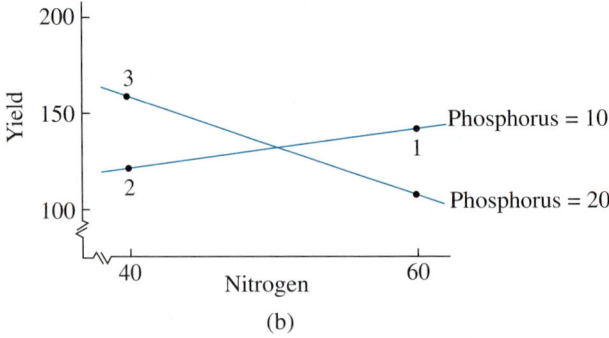

(b)

We have seen that the one-at-a-time approach to investigating the effect of two factors on a response is suitable only for situations in which the two factors do not interact. Although this was illustrated for the simple case in which two factors were to be investigated at each of two levels, the inadequacies of a one-at-a-time approach are even more salient when trying to investigate the effects of more than two factors on a response.

factorial experiment

Factorial experiments are useful for examining the effects of two or more factors on a response y, whether or not interaction exists. As before, the choice of the number of levels of each variable and the actual settings of these variables is important. However, assuming that we have made these selections with help from an investigator knowledgeable in the area being examined, we must decide at what factor–level combinations we will observe y.

Classically, factorial experiments have not been referred to as designs because they deal with the choice of levels and the selection of factor–level combinations (treatments) rather than with how the treatments are assigned to experimental units. Unless otherwise specified, we will assume that treatments are assigned to experimental units at random. The factorial–level combinations will then correspond to the "treatments" of a completely randomized design.

DEFINITION 15.3

A **factorial experiment** is an experiment in which the response y is observed at all factor–level combinations of the independent variables.

Using our previous example, if we are interested in examining the effect of two levels of nitrogen x_1 at 40 and 60 pounds per plot and two levels of phosphorus x_2 at 10 and 20 pounds per plot on the yield of a crop, we could use a completely randomized design where the four factor–level combinations (treatments) of Table 15.22 are assigned at random to the experimental units.

TABLE 15.22

2×2 factorial experiment for crop yield

Factor–Level	Combinations
x_1	x_2
40	10
40	20
60	10
60	20

Similarly, if we wished to examine x_1 at the two levels 40 and 60 and x_2 at the three levels 10, 15, and 20, we could use the six factor–level combinations of Table 15.23 as treatments in a completely randomized design.

TABLE 15.23

2×3 factorial experiment for crop yield

Factor–Level	Combinations
x_1	x_2
40	10
40	15
40	20
60	10
60	15
60	20

EXAMPLE 15.7

An auto manufacturer is interested in examining the effect of engine speed x_1, measured in revolutions per minute, and ground speed x_2, measured in miles per hour, on gasoline mileage. The investigators, in consultation with company mechanics and other personnel, decided to consider settings of x_1 at 800, 1,000, and 1,200 and settings of x_2 at 30, 50, and 70. Give the factor–level combinations to be used in a 3×3 factorial experiment.

Solution   Using the definition of factorial experiment, we would observe gasoline mileage at the following settings of x_1 and x_2:

x_1	800	800	800	1,000	1,000	1,000	1,200	1,200	1,200
x_2	30	50	70	30	50	70	30	50	70

The examples of factorial experiments presented in this section have concerned two independent variables. However, the procedure applies to any number of factors and levels per factor. Thus, if we had four different factors x_1, x_2, x_3, and x_4 at two, three, three, and four levels, respectively, we could formulate a $2 \times 3 \times 3 \times 4$ factorial experiment by considering all $2 \cdot 3 \cdot 3 \cdot 4 = 72$ factor–level combinations.

One final comparison should be made between the one-at-a-time approach and a factorial experiment. Not only do we get information concerning factor interactions using a factorial experiment, but also, when there are no interactions, we get at least the same amount of information about the effects of each individual factor using fewer observations. To illustrate this idea, let us consider the 2×2 factorial experiment with nitrogen and phosphorus. If there is no interaction between the two factors, the data appear as shown in Figure 15.6(a). For convenience, the data are reproduced in Table 15.24, with the four sample combinations designated by the numbers 1 through 4. If a 2×2 factorial experiment is used and no interaction exists between the two factors, we can obtain two independent differences to examine the effects of each of the factors on the response. Thus, from Table 15.24, the differences between observations 1 and 4 and the difference between observations 2 and 3 would be used to measure the effect of phosphorus. Similarly, the difference between observations 4 and 3 and the difference between observations 2 and 1 would be used to measure the effect of the two levels of nitrogen on plot yield.

FIGURE 15.6

Illustrations of the absence and presence of interaction in a 2×2 factorial experiment: (a) factors A and B do not interact; (b) factors A and B interact; (c) factors A and B interact

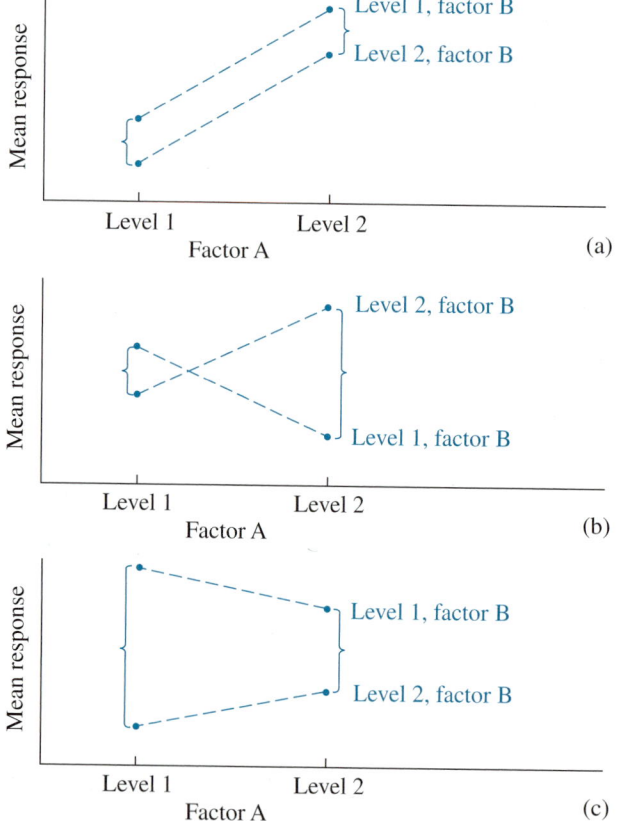

TABLE 15.24

Factor–level combinations for a 2×2 factorial experiment

Combination	Yield	Nitrogen	Phosphorus
1	145	60	10
2	125	40	10
3	165	40	20
4	180	60	20

If we employed a one-at-a-time approach for the same experimental situation, it would take six observations (two observations at each of the three initial factor–level combinations shown in Table 15.24) to obtain the same number of independent differences for examining the separate effects of nitrogen and phosphorus when no interaction is present.

The model for an observation in a completely randomized design with a two-factor factorial treatment structure and $n > 1$ replications can be written in the form

$$y_{ijk} = \mu + \alpha_i + \beta_j + \alpha\beta_{ij} + \varepsilon_{ijk}$$

where the terms of the model are defined as follows:

y_{ijk}: The response from the kth experimental unit receiving the ith level of factor A and the jth level of factor B.

μ: Overall mean, an unknown constant.

α_i: An effect due to the ith level of factor A, an unknown constant.

β_j: An effect due to the jth level of factor B, an unknown constant.

$\alpha\beta_{ij}$: An interaction effect of the ith level of factor A with the jth level of factor B, an unknown constant.

ε_{ijk}: A random error associated with the response from the kth experimental unit receiving the ith level of factor A combined with the jth level of factor B. We require that the ε_{ij}s have a normal distribution with mean 0 and common variance σ_ε^2. In addition, the errors must be independent.

The conditions given above for our model can be shown to imply that the recorded response from the kth experimental unit receiving the ith level of factor A combined with the jth level of factor B is normally distributed with mean

$$E(y_{ijk}) = \mu + \alpha_i + \beta_j + \alpha\beta_{ij}$$

and variance σ_ε^2.

To illustrate this model, consider the model for a two-factor factorial experiment with no interaction, such as the 2×2 factorial experiment with nitrogen and phosphorus:

$$y_{ijk} = \mu + \alpha_i + \beta_j + \varepsilon_{ijk}$$

Expected values for a 2×2 factorial experiment are shown in Table 15.25.

This model assumes that difference in population means (expected values) for any two levels of factor A is the same no matter what level of B we are considering. The same property holds when comparing two levels of factor B. For example, the difference in mean response for levels 1 and 2 of factor A is the same value, $\alpha_1 - \alpha_2$, no matter what level of factor B we are considering. Thus,

TABLE 15.25

Expected values for a 2×2 factorial experiment

| Factor A | Factor B | |
	Level 1	Level 2
Level 1	$\mu + \alpha_1 + \beta_1$	$\mu + \alpha_1 + \beta_2$
Level 2	$\mu + \alpha_2 + \beta_1$	$\mu + \alpha_2 + \beta_2$

a test for no differences among the two levels of factor A would be of the form $H_0: \alpha_1 - \alpha_2 = 0$. Similarly, the difference between levels of factor B is $\beta_1 - \beta_2$ for either level of factor A, and a test of no difference between the factor B means is $H_0: \beta_1 - \beta_2 = 0$. This phenomenon was also noted for the randomized block design.

interaction If the assumption of additivity of terms in the model does not hold, then we need a model that employs terms to account for **interaction.**

The expected values for a 2×2 factorial experiment with n observations per cell are presented in Table 15.26.

TABLE 15.26

Expected values for a 2×2 factorial experiment, with replications

| Factor A | Factor B | |
	Level 1	Level 2
Level 1	$\mu + \alpha_1 + \beta_1 + \alpha\beta_{11}$	$\mu + \alpha_1 + \beta_2 + \alpha\beta_{12}$
Level 2	$\mu + \alpha_2 + \beta_1 + \alpha\beta_{21}$	$\mu + \alpha_2 + \beta_2 + \alpha\beta_{22}$

As can be seen from Table 15.26, the difference in mean response for levels 1 and 2 of factor A on level 1 of factor B is

$$(\alpha_1 - \alpha_2) + (\alpha\beta_{11} - \alpha\beta_{21})$$

but for level 2 of factor B this difference is

$$(\alpha_1 - \alpha_2) + (\alpha\beta_{12} - \alpha\beta_{22})$$

Because the difference in mean response for levels 1 and 2 of factor A is *not* the same for different levels of factor B, the model is no longer additive, and we say that the two factors interact.

DEFINITION 15.4 Two factors A and B are said to **interact** if the difference in mean responses for two levels of one factor is not constant across levels of the second factor.

In measuring the octane rating of gasoline, interaction can occur when two components of the blend are combined to form a gasoline mixture. The octane properties of the blended mixture may be quite different than would be expected by examining each component of the mixture. Interaction in this situation could have a positive or negative effect on the performance of the blend, in which case the components are said to potentiate, or antagonize, one another.

profile plot We can amplify the notion of an interaction with the **profile plots** shown previously in Figure 15.6. As we see from Figure 15.6(a), when no interaction is present, the difference in the mean response between levels 1 and 2 of factor B (as indicated by the braces) is the same for both levels of factor A. However, for the two illustrations in Figure 15.6(b) and (c), we see that the difference between the levels of factor B changes from level 1 to level 2 of factor A. For these cases, we have an interaction between the two factors.

Note that an interaction is not restricted to two factors. With three factors A, B, and C, we might have an interaction between factors A and B, A and C, and B and C, and the two-factor interactions would have interpretations that follow immediately from Definition 15.4. Thus, the presence of an AC interaction indicates that the difference in mean response for levels of factor A varies across levels of factor C. A three-way interaction between factors A, B, and C might indicate that the difference in mean response for levels of C changes across combinations of levels for factors A and B.

The analysis of variance for a factorial experiment with an interaction between the factors requires that we have $n > 1$ observations on each of the treatments (factor–level combinations). We will construct the analysis of variance table for a completely randomized two-factor experiment with a levels of factor A, b levels of factor B, and n observations on each of the ab treatments. Before partitioning the total sum of squares into its components we need the notation defined here.

y_{ijk}: Observation on the kth experimental unit receiving the ith level of factor A and jth level of factor B.

$\bar{y}_{i..}$: Sample mean for observations at the ith level of factor A, $\bar{y}_{i..} = \frac{1}{bn}\Sigma_{jk}\, y_{ijk}$.

$\bar{y}_{.j.}$: Sample mean for observations at the jth level of factor B, $\bar{y}_{.j.} = \frac{1}{an}\Sigma_{ik}\, y_{ijk}$.

$\bar{y}_{ij.}$: Sample mean for observations at the ith level of factor A and the jth level of factor B, $\bar{y}_{ij.} = \frac{1}{n}\Sigma_{k}\, y_{ijk}$

$\bar{y}_{...}$: Overall sample mean, $\bar{y}_{...} = \frac{1}{abn}\Sigma_{ijk}\, y_{ijk}$.

total sum of squares　　The **total sum of squares** of the measurements about their mean $\bar{y}_{...}$ is defined as before:

$$\text{TSS} = \sum_{ijk} (y_{ijk} - \bar{y}_{...})^2$$

This sum of squares will be partitioned into four sources of variability: two due to the main effects of factors A and B, one due to the interaction between factors A and B, and one due to the variability from all sources not accounted for by the main effects and interaction. We call this source of variability **error**. The **partition of TSS** follows from an examination of the model:

$$y_{ijk} = \mu + \alpha_i + \beta_j + \alpha\beta_{ij} + \varepsilon_{ijk}$$

The parameters in the model have sample estimates:

$$\hat{\mu} = \bar{y}_{...} \qquad \hat{\alpha}_i = \bar{y}_{i..} - \bar{y}_{...} \qquad \hat{\beta}_j = \bar{y}_{.j.} - \bar{y}_{...} \qquad \widehat{\alpha\beta}_{ij} = (\bar{y}_{ij.} - \bar{y}_{...}) - \hat{\alpha}_i - \hat{\beta}_j$$

$$= \bar{y}_{ij.} - \bar{y}_{i..} - \bar{y}_{.j.} + \bar{y}_{...}$$

It can be shown algebraically that TSS takes the following form:

$$\sum_{ijk} (y_{ijk} - \bar{y}_{...})^2 = bn \sum_{i} (\bar{y}_{i..} - \bar{y}_{...})^2 + an \sum_{j} (\bar{y}_{.j.} - \bar{y}_{...})^2$$

$$+ n \sum_{ij} (\bar{y}_{ij.} - \bar{y}_{i..} - \bar{y}_{.j.} + \bar{y}_{...})^2 + \sum_{ijk} (y_{ijk} - \bar{y}_{ij.})^2$$

main effect of factor A

We will interpret the terms in the partition using the parameter estimates. The first quantity on the right-hand side of the equal sign measures the **main effect of factor A** and can be written as

$$\text{SSA} = bn \sum_{i} (\bar{y}_{i..} - \bar{y}_{...})^2 = bn \sum_{i} (\hat{\alpha}_i)^2$$

main effect of factor B

Similarly, the second quantity on the right-hand side of the equal sign measures the **main effect of factor B** and can be written as

$$\text{SSB} = an \sum_{j} (\bar{y}_{.j.} - \bar{y}_{...})^2 = an \sum_{j} (\hat{\beta}_j)^2$$

interaction effect of factors A and B

The third quantity measures the **interaction effect of factors A and B** and can be written as

$$\text{SSAB} = n \sum_{ij} (\bar{y}_{ij.} - \bar{y}_{i..} - \bar{y}_{.j.} + \bar{y}_{...})^2 = n \sum_{ij} (\widehat{\alpha\beta}_{ij})^2$$

sum of squares for error

The final term is the **sum of squares for error,** SSE, and represents the variability in the y'_{ijk}s not accounted for by the main effects and interaction effects. There are several forms for this term. Defining the residuals from the model as before, we have $e_{ijk} = y_{ijk} - \hat{\mu} - \hat{\alpha}_i - \hat{\beta}_j - \widehat{\alpha\beta}_{ij} = y_{ijk} - \bar{y}_{ij.}$. Therefore,

$$\text{SSE} = \sum_{ijk} (y_{ijk} - \bar{y}_{ij.})^2 = \sum_{ijk} (e_{ijk})^2$$

Alternatively, SSE = TSS − SST − SSB − SSAB. We summarize the partition of the sum of squares in the AOV table as given in Table 15.27.

TABLE 15.27

AOV table for a completely randomized two-factor factorial experiment

Source	SS	df	MS	F
Main Effect				
A	SSA	$a - 1$	$\text{MSA} = \text{SSA}/(a - 1)$	MSA/MSE
B	SSB	$b - 1$	$\text{MSB} = \text{SSB}/(b - 1)$	MSB/MSE
Interaction				
AB	SSAB	$(a - 1)(b - 1)$	$\text{MSAB} = \text{SSAB}/(a - 1)(b - 1)$	MSAB/MSE
Error	SSE	$ab(n - 1)$	$\text{MSE} = \text{SSE}/ab(n - 1)$	
Total	TSS	$abn - 1$		

From the AOV table we observe that if we have only one observation on each treatment, $n = 1$, then there are 0 degrees of freedom for error. Thus, if factors A and B interact and $n = 1$, then there are no valid tests for interactions or main effects. However, if the factors do not interact, then the interaction term can be used as the error term and we replace SSE with SSAB. However, it would be an exceedingly rare situation to run experiments with $n = 1$ since in most cases the researcher would not know prior to running the experiment whether or not factors A and B interact. Hence, in order to have valid tests for main effects and interactions, we need $n > 1$.

EXAMPLE 15.8

An experiment was conducted to determine the effects of four different pesticides on the yield of fruit from three different varieties (B_1, B_2, B_3) of a citrus tree. Eight trees from each variety were randomly selected from an orchard. The four pesticides were then randomly assigned to two trees of each variety and applications were made according to recommended levels. Yields of fruit (in bushels per tree) were obtained after the test period. The data appear in Table 15.28.

TABLE 15.28

Data for the 3 × 4 factorial experiment of fruit tree yield, $n = 2$ observations per treatment

Variety, B	Pesticide, A			
	1	2	3	4
1	49	50	43	53
	39	55	38	48
2	55	67	53	85
	41	58	42	73
3	66	85	69	85
	68	92	62	99

   a. Write an appropriate model for this experiment.

   b. Set up an analysis of variance table and conduct the appropriate F-tests of main effects and interactions using $\alpha = .05$.

profile plot   **c.** Construct a plot of the treatment means, called a **profile plot.**

Solution The experiment described is a completely randomized 3 × 4 factorial experiment with factor A, pesticides having $a = 3$ levels and factor B, variety having $b = 4$ levels. There are $n = 2$ replications of the 12 factor–level combinations of the two factors.

   a. The model for a 3 × 4 factorial experiment with interaction between the two factors is

$$y_{ijk} = \mu + \alpha_i + \beta_j + \alpha\beta_{ij} + \varepsilon_{ijk}, \quad \text{for } i = 1, 2, 3, 4; j = 1, 2, 3; k = 1, 2$$

where μ is the overall mean yield per tree, α_i is the effect of the ith level of the pesticide, β_j is the effect of the jth variety of citrus tree, and $\alpha\beta_{ij}$ is the interaction effect of the ith level of pesticide with jth variety of citrus tree.

   b. In most experiments we would strongly recommend using a computer software program to obtain the AOV table, but to illustrate the calculations we will construct the AOV for this example using the definitions of the individual sum of squares. First, we estimate the parameters in the model. To accomplish this we use the treatment means given in Table 15.29.

TABLE 15.29

Sample means for factor–level combinations (treatments) of A and B

Variety, B	Pesticide, A				Variety Means
	1	2	3	4	
1	44	52.5	40.5	50.5	46.875
2	48	62.5	47.5	79	59.25
3	67	88.5	65.5	92	78.25
Pesticide Means	53	67.83	51.17	73.83	61.46

Next, we obtain the parameter estimates:

Main Effects

$\hat{\alpha}_1 = \bar{y}_{1..} - \bar{y}_{...} = 53 - 61.46 = -8.46$
$\hat{\alpha}_2 = \bar{y}_{2..} - \bar{y}_{...} = 67.83 - 61.46 = 6.37$
$\hat{\alpha}_3 = \bar{y}_{3..} - \bar{y}_{...} = 51.17 - 61.46 = -10.29$
$\hat{\alpha}_4 = \bar{y}_{4..} - \bar{y}_{...} = 73.83 - 61.46 = 12.37$

Factor B

$\hat{\beta}_1 = \bar{y}_{.1.} - \bar{y}_{...} = 46.875 - 61.46 = -14.585$
$\hat{\beta}_2 = \bar{y}_{.2.} - \bar{y}_{...} = 59.25 - 61.46 = -2.21$
$\hat{\beta}_3 = \bar{y}_{.3.} - \bar{y}_{...} = 78.25 - 61.46 = 16.79$

Interaction Effects

$\widehat{\alpha\beta}_{11} = \bar{y}_{11.} - \bar{y}_{1..} - \bar{y}_{.1.} + \bar{y}_{...} = 44 - 53 - 46.875 + 61.46 = 5.585$
$\widehat{\alpha\beta}_{12} = \bar{y}_{12.} - \bar{y}_{1..} - \bar{y}_{.2.} + \bar{y}_{...} = 48 - 53 - 59.25 + 61.46 = -2.79$
$\widehat{\alpha\beta}_{13} = \bar{y}_{13.} - \bar{y}_{1..} - \bar{y}_{.3.} + \bar{y}_{...} = 67 - 53 - 78.25 + 61.46 = -2.79$
$\widehat{\alpha\beta}_{21} = \bar{y}_{21.} - \bar{y}_{2..} - \bar{y}_{.1.} + \bar{y}_{...} = 52.5 - 67.83 - 46.875 + 61.46 = -.745$
$\widehat{\alpha\beta}_{22} = \bar{y}_{22.} - \bar{y}_{2..} - \bar{y}_{.2.} + \bar{y}_{...} = 62.5 - 67.83 - 59.25 + 61.46 = -3.12$
$\widehat{\alpha\beta}_{23} = \bar{y}_{23.} - \bar{y}_{2..} - \bar{y}_{.3.} + \bar{y}_{...} = 88.5 - 67.83 - 78.25 + 61.46 = 3.88$
$\widehat{\alpha\beta}_{31} = \bar{y}_{31.} - \bar{y}_{3..} - \bar{y}_{.1.} + \bar{y}_{...} = 40.5 - 51.17 - 46.875 + 61.46 = 3.915$
$\widehat{\alpha\beta}_{32} = \bar{y}_{32.} - \bar{y}_{3..} - \bar{y}_{.2.} + \bar{y}_{...} = 47.5 - 51.17 - 59.25 + 61.46 = -1.46$
$\widehat{\alpha\beta}_{33} = \bar{y}_{33.} - \bar{y}_{3..} - \bar{y}_{.3.} + \bar{y}_{...} = 65.5 - 51.17 - 78.25 + 61.46 = -2.46$
$\widehat{\alpha\beta}_{41} = \bar{y}_{41.} - \bar{y}_{4..} - \bar{y}_{.1.} + \bar{y}_{...} = 50.5 - 73.83 - 46.875 + 61.46 = -8.745$
$\widehat{\alpha\beta}_{42} = \bar{y}_{42.} - \bar{y}_{4..} - \bar{y}_{.2.} + \bar{y}_{...} = 79 - 73.83 - 59.25 + 61.46 = 7.38$
$\widehat{\alpha\beta}_{43} = \bar{y}_{43.} - \bar{y}_{4..} - \bar{y}_{.3.} + \bar{y}_{...} = 92 - 73.83 - 78.25 + 61.46 = 1.38$

We next calculate the total sum of squares.

$$\text{TSS} = \sum_{ijk} (y_{ijk} - \bar{y}_{...})^2 = (49 - 61.46)^2 + (50 - 61.46)^2 + \cdots + (99 - 61.46)^2$$

$$= 7{,}187.96$$

The main effect sum of squares is

$$\text{SSA} = bn \sum_i \hat{\alpha}_i^2 = (3)(2)[(-8.46)^2 + (6.37)^2 + (-10.29)^2 + (12.37)^2]$$

$$= 2{,}226.29$$

$$\text{SSB} = an \sum_i \hat{\beta}_j^2 = (4)(2)[(-14.585)^2 + (-2.21)^2 + (16.79)^2]$$

$$= 3{,}996.08$$

The interaction sum of squares is

$$\text{SSAB} = n \sum_i (\widehat{\alpha\beta}_{ij})^2 = (2)[(5.585)^2 + (-2.79)^2 + \cdots + (1.38)^2] = 456.92$$

The sum of squares error is obtained as

$$\text{SSE} = \text{TSS} - \text{SSA} - \text{SSB} - \text{SSAB} = 7{,}187.96 - 2{,}226.29 - 3{,}996.08$$
$$- 456.92 = 508.67$$

The analysis of variance table for this completely randomized 3×4 factorial experiment with $n = 2$ replications per treatment is given in Table 15.30.

The first test of significance *must* be to test for an interaction between factors A and B, because if the interaction is significant then the main effects *may have*

TABLE 15.30

AOV table for fruit yield experiment of Example 15.8

Source	SS	df	MS	F
Pesticide, A	2,226.29	3	742.10	17.51
Variety, B	3,996.08	2	1,998.04	47.13
Interaction, AB	456.92	6	76.15	1.80
Error	508.67	12	42.39	
Total	7,187.96	23		

no interpretation. The *F* statistic is

$$F = \frac{\text{MSAB}}{\text{MSE}} = \frac{76.15}{42.39} = 1.80$$

The computed value of *F* does not exceed the tabulated value of 3.00 for $a = .05$, $df_1 = 6$, $df_2 = 12$ in the *F* tables. Hence, we have insufficient evidence to indicate an interaction between pesticide levels and variety of trees levels. We can observe this lack of interaction by constructing a profile plot. Figure 15.7 contains a plot of the sample treatment means for this experiment.

FIGURE 15.7

Profile plot for fruit yield experiment of Example 15.8

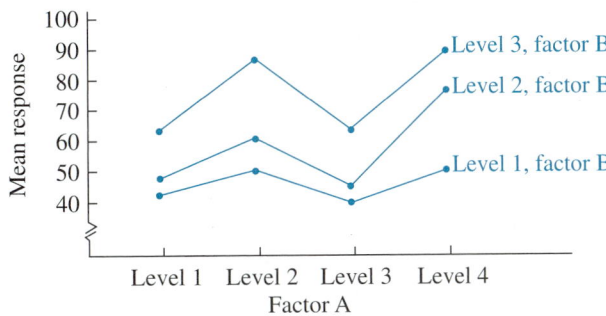

From the profile plot we can observe that the differences in mean yields between the three varieties of citrus trees remain nearly constant across the four pesticide levels. That is, the three lines for the three varieties are nearly parallel lines and hence the interaction between the levels of variety and pesticide is not significant. Because the interaction is not significant, we can next test the main effects of the two factors. These tests separately examine the differences among the levels of variety and the levels of pecticides. For pesticides, the *F*-statistic is

$$F = \frac{\text{MSA}}{\text{MSE}} = \frac{742.10}{42.39} = 17.51$$

The computed value of *F* does exceed the tabulated value of 3.49 for $a = .05$, $df_1 = 3$, $df_2 = 12$ in the *F* tables. Hence, we have sufficient evidence to indicate a significant difference in the mean yields among the four pesticide levels. For varieties, the *F*-statistic is

$$F = \frac{\text{MSB}}{\text{MSE}} = \frac{1,998.04}{42.39} = 47.13$$

The computed value of *F* does exceed the tabulated value of 3.89 for $a = .05$, $df_1 = 2$, $df_2 = 12$ in the *F* tables. Hence, we have sufficient evidence to indicate a significant difference in the mean yields among the three varieties of citrus trees.

In Section 15.7, we will discuss how to explore which pairs of levels differ for both factors A and B.

The results of an *F* test for main effects for factors A or B must be interpreted very carefully in the presence of a **significant interaction.** The first thing we would do is to construct a profile plot using the sample treatment means, $\bar{y}_{ij.}$. If the profile plot for Example 15.8 had appeared as shown in Figure 15.8, there would have been an indication of an interaction between factors A and B (pesticides and varieties). Provided that the MSE was not too large relative to MSAB, the *F* test for interaction would undoubtedly have been significant.

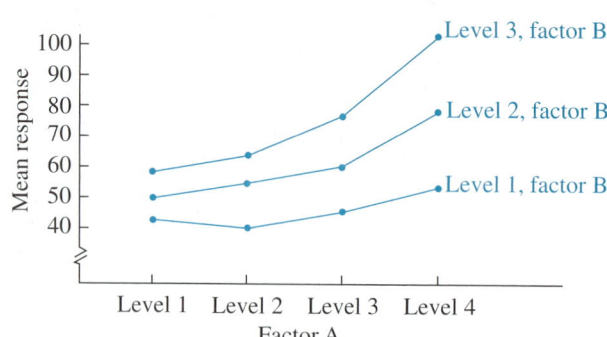

Would *F* tests for main effects have been appropriate for the profile plot of Figure 15.8? The answer is yes, because there is an *orderly* interaction; the *order* of the means for levels of factor B is always the same even though the *magnitude* of the differences between levels of factor B may change from level to level of factor A. Clearly, the profile plot in Figure 15.8 shows that the level 3 mean of factor B is always larger than the means for levels 1 and 2. Similarly, the level 2 mean for factor B is always larger than the mean for level 1 for factor B, no matter which level of factor A that we examine. However, we must be very careful in the conclusions we obtain from such a situation. If we find a significant difference in the levels of factor B, with mean response at level 3 larger than levels 1 and 2 of factor B across all levels of factor A, we may be led to conclude that level 3 of factor B produces significantly larger mean values than the other two levels of factor B. However, note that at level 1 of factor A, there is very little difference in the mean responses of the three levels of factor B. Thus, if we were to use level 1 of factor A, the three levels of factor B would produce equivalent mean responses. Thus, our conclusions about the differences in the mean responses

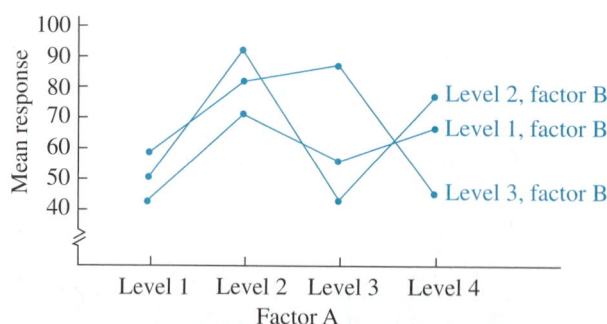

among the levels of factor B are not consistent across the levels of factor A and may contradict the test for main effects of factor B at certain levels of factor A.

When the interaction is orderly, a test on main effects can be meaningful; however, not all interactions are orderly. The profile plot in Figure 15.9 shows a situation in which a test of main effects in the presence of a significant interaction might be misleading. A *disorderly* interaction, such as in Figure 15.9, can obscure the main effects. It is not that the tests are statistically incorrect; it is that they may lead to a misinterpretation of the results of the experiment. At level 1 of factor A, there is very little difference in the mean responses of the three levels of factor B. At level 3 of factor A, level 3 of factor B produces a much larger response than level 2 of factor B. In contradiction to this result, we have at level 4 of factor A, level 2 of factor B produces a much large mean response than level 3 of factor B. Thus, when the two factors have significant interactions, conclusions about the differences in the mean responses among the levels of factor B must be made separately at *each level* of factor A. That is, a single conclusion about the levels of factor B does not hold for all levels of factor A.

When our experiment involves three factors, the calculations become considerably more complex. However, interpretations about main effects and interactions are similar to the interpretations when we have only two factors. With three factors A, B, and C, we might have an interaction between factors A and B, A and C, and B and C. The interpretations for these two-way interactions would follow immediately from Definition 15.4. Thus, the presence of an AC interaction indicates that the differences in mean responses among the levels of factor A vary across the levels of factor C. The same care must be taken in making interpretations among main effects as we discussed previously. A three-way interaction between factors A, B, and C might indicate that the differences in mean responses for levels of factor C change across combinations of levels for factors A and B. A second interpretation of a three-way interaction is that the pattern in the interactions between factors A and B changes across the levels of factor C. Thus, if a three-way interaction were present, and we plotted a separate profile plot for the two-way interaction between factors A and B at each level of factor C, we would see decidedly different patterns in several of the profile plots.

The model for an observation in a completely randomized design with a three-factor factorial treatment structure and $n > 1$ replications can be written in the form

$$y_{ijkm} = \mu + \alpha_i + \beta_j + \gamma_k + \alpha\beta_{ij} + \alpha\gamma_{ik} + \beta\gamma_{jk} + \alpha\beta\gamma_{ijk} + \varepsilon_{ijkm}$$

where the terms of the model are defined as follows:

y_{ijkm}: The response from the mth experimental unit receiving the ith level of factor A, the jth level of factor B, and the kth level of factor C.

μ: Overall mean, an unknown constant.

α_i: An effect due to the ith level of factor A, an unknown constant.

β_j: An effect due to the jth level of factor B, an unknown constant.

γ_k: An effect due to the kth level of factor C, an unknown constant.

$\alpha\beta_{ij}$: A two-way interaction effect of the ith level of factor A with the jth level of factor B, an unknown constant.

TABLE 15.32

AOV table for data in case study, a three-factor factorial experiment

```
General Linear Models Procedure

Dependent Variable: Texture of Meat:
```

| | | Sum of | Mean | | |
Source	DF	Squares	Square	F Value	Pr > F
Model	11	2080.28750	189.11705	62.40	0.0001
Error	24	72.74000	3.03083		
Corrected Total	35	2153.02750			

	R-Square	C.V.	Root MSE	Y Mean
	0.966215	1.769387	1.74093	98.3917

Source	DF	Type I SS	Mean Square	F Value	Pr > F
Main Effects:					
L	1	526.70250	526.70250	173.78	0.0001
B	1	113.42250	113.42250	37.42	0.0001
P	2	1090.11500	545.05750	179.84	0.0001
Interactions:					
L*B	1	44.22250	44.22250	14.59	0.0008
L*P	2	182.53500	91.26750	30.11	0.0001
B*P	2	115.84500	57.92250	19.11	0.0001
L*B*P	2	7.44500	3.72250	1.23	0.3106

TABLE 15.33

Table of means for data in case study

Level (%)	Blend	SPI (%)	Two-Way Means
.5	KSS	*	99.3
.5	KNC	*	105.1
1	KSS	*	93.9
1	KNC	*	95.2
.5	*	1.1	110.0
.5	*	2.2	105.0
.5	*	4.4	91.7
1	*	1.1	97.5
1	*	2.2	96.2
1	*	4.4	90.0
*	KSS	1.1	104.5
*	KSS	2.2	97.4
*	KSS	4.4	88.0
*	KNC	1.1	103.0
*	KNC	2.2	103.9
*	KNC	4.4	93.7

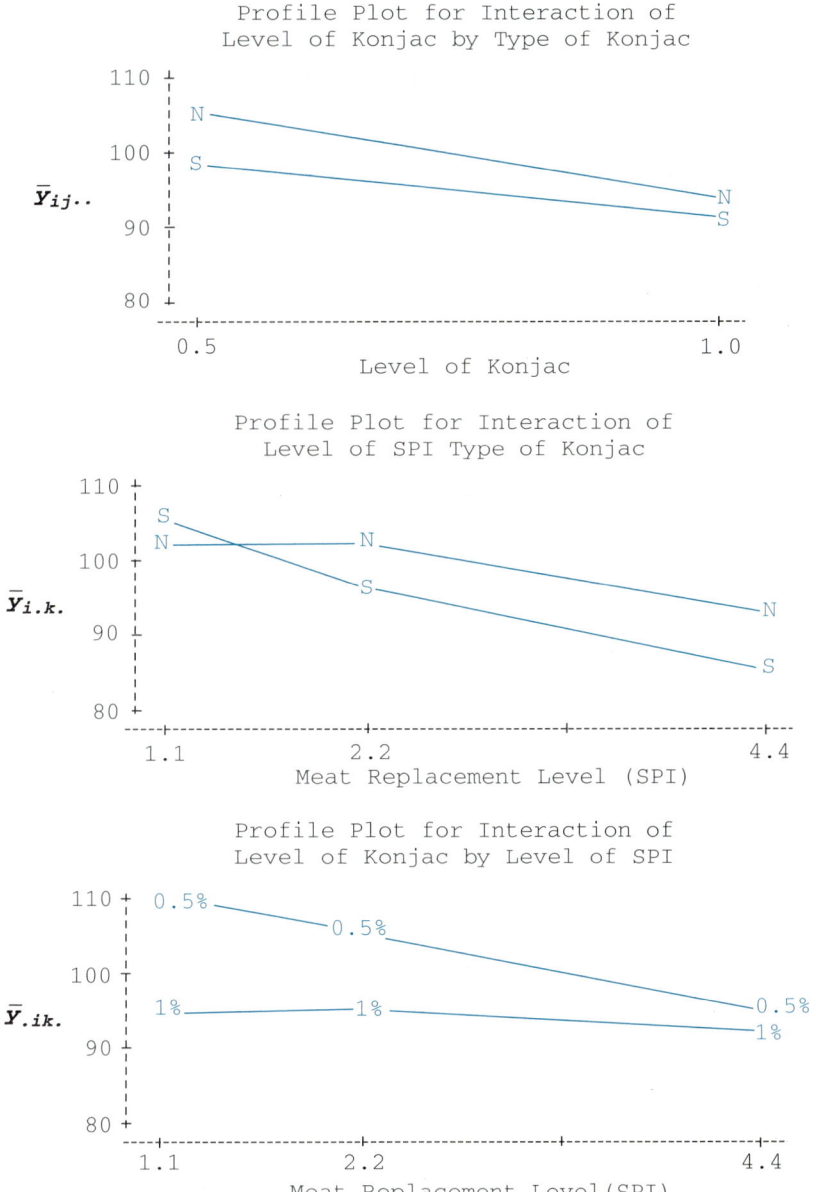

FIGURE 15.11

Profile plots of the two-way interactions

From Figure 15.11, we can observe that there are considerable differences in the mean texture of the meat product depending on the type of Konjac, level of Konjac, and level of SPI in the meat product. When the level of Konjac is 1%, there is very little difference in the mean texture of the meat; however, at the .5% level, KNC Konjac produced a product with a higher mean texture than did the KSS blend of Konjac. When considering the effect of level of SPI on the mean texture of the bologna, we can observe that at a level of 1.1% SPI there was a sizeable difference between using .5% Konjac and 1% Konjac. As the level of SPI increased, the size of the difference decreased markedly. Furthermore, at a 1.1% level of SPI there was essentially no difference between the two blends of Konjac, but as the level of SPI increased, the KNC blend produced a meat product having

Supplementary Exercises

Engin. **15.17** An experimenter is interested in examining the bond strength of a new adhesive product prepared under three different temperature settings (280°F, 300°F, and 320°F) and four different pressure settings (100, 150, 200, and 250 psi). The experimenter will prepare a sufficient amount of the adhesive so that each temperature–pressure setting combination is tested on three samples of the adhesive. Suppose that the experimenter can only test 12 samples per day and the conditions in the laboratory are somewhat variable from day-to-day. Describe an experimental design that takes into account the day-to-day variation in the laboratory. Include a diagram that displays the assignment of the temperature–pressure setting combinations to adhesive samples.

Edu. **15.18** A study was conducted to study the impact of child abuse on performance in school. Three categories of child abuse were defined as follows:

> "Abused child"—a child who is physically abused.

> "Neglected child"—a child receiving inadequate care.

> "Nonabuse"—a child receiving normal care and not physically abused.

The researchers randomly selected 30 boys and 30 girls from each of the three categories using the records of the state child-welfare agency for the abused and neglected children and a local school for the nonabused children. The scores on a standard grade-level assessment test of reading, mathematics, and general science were recorded for all the selected children.
  a. Suppose the students were all in the seventh grade. Identify the design.
  b. Suppose the children were equally divided between the third, fifth, and seventh grades. Identify the design.

Gov. **15.19** The city manager of a large midwestern city was negotiating with the three unions that represented the policemen, firemen, and building inspectors over the salaries for these groups of employees. The three unions claimed that the starting salaries were substantially different between the three groups, whereas in most cities there was not a significant difference in starting salaries between the three groups. To obtain information on starting salaries across the nation, the city manager decided to randomly select one city in each of eight geographical regions. The starting yearly salaries (in thousands of dollars) were obtained for each of the three groups in each of the eight regions. The data appear here.

Region	1	2	3	4	5	6	7	8	Mean
Policemen	32.3	33.2	30.8	30.5	30.1	30.2	28.4	27.9	30.42
Fireman	31.9	32.8	31.6	31.2	30.8	30.6	28.7	27.5	30.64
Inspectors	27.9	27.8	26.5	26.8	26.4	26.8	25.3	25.9	26.68
Region Mean	30.7	31.3	29.6	29.5	29.1	29.2	27.5	27.1	29.25

  a. Write a model for this study, identifying all the terms in the model.
  b. Using the analysis of variance from the Minitab computer output shown here, do the data suggest a difference in mean starting salary for the three groups of employees? Use $\alpha = .05$.
  c. Give the level of significance for your test.
  d. Which pairs of jobs types have significantly different starting salaries?

```
Two-Way Analysis of Variance for Salary

Source        DF        SS          MS          F          P
Job            2     62.8712     31.4356     955.23     0.000
Region         7      6.5213      0.9316      28.31     0.000
Error         14      0.4607      0.0329
Total         23     69.8532
```

15.20 Refer to Exercise 15.19.
  a. Plot the data in a profile plot with factors job type and region. Does there appear to be an interaction between the two factors? If there was an interaction, would you be able to test for it using the given data? If not, why not?
  b. Did the geographical region variable increase the efficiency of the design over conducting the study as a completely randomized design where the city manager would have just randomly selected eight cities regardless of their location?
  c. Identify additional sources of variability that may need to be included in future studies.

Ag. **15.21** A study was conducted to compare the effect of four manganese rates (from $MnSO_4$) and four copper rates (from $CuSO_4$ $5H_2O$) on the yield of soybeans. A large field was subdivided into 32 separate plots. Two plots were randomly assigned to each of the 16 factor–level combinations (treatments) and the treatments were applied to the designated plot. Soybeans were then planted over the entire field in rows 3 feet apart. The yields from the 32 plots are given here (in kilograms/hectare).

Cu	Mn 20	Mn 50	Mn 80	Mn 110	Cu Mean
1	1,558 1,578	2,003 2,033	2,490 2,470	2,830 2,810	2,221.5
3	1,590 1,610	2,020 2,051	2,620 2,632	2,860 2,841	2,278.0
5	1,558 1,550	2,003 2,010	2,490 2,690	2,830 2,910	2,255.1
7	1,328 1,427	2,010 2,031	2,887 2,832	2,960 2,941	2,302.0
Mn Mean	1,524.9	2,020.1	2,638.9	2,872.8	2,264.2

  a. Identify the design for this experiment.
  b. Write an appropriate statistical model for this experiment.
  c. Construct a profile plot and describe what this plot says about the effect of Mn and Cu on soybean yield.

15.22 Refer to Exercise 15.21.
  a. Using the computer printout given here, test for an interaction between the effect of Mn and Cu on soybean yield. Use $\alpha = .05$.
  b. What level of Mn appears to produce the highest yield?
  c. What level of Cu appears to produce the highest yield?
  d. What combination of Cu-Mn appears to produce the highest yield?

```
General Linear Models Procedure for Exercise 15.22

Dependent Variable: SOYBEAN YIELD

                                 Sum of          Mean
Source                   DF      Squares        Square    F Value    Pr > F
Model                    15    9167706.7      611180.4    305.08     0.0001
Error                    16      32053.5        2003.3
Corrected Total          31    9199760.2

               R-Square              C.V.      Root MSE              Y Mean
               0.996516          1.976839        44.759              2264.2

Source                   DF   Type III SS   Mean Square   F Value    Pr > F
CU                        3      28199.3        9399.8       4.69     0.0155
MN                        3    8935108.1     2978369.4    1486.70     0.0001
CU*MN                     9     204399.3       22711.0      11.34     0.0001
```

15.23 Suppose we have a completely randomized three-factor factorial experiment with levels $3 \times 4 \times 6$, with three replications of each of the 72 treatments. Assume that the three-way interaction is not significant.

    a. Write a model to describe the response y_{ijkm} for this type of experiment.

    b. Provide a complete AOV table for this type of experiment.

    c. Sketch three profile plots to depict the following three two-way interactions: F_1*F_2 significant but orderly, F_2*F_3 nonsignificant, F_1*F_3 significant and disorderly.

Ag.   **15.24** An experiment was set up to compare the effect of different soil pH and calcium additives on the increase in trunk diameters for orange trees. Annual applications of elemental sulfur, gypsum, soda ash, and other ingredients were applied to provide pH value levels of 4, 5, 6, and 7. Three levels of a calcium supplement (100, 200, and 300 pounds per acre) were also applied. All factor–level combinations of these two variables were used in the experiment. At the end of a 2-year period, three diameters were examined at each factor–level combination. The data appear next.

pH Value	Calcium 100	Calcium 200	Calcium 300
4.0	5.2	7.4	6.3
	5.9	7.0	6.7
	6.3	7.6	6.1
5.0	7.1	7.4	7.3
	7.4	7.3	7.5
	7.5	7.1	7.2
6.0	7.6	7.6	7.2
	7.2	7.5	7.3
	7.4	7.8	7.0
7.0	7.2	7.4	6.8
	7.5	7.0	6.6
	7.2	6.9	6.4

a. Construct a profile plot. What do the data suggest?
b. Write an appropriate statistical model.
c. Perform an analysis of variance and identify the experimental design. Use $\alpha = .05$.

15.25 Refer to Exercise 15.24.
a. Use the computer output given here to test for interactions and main effects. Use $\alpha = .05$.
b. What can you conclude about the effects of pH and calcium on the increase in the mean trunk diameters for orange trees?

```
General Linear Models Procedure for Exercise 15.25


Dependent Variable: Increase in Tree Diameter


                            Sum of           Mean
Source              DF      Squares         Square     F Value    Pr > F
Model               11      9.1830556      0.8348232    12.32     0.0001
Error               24      1.6266667      0.0677778
Corrected Total     35     10.8097222


             R-Square           C.V.        Root MSE              Y Mean
             0.849518        3.691335         0.2603              7.0528


Source              DF   Type III SS    Mean Square    F Value    Pr > F
PH                   3    4.4608333      1.4869444      21.94     0.0001
CA                   2    1.4672222      0.7336111      10.82     0.0004
PH*CA                6    3.2550000      0.5425000       8.00     0.0001
```

15.26 Refer to Exercise 15.24.
a. Use Tukey's W procedure to determine differences in mean increase in trunk diameters among the three calcium rates. Use $\alpha = .05$.
b. Are your conclusions about the differences in mean increase in diameters among the three calcium rates the same for all four pH values?

```
         Level of               --------------Y--------------
           PH          N            Mean                SD
            4          9         6.50000000          0.75828754
            5          9         7.31111111          0.15365907
            6          9         7.40000000          0.25000000
            7          9         7.00000000          0.36400549


         Level of               --------------Y--------------
           CA          N            Mean                SD
           100         12        6.95833333          0.75252102
           200         12        7.33333333          0.28069179
           300         12        6.86666667          0.45193188
```

Level of PH	Level of CA	N	Y Mean	Y SD
4	100	3	5.80000000	0.55677644
4	200	3	7.33333333	0.30550505
4	300	3	6.36666667	0.30550505
5	100	3	7.33333333	0.20816660
5	200	3	7.26666667	0.15275252
5	300	3	7.33333333	0.15275252
6	100	3	7.40000000	0.20000000
6	200	3	7.63333333	0.15275252
6	300	3	7.16666667	0.15275252
7	100	3	7.30000000	0.17320508
7	200	3	7.10000000	0.26457513
7	300	3	6.60000000	0.20000000

15.27   Refer to Exercise 15.24.

a. Use the residual analysis contained in the computer output given here to determine whether any of the conditions required to conduct an appropriate F-test have been violated.

b. If any of the conditions have been violated, suggest ways to overcome these difficulties.

```
Variable-RESIDUALS
                       Moments
        N                  36   Sum Wgts           36
        Mean                0   Sum                 0
        Std Dev      0.215583   Variance     0.046476
        Skewness     -0.22276   Kurtosis     0.699897

        W:Normal     0.986851   Pr<W           0.9562

Variable-RESIDUALS
        Stem Leaf                 #      Boxplot
          5 0                     1         |
          4                                 |
          3 03                    2         |
          2 0007                  4      +-----+
          1 033777                6      |     |
          0 003377                6      *--+--*
         -0 733                   3      |     |
         -1 7733000               7      +-----+
         -2 73000                 5         |
         -3 3                     1         |
         -4                                 |
         -5                                 |
         -6 0                     1         |
            ---+---+---+---+
   Multiply Stem.Leaf by 10**-1
```

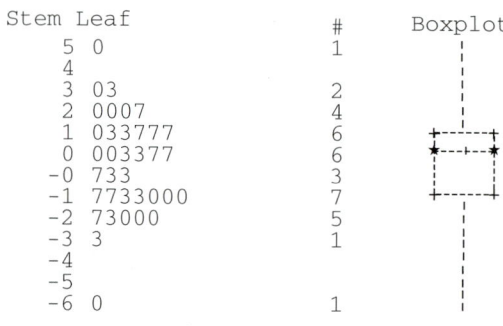

Normal Probability Plot

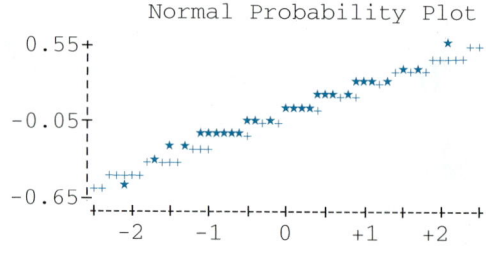

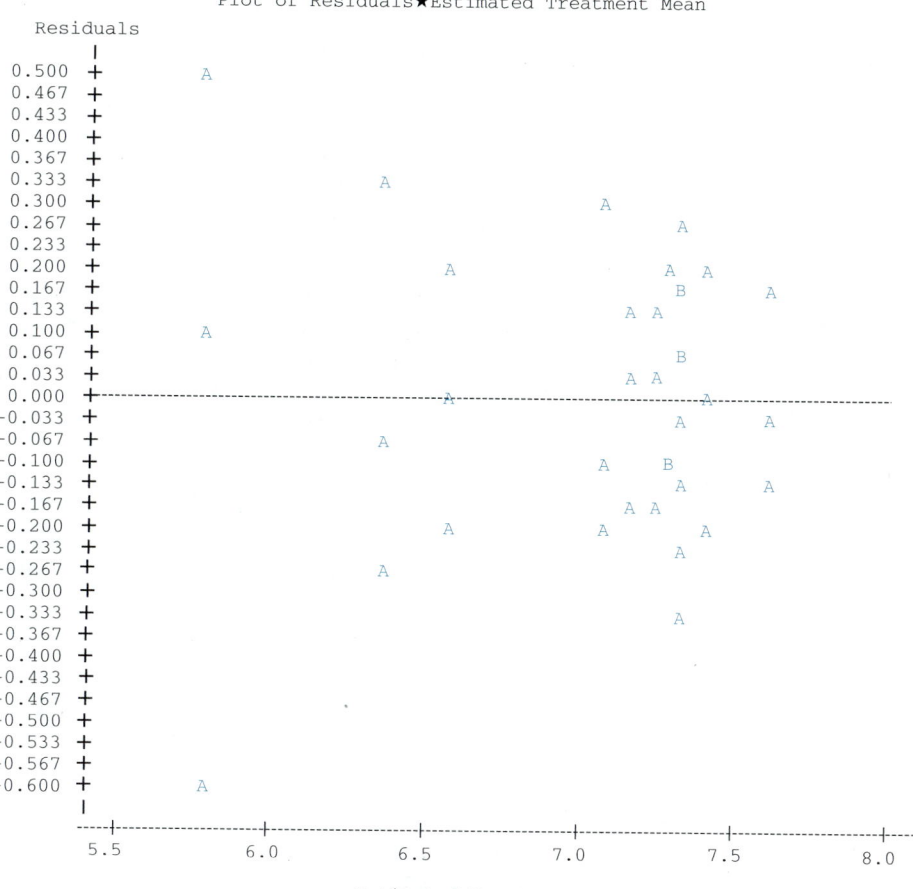

Plot of Residuals★Estimated Treatment Mean

Med.   **15.28**   Researchers conducted an experiment to compare the average oral body temperature for persons taking one of nine different medications often prescribed for high blood
pressure. The researchers were concerned that the effect of the drug may be different
depending on the severity of the patient's high blood pressure disorder. Patients with high
blood pressure who satisfy the study's entrance criteria were classified into one of the three
levels of severity of the blood pressure disorder. The patients were then randomly assigned
to receive one of the nine medications. Each patient in the study was given the assigned
medication at 6:00 A.M. of the designated study day. Temperatures were taken at hourly
intervals beginning at 8:00 A.M. and continuing for 10 hours. During this time, the patients
were not allowed to do any physical activity and had to lie in bed. To eliminate the variability
of temperature readings within a day, the average of the hourly determinations was the
recorded response for each patient. These data are given in the accompanying table.

    a. Identify the design for this experiment.

    b. Write an appropriate statistical model and identify the parameters of the model.

Severity	Medication								
	A	**B**	**C**	**D**	**E**	**F**	**G**	**H**	**I**
	97.8	98.1	98.0	97.3	97.9	97.9	97.1	98.0	97.8
	97.2	98.1	97.8	97.3	97.8	97.9	97.6	97.8	98.0
1	97.6	98.0	98.1	97.5	97.8	97.8	97.3	98.0	97.7
	97.2	97.7	97.8	97.5	97.7	97.8	97.7	97.9	97.9
	97.6	97.7	97.9	97.6	97.8	97.6	97.5	98.0	97.8
	97.6	97.8	97.9	97.5	97.8	98.0	97.6	97.9	98.0
	97.4	97.7	98.1	97.4	97.8	97.7	97.5	98.0	97.6
2	97.3	97.6	97.8	97.5	97.7	97.8	97.6	97.9	98.0
	97.5	97.7	97.8	97.6	97.7	97.9	97.5	97.9	97.9
	97.5	97.7	97.6	97.7	97.8	97.8	97.3	97.8	97.9
	97.5	97.6	98.0	97.9	97.7	97.9	97.4	97.8	98.0
	97.9	97.7	97.8	97.8	97.8	98.0	97.8	97.8	98.1
3	97.6	97.9	98.1	97.8	97.9	97.7	97.4	98.0	97.9
	97.6	97.9	97.7	97.8	98.0	97.9	97.6	97.9	98.1
	97.7	97.8	98.7	97.6	98.1	97.9	97.6	97.8	97.9

15.29 Refer to Exercise 15.28.
   a. Complete the AOV table for the experiment given here.

Source	SS	df	MS	*F*
Severity	0.3628			
Medication	3.5117			
Interaction	0.5012			
Error	2.6520			
Total	7.0277			

   b. Are the differences in mean temperatures for the nine medications the same for all three severities of the blood pressure disorders? Use $\alpha = .05$.
   c. Is there a significant difference in mean temperatures for medications and severity of the disorder? Use $\alpha = .05$.
   d. Use a profile plot to assist in discussing your conclusions concerning the effect of medication and severity on the mean temperatures of the patients.

Med. **15.30** A physician was interested in examining the relationship between work performed by individuals in an exercise tolerance test and the excess weight (as determined by standard weight–height tables) they carried. To do this, a random sample of 28 healthy adult females, ranging in age from 25 to 40, was selected from the community clinic during routine visits for physical examinations. The selection process was restricted so that seven persons were selected from each of the following weight classifications.

   Normal weight (less than 10% underweight)

   1%–10% overweight

   11%–20% overweight

   More than 20% overweight

As part of the physical examination, each person was required to exercise on a bicycle ergometer until the onset of fatigue. The time to fatigue (in minutes) was recorded for each person. The data are given next.

Classification	Fatigue Time
Normal	25, 28, 19, 27, 23, 30, 35
1%–10% overweight	24, 26, 18, 16, 14, 12, 17
11%–20% overweight	15, 18, 17, 25, 12, 10, 23
More than 20% overweight	10, 9, 18, 14, 6, 4, 15

a. Identify the experimental design and write an appropriate statistical model.
b. Use $\alpha = .05$ and perform an analysis of variance.

15.31 Refer to Exercise 15.30.
a. How would you design an experiment to investigate the effects of age, gender, and excess weight on fatigue time?
b. Suppose the physician wanted to investigate the relationship among the quantitative variables percentage overweight, age, and fatigue time. Write a possible model.

Env. **15.32** An experiment was conducted to investigate the heat loss for five different designs for commercial thermal panes. The researcher in order to obtain results that would be applicable throughout most regions of the country decided to evaluate the panes at five temperatures, 0°F, 20°F, 40°F, 60°F, and 80°F. A sample of 10 panes of each design was obtained. Two panes of each design were randomly assigned to each of the five exterior temperature settings. The interior temperature of the test was controlled at 70°F for all five exterior temperatures. The heat losses associated with the five pane designs are given here.

| Exterior Temperature Setting (°F) | Pane Design | | | | |
	A	B	C	D	E
80	7.2, 7.8	7.1, 7.9	8.1, 8.8	8.3, 8.9	9.3, 9.8
60	8.1, 8.1	8.0, 8.9	8.2, 8.9	8.1, 8.8	9.2, 9.9
40	9.0, 9.9	9.2, 9.8	10.0, 10.8	10.2, 10.7	9.9, 9.0
20	9.2, 9.8	9.1, 9.9	10.1, 10.8	10.3, 10.9	9.3, 9.8
0	10.2, 10.8	10.1, 10.9	11.1, 11.8	11.3, 11.9	9.3, 9.9

a. Identify the experimental design and write an appropriate statistical model.
b. Is there a significant difference in the mean heat loss of the five pane designs? Use $\alpha = .05$. An AOV table for the data is given here.
c. Are the differences in the five designs consistent across the five temperatures? Use $\alpha = .05$ and a profile plot in reaching your conclusion.
d. Use Tukey's W procedure at an $\alpha = .05$ level to compare the mean heat loss for the five pane designs.

```
General Linear Models Procedure for Exercise 15.32

Dependent Variable: HEAT LOSS

Source                     DF    Sum of Squares    F Value    Pr > F

Model                      24      58.07280000      10.47     0.0001
Error                      25       5.78000000
Corrected Total            49      63.85280000


                R-Square              C.V.              Y Mean
                0.909479           5.067797          9.48800000


Source                     DF    Type III SS    F Value    Pr > F
T                           4    39.77880000     43.01     0.0001
D                           4     7.32280000      7.92     0.0003
T*D                        16    10.97120000      2.97     0.0073


T     D    N      Mean        T     N       Mean        D     N       Mean
0     a    2    10.5000000                               a    10    9.01000000
0     b    2    10.5000000    0    10    10.7300000      b    10    9.09000000
0     c    2    11.4500000    20   10     9.9200000      c    10    9.86000000
0     d    2    11.6000000    40   10     9.8500000      d    10    9.94000000
0     e    2     9.5000000    60   10     8.6200000      e    10    9.54000000
20    a    2     9.5000000    80   10     8.3200000
20    b    2     9.5000000
20    c    2    10.4500000
20    d    2    10.6000000
```

15.33 Refer to Exercise 15.24. In the description of this experiment, the researchers failed to note that the experiment in fact had been conducted at four different orange groves, which were located in different states. Grove 1 had a soil pH of 4.0, grove 2 had a soil pH of 5.0, grove 3 had a soil pH of 6.0, and grove 4 had a soil pH of 7.0. At each of the groves, three trees were randomly assigned to one of the calcium levels 100, 200, or 300 pounds per acre. The data are given here.

		Calcium		
Grove	pH Value	100	200	300
1	4.0	5.2, 5.9, 6.3	7.4, 7.0, 7.6	6.3, 6.7, 6.1
2	5.0	7.1, 7.4, 7.5	7.4, 7.3, 7.1	7.3, 7.5, 7.2
3	6.0	7.6, 7.2, 7.4	7.6, 7.5, 7.8	7.2, 7.3, 7.0
4	7.0	7.2, 7.5, 7.2	7.4, 7.0, 6.9	6.8, 6.6, 6.4

a. How would this new information alter the conclusions reached in Exercise 15.24 concerning the effect of soil pH and calcium on the mean increase in tree diameter?

b. Design a new experiment in which the effects of soil pH and calcium on the mean increase in tree diameter could be validly evaluated. All four groves must be used in your design along with the four levels of pH and three levels of calcium.

Psy.   **15.34**   An experiment was conducted to examine the effects of different levels of reinforcement and different levels of isolation on children's ability to recall. A single analyst was to work with a random sample of 36 children selected from a relatively homogeneous group of fourth-grade students. Two levels of reinforcement (none and verbal) and three levels of isolation (20, 40, and 60 minutes) were to be used. Students were randomly assigned to the six treatment groups, with a total of six students being assigned to each group.

Each student was to spend a 30-minute session with the analyst. During this time, the student was to memorize a specific passage, with reinforcement provided as dictated by the group to which the student was assigned. Following the 30-minute session, the student was isolated for the time specified for his or her group and then tested for recall of the memorized passage. The data appear next.

Level of Reinforcement	Time of Isolation (minutes)					
	20		**40**		**60**	
None	26	19	30	36	6	10
	23	18	25	28	11	14
	28	25	27	24	17	19
Verbal	15	16	24	26	31	38
	24	22	29	27	29	34
	25	21	23	21	35	30

Use the computer output shown here to draw your conclusions.

```
General Linear Models Procedure for Exercise 15.34

Dependent Variable: TEST SCORE

                            Sum of          Mean
Source            DF        Squares         Square      F Value    Pr > F
Model              5       1410.8889      282.1778       17.88     0.0001
Error             30        473.3333       15.7778
Corrected Total   35       1884.2222

            R-Square          C.V.        Root MSE               Y Mean
            0.748791       16.70520        3.9721               23.778

Source            DF    Type III SS    Mean Square    F Value    Pr > F
REINFORCE          1      196.0000       196.0000       12.42     0.0014
TIME               2      156.2222        78.1111        4.95     0.0139
INTERACTION        2     1058.6667       529.3333       33.55     0.0001
```

Bus.   **15.35**   A food-processing plant has tested several different formulations of a new breakfast drink. Each of six panels rated the 12 different formulations obtained from combining one of three levels of sweetness, one of two levels of caloric content, and one of two colors.
  a. Identify the design.
  b. Write an appropriate model.
  c. Give the analysis of variance table for this design.

```
              MEAN SPEED FOR TREATMENTS AND EACH FACTOR

      TIME      FIRM      MEAN  |  TIME  MEAN      FIRM    MEAN
--------------------------------|-------------------------------------
      A         1         9.33  |   A    9.167      1      9.611
      A         2        11.00  |   M    8.000      2      9.888
      A         3         7.16  |   N   10.056      3      7.722
      M         1         7.83  |
      M         2        10.00  |
      M         3         6.16  |
      N         1        11.66  |
      N         2         8.66  |
      N         3         9.83  |
```

Engin.   **15.45**   Three dye formulas for a certain synthetic fiber are under consideration by a textile manufacturer who wishes to know whether the three are in fact different in quality. To aid in this decision, the manufacturer conducts an experiment in which five specimens of fabric are cut into thirds, and one third is randomly assigned to be dyed by each of the thre dyes. Each piece of fabric is later graded and assigned a score measuring the quality of the dye. The results are as follows.

	Fabric Specimen				
Dye	**1**	**2**	**3**	**4**	**5**
A	74	78	76	82	77
B	81	86	90	93	73
C	95	99	90	87	93

a. Identify the design.
b. Run an analysis of variance and draw conclusions about the dyes. Use $\alpha = .05$.
c. Give a measure of the efficiency of this design to one not blocking on fabric specimens.

```
General Linear Models Procedure for Exercise 15.45

Dependent Variable: QUALITY
                                Sum of          Mean
Source                   DF     Squares         Square      F Value    Pr > F

Model                     6    688.00000      114.66667       3.34     0.0596
Error                     8    274.93333       34.36667
Corrected Total          14    962.93333

                    R-Square         C.V.       ROOT MSE            Y Mean
                    0.714484       6.902248       5.8623            84.933

Source                   DF  Type III SS    Mean Square      F Value    Pr > F

DYE                       2    593.73333      296.86667        8.64     0.0100
SPECIMEN                  4     94.26667       23.56667        0.69     0.6216
```

DYE	Mean	SPECIMEN	Mean
1	77.40	1	83.33
2	84.60	2	87.67
3	92.80	3	85.33
		4	87.33
		5	81.00

Psy.   **15.46**   An experiment tested the effect of factory music on workers' production. Four music programs (A, B, C, D) were compared with no music (E). Each program was played for one entire day, and five replications for each program were desired. The length of the experiment was thus 5 weeks. To control for variation in week and day of week, a Latin square design was adopted for the 25 days of the experiment. Each program was played once on each day of the week and once each week.

Week	Monday	Tuesday	Wednesday	Thursday	Friday
1	133 (E)	139 (B)	140 (C)	140 (D)	145 (A)
2	139 (A)	136 (E)	141 (B)	143 (C)	146 (D)
3	138 (B)	139 (D)	140 (E)	139 (A)	142 (C)
4	137 (C)	140 (A)	136 (D)	129 (E)	132 (B)
5	142 (D)	143 (C)	142 (A)	144 (B)	132 (E)

a. Does there appear to be a difference in mean workers' production between the five types of music? Use $\alpha = .05$.

b. If there is a difference in worker's production, which of the four music programs appear to be associated with higher worker production in comparison to no music?

```
General Linear Models Procedure for Exercise 15.45

Dependent Variable: Y    OUTPUT
                              Sum of          Mean
Source                 DF     Squares        Square     F Value    Pr > F
Model                  12    313.12000      26.09333      2.59     0.0561
Error                  12    120.72000      10.06000
Corrected Total        24    433.84000


           R-Square              C.V.        Root MSE              Y Mean
           0.721741          2.280522          3.1718             139.08


Source                 DF    Type III SS   Mean Square    F Value    Pr > F


WEEK                    4    123.44000      30.86000        3.07     0.0589
DAY                     4     11.84000       2.96000        0.29     0.8761
MUSIC                   4    177.84000      44.46000        4.42     0.0200
```

```
WEEK      Mean      DAY      Mean      MUSIC      Mean
 1       139.4       1      137.8       A        141.0
 2       141.0       2      139.4       B        138.8
 3       139.6       3      139.8       C        141.0
 4       134.8       4      139.0       D        140.6
 5       140.6       5      139.4       E        134.0
```

Ag.   **15.47** The yields of wheat (in pounds) are shown here for 5 farms. Five plots are selected based on their soil fertility at each farm with the most fertile plots designated as 1. The treatments (fertilizers) applied to each plot are shown in parentheses.

	Fertility				
Farm	**1**	**2**	**3**	**4**	**5**
1	(D) 10.3	(E) 8.6	(A) 6.7	(C) 7.6	(B) 5.8
2	(E) 8.8	(B) 6.7	(C) 6.7	(A) 4.8	(D) 6.0
3	(A) 6.3	(C) 8.3	(B) 6.8	(D) 8.0	(E) 8.8
4	(C) 8.9	(D) 7.4	(E) 8.2	(B) 6.2	(A) 4.4
5	(B) 7.3	(A) 4.4	(D) 7.7	(E) 6.8	(C) 6.7

a. Identify the designs.
b. Do an analysis of variance and draw conclusions concerning the five fertilizers. Use $\alpha = .01$.

15.48 Refer to Exercise 15.47. Run a multiple-comparison procedure to make all pairwise comparisons of the treatment means. Identify which error rate was controlled.

```
General Linear Models Procedure for Exercises 15.47 and 15.48

Dependent Variable: YIELD
                                Sum of           Mean
Source               DF         Squares          Square      F Value    Pr > F
Model                12        46.067200        3.838933       9.88      0.0002
Error                12         4.663200        0.388600
Corrected Total      24        50.730400

                 R-Square              C.V.        Root MSE              Y Mean
                 0.908079           8.745481         0.6234              7.1280

Source               DF     Type III SS    Mean Square    F Value    Pr > F

FARM                  4       6.522400       1.630600       4.20      0.0236
PLOT                  4      11.266400       2.816600       7.25      0.0033
FERT                  4      28.278400       7.069600      18.19      0.0001


     FARM     Mean   |    PLOT     Mean   |   FERTILIZER     Mean
----------------------|--------------------|------------------------
      1       7.80   |     1       8.32   |       A          5.32
      2       6.60   |     2       7.08   |       B          6.56
      3       7.64   |     3       7.22   |       C          7.64
      4       7.02   |     4       6.68   |       D          7.88
      5       6.58   |     5       6.34   |       E          8.24
```

CHAPTER 16

The Analysis of Covariance

16.1 Introduction and Case Study

16.2 A Completely Randomized Design with One Covariate

16.3 The Extrapolation Problem

16.4 Multiple Covariates and More Complicated Designs

16.5 Summary

16.1 Introduction and Case Study

In some experiments, the experimental units are nonhomogeneous or there is variation in the experimental conditions that are not due to the treatments. For example, a study is designed to evaluate different methods of teaching reading to 8-year-old children. The response variable is final scores of the children after participating in the reading program. However, the children participating in the study will have different reading ability prior to entering the program. Also, there will be many factors outside the school that may have an influence on the reading score of a child, such as socioeconomic variables associated with the child's family. The variables that describe the differences in experimental units or experimental conditions are called **covariates.** The analysis of covariance is a method by which the influence of the covariates on the treatment means is reduced. This will often result in increased precision for parameter estimates and increased power for tests of hypotheses.

covariates

In Chapter 15, we addressed this problem through the use of randomized complete block and Latin square designs. The experimental units were grouped into blocks of experimental units, which provided for greater homogeneity of the experimental units within each block than was present in the collection of experimental units as a whole. Thus, we achieved a reduction in the variation of the responses due to factors other than the treatments.

In many experiments it may be difficult or impossible to block the experimental units. The characteristics that differentiate the experimental units may not be known prior to running the experiment, or the variables that affect the response may not surface until after the experiments have started. In some cases, there may be too few experimental units in each block to examine all the treatments. Several examples of these types of experiments include the following:

- A clinical trial is run to evaluate the several traditional methods for treating chronic pain and some new alternative approaches. The patients included in the trial would have different levels of pain depending on the length of time they have been inflicted with the syn-

943

drome, their ages, physical condition, and many other factors that can affect the performance of the treatment. Researchers could block on several of these factors but the influence of the other covariates may have an undue influence on the outcome of the trial.

- The aerial application of insecticides to control fire ants is proposed for large pasture lands in Texas. There are a number of possible methods for applying the insecticide to the pastures. Because the EPA is concerned about the spray drifting off the target areas, a study is designed to evaluate the accuracy of the spraying techniques. The amount of the insecticide, y, landing within the target areas is recorded for each of the four methods of applying the insecticide. The testing is to be conducted only on those days in which there is little or no wind. However, in Texas there are always wind gusts that may affect the accuracy of the spraying. Thus, an important covariate is the wind speed at the target area during the spraying.

- A fiber-optic cable manufacturer is investigating three new machines used in coating the cable. The response of interest is the tensile strength, y, of the cable after the coating is applied. Although the coating thickness is set at a uniform thickness of 1.5 mm, there is some variation in thickness along the length of a 100-meter cable. This variation in thickness may affect the tensile strength of the cable. The testing is conducted in a laboratory with a constant temperature. The experiments are run over a 5-day period of time. Because there are some environmental and technician differences in the laboratory from day to day, the researchers decide to block on day and to record the thickness of the coating at the break point in the cable. Thus, both a blocking variable and a covariate will be involved in the experiment.

The experiments described in the following case study involves an experiment in which the measured response is related not only to the assigned treatment but also to a covariate, which was measured on the experimental unit during the study.

Case Study: Evaluation of Cool-Season Grasses for Putting Greens

A problem confronting greenskeepers on golf courses is the prevalence of viral diseases, which damage putting greens. The diseases are particularly dangerous during the early spring when the weather is cool and wet and the grasses on the greens have not completely recovered from winter dormancy. Several new cultivars of turfgrass for use on golf course greens have been developed. These cultivars are resistant to the type of viral diseases that are of concern to the greenskeepers. Prior to adopting the grasses for use on golf course greens, it is necessary to evaluate the cultivars with respect to their appropriateness for use on the putting surfaces.

Designing the Data Collection The researchers considered the following issues in designing an appropriate experiment to evaluate the cultivars:

1. What performance measures should be used to evaluate the cultivars?
2. Does the geographical region of the country affect the performance of the cultivar?

3. Do the cultivars perform differently during different times of the golf season?
4. What soil factors affect the performance characteristics of the cultivars?
5. How many replications per cultivar are needed to obtain a reliable estimate of cultivar performance?
6. What environmental factors may affect the performance of the cultivars during the test period?
7. What are the valid statistical procedures for evaluating differences in the cultivars?
8. What type of information should be included in a final report to document the differences in the suitability of the cultivars for use on golf course putting greens?

From previous studies, three cultivars (C_1, C_2, and C_3) were found to have the greatest resistance to the early spring viral diseases. Next the researchers determined from discussions with golf course superintendents that the performance measure of greatest interest was the speed that a ball rolls on the green after being struck by a putter. The U.S. Golf Association (USGA) has developed a device called the Stimpmeter to evaluate the speed of the greens. The Stimpmeter is a 36-inch extruded aluminum bar with a grooved runway on one side. A notch in the runway is used to support a golf ball until one end of the Stimpmeter is lifted to an angle of roughly 20 degrees. The average distance the golf ball travels after two opposing rolls down the Stimpmeter is referred to as the speed of the green. The farther the ball rolls, the faster the green. Important factors that affect speed are length of grass, hardness of surface, and slope of the surface.

The researchers decided to study eight different regions of the country. In each region, they selected a golf course and three putting greens were constructed. The three greens had the same soil composition and slope. The three cultivars were randomly assigned to a single green at each of the eight golf courses. Thus, the factors affecting green speed that are associated with geographical location were controlled through the use of blocking. A factor that was considered to be important but that the researchers were not able to control was the humidity during the testing period. Thus, it was decided to record humidity and use it as a covariate. The measurements of green speed (in feet) and humidity at the six locations are given in Table 16.1.

TABLE 16.1
Greens speed of three cultivars

	C_1		C_2		C_3	
Region	**Humidity**	**Speed**	**Humidity**	**Speed**	**Humidity**	**Speed**
1	31.60	7.56	29.42	8.88	89.60	8.20
2	54.12	7.41	44.44	8.20	37.17	9.15
3	42.34	7.64	84.38	7.20	37.32	9.24
4	53.82	6.81	88.42	7.12	89.21	8.31
5	86.70	6.86	71.33	8.16	58.57	9.42
6	76.27	6.86	45.50	8.68	66.68	9.26
7	68.66	7.22	66.79	8.25	82.78	8.93
8	47.27	7.64	58.34	8.22	29.52	9.89

The measurement of speed for each of the greens versus the humidity readings during the testing period are plotted in Figure 16.1. The plotted points suggest a negative relationship between speed and humidity level with the relationship similar for all three cultivars. However, cultivar (C3) appears to yield a uniformly greater speed value than the other two cultivars. We will develop models and analysis techniques in the remainder of this chapter to enable us to adjust the speed readings for the cultivars for both the region of the country in which the greens were located and the humidity during the time in which the tests were conducted. Since the analysis of covariance combines features of the analysis of variance and regression analysis, we will make use of a general linear model formulation for the analysis of this type of data. By referring to and building on our work with general linear models in preceding chapters, we can more easily understand the blending of analysis of variance with regression modeling. We begin our presentation with a single covariate in a completely randomized design.

FIGURE 16.1

Speed of golf greens for three cultivars with humidity readings

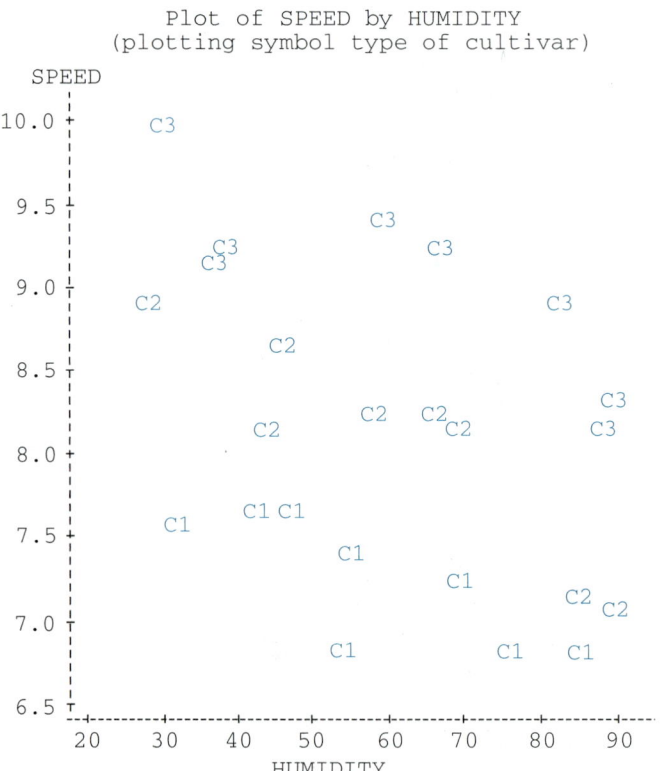

16.2 A Completely Randomized Design with One Covariate

A completely randomized design is used to compare t population means. To do this, we obtain a random sample of n_i observations on the variable y in the ith population ($i = 1, 2, \ldots, t$). Now, in addition to measuring the response variable

y on each experimental unit, we measure a second variable x, often called a *covariable*, or a *covariate*. For example, in studying the effects of different methods of reinforcement on the reading achievement levels of 8-year-old children, we could measure not only the final achievement level y for each child but also the prestudy reading performance level x. Ultimately, we would want to make comparisons among the different methods while taking into account information on both y and x.

Note that x can be thought of as an independent variable, but unlike most situations discussed in previous chapters, here we cannot control the value of x (as we controlled settings of temperature or pressure) prior to observing the variable. In spite of this, we may still write a model for the completely randomized design treating the covariate as an independent variable.

We will examine an experiment comparing $t = 3$ treatments from a completely randomized experiment with one covariate to illustrate the analysis of covariance procedures.

EXAMPLE 16.1

In this study, the effect of two treatments, a slow-release fertilizer (S) and a fast-release fertilizer (F), on seed yield (grams) of peanut plants were compared with a control (C), a standard fertilizer. Ten replications of each treatment were to be grown in a greenhouse study. When setting up the experiment, the researcher recognized that the 30 peanut plants were not exactly at the same level of development or health. Consequently, the researcher recorded the height (cm) of the plant, a measure of plant development and health, at the start of the experiment, as shown in the following table.

Control (C)		Slow Release (S)		Fast Release (F)	
Yield	**Height**	**Yield**	**Height**	**Yield**	**Height**
12.2	45	16.6	63	9.5	52
12.4	52	15.8	50	9.5	54
11.9	42	16.5	63	9.6	58
11.3	35	15.0	33	8.8	45
11.8	40	15.4	38	9.5	57
12.1	48	15.6	45	9.8	62
13.1	60	15.8	50	9.1	52
12.7	61	15.8	48	10.3	67
12.4	50	16.0	50	9.5	55
11.4	33	15.8	49	8.5	40

A plot of the yields for each treatment is shown in Figure 16.2 with the covariate, plant height, given on the horizontal axis.

FIGURE 16.2

Seed yield for three
treatments with covariate
plant height

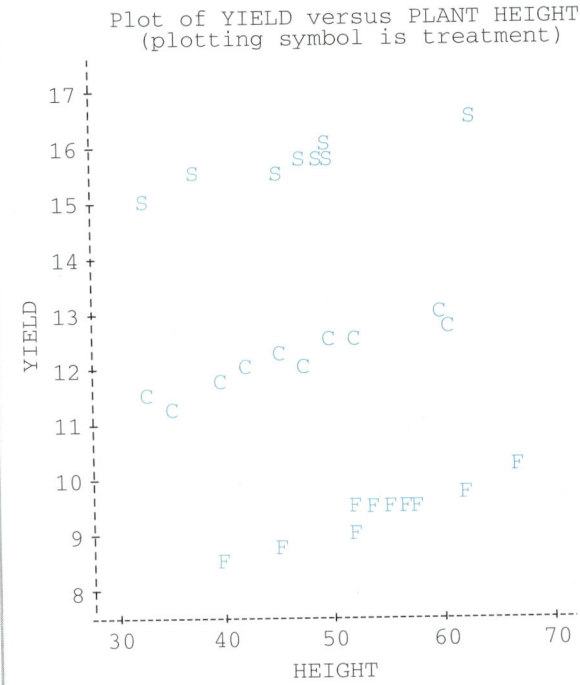

The experiment described in Example 16.1 was conducted in a completely randomized design with three treatment groups and a single covariate. If we assume a straight-line relationship between seed yield, y_{ij}, and the covariate, plant height, x_{ij}, the model for the completely randomized design with a single covariate is given by

$$y_{ij} = \mu_i + \beta_1(x_{ij} - \bar{x}_{..}) + \varepsilon_{ij}$$

or

$$y_{ij} = \beta_0 + \alpha_i + \beta_1 x_{ij} + \varepsilon_{ij}$$

with $i = 1, 2, \ldots, t$ and $j = 1, 2, \ldots, n$, where μ_i is the ith treatment mean, β_1 is the slope of the regression of y_{ij} on x_{ij}, β_0 is intercept of the regression of y_{ij} on x_{ij}, α_i is the ith treatment effect, and ε_{ij} are random independent normally distributed experimental errors with mean 0 and variance σ_{ε}^2. The other major conditions imposed on the model in an analysis of covariance are as follows.

1. The relationship between the response y and the covariate x is linear.
2. The regression coefficient β_1 is the same for all treatments.
3. The treatments do not affect the covariate, x_{ij}.

adjusted treatment means

The analysis of covariance involves fitting a number of models to the response variable, y. First, we evaluate whether the covariate, x, provides a significant reduction in the experimental error. If the reduction is significant, then we replace the observed treatment means, $\bar{y}_{i.}$, with estimated **adjusted treatment means**, $\hat{\mu}_{\text{Adj},i}$, which are adjusted for the effect of the covariate on the response variable. The significance of the treatment differences is then made on the basis of the adjusted means and not on the observed means.

We will formulate the required models needed in the analysis of covariance. The model relating y_{ij} to the t treatments and the covariate can be written in the form of analysis of variance models and then reformulated in regression form.

Full Model: $\quad y_{ij} = \beta_0 + \alpha_i + \beta_1 x_{ij} + \varepsilon_{ij}$

Next, we will formulate two reduced models, one without the covariate and then a model without treatment differences but with the covariate.

Reduced Model I: $\quad y_{ij} = \beta_0 + \alpha_i + \varepsilon_{ij}$

Reduced Model II: $\quad y_{ij} = \beta_0 + \beta_1 x_{ij} + \varepsilon_{ij}$

These three models also can be written in the form of the regression (general linear) models of Chapter 12. We make this transition to regression models because it facilitates analysis using various statistical software packages.

Full Model: $\quad y = \beta_0 + \beta_1 x_1 + \beta_2 x_2 + \cdots + \beta_t x_t + \varepsilon$

where

x_1 = covariate

$x_2 = 1$ if treatment 2 is used $\qquad x_2 = 0$ otherwise

$x_3 = 1$ if treatment 3 is used $\qquad x_3 = 0$ otherwise

$\cdots$

$x_t = 1$ if treatment t is used $\qquad x_t = 0$ otherwise

It is helpful with these models to refer to a table of expected values, as shown in Table 16.2, based on the full model. Note that the treatments have the same slope (β_1) but different intercepts.

TABLE 16.2

Expected values for the full model

Treatment	Expected Value
1	$\beta_0 + \beta_1 x_1$
2	$(\beta_0 + \beta_2) + \beta_1 x_1$
.	.
.	.
.	.
t	$(\beta_0 + \beta_t) + \beta_1 x_1$

We next fit a reduced model in which the covariate is removed in order to determine the influence of the covariate.

Reduced Model I: $\quad y = \beta_0 + \beta_2 x_2 + \beta_3 x_3 + \cdots + \beta_t x_t + \varepsilon$

A second reduced model is fit in which the treatment effects are removed but the covariate remains in the model.

Reduced Model II: $\quad y = \beta_0 + \beta_1 x_1 + \varepsilon$

From each of these models we obtain the sum of squares error, which we will denote as follows:

SSE_F = sum of squares error from the full model

SSE_{RI} = sum of squares error from reduced model I

SSE_{RII} = sum of squares error from reduced model II

The significance of the influence of the covariate on the response variable is determined by testing the hypothesis that the regression lines for the treatments have a slope of zero. This hypothesis is

$$H_0: \quad \beta_1 = 0 \text{ versus } H_a: \quad \beta_1 \neq 0$$

for the full model. Our test statistic is based on the sum of squares reduction due to the addition of the covariate x to the model and is given as

$$SS_{Cov} = SSE_{RI} - SSE_F$$

We then form the F test

$$F = \frac{SS_{Cov}}{SSE_F/(N - t - 1)}$$

where N is the number of observations in the experiment. Our decision rule is then given by

$$\text{Reject } H_0: \quad \beta_1 = 0 \text{ if } F \geq F_{\alpha,1,N-t-1}$$

If we determine that the covariate does have a significant linear relationship with the response variable, we would next test for a significant treatment effect using the adjusted treatment means. That is, we want to test the hypotheses

$$H_0: \quad \alpha_1 = \alpha_2 = \cdots = \alpha_t = 0 \text{ versus } H_a: \quad \text{Not all } \alpha_i\text{s are } 0$$

which in the regression model is equivalent to testing that the regression lines have the same intercept (β_0). Thus, from Table 16.2 we are testing

$$H_0: \quad \beta_2 = \beta_3 = \cdots = \beta_t = 0 \text{ versus } H_a: \quad \text{Not all of } \beta_2, \beta_3, \ldots, \beta_t \text{ are } 0$$

Our test statistic is based on the sum of squares reduction due to the addition of the differences in the treatment means to the model and is given

$$SS_{Trt} = SSE_{RII} - SSE_F$$

We then form the F test

$$F = \frac{SS_{Trt}/(t - 1)}{SSE_F/(N - t - 1)}$$

Our decision rule is then given by

$$\text{Reject } H_0: \quad \beta_2 = \beta_3 = \cdots \beta_t = 0 \text{ if } F \geq F_{\alpha,t-1,N-t-1}$$

If we reject H_0, then we can evaluate treatment differences by examining the estimated adjusted treatment means using the formula

$$\hat{\mu}_{Adj,i} = \bar{y}_{i.} - \hat{\beta}_1(\bar{x}_{i.} - \bar{x}_{..})$$

which adjusts the observed treatment means for the effect of the covariate. This effect is estimated by considering how large a difference exists between the mean value of the covariate observed of the experimental units receiving treatment i and the average value on the covariate over all treatments.

We can also estimate the adjusted treatment means using the regression model. From Table 16.2 we have that for treatments $i = 2, 3, \ldots, t$

$$\mu_i = E(y) = \beta_0 + \beta_i + \beta_1 x_1$$

and for $i = 1$,

$$\mu_1 = E(y) = \beta_0 + \beta_1 x_1$$

The estimated adjusted treatment means are obtained by estimating the mean value of y for each treatment group corresponding to the overall mean value of the covariate, $x_1 = \bar{x}_{..}$. It follows that

$$\hat{\mu}_{\text{Adj},i} = \hat{\beta}_0 + \hat{\beta}_i + \hat{\beta}_1 \bar{x}_{..}$$

for treatments $i = 2, 3, \ldots, t$ and

$$\hat{\mu}_{\text{Adj},1} = \hat{\beta}_0 + \hat{\beta}_1 \bar{x}_{..}$$

for treatment 1. The estimated standard error of the estimated ith treatment mean, $\hat{\mu}_{\text{Adj},i}$ is given by

$$SE(\hat{\mu}_{\text{Adj},i}) = \sqrt{\text{MSE}_F \left(\frac{1}{n} + \frac{(\bar{x}_{i.} - \bar{x}_{..})^2}{E_{xx}} \right)}$$

where $E_{xx} = \sum\sum_{ij}(x_{ij} - \bar{x}_{i.})^2$. The estimated standard error of the difference between two adjusted treatment means $\hat{\mu}_{\text{Adj},i} - \hat{\mu}_{\text{Adj},h}$ is given by

$$SE(\hat{\mu}_{\text{Adj},i} - \hat{\mu}_{\text{Adj},h}) = \sqrt{\text{MSE}_F \left(\frac{2}{n} + \frac{(\bar{x}_{i.} - \bar{x}_{h.})^2}{E_{xx}} \right)}$$

where MSE_F is the MSE from the full model. These estimated standard errors can now be used to place confidence intervals on the adjusted treatment means and their differences.

The following example will illustrate the ideas of analysis of covariance.

EXAMPLE 16.2

Refer to Example 16.1, where we had three treatments—a control (C), a slow-release fertilizer (S), and a fast-release fertilizer (F)—and we used plant height at the beginning of the study as a covariate. Our response variable was the seed yield of peanut plants and we had ten replicates.

a. Write the model for an analysis of covariance.
b. Use the computer output shown here to test whether the covariate provides a significant reduction in experimental error.
c. Give the linear regression equations for the three treatment groups.
d. Compute the observed and adjusted treatment means for the three treatment groups.
e. Does there appear to be a significant difference between the three treatments after adjusting for the covariate?

The computer printout for the analysis is given here.

```
FULL MODEL

General Linear Models Procedure

Dependent Variable: Y    YIELD
```

Source	DF	Sum of Squares	Mean Square	F Value	Pr > F
Model	3	214.37595	71.45865	4447.85	0.0001
Error	26	0.41771	0.01607		
Corrected Total	29	214.79367			

```
                              T for H0:    Pr > |T|    Std Error of
INTERCEPT      9.529256364       71.34      0.0001       0.13357349
X1 (COV)       0.055809949       20.41      0.0001       0.00273429
X2 (S)         3.571637117       62.62      0.0001       0.05703267
X3 (F)        -3.144155615      -52.08      0.0001       0.06037390
-------------------------------------------------------------------
```

REDUCED MODEL I

General Linear Models Procedure

Dependent Variable: Y   YIELD

Source	DF	Sum of Squares	Mean Square	F Value	Pr > F
Model	2	207.68267	103.84133	394.28	0.0001
Error	27	7.11100	0.26337		
Corrected Total	29	214.79367			

Parameter	Estimate	T for H0: Parameter = 0	Pr > |T|	Std Error of Estimate
INTERCEPT	12.13000000	74.74	0.0001	0.16228690
X2 (S)	3.70000000	16.12	0.0001	0.22950833
X3 (F)	-2.72000000	-11.85	0.0001	0.22950833

REDUCED MODEL II

General Linear Models Procedure

Dependent Variable: Y   YIELD

Source	DF	Sum of Squares	Mean Square	F Value	Pr > F
Model	1	0.4721494	0.4721494	0.06	0.8057
Error	28	214.3215172	7.6543399		
Corrected Total	29	214.7936667			

Parameter	Estimate	T for H0: Parameter = 0	Pr > |T|	Std Error of Estimate
INTERCEPT	13.14900450	4.64	0.0001	2.83300563
X1 (COV)	-0.01387451	-0.25	0.8057	0.05586395

Solution

a. We have a completely randomized design with three treatments, ten replications per treatment, and a single covariate. The model is thus given by $y_{ij} = \mu_i + \beta_1(x_{ij} - \bar{x}_{..}) + \varepsilon_{ij}$, for $i = 1, 2, 3$ and $j = 1, \ldots, 10$.

The full model using regression notation is

Full Model (in which the regression lines have different intercepts but a common slope):

$$y = \beta_0 + \beta_1 x_1 + \beta_2 x_2 + \beta_3 x_3 + \varepsilon$$

where

y = yield

x_1 = plant height

x_2 = 1 if treatment is S x_2 = 0 otherwise

x_3 = 1 if treatment is F x_3 = 0 otherwise

The expected values of the response for the three treatments are shown here.

Treatment	Expected Value
C	$\beta_0 + \beta_1 x_1$
S	$(\beta_0 + \beta_2) + \beta_1 x_1$
F	$(\beta_0 + \beta_3) + \beta_1 x_1$

The corresponding reduced models are:
Reduced Model I (in which the regression lines have a slope equal to zero; that is, the covariate is unrelated to the response variable):

$$y = \beta_0 + \beta_2 x_2 + \beta_3 x_3 + \varepsilon$$

and

Reduced Model II (in which the regression lines have a common intercept. β_0, and common slope, β_1):

$$y = \beta_0 + \beta_1 x_1 + \varepsilon$$

b. We want to test whether the covariate provides a reduction in the experimental error. That is, we need to test that the common slope (β_1) is zero.

$$H_0: \quad \beta_1 = 0 \text{ versus } H_a: \quad \beta_1 \neq 0$$

From the computer output,

$$SSE_F = .41771 \qquad SSE_{RI} = 7.11100$$

Thus, we have

$$SS_{Cov} = SSE_{RI} - SSE_F = 7.111 - .41771 = 6.69329$$

Our F test is

$$F = \frac{6.69329}{.41771/(30 - 3 - 1)} = 416.62 \quad \text{and} \quad F_{.05,1,26} = 4.23$$

Because 416.62 is greater than 4.23, we reject H_0 and conclude that the plant height (the covariate) is significantly related to plant seed yield (i.e., the slope β_1 is different from zero).

c. From the output for the full model we obtain the least-squares estimates:

$$\hat{\beta}_0 = 9.53, \qquad \hat{\beta}_1 = 0.0558, \qquad \hat{\beta}_2 = 3.57, \qquad \hat{\beta}_3 = -3.14$$

The estimated seed yields, with adjustments for initial plant height for the three treatments, are

Control: $\qquad\qquad \hat{y} = \hat{\beta}_0 + \hat{\beta}_1 x_1 = 9.53 + .0558x_1$

Slow Release: $\quad \hat{y} = (\hat{\beta}_0 + \hat{\beta}_2) + \hat{\beta}_1 x_1 = (9.53 + 3.57) + .0558x_1$
$$= 13.1 + .0558x_1$$

Fast Release: $\quad \hat{y} = (\hat{\beta}_0 + \hat{\beta}_3) + \hat{\beta}_1 x_1 = (9.53 - 3.14) + .0558x_1$
$$= 6.39 + .0558x_1$$

d. The observed sample means are given here.

	Control	Slow Release	Fast Release	Overall
y	12.13	15.83	9.41	12.457
x	46.60	48.90	54.20	49.900

We can obtain the estimated adjusted means by substituting the overall mean plant height for x_1 in the separate regression equations:

Control: $\hat{\mu}_{\text{Adj},1} = 9.53 + .0558(49.90) = 12.31$

Slow Release: $\hat{\mu}_{\text{Adj},2} = 13.1 + .0558(49.90) = 15.88$

Fast Release: $\hat{\mu}_{\text{Adj},3} = 6.39 + .0558(49.90) = 9.17$

Alternatively, we could obtain the estimated adjusted means, using the formula

$$\hat{\mu}_{\text{Adj},i} = \bar{y}_{i.} - \hat{\beta}_1(\bar{x}_{i.} - \bar{x}_{..})$$

Control: $\hat{\mu}_{\text{Adj},1} = 12.13 - .0558(46.60 - 49.90) = 12.31$

Slow Release: $\hat{\mu}_{\text{Adj},2} = 15.83 - .0558(48.9 - 49.90) = 15.88$

Fast Release: $\hat{\mu}_{\text{Adj},3} = 9.41 - .0558(54.20 - 49.90) = 9.17$

Because the slow-release fertilizer plants had an average plant height less than the overall average height, the observed average seed yield was adjusted upward from 15.83 to 15.88, whereas the fast-release fertilizer's average seed yield was adjusted downward from 9.41 to 9.17.

e. We can test for a difference in the average seed yields of the three treatments by examining the sum of squares error in reduced model II. We want to test the following hypotheses:

H_0: $\mu_{\text{Adj},1} = \mu_{\text{Adj},2} = \ldots = \mu_{\text{Adj},t}$ versus H_a: Not all $\mu'_{\text{Adj},i}$s are equal.

This is equivalent to testing the null hypothesis that the regression lines have a common intercept (β_0); that is, we want to test

H_0: $\beta_2 = \beta_3 = 0$ versus H_a: $\beta_2 \neq 0$ or $\beta_3 \neq 0$

From the computer output,

$$\text{SSE}_F = .41771 \qquad \text{SSE}_{\text{RII}} = 214.3215$$

Thus, we have

$$SS_{Trt} = SSE_{RII} - SSE_F = 214.3215 - .41771 = 213.90$$

Our F test thus is

$$F = \frac{213.90/(3 - 1)}{.41771/(30 - 3 - 1)} = 6{,}657.13 \quad \text{and} \quad F_{.05,2,26} = 3.37$$

Because 6,657.13 is greater than 3.37, we reject H_0 and conclude that the intercepts are not equal and hence there is a significant difference in the adjusted plant seed yields for the three types of fertilizers.

The conclusions we reached in Example 16.2 are dependent on the validity of the conditions we placed on the model. We can evaluate the condition of independent and homogeneous normally distributed error terms by examining the residuals from the fitted model:

$$e_{ij} = y_{ij} - \hat{\beta}_0 - \hat{\beta}_1 x_{1ij} - \hat{\beta}_2 x_{2ij} - \cdots - \hat{\beta}_t x_{tij}$$

We can then apply plots and tests of normality to the e_{ij}s to evaluate the equal variance and normality conditions.

The three added conditions for the analysis of covariance are evaluated in the following manner.

The Relationship between the Response and Covariate Is Linear. We can evaluate this condition as we did in regression analysis through the use of plots and tests of hypotheses. We can plot y versus x separately for each treatment and assess whether the plotted points follow a straight line. A separate regression line can be fitted for each treatment using the methods of Chapter 11. We then would assess the residuals from the t fitted lines and conduct tests of lack of fit to determine whether any of the t fitted lines needed higher-order terms in the covariate x_{ij}.

The Regression (Slope) Coefficient Is the Same for all t Treatments. Consider the following model:

$$\text{Model A:} \quad y = \beta_0 + \beta_1 x_1 + \beta_2 x_2 + \beta_3 x_3 + \cdots + \beta_t x_t + \beta_{t+1} x_1 x_2$$
$$+ \beta_{t+2} x_1 x_3 + \cdots + \beta_{2t-1} x_1 x_t + \varepsilon$$

where $x_2, \ldots, x_t$ are the indicator variables for the treatments and x_1 is the covariate. This regression model yields separate regression lines, possibly different slopes and different intercepts, for each treatment. (See the expected values for model A shown in Table 16.3.)

TABLE 16.3
Expected values for model A

Treatment	Expected Value
1	$\beta_0 + \beta_1 x_1$
2	$(\beta_0 + \beta_2) + (\beta_1 + \beta_{t+1})x_1$
3	$(\beta_0 + \beta_3) + (\beta_1 + \beta_{t+2})x_1$
.	.
.	.
.	.
t	$(\beta_0 + \beta_t) + (\beta_1 + \beta_{2t-1})x_1$

We next consider a reduced model, in which we require the slopes to be the same for all treatments but allow for different intercepts.

Model B: $\quad y = \beta_0 + \beta_2 x_2 + \beta_3 x_3 + \cdots + \beta_t x_t + \varepsilon$

The test for equal slopes would involve testing

$$H_0: \quad \beta_{t+1} = \beta_{t+2} = \cdots = \beta_{2t-1} = 0$$

$$H_a: \quad \text{At least one of } \beta_{t+1}, \beta_{t+2}, \ldots, \beta_{2t-1} \text{ is not } 0$$

The test statistic would be obtained by fitting models A and B.

$$F = \frac{(\text{SSE}_B - \text{SSE}_A)/(t-1)}{\text{SSE}_A/(N-2t)} \qquad \text{with df}_1 = t - 1, \text{df}_2 = N - 2t$$

This would determine whether the set of regression lines relating the response to the covariate have the same slope. This is a crucial assumption because if the slopes are different, then the difference in the adjusted treatment means is highly dependent on the level of the covariate chosen for adjustment. This situation is similar to experiments in which we had two factors with significant interactions and inferences about one factor depending on the level of the second factor. The situation in which the lines relating the response to the covariate have different slopes is displayed in Figure 16.3. From this figure we can observe that amount of adjustment varies greatly depending on which treatment and which value of the covariate are selected for adjustment.

FIGURE 16.3

Regression lines relating the response and covariate with different slopes

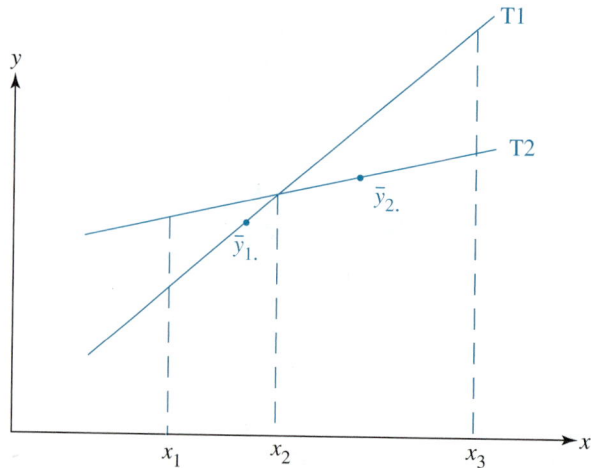

The Treatments Do Not Affect the Covariate, x_{ij}. In experiments where both the covariate x as well as the response variable y are affected by the treatments, then we cannot validly apply the methods of analysis of covariance. The appropriate method of analysis would involve multivariate analysis where we treat the response as a bivariate variable (x, y). When the covariate is measured prior to the random assignment of treatments to the experimental units, the analysis of covariance model would be appropriate, because it would be impossible for the treatment to affect the covariate. When the covariate is measuring conditions in the experimental setting—that is, the covariate is measured during the running of the experiment—the experimenter must decide whether the treatments have an affect on the covariate. Only after the experimenter determines that the treatments have not affected the covariate can we correctly adjust the treatment means for the covariate.

EXAMPLE 16.3

Refer to Example 16.1. Evaluate the necessary conditions in the analysis of covariance model using the computer output given here.

```
MODEL A: DIFFERENT SLOPES FOR EACH TREATMENT


General Linear Models Procedure
Number of observations in data set = 30


Dependent Variable: Y    YIELD
                                Sum of          Mean
Source                  DF      Squares        Square     F Value    Pr > F
Model                    5    214.43722      42.88744    2887.70     0.0001
Error                   24      0.35644       0.01485
Corrected Total         29    214.79367


Source                  DF    Type III SS   Mean Square   F Value    Pr > F


X1                       1     2.6167178     2.6167178    176.19     0.0001
X2                       1     2.5905994     2.5905994    174.43     0.0001
X3                       1     1.4990044     1.4990044    100.93     0.0001
X2*X1                    1     0.0190292     0.0190292      1.28     0.2688
X3*X1                    1     0.0151538     0.0151538      1.02     0.3225


                                  T for H0:   Pr > |T|   Std Error of
Parameter       Estimate      Parameter = 0               Estimate
INTERCEPT      9.491768741         46.88      0.0001     0.20245904
X1             0.056614405         13.27      0.0001     0.00426518
X2             3.906558043         13.21      0.0001     0.29578964
X3            -3.519620102        -10.05      0.0001     0.35033468
X2*X1         -0.006886936         -1.13      0.2688     0.00608421
X3*X1          0.006814587          1.01      0.3225     0.00674632
-----------------------------------------------------------------------
MODEL B: SAME SLOPE FOR ALL TREATMENTS


General Linear Models Procedure
Number of observations in data set = 30


Dependent Variable: Y    YIELD
                                Sum of          Mean
Source                  DF      Squares        Square     F Value    Pr > F
Model                    3    214.37595      71.45865    4447.85     0.0001
Error                   26      0.41771       0.01607
Corrected Total         29    214.79367
```

Source	DF	Type III SS	Mean Square	F Value	Pr > F
X1	1	6.693287	6.693287	416.62	0.0001
X2	1	63.007424	63.007424	3921.82	0.0001
X3	1	43.572654	43.572654	2712.12	0.0001

Parameter	Estimate	T for H0: Parameter = 0	Pr > |T|	Std Error of Estimate
INTERCEPT	9.529256364	71.34	0.0001	0.13357349
X1	0.055809949	20.41	0.0001	0.00273429
X2	3.571637117	62.62	0.0001	0.05703267
X3	-3.144155615	-52.08	0.0001	0.06037390

Solution From Figure 16.2, we can see that the lines relating seed yield to plant height for the three treatments appear to be adequately fit by a straight line and the three slopes appear to be the same; that is, we have three parallel lines with possibly different intercepts. The following computer output is obtained by fitting model A (different slopes and different intercepts) and model B (same slopes but different intercepts) to the plant seed yield data.
From the output we can compute

$$F = \frac{(\text{SSE}_B - \text{SSE}_A)/(t - 1)}{\text{SSE}_A/(N - 2t)} = \frac{(.41771 - .35644)/(3 - 1)}{.35644/(30 - 6)} = 2.06$$

with $df_1 = 2$, $df_2 = 24$. Because $F_{.05,2,24} = 3.40$, we fail to reject H_0 and conclude that there is not significant evidence of a difference in the slopes of the three lines. Because the covariate, plant height, was measured prior to assigning the type of fertilizer to the plants, the treatments cannot have an affect on the covariate. The remaining conditions of equal variance and normality can be assessed using a residual analysis.

EXERCISES   **Basic Techniques**

16.1   Consider a completely randomized design for $t = 5$ treatments, with a single covariate x_1 and six observations per treatment. Write the complete general linear model under the assumption that the response y is linearly related to the covariate x_1 for each treatment Identify the parameters in your model.

16.2   Refer to Exercise 16.1. Indicate the relationships among the parameters of the model for the following cases; show a graph for each case.
  a. The lines are not parallel.
  b. The lines are parallel, but do not coincide.
  c. The lines are coincident.

16.3   Refer to Exercise 16.1. How would you test for parallelism among the straight lines for the five treatment groups? Identify how you would obtain the test statistic. What are the degrees of freedom associated with the test statistics?

16.4   Refer to Exercise 16.1. Assume the lines are parallel. Give the test for adjusted treatment means. How would you estimate the mean response for treatment 1 with $x_1 = 5$?

16.5   Perform an analysis of variance on the following experiment. A researcher wants to evaluate the difference in the mean film thickness of a coating placed on silicon wafers using three different processes. Six wafers are randomly assigned to each of the processes. The film thickness (Y) and the temperature (X) in the lab during the coating process are recorded on each wafer. The researcher is concerned that fluctuations in temperature may affect the thickness of the coating. Test whether the processes have a difference in mean film thickness.

	Process				
1		**2**		**3**	
x	*y*	*x*	*y*	*x*	*y*
26	100	24	118	37	124
35	150	28	134	31	95
28	106	29	138	34	120
31	95	32	147	27	86
29	113	36	165	28	98
34	144	35	159	25	81

16.3 The Extrapolation Problem

In the previous section, we discussed how to compare two (or more) treatments from a completely randomized design with one covariable. If the regression equations for the treatments are linear in the covariable and parallel, we said we could compare the treatments using the adjusted treatment means. However, as with most methods, the analysis of covariance methods should not be used blindly. Even if the linearity and parallelism assumptions hold, we can have problems if the values of the covariable do not have considerable overlap for the treatment groups. We will illustrate this with an example.

Suppose that we were interested in comparing self-esteem scores for alcoholics and drug addicts. We collected a sample of nine alcoholics and a sample of nine drug addicts and for each individual, we obtained his or her self-esteem score and age. The data are shown in Table 16.4.

TABLE 16.4

Self-esteem scores and ages for a sample of alcoholics and drug addicts

Alcoholics		Drug Addicts	
Self-Esteem	**Age**	**Self-Esteem**	**Age**
25	15	20	30
22	17	17	31
24	18	18	33
20	19	15	35
21	21	14	36
17	22	15	37
14	23	12	38
16	24	10	40
15	25	11	41

If we blindly followed the analysis of covariance procedures without looking at the data, we would find the regression equations for alcoholics and drug addicts to be reasonably linear and parallel. From the computer output displayed in Figure 16.4, we note from the plotted data that the data values for alcoholics (A) would fall near a straight line, as would the points for drug addicts (D). If we used the

FIGURE 16.4   Plot of SELF-ESTEEM versus AGE

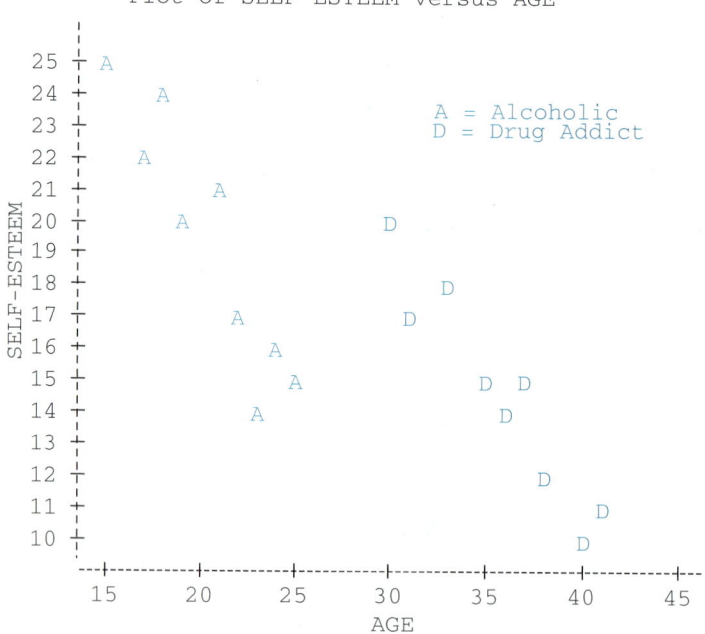

sum of squares error for the two models, we would obtain

$$F = \frac{(30.88 - 27.39)/(2 - 1)}{1.9567} = 1.78$$

with $df_1 = 1$, $df_2 = 14$. The p-value for the observed F value is Pr $(F \geq 1.78) = 0.2035$. Thus, we would accept the hypothesis that the slopes of the lines relating self-esteem to age are the same for the alcoholics and the drug addicts. Furthermore, from the computer output for model B, we have that the p-value for testing a difference in the adjusted mean self-esteem scores is Pr $(F \geq 34.14) < 0.0001$. The two groups of addicts appear to have significantly different adjusted mean self-esteem scores.

MODEL A: DIFFERENT SLOPES AND TREATMENT DIFFERENCES

Dependent Variable: Y   SELF-ESTEEM

Source	DF	Sum of Squares	Mean Square	F Value	Pr > F
Model	3	286.60611	95.53537	48.82	0.0001
Error	14	27.39389	1.95671		
Corrected Total	17	314.00000			

Source	DF	Type III SS	Mean Square	F Value	Pr > F
X1	1	188.51593	188.51593	96.34	0.0001
X2	1	0.43265	0.43265	0.22	0.6454
X2*X1	1	3.48284	3.48284	1.78	0.2035

Parameter	Estimate	T for H0: Parameter = 0	Pr > |T|	Std Error of Estimate
INTERCEPT	44.18390805 B	9.49	0.0001	4.65570471
X1	-0.82758621 B	-6.37	0.0001	0.12987748
X2	-2.60800443 B	-0.47	0.6454	5.54628759
X2*X1	-0.26036560 B	-1.33	0.2035	0.19515497

MODEL B: SAME SLOPES AND TREATMENT DIFFERENCES

Dependent Variable: Y   SELF-ESTEEM

Source	DF	Sum of Squares	Mean Square	F Value	Pr > F
Model	2	283.12327	141.56163	68.77	0.0001
Error	15	30.87673	2.05845		
Corrected Total	17	314.00000			

Source	DF	Type III SS	Mean Square	F Value	Pr > F
X1	1	185.12327	185.12327	89.93	0.0001
X2	1	70.27928	70.27928	34.14	0.0001

Parameter	Estimate	T for H0: Parameter = 0	Pr > |T|	Std Error of Estimate
INTERCEPT	48.29686944 B	13.50	0.0001	3.57834982
X1	-0.94290288	-9.48	0.0001	0.09942750
X2	-9.68641053 B	-5.84	0.0001	1.65775088

REDUCED MODEL I: TREATMENT DIFFERENCES WITH NO COVARIATE

General Linear Models Procedure

Dependent Variable: Y   SELF-ESTEEM

Source	DF	Sum of Squares	Mean Square	F Value	Pr > F
Model	1	98.000000	98.000000	7.26	0.0160
Error	16	216.000000	13.500000		
Corrected Total	17	314.000000			

Source	DF	Type III SS	Mean Square	F Value	Pr > F
X2	1	98.000000	98.000000	7.26	0.0160

REDUCED MODEL II: COVARIATE BUT NO TREATMENT DIFFERENCES   51

General Linear Models Procedure

Dependent Variable: Y   SELF-ESTEEM

Source	DF	Sum of Squares	Mean Square	F Value	Pr > F
Model	1	212.84398	212.84398	33.67	0.0001
Error	16	101.15602	6.32225		
Corrected Total	17	314.00000			

The model we formulated in Example 16.4 not only provides for a linear relationship between y and x_1 for each of the treatments in each block, but it also allows for differences among intercepts and slopes. If we wanted to test for the equality of the slopes across treatments and blocks, we would use the null hypothesis

$$H_0: \quad \beta_5 = \beta_6 = \beta_7 = 0$$

If there is insufficient evidence to reject H_0, we would proceed with the reduced model (obtained by setting $\beta_5 = \beta_6 = \beta_7 = 0$ in our model)

$$y = \beta_0 + \beta_1 x_1 + \beta_2 x_2 + \beta_3 x_3 + \beta_4 x_4 + \varepsilon$$

A test for differences among treatments adjusted for the covariate could be obtained by fitting a complete and a reduced model for the null hypothesis

$$H_0: \quad \beta_3 = \beta_4 = 0$$

Analyzing Data for Case Study, Speed of Golf Course Greens

The objective of the study was to compare the mean speed of putted golf balls for three cultivars used on golf course greens. A plot of the data was presented in Figure 16.1. From this plot it would appear that the response variable, speed of putted ball, was linearly related to relative humidity with a similar slope coefficient for the three cultivars. We will model the data, evaluate the model conditions, and then test for differences in the adjusted mean speeds for the three cultivars. Because there were regional differences in soil characteristics and climatic conditions, eight different regions of the country were selected for testing sites. At each site, there was a single green for each of the three cultivars. A covariate, relative humidity, was recorded during the time at which the speed measurements were obtained on each green. Thus, we have a randomized complete block design with eight blocks (region of country), three treatments (cultivars), and a single covariate (relative humidity). We'll assume a model that relates the response variable (speed of green) to the blocks, treatments, and covariate and allows for different slopes for the treatment (cultivars) within a region, but assumes that a green treatment has the same slope across regions.

Model I: The full model for this situation in general linear model notation is shown here.

$$y = \beta_0 + \beta_1 x_1 + \beta_2 x_2 + \beta_3 x_3 + \beta_4 x_4 + \beta_5 x_5 + \beta_6 x_6 + \beta_7 x_7 + \beta_8 x_8 + \beta_9 x_9$$
$$+ \beta_{10} x_{10} + \beta_{11} x_9 x_1 + \beta_{12} x_{10} x_1 + \varepsilon$$

where

x_1 = relative humidity (covariate)

$x_2 = 1$ if region 1 is used $x_2 = 0$ otherwise

$x_3 = 1$ if region 2 is used $x_3 = 0$ otherwise

$x_4 = 1$ if region 3 is used $x_4 = 0$ otherwise

$$x_5 = 1 \text{ if region 4 is used} \qquad x_5 = 0 \text{ otherwise}$$

$$x_6 = 1 \text{ if region 5 is used} \qquad x_6 = 0 \text{ otherwise}$$

$$x_7 = 1 \text{ if region 6 is used} \qquad x_7 = 0 \text{ otherwise}$$

$$x_8 = 1 \text{ if region 7 is used} \qquad x_8 = 0 \text{ otherwise}$$

$$x_9 = 1 \text{ if cultivar 1 is used} \qquad x_9 = 0 \text{ otherwise}$$

$$x_{10} = 1 \text{ if cultivar 2 is used} \qquad x_{10} = 0 \text{ otherwise}$$

The expected values for model I are shown in Table 16.6.

TABLE 16.6

Expected values for model I in the case study

	Cultivar		
Region	**1**	**2**	**3**
1	$(\beta_0 + \beta_2) + (\beta_1 + \beta_{11})x_1$	$(\beta_0 + \beta_2) + (\beta_1 + \beta_{12})x_1$	$(\beta_0 + \beta_2) + \beta_1 x_1$
2	$(\beta_0 + \beta_3) + (\beta_1 + \beta_{11})x_1$	$(\beta_0 + \beta_3) + (\beta_1 + \beta_{12})x_1$	$(\beta_0 + \beta_3) + \beta_1 x_1$
.	.	.	.
.	.	.	.
.	.	.	.
7	$(\beta_0 + \beta_8) + (\beta_1 + \beta_{11})x_1$	$(\beta_0 + \beta_8) + (\beta_1 + \beta_{12})x_1$	$(\beta_0 + \beta_8) + \beta_1 x_1$
8	$\beta_0 + (\beta_1 + \beta_{11})x_1$	$\beta_0 + (\beta_1 + \beta_{12})x_1$	$\beta_0 + \beta_1 x_1$

Note that the cultivars have different slopes, but that each cultivar has the same slope across regions.

To test whether the linear relationship between speed and relative humidity is the same for the three cultivars—that is, whether the three lines have equal slopes—we would fit a model to the data in which the three lines have the same slope, but different intercepts.

Model II: Region and cultivar differences with covariate having equal slopes

$$y = \beta_0 + \beta_1 x_1 + \beta_2 x_2 + \beta_3 x_3 + \beta_4 x_4 + \beta_5 x_5 + \beta_6 x_6 + \beta_7 x_7 + \beta_8 x_8 + \beta_9 x_9$$
$$+ \beta_{10} x_{10} + \varepsilon$$

The computer output from fitting these two models is given here.

```
MODEL I: REGION AND TREATMENT DIFFERENCES WITH COVARIATE HAVING UNEQUAL SLOPES

General Linear Models Procedure

Dependent Variable: S    SPEED
                                 Sum of        Mean
Source                 DF       Squares      Square    F Value    Pr > F
Model                  12     18.592940    1.549412      54.78    0.0001
Error                  11      0.311125    0.028284
Corrected Total        23     18.904065
```

Source	DF	Type III SS	Mean Square	F Value	Pr > F
X1	1	0.8535690	0.8535690	30.18	0.0002
X2	1	0.2203698	0.2203698	7.79	0.0175
X3	1	0.1929255	0.1929255	6.82	0.0242
X4	1	0.1384441	0.1384441	4.89	0.0490
X5	1	0.2766251	0.2766251	9.78	0.0096
X6	1	0.0476899	0.0476899	1.69	0.2207
X7	1	0.0011902	0.0011902	0.04	0.8412
X8	1	0.0189954	0.0189954	0.67	0.4299
X9	1	0.4863322	0.4863322	17.19	0.0016
X10	1	0.0252386	0.0252386	0.89	0.3651
X1*X9	1	0.0902496	0.0902496	3.19	0.1016
X1*X10	1	0.1332566	0.1332566	4.71	0.0527

MODEL II: REGION AND TREATMENT DIFFERENCES WITH COVARIATE HAVING EQUAL SLOPES

General Linear Models Procedure

Dependent Variable: S   SPEED

Source	DF	Sum of Squares	Mean Square	F Value	Pr > F
Model	10	18.435323	1.843532	51.13	0.0001
Error	13	0.468741	0.036057		
Corrected Total	23	18.904065			

Source	DF	Type III SS	Mean Square	F Value	Pr > F
X1	1	3.135616	3.135616	86.96	0.0001
X2	1	0.099813	0.099813	2.77	0.1201
X3	1	0.166114	0.166114	4.61	0.0513
X4	1	0.169050	0.169050	4.69	0.0496
X5	1	0.234314	0.234314	6.50	0.0242
X6	1	0.039421	0.039421	1.09	0.3148
X7	1	0.009653	0.009653	0.27	0.6136
X8	1	0.039862	0.039862	1.11	0.3122
X9	1	14.089314	14.089314	390.75	0.0001
X10	1	3.730729	3.730729	103.47	0.0001

A test for equal slopes is obtained by testing in model I the hypotheses

$$H_0: \quad \beta_{11} = \beta_{12} = 0 \text{ versus } H_a: \quad \beta_{11} \neq 0 \text{ and/or } \beta_{12} \neq 0$$

The test statistic for H_0 versus H_a is

$$F = \frac{(\text{SSE}_{II} - \text{SSE}_I)/(\text{df}_{EII} - \text{df}_{EI})}{\text{MSE}_I} = \frac{(.4687 - .3111)/(13 - 11)}{.0283} = 2.79$$

The p-value is given by Pr $(F_{2,11} \geq 2.79) = .1050$. Thus, the data support the hypothesis that the three cultivars have the same slope. Next, we can test for differences in adjusted means of the three cultivars. We would fit a model in which the covariate has equal slopes for the three cultivars but would remove any differences in the cultivars and retain differences due to the blocking variable, regions.

Model III: Covariate with equal slopes, region but no cultivar differences

$$y = \beta_0 + \beta_1 x_1 + \beta_2 x_2 + \beta_3 x_3 + \beta_4 x_4 + \beta_5 x_5 + \beta_6 x_6 + \beta_7 x_7 + \beta_8 x_8 + \varepsilon$$

The computer output from fitting this model is given here.

```
MODEL III: COVARIATE WITH EQUAL SLOPES, REGION BUT NO
CULTIVAR DIFFERENCES

General Linear Models Procedure

Dependent Variable: S    SPEED
                              Sum of            Mean
Source                DF     Squares          Square    F Value    Pr > F
Model                  8   4.3410695       0.5426337       0.56    0.7950
Error                 15  14.5629951       0.9708663
Corrected Total       23  18.9040646


Source                DF   Type III SS   Mean Square    F Value    Pr > F
X1                     1   2.0172033       2.0172033       2.08    0.1700
X2                     1   0.1190733       0.1190733       0.12    0.7311
X3                     1   0.1670196       0.1670196       0.17    0.6842
X4                     1   0.2163489       0.2163489       0.22    0.6437
X5                     1   0.4184864       0.4184864       0.43    0.5214
X6                     1   0.0032885       0.0032885       0.00    0.9544
X7                     1   0.0000020       0.0000020       0.00    0.9989
X8                     1   0.0031418       0.0031418       0.00    0.9554
```

A test for differences in the adjusted cultivar means is a test of

$$H_0: \mu_{Adj,C1} = \mu_{Adj,C2} = \mu_{Adj,C3} \quad \text{versus} \quad H_a: \mu_{Adj,C} \text{ is not all equal}$$

This set of hypotheses is equivalent to testing in model II the hypotheses

$$H_0: \quad \beta_9 = \beta_{10} = 0 \quad \text{versus} \quad H_a: \quad \beta_9 \neq 0 \text{ and/or } \beta_{10} \neq 0$$

The test statistic for H_0 versus H_a is

$$F = \frac{(\text{SSE}_{\text{III}} - \text{SSE}_{\text{II}})/(\text{df}_{\text{EIII}} - \text{df}_{\text{EII}})}{\text{MSE}_{\text{II}}} = \frac{(14.5630 - .4687)/(15 - 13)}{.0361} = 195.21$$

The p-value is given by $\text{Pr}\,(F_{2,13} \geq 195.21) < .0001$. Thus, the data strongly support the research hypothesis that there is a significant difference in the adjusted mean speeds for the three cultivars. We can further investigate what type of differences exist in the three cultivars by examining the plot of the speed and relative humidity data values in Figure 16.5. The lines drawn through the data values were obtained from the parameter estimates in model II. We can observe that cultivar C3 consistently yields higher speeds than the other two cultivars, with cultivar C2 yielding higher speeds than cultivar C1.

The estimated adjusted mean speeds are given in Table 16.7 along with their estimated standard errors, which were used to construct 95% confidence intervals on the mean speeds. From the results in Table 16.7, cultivar C3 has an adjusted mean speed about 1 unit larger than cultivar C2, which has an adjusted mean

FIGURE 16.5

Cultivar speeds plotted versus relative humidity readings along with fitted lines from regression model

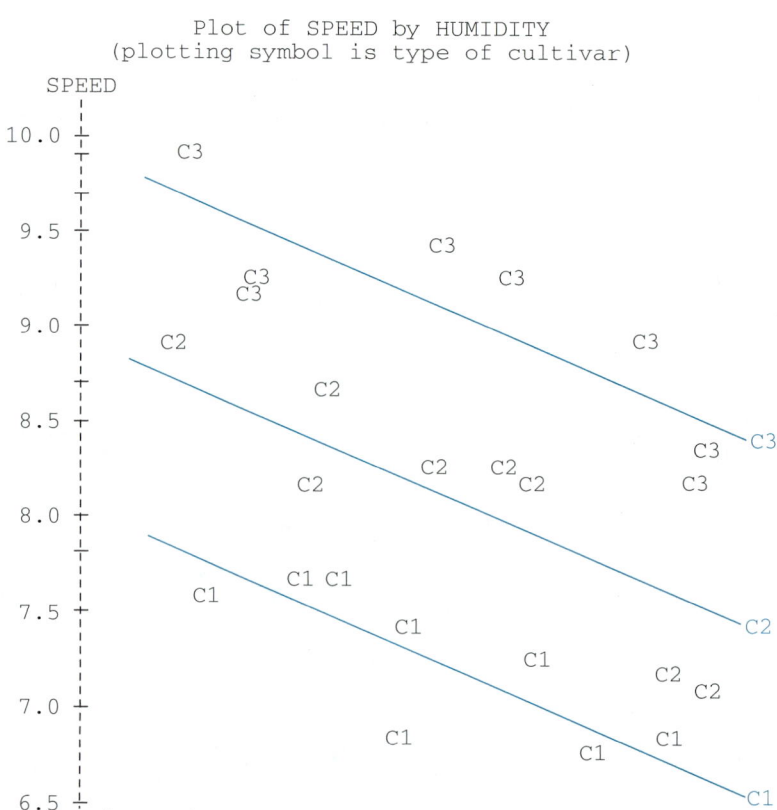

speed about 1 unit larger than the adjusted mean speed for cultivar C1. This size of differences in the mean speed is considered to be a practical difference and will greatly assist golf course designers in selecting the proper cultivar for their course.

Prior to using the results obtained above, the researcher must check whether the conditions placed on the analysis of covariance model are satisfied in this experiment. An examination of the following plots of the residuals and plots of the observed data will assist in checking on the validity of the model conditions. The computer printouts of the analysis of the residuals from model II are given in Figure 16.6.

The boxplot and stem-and-leaf plot of the residuals do not indicate any extreme values. The normal probability plot indicates that a few residuals are somewhat deviant from the fitted line. However, the test of normality yields a p-value of .3405, so there is strong support for the normality of the residuals. The plot of the residuals versus predicted values does not indicate a violation of the

TABLE 16.7

Estimated adjusted cultivar speeds with 95% confidence intervals

Cultivar	$\hat{\mu}_{Adj}$	SE($\hat{\mu}_{Adj}$)	95% Confidence Interval
C1	7.20	.0674	(7.05, 7.35)
C2	8.12	.0672	(7.98, 8.27)
C3	9.08	.0672	(8.94, 9.23)

FIGURE 16.6

```
Univariate Procedure
Variable-RESIDUALS
                    Moments
   N                    24  Sum Wgts           24
   Mean                  0  Sum                 0
   Std Dev        0.142759  Variance      0.02038
   Skewness       0.522974  Kurtosis     -0.22996

   W:Normal       0.954191  Pr<W           0.3405

Variable-RESIDUALS
      Stem Leaf                    #    Boxplot
         2 79                      2       |
         2 12                      2       |
         1                                 |
         1 3                       1       |
         0 8                       1    +-----+
         0 11134                   5    |  +  |
        -0 220                     3    *-----*
        -0 8755                    4    +-----+
        -1 433                     3       |
        -1 6                       1       |
        -2 40                      2       |
           ---+---+---+---+
   Multiply Stem.Leaf by 10**-1
```

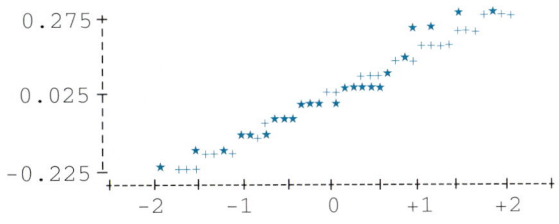

Normal Probability Plot

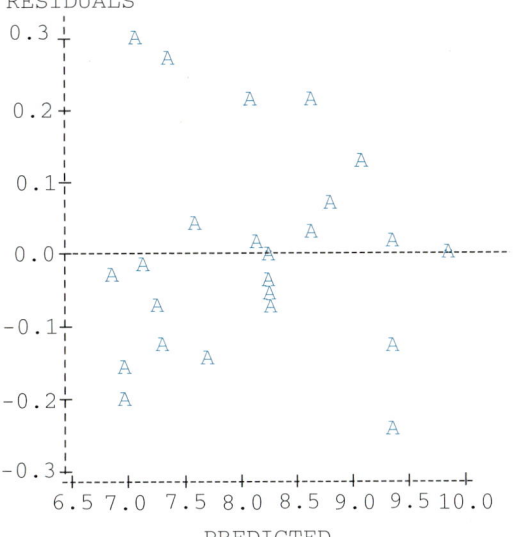

Plot of RESIDUALS versus PREDICTED

ment are randomly selected from a population of possible levels, the model used to relate the response variable to the levels of the factors is referred to as a **random-effects** model. The inferences from these models are generalized to the population of levels from the levels used in the experiment, which were randomly selected. In a product improvement study, one of the common sources of variability is the operator of the process. The company may have hundreds of operators but only five or six will be randomly selected to participate in the study. However, the quality engineer is interested in the performance of all operators, not only the operators that are involved in the study.

DEFINITION 17.2

In a **random effects model** for an experiment, the levels of factors used in the experiment are randomly selected from a population of possible levels. The inferences from the data in the experiment are for all levels of the factors in the population from which the levels were selected and not only the levels used in the experiment.

Many studies will involve factors having a predetermined set of levels and factors in which the levels used in the study are randomly selected from a population of levels. The blocks in a randomized block design might represent a random sample of b plots of land taken from a population of plots in an agricultural research facility. Then the effects due to the blocks are considered to be random effects. Suppose the treatments are four new varieties of soybeans that have been developed to be resistant to a specific virus. The levels of the treatment are fixed because these are the only varieties of interest to the researchers, whereas the levels of the plots of land are random because the researchers are not interested in only these plots of land but are interested in the effects of these treatments on a wide range of plots of land. When some of the factors to be used in the experiment have levels randomly selected from a population of possible levels and other factors have predetermined levels, the model used to relate the response variable to the levels of the factors is referred to as a **mixed-effects** model.

DEFINITION 17.3

In a **mixed-effects model** for an experiment, the levels of some of the factors used in the experiment are randomly selected from a population of possible levels, whereas the levels of the other factors in the experiment are predetermined. The inferences from the data in the experiment concerning factors with fixed levels are only for the levels of the factors used in the experiment, whereas inferences concerning factors with randomly selected levels are for all levels of the factors in the population from which the levels were selected.

In this chapter, we will consider various random-effects and mixed-effects models. For each model, we will indicate the appropriate analysis of variance and show how to estimate all relevant components of variance. The following case study will describe a mixed-effects experiment.

Case Study: Evaluation of Pressure Drop Across Expansion Joints

A major problem in power plants is pressure drops across expansion joints in electric turbines. The design engineer wants to design a study to identify the factors that are most likely to influence the pressure drop readings. Once these factors are identified and the most crucial factors determined by the size of their contribution to the pressure drops across the expansion joint during the study, the engineer can make design changes in the process or alter the method by which the operators of the process are trained. These types of changes may be expensive or time-consuming so the engineer wants to be certain which factors will have the greatest impact on reducing the pressure drop.

Designing the Data Collection The design engineer considered the following issues in designing an appropriate experiment to evaluate pressure drop:

1. Which factors should be used in the study?
2. Which levels of the factors are of interest?
3. How many levels are needed to adequately identify the important sources of variation?
4. How many replications per factor–level combinations are needed to obtain a reliable estimate of the variance components?
5. What environmental factors may affect the performance of the pressure gauge during the test period?
6. What are the valid statistical procedures for evaluating the causes of the variability in pressure drops across the expansion joints?
7. What type of information should be included in a final report to document that all importance sources of variability have been identified?

The factors selected for study are the gas temperature on the inlet side of the joint and the type of pressure gauge used by the operator. The engineer decides that a factorial experiment is required to determine which of these factors has the greatest effect on the pressure drop. Three temperatures that cover the feasible range for operation of the turbine are 15°C, 25°C, and 35°C. From the hundreds of different types of pressure gauges used to monitor pressure in the lines, four types of gauges are randomly selected for use in the study. To obtain precise estimates of the mean pressure drop for each of the 12 factor–level combinations, it was decided to obtain 6 replications of each of 12 treatments. The data from the 72 experimental runs are given in Table 17.1.

A profile plot of the 12 sample treatment means is presented in Figure 17.1. From the plot, gauge type G1 has a more variable mean pressure drop than do the other three gauge types. To determine whether this observed difference is more

TABLE 17.1

Pressure drop across expansion joints

	15°C				25°C				35°C			
	G1	**G2**	**G3**	**G4**	**G1**	**G2**	**G3**	**G4**	**G1**	**G2**	**G3**	**G4**
	40	43	42	47	57	49	44	36	35	41	42	41
	40	34	35	47	57	43	45	49	35	43	41	44
	37	38	35	40	65	51	49	38	35	44	34	35
	47	42	41	36	67	49	45	45	46	36	35	46
	42	39	43	41	63	45	46	38	41	42	39	44
	41	35	36	47	59	43	43	42	42	41	36	46
Mean	41.17	38.50	38.67	43.00	61.33	46.67	45.33	41.33	39.00	41.17	37.83	42.67

from a normal population with mean 0 and variance 0; that is, all α values in the population are equal to 0.

Thus, although the forms of the null hypotheses are different for the two models, the meanings attached to them are very similar. For the fixed-effects model, we are assuming that the sampled αs (which are the only αs) are identically 0, whereas in the random-effects model, the null hypothesis leads us to assume that the sampled αs, as well as all other αs in the population, are 0.

The alternative hypotheses are also similar. In the fixed-effects model, we are assuming that at least one of the αs is different from the rest; that is, there is some variability among the set of αs. For the random-effects model, the alternative hypothesis is that $\sigma_\alpha^2 > 0$; that is, not all α values are the same in the population.

In a random-effects model with a single factor, we have that the response variable has mean value and variance given by

$$E(y_{ij}) = \mu \quad \text{and} \quad \sigma_y^2 = \text{Var}(y_{ij}) = \sigma_\alpha^2 + \sigma_\varepsilon^2$$

Thus, in many random-effects experiments, we want to determine the relative size of σ_α^2 to σ_ε^2 in order to assess the size of the treatment effect to the overall variability in the response variable. Because we do not know σ_α^2 or σ_ε^2, we can form estimates of these terms by using the idea of **AOV moment matching** estimators. From the AOV table shown next, we see that MST has expected mean square of $\sigma_\alpha^2 + n\sigma_\varepsilon^2$ and MSE has expected mean square of σ_ε^2.

Source	MS	EMS
Treatments	MST	$\sigma_\alpha^2 + n\sigma_\varepsilon^2$
Error	MSE	σ_ε^2

We equate sample mean square to its expected value and solve for the population variance, to get

$$\hat{\sigma}_\varepsilon^2 = \text{MSE} \quad \text{and} \quad \hat{\sigma}_\alpha^2 = (\text{MST} - \text{MSE})/n.$$

Thus, we have $\hat{\sigma}_y^2 = \hat{\sigma}_\alpha^2 + \hat{\sigma}_\varepsilon^2$. The variance in the response variable can thus be proportionally allocated to the two sources of variability, the treatment and experimental error, shown in Table 17.3.

TABLE 17.3
Proportional allocation of total variability in the response variable

Source of Variance	Estimator	Proportion of Total
Treatment	$\hat{\sigma}_\alpha^2 = (\text{MST} - \text{MSE})/n$	$\hat{\sigma}_\alpha^2/\hat{\sigma}_y^2$
Error	$\hat{\sigma}_\varepsilon^2 = \text{MSE}$	$\hat{\sigma}_\varepsilon^2/\hat{\sigma}_y^2$
Total	$\hat{\sigma}_y^2 = \hat{\sigma}_\alpha^2 + \hat{\sigma}_\varepsilon^2$	1.0

It might also be of interest to the researchers to estimate the mean value for the response variable, μ. We have that the point estimator of μ and its estimated standard error are given by

$$\hat{\mu} = \bar{y}_{..} \quad \text{and} \quad \text{SE}(\hat{\mu}) = \sqrt{\text{MST}/tn}$$

We can then construct a $100(1 - \alpha)\%$ confidence interval for μ as given here.

$$\hat{\mu} \pm t_{\alpha/2, \text{df}_{\text{Error}}} \text{SE}(\hat{\mu}) \quad \text{or} \quad \bar{y}_{..} \pm t_{\alpha/2, t(n-1)} \sqrt{\text{MST}/tn}$$

EXAMPLE 17.1

Consider the problem we used to illustrate a one-factor experiment with random treatment effects. Two graduate students working for a professor in electrical engineering have been funded to record lightning discharge intensities (intensities of the electrical field) at three tracking stations. Because of the high frequency of thunderstorms in the summer months (in Florida, storms occur on 80 or more days per year), the graduate students were to choose a point at random on a map of the 20-mile-radius region and assemble their tracking equipment (provided they could get permission of the property owners). Each day during the hours from 8 A.M. to 5 P.M., they were to monitor their instruments until the maximum intensity had been recorded for five separate storms. They then repeated the process separately at the two other locations chosen at random. The sample data (in volts per meter) appear in Table 17.4.

TABLE 17.4
Lightning discharge intensities (in volts per meter)

Tracking Station	Intensities					Mean
1	20	1,050	3,200	5,600	50	1,984
2	4,300	70	2,560	3,650	80	2,132
3	100	7,700	8,500	2,960	3,340	4,520
Overall Mean						2,878.67

a. Write an appropriate statistical model, defining all terms.
b. Perform an analysis of variance and interpret your results. Use $\alpha = .05$.
c. Estimate the variance components and their proportional allocation of the total variability.
d. Estimate the mean maximum daily lightning discharge intensity and place a 95% confidence on this mean.

Solution Because the tracking stations were selected at random, we can use a single-factor random-effects model to relate maximum lightning discharge intensity, y_{ij}, to the ith station and jth day.

$$y_{ij} = \mu + \alpha_i + \varepsilon_{ij} \qquad (i = 1, 2, 3; \quad j = 1, 2, \ldots, 5)$$

where μ is the mean maximum daily lightning discharge intensity, α_i is the random effect of the ith randomly selected station, and ε_{ij} is random effect due to all other sources of variability. The formulas for computing the sum of squares for the random-effects analysis of variance are identical to the formulas in the fixed-effects analysis of variance. Thus, we have

$$\text{SST} = n \sum_i (\bar{y}_{i.} - \bar{y}_{..})^2 = 5\{(1,984 - 2,878.67)^2 + (2,132 - 2,878.67)^2$$

$$+ (4,520 - 2,878.67)^2\} = 20,259,573.3$$

$$\text{TSS} = \sum_{ij} (y_{ij} - \bar{y}_{..})^2 = (20 - 2,878.67)^2 + (1,050 - 2,878.67)^2$$

$$+ \cdots + (3,340 - 2,878.67)^2 = 108,249,173.3$$

By subtraction,

$$\text{SSE} = \text{TSS} - \text{SST} = 108,249,173.3 - 20,259,573.3 = 87,989,600$$

We can use these calculations to construct an AOV table, as shown in Table 17.5.

TABLE 17.5
AOV table for the data of Example 17.1

Source	SS	df	MS	EMS	F
Tracking stations	20,259,573.3	2	10,129,786.65	$\sigma_\varepsilon^2 + 5\sigma_\alpha^2$	1.38
Error	87,989,600.0	12	7,332,466.67	σ_ε^2	
Totals	108,249,173.3	14			

The F test for H_0: $\sigma_\alpha^2 = 0$ is based on $df_1 = 2$ and $df_2 = 12$ degrees of freedom. Because the computed value of F, 1.38, does not exceed 3.89, the value in Appendix Table 8 for $a = .05$, $df_1 = 2$, and $df_2 = 12$, we have insufficient evidence to indicate that there is a significant random component due to variability in intensities from tracking station to tracking station. Rather, as an electrical engineer postulated, it is probably best to work with a single tracking station, because most of the variability in intensities is related to the distance of the tracking station from the point of discharge, and we have no control of this source. In fact, we can compute estimates of the variance components and obtain

$$\hat{\sigma}_\varepsilon^2 = 7,332,466.67 \qquad \hat{\sigma}_\alpha^2 = (10,129,786.65 - 7,332,466.67)/5 = 559,464$$

which yields

$$\hat{\sigma}_y^2 = 7,332,466.67 + 559,464 = 7,891,930.67$$

We have that the proportion of the total variability due to station differences is $559,464/7,891,930.67 = .0709$. Only 7.1% of the variability in maximum daily lightning intensity is due to station differences. We can place a 95% confidence interval on the mean maximum daily lightning intensity as given here.

$$\bar{y}_{..} \pm t_{0.25,12}SE(\hat{\mu})$$
$$2878.67 \pm (2.179)\sqrt{10,129,786.65/15} \quad \text{or} \quad 2878.67 \pm 1790.65$$

Thus, we are 95% confident that the mean daily maximum lightning intensity is within (1088, 4669).

EXERCISES

Applications

Med.

17.1 A pharmaceutical company would like to examine the potency of a liquid medication mixed in large vats. To do this, a random sample of five vats from a month's production was obtained, and four separate samples were selected from each vat.
   a. Write a random-effects model for this experimental situation, identifying all terms in the model.
   b. Run an analysis of variance for the sample data given here. Use $\alpha = .05$.

Vat 1	Vat 2	Vat 3	Vat 4	Vat 5
3.2	2.6	3.4	4.2	1.8
3.8	2.9	3.9	4.4	2.3
3.5	2.8	3.3	4.3	1.9
3.0	2.0	3.1	4.2	2.1

17.2 Suppose that the pharmaceutical company of Exercise 17.1 wishes to estimate the expected potency for a measurement made on a vat selected at random from a month's production of a liquid medication.

     a. Using the sample data of Exercise 17.1, form a point estimate of the average potency for a measurement made on a randomly selected vat.

     b. Place a 95% confidence interval on the average potency for a measurement made on a randomly selected vat.

17.3   Extensions of Random-Effects Models

The ideas presented for a random-effects model in a one-factor experiment can be extended to any of the block designs and factorial experiments covered in Chapter 15. Although we will not have time to cover all such situations, we will consider first a randomized block design in which the block effects and the treatment effects are random.

     Consider an experiment to examine the effects of different analysts and subjects in chemical analyses for the DNA content of plaque. Three female subjects (ages 18–20 years) were chosen for the study. Each subject was allowed to maintain her usual diet, supplemented with 30 mg (15 tablets) of sucrose per day. No toothbrushing or mouthwashing was allowed during the study. At the end of the week, plaque was scraped from the entire dentition of each subject and divided into three samples. Each of three analysts chosen at random was then given an unmarked sample of plaque from each of the subjects and asked to perform an analysis for the DNA content (in micrograms). The two-factor experiment of sample data could then be organized as shown in Table 17.6.

TABLE 17.6
DNA concentrations for samples of plaque

Analyst	Subject 1	Subject 2	Subject 3	Means
1	13.2	10.6	8.5	10.77
2	12.5	9.6	7.9	10.00
3	13.0	9.9	8.3	10.40
Means	12.9	10.03	8.23	10.39

randomized block design

This experimental design is recognized as a **randomized block design,** with subjects representing blocks and analysts being the treatments. The experimental units are samples of plaque scraped from the dentition of subjects. If we assume that the three subjects represent a random sample from a large population of possible subjects, and, similarly, that the three analysts represent a random sample from a large population of possible analysts, we can write the following random-effects model relating DNA concentration to the two factors "analysts" and "subjects":

random-effects model

$$y_{ij} = \mu + \alpha_i + \beta_j + \varepsilon_{ij}$$

We assume the following:

assumptions

1. μ is an overall unknown concentration mean.

2. α_i is a random effect due to the ith analyst. α_i is normally distributed, with mean 0 and variance σ_α^2.

3. The α_is are independent.
4. β_j is a random effect due to the jth subject. β_j is a normally distributed random variable, with mean 0 and variance σ_β^2.
5. The β_js are independent.
6. The α_is, β_js, and ε_{ij}s are independent.

Again note the difference between assuming that the treatments and blocks are random rather than fixed effects. If, for example, the three analysts chosen for the study were the only analysts of interest, we would be concerned with differences in mean DNA concentrations for these specific analysts. Now, however, treating the effect due to an analyst as a random variable, our inference will be about the population of analysts' effects. Because the mean of this normal population is assumed to be 0, we want to determine whether the variance σ_α^2 is greater than 0.

The AOV table for a general two-factor completely randomized design of a levels of factor A and b levels of factor B and no replication is shown in Table 17.7. This same AOV can apply to a randomized complete block design where A denotes treatment and B denotes blocks. As with the random-effects model for a one-factor experiment, the analysis of variance tables for a fixed- and random-effects models of a two-factor experiment are identical, except for the expected mean squares.

TABLE 17.7

AOV table for a two-factor experiment, a Levels of factor A and b Levels of factor B

				EMS	
Source	**SS**	**df**	**MS**	**Fixed Effects**	**Random Effects**
A	SSA	$a-1$	MSA	$\sigma_\varepsilon^2 + b\theta_A$	$\sigma_\varepsilon^2 + b\sigma_\alpha^2$
B	SSB	$b-1$	MSB	$\sigma_\varepsilon^2 + a\theta_B$	$\sigma_\varepsilon^2 + a\sigma_\beta^2$
Error	SSE	$(a-1)(b-1)$	MSE	σ_ε^2	σ_ε^2
Totals	TSS	$ab-1$			

tests

The computation of sums of squares and mean squares would proceed exactly as shown in Chapter 15. The difference in test procedures is illustrated in Table 17.8 for factor A. Similar results would also apply to factor B.

TABLE 17.8

Difference in test procedures for factor A

Fixed-Effects Model	**Random-Effects Model**
H_0: $\alpha_1 = \alpha_2 = \cdots = \alpha_a = 0$	H_0: $\sigma_\alpha^2 = 0$
H_a: At least one of the αs differs from the rest	H_a: $\sigma_\alpha^2 > 0$
T.S.: $F = \dfrac{\text{MSA}}{\text{MSE}}$	T.S.: $F = \dfrac{\text{MSA}}{\text{MSE}}$
R.R.: Based on $df_1 = a-1$, $df_2 = (a-1)(b-1)$	R.R.: Same

Rather than proceed with an example at this point, we will discuss a random-effects model for a factorial experiment with $n > 1$ observations at each factor–level combination. Then we will illustrate the test procedure.

In Chapter 15, we considered the fixed-effects model for a $a \times b$ factorial experiment in a completely randomized design with $n > 1$ observations per cell.

$a \times b$ factorial, $n > 1$

The random-effects model for an $a \times b$ **factorial** experiment would be of the same form as the corresponding fixed-effects experiment, but with different assumptions.

model

$$y_{ijk} = \mu + \alpha_i + \beta_j + \alpha\beta_{ij} + \varepsilon_{ijk}$$

where y_{ijk} is the response for the kth observation at the ith level of factor A and the jth level of factor B; μ, α_i, β_j, and ε_{ijk} are defined as before for the random-effects model without replication. In addition, we assume the following:

assumptions

1. $\alpha\beta_{ij}$ is a random effect due to the ith level of factor A and the jth level of factor B. $\alpha\beta_{ij}$ is normally distributed, with mean 0 and variance $\sigma_{\alpha\beta}^2$.
2. The $\alpha\beta_{ij}$s are independent.
3. The α_is, β_js, $\alpha\beta_{ij}$s, and ε_{ijk}s are independent.

AOV tables

The appropriate **AOV tables** for fixed- and random-effects models are shown in Table 17.9.

TABLE 17.9
AOV table for an $a \times b$ factorial experiment, with n observations per cell

				EMS	
Source	**SS**	**df**	**MS**	**Fixed Effects**	**Random Effects**
A	SSA	$a - 1$	MSA	$\sigma_\varepsilon^2 + nb\theta_A$	$\sigma_\varepsilon^2 + n\sigma_{\alpha\beta}^2 + bn\sigma_\alpha^2$
B	SSB	$b - 1$	MSB	$\sigma_\varepsilon^2 + na\theta_B$	$\sigma_\varepsilon^2 + n\sigma_{\alpha\beta}^2 + an\sigma_\beta^2$
AB	SSAB	$(a - 1)(b - 1)$	MSAB	$\sigma_\varepsilon^2 + n\theta_{AB}$	$\sigma_\varepsilon^2 + n\sigma_{\alpha\beta}^2$
Error	SSE	$ab(n - 1)$	MSE	σ_ε^2	σ_ε^2
Totals	TSS	$abn - 1$			

The appropriate tests using the AB interaction sum of squares are illustrated in Table 17.10 for the two models.

TABLE 17.10
A comparison of appropriate interaction tests for fixed- and random-effects models

Fixed-Effects Model	**Random-Effects Model**
H_0: $\alpha\beta_{11} = \alpha\beta_{12} = \cdots = \alpha\beta_{ab} = 0$	H_0: $\sigma_{\alpha\beta}^2 = 0$
H_a: At least one $\alpha\beta_{ij}$ differs from the rest	H_a: $\sigma_{\alpha\beta}^2 > 0$
T.S.: $F = \dfrac{\text{MSAB}}{\text{MSE}}$	T.S.: $F = \dfrac{\text{MSAB}}{\text{MSE}}$
R.R.: Based on $df_1 = (a - 1)(b - 1)$, $df_2 = ab(n - 1)$	R.R.: Same

Now, unlike the one-factor experiment and the two-factor experiment without replication, the test statistic for main effects are different for the fixed- and random-effects models. In addition, for the random-effects model, the tests for σ_α^2 and σ_β^2 can proceed even when the test on the AB interaction ($\sigma_{\alpha\beta}^2$) is significant. We have seen previously that for fixed-effects models, a test for main effects in the presence of a significant interaction only seems to make sense when the profile plot suggests that the interaction is "orderly." For random-effects models, we are interested in identifying the various sources of variability (e.g., $\sigma_{\alpha\beta}^2$, σ_α^2, and σ_β^2) that affect the response y. Tests for σ_α^2 and σ_β^2 do make sense even when $\sigma_{\alpha\beta}^2$ has been shown to be greater than zero.

For the fixed-effects model following a nonsignificant test on the AB interaction, we can test for main effects due to factors A and B by using

$$F = \frac{MSA}{MSE} \quad \text{and} \quad F = \frac{MSB}{MSE}$$

respectively. As we see from the expected mean squares column of Table 17.9, no matter what the results of the test $H_0: \sigma_{\alpha\beta}^2 = 0$, we can form an F test for the components σ_α^2 and σ_β^2 using the test procedures shown in Table 17.11. Note that the test statistics differ from those used in the fixed-effects case, where the denominator of all F statistics is MSE.

tests, main effects

TABLE 17.11

Tests for an $a \times b$ factorial experiment with replication: random-effects model

Factor A	Factor B
H_0: $\sigma_\alpha^2 = 0$	H_0: $\sigma_\beta^2 = 0$
H_a: $\sigma_\alpha^2 > 0$	H_a: $\sigma_\beta^2 > 0$
T.S.: $F = \dfrac{MSA}{MSAB}$	T.S.: $F = \dfrac{MSB}{MSAB}$
R.R.: Based on $df_1 = (a-1)$, $df_2 = (a-1)(b-1)$	R.R.: Based on $df_1 = (b-1)$, $df_2 = (a-1)(b-1)$

In many experiments involving factors having random effects, we will want to estimate the **variance components** σ_α^2, σ_β^2, $\sigma_{\alpha\beta}^2$, and σ_ε^2. We can once again use the AOV moment matching estimators, which are obtained by matching the sample mean squares with the expected mean squares in the AOV and then solving for the individual variance components. Using the MSs and EMSs in Table 17.9, we obtain

$$\hat{\sigma}_\beta^2 = MSE$$

$$\hat{\sigma}_{\alpha\beta}^2 = (MSAB - MSE)/n$$

$$\hat{\sigma}_\beta^2 = (MSB - MSAB)/an$$

and

$$\hat{\sigma}_\alpha^2 = (MSA - MSAB)/bn$$

Also, from the random-effects model for two factors having randomly selected levels, we have

$$E(y_{ijk}) = \mu \quad \text{and} \quad \sigma_y^2 = \sigma_\alpha^2 + \sigma_\beta^2 + \sigma_{\alpha\beta}^2 + \sigma_\varepsilon^2$$

Thus, we have $\hat{\sigma}_y^2 = \hat{\sigma}_\alpha^2 + \hat{\sigma}_\beta^2 + \hat{\sigma}_{\alpha\beta}^2 + \hat{\sigma}_\varepsilon^2$. We can then proportionally allocate the total variability $\hat{\sigma}_y^2$ into the four sources of variability: factor A, factor B, the interaction, and experimental error. See Table 17.12.

TABLE 17.12

Proportional allocation of total variability in the response variable

Source of Variance	Estimator	Proportion of Total
Factor A	$\hat{\sigma}_\alpha^2 = (MSA - MSAB)/bn$	$\hat{\sigma}_\alpha^2/\hat{\sigma}_y^2$
Factor B	$\hat{\sigma}_\beta^2 = (MSB - MSAB)/an$	$\hat{\sigma}_\beta^2/\hat{\sigma}_y^2$
Interaction AB	$\hat{\sigma}_{\alpha\beta}^2 = (MSAB - MSE)/n$	$\hat{\sigma}_{\alpha\beta}^2/\hat{\sigma}_y^2$
Error	$\hat{\sigma}_\varepsilon^2 = MSE$	$\hat{\sigma}_\varepsilon^2/\hat{\sigma}_y^2$
Total	$\hat{\sigma}_y^2 = \hat{\sigma}_\alpha^2 + \hat{\beta}_\alpha^2 + \hat{\sigma}_{\alpha\beta}^2 + \hat{\sigma}_\varepsilon^2$	1.0

The researchers might also be interested in estimating the mean value for the response variable, μ. We have that the point estimator of μ and its estimated standard error are given by

$$\hat{\mu} = \bar{y}_{...} \quad \text{and} \quad SE(\hat{\mu}) = \sqrt{(MSA + MSB - MSAB)/abn}$$

We can then construct a $100(1 - \alpha)\%$ confidence interval for μ as given here.

$$\bar{y}_{...} \pm t_{\alpha/2, df_{Approx.}} \sqrt{(MSA + MSB - MSAB)/abn}$$

where the degrees of freedom for the t tables is obtained from the Satterthwaite approximation,

$$df_{Approx.} = \frac{(MSA + MSB - MSAB)^2}{(MSA)^2/(a-1) + (MSB)^2/(b-1) + (MSAB)^2/(a-1)(b-1)}$$

Because in most cases this value is not an integer, we take the largest integer less than or equal to $df_{Approx.}$.

In some experiments the estimates of some of the variance components may result in a negative number. Of course by definition a variance component must be a nonnegative number; thus we must consider alternatives whenever the sample estimator is negative.

A1. We can set the estimator equal to zero and use zero as the estimator of the variance component. However, the estimator will no longer be an unbiased estimator of the variance component.

A2. A negative estimator of a variance component may be an indication that we have elements in our model that are not appropriate for this experiment. A more complex model may be needed for this experiment.

A3. There are alternative estimators of variance components that are mathematically beyond the level of this book. Such methods as REML or MINIQUE are currently available in SAS. However, we should still carefully examine the data, because a negative variance component estimator is often an indicator of an inadequate model.

EXAMPLE 17.2

A consumer product agency wants to evaluate the accuracy of determining the level of calcium in a food supplement. There are a large number of possible testing laboratories and a large number of chemical assays for calcium. The agency randomly selects three laboratories and three assays for use in the study. Each laboratory will use all three assays in the study. Eighteen samples containing 10 mg of calcium are prepared and each assay–laboratory combination is randomly assigned to two samples. The determinations of calcium content are given here (numbers in parentheses are averages for the assay–laboratory combinations).

Assay	Laboratory			Assay Mean
	1	**2**	**3**	
1	10.9	10.5	9.7	10.3
	10.9	9.8	10.0	
	(10.9)	(10.15)	(9.85)	
2	11.3	9.4	8.8	10.1
	11.7	10.2	9.2	
	(11.5)	(9.8)	(9.0)	
3	11.8	10.0	10.4	10.8
	11.2	10.7	10.7	
	(11.5)	(10.35)	(10.55)	
Lab Mean	11.3	10.1	9.8	10.4

a. Perform an analysis of variance for this experiment. Conduct all tests with $\alpha = .05$.
b. Estimate all variance components and determine their proportional allocation to the total variability.
c. Estimate the average calcium level over all laboratories and assays.

Solution   Using the formulas from Chapter 15, we obtain the sum of squares as follows:

$$\text{TSS} = \sum_{ijk} (y_{ijk} - \bar{y}_{...})^2 = (10.9 - 10.4)^2 + (10.9 - 10.4)^2 + \cdots + (10.7 - 10.4)^2$$

$$= 12.00$$

$$\text{SSA} = \sum_{i} 6(\bar{y}_{i..} - \bar{y}_{...})^2 = 6\{(10.3 - 10.4)^2 + (10.1 - 10.4)^2 + (10.8 - 10.4)^2\}$$

$$= 1.56$$

$$\text{SSL} = \sum_{j} 6(\bar{y}_{.j.} - \bar{y}_{...})^2 = 6\{(11.3 - 10.4)^2 + (10.1 - 10.4)^2 + (9.8 - 10.4)^2\}$$

$$= 7.56$$

$$\text{SSAL} = \sum_{ij} 2(\bar{y}_{ij.} - \bar{y}_{...})^2 = 2\{(10.9 - 10.4)^2 + (10.15 - 10.4)^2 + (9.85 - 10.4)^2$$

$$+ \cdots + (10.55 - 10.4)^2\} = 1.64$$

$$\text{SSE} = \text{TSS} - \text{SSA} - \text{SSL} - \text{SSAL} = 12.00 - 1.56 - 7.56 - 1.64 = 1.24$$

Our results are summarized in an analysis of variance table.

Source	SS	df	MS	EMS
Assay	1.56	2	.78	$\sigma_\varepsilon^2 + 2\sigma_{\alpha\beta}^2 + 6\sigma_\alpha^2$
Lab	7.56	2	3.78	$\sigma_\varepsilon^2 + 2\sigma_{\alpha\beta}^2 + 6\sigma_\beta^2$
Assay*Lab	1.64	4	.41	$\sigma_\varepsilon^2 + 2\sigma_{\alpha\beta}^2$
Error	1.24	9	.1378	σ_ε^2
Total	12.00	17		

We can proceed with appropriate statistical tests, using the results presented in the AOV table. For the AB interaction we have

H_0: $\sigma_{\alpha\beta}^2 = 0$

H_a: $\sigma_{\alpha\beta}^2 > 0$

T.S.: $F = \dfrac{\text{MSAB}}{\text{MSE}} = \dfrac{.41}{.1378} = 2.98$

R.R.: For $\alpha = .05$, we will reject H_0 if F exceeds 3.63, the critical value for $a = .05$, $df_1 = 4$, and $df_2 = 9$.

Conclusion: There is insufficient evidence to reject H_0. There does not appear to be a significant interaction between the levels of factors A and B.

For factor B we have

H_0: $\sigma_{\beta}^2 = 0$

H_a: $\sigma_{\beta}^2 > 0$

T.S.: $F = \dfrac{\text{MSB}}{\text{MSAB}} = \dfrac{3.78}{.41} = 9.22$

R.R.: For $\alpha = .05$, we will reject H_0 if F exceeds 6.94, the critical value based on $a = .05$, $df_1 = 2$, and $df_2 = 4$.

Conclusion: Because the observed value of F is much larger than 6.94, we reject H_0 and conclude that there is a significant variability in calcium concentrations from lab to lab.

The test for factor A follows:

H_0: $\sigma_{\alpha}^2 = 0$.

H_a: $\sigma_{\alpha}^2 > 0$

T.S.: $F = \dfrac{\text{MSA}}{\text{MSAB}} = \dfrac{.78}{.41} = 1.90$

R.R.: For $\alpha = .05$, we will reject H_0 if F exceeds 6.94, the critical value for $a = .05$, $df_1 = 2$, and $df_2 = 4$.

Conclusion: There is insufficient evidence to indicate a significant variability in calcium determinations from assay to assay.

We will next estimate the variance components. Using the MSs and EMSs in analysis of variance table, we obtain

$\hat{\sigma}_{\varepsilon}^2 = \text{MSE} = .1378$

$\hat{\sigma}_{\alpha\beta}^2 = (\text{MSAB} - \text{MSE})/n = (.41 - .1378)/2 = .1361$

$\hat{\sigma}_{\beta}^2 = (\text{MSB} - \text{MSAB})/an = (3.78 - .41)/6 = .5617$

and

$\hat{\sigma}_{\alpha}^2 = (\text{MSA} - \text{MSAB})/bn = (.78 - .41)/6 = .0617$

a. Write an appropriate linear statistical model. List the assumptions and identify terms.

b. Perform an analysis of variance, showing expected mean squares. Use $\alpha = .05$.

c. Allocate the total variability in sales to the various sources of variability.

17.4 Mixed-Effects Models

In Section 17.3, we compared the analysis of variance tables for fixed- and random-effects models for a randomized block design and for a general $a \times b$ factorial laid out in a completely randomized design. Suppose, however, that we have a **mixed-effects model** for these same experimental designs where one effect is fixed and the other is random. For example, in Section 17.3, we considered an experiment to examine the effects of different subjects and different analysts on the DNA content of plaque. If the three subjects were selected at random and if the three analysts chosen were the only analysts of interest, we would have a mixed model for a randomized block design with fixed analysts and random subjects.

mixed-effects model

Let us consider a mixed model for a general $a \times b$ factorial in a completely randomized design. The model is the same as that given in Section 17.3 except that there are different assumptions.

$$y_{ijk} = \mu + \alpha_i + \beta_j + \alpha\beta_{ij} + \varepsilon_{ijk}$$

where we use the following conditions with the levels of factor A fixed and the levels of factor B randomly selected:

assumptions

1. μ is the unknown overall mean response.

2. α_i is a fixed effect corresponding to the ith level of factor A with $\sum_i \alpha_i = 0$.

3. β_j is a random effect due to the jth level of factor B. The β_js have independent normal distributions, with mean 0 and variance σ_β^2.

4. $\alpha\beta_{ij}$ is a random effect due to the interaction of the ith level of factor A with the jth level of factor B with $\sum_i \alpha\beta_{ij} = 0$, for each $j = 1, \ldots, b$. The $\alpha\beta_{ij}$s have normal distributions with mean 0 and variance $\sigma_{\alpha\beta}^2$. For all $j \neq j'$, $\alpha\beta_{ij}$ and $\alpha\beta_{i'j'}$ are independent.

5. The β_js, $\alpha\beta_{ij}$s, and ε_{ijk}s are independent.

Since $\sum_i \alpha\beta_{ij} = 0$, for each $j = 1, \ldots, b$, we have that for each j, $\alpha\beta_{1j}, \alpha\beta_{2j}, \ldots, \alpha\beta_{aj}$, are correlated, whereas, $\alpha\beta_{ij}$ and $\alpha\beta_{i'j'}$ are independent for all $j \neq j'$. That is, the interaction effects from the same level of factor B are correlated, whereas the interaction effects from different levels of factor B are independent.

Using these assumptions, the analysis of variance table for a fixed, random, or mixed model in a two-factor experiment with replication is as shown in Table 17.13.

The expected mean squares column of Table 17.13 can be helpful in determining appropriate tests of significance. The test for $\sigma_{\alpha\beta}^2$ is the same in the mixed model as in the random-effects model.

test for $\sigma_{\alpha\beta}^2$

H_0: $\sigma_{\alpha\beta}^2 = 0$

H_a: $\sigma_{\alpha\beta}^2 > 0$

T.S.: $F = \dfrac{\text{MSAB}}{\text{MSE}}$

R.R.: Based on $\text{df}_1 = (a-1)(b-1)$ and $\text{df}_2 = ab(n-1)$

TABLE 17.13
AOV table for an $a \times b$ factorial experiment, with n observations per cell

Source	SS	df	MS	EMS Fixed Effects	EMS Random Effects	EMS Mixed Effects A Fixed, B Random
A	SSA	$a - 1$	MSA	$\sigma_\varepsilon^2 + bn\theta_A$	$\sigma_\varepsilon^2 + n\sigma_{\alpha\beta}^2 + bn\sigma_\alpha^2$	$\sigma_\varepsilon^2 + n\sigma_{\alpha\beta}^2 + bn\theta_A$
B	SSB	$b - 1$	MSB	$\sigma_\varepsilon^2 + an\theta_B$	$\sigma_\varepsilon^2 + n\sigma_{\alpha\beta}^2 + an\sigma_\beta^2$	$\sigma_\varepsilon^2 + an\sigma_\beta^2$
AB	SSAB	$(a - 1)(b - 1)$	MSAB	$\sigma_\varepsilon^2 + n\theta_{AB}$	$\sigma_\varepsilon^2 + n\sigma_{\alpha\beta}^2$	$\sigma_\varepsilon^2 + n\sigma_{\alpha\beta}^2$
Error	SSE	$ab(n - 1)$	MSE	σ_ε^2	σ_ε^2	σ_ε^2
Totals	TSS	$nab - 1$				

No matter what the results of our tests for $\sigma_{\alpha\beta}^2$, we could proceed to use the following tests for factors A and B, which follow from entries in the expected mean squares column of Table 17.13. For factor A we have

test, factor A

H_0: $\alpha_1 = \alpha_2 = \cdots = \alpha_a = 0$

H_a: At least one of the αs differs from the rest

T.S.: $F = \dfrac{\text{MSA}}{\text{MSAB}}$

R.R.: Based on $df_1 = (a - 1)$ and $df_2 = (a - 1)(b - 1)$

For factor B we have

test, factor B

H_0: $\sigma_\beta^2 = 0$

H_a: $\sigma_\beta^2 > 0$

T.S.: $F = \dfrac{\text{MSB}}{\text{MSE}}$

R.R.: Based on $df_1 = (b - 1)$ and $df_2 = ab(n - 1)$

The analysis of variance procedure outlined for a mixed-effects model from an $a \times b$ factorial experiment can be used as well for a randomized block design, where treatments are fixed, blocks are assumed to be random, and there are n observations for each block and treatment. We will illustrate a mixed model in the following example.

EXAMPLE 17.3

A study was designed to evaluate the effectiveness of two different sunscreens (s_1 and s_2) for protecting the skin of persons who want to avoid burning or additional tanning while exposed to the sun. A random sample of 40 subjects (ages 20–25) agreed to participate in the study. For each subject a 1-inch square was marked off on their back, under the shoulder but above the small of the back. Twenty subjects were randomly assigned to each of the two types of sunscreen. A reading based on the color of the skin in the designated square was made prior to the application of a fixed amount of the assigned sunscreen, and then again after application and exposure to the sun for a 2-hour period. The company was concerned that the measurement of color is extremely variable, and wanted to assess the variability in the readings due to the technician taking the readings. Thus, the company randomly selected ten technicians from their worldwide staff to participate in the study. Four subjects, two having s_1 and two having s_2, were randomly assigned to each technician for evaluation. The data recorded in Table 17.14 are differences (postexposure minus preexposure) for the subjects in the study. A high response indicates a greater degree of burning.

TABLE 17.14

Data for sunscreen experiment in Example 17.3

Sunscreen (A)	Technician (B)										Mean
	1	2	3	4	5	6	7	8	9	10	
s_1	8.2	3.6	10.7	3.9	12.9	5.5	9.1	13.7	8.1	2.5	7.82
	7.6	3.5	10.3	4.4	12.1	5.9	9.7	13.2	8.7	2.8	
Mean	(7.9)	(3.55)	(10.5)	(4.15)	(12.5)	(5.7)	(9.4)	(13.45)	(8.4)	(2.65)	
s_2	6.1	4.3	9.6	2.3	12.4	4.8	8.3	12.9	8.0	2.1	7.15
	6.8	4.7	9.2	2.5	12.8	4.0	8.6	13.6	7.5	2.5	
Mean	(6.45)	(4.5)	(9.4)	(2.4)	(12.6)	(4.4)	(8.45)	(13.25)	(7.75)	(2.3)	
Mean	(7.175)	(4.025)	(9.95)	(3.275)	(12.55)	(5.05)	(8.925)	(13.35)	(8.075)	(2.475)	7.485

The experiment is a completely randomized design with two factors, sunscreen type (A) with two fixed levels and technician (B) with ten randomly selected levels. There are two subjects for each sunscreen–technician combination. Analyze the data to determine any differences in sunscreens and technicians.

Solution We can compute the sums of squares for the sources of variability in the AOV table using the following formulas.

$$\text{TSS} = \sum_{ijk} (y_{ijk} - \bar{y}_{...})^2 = (8.2 - 7.485)^2 + (7.6 - 7.485)^2 + \cdots$$

$$+ (2.5 - 7.485)^2 = 530.59$$

$$\text{SSA} = \sum_{i} 20(\bar{y}_{i..} - \bar{y}_{...})^2 = 20\{(7.82 - 7.485)^2 + (7.15 - 7.485)^2\} = 4.49$$

$$\text{SSB} = \sum_{j} 4(\bar{y}_{.j.} - \bar{y}_{...})^2 = 4\{(7.175 - 7.485)^2 + (4.025 - 7.485)^2 + \cdots$$

$$+ (2.475 - 7.485)^2\} = 517.49$$

$$\text{SSAB} = \sum_{ij} 2(\bar{y}_{ij.} - \bar{y}_{...})^2 - \text{SSA} - \text{SSB} = 2\{(7.9 - 7.485)^2 + (3.55 - 7.485)^2$$

$$+ (10.5 - 7.485)^2 + \cdots + (2.3 - 7.485)^2\} - 4.49 - 517.49 = 5.97$$

$$\text{SSE} = \text{TSS} - \text{SSA} - \text{SSB} - \text{SSAB} = 530.59 - 4.49 - 517.49 - 5.97 = 2.64$$

Substituting $a = 2$, $b = 10$, and $n = 2$ into an AOV table similar to that shown in Table 17.13, we have the results shown in Table 17.15.

A test for the random component $\alpha\beta_{ij}$ is as follows:

$$H_0: \quad \sigma_{\alpha\beta}^2 = 0$$

$$H_a: \quad \sigma_{\alpha\beta}^2 > 0$$

TABLE 17.15

AOV table for the data of Example 17.3

Source	SS	df	MS	EMS Mixed Model
A	4.49	1	4.49	$\sigma_\varepsilon^2 + 2\sigma_{\alpha\beta}^2 + 20\theta_A$
B	517.49	9	57.50	$\sigma_\varepsilon^2 + 4\sigma_\beta^2$
AB	5.97	9	.66	$\sigma_\varepsilon^2 + 2\sigma_{\alpha\beta}^2$
Error	2.64	20	.13	σ_ε^2
Totals	530.59	39		

T.S.: $F = \dfrac{\text{MSAB}}{\text{MSE}} = \dfrac{.66}{.13} = 5.08$

R.R.: For $\alpha = .05$, we will reject H_0 if the computed value of F exceeds 2.39, the value in Table 8 for $a = .05$, $df_1 = 9$, and $df_2 = 20$.

Conclusion: Because 5.08 exceeds 2.39, we reject H_0 and conclude that $\alpha_{\alpha\beta}^2 > 0$; that is, there is a significant source of random variation due to the combination of the ith level of A (sunscreens) and the jth level of B (technician). We would infer from this that the variation in the determination of skin color due to technician differences is different for the two types of sunscreen.

We next proceed to evaluate the effects due to the technicians.

H_0: $\sigma_\beta^2 = 0$

H_a: $\sigma_\beta^2 > 0$

T.S.: $F = \dfrac{\text{MSB}}{\text{MSE}} = \dfrac{57.50}{.13} = 442.31$

R.R.: For $\alpha = .05$, we will reject H_0 if F exceeds 2.39, the value in Appendix Table 8 for $a = .05$, $df_1 = 9$, and $df_2 = 20$.

Conclusion: Because 442.31 exceeds 2.39, we reject H_0 and conclude that $\sigma_\beta^2 > 0$. Thus there is a significant source of random variation due to variability from technician to technician.

For Factor A we have

H_0: $\alpha_1 = \alpha_2 = 0$

H_a: $\alpha_1 \neq \alpha_2$

T.S.: $F = \dfrac{\text{MSA}}{\text{MSAB}} = \dfrac{4.49}{.66} = 6.80$

R.R.: For $\alpha = .05$, we will reject H_0 if F exceeds 5.12, the value in Appendix Table 8 for $a = .05$, $df_1 = 1$, and $df_2 = 9$.

Conclusion: Because $6.80 > 5.12$, we reject H_0 and conclude that the mean response (post minus pre) differs for the two sunscreens. Because $\bar{y}_{s1} = 7.82$ and $\bar{y}_{s2} = 7.15$, we would conclude that s_2 offers more protection on the average than s_1. However, as noted previously, there are significant sources of variability due to technicians and the combination of technicians with sunscreens.

We will next analyze the data in the case study.

Analyzing Data for the Case Study: Pressure Drop across Expansion Joints

The objective of the study was to determine whether the pressure drop across the expansion joint in electric turbines was related to gas temperature. Also, the researchers wanted to assess the variation in readings from the various types of pressure gauges and find out whether variation in readings was consistent across different gas temperatures. In Figure 17.1, we observed that there was a slight increase in pressure drop as the temperature increased from 15°C to 25°C but a subsequent decrease in pressure drop when the temperature was further increased from 25°C to 35°C. The pressure drops recorded by the four gauges were fairly consistent over the three temperatures, with the exception that gauge G1 recorded a much higher mean pressure drop than the other three gauges at 25°C. The following table of means and standard deviations for the 12 temperature–gauge combinations reveals a fairly constant standard deviation, but gauge G1 has a much higher mean pressure drop at 25°C than the mean pressure drops of the other 11 temperature–gauge treatments.

| | Mean and Standard Deviations in Pressure Drop Readings | | | | | | | |
| | Mean | | | | Standard Deviation | | | |
Temperature	G1	G2	G3	G4	G1	G2	G3	G4
15	41.17	38.50	38.67	43.00	3.31	3.62	3.72	4.69
25	61.33	46.67	45.33	41.33	4.27	3.44	2.07	4.97
35	39.00	41.17	37.83	42.67	4.69	2.79	3.31	4.18

Because the four gauges were a random sample from a population of gauges that are used in the company, we want to assess whether the pattern observed in the above table and in Figure 17.1 were significant differences relative to the population from which the gauges were selected. Also, we want to determine whether there are significant differences in mean pressure drop across the temperature range 15°C to 35°C. The temperature factor is fixed and the gauge factor is random. The following model will be fit to the data:

$$y_{ijk} = \mu + \alpha_i + \beta_j + \alpha\beta_{ij} + \varepsilon_{ijk}$$

where y_{ijk} is the pressure drop during the kth replication using gauge j with temperature i. Prior to running tests of hypotheses or constructing confidence intervals, we will evaluate the conditions that the experiment must satisfy for inferences to be appropriate. An examination of the following plots of the residuals will assist us in checking on the validity of the model conditions. The computer printouts of the analysis of the residuals from the fitted model are given here.

```
Residual Analysis for Case Study

Univariate Procedure

Variable=RESIDUALS

                                            Moments

                                N        72    Sum Wgts          72
                              Mean        0    Sum                0
                              Std Dev   3.532545  Variance   12.47887
                              Skewness  0.014971  Kurtosis   -0.87963

                              W:Normal  0.959963  Pr<W          0.0655
```

```
Variable=RESIDUALS

          Stem Leaf                        #          Boxplot
            6 07                           2             |
            4 000233578                    9             |
            2 03338023335777              14          +-----+
            0 5778823378                  10          |  +  |
           -0 877322533222                12          *-----*
           -2 87777533087330              14          +-----+
           -4 325332000                    9             |
           -6 70                           2             |
             ---+---+---+---+
```

```
Variable=RESIDUALS
```

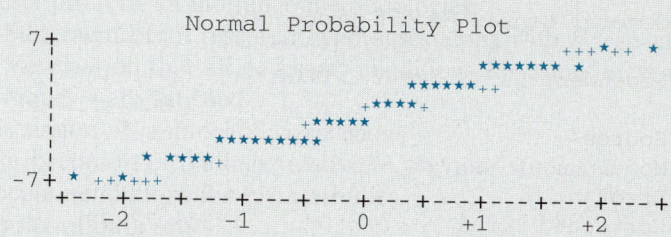

```
                        Normal Probability Plot
          7+                                           +++*++ *
           |                                      *******++
           |                                 ******++
           |                             +****+
           |                         +*****
           |                  *********
           |              * ****+
          -7+ *  ++*+++
             +---+---+---+---+---+---+---+---+---+---+
                -2       -1        0       +1       +2
```

Plot of RESIDUALS★PREDICTED   Legend: A = 1 obs, B = 2 obs, etc.

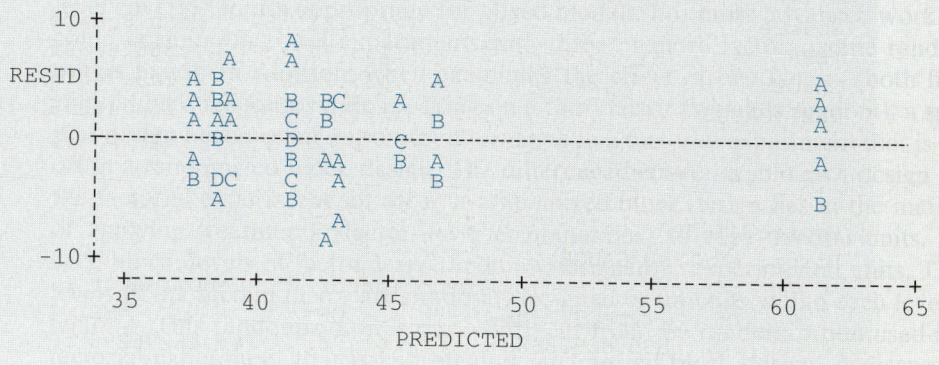

EXERCISES   **Applications**

Env.   **17.6**   The following study was designed to evaluate the effectiveness of four chemicals developed to control fire ants. The type of environmental conditions in which the chemical is placed might have an effect on the effectiveness of the treatment to kill fire ants. Thus, the researcher randomly selected five locations from a large selection of location; each location representing a randomly selected environment. To reduce the effect of different colonies of fire ants and the type of mounds they inhabit, the researcher created 40 artificial fire ant mounds and populated them with 50,000 ants having similar ancestry. The researcher randomly assigned two mounds to each of the 20 treatment–location combinations. The number of fire ants killed during a 1-week period was recorded. The number of fire ants killed (in thousands) are given here.

   a. Write an appropriate linear statistical model for this study. Identify all terms in your model.
   b. Compute the sum of squares for this experiment and report this value in an AOV table. Be sure to include the expected mean squares column in the AOV table.

	Chemicals			
Locations	**1**	**2**	**3**	**4**
1	7.2	4.2	9.5	5.4
	9.6	3.5	9.3	3.9
2	8.5	2.9	8.8	6.3
	9.6	3.3	9.2	6.0
3	9.1	1.8	7.6	6.1
	8.6	2.4	7.1	5.6
4	8.2	3.6	7.3	5.0
	9.0	4.4	7.0	5.4
5	7.8	3.7	9.2	6.5
	8.0	3.9	8.3	6.9

17.7   Refer to Exercise 17.6. Perform an analysis of variance. Draw your conclusions, using $\alpha = .05$.

17.5   Rules for Obtaining Expected Mean Squares

We discussed the AOVs for one- and two-factor experiments for fixed-effects models in Chapter 15 and for random or mixed models earlier in this chapter. We will see in this section that for any k-factors experiment of data, with n observations per factor–level combination, it is possible to write expected mean squares for all main effects and interactions for fixed, random, or mixed models using some rather simple rules. *The importance of these rules is that, having written down the expected mean squares for an unfamiliar experimental design, we often can construct appropriate F tests.* The assumptions for the fixed and random models will be the same as we have used in describing fixed, random, and mixed models in previous sections.

classifying interactions   Two rules for **classifying interactions** as fixed or random effects are needed before we can proceed with the rules for obtaining expected mean squares.

Rules for the Classification of Interactions

1. If a fixed effect interacts with another fixed effect, the resulting interaction term is a fixed effect.
2. If a random effect interacts with another effect (fixed or random), the resulting interaction term is a random component.

EXAMPLE 17.4

Consider a 3×6 factorial with two observations per factor–level combination. Classify the AB interaction as fixed or random for the following situations:

    a. A and B are both fixed effects.
    b. A is fixed and B is random.
    c. A and B are both random.

Solution   We apply the rules for classifying interactions.

    a. AB is a fixed effect because A (fixed) interacts with B (fixed).
    b. AB is a random component because A (fixed) interacts with B (random).
    c. AB is random because A (random) interacts with B (random).

EXAMPLE 17.5

Consider a factorial experiment in the factors A, B, and C. Classify the AB, AC, BC, and ABC interactions as fixed or random when A and B are fixed effects and C is random.

Solution   We apply the classification rules.

    AB is fixed; A (fixed) interacts with B (fixed).

    AC is random; A (fixed) interacts with C (random).

    BC is random; B (fixed) interacts with C (random).

    ABC is random; A (fixed) interacts with BC (random).

mean square table   Before we state the rules for determining expected mean squares, it is convenient to construct a **mean square table.** These steps are summarized here and illustrated for an $a \times b$ factorial experiment with factor A random, B fixed, and n observations for each factor–level combination of A and B.

Steps in Constructing a Mean Square Table

1. Write the model for the experiment. For an $a \times b$ factorial experiment, the model is

$$y_{ijk} = \mu + \alpha_i + \beta_j + \alpha\beta_{ij} + \varepsilon_{k[ij]}$$

Note: We use brackets in the ε-term to indicate that there are $k = 1, 2, \ldots, n$ observations for each factor–level combination of factors A and B (i.e., for each choice of i, j).

2. Construct a table with each term in the model (except μ) forming a row heading. This table takes the following form for our example.

α_j
β_j
$\alpha\beta_{ij}$
$\varepsilon_{k[ij]}$

3. Form a column in the table for each subscript in the model.

	i	j	k
α_j			
β_j			
$\alpha\beta_{ij}$			
$\varepsilon_{k[ij]}$			

4. Above each column heading indicate whether the subscript corresponds to a fixed (F) or random (R) effect.

5. Also indicate the number levels for each subscript. Additions to the table from steps 4 and 5 are shown here for our example.

	a R i	b F j	n R k
α_j			
β_j			
$\alpha\beta_{ij}$			
$\varepsilon_{k[ij]}$			

6. For each row in a given column, enter the number of levels associated with the column subscript, unless the row term contains the column subscript.

	a R i	b F j	n R k
α_i		b	n
β_j	a		n
$\alpha\beta_{ij}$			n
$\varepsilon_{k[ij]}$			

7. Examine the terms of the model, which are listed in the first column of the table. For each term with brackets in the subscript, place a 1 under the column(s) with a subscript included in the brackets.

	a	b	n
	R	F	R
	i	j	k
α_i		b	n
β_j	a		n
$\alpha\beta_{ij}$			n
$\varepsilon_{k[ij]}$	1	1	

8. Fill in the remaining cells of a column with a 0 if the column is headed by an F; fill in the remaining cells of a column headed by an R with a 1.

	a	b	n
	R	F	R
	i	j	k
α_i	1	b	n
β_j	a	0	n
$\alpha\beta_{ij}$	1	0	n
$\varepsilon_{k[ij]}$	1	1	1

This is the *mean square table* used for computing the expected mean squares for a two factor experiment with factor A random, factor B, fixed and n observations per factor–level combination of A and B.

It is easy to compute expected mean squares once you have the mean square table. These rules are listed here.

Rules for Obtaining an EMS Using a Mean Square Table

1. Examine the subscript(s) of the term.
2. Eliminate any row in the mean square table that does not have the subscript(s).
3. Cover each column of the table headed by a nonbracketed subscript of the term.
4. Multiply the remaining, uncovered entries in each row to obtain the coefficients of terms in the expected mean square.

EXAMPLE 17.6

Compute $E(\text{MSA})$ for a two-factor experiment with a levels of factor A (random), b levels of factor B (fixed), and n observations per factor–level combination.

Solution   Refer to the mean square table just given. We note that α_i has the subscript i; thus, we eliminate the second row of the table. Covering column 1 (which is headed by i), we multiply the remaining entries. Table 17.16 shows these steps, the multiplication, and the terms of the EMS.

TABLE 17.16
Mean square table for computing $E(MSA)$

	a R i	b F j	n R k	Product of Remaining Entries	Term EMS
α_i	1	b	n	bn	$bn\sigma_\alpha^2$
β_i	a	0	n	—	
$\alpha\beta_{ij}$	1	0	n	0	
$\varepsilon_{k[ij]}$	1	1	1	1	σ_ε^2

Using the last column of Table 17.16, we have

$$E(MSA) = \sigma_\varepsilon^2 + bn\sigma_\alpha^2$$

The computation $E(MSB)$ follows in a similar manner. The term β_j has a subscript j, so we eliminate the second column (which is headed by j) and the first row (which contains no j) of the mean square table. The remaining entries are multiplied to obtain the coefficients of the expected mean square (see Table 17.17).

TABLE 17.17
Mean square table for computing $E(MSB)$

	a R i	b F j	n R k	Product of Remaining Entries	Term EMS
α_i	1	b	n	—	
β_i	a	0	n	an	$an\theta_\beta$
$\alpha\beta_{ij}$	1	0	n	n	$n\sigma_{\alpha\beta}^2$
$\varepsilon_{k[ij]}$	1	1	1	1	σ_ε^2

Hence,

$$E(MSB) = \sigma_\varepsilon^2 + n\sigma_{\alpha\alpha}^2 + an\theta_B$$

where θ_B is a constant of the form

$$\theta_B = \frac{\sum_j \beta_j^2}{b - 1}$$

EXAMPLE 17.7

a. Set up the mean square table for computing expected mean squares for a two-factor experiment with factor A fixed, factor B random, and n observations for each factor–level combination of A and B.

b. Compute $E(MSA)$.

Solution

a. The model for this experiment situation is

$$y_{ijk} = \mu + \alpha_i + \beta_j + \alpha\beta_{ij} + \varepsilon_{k[ij]}$$

The corresponding mean square table is shown in Table 17.18.

b. $E(MSA)$ is found as shown in Table 17.19.

<table>
<tr><td colspan="4">TABLE 17.18
Mean square table for
Example 17.7</td></tr>
</table>

	a F i	b R j	n R k
α_i	0	b	n
β_i	a	1	n
$\alpha\beta_{ij}$	0	1	n
$\varepsilon_{k[ij]}$	1	1	1

TABLE 17.19

Computations for $E(MSA)$

	a F i	b R j	n R k	**Product of Remaining Entries**	**Term EMS**
α_i	0	b	n	bn	$bn\theta_\alpha$
β_i	a	1	n	—	
$\alpha\beta_{ij}$	0	1	n	n	$n\sigma^2_{\alpha\beta}$
$\varepsilon_{k[ij]}$	1	1	1	1	σ^2_ε

Thus, $E(MSA) = \sigma^2_\varepsilon + n\sigma^2_{\alpha\beta} + bn\theta_A$, where $\theta_A = \sum_i \alpha_i^2/(a-1)$. This agrees with what we obtained in the previous example for the fixed factor (B) with appropriate changes in notation.

Previously, we have been concerned with only fixed-effects models. For these models the test statistics are always formed using the affected mean square in the numerator divided by MSE. However, for random and mixed models the test statistics are not always the same. The test statistic for interaction, F equals MSAB/MSE, is the same for the fixed, random, and mixed models, but the F tests for factors A and B change depending on the assumptions for α_i and β_j. For example, the F test for factor A is MSA/MSE for A fixed, B fixed, and for A random, B fixed. In contrast, the F test for factor A is MSA/MSAB for A fixed, B random, and for A random, B random. Thus you can see the importance of knowing the expected mean squares for random and mixed models.

A special case of the two-factor experiment should be mentioned: when there is only one observation ($n = 1$) per factor–level combination of A and B, and we have the model

$$y_{ij} = \mu + \alpha_i + \beta_j + \alpha\beta_{ij} + \varepsilon_{ij}$$

You can see from the abbreviated AOV table of Table 17.20 that there are no degrees of freedom for error, so there is no test for interaction and, depending on the model, there may not be a valid test for the main effects. The only possible remedy is for the situation in which one can assume that there is no interaction between factors A and B, in which case all main effects can be tested using the mean square error from the model

$$y_{ij} = \mu + \alpha_i + \beta_j + \varepsilon_{ij}$$

TABLE 17.20

Abbreviated AOV table, two-factor experiment ($n = 1$)

Source	df
A	$a - 1$
B	$b - 1$
AB	$(a - 1)(b - 1)$
Error	—
Total	$ab - 1$

regardless of whether we have a fixed, random, or mixed model. When the no-interaction assumption is not reasonable, the experiment must allow for replication ($n > 1$) at the factor–level combinations of A and B in order to obtain valid tests of the interaction and main effects.

The same rules used for the two-factor experiment can also be used for more complicated experiments and, although the rules may seem a bit cumbersome, with practice they are quite easy to use. We will give one more example using a three-factor experiment. For additional details regarding assumptions, derivations, and more complicated applications, see Hicks and Turner (1999) and Kuchl (1999).

EXAMPLE 17.8

Give the expected mean squares for a $3 \times 5 \times 2$ factorial experiment with $n = 4$ observations per factor–level combination. Treat factors A and B as fixed and factor C as random.

Solution   The complete model for this experiment is given here along with the corresponding mean square table:

$$y_{ijkl} = \mu + \alpha_i + \beta_j + \gamma_k + \alpha\beta_{ij} + \alpha\gamma_{ik} + \beta\gamma_{ik} + \alpha\beta\gamma_{ijk} + \varepsilon_{l[ijk]}$$

We will set up the mean square table for general values of a, b, c, and n, and then substitute later. The mean square table using the rules discussed previously is shown in Table 17.21. The expected mean squares for the model terms can be obtained by applying the EMS rules to this mean square table. For example, for $E(\text{MSA})$, we have the uncovered entries in Table 17.22.

TABLE 17.21
Mean square table for
Example 17.8

	a F i	b F j	c R k	n R j
α_i	0	b	c	n
β_j	a	0	c	n
γ_k	a	b	1	n
$\alpha\beta_{ij}$	0	0	c	n
$\alpha\gamma_{ik}$	0	b	1	n
$\beta\gamma_{jk}$	a	0	1	n
$\alpha\beta\gamma_{ijk}$	0	0	1	n
$\varepsilon_{l[ijk]}$	1	1	1	1

TABLE 17.22
Computations for $E(\text{MSA})$
Example 17.8

	a F i	b F j	c R k	n R l	**Product of Remaining Entries**	**Term EMS**
α_i	0	b	c	n	bcn	$bcn\theta_A$
β_j	a	0	c	n	—	
γ_k	a	b	1	n	—	
$\alpha\beta_{ij}$	0	0	c	n	0	
$\alpha\gamma_{ik}$	0	b	1	n	bn	$bn\sigma^2_{\alpha\gamma}$
$\beta\gamma_{jk}$	a	0	1	n	—	
$\alpha\beta\gamma_{ijk}$	0	0	1	n	0	
$\varepsilon_{l[ijk]}$	1	1	1	1	1	σ^2_ε

From the last column of this table we have

$$E(\text{MSA}) = \sigma_\varepsilon^2 + bn\sigma_{\alpha\gamma}^2 + bcn\theta_A$$

Substituting $a = 3$, $b = 5$, $c = 2$, and $n = 4$, this becomes

$$E(\text{MSA}) = \sigma_\varepsilon^2 + 20\sigma_{\alpha\gamma}^2 + 40\theta_A$$

where

$$\theta_A = \sum_i \alpha_i^2 / 2$$

Similarly, the expected mean squares for factors B and C can be shown to be

$$E(\text{MSB}) = \sigma_\varepsilon^2 + an\sigma_{\beta\gamma}^2 + acn\theta_B$$
$$= \sigma_\varepsilon^2 + 12\sigma_{\beta\gamma}^2 + 24\theta_B$$

and

$$E(\text{MSC}) = \sigma_\varepsilon^2 + abn\sigma_\gamma^2$$
$$= \sigma_\varepsilon^2 + 60\sigma_{\beta\gamma}^2$$

The table for computing the expected mean square for the AB interaction is shown in Table 17.23. The expected mean square for MSAB is

$$E(\text{MSAB}) = \sigma_\varepsilon^2 + n\sigma_{\alpha\beta\gamma}^2 + cn\theta_{AB}$$
$$= \sigma_\varepsilon^2 + 4\sigma_{\alpha\beta\gamma}^2 + 8\theta_{AB}$$

where

$$\theta_{AB} = \sum_i \frac{\alpha\beta_{ij}^2}{8}$$

TABLE 17.23
Computations for $E(\text{MSAB})$
Example 17.8

	a F i	b F j	c R k	n R l	Product of Remaining Entries	Term EMS
α_i	0	b	c	n	—	
β_j	a	0	c	n	—	
γ_k	a	b	1	n	—	
$\alpha\beta_{ij}$	0	0	c	n	cn	$cn\theta_{AB}$
$\alpha\gamma_{ik}$	0	b	1	n	—	
$\beta\gamma_{jk}$	a	0	1	n	—	
$\alpha\beta\gamma_{ijk}$	0	0	1	n	n	$n\sigma_{\alpha\beta\gamma}^2$
$\varepsilon_{kl[ijk]}$	1	1	1	1	1	σ_ε^2

After application of the EMS rules to the AC and BC interactions, we obtain

$$E(\text{MSAC}) = \sigma_\varepsilon^2 + bn\sigma_{\alpha\gamma}^2$$
$$= \sigma_\varepsilon^2 + 20\sigma_{\alpha\gamma}^2$$

and

$$E(\text{MSBC}) = \sigma_\varepsilon^2 + an\sigma_{\beta\gamma}^2$$
$$= \sigma_\varepsilon^2 + 12\sigma_{\beta\gamma}^2$$

Similarly, it can be shown that MSABC and MSE have expectations

$$E(\text{MSABC}) = \sigma_\varepsilon^2 + n\sigma_{\alpha\beta\gamma}^2$$

$$= \sigma_\varepsilon^2 + 4\sigma_{\alpha\beta\gamma}^2$$

and

$$E(\text{MSE}) = \sigma_\varepsilon^2$$

A summary of the expected mean squares, which we have computed using the EMS rules of this chapter, for the $3 \times 5 \times 2$ factorial with $n = 4$ observations per cell, and factors A and B fixed but factor C random, is shown in Table 17.24. We have included the denominator of the valid F test for testing whether this source of variation is significant.

TABLE 17.24
Partial AOV for
Example 17.8

Source	EMS	Denominator of F
A	$\sigma_\varepsilon^2 + 20\sigma_{\alpha\gamma}^2 + 40\theta_A$	MSAC
B	$\sigma_\varepsilon^2 + 12\sigma_{\beta\gamma}^2 + 24\theta_B$	MSBC
C	$\sigma_\varepsilon^2 + 60\sigma_{\alpha\gamma}^2$	MSE
AB	$\sigma_\varepsilon^2 + 4\sigma_{\alpha\beta\gamma}^2 + 8\theta_{\alpha\beta}$	MSABC
AC	$\sigma_\varepsilon^2 + 20\sigma_{\alpha\gamma}$	MSE
BC	$\sigma_\varepsilon^2 + 12\sigma_{\beta\gamma}^2$	MSE
ABC	$\sigma_\varepsilon^2 + 4\sigma_{\alpha\beta\gamma}^2$	MSE
Error	σ_ε^2	*

EXAMPLE 17.9

Refer to Example 17.8. Give an appropriate F statistic for

$$H_0: \quad \theta_A = 0 \quad \text{and} \quad H_0: \quad \sigma_{\beta\gamma}^2 = 0$$

Solution   Using the expected mean squares listed in Table 17.24, it is clear that the test statistic for $H_0: \theta_A = 0$ is $F = \text{MSA}/\text{MSAC}$; the test statistic for $H_0: \sigma_{\beta\gamma}^2 = 0$ is $F = \text{MSBC}/\text{MSE}$.

We can always obtain valid tests for all sources of variability in fixed-effects models, but this is not true for some random-effects and mixed-effects models. Tables 17.25, 17.26, 17.27, and 17.28 display the EMS for several three-factor experiments. In these tables, we provide the denominator of the F test for those

TABLE 17.25
Three-factor $a \times b \times c$
design with all factors fixed
and n replications

Source	EMS	Denominator of F
All Factors Fixed		
A	$\sigma_\varepsilon^2 + bcn\theta_A$	MSE
B	$\sigma_\varepsilon^2 + acn\theta_B$	MSE
C	$\sigma_\varepsilon^2 + abn\theta_C$	MSE
AB	$\sigma_\varepsilon^2 + cn\theta_{AB}$	MSE
AC	$\sigma_\varepsilon^2 + bn\theta_{AC}$	MSE
BC	$\sigma_\varepsilon^2 + an\theta_{BC}$	MSE
ABC	$\sigma_\varepsilon^2 + n\theta_{ABC}$	MSE
Error	σ_ε^2	*

TABLE 17.26

Three-factor $a \times b \times c$ design with all factors random and n replications

	All Factors Random	
Source	**EMS**	**Denominator of F**
A	$\sigma_\varepsilon^2 + n\sigma_{\alpha\beta\gamma}^2 + cn\sigma_{\alpha\beta}^2 + bn\sigma_{\alpha\gamma}^2 + bcn\sigma_\alpha^2$	*
B	$\sigma_\varepsilon^2 + n\sigma_{\alpha\beta\gamma}^2 + cn\sigma_{\alpha\beta}^2 + an\sigma_{\beta\gamma}^2 + acn\sigma_\beta^2$	*
C	$\sigma_\varepsilon^2 + n\sigma_{\alpha\beta\gamma}^2 + bn\sigma_{\alpha\gamma}^2 + an\sigma_{\beta\gamma}^2 + abn\sigma_\gamma^2$	*
AB	$\sigma_\varepsilon^2 + n\sigma_{\alpha\beta\gamma}^2 + cn\sigma_{\alpha\beta}^2$	MSABC
AC	$\sigma_\varepsilon^2 + n\sigma_{\alpha\beta\gamma}^2 + bn\sigma_{\alpha\gamma}^2$	MSABC
BC	$\sigma_\varepsilon^2 + n\sigma_{\alpha\beta\gamma}^2 + an\sigma_{\beta\gamma}^2$	MSABC
ABC	$\sigma_\varepsilon^2 + n\sigma_{\alpha\beta\gamma}^2$	MSE
Error	σ_ε^2	*

TABLE 17.27

Three-factor $a \times b \times c$ design with A and B random, C fixed and n replications

	A and B Random, C Fixed	
Source	**EMS**	**Denominator of F**
A	$\sigma_\varepsilon^2 + cn\sigma_{\alpha\beta}^2 + bcn\sigma_\alpha^2$	MSAB
B	$\sigma_\varepsilon^2 + cn\sigma_{\alpha\beta}^2 + acn\sigma_\beta^2$	MSAB
C	$\sigma_\varepsilon^2 + n\sigma_{\alpha\beta\gamma}^2 + bn\sigma_{\alpha\gamma}^2 + an\sigma_{\beta\gamma}^2 + abn\theta_C$	*
AB	$\sigma_\varepsilon^2 + cn\sigma_{\alpha\beta}^2$	MSE
AC	$\sigma_\varepsilon^2 + n\sigma_{\alpha\beta\gamma}^2 + bn\sigma_{\alpha\gamma}^2$	MSABC
BC	$\sigma_\varepsilon^2 + n\sigma_{\alpha\beta\gamma}^2 + an\sigma_{\beta\gamma}^2$	MSABC
ABC	$\sigma_\varepsilon^2 + n\sigma_{\alpha\beta\gamma}^2$	MSE
Error	σ_ε^2	*

TABLE 17.28

Three-factor $a \times b \times c$ design with A random, B and C fixed and n replications

	A Random, B and C Fixed	
Source	**EMS**	**Denominator of F**
A	$\sigma_\varepsilon^2 + bcn\sigma_\alpha^2$	MSE
B	$\sigma_\varepsilon^2 + cn\sigma_{\alpha\beta}^2 + acn\theta_B$	MSAB
C	$\sigma_\varepsilon^2 + bn\sigma_{\alpha\gamma}^2 + abn\theta_C$	MSAC
AB	$\sigma_\varepsilon^2 + cn\sigma_{\alpha\beta}^2$	MSE
AC	$\sigma_\varepsilon^2 + bn\sigma_{\alpha\gamma}^2$	MSE
BC	$\sigma_\varepsilon^2 + n\sigma_{\alpha\beta\gamma}^2 + an\theta_{BC}$	MSABC
ABC	$\sigma_\varepsilon^2 + n\sigma_{\alpha\beta\gamma}^2$	MSE
Error	σ_ε^2	*

variance components having valid F tests. An * indicates those variance components for which there is not a valid F test. Approximate F tests can be constructed for sources of variability in random-effects and mixed-effects models where no valid F test is available. These tests are available in some of the computer software programs—for example, SAS and SPSS. A discussion of these tests can be found in Hicks and Turner (1999) and Kuehl (1999).

The estimation of variance components was illustrated in Sections 17.2 and 17.3. This procedure of equating mean squares to expected mean squares can be used for obtaining estimates of variance components in random-effects and mixed-effects models for balanced designs following the procedure that we introduced in these earlier sections. Many computer software programs will carry out these calculations—for example, SAS and SPSS. However, in some of these programs, the conditions placed on the model are different from the conditions that we have used, the so-called classical conditions. Under these alternative conditions that the software programs impose on the model, the expected mean squares are different for some of the mixed models. Hence the estimators of the variance components and the appropriate F tests may be different from the results obtained under the classical conditions. The problem of variance components estimation for unbalanced designs is a complex one and is beyond the scope of this text. A detailed discussion of this topic can be found in Searle, Casella, and McCulloch (1992).

17.6   Nested Sampling and the Split-Plot Design

Sometimes in an experiment one factor is "nested" within another. This can be illustrated with the following example. A pharmaceutical company conducted tests to determine the stability of its product (under room-temperature conditions) at a specific point in time. Two manufacturing sites were used. At each site, a random sample of three batches of the product was obtained and additional random samples of ten different tablets were obtained from each batch. The design can be represented as shown in Figure 17.2.

FIGURE 17.2

Two-factor experiment with batches nested in sites

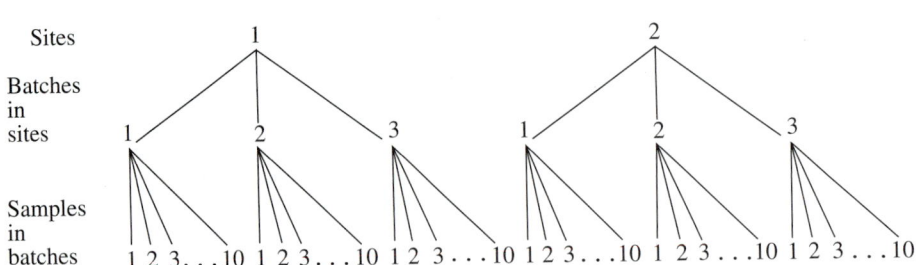

Although this might look like the usual two-factor experiment with sites (factor A) and batches (factor B), note that the three batches taken from site 1 are different from the three batches taken from site 2. In this sense, factor B (batches) is said to be *nested* in factor A (sites). For this experimental situation, it will be impossible to evaluate the effect of the interaction of factor B with factor A, because each level of factor B does not appear with each level of factor A, as would happen with a factorial arrangement of factors A and B. Here, the three batches within a site are unique to that site.

The general model for a two-factor experiment (n observations per cell) where factor B is nested in factor A can be written as

$$y_{ijk} = \mu + \alpha_i + \beta_{j(i)} + \varepsilon_{ijk} \qquad \begin{matrix} i = 1, 2, \ldots, a \\ j = 1, 2, \ldots, b \\ k = 1, 2, \ldots, n \end{matrix}$$

Note that this model is similar to the model for the two-factor experiment of Section 17.3, except that there is no interaction term $\alpha\beta_{ij}$ and the term for factor B, $\beta_{j(i)}$, is subscripted to denote the jth level of factor B is nested in the ith level of factor A. The analysis of variance table for this design is shown in Table 17.29.

TABLE 17.29
AOV table for a two-factor experiment (n observations per cell) with factor B nested in factor A

					EMS	
Source	SS	df	MS	Fixed	Mixed (A Fixed)	Random
A	SSA	$a - 1$	MSA	$\sigma_\varepsilon^2 + bn\theta_A$	$\sigma_\varepsilon^2 + n\sigma_\beta^2 + bn\theta_A$	$\sigma_\varepsilon^2 + n\sigma_\beta^2 + bn\sigma_\alpha^2$
$B(A)$	SSB(A)	$a(b - 1)$	MSB(A)	$\sigma_\varepsilon^2 + n\theta_B$	$\sigma_\varepsilon^2 + n\sigma_\beta^2$	$\sigma_\varepsilon^2 + n\sigma_\beta^2$
Error	SSE	$ab(n - 1)$	MSE	σ_ε^2	σ_ε^2	σ_ε^2
Total	TSS	$abn - 1$				

The sum of squares in the AOV table are computed using the formulas given here.

$$TSS = \sum_{ijk} (y_{ijk} - \bar{y}_{...})^2$$

$$SSA = \sum_{i} bn(\bar{y}_{i..} - \bar{y}_{...})^2$$

$$SSB(A) = \sum_{i} \sum_{j} n(\bar{y}_{ij.} - \bar{y}_{i..})^2$$

$$SSE = TSS - SSA - SSB(A)$$

Three of the more common situations are shown in Table 17.29 with the expected mean squares. Note the following in particular:

1. The F test for factor B is always

$$F = \frac{MSB(A)}{MSE}$$

2. The F test for factor A in the fixed-effects model is

$$F = \frac{MSA}{MSE}$$

For the random- and mixed-effects model, however, the corresponding test for factor A is

$$F = \frac{MSA}{MSB(A)}$$

3. When $n = 1$, there is no test for factor B, but we can test for factor A in the random- and mixed-effects model using

$$F = \frac{MSA}{MSB(A)}$$

EXAMPLE 17.10

Researchers conducted an experiment to determine the content uniformity of film-coated tablets produced for a cardiovascular drug used to lower blood pressure. They obtained a random sample of three batches from each of two blending sites; within each batch they assayed a random sample of five tablets to determine content uniformity. The data are shown here:

	Site		1			2		
Batches within each site			1	2	3	1	2	3
Tablets within each batch			5.03	4.64	5.10	5.05	5.46	4.90
			5.10	4.73	5.15	4.96	5.15	4.95
			5.25	4.82	5.20	5.12	5.18	4.86
			4.98	4.95	5.08	5.12	5.18	4.86
			5.05	5.06	5.14	5.05	5.11	5.07

a. Run an analysis of variance. Use $\alpha = .05$.

b. Is there evidence to indicate batch-to-batch variability in content uniformity? Does the F test run depend on whether we assume batches are fixed or random?

c. Draw conclusions about batch.

Solution

a. For these data we have $a = 2$ blending sites, $b = 3$ batches within each blending site, and $n = 5$ tablets per batch. From the data we compute the following sample means:

		Batch			
Site	**1**	**2**	**3**	**Site Mean**	
1	5.082	4.84	5.134	5.01867	
2	5.06	5.216	4.928	5.068	
Overall				5.04333	

From the data we compute the following sum of squares:

$$\text{TSS} = (5.03 - 5.04333)^2 + (5.10 - 5.04333)^2 + \cdots + (5.07 - 5.04333)^2$$

$$= .76348$$

$$\text{SSA} = 15\{(5.01867 - 5.04333)^2 + (5.068 - 5.04333)^2\} = .01824$$

$$\text{SSB}(A) = 5\{(5.082 - 5.01867)^2 + (4.84 - 5.01867)^2 + (5.134 - 5.01867)^2$$

$$+ (5.06 - 5.068)^2 + (5.216 - 5.068)^2 + (4.928 - 5.068)^2\} = .45401$$

$$\text{SSE} = \text{TSS} - \text{SSA} - \text{SSB}(A) = .76348 - .01824 - .45401 = .29123$$

The computer output for the analysis of this data set is given here. Note that the sum of squares differ slightly from our calculations. This is due to round-off error because we are dealing with very small deviations. We will use the sum of squares from the computer output in the analysis of variance table for this experiment, which is given here.

Source	SS	df	MS	F
A	.01825	1	.01825	.16
$B(A)$	.45401	4	.11350	9.39
Error	.29020	24	.01209	
Total	.76246	29		

```
CONTENT UNIFORMITY OF FILM-COATED TABLETS

General Linear Models Procedure

Dependent Variable: Y   CONTENT
```

Source	DF	Sum of Squares	Mean Square	F Value	Pr > F
Model	5	0.47226667	0.09445333	7.81	0.0002
Error	24	0.29020000	0.01209167		
Corrected Total	29	0.76246667			

R-Square	C.V.	Root MSE	Y Mean
0.619393	2.180346	0.10996	5.04333

Source	DF	Type III SS	Mean Square	F Value	Pr > F
SITE	1	0.01825333	0.01825333	1.51	0.2311
BATH (SITE)	4	0.45401333	0.11350333	9.39	0.0001

```
Tests of Hypotheses for Mixed Model Analysis of Variance

Dependent Variable: Y   CONTENT

Source: SITE
Error: MS(BATCH(SITE))
```

DF	Type III MS	Denominator DF	Denominator MS	F Value	Pr > F
1	0.0182533333	4	0.1135033333	0.1608	0.7089

```
Source: BATCH(SITE)
Error: MS(Error)
```

DF	Type III MS	Denominator DF	Denominator MS	F Value	Pr > F
4	0.1135033333	24	0.0120916667	9.3869	0.0001

b., c. The F test for batches is

$$F = \frac{\text{MSB(A)}}{\text{MSE}} = 9.39$$

based on $df_1 = 4$ and $df_2 = 24$ degrees of freedom. Because the observed value of F, 9.39, exceeds the tabled value of F for $\alpha = .05$, we conclude that there is considerable batch-to-batch variability in content uniformity of tablets. This test does not depend on whether the batches are random.

By now you may have realized that a whole new series of experimental designs have opened up with the introduction of nested effects. Thinking beyond the two-factor design, one could imagine a general multifactor design with factor A, factor B nested in levels of factor A, factor C nested in levels of A, and B, and so on. The analysis of variance table for a three-factor nested design with all factors random is shown in Table 17.30.

Other extensions of these designs are possible as well. For example, one could have a three-factor experiment, where factors A and B are cross-classified but factor C is nested within levels of factors A and B. This would be an example of a *partially nested design*.

TABLE 17.30

AOV table for a three-factor
nested design—all
factors random (n
observations per cell)

Source	SS	df	MS	EMS
A	SSA	$a - 1$	MSA	$\sigma_\varepsilon^2 + n\sigma_\gamma^2 + cn\sigma_\beta^2 + bcn\sigma_\alpha^2$
$B(A)$	SSB(A)	$a(b - 1)$	MSB(A)	$\sigma_\varepsilon^2 + n\sigma_\gamma^2 + cn\sigma_\beta^2$
$C(A, B)$	SSC(A, B)	$ab(c - 1)$	MSC(A, B)	$\sigma_\varepsilon^2 + n\sigma_\gamma^2$
Error	SSE	$abc(n - 1)$	MSE	σ_ε^2
Total	TSS	$abcn - 1$		

Suppose that a marketing research firm is responsible for sampling potential customers to obtain their opinions on two products (A_1 and A_2) in four geographic areas of the country ($B_1, \ldots, B_4$). A random sample of six stores selling product A_i is obtained in each geographic area. For each store selected for product A_i in geographic area B_j, ten people are interviewed concerning product i. For this design, factor C (stores) would be nested in levels of factors A (products) and B (geographic areas) and there would be $n = 10$ observations (opinions) for each level of factor C (stores) nested in levels of factors A and B.

The possibilities of nested and partial nested designs are seemingly endless, but, unfortunately, we will not have an opportunity to examine them here. The interested reader should refer to Kuehl (1999) and Montgomery (1997) for a more extensive treatment of this topic. We will, however, consider one very popular **split-plot design** design that is similar to a partially nested design. It is called the **split-plot design** because it had its origin in agriculture experimentation. We will illustrate its use with an example.

The yields of three different varieties of soybeans are to be compared under two different levels of fertilizer application. If we were interested in getting (say) $n = 2$ observations at each combination of fertilizer and variety of soybeans, we would need 12 equal-sized plots. Taking fertilizers as factor A and varieties as a treatment factor T, one possible design would be the standard 2×3 factorial experiment in a completely randomized design with $n = 2$ observations per factor–level combination. However, since the application of fertilizer to a plot occurs when the soil is being prepared for planting, it would be difficult (logistically) to first apply fertilizer A_1 to six of the plots dictated by the factorial arrangement of factors A and T and then fertilizer A_2 to the other six plots before planting the required varieties of soybeans in each plot.

An easier design to execute would have each fertilizer applied to two larger "wholeplots" and then the varieties of soybeans planted in three "subplots" (equal in size to the plots of the previous design) within each wholeplot. A design of this type appears in Figure 17.3.

FIGURE 17.3
Split-plot design

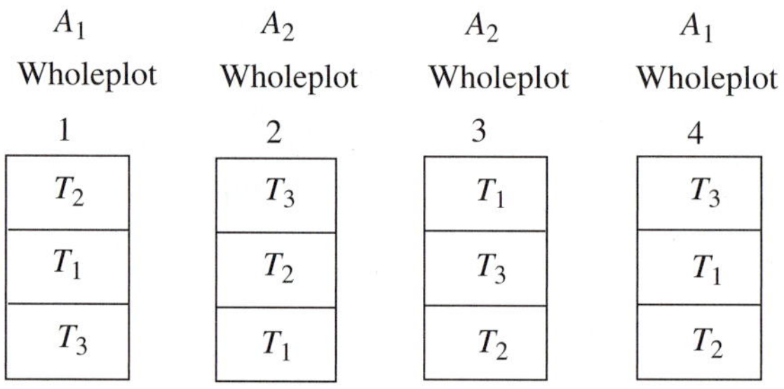

A_1	A_2	A_2	A_1
Wholeplot	Wholeplot	Wholeplot	Wholeplot
1	2	3	4
T_2	T_3	T_1	T_3
T_1	T_2	T_3	T_1
T_3	T_1	T_2	T_2

This design is called a split-plot design, and with this design there is a two-stage randomization. First, levels of factor A (fertilizers) are randomly assigned to the wholeplots; second, the levels of factor T (soybeans) are randomly assigned to the subplots within a wholeplot (see Figure 17.4). Using this design, it would be much easier to prepare the soil and to apply the appropriate fertilizer to the larger wholeplots and then to plant varieties of soybeans in the subplots, rather than to prepare the soil and to apply fertilizer to the subplots and then to plant soybeans in the subplots, as would be the case for a standard 2×3 factorial experiment.

FIGURE 17.4

Two-stage randomization for a completely randomized split-plot design

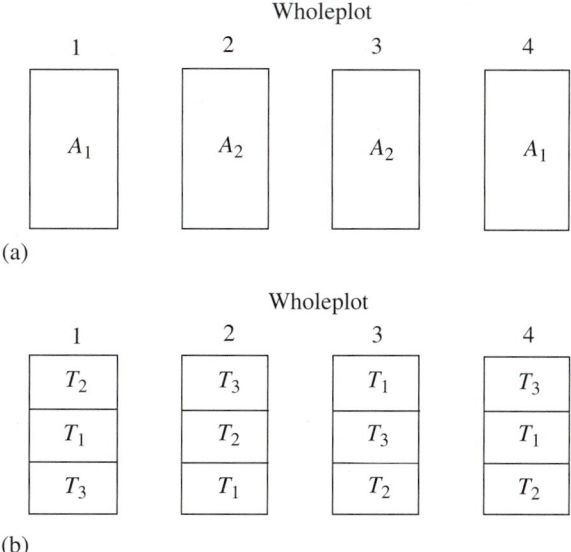

Because the randomization at the wholeplot level and at the subplot level is according to a completely randomized design, the design is often referred to as a completely randomized split-plot design.

Consider the model for the completely randomized split-plot design with a levels of factor A, t levels of factor T, and n repetitions of the ith level of factor A. If y_{ijk} denotes the kth response for the ith level of factor A, jth level of factor T, then

$$y_{ijk} = \mu + \alpha_i + \tau_j + \alpha\tau_{ij} + \delta_{ik} + \varepsilon_{ijk}$$

where

α_i: Fixed effect for ith level of A

τ_j: Fixed effect for jth level of T

$\alpha\tau_{ij}$:   Fixed effect for ith level of A, jth level of T

δ_{ik}:   Random effect for the kth wholeplot receiving the ith level of A. The δ_{ik} are independent normal with mean 0 and variance σ_δ^2.

ε_{ijk}:   Random error. The ε_{ijk} are independent normal with mean 0 and variance σ_ε^2.

The δ_{ik} and ε_{ijk} are mutually independent.

The AOV for this model and design is shown in Table 17.31.

You could compute the sums of square for the AOV using our standard formulas, but we suggest going to computer output to get them. It follows from the expected mean square that we have the following analyses:

TABLE 17.31
AOV for a completely randomized split-plot design

Source	SS	df	EMS
Between wholeplots			
A	SSA	$a - 1$	$\sigma_\varepsilon^2 + t\sigma_\delta^2 + tn\theta_A$
Wholeplot Error	SS(A)	$a(n - 1)$	$\sigma_\varepsilon^2 + t\sigma_\delta^2$
Within wholeplots			
T	SST	$t - 1$	$\sigma_\varepsilon^2 + an\theta_T$
AT	SSAT	$(a - 1)(t - 1)$	$\sigma_\varepsilon^2 + n\theta_{AT}$
Subplot Error	SSE	$a(n - 1)(t - 1)$	σ_ε^2
Total	TSS	$atn - 1$	

Wholeplot Analysis

$$H_0: \quad \theta_A = 0 \text{ (or, equivalently, } H_0: \text{All } \alpha_i = 0), \quad F = \frac{\text{MSA}}{\text{MS(A)}}$$

Subplot Analysis

$$H_0: \quad \theta_{AT} = 0 \text{ (or, equivalently, } H_0: \text{All } \alpha\tau_{ij} = 0), \quad F = \frac{\text{MSAT}}{\text{MSE}}$$

$$H_0: \quad \theta_T = 0 \text{ (or, equivalently, } H_0: \text{All } \tau_k = 0), \quad F = \frac{\text{MST}}{\text{MSE}}$$

A variation on this design introduces a *blocking factor* (such as farms). Thus for our example, there may be $b = 2$ farms with $a = 2$ wholeplots per farm and $t = 3$ subplots per wholeplot. This design is shown in Figure 17.5. Because the randomization to the wholeplots is done according to a randomized block design and the randomization to the subplot units within a wholeplot occurs according to a completely randomized design, the design is often referred to as a randomized block split-plot design.

FIGURE 17.5
Randomized block split-plot design

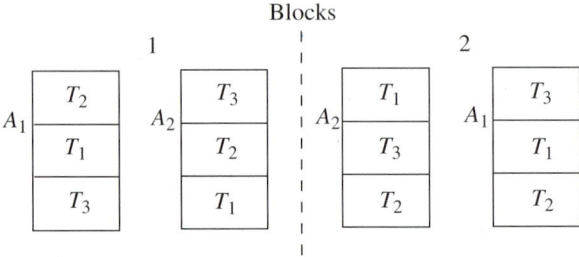

The model for this more general two-factor split-plot design laid off in b blocks is as follows:

$$y_{ijk} = \mu + \alpha_i + \beta_j + \alpha\beta_{ij} + \tau_k + \alpha\tau_{ik} + \varepsilon_{ijk}$$

where y_{ijk} denotes the measurement receiving the ith level of factor A and the kth level of factor T in the jth block. The parameters α_i, τ_k, and $\alpha\tau_{ik}$ are the usual main effects and interaction parameters for a two-factor experiment, whereas β_j is the effect due to block j and $\alpha\beta_{ij}$ is the interaction between the ith level of factor A and the jth block. The analysis corresponding to this model is shown in Table 17.32. Here we assume factors A and T are fixed effects, whereas blocks are random.

TABLE 17.32
AOV for a randomized block split-plot design (A, τ fixed; blocks random)

Source	SS	df	EMS
Between wholeplots			
Blocks	SSB	$b - 1$	$\sigma_\varepsilon^2 + at\sigma_\beta^2$
A	SSA	$a - 1$	$\sigma_\varepsilon^2 + t\sigma_{\alpha\beta}^2 + bt\theta_A$
AB (wholeplot error)	SSAB	$(a-1)(b-1)$	$\sigma_\varepsilon^2 + t\sigma_{\alpha\beta}^2$
Within wholeplots			
T	SST	$(t-1)$	$\sigma_\varepsilon^2 + ab\theta_T$
AT	SSAT	$(a-1)(t-1)$	$\sigma_\varepsilon^2 + b\theta_{AT}$
Subplot Error	SSE	$a(b-1)(t-1)$	σ_ε^2
Totals	TSS	$abt - 1$	

The sums of squares for the sources of variability listed in Table 17.32 can be obtained using the general formulas for main effects and interactions in a factorial experiment or from appropriate software packages. Using these expected mean squares, we can obtain a valid F test for factor A in the wholeplot portion of the analysis and for factor T and the AT interaction in the subplot portion. These are shown here. Note that no test is made for the variability due to blocks.

Wholeplot Analysis

$$H_0: \quad \theta_A = 0 \text{ (or, equivalently, } H_0\text{: all } \alpha_i = 0), \quad F = \frac{\text{MSA}}{\text{MSAB}}$$

Subplot Analysis

$$H_0: \quad \theta_{AT} = 0 \text{ (or, equivalently, } H_0\text{: all } \alpha\tau_{ik} = 0), \quad F = \frac{\text{MSAT}}{\text{MSE}}$$

$$H_0: \quad \theta_T = 0 \text{ (or, equivalently, } H_0\text{: all } \tau_k = 0), \quad F = \frac{\text{MST}}{\text{MSE}}$$

EXAMPLE 17.11

Soybean yields (in bushels per subplot unit) are shown here for a two-factor split-plot design laid off in $b = 3$ blocks. Fertilizers (factor A) were applied at random to the wholeplot units within each farm. Soybean varieties (factor T) were then randomly allocated to the subplots within each wholeplot. Conduct an analysis of variance using these sample data. Give an approximate p-value for each test.

	1			2			3	
	Fertilizers			**Fertilizers**			**Fertilizers**	
Varieties	**1**	**2**	**Varieties**	**2**	**1**	**Varieties**	**1**	**2**
1	10.6	10.9	2	11.9	11.5	3	9.5	9.8
2	11.4	11.7	3	12.6	12.1	1	8.1	8.2
3	11.8	12.4	1	11.6	10.8	2	8.7	9.3

Solution   For these data with $a = 2$, $b = 3$, $t = 3$, and $n = 1$, the sum of squares are as shown (see the Type III SS column in the following computer output):

$$\text{SSA} = 0.845 \qquad \text{SSAT} = 0.00333$$
$$\text{SSB} = 28.863 \qquad \text{SSE} = 0.227$$
$$\text{SSAB} = 0.0433 \qquad \text{TSS} = 35.325$$
$$\text{SST} = 5.343$$

```
SPLIT-PLOT DESIGN: WHOLE PLOT TRT-FERT AND SPLIT-PLOT TRT-VARIETY

General Linear Models Procedure

Dependent Variable: Y   YIELD
                            Sum of            Mean
Source              DF      Squares          Square    F Value    Pr > F
Model                9   35.0983333       3.8998148     137.64    0.0001
Error                8    0.2266667       0.0283333
Corrected Total     17   35.3250000

              R-Square            C.V.        Root MSE              Y Mean
              0.993583        1.570685         0.16833             10.7167

Source              DF    Type III SS     Mean Square    F Value    Pr > F
F                    1      0.8450000       0.8450000      29.82    0.0006
B                    2     28.8633333      14.4316667     509.35    0.0001
B*F                  2      0.0433333       0.0216667       0.76    0.4967
V                    2      5.3433333       2.6716667      94.29    0.0001
V*F                  2      0.0033333       0.0016667       0.06    0.9433

Dependent Variable: Y   YIELD

Tests of Hypotheses using the Type III MS for B*F as an error term

Source              DF    Type III SS     Mean Square    F Value    Pr > F
F                    1     0.84500000      0.84500000      39.00    0.0247
```

The analysis of variance table is shown here.

Source	SS	df	MS	F	p-value
Between Wholeplots					
Blocks	28.863	2	14.431	—	—
A	.845	1	0.845	39.00	.0247
AB (Wholeplot Error)	.0433	2	.0217	—	—
Within Wholeplots					
T	5.343	2	2.672	94.29	.0001
AT	.00333	2	.00167	.060	.9433
Subplot Error	.227	8	.0283	—	—
Total	35.325	17			

The distinction between this two-factor split-plot design and the standard two-factor experiments discussed in Chapter 15 lies in the randomization. In a split-plot design, there are two stages to the randomization process; first levels of factor A are randomized to the wholeplots within each block, and then levels of factor B are randomized to the subplot units within each wholeplot of every block. In contrast, for a two-factor experiment laid off in a randomized block design (see Section 15.3), the randomization is a one-step procedure; treatments (factor–level combinations of the two factors) are randomized to the experimental units in each block. The post-AOV analysis involving mean separations, contrasts, estimated treatment means, and confidence intervals are somewhat more complex for the split-plot design than for the designs that we have discussed previously. Excellent references for further reading on this topic are Kuehl (1999). Snedecor and Cochran (1980), and Lentner and Bishop (1993).

17.7 Summary

Fixed, random, and mixed models are easily distinguished if we think in terms of the general linear model. The fixed-effects model relates a response to $k \geq 1$ independent variables and one random component, whereas a random-effects model is a general linear model with $k = 0$ and more than one random component. The mixed model, a combination of the fixed- and the random-effects models, relates a response to $k \geq 1$ independent variables and more than one random component.

We illustrated the application of random-effects models to experimental situations for the completely randomized design and for the $a \times b$ factorial experiment laid off in a completely randomized design. We noted similarities between tests of significance in an analysis of variance for a random-effects model and for the corresponding fixed-effects model. Inferences resulting from an analysis of variance for a mixed model were illustrated using the $a \times b$ factorial experiment.

Unfortunately, in an introductory course, only a limited amount of time can be devoted to a discussion of random- and mixed-effects models. To expand our discussion in the text, the results of Section 17.5 are useful in developing the expected mean squares for sources of variability in the analysis of variance table for balanced designs. Using these expectations we can then attempt to construct appropriate test statistics for evaluating the significance of any of the fixed or random effects in the model.

The hardest part in any of these problems involving random- or mixed-effects models arises from trying to estimate $E(y)$, with an appropriate confidence interval for a random-effects model and the average value of y at some level or combination of levels for fixed effects in a mixed model. We illustrated how to obtain an estimate of $E(y)$ for a random-effects model and how to construct an approximate confidence interval. The problem becomes even more complicated for mixed models.

The final topics covered in this chapter were nested designs and split-plot designs. A brief introduction showed several variations on the basic factorial experiments discussed in Chapter 15 and in earlier sections of this chapter. The designs presented are only a few of the more common designs possible when considering nested effects in a multifactor experimental setting. The interested reader should consult the references at the end of this book to pursue these topics in more detail; in particular, Kuehl (1999) is an excellent reference.

Supplementary Exercises

17.8 Distinguish between inferences related to θ_A (when factor A is fixed) and σ_α^2 (when factor A is random).

17.9 Consider a $2 \times 3 \times 4$ factorial experiment with $n = 5$ replications. Suppose that the levels of factor A are fixed but the levels of factors B and C are randomly selected from a population of levels.

    a. Write down a model for this experiment. Identify all terms in the model and state all conditions that are placed on the terms in the model.

    b. Construct a partial AOV including df and expected mean squares for all sources of variability.

    c. Provide the ratio of mean squares for all appropriate F tests for the significance of sources of variability.

17.10 Consider a $3 \times 4 \times 2$ factorial experiment with $n = 6$ replications. Suppose that the levels of factors A and B are fixed but the levels of factor C are randomly selected from a population of levels.

    a. Write down a model for this experiment. Identify all terms in the model and state all conditions that are placed on the terms in the model.

    b. Construct a partial AOV including df and expected mean squares for all sources of variability.

    c. Provide the ratio of mean squares for all appropriate F tests for the significance of sources of variability.

17.11 Consider a $3 \times 5 \times 5$ factorial experiment with $n = 4$ replications. Suppose that the levels of factor B are fixed but the levels of factors A and C are randomly selected from a population of levels.

    a. Write down a model for this experiment. Identify all terms in the model and state all conditions that are placed on the terms in the model.

    b. Construct a partial AOV including df and expected mean squares for all sources of variability.

    c. Provide the ratio of mean squares for all appropriate F tests for the significance of sources of variability.

Env. **17.12** Refer to Exercise 17.6. Suppose the four chemicals were randomly selected from the hundreds of different chemicals used to control fire ants. The researchers were interested in whether the effectiveness of a chemical to control fire ants varied across different environments.

    a. Write an appropriate model for this situation. Indicate how the conditions placed on the terms in the model differ from the conditions placed on the model used when the chemicals were the only chemicals of interest to the researchers.

    b. Construct the AOV table and test all relevant hypotheses.

    c. Compare the conclusions and inferences in this problem to those of Exercise 17.6.

17.13 Refer to Exercise 17.12.

    a. Which model and analysis seem to be more appropriate? Explain your answer.

    b. Under what circumstances would a fixed-effects model be appropriate?

    b. Indicate the test statistics for those sources of variability that can be tested using an exact F test.

Engin. **17.14** The civil engineering department at a university was awarded a large grant to study the campus traffic problems and to recommend alternative solutions. One small phase of the study involved obtaining daily counts on the number of cars crossing, but not making use of, the campus facilities. To do this, a team of volunteers was stationed at each entrance to monitor simultaneously the license number and the time of entrance or exit for each car passing through the checkpoint. By comparing lists for all checkpoints and allowing a reasonable time for cars to traverse the campus, the teams were able to determine the number of cars crossing but not using the campus facilities during the 8:00 A.M. to 5:00 P.M. time period. A random sample of 6 weeks throughout the academic year was used, with 2 midweek days selected for study in the weeks sampled. The traffic volume data appear next.

Week 1	Week 2	Week 3	Week 4	Week 5	Week 6
680	438	539	264	693	530
618	520	600	198	646	575

a. Write an appropriate linear statistical model. Identify all terms in the model.
b. Perform an analysis of variance, indicating expected mean squares. Use $\alpha = .05$.

17.15 Refer to Exercise 17.14. Estimate the average number of cars crossing but not using the campus facilities for a midweek day of a randomly selected week and give an approximate confidence interval. (*Hint:* Refer to Example 17.1.)

Med.   **17.16** A study was designed to evaluate the effectiveness of new treatments to reduce the systolic blood pressure of patients determined to have high blood pressure. Three drugs were selected for evaluation (D1, D2, D3). There are numerous nondrug treatments for reducing blood pressure, including various combinations of a controlled diet, exercise programs, biofeedback, and so on. The researchers randomly selected three nondrug treatments (ND1, ND2, ND3) for examination in the study. The age of the patient often may hinder the effectiveness of any treatment. Thus, patients with high blood pressure were divided into two age groups (A1, A2). A group of 54 patients was divided into the two age groups and then randomly assigned to a combination of one of the three drugs and one of the three nondrug treatments. After participating in the program for 2 months, the reduction in systolic blood pressure from the blood pressure readings at the beginning of the program was recorded for each patient. These values are given in the table.

	Age A1 Nondrug			Age A2 Nondrug		
	ND1	ND2	ND3	ND1	ND2	ND3
Drug	33	37	41	34	48	44
D1	34	38	42	33	46	46
	35	36	39	38	45	49
Drug	46	44	43	47	44	44
D2	45	48	44	49	48	46
	46	49	45	45	46	41
Drug	38	45	36	36	46	38
D3	34	45	37	39	47	36
	37	44	35	35	44	35

a. Write a model for this study. Identify all terms in your model and state all necessary conditions placed on the terms in the model.
b. Construct the AOV table for the study, including the expected mean squares.
c. Test the significance of all relevant sources of variation. Use $\alpha = .05$.
d. What conclusions do you draw about the difference in the effectiveness of the combinations of nondrug and drug treatments for high blood pressure?

17.17 Refer to Exercise 15.3. Suppose that we consider the five investigators as a random sample from a population of all possible investigators for the rocket propellant experiment.
a. Write an appropriate linear statistical model, identifying all terms and listing your assumptions.
b. Perform an analysis of variance. Include an expected mean squares column in the analysis of variance table.

17.18 Refer to Exercise 17.17. Indicate the differences in the hypothesis under test and differences in the conclusions drawn for the fixed and random effects.

17.19 Refer to Exercise 15.42. Suppose that the two laboratories were randomly selected from a population of laboratories for participation in the study, which also included time and temperature as possible sources of variability.
  a. Obtain the expected mean squares for all sources of variability.
  b. Test all relevant sources of variability for significance. Use $\alpha = .05$.
  c. Compare the results obtained here to the results obtained in Exercise 15.42.
  d. Does considering the laboratory effects to be random effects seem more relevant than considering them as fixed effects? Explain your answer.

17.20 Refer to Exercise 15.32. Suppose that the five pane designs were randomly selected from a population of pane designs for participation in the study.
  a. Obtain the expected mean squares for all sources of variability.
  b. Test all relevant sources of variability for significance. Use $\alpha = .05$.
  c. Compare the results obtained here to the results obtained in Exercise 15.32.
  d. Does considering the pane design effects to be random effects seem more relevant than considering them as fixed effects? Explain your answer.

17.21 Refer to the study described in Exercise 15.28.
  a. Considering the nine medications to be randomly selected from a population of possible medications, write a model for the study.
  b. Give the expected mean squares for all sources of variability.
  c. Indicate how your analysis and conclusions would change from those of Exercise 15.28.

Engin.　**17.22** The two most crucial factors that influence the strength of solders used in cementing computer chips into the mother board of the guidance system of an airplane are identified as the machine used to insert the solder and the operator of the machine. Four solder machines and three operators were randomly selected from the many machines and operators available at the company's plants. Each operator made two solders on each of the four machines. The resulting strength determinations of the solders are given here.

		Machine		
Operator	1	2	3	4
1	204	205	203	205
	205	210	204	203
2	205	205	206	209
	207	206	204	207
3	211	207	209	215
	209	210	214	212

  a. Write a model for this study. Include all terms and conditions placed on the terms in the model.
  b. Present the AOV table for this study and include the expected mean squares.
  c. What conclusions can you make about the effect of machine and operator on the variability in solder strength?

17.23 Refer to Exercise 17.22.
  a. Estimate the variance components in this study.
  b. Proportionally allocate the sources of variability with respect to the the total variability in solder strength.
  c. Place a 95% confidence interval on the average solder strength.

Env.   **17.24**   Core soil samples are taken in each of six locations within a territory being investigated for surface mining of bituminous coal. Each of the core samples is divided into four sub-samples for separate analyses of the sulfur content of the sample.
  a. Identify the design and give a model for this experimental setting.
  b. Give the sources of variability and degrees of freedom for an AOV.

17.25   The sample data for Exercise 17.24 are shown here. Run an AOV and draw conclusions. Use $\alpha = .05$.

	Analyses			
Location	1	2	3	4
1	15.2	16.8	17.5	16.2
2	13.1	13.8	12.6	12.9
3	17.5	17.1	16.7	16.5
4	18.3	18.4	18.6	17.9
5	12.8	13.6	14.2	14.0
6	13.5	13.9	13.6	14.1

Engin.   **17.26**   Tablet hardness is one comparative measure for different formulations of the same drug product; some combinations of ingredients (in addition to the active drug) in a formulation give rise to harder tablets than do other combinations. Suppose that three batches of a formulation are randomly selected for examination. Three different 1-kg samples of tablets are randomly selected from each batch and seven tablets are randomly selected for testing from each of the 1-kg samples. The hardness readings are given here.

	Batch 1			Batch 2			Batch 3		
Sample	1	2	3	1	2	3	1	2	3
	85	76	95	108	117	101	71	81	72
	94	87	98	100	106	108	85	70	68
	91	90	94	105	103	100	78	84	80
	98	91	96	109	109	99	68	83	72
	85	88	99	104	100	117	85	72	75
	96	94	100	102	104	109	67	81	79
	93	96	93	108	102	105	76	78	74

  a. Identify the design.
  b. Give an appropriate model with assumptions.
  c. Give the sources of variability and degrees of freedom for an AOV.
  d. Perform an analysis of variance and draw conclusions about the tablet hardness data for the formulation under study. Use $\alpha = .05$.

Psy.   **17.27**   An experimenter is designing an experiment in which she plans to compare nine different formulations of a meat product. One factor, F, is percent fat (10%, 15%, 20%) in the meat. The other factor, C, is cooking method (broil, bake, fry). She will prepare samples of each of the nine combinations and present them to tasters who will score the samples based on various criteria. Four tasters are available for the study. Each taster will taste nine samples. There are taster-to-taster differences, but the order in which the samples are tasted will not influence the taste scores. The samples will be prepared in the following manner so that the meat samples can be prepared and kept warm for the tasters. A portion of meat containing 15% fat will be divided into three equal portions. Each of the three methods of cooking will then be randomly assigned to one of the three portions. This procedure will be repeated for meat samples having 15% and 20% fat. The nine meat samples will then be tasted and scored by the taster. The whole process is repeated for the other three tasters. The taste scores (0 to 100) are given here.

	10% Fat			15% Fat			20% Fat		
	Broil	**Bake**	**Fry**	**Broil**	**Bake**	**Fry**	**Broil**	**Bake**	**Fry**
Taster 1	75	79	82	78	82	81	81	85	87
Taster 2	61	65	74	65	73	80	75	81	86
Taster 3	75	78	79	80	82	83	87	88	92
Taster 4	73	76	78	75	81	90	86	92	93

a. Identify the design.
b. Give an appropriate model with assumptions.
c. Give the sources of variability and degrees of freedom for an AOV.
d. Perform an analysis of variance and draw conclusions about the effect of fat percentage and method of cooking on the taste of the meat product. Use $\alpha = .05$. A computer output for this data set is given here.

```
General Linear Models Procedure

                    Class      Levels     Values
                    TASTER        4        1 2 3 4
                    FAT           3        10 15 20
                    METHOD        3        BA BR F

Dependent Variable: Y   SCORE
                              Sum of             Mean
Source               DF      Squares           Square      F Value     Pr > F
Model                17     931.666667        54.803922      5.97       0.0002
Error                18     165.333333         9.185185
Corrected Total      35    1097.000000

               R-Square           C.V.         Root MSE              Y Mean
               0.849286         3.780508        3.03071              80.1667
Source               DF     Type III SS    Mean Square      F Value     Pr > F
T                     3     291.888889      97.296296       10.59       0.0003
F                     2     146.000000      73.000000        7.95       0.0034
T*F                   6     451.777778      75.296296        8.20       0.0002
M                     2      32.166667      16.083333        1.75       0.2019
F*M                   4       9.833333       2.458333        0.27       0.8949

        Level of  Level of
        FAT       METHOD     N       MEAN          FAT  MEAN      METHOD MEAN
                                     SCORE              SCORE            SCORE
        10        BA         4       80.75         10   80.25     BAKE   81.75
        10        BR         4       78.75         15   79.75     BROIL  80.33
        10        F          4       81.25         20   84.25     FRY    82.17
        15        BA         4       80.25
        15        BR         4       79.00
        15        F          4       80.00
        20        BA         4       84.25
        20        BR         4       83.25
        20        F          4       85.25
```

CHAPTER 18

18.1 Introduction and Case Study

18.2 Single-Factor Experiments with Repeated Measures

18.3 Two-Factor Experiments with Repeated Measures on One of the Factors

18.4 Crossover Designs

18.5 Summary

Repeated Measures and Crossover Designs

18.1 Introduction and Case Study

In all of the experimental situations discussed so far in this text (except for the paired difference experiment), we have assumed that only one observation is taken on each experimental unit. For example, in an experiment to compare the effects of three different cardiovascular compounds on blood pressure, we could use a single-factor design where n_1 patients are assigned to compound 1, n_2 to compound 2, and n_3 to compound 3. Then the model would be

$$y_{ij} = \mu + \alpha_i + \varepsilon_{ij}$$

where α_i is the (fixed or random) effect due to compound i and ε_{ij} is the random effect associated with patient j treated with compound i. For this design, we would get one measurement (y_{ij}) for each patient.

The practicalities of many applied research settings make it mandatory from a cost and efficiency standpoint to obtain more than one observation per experimental unit. For example, in conducting clinical research, it is often difficult to find patients who have the condition to be studied *and* who are willing to participate in a clinical trial. Hence, it is important to obtain as much information as possible once a suitable number of patients have been located. In this chapter, we will consider several different experimental settings involving one or more factors and repeated measures.

In a **crossover designed experiment,** each subject receives all treatments. The individual subjects in the study are serving as blocks and hence decreasing the experimental error. This provides an increased precision of the treatment comparisons when compared to the design in which each subject receives a single treatment. In the **repeated measures designed experiment,** we obtain t different measurements corresponding to t different time points following administration of the assigned treatment. This experimental setting is shown in Table 18.1. In Table 18.1, y_{ijk} denotes the observation at the time k for the jth patient on compound i. Note that we are getting $t > 1$ observations per patient, rather than only 1.

The multiple observations over time on the same subject often yield a more efficient use of experimental resources than using a different subject for each observation time. Thus, fewer subjects are required, with a subsequent reduction in cost. Also, the estimation of time trends will be measured with a greater degree

TABLE 18.1

Repeated time points for each patient

Compound	Time Period			
	1	**2**	$\cdots$	**t**
1	y_{111}	y_{112}	$\cdots$	y_{11t}
	$\vdots$	$\vdots$		$\vdots$
	$y_{1n_1 1}$	$y_{1n_1 2}$	$\cdots$	$y_{1n_1 t}$
2	y_{211}	y_{212}	$\cdots$	y_{21t}
	$\vdots$	$\vdots$		$\vdots$
	$y_{2n_2 1}$	$y_{2n_2 2}$	$\cdots$	$y_{2n_2 t}$
3	y_{311}	y_{312}	$\cdots$	y_{31t}
	$\vdots$	$\vdots$		$\vdots$
	$y_{3n_3 1}$	$y_{3n_3 2}$	$\cdots$	$y_{3n_3 t}$

of precision. The methods of this chapter can be used to analyze data from both crossover studies and repeated measures studies. The application of both of these designs is broad based. Applications abound in the pharmaceutical industry and in the R & D and manufacturing operations of most industries. Medical researchers, ecological studies, and numerous other areas of research involve the evaluation of time trends and hence may find the repeated measures design useful. An extension of these designs may also be appropriate for studies in which the data have a spatial relationship in place of the time trend. Examples include the reclaimation of strip-mined coal fields, evaluation of the effects of an oil spill, and air pollution around an industrial facility. Studies involving spatially repeated measures are generally more complex to model than the time trends we will address in this chapter. Further reading on the modeling of spatial data can be found in Ripley (1976), Haining (1990), and Cressie (1993).

The following case study will illustrate the evaluation of time trends in a repeated measures design.

Case Study: Effects of Oil Spill on Plant Growth

We examined a small portion of this case study in Chapter 6. On January 7, 1992, an underground oil pipeline ruptured and caused the contamination of a marsh along the Chiltipin Creek in San Patricio County, Texas. The cleanup process consisted of burning the contaminated regions in the marsh. To evaluate the influence of the oil spill on the flora, the researchers designed a study of plant growth after the burning was designed. In an unpublished Texas A&M University dissertation, Newman (1997) describes the researchers' findings with respect to *Distichlis spicata,* a flora of particular importance to the area of the spill.

Designing the Data Collection   Two questions of importance to the researchers were as follows.

1. Did the oil site recover after the spill and burning?
2. How long did it take for the recovery?

To answer these questions, the researchers needed to have a baseline from which they could compare the *Distichlis spicata* density in the months after the burning of the site. The density of the flora depended on soil characteristics, slope of the land, environmental conditions, weather, and many other factors. The researchers selected a nearby section of land, designated the control site, which was not affected by the oil spill but had similar soil and environmental properties as the spill site. At both the oil spill site and and the control site, 20 tracts were randomly chosen. After a 9-month transition period, measurements were taken at approximately 3-month intervals for a total of eight time periods. During each time period, the number of *Distichlis spicata* within each of the 40 tracts was recorded.

The experimental design is a repeated measures design with two treatments, the oil spill and the control region, and eight measurements taken over time on each of the tracts over a 2-year period. To answer the researchers' questions, we will state them in terms of the *Distichlis spicata* counts. Thus, our research hypotheses are stated as follows.

Research Hypothesis 1 Was there a difference in the average density of *Distichlis spicata* between the oil spill tracts and the control tracts during the study period?

Research Hypothesis 2 Were there significant trends in average density of *Distichlis spicata* during the study period?

Research Hypothesis 3 Were the trends for the oil spill and control tracts different?

Managing the Data The data consisted of the number of *Distichlis spicata* plants found on each tract during the eight observation periods on both the control and the burned (oil spill) sites. There were a total of 320 data values. The researchers examined the data to determine whether any obvious errors were present in the field records. The researchers initially decided to eliminate tracts 1 and 20 in the control sites because there were so few plants found during the October 1992 measurement period. However, they kept both tracts because they did not want to bias the sampling plan. The data were then transferred to computer files and prepared for analysis following the steps outlined in Section 2.5. The data are given in Table 18.2.

Analyzing the Data The next step in the study is for the researchers to summarize and plot the data. The mean flora counts by treatment and date are given in Table 18.3.

The flora counts are plotted in Figure 18.1 using boxplots for each date and treatment. The boxplots reveal that the control plots have higher median flora counts than the oil spill plots. The control plots, however, are somewhat more variable than the oil spill plots. This may be due to the burning treatment used on the oil spill plots, which often results in more homogeneous tracts than the conditions that were present on the tracts prior to the burning. The extension of these observations to the population of tracts and not only the observed tracts in the study will require modeling of the data and testing of the relevant statistical research hypotheses. We will provide this analysis at the end of the chapter after introducing the methods of analyzing repeated measures designs.

TABLE 18.2

Number of *Distichlis spicata* under two treatments

Treatment	Tract	Oct. 92	Jul. 93	Oct. 93	Jan. 94	Apr. 94	Jul. 94	Oct. 94	Jan. 95
Burned	1	27	25	18	21	26	22	20	27
	2	5	15	10	12	10	11	12	9
	3	17	26	26	25	15	10	14	17
	4	41	41	42	38	34	26	26	25
	5	25	28	22	27	24	16	18	23
	6	11	24	13	20	16	13	10	14
	7	37	40	33	31	32	30	25	31
	8	38	38	33	38	39	35	32	38
	9	31	33	25	30	28	21	17	19
	10	24	25	21	24	24	19	17	22
	11	22	27	31	30	32	30	25	34
	12	26	45	39	35	35	36	30	27
	13	32	38	34	45	41	28	31	31
	14	35	37	35	42	35	32	27	29
	15	26	23	19	18	21	13	11	19
	16	22	29	24	24	20	16	18	24
	17	50	54	56	60	51	52	49	52
	18	17	29	23	39	31	24	26	34
	19	25	37	29	32	28	14	13	24
	20	33	39	39	48	36	34	30	34

Treatment	Tract	Oct. 92	Jul. 93	Oct. 93	Jan. 94	Apr. 94	Jul. 94	Oct. 94	Jan. 95
Control	1	7	0	0	1	0	0	0	0
	2	57	46	49	51	48	43	40	40
	3	43	59	59	60	58	53	55	58
	4	43	53	52	53	53	53	52	54
	5	59	55	59	60	54	47	54	53
	6	42	48	50	48	43	37	38	38
	7	35	42	50	55	41	40	44	45
	8	40	51	53	57	53	38	43	36
	9	24	52	54	59	57	55	57	39
	10	42	49	50	54	51	44	39	41
	11	16	31	39	47	24	22	33	35
	12	54	58	60	60	54	51	48	51
	13	30	43	43	47	39	36	49	56
	14	47	50	60	60	54	52	57	57
	15	40	40	47	49	43	41	48	52
	16	11	23	27	31	17	19	24	29
	17	41	45	42	44	41	33	31	42
	18	50	52	55	53	45	42	35	51
	19	8	8	7	12	6	5	8	10
	20	0	0	0	1	0	0	0	0

TABLE 18.3

Flora count means by treatment and date

Treatment	Oct. 92	Jul. 93	Oct. 93	Jan. 94	Apr. 94	Jul. 94	Oct. 94	Jan. 95
Burned	27.20	32.65	28.60	31.95	28.90	24.10	22.55	26.65
Control	34.45	40.25	42.80	45.10	39.05	35.55	37.75	39.35

FIGURE 18.1

Boxplots of flora counts by treatment and date

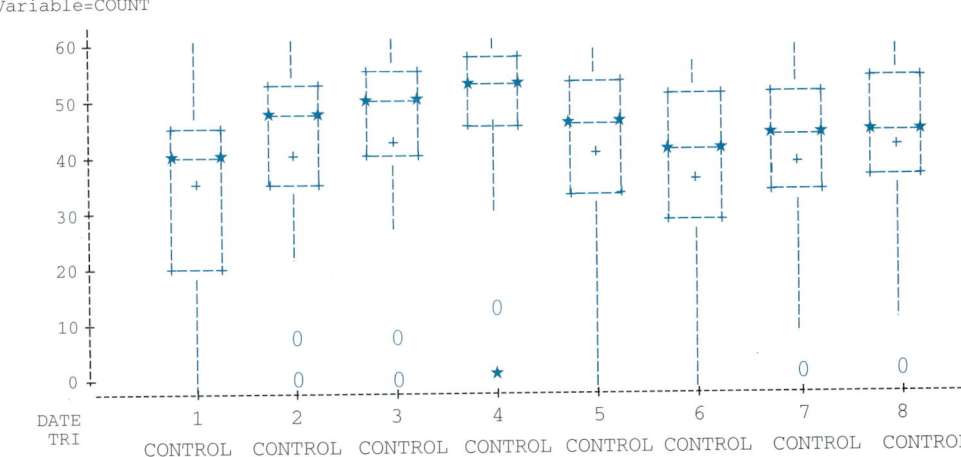

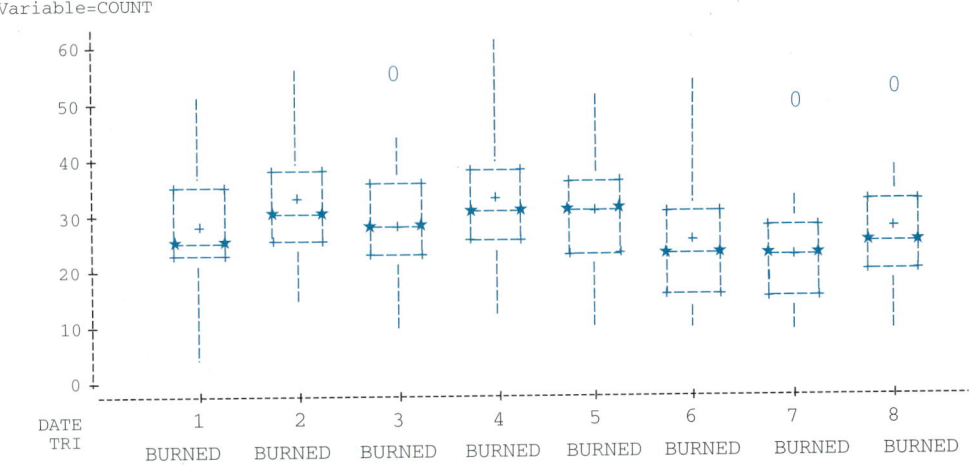

18.2 Single-Factor Experiments with Repeated Measures

In the previous section, we discussed some reasons why one might want to get more than one observation per patient. Another reason for obtaining more than one observation per patient is that frequently the variability *among* or *between* patients is much greater than the variability *within* a patient. We observed this in the paired *t*-test example of Section 6.5. If this is the case, it might be better to block on patients and to give each patient each treatment. Then the comparison among compounds is a within-patient comparison rather than a comparison between patients, as would be the case with the single-factor experiment with n_i different patients assigned to compound i. A single-factor design that reflects this within-patient emphasis is shown in Table 18.4.

TABLE 18.4

A within-patient comparison of compounds 1, 2, and 3

	Patient			
Compound	**1**	**2**	$\cdots$	n
1	y_{11}	y_{12}	$\cdots$	y_{1n}
2	y_{21}	y_{22}	$\cdots$	y_{2n}
3	y_{31}	y_{32}	$\cdots$	y_{3n}

With this design, the three compounds are administered in sequence to each of the n patients. A compound is administered to a patient during a given treatment period. After a sufficiently long "washout" period, another compound is given to the same patient. This procedure is repeated until the patient has been treated with all three compounds. The order in which the compounds are administered would be randomized. In this design, it is crucial that the washout period between treatments is sufficiently long that the results from one compound would not affect the results for another compound.

Here again, we are obtaining more than one observation per patient and presumably getting more useful information about the three drug products in question. One model for this experimental setting is

$$y_{ij} = \mu + \alpha_i + \pi_j + \varepsilon_{ij}$$

where μ is the overall mean response, α_i is the effect of the ith compound, π_j is the effect of the jth patient, and ε_{ij} is the experimental error for the jth patient receiving the ith compound.

Note that this model looks like any other single-factor experimental setting with a compounds and n patients. However, the assumptions are different because we are obtaining more than one observation per patient. For this model, we make the following assumptions.

1. α_i is a constant and $\Sigma\, \alpha_i = 0$.
2. The π_j are independent and normally distributed $(0, \sigma_\pi^2)$.
3. The ε_{ij} are independent of the π_j.
4. The ε_{ij} are normally distributed $(0, \sigma_\varepsilon^2)$.
5. The ε_{ij}s have the following correlation relationship:
   ε_{ij} and $\varepsilon_{i'j}$ are correlated for $i \neq i'$.
   ε_{ij} and $\varepsilon_{i'j'}$ are independent for $j \neq j'$.

That is, two observations from the same patient are correlated but observations from different patients are independent. From these assumptions it can be shown that the variance of y_{ij} is $\sigma_\pi^2 + \sigma_\varepsilon^2$. A further assumption is that the covariance for any two observations from patient j, y_{ij} and $y_{i'j}$, is constant. These assumptions give rise to a variance–covariance matrix for the observations, which exhibits *compound symmetry*. The discussion of correlated observations is beyond the scope of this book and we refer the interested reader to Kuehl (1999) and Vonesh and Chinchilli (1997).

The analysis of variance for the experimental design being discussed and this set of assumptions is shown in Table 18.5. This AOV should be familiar.

TABLE 18.5
AOV for the experimental
setting depicted in
Table 18.4

Source	SS	df	EMS (A fixed, patients random)
Between patients	SSP	$n - 1$	$\sigma_\varepsilon^2 + a\sigma_\pi^2$
Within patients			
A	SSA	$a - 1$	$\sigma_\varepsilon^2 + n\theta_A$
Error	SSE	$(a - 1)(n - 1)$	σ_ε^2
Totals	TSS	$an - 1$	

When the assumptions hold, and hence when compound symmetry holds, the statistical test on factor A ($F = MSA/MSE$) is appropriate. However, there are some other more general conditions that also lead to a valid F test for factor A using $F = MSA/MSE$. How restrictive are these assumptions and how can we tell when the test is appropriate?

There are no easy answers to these questions because there are no simple tests to check for compound symmetry. The general conditions (called the Huynh–Feldt conditions) under which the F test for factor A is valid are often not met because observations on the same patient taken closely in time are more highly correlated than are observations taken farther apart in time. So be careful about this. In general, when the variance–covariance matrix does not follow a pattern of compound symmetry, the F test for factor A has a positive bias, which allows rejection of H_0: all $\alpha_i = 0$ more often than is indicated by the critical F-values.

From a practical standpoint, the best thing to do in a given experimental setting is to make certain that there is sufficient time between applications of the treatment to allow washout (or elimination) of the previous treatment and to make certain that the design is applied in only those situations where the disease is relatively stable, so that following treatment and washout, each patient (or experimental unit) is essentially the same as prior to receiving treatment. For example, even when studying the effect of blood-pressure-lowering drugs, we would expect the hypertension to be stable enough that the patients would return to their predrug level blood pressures after washout of the first assigned compound before receiving the second assigned compound, and so on.

In Section 18.3, more will be said about how to judge whether the underlying assumptions for the test hold, and if they do not, how to proceed. For further information on this topic, refer to higher-level textbooks covering repeated measures experiments in detail [for example, Kuehl (1999) and Vonesh and Chinchilli (1997)].

18.3 Two-Factor Experiments with Repeated Measures on One of the Factors

We can extend our discussion of repeated measures experiments to two-factor settings. For example, in comparing the blood-pressure-lowering effects of cardiovascular compounds, we could randomize the patients so that n different patients receive each of the three compounds. However, rather than having each patient receive each compound, we could take multiple measurements across time for each patient. For example, we might be interested in obtaining blood pressure readings immediately prior to receiving a single dose of the assigned compound and then every 15 minutes for the first hour and hourly thereafter for the next 6 hours. The data for this type of setting are depicted in Table 18.1. Note that this is a two-factor experiment (compounds and time) with repeated measures taken over one of the two factors (time).

The model that we will use for a two-factor experiment comparing a levels of factor A (compounds), having n patients per level of factor A and b levels of factor B (time) is

$$y_{ijk} = \mu + \alpha_i + \pi_{j(i)} + \beta_k + \alpha\beta_{ik} + \varepsilon_{ijk}$$

where α_i, β_k, and $\alpha\beta_{ik}$ are fixed effects corresponding to main effects for factor A (compounds), factor B (times), and their interaction, respectively. The term $\pi_{j(i)}$ denotes a random effect due to the jth patient in the ith level of factor A. We assume that the $\pi_{j(i)}$ are independent and normally distributed $(0, \sigma_\pi^2)$.

Based on these assumptions, we have the analysis of variance shown in Table 18.6. Based on Table 18.6, it is clear that the following tests can be performed:

1. $H_0: \theta_{AB} = 0$

$$F = \frac{\text{MSAB}}{\text{MSE}}$$

2. $H_0: \theta_B = 0$

$$F = \frac{\text{MSB}}{\text{MSE}}$$

3. $H_0: \theta_A = 0$

$$F = \frac{\text{MSA}}{\text{MSP(A)}}$$

TABLE 18.6

Analysis of variance for a two-factor experiment, repeated measures on one factor

Source	SS	df	EMS (A, B fixed; patients random)
Between patients			
A	SSA	$a - 1$	$\sigma_\varepsilon^2 + b\sigma_\pi^2 + bn\theta_A$
Patients in A	SSP(A)	$a(n - 1)$	$\sigma_\varepsilon^2 + b\sigma_\pi^2$
Within patients			
B	SSB	$b - 1$	$\sigma_\varepsilon^2 + an\theta_B$
AB	SSAB	$(a - 1)(b - 1)$	$\sigma_\varepsilon^2 + n\theta_{AB}$
Error	SSE	$a(b - 1)(n - 1)$	σ_ε^2
Totals	TSS	$abn - 1$	

EXAMPLE 18.1

Ten subjects agreed to participate in a study to examine the concentration of drug in the bloodstream for two different dosage forms (capsule and tablet) of the same product following a single dose. Presumably, within limits, the higher the concentration, the more effective the drug product. Five subjects were allocated at random to the capsule form and the other five to the tablet form. Subjects fasted from 8:00 P.M. of the night prior to starting the study until 4 hours following ingestion of the assigned dose form (at 8:00 A.M. of the next day). Blood samples (15 mL) were obtained at .5, 1, 2, 3, and 4 hours after dosing, and were analyzed for the concentration of the drug product in the bloodstream. The data (in ng/mL) are shown here.

	Tablet Time						Capsule Time				
Subject	.5	1	2	3	4	Subject	.5	1	2	3	4
1	50	75	120	60	30	1	30	55	80	130	65
2	40	80	135	70	40	2	25	50	75	125	60
3	55	75	125	85	50	3	35	65	85	140	85
4	70	85	140	90	40	4	45	70	90	145	80
5	60	90	150	95	50	5	50	75	95	160	90

a. Plot the mean sample data (response versus time) for each of the dose forms. Do the dose forms seem to have different availability patterns across time?

b. Run a repeated measures analysis of variance.

Solution

a. The mean plasma concentrations by formulation and time are given here.

	Time				
Form	.5	1	2	3	4
C	37	63	85	140	76
T	55	81	134	80	42

A profile plot of the plasma concentrations is given here.

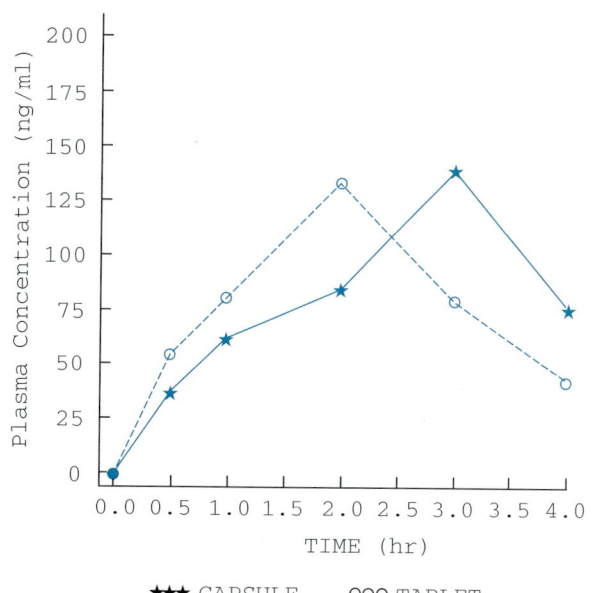

Plot of Plasma Concentration by Time

★★★ CAPSULE   ooo TABLET

b.

Source	SS	df	MS	F	p-value
Between patients					
Formulations	40.5	1	40.5	.083	.7810
Patients in formulation	3,920	8	490		
Within patients					
Time	34,288	4	8,572	279.90	.0001
Time × formulation	19,472	4	4,868	159.00	.0001
Error	980	32	30.625		

Conclusion: There is evidence (based on the significant time × formulation interaction) ($p < .0001$) that the two formulations have different availability (concentration) patterns across time.

    The F test for factor A is based on between-subject effects and hence is *not* affected by the repeated measures on factor B. However, the F-ratios for the within-patients effects are affected and, as with the one-factor experiment with repeated measures, we must worry about the conditions under which these F tests are appropriate. If compound symmetry of the variance–covariance matrix for the y_{ijk}s holds, then we can apply these tests; also if the Huynh–Feldt conditions alluded to previously hold, then we can apply these F tests. Some have suggested [Greenhouse and Geisser (1959); Huynh and Feldt (1970)] that "adjusted" F-values be used to determine the statistical significance of a repeated measures F test when there is some departure from the underlying conditions for that test. The adjustments recommended by the various authors follow the same pattern. A quantity epsilon is defined as a multiplicative adjustment factor for the numerator and denominator degrees of freedom for the F test in question. This epsilon (which we will denote by e) is not to be confused with the random error term ε in our models. For most of these adjustments, the multiplicative factor e ranges between 0 and 1, taking on a value of 1 when the underlying conditions for a valid F test are met and smaller values as the degree of departure from those conditions increases. A value of e having been determined for a given situation, the computed F statistic is compared to the critical value for an F distribution with numerator and denominator degrees of freedom multiplied by e.

    The ideas behind the adjustment can be seen if we use the experimental setting for Table 18.6 as the basis for discussion. Here we have a two-factor experiment with repeated measures on the second factor (B). The F tests for the within-patient effects, B and AB shown in Table 18.6, are valid provided the Huynh–Feldt conditions hold.

    For a given experiment, we compute a value of e and adjust the degrees of freedom for the F test by multiplying df_1 and df_2 by e. So, to run a test of H_0: $\theta_{AB} = 0$, a value of e is computed from the sample data and the computed F statistic

$$F = \frac{\text{MSAB}}{\text{MSE}}$$

is compared to a critical value of F_α based on $df_1 = e(a - 1)(b - 1)$ and $df_2 = ea(b - 1)(n - 1)$. Note that when $e = 1$, the underlying conditions hold and we have the original, recommended degrees of freedom, $df_1 = (a - 1)(b - 1)$ and $df_2 = a(b - 1)(n - 1)$.

In experimental situations where repeated measures data are to be analyzed and where you have access to SAS, you can use PROC GLM to compute revised p-values for two different adjustments to the degrees of freedom. The first adjustment, proposed by Greenhouse and Geisser (1959), uses a sample estimate of e. This adjustment, labeled "G-G" in the SAS output, has been shown, in simulation studies, to be ultraconservative, because the actual p-value may be much smaller than that indicated by the p-value using the G-G adjustment. The second adjustment factor [proposed by Huynh and Feldt (1970)] is based on a different formula for e. Once again, however, an estimate of this adjustment factor is computed from the sample data. The degrees of freedom for critical values of the F statistics are then adjusted using the estimate of e. This adjustment is labeled "H-F" in the PROC GLM output. Although the Greenhouse–Geisser e and Huynh–Feldt e both must be in the interval $0 < e \leq 1$, the H-F estimate of e can sometimes be greater than 1. In these situations, a value of $e = 1$ is used in determining the appropriate degrees of freedom for the F test.

```
General Linear Models Procedure

Dependent Variable: CONC

Source              DF    Sum of Squares    F Value    Pr > F
Model               17     57720.5000000    110.87     0.0001
Error               32       980.0000000
Corrected Total     49     58700.5000000


Source              DF       Type I SS    F Value    Pr > F
FORM                 1      40.5000000       1.32     0.2587
PATIENT(FORM)        8    3920.0000000      16.00     0.0001
TIME                 4   34288.0000000     279.90     0.0001
FORM*TIME            4   19472.0000000     158.96     0.0001

Source: FORM
Error: MS(PATIENT(FORM))
                   Denominator    Denominator
    DF   Type III MS       DF         MS    F Value  Pr > F
     1          40.5        8        490     0.0827  0.7810


Repeated Measures Analysis of Variance
Tests of Hypotheses for Between Subjects Effects

Source              DF      Type III SS    F Value    Pr > F
FORM                 1       40.5000000       0.08     0.7810
Error                8     3920.0000000


Repeated Measures Analysis of Variance
Univariate Tests of Hypotheses for Within Subject Effects

Source: TIME
                                                        Adj  Pr > F
    DF   Type III SS  Mean Square  F Value   Pr > F    G - G    H - F
     4   34288.00000   8572.00000   279.90   0.0001   0.0001   0.0001
```

```
Source: TIME*FORM
                                                            Adj  Pr > F
     DF  Type III SS  Mean Square   F Value   Pr > F    G - G      H - F
     4   19472.00000  4868.00000    158.96    0.0001    0.0001     0.0001


Source: Error(TIME)

     DF  Type III SS  Mean Square
     32    980.00000    30.62500


Greenhouse-Geisser Epsilon = 0.7374
       Huynh-Feldt Epsilon = 1.3610
```

EXAMPLE 18.2

Refer to the output for Example 18.1.

    a. Locate the estimated values for the Greenhouse–Geisser adjustment factor and the Huynh–Feldt adjustment factor.
    b. Are the conclusions for the tests on time effects and the time formulation interaction affected by these adjustments?

Solution

    a. The Greenhouse–Geisser estimate of e is .7374 and the Huynh–Feldt estimate of e is 1.3610.
    b. Time effects: F tests based on the G-G adjustment and on the H-F adjustment yield p-values of .0001 and .0001, respectively, the same as the original F tests; hence, the adjustments do not change the original conclusion.

Time × formulation interactions: As with the F test on time, the adjustments did not change the p-values or the conclusions drawn from the original tests.

In conclusion, if you have access to SAS when doing an analysis of variance for a repeated measures experiment, it would be wise to check the effects of adjustments to F tests on the factors affected by repeated measures. If the conclusions based on the original ($e = 1$) test differ from those based on the H-F or G-G adjustment, we recommend adhering to the conclusions based on the less conservative H-F adjustment.

Next we will complete the analysis of the data in the case study.

Analyzing Data for Case Study: Effects of Oil Spill on Plant Growth

The objective of the study was to examine the effects of the oil spill and subsequent burning of the tracts on which the oil spill occurred on the density of the flora, *Distichlis spicata*. Because baseline density of the flora prior to the oil spill and burning did not exist, a comparison will be made with tracts that were not involved in the oil spill. In Figure 18.2, a profile plot of the flora densities is displayed for the control and burned tracts across the eight observation dates. The mean densities for the control (C) tracts are consistently higher than the mean densities for the burned (B) tracts. The changes in mean densities have similar trends except on

FIGURE 18.2

Profile plot of flora densities by type and date

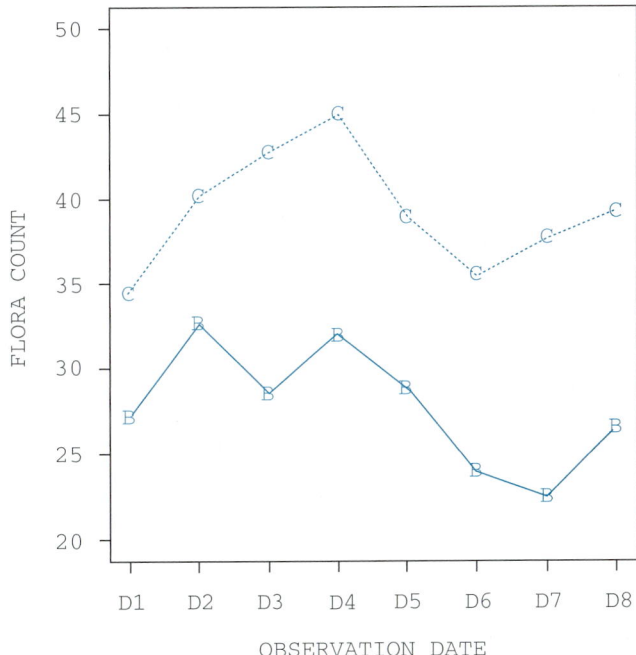

two of the observation dates (D2 and D7). On these two dates, the mean density of the flora on the burned tracts decreased from the previous date, whereas the mean densities for the control plot increased. We will next construct the repeated measures AOV to confirm these observations.

The analysis of variance of the flora density data is given here.

Source	SS	df	MS	F	p-value	Adjusted p-value G-G	Adjusted p-value H-F
Between tracts							
Treatment	10,511.11	1	10,511.11	6.56	0.0145		
Tracts in treatment	60,844.63	38	1,601.17				
Within tracts							
Date	2,845.09	7	406.44	19.35	0.0001	0.0001	0.0001
Date × treatment	602.29	7	86.04	4.10	0.0003	0.0046	0.0029
Error	5,587.88	266	21.01				

Greenhouse–Geisser Epsilon = 0.5269
Huynh–Feldt Epsilon = 0.5355

There is a highly significant date by treatment interaction that confirms the observations we had made from examining the profile plot. Furthermore, there is a significant difference between the mean densities of the burned and control plots. The control plots had a larger mean flora density than the burned plots; see Table 18.3. This difference was 7.25 at the first observation date and increased to a final difference of 12.70 on the final observation date, slightly more than 2 years later. Thus, the mean flora density for the tracts on which the oil spill occurred showed no recovery in flora density, 27.20 on October 1992 to 26.65 on January 1995. Because the flora density on the control tracts, which had similar soil conditions and environmental exposures during the study period, had an increase in flora density, 34.45 to 39.35, we would conclude that the oil spill and subsequent burning has resulted in reduced flora density on these tracts.

Reporting Conclusions   We need to include the following items in a report summarizing our findings of this study.

1. Statement of objectives of study
2. Description of study design, how the tracts were selected, how the measurements of flora density were made, how the dates were selected for observations, and justification of the similarity of tracts with respect to environmental and soil conditions
3. Discussion of how general the conclusions of this study are to other oil spill sites
4. Numerical and graphical representations of the data
5. Description of all inference methodologies:
   - Statement of research hypotheses
   - Model that represents experimental conditions
   - Verification of model conditions
   - AOV table, including *p*-values
6. Discussion of results and conclusions
7. Interpretation of findings relative to previous studies
8. Inferences relative to similar types of treatments for oil spills
9. Listing of data

EXERCISES   **Applications**

Env.

18.1   The cayenne tick is recognized as a pest of wildlife, livestock, and humans. It is distributed in the western hemisphere between 30°N and 30°S latitudes. This tick has been identified as a potential vector of several diseases, but the ecology of the cayenne tick is poorly understood. The following study was conducted to examine the survival potential of this tick as a function of the saturation deficit (SD) of the environment. Saturation deficit is an index of environmental conditions that combines both temperature and relative humidity with SD increasing with temperature but decreasing with relative humidity. Thus, high values of SD are associated with high temperatures and low relative humidities, conditions that cause ticks to experience maximum water loss. Five values were selected for SD (2.98, 4.83, 5.80, 8.88, and 13.38 mm of Hg) for use in the study. The conditions were established in an artificial environment, with five ticks randomly assigned to each of these conditions. The whole-body water loss of the ticks was recorded every 2 days over approximately a 3-week study period. The water losses (mg) of the ticks are given here.

		Days of Exposure										
SD	Tick	1	2	3	4	5	6	7	8	9	10	11
2.98	1	.54	.59	.64	.73	.76	.89	.93	1.01	1.08	1.15	1.23
	2	.69	.75	.81	.90	.97	1.20	1.14	1.19	1.26	1.38	1.43
	3	.77	.80	.87	.94	1.01	1.10	1.17	1.24	1.34	1.41	1.51
	4	.64	.69	.77	.83	.88	.96	1.04	1.09	1.20	1.23	1.31
	5	.51	.58	.62	.71	.74	.81	.88	.93	.99	1.03	1.13

(continues)

(*continued*)

SD	Tick	1	2	3	4	5	6	7	8	9	10	11
						Days of Exposure						
4.83	1	.64	.71	.77	.89	.90	1.00	1.06	1.14	1.22	1.34	1.39
	2	.80	.91	.97	1.01	1.11	1.19	1.29	1.31	1.37	1.47	1.54
	3	.79	.85	.89	.99	1.04	1.05	1.16	1.21	1.32	1.39	1.47
	4	.77	.82	.88	.92	1.01	1.09	1.19	1.27	1.35	1.44	1.58
	5	.79	.84	.91	.98	1.07	1.14	1.19	1.31	1.37	1.46	1.55
5.80	1	.72	.79	.83	.94	.98	1.09	1.12	1.21	1.28	1.34	1.41
	2	.89	.94	1.01	1.21	1.27	1.40	1.44	1.49	1.49	1.58	1.63
	3	.97	.99	1.07	1.09	1.21	1.30	1.37	1.44	1.54	1.61	1.73
	4	.85	.88	.97	1.05	1.09	1.17	1.24	1.29	1.30	1.23	1.51
	5	.71	.78	.82	.91	.94	1.11	1.19	1.23	1.29	1.33	1.43
8.88	1	.93	.99	1.03	1.14	1.18	1.29	1.33	1.36	1.38	1.54	1.62
	2	1.09	1.14	1.21	1.41	1.47	1.55	1.64	1.69	1.71	1.78	1.83
	3	1.19	1.20	1.07	1.29	1.31	1.50	1.57	1.64	1.74	1.81	1.93
	4	1.05	1.08	1.17	1.25	1.29	1.37	1.44	1.49	1.50	1.53	1.71
	5	1.01	1.09	1.18	1.21	1.29	1.31	1.39	1.43	1.49	1.53	1.63
13.38	1	1.05	1.09	1.13	1.24	1.28	1.39	1.43	1.56	1.68	1.74	1.82
	2	1.29	1.34	1.41	1.51	1.57	1.65	1.74	1.79	1.83	1.88	1.93
	3	1.38	1.40	1.47	1.49	1.51	1.60	1.69	1.74	1.79	1.87	2.03
	4	1.23	1.28	1.37	1.45	1.49	1.57	1.64	1.69	1.70	1.73	1.81
	5	1.23	1.29	1.38	1.41	1.49	1.52	1.48	1.53	1.59	1.63	1.78

    a. Display the profile plot for these data showing mean whole-body weight loss by time period for each value of SD.
    b. Does an increase in saturation deficit appear to increase the whole-body weight loss for the cayenne tick?

18.2 Refer to the data in Exercise 18.1.
    a. Provide a model for this design.
    b. Construct an AOV table for the study.
    c. Does an increase in saturation deficit appear to increase the whole-body weight loss for the cayenne tick? Use $\alpha = .05$.
    d. Is the increase in whole-body weight loss for the cayenne tick over the study the same for all levels of SD?

Med.   **18.3** An antihistamine is frequently studied using a model to examine its effectiveness (compared to a placebo) in inhibiting a positive skin reaction to a known allergen. Consider the following situation. Individuals are screened to find 20 subjects who demonstrate sensitivity to the allergen to be used in the study. The 20 subjects are then randomly assigned to one of two treatment groups (the known antihistamine and an identical-appearing placebo), with 10 subjects per group. At the start of the study, a baseline (predrug) sensitivity reading is obtained, and then each patient begins taking the assigned medication for 3 days. Skin sensitivity readings are taken at 1, 2, 3, 4, and 8 hours following the first dose. The percentage inhibition of skin sensitivity reaction (reduction in swelling area where the allergen is applied, compared to baseline) is shown here for each of the 20 patients.

		Percentage Inhibition Time (hours)				
Treatment	Patient	1	2	3	4	8
1	1	10.5	28.2	15.3	43.0	29.0
	2	41.2	25.3	27.8	28.0	53.2
	3	43.0	20.8	29.3	5.2	26.5
	4	61.4	61.6	62.8	43.8	19.6
	5	5.0	28.2	31.6	19.5	2.3
	6	−10.2	27.2	38.1	35.5	18.0
	7	−12.9	22.1	34.0	43.4	34.2
	8	27.1	26.5	38.8	28.5	17.4
	9	13.0	19.7	23.5	29.4	39.6
	10	28.9	26.1	11.2	18.1	16.5
2	1	3.0	9.3	1.0	15.0	3.0
	2	−1.5	−10.1	20.2	18.3	13.5
	3	10.8	20.6	28.3	25.2	15.8
	4	15.3	19.8	25.4	31.3	21.7
	5	8.7	8.0	17.5	26.6	16.4
	6	−4.6	5.8	12.7	15.6	29.6
	7	−16.6	28.4	32.7	34.4	15.8
	8	9.4	15.7	22.7	29.8	23.2
	9	−19.3	15.7	21.7	30.4	26.1
	10	−12.8	12.3	0.1	21.3	10.6

(A negative value means there was an increase in swelling, compared to baseline.)

    a. Compare means and standard deviations by time period for each group.

    b. Plot these data showing mean percentage inhibition by time for each treatment group. Does the antihistamine group appear to differ from the placebo group?

18.4   Refer to the data from Exercise 18.3. Give a model for this design and run a repeated measures analysis of variance to compare the two treatment groups. Do the analysis of variance results agree with your intuition based on the plot of Exercise 18.3?

18.5   Another question that may be asked relates to the onset of antihistaminic activity. How might you define onset? For each of the treatment groups, use a t test to determine the test time at which there is a significant reduction from baseline. What do these results suggest?

18.4   Crossover Designs

We will now consider an extension to the single-factor experiment discussed in Section 18.2. Recall that in Table 18.4 we presented data for an experimental situation where each of n patients received the same three compounds in a random order. A Latin square arrangement of the compounds is an experimental design that provides the same advantages as the single-factor experiment with repeated measures (namely, multiple observations per patient and a within-patient comparison of the treatments) while offering some protection that patients or conditions did not change with time.

    A 3×3 Latin square design for this experimental situation is shown in Table 18.7. The design itself is called a three-period crossover design.

    With this design, $3n$ patients are randomly assigned to the sequences (rows) of the design, n to each sequence. The periods correspond to the order in which

TABLE 18.7

A 3 × 3 Latin square design

Sequence	Patient	Factor B (periods) 1	2	3
1	n	A_1	A_2	A_3
2	n	A_2	A_3	A_1
3	n	A_3	A_1	A_2

the compounds are taken. The model for this design is

$$y_{ijkl} = \mu + \delta_k + \pi_{l(k)} + \alpha_i + \beta_j + \alpha\beta_{ij}^* + \varepsilon_{ijkl}$$

where δ_k is the fixed effect for the kth sequence and $\pi_{l(k)}$ is the random effect of the lth patient in sequence k; α_i and β_j are the fixed effects for compounds (factor A) and periods (factor B). The reason for the asterisk on the interaction term will be discussed later. The analysis of variance for this design (three-period crossover design) is shown in Table 18.8.

TABLE 18.8

Analysis of variance for a three-period crossover design

Source	SS	df	EMS (A, B fixed; patients random)
Between patients			
Sequence	SSSeq	2	$\sigma_\varepsilon^2 + 3\sigma_\pi^2 + 3n\theta_{Seq}$
Patients in sequences	SSP(Seq)	$3(n-1)$	$\sigma_\varepsilon^2 + 3\sigma_\pi^2$
Within patients			
A (compounds)	SSA	2	$\sigma_\varepsilon^2 + 3n\theta_A$
B (periods)	SSB	2	$\sigma_\varepsilon^2 + 3n\theta_B$
AB^*	SSAB*	2	$\sigma_\varepsilon^2 + n\theta_{AB}$
Error	SSE	$3(2)(n-1)$	σ_ε^2
Totals	TSS	$9n-1$	

The sums of squares, degrees of freedom, and expected mean squares for the between-patient effects for factors A and B in the within-patient portion of the analysis of variance are straightforward, but note that there is an asterisk on the AB interaction term and that this interaction term has only two, rather than four, degrees of freedom. Actually, the missing two degrees of freedom are in the sum of squares due to sequences. In fact, it can be shown that the AB interaction sums of squares is equal to

$$\text{SSAB} = \text{SSSeq} + \text{SSAB}^*$$

We will use this identity to compute SSAB*.

The test that we run in the analysis of variance will give us partial information about the AB interaction; actually, we are testing the within-patient portion of that interaction.

EXAMPLE 18.3

Twelve males volunteered to participate in a study to compare the durations of effect of three different formulations of a drug product. Formulation 1 was a 50-mg tablet, formulation 2 was a 100-mg tablet, and formulation 3 was a sustained-release formulation capsule. A three-period crossover design was used, with four volunteers assigned to each of the three treatment sequences. On each treatment day, volunteers were given their assigned formulation and were observed to determine the duration of effect (blood pressure lowering). There was a 1-week washout between each treatment period of the study. The sample data are shown here.

Sequence	Patient	Period 1	2	3
1	$n = 4$	A_1	A_2	A_3
2	$n = 4$	A_2	A_3	A_1
3	$n = 4$	A_3	A_1	A_2

Sequence	Patient	Period 1	2	3
1	1	1.5	2.2	3.4
	2	2.0	2.6	3.1
	3	1.6	2.7	3.2
	4	1.1	2.3	2.9
2	1	2.5	3.5	1.9
	2	2.8	3.1	1.5
	3	2.7	2.9	2.4
	4	2.4	2.6	2.3
3	1	3.3	1.9	2.7
	2	3.1	1.6	2.5
	3	3.6	2.3	2.2
	4	3.0	2.5	2.0

Solution   Based on the analysis of variance shown in the accompanying computer output, there is a hint of a period by treatment interaction ($p = .0853$), which appears negligible in the presence of a highly significant treatment effect ($p = .0001$). This is borne out in the plot of mean durations versus period for the three sequences shown here. As can be seen, the longest durations on the average were observed with formulation 3, followed by formulation 2 and then 1.

```
General Linear Models Procedure

Dependent Variable: DURATION

Source            DF    Sum of Squares    F Value    Pr > F
Model             17       11.08638889       5.69    0.0003
Error             18        2.06333333
Corrected Total   35       13.14972222

Source            DF         Type I SS    F Value    Pr > F
SEQ                2        0.23388889       1.02    0.3804
PATIENT(SEQ)       9        0.66916667       0.65    0.7425
TREAT              2        9.51722222      41.51    0.0001
PERIOD             2        0.01722222       0.08    0.9279
TREAT*PERIOD       2        0.64888889       2.83    0.0853

Tests of Hypotheses for Mixed Model Analysis of Variance

Dependent Variable: DURATION

Source: SEQ
Error: MS(PATIENT(SEQ))
                         Denominator      Denominator
   DF      Type I MS             DF               MS    F Value    Pr > F
    2    0.1169444444              9     0.0743518519     1.5729    0.2595
```

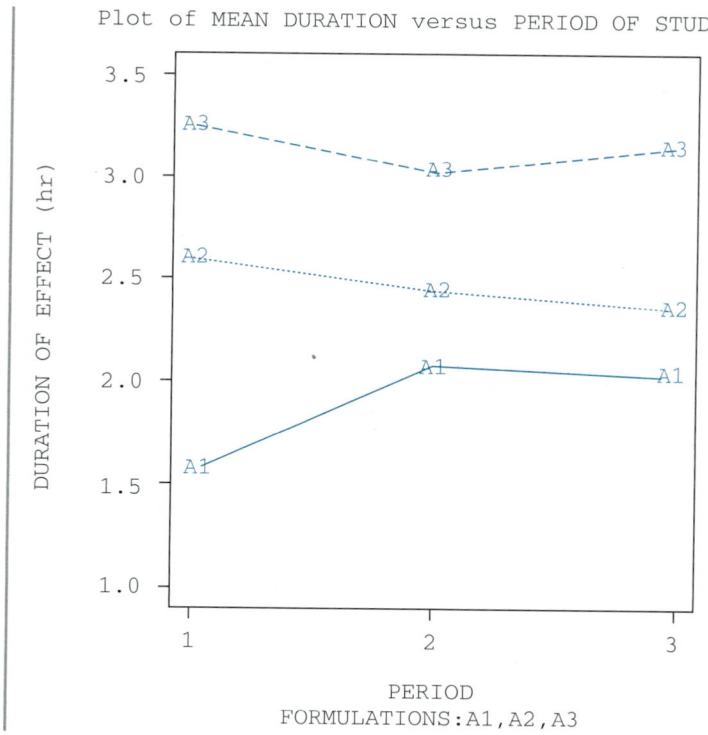

When there are only two compounds to be examined, the Latin square arrangement, called a two-period crossover design, would have $2n$ patients randomly assigned to the two sequences, n to each sequence. The two-period crossover design is shown in Table 18.9.

TABLE 18.9

Layout for a two-period crossover design

Sequence	Patient	Factor B (periods) 1	2
1	n	A_1	A_2
2	n	A_2	A_1

The corresponding analysis of variance for the model is

$$y_{ijkl} = \mu + \delta_k + \pi_{l(k)} + \alpha_i + \beta_j + \varepsilon_{ijkl}$$

where δ_k is the fixed effect due to sequence k, and α_i and β_j are the fixed effects due to treatment i and period j. As before, $\pi_{l(k)}$ represents the lth person in sequence k.

Note there is no AB interaction term in this model. We must assume this interaction is negligible; otherwise the design is inappropriate because there are no degrees of freedom available for testing the significance of the AB interaction. The AOV table for a two-period crossover design is shown in Table 18.10.

There are many other extensions to the repeated measures designs discussed in this chapter. For example, one could combine the concept of repeated measures on the same factor illustrated in Table 18.4 with the crossover design. Such a plan

TABLE 18.10

AOV table for a two-period crossover design

Source	SS	df	EMS (A, B fixed; patients random)
Between patients			
Sequences	SSSeq	1	$\sigma_\varepsilon^2 + 2\sigma_\pi^2 + 2n\theta_{Seq}$
Patients in sequences	SSP(Seq)	$2(n-1)$	$\sigma_\varepsilon^2 + 2\sigma_\pi^2$
Within patients			
A	SSA	1	$\sigma_\varepsilon^2 + 2n\theta_A$
B	SSB	1	$\sigma_\varepsilon^2 + 2n\theta_B$
Error	SSE	$2(n-1)$	σ_ε^2
Totals	TSS	$4n-1$	

is illustrated in Table 18.11. Thus, rather than taking one observation per patient within each period, we would take observations at t different time points. For example, we could measure blood pressure every 15 minutes for the first hour following treatment with compound i, and then hourly for the next 7 hours. This would be done in each of the periods for a total of 10 blood pressure measurements on each patient in each time period.

TABLE 18.11

Two-period crossover design with repeated measures

	Period	
	1	**2**
	Time	**Time**
Sequence	**1 2 ⋯ t**	**1 2 ⋯ t**
1	A_1	A_2
2	A_2	A_1

Although we will not give the analysis of variance for this extension to the repeated measures experiments discussed in this chapter, and will not cover other more complicated repeated measures designs, we want you to be aware of the wealth of possible designs that are available if you are willing to take more than one observation per experimental unit. The interested reader is referred to Vonesh and Chinchilli (1997); Crowder and Hand (1990); Jones and Kenward (1994); Diggle, Liang, and Zeger (1996).

18.5 Summary

In this chapter, we have discussed some of the initial concepts and designs associated with repeated measures experiments. We introduced single- and two-factor experiments, analyses for these experiments, and the special topics of two- and three-period crossover designs. These methods are only a beginning, however. Rather than presenting an exhaustive, detailed account of the subject, we have looked at these few situations to see the applicability and utility of some of the repeated measures designs and procedures. Facility in designing and analyzing such experiments can be gained only after more detailed coverage of repeated measures topics through additional reading and course work.

Supplementary Exercises

Psy.   **18.6**   An investigational drug product was studied under sleep laboratory conditions to determine its effect on duration of sleep. A group of 16 patients willing to participate in the study were randomly assigned to one of two drug sequences; 8 were to receive the investigational drug in period 1 and an identical-appearing placebo in period 2, and the remaining 8 patients were to receive the treatment in the reverse order.

    a. Identify the design.

    b. Give a model for this design.

    c. State the assumptions that might affect the appropriateness of this design.

18.7   Sleep duration data (in hours/night) are shown for the patients of Exercise 18.6.

		Period	
Sequence	Patient	1	2
1	1	8.6	8.0
	2	7.5	7.1
	3	8.3	7.4
	4	8.4	7.3
	5	6.4	6.4
	6	6.9	6.8
	7	6.5	6.1
	8	6.0	5.7
2	9	7.3	7.9
	10	7.5	7.6
	11	6.4	6.3
	12	6.8	7.5
	13	7.1	7.7
	14	8.2	8.6
	15	7.2	7.8
	16	6.7	6.9

Sequence 1 received the investigational drug first and placebo second; the reverse order applied to sequence 2.

    a. Compute means and standard errors per sequence, per period.

    b. Plot these data to show what happened during the study. Does the investigational drug appear to affect sleep duration? In what way? Use $\alpha = .05$.

    c. Run a repeated measures analysis of variance for this design. Draw conclusions. Does the analysis of variance confirm your impressions in part (b)?

18.8   Refer to Exercise 18.6. Suppose we ignore the order in which the patients received the treatments. Count the number of patients who had higher sleep duration on the investigational drug than on placebo.

    a. Suggest another simple test for assessing the effectiveness of the investigational drug.

    b. Give a p-value for the test of part (a).

18.9   Refer to Exercise 18.6. Suppose the sleep durations for period 2 of sequence 1 were as follows:

    8.5   7.6   8.5   8.3   7.2   7.0   6.4   6.1

    a. Plot the study data for both sequences.

    b. Does the design still seem to be appropriate? Is there a possible explanation for what happened?

18.10   Refer to Exercise 18.9. In spite of the results from period 2, we can still get a between-patient comparison of the treatment groups if we use the period 1 results only. Suggest an appropriate test, run the test, and give the p-value for your test. Draw a conclusion.

Med.

18.11   Many of us have been exposed to advertising related to the "bioavailability" of generic and brand-name formulations of the same drug product. One way to compare the bioavailability of two formulations of a drug product is to compare areas under the concentration curve (AUC) for subjects treated with both formulations. For example, the shaded area in the figure represents the AUC for a patient treated with a single dose of a drug.

AUC for a patient treated
with a single dose of drug,
Exercise 18.11

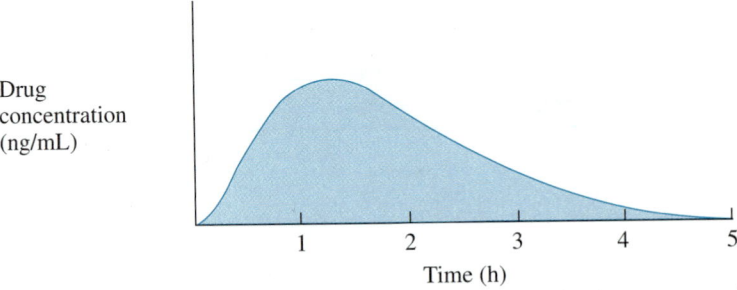

A three-period crossover design was used to compare the bioavailability of two brand-name (A_1, A_2) and one generic version (A_3) of weight-reducing agents. Three sequences of administering the drugs were used in the study:

    Sequence 1: A_1, A_2, A_3

    Sequence 2: A_2, A_3, A_1

    Sequence 3: A_3, A_1, A_2

A random sample of five subjects was assigned to each of the three sequences. The AUCs for these 15 patients are shown here.

			Period	
Sequence	**Patient**	**1**	**2**	**3**
1	1	80.2	40.4	38.4
	2	79.1	38.5	36.1
	3	108.4	78.3	56.5
	4	41.2	38.2	26.2
	5	72.7	58.5	36.3
2	1	74.6	51.2	48.6
	2	125.3	100.5	86.4
	3	145.5	108.5	96.4
	4	86.7	68.8	58.2
	5	107.8	78.5	53.1
3	1	79.7	40.4	37.2
	2	89.2	68.8	56.2
	3	99.1	76.5	43.9
	4	102.4	88.1	53.4
	5	109.3	98.5	76.8

a. Plot the formulation means (AUC) by period for each sequence.
b. Is there evidence of a period effect?
c. Do the formulations appear to differ relative to AUC?

18.12 Refer to Exercise 18.11. Run an analysis of variance for a three-period crossover design. Does your analysis confirm the intuition you expressed in Exercise 18.11? Use $\alpha = .05$.

18.13 Refer to Exercise 18.11. Compare the mean AUCs for the three formulations using *only* the period 1 data. Does this analysis confirm the analysis of Exercise 18.12? Why or why not might the analysis of Exercise 18.12 be more suitable than the "parallel" analysis of this exercise?

Med. **18.14** A study was conducted to demonstrate the effectiveness of an investigational drug product in reducing the number of epileptic seizures in patients who have not been helped by standard therapy. Thirty patients participated in the study, with 15 randomized to the drug treatment group and 15 to the placebo group. Patient demographic data are displayed here.

		Group	
		Investigational Drug ($n_1 = 15$)	**Placebo ($n_2 = 15$)**
Age (yr)	Mean ($\pm$SD)	37.2 ($\pm$10.5)	39.5 ($\pm$9.6)
	Range	19–68	21–65
Gender	M	20	16
	F	10	14
Duration of illness (yr)	Mean ($\pm$SD)	10.7 ($\pm$6.5)	11.5 ($\pm$7.3)
	Range	1–18	1–26

a. Do the groups appear to be comparable related to these demographic variables?
b. Are the mean ages or durations of illness different? How would you make this comparison?
c. How might you compare the sex distributions of the two groups?

18.15 The seizure data for the study of Exercise 18.14 are shown here. Note that we have baseline seizure rates as well as seizure rates for 5 months while on therapy.
a. Plot the mean seizure rates by month for the two groups. Does the investigational drug appear to work?
b. Run a repeated measures AOV and draw conclusions based on $\alpha = .01$.

			Time (months)				
Group	**Patient**	**Baseline**	**1**	**2**	**3**	**4**	**5**
Drug	1	15	11	10	6	5	3
	2	13	6	5	1	2	1
	3	12	8	3	0	3	0
	4	18	4	2	3	1	2
	5	30	15	14	10	8	20
	6	14	7	9	3	4	1
	7	25	12	18	13	10	6
	8	22	21	18	16	17	25

(continues)

(continued)

Group	Patient	Baseline	Time (months) 1	2	3	4	5
	9	23	17	14	10	7	1
	10	14	2	1	0	0	0
	11	15	4	5	6	3	2
	12	17	8	7	8	2	6
	13	26	13	10	9	7	4
	14	28	2	1	3	1	3
	15	29	27	29	25	24	22
Placebo	1	16	15	18	14	13	12
	2	18	14	13	12	10	15
	3	14	10	5	4	6	7
	4	19	15	16	9	12	15
	5	12	10	14	16	17	12
	6	11	13	8	7	6	11
	7	31	32	30	21	24	20
	8	32	35	34	31	20	24
	9	21	20	18	15	16	18
	10	26	22	23	21	15	14
	11	13	10	14	12	8	6
	12	17	15	10	3	2	3
	13	18	16	12	14	13	11
	14	23	15	14	18	19	20
	15	10	8	11	10	9	6

18.16   Refer to the data of Exercise 18.15.
  a. Consider the change in seizure rate from baseline to the 5-month reading. Compare the two groups using these data. Do you reach a similar conclusion?
  b. Because seizure rates can be quite variable, some people might compare the maximum change for patients in the two groups. Do these data support your previous conclusions?

Env.   **18.17**   Gasoline efficiency ratings were obtained on a random sample of 12 automobiles, 6 each of two different models. These ratings were taken at five different times for each of the 12 automobiles.
  a. Compute the mean efficiencies for each model at each time point, and plot these data.
  b. Draw conclusions from the analysis of variance. Use $\alpha = .05$.
  c. What effects, if any, do the correction factors have on the within-model comparisons in the analysis of variance shown here?

Model	Car	Time 1	Time 2	Time 3	Time 4	Time 5
1	1	1.43	1.47	1.39	1.40	1.44
1	2	1.50	1.41	1.51	1.53	1.41
1	3	1.79	1.88	1.89	2.00	1.90
1	4	1.87	1.78	2.00	2.00	2.11
1	5	1.85	1.89	1.93	1.86	1.81
1	6	1.89	1.66	1.78	1.77	1.67

(continues)

(*continued*)

Model	Car	Time 1	Time 2	Time 3	Time 4	Time 5
2	1	1.63	1.62	1.64	1.63	1.53
2	2	1.81	1.83	1.84	1.83	1.86
2	3	2.25	2.10	2.34	2.27	2.32
2	4	1.79	1.80	1.92	2.03	2.02
2	5	2.11	2.00	2.33	2.46	2.35
2	6	2.10	2.03	2.00	2.09	1.87

```
General Linear Models Procedure

Repeated Measures Analysis of Variance

Tests of Hypotheses for Between Subjects Effects

Source    DF    Type III SS    F Value    Pr > F

MODEL      1     0.95760667       3.38     0.0960
Error     10     2.83722667

General Linear Models Procedure

Repeated Measures Analysis of Variance

Tests of Hypotheses for Within Subject Effects

Source: TIME

                                                        Adj Pr > F
    DF    Type III SS    Mean Square   F Value   Pr > F   G - G     H - F
     4    0.09579333     0.02394833      3.03    0.0285   0.0719    0.0512

Source: TIME*MODEL

                                                        Adj Pr > F
    DF    Type III SS    Mean Square   F Value   Pr > F   G - G     H - F
     4    0.01182667     0.00295667      0.37    0.8260   0.6906    0.7528

Source: Error (TIME)

    DF    Type III SS    Mean Square
    40    0.31654000     0.00791350

Greenhouse-Geisser Epsilon = 0.4943
      Huynh-Feldt Epsilon = 0.6770
```

Plot of EFFICIENCY RATING versus TIME

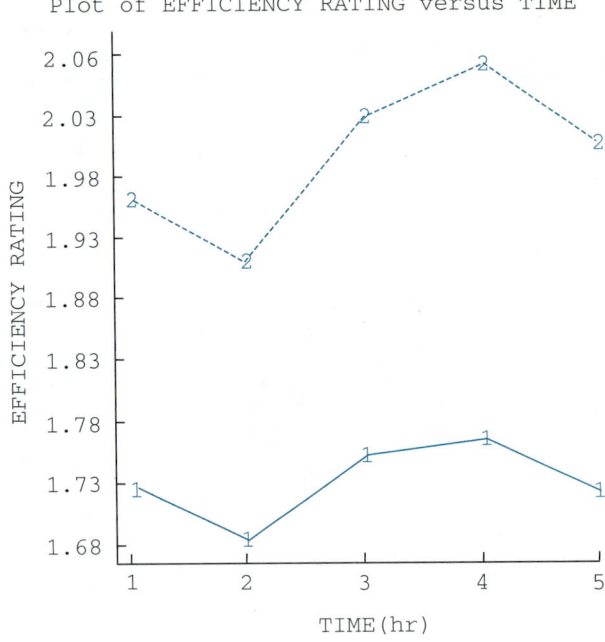

MODEL ~~111~~ 1   ~~222~~ 2

CHAPTER 19

Analysis of Variance for Some Unbalanced Designs

19.1 Introduction and Case Study

19.2 A Randomized Block Design with One or More Missing Observations

19.3 A Latin Square Design with Missing Data

19.4 Balanced Incomplete Block (BIB) Designs

19.5 Summary

19.1 Introduction and Case Study

We examined the analysis of variance for balanced designs in Chapters 8 and 15, where we used appropriate formulas (and corresponding computer solutions) to construct AOV tables and set up hypothesis tests. We also considered another way of performing an analysis of variance. We saw that the sum of squares associated with a source of variability in the analysis of variance table can be found as the drop in the sum of squares for error obtained from fitting reduced and complete models. Although we did not advocate the use of complete and reduced models for obtaining the sums of squares for sources of variability in balanced designs, we did indicate that the procedure was completely general and could be used for any experimental design. In particular, in this chapter, we will make use of complete and reduced models for obtaining the sums of squares in the analysis for *unbalanced designs,* where formulas are no longer readily available and easy to apply.

You might ask why an experimenter would run a study using an unbalanced design, especially since unbalanced designs seem to be more difficult to analyze. In point of fact, most studies do begin by using a balanced design, but for any one of many different reasons, the experimenter is unable to obtain the same number of observations per cell as dictated by the balanced design being employed. Consider a study of three different weight-reducing agents in which five different clinics (blocks) are employed and patients are to be randomly assigned to the three treatment groups according to a randomized block design. Even if the experimenter plans to have five overweight persons assigned to each treatment at each clinic, the final count will almost certainly show an imbalance of persons assigned to each treatment group. Almost every clinic could be expected to have a few people who would not complete the study. Some people might move from the community, others might drop out due to a lack of efficacy in the program, and so on. In addition, the experimenter might find it impossible to locate 15 overweight people at each clinic who are willing to participate in the study. Because an unbalanced design at the end of a study occurs quite often, we must learn how to analyze data arising from unbalanced designs.

We will next consider a case study in which we are aware of the unbalanced nature of the design prior to running the experiment and hence can design the study to partially accommodate the imbalance so as to minimize any bias with respect to estimating the treatment effects.

Case Study: Evaluation of the Consistency of Property Assessors

The county in which a large southwestern city is located received over the past year a large number of complaints concerning the assessed valuation of residential homes. Some of the county residents stated that there was wide variation in the valuation of residential property values depending on which county property assessor determined the property's value. The county employs hundreds of assessors who determine the value of residential property for the purposes of computing property taxes due from each property owner in the county. The county manager decided to design a study to see whether the assessors differ systematically in their determination of property values.

Designing the Data Collection   The manager needed to determine how to evaluate the consistency in the assessors' determinations of property values. Because the county assessor's office is generally understaffed and the assessors have a complete work schedule, it was decided to randomly select 16 assessors for participation in the study. To determine consistency, it would be necessary to have the assessors evaluate the same properties. However, there is a wide variety in the types of homes and extent of landscaping in the properties throughout the county. This variation in values and styles is thought to be one of the sources of deviations in the assessed valuations of the properties. Thus, the manager carefully selected 16 properties that would represent the wide diversity of properties in the county but all within the midpriced range of homes. Initially, the study was to have each of the 16 assessors determine a value for each of the 16 properties. This would require a total of 256 valuations to be done by the 16 assessors. However, this would be too time-consuming. Thus, each assessor was assigned to evaluate 6 of the 16 properties. The necessary number of valuations would be reduced from 256 to 96. The design is a randomized block design with the blocking variable being the 16 properties and the treatment variable being the 16 assessors. Note that the design is no longer a randomized complete block design because each assessor only valuated 6 of the 16 properties. The county statistician was concerned about the incomplete nature of the block design because some of the properties may be more difficult to evaluate than others. Although it would not be possible to have a complete block design, the statistician decided on the following method of assigning the properties to the assessors. We will demonstrate that the design is in fact a **balanced incomplete block design** when we provide the analysis of the case study in Section 19.4. The assessed valuations of the properties (in thousands of dollars) are provided here.

Property	\multicolumn{16}{c}{Assessor}

Property	1	2	3	4	5	6	7	8	9	10	11	12	13	14	15	16
1				125	120					112		115	118		110	
2				126		118		110		128	125			125		
3	110								125	118	138	110				126
4		131	150	157			125				150	156				
5	150	154	152		125			157		139						
6		138			118	110		120			124		129			
7	134				144	146	130					130		145		
8	157	159		150		134									120	158
9			156		155						150			138	124	156
10			156			128		155				153	155			122
11	155		158	157					142				123	155		
12		118					110			113			118	125		111
13			152			111	150		112	128					130	
14	115							112	110		135		130		128	
15		115						110	145			135		124	120	
16				157	120		150	135	120							132

Managing the Data The county manager's staff would next prepare the data for a statistical analysis following the procedures outlined in Section 2.5. The staff would need to verify that the assessors independently valuated each property without consulting with any other assessor. The data would then be checked for transmission errors and a computer file would be prepared for a statistical analysis of the data.

Analyzing the Data Because the design is not a complete block design—only 96 of the 256 possible block–treatment combinations were observed—we cannot use the models and analysis techniques from Chapter 15. The analysis of the case study will be provided at the end of Section 19.4.

19.2 A Randomized Block Design with One or More Missing Observations

unbalanced design

Any time the number of observations is not the same for all factor–level combinations, we call the design **unbalanced.** Thus, a randomized block design or a Latin square design with one or more missing observations is an unbalanced design. We will begin our examination by considering a simple case, a randomized block design with one missing observation.

The analysis of variance for a randomized block design with one missing observation can be performed rather easily by using the formulas for a randomized complete block design, after we have estimated the **value of the missing observation** and corrected for the **estimation bias.** The formula for estimating the missing observation M is given by

value of missing observation

estimation bias

$$\hat{M} = \frac{t y_{i\cdot} + b y_{\cdot j} - y_{\cdot\cdot}}{(t-1)(b-1)}$$

where t is the number of treatments, b is the number of blocks, $y_i.$ is the sum of all the observations on the treatment with the missing observation, $y_{.j}$ is the sum of all measurements in the block with the missing observation, and $y_{..}$ is the sum of all the measurements. After calculating the sum of squares for treatment, SST, using the formulas for a balanced design with the missing value replaced with its estimated value, we must correct SST for the bias in its estimation by subtracting,

$$\text{Bias} = \frac{\{y_{.j} - (t-1)\hat{M}\}^2}{t(t-1)}$$

That is, the corrected treatment sum of squares is $\text{SST}_C = \text{SST} - \text{Bias}$. The corrected mean squares treatment, $\text{MST}_C = \text{SST}_C/(t-1)$, has expected value $\sigma_\varepsilon^2 + b\theta_T$, the same as for a complete block design. An exact F can then be performed for the treatment effects.

We illustrate the analysis of variance for this design with an example.

EXAMPLE 19.1

An experiment was conducted to determine the nutritional value of diets for cows that are supplemented by whey. Five dairies were involved in the study. Each cow in a sample of four cows from a dairy was randomly assigned to one of the four treatment groups, so that a total of five cows were in each treatment group.

      Treatment 1:   water only

      Treatment 2:   whey plus 30.2 L of water/day

      Treatment 3:   whey plus 15.1 L of water/day

      Treatment 4:   whey only

In addition to the liquid portion of the diet listed for each treatment group, each cow was fed 7.5 kg of grain per day.

One response of interest was the amount of hay consumed per day. These data (in kilograms per animal) are listed in Table 19.1. Unfortunately, as can be seen from the data, the cow on diet 4 from dairy 2 was dropped from the study and no replacement was made. The cow developed an infection (unrelated to the treatment) and was dropped from the study for safety reasons.

TABLE 19.1

Consumption of hay for cows

Dairy	Treatment 1	2	3	4
1	15.4	9.6	9.5	8.4
2	14.8	9.3	9.4	M
3	15.9	9.8	9.7	9.3
4	15.5	9.4	9.2	8.1
5	14.7	9.2	9.0	7.9

Estimate the missing value and then perform an analysis of variance. Use $\alpha = .01$.

Solution   For this randomized block design with $b = 5$ and $t = 4$, the quantities $y_i.$, $y_{.j}$, and $y_{..}$ are defined as follows:

$y_{i\cdot}$ = sum of all observations on treatment 4
$$= 8.4 + 9.3 + 8.1 + 7.9 = 33.7$$

$y_{\cdot j}$ = sum of all observations in block 2
$$= 14.8 + 9.3 + 9.4 = 33.5$$

$y_{\cdot\cdot}$ = sum of all measurements
$$= 15.4 + 9.6 + \cdots + 7.9 = 204.1$$

The estimate of the missing value is

$$\hat{M} = \frac{ty_{i\cdot} + by_{\cdot j} - y_{\cdot\cdot}}{(t-1)(b-1)} = \frac{4(33.7) + 5(33.5) - 204.1}{3(4)}$$

$$= \frac{98.2}{12} = 8.183$$

Having estimated the missing value, we can compute sums of squares for our analysis of variance by using the formulas of Chapter 15. The treatment and block means are given by

$$\bar{y}_{1\cdot} = 15.26 \qquad \bar{y}_{\cdot1} = 10.725$$

$$\bar{y}_{2\cdot} = 9.46 \qquad \bar{y}_{\cdot2} = 10.425$$

$$\bar{y}_{3\cdot} = 9.36 \qquad \bar{y}_{\cdot3} = 11.175$$

$$\bar{y}_{4\cdot} = 8.377 \qquad \bar{y}_{\cdot4} = 10.55$$

$$\bar{y}_{\cdot5} = 10.20$$

$$\bar{y}_{\cdot\cdot} = 10.615$$

Note that the new means for treatment 4 and for block 2 incorporate the estimated missing observation. Similarly, the mean of all measurements includes the estimated missing value.

$$\text{TSS} = (15.4 - 10.615)^2 + (9.6 - 10.615)^2 + \cdots + (7.9 - 10.615)^2 = 150.21$$

$$\text{SSB} = 4\{(10.725 - 10.615)^2 + (10.425 - 10.615)^2 + (11.175 - 10.615)^2$$
$$+ (10.55 - 10.615)^2 + (10.2 - 10.615)^2\} = 2.16$$

$$\text{SST} = 5\{(15.26 - 10.615)^2 + (9.46 - 10.615)^2 + (9.36 - 10.615)^2$$

$$+ (8.377 - 10.615)^2\} = 147.48$$

$$\text{SSE} = 150.21 - 2.16 - 147.48 = 0.57$$

$$\text{Correction for bias} = \frac{\{33.7 - (4-1)8.183\}^2}{4(4-1)} = 6.98$$

$$\text{Corrected treatment SS} = 147.48 - 6.98 = 140.50$$

The only difference in the analysis of variance table for unbalanced and balanced randomized block designs is that since n refers to the number of actual observations, the error for an unbalanced design loses one degree of freedom for each missing observation when compared to the corresponding balanced design. The AOV table for our example is shown in Table 19.2.

The F tests for treatments and blocks are both significant, using $\alpha = .01$ (the critical values of F are 6.22 and 5.67, respectively). As can be seen from the data, the cows on treatment 1 (water only) consumed much more hay than cows on any of the diets supplemented with whey.

TABLE 19.2

AOV table for the data
of Example 19.1

Source	SS	df	MS	F	p-value
Diary	2.16	4	.54		
Treatment	140.50	3	46.83	904.0	.0001
Error	0.57	11	.0518		
Totals	150.21	18			

comparisons among treatment means

Having seen an analysis of variance, we may wish to make certain **comparisons among the treatment means.** We'll run pairwise comparisons using Fisher's least significant difference. The least significant difference between the treatment with a missing observation and any other treatment mean is

LSD

$$\text{LSD} = t_{\alpha/2} \sqrt{\text{MSE}\left(\frac{2}{b} + \frac{t}{b(b-1)(t-1)}\right)}$$

For any pair of treatments with no missing value, the least significant difference is as before; namely,

$$\text{LSD} = t_{\alpha/2} \sqrt{\frac{2\text{MSE}}{b}}$$

fitting complete and reduced models

The formulas for estimating missing observations in a randomized block design become more complicated with more missing data, as do the formulas for least significant differences. Because of this, we will consider **fitting complete and reduced models** to analyze unbalanced designs. We will illustrate the procedure first by examining an unbalanced randomized block design.

Because it would require more data input for a computer solution using the general linear model format with dummy variables presented in Chapter 12, we will represent the complete and reduced models for testing treatments as follows:

models

complete model (model 1): $y_{ij} = \mu + \beta_j + \alpha_i + \varepsilon_{ij}$

reduced model (model 2): $y_{ij} = \mu + \beta_j + \varepsilon_{ij}$

where β_j is the jth block effect and α_i is the ith treatment effect.

By fitting model 1 (using SAS or other computer software), we obtain SSE_1. Similarly, a fit of model 2 yields SSE_2. The difference in the two sums of squares for error, $\text{SSE}_2 - \text{SSE}_1$, gives the drop in the sum of squares due to treatments. Because this is an unbalanced design, the block effects do not cancel out when comparing treatment means as they do in a balanced randomized block design (see Chapter 15). The difference in the sums of squares, $\text{SSE}_2 - \text{SSE}_1$, has been adjusted for any effects due to blocks caused by the imbalance in the design. This difference is called the **sum of squares due to treatments adjusted for blocks.**

SST_adj

$$\text{SSE}_2 - \text{SSE}_1 = \text{SST}_{\text{adj}}$$

The sum of squares due to blocks **unadjusted for any treatment differences** is obtained by subtraction:

$$\text{SSB} = \text{TSS} - \text{SST}_{\text{adj}} - \text{SSE}$$

where SSE and TSS are sums of squares from the complete model. (*Note:* We could also obtain SSB, the uncorrected sum of squares for blocks, using the formula of Section 15.3).

AOV table, treatments

The **analysis of variance table for testing the effect of treatments** is shown in Table 19.3. In the table, n is the number of actual observations.

TABLE 19.3
AOV table for testing the effects of treatments, unbalanced randomized block design

Source	SS	df	MS	F
Blocks	SSB	$b - 1$	—	—
Treatments$_{adj}$	SST$_{adj}$	$t - 1$	MST$_{adj}$	MST$_{adj}$/MSE
Error	SSE	$n - b - t + 1$	MSE	
Totals	TSS	$n - 1$		

The corresponding sum of squares for testing the effect of blocks has the same complete model (model 1) as before, and

$$y_{ij} = \mu + \alpha_i + \varepsilon_{ij}$$

is the reduced model (model 2). The sum of squares drop, $SSE_2 - SSE_1$, is the **SSB$_{adj}$**   **sum of squares due to blocks after adjusting for the effects of treatments.** By subtraction, we obtain

$$SST = TSS - SSB_{adj} - SSE$$

AOV table, blocks   The **AOV table** is shown in Table 19.4.

TABLE 19.4
AOV table for testing effects of blocks, unbalanced randomized block design

Source	SS	df	MS	F
Blocks$_{adj}$	SSB$_{adj}$	$b - 1$	MSB$_{adj}$	MSB$_{adj}$/MSE
Treatments	SST	$t - 1$	—	—
Error	SSE	$n - t - b + 1$	MSE	—
Totals	TSS	$n - 1$		

Note that SST and SST$_{adj}$ are not the same quantity in an unbalanced design; they will be the same only for a balanced design. Similarly, SSB and SSB$_{adj}$ are different quantities in an unbalanced design. For an unbalanced design, we have the following identities:

$$TSS = SST_{adj} + SSB + SSE = SST + SSB_{adj} + SSE$$

but

$$TSS \neq SST_{adj} + SSB_{adj} + SSE$$

EXERCISES   **Basic Techniques**

19.1   Refer to the data of Example 19.1 and the SAS computer output shown here.

```
General Linear Models Procedure: FULL MODEL

Dependent Variable: CONS

Source              DF    Sum of Squares    F Value    Pr > F
Model                7      143.41548246     394.80    0.0001
Error               11        0.57083333
Corrected Total     18      143.98631579
```

Source	DF	Type III SS	F Value	Pr > F
DAIRY	4	2.11266667	10.18	0.0011
TREAT	3	140.80083333	904.41	0.0001

General Linear Models Procedure: REDUCED MODEL WITHOUT DIARY

Dependent Variable: CONS

Source	DF	Sum of Squares	F Value	Pr > F
Model	3	141.30281579	263.28	0.0001
Error	15	2.68350000		
Corrected Total	18	143.98631579		

Source	DF	Type III SS	F Value	Pr > F
TREAT	3	141.30281579	263.28	0.0001

General Linear Models Procedure: REDUCED MODEL WITHOUT TREAT

Dependent Variable: CONS

Source	DF	Sum of Squares	F Value	Pr > F
Model	4	2.61464912	0.06	0.9914
Error	14	141.37166667		
Corrected Total	18	143.98631579		

Source	DF	Type III SS	F Value	Pr > F
DAIRY	4	2.61464912	0.06	0.9914

a. Indicate the complete and reduced models for testing treatments.

b. Construct an analysis of variance table for testing treatments. Give the level of significance for your test and draw conclusions.

19.2   Refer to Example 19.1. Use the least significant difference criterion for identifying which treatments differ from the others. Use $\alpha = .05$.

19.3   Refer to Example 15.2. Suppose that the first observation in block 1 (plot 1) is missing. Analyze the data by estimating the missing value and then performing an analysis of variance. Use $\alpha = .05$.

19.4   Refer to Exercise 19.3. Perform the corresponding analysis by fitting complete and reduced models. Compare your conclusions to those in Exercise 19.3.

19.5   Refer to Exercise 19.1. Fit the reduced model $y_{ij} = \mu + \alpha_i + \varepsilon_{ij}$ to obtain SSE_2. The sum of squares drop will be the sum of squares due to blocks, adjusted for treatments. Verify that this computer value for SSB_{adj} is the same as that shown in the TYPE III SS column of the computer output in Exercise 19.1.

19.6   Refer to the data of Exercise 15.3. Suppose that in the rocket propellant test for the second mixture to be analyzed by investigator 3, a piece of equipment malfunctioned. Instead of going back to the laboratories to prepare a duplicate mixture, the investigators proceeded to obtain the remaining propellent thrust data.

a. Estimate the missing value.

b. Perform an analysis of variance, using $\alpha = .05$.

19.7   Refer to Exercise 19.6.

a. Use complete and reduced models to obtain an analysis of variance. Compare your results to those in Exercise 19.6.

b. How would you analyze the data if the response for mixture 4 and investigator 1 were also missing?

19.3 A Latin Square Design with Missing Data

Recall that a $t \times t$ Latin square design can be used to compare t treatment means while filtering out two additional sources of variability (rows and columns). The treatments are randomly assigned in such a way that each treatment appears in every row and in every column. In this section, we will illustrate the method for performing an analysis of variance in a Latin square design when one observation is missing. Then we will use the general method of fitting complete and reduced models with missing observations, described for the randomized block design in Section 19.2, for more complicated designs.

estimating missing value

The formula for **estimating a single missing value** in a Latin square design is

$$\hat{M} = \frac{t(y_{i.} + y_{.j} + y_{k}) - 2y_{..}}{(t-1)(t-2)}$$

where $y_{i.}$, $y_{.j}$, and y_{k} represent the row, column, and treatment totals, respectively, corresponding to the missing observation. $y_{..}$ *is the sum of all observations in the experiment and t is the number of treatments in the Latin square design.*

The total sum of squares and the sum of squares for rows, columns, treatments, and error are computed using the formulas from Chapter 15 for a complete Latin square design, with the missing observation replaced with $\hat{M}$. The mean squares for treatment is a biased estimator for the expected mean square treatment in a balanced Latin square. This bias is estimated by

$$\text{bias} = \left(\frac{y_{..} - y_{i.} - y_{.j} - (t-1)y_{k}}{(t-1)(t-2)} \right)^2$$

The corrected treatment sum of squares is then given by

$$\text{SST}_C = \text{SST} - \text{Bias}$$

Then $\text{MST}_C = \text{SST}_C/(t-1)$ is an unbiased estimator of $\sigma_\varepsilon + t\theta_T$. With $n = t^2 - 1$, the number of observed data values in the Latin square design, we obtain the following AOV table for the Latin square design with one missing observation estimated by $\hat{M}$.

TABLE 19.5
AOV table for a Latin square design with one missing value

Source	SS	df	MS	F
Row	SSR	$t - 1$	MSR	—
Column	SSC	$t - 1$	MSC	—
Treatment	SST_C	$t - 1$	MST_C	MST_C/MSE
Error	SSE	$n - 3t + 2$	MSE	
Totals	TSS	$n - 1$		

EXAMPLE 19.2

A company has considered the properties (such as strength, elongation, and so on) of many different variations of nylon stocking in trying to select the experimental stockings to be placed in extensive consumer acceptance surveys.

Five versions (A, B, C, D, and E) of the stockings have passed the preliminary screening and are scheduled for more extensive testing. As part of the testing, five samples of each type are to be examined for elongation under constant stress by each of five investigators on five separate days. The analyses are to be performed

TABLE 19.6
Elongation data for
Example 19.2

	Day				
Investigator	**1**	**2**	**3**	**4**	**5**
1	B 22.1	A 18.6	C 23.0	E 24.3	D 17.1
2	C 23.5	D 16.5	A 18.7	B 22.0	E M
3	D 17.4	E 23.8	B 22.8	C 23.9	A 20.0
4	A 20.3	B 23.4	E 25.9	D 18.7	C 24.2
5	E 25.7	C 24.8	D 18.9	A 20.6	B 24.6

following the random assignment of a Latin square. The elongation data (in centimeters) are displayed in Table 19.6.

Note that the measurement on variety E stockings for investigator 2 is missing and that the experiment was not rerun to obtain an observation. Use the methods of this section to estimate the missing value.

Solution   For our data the treatment, row, and column totals corresponding to the missing observations are

$$y_5 = 99.70 \qquad y_{2.} = 80.70 \qquad y_{.5} = 85.90 \qquad y_{..} = 520.80$$

Then with $t = r = c = 5$, we find

$$\hat{M} = \frac{5(80.70 + 85.90 + 99.70) - 2(520.80)}{(5 - 1)(5 - 2)} = 24.1583$$

We will replace the missing observation with its least-squares estimate, $\hat{M}$, and compute sum of squares using the formulas from Chapter 15.

Investigator	Day	Version	Overall
$\bar{y}_{1.} = 21.020$	$\bar{y}_{.1} = 21.800$	$\bar{y}_1 = 19.640$	$\bar{y}_{..} = 21.79833$
$\bar{y}_{2.} = 20.97166$	$\bar{y}_{.2} = 21.420$	$\bar{y}_2 = 22.980$	
$\bar{y}_{3.} = 21.580$	$\bar{y}_{.3} = 21.860$	$\bar{y}_3 = 23.880$	
$\bar{y}_{4.} = 22.500$	$\bar{y}_{.4} = 21.900$	$\bar{y}_4 = 17.720$	
$\bar{y}_{5.} = 22.920$	$\bar{y}_{.5} = 22.01166$	$\bar{y}_5 = 24.77166$	

$$\begin{aligned}
\text{TSS} &= (22.1 - 21.79833)^2 + (18.6 - 21.79833)^2 + \cdots + (24.6 - 21.79833)^2 \\
&= 197.20
\end{aligned}$$

$$\begin{aligned}
\text{SSR} &= 5\{(21.020 - 21.79833)^2 + (20.97166 - 21.79833)^2 + \cdots + (22.920 \\
&\quad - 21.79833)^2\} = 15.44
\end{aligned}$$

$$\begin{aligned}
\text{SSC} &= 5\{(21.8 - 21.79833)^2 + (21.42 - 21.79833)^2 + \cdots + (22.01166 \\
&\quad - 21.79833)^2\} = 1.01
\end{aligned}$$

$$\begin{aligned}
\text{SST} &= 5\{(19.64 - 21.79833)^2 + (22.98 - 21.79833)^2 + \cdots + (24.77166 \\
&\quad - 21.79833)^2\} = 179.31
\end{aligned}$$

$$\text{SSE} = 197.20 - 15.44 - 1.01 - 179.31 = 1.44$$

$$\text{Bias} = \left(\frac{520.80 - 80.70 - 85.90 - (5 - 1)99.70}{(5 - 1)(5 - 2)}\right)^2 = 13.82$$

Corrected treatment $= \text{SST}_C = 179.31 - 13.82 = 165.49$

The analysis of variance table for this study is given here.

Source	SS	df	MS	F
Investigator	15.44	4	3.86	—
Day	1.01	4	.25	—
Version	165.49	4	41.37	316.04
Error	1.44	11	.13	
Totals	197.20	23		

Having located a significant effect due to treatments, we can make pairwise treatment comparisons using the following formulas. The least significant difference between the treatment with the missing value and any other treatment is

LSD
$$\text{LSD} = t_{\alpha/2} \sqrt{\text{MSE} \left(\frac{2}{t} + \frac{1}{(t-1)(t-2)} \right)}$$

For any other pair of treatments, the LSD is as before:

$$\text{LSD} = t_{\alpha/2} \sqrt{\frac{2\text{MSE}}{t}}$$

The value for MSE is taken from the analysis of variance table.

fitting full and reduced models
For Latin square designs with more than one missing observation, it is easier to use the method of **fitting full and reduced models** to adjust the treatment sum of squares for imbalances in the design due to missing observations. The complete model is given by

$$\text{model 1:} \quad y_{ij} = \mu + \alpha_k + \beta_i + \gamma_j + \varepsilon_{ij}$$

where y_{ij} is the observation in the ith row and jth column on treatment k. This model is fit to the observed data without estimating the missing values. We obtain the error sum of squares, which we will denote as SSE_1. Next, we fit the reduced model without the treatment effect,

$$\text{model 2:} \quad y_{ij} = \mu + \beta_i + \gamma_j + \varepsilon_{ij}$$

to the observed data without estimating the missing values. We will again obtain an error sum of squares, which we will denote as SSE_2. The difference in these two error sum of squares is the corrected sum of squares for treatments,

$$\text{SST}_C = \text{SSE}_2 - \text{SSE}_1$$

The test for treatment effects is the F test given in Table 19.6,

$$F = \frac{\text{SST}_C/(t-1)}{\text{SSE}_1/(n-3t+2)}$$

where n is the number of observed data values. We could obtain the corrected sum of squares for row and column effects in a similar fashion. By fitting a reduced model including the treatment effect and row effect but removing the column effect, we could obtain the sum of squares error needed to obtain the adjusted column effect. Similarly, we could obtain the adjusted row effect. In most cases the test for significant column or row effects is not of interest.

EXERCISES

Basic Techniques

19.8 Refer to Example 19.2 and the computer output shown here to compute the sum of squares for treatment and error. Compare these values to the values computed using the estimated missing value formulas.

```
General Linear Models Procedure: FULL MODEL FOR EXAMPLE 19.2

Dependent Variable: ELONG

Source                    DF    Sum of Squares    F Value    Pr > F
Model                     12     189.95683333     120.66     0.0001
Error                     11       1.44316667
Corrected Total           23     191.40000000


Source                    DF       Type I SS       F Value    Pr > F
INVEST                     4      22.32850000      42.55      0.0001
DAY                        4       2.13400000       4.07      0.0291
VERSION                    4     165.49433333      315.35      0.0001


Source                    DF      Type III SS      F Value    Pr > F
INVEST                     4      14.36883333      27.38      0.0001
DAY                        4       0.94283333       1.80      0.1998
VERSION                    4     165.49433333      315.35      0.0001



General Linear Models Procedure: REDUCED MODEL WITHOUT TREATMENT

Dependent Variable: ELONG

Source                    DF    Sum of Squares    F Value    Pr > F
Model                      8      24.46250000      0.27      0.9646
Error                     15     166.93750000
Corrected Total           23     191.40000000


Source                    DF       Type I SS       F Value    Pr > F
INVEST                     4      22.32850000       0.50      0.7352
DAY                        4       2.13400000       0.05      0.9952


Source                    DF      Type III SS      F Value    Pr > F
INVEST                     4      23.49000000       0.53      0.7172
DAY                        4       2.13400000       0.05      0.9952
```

19.9 Refer to Exercise 19.8.

    a. Test for a significant difference in the mean elongation of the versions of the stockings.

    b. Use the least significant difference criterion for determining which pairs of versions of the stockings are significantly different.

Applications

Env.   **19.10** A petroleum company was interested in comparing the miles per gallon achieved by four different gasoline blends (I, II, III, IV). Because there can be considerable variability due to differences in drivers and car models, these two extraneous sources of variability were included as "blocking" variables in the following Latin square design. Each driver drove each car model over a standard course with the assigned gasoline blend using a Latin square design. However, when driver 3 was operating a model 4 car using blend II gasoline, there was a malfunction of the car's carburetor that invalidated the data. This malfunction was not discovered until well after the completion of the study, and hence the data could not be replaced. The miles per gallon data are given here.

	Car Model			
Driver	**1**	**2**	**3**	**4**
1	IV 15.5	II 33.9	III 13.2	I 29.1
2	II 16.3	III 26.6	I 19.4	IV 22.8
3	III 10.8	I 31.1	IV 17.1	II —
4	I 14.7	IV 34.0	II 19.7	III 21.6

    a. Run an analysis of variance by estimating the missing value. Use $\alpha = .05$.
    b. Make treatment comparisons by using Fisher's least significant difference, with $\alpha = .05$.

19.11 Use the method of fitting complete and reduced models to obtain an analysis of variance for the data in Exercise 19.10.

19.4 Balanced Incomplete Block (BIB) Designs

The designs we have discussed thus far in this chapter were unbalanced due to unforeseen circumstances caused by some accident while conducting the experiment or during data processing. Sometimes, however, we may be forced to design an experiment in which we must sacrifice some balance in order to perform the experiment. This often occurs when the number of experimental units per block is fewer than the number of treatments under consideration. Consider the following example.

EXAMPLE 19.3

Suppose the quality control laboratory of a chemical company needs to evaluate five different formulations (A, B, C, D, E) of a paint for consistency of color. Four samples of each formulation are evaluated on a daily basis. The laboratory has five technicians available for running the tests, and each technician can evaluate at most four samples per day. Thus, it is not possible to conduct a randomized complete block design because every formulation cannot be evaluated by every technician. However, it may be possible to achieve a partial balance in the design by having each pair of formulations evaluated by the same number of technicians. One such design is listed in Table 19.7.

TABLE 19.7
Assignment of formulations to quality control technicians

Technician	Formulations			
1	D	B	A	E
2	E	A	A	D
3	A	C	D	B
4	C	E	B	A
5	B	D	E	C

Note that each pair of formulations is evaluated by three technicians.

Any randomized block design in which the number of treatments t to be investigated is larger than the number of experimental units available per block is called an **incomplete block design.** Thus, whenever homogeneous blocks of $k < t$ experimental units exist or can be constructed, an incomplete block design cannot be avoided. However, it may be possible to achieve partial balance in the design. One such incomplete block design is defined here.

DEFINITION 19.1

A **balanced incomplete block (BIB) design** is an experimental design in which there are t treatments assigned to b blocks such that

1. Each block contains $k < t$ experimental units.
2. Each treatment appears at most once in each block.
3. Each block contains k treatments.
4. Every treatment appears in exactly r blocks.
5. Every pair of treatments occurs together in λ blocks.

From Definition 19.1, we can conclude that for a design to be a BIB design,

- Every pair of treatments appear together in the same block equally often.
- Each treatment is observed r times.
- The number of observations, n, must satisfy $n = rt = kb$.
- $\lambda < r < b$
- $\lambda = r(k - 1)/(t - 1)$ must be an integer.

EXAMPLE 19.4

Refer to Example 19.3. We had $b = 5$ blocks (technicians) and $t = 5$ treatments (formulations). There were $k = 4$ treatments per block, hence $k = 4 < 5 = t$, which results in an incomplete block design. Now, each formulation appeared in exactly $r = 4$ blocks. For the design to be a BIB design, we would need to have every pair of formulations evaluated by $\lambda = r(k - 1)/(t - 1) = 4(4 - 1)/(5 - 1) = 3$ technicians. Examining the assignment of technicians to formulations in Table 19.7, we find that each pair of formulations is evaluated by three technicians. Thus, the design given in Table 19.7 is a BIB design.

In many situations we do not have complete flexibility in designing an experiment because a BIB design does not exist for all possible choices of t, k, b, and r. For example, suppose we have $t = 6$ treatments to be investigated and $b = 4$ blocks, each containing $k = 3$ experimental units. Thus, each treatment could be observed $r = 2$ times. However, for the design to be a BIB design, $\lambda = r(k - 1)/(t - 1)$ would have to be an integer. In fact, however, $\lambda = 2(3 - 1)/(6 - 1) = 4/5$, which is obviously not an integer. Thus, a BIB design cannot be constructed for this combination of treatments and blocks. There are procedures for constructing BIB designs and more complicated incomplete block designs. The books by Cochran and Cox (1957), Lentner and Bishop (1993), and Kuehl (1999) contain tables of BIB designs and methods for constructing such designs. Several statistical software programs (SAS and Minitab, for example) will construct BIB designs for specified values of t, k, b, and r.

The analysis of variance for a balanced incomplete block design can be performed either by using specifically developed formulas or by using the method of fitting complete and reduced models as discussed for unbalanced designs. We will present the shortcut formulas for the analysis of variance table shown in Table 19.8.

TABLE 19.8

Analysis of variance table for a balanced incomplete block design

Source	SS	df	MS	F
Blocks	SSB	$b - 1$	—	—
Treatments$_{adj}$	SST$_{adj}$	$t - 1$	MST$_{adj}$	MST$_{adj}$/MSE
Error	SSE	$n - t - b + 1$	MSE	
Totals	TSS	$n - 1$		

The model for a BIB design is given here.

$$y_{ijg} = \mu + \alpha_i + \beta_j + \varepsilon_{ijg} \qquad \text{for } i = 1, \ldots, t; \quad j = 1, \ldots, b; \quad g = n_{ij}$$

where $n_{ij} = 1$ if the ith treatment appears in the jth block, and equals zero otherwise. The terms in the model are μ, the overall mean; α_i, the ith treatment effect; β_j, the jth block effect; and ε_{ijg}s are independent and normally distributed with mean 0 and variance σ_ε^2. From this model, we compute the quantities SSB (the sum of squares unadjusted for treatments) and the total sum of squares are computed as previously:

$$\text{TSS} = \sum_{ij} (y_{ij} - \bar{y}_{..})^2$$

where $n = rt = bk$ is the actual number of data values and

$$\text{SSB} = k \sum_j (\bar{y}_{.j} - \bar{y}_{..})^2$$

where $\bar{y}_{.j}$ is the mean of all observations in the jth block and $\bar{y}_{..}$ is the overall mean. Then if we define

$$y_{i.} = \text{sum of all observations on treatment } i$$

$$B_{(i)} = \text{sum of all measurements for blocks that contain treatment } i$$

the sum of squares for treatments adjusted for blocks is

$$\text{SST}_{adj} = \frac{t - 1}{nk(k - 1)} \sum_i (ky_{i.} - B_{(i)})^2$$

The sum of squares for error is found by subtraction:

$$\text{SSE} = \text{TSS} - \text{SSB} - \text{SST}_{adj}$$

As indicated in the analysis of variance table, the test statistic for testing the hypothesis of no difference among the treatment means is MST$_{adj}$/MSE.

EXAMPLE 19.5

A large company enlisted the help of a random sample of 12 potential consumers in a given geographical location to compare the physical characteristics (such as firmness and rebound) of eight experimental pillows and one presently marketed pillow. Because the company knew from previous studies that most people's attention span allowed for them to evaluate at most three pillows at a given time, it decided to employ the design shown in Table 19.9.

After the pillow types were randomly assigned the letters from A to I, tables were prepared with the appropriate pillow types assigned to each table. Each pillow was sealed in an identical white pillowcase and hence could not be distinguished from the others by color. The only marking on the pillowcase was a four-digit number, which provided the investigators with an identification code. With all tables in place, the 12 potential consumers were randomly assigned to a table to compare the three pillows. The consumers were to rate each pillow with a comfort score, based on a 1 to 100 point scale (higher score indicates greater comfort). The scores for each pillow are recorded in Table 19.9 (letters identify the pillow type with A being the presently marketed pillow).

TABLE 19.9
Comfort scores for Example 19.5

Block (Consumers)	Treatment (Pillow)						Block Totals	Block Means
1	A	59	B	26	C	38	123	41
2	D	85	E	92	F	69	246	82
3	G	74	H	52	I	27	153	51
4	A	63	D	70	G	68	201	67
5	B	26	E	98	H	59	183	61
6	C	31	F	60	I	35	126	42
7	A	62	E	85	I	30	177	59
8	B	23	F	73	G	75	171	57
9	C	49	D	74	H	51	174	58
10	A	52	F	76	H	43	171	57
11	B	18	D	79	I	41	138	46
12	C	42	E	84	G	81	207	69
							2,070	57.5

Verify that the design used is a BIB design. Use the formulas of this section to perform an analysis of variance. Use $\alpha = .05$ to test for a difference in mean comfort score among the nine pillow types.

Solution We need to verify that all the conditions required for a BIB design have been satisfied. We note that there were nine treatments (pillows), twelve blocks (consumers), three observations per block (pillows per consumer), and each pillow was rated by four consumers, with a consumer rating at most one pillow of each type. That is, $t = 9$, $b = 12$, $k = 3$, $r = 4$, which yields $n = (9)(4) = (12)(3) = 36$.

We next compute $\lambda = r(k - 1)/(t - 1) = 4(3 - 1)/(9 - 1) = 1$. That is, each pair of pillows was rated by exactly one consumer. We confirm this by examining Table 19.9. Thus, we have that the design used in the study was a BIB design. For an analysis using the formulas given in this section, it is convenient to construct a table of totals and means, as shown in Table 19.10.

To illustrate the values in Table 19.10, let us consider the elements for treatment A:

$y_{1.} = $ sum of values for treatment A $ = 59 + 63 + 62 + 52 = 236$

$B_{(1)} = $ sum of block totals for blocks containing A $ = 123 + 201 + 177 + 171$
$ = 672$

$ky_i - B_{(i)} = (3)(236) - 672 = 36$

TABLE 19.10
Totals for the data
of Table 19.9

Treatment	$y_{i.}$	$B_{(i)}$	$ky_{i.} - B_{(i)}$
A	236	672	36
B	93	615	−336
C	160	630	−150
D	308	759	165
E	359	813	264
F	278	714	120
G	298	732	162
H	205	681	−66
I	133	594	−195
Total	2,070		0

To compute the sum of squares, using the values in Tables 19.9 and 19.10 we have

$$\text{SST}_{\text{Adj}} = \frac{(t-1)}{nk(k-1)} \sum_i (ky_{i.} - B_{(i)})^2 = \frac{(9-1)(316{,}638)}{(36)(3)(3-1)} = 11{,}727.33$$

Similarly, using the block means from Table 19.9, we obtain

$$\text{SSB} = k \sum_j (\bar{y}_{.j} - \bar{y}_{..})^2 = 3\{(41 - 57.5)^2 + \cdots + (69 - 57.5)^2\} = 4{,}575$$

Using the values from Table 19.9, we obtain the total sum of squares

$$\text{TSS} = \sum_{ij} (y_{ij} - \bar{y}_{..})^2 = \{(59 - 57.5)^2 + \cdots + (81 - 57.5)^2\} = 16{,}861$$

and the sum of squares error

$$\text{SSE} = \text{TSS} - \text{SST}_{\text{Adj}} - \text{SSB} = 16{,}861 - 11{,}727.33 - 4{,}575 = 558.67$$

The analysis of variance table for testing for differences in the mean comfort values among the nine types of pillows is shown in Table 19.11. Since the computed value of F, 41.98, exceeds the table value, 2.59, for $\text{df}_1 = 8$, $\text{df}_2 = 16$, and $\alpha = .05$, we conclude that there are significant (p-value < 0.0001) differences in the mean comfort rating among the nine types of pillows.

TABLE 19.11
AOV table for the data
of Example 19.5

Source	SS	df	MS	F	p-value
Consumer	4,575	11	415.91	—	—
Treatment	11,727.33	8	1,465.92	41.98	0.0001
Error	558.67	16	34.92	—	—
Totals	16,861	35	—	—	—

comparison among treatment means

Following the observation of a significant F test concerning differences among treatment means, we naturally might like to determine which treatment means are significantly different from others. To do this, we make use of the following notation: $\hat{\mu}_i$, an estimate of the mean for treatment i, given by

$$\hat{\mu}_i = \bar{y}_{..} + \frac{ky_{i.} - B_{(i)}}{t\lambda}$$

where $\bar{y}_{..}$ is the overall sample mean. An estimate of the difference between two treatment means i and i' is then

$$\hat{\mu}_i - \hat{\mu}_{i'.} = \frac{[ky_{i.} - B_{(i)}] - [ky_{i'.} - B_{(i')}]}{t\lambda}$$

The least significant difference between any pair of treatment means is

LSD
$$LSD = t_{\alpha/2} \sqrt{\frac{2k\text{MSE}}{t\lambda}}$$

EXAMPLE 19.6

Compute the estimated treatment means and determine all pairwise differences, using $\alpha = .05$.

Solution For the BIB design of Example 19.5, we have $\bar{y}_{..} = 57.5$, $t = 9$, and $\lambda = 1$. Thus, using the $ky_{i.} - B_{(i)}$ column in Table 19.10, we have the following estimated treatment means using

$$\hat{\mu}_i = \bar{y}_{..} + \frac{ky_{i.} - B_{(i)}}{t\lambda} = 57.5 + \frac{ky_{i.} - B_{(i)}}{(9)(1)}$$

Treatment	$\bar{y}_{i.}$	$ky_{i.} - B_{(i)}$	$\hat{\mu}_i$
A	59.00	36	61.50
B	23.25	−336	20.17
C	40.00	−150	40.83
D	77.00	165	75.83
E	89.75	264	86.83
F	69.50	120	70.83
G	74.50	162	75.50
H	51.25	−66	50.17
I	33.25	−195	35.83

Note that when comparing the raw treatment means $\bar{y}_{i.}$ to the least-squares estimated means $\hat{\mu}_i$, some of the raw means are increased, whereas some are decreased depending on the relative sizes of the block totals in which the treatment appears.

Using MSE = 34.92, based on $df_{\text{Error}} = 16$, we obtain

$$LSD = t_{\alpha/2} \sqrt{\frac{2k\text{MSE}}{t\lambda}} = 2.12 \sqrt{\frac{2(3)(34.92)}{(9)(1)}} = 10.23$$

The nine least-squares estimated treatment means are arranged in ascending order, with a summary of the significant results. Those treatments underlined by a common line are not significantly different from each other, using the value of LSD to declare pairs significantly different.

B	I	C	H	A	F	G	D	E
20.17	35.83	40.83	50.17	61.50	70.83	75.50	75.83	86.83

Alternatively, the computation of the adjusted sum of squares for treatments and the corresponding F test for testing differences in the treatment means can be accomplished by fitting two models. First, fit a full model with both block and treatment effects obtaining SSE_1. Next, fit a reduced model without block effects obtaining SSE_2. The adjusted sum of squares for treatments, SST_{Adj}, is then obtained by

$$SST_{Adj} = SSE_2 - SSE_1$$

with $df_{Trt} = df_{E2} - df_{E1}$. The F test for treatment effects is then $F = MST_{Adj}/MSE_1$.

Analyzing the Property Valuation Case Study The objective of the study was to determine whether the county assessors provided a consistent valuation of residential property values. The factors in the study were the blocking factor, 16 residential properties, and the treatment factor, 16 county property assessors. The treatment effects are random because the assessors were randomly selected from the population of county assessors and the county manager was interested in the results not only for the 16 assessors in the study but for all county assessors. The design was an incomplete block design because each treatment (assessor) was observed in only 6 of the 16 blocks (properties). We will next verify that the design was a BIB design.

First we identify the parameters in a BIB:

$$t = 16 \qquad r = 6 \qquad b = 16 \qquad k = 6$$

This would require that $n = (16)(6) = 96$ observations and $\lambda = 6(6 - 1)/(16 - 1) = 2$. From this we would conclude that for the study to be a BIB design, it is necessary for every pair of assessors to valuate two of the same properties, each assessor must valuate 6 of the 16 properties, and we have a total of 96 valuations. An examination of the data reveals that all these conditions have been satisfied. We will next fit the models necessary for an evaluation of the data. The model for relating the variation in valuations to assessor effects, property effects, and all other sources is given by

$$\text{Full model: } y_{ijg} = \mu + \alpha_i + \beta_j + \varepsilon_{ijg}$$

where μ is the overall mean valuation across all assessors, α_i is the random effect on the valuation due to assessor i, β_j is the random effect on the valuation due to property j, and ε_{ijg} represents the random effect of all other sources of variation on the valuation. Next, we fit the reduced models. First is the model without the assessor effect.

$$\text{Reduced model I: } y_{ijg} = \mu + \beta_j + \varepsilon_{ijg}$$

From this model we would obtain the adjusted sum of squares for assessors. Next, we fit the model without the property effect.

$$\text{Reduced model II: } y_{ijg} = \mu + \alpha_i + \varepsilon_{ijg}$$

From this model we would obtain the adjusted sum of squares for properties.

The computer output given here provides us with the sum of squares errors from the three fitted models, SSE_{Full}, $SSE_{Red\,I}$, and $SSE_{Red\,II}$.

```
General Linear Models: FULL MODEL

Dependent Variable: VALUATION

Source                DF    Sum of Squares    F Value    Pr > F
Model                 30       16976.0919219     4.51     0.0001
Error                 65        8161.2414114
Corrected Total       95       25137.3333333

Source                DF       Type III SS      F Value    Pr > F
ASR                   15        3759.0919219     2.00      0.0291
P                     15       10343.8800172     5.49      0.0001


General Linear Models: REDUCED MODEL WITHOUT TREATMENT VARIABLE (ASSESSOR)

Dependent Variable: VAL

Source                DF    Sum of Squares    F Value    Pr > F
Model                 15       13217.0000000     5.91     0.0001
Error                 80       11920.3333333
Corrected Total       95       25137.3333333

Source                DF       Type III SS      F Value    Pr > F
P                     15       13217.0000000     5.91      0.0001


General Linear Models: REDUCED MODEL WITHOUT BLOCK VARIABLE (PROPERTY)

Source                DF    Sum of Squares    F Value    Pr > F
Model                 15        6632.21190476    1.91     0.0339
Error                 80       18505.12142857
Corrected Total       95       25137.3333333

Source                DF       Type III SS      F Value    Pr > F
ASR                   15        6632.21190476    1.91      0.0339
```

The test for statistically significant differences in the mean valuations due to assessor differences is obtained as follows.

$$\text{SST}_{\text{Adj}} = \text{SSE}_{\text{Red I}} - \text{SSE}_{\text{Full}} = 11{,}920.33 - 8{,}161.24 = 3{,}759.09$$

with $\text{df}_{\text{Trt}} = \text{df}_{\text{ERed I}} - \text{df}_{\text{EFull}} = 80 - 65 = 15$. We can then test whether there is a significant variation in the valuation due to differences in the assessors. Since assessor is a random source of variation, we want to test

$$H_0: \sigma_\alpha^2 = 0 \quad \text{versus} \quad H_a: \sigma_\alpha^2 \neq 0$$

We compute the value of the test statistic

$$F = \frac{\text{SST}_{\text{Adj}}/\text{df}_{\text{Trt}}}{\text{SSE}_{\text{Full}}/\text{df}_{\text{EFull}}} = \frac{3{,}759.09/15}{8{,}161.24/65} = 2.00$$

with p-value $= .0291$. We can compare the F-value to the tabled .05 percentile from an F distribution with $\text{df}_1 = 15$, $\text{df}_2 = 65$, 1.82, and conclude that there is significant (p-value $= .0291$) variation due to the differences in the assessors. Similarly, we obtain the adjusted sum of squares due to the differences in the properties.

$$\text{SSB}_{\text{Adj}} = \text{SSE}_{\text{Red II}} - \text{SSE}_{\text{Full}} = 18{,}505.12 - 8{,}161.24 = 10{,}343.88$$

with $\text{df}_{\text{Block}} = \text{df}_{\text{ERed II}} - \text{df}_{\text{EFull}} = 80 - 65 = 15$. We can summarize our findings in an AOV table given here.

Source	df	SS	EMS	F	p-value
Property	15	10,343.88	$\sigma_\varepsilon^2 + 5.33\sigma_{\text{Prop.}}^2$	—	—
Assessor	15	3,759.09	$\sigma_\varepsilon^2 + 5.33\sigma_{\text{Asr.}}^2$	2.00	.0291
Error	65	8,161.24	σ_ε^2	—	—

Note that the multipliers for the variances from property and assessor effects are not 16, as they would be in a randomized complete design. Because of the incompleteness of the design, we have the following values for the expected mean squares:

$$\text{Expected mean square for blocks:} \quad \text{EMS}_{\text{Block}} = \sigma_\varepsilon^2 + \frac{bk - t}{b - 1}\sigma_{\text{Block}}^2$$

and

$$\text{Expected mean square for treatment:} \quad \text{EMS}_{\text{Trt}} = \sigma_\varepsilon^2 + \frac{\lambda t}{k}\sigma_{\text{Trt}}^2$$

From the AOV table, we can obtain the following estimates of the variance components.

$$\hat{\sigma}_\varepsilon^2 = 8,161.24/65 = 125.56$$

$$\hat{\sigma}_{\text{Prop.}}^2 = (10,343.88/15 - 125.56)/5.33 = 105.82$$

$$\hat{\sigma}_{\text{Asr.}}^2 = (3,759.09/15 - 125.56)/5.33 = 23.46$$

Thus, we have the following proportional allocation of the total variability in the valuations.

Source of Variation	Estimated Variance	Proportion of Total Variation (%)
Properties	105.82	41.5
Assessors	23.46	9.2
Exp. Error	125.56	49.3
Totals	254.84	100

Although we found that there was significant (p-value = .0291) variability due to the assessors, less than 10% of the variability in the assessed valuations of the properties was due to assessors. Thus, we have determined that the assessors are reasonably consistent in their valuations of midpriced residental properties in the county.

Reporting Conclusions The report from the county staff personnel to the county manager should include the following items.

1. Statement of objectives of study
2. Description of study design, how the properties used in the study were selected, how the assessors were selected, and the manner in which the valuations were conducted

3. Discussion of the relevance of the conclusions of this study to valuations throughout the county

4. Numerical and graphical representations of the data

5. Description of all inference methodologies:
- Statement of research hypotheses
- Model that represents experimental conditions
- Verification of model conditions
- AOV table, including p-values

6. Discussion of results and conclusions

7. Interpretation of findings relative to residential complaints about the biases in property valuations

8. Listing of data

19.5 Summary

In this chapter, we discussed the analysis of variance for some unbalanced designs, beginning with a discussion of the analysis for a randomized block design with one missing observation. Two possible analysis were proposed. The first required that we estimate the missing value and then proceed with the usual formulas developed in Chapter 15. Although estimating a single missing value is quite easy to do, the procedure becomes more difficult when there is more than one missing value. The second procedure, that of fitting complete and reduced models to obtain adjusted sums of squares, can be used for one or more missing observations.

With the Latin square design, we again showed how to estimate a single missing observation and proceed with the usual analysis. However, as with the randomized block design, the method of analysis by fitting complete and reduced models is more appropriate when there is more than one missing value.

Finally, we considered another class of unbalanced designs, incomplete block designs. The particular designs that we discussed were incomplete randomized block designs in which not all treatments appear in each block. These incomplete block designs retain a certain amount of balance, because all pairs of treatments appear together in a block the same number of times. We illustrated the analysis for balanced incomplete block designs using appropriate formulas. The method of analysis for BIB designs can be accomplished by fitting full and reduced models as was done in the case of missing values in the randomized block design and Latin square design.

Key Formulas

1. Missing observation, randomized block design

a.
$$\hat{M} = \frac{ty_{i\cdot} + by_{\cdot j} - y_{\cdot\cdot}}{(t-1)(b-1)}$$

b. Bias correction for sum of squares treatment
$$\text{Bias} = \frac{\{y_{\cdot j} - (t-1)\hat{M}\}^2}{t(t-1)}$$

The corrected treatment sum of squares is then $\text{SST}_C = \text{SST} - \text{Bias}$.

2. Fisher's LSD for a randomized block design
a. For any pair of treatments with no missing value

$$\text{LSD} = t_{\alpha/2} \sqrt{\frac{2\text{MSE}}{b}}$$

b. Between the treatment with a missing value and any other treatment

$$\text{LSD} = t_{\alpha/2} \sqrt{\text{MSE}\left(\frac{2}{b} + \frac{t}{b(b-1)(t-1)}\right)}$$

3. Equalities for randomized block design

$$\text{SSB} = \text{TSS} - \text{SST}_{\text{Adj}} - \text{SSE}$$

$$\text{SST} = \text{TSS} - \text{SSB}_{\text{Adj}} - \text{SSE}$$

4. Missing observation, Latin square design

a. $$\hat{M} = \frac{t(y_{i..} + y_{.j.} + y_{..k}) - 2y_{...}}{(t-1)(t-2)}$$

b. Bias correction for sum of squares treatment

$$\text{Bias} = \frac{\{y_{.j} - (t-1)\hat{M}\}^2}{t(t-1)}$$

The corrected treatment sum of squares is then $\text{SST}_C = \text{SST} - \text{Bias}$.

5. Fisher's LSD for a Latin square design
a. For any pair of treatments with no missing value

$$\text{LSD} = t_{\alpha/2} \sqrt{\frac{2\text{MSE}}{t}}$$

b. Between the treatment with the missing value and any other treatment

$$\text{LSD} = t_{\alpha/2} \sqrt{\text{MSE}\left(\frac{2}{t} + \frac{1}{(t-1)(t-2)}\right)}$$

6. Sums of squares for an incomplete block design

$$\text{SST}_{\text{Adj}} = \frac{t-1}{n(k)(k-1)} \sum_i (k_{y_{i.}} - B_{(i)})^2$$

$$\text{SSE} = \text{TSS} - \text{SSB} - \text{SST}_{\text{Adj}}$$

7. Pairwise comparisons of treatment means, incomplete block design

$$\hat{\mu}_i - \hat{\mu}_{i'} = \frac{[ky_{i.} - B_{(i)}] - [ky_{i'.} - B_{(i')}]}{t\lambda}$$

$$\text{LSD} = t_{\alpha/2} \sqrt{\frac{2k\text{MSE}}{t\lambda}}$$

Source	DF	Type III SS	F Value	Pr > F
INSPECT	1	0.66666667	0.08	0.7872
TRAIN	1	130.66666667	14.93	0.0023
INSPECT*TRAIN	1	1.50000000	0.17	0.6861
LINE (TRAIN)	4	56.66666667	1.62	0.2329
LINE*INSPECT (TRAIN)	4	1.33333333	0.04	0.9968

19.23 Refer to Exercise 19.21. Suppose that inspector 2 was unable to evaluate the second car from assembly line 4 and that inspector 1 missed car 1 from assembly line 3.

a. Does the model change? Suggest a method for analyzing the data.

b. Use the computer output shown here to draw conclusions.

General Linear Models Procedure: FULL MODEL FOR EXERCISE 19.23

Dependent Variable: DEFECTS

Source	DF	Sum of Squares	F Value	Pr > F
Model	11	180.81818182	1.60	0.2326
Error	10	102.50000000		
Corrected Total	21	283.31818182		

Source	DF	Type III SS	F Value	Pr > F
INSPECT	1	0.03571429	0.00	0.9541
TRAIN	1	124.32142857	12.13	0.0059
INSPECT*TRAIN	1	0.89285714	0.09	0.7739
LINE (TRAIN)	4	48.75000000	1.19	0.3733
LINE*INSPECT (TRAIN)	4	2.62500000	0.06	0.9913

General Linear Models Procedure: REDUCED MODEL I FOR EXERCISE 19.23

Dependent Variable: DEFECTS

Source	DF	Sum of Squares	F Value	Pr > F
Model	11	180.81818182	1.60	0.2326
Error	10	102.50000000		
Corrected Total	21	283.31818182		

Source	DF	Type III SS	F Value	Pr > F
INSPECT	1	0.03571429	0.00	0.9541
INSPECT*TRAIN	1	0.89285714	0.09	0.7739
LINE (TRAIN)	4	48.75000000	1.19	0.3733
LINE*INSPECT (TRAIN)	4	2.62500000	0.06	0.9913

General Linear Models Procedure: REDUCED MODEL II FOR EXERCISE 19.23

Dependent Variable: DEFECTS

Source	DF	Sum of Squares	F Value	Pr > F
Model	11	180.81818182	1.60	0.2326
Error	10	102.50000000		
Corrected Total	21	283.31818182		

Source	DF	Type III SS	F Value	Pr > F
INSPECT	1	0.03571429	0.00	0.9541
TRAIN	1	124.32142857	12.13	0.0059
INSPECT*TRAIN	1	0.89285714	0.09	0.7739
LINE*INSPECT (TRAIN)	8	52.76666667	0.64	0.7279

CHAPTER 20

Communicating and Documenting the Results of Analyses

20.1 Introduction

20.2 The Difficulty of Good Communication

20.3 Communication Hurdles: Graphical Distortions

20.4 Communication Hurdles: Biased Samples

20.5 Communication Hurdles: Sample Size

20.6 Preparing Data for Statistical Analysis

20.7 Guideline for a Statistical Analysis and Report

20.8 Documentation and Storage of Results

20.9 Summary

20.1 Introduction

In Chapter 1, we introduced the subject of statistics as the science of learning from data. The four steps in learning from data—(1) designing the data collection, (2) preparing data for analysis and summarization, (3) analyzing the data, and (4) reporting the conclusions—form the major outline for this book. We have tried to keep these four steps in focus as we progressed through the book by showing these steps, however briefly, in our case studies.

In the previous chapters, we have discussed particular statistical methods, how those methods are applied to specific data sets, and how findings from statistical analyses in the form of computer output are interpreted. We have not concentrated on the processing steps that one follows between the time the data are received and the time they are available in computer-readable form for analysis, nor have we discussed the form and content of the report that summarizes the results of a statistical analysis. In this chapter, we consider the data-processing steps and statistical report writing. This chapter is not a complete manual with all the tools required; rather, it is an overview—what a manager or researcher should know about these steps. As an example, the chapter reflects standard procedures in the pharmaceutical industry, which is highly regulated. Procedures differ somewhat in other industries and organizations. We begin this chapter with some general pitfalls and difficulties you might encounter in communicating effectively.

20.2   The Difficulty of Good Communication

We have spent time throughout the text making sense of data; the final step in this process is the communication of results. How might you communicate the results of a study or survey? The list of possibilities is almost endless, including all forms of verbal and written communication. There is quite a range of possibilities for verbal and written communication. For example, written communication within a company or organization can vary from an informal short note or memo to a formal project report (Figure 20.1).

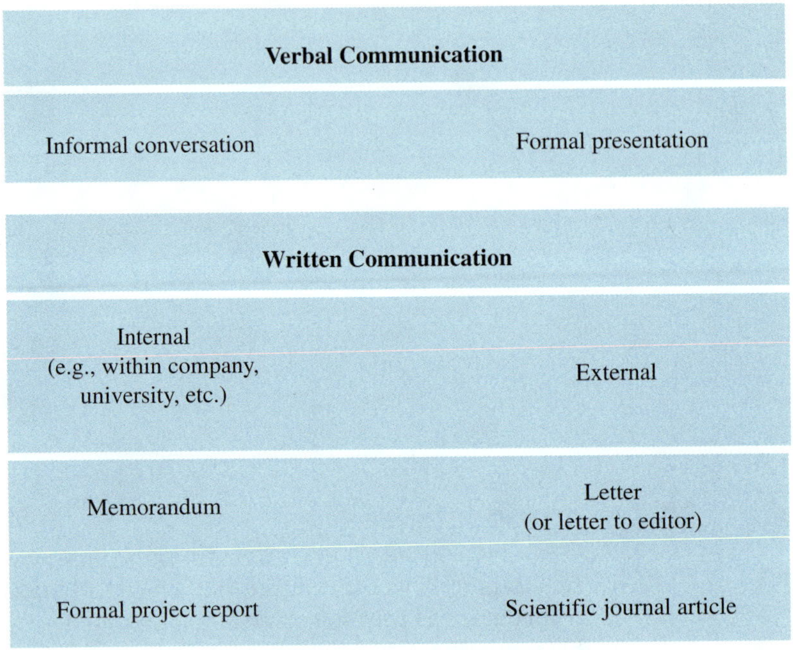

FIGURE 20.1

Forms of written and verbal communication

Verbal Communication	
Informal conversation	Formal presentation

Written Communication	
Internal (e.g., within company, university, etc.)	External
Memorandum	Letter (or letter to editor)
Formal project report	Scientific journal article

Communicating the results of a statistical analysis in concise, unambiguous terms is difficult. In fact, descriptions of most things are difficult. For example, try to describe the person sitting next to you so precisely that a stranger could select the individual from a group of others having similar physical characteristics. It is not an easy task. Fingerprints, voiceprints, and photographs—all pictorial descriptions—are the most precise methods of human identification. The description of a set of measurements is also a difficult task. However, like the description of a person, it can be accomplished more easily by using graphics or pictorial methods.

Cave drawings convey to us scattered bits of information about the life of prehistoric people. Similarly, vast quantities of knowledge about the ancient lives and cultures of the Babylonians, Egyptians, Greeks, and Romans are brought to life by means of drawings and sculpture. Art has been used to convey a picture of various lifestyles, history, and culture in all ages. Not surprisingly, use of graphs and tables along with a written description can help to convey the meaning of a statistical analysis.

In reading the results of a statistical analysis and in communicating the results of our own analyses, we must be careful not to distort them because of the way we present the data and results. You have all heard the expression, "It is easy to lie with statistics." The idea is *not* new. The famous British statesman Disraeli is quoted as saying, "There are three kinds of lies: lies, damned lies, and statistics." Where do things go wrong?

First of all, distortion of truth can occur only when we communicate. And since communication can be accomplished with graphs, pictures, sound, aroma, taste, words, numbers, or any other means devised to reach our senses, distortions can occur using any one or any combination of these methods of communication.

In this respect, statements that we make could be misleading to others because we might have omitted something in the explanation of the data-gathering stage or with the analyses done. For example, we might unintentionally fail to clearly explain the meaning of a numerical statement, or we might omit some background information that is necessary for a clear interpretation of the results. Even a correct statement may appear to be distorted if the reader lacks knowledge of elementary statistics. Thus a very clear expression of an inference using a 95% confidence interval is meaningless to a person who has not been exposed to the introductory concepts of statistics.

Now we will look at some potential hurdles to effective communication that we must carefully consider when we present the results of a statistical analysis—or when we try to interpret what someone else has presented.

20.3   Communication Hurdles: Graphical Distortions

Pictures can easily distort the truth. The marketing of many products, including soft drinks, beers, cosmetics, clothing, automobiles, and many more, involves the use of attractive, youthful models. The not-so-subtle impression we are left with is that (somehow) by using the product, we, too, will look like these models. Have you ever "stepped back" from one of these commercials and wondered how the commercial message relates to the quality and usefulness of the product? Have you thought about how you are being misled by a commercial? The use of sex appeal to sell products is very prominent, and we seem to accept this type of distortion. The beer ad article shown here, which appeared in *USA Today,* March 15, 2000, illustrates how we are "manipulated" through these types of ads.

Sex Appeal Slipping Back into Beer Ads

**'Risqué' TV
leads to more
liberal promotions**

By Michael McCarthy
USA TODAY

NEW YORK—The "babes" in socalled Beer & Babes advertising are back.

Nearly a decade ago national soul-searching about sexual harrassment in the wake of the Anita Hill controversy forced beer marketers to purge traditional and what were seen as sexist symbols, such as "The Swedish Bikini Team," from their ads.

Now, brewers including Miller Brewing, Anheuser-Busch and Heineken are injecting sex appeal back into TV ads.

"We're in a new decade, and tastes are changing," says marketing consultant Laura Ries.

As TV is more "risque," beer ads reflect a "more liberal environment," she says. "For a while, it was dogs, cats and penguins. Now, they're reverting back to sex sells. It's an attention grabber."

Others still see it as sexist sells. "We definitely see a trend," say Sonia Ossorio, chairwoman of the media watch committee of the National Organization for Women (NOW). And she warns that the new spots are "more insidious" because they take a humorous rather than leering approach.

Recent ads illustrating the trend:

▶**Miller.** The brewer will begin a campaign for Miller Lite during NCAA basketball tournament games starting this week. The ads resurrect their classic slogan, "It's Miller Time," and classic images. In one, three guys in a deserted bar on a rainy day pass the time rubbing their eyes. They completely miss two scantily clad and soaked, beauties who duck in briefly to dry off.

In another, an unlucky buddy who draws the designated-driver straw ends up with the real prize. Three beautiful exchange students looking for a ride home. His pals look on enviously.

"They're done in a humorous way," says Miller's senior vice president of marketing Bob Mikulay. The commercials are "about the connection that guys make when they are hanging out over a beer."

Miller Lite could use a jump-start. Its 8% share of the U.S. market is down from 10% a decade ago, says Eric Shepard, executive editor of *Beer Marketer's Insights.*"

▶**Heineken.** The brewer's "It's all about the Beer" campaign uses frank sexual imagery.

In one literally over-the-top TV spot, called "The Premature Pour," a guy spies a beautiful woman across the room in a bar. She slowly and seductively pours her Heineken into a tall glass. When he tries to do the same, he fumbles, pouring too fast and spilling foam all over the place.

In a spot called "The Wrong Bar Car," a guy on a train sits drinking the wrong beer in a bar car full of weirdos and losers also drinking mundane beer. Out the window he spies on the next track a beautiful blonde in a bar car full of partying Heineken drinkers. To his chagrin she waves as his train pulls out.

▶**Anheuser-Busch.** You don't have to be a Fellini fan to get the symbolism in its recent "Stranger on a Train" spot for Michelob Light.

A woman enters a train compartment and sits next to one of two guys sitting opposite each other. The train goes into a tunnel. When it emerges, she has switched seats and is cuddling the guy drinking a Michelob Light. He grins triumphantly at his rival.

Ries believes the new round of ads have brought "beercake" full circle. "The ads show equal opportunity sexism. The women are in charge, and the guys are the butt of the jokes," she says.

Cheryl Berman, chief creative officer for Leo Burnett, doesn't see this as a return to old-style beer ads. "The ads are more about male bonding than anything else," she says.

And she thinks this work is an improvement on some of the odd beer campaigns that fell flat in recent years, such as Miller's fictional ad man "Dick."

"There were a lot of awards-show ads," she says. They won really big at (industry) creative shows—but the public didn't buy into them."

Rick Boyko of Miller Lite ad agency Ogilvy & Mather says the campaign will show female friends enjoying "Miller Time," too. "It resonates with today's consumers. I have a 21-year old daughter who says 'It's Miller Time.' "

But the poster boy for sexist beer advertising knows the pitfalls firsthand.

Ad agency executive Patrick Scullin lost his job and nearly his career after Stroh yanked his infamous "Swedish Bikini Team" ads for Old Milwaukee beer from the airwaves in 1991. Stroh beat a hasty retreat after a group of women employees sued, alleging the ad campaign helped create an atmosphere encouraging sexual harassment.

"My ZIP code became Siberia," says Scullin, now creative director at Ames Scullin O'Haire in Atlanta. "They were intended as a spoof. But they turned into a media circus."

Scullin doesn't see a big change in the new ads. "The women just wear tight jeans instead of bikinis," he says. "People don't like to admit it. But as long as there's biology, there will be sexism in beer advertising."

Television, Internet, and catalog pictures of products are frequently more attractive than the real thing, but we usually take this type of distortion for granted.

Statistical pictures are the histograms, frequency polygons, pie charts, and bar graphs of Chapter 3. These drawings or displays of numerical results are difficult to combine with sketches of lovely women or handsome men and hence are secure from the most common form of graphic distortion. However, other distortions are possible. One could shrink or stretch the axes, thus distorting the actual results. The idea behind these distortions is that shallow and steep slopes are commonly associated with small and large increases, respectively.

For example, suppose that the values of a leading consumer price index over the first 6 months of the year were 160, 165, 178, 189, 196, and 210. We might show the upward movement of this consumer price index by using the frequency polygon of Figure 20.2. In this graph, the increase in the index is apparent, but it does not appear to be very great. On the other hand, we could present the sample data in a much different light, as shown in Figure 20.3. For this graph, the vertical axis is stretched and does not include 0. Note the impression of a substantial rise that is indicated by the steeper slope. Another way to achieve the same effect—to decrease or include a slope—is to stretch or shrink the horizontal axis.

FIGURE 20.2

Changes in a consumer price index

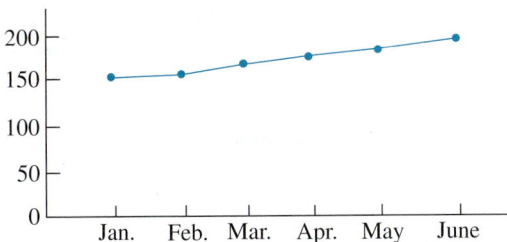

FIGURE 20.3

Changes in a consumer price index

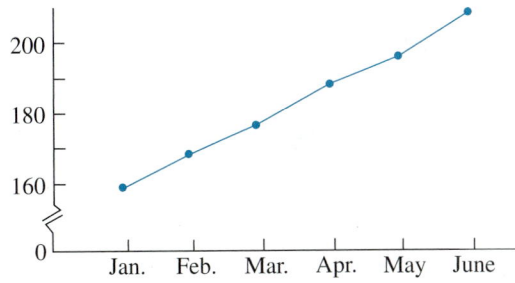

When we present data in the form of bar graphs, histograms, frequency polygons, or other figures, we must be careful not to shrink or stretch axes because doing so will catch most readers off guard. Increases or decreases in responses should be judged large or small depending on the arbitrary importance to the observer of the change, not on the slopes shown in graphic representations. In reality, most people look only at the slopes in the "pictures."

20.4   Communication Hurdles: Biased Samples

One of the most common statistical distortions occurs because the experimenter unwittingly (or sometimes knowingly) samples the wrong population. That is, he or she draws the sample from a set of measurements that is not the proper population of interest.

For example, suppose that we want to assess the reaction of taxpayers to a proposed park and recreation center for children. A random sample of households is selected, and interviewers are sent to those households in the sample. Unfortunately, no one is at home in 40% of the sample households, so we randomly select and substitute other households in the city to make up the deficit. The resulting sample is selected from the wrong population, and the sample is therefore said to be biased.

The specified population of interest in the household survey is the collection of opinions that would be obtained from the complete set of all households in the city. In contrast, the sample was drawn from a much smaller population or subset of this group—the set of opinions from householders who were at home when the sample was taken. It is possible that the fractions of householders favoring the park in these two populations are equal, and no damage was done by confining the sampling to those at home. However, it is much more likely that those at home had small children and that this group would yield a higher fraction in favor of the park than would the city as a whole. Thus, we have a biased sample because it is loaded in favor of families with small children. Perhaps a better way to see the difficulty is to note that we unwittingly selected the sample only from a special subset of the population of interest.

Biased samples frequently result from surveys that utilize mailed questionnaires. In a sense, the investigator lets the selection and number of the sampling units depend on the interests, available time, and various other personal characteristics of the individuals who receive the questionnaires. Extremely busy and energetic people may drop the questionnaires in the nearest wastebasket; you rarely hear from those low-energy folk who are uninterested or who are engrossed with other activities. Most often, the respondents are activists—those who are highly in favor, those who are opposed, or those who have something to gain from a certain outcome of the survey.

Although numerous newscasters and analysts utilize election results as an expression of public opinion on major issues, it is a well-known fact that voting results represent a biased sample of public opinion. Those who vote represent much less than half of the eligible voters; they are individuals who desire to exercise their rights and responsibilities as citizens or are individuals who have been specially motivated to participate. The resultant subset of voters is not representative of the interests and opinions of all eligible voters in the country.

Sampling the wrong population also occurs when people attempt to extrapolate experimental results from one population to another. Numerous experimental results have been published about the effect of various products (e.g., saccharin) in inducing cancer in moles, rats, the breasts of beagles, and so forth. These results are often used to imply that humans have a high risk of developing cancer after frequent or extended exposure to the product. These inferences are not always justified, because the experimental results were not obtained on humans. It is quite possible that humans are capable of resisting much higher doses than rats, or perhaps humans may be completely resistant for some reason. Drug induction of cancer in small mammals *does* indicate a need for concern and caution by humans, but it does not prove that the drug is definitely harmful to humans. Note that we are not criticizing experimentation in various species of animals, because it is frequently the only way we can obtain any information about potential toxicity in human beings. We simply point out that the experimenter is knowingly sampling a population that is only similar (and quite likely not too similar) to the one of interest.

Engineers also test "rats" instead of "humans." Rats, in this context, are miniature models or pilot plants of a new engineering system. Experiments on the models occasionally yield results that differ substantially from the results of the larger, real systems. So again, we see a sampling from the wrong population, but it is the best the engineer can do because of the economics of the situation. Funds are not usually available to test a number of full-scale models prior to production.

Many other examples could be given of biased samples or of sampling from the wrong populations. The point is that when we communicate the results of a study or survey we should be clear about how the sample was drawn and whether it was *randomly* selected from the population of interest. If this information is not given in the published results of a survey or experiment, the reader should take the inferences with a grain of salt.

20.5   Communication Hurdles: Sample Size

Distortions can occur when the sample size is not discussed. For example, suppose you read that a survey indicates that approximately 75% of a sample favor a new high-rise building complex. Further investigation might reveal that the investigator sampled only four people. When three out of the four favored the project, the investigator decided to stop the survey. Of course, we exaggerate with this example; but we could also have revealed inconclusive results based on a sample of 25, even though many buyers would consider this sample size to be large enough. As you well know, very large samples are required to achieve adequate information in sampling binomial populations.

Fortunately, many publications now provide more information about the sample size and how opinion surveys are conducted. Ten years ago, it was rare to find how many people were sampled, much less how they were sampled. The situation is different now. In fact, sometimes the media have gone too far in an attempt to be completely open about how a survey was done. A case in point is the following article from the *Wall Street Journal*. How many of us understand much more than the number of persons sampled and the approximate plus or minus (confidence interval)? It would take a person well trained in statistics and survey sampling to interpret what was done. Again, the moral of the story is simple: Try to communicate in unambiguous terms.

How Poll Was Conducted

The Wall Street Journal/NBC News poll was based on nationwide telephone interviews conducted last Friday through Mon- day with 4,159 adults age 18 or older. There were 2,630 likely voters.

The sample was drawn from a complete list of telephone exchanges, chosen so that each region of the country was represented in proportion to its population. Households were selected by a method that gave all telephone numbers, listed and unlisted, a proportionate chance of being included. The results of the survey were weighted to adjust for variations in the sample relating to education, age, race, gender, and religion.

Chances are 19 of 20 that if all adults in the United States had been surveyed using the same questionnaire, the findings would differ from these poll results by no more than two percentage points in either direction. The margin of error for subgroups may be larger.

20.6 Preparing Data for Statistical Analysis

We begin with a discussion of the steps involved in processing data from a study. In practice, these steps may consume 75% of the total effort from the receipt of the raw data to the presentation of results from the analysis. What are these steps, why are they so important, and why are they so time-consuming?

To answer these questions, let's list the major data-processing steps in the cycle, which begin with receipt of the data and end when the statistical analysis begins. Then we'll discuss each step separately.

DEFINITION 20.1

Steps in Preparing Data for Analysis

1. Receiving the raw data source
2. Creating the database from the raw data source
3. Editing the database
4. Correcting and clarifying the raw data source
5. Finalizing the database
6. Creating data files from the database

raw data source

1. Receiving the raw data source. For each study that is to be summarized and analyzed, the data arrive in some form, which we'll refer to as the **raw data source.** For a clinical trial, the raw data source is usually case report forms, sheets of $8\frac{1}{2}'' \times 11''$ paper that have been used to record study data for each patient entered into the study. For other types of studies, the raw data source may be sheets of paper from a laboratory notebook, a magnetic tape (or any other form of machine-readable data), hand tabulations, and so on.

data trail

It is important to retain the raw data source because it is the beginning of the **data trail,** which leads from the raw data to the conclusions drawn from a study. Many consulting operations involved with the analysis and summarization of many different studies keep a log that contains vital information related to the study and raw data source. General information contained in a study log is shown next.

DEFINITION 20.2

> **Log for Study Data**
>
> 1. Data received, and from whom
> 2. Study investigator
> 3. Statistician (and others) assigned
> 4. Brief description of study
> 5. Treatments (compounds, preparations, and so on) studied
> 6. Raw data source
> 7. Response(s) measured
> 8. Reference number for study
> 9. Estimated (actual) completion date
> 10. Other pertinent information

Later, when the study has been analyzed and results have been communicated, additional information can be added to the log on how the study results were communicated, where these results are recorded, what data files have been saved, and where these files are stored.

2. Creating the database from the raw data source. For most studies that are scheduled for a statistical analysis, a machine-readable database is created. The steps taken to create the database and the eventual form of the database vary from one operation to another, depending on the software systems to be used in the statistical analysis. However, we can give a few guidelines based on the form of the entry system.

When the data are to be *key-entered* at a terminal, the raw data are first checked for legibility. Any illegible numbers or letters or other problems should be brought to the attention of the study coordinator. Then a coding guide that assigns column numbers and variable names to the data is filled out. Certain codes for missing values (for example, those not available) are also defined here. Also, it is helpful to give a brief description of each variable. The data file keyed in at the terminal is referred to as the **machine-readable database.** A listing of the contents of the database should be obtained and checked carefully against the raw data source. Any errors should be corrected at the terminal and verified against an updated listing.

machine-readable database

Sometimes data are received in machine-readable form. In these situations, the magnetic tape or disk file is considered to be the database. You must, however, have a coding guide to "read" the database. Using the coding guide, obtain a listing of the contents of the database and check it *carefully* to see that all numbers and characters look reasonable and that proper formats were used to create the file. Any problems that arise must be resolved before proceeding further.

Some data sets are so small that it is not necessary to create a machine-readable data file from the raw data source. Instead, calculations can be performed by hand or the data entered into an electronic calculator. In these situations, check any calculations to see that they make sense. Don't believe everything you see; redoing the calculations is not a bad idea.

3. Editing the database. The types of edits done and the completeness of the editing process really depend on the type of study and how concerned you are about the accuracy and completeness of the data prior to analysis. For example, in using SAS files it is wise to examine the minimum, maximum, and frequency distribution for each variable to make certain nothing looks unreasonable.

logic checks

Certain other checks should be made. Plot the data and look for problems. Also, certain **logic checks** should be done, depending on the structure of the data. For example, if data are recorded for patients during several different visits, then the data recorded for visit 2 cannot be earlier than the data for visit 1; similarly, if a patient is lost to follow-up after visit 2, we cannot have any data for that patient at later visits.

For small data sets, we can do these data edits by hand, but for large data sets the job may be too time-consuming and tedious. If machine editing is required, look for a software system that allows the user to specify certain data edits. Even so, for more complicated edits and logic checks, it may be necessary to have a customized edit program written in order to machine-edit the data. This programming chore can be a time-consuming step; plan for this well in advance of receipt of the data.

4. Correcting and clarifying the raw data source. Questions frequently arise concerning the legibility or accuracy of the raw data during any one of the steps from the receipt of the raw data to the communication of the results from the statistical analysis. We have found it helpful to keep a list of these problems or discrepancies in order to define the data trail for a study. If a correction (or clarification) is required to the raw data source, this should be indicated on the form and the appropriate change made to the raw data source. If no correction is required, this should be indicated on the form as well. Keep in mind that the machine-readable database should be changed to reflect any changes made to the raw data source.

5. Finalizing the database. You may have been led to believe that all data for a study arrive at one time. This, of course, is not always the case. For example, with a marketing survey, different geographic locations may be surveyed at different times, and hence those responsible for data processing do not receive all the data at once. All these subsets of data, however, must be processed through the cycles required to create, edit, and correct the database. Eventually, the study is declared complete and the data are processed into the database. At this time, the database should be reviewed again and final corrections made before beginning the analysis. This is because, for large data sets, the analysis and summarization chores take considerable staff and computer time. It's better to agree on a final database analysis than to have to repeat all analyses on a changed database at a later date.

6. Creating data files from the database. Generally, one or two sets of data files are created from the machine-readable database. The first set, referred to as

original files

original files, reflects the basic structure of the database. A listing of the files is checked against the database listing to verify that the variables have been read with correct formats and missing value codes have been retained. For some studies, the original files are actually used for editing the database.

work files

A second set of data files, called **work files,** may be created from the original files. Work files are designed to facilitate the analysis. They may require restructuring of the original files, a selection of important variables, or the creation or addition of new variables by insertion, computation, or transformation. A listing of the work files is checked against that of the original files to ensure proper restructuring and variable selection. Computed and transformed variables are checked by hand calculations to verify the program code.

If original and work files are SAS data sets, you should utilize the documentation features provided by SAS. At the time an SAS data set is created, a descriptive label for the data set of up to 40 characters should be assigned. The label can be stored with the data set, imprinted wherever the contents procedure is used to print the data set's contents. All variables can be given descriptive names, up to 8 characters in length, which are meaningful to those involved in the project. In addition, variable labels up to 40 characters in length can be used to provide additional information. Title statements can be included in the SAS code to identify the project and describe each job. For each file, a listing (proc print) and a dictionary (proc contents) can be retained.

For files created from the database using other software packages, use the labeling and documentation features available in the computer program.

Even if appropriate statistical methods are applied to data, the conclusions drawn from the study are only as good as the data on which they are based. So you be the judge. The amount of time spent on these data-processing chores before analysis really depends on the nature of the study, the quality of the raw data source, and how confident you want to be about the completeness and accuracy of the data.

20.7 Guidelines for a Statistical Analysis and Report

In this section, we briefly discuss a few guidelines for performing a statistical analysis and list some important elements of a statistical report used to communicate results. The statistical analysis of a large study can usually be broken down into three types of analyses; (1) preliminary analyses, (2) primary analyses, and (3) backup analyses.

preliminary analyses

The **preliminary analyses,** which are often descriptive or graphic, familiarize the statistician with the data and provide a foundation for all subsequent analyses. These analyses may include frequency distributions, histograms, descriptive statistics, an examination of comparability of the treatment groups, correlations, or univariate and bivariate plots.

primary analyses
backup analyses

Primary analyses address the objectives of the study and the analyses on which conclusions are drawn. **Backup analyses** include alternate methods for examining the data that confirm the results of the primary analyses; they may also include new statistical methods that are not as readily accepted as the more standard methods. Several guidelines for analyses follow.

DEFINITION 20.3

Preliminary, Primary, and Backup Analyses

1. Perform the analyses with software that has been extensively tested.
2. Label the computer output to reflect which study is analyzed, what subjects (animals, patients, and so on) are used in the analysis, and a brief description of the analysis preferred. For example, TITLE statements in SAS are very helpful.
3. Use variable labels and value labels (for example, 0 = none, 1 = mild) on the output.
4. Provide a list of the data used in each analysis.
5. Check the output *carefully* for all analyses. Did the job run successfully? Are the sample sizes, means, and degrees of freedom correct? Other checks may be necessary as well.
6. Save all preliminary, primary, and backup analyses that provide the informational base from which study conclusions are drawn.

After the statistical analysis is completed, conclusions must be drawn and the results communicated to the intended audience. Sometimes it is necessary to communicate these results as a formal, written, statistical report. A general outline for a statistical report that we have found useful and informative follows.

DEFINITION 20.4

General Outline for a Statistical Report

1. Summary
2. Introduction
3. Experimental design and study procedures
4. Descriptive statistics
5. Statistical methodology
6. Results and conclusions
7. Discussion
8. Data listings

20.8 Documentation and Storage of Results

The final part of this cycle of data processing, analysis, and summarization concerns the documentation and storage of results. For formal statistical analyses that are subject to careful scrutiny by others, it is important to provide detailed documentation for all data processing and the statistical analyses so the data trail is clear and the database or work files readily accessible. Then the reviewer can follow what has been done, redo it, or extend the analyses. The elements of a documentation and storage file depend on the particular setting in which you work. The contents for a general documentation storage file are as follows.

DEFINITION 20.5

Study Documentation and Storage File

1. Statistical report
2. Study description
3. Random code (used to assign subjects to treatment groups)
4. Important correspondence
5. File creation information
6. Preliminary, primary, and backup analyses
7. Raw data source
8. A data management sheet, which includes the log, as well as information on the storage of the data files

The major thrust behind the documentation and storage file is that we want to provide a clear data and analysis "trail" for our own use or for someone else's use, should there be a need to revisit the data. For any given situation, ask yourself whether such documentation is necessary and, if so, how detailed it must be. A good test of the completeness and understandability of your documentation is to ask a colleague, who is unfamiliar with your project but knowledgeable in your field, to try to reconstruct and even redo the primary analyses you did. If he or she can navigate through your documentation trail, you have done the job.

20.9 Summary

In this chapter, we have discussed how to present the results of a statistical analysis of data and some of the problems with effectively communicating these results to the intended audience. The task is not easy. Some of the obstacles or hurdles standing in the way of effectively communicating the results of statistical analyses include graphical distortions, biased sampling, and omitting a discussion of the sample size and sampling technique. With some understanding of these obstacles, we can better critique and understand communications aimed at us and also do a better job communicating the results of our analyses to others.

The final topic in this chapter dealt with the documentation and storage of results. Having completed your analyses, drawn conclusions, and communicated these results to the intended audience, the temptation is to postpone or eliminate the documentation and storage of results. However, it is worth your time to assess the potential for revisiting your analyses in the future and determine what steps should be taken to facilitate this process (for you or for others).

Finally, we should put our statistical analyses in the context of the practical problem(s) being addressed. The report of statistical analyses will not necessarily be the answer to an important question; it is only *part* of the answer. For example, we may demonstrate that a tablet delivers a drug more quickly than a capsule, but this is not the only consideration in the decision to market the capsule or tablet form. Such factors as cost, palatability, and stability will also be considered. Some of the relevant analyses addressing these considerations are not statistical.

Supplementary Exercise

Class Project

Each person in the class chooses a commercial he or she has heard or seen lately and critiques it for possible distortions, as well as make suggestions for improvement with regard to clarity of message, etc. (*Note:* We didn't ask you to improve it from a commercial standpoint.) Present these findings.

APPENDIX

Statistical Tables

Table 1   Standard Normal Curve Areas

Table 2   Percentage Points of Student's t Distribution

Table 3 t Test Type II Error Curves

Table 4   Percentage Points of Sign Test: $C_{\alpha,n}$

Table 5   Percentage Points of Wilcoxon Rank Sum Test: T_L and T_U

Table 6   Percentage Points of Wilcoxon Signed-Rank Test

Table 7   Percentage Points of Chi-Square Distribution: χ_α^2

Table 8   Percentage Points of F Distribution: F_α

Table 9   Values of $2 \, \text{Arcsin} \sqrt{\hat{\pi}}$

Table 10   Percentage Points of Studentized Range Distribution: $q_\alpha(t, v)$

Table 11   Percentage Points for Dunnett's Test: $d_\alpha(k, v)$

Table 12   Percentage Points for Hartley's $F_{\max}$ Test: $F_{\max,\alpha}$

Table 13   Random Numbers

Table 14 F Test Power Curves for AOV

Table 15   Poisson Probabilities: $Pr(Y = y)$

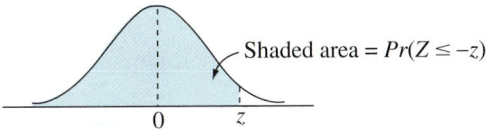

Shaded area = $Pr(Z \le -z)$

TABLE 1

Standard normal curve areas

z	0.00	0.01	0.02	0.03	0.04	0.05	0.06	0.07	0.08	0.09
−3.4	0.0003	0.0003	0.0003	0.0003	0.0003	0.0003	0.0003	0.0003	0.0003	0.0002
−3.3	0.0005	0.0005	0.0005	0.0004	0.0004	0.0004	0.0004	0.0004	0.0004	0.0003
−3.2	0.0007	0.0007	0.0006	0.0006	0.0006	0.0006	0.0006	0.0005	0.0005	0.0005
−3.1	0.0010	0.0009	0.0009	0.0009	0.0008	0.0008	0.0008	0.0008	0.0007	0.0007
−3.0	0.0013	0.0013	0.0013	0.0012	0.0012	0.0011	0.0011	0.0011	0.0010	0.0010
−2.9	0.0019	0.0018	0.0018	0.0017	0.0016	0.0016	0.0015	0.0015	0.0014	0.0014
−2.8	0.0026	0.0025	0.0024	0.0023	0.0023	0.0022	0.0021	0.0021	0.0020	0.0019
−2.7	0.0035	0.0034	0.0033	0.0032	0.0031	0.0030	0.0029	0.0028	0.0027	0.0026
−2.6	0.0047	0.0045	0.0044	0.0043	0.0041	0.0040	0.0039	0.0038	0.0037	0.0036
−2.5	0.0062	0.0060	0.0059	0.0057	0.0055	0.0054	0.0052	0.0051	0.0049	0.0048
−2.4	0.0082	0.0080	0.0078	0.0075	0.0073	0.0071	0.0069	0.0068	0.0066	0.0064
−2.3	0.0107	0.0104	0.0102	0.0099	0.0096	0.0094	0.0091	0.0089	0.0087	0.0084
−2.2	0.0139	0.0136	0.0132	0.0129	0.0125	0.0122	0.0119	0.0116	0.0113	0.0110
−2.1	0.0179	0.0174	0.0170	0.0166	0.0162	0.0158	0.0154	0.0150	0.0146	0.0143
−2.0	0.0228	0.0222	0.0217	0.0212	0.0207	0.0202	0.0197	0.0192	0.0188	0.0183
−1.9	0.0287	0.0281	0.0274	0.0268	0.0262	0.0256	0.0250	0.0244	0.0239	0.0233
−1.8	0.0359	0.0351	0.0344	0.0336	0.0329	0.0322	0.0314	0.0307	0.0301	0.0294
−1.7	0.0446	0.0436	0.0427	0.0418	0.0409	0.0401	0.0392	0.0384	0.0375	0.0367
−1.6	0.0548	0.0537	0.0526	0.0516	0.0505	0.0495	0.0485	0.0475	0.0465	0.0455
−1.5	0.0668	0.0655	0.0643	0.0630	0.0618	0.0606	0.0594	0.0582	0.0571	0.0559
−1.4	0.0808	0.0793	0.0778	0.0764	0.0749	0.0735	0.0721	0.0708	0.0694	0.0681
−1.3	0.0968	0.0951	0.0934	0.0918	0.0901	0.0885	0.0869	0.0853	0.0838	0.0823
−1.2	0.1151	0.1131	0.1112	0.1093	0.1075	0.1056	0.1038	0.1020	0.1003	0.0985
−1.1	0.1357	0.1335	0.1314	0.1292	0.1271	0.1251	0.1230	0.1210	0.1190	0.1170
−1.0	0.1587	0.1562	0.1539	0.1515	0.1492	0.1469	0.1446	0.1423	0.1401	0.1379
−0.9	0.1841	0.1814	0.1788	0.1762	0.1736	0.1711	0.1685	0.1660	0.1635	0.1611
−0.8	0.2119	0.2090	0.2061	0.2033	0.2005	0.1977	0.1949	0.1922	0.1894	0.1867
−0.7	0.2420	0.2389	0.2358	0.2327	0.2296	0.2266	0.2236	0.2206	0.2177	0.2148
−0.6	0.2743	0.2709	0.2676	0.2643	0.2611	0.2578	0.2546	0.2514	0.2483	0.2451
−0.5	0.3085	0.3050	0.3015	0.2981	0.2946	0.2912	0.2877	0.2843	0.2810	0.2776
−0.4	0.3446	0.3409	0.3372	0.3336	0.3300	0.3264	0.3228	0.3192	0.3156	0.3121
−0.3	0.3821	0.3783	0.3745	0.3707	0.3669	0.3632	0.3594	0.3557	0.3520	0.3483
−0.2	0.4207	0.4168	0.4129	0.4090	0.4052	0.4013	0.3974	0.3936	0.3897	0.3859
−0.1	0.4602	0.4602	0.4602	0.4602	0.4602	0.4602	0.4602	0.4602	0.4602	0.4602
−0.0	0.5000	0.4960	0.4920	0.4880	0.4840	0.4801	0.4761	0.4721	0.4681	0.4641

z	Area
−3.50	0.00023263
−4.00	0.00003167
−4.50	0.00000340
−5.00	0.00000029

Source: Computed by M. Longnecker using Splus.

TABLE 1
Standard normal curve areas

z	0.00	0.01	0.02	0.03	0.04	0.05	0.06	0.07	0.08	0.09
0.0	0.5000	0.5040	0.5080	0.5120	0.5160	0.5199	0.5239	0.5279	0.5319	0.5359
0.1	0.5398	0.5398	0.5398	0.5398	0.5398	0.5398	0.5398	0.5398	0.5398	0.5398
0.2	0.5793	0.5832	0.5871	0.5910	0.5948	0.5987	0.6026	0.6064	0.6103	0.6141
0.3	0.6179	0.6217	0.6255	0.6293	0.6331	0.6368	0.6406	0.6443	0.6480	0.6517
0.4	0.6554	0.6591	0.6628	0.6664	0.6700	0.6736	0.6772	0.6808	0.6844	0.6879
0.5	0.6915	0.6950	0.6985	0.7019	0.7054	0.7088	0.7123	0.7157	0.7190	0.7224
0.6	0.7257	0.7291	0.7324	0.7357	0.7389	0.7422	0.7454	0.7486	0.7517	0.7549
0.7	0.7580	0.7611	0.7642	0.7673	0.7704	0.7734	0.7764	0.7794	0.7823	0.7852
0.8	0.7881	0.7910	0.7939	0.7967	0.7995	0.8023	0.8051	0.8078	0.8106	0.8133
0.9	0.8159	0.8186	0.8212	0.8238	0.8264	0.8289	0.8315	0.8340	0.8365	0.8389
1.0	0.8413	0.8438	0.8461	0.8485	0.8508	0.8531	0.8554	0.8577	0.8599	0.8621
1.1	0.8643	0.8665	0.8686	0.8708	0.8729	0.8749	0.8770	0.8790	0.8810	0.8830
1.2	0.8849	0.8869	0.8888	0.8907	0.8925	0.8944	0.8962	0.8980	0.8997	0.9015
1.3	0.9032	0.9049	0.9066	0.9082	0.9099	0.9115	0.9131	0.9147	0.9162	0.9177
1.4	0.9192	0.9207	0.9222	0.9236	0.9251	0.9265	0.9279	0.9292	0.9306	0.9319
1.5	0.9332	0.9345	0.9357	0.9370	0.9382	0.9394	0.9406	0.9418	0.9429	0.9441
1.6	0.9452	0.9463	0.9474	0.9484	0.9495	0.9505	0.9515	0.9525	0.9535	0.9545
1.7	0.9554	0.9564	0.9573	0.9582	0.9591	0.9599	0.9608	0.9616	0.9625	0.9633
1.8	0.9641	0.9649	0.9656	0.9664	0.9671	0.9678	0.9686	0.9693	0.9699	0.9706
1.9	0.9713	0.9719	0.9726	0.9732	0.9738	0.9744	0.9750	0.9756	0.9761	0.9767
2.0	0.9772	0.9778	0.9783	0.9788	0.9793	0.9798	0.9803	0.9808	0.9812	0.9817
2.1	0.9821	0.9826	0.9830	0.9834	0.9838	0.9842	0.9846	0.9850	0.9854	0.9857
2.2	0.9861	0.9864	0.9868	0.9871	0.9875	0.9878	0.9881	0.9884	0.9887	0.9890
2.3	0.9893	0.9896	0.9898	0.9901	0.9904	0.9906	0.9909	0.9911	0.9913	0.9916
2.4	0.9918	0.9920	0.9922	0.9925	0.9927	0.9929	0.9931	0.9932	0.9934	0.9936
2.5	0.9938	0.9940	0.9941	0.9943	0.9945	0.9946	0.9948	0.9949	0.9951	0.9952
2.6	0.9953	0.9955	0.9956	0.9957	0.9959	0.9960	0.9961	0.9962	0.9963	0.9964
2.7	0.9965	0.9966	0.9967	0.9968	0.9969	0.9970	0.9971	0.9972	0.9973	0.9974
2.8	0.9974	0.9975	0.9976	0.9977	0.9977	0.9978	0.9979	0.9979	0.9980	0.9981
2.9	0.9981	0.9982	0.9982	0.9983	0.9984	0.9984	0.9985	0.9985	0.9986	0.9986
3.0	0.9987	0.9987	0.9987	0.9988	0.9988	0.9989	0.9989	0.9989	0.9990	0.9990
3.1	0.9990	0.9991	0.9991	0.9991	0.9992	0.9992	0.9992	0.9992	0.9993	0.9993
3.2	0.9993	0.9993	0.9994	0.9994	0.9994	0.9994	0.9994	0.9995	0.9995	0.9995
3.3	0.9995	0.9995	0.9995	0.9996	0.9996	0.9996	0.9996	0.9996	0.9996	0.9997
3.4	0.9997	0.9997	0.9997	0.9997	0.9997	0.9997	0.9997	0.9997	0.9997	0.9998

z	Area
3.50	0.99976737
4.00	0.99996833
4.50	0.99999660
5.00	0.99999971

Shaded area = α

TABLE 2
Percentage points of Student's t distribution

df/α =	.40	.25	.10	.05	.025	.01	.005	.001	.0005
1	0.325	1.000	3.078	6.314	12.706	31.821	63.657	318.309	636.619
2	0.289	0.816	1.886	2.920	4.303	6.965	9.925	22.327	31.599
3	0.277	0.765	1.638	2.353	3.182	4.541	5.841	10.215	12.924
4	0.271	0.741	1.533	2.132	2.776	3.747	4.604	7.173	8.610
5	0.267	0.727	1.476	2.015	2.571	3.365	4.032	5.893	6.869
6	0.265	0.718	1.440	1.943	2.447	3.143	3.707	5.208	5.959
7	0.263	0.711	1.415	1.895	2.365	2.998	3.499	4.785	5.408
8	0.262	0.706	1.397	1.860	2.306	2.896	3.355	4.501	5.041
9	0.261	0.703	1.383	1.833	2.262	2.821	3.250	4.297	4.781
10	0.260	0.700	1.372	1.812	2.228	2.764	3.169	4.144	4.587
11	0.260	0.697	1.363	1.796	2.201	2.718	3.106	4.025	4.437
12	0.259	0.695	1.356	1.782	2.179	2.681	3.055	3.930	4.318
13	0.259	0.694	1.350	1.771	2.160	2.650	3.012	3.852	4.221
14	0.258	0.692	1.345	1.761	2.145	2.624	2.977	3.787	4.140
15	0.258	0.691	1.341	1.753	2.131	2.602	2.947	3.733	4.073
16	0.258	0.690	1.337	1.746	2.120	2.583	2.921	3.686	4.015
17	0.257	0.689	1.333	1.740	2.110	2.567	2.898	3.646	3.965
18	0.257	0.688	1.330	1.734	2.101	2.552	2.878	3.610	3.922
19	0.257	0.688	1.328	1.729	2.093	2.539	2.861	3.579	3.883
20	0.257	0.687	1.325	1.725	2.086	2.528	2.845	3.552	3.850
21	0.257	0.686	1.323	1.721	2.080	2.518	2.831	3.527	3.819
22	0.256	0.686	1.321	1.717	2.074	2.508	2.819	3.505	3.792
23	0.256	0.685	1.319	1.714	2.069	2.500	2.807	3.485	3.768
24	0.256	0.685	1.318	1.711	2.064	2.492	2.797	3.467	3.745
25	0.256	0.684	1.316	1.708	2.060	2.485	2.787	3.450	3.725
26	0.256	0.684	1.315	1.706	2.056	2.479	2.779	3.435	3.707
27	0.256	0.684	1.314	1.703	2.052	2.473	2.771	3.421	3.690
28	0.256	0.683	1.313	1.701	2.048	2.467	2.763	3.408	3.674
29	0.256	0.683	1.311	1.699	2.045	2.462	2.756	3.396	3.659
30	0.256	0.683	1.310	1.697	2.042	2.457	2.750	3.385	3.646
35	0.255	0.682	1.306	1.690	2.030	2.438	2.724	3.340	3.591
40	0.255	0.681	1.303	1.684	2.021	2.423	2.704	3.307	3.551
50	0.255	0.679	1.299	1.676	2.009	2.403	2.678	3.261	3.496
60	0.254	0.679	1.296	1.671	2.000	2.390	2.660	3.232	3.460
120	0.254	0.677	1.289	1.658	1.980	2.358	2.617	3.160	3.373
inf.	0.253	0.674	1.282	1.645	1.960	2.326	2.576	3.090	3.291

Source: Computed by M. Longnecker using Splus.

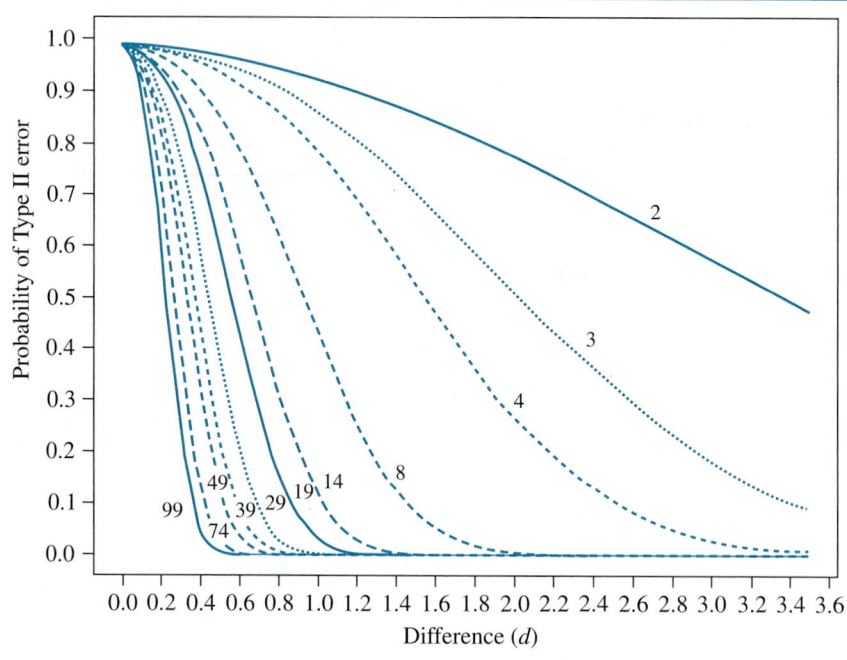

Source: Computed by M. Longnecker using SAS.

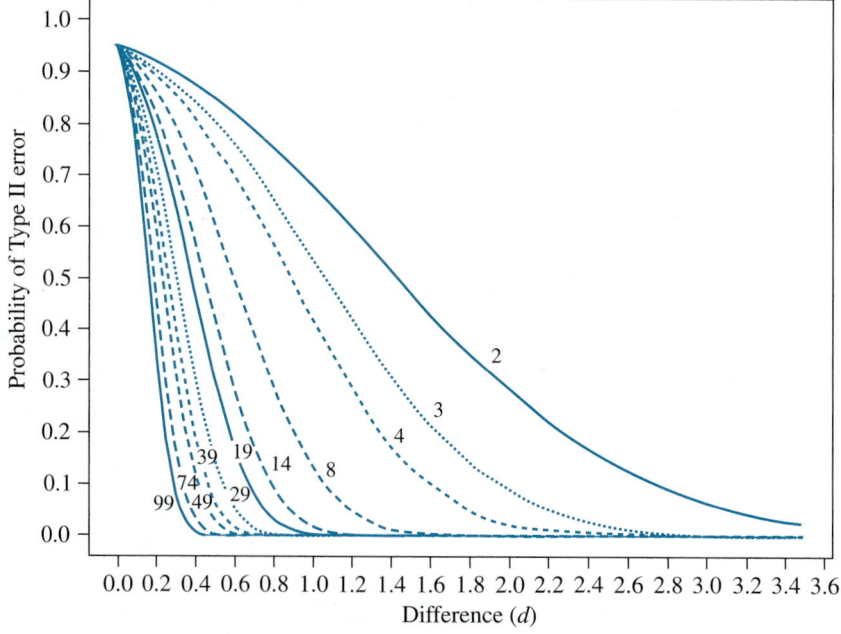

Source: Computed by M. Longnecker using SAS.

TABLE 3

Probability of Type II error curves for $\alpha = .01$ (two-sided)

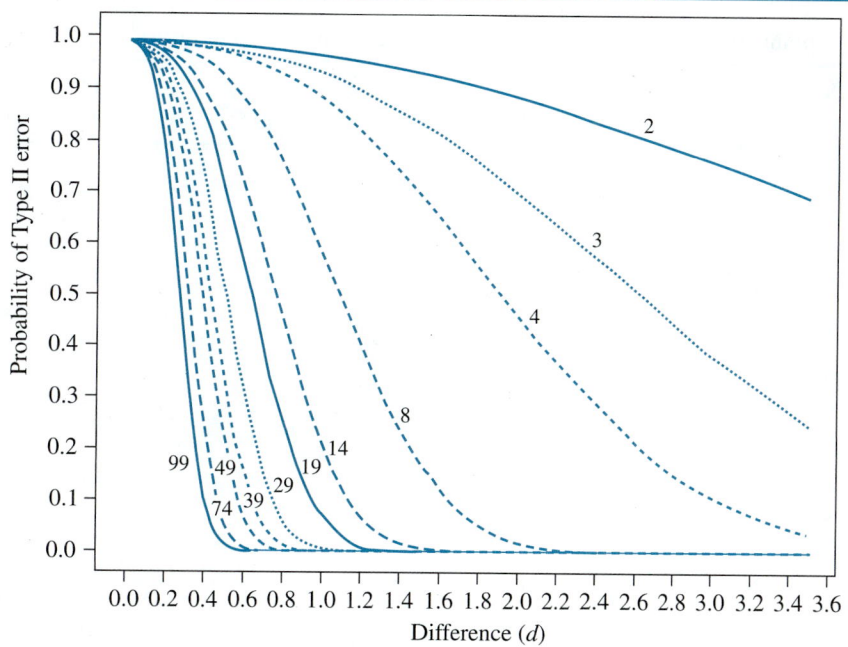

Source: Computed by M. Longnecker using SAS.

TABLE 3

Probability of Type II error curves for $\alpha = .05$ (two-sided)

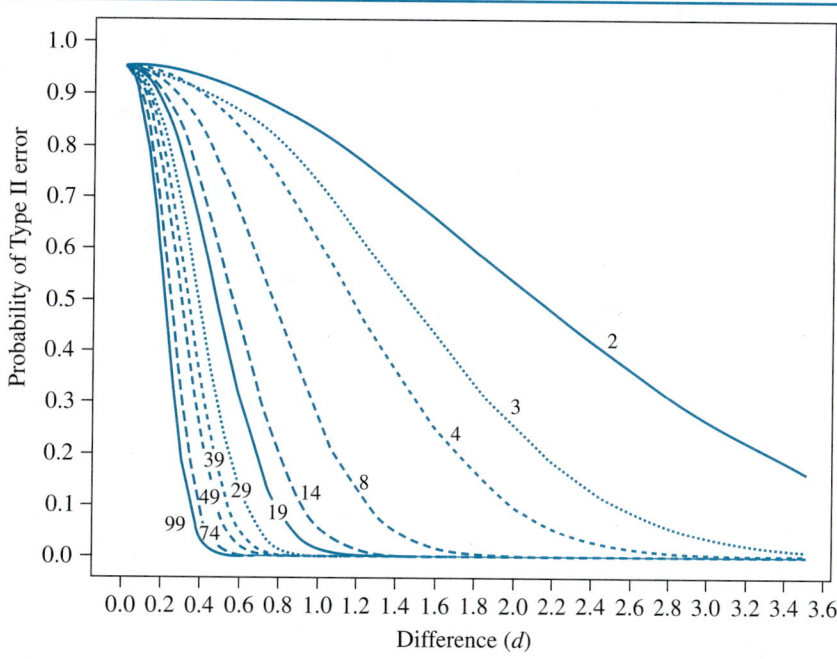

Source: Computed by M. Longnecker using SAS.

TABLE 4

Percentage points for confidence intervals on the median and the sign test: $C_{\alpha,n}$

$\alpha(2)$ $\alpha(1)$	.20 .10	.10 .05	.05 .025	.02 .01	.01 .005	.005 .0025	.002 .001	$\alpha(2)$ $\alpha(1)$	.20 .10	.10 .05	.05 .025	.02 .01	.01 .005	.005 .0025	.002 .001
n								n							
1	*	*	*	*	*	*	*	26	9	8	7	6	6	5	4
2	*	*	*	*	*	*	*	27	9	8	7	7	6	5	5
3	*	*	*	*	*	*	*	28	10	9	8	7	6	6	5
4	0	*	*	*	*	*	*	29	10	9	8	7	7	6	5
5	0	0	*	*	*	*	*	30	10	10	9	8	7	6	6
6	0	0	0	*	*	*	*	31	11	10	9	8	7	7	6
7	1	0	0	0	*	*	*	32	11	10	9	8	8	7	6
8	1	1	0	0	0	*	*	33	12	11	10	9	8	8	7
9	2	1	1	0	0	0	*	34	12	11	10	9	9	8	7
10	2	1	1	0	0	0	0	35	13	12	11	10	9	8	8
11	2	2	1	1	0	0	0	36	13	12	11	10	9	9	8
12	3	2	2	1	1	0	0	37	14	13	12	10	10	9	8
13	3	3	2	1	1	1	0	38	14	13	12	11	10	9	9
14	4	3	2	2	1	1	1	39	15	13	12	11	11	10	9
15	4	3	3	2	2	1	1	40	15	14	13	12	11	10	9
16	4	4	3	2	2	2	1	41	15	14	13	12	11	11	10
17	5	4	4	3	2	2	1	42	16	15	14	13	12	11	10
18	5	5	4	3	3	2	2	43	16	15	14	13	12	11	11
19	6	5	4	4	3	3	2	44	17	16	15	13	13	12	11
20	6	5	5	4	3	3	2	45	17	16	15	14	13	12	11
21	7	6	5	4	4	3	3	46	18	16	15	14	13	13	12
22	7	6	5	5	4	4	3	47	18	17	16	15	14	13	12
23	7	7	6	5	4	4	3	48	19	17	16	15	14	13	12
24	8	7	6	5	5	4	4	49	19	18	17	15	15	14	13
25	8	7	7	6	5	5	4	50	19	18	17	16	15	14	13

Note: An * means that no test or confidence interval of this level exists.

Source: Computed by M. Longnecker using Splus.

TABLE 5

Critical values of T_L and T_U for the Wilcoxon rank sum test: independent samples. Test statistic is rank sum associated with smaller sample (if equal sample sizes, either rank sum can be used).

a. $\alpha = .025$ one-tailed; $\alpha = .05$ two-tailed

n_2		n_1															
		3		**4**		**5**		**6**		**7**		**8**		**9**		**10**	
		T_L	T_U	T_L	T_U	T_L	T_U	T_L	T_U	T_L	T_U	T_L	T_U	T_L	T_U	T_L	T_U
3		5	16	6	18	6	21	7	23	7	26	8	28	8	31	9	33
4		6	18	11	25	12	28	12	32	13	35	14	38	15	41	16	44
5		6	21	12	28	18	37	19	41	20	45	21	49	22	53	24	56
6		7	23	12	32	19	41	26	52	28	56	29	61	31	65	32	70
7		7	26	13	35	20	45	28	56	37	68	39	73	41	78	43	83
8		8	28	14	38	21	49	29	61	39	73	49	87	51	93	54	98
9		8	31	15	41	22	53	31	65	41	78	51	93	63	108	66	114
10		9	33	16	44	24	56	32	70	43	83	54	98	66	114	79	131

b. $\alpha = .05$ one-tailed; $\alpha = .10$ two-tailed

n_2		n_1															
		3		**4**		**5**		**6**		**7**		**8**		**9**		**10**	
		T_L	T_U	T_L	T_U	T_L	T_U	T_L	T_U	T_L	T_U	T_L	T_U	T_L	T_U	T_L	T_U
3		6	15	7	17	7	20	8	22	9	24	9	27	10	29	11	31
4		7	17	12	24	13	27	14	30	15	33	16	36	17	39	18	42
5		7	20	13	27	19	36	20	40	22	43	24	46	25	50	26	54
6		8	22	14	30	20	40	28	50	30	54	32	58	33	63	35	67
7		9	24	15	33	22	43	30	54	39	66	41	71	43	76	46	80
8		9	27	16	36	24	46	32	58	41	71	52	84	54	90	57	95
9		10	29	17	39	25	50	33	63	43	76	54	90	66	105	69	111
10		11	31	18	42	26	54	35	67	46	80	57	95	69	111	83	127

Source: From F. Wilcoxon and R. A. Wilcox, *Some Rapid Approximate Statistical Procedures* (Pearl River, N.Y. Lederle Laboratories, 1964), pp. 20–23. Reproduced with the permission of American Cyanamid Company.

TABLE 6
Critical values for the
Wilcoxon signed-rank test
$[n = 5(1)54]$

One-Sided	Two-Sided	n = 5	n = 6	n = 7	n = 8	n = 9
p = .1	p = .2	2	3	5	8	10
p = .05	p = .1	0	2	3	5	8
p = .025	p = .05		0	2	3	5
p = .01	p = .02			0	1	3
p = .005	p = .01				0	1
p = .0025	p = .005					0
p = .001	p = .002					

One-Sided	Two-Sided	n = 15	n = 16	n = 17	n = 18	n = 19
p = .1	p = .2	36	42	48	55	62
p = .05	p = .1	30	35	41	47	53
p = .025	p = .05	25	29	34	40	46
p = .01	p = .02	19	23	27	32	37
p = .005	p = .01	15	19	23	27	32
p = .0025	p = .005	12	15	19	23	27
p = .001	p = .002	8	11	14	18	21

One-Sided	Two-Sided	n = 25	n = 26	n = 27	n = 28	n = 29
p = .1	p = .2	113	124	134	145	157
p = .05	p = .1	100	110	119	130	140
p = .025	p = .05	89	98	107	116	126
p = .01	p = .02	76	84	92	101	110
p = .005	p = .01	68	75	83	91	100
p = .0025	p = .005	60	67	74	82	90
p = .001	p = .002	51	58	64	71	79

One-Sided	Two-Sided	n = 35	n = 36	n = 37	n = 38	n = 39
p = .1	p = .2	235	250	265	281	297
p = .05	p = .1	213	227	241	256	271
p = .025	p = .05	195	208	221	235	249
p = .01	p = .02	173	185	198	211	224
p = .005	p = .01	159	171	182	194	207
p = .0025	p = .005	146	157	168	180	192
p = .001	p = .002	131	141	151	162	173

One-Sided	Two-Sided	n = 45	n = 46	n = 47	n = 48	n = 49
p = .1	p = .2	402	422	441	462	482
p = .05	p = .1	371	389	407	426	446
p = .025	p = .05	343	361	378	396	415
p = .01	p = .02	312	328	345	362	379
p = .005	p = .01	291	307	322	339	355
p = .0025	p = .005	272	287	302	318	334
p = .001	p = .002	249	263	277	292	307

Source: Computed by P. J. Hildebrand.

TABLE 6
(*continued*)

One-Sided	Two-Sided	$n = 10$	$n = 11$	$n = 12$	$n = 13$	$n = 14$
$p = .1$	$p = .2$	14	17	21	26	31
$p = .05$	$p = .1$	10	13	17	21	25
$p = .025$	$p = .05$	8	10	13	17	21
$p = .01$	$p = .02$	5	7	9	12	15
$p = .005$	$p = .01$	3	5	7	9	12
$p = .0025$	$p = .005$	1	3	5	7	9
$p = .001$	$p = .002$	0	1	2	4	6

One-Sided	Two-Sided	$n = 20$	$n = 21$	$n = 22$	$n = 23$	$n = 24$
$p = .1$	$p = .2$	69	77	86	94	104
$p = .05$	$p = .1$	60	67	75	83	91
$p = .025$	$p = .05$	52	58	65	73	81
$p = .01$	$p = .02$	43	49	55	62	69
$p = .005$	$p = .01$	37	42	48	54	61
$p = .0025$	$p = .005$	32	37	42	48	54
$p = .001$	$p = .002$	26	30	35	40	45

One-Sided	Two-Sided	$n = 30$	$n = 31$	$n = 32$	$n = 33$	$n = 34$
$p = .1$	$p = .2$	169	181	194	207	221
$p = .05$	$p = .1$	151	163	175	187	200
$p = .025$	$p = .05$	137	147	159	170	182
$p = .01$	$p = .02$	120	130	140	151	162
$p = .005$	$p = .01$	109	118	128	138	148
$p = .0025$	$p = .005$	98	107	116	126	136
$p = .001$	$p = .002$	86	94	103	112	121

One-Sided	Two-Sided	$n = 40$	$n = 41$	$n = 42$	$n = 43$	$n = 44$
$p = .1$	$p = .2$	313	330	348	365	384
$p = .05$	$p = .1$	286	302	319	336	353
$p = .025$	$p = .05$	264	279	294	310	327
$p = .01$	$p = .02$	238	252	266	281	296
$p = .005$	$p = .01$	220	233	247	261	276
$p = .0025$	$p = .005$	204	217	230	244	258
$p = .001$	$p = .002$	185	197	209	222	235

One-Sided	Two-Sided	$n = 50$	$n = 51$	$n = 52$	$n = 53$	$n = 54$
$p = .1$	$p = .2$	503	525	547	569	592
$p = .05$	$p = .1$	466	486	507	529	550
$p = .025$	$p = .05$	434	453	473	494	514
$p = .01$	$p = .02$	397	416	434	454	473
$p = .005$	$p = .01$	373	390	408	427	445
$p = .0025$	$p = .005$	350	367	384	402	420
$p = .001$	$p = .002$	323	339	355	372	389

TABLE 7

Percentage points of the chi-square distribution

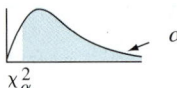

df/α =	.999	.995	.99	.975	.95	.90
1	.000002	.000039	.000157	.000982	.003932	.01579
2	.002001	.01003	.02010	.05064	.1026	.2107
3	.02430	.07172	.1148	.2158	.3518	.5844
4	.09080	.2070	.2971	.4844	.7107	1.064
5	.2102	.4117	.5543	.8312	1.145	1.610
6	.3811	.6757	.8721	1.237	1.635	2.204
7	.5985	.9893	1.239	1.690	2.167	2.833
8	.8571	1.344	1.646	2.180	2.733	3.490
9	1.152	1.735	2.088	2.700	3.325	4.168
10	1.479	2.156	2.558	3.247	3.940	4.865
11	1.834	2.603	3.053	3.816	4.575	5.578
12	2.214	3.074	3.571	4.404	5.226	6.304
13	2.617	3.565	4.107	5.009	5.892	7.042
14	3.041	4.075	4.660	5.629	6.571	7.790
15	3.483	4.601	5.229	6.262	7.261	8.547
16	3.942	5.142	5.812	6.908	7.962	9.312
17	4.416	5.697	6.408	7.564	8.672	10.09
18	4.905	6.265	7.015	8.231	9.390	10.86
19	5.407	6.844	7.633	8.907	10.12	11.65
20	5.921	7.434	8.260	9.591	10.85	12.44
21	6.447	8.034	8.897	10.28	11.59	13.24
22	6.983	8.643	9.542	10.98	12.34	14.04
23	7.529	9.260	10.20	11.69	13.09	14.85
24	8.085	9.886	10.86	12.40	13.85	15.66
25	8.649	10.52	11.52	13.12	14.61	16.47
26	9.222	11.16	12.20	13.84	15.38	17.29
27	9.803	11.81	12.88	14.57	16.15	18.11
28	10.39	12.46	13.56	15.31	16.93	18.94
29	10.99	13.12	14.26	16.06	17.71	19.77
30	11.59	13.79	14.95	16.79	18.49	20.60
40	17.92	20.71	22.16	24.43	26.51	29.05
50	24.67	27.99	29.71	32.36	34.76	37.69
60	31.74	35.53	37.48	40.48	43.19	46.46
70	39.04	43.28	45.44	48.76	51.74	55.33
80	46.52	51.17	53.54	57.15	60.39	64.28
90	54.16	59.20	61.75	65.65	69.13	73.29
100	61.92	67.33	70.06	74.22	77.93	82.36
120	77.76	83.85	86.92	91.57	95.70	100.62
240	177.95	187.32	191.99	198.98	205.14	212.39

TABLE 7
(*continued*)

α = .10	.05	.025	.01	.005	.001	df
2.706	3.841	5.024	6.635	7.879	10.83	1
4.605	5.991	7.378	9.210	10.60	13.82	2
6.251	7.815	9.348	11.34	12.84	16.27	3
7.779	9.488	11.14	13.28	14.86	18.47	4
9.236	11.07	12.83	15.09	16.75	20.52	5
10.64	12.59	14.45	16.81	18.55	22.46	6
12.02	14.07	16.01	18.48	20.28	24.32	7
13.36	15.51	17.53	20.09	21.95	26.12	8
14.68	16.92	19.02	21.67	23.59	27.88	9
15.99	18.31	20.48	23.21	25.19	29.59	10
17.28	19.68	21.92	24.72	26.76	31.27	11
18.55	21.03	23.34	26.22	28.30	32.91	12
19.81	22.36	24.74	27.69	29.82	34.53	13
21.06	23.68	26.12	29.14	31.32	36.12	14
22.31	25.00	27.49	30.58	32.80	37.70	15
23.54	26.30	28.85	32.00	34.27	39.25	16
24.77	27.59	30.19	33.41	35.72	40.79	17
25.99	28.87	31.53	34.81	37.16	42.31	18
27.20	30.14	32.85	36.19	38.58	43.82	19
28.41	31.41	34.17	37.57	40.00	45.31	20
29.62	32.67	35.48	38.93	41.40	46.80	21
30.81	33.92	36.78	40.29	42.80	48.27	22
32.01	35.17	38.08	41.64	44.18	49.73	23
33.20	36.42	39.36	42.98	45.56	51.18	24
34.38	37.65	40.65	44.31	46.93	52.62	25
35.56	38.89	41.92	45.64	48.29	54.05	26
36.74	40.11	43.19	46.96	49.65	55.48	27
37.92	41.34	44.46	48.28	50.99	56.89	28
39.09	42.56	45.72	49.59	52.34	58.30	29
40.26	43.77	46.98	50.89	53.67	59.70	30
51.81	55.76	59.34	63.69	66.77	73.40	40
63.17	67.50	71.42	76.15	79.49	86.66	50
74.40	79.08	83.30	88.38	91.95	99.61	60
85.53	90.53	95.02	100.43	104.21	112.32	70
96.58	101.88	106.63	112.33	116.32	124.84	80
107.57	113.15	118.14	124.12	128.30	137.21	90
118.50	124.34	129.56	135.81	140.17	149.45	100
140.23	146.57	152.21	158.95	163.65	173.62	120
268.47	277.14	284.80	293.89	300.18	313.44	240

Source: Computed by P. J. Hildebrand.

TABLE 8
Percentage points of the F distribution (df_2 between 1 and 6)

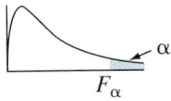

df_2	α	1	2	3	4	5	6	7	8	9	10
1	.25	5.83	7.50	8.20	8.58	8.82	8.98	9.10	9.19	9.26	9.32
	.10	39.86	49.50	53.59	55.83	57.24	58.20	58.91	59.44	59.86	60.19
	.05	161.4	199.5	215.7	224.6	230.2	234.0	236.8	238.9	240.5	241.9
	.025	647.8	799.5	864.2	899.6	921.8	937.1	948.2	956.7	963.3	968.6
	.01	4052	5000	5403	5625	5764	5859	5928	5981	6022	6056
2	.25	2.57	3.00	3.15	3.23	3.28	3.31	3.34	3.35	3.37	3.38
	.10	8.53	9.00	9.16	9.24	9.29	9.33	9.35	9.37	9.38	9.39
	.05	18.51	19.00	19.16	19.25	19.30	19.33	19.35	19.37	19.38	19.40
	.025	38.51	39.00	39.17	39.25	39.30	39.33	39.36	39.37	39.39	39.40
	.01	98.50	99.00	99.17	99.25	99.30	99.33	99.36	99.37	99.39	99.40
	.005	198.5	199.0	199.2	199.2	199.3	199.3	199.4	199.4	199.4	199.4
	.001	998.5	999.0	999.2	999.2	999.3	999.3	999.4	999.4	999.4	999.4
3	.25	2.02	2.28	2.36	2.39	2.41	2.42	2.43	2.44	2.44	2.44
	.10	5.54	5.46	5.39	5.34	5.31	5.28	5.27	5.25	5.24	5.23
	.05	10.13	9.55	9.28	9.12	9.01	8.94	8.89	8.85	8.81	8.79
	.025	17.44	16.04	15.44	15.10	14.88	14.73	14.62	14.54	14.47	14.42
	.01	34.12	30.82	29.46	28.71	28.24	27.91	27.67	27.49	27.35	27.23
	.005	55.55	49.80	47.47	46.19	45.39	44.84	44.43	44.13	43.88	43.69
	.001	167.0	148.5	141.1	137.1	134.6	132.8	131.6	130.6	129.9	129.2
4	.25	1.81	2.00	2.05	2.06	2.07	2.08	2.08	2.08	2.08	2.08
	.10	4.54	4.32	4.19	4.11	4.05	4.01	3.98	3.95	3.94	3.92
	.05	7.71	6.94	6.59	6.39	6.26	6.16	6.09	6.04	6.00	5.96
	.025	12.22	10.65	9.98	9.60	9.36	9.20	9.07	8.98	8.90	8.84
	.01	21.20	18.00	16.69	15.98	15.52	15.21	14.98	14.80	14.66	14.55
	.005	31.33	26.28	24.26	23.15	22.46	21.97	21.62	21.35	21.14	20.97
	.001	74.14	61.25	56.18	53.44	51.71	50.53	49.66	49.00	48.47	48.05
5	.25	1.69	1.85	1.88	1.89	1.89	1.89	1.89	1.89	1.89	1.89
	.10	4.06	3.78	3.62	3.52	3.45	3.40	3.37	3.34	3.32	3.30
	.05	6.61	5.79	5.41	5.19	5.05	4.95	4.88	4.82	4.77	4.74
	.025	10.01	8.43	7.76	7.39	7.15	6.98	6.85	6.76	6.68	6.62
	.01	16.26	13.27	12.06	11.39	10.97	10.67	10.46	10.29	10.16	10.05
	.005	22.78	18.31	16.53	15.56	14.94	14.51	14.20	13.96	13.77	13.62
	.001	47.18	37.12	33.20	31.09	29.75	28.83	28.16	27.65	27.24	26.92
6	.25	1.62	1.76	1.78	1.79	1.79	1.78	1.78	1.78	1.77	1.77
	.10	3.78	3.46	3.29	3.18	3.11	3.05	3.01	2.98	2.96	2.94
	.05	5.99	5.14	4.76	4.53	4.39	4.28	4.21	4.15	4.10	4.06
	.025	8.81	7.26	6.60	6.23	5.99	5.82	5.70	5.60	5.52	5.46
	.01	13.75	10.92	9.78	9.15	8.75	8.47	8.26	8.10	7.98	7.87
	.005	18.63	14.54	12.92	12.03	11.46	11.07	10.79	10.57	10.39	10.25
	.001	35.51	27.00	23.70	21.92	20.80	20.03	19.46	19.03	18.69	18.41

TABLE 8

(*continued*)

					df$_1$							
12	**15**	**20**	**24**	**30**	**40**	**60**	**120**	**240**	**inf.**	α	df$_2$
9.41	9.49	9.58	9.63	9.67	9.71	9.76	9.80	9.83	9.85	.25	**1**
60.71	61.22	61.74	62.00	62.26	62.53	62.79	63.06	63.19	63.33	.10	
243.9	245.9	248.0	249.1	250.1	251.1	252.2	253.3	253.8	254.3	.05	
976.7	984.9	993.1	997.2	1001	1006	1010	1014	1016	1018	.025	
6106	6157	6209	6235	6261	6287	6313	6339	6353	6366	.01	
3.39	3.41	3.43	3.43	3.44	3.45	3.46	3.47	3.47	3.48	.25	**2**
9.41	9.42	9.44	9.45	9.46	9.47	9.47	9.48	9.49	9.49	.10	
19.41	19.43	19.45	19.45	19.46	19.47	19.48	19.49	19.49	19.50	.05	
39.41	39.43	39.45	39.46	39.46	39.47	39.48	39.49	39.49	39.50	.025	
99.42	99.43	99.45	99.46	99.47	99.47	99.48	99.49	99.50	99.50	.01	
199.4	199.4	199.4	199.5	199.5	199.5	199.5	199.5	199.5	199.5	.005	
999.4	999.4	999.4	999.5	999.5	999.5	999.5	999.5	999.5	999.5	.001	
2.45	2.46	2.46	2.46	2.47	2.47	2.47	2.47	2.47	2.47	.25	**3**
5.22	5.20	5.18	5.18	5.17	5.16	5.15	5.14	5.14	5.13	.10	
8.74	8.70	8.66	8.64	8.62	8.59	8.57	8.55	8.54	8.53	.05	
14.34	14.25	14.17	14.12	14.08	14.04	13.99	13.95	13.92	13.90	.025	
27.05	26.87	26.69	26.60	26.50	26.41	26.32	26.22	26.17	26.13	.01	
43.39	43.08	42.78	42.62	42.47	42.31	42.15	41.99	41.91	41.83	.005	
128.3	127.4	126.4	125.9	125.4	125.0	124.5	124.0	123.7	123.5	.001	
2.08	2.08	2.08	2.08	2.08	2.08	2.08	2.08	2.08	2.08	.25	**4**
3.90	3.87	3.84	3.83	3.82	3.80	3.79	3.78	3.77	3.76	.10	
5.91	5.86	5.80	5.77	5.75	5.72	5.69	5.66	5.64	5.63	.05	
8.75	8.66	8.56	8.51	8.46	8.41	8.36	8.31	8.28	8.26	.025	
14.37	14.20	14.02	13.93	13.84	13.75	13.65	13.56	13.51	13.46	.01	
20.70	20.44	20.17	20.03	19.89	19.75	19.61	19.47	19.40	19.32	.005	
47.41	46.76	46.10	45.77	45.43	45.09	44.75	44.40	44.23	44.05	.001	
1.89	1.89	1.88	1.88	1.88	1.88	1.87	1.87	1.87	1.87	.25	**5**
3.27	3.24	3.21	3.19	3.17	3.16	3.14	3.12	3.11	3.10	.10	
4.68	4.62	4.56	4.53	4.50	4.46	4.43	4.40	4.38	4.36	.05	
6.52	6.43	6.33	6.28	6.23	6.18	6.12	6.07	6.04	6.02	.025	
9.89	9.72	9.55	9.47	9.38	9.29	9.20	9.11	9.07	9.02	.01	
13.38	13.15	12.90	12.78	12.66	12.53	12.40	12.27	12.21	12.14	.005	
26.42	25.91	25.39	25.13	24.87	24.60	24.33	24.06	23.92	23.79	.001	
1.77	1.76	1.76	1.75	1.75	1.75	1.74	1.74	1.74	1.74	.25	**6**
2.90	2.87	2.84	2.82	2.80	2.78	2.76	2.74	2.73	2.72	.10	
4.00	3.94	3.87	3.84	3.81	3.77	3.74	3.70	3.69	3.67	.05	
5.37	5.27	5.17	5.12	5.07	5.01	4.96	4.90	4.88	4.85	.025	
7.72	7.56	7.40	7.31	7.23	7.14	7.06	6.97	6.92	6.88	.01	
10.03	9.81	9.59	9.47	9.36	9.24	9.12	9.00	8.94	8.88	.005	
17.99	17.56	17.12	16.90	16.67	16.44	16.21	15.98	15.86	15.75	.001	

TABLE 8

Percentage points of the F distribution (df_2 between 7 and 12)

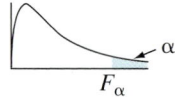

df_2	α	df_1 1	2	3	4	5	6	7	8	9	10
7	.25	1.57	1.70	1.72	1.72	1.71	1.71	1.70	1.70	1.69	1.69
	.10	3.59	3.26	3.07	2.96	2.88	2.83	2.78	2.75	2.72	2.70
	.05	5.59	4.74	4.35	4.12	3.97	3.87	3.79	3.73	3.68	3.64
	.025	8.07	6.54	5.89	5.52	5.29	5.12	4.99	4.90	4.82	4.76
	.01	12.25	9.55	8.45	7.85	7.46	7.19	6.99	6.84	6.72	6.62
	.005	16.24	12.40	10.88	10.05	9.52	9.16	8.89	8.68	8.51	8.38
	.001	29.25	21.69	18.77	17.20	16.21	15.52	15.02	14.63	14.33	14.08
8	.25	1.54	1.66	1.67	1.66	1.66	1.65	1.64	1.64	1.63	1.63
	.10	3.46	3.11	2.92	2.81	2.73	2.67	2.62	2.59	2.56	2.54
	.05	5.32	4.46	4.07	3.84	3.69	3.58	3.50	3.44	3.39	3.35
	.025	7.57	6.06	5.42	5.05	4.82	4.65	4.53	4.43	4.36	4.30
	.01	11.26	8.65	7.59	7.01	6.63	6.37	6.18	6.03	5.91	5.81
	.005	14.69	11.04	9.60	8.81	8.30	7.95	7.69	7.50	7.34	7.21
	.001	25.41	18.49	15.83	14.39	13.48	12.86	12.40	12.05	11.77	11.54
9	.25	1.51	1.62	1.63	1.63	1.62	1.61	1.60	1.60	1.59	1.59
	.10	3.36	3.01	2.81	2.69	2.61	2.55	2.51	2.47	2.44	2.42
	.05	5.12	4.26	3.86	3.63	3.48	3.37	3.29	3.23	3.18	3.14
	.025	7.21	5.71	5.08	4.72	4.48	4.32	4.20	4.10	4.03	3.96
	.01	10.56	8.02	6.99	6.42	6.06	5.80	5.61	5.47	5.35	5.26
	.005	13.61	10.11	8.72	7.96	7.47	7.13	6.88	6.69	6.54	6.42
	.001	22.86	16.39	13.90	12.56	11.71	11.13	10.70	10.37	10.11	9.89
10	.25	1.49	1.60	1.60	1.59	1.59	1.58	1.57	1.56	1.56	1.55
	.10	3.29	2.92	2.73	2.61	2.52	2.46	2.41	2.38	2.35	2.32
	.05	4.96	4.10	3.71	3.48	3.33	3.22	3.14	3.07	3.02	2.98
	.025	6.94	5.46	4.83	4.47	4.24	4.07	3.95	3.85	3.78	3.72
	.01	10.04	7.56	6.55	5.99	5.64	5.39	5.20	5.06	4.94	4.85
	.005	12.83	9.43	8.08	7.34	6.87	6.54	6.30	6.12	5.97	5.85
	.001	21.04	14.91	12.55	11.28	10.48	9.93	9.52	9.20	8.96	8.75
11	.25	1.47	1.58	1.58	1.57	1.56	1.55	1.54	1.53	1.53	1.52
	.10	3.23	2.86	2.66	2.54	2.45	2.39	2.34	2.30	2.27	2.25
	.05	4.84	3.98	3.59	3.36	3.20	3.09	3.01	2.95	2.90	2.85
	.025	6.72	5.26	4.63	4.28	4.04	3.88	3.76	3.66	3.59	3.53
	.01	9.65	7.21	6.22	5.67	5.32	5.07	4.89	4.74	4.63	4.54
	.005	12.23	8.91	7.60	6.88	6.42	6.10	5.86	5.68	5.54	5.42
	.001	19.69	13.81	11.56	10.35	9.58	9.05	8.66	8.35	8.12	7.92
12	.25	1.46	1.56	1.56	1.55	1.54	1.53	1.52	1.51	1.51	1.50
	.10	3.18	2.81	2.61	2.48	2.39	2.33	2.28	2.24	2.21	2.19
	.05	4.75	3.89	3.49	3.26	3.11	3.00	2.91	2.85	2.80	2.75
	.025	6.55	5.10	4.47	4.12	3.89	3.73	3.61	3.51	3.44	3.37
	.01	9.33	6.93	5.95	5.41	5.06	4.82	4.64	4.50	4.39	4.30
	.005	11.75	8.51	7.23	6.52	6.07	5.76	5.52	5.35	5.20	5.09
	.001	18.64	12.97	10.80	9.63	8.89	8.38	8.00	7.71	7.48	7.29

TABLE 8
(*continued*)

					df$_1$							
12	**15**	**20**	**24**	**30**	**40**	**60**	**120**	**240**	**inf.**	α	**df$_2$**
1.68	1.68	1.67	1.67	1.66	1.66	1.65	1.65	1.65	1.65	.25	**7**
2.67	2.63	2.59	2.58	2.56	2.54	2.51	2.49	2.48	2.47	.10	
3.57	3.51	3.44	3.41	3.38	3.34	3.30	3.27	3.25	3.23	.05	
4.67	4.57	4.47	4.41	4.36	4.31	4.25	4.20	4.17	4.14	.025	
6.47	6.31	6.16	6.07	5.99	5.91	5.82	5.74	5.69	5.65	.01	
8.18	7.97	7.75	7.64	7.53	7.42	7.31	7.19	7.13	7.08	.005	
13.71	13.32	12.93	12.73	12.53	12.33	12.12	11.91	11.80	11.70	.001	
1.62	1.62	1.61	1.60	1.60	1.59	1.59	1.58	1.58	1.58	.25	**8**
2.50	2.46	2.42	2.40	2.38	2.36	2.34	2.32	2.30	2.29	.10	
3.28	3.22	3.15	3.12	3.08	3.04	3.01	2.97	2.95	2.93	.05	
4.20	4.10	4.00	3.95	3.89	3.84	3.78	3.73	3.70	3.67	.025	
5.67	5.52	5.36	5.28	5.20	5.12	5.03	4.95	4.90	4.86	.01	
7.01	6.81	6.61	6.50	6.40	6.29	6.18	6.06	6.01	5.95	.005	
11.19	10.84	10.48	10.30	10.11	9.92	9.73	9.53	9.43	9.33	.001	
1.58	1.57	1.56	1.56	1.55	1.54	1.64	1.53	1.53	1.53	.25	**9**
2.38	2.34	2.30	2.28	2.25	2.23	2.21	2.18	2.17	2.16	.10	
3.07	3.01	2.94	2.90	2.86	2.83	2.79	2.75	2.73	2.71	.05	
3.87	3.77	3.67	3.61	3.56	3.51	3.45	3.39	3.36	3.33	.025	
5.11	4.96	4.81	4.73	4.65	4.57	4.48	4.40	4.35	4.31	.01	
6.23	6.03	5.83	5.73	5.62	5.52	5.41	5.30	5.24	5.19	.005	
9.57	9.24	8.90	8.72	8.55	8.37	8.19	8.00	7.91	7.81	.001	
1.54	1.53	1.52	1.52	1.51	1.51	1.50	1.49	1.49	1.48	.25	**10**
2.28	2.24	2.20	2.18	2.16	2.13	2.11	2.08	2.07	2.06	.10	
2.91	2.85	2.77	2.74	2.70	2.66	2.62	2.58	2.56	2.54	.05	
3.62	3.52	3.42	3.37	3.31	3.26	3.20	3.14	3.11	3.08	.025	
4.71	4.56	4.41	4.33	4.25	4.17	4.08	4.00	3.95	3.91	.01	
5.66	5.47	5.27	5.17	5.07	4.97	4.86	4.75	4.69	4.64	.005	
8.45	8.13	7.80	7.64	7.47	7.30	7.12	6.94	6.85	6.76	.001	
1.51	1.50	1.49	1.49	1.48	1.47	1.47	1.46	1.45	1.45	.25	**11**
2.21	2.17	2.12	2.10	2.08	2.05	2.03	2.00	1.99	1.97	.10	
2.79	2.72	2.65	2.61	2.57	2.53	2.49	2.45	2.43	2.40	.05	
3.43	3.33	3.23	3.17	3.12	3.06	3.00	2.94	2.91	2.88	.025	
4.40	4.25	4.10	4.02	3.94	3.86	3.78	3.69	3.65	3.60	.01	
5.24	5.05	4.86	4.76	4.65	4.55	4.45	4.34	4.28	4.23	.005	
7.63	7.32	7.01	6.85	6.68	6.52	6.35	6.18	6.09	6.00	.001	
1.49	1.48	1.47	1.46	1.45	1.45	1.44	1.43	1.43	1.42	.25	**12**
2.15	2.10	2.06	2.04	2.01	1.99	1.96	1.93	1.92	1.90	.10	
2.69	2.62	2.54	2.51	2.47	2.43	2.38	2.34	2.32	2.30	.05	
3.28	3.18	3.07	3.02	2.96	2.91	2.85	2.79	2.76	2.72	.025	
4.16	4.01	3.86	3.78	3.70	3.62	3.54	3.45	3.41	3.36	.01	
4.91	4.72	4.53	4.43	4.33	4.23	4.12	4.01	3.96	3.90	.005	
7.00	6.71	6.40	6.25	6.09	5.93	5.76	5.59	5.51	5.42	.001	

TABLE 8

Percentage points of the F distribution (df$_2$ between 13 and 18)

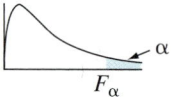

df$_2$	α	1	2	3	4	5	6	7	8	9	10
						df$_1$					
13	.25	1.45	1.55	1.55	1.53	1.52	1.51	1.50	1.49	1.49	1.48
	.10	3.14	2.76	2.56	2.43	2.35	2.28	2.23	2.20	2.16	2.14
	.05	4.67	3.81	3.41	3.18	3.03	2.92	2.83	2.77	2.71	2.67
	.025	6.41	4.97	4.35	4.00	3.77	3.60	3.48	3.39	3.31	3.25
	.01	9.07	6.70	5.74	5.21	4.86	4.62	4.44	4.30	4.19	4.10
	.005	11.37	8.19	6.93	6.23	5.79	5.48	5.25	5.08	4.94	4.82
	.001	17.82	12.31	10.21	9.07	8.35	7.86	7.49	7.21	6.98	6.80
14	.25	1.44	1.53	1.53	1.52	1.51	1.50	1.49	1.48	1.47	1.46
	.10	3.10	2.73	2.52	2.39	2.31	2.24	2.19	2.15	2.12	2.10
	.05	4.60	3.74	3.34	3.11	2.96	2.85	2.76	2.70	2.65	2.60
	.025	6.30	4.86	4.24	3.89	3.66	3.50	3.38	3.29	3.21	3.15
	.01	8.86	6.51	5.56	5.04	4.69	4.46	4.28	4.14	4.03	3.94
	.005	11.06	7.92	6.68	6.00	5.56	5.26	5.03	4.86	4.72	4.60
	.001	17.14	11.78	9.73	8.62	7.92	7.44	7.08	6.80	6.58	6.40
15	.25	1.43	1.52	1.52	1.51	1.49	1.48	1.47	1.46	1.46	1.45
	.10	3.07	2.70	2.49	2.36	2.27	2.21	2.16	2.12	2.09	2.06
	.05	4.54	3.68	3.29	3.06	2.90	2.79	2.71	2.64	2.59	2.54
	.025	6.20	4.77	4.15	3.80	3.58	3.41	3.29	3.20	3.12	3.06
	.01	8.68	6.36	5.42	4.89	4.56	4.32	4.14	4.00	3.89	3.80
	.005	10.80	7.70	6.48	5.80	5.37	5.07	4.85	4.67	4.54	4.42
	.001	16.59	11.34	9.34	8.25	7.57	7.09	6.74	6.47	6.26	6.08
16	.25	1.42	1.51	1.51	1.50	1.48	1.47	1.46	1.45	1.44	1.44
	.10	3.05	2.67	2.46	2.33	2.24	2.18	2.13	2.09	2.06	2.03
	.05	4.49	3.63	3.24	3.01	2.85	2.74	2.66	2.59	2.54	2.49
	.025	6.12	4.69	4.08	3.73	3.50	3.34	3.22	3.12	3.05	2.99
	.01	8.53	6.23	5.29	4.77	4.44	4.20	4.03	3.89	3.78	3.69
	.005	10.58	7.51	6.30	5.64	5.21	4.91	4.69	4.52	4.38	4.27
	.001	16.12	10.97	9.01	7.94	7.27	6.80	6.46	6.19	5.98	5.81
17	.25	1.42	1.51	1.50	1.49	1.47	1.46	1.45	1.44	1.43	1.43
	.10	3.03	2.64	2.44	2.31	2.22	2.15	2.10	2.06	2.03	2.00
	.05	4.45	3.59	3.20	2.96	2.81	2.70	2.61	2.55	2.49	2.45
	.025	6.04	4.62	4.01	3.66	3.44	3.28	3.16	3.06	2.98	2.92
	.01	8.40	6.11	5.18	4.67	4.34	4.10	3.93	3.79	3.68	3.59
	.005	10.38	7.35	6.16	5.50	5.07	4.78	4.56	4.39	4.25	4.14
	.001	15.72	10.66	8.73	7.68	7.02	6.56	6.22	5.96	5.75	5.58
18	.25	1.41	1.50	1.49	1.48	1.46	1.45	1.44	1.43	1.42	1.42
	.10	3.01	2.62	2.42	2.29	2.20	2.13	2.08	2.04	2.00	1.98
	.05	4.41	3.55	3.16	2.93	2.77	2.66	2.58	2.51	2.46	2.41
	.025	5.98	4.56	3.95	3.61	3.38	3.22	3.10	3.01	2.93	2.87
	.01	8.29	6.01	5.09	4.58	4.25	4.01	3.84	3.71	3.60	3.51
	.005	10.22	7.21	6.03	5.37	4.96	4.66	4.44	4.28	4.14	4.03
	.001	15.38	10.39	8.49	7.46	6.81	6.35	6.02	5.76	5.56	5.39

TABLE 8
(*continued*)

12	15	20	24	30	40	60	120	240	inf.	α	df$_2$
1.47	1.46	1.45	1.44	1.43	1.42	1.42	1.41	1.40	1.40	.25	**13**
2.10	2.05	2.01	1.98	1.96	1.93	1.90	1.88	1.86	1.85	.10	
2.60	2.53	2.46	2.42	2.38	2.34	2.30	2.25	2.23	2.21	.05	
3.15	3.05	2.95	2.89	2.84	2.78	2.72	2.66	2.63	2.60	.025	
3.96	3.82	3.66	3.59	3.51	3.43	3.34	3.25	3.21	3.17	.01	
4.64	4.46	4.27	4.17	4.07	3.97	3.87	3.76	3.70	3.65	.005	
6.52	6.23	5.93	5.78	5.63	5.47	5.30	5.14	5.05	4.97	.001	
1.45	1.44	1.43	1.42	1.41	1.41	1.40	1.39	1.38	1.38	.25	**14**
2.05	2.01	1.96	1.94	1.91	1.89	1.86	1.83	1.81	1.80	.10	
2.53	2.46	2.39	2.35	2.31	2.27	2.22	2.18	2.15	2.13	.05	
3.05	2.95	2.84	2.79	2.73	2.67	2.61	2.55	2.52	2.49	.025	
3.80	3.66	3.51	3.43	3.35	3.27	3.18	3.09	3.05	3.00	.01	
4.43	4.25	4.06	3.96	3.86	3.76	3.66	3.55	3.49	3.44	.005	
6.13	5.85	5.56	5.41	5.25	5.10	4.94	4.77	4.69	4.60	.001	
1.44	1.43	1.41	1.41	1.40	1.39	1.38	1.37	1.36	1.36	.25	**15**
2.02	1.97	1.92	1.90	1.87	1.85	1.82	1.79	1.77	1.76	.10	
2.48	2.40	2.33	2.29	2.25	2.20	2.16	2.11	2.09	2.07	.05	
2.96	2.86	2.76	2.70	2.64	2.59	2.52	2.46	2.43	2.40	.025	
3.67	3.52	3.37	3.29	3.21	3.13	3.05	2.96	2.91	2.87	.01	
4.25	4.07	3.88	3.79	3.69	3.58	3.48	3.37	3.32	3.26	.005	
5.81	5.54	5.25	5.10	4.95	4.80	4.64	4.47	4.39	4.31	.001	
1.43	1.41	1.40	1.39	1.38	1.37	1.36	1.35	1.35	1.34	.25	**16**
1.99	1.94	1.89	1.87	1.84	1.81	1.78	1.75	1.73	1.72	.10	
2.42	2.35	2.28	2.24	2.19	2.15	2.11	2.06	2.03	2.01	.05	
2.89	2.79	2.68	2.63	2.57	2.51	2.45	2.38	2.35	2.32	.025	
3.55	3.41	3.26	3.18	3.10	3.02	2.93	2.84	2.80	2.75	.01	
4.10	3.92	3.73	3.64	3.54	3.44	3.33	3.22	3.17	3.11	.005	
5.55	5.27	4.99	4.85	4.70	4.54	4.39	4.23	4.14	4.06	.001	
1.41	1.40	1.39	1.38	1.37	1.36	1.35	1.34	1.33	1.33	.25	**17**
1.96	1.91	1.86	1.84	1.81	1.78	1.75	1.72	1.70	1.69	.10	
2.38	2.31	2.23	2.19	2.15	2.10	2.06	2.01	1.99	1.96	.05	
2.82	2.72	2.62	2.56	2.50	2.44	2.38	2.32	2.28	2.25	.025	
3.46	3.31	3.16	3.08	3.00	2.92	2.83	2.75	2.70	2.65	.01	
3.97	3.79	3.61	3.51	3.41	3.31	3.21	3.10	3.04	2.98	.005	
5.32	5.05	4.78	4.63	4.48	4.33	4.18	4.02	3.93	3.85	.001	
1.40	1.39	1.38	1.37	1.36	1.35	1.34	1.33	1.32	1.32	.25	**18**
1.93	1.89	1.84	1.81	1.78	1.75	1.72	1.69	1.67	1.66	.10	
2.34	2.27	2.19	2.15	2.11	2.06	2.02	1.97	1.94	1.92	.05	
2.77	2.67	2.56	2.50	2.44	2.38	2.32	2.26	2.22	2.19	.025	
3.37	3.23	3.08	3.00	2.92	2.84	2.75	2.66	2.61	2.57	.01	
3.86	3.68	3.50	3.40	3.30	3.20	3.10	2.99	2.93	2.87	.005	
5.13	4.87	4.59	4.45	4.30	4.15	4.00	3.84	3.75	3.67	.001	

df$_1$

TABLE 8

Percentage points of the F distribution (df$_2$ between 19 and 24)

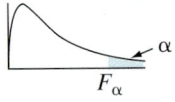

df$_2$	α	1	2	3	4	5	6	7	8	9	10
						df$_1$					
19	.25	1.41	1.49	1.49	1.47	1.46	1.44	1.43	1.42	1.41	1.41
	.10	2.99	2.61	2.40	2.27	2.18	2.11	2.06	2.02	1.98	1.96
	.05	4.38	3.52	3.13	2.90	2.74	2.63	2.54	2.48	2.42	2.38
	.025	5.92	4.51	3.90	3.56	3.33	3.17	3.05	2.96	2.88	2.82
	.01	8.18	5.93	5.01	4.50	4.17	3.94	3.77	3.63	3.52	3.43
	.005	10.07	7.09	5.92	5.27	4.85	4.56	4.34	4.18	4.04	3.93
	.001	15.08	10.16	8.28	7.27	6.62	6.18	5.85	5.59	5.39	5.22
20	.25	1.40	1.49	1.48	1.47	1.45	1.44	1.43	1.42	1.41	1.40
	.10	2.97	2.59	2.38	2.25	2.16	2.09	2.04	2.00	1.96	1.94
	.05	4.35	3.49	3.10	2.87	2.71	2.60	2.51	2.45	2.39	2.35
	.025	5.87	4.46	3.86	3.51	3.29	3.13	3.01	2.91	2.84	2.77
	.01	8.10	5.85	4.94	4.43	4.10	3.87	3.70	3.56	3.46	3.37
	.005	9.94	6.99	5.82	5.17	4.76	4.47	4.26	4.09	3.96	3.85
	.001	14.82	9.95	8.10	7.10	6.46	6.02	5.69	5.44	5.24	5.08
21	.25	1.40	1.48	1.48	1.46	1.44	1.43	1.42	1.41	1.40	1.39
	.10	2.96	2.57	2.36	2.23	2.14	2.08	2.02	1.98	1.95	1.92
	.05	4.32	3.47	3.07	2.84	2.68	2.57	2.49	2.42	2.37	2.32
	.025	5.83	4.42	3.82	3.48	3.25	3.09	2.97	2.87	2.80	2.73
	.01	8.02	5.78	4.87	4.37	4.04	3.81	3.64	3.51	3.40	3.31
	.005	9.83	6.89	5.73	5.09	4.68	4.39	4.18	4.01	3.88	3.77
	.001	14.59	9.77	7.94	6.95	6.32	5.88	5.56	5.31	5.11	4.95
22	.25	1.40	1.48	1.47	1.45	1.44	1.42	1.41	1.40	1.39	1.39
	.10	2.95	2.56	2.35	2.22	2.13	2.06	2.01	1.97	1.93	1.90
	.05	4.30	3.44	3.05	2.82	2.66	2.55	2.46	2.40	2.34	2.30
	.025	5.79	4.38	3.78	3.44	3.22	3.05	2.93	2.84	2.76	2.70
	.01	7.95	5.72	4.82	4.31	3.99	3.76	3.59	3.45	3.35	3.26
	.005	9.73	6.81	5.65	5.02	4.61	4.32	4.11	3.94	3.81	3.70
	.001	14.38	9.61	7.80	6.81	6.19	5.76	5.44	5.19	4.99	4.83
23	.25	1.39	1.47	1.47	1.45	1.43	1.42	1.41	1.40	1.39	1.38
	.10	2.94	2.55	2.34	2.21	2.11	2.05	1.99	1.95	1.92	1.89
	.05	4.28	3.42	3.03	2.80	2.64	2.53	2.44	2.37	2.32	2.27
	.025	5.75	4.35	3.75	3.41	3.18	3.02	2.90	2.81	2.73	2.67
	.01	7.88	5.66	4.76	4.26	3.94	3.71	3.54	3.41	3.30	3.21
	.005	9.63	6.73	5.58	4.95	4.54	4.26	4.05	3.88	3.75	3.64
	.001	14.20	9.47	7.67	6.70	6.08	5.65	5.33	5.09	4.89	4.73
24	.25	1.39	1.47	1.46	1.44	1.43	1.41	1.40	1.39	1.38	1.38
	.10	2.93	2.54	2.33	2.19	2.10	2.04	1.98	1.94	1.91	1.88
	.05	4.26	3.40	3.01	2.78	2.62	2.51	2.42	2.36	2.30	2.25
	.025	5.72	4.32	3.72	3.38	3.15	2.99	2.87	2.78	2.70	2.64
	.01	7.82	5.61	4.72	4.22	3.90	3.67	3.50	3.36	3.26	3.17
	.005	9.55	6.66	5.52	4.89	4.49	4.20	3.99	3.83	3.69	3.59
	.001	14.03	9.34	7.55	6.59	5.98	5.55	5.23	4.99	4.80	4.64

TABLE 8
(*continued*)

12	15	20	24	30	40	60	120	240	inf.	α	df_2
					df_1						
1.40	1.38	1.37	1.36	1.35	1.34	1.33	1.32	1.31	1.30	.25	**19**
1.91	1.86	1.81	1.79	1.76	1.73	1.70	1.67	1.65	1.63	.10	
2.31	2.23	2.16	2.11	2.07	2.03	1.98	1.93	1.90	1.88	.05	
2.72	2.62	2.51	2.45	2.39	2.33	2.27	2.20	2.17	2.13	.025	
3.30	3.15	3.00	2.92	2.84	2.76	2.67	2.58	2.54	2.49	.01	
3.76	3.59	3.40	3.31	3.21	3.11	3.00	2.89	2.83	2.78	.005	
4.97	4.70	4.43	4.29	4.14	3.99	3.84	3.68	3.60	3.51	.001	
1.39	1.37	1.36	1.35	1.34	1.33	1.32	1.31	1.30	1.29	.25	**20**
1.89	1.84	1.79	1.77	1.74	1.71	1.68	1.64	1.63	1.61	.10	
2.28	2.20	2.12	2.08	2.04	1.99	1.95	1.90	1.87	1.84	.05	
2.68	2.57	2.46	2.41	2.35	2.29	2.22	2.16	2.12	2.09	.025	
3.23	3.09	2.94	2.86	2.78	2.69	2.61	2.52	2.47	2.42	.01	
3.68	3.50	3.32	3.22	3.12	3.02	2.92	2.81	2.75	2.69	.005	
4.82	4.56	4.29	4.15	4.00	3.86	3.70	3.54	3.46	3.38	.001	
1.38	1.37	1.35	1.34	1.33	1.32	1.31	1.30	1.29	1.28	.25	**21**
1.87	1.83	1.78	1.75	1.72	1.69	1.66	1.62	1.60	1.59	.10	
2.25	2.18	2.10	2.05	2.01	1.96	1.92	1.87	1.84	1.81	.05	
2.64	2.53	2.42	2.37	2.31	2.25	2.18	2.11	2.08	2.04	.025	
3.17	3.03	2.88	2.80	2.72	2.64	2.55	2.46	2.41	2.36	.01	
3.60	3.43	3.24	3.15	3.05	2.95	2.84	2.73	2.67	2.61	.005	
4.70	4.44	4.17	4.03	3.88	3.74	3.58	3.42	3.34	3.26	.001	
1.37	1.36	1.34	1.33	1.32	1.31	1.30	1.29	1.28	1.28	.25	**22**
1.86	1.81	1.76	1.73	1.70	1.67	1.64	1.60	1.59	1.57	.10	
2.23	2.15	2.07	2.03	1.98	1.94	1.89	1.84	1.81	1.78	.05	
2.60	2.50	2.39	2.33	2.27	2.21	2.14	2.08	2.04	2.00	.025	
3.12	2.98	2.83	2.75	2.67	2.58	2.50	2.40	2.35	2.31	.01	
3.54	3.36	3.18	3.08	2.98	2.88	2.77	2.66	2.60	2.55	.005	
4.58	4.33	4.06	3.92	3.78	3.63	3.48	3.32	3.23	3.15	.001	
1.37	1.35	1.34	1.33	1.32	1.31	1.30	1.28	1.28	1.27	.25	**23**
1.84	1.80	1.74	1.72	1.69	1.66	1.62	1.59	1.57	1.55	.10	
2.20	2.13	2.05	2.01	1.96	1.91	1.86	1.81	1.79	1.76	.05	
2.57	2.47	2.36	2.30	2.24	2.18	2.11	2.04	2.01	1.97	.025	
3.07	2.93	2.78	2.70	2.62	2.54	2.45	2.35	2.31	2.26	.01	
3.47	3.30	3.12	3.02	2.92	2.82	2.71	2.60	2.54	2.48	.005	
4.48	4.23	3.96	3.82	3.68	3.53	3.38	3.22	3.14	3.05	.001	
1.36	1.35	1.33	1.32	1.31	1.30	1.29	1.28	1.27	1.26	.25	**24**
1.83	1.78	1.73	1.70	1.67	1.64	1.61	1.57	1.55	1.53	.10	
2.18	2.11	2.03	1.98	1.94	1.89	1.84	1.79	1.76	1.73	.05	
2.54	2.44	2.33	2.27	2.21	2.15	2.08	2.01	1.97	1.94	.025	
3.03	2.89	2.74	2.66	2.58	2.49	2.40	2.31	2.26	2.21	.01	
3.42	3.25	3.06	2.97	2.87	2.77	2.66	2.55	2.49	2.43	.005	
4.39	4.14	3.87	3.74	3.59	3.45	3.29	3.14	3.05	2.97	.001	

TABLE 8

Percentage points of the F distribution (df$_2$ between 25 and 30)

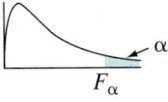

F_α

df$_2$	α	1	2	3	4	5	6	7	8	9	10
							df$_1$				
25	.25	1.39	1.47	1.46	1.44	1.42	1.41	1.40	1.39	1.38	1.37
	.10	2.92	2.53	2.32	2.18	2.09	2.02	1.97	1.93	1.89	1.87
	.05	4.24	3.39	2.99	2.76	2.60	2.49	2.40	2.34	2.28	2.24
	.025	5.69	4.29	3.69	3.35	3.13	2.97	2.85	2.75	2.68	2.61
	.01	7.77	5.57	4.68	4.18	3.85	3.63	3.46	3.32	3.22	3.13
	.005	9.48	6.60	5.46	4.84	4.43	4.15	3.94	3.78	3.64	3.54
	.001	13.88	9.22	7.45	6.49	5.89	5.46	5.15	4.91	4.71	4.56
26	.25	1.38	1.46	1.45	1.44	1.42	1.41	1.39	1.38	1.37	1.37
	.10	2.91	2.52	2.31	2.17	2.08	2.01	1.96	1.92	1.88	1.86
	.05	4.23	3.37	2.98	2.74	2.59	2.47	2.39	2.32	2.27	2.22
	.025	5.66	4.27	3.67	3.33	3.10	2.94	2.82	2.73	2.65	2.59
	.01	7.72	5.53	4.64	4.14	3.82	3.59	3.42	3.29	3.18	3.09
	.005	9.41	6.54	5.41	4.79	4.38	4.10	3.89	3.73	3.60	3.49
	.001	13.74	9.12	7.36	6.41	5.80	5.38	5.07	4.83	4.64	4.48
27	.25	1.38	1.46	1.45	1.43	1.42	1.40	1.39	1.38	1.37	1.36
	.10	2.90	2.51	2.30	2.17	2.07	2.00	1.95	1.91	1.87	1.85
	.05	4.21	3.35	2.96	2.73	2.57	2.46	2.37	2.31	2.25	2.20
	.025	5.63	4.24	3.65	3.31	3.08	2.92	2.80	2.71	2.63	2.57
	.01	7.68	5.49	4.60	4.11	3.78	3.56	3.39	3.26	3.15	3.06
	.005	9.34	6.49	5.36	4.74	4.34	4.06	3.85	3.69	3.56	3.45
	.001	13.61	9.02	7.27	6.33	5.73	5.31	5.00	4.76	4.57	4.41
28	.25	1.38	1.46	1.45	1.43	1.41	1.40	1.39	1.38	1.37	1.36
	.10	2.89	2.50	2.29	2.16	2.06	2.00	1.94	1.90	1.87	1.84
	.05	4.20	3.34	2.95	2.71	2.56	2.45	2.36	2.29	2.24	2.19
	.025	5.61	4.22	3.63	3.29	3.06	2.90	2.78	2.69	2.61	2.55
	.01	7.64	5.45	4.57	4.07	3.75	3.53	3.36	3.23	3.12	3.03
	.005	9.28	6.44	5.32	4.70	4.30	4.02	3.81	3.65	3.52	3.41
	.001	13.50	8.93	7.19	6.25	5.66	5.24	4.93	4.69	4.50	4.35
29	.25	1.38	1.45	1.45	1.43	1.41	1.40	1.38	1.37	1.36	1.35
	.10	2.89	2.50	2.28	2.15	2.06	1.99	1.93	1.89	1.86	1.83
	.05	4.18	3.33	2.93	2.70	2.55	2.43	2.35	2.28	2.22	2.18
	.025	5.59	4.20	3.61	3.27	3.04	2.88	2.76	2.67	2.59	2.53
	.01	7.60	5.42	4.54	4.04	3.73	3.50	3.33	3.20	3.09	3.00
	.005	9.23	6.40	5.28	4.66	4.26	3.98	3.77	3.61	3.48	3.38
	.001	13.39	8.85	7.12	6.19	5.59	5.18	4.87	4.64	4.45	4.29
30	.25	1.38	1.45	1.44	1.42	1.41	1.39	1.38	1.37	1.36	1.35
	.10	2.88	2.49	2.28	2.14	2.05	1.98	1.93	1.88	1.85	1.82
	.05	4.17	3.32	2.92	2.69	2.53	2.42	2.33	2.27	2.21	2.16
	.025	5.57	4.18	3.59	3.25	3.03	2.87	2.75	2.65	2.57	2.51
	.01	7.56	5.39	4.51	4.02	3.70	3.47	3.30	3.17	3.07	2.98
	.005	9.18	6.35	5.24	4.62	4.23	3.95	3.74	3.58	3.45	3.34
	.001	13.29	8.77	7.05	6.12	5.53	5.12	4.82	4.58	4.39	4.24

TABLE 8

(*continued*)

							df$_1$				
12	**15**	**20**	**24**	**30**	**40**	**60**	**120**	**240**	**inf.**	**α**	**df$_2$**
1.36	1.34	1.33	1.32	1.31	1.29	1.28	1.27	1.26	1.25	.25	**25**
1.82	1.77	1.72	1.69	1.66	1.63	1.59	1.56	1.54	1.52	.10	
2.16	2.09	2.01	1.96	1.92	1.87	1.82	1.77	1.74	1.71	.05	
2.51	2.41	2.30	2.24	2.18	2.12	2.05	1.98	1.94	1.91	.025	
2.99	2.85	2.70	2.62	2.54	2.45	2.36	2.27	2.22	2.17	.01	
3.37	3.20	3.01	2.92	2.82	2.72	2.61	2.50	2.44	2.38	.005	
4.31	4.06	3.79	3.66	3.52	3.37	3.22	3.06	2.98	2.89	.001	
1.35	1.34	1.32	1.31	1.30	1.29	1.28	1.26	1.26	1.25	.25	**26**
1.81	1.76	1.71	1.68	1.65	1.61	1.58	1.54	1.52	1.50	.10	
2.15	2.07	1.99	1.95	1.90	1.85	1.80	1.75	1.72	1.69	.05	
2.49	2.39	2.28	2.22	2.16	2.09	2.03	1.95	1.92	1.88	.025	
2.96	2.81	2.66	2.58	2.50	2.42	2.33	2.23	2.18	2.13	.01	
3.33	3.15	2.97	2.87	2.77	2.67	2.56	2.45	2.39	2.33	.005	
4.24	3.99	3.72	3.59	3.44	3.30	3.15	2.99	2.90	2.82	.001	
1.35	1.33	1.32	1.31	1.30	1.28	1.27	1.26	1.25	1.24	.25	**27**
1.80	1.75	1.70	1.67	1.64	1.60	1.57	1.53	1.51	1.49	.10	
2.13	2.06	1.97	1.93	1.88	1.84	1.79	1.73	1.70	1.67	.05	
2.47	2.36	2.25	2.19	2.13	2.07	2.00	1.93	1.89	1.85	.025	
2.93	2.78	2.63	2.55	2.47	2.38	2.29	2.20	2.15	2.10	.01	
3.28	3.11	2.93	2.83	2.73	2.63	2.52	2.41	2.35	2.29	.005	
4.17	3.92	3.66	3.52	3.38	3.23	3.08	2.92	2.84	2.75	.001	
1.34	1.33	1.31	1.30	1.29	1.28	1.27	1.25	1.24	1.24	.25	**28**
1.79	1.74	1.69	1.66	1.63	1.59	1.56	1.52	1.50	1.48	.10	
2.12	2.04	1.96	1.91	1.87	1.82	1.77	1.71	1.68	1.65	.05	
2.45	2.34	2.23	2.17	2.11	2.05	1.98	1.91	1.87	1.83	.025	
2.90	2.75	2.60	2.52	2.44	2.35	2.26	2.17	2.12	2.06	.01	
3.25	3.07	2.89	2.79	2.69	2.59	2.48	2.37	2.31	2.25	.005	
4.11	3.86	3.60	3.46	3.32	3.18	3.02	2.86	2.78	2.69	.001	
1.34	1.32	1.31	1.30	1.29	1.27	1.26	1.25	1.24	1.23	.25	**29**
1.78	1.73	1.68	1.65	1.62	1.58	1.55	1.51	1.49	1.47	.10	
2.10	2.03	1.94	1.90	1.85	1.81	1.75	1.70	1.67	1.64	.05	
2.43	2.32	2.21	2.15	2.09	2.03	1.96	1.89	1.85	1.81	.025	
2.87	2.73	2.57	2.49	2.41	2.33	2.23	2.14	2.09	2.03	.01	
3.21	3.04	2.86	2.76	2.66	2.56	2.45	2.33	2.27	2.21	.005	
4.05	3.80	3.54	3.41	3.27	3.12	2.97	2.81	2.73	2.64	.001	
1.34	1.32	1.30	1.29	1.28	1.27	1.26	1.24	1.23	1.23	.25	**30**
1.77	1.72	1.67	1.64	1.61	1.57	1.54	1.50	1.48	1.46	.10	
2.09	2.01	1.93	1.89	1.84	1.79	1.74	1.68	1.65	1.62	.05	
2.41	2.31	2.20	2.14	2.07	2.01	1.94	1.87	1.83	1.79	.025	
2.84	2.70	2.55	2.47	2.39	2.30	2.21	2.11	2.06	2.01	.01	
3.18	3.01	2.82	2.73	2.63	2.52	2.42	2.30	2.24	2.18	.005	
4.00	3.75	3.49	3.36	3.22	3.07	2.92	2.76	2.68	2.59	.001	

TABLE 8
Percentage points of the F distribution (df_2 at least 40)

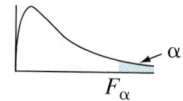

F_α

| df_2 | α | \multicolumn{10}{c}{df_1} |
		1	2	3	4	5	6	7	8	9	10
40	.25	1.36	1.44	1.42	1.40	1.39	1.37	1.36	1.35	1.34	1.33
	.10	2.84	2.44	2.23	2.09	2.00	1.93	1.87	1.83	1.79	1.76
	.05	4.08	3.23	2.84	2.61	2.45	2.34	2.25	2.18	2.12	2.08
	.025	5.42	4.05	3.46	3.13	2.90	2.74	2.62	2.53	2.45	2.39
	.01	7.31	5.18	4.31	3.83	3.51	3.29	3.12	2.99	2.89	2.80
	.005	8.83	6.07	4.98	4.37	3.99	3.71	3.51	3.35	3.22	3.12
	.001	12.61	8.25	6.59	5.70	5.13	4.73	4.44	4.21	4.02	3.87
60	.25	1.35	1.42	1.41	1.38	1.37	1.35	1.33	1.32	1.31	1.30
	.10	2.79	2.39	2.18	2.04	1.95	1.87	1.82	1.77	1.74	1.71
	.05	4.00	3.15	2.76	2.53	2.37	2.25	2.17	2.10	2.04	1.99
	.025	5.29	3.93	3.34	3.01	2.79	2.63	2.51	2.41	2.33	2.27
	.01	7.08	4.98	4.13	3.65	3.34	3.12	2.95	2.82	2.72	2.63
	.005	8.49	5.79	4.73	4.14	3.76	3.49	3.29	3.13	3.01	2.90
	.001	11.97	7.77	6.17	5.31	4.76	4.37	4.09	3.86	3.69	3.54
90	.25	1.34	1.41	1.39	1.37	1.35	1.33	1.32	1.31	1.30	1.29
	.10	2.76	2.36	2.15	2.01	1.91	1.84	1.78	1.74	1.70	1.67
	.05	3.95	3.10	2.71	2.47	2.32	2.20	2.11	2.04	1.99	1.94
	.025	5.20	3.84	3.26	2.93	2.71	2.55	2.43	2.34	2.26	2.19
	.01	6.93	4.85	4.01	3.53	3.23	3.01	2.84	2.72	2.61	2.52
	.005	8.28	5.62	4.57	3.99	3.62	3.35	3.15	3.00	2.87	2.77
	.001	11.57	7.47	5.91	5.06	4.53	4.15	3.87	3.65	3.48	3.34
120	.25	1.34	1.40	1.39	1.37	1.35	1.33	1.31	1.30	1.29	1.28
	.10	2.75	2.35	2.13	1.99	1.90	1.82	1.77	1.72	1.68	1.65
	.05	3.92	3.07	2.68	2.45	2.29	2.18	2.09	2.02	1.96	1.91
	.025	5.15	3.80	3.23	2.89	2.67	2.52	2.39	2.30	2.22	2.16
	.01	6.85	4.79	3.95	3.48	3.17	2.96	2.79	2.66	2.56	2.47
	.005	8.18	5.54	4.50	3.92	3.55	3.28	3.09	2.93	2.81	2.71
	.001	11.38	7.32	5.78	4.95	4.42	4.04	3.77	3.55	3.38	3.24
240	.25	1.33	1.39	1.38	1.36	1.34	1.32	1.30	1.29	1.27	1.27
	.10	2.73	2.32	2.10	1.97	1.87	1.80	1.74	1.70	1.65	1.63
	.05	3.88	3.03	2.64	2.41	2.25	2.14	2.04	1.98	1.92	1.87
	.025	5.09	3.75	3.17	2.84	2.62	2.46	2.34	2.25	2.17	2.10
	.01	6.74	4.69	3.86	3.40	3.09	2.88	2.71	2.59	2.48	2.40
	.005	8.03	5.42	4.38	3.82	3.45	3.19	2.99	2.84	2.71	2.61
	.001	11.10	7.11	5.60	4.78	4.25	3.89	3.62	3.41	3.24	3.09
inf.	.25	1.32	1.39	1.37	1.35	1.33	1.31	1.29	1.28	1.27	1.25
	.10	2.71	2.30	2.08	1.94	1.85	1.77	1.72	1.67	1.63	1.60
	.05	3.84	3.00	2.60	2.37	2.21	2.10	2.01	1.94	1.88	1.83
	.025	5.02	3.69	3.12	2.79	2.57	2.41	2.29	2.19	2.11	2.05
	.01	6.63	4.61	3.78	3.32	3.02	2.80	2.64	2.51	2.41	2.32
	.005	7.88	5.30	4.28	3.72	3.35	3.09	2.90	2.74	2.62	2.52
	.001	10.83	6.91	5.42	4.62	4.10	3.74	3.47	3.27	3.10	2.96

TABLE 8
(*continued*)

					df$_1$							
12	15	20	24	30	40	60	120	240	inf.	α	df$_2$
1.31	1.30	1.28	1.26	1.25	1.24	1.22	1.21	1.20	1.19	.25	40
1.71	1.66	1.61	1.57	1.54	1.51	1.47	1.42	1.40	1.38	.10	
2.00	1.92	1.84	1.79	1.74	1.69	1.64	1.58	1.54	1.51	.05	
2.29	2.18	2.07	2.01	1.94	1.88	1.80	1.72	1.68	1.64	.025	
2.66	2.52	2.37	2.29	2.20	2.11	2.02	1.92	1.86	1.80	.01	
2.95	2.78	2.60	2.50	2.40	2.30	2.18	2.06	2.00	1.93	.005	
3.64	3.40	3.14	3.01	2.87	2.73	2.57	2.41	2.32	2.23	.001	
1.29	1.27	1.25	1.24	1.22	1.21	1.19	1.17	1.16	1.15	.25	60
1.66	1.60	1.54	1.51	1.48	1.44	1.40	1.35	1.32	1.29	.10	
1.92	1.84	1.75	1.70	1.65	1.59	1.53	1.47	1.43	1.39	.05	
2.17	2.06	1.94	1.88	1.82	1.74	1.67	1.58	1.53	1.48	.025	
2.50	2.35	2.20	2.12	2.03	1.94	1.84	1.73	1.67	1.60	.01	
2.74	2.57	2.39	2.29	2.19	2.08	1.96	1.83	1.76	1.69	.005	
3.32	3.08	2.83	2.69	2.55	2.41	2.25	2.08	1.99	1.89	.001	
1.27	1.25	1.23	1.22	1.20	1.19	1.17	1.15	1.13	1.12	.25	90
1.62	1.56	1.50	1.47	1.43	1.39	1.35	1.29	1.26	1.23	.10	
1.86	1.78	1.69	1.64	1.59	1.53	1.46	1.39	1.35	1.30	.05	
2.09	1.98	1.86	1.80	1.73	1.66	1.58	1.48	1.43	1.37	.025	
2.39	2.24	2.09	2.00	1.92	1.82	1.72	1.60	1.53	1.46	.01	
2.61	2.44	2.25	2.15	2.05	1.94	1.82	1.68	1.61	1.52	.005	
3.11	2.88	2.63	2.50	2.36	2.21	2.05	1.87	1.77	1.66	.001	
1.26	1.24	1.22	1.21	1.19	1.18	1.16	1.13	1.12	1.10	.25	120
1.60	1.55	1.48	1.45	1.41	1.37	1.32	1.26	1.23	1.19	10	
1.83	1.75	1.66	1.61	1.55	1.50	1.43	1.35	1.31	1.25	.05	
2.05	1.94	1.82	1.76	1.69	1.61	1.53	1.43	1.38	1.31	.025	
2.34	2.19	2.03	1.95	1.86	1.76	1.66	1.53	1.46	1.38	.01	
2.54	2.37	2.19	2.09	1.98	1.87	1.75	1.61	1.52	1.43	.005	
3.02	2.78	2.53	2.40	2.26	2.11	1.95	1.77	1.66	1.54	.001	
1.25	1.23	1.21	1.19	1.18	1.16	1.14	1.11	1.09	1.07	.25	240
1.57	1.52	1.45	1.42	1.38	1.33	1.28	1.22	1.18	1.13	10	
1.79	1.71	1.61	1.56	1.51	1.44	1.37	1.29	1.24	1.17	.05	
2.00	1.89	1.77	1.70	1.63	1.55	1.46	1.35	1.29	1.21	.025	
2.26	2.11	1.96	1.87	1.78	1.68	1.57	1.43	1.35	1.25	.01	
2.45	2.28	2.09	1.99	1.89	1.77	1.64	1.49	1.40	1.28	.005	
2.88	2.65	2.40	2.26	2.12	1.97	1.80	1.61	1.49	1.35	.001	
1.24	1.22	1.19	1.18	1.16	1.14	1.12	1.08	1.06	1.00	.25	inf.
1.55	1.49	1.42	1.38	1.34	1.30	1.24	1.17	1.12	1.00	10	
1.75	1.67	1.57	1.52	1.46	1.39	1.32	1.22	1.15	1.00	.05	
1.94	1.83	1.71	1.64	1.57	1.48	1.39	1.27	1.19	1.00	.025	
2.18	2.04	1.88	1.79	1.70	1.59	1.47	1.32	1.22	1.00	.01	
2.36	2.19	2.00	1.90	1.79	1.67	1.53	1.36	1.25	1.00	.005	
2.74	2.51	2.27	2.13	1.99	1.84	1.66	1.45	1.31	1.00	.001	

Source: Computed by P. J. Hildebrand.

TABLE 9
Values of 2 arcsin $\sqrt{\hat{\pi}}$

$\hat{\pi}$		$\hat{\pi}$		$\hat{\pi}$		$\hat{\pi}$		$\hat{\pi}$	
.001	.0633	.041	.4078	.36	1.2870	.76	2.1177	.971	2.7993
.002	.0895	.042	.4128	.37	1.3078	.77	2.1412	.972	2.8053
.003	.1096	.043	.4178	.38	1.3284	.78	2.1652	.973	2.8115
.004	.1266	.044	.4227	.39	1.3490	.79	2.1895	.974	2.8177
.005	.1415	.045	.4275	.40	1.3694	.80	2.2143	.975	2.8240
.006	.1551	.046	.4323	.41	1.3898	.81	2.2395	.976	2.8305
.007	.1675	.047	.4371	.42	1.4101	.82	2.2653	.977	2.8371
.008	.1791	.048	.4418	.43	1.4303	.83	2.2916	.978	2.8438
.009	.1900	.049	.4464	.44	1.4505	.84	2.3186	.979	2.8507
.010	.2003	.050	.4510	.45	1.4706	.85	2.3462	.980	2.8578
.011	.2101	.06	.4949	.46	1.4907	.86	2.3746	.981	2.8650
.012	.2195	.07	.5355	.47	1.5108	.87	2.4039	.982	2.8725
.013	.2285	.08	.5735	.48	1.5308	.88	2.4341	.983	2.8801
.014	.2372	.09	.6094	.49	1.5508	.89	2.4655	.984	2.8879
.015	.2456	.10	.6435	.50	1.5708	.90	2.4981	.985	2.8960
.016	.2537	.11	.6761	.51	1.5908	.91	2.5322	.986	2.9044
.017	.2615	.12	.7075	.52	1.6108	.92	2.5681	.987	2.9131
.018	.2691	.13	.7377	.53	1.6308	.93	2.6062	.988	2.9221
.019	.2766	.14	.7670	.54	1.6509	.94	2.6467	.989	2.9315
.020	.2838	.15	.7954	.55	1.6710	.95	2.6906	.990	2.9413
.021	.2909	.16	.8230	.56	1.6911	.951	2.6952	.991	2.9516
.022	.2978	.17	.8500	.57	1.7113	.952	2.6998	.992	2.9625
.023	.3045	.18	.8763	.58	1.7315	.953	2.7045	.993	2.9741
.024	.3111	.19	.9021	.59	1.7518	.954	2.7093	.994	2.9865
.025	.3176	.20	.9273	.60	1.7722	.955	2.7141	.995	3.0001
.026	.3239	.21	.9521	.61	1.7926	.956	2.7189	.996	3.0150
.027	.3301	.22	.9764	.62	1.8132	.957	2.7238	.997	3.0320
.028	.3363	.23	1.0004	.63	1.8338	.958	2.7288	.998	3.0521
.029	.3423	.24	1.0239	.64	1.8338	.959	2.7338	.999	3.0783
.030	.3482	.25	1.0472	.65	1.8546	.960	2.7389		
.031	.3540	.26	1.0701	.66	1.8965	.961	2.7440		
.032	.3597	.27	1.0928	.67	1.9177	.962	2.7492		
.033	.3654	.28	1.1152	.68	1.9391	.963	2.7545		
.034	.3709	.29	1.1374	.69	1.9606	.964	2.7598		
.035	.3764	.30	1.1593	.70	1.9823	.965	2.7652		
.036	.3818	.31	1.1810	.71	2.0042	.966	2.7707		
.037	.3871	.32	1.2025	.72	2.0264	.967	2.7762		
.038	.3924	.33	1.2239	.73	2.0488	.968	2.7819		
.039	.3976	.34	1.2451	.74	2.0715	.969	2.7876		
.040	.4027	.35	1.2661	.75	2.0944	.970	2.7934		

From *Experimental Design: Procedures for the Behavioral Sciences,* by
Roger E. Kirk. Copyright © 1968 by Wadsworth Publishing Company, Inc.
Reprinted by permission of the publisher, Brooks/Cole, Pacific Grove, Calif.

TABLE 10

Percentage points of the Studentized range

Error df	α	2	3	4	5	6	7	8	9	10	11
5	.05	3.64	4.60	5.22	5.67	6.03	6.33	6.58	6.80	6.99	7.17
	.01	5.70	6.98	7.80	8.42	8.91	9.32	9.67	9.97	10.24	10.48
6	.05	3.46	4.34	4.90	5.30	5.63	5.90	6.12	6.32	6.49	6.65
	.01	5.24	6.33	7.03	7.56	7.97	8.32	8.61	8.87	9.10	9.30
7	.05	3.34	4.16	4.68	5.06	5.36	5.61	5.82	6.00	6.16	6.30
	.01	4.95	5.92	6.54	7.01	7.37	7.68	7.94	8.17	8.37	8.55
8	.05	3.26	4.04	4.53	4.89	5.17	5.40	5.60	5.77	5.92	6.05
	.01	4.75	5.64	6.20	6.62	6.96	7.24	7.47	7.68	7.86	8.03
9	.05	3.20	3.95	4.41	4.76	5.02	5.24	5.43	5.59	5.74	5.87
	.01	4.60	5.43	5.96	6.35	6.66	6.91	7.13	7.33	7.49	7.65
10	.05	3.15	3.88	4.33	4.65	4.91	5.12	5.30	5.46	5.60	5.72
	.01	4.48	5.27	5.77	6.14	6.43	6.67	6.87	7.05	7.21	7.36
11	.05	3.11	3.82	4.26	4.57	4.82	5.03	5.30	5.35	5.49	5.61
	.01	4.39	5.15	5.62	5.97	6.25	6.48	6.67	6.84	6.99	7.13
12	.05	3.08	3.77	4.20	4.52	4.75	4.95	5.12	5.27	5.39	5.51
	.01	4.32	5.05	5.50	5.84	6.10	6.32	6.51	6.67	6.81	6.94
13	.05	3.06	3.73	4.15	4.45	4.69	4.88	5.05	5.19	5.32	5.43
	.01	4.26	4.96	5.40	5.73	5.98	6.19	6.37	6.53	6.67	6.79
14	.05	3.03	3.70	4.11	4.41	4.64	4.83	4.99	5.13	5.25	5.36
	.01	4.21	4.89	5.32	5.63	5.88	6.08	6.26	6.41	6.54	6.66
15	.05	3.01	3.67	4.08	4.37	4.59	4.78	4.94	5.08	5.20	5.31
	.01	4.17	4.84	5.25	5.56	5.80	5.99	6.16	6.31	6.44	6.55
16	.05	3.00	3.65	4.05	4.33	4.56	4.74	4.90	5.03	5.15	5.26
	.01	4.13	4.79	5.19	5.49	5.72	5.92	6.08	6.22	6.35	6.46
17	.05	2.98	3.63	4.02	4.30	4.52	4.70	4.86	4.99	5.11	5.21
	.01	4.10	4.74	5.14	5.43	5.66	5.85	6.01	6.15	6.27	6.38
18	.05	2.97	3.61	4.00	4.28	4.49	4.67	4.82	4.96	5.07	5.17
	.01	4.07	4.70	5.09	5.38	5.60	5.79	5.94	6.08	6.20	6.31
19	.05	2.96	3.59	3.98	4.25	4.47	4.65	4.79	4.92	5.04	5.14
	.01	4.05	4.67	5.05	5.33	5.55	5.73	5.89	6.02	6.14	6.25
20	.05	2.95	3.58	3.96	4.23	4.45	4.62	4.77	4.90	5.01	5.11
	.01	4.02	4.64	5.02	5.29	5.51	5.69	5.84	5.97	6.09	6.19
24	.05	2.92	3.53	3.90	4.17	4.37	4.54	4.68	4.81	3.92	5.01
	.01	3.96	4.55	4.91	5.17	5.37	5.54	5.69	5.81	5.92	6.02
30	.05	2.89	3.49	3.85	4.10	4.30	4.46	4.60	4.72	4.82	4.92
	.01	3.89	4.45	4.80	5.05	5.24	5.40	5.54	5.65	5.76	5.85
40	.05	2.86	3.44	3.79	4.04	4.23	4.39	4.52	4.63	4.73	4.82
	.01	3.82	4.37	4.70	4.93	5.11	5.26	5.39	5.50	5.60	5.69
60	.05	2.83	3.40	3.74	3.98	4.16	4.31	4.44	4.55	4.65	4.73
	.01	3.76	4.28	4.59	4.82	4.99	5.13	5.25	5.36	5.45	5.53
120	.05	2.80	3.36	3.68	3.92	4.10	4.24	4.36	4.47	4.56	4.64
	.01	3.70	4.20	4.50	4.71	4.87	5.01	5.12	5.21	5.30	5.37
∞	.05	2.77	3.31	3.63	3.86	4.03	4.17	4.29	4.39	4.47	4.55
	.01	3.64	4.12	4.40	4.60	4.76	4.88	4.99	5.08	5.16	5.23

TABLE 10

(*continued*)

| Error df | \multicolumn{9}{c}{t = Number of Treatment Means} | | | | | | | | | α |
	12	13	14	15	16	17	18	19	20	
5	7.32	7.47	7.60	7.72	7.83	7.93	8.03	8.12	8.21	.05
	10.70	10.89	11.08	11.24	11.40	11.55	11.68	11.81	11.93	.01
6	6.79	6.92	7.03	7.14	7.24	7.34	7.43	7.51	7.59	.05
	9.48	9.65	9.81	9.95	10.08	10.21	10.32	10.43	10.54	.01
7	6.43	6.55	6.66	6.76	6.85	6.94	7.02	7.10	7.17	.05
	8.71	8.86	9.00	9.12	9.24	9.35	9.46	9.55	9.65	.01
8	6.18	6.29	6.39	6.48	6.57	6.65	6.73	6.80	6.87	.05
	8.18	8.31	8.44	8.55	8.66	8.76	8.85	8.94	9.03	.01
9	5.98	6.09	6.19	6.28	6.36	6.44	6.51	6.58	6.64	.05
	7.78	7.91	8.03	8.13	8.23	8.33	8.41	8.49	8.57	.01
10	5.83	5.93	6.03	6.11	6.19	6.27	6.34	6.40	6.47	.05
	7.49	7.60	7.71	7.81	7.91	7.99	8.08	8.15	8.23	.01
11	5.71	5.81	5.90	5.98	6.06	6.13	6.20	6.27	6.33	.05
	7.25	7.36	7.46	7.56	7.65	7.73	7.81	7.88	7.95	.01
12	5.61	5.71	5.80	5.86	5.95	6.02	6.09	6.15	6.21	.05
	7.06	7.17	7.26	7.36	7.44	7.52	7.59	7.66	7.73	.01
13	5.53	5.63	5.71	5.79	5.86	5.93	5.99	6.05	6.11	.05
	6.90	7.01	7.10	7.19	7.27	7.35	7.42	7.48	7.55	.01
14	5.46	5.55	5.64	5.71	5.79	5.85	5.91	5.97	6.03	.05
	6.77	6.87	6.96	7.05	7.13	7.20	7.27	7.33	7.39	.01
15	5.40	5.49	5.57	5.65	5.72	5.78	5.85	5.90	5.96	.05
	6.66	6.76	6.84	6.93	7.00	7.07	7.14	7.20	7.26	.01
16	5.35	5.44	5.52	5.59	5.66	5.73	5.79	5.84	5.90	.05
	6.56	6.66	6.74	6.82	6.90	6.97	7.03	7.09	7.15	.01
17	5.31	5.39	5.47	5.54	5.61	5.67	5.73	5.79	5.84	.05
	6.48	6.57	6.66	6.73	6.81	6.87	6.94	7.00	7.05	.01
18	5.27	5.35	5.43	5.50	5.57	5.63	5.69	5.74	5.79	.05
	6.41	6.50	6.58	6.65	6.73	6.79	6.85	6.91	6.97	.01
19	5.23	5.31	5.39	5.46	5.53	5.59	5.65	5.70	5.75	.05
	6.34	6.43	6.51	6.58	6.65	6.72	6.78	6.84	6.89	.01
20	5.20	5.28	5.36	5.43	5.49	5.55	5.61	5.66	5.71	.05
	6.28	6.37	6.45	6.52	6.59	6.65	6.71	6.77	6.82	.01
24	5.10	5.18	5.25	5.32	5.38	5.44	5.49	5.55	5.59	.05
	6.11	6.19	6.26	6.33	6.39	6.45	6.51	6.56	6.61	.01
30	5.00	5.08	5.15	5.21	5.27	5.33	5.38	5.43	5.47	.05
	5.93	6.01	6.08	6.14	6.20	6.26	6.31	6.36	6.41	.01
40	4.90	4.98	5.04	5.11	5.16	5.22	5.27	5.31	5.36	.05
	5.76	5.83	5.90	5.96	6.02	6.07	6.12	6.16	6.21	.01
60	4.81	4.88	4.94	5.00	5.06	5.11	5.15	5.20	5.24	.05
	5.60	5.67	5.73	5.78	5.84	5.89	5.93	5.97	6.01	.01
120	4.71	4.78	4.84	4.90	4.95	5.00	5.04	5.09	5.13	.05
	5.44	5.50	5.56	5.61	5.66	5.71	5.75	5.79	5.83	.01
∞	4.62	4.68	4.74	4.80	4.85	4.89	4.93	4.97	5.01	.05
	5.29	5.35	5.40	5.45	5.49	5.54	5.57	5.61	5.65	.01

TABLE 11

Percentage points for Dunnett's test: $d_\alpha(k, \nu)$

						$\alpha = 0.05$ (one-sided)							
ν	$k = 2$	3	4	5	6	7	8	9	10	11	12	15	20
5	2.44	2.68	2.85	2.98	3.08	3.16	3.24	3.30	3.36	3.41	3.45	3.57	3.72
6	2.34	2.56	2.71	2.83	2.92	3.00	3.07	3.12	3.17	3.22	3.26	3.37	3.50
7	2.27	2.48	2.62	2.73	2.82	2.89	2.95	3.01	3.05	3.10	3.13	3.23	3.36
8	2.22	2.42	2.55	2.66	2.74	2.81	2.87	2.92	2.96	3.01	3.04	3.14	3.25
9	2.18	2.37	2.50	2.60	2.68	2.75	2.81	2.86	2.90	2.94	2.97	3.06	3.18
10	2.15	2.34	2.47	2.56	2.64	2.70	2.76	2.81	2.85	2.89	2.92	3.01	3.12
11	2.13	2.31	2.44	2.53	2.60	2.67	2.72	2.77	2.81	2.85	2.88	2.96	3.07
12	2.11	2.29	2.41	2.50	2.58	2.64	2.69	2.74	2.78	2.81	2.84	2.93	3.03
13	2.09	2.27	2.39	2.48	2.55	2.61	2.66	2.71	2.75	2.78	2.82	2.90	3.00
14	2.08	2.25	2.37	2.46	2.53	2.59	2.64	2.69	2.72	2.76	2.79	2.87	2.97
15	2.07	2.24	2.36	2.44	2.51	2.57	2.62	2.67	2.70	2.74	2.77	2.85	2.95
16	2.06	2.23	2.34	2.43	2.50	2.56	2.61	2.65	2.69	2.72	2.75	2.83	2.93
17	2.05	2.22	2.33	2.42	2.49	2.54	2.59	2.64	2.67	2.71	2.74	2.81	2.91
18	2.04	2.21	2.32	2.41	2.48	2.53	2.58	2.62	2.66	2.69	2.72	2.80	2.89
19	2.03	2.20	2.31	2.40	2.47	2.52	2.57	2.61	2.65	2.68	2.71	2.79	2.88
20	2.03	2.19	2.30	2.39	2.46	2.51	2.56	2.60	2.64	2.67	2.70	2.77	2.87
24	2.01	2.17	2.28	2.36	2.43	2.48	2.53	2.57	2.60	2.64	2.66	2.74	2.83
30	1.99	2.15	2.25	2.33	2.40	2.45	2.50	2.54	2.57	2.60	2.63	2.70	2.79
40	1.97	2.13	2.23	2.31	2.37	2.42	2.47	2.51	2.54	2.57	2.60	2.67	2.75
60	1.95	2.10	2.21	2.28	2.35	2.39	2.44	2.48	2.51	2.54	2.56	2.63	2.72
120	1.93	2.08	2.18	2.26	2.32	2.37	2.41	2.45	2.48	2.51	2.53	2.60	2.68
∞	1.92	2.06	2.16	2.23	2.29	2.34	2.38	2.42	2.45	2.48	2.50	2.56	2.64

From C. W. Dunnett (1955), "A Multiple Comparison Procedure for Comparing Several Treatments with a Control," *Journal of the American Statistical Association* 50, 1112–1118. Reprinted with permission from *Journal of the American Statistical Association.* Copyright 1955 by the American Statistical Association. All rights reserved. C. W. Dunnett (1964), "New Tables for Multiple Comparisons with a Control," *Biometrics* 20, 482–491. Also additional tables produced by C. W. Dunnett in 1980.

TABLE 11

Percentage points for Dunnett's test: $d_\alpha(k, \nu)$

ν	$k = 2$	3	4	5	6	7	8	9	10	11	12	15	20
					$\alpha = 0.01$ (one-sided)								
5	3.90	4.21	4.43	4.60	4.73	4.85	4.94	5.03	5.11	5.17	5.24	5.39	5.59
6	3.61	3.88	4.07	4.21	4.33	4.43	4.51	4.59	4.64	4.70	4.76	4.89	5.06
7	3.42	3.66	3.83	3.96	4.07	4.15	4.23	4.30	4.35	4.40	4.45	4.57	4.72
8	3.29	3.51	3.67	3.79	3.88	3.96	4.03	4.09	4.14	4.19	4.23	4.34	4.48
9	3.19	3.40	3.55	3.66	3.75	3.82	3.89	3.94	3.99	4.04	4.08	4.18	4.31
10	3.11	3.31	3.45	3.56	3.64	3.71	3.78	3.83	3.88	3.92	3.96	4.06	4.18
11	3.06	3.25	3.38	3.48	3.56	3.63	3.69	3.74	3.79	3.83	3.86	3.96	4.08
12	3.01	3.19	3.32	3.42	3.50	3.56	3.62	3.67	3.71	3.75	3.79	3.88	3.99
13	2.97	3.15	3.27	3.37	3.44	3.51	3.56	3.61	3.65	3.69	3.73	3.81	3.92
14	2.94	3.11	3.23	3.32	3.40	3.46	3.51	3.56	3.60	3.64	3.67	3.76	3.87
15	2.91	3.08	3.20	3.29	3.36	3.42	3.47	3.52	3.56	3.60	3.63	3.71	3.82
16	2.88	3.05	3.17	3.26	3.33	3.39	3.44	3.48	3.52	3.56	3.59	3.67	3.78
17	2.86	3.03	3.14	3.23	3.30	3.36	3.41	3.45	3.49	3.53	3.56	3.64	3.74
18	2.84	3.01	3.12	3.21	3.27	3.33	3.38	3.42	3.46	3.50	3.53	3.61	3.71
19	2.83	2.99	3.10	3.18	3.25	3.31	3.36	3.40	3.44	3.47	3.50	3.58	3.68
20	2.81	2.97	3.08	3.17	3.23	3.29	3.34	3.38	3.42	3.45	3.48	3.56	3.65
24	2.77	2.92	3.03	3.11	3.17	3.22	3.27	3.31	3.35	3.38	3.41	3.48	3.57
30	2.72	2.87	2.97	3.05	3.11	3.16	3.21	3.24	3.28	3.31	3.34	3.41	3.50
40	2.68	2.82	2.92	2.99	3.05	3.10	3.14	3.18	3.21	3.24	3.27	3.34	3.42
60	2.64	2.78	2.87	2.94	3.00	3.04	3.08	3.12	3.15	3.18	3.20	3.27	3.35
120	2.60	2.73	2.82	2.89	2.94	2.99	3.03	3.06	3.09	3.12	3.14	3.20	3.28
∞	2.56	2.68	2.77	2.84	2.89	2.93	2.97	3.00	3.03	3.06	3.08	3.14	3.21

TABLE 11

Percentage points for Dunnett's test: $d_\alpha(k, \nu)$ (*continued*)

ν	$k = 2$	3	4	5	6	7	8	9	10	11	12	15	20
				$\alpha = 0.05$ (two-sided)								
5	3.03	3.29	3.48	3.62	3.73	3.82	3.90	3.97	4.03	4.09	4.14	4.26	4.42
6	2.86	3.10	3.26	3.39	3.49	3.57	3.64	3.71	3.76	3.81	3.86	3.97	4.11
7	2.75	2.97	3.12	3.24	3.33	3.41	3.47	3.53	3.58	3.63	3.67	3.78	3.91
8	2.67	2.88	3.02	3.13	3.22	3.29	3.35	3.41	3.46	3.50	3.54	3.64	3.76
9	2.61	2.81	2.95	3.05	3.14	3.20	3.26	3.32	3.36	3.40	3.44	3.53	3.65
10	2.57	2.76	2.89	2.99	3.07	3.14	3.19	3.24	3.29	3.33	3.36	3.45	3.57
11	2.53	2.72	2.84	2.94	3.02	3.08	3.14	3.19	3.23	3.27	3.30	3.39	3.50
12	2.50	2.68	2.81	2.90	2.98	3.04	3.09	3.14	3.18	3.22	3.25	3.34	3.45
13	2.48	2.65	2.78	2.87	2.94	3.00	3.06	3.10	3.14	3.18	3.21	3.29	3.40
14	2.46	2.63	2.75	2.84	2.91	2.97	3.02	3.07	3.11	3.14	3.18	3.26	3.36
15	2.44	2.61	2.73	2.82	2.89	2.95	3.00	3.04	3.08	3.12	3.15	3.23	3.33
16	2.42	2.59	2.71	2.80	2.87	2.92	2.97	3.02	3.06	3.09	3.12	3.20	3.30
17	2.41	2.58	2.69	2.78	2.85	2.90	2.95	3.00	3.03	3.07	3.10	3.18	3.27
18	2.40	2.56	2.68	2.76	2.83	2.89	2.94	2.98	3.01	3.05	3.08	3.16	3.25
19	2.39	2.55	2.66	2.75	2.81	2.87	2.92	2.96	3.00	3.03	3.06	3.14	3.23
20	2.38	2.54	2.65	2.73	2.80	2.86	2.90	2.95	2.98	3.02	3.05	3.12	3.22
24	2.35	2.51	2.61	2.70	2.76	2.81	2.86	2.90	2.94	2.97	3.00	3.07	3.16
30	2.32	2.47	2.58	2.66	2.72	2.77	2.82	2.86	2.89	2.92	2.95	3.02	3.11
40	2.29	2.44	2.54	2.62	2.68	2.73	2.77	2.81	2.85	2.87	2.90	2.97	3.06
60	2.27	2.41	2.51	2.58	2.64	2.69	2.73	2.77	2.80	2.83	2.86	2.92	3.00
120	2.24	2.38	2.47	2.55	2.60	2.65	2.69	2.73	2.76	2.79	2.81	2.87	2.95
∞	2.21	2.35	2.44	2.51	2.57	2.61	2.65	2.69	2.72	2.74	2.77	2.83	2.91

TABLE 11

Percentage points for Dunnett's test: $d_\alpha(k, \nu)$ (*continued*)

ν	$k = 2$	3	4	5	6	7	8	9	10	11	12	15	20
					$\alpha = 0.01$ (two-sided)								
5	4.63	4.98	5.22	5.41	5.56	5.69	5.80	5.89	5.98	6.05	6.12	6.30	6.52
6	4.21	4.51	4.71	4.87	5.00	5.10	5.20	5.28	5.35	5.41	5.47	5.62	5.81
7	3.95	4.21	4.39	4.53	4.64	4.74	4.82	4.89	4.95	5.01	5.06	5.19	5.36
8	3.77	4.00	4.17	4.29	4.40	4.48	4.56	4.62	4.68	4.73	4.78	4.90	5.05
9	3.63	3.85	4.01	4.12	4.22	4.30	4.37	4.43	4.48	4.53	4.57	4.68	4.82
10	3.53	3.74	3.88	3.99	4.08	4.16	4.22	4.28	4.33	4.37	4.42	4.52	4.65
11	3.45	3.65	3.79	3.89	3.98	4.05	4.11	4.16	4.21	4.25	4.29	4.39	4.52
12	3.39	3.58	3.71	3.81	3.89	3.96	4.02	4.07	4.12	4.16	4.19	4.29	4.41
13	3.33	3.52	3.65	3.74	3.82	3.89	3.94	3.99	4.04	4.08	4.11	4.20	4.32
14	3.29	3.47	3.59	3.69	3.76	3.83	3.88	3.93	3.97	4.01	4.05	4.13	4.24
15	3.25	3.43	3.55	3.64	3.71	3.78	3.83	3.88	3.92	3.95	3.99	4.07	4.18
16	3.22	3.39	3.51	3.60	3.67	3.73	3.78	3.83	3.87	3.91	3.94	4.02	4.13
17	3.19	3.36	3.47	3.56	3.63	3.69	3.74	3.79	3.83	3.86	3.90	3.98	4.08
18	3.17	3.33	3.44	3.53	3.60	3.66	3.71	3.75	3.79	3.83	3.86	3.94	4.04
19	3.15	3.31	3.42	3.50	3.57	3.63	3.68	3.72	3.76	3.79	3.83	3.90	4.00
20	3.13	3.29	3.40	3.48	3.55	3.60	3.65	3.69	3.73	3.77	3.80	3.87	3.97
24	3.07	3.22	3.32	3.40	3.47	3.52	3.57	3.61	3.64	3.68	3.70	3.78	3.87
30	3.01	3.15	3.25	3.33	3.39	3.44	3.49	3.52	3.56	3.59	3.62	3.69	3.78
40	2.95	3.09	3.19	3.26	3.32	3.37	3.41	3.44	3.48	3.51	3.53	3.60	3.68
60	2.90	3.03	3.12	3.19	3.25	3.29	3.33	3.37	3.40	3.42	3.45	3.51	3.59
120	2.85	2.97	3.06	3.12	3.18	3.22	3.26	3.29	3.32	3.35	3.37	3.43	3.51
∞	2.79	2.92	3.00	3.06	3.11	3.15	3.19	3.22	3.25	3.27	3.29	3.35	3.42

TABLE 12

Percentage points of $F_{max} = s^2_{max}/s^2_{min}$

Upper 5% Points

df$_2$ \ t	2	3	4	5	6	7	8	9	10	11	12
2	39.0	87.5	142	202	266	333	403	475	550	626	704
3	15.4	27.8	39.2	50.7	62.0	72.9	83.5	93.9	104	114	124
4	9.60	15.5	20.6	25.2	29.5	33.6	37.5	41.1	44.6	48.0	51.4
5	7.15	10.8	13.7	16.3	18.7	20.8	22.9	24.7	26.5	28.2	29.9
6	5.82	8.38	10.4	12.1	13.7	15.0	16.3	17.5	18.6	19.7	20.7
7	4.99	6.94	8.44	9.70	10.8	11.8	12.7	13.5	14.3	15.1	15.8
8	4.43	6.00	7.18	8.12	9.03	9.78	10.5	11.1	11.7	12.2	12.7
9	4.03	5.34	6.31	7.11	7.80	8.41	8.95	9.45	9.91	10.3	10.7
10	3.72	4.85	5.67	6.34	6.92	7.42	7.87	8.28	8.66	9.01	9.34
12	3.28	4.16	4.79	5.30	5.72	6.09	6.42	6.72	7.00	7.25	7.48
15	2.86	3.54	4.01	4.37	4.68	4.95	5.19	5.40	5.59	5.77	5.93
20	2.46	2.95	3.29	3.54	3.76	3.94	4.10	4.24	4.37	4.49	4.59
30	2.07	2.40	2.61	2.78	2.91	3.02	3.12	3.21	3.29	3.36	3.39
60	1.67	1.85	1.96	2.04	2.11	2.17	2.22	2.26	2.30	2.33	2.36
∞	1.00	1.00	1.00	1.00	1.00	1.00	1.00	1.00	1.00	1.00	1.00

Upper 1% Points

df$_2$ \ t	2	3	4	5	6	7	8	9	10	11	12
2	199	448	729	1036	1362	1705	2063	2432	2813	3204	3605
3	47.5	85	120	151	184	21(6)	24(9)	28(1)	31(0)	33(7)	36(1)
4	23.2	37	49	59	69	79	89	97	106	113	120
5	14.9	22	28	33	38	42	46	50	54	57	60
6	11.1	15.5	19.1	22	25	27	30	32	34	36	37
7	8.89	12.1	14.5	16.5	18.4	20	22	23	24	26	27
8	7.50	9.9	11.7	13.2	14.5	15.8	16.6	17.9	18.9	19.8	21
9	6.54	8.5	9.9	11.1	12.1	13.1	13.9	14.7	15.3	16.0	16.6
10	5.85	7.4	8.6	9.6	10.4	11.1	11.8	12.4	12.9	13.4	13.9
12	4.91	6.1	6.9	7.6	8.2	8.7	9.1	9.5	9.9	10.2	10.6
15	4.07	4.9	5.5	6.0	6.4	6.7	7.1	7.3	7.5	7.8	8.0
20	3.32	3.8	4.3	4.6	4.9	5.1	5.3	5.5	5.6	5.8	5.9
30	2.63	3.0	3.3	3.4	3.6	3.7	3.8	3.9	4.0	4.1	4.2
60	1.96	2.2	2.3	2.4	2.4	2.5	2.5	2.6	2.6	2.7	2.7
∞	1.00	1.0	1.0	1.0	1.0	1.0	1.0	1.0	1.0	1.0	1.0

s^2_{max} is the largest and s^2_{min} the smallest in a set of t independent mean squares, each based on df$_2 = n - 1$ degrees of freedom. Values in the column $t = 2$ and in the rows df$_2 = 2$ and ∞ are exact. Elsewhere, the third digit may be in error by a few units for the 5% points and several units for the 1% points. The third-digit figures in parentheses for df$_2 = 3$ are the most uncertain. From *Biometrika Tables for Statisticians*, 3rd ed., Vol. 1, edited by E. S. Pearson and H. O. Hartley (New York: Cambridge University Press, 1966), Table, p. 202. Reproduced by permission of the *Biometrika Trustees*.

TABLE 13
Random numbers

Line/ Col.	(1)	(2)	(3)	(4)	(5)	(6)	(7)	(8)	(9)	(10)	(11)	(12)	(13)	(14)
1	10480	15011	01536	02011	81647	91646	69179	14194	62590	36207	20969	99570	91291	90700
2	22368	46573	25595	85393	30995	89198	27982	53402	93965	34095	52666	19174	39615	99505
3	24130	48360	22527	97265	76393	64809	15179	24830	49340	32081	30680	19655	63348	58629
4	42167	93093	06243	61680	07856	16376	39440	53537	71341	57004	00849	74917	97758	16379
5	37570	39975	81837	16656	06121	91782	60468	81305	49684	60672	14110	06927	01263	54613
6	77921	06907	11008	42751	27756	53498	18602	70659	90655	15053	21916	81825	44394	42880
7	99562	72905	56420	69994	98872	31016	71194	18738	44013	48840	63213	21069	10634	12952
8	96301	91977	05463	07972	18876	20922	94595	56869	69014	60045	18425	84903	42508	32307
9	89579	14342	63661	10281	17453	18103	57740	84378	25331	12566	58678	44947	05585	56941
10	85475	36857	53342	53988	53060	59533	38867	62300	08158	17983	16439	11458	18593	64952
11	28918	69578	88231	33276	70997	79936	56865	05859	90106	31595	01547	85590	91610	78188
12	63553	40961	48235	03427	49626	69445	18663	72695	52180	20847	12234	90511	33703	90322
13	09429	93969	52636	92737	88974	33488	36320	17617	30015	08272	84115	27156	30613	74952
14	10365	61129	87529	85689	48237	52267	67689	93394	01511	26358	85104	20285	29975	89868
15	07119	97336	71048	08178	77233	13916	47564	81056	97735	85977	29372	74461	28551	90707
16	51085	12765	51821	51259	77452	16308	60756	92144	49442	53900	70960	63990	75601	40719
17	02368	21382	52404	60268	89368	19885	55322	44819	01188	65255	64835	44919	05944	55157
18	01011	54092	33362	94904	31273	04146	18594	29852	71585	85030	51132	01915	92747	64951
19	52162	53916	46369	58586	23216	14513	83149	98736	23495	64350	94738	17752	35156	35749
20	07056	97628	33787	09998	42698	06691	76988	13602	51851	46104	88916	19509	25625	58104
21	48663	91245	85828	14346	09172	30168	90229	04734	59193	22178	30421	61666	99904	32812
22	54164	58492	22421	74103	47070	25306	76468	26384	58151	06646	21524	15227	96909	44592
23	32639	32363	05597	24200	13363	38005	94342	28728	35806	06912	17012	64161	18296	22851
24	29334	27001	87637	87308	58731	00256	45834	15398	46557	41135	10367	07684	36188	18510
25	02488	33062	28834	07351	19731	92420	60952	61280	50001	67658	32586	86679	50720	94953

Abridged from William H. Beyer, ed., *Handbook of Tables for Probability and Statistics*, 2nd ed. © The Chemical Rubber Co., 1968. Used by permission of CRC Press, Inc.

TABLE 14

Power of the analysis
of variance test
($\alpha = .05$, $t = 3$)

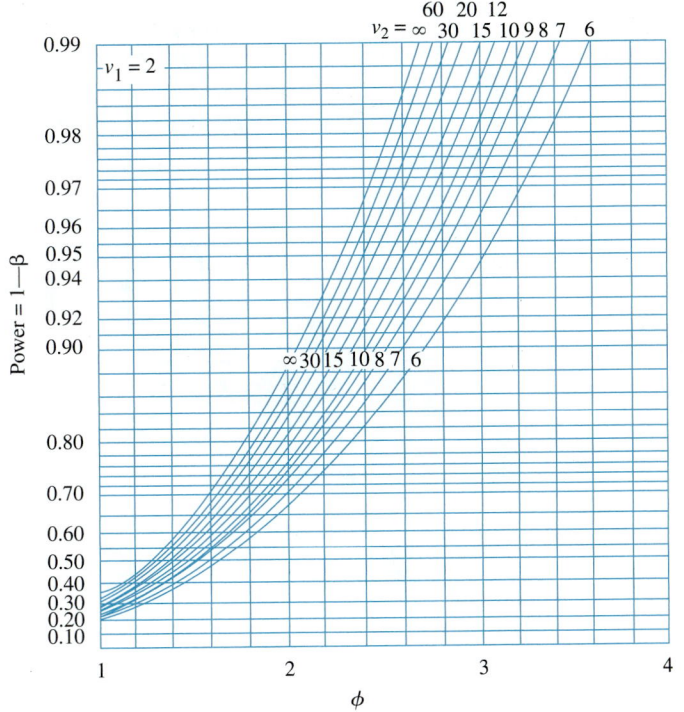

Power of the analysis
of variance test
($\alpha = .05$, $t = 4$)

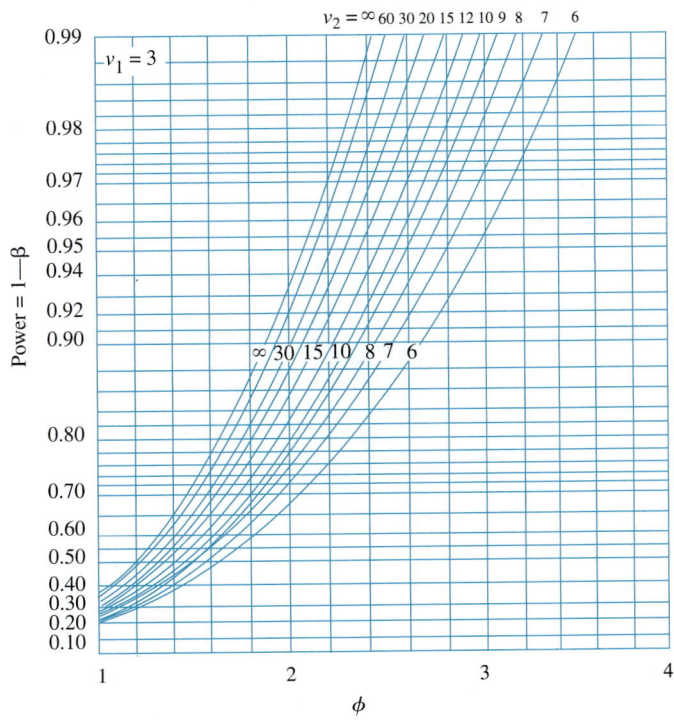

TABLE 14

Power of the analysis
of variance test
($\alpha = .05$, $t = 5$)

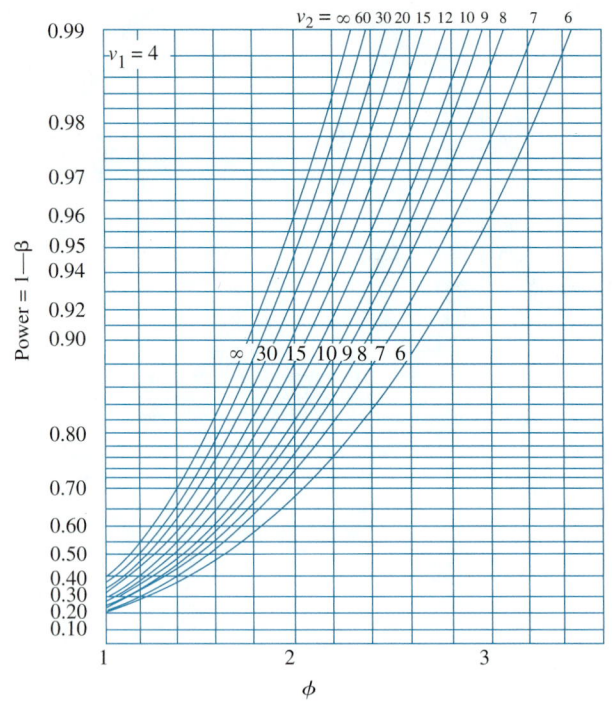

Power of the analysis
of variance test
($\alpha = .05$, $t = 6$)

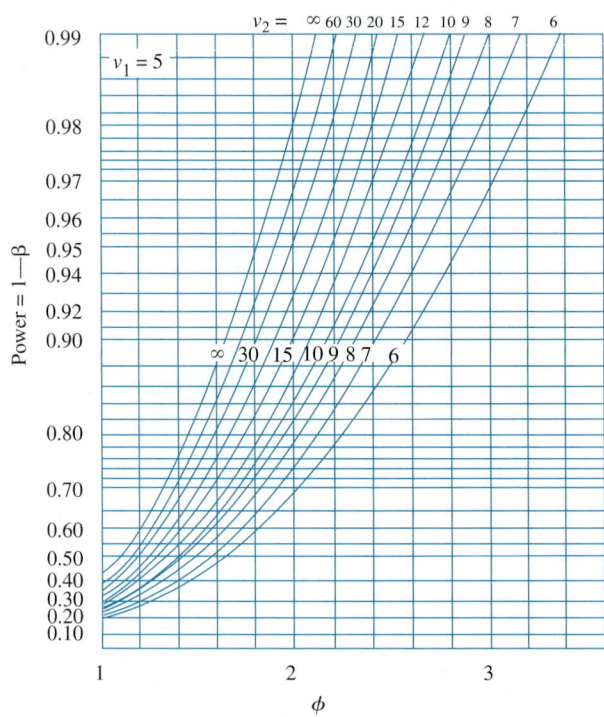

TABLE 14

Power of the analysis
of variance test
($\alpha = .05$, $t = 7$)

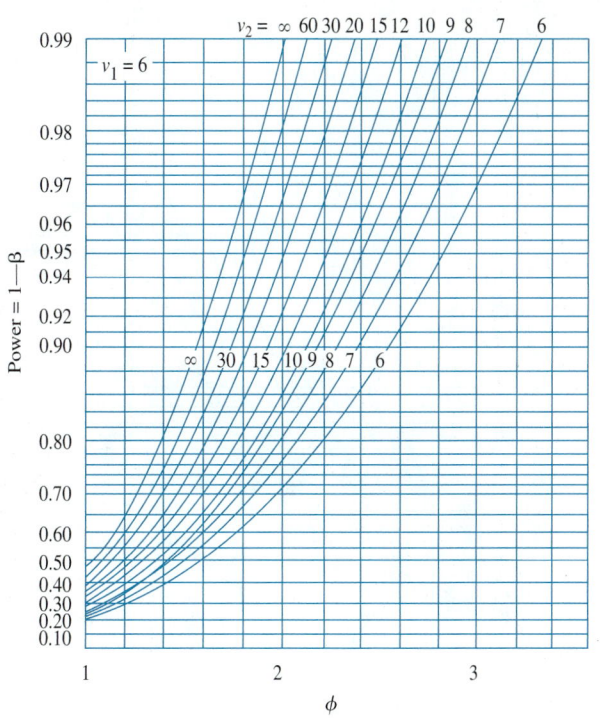

Power of the analysis
of variance test
($\alpha = .05$, $t = 8$)

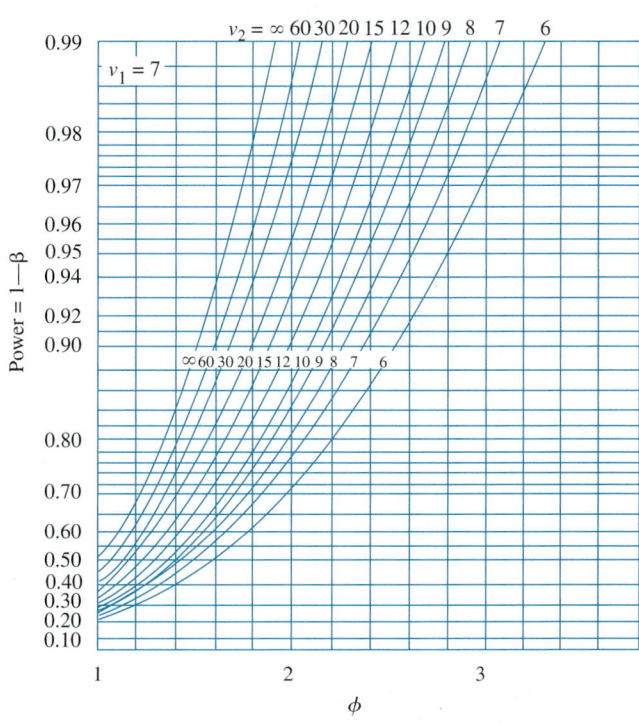

TABLE 14

Power of the analysis
of variance test
$(\alpha = .05, t = 9)$

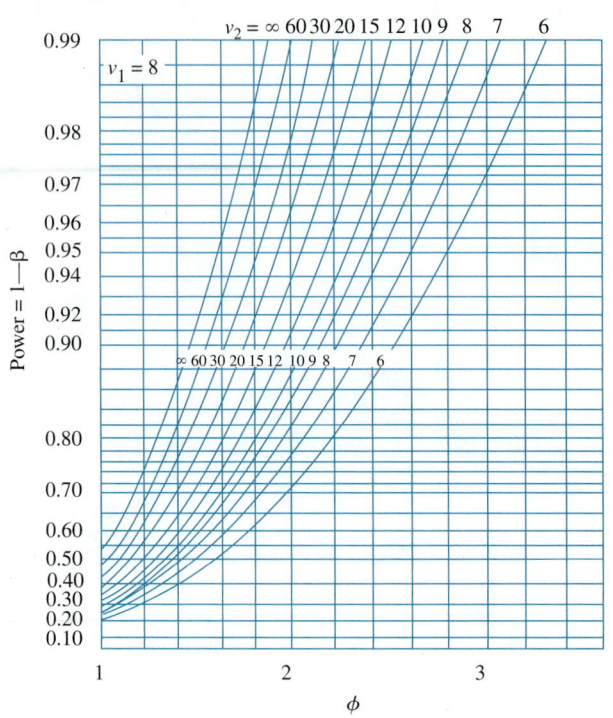

TABLE 15
Poisson probabilities (μ between .1 and 4.0)

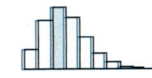

y	.1	.2	.3	.4	μ .5	.6	.7	.8	.9	1.0
0	.9048	.8187	.7408	.6703	.6065	.5488	.4966	.4493	.4066	.3679
1	.0905	.1637	.2222	.2681	.3033	.3293	.3476	.3595	.3659	.3679
2	.0045	.0164	.0333	.0536	.0758	.0988	.1217	.1438	.1647	.1839
3	.0002	.0011	.0033	.0072	.0126	.0198	.0284	.0383	.0494	.0613
4	.0000	.0001	.0003	.0007	.0016	.0030	.0050	.0077	.0111	.0153
5	.0000	.0000	.0000	.0001	.0002	.0004	.0007	.0012	.0020	.0031
6	.0000	.0000	.0000	.0000	.0000	.0000	.0001	.0002	.0003	.0005

y	1.1	1.2	1.3	1.4	μ 1.5	1.6	1.7	1.8	1.9	2.0
0	.3329	.3012	.2725	.2466	.2231	.2019	.1827	.1653	.1496	.1353
1	.3662	.3614	.3543	.3452	.3347	.3230	.3106	.2975	.2842	.2707
2	.2014	.2169	.2303	.2417	.2510	.2584	.2640	.2678	.2700	.2707
3	.0738	.0867	.0998	.1128	.1255	.1378	.1496	.1607	.1710	.1804
4	.0203	.0260	.0324	.0395	.0471	.0551	.0636	.0723	.0812	.0902
5	.0045	.0062	.0084	.0111	.0141	.0176	.0216	.0260	.0309	.0361
6	.0008	.0012	.0018	.0026	.0035	.0047	.0061	.0078	.0098	.0120
7	.0001	.0002	.0003	.0005	.0008	.0011	.0015	.0020	.0027	.0034
8	.0000	.0000	.0001	.0001	.0001	.0002	.0003	.0005	.0006	.0009

y	2.1	2.2	2.3	2.4	μ 2.5	2.6	2.7	2.8	2.9	3.0
0	.1225	.1108	.1003	.0907	.0821	.0743	.0672	.0608	.0550	.0498
1	.2572	.2438	.2306	.2177	.2052	.1931	.1815	.1703	.1596	.1494
2	.2700	.2681	.2652	.2613	.2565	.2510	.2450	.2384	.2314	.2240
3	.1890	.1966	.2033	.2090	.2138	.2176	.2205	.2225	.2237	.2240
4	.0992	.1082	.1169	.1254	.1336	.1414	.1488	.1557	.1622	.1680
5	.0417	.0476	.0538	.0602	.0668	.0735	.0804	.0872	.0940	.1008
6	.0146	.0174	.0206	.0241	.0278	.0319	.0362	.0407	.0455	.0504
7	.0044	.0055	.0068	.0083	.0099	.0118	.0139	.0163	.0188	.0216
8	.0011	.0015	.0019	.0025	.0031	.0038	.0047	.0057	.0068	.0081
9	.0003	.0004	.0005	.0007	.0009	.0011	.0014	.0018	.0022	.0027
10	.0001	.0001	.0001	.0002	.0002	.0003	.0004	.0005	.0006	.0008
11	.0000	.0000	.0000	.0000	.0000	.0001	.0001	.0001	.0002	.0002

y	3.1	3.2	3.3	3.4	μ 3.5	3.6	3.7	3.8	3.9	4.0
0	.0450	.0408	.0369	.0334	.0302	.0273	.0247	.0224	.0202	.0183
1	.1397	.1304	.1217	.1135	.1057	.0984	.0915	.0850	.0789	.0733
2	.2165	.2087	.2008	.1929	.1850	.1771	.1692	.1615	.1539	.1465
3	.2237	.2226	.2209	.2186	.2158	.2125	.2087	.2046	.2001	.1954
4	.1733	.1781	.1823	.1858	.1888	.1912	.1931	.1944	.1951	.1954
5	.1075	.1140	.1203	.1264	.1322	.1377	.1429	.1477	.1522	.1563
6	.0555	.0608	.0662	.0716	.0771	.0826	.0881	.0936	.0989	.1042

TABLE 15
Poisson probabilities (μ between 3.1 and 10.0)

y	3.1	3.2	3.3	3.4	3.5	3.6	3.7	3.8	3.9	4.0
7	.0246	.0278	.0312	.0348	.0385	.0425	.0466	.0508	.0551	.0595
8	.0095	.0111	.0129	.0148	.0169	.0191	.0215	.0241	.0269	.0298
9	.0033	.0040	.0047	.0056	.0066	.0076	.0089	.0102	.0116	.0132
10	.0010	.0013	.0016	.0019	.0023	.0028	.0033	.0039	.0045	.0053
11	.0003	.0004	.0005	.0006	.0007	.0009	.0011	.0013	.0016	.0019
12	.0001	.0001	.0001	.0002	.0002	.0003	.0003	.0004	.0005	.0006
13	.0000	.0000	.0000	.0000	.0001	.0001	.0001	.0001	.0002	.0002

y	4.1	4.2	4.3	4.4	4.5	4.6	4.7	4.8	4.9	5.0
0	.0166	.0150	.0136	.0123	.0111	.0101	.0091	.0082	.0074	.0067
1	.0679	.0630	.0583	.0540	.0500	.0462	.0427	.0395	.0365	.0337
2	.1393	.1323	.1254	.1188	.1125	.1063	.1005	.0948	.0894	.0842
3	.1904	.1852	.1798	.1743	.1687	.1631	.1574	.1517	.1460	.1404
4	.1951	.1944	.1933	.1917	.1898	.1875	.1849	.1820	.1789	.1755
5	.1600	.1633	.1662	.1687	.1708	.1725	.1738	.1747	.1753	.1755
6	.1093	.1143	.1191	.1237	.1281	.1323	.1362	.1398	.1432	.1462
7	.0640	.0686	.0732	.0778	.0824	.0869	.0914	.0959	.1002	.1044
8	.0328	.0360	.0393	.0428	.0463	.0500	.0537	.0575	.0614	.0653
9	.0150	.0168	.0188	.0209	.0232	.0255	.0281	.0307	.0334	.0363
10	.0061	.0071	.0081	.0092	.0104	.0118	.0132	.0147	.0164	.0181
11	.0023	.0027	.0032	.0037	.0043	.0049	.0056	.0064	.0073	.0082
12	.0008	.0009	.0011	.0013	.0016	.0019	.0022	.0026	.0030	.0034
13	.0002	.0003	.0004	.0005	.0006	.0007	.0008	.0009	.0011	.0013
14	.0001	.0001	.0001	.0001	.0002	.0002	.0003	.0003	.0004	.0005
15	.0000	.0000	.0000	.0000	.0001	.0001	.0001	.0001	.0001	.0002

y	5.5	6.0	6.5	7.0	7.5	8.0	8.5	9.0	9.5	10.0
0	.0041	.0025	.0015	.0009	.0006	.0003	.0002	.0001	.0001	.0000
1	.0225	.0149	.0098	.0064	.0041	.0027	.0017	.0011	.0007	.0005
2	.0618	.0446	.0318	.0223	.0156	.0107	.0074	.0050	.0034	.0023
3	.1133	.0892	.0688	.0521	.0389	.0286	.0208	.0150	.0107	.0076
4	.1558	.1339	.1118	.0912	.0729	.0573	.0443	.0337	.0254	.0189
5	.1714	.1606	.1454	.1277	.1094	.0916	.0752	.0607	.0483	.0378
6	.1571	.1606	.1575	.1490	.1367	.1221	.1066	.0911	.0764	.0631
7	.1234	.1377	.1462	.1490	.1465	.1396	.1294	.1171	.1037	.0901
8	.0849	.1033	.1188	.1304	.1373	.1396	.1375	.1318	.1232	.1126
9	.0519	.0688	.0858	.1014	.1144	.1241	.1299	.1318	.1300	.1251
10	.0285	.0413	.0558	.0710	.0858	.0993	.1104	.1186	.1235	.1251
11	.0143	.0225	.0330	.0452	.0585	.0722	.0853	.0970	.1067	.1137
12	.0065	.0113	.0179	.0263	.0366	.0481	.0604	.0728	.0844	.0948
13	.0028	.0052	.0089	.0142	.0211	.0296	.0395	.0504	.0617	.0729
14	.0011	.0022	.0041	.0071	.0113	.0169	.0240	.0324	.0419	.0521
15	.0004	.0009	.0018	.0033	.0057	.0090	.0136	.0194	.0265	.0347

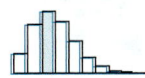

TABLE 15
Poisson probabilities (μ between 5.5 and 20.0)

					μ					
y	5.5	6.0	6.5	7.0	7.5	8.0	8.5	9.0	9.5	10.0
16	.0001	.0003	.0007	.0014	.0026	.0045	.0072	.0109	.0157	.0217
17	.0000	.0001	.0003	.0006	.0012	.0021	.0036	.0058	.0088	.0128
18	.0000	.0000	.0001	.0002	.0005	.0009	.0017	.0029	.0046	.0071
19	.0000	.0000	.0000	.0001	.0002	.0004	.0008	.0014	.0023	.0037
20	.0000	.0000	.0000	.0000	.0001	.0002	.0003	.0006	.0011	.0019
21	.0000	.0000	.0000	.0000	.0000	.0001	.0001	.0003	.0005	.0009
22	.0000	.0000	.0000	.0000	.0000	.0000	.0001	.0001	.0002	.0004
23	.0000	.0000	.0000	.0000	.0000	.0000	.0000	.0000	.0001	.0002

					μ					
y	11.0	12.0	13.0	14.0	15.0	16.0	17.0	18.0	19.0	20.0
0	.0000	.0000	.0000	.0000	.0000	.0000	.0000	.0000	.0000	.0000
1	.0002	.0001	.0000	.0000	.0000	.0000	.0000	.0000	.0000	.0000
2	.0010	.0004	.0002	.0001	.0000	.0000	.0000	.0000	.0000	.0000
3	.0037	.0018	.0008	.0004	.0002	.0001	.0000	.0000	.0000	.0000
4	.0102	.0053	.0027	.0013	.0006	.0003	.0001	.0001	.0000	.0000
5	.0224	.0127	.0070	.0037	.0019	.0010	.0005	.0002	.0001	.0001
6	.0411	.0255	.0152	.0087	.0048	.0026	.0014	.0007	.0004	.0002
7	.0646	.0437	.0281	.0174	.0104	.0060	.0034	.0019	.0010	.0005
8	.0888	.0655	.0457	.0304	.0194	.0120	.0072	.0042	.0024	.0013
9	.1085	.0874	.0661	.0473	.0324	.0213	.0135	.0083	.0050	.0029
10	.1194	.1048	.0859	.0663	.0486	.0341	.0230	.0150	.0095	.0058
11	.1194	.1144	.1015	.0844	.0663	.0496	.0355	.0245	.0164	.0106
12	.1094	.1144	.1099	.0984	.0829	.0661	.0504	.0368	.0259	.0176
13	.0926	.1056	.1099	.1060	.0956	.0814	.0658	.0509	.0378	.0271
14	.0728	.0905	.1021	.1060	.1024	.0930	.0800	.0655	.0514	.0387
15	.0534	.0724	.0885	.0989	.1024	.0992	.0906	.0786	.0650	.0516
16	.0367	.0543	.0719	.0866	.0960	.0992	.0963	.0884	.0772	.0646
17	.0237	.0383	.0550	.0713	.0847	.0934	.0963	.0936	.0863	.0760
18	.0145	.0255	.0397	.0554	.0706	.0830	.0909	.0936	.0911	.0844
19	.0084	.0161	.0272	.0409	.0557	.0699	.0814	.0887	.0911	.0888
20	.0046	.0097	.0177	.0286	.0418	.0559	.0692	.0798	.0866	.0888
21	.0024	.0055	.0109	.0191	.0299	.0426	.0560	.0684	.0783	.0846
22	.0012	.0030	.0065	.0121	.0204	.0310	.0433	.0560	.0676	.0769
23	.0006	.0016	.0037	.0074	.0133	.0216	.0320	.0438	.0559	.0669
24	.0003	.0008	.0020	.0043	.0083	.0144	.0226	.0328	.0442	.0557
25	.0001	.0004	.0010	.0024	.0050	.0092	.0154	.0237	.0336	.0446
26	.0000	.0002	.0005	.0013	.0029	.0057	.0101	.0164	.0246	.0343
27	.0000	.0001	.0002	.0007	.0016	.0034	.0063	.0109	.0173	.0254
28	.0000	.0000	.0001	.0003	.0009	.0019	.0038	.0070	.0117	.0181
29	.0000	.0000	.0001	.0002	.0004	.0011	.0023	.0044	.0077	.0125
30	.0000	.0000	.0000	.0001	.0002	.0006	.0013	.0026	.0049	.0083
31	.0000	.0000	.0000	.0000	.0001	.0003	.0007	.0015	.0030	.0054
32	.0000	.0000	.0000	.0000	.0001	.0001	.0004	.0009	.0018	.0034
33	.0000	.0000	.0000	.0000	0000	.0001	.0002	.0005	.0010	.0020

Source: Computed by D. K. Hildebrand.

REFERENCES

Agresti, A. (1990), *Categorical Data Analysis.* New York: Wiley.

Brown, S., M. Healy, and M. Kearns (1981), "Report on the interlaboratory trial of the reference method for the determination of total calcium in serum." *Journal of Clinical Chemistry and Clinical Biochemistry* 19, 395–426.

Carmer, S., and M. Swanson (1973), "An evaluation of ten pairwise multiple comparison procedures by Monte Carlo methods." *Journal of the American Statistical Association* 68, 66–74.

Carroll, R., R. Chen, E. George, T. Li, H. Newton, H. Schmiedliche, and N. Wang (1997), "Ozone exposure and population density in Harris County, Texas." *Journal of the American Statistical Association* 92, 392–415.

Carter, R. (1981), "Restricted maximum likelihood estimation of bias and reliability in the comparison of several measuring methods." *Biometrics* 37, 733–741.

Cochran, W. (1954), "Some methods for strengthening the common χ^2 test." *Biometrics* 10, 417–451.

Cochran, W., and G. Cox (1957), *Experimental Design.* 2nd ed. New York: Wiley.

Conover, J. (1998), *Practical Nonparametric Statistics.* 3rd ed. New York: Wiley.

Cressie, N. (1993), *Statistics for Spatial Data.* New York: Wiley.

Crowder, M., and D. Hand (1990), *Analysis of Repeated Measures.* London: Chapman and Hall.

Cryer, J., and R. Miller (1991), *Statistics for Business: Data Analysis and Modeling.* Boston, PWS-Kent.

Deming, W. (1981), *Quality, Productivity, and Competitive Position.* Cambridge, Mass.: MIT-CAES.

Devore, J. (2000), *Probability and Statistics for Engineering and the Sciences.* 5th ed. Pacific Grove, Cal.: Duxbury Press.

Diggle, P., K. Liang, and S. Zeger (1996), *Analysis of Longitudinal Data.* New York: Oxford University Press.

Dunnett, C. (1955), "A multiple comparison procedure for comparing several treatments with a control." *Journal of the American Statistical Association* 50, 1096–1121.

Dunnett, C. (1964), "New tables for multiple comparisons with a control." *Biometrics* 20, 482–491.

Durbin, J., and G. Watson (1951), "Testing for serial correlation in least squares, II." *Biometrika* 38, 159–178.

Effron, B., and R. Tibshirani (1993), *An Introduction to the Bootstrap.* London: Chapman and Hall.

Fisher, R. A. (1949), *The Design of Experiments.* Edinburgh: Oliver and Boyd.

Greenhouse, S., and S. Geisser (1959), "On methods in the analysis of profile data." *Psychometrika* 24, 95–112.

Haining, R. (1990), *Spatial Data Analysis in the Social and Environmental Sciences.* Cambridge: Cambridge University Press.

Hammer, M. (1996), *Beyond Reengineering.* New York: Harper Collins.

Hammer, M., and J. Champy (1993), *Reengineering the Corporation: A Manifesto for Business Revolution.* New York: Harper Collins.

Hicks, C., and K. Turner (1999), *Fundamental Concepts in the Design of Experiments.* 5th ed. New York: Oxford University Press.

Hildebrand, D., and L. Ott (1998), *Statistical Thinking for Managers.* 4th ed. Pacific Grove, Cal.: Duxbury Press.

Huynh, H., and L. Feldt (1970), "Conditions under which mean square ratios in repeated measurement designs have fixed F-distributions." *Journal of the American Statistical Association* 65, 1582–1589.

Jones, B., and M. Kenard (1989), *Design and Analysis of Cross-Over Trials.* London: Chapman and Hall.

Koehler, K. (1986), "Goodness-of-fit tests for log-linear models in sparse contingency tables." *Journal of the American Statistical Association* 81, 483–493.

Kuehl, R. (1999), *Design of Experiments: Statistical Principles of Research Design and Analysis.* 2nd ed. Pacific Grove, Cal.: Duxbury Press.

Larntz, K. (1978), "Small-sample comparison of exact levels for chi-squared goodness-of-fit statistics." *Journal of the American Statistical Association* 73, 253–263.

Lentner, M., and T. Bishop (1993), *Experimental Design and Analysis.* 2nd ed. Blacksburg, Va.: Valley Book Company.

Mallows, C. (1973), "Some comments on C_p." *Technometrics* 15, 661–675.

Manly, B. (1998), *Randomization, Bootstrap and Monte Carlo Methods in Biology.* 2nd ed. London: Chapman and Hall.

Meyer, C. (1993), *Fast Cycle Time.* New York: The Free Press.

Miller, R. (1981), *Simultaneous Statistical Inference.* 2nd ed. New York: Springer-Verlag.

Montgomery, D. (1997), *Design and Analysis of Experiments.* 5th ed. New York: Wiley.

Neter, J., M. Kutner, C. Nachtsheim, and W. Wasserman (1996), *Applied Linear Statistical Models.* 4th ed. Boston: WCB McGraw-Hill.

Newman, R. (1998), "Testing parallelism among the profiles after a certain time period." Unpublished PhD dissertation. Texas A&M University.

Raftery, A., and J. Zeh (1998), "Estimating bowhead whale population size and rate of increase from the 1993 census." *Journal of the American Statistical Association* 93, 451–462.

Randles, R., and D. Wolfe (1979), *Introduction to the Theory of Nonparametric Statistics.* New York: Wiley.

Ripley, B. (1988), *Statistical Inference for Spatial Processes.* Cambridge: Cambridge University Press.

Rowe, N., R. Anderson, and L. Wanninger (1974), "Effects of ready-to-eat breakfast cereals on dental caries experience in adolescent children: A three-year study." *Journal of Dental Research* 53, 33.

Scheaffer, R. L., W. Mendenhall, and L. Ott (1996), *Elementary Survey Sampling.* 5th ed. Pacific Grove, Cal.: Duxbury Press.

Scheffé, H. (1953), "A method for judging all contrasts in the analysis of variance." *Biometrika* 40, 87–104.

Searle, S., G. Casella, and C. McCulloch (1992), *Variance Components.* New York: Wiley.

Snedecor, G., and W. Cochran (1980), *Statistical Methods.* 7th ed. Ames, Iowa: Iowa State University Press.

Tukey, J. (1977), *Exploratory Data Analysis.* Reading, Mass.: Addison-Wesley.

U.S. Bureau of Labor Statistics, *Handbook of Methods,* Vols. I and II (1982). Washington, D.C.: U.S. Department of Labor.

Vonesh, E., and V. Chinchilli (1997), *Linear and Nonlinear Models for the Analysis of Repeated Measurements.* New York: Marcel Dekker.

Welch, B. (1938), "The significance of the difference between two means when the population variances are unequal," *Biometrika* 29, 350–362.

additive effects, 621
adjusted treatment means, 948, 950
American Statistical Association
        (ASA), 22
analysis of covariance, 943–974
    case study, 944–946
    completely randomized design,
            multiple covariates, 962–970
        checking results with plots of
                residuals, 968–970
        example of, 963
        exercises for, 970
        models for, 964–968
        overview of, 962–963
        reporting conclusions of, 970
    completely randomized design,
            one covariate, 946–958
        adjusted treatment means and,
                948, 950
        examples of, 947, 951–952,
                957–958
        exercises for, 958–959
        model conditions, 948, 955–957
        models and tests of hypotheses,
                949–951
        overview of, 946–947
    exercises for, 971–974
    extrapolation problems, 959–962
    introduction, 943–944
        covariates defined, 943
        examples of covariate applica-
                tions, 943–944
analysis of variance (AOV), 383–384
    alternatives to
        Kruskal–Wallis test, 410–414
        transformations of data,
                403–410
    checking conditions for, 396–403
        examples of, 397–403
        overview of, 396–397
        reporting conclusions of, 403
        using residuals analysis for, 397
    crossover designs and, 1041
    model of, 394–396
    overview of, 379
    power of, 848
    sample variation and, 383
    statistical tests for, 384–393
        AOV table summarizing, 389

completely randomized design,
        387–388
    developing test statistic,
            386–387
    examples of, 389–392
    exercises for, 392–393
    mean square and, 388
    multiple t tests, 385
    sum of squares and, 388
for two-factor experiment, re-
        peated measures, 1032
analysis of variance (AOV), fixed-,
        random-, and mixed-effects,
        975–1024
case study, 976–978
exercises for, 1020–1024
fixed-effects model
    defined, 975
    vs. random-effects model,
            978–979
introduction, 975–976
mixed-effects model, 992–1000
    assumptions for, 992
    defined, 976
    example of, 993–996
    exercises for, 1000
    tests of significance for,
            992–993
nested sampling and split-plot de-
        sign, 1010–1020
    AOV table for, 1014
    examples of, 1011–1014,
            1018–1019
    overview of, 1010–1011
    split-plot design, 1014–1018
random-effects model, 978–983
    AOV moment matching and,
            980
    assumptions of, 978–979
    defined, 976
    example of, 981–982
    exercises for, 982–983
    expected mean squares and,
            979
    extensions of, 983–992
    hypothesis tests for, 979–980
    vs. fixed-effects model, 978–979
random-effects model, block
    design

assumptions for, 983–984
estimating variance compo-
        nents, 986–987
example of, 987–989
exercises for, 991–992
hypothesis testing for, 989–990
model and AOV tables for,
        985
nested sampling and, 990–991
rules for obtaining expected mean
        squares, 1000–1010
    classifying interactions,
            1000–1001
    examples of, 1001, 1003–1005,
            1006–1010
    mean square tables, 1001–1003
analysis of variance (AOV), for stan-
        dard designs, 853–942
case study, 854–855
comparing treatment means,
        916–922
    applying multiple comparison
            procedures, 920
    case study, 921–922
    examples of, 917–921
    using Fisher's LSD procedure,
            916–917
completely randomized design, sin-
        gle factor, 855–859
    advantages/disadvantage of,
            858
    example of, 858–859
    model for, 856
    overview of, 855–856
    using total sum of squares as
            test statistic, 856–857
exercises for, 924–942
factorial treatment in completely
        randomized design, 891–914
    accounting for interaction in,
            898–899
    AOV table for, 908
    constructing AOV table for,
            899–900
    example of, 901
    exercises for, 913–914
    first test of significance for,
            902–903
    interaction of factors and, 894

analysis of variance (AOV) (*contd.*)
 interpreting *F* test for main effects, 904–905
 models for, 897–898, 905–907
 one-at-a-time approach, 892
 overview of factorial experiments, 894–897
 sum of squares calculations for, 907–908
 summary of method of analysis, 909
 factorial treatment in randomized complete block design, 914–916
  AOV table for, 916
  example of, 915–916
  exercise for, 916
  overview of, 914–915
 introduction, 853–854
 key formulas for, 923
 Latin square design, 879–891
  advantages/disadvantages of, 880
  defined, 881
  examples of, 879–880, 884–888
  exercises for, 888–891
  filtering with, 881–882
  model for, 881
  overview of, 879–880
  relative efficiency of, 884
  testing for treatment effects, 882–884
 randomized complete block design, 859–878
  advantages/disadvantages of, 861
  defined, 861
  examples of, 868–871
  exercises for, 872–878
  filtering and, 863–864
  model for, 863–864
  overview of, 859–861
  relative efficiency of, 867–868
  testing difference among treatment means, 865–867
analysis of variance (AOV), unbalanced designs, 1051–1076
 balanced incomplete block designs, 1063–1072
  AOV table for, 1065
  comparing treatment means, 1067–1068
  defined, 1064
  examples of, 1063–1066, 1068
  model for, 1065

overview of, 1063
 reporting conclusions of, 1071–1072
 sum of squares for, 1066
 case study, 1052–1053
 exercises for, 1074–1076
 introduction, 1051–1052
 key formulas for, 1072–1073
 Latin square design, missing data, 1059–1063
  comparing treatment means, 1061
  estimating missing value, 1059
  example of, 1059–1060
  exercises for, 1062–1063
  fitting complete and reduced models for, 1061
 randomized block design, one or more missing observations, 1053–1058
  comparing treatment means, 1056
  estimating value of missing observation, 1053
  estimation bias and, 1053–1054
  examples of, 1054–1055
  exercises for, 1057–1058
  fitting complete and reduced models for, 1056–1057
  testing effect of treatments, 1057
AOV moment matching, 980
AOV tables
 balanced incomplete block designs, 1065
 factorial experiment ($a \times b$), 985
 factorial treatment in completely randomized design, 899–900, 908, 916
 fixed or random-effects models, 979
 mixed-effects model, 998
 nested sampling and split-plot design, 1014
 random-effects model, block design, 985
 split-plot design, 1016, 1019
 statistical tests for AOV, 389
 two-factor experiment, 984
 two-period crossover design, 1044
arithmetic mean (*See* mean)
associations
 strength of association
  contingency tables, 505
  measuring, 510–516

unit of association, 532
 vs. causal relationships, 830

backup analyses, 1087
backward elimination, 717–720
balanced incomplete block (BIB) designs, 1063–1072
 AOV table for, 1065
 comparing treatment means, 1067–1068
 defined, 1064
 examples of, 1063–1066, 1068
 model for, 1065
 overview of, 1063
 sum of squares for, 1066
bar charts
 accuracy of data presentation with, 1081
 cluster bar graphs, 102–103
 description of, 45
 examples of, 46
 guideline for constructing, 46
 stacked bar graphs, 102
 uses of, 50
Bayes' formula, 136–140
 equation for, 138
 false positive/false negative test results, 136
 likelihoods in, 138
 observable events in, 138
 prior/posterior probabilities in, 138
 sensitivity/specificity of diagnostic tests, 137
 states of nature in, 138
benchmarks (*See* control treatments)
best-guess estimation (*See* point estimation)
best subset regression
 multiple regression, 730–731
 selecting multiple regression variables, 716–717
biased samples, 1082–1083
binomials, 144–154
 defining binomial experiments, 144
 formula for computation of, 146–149
 mean and standard deviation of, 149–151
 normal approximation to, 182–185
 properties of, 145
 public opinion polls as examples of, 144
block design (*See* randomized complete block design)

blocking variables (*See also* randomized complete block design)
  split-plot design and, 1016
  use of, 891
Bonferroni procedure
  error rate control and, 439–440
  testing hypothesis with, 457
bootstrap techniques, for sampling distribution, 349–350
box-and-whiskers plot, 97
boxplots, 96–101
  examples of, 99, 106–107
  overview of, 96–97
  skeletal boxplot, 96–97
  steps in construction of, 98
  strengths of, 99–100
  using for residuals analysis, 968
  using to check normality, 396–397
  vs. stem-and-leaf plot, 96

calculators, data analysis with, 41–43
calibration, inverse regression and, 582
categorical data, 469–530
  case study, 470–471
  contingency tables and, 501–510
    cross tabulations with, 502
    dependence and, 502
    estimated expected value, 503
    examples of, 503–508
    exercises for, 508–510
    likelihood ratio statistic, 505
    strength of association, 505
    tests of homogeneity, 505–506
    tests of independence, 504
  exercises for, 521–530
  inferences about a population proportion, 471–482
    confidence interval for, 472–474
    examples of, 472–476
    exercises for, 477–482
    mean and standard error, 472
    sample size requirement, 477
    statistical test for, 475
  inferences about difference between two population proportions, 482–488
    confidence interval for, 483
    examples of, 483–485
    exercises for, 486
    notation for comparing, 482
    rule for sample size, 484
    statistical test for, 484–485

inferences about several proportions, 488–497
  examples of, 489, 491–493
  exercises for, 494–497
  expected number of outcomes, 490
  multinomial distribution, 488–489
  multinomial experiments, 488
  statistical test for, 490–494
introduction, 469
key formulas for, 520–521
measuring strength of relation, 510–516
  dependent and independent variables and, 512
  examples of, 513–516
  percentage analysis, 510–511
  predictability analysis, 511–512
odds and odds ratios, 516–519
  example of, 517
  formula for, 516–517
  odds ratio defined, 518–519
Poisson distribution, 497–501
  assumptions for, 498
  defined, 497
  examples of, 497–500
  formula for, 497
  tests using, 498–499
causal relationships, compared with associations, 830
cell probabilities, 490
census polls, 19–20
Central Limit Theorem, 175, 179–180, 349
central tendency, measures of
  data skew and, 76–77
  exercises for, 77–81
  mean, 73–76, 77
  median, 71–73, 77
  mode, 70, 77
  overview of, 69–70
chi-square distribution, 344–346
  approximating, 490
  critical values of, 345
  data normality and, 349, 351
  densities of, 345
  overview of, 344–345
  statistical table of, 1100–1101
  upper-tail and lower-tail values of, 346
chi-square goodness-of-fit test, 490–497
class frequency, 48

class intervals, constructing, 47–48
classical interpretation of probability
  defined, 124
  determining probability with, 126
clinical trials, 7
cluster bar graphs, 102–103
cluster effect, of data, 274
cluster sampling, 20
Cobb–Douglas production function, 738
coefficient of determination
  for multiple regression, 646, 712
  for simple linear regression, 592–593
coefficient of variation
  as measure of variability, 96
  transformations of data and, 406
coefficients, estimating mean, 197–200
coefficients, linear regression, 591
coefficients, multiple regression, 627–646
  examples for, 628–633
  exercises for, 633–646
  least-squares estimates, 630
  normal equations for, 627–628
  residual standard deviation and, 632–633
  testing subset of regression coefficients, 657–665
collinearity
  assessing, 708–710
  avoiding, 710
  effect of, 652
  overspecification and, 715
  problems of, 646–647
communication and documentation, 1077–1089
  communication hurdles
    biased samples, 1082–1083
    graphical distortions, 1079–1082
    sample size, 1083–1084
  data preparation, steps in, 1084–1087
  difficulty of good communication, 1078–1079
  documenting and storing results, 1088–1089
  introduction, 1077
  report guidelines, 1087–1088
  types of, 1079
complements of events, 129–130 (*See also* events)

completely randomized design
    analysis of variance for, 387–388
    experimental designs and, 28–29
    model for observations in,
        394–396
        assumptions for, 395
        conditions for, 394
        sum of terms and, 394–395
completely randomized design, multiple covariates, 962–970
    checking results with plots of residuals, 968–970
    example of, 963
    exercises for, 970
    models for, 964–968
    overview of, 962–963
    reporting conclusions of, 970
    testing hypotheses, 964
completely randomized design, one covariate, 947
    adjusted treatment means and, 948, 950
    examples of, 947, 951–952, 957–958
    exercises for, 958–959
    model assumptions
        reviewed, 955–957
        stated, 948
    models and sum of squares for, 949–951
    overview of, 946–947
completely randomized design, single factor, 855–859
    advantages/disadvantages of, 858
    example of, 858–859
    model for, 856
    overview of, 855–856
    using total sum of squares as test statistic, 856–857
compound symmetry of observations, 1030–1031
computers, data analysis with, 41–43
conditional probability, 131–136
    defined, 132
    independent vs. dependent events, 133–134
    multiplication law and, 132
    vs. unconditional probability, 132
confidence coefficient
    estimating population mean, 197–200
    estimating population variance, 346
    population proportion inferences and, 472

confidence interval
    defined, 666
    estimating standard error in multiple regression, 653
    Fisher's least significant difference and, 444
    linear regression forecasting and, 567–568
    mean and, 196, 200–201
    mean, unknown variance and, 234
    median and, 243–245
    population proportions and, 472–474, 483
    population variance and
        comparing two, 359
        single, 346
    sample size and, 205, 316
    Scheffé's S method and, 453–454
    slope and, 559–560
    Tukey's W procedure and, 447
confounded factors, 860
constant variance
    regression model assumptions, 759, 760–765
    using weighted least-squares, 762–763
contingency tables, 501–510
    constructing, 101–102
    cross tabulations with, 502
    dependence and, 502
    examples of, 503–508
    exercises for, 508–510
    expected value and, 503
    likelihood ratio and, 505
    strength of association in, 505
    tests of homogeneity, 505–506
    tests of independence, 504
continuity correction, normal approximation and, 184
continuous random variables, 154–157
    defined, 142
    probability distribution of, 155–157
    vs. discrete random variables, 154
contrasts
    linear contrasts, 431–438
        defined, 432
        examples of, 432–437
        exercises for, 437–438
        F test for, 436
        notation for, 431
    orthogonal contrasts
        defined, 432
        overview of, 432–433

control treatments
    defined, 833
    use in experiments, 450
correlation, 590–600
    assumptions for, 593–595
    coefficient of determination and, 592–593
    correlation coefficient and, 591
    examples of, 592–597
    exercises for, 597–600
    measuring predictions with, 590–591
correlation coefficient, 591
correlation matrix, 708
count data (*See* categorical data)
covariance, analysis of (*See* analysis of covariance)
covariates (*See also* analysis of covariance)
    defined, 947
    using to reduce variability, 839–840
C_p statistic, 716
cross-product term, in multiple regression, 619
cross tabulations, with contingency tables, 502
crossover designs, 1040–1044
    analysis of variance for, 1041
    defined, 1025
    example of, 1041–1043
    layout for two-period crossover design, 1043
    model for, 1040
    using Latin square design, 1040

data
    analysis, 2
    cluster effect of, 274
    collection
        design for, 18
        processes included in, 19
        role in statistics, 2
        statisticians and, 10
    collection techniques, 24–25
        direct observation, 25
        personal interviews, 24
        self-administered questionnaire, 25
        telephone interviews, 24–25
    evaluating normality of, 351
    management
        procedures for, 35–38
        steps in process of, 35–37
    preparation, role in statistics, 2

preparation, steps in
process of learning from, 2–5
spatial correlation of, 274
data description, 40–120
  boxplot, 96–101
    construction of, 98
    overview of, 96–97
    skeletal boxplot, 96–97
    strengths of, 99–100
  exercises for, 110–120
  graphical methods, 43–69
    bar charts, 45–46
    data organization guidelines, 43
    exercises for, 63–69
    frequency histogram, 46–54
    general guidelines, 62
    pie charts, 43–45
    stem-and-leaf plots, 54–57
    time series, 57–62
  introduction, 40–41
  key formulas, 110
  measures of central tendency, 69–81, 81–96
    data skew and, 76–77
    exercises for, 77–81
    mean, 73–76, 77
    median, 71–73, 77
    mode, 70, 77
    overview of, 69–70
  measures of variability
    coefficient of variation, 96
    deviation, 86–87
    Empirical Rule for, 89–93
    exercises for, 93–96
    interquartile range, 86
    overview of, 81–82
    percentiles, 82–86
    range, 82
    standard deviation, 88–89
    variance, 87–88
  multiple variables, 101–109
    cluster bar graphs and, 102–103
    contingency tables and, 101–102
    exercises for, 107–109
    scatterplots and, 102–103
    side-by-side boxplots and, 104–107
    stacked bar graphs and, 102
  using calculators and computers for, 41–43
data dredging, 427–428
data snooping, 427–428

data-splitting approach, multiple regression, 714
data trail
  defined, 1084
  overview of, 36
data variation (*See* variability, measures of)
databases
  creating, 1085
  creating data files from, 1086–1087
  editing, 1085–1086
  finalizing, 1086
dependent events, 133–134
dependent variables
  contingency tables and, 502
  measuring strength of relations in, 512
  vs. independent variables, 512
descriptive statistics, 40
designed experiments, 829–852 (*See also* experimental design)
  comparing observation studies with, 830–831
  controlling experimental error, 836–840
    experimental procedures and, 836–837
    selecting experimental and measurement units, 837–838
    using blocking, 838–839
    using covariates, 839–840
  determining number of replications, 845–848
    accuracy of estimator specifications and, 845–846
    examples of, 846–847
    using *F* test for, 846–848
  elements of plan for, 831
  exercises for, 848–852
  introduction, 829
  randomization of experimental units to treatments, 840–844
    examples of, 841–842, 844
    overview of, 840–841
    randomized complete block design and, 843–845
    steps in random assignment process, 842–843
  terminology, 831–836
    control treatment, 833
    experimental error, 834–835
    experimental unit, 833
    factorial treatment design, 832
    factors, 832

    fractional factorial experiments, 833
    measurement unit, 834
    measurements or observations, 832
    one-way classification, 832
    replication, 833–834
    treatment design, 832
    treatments, 832
    variance of experimental error, 835–836
  unbalanced designs, 1051
deviation, 86–87 (*See also* residual standard deviation; standard deviation)
df (degree of freedom), 504
diagnostic measures, leverage and influence, 546
direct observation, data collection, 25
discrete random variables (*See also* binomials)
  probability distribution for, 142
  properties of, 143
documentation, 1088–1089
dummy variables
  defined, 623
  lack of fit and, 730
Dunnett's procedure, 450–452
  example of, 451–452
  statistical table of, 1117–1120
  steps in, 450–451
  use of controls, 450

Empirical Rule, 89–93
  approximating standard deviation, 91–92, 632
  statement of, 89
  utility of, 90–91
equal variance condition, evaluating, 355
error rate control, 438–440
  Bonferroni inequality and, 439–440
  examples of, 440
  Type I error rate, 438–440
error terms, 395
error variance (*See also* residual standard deviation)
  estimating based on residuals, 546
  estimating for linear regression, 540
  estimating for multiple regression, 712
  mean square estimates of, 580
estimated expected value, 503

estimation
  comparing with hypothesis testing, 193
  estimation bias
    in Latin square design, 1059
    in randomized block design, 1053–1054
  inferences and, 193
  of mean, 196–204
  ratio estimation, 20
event relations, probability, 128–131
  complements of events, 129–130
  intersection of events, 130
  mutually exclusive events, 128–129
  union of events, 130
events, defined, 124
expected cell counts, 490
expected mean squares (EMS), 1000–1010 (*See also* mean squares)
  classifying interactions, 1000–1001
  comparing fixed- and random-effects model, 979
  in completely randomized design with single factor, 857
  constructing mean square tables, 1001–1003
  examples of, 1001, 1003–1005, 1006–1010
  in randomized complete block design, 866
  rules for obtaining, 1003
expected number of outcomes, in multinomial experiments, 490
expected values, defined, 619
experimental design, 28–30 (*See also* designed experiments)
  completely randomized design, 28–29
  Latin square design, 29
  randomized block design, 29
experimental error
  causes of, 834–835
  controlling, 836–840
    experimental procedures and, 836–837
    selecting experimental and measurement units, 837–838
    using blocking, 838–839
    using covariates, 839–840
  example of, 834–835
  lack of fit and, 579–580
  sources of, 836
  variance of, 835–836

experimental units
  defined, 833
  randomization of, 840–841
  selecting, 837–838
experiments (*See* designed experiments)
experimentwise error rate
  Bonferroni inequality and, 439
  controlling, 438–440
  Dunnett's procedure for controlling, 450
  in Scheffé's method, 450
  in Tukey's W procedure, 445
explained sum of squares (*See* sum of squares regression-SS(Regression))
explanation, compared with prediction, 531–532
exploratory data analysis (EDA)
  graphical methods of, 54
  quartiles of distribution in, 96
extrapolation
  analysis of covariance and, 959–962
  extrapolation penalty, 569
  linear regression forecasting and, 568–569
  multiple regression and, 667

F distribution, 356–357
  critical values of, 357
  defined, 356
  densities of, 356
  properties of, 356
  statistical table of, 1102–1113
F tests
  analysis of covariance and, 950
  analysis of variance and, 386–387
  contrasts and, 436
  correlated variables and, 708
  determining number of replications, 846–848
  expected mean square rules and, 1000
  factorial treatment, completely randomized design, 904–905
  null hypothesis and, 560
  one-factor experiments, repeated measures, 1031
  power of, 846–848
  statistical table of power curves for AOV, 1123–1126
  testing regression coefficients, 648–651, 657–658
  two-factor experiments, repeated measures, 1034

factorial experiments, 30–33
  applications of, 32
  defined, 32, 832, 895
  extending random-effects model to, 983
  interaction of factors in, 31
  one-at-a-time approach and, 30–31, 892–894, 896
  overview of, 894–897
  random-effects model for, 985–986
  use of factors in, 30
  vs. split-plot design, 999
factorial treatment, completely randomized design, 891–914
  accounting for interaction in, 898–899
  AOV table for, 908
  constructing AOV table for, 899–900
  example of, 901
  exercises for, 913–914
  first test of significance for, 902–903
  interaction of factors and, 894
  interpreting F test for main effects, 904–905
  models for, 897–898
  models for three factors, 905–907
  one-at-a-time approach, 892
  sum of squares calculations for, 907–908
  summary of method of analysis, 909
factorial treatment, randomized complete block design, 914–916
  AOV table for, 916
  example of, 915–916
  exercise for, 916
  overview of, 914–915
factors (*See also* variables)
  confounded factors, 860
  in designed experiments, 30, 832
  measuring main effects of, 900
  nested sampling and, 1010
  in observational studies, 830
  three-factor experiment measuring, 909
    random and fixed effects, 1008–1009
  treatment design and, 892
false negative/false positive, in Bayes' formula, 136
filtering
  Latin square design, 881–882

randomized complete block design, 863–864

first-order models, multiple regression, 619

Fisher's least significant difference (LSD) (*See* least significant difference (LSD))

fixed-effects model
  defined, 975
  formula for, 978
  vs. random-effects model, 978–979, 1005

forecasting
  linear regression, 567–576
    confidence interval for, 567–568
    dangers of extrapolation, 568–569
    examples of, 569–572
    exercises for, 572–576
    overview of, 567
    prediction interval in, 570–571
    prediction vs. explanation, 531–532
  multiple regression, 666–670
    examples of, 666–668
    exercises for, 668–670
    extrapolation and, 667
    overview of, 666

formulas
  analysis of variance (AOV), 923
  analysis of variance (AOV), unbalanced designs, 1072–1073
  categorical data, 520–521
  data description, 110
  linear regression, 602–603
  multiple-comparison procedures, 459
  multiple regression, 620, 688, 783
  population central values, 251–253
  population variances, 373–374, 415
  probability distributions, 187

forward selection (*See* stepwise regression)

fractional factorial experiments, 833

frequency histogram (*See also* histograms)
  accuracy of data presentation with, 1081
  examples of, 49

frequency tables, 47

general linear model
  matrix notation for, 684

multiple regression and, 625–627
  exercises using, 626–627
  formula for, 626

graphical methods
  bar charts, 45–46
  communication distortions with, 1079–1082
  exercises for, 63–69
  frequency histogram, 46–54
  guidelines for, 43, 62
  pie charts, 43–45
  representing lack of fit with, 576–577
  stem-and-leaf plots, 54–57
  time series, 57–62

grouped data
  median for, 72
  range for, 82

Hartley's F_{max} test
  examples, 366–371
  overview of, 366
  recommended use of, 396
  statistical table of, 1121
  when to use, 368

heavy-tailed distributions, 236

high influence point
  effect on regression equations, 767
  estimating regression slope and, 545

high leverage point, 545–546

histograms (*See also* frequency histogram)
  accuracy of data presentation with, 1081
  creating frequency table for, 47–48
  Empirical Rule and, 89
  examples of, 49, 51–54
  measuring probability in, 50
  skewed right or left, 54
  symmetric, 54
  uniform, 54
  unimodal and bimodal, 52–54
  uses of, 50
  variability in, 81
  vs. bar and pie charts, 50

homogeneity, tests of, 505–506

hypothesis generation, data snooping and, 428

hypothesis testing
  comparing with estimation, 193
  decision rule for, 224–225
  inferences and, 193
  null hypotheses and, 227
  research hypotheses and, 207

incomplete block design, 1064 (*See also* balanced incomplete block (BIB) design)

independence
  conditional probability and, 131–136
  multiple regression and, 773–781
    assumptions for, 759
    diagnosing problems with residual plots, 776
    Durbin-Watson test statistic, 773
    first differences alternative, 776
    serial correlation and, 773–776
    time series data and, 773
  tests of, 503, 504

independent events, 133–134

independent samples, 134

independent variables
  measuring strength of relations in, 512
  selecting, 708
  too many, 711
  vs. dependent variables, 512

individual error rate
  Bonferroni inequality and, 439
  Type I errors, 438

inferential statistics (*See* statistical inference)

influence (outliers), 767

interaction of factors, 622
  accounting for, 898–899
  classifying as fixed or random effect, 1000–1001
  defined, 898
  disorderly interactions, 905
  in factorial experiments, 31
  measuring, 900
  more than two factors, 899
  one-at-a-time approach to, 894
  orderly interactions, 904

intercept
  comparing slopes and, 671
  defined, 532
  least-squares estimates of, 542
  in linear regression, 540
  in multiple regression, 630, 632
  random error and, 557

interquartile range (IQR), 86

intersecting lines, compared with parallel lines, 671

intersection of events, probability laws and, 130

interval estimate, 196

inverse regression problems, 582–590
calibration as, 582
examples of, 583–584
exercises for, 588–590
predictions, 583, 585
inverse transformations, 739

jackknife method, for outlier detection, 769–770

key formulas (*See* formulas)
Kruskal–Wallis test, 410–414
example of, 410–412
exercises for, 413–414
testing more than two populations, 410

lack of fit
checking with residual plots, 759
dummy variables and, 730
in linear regression, 576–582
examples of, 577–579, 581–582
exercises for, 582
mean square estimates and, 580
partitioning residuals, 579–580
process for, 580
using graphs for, 576–577
in multiple regression, 727
nonlinearity and, 737
large samples, median, 245–246, 247–248
Latin square design, 879–891
advantages/disadvantages of, 880
crossover designs and, 1040
defined, 881
examples of, 879–880, 884–888
exercises for, 888–891
filtering with, 881–882
model for, 881
overview of, 879–880
relative efficiency of, 884
testing for treatment effects, 882–884
types of experimental designs, 29
Latin square design, missing data, 1059–1063
comparing treatment means, 1061
estimating missing value, 1059
example of, 1059–1060
exercises for, 1062–1063
fitting complete and reduced models for, 1061
least significant difference (LSD), 440–444

AOV for standard designs and, 920
balanced incomplete block designs, 1068
comparing treatment means
standard designs, 916–917
unbalanced designs, 1056
comparing with Tukey's W procedure, 446
confidence interval for, 444
examples of, 441–444
Fisher's protected LSD and, 440–441
sample size and, 444
steps in, 441, 442–443
using with Latin square design, 1061
least-squares method
estimating multiple regression coefficients, 627, 630
linear regression, 541–544
multiple regression, 728
level of confidence, estimating mean and, 196, 204–205
level of significance (*p*-value), 224–228
defined, 224
hypothesis testing using, 224–225
null hypotheses and, 227
Levine's test
confidence interval and, 454
examples of, 368–371
exercises for, 371–372
overview of, 368
use of, 368, 396
leverage (outliers), 767
likelihoods
defined, 138
ratio statistic for, 505
linear contrasts, 431–438
defined, 432
examples of, 432–437
exercises for, 437–438
F test for contrasts, 436
mutually orthogonal contrasts, 432–433
notation for, 431
orthogonal contrasts, 432–433
linear regression, 531–616
analyzing, 532
case study, 538–540
comparing slopes, 671
correlation and, 590–600
assumptions for correlation inference, 593–595

coefficient of determination and, 592–593
correlation coefficient and, 591
examples of, 592–597
exercises for, 597–600
measuring accuracy of predictions with, 590–591
estimating parameters, 540–557
creating scatterplot for, 540–541
examples of, 542–544, 547–548
exercises for, 548–557
high leverage points and, 545–546
measures of leverage and influence, 546
using least-squares method, 541–542
using residuals analysis, 546–547
exercises for, 603–616
forecasting, 567–576
confidence interval for, 567–568
dangers of extrapolation, 568–569
examples of, 569–572
exercises for, 572–576
overview of, 567
prediction interval in, 570–571
inferences about parameters, 557–566
accounting for random error, 557
confidence interval for slope, 559–560
examples of, 558–561
exercises for, 561–566
using F test for null hypothesis, 560
using t test for slope, 557–558
introduction
analyzing simple regression, 532
assumptions, 532–535
checking assumptions, 534–537
choosing transformations, 537–538
comparing prediction and explanation, 531–532
use of random error term, 533
inverse regression problems, 582–590
calibration as, 582
examples of, 583–584

exercises for, 588–590
predictions, 583, 585
key formulas for, 602–603
lack of fit, 576–582
   examples of, 577–579, 581–582
   exercises for, 582
   mean square estimates and, 580
   partitioning residuals and, 579–580
   process for, 580
   using graphs for, 576–577
linear regression variables vs. multiple regression variables, 631
linearity, assumption of, 532–533
logarithmic transformation, 738–739, 759–760
logic checks, raw data source, 37, 1086
logistic regression, 675–683
   example of, 677–679
   exercises for, 679–683
   functions of, 676
   modeling associations, 675
   multiple logistic regression model, 677
   simple logistic regression model, 675–677
LOWESS curve
   as a smoother, 534–536
   using with regression models, 739
   using with residual plots, 759

machine-readable database
   defined, 36
   key-entered data and, 1085
Mann-Whitney U test, 289 (*See also* Wilcoxon rank sum test)
mean (μ)
   binomial probability distribution and, 149–151
   characteristics of, 77
   comparing with variability, 81
   defined, 73
   estimating, 196–204
      confidence coefficient, 197–200
      confidence interval, 200–201
      interval estimate and level of confidence, 196
      sampling distribution, 196
      standard deviation, 201–202
   measures of central tendency and, 73–76
   population central values and, 193

population proportion inferences and, 472
sample size for estimating, 204–207
   formula for computation of, 205
   level of confidence, 204–205
   tolerable error, 204–205
sample size for testing, 219–224
   for two-sided test, 221
   Type I/Type II errors and, 219–220
statistical test for
   computing probability of Type II error, 214, 214–216, 218
   guidelines for, 208
   null hypothesis, 208, 209
   one-tailed tests, 211, 216
   rejection region, 208–210
   research hypothesis, 207
   summary of, 212–213
   test statistic and, 208
   two-tailed tests, 211, 216
   Type I/Type II errors, 209, 218–219
statistical tests for, 207–219
transformations of data and, 407
mean (μ), multiple comparisons (*See* multiple-comparison procedures)
mean (μ), unknown variance, 228–243
   accuracy considerations for t procedures, 237–238
   adjusting for nonnormal distributions, 236–237
   case study, 238–239
   confidence interval for, 234
   robust methods for, 238
   Student's t distribution and, properties of, 230
   vs. normal distribution, 229
   summary of statistical test for, 231
mean (μ_1) – mean (μ_2)
   independent samples, 267–287
      cluster effect of data and, 274
      confidence interval for, 267–268
      confidence interval for (when variance is unequal), 276
      effect of unequal population variances, 274–275
      exercises for, 280–287
      sample size for, 314–316

spatial correlation and, 274
statistical test for, 271–274
t test for, 275–276
Type I errors and, 279
weighted average of sampling variances, 268
paired data
   confidence interval for, 303
   estimating, 299–308
   overview of, 299–300
mean squares (*See also* expected mean squares (EMS))
   constructing mean square tables, 1001–1003
   defined, 388
   estimating error variance with, 580
measurements
   measurement units
      defined, 834
      example of, 834
      selecting, 837–838
   remedies for measurement problems, 23
   as response variables, 832
median (M)
   characteristics of, 77
   defined, 71
   for grouped data, 72
   inferences about, 243–250
      approximating for large samples, 245–248
      confidence interval for, 243–245
      sign test vs. t test, 248
      summarizing statistical test for, 246–247
      using sign test, 246
   population central values and, 193
Minitab
   probability and, 185–186
      calculating binomial probabilities, 185
      calculating normal probabilities, 185
      generating random numbers, 185
      generating sampling distribution, 186
   scatterplot matrix, 708
   statistics software and, 42
mixed-effects model, 992–1000
   assumptions for, 992
   defined, 976
   example of, 993–996

mixed-effects model (*contd.*)
  exercises for, 1000
  reporting conclusions of, 999
  residual analysis for, 997
  tests of significance for, 992–993
mode
  characteristics of, 77
  defined, 70
models
  balanced incomplete block designs, 1065
  completely randomized design, 394–396
    assumptions for, 395
    conditions for, 394
    multiple covariates, 964–968
    one covariate, 949–951
    single factor, 856
    sum of terms and, 394–395
  crossover designs, 1040
  factorial treatment, completely randomized design, 897–898, 905–907
  Latin square design, 881
  linear regression, 540–557
    creating scatterplot for, 540–541
    examples of, 542–544, 547–548
    exercises for, 548–557
    high leverage points and, 545–546
    measures of leverage and influence, 546
    using least-squares method, 541–542
    using residuals analysis, 546–547
  multiple regression, 727–782
    checking assumptions, 731–737, 758–759
    checking constant variance, 760–765
    checking independence, 773–781
    checking lack of fit, 727–730
    checking normality, 765–773
    checking zero expectation, 759–760
    exercises for, 745–758
    logarithmic transformation and, 738–739
    nonlinear least-squares, 740–744
    using best subset regression, 730–731

    using nonlinear models, 737–744
    using scatterplots, 727–730
  randomized complete block design, 862–864
  two-factor experiments, repeated measures, 1031
multicollinearity (*See* collinearity)
multinomial distribution, 488–489
multinomial experiments, 488
multiple-comparison procedures, 427–468
  case study, 428–431
  comparing treatment means, 920
  Dunnett's procedure, 450–452
    example of, 451–452
    steps in, 450–451
    use of controls, 450
  error rate control and, 438–440
    Bonferroni inequality and, 439–440
    examples of, 440
    experimentwise Type I error rate, 438–440
    individual comparison Type I error rate, 438
  exercises for, 459–468
  Fisher's least significant difference and, 440–444
    confidence interval for, 444
    for equal sample sizes, 444
    examples of, 441–444
    Fisher's protected LSD and, 440–441
    steps in, 441
  introduction, 427–428
  key formulas for, 459
  linear contrasts and, 431–438
    defined, 432
    examples of, 432–437
    exercises for, 437–438
    F test for, 436
    notation for, 431
    orthogonal contrasts and, 432–433
  reporting conclusions of, 458
  Scheffé's S method, 452–458
    confidence interval for, 453–454
    example of, 453
    overview of, 452
    steps in, 452
  Student–Newman–Keuls procedure, 447–450
    comparing with Tukey's procedure, 447–448

    example of, 448–449
    steps in, 448
  Tukey's W procedure, 444–447
    comparing with LSD procedure, 446
    confidence interval for, 447
    example of, 445–446
    experimentwise error rate in, 445
    steps in, 445
    use of Studentized range distribution in, 444
multiple logistic regression models, 677
multiple regression, 617–709
  case study, 617–620
  comparing slopes, 670–675
    examples of, 670–674
    exercises for, 674–675
    extending to three or more lines, 674
    intersecting lines vs. parallel lines, 671
  estimating coefficients, 627–646
    examples for, 628–633
    exercises for, 633–646
    least-squares estimates, 630
    linear regression variables vs. multiple regression variables, 631
    normal equations for, 627–628
    residual standard deviation and, 632–633
  exercises for, 689–709
  forecasting, 666–670
    examples of, 666–668
    exercises for, 668–670
    extrapolation and, 667
    overview of, 666
  general linear model for
    exercises using, 626–627
    formula for, 626
  inferences in, 646–657
    coefficient of determination and, 646
    colinearity and, 646–647
    examples of, 647–651, 653–655
    exercises for, 655–657
    F test of all coefficients, 648–651, 655
    t test of individual coefficients, 652–655
  introduction, 620–625
    assumptions for, 621
    examples of, 621–622, 624–625

first-order models, 620–622
formula for multiple regression
model, 620
higher-order models, 622–625
parameters of, 620–621
key formulas for, 688
logistic regression and, 675–683
example of, 677–679
exercises for, 679–683
modeling associations, 675
multiple logistic regression
model, 677
simple logistic regression
model, 675–677
testing coefficients, 657–665
complete and reduced models
for, 658
example of, 658–661
exercises for, 661–665
F test of predictors, 657–658
theory, 683–686
computing estimated standard
error, 686
computing inverse of the $\mathbf{X'X}$
matrix, 685
computing SS(Regression) and
SS(Total), 686
example of, 684–686
general linear model, 684
normal equations in matrix notation, 684
using general linear model for,
625–627
multiple regression, applying,
705–828
case study, 706–707
checking model assumptions,
758–782
checking constant variance,
760–765
checking independence,
773–781
checking normality, 765–773
checking zero expectation,
759–760
overview of, 758–759
exercises for, 783–828
introduction, 705
key formulas for, 783
model formation, 727–758
exercises for, 745–758
logarithmic transformation and,
738–739
nonlinear least squares,
740–744

testing lack of fit, 727–730
testing with best subset regression, 730–731
testing with residual plots,
731–737
using nonlinear models,
737–744
using scatterplots, 727–730
selecting variables, 707–727
backward elimination, 717–720
best subset regression, 716–717
collinearity and, 708–710
data-splitting approach, 714
examples for, 709–722
exercises for, 722–727
overview of, 707–708
performing all possible regressions, 711–714
PRESS statistic and, 714–715
stepwise regression, 718,
720–722
underspecification/over-
specification, 715
multiple regression coefficients,
627–646
examples for, 628–633
exercises for, 633–646
least-squares estimates, 630
linear regression variables vs. multiple regression variables,
631
normal equations for, 627–628
residual standard deviation,
632–633
testing, 657–665
multiple variables (*See* variables, multiple)
multiplication law, conditional probability and, 132
mutually exclusive events, probability
laws and, 128–129
mutually orthogonal contrasts,
432–433

natural logarithms, 407, 739
nested sampling, 1010–1020
AOV table for, 1014
examples of, 1011–1014,
1018–1019
F tests for, 1012–1014
factors of, 1010
overview of, 1010–1011
partially nested design, 1014
random-effects model, 990–991
split-plot design, 1014–1018

nonlinearity
least squares and, 740–744
multiple regression and, 737–744
using transformations with,
759–760
nonnormal distributions, adjusting,
236–237
normal approximation, 292–295
normal curve, 157–166, 765
100pth percentile of distribution,
162–166
approximating binomial, 182–185
area under, 158–162
assumptions
checking, 765–773
stating, 759
calculating normal probabilities,
185
continuity correction and, 184
overview of, 157–158
standard normal curve and,
1091–1092
using probability plots with, 766
using scatterplots and residual
plots with, 765
normal equations
estimating regression coefficients,
627–628
for multiple regression, 684
normal population (*See* population
mean)
normal probability distribution (*See*
normal curve)
null hypothesis
accepting/rejecting, 227
estimating standard error,
653–654
multiple t tests and, 385
randomized complete block design
and, 865
statistical tests for, 208, 209
testing regression coefficients, 658
numerical descriptive measures
comparing parameters and statistics, 69–70
overview of, 69
value of, 109

observational studies
vs. designed experiments, 830–831
vs. scientific studies, 34–35
observations
compound symmetry of,
1030–1031
costs of, 714

observations (*contd.*)
  defined, 138
  direct observation, 25
  estimating value of missing obser-
    vations, 1053
  reasons for multiple observations,
    1027–1029
  as response variables, 832
observed cell counts, 490
OC curve, 214
odds, 516–519 (*See also* probability)
  example of, 517
  formula for, 516–517
  odds ratio defined, 518–519
one-at-a-time approach
  factorial experiments and, 30–31,
    892, 896
  problems of interaction and, 894
one-tailed tests
  computing probability for, 216
  overview of, 211
one-way classification (*See also* com-
    pletely randomized design)
  defined, 394
  population mean and, 431
  treatment design and, 832
opinion polls
  applying statistics to, 9–10
  Gallup and Harris polls, 20
original files
  machine-readable database and,
    37–38
  sets of, 1086
orthogonal contrasts, 432–433
outcome, defined, 124
outliers
  detecting
    jackknife method, 769–770
    scatterplots or residual plots,
      765
  how to deal with, 771
  leverage and influence types, 767
  problems in identifying, 769
overspecification, multiple regression
    variables, 715

p-value (*See* level of significance (*p*-
    value))
paired data (*See* mean (μ_1) − mean
    (μ_2), paired data)
parallel lines vs. intersecting lines,
    671
parameters
  estimating with scatterplots,
    540–541

linear regression
  estimating model parameters,
    540–557
  inferences about parameters,
    557–566
  multiple regression, 620–621
  numerical descriptive measures
    and, 69–70
  population parameters, 193, 341
  regression parameters
    accounting for random error,
      557
    confidence interval for slope,
      559–560
    examples of, 558–561
    exercises for, 561–566
    inferences about, 557–566
    using F test for null hypothesis,
      560
    using t test for slope, 557–558
partial slopes, 621
partitions of TSS
  completely randomized design, sin-
    gle factor, 857
  factorial treatment, completely
    randomized design, 899
  Latin square design, 882
  randomized complete block de-
    sign, 865
Pearson correlation matrix (*See* corre-
    lation matrix)
percentiles
  defined, 82
  examples of, 82–86
  percentage analysis, 510–511
  percentile of distribution (100*p*th),
    162–166
  quartiles of distribution, 83
personal interviews, data collection,
    24
pi (*See* proportion)
pie charts
  constructing, 45
  description of, 43
  examples of, 44–45
  uses of, 50
placebo effect
  controls and, 450, 833
  use of, 7
plots (*See* by individual type)
point estimation, 194, 648
Poisson distribution, 497–501
  assumptions for, 498
  defined, 497
  examples of, 497–500

formula for, 497
  tests using, 498–499
Poisson probabilities, statistical table,
    1127–1129
Poisson random variable, 403–404
pooled estimate, population variance,
    384
population central values, more than
    two populations, 379–425
  case study, 379–384
  checking AOV conditions,
    396–403
    examples of, 397–403
    overview of, 396–397
    reporting conclusions of, 403
    using residuals analysis for,
      397
  exercises for, 416–425
  introduction, 379
  key formulas for, 415
  Kruskal–Wallis test for, 410–414
    example of, 410–412
    exercises for, 413–414
    testing more than two popula-
      tions, 410
  observation model, completely ran-
    domized design, 394–396
    assumptions for, 395
    conditions for, 394
    sum of terms and, 394–395
  testing equality, 384–393
    AOV table of results, 389
    completely randomized design,
      387–388
    developing test statistic for,
      386–387
    examples of, 389–392
    exercises for, 392–393
    mean square and, 388
    sum of squares and, 388
    using multiple t tests, 385
  transformations, 403–409
    defined, 403
    examples of, 404–409
    exercises for, 409
    selecting new variables,
      403–404
population central values, single pop-
    ulation
  case study, 193–195
  estimating mean, 196–204
    confidence coefficient, 197–200
    confidence interval, 200–201
    interval estimate and level of
      confidence, 196

sampling distribution, 196
standard deviation, 201–202
estimating mean, unknown variance, 228–243
accuracy of *t* procedures,
237–238
adjusting for nonnormal distributions, 236–237
applying to case study, 238–239
confidence interval for, 234
robust methods for, 238
Student's *t* distribution vs. normal distribution, 229
Student's *t* properties, 230
summary of statistical test for,
231
estimating median, 243–250
approximating for large samples, 245–248
confidence interval, 243–245
sign test vs. *t* test, 248
statistical test for, 246–247
using sign test, 246
exercises for, 253–262
introduction, 192–193
estimation vs. hypothesis testing, 193
parameters, 193
key formulas, 251–253
level of significance (*p*-value),
224–228
accepting/rejecting null hypotheses, 227
decision rule for hypothesis
testing, 224–225
sample size, 204–207, 219–224
formula for computation of,
205
level of confidence and,
204–205
tolerable error and, 204–205
for two-sided test, 221
Type I/Type II errors and,
219–220
statistical test for mean, 207–219
computing probability of Type
II error, 214–216, 218
guidelines for, 208
null hypothesis, 208, 209
one-tailed tests, 211, 216
rejection region and, 208–210
research hypothesis and, 207
summary of, 212–213
test statistic and, 208
two-tailed tests, 211, 216

Type I/Type II errors, 209,
218–219
population central values, two populations, 263–339
estimating mean (μ_1) − mean (μ_2)
paired data, 299–308
confidence interval, 303
examples of, 300–304
exercises for, 305–308
overview of, 299–300
estimating mean (μ_1) − mean (μ_2)
independent samples,
267–287
cluster effect of data and, 274
confidence interval, 267–268
confidence interval, variance
unequal, 276
effect of unequal variances,
274–275
examples of, 268–274
exercises for, 280–287
spatial correlation and, 274
statistical test for, 271–274
t test for, 275–276
Type I errors and, 279
weighted average of sampling
variances, 268
exercises for, 319–339
introduction to, 263–267
sampling distribution properties, 266–267
theorem for sampling distribution, 266
sample size, 314–316
confidence interval, independent samples, 314
confidence interval, paired samples, 316
examples of, 315
testing independent samples,
315
testing paired samples, 316
Wilcoxon rank sum test, 287–299
calculating rank sum statistics,
288–289
examples of, 289–296
exercises for, 296–299
normal approximation and,
292–295
overview of, 288
summary of, 289
vs. *t* test, 295–296
Wilcoxon signed-rank test, 308–314
computing, 308
example of, 309–312

exercises for, 312–314
g groups and, 308–309
overview of, 309
population mean (*See* mean)
population mean, multiple comparisons (*See* multiple-comparison
procedures)
population median (*See* median)
population parameters
central values and, 193
statistical inference and, 341
population proportion, inferences
about
confidence interval for, 472–474
examples of, 472–476
exercises for, 477–482
mean and standard error, 472
sample size requirement, 477
statistical test for, 475
population proportion, inferences
about difference between two
populations
confidence interval for, 483
examples of, 483–485
exercises for, 486
notation for comparing, 482
rule for sample size, 484
statistical test for, 484–485
population proportion, inferences
about several proportions
examples of, 489, 491–493
exercises for, 494–497
expected number of outcomes, 490
multinomial distribution, 488–489
multinomial experiments, 488
statistical test for, 490–494
population variances, 341–378 (*See
also* variance)
case study, 341–344
introduction, 341
key formulas for, 373–374
for single population, 344–355
chi-square distribution and,
344–346
confidence interval, 346
examples of, 346–351
exercises for, 351–355
statistical test for, 348
Type I errors and, 350–351
unbiased estimators and, 344
for than two populations, 365–372
examples of, 366–371
exercises for, 371–372
Hartley's F_{max} test, 366–368
Levine's test, 368–371

population variances (*contd.*)
  for two populations, 355–365
    confidence interval, 359
    evaluating equal variance condi-
      tion, 355
    examples of, 357–363
    exercises for, 363–365
    F distribution and, 356–357
    reporting conclusions of,
      362–363
    statistical test for, 358
posterior probabilities, 138
power curves, 214
power values
  t test and, 237
  t test vs. sign test, 248
predictability analysis
  measuring strength of relation,
    511–512
prediction (*See* forecasting, linear re-
    gression)
prediction interval
  defined, 666
  for linear regression forecasting,
    570–571
preliminary analyses, 1087
PRESS statistic, 714–715
primary analyses, 1087
probability (*See also* odds)
  basing inferences on, 122–125
  Bayes' formula, 136–140
    false positive/false negative test
      results, 136
    likelihoods in, 138
    observable events in, 138
    prior/posterior probabilities in,
      138
    sensitivity/specificity of diagnos-
      tic tests, 137
    statement of, 138
    states of nature in, 138
  classical interpretation
    defined, 124
    determining probability with,
      126
  computation of probability of
    Type II error, 214, 216
  conditional probability, 131–136
    conditional vs. unconditional
      probability, 132
    defined, 132
    independent vs. dependent
      events, 133–134
    multiplication law and, 132
    probability of the intersection,
      132–133

definitions of
  classical interpretation, 124
  relative frequency interpreta-
    tion, 124
  subjective interpretation,
    124–125
determining, 125–128
  with classical interpretation,
    126
  with relative frequency interpre-
    tation, 126–128
event relations and, 128–131
  complements of events,
    129–130
  intersection of events, 130
  mutually exclusive events,
    128–129
  union of events, 130
exercises for, 187–190. *See also*
    probability distributions
measuring with histograms, 50
Minitab instructions for, 185–186
  calculating binomial probabili-
    ties, 185
  calculating normal probabili-
    ties, 185
  generating random numbers,
    185
  generating sampling distribu-
    tion, 186
probability of the intersection,
    132–133
random variables and, 141–142
probability distributions
  binomials and, 144–154
    defining binomial experiments,
      144
    formula for computation of,
      146–149
    mean and standard deviation
      of, 149–151
    properties of, 145
  chi-square distribution, 344–346
  for continuous random variables,
    154–157
  for discrete random variables,
    142–143
  F distribution and, 356–357
  key formulas for, 187
  normal curve and, 157–166
    100*p*th percentile of distribu-
      tion, 162–166
    approximating binomial,
      182–185
    area under, 158–162
    overview of, 157–158

random sampling and, 166–171
  defined, 167
  random number tables and,
    168
  sampling distribution and,
    171–182
    Central Limit Theorem, 175,
      179–180
    contrasting sample histogram
      with, 180
    defined, 171
    examples of, 171–173
    interpretations of, 180
    standard of error, 175
probability plots
  checking normality, 766
  identifying outliers, 770
profile plots
  constructing, 901
  two-way interactions, 911
  using with interactions, 898–899
proportions
  defined, 193
  transformations of data and, 409
pure experimental error, lack of fit
    and, 579–580

quadratic transformations, 739
qualitative random variables, 141,
    623
quality improvement process, 13–14
quantitative random variables, 623
  (*See also* random variables)
quartiles of distribution, 96

random-effects model, 978–983
  AOV moment matching and, 980
  assumptions of, 978–979
  defined, 976
  example of, 981–982
  exercises for, 982–983
  expected mean squares and, 979
  formula for, 978
  hypothesis tests for, 979–980
  vs. fixed-effects model, 978–979,
    1005
random-effects model, block design,
    983–992
  assumptions for, 983–984
  estimating variance components,
    986–987
  example of, 987–989
  exercises for, 991–992
  hypothesis testing for, 989–990
  model and AOV tables for, 985
  nested sampling and, 990–991

random error
  accounting for in linear regression, 557
  use of random error term, 533
random numbers
  generating, 185
  statistical table of, 1122
  tables of, 168
random sampling, 166–171
  defined, 167
  random number tables and, 168
random variables, 141–142, 623
randomization, experimental units to treatments, 840–844
  examples of, 841–842, 844
  overview of, 840–841
  randomized complete block design and, 843–845
  steps in random assignment process, 842–843
randomized complete block design, 859–878
  advantages/disadvantages of, 861
  analysis of covariance and, 943
  comparing treatment means, 865–867
  criteria for, 838–839
  defined, 33, 861
  examples of, 868–871
  exercises for, 872–878
  experimental randomization and, 843–845
  extending random-effects model to, 983
  filtering and, 863–864
  model for, 863–864
  overview of, 859–861
  reducing experimental error with, 834–835
  relative efficiency of, 867–868
  types of experimental designs, 29
randomized complete block design, one or more missing observations, 1053–1058
  comparing treatment means, 1056
  estimating value of missing observation, 1053
  estimation bias and, 1053–1054
  examples of, 1054–1055
  exercises for, 1057–1058
  fitting complete and reduced models for, 1056–1057
  testing effect of treatments, 1057
range
  defined, 82
  interquartile range, 86

Studentized range distribution, 444, 1115–1116
rank sum statistics
  calculating, 288–289
  for more than two population variances, 410
ranks, defined, 288
ratio estimation, 20
raw data source
  correcting and clarifying, 37, 1086
  creating database from, 36–37
  data preparation and, 1084–1085
  defined, 1084
  receiving, 36
regression analysis (*See also* linear regression; multiple regression)
  analysis of covariance and, 949–950
  diagnostic measures of leverage and influence, 546
  formal assumptions of, 534–535
  logistic regression and, 675–677
  straight-line prediction and, 541
regression parameters
  accounting for random error, 557
  confidence interval for slope, 559–560
  examples of, 558–561
  exercises for, 561–566
  inferences about, 557–566
  using F test for null hypothesis, 560
  using t test for slope, 557–558
rejection region, normal curve, 208–210
relative efficiency
  Latin square design, 884
  randomized complete block design, 867–868
relative frequency, defined, 48
relative frequency histogram (*See also* histograms)
  different variability, same mean in, 81
  examples of, 49–51
relative frequency interpretation of probability
  defined, 124
  determining probability with, 126–128
repeated measures, 1025–1050
  case study, 1026–1027
  crossover designs and, 1040–1044
  analysis of variance for, 1041
  example of, 1041–1043
  model for, 1040

using Latin square design, 1040
  defined, 1025
  exercises for, 1045–1050
  introduction, 1025–1026
  crossover designs defined, 1025
  repeated measure designs defined, 1025
  single-factor experiments, 1027–1031
  assumptions for, 1029–1030
  compound symmetry of observations, 1030–1031
  reasons for multiple observations, 1027–1029
  two-factor experiments, 1031–1040
  examples of, 1032–1034, 1036
  exercises for, 1038–1040
  F tests for, 1034
  model for, 1031
  reporting conclusions of, 1038
  tests for, 1032
replications
  defined, 833–834
  determining number of, 845–848
  accuracy of estimator specifications and, 845–846
  examples of, 846–847
  using F test for, 846–848
report guidelines, 1087–1088
research hypotheses, 207
residual plots
  checking analysis of covariance with, 968–970
  checking constant variance with, 760
  checking lack of fit with, 759
  checking multiple regression assumptions with, 731–737
  comparing with scatterplots and probability plots, 770
  detecting outliers with, 765, 770
  diagnosing collinearity with, 776
residual standard deviation
  analysis of covariance and, 951
  defined, 547
  measuring prediction accuracy, 590, 632–633
  random error and, 557
residuals analysis
  checking AOV conditions with, 397
  estimating true error variance with, 546–547
  using boxplots for, 968

response variables, 832, 839–840
robust methods, 238
root MSE (*See* residual standard deviation)
round-off error, 909

sample histogram vs. sampling distribution, 180
sample size
  communication hurdles and, 1083–1084
  for mean, 204–207
    computation of, 205
    level of confidence, 204–205
    tolerable error, 204–205
  for mean (μ_1) − mean (μ_2)
    confidence interval, independent samples, 314
    confidence interval, paired samples, 316
    examples of, 315
    testing independent samples, 315
    testing paired samples, 316
  population proportion inferences and, 477, 484
  as statistical problem, 7
  testing mean and, 219–224
    two-sided test and, 221
    Type I/Type II errors and, 219–220
samples
  biased samples, 1082–1083
  standard deviation, 547
  variation, 383–384
sampling distribution, 171–182
  bootstrap techniques and, 349–350
  Central Limit Theorem, 175, 179–180
  defined, 171
  estimating mean, 196
  estimating mean (μ_1) − mean (μ_2), 266–267
  examples of, 171–173
  generating, 186
  interpretations of, 180
  standard of error, 175
  theorem for, 266
  vs. sample histogram, 180
sampling techniques, 21–22
  cluster sampling, 20
  ratio estimation, 20
  single random sampling, 21
  stratified random sampling, 21–22
  systematic sampling, 22

SAS
  checking models with, 742
  generating residual plots with, 760
  statistics software types and, 42
scatterplots
  checking regression analysis assumptions, 534–537
  detecting outliers (nonlinear relations), 709, 765, 770
  displaying data with multiple variables, 102–103
  estimating model parameters with, 540–541
  key features of, 536–537
  limitations of, 730
  testing multiple regression models for lack of fit, 727–730
  testing variables with, 708–709
  use of smoothers with, 534–535
  using transformation variables with, 535
  vs. residual plots and probability plots, 770
Scheffé's *S* method, 452–458
  AOV for standard designs and, 920
  confidence interval for, 453–454
  example of, 453
  overview of, 452
  steps in, 452
scientific studies, 27–34
  complicated designs, 30–33
  data collection and, 19
  experimental designs for, 28–30
    completely randomized design, 28–29
    Latin square design, 29
    randomized block design, 29
  factorial experiments, 30–33
    applications of, 32
    defined, 32
    interaction of factors in, 31
    one-at-a-time approach, 30–31
    use of factors, 30
  overview of, 27–28
  vs. observational studies, 34–35
self-administered questionnaires, data collection, 25
sensitivity, of diagnostic tests, 137
separate-variance *t* tests, 276
sequential sums of squares (SS), 647
serial correlation
  of data, 274
  Durbin–Watson test statistic for, 773

first differences approach and, 776
  positive/negative correlation, 773
  time series data and, 773
side-by-side boxplots, 104–107
sign test
  comparing with *t* test, 248
  statistical table of, 1096
  testing one-sided hypotheses with, 246
simple linear regression (*See* linear regression)
simple logistic regression model, 675–677 (*See also* logistic regression)
single-factor experiments, repeated measures, 1027–1031
  assumptions for, 1029–1030
  compound symmetry of observations, 1030–1031
  reasons for multiple observations, 1027–1029
single random sampling, 21
skeletal boxplot, 96–97
skewness (*See also* normal curve)
  checking normality, 765
  measures of central tendency and, 76–77
  skewed distributions, 236
  skewed right or left histograms
    defined, 54
    example of, 53
slope
  comparing, 670–675
    examples of, 670–674
    exercises for, 674–675
    extending comparison to three or more lines, 674
    intersecting lines vs. parallel lines, 671
  confidence interval for, 559–560
  defined, 532
  least-squares estimates of, 542
  in linear regression, 540
  in multiple regression, 630, 632
  random error and, 557
smoothers, 535, 759
software packages, for statistics, 42–43
spatial correlation, of data, 274
specificity of diagnostic tests, in Bayes' formula, 137
spline fit, 535
split-plot design
  nested sampling and split-plot design, 1014–1018
  subplot analysis, 1018

vs. factorial experiments, 999
vs. standard two-factor experiments, 1019
wholeplot analysis, 1017
SPSS, statistics software, 42
squared prediction error, 541
SS(Error) (*See* sum of squares residuals-SS(Residual))
SS(Model) (*See* sum of squares regression-SS(Regression))
SS(Regression) (*See* sum of squares regression-SS(Regression))
SS(Residual) (*See* sum of squares residuals-SS(Residual))
SS(Total) (*See* sum of squares, total (TSS))
stacked bar graphs, 102
standard deviation, 88–89 (*See also* population variances)
    applications of, 341
    approximating, 91–92
    binomial probability distribution and, 149–151
    defined, 88
    estimating mean and, 201–202
    in multiple regression, 652, 686
    multiple regression forecasting and, 666
    population parameters, 193
    transformations of data and, 407
standard error, 175, 472 (*See also* residual standard deviation)
standard normal curve, statistical table of, 1091–1092
standardized residuals, zero expectation and, 759
states of nature, 138
statistical inference
    based on probability, 122–125
    data description and, 40–41
    guidelines for, 1087–1088
    hypothesis testing and, 193
    linear regression, 557–566
        accounting for random error, 557
        confidence interval for slope, 559–560
        examples of, 558–561
        exercises for, 561–566
        using F test for null hypothesis, 560
        using t test for slope, 557–558
    median (M), 243–250
        approximating for large samples, 245–248

confidence interval for, 243–245
    sign test vs. t test, 248
    summarizing statistical test for, 246–247
    using sign test, 246
multiple regression
    coefficient of determination and, 646
    colinearity and, 646–647
    examples of, 647–651, 653–655
    exercises for, 655–657
    F test of all coefficients, 648–651, 655
    t test of individual coefficient, 652–655
population central values, single population, 251–253
population central values, two populations
    calculating rank sum statistics, 288–289
    computing, 308
    examples of, 289–292, 309–312
    exercises for, 296–299, 312–314
    g groups and, 308–309
    normal approximation and, 292–295
    overview of, 288, 309
    summary of, 289
    using t test for, 295–296
population parameters and, 341
population proportions, difference between two, 482–488
    confidence interval for, 483
    examples of, 483–485
    exercises for, 486
    notation for comparing, 482
    rule for sample size, 484
    statistical test for, 484–485
population proportions, several, 488–497
    examples of, 489, 491–493
    exercises for, 494–497
    expected number of outcomes, 490
    multinomial distribution, 488–489
    multinomial experiments, 488
    statistical test for, 490–494
population proportions, single, 471–482
    confidence interval for, 472–474
    examples of, 472–476

exercises for, 477–482
    mean and standard error, 472
    sample size requirement, 477
    statistical test for, 475
statistical reports, 1088
statistical tables, 1091–1129
    F test power curves for AOV, 1123–1126
    percentage points for Dunnett's test, 1117–1120
    percentage points for Hartley's F_{max} test, 1121
    percentage points of chi-square distribution, 1100–1101
    percentage points of F distribution, 1102–1113
    percentage points of sign test, 1096
    percentage points of Student's t distribution, 1093
    percentage points of the Studentized range, 1115–1116
    percentage points of Wilcoxon rank sum test, 1097
    percentage points of Wilcoxon signed-rank test, 1098–1099
    Poisson probabilities, 1127–1129
    random numbers, 1122
    standard normal curve, 1091–1092
    values of 2 arcsin, 1114
statistical tests (*See also* hypothesis testing; t test)
    five parts of, 208
    for mean (μ), 231
    for mean (μ_1) − mean (μ_2) independent samples, 271–274
    for median (M), 246–247
    for more than two population variances, 384–393
        completely randomized design for, 387–388
        developing test statistic for, 386–387
        examples of, 389–392
        exercises for, 392–393
        mean square and, 388
        sum of squares and, 388
        summarizing results in AOV table, 389
        using multiple t tests, 385
    population proportion inferences and, 475
    for single population variance, 348
    t test power curves and, 1094–1095

statistical tests (*contd.*)
  for two population proportions, 484–485, 490–494
  for two population variances, 358
statisticians, 2, 10–11
statistics
  application examples, 6–10
    acid rain, 6–7
    bowhead whale population, 8–9
    libel suits, 8
    new drug testing, 7–8
    ozone exposure, 9
  introduction to, 2–6
  numerical descriptive measures of, 69–70
  reasons for study of, 6
  role statisticians and, 10–11
  statement of objective of, 10
  use in quality and process improvement, 12–14
stem-and-leaf plots, 54–57
  examples of, 57
  guideline for constructing, 56
  overview of, 54
  using for residuals analysis, 968
  vs. boxplot, 96
stepwise regression, 718, 720–722
storage files, 1088
storing, results, 1088–1089
stratified random sampling, 21–22
strength of association
  contingency tables, 505
  measuring, 510–516
    dependent and independent variables and, 512
    examples of, 513–516
    percentage analysis, 510–511
    predictability analysis, 511–512
Student–Newman–Keuls procedure (SNK), 447–450
  AOV for standard designs and, 920
  example of, 448–449
  steps in, 448
  vs. Tukey's procedure, 447–448
Studentized range distribution
  statistical table of, 1115–1116
  use of, 444
Studentized residual, 776
Student's *t* distribution
  properties of, 230
  statistical table of, 1093
  vs. normal distribution, 229
study log for data, 36, 1085

subjective interpretation of probability, 124–125
sum, in notation, 73
sum of squares
  balanced incomplete block designs, 1066
  completely randomized design, one covariate, 949–951
  factorial treatment, completely randomized design, 907–908
    main effects, 907
    three-way interactions, 907
    two-way interactions, 907
  Latin square design
    between-columns, 883
    between-rows, 883
  mean square, 388–389
sum of squares between samples (SSB)
  balanced incomplete block designs, 1065, 1067
  defined, 388
  randomized complete block design, one or more missing observations, 1057
sum of squares between treatment (SST)
  balanced incomplete block designs, 1065, 1067
  completely randomized design with single factor, 857
  Latin square design, 883
  randomized complete block design, 866
  randomized complete block design, one or more missing observations, 1056
sum of squares for error (SSE)
  analysis of covariance and, 949
  balanced incomplete block designs, 1065, 1067
  completely randomized design with single factor, 857
  factorial treatment, completely randomized design, 900, 907
  Latin square design, 883
  randomized complete block design, 866
sum of squares regression-SS(Regression)
  complete and reduced models for, 658
  computing for multiple regression, 686

defined, 649
  vs. SS(Residual) and SS(Total), 649
sum of squares residuals-SS(Residual)
  defined, 632, 740
  measuring accuracy of predictions with, 590
  vs. SS(Regression), 649
sum of squares sequential (SS), 647
sum of squares, total (TSS)
  completely randomized design with single factor, 856–857
  computing for multiple regression, 686
  defined, 388
  factorial treatment, completely randomized design, 899
  Latin square design, 882
  randomized complete block design, 865
  vs. SS(Regression), 649
sum of squares with contrasts (SSC), 433–434
sum of squares within samples (SSW), 388
surveys, 19–27
  conducted by, 19–20
    Gallup and Harris opinion polls, 20
    U.S. Bureau of Labor Statistics, 20
    U.S. Bureau of the Census, 19–20
  data collection techniques, 24–25
    direct observation, 25
    personal interviews, 24
    self-administered questionnaire, 25
    telephone interviews, 24–25
  influence of, 19
  problems with, 22–23
    measurement problems, 23
    survey nonresponse, 22–23
  sampling techniques for, 21–22
    cluster sampling, 21
    ratio estimation, 21
    single random sampling, 21
    stratified random sampling, 21–22
    systematic sampling, 22
  stages of, 24
symmetric histogram
  defined, 54
  example of, 52

symmetry of observations, compound, 1030–1031
systematic sampling, 22

t statistic (*See* t tests)
t tests
  accuracy of, 237–238
  evaluating equal variance, 355
  for mean (μ), 208
  for mean (μ_1) − mean (μ_2)
    independent samples, 275–276
    paired data, 303
  for more than two population variances, 385
  for slope, 557–558
  statistical table of power curves and, 1094–1095
  vs. sign test, 248
  vs. Wilcoxon rank sum test, 295–296
  vs. Wilcoxon signed-rank test, 311
telephone interviews, data collection, 24–25
terms, in completely randomized design, 394–395
test statistic (*See* t tests)
tests
  Poisson distribution, 498–499
  regression coefficients
    complete and reduced models for, 658
    example of, 658–661
    exercises for, 661–665
    F test of predictors, 657–658
  tests of significance
    factorial treatment in completely randomized design, 902–903
    mixed-effects model, 992–993
theorems
  Central Limit Theorem, 175, 179–180
  sampling distribution, 266
time series
  overview of, 57–58
  serial correlation and, 773
  uses of, 59
tolerable error, 204–205
total sum of squares (TSS) (*See* sum of squares, total (TSS))
transformations
  for central values, more than two populations
    defined, 403
    examples of, 404–409

    exercises for, 409
    selecting new variables, 403–404
  choosing, 537–538
  inverse transformations and, 739
  logarithmic, 538, 738
  for more than two population variances, 403–409
  quadratic, 739
  straightening scatterplot graphs with, 535–536
treatment means
  adjusted treatment means, 948, 950–951
  balanced incomplete block designs, 1067–1068
  comparing, 916–922
    case study, 921–922
    examples of, 917–921
    multiple comparison procedures and, 920
    using Fisher's LSD procedure, 916–917
  for Latin square design, 1061
  for randomized block design, 1056
treatments
  control treatments, 833
  defined, 623, 832
  design of, 892
  factorial treatment design, 832
  observational studies and, 830
  randomly assigning experimental units to, 840–841
trial-and-error, selecting multiple regression models, 730
true error variance, linear regression, 546–547
Tukey's W procedure, 444–447
  AOV for standard designs and, 920
  confidence interval for, 447
  example of, 445–446
  experimentwise error rate in, 445
  steps in, 445
  use of Studentized range distribution in, 444
  vs. LSD procedure, 446
  vs. SNK procedure, 447–448
two-factor experiments, repeated measures, 1031–1040
  examples of, 1032–1034, 1036
  exercises for, 1038–1040
  F tests for, 1034
  model for, 1031

  reporting conclusions of, 1038
  tests for, 1032
two-tailed tests
  computing probability for, 216
  overview of, 211
  sample size for, 221
two-way decision process, 209
Type I errors
  Bonferroni inequality and, 438
  defined, 209
  mean (μ) and, 219–220
  mean (μ_1) − mean (μ_2) and, 279
  multiple t tests and, 385
  probability of, 218–219
  for single population variance, 349–351
  for two population variances, 363
Type II errors
  Bonferroni inequality and, 438
  defined, 209
  mean (μ) and, 219–220
  multiple t tests and, 385
  probability of, 214–216, 218–219, 234
  for two population variances, 363

unbalanced designs, 1053
unbiased estimators
  for completely randomized design, 857
  population variances and, 344
  for randomized complete block design, 866
  variance and, 87
unconditional probability, 132
underspecification, 715
unequal population variances, 274–275
uniform histogram
  defined, 54
  example of, 52
unimodal/bimodal histograms, 52–54
  defined, 52
  examples of, 52–54
union of events, 130
unique predictive value, 647
unit of association, in prediction, 532

variability, measures of
  deviation, 86–87
  Empirical Rule for, 89–93
  exercises for, 93–96
  interquartile range, 86
  overview of, 81–82
  percentiles, 82–86

variability, measures of (*contd.*)
  range, 82
  standard deviation, 88–89
  variance, 87–88
  vs. measures of central tendency,
    81
variables (*See also* factors)
  covariates and, 839–840
  in designed experiments, 832
  multiple, 101–109
    cluster bar graphs, 102–103
    contingency tables and, 101–102
    exercises for, 107–109
    scatterplots, 102–103
    side-by-side boxplots, 104–107
    stacked bar graphs, 102
  selecting for multiple regression,
    707–727
    backward elimination, 717–720
    best subset regression, 716–717
    collinearity and, 708–710
    data-splitting approach, 714
    examples for, 709–722
    exercises for, 722–727
    overview of, 707–708
    performing all possible regres-
      sions, 711–714

PRESS statistic and, 714–715
    stepwise regression, 718,
      720–722
    underspecification/over-
      specification, 715
    too many independent variables,
      711
  types, 141–142
    continuous random variables,
      142
    discrete random variables, 142
    qualitative random variables,
      141
    quantitative random variables,
      141–142
variance (*See also* analysis of vari-
    ance (AOV); population vari-
    ances)
  components
    defined, 975
    estimating, 986–989
  defined, 87–88
  in designed experiments, 835–836
  reducing through blocking, 838–839
variance inflation factor (VIF)
  defined, 652
  diagnosing collinearity with, 709

weighted average, sampling vari-
    ances, 268
weighted least-squares, 762–763
Wilcoxon rank sum test, 287–299
  calculating rank sum statistics,
    288–289
  normal approximation and,
    292–295
  overview of, 288
  statistical table of, 1097
  summary of, 289
  vs. *t* test, 295–296
Wilcoxon signed-rank test, 308–314
  computing, 308
  example of, 309–312
  exercises for, 312–314
  g groups and, 308–309
  overview of, 309
  statistical table of, 1098–1099
  vs. paired *t* test, 311
work files
  from machine-readable database,
    38
  sets of data files and, 1086

zero expectation, multiple regression,
    759–760